AA002282

2022 24th European Conference on Power Electronics and Applications (EPE'22 ECCE Europe)

Hanover, Germany
5-9 September 2022

Pages 2021-2688

IEEE Catalog Number: CFP22850-POD
ISBN: 978-1-6654-8700-9

Copyright © 2022, The European Power Electronics and Drives Association
All Rights Reserved

*** This is a print representation of what appears in the IEEE Digital Library. Some format issues inherent in the e-media version may also appear in this print version.

IEEE Catalog Number: CFP22850-POD
ISBN (Print-On-Demand):   978-1-6654-8700-9
ISBN (Online):   978-9-0758-1539-9

Additional Copies of This Publication Are Available From:

Curran Associates, Inc
57 Morehouse Lane
Red Hook, NY 12571 USA
Phone: (845) 758-0400
Fax: (845) 758-2633
E-mail: curran@proceedings.com
Web: www.proceedings.com

2022 24th European Conference on Power Electronics and Applications (EPE'22 ECCE Europe)

Hanover, Germany
5-9 September 2022

Pages 2021-2688

IEEE Catalog Number: CFP22850-POD
ISBN: 978-1-6654-8700-9

TABLE OF CONTENTS

Dynamic Power Analysis of Inverter-Fed Drives Based on the Switching Period of the Power Electronics 1
Alexander Stock

Stability Analysis in an Inverter-Dominant Microgrid Facing In-Rush Current of an Induction Machine 11
Nastaran Fazli, David Hammes, Sidney Gierschner, Hans-Gunter Eckel

Self-Oscillating Capacitive Power Transfer with Multiple Receiver Capability and Coupling Path Adaption 22
Norbert Seliger

An Electrically Driven Gas Compressor for Hydrogen Refueling Stations with Active Power Smoothing 30
Alfred Rufer

Unsymmetrical Fault Behavior of PLL Based Grid-Connected Converters 39
Philipp Hackl, Ziqian Zhang, Robert Schuerhuber

Stability Assessment and Optimization of MMC Energy Balancing for Drive Applications at Standstill using an Averaging Approach 49
Qiuye Gui, Hendrik Fehr, Albrecht Gensior

Turn-On Losses Optimization for Medium Power SiC MOSFET Half-Bridge Module 59
Pham Ha Trieu To, Felix Kayser, Hans-Günter Eckel

Oscillation Damping in a 500kW Hybrid Si/SiC Three-Level ANPC Inverter with Decoupling Capacitor 70
Pham Ha Trieu To, Hans-Günter Eckel

Multi Busbar Sub-Module Modular Multilevel STATCOM with Partially Rated Energy Storage Configured in Sub-Stacks 80
Chuantong Hao, Wenhao Ma, Michael Merlin, Paul Judge, Stephen Finney

Three-Phase ZVS Inverter with Variable and Fixed Frequency Operation Based on GaN Semiconductors 88
Benedikt Kohlhepp, Michael Lutsch, Thomas Dürbaum

Influences of Conductor Positions and Fast Rising Impulse Voltages on the Line-End Coil Based on a Three-Phase High-Frequency Model 97
Ting Helmholdt-Zhu, Volker Grabs

Simulation Tool for Optimization of Digital Active Gate Drive Sequence using Genetic Algorithm 108
Hajime Takayama, Shuhei Fukunaga, Takashi Hikihara

Analysis of Balancing Algorithms for Quasi- Two/Three-Level Single Phase Operation of a Flying Capacitor Converter 115
Stefan Mersche, Markus Bayer, Kai Rickert, Marc Hiller

Instability in Active Balancing Control of Dc Bus Voltages in VSC Converters Interconnected via Multi-Winding Transformers 125
Duro Basic, Sami Siala

Online Learning-Based Islanding Detection Scheme for Grid-Connected Systems....................................... 135
Mohammed Ali Khan, V S Bharath Kurukuru, Rupam Singh

Difference in the Design Process of LCL Filters for Grid Connected VSI When using SiC/GaN
Instead of Si Semiconductors 145
Dennis Kampen, Lukas Fräger, Niklas Badenhop, Arthur Mambetow

Analysis and Design of a Resonant DC/DC Transformer in Modular Operation.................................. 152
Abraham López, Manuel Arias, Pablo F. Miaja, Arturo Fernández

Predictive Braking Algorithm for Soft Starter Driven Induction Motors.. 160
Hauke Nannen, Heiko Zatocil, Gerd Griepentrog

Ambient Electromagnetic Energy Harvesting Circuit using Rectennas Manufactured with
Stereolithography Resin 169
Xuan Viet Linh Nguyen, Tony Gerges, Jacques Verdier, Philippe Lombard, Michel Cabrera,
Bruno Allard, Jean-Marc Duchamp, Philippe Benech

Boost/Buck-Boost Based Grid Connected Solar PV Micro-Inverter with Reduced Number of
Switches and Having Power Decoupling Capability 178
Arup Ratan Paul, Arghyadip Bhattacharya, Kishore Chatterjee

Operation and Selection of Multilevel Power Converters for Doubly Fed Induction Generator-
Based Wind Turbines 187
Kapil Jha, Joseph Banda, Hridya I, Arvind Tiwari

A Detailed View on the Trapezoidal Operation for MMC Type Braking Chopper in Medium
Voltage Application.. 195
Patrick Hofstetter, Viktor Hofmann, Dennis Karwatzki

Influence of Operating Frequency on High-Power Medium-Voltage Medium-Frequency
Transformers 203
Thomas B. Gradinger, Ralph M. Burkart, Marko Mogorovic

Output Power Characteristics of Isolated Secondary-Resonant SAB DC-DC Converter for Output
Voltage Variation 213
Shota Yamashita, Kohei Budo, Takaharu Takeshita

Hardware and Control Design of a High Precision Modular Power Converter Based on GaN
Technology for Particle Accelerator Magnets 223
Thomas Margreiter, Ivan De Cesaris, Maurizio Incurvati, Sebastien Pelletier, Martin
Schiestl, Ronald Stärz

Battery Cycler to Generate Open Li-Ion Cell Aging Data and Models.. 232
Matthias Luh, Thomas Blank

Function Blocks of a Highly-Integrated All-In-GaN Power IC for DC-DC Conversion 242
Michael Basler, Richard Reiner, Stefan Moench, Patrick Waltereit, Rüdiger Quay

Comparison of Redundancy Requirements for Modular Multilevel Converter Considering
Manufacturer Reliability Inputs and Mission Profile 251
Diego Velazco, Guy Clerc, Emmanuel Boutleux, Francois Wallart

Impact of Insulation and Cooling on Performance Due to Reliability-Oriented Design of Electrical
Machines 261
Lucas Vincent Hanisch, Jonas Franzki, Markus Henke

Long Switching Horizon Model Predictive Controller for High-Speed Integrated Modular Motor Drives 268

Martin Schiestl, Maurizio Incurvati, Ronald Starz, Markus Schmid

Standalone Power Management System for Flexible Piezo Electric Nano Generators (PENG) Based on the Co-Polymer P(VDF:TrFE) 279

Alexander Wölk, Mahmoud Shousha, Shashank Shekhawat Singh, Martin Haug, Lorandt Fölkel, Michael Brooks, Asier Alvarez, Andreas Petritz, Philipp Schäffner, Jonas Groten, Andreas Tschepp, Barbara Stadlober

Analysis and Estimation of Neutral-Point Voltage Balancing Ability of an Optimized Balancing Algorithm for Grid Connected Active-NPC Converter 289

Joseph Banda, Kapil Jha, Hridya Ittamveettil, Arvind Kumar Tiwari, Fernando Ramirez

A Direct Model Predictive Control Strategy of Back-To-Back Modular Multilevel Converters using Arm Energy Estimation 297

Akseli Hakkila, Antonios Antonopoulos, Petros Karamanakos

Study on Commutation Loop Inductance and Current Distribution to DC-Link Capacitors in a GaN Half-Bridge 307

Benedikt Kohlhepp, Samuel Faber, Jeremias Kaiser, Thomas Dürbaum

Cooperative Control of Online Impedance Spectroscopy Monitoring Method and Maximum Power Point Tracking Method for Photovoltaic Panels 315

Xin Wang, Zhixue Zheng, Michel Aillerie, Alexandre De Bernardinis, Jean–paul Sawicki, Marie-Cécile Péra, Daniel Hissel

Benefits of Switching from Si to SiC Modules with Further Converter Optimization 325

Antxon Arrizabalaga, Mikel Mazuela, Iosu Aizpuru, June Urkizu, Jon Aztiria

On the Reduction of Output Capacitance in Two-Level Three Phase PFC Boost Rectifier for Pulsating Loads 335

Tania C. Cano, Douglas Pedroso, Alberto Rodríguez, Ignacio Castro, Diego G. Lamar

Cognitive Insights into Metaheuristic Digital Twin Based Health Monitoring of DC-DC Converters 344

Abdul Basit Mirza, Kushan Choksi, Sama Salehi Vala, Krishna Moorthy Radha, Madhu Sudhan Chinthavali, Fang Luo

A Three-Phase Isolated Secondary-Resonant Single-Active-Bridge DC-DC Converter with a Delta-Star Connected Transformer 351

Atsushi Nishio, Kohei Budo, Mai Van Tuan, Takaharu Takeshita

A Novel Concept to Optimize Core Loss in Planar Magnetic Based on an Unbalanced-Flux-Approach 361

Sobhi Barg, Kent Bertilsson, Grover Torrico

Model Reduction using Singular Perturbation Methods for a Microgrid Application 370

Lasse Gnärig, Albrecht Gensior, Saioa Burutxaga Laza, Miguel Carrasco, Carsten Reincke-Collon

Drive Level Parameter Identification of an Induction Motor 380

Andreas Bünte, Alex Hald, Andreas Kirsch

Impedance Stability of Single-Phase LCL Grid-Connected Voltage Source Inverters with Wideband Gap Devices Under Different Control Approaches 390

Ramy Ali, Terence O'Donnell

Design and Modulation Optimization of an MMC Based Braking Chopper.. 400
   Viktor Hofmann, Patrick Hofstetter

Modeling the Arrangement of Drill Holes for Orthogonal Biasing in Controllable Inductors for
Power Electronic Converters.. 411
   Jonas Pfeiffer, Christoph Drexler, Pierre Küster, Peter Zacharias, Michael Schmidhuber

A Sectorized FCS-MPC Transformerless SST for Power Transmission Application 421
   *Gabriel Gaburro Bacheti, Renner Sartório Camargo, Emilio José Bueno, Marco Liserre,
   Lucas Frizera Encarnação*

Inductance Estimation for Square-Shaped Multilayer Planar Windings 432
   Theofilos Papadopoulos, Antonios Antonopoulos

Cost and Efficiency Considerations in On-Board Chargers 442
   *Marija Jankovic, Christian Felgemacher, Kevin Lenz, Aly Mashaly, Abdelmouneim
   Charkaoui*

A Novel Combined Control of Ground Current and DC-Pole-To-Ground Voltage in Symmetrical
Monopole Modular Multilevel Converters for HVDC Applications.. 451
   Pablo Briff, Amit Kumar

A PFC Boost Converter with Reduced Switching Losses Operating at a Fixed Switching Frequency............ 459
   Burkhard Ulrich

Predictive Control of Power Electronics Autotransformer for Mitigating Three-Phase Grid Current
Unbalance in Railway Supply Systems... 468
   Tabish Nazir Mir, Faysal Hardan, Masood Hajian, Tamer Kamel, Pietro Tricoli

Parameter Sensitivity of a MRAS-Based Sensorless Control for AFPMSM Considering Speed
Accuracy and Dynamic Response at Multiple Parameter Variations.. 474
   Michael Brüns, Christian Rudolph, Tankred Müller

Synchronization Stability of a Grid Forming Converter Under the Effect of Current Limit in
Voltage Dips with VI Based Current Limiting Method: Analysis and Solution 484
   *Siam Hasan Khan, Markel Zubiaga Lazkano, Pedro Izurza, Alain Sanchez-Ruiz, Javier Cañas
   Aceña, Joseba Arza*

Analytic Calculation of Touch and Leakage Currents of Non-Isolated EV Chargers using a Fast
Common Mode Calculation Method and Non-Ideal Passive Component Models 493
   Christian Stutz, Sebastian Nielebock, Martin März

Triple-Phase-Shift Controlled Dual Active Bridge Converter with Variable Input Voltage in
Auxiliary Railway Supply 504
   Martin Scohier, Olivier Deblecker, Carlos Valderrama

Loss Characterization Methodology for Soft Magnetic Nano-Crystalline Tape Materials in Coupled
Inductors... 514
   David Bohne, Valentin Wagner, Patrick Deck, Christian P. Dick

Substitution of Nanocrystalline Toroid by Laminated Ferrite Toroid in the Application of a
Common-Mode Choke 525
   Lukas Reißenweber, Fritz Wohlrath, Alexander Stadler

Direct Active Stabilization of the DC-Link in Voltage-Source Converters 534
   Matthieu Bertin, Mohamad Koteich

Hardware-In-The-Loop Control of a Modular Induction Motor Drive in Power Electronics Education.. 544
Jens Peter Kaerst

Design and Efficiency Analysis of an LCL Capacitive Power Transfer System with Load-Independent ZPA.. 554
Francesco Musolino, Ahmed Abdullah, Mario Pavone, Fabio Ferreyra, Paolo Crovetti

A Pulse Generator Based on Transmission Line Transformer for Insulation Aging Test 562
Xiao Yu, Khanh-Hung Nguyen, Peter Zacharias

Design of a Single-Phase Common Mode and Differential Mode Inductor for Interleaved Converters 572
Jonathan Robinson, Gopal Mondal, Stefan Hänsel, Matthias Neumeister

Steady-State Analysis and Comparison of SSFB, SDFB and DSFB MMC-Based STATCOM 582
Mohamed Moez Belhaouane, Pierre Vermeerch, François Gruson, Pierre Rault, Sébastien Dennetiere, Xavier Guillaud

Current Distribution Control in Parallel Connected Power Converters with Continuous Output Voltage 593
Sabrina Ulmer, Andreas Brunner, Philipp Czerwenka, Gernot Schullerus, Ertugrul Sönmez

Optimized Pulse Pattern with Half-Wave Symmetry for 5-Level Converter 604
Jonas Weires, Pedro Leal Dos Santos, Steven Liu

Characterization of Si-IGBT Crosstalk with a Concentration on Power Circuit Parasitic Elements and the Device Operation Point.. 614
Amir Azam Rajabian, Sadegh Mohsenzade, Javad Naghibi, Kamyar Mehran

Impact of Higher Current Harmonics on Component Current Stress and Conduction Losses of Half-Bridge-Series-Resonant-Converters in Discontinuous Conduction Mode for High-Power Applications.. 624
Daniel Haake, Anton Grodnichev, Fabian Schnabel, Marco Jung

Control of a Zero-Voltage Switching Isolated Series-Resonant Power Circuit for Direct 3-Phase AC to DC Conversion 634
Yusuf Kosesoy, Remco Bonten, Henk Huisman, Jan Schellekens

Design of a Robust Voltage Control for Inverters with LC Filter Based on the Internal Model Control.. 641
Frederik Stallmann, Axel Mertens, Lukas Fräger

Influence of Power Semiconductor Device Variations on Pulse Shape of Nanosecond Pulses in a Solid-State Linear Transformer Driver.. 651
Raffael Risch, Anliang Hu, Jürgen Biela

Optimal Design of Integrated Motor Drives - Comparison of Topologies (2L/3L/Modular), PWM Variants, and Switch Technologies (Si/SiC/GaN).. 662
Thilo Bringezu, Jürgen Biela

Distribution Transformer Voltage Control using a Single-Phase Matrix Converter 673
Rui Wang, Henk Huisman, Korneel Wijnands

Influence of Carrier-Based PWM Techniques on the Common-Mode Voltage and Common-Mode Current of Six-Phase Full-Bridge Inverters... 681
Juris Arrozy, Esin Ilhan Caarls, Henk Huisman, Jorge L. Duarte, Lorenzo Ceccarelli

Mitigation of Dead-Time Effects on Transient DC Bias Elimination in Dual Active Bridge Link Current 689
    MK Kharabela Mohanta, Dipankar De, Silpashree Sahu, Alberto Castellazzi

Generalized Automated Tool for Analysis and Design of Multiphase Coupled Inductor Buck Converters 698
    Rana Asad Ali, Mahmoud Shousha, Martin Haug

Experimental Study of a Directly Oil-Cooled Electrical Machine for a Full-Electric Vehicle by using Low Viscosity Oil 709
    Huihui Xu, Georg Tobias Götz, Shimin Zhang, Rik W. De Doncker

Development of a Family of High Voltage Gain Step-Up Multi-Port DC-DC Converters for Fuel Cell-Based Hybrid Vehicular Power Systems 719
    Pouya Zolfi, Sina Vahid, Ayman El-Refaie

Bidirectional DC Circuit Breaker with Improved Performance During Commissioning and Reclosing 730
    Aditya Pogulaguntla, Venkata Raghavendra I, Satish Naik Banavath, Andrii Chub, T Sreekanth, Harish Sarma Krishnamoorthy

Modeling Method for Conducted Noise Flowing in Power Lines of DC/DC Converter 739
    Takato Hattori, Wataru Kitagawa, Takaharu Takeshita

High-Bandwidth Power Hardware-In-The-Loop for Motor and Battery Emulation at High Voltage Levels 749
    Manuel Fischer, Philipp Kemper, Johannes Herbold, Daniel Epping, Frank Puschmann

Analysis and Discussion of Different Three-Phase dv/dt Filter Topologies and the Influences of Their Filter Parameters on Losses and EMC 758
    Eric Fritze, Michael Meissner, Klaus F. Hoffmann, Kai-Uwe Rathjen, Stefan Dickmann, Oliver Woywode

State of Charge Prediction of Lithium-Ion Batteries Based on Artificial Neural Networks and Reduced Data 767
    Sebastian Pohlmann, Ali Mashayekh, Dominic Karnehm, Manuel Kuder, Antje Gieraths, Thomas Weyh

Investigation for Condensation Test Condition of HVIGBT Modules.. 777
    Kenji Hatori, Keiichi Nakamura, Wakana Noboru, Nils Soltau, Eugen Wiesner

Three Phase PV Inverter LCOE Optimization Considering Technological Choice 787
    Morteza Tadbiri Nooshabadi, Jean-Luc Schanen, Shahrokh Farhangi, Hossein Iman-Eini

Square Wave Operation to Reduce Pulsating Power in Isolated MMC-Based Ultrafast Chargers 798
    Ygor Pereira Marca, Maurice G. L. Roes, Jorge L. Duarte, Korneel Wijnands

Surge Current Protection for Railway Traction Applications... 805
    Michael Gleissner, Mark-M. Bakran

Impedance-Based Analysis of HVDC Converter Control for Robust Stability in AC Power Systems............ 814
    André Schön, Andreas Lorenz, Rodrigo Alonso Alvarez Valenzuela

Class-E Push-Pull Resonance Converter with Load Variation Robustness for Industrial Induction Heating 825
    Janus Dybdahl Meinert, Benjamin Futtrup Kjærsgaard, Thore Stig Aunsborg, Asger Bjorn Jorgensen, Stig Munk-Nielsen, Sune Bro Duun

Review of Power Converter Topologies for Electrochemical Impedance Spectroscopy of Lithium-Ion Batteries 833

Hamzeh Beiranvand, Julius M. Placzek, Marco Liserre, Giorgia Zampardi, Doriano Constantino Brogioli, Fabio La Mantia

Design and Experimental Validation of a Voltage Sensing-Current Cancellation Common Mode Linear Active Filter 843

B. Mohamed Nassurdine, PE Lévy, D. Labrousse, JL Schanen, X. Maynard, S. Carcouet

Partial Discharges of Insulated Wires Under Impulses from Wide Bandgap Power Electronics 854

Ting Helmholdt-Zhu, Vivien Grau, Urs Obernolte

Analysis of a Droop-Based Power Controller for Three-Phase Microgrids 865

Andrea Lauri, Hossein Abedini, Davide Biadene, Tommaso Caldognetto, Paolo Mattavelli

Efficiently Paralleling GaN-Transistors for High Current and High Frequency Applications using a Butterfly Layout 873

Martin Wattenberg, Oscar Lorenz, Juan Sanchez

Data-Driven Decentralized Volt/Var Control for Smart PV Inverters in Distribution Systems 883

Yizhou Lu, Qianwen Xu, Lars Nordström

Study of Current Ripple Generators for Accelerated Ageing of Capacitors 891

Robert Keilmann, Hendrik Schefer, Regine Mallwitz

Intra-Arm Balancing Control of Cascaded Multi-Port Converter for Whole Power Unbalance Conditions 902

Takumi Yasuda, Jun-Ichi Itoh

Investigation of Creepage Distances on Printed Circuit Boards for Avionic Applications 912

Hendrik Schefer, Zhongqing Xu, Tobias Kopp, Regine Mallwitz, Michael Kurrat

A 20 kW, 3-Level Flying Capacitor 1500 V Inverter with Characterized GaN Devices for Grid-Tie Applications 922

Van Sang Nguyen, Anthony Bier, Hajar Es-Seghier, Ulrich Soupremanien, Gérard Delette, Stephane Catellani

New Analytical Model for Calculating HF-Losses in Litz Wire Regions Located Outside the E/U-CoreWindow of Transformers 933

Qingchao Meng, Jürgen Biela

Fast and Accurate Soft-Switching and Hard-Switching Losses Estimation for Power Converter, Application to the Dual Active Bridge (DAB) Converter 944

Francois Boige, Nicolas Videau, Adel Ziani, Bruno Guerrero, Julien Laclaverie

Influence of an Electrical Machine on the Dimension and Packaging of Multi-Machine Systems 952

Thomas Stöckl, Hans-Georg Herzog

Design of a Serial Impingement Cooling Heatsink for a 30 kW PV String Inverter 960

Paul Bruyere, Guillaume Piquet Boisson, Gaëtan Perez

Online Junction Temperature Measurement of SiC-MOSFETs via Gate Impedance using the Gate-Signal Injection Method 971

David Hirning, Luca Bauer, Johannes Ruthardt, Jörg Haarer, Philipp Ziegler, Jörg Roth-Stielow

Powercycling Test Bench with Realistic Loss Distribution and Temperature Ripples 980
Till-Mathis Plötz, Jan Fuhrmann, Hans-Günter Eckel

Design, Implementation and Characterization of an Integrated Current Sensing in GaN HEMT
Device by using the Current-Mirroring Technique 990
Van-Sang Nguyen, René Escoffier, Stéphane Catellani, Murielle Fayolle-Lecocq, Jérémy
Martin

GaN-Based Modular Multilevel Converter for Low-Voltage Grid Enables High Efficiency 999
Philip Kiehnle, Patrick Himmelmann, Marc Hiller

Energy Management of Smart Homes with Electric Vehicles using Deep Reinforcement Learning............. 1006
Xavier Weiss, Qianwen Xu, Lars Nordström

Simple and Low-Computational Losses Modeling for Efficiency Enhancement of Differential
Inverters with High Accuracy at Different Modulation Schemes.. 1015
Ahmed Shawky, Mokhtar Aly, Emad M. Ahmed, Samir Kouro, José Rodriguez

Estimation of Battery Parameters in Cascaded Half-Bridge Converters with Reduced Voltage
Sensors 1025
Nima Tashakor, Bita Arabsalmanabadi, Elham Hosseini, Kamal Al-Haddad, Stefan Goetz

Method to Analyze the Influence of Switching Behavior in Hard Switching Half Bridge Topologies
for Traction Application.. 1036
Dominik Nehmer, Michael Gleissner, Lukas Bergmann, Mark-M. Bakran

Impact of Aluminum Casing on High-Frequency Transformer Leakage Inductance and AC
Resistance... 1046
Reda Bakri, Xavier Margueron, Wendell Da Cunha Alves, Xavier Cimetiere, Frédéric Gillon,
Antoine Bruyere, Lucian Vatamanu

Neural Networks-Generalized Predictive Control for MIMO Grid-Connected Z-Source Inverter
Model 1056
Navid Salehi, Herminio Martinez-Garcia, Guillermo Velasco-Quesada

Voltage Estimation for Diode-Clamped MMCs Based on a Simplified Neural Network 1064
Nima Tashakor, Davood Keshavarzi, Shady Banana, Stefan Goetz

A Non-Cooperative Game-Theoretic Distributed Control Approach for Power Quality
Compensators 1074
Claudio Burgos-Mellado, Victor Bucarey, Helmo K. Morales-Paredes, Diego Muñoz-
Carpintero

A Comparative Analysis of Power Converter Topologies for Integration of Modular Batteries in
Electric Vehicles.. 1083
Alberto Cárcamo, Aitor Vázquez, Alberto Rodriguez, Diego G. Lamar, Marta M. Hernando,
Daniel Remón

Design of a High-Dynamic Test Bench for Accelerated Dielectric Lifetime Testing with Adjustable
Voltage Slopes and Temperatures 1094
Hendrik Schefer, Lucas Hanisch, Tim-Hendrik Dietrich, Regine Mallwitz, Markus Henke

Novel Modulation Method for Common-Mode Noise Reduction in Solid-State Transformer Based
on ISOP Configuration 1104
Naoto Kikuchi, Hiroki Watanabe, Keisuke Kusaka, Jun-Ichi Itoh

Modular STATCOM for Compensation of Reactive Power and Voltage Asymmetry in Medium-Voltage Distribution Power Grids 1114
Josef Štengl, Tomáš Kormska, Jakub Talla, Zdenek Peroutka

Novel Method for Active Short Circuit (ASC) Tests of Power Module in Automotive Traction Application 1121
Tobias Appel, Arne Bieler

Short Circuit Performance and Current Limiting Mode of a Monolithically Integrated SiC Circuit Breaker for DC Applications Up to 800 V 1128
Norman Boettcher, Taro Takamori, Keiji Wada, Wataru Saito, Shin-Ichi Nishizawa, Tobias Erlbacher

Application of a HV Bipolar Square-Wave Voltage Generator for Qualification and Assessment of Energy Equipment 1137
Rico Fischer-Baeumer, Kai Gohrmann, Konrad Domes, Benjamin Sahan, Christian Staubach

A Decentralized and Communication-Free Control Algorithm of DC Microgrids for the Electrification of Rural Africa 1147
Lucas Richard, David Frey, Marie-Cecile Alvarez-Herault, Bertrand Raison

Universal Real-Time Model for Active Rectifiers in Versatile Totem-Pole PFC Configurations 1157
Axel Kiffe, Thorben Hoffstadt

Investigation of Core-Loss Mechanisms in Large-Scale Ferrite Cores for High-Frequency Applications 1167
Michael Baumann, Christoph Drexler, Jonas Pfeiffer, Jens Schueltzke, Erwin Lorenz, Michael Schmidhuber

Generation of Methodology for Making Benchmark Microgrids and Application in ESUSCON Microgrid 1177
Oscar Dorner, Patricio Mendoza-Araya

An Overview of Grid-Connection Requirements for Converters and Their Impact on Grid-Forming Control 1187
Paul Imgart, Mebtu Beza, Massimo Bongiorno, Jan R. Svensson

Modular Battery-Integrated Power Electronics-Modelling, Advantages, and Challenges 1197
Nima Tashakor, Jan Kacetl, Tomas Kacetl, Stefan Goetz

Design of Triple-Active Bridge Converter with Inherently Decoupled Power Flows 1207
Dong-Uk Kim, Byengjoo Byen, Byunghwang Jeong, Sungmin Kim

Application of a Multi-Winding Magnetic Component Characterization Method to Optimize Cross-Regulation Performances in DCM Flyback Converters 1216
Denis Motte-Michellon, Brahim Ramdane, Yves Lembeye, Bruno Cogitore

Application of an Electrostatic Machine in a Low-Voltage Microgrid 1226
Gabriel Ramos Huerta, Patricio Mendoza-Araya

Influences of Parasitic Capacitances in Wide Bandwidth Rogowski Coils for Commutation Current Measurement 1237
Philipp Ziegler, Tobias Festerling, Jorg Haarer, Philipp Marx, David Hirning, Jorg Roth-Stielow

Systematic Analysis of Oscillations in DC-Links of Fast Switching Power Electronics 1247
Tobias Fricke, Regine Mallwitz

EMI Mitigation Induced by an IGBT Driver Based on a Controlled Gate Current Profile 1256
Daniel S. Martinez-Padron, Nicolas Patin, Eric Monmasson

An Accurate and Fast Model of Three-Level Three-Phase Dual-Active Bridge Converters in Real-Time Simulation 1266
Ming Jia, Philipp Joebges, Rik W. De Doncker

A Calorimetric and Electrical Method for Measuring Loss Energies of Half-Bridges 1277
Jörg Haarer, Mattea Eckstein, Philipp Ziegler, Philipp Marx, David Hirning, Jörg Roth-Stielow

Condition Monitoring Approach of a SiC Power Semiconductor using Turn-Off Delay with an Integration in a SiC Driver 1286
Victor Golev, Ulf Schümann, Rando Raßmann, Jan Bockholt

Measurement Results of Multilevel Hysteresis Control for Paralleled Two-Level Converters 1294
Magdalena Gierschner, Yves Hein, Hans-Günter Eckel, Christian Heien

Design and Development of a Short-Circuit Test Bench for Low-Voltage Direct Current Protection Devices 1300
Simon Ravyts, Thomas Vandenbussche, Koen Stul, Jan Cappelle

A Novel Modified-TOGI Based PLL for the Three-Phase Unbalanced and Distorted Grid Conditions 1309
Khanh-Hung Nguyen, Ahmad Ali Nazeri, Xiao Yu, Peter Zacharias

Comparison of Two and Three-Level AC-DC Rectifier Semiconductor Losses with SiC MOSFETs Considering Reverse Conduction 1319
Guangyao Yu, Thiago Batista Soeiro, Jianning Dong, Pavol Bauer

Measurement Method for Simple Determination of Sinusoidal Large Signal Losses in Inductive Components 1328
Peter Zacharias, Alejandro Aganza-Torres

A Novel Technique for the Suppression of the Displacement Current Through Power Module Base-Plate Capacitance 1336
Mahmoud Saeidi, Ahmad Ali Nazeri, Rufad Zilic, Peter Zacharias

Analysis and Implementation of Effective Placement of EMC Capacitors for WBG Modules 1343
Mahmoud Saeidi, Ahmad Ali Nazeri, Firas Jenhani, Peter Zacharias

Power Hardware-In-The-Loop Verification of a Cold Load Pickup Scenario for a Bottom-Up Black Start of an Inverter-Dominated Microgrid 1350
Mina Mirzadeh, Robin Strunk, Tobias Erckrath, Axel Mertens

Detection of Incipient Inter-Turn Short-Circuit Faults by Artificial Intelligence Classifiers 1361
Osman Örgüt, Ilker Sahin, Ece Olcay Günes

Modeling the Impact of Grid-Forming E-STATCOMs on Inter-Area System Oscillations 1371
A. Bolzoni, N. Johansson, J. P. Hasler

Combining Schwarz-Christoffel Mappings and Biot-Savart Law to Calculate the High-Frequency Current Distribution Inside a Single Slot 1381
Torben Fricke, Phil Leon Pickert, Babette Schwarz, Bernd Ponick

Standardised Switching Cell Building Block for Converter Design Optimisation with Detailed Electro-Thermal Model 1391
Georgios Papadopoulos, Jürgen Biela

Design Procedure for Transformer-Based Solid-State Pulse Modulators with Damping Network 1402
Spyridon Stathis, Juergen Biela

DC Bias Impact on Magnetic Core Losses at High Frequency 1413
Bima Nugraha Sanusi, Ziwei Ouyang

Investigation of the Short-Circuit Type II Safe Operating Area of IGBTs 1424
Madhu Lakshman Mysore, Mohamed Alaluss, Abhishek Maitra, Thomas Basler, Roman Baburske, Franz-Josef Niedernostheide, Hans-Joachim Schulze

Single Transformer, MMC Based MV Power Electronic Traction Transformer 1434
Simon Fuchs, Simon Beck, Jürgen Biela

A New Power MOSFET Technology Achieves a Further Milestone in Efficiency 1445
Ralf Siemieniec, Michael Hutzler, Cesar Braz, Tomasz Naeve, Elias Pree, Heimo Hofer, Ingmar Neumann, David Laforet

Experimental Evaluation of Battery Impedance and Submodule Loss Distribution for Battery Integrated Modular Multilevel Converters 1456
Arvind Balachandran, Tomas Jonsson, Lars Eriksson, Anders Larsson

Constant DC Power Infeed Grid Forming with Improved Ability to Ride-Through Unbalanced Low-Voltage Faults 1466
Tayssir Hassan, Malte Eggers, Huoming Yang, Peter Teske, Sibylle Dieckerhoff

Constrained Long-Horizon Direct Model Predictive Control for Grid-Connected Converters with LCL Filters 1476
Mattia Rossi, Petros Karamanakos, Francesco Castelli-Dezza

Performance Evaluation of SiC-Based Isolated Bidirectional DC/DC Converters for Electric Vehicle Charging 1486
Kaushik Naresh Kumar, Rafal Miskiewicz, Przemyslaw Trochimiuk, Jacek Rabkowski, Dimosthenis Peftitsis

Impact of Threshold Voltage Shifting on Junction Temperature Sensing in GaN HEMTs 1497
Burhan Etoz, Jose Ortiz Gonzalez, Arkadeep Deb, Saeed Jahdi, Olayiwola Alatise

Comparison of Power Cycling Results of Discrete GaN Cascodes for Automotive Power Electronics with High Temperature Swings 1506
Florian Lippold, Philipp Hauenschild, Regine Mallwitz

Current Distortion Study for Hybrid Multi-Level Grid Inverter with Active Neutral-Point-Clamped 4-Leg Topology 1515
Jonas Steffen, Matthias Klee, Fabian Schnabel, Axel Seibel, Marco Jung

Dynamic Maximum Power Point Tracking Method Including Detection of Varying Partial Shading Conditions for Photovoltaic Systems 1525
Rosalie Rouphael, Nezha Maamri, Jean-Paul Gaubert

Novel Operation Mode of the Modular Multilevel Matrix Converter Based on a Dimensioning Algorithm 1533
Rebecca Dierks, Axel Mertens

On the Cosmic Ray Influence on the Electronics Design of a High Altitude Electric Aircraft 1543
  Philippe Morey, Mauro Carpita

DC-Bus Control Considerations of Asymmetrical Multilevel Inverters with Embedded Buck-Boost
Converter 1551
  Theodoros P. Mouselinos, Emmanuel C. Tatakis

A Seamless Modulation Strategy for Step-Up/Down Partial Power Processing Converter (SUD-
P3C) 1561
  Chao Liu, Zhe Zhang, Ziwei Ouyang, Jiasheng Huang, Michael A. E. Andersen, Tiberiu
  Gabriel Zsurzsan

Performances Analysis of Non-Model-Based Speed Estimation Algorithms for Motor Drives 1569
  Gaetano Turrisi, Luigi Danilo Tornello, Giacomo Scelba, Giulio De Donato, Giuseppe
  Scarcella

A Method to Design Power Control System of Wayside Energy Storage System for Energy Saving
in DC-Electrified Railway 1580
  Kota Sato, Keiichiro Kondo, Hiroyasu Kobayashi, Makoto Chida

A Reconfigurable Single-Stage Three-Phase Electric Vehicle DC Fast Charger Compatible with
Both 400V and 800V Automotive Battery Packs 1590
  Mojtaba Forouzesh, Yan-Fei Liu, Paresh C. Sen

Efficiency Improvement of Single-Stage AC-DC LLC Converter using a Line Cycle Synchronous
Rectifier (SR) Driving Strategy 1601
  Mojtaba Forouzesh, Yan-Fei Liu, Paresh C. Sen

Influence of DC Supply Voltage Unbalances on the Performance of ARCP Inverters 1611
  Gholamreza Tabrizi, Sebastian Sprunck, Marco Jung

Grid-Forming Control for Enhanced Microgrid Interconnection 1620
  Tobias Erckrath, Christian Bendfeld, Peter Unruh, Axel Seibel, Marco Jung

Low Phase Shift Filter for Current Sensing Based on the Difference Between AC Machine Models
with and Without Iron Losses 1631
  Niklas Himker, Marcel Krümpelmann, Axel Mertens

Design and Analysis of a Voltage Clamping Active Delay Control Method for Series Connected
SiC MOSFETs 1641
  Rui Wang, Asger Bjørn Jørgensen, Hongbo Zhao, Stig Munk-Nielsen

Practical Implementation of a Concept for In-Situ Detection of Humidity-Related Degradation of
IGBT Modules 1649
  Benedikt Kostka, Axel Mertens

Design for Enhanced Noise Immunity of PCB Coils Used for Sensing Current Through Power
Devices 1658
  Aamir Rafiq, Sumit Pramanick

Measurement Principle for Measuring High Frequency Bearing Currents in Electric Machines and
Drive Systems 1665
  Benjamin Knebusch, Lennart Junemann, Pauline Holtje, Axel Mertens, Bernd Ponick

Climatically Induced Insulation Degradation in Power Semiconductor Modules of Wind Turbines 1674
  Timo Lichtenstein, Sören Fröhling, Bernd Tegtmeier, Katharina Fischer

Comparison of Magnetic Noise Compensation Techniques for Dual Three-Phase Electrically Excited Synchronous Machines.. 1684

Jonas Henkenjohann, Jan Andresen, Axel Mertens

PCB Technology Comparison Enabling a 900V SiC MOSFET Half Bridge Design for Automotive Traction Inverters 1692

Matthias Spieler, Che-Wei Chang, Ayman El-Refaie, Muhammad H Alvi, Dong Dong, Rolando Burgos

Desaturated Turn-Off of Low-Saturation IGBTs with Clamping Method to Reduce Turn-Off Energy Losses.. 1703

Vishwas Acharya Nayampalli, Hans-Günter Eckel

Impact of Bond Wire Configuration on the Power Cycling Capability of Discrete SiC-MOSFET Devices 1713

Patrick Heimler, Nick Thönelt, Josef Lutz, Thomas Basler

A Low-Leakage, Low-Loss Magnetic Transformer Structure for High-Frequency Applications................. 1722

Allen Nguyen, Ajinkya Phanse, Michael Solomentsev, Alex J. Hanson

Temperature Distribution of an IGBT Chip During Repetitive Switching Events Under Consideration of Front-Side Ageing... 1733

Christian Bäumler, Bo Zhang, Maximilian Goller, Xing Liu, Thomas Basler

Boosting Pilot-Diode Reverse-Conducting IGBTs Turn-ON and Reverse-Recovery Losses with a Simple Gate-Control Technique.. 1744

Daniel Lexow, Hans-Günter Eckel

Modeling of an Interleaved DC-DC Boost Converter for a Direct Model Predictive Control Strategy.. 1754

Thomas Effenberger, Hannes Böorngen, Eyke Liegmann, Michael Hoerner, Petros Karamanakos, Ralph Kennel

Static Analysis and Control Strategies of the Single Active Bridge Converter 1765

Alexis A. Gómez, Alberto Rodríguez, Marta M. Hernando, Diego G. Lamar, Javier Sebastián, Ibán Ayarzaguena, Jose Manuel Bermejo, Igor Larrazabal, David Ortega, Francisco Vázquez

Multi-Port Inductive Power Transfer System Considering Charging Auxiliary Battery in EVs................... 1776

Zhuoqi Zhang, Ryosuke Ota, Ryohei Okada, Nobukazu Hoshi

Influence of IGBT and Diode Parameters on the Current Sharing and Switching-Waveform Characteristics of Parallel-Connected Power Modules.. 1785

Y. Ando, J. Sakai, K. Hatori, N. Soltau, E. Wiesner

Innovative Driving Scheme for Electrical Generators in More Electric Aircrafts Employing Series Active Filtering.. 1796

Nena Apostolidou, Nick Papanikolaou

Field-Measurement Based Hygrothermal Modelling of the Converter-Cabinet Climate in Wind Turbines... 1804

Katharina Fischer, Katherina Gohler

A Multi-Mode Control Based Asymmetrical Dual-Active-Bridge Series-Resonant DC-DC Converter (DABSRC) 1815

M. Yaqoob, Grover Torrico, Wang Shuqin

Extended Balancing and Dimensioning of Capacitors in MMC Double Submodules 1824
Ali Sharaf Addin, Christopher Dahmen, Thomas Brückner

Saliency Extraction and Torque Sharing Estimation of Dual Motor Drive using Special Current Sensor Configuration.. 1834
E. Rodriguez Montero, M. Vogelsberger, T. Wolbank

Soft-Switching Converter for Inductive Power Transfer System with Double-Sided LCC Resonant Network 1844
Ryohei Okada, Ryosuke Ota, Nobukazu Hoshi

Ultra Low Loss - MMC Submodules Favorable for SiC-FET Enabling High Functional Safety 1855
Christopher Dahmen, Rainer Marquardt

Control of an Active Gate Driver for an Electric Vehicle Traction Inverter using Artificial Neural Networks 1865
Julius Wiesemann, Jacob Dumtzlaff, Axel Mertens

Cascaded H-Bridge Converter Designs for Future Short-Range All-Electric Aircraft Propulsion 1875
Maximilian Hagedorn, Malte Lorenz, Axel Mertens

Overview and Evaluation of Energy Balancing Techniques for MMCs with Various Input and Output Frequencies.. 1885
Gyanendra Kumar Sah, Michael Schütt, Hans-Günter Eckel

Comparative Lifetime Estimations for IGBT Modules in Wind Turbine Converters 1895
Christian Neumann, Hans-Gunter Eckel

Single-Phase, Five-Level Inverter with SPWM-Based Neutral Point Voltage Balancing Scheme................ 1906
Dmytro Kondratenko, Arkadiusz Lewicki, Charles Odeh

Magnetic Core Evaluation Kit for the Comparison of Core Losses 1914
Wilmar Martinez, Xiaobing Shen, Siqi Lin, Jens Friebe

Multi-Objective Optimization of Modular Multilevel Converter Systems... 1923
Nikolaus Patzelt, Christian Schlegel, Michail Vasiladiotis

Sizing of Hybrid Energy Storage System for Residential PV Applications 1933
Xiangqiang Wu, Zhongting Tang, Tamas Kerekes

DC Bias Currents in Full-Bridge DC-DC Converters in Context of WBG Semiconductors and High Switching Frequencies.. 1939
Niklas Badenhop, Lukas Fräger, Dennis Kampen, Sascha Langfermann, Michael Owzareck

Parameter Tuning Method for Class Φ_2 Converters for High-Frequency Wireless Power Transfer Applications.. 1947
Yining Liu, Prasad Jayathurathnage, Jorma Kyyrä

Inductor Design Optimization using FEA Supervised Machine Learning 1955
D. Cajander, I. Viarouge, P. Viarouge, D. Aguglia

Enabling Large-Scaled MMC EMT-RMS Co-Simulation by Data Exchange in the Loop (DXiL)............... 1966
Xiong Xiao, Soham Choudhury, Martin Coumont, Jutta Hanson

Advanced Low-Voltage System-In-Package Half-Bridge MOSFET with Added Protection Features.......... 1975
S. Musumeci, V. Barba, F. Scrimizzi, C. Mistretta

Evaluation of Common-Mode Leakage Current of Aalborg-Type Transformerless PV Inverters 1985
Georgios I. Orfanoudakis, Eftychios Koutroulis, Georgios Foteinopoulos, Weimin Wu

Multi-Frequency Traction-To-Auxiliary Integrated EV Drivetrain: Eliminating the Need for an
Auxiliary Power Module 1995
Caniggia Viana, Mehanathan Pathmanathan, Peter W. Lehn

Potentials to Improve the Post-Fault Performance of a Fault-Tolerant Inverter System in Electrified
Aircraft Propulsion System 2003
Yongtao Cao, Leon Fauth, Jens Friebe, Axel Mertens

Model Predictive Control-Enabled Fault Ride Through Operation Strategy for High Power Wind
Turbine 2011
Pedro Catalán, Yanbo Wang, Zhe Chen, Joseba Arza

A Theoretical Comparison of Different Virtual Synchronous Generator Implementations on
Inverters 2021
Patrick Körner, Andrea Reindl, Hans Meier, Michael Niemetz

Linear Flux-Switching Machine Design - A Multiobjective Optimization 2030
Hendrik Marks, Henning Schillingmann, Sridhar Balasubramanian, Markus Henke

Single-Arm MMC-Based Converter for Transformerless Rail Interties.. 2038
Simon Beck, Simon Fuchs, Jürgen Biela

Medium Voltage Diode Rectifier Design for High Step-Up DC-DC Converter 2049
Pierre Le Métayer, Cyril Buttay, Drazen Dujic, Piotr Dworakowski

Fast Switching Planar Inductance Current Source ZETA Converter with Integrated Common Mode
Filter 2058
Benjamin H. Zacher, Christian Schumann

System Level Simulation of Moisture Propagation and Effects in Wind Power Converters.............. 2066
Johannes C. Wenzel, Axel Mertens

PWM-Based Optimization-Free Active Voltage-Balancing Control of 7-Level Active Neutral-
Point-Clamped Flying-Capacitor Multicell Inverters 2073
Vahid Dargahi

Model Predictive Power Sharing Algorithm for Fuel Cell Integration in a Dual Inverter Electric
Vehicle Drivetrain 2084
Mehanathan Pathmanathan, Caniggia Viana, Sukhjit Singh, Peter W. Lehn

Comparative Evaluation of the 5-Phase Vienna and the 5-Phase PWM Rectifiers Under DC
Voltage Control 2092
A. Dieng

Modelling and Control of a 50kW SiC-Based Isolated DAB Converter for Off-Board Chargers of
Electric Vehicles... 2101
*Haaris Rasool, Manh Tuan Tran, Sajib Chakraborty, Joeri Van Mierlo, Thomas Geury,
Mohamed El Baghdadi, Omar Hegazy*

Impact of Cyber Attacks on Cost Oriented Power Routing Schemes in Microgrids 2110
Kirti Gupta, Subham Sahoo, Bijaya Ketan Panigrahi, Frede Blaabjerg

Response of IGBT Chip Characteristics Due to Critical Stress... 2119
Kohei Yamauchi, Rik W. De Doncker

Mega-Hertz High-Power WPT System with Parallel-Connected Inverters using Current Balance Circuit... 2127
Masamichi Yamaguchi, Keisuke Kusaka, Jun-Ichi Itoh

Investigation and Mitigation of Common-Mode Voltage in Four-Level NPC Converters Modulated by Redundant Level Modulation 2136
Jun Wang, Wei Xu, Xibo Yuan, Lihong Xie

Ferrite Optimization for a Three-Phase Wireless Power Transfer System for Electric Vehicles 2145
Shuang Nie, Mehanathan Pathmanathan, Peter W. Lehn

Frequency and Modulation Index Related Effects in Continuous and Discontinuous Modulated Y-Inverter for Motor-Drive Applications 2156
Hamzeh J. Jaber, Alberto Castellazzi

Performance Evaluation of Sinusoidal-Flux Reluctance Machine for Improving Power Density with Reduced Torque and Input-Current Ripples 2164
Kiwa Nagayasu, Masaki Iida, Kazuhiro Umetani, Mastaka Ishihara, Eiji Hiraki

Power Hardware-In-The-Loop Test of Low-Voltage Battery for a Plug-In Hybrid Electric Vehicle 2175
Ronan German, Florian Tournez, Alain Bouscayrol, Aurelien Lievre, Betty Lemaire-Semail

Stability Analysis of DFIG System Connected with High-Frequency Capacitive Grid Based on Closed-Loop Current Control and Direct Power Control 2182
Bin Hu, Heng Nian, Subham Sahoo, Frede Blaabjerg, Yaqian Zhang, Zixiao Xu

Full-Bridge Modular Multilevel Converter for the Four-Quadrant Supply of High Power Magnets in Particle Accelerators... 2189
Manuel Colmenero, Ricardo Vidal-Albalate, Francisco R. Blanquez, Ramon Blasco-Gimenez

Deep Neural Network for Magnetic Core Loss Estimation using the MagNet Experimental Database 2197
Xiaobing Shen, Hans Wouters, Wilmar Martinez

Hybrid Circuit Board Structure for Power Electronics... 2205
Gerrit Braun, Deniz-Heinz Moldenhauer

Active Control of Gear Mesh Vibration using a Permanent-Magnet Synchronous Motor and Simultaneous Equation Method.. 2211
Dominik Reitmeier

Research Laboratory for Testing Grid Connected Devices Under Grid Voltage / Grid Impedance Variations and Microgrid Conditions 2219
Swen Bosch, Jochen Staiger, Heinrich Steinhart

Reducing the Impact of Skin Effect Induced Measurement Errors in M-Shunts by Deliberate Field Coupling 2230
Hauke Lutzen, Jonas Müller, Vladimir Polezhaev, Till Huesgen, Nando Kaminski

Grid Forming Control for HVDC Systems: Opportunities and Challenges 2241
Adil Abdalrahman, Ying-Jiang Häfner, Malaya Kumar Sahu, Khirod Kumar Nayak, Ashkan Nami

A Highly Integrated and Modular High Speed Electric Drive for Lightweight Electric Mountain Bikes.. 2251
Matthias Hofer, Mario Nikowitz, Manfred Schrödl

Performance Enhancement of Power Conditioning Systems in More Electric Aircrafts 2257
Nick Rigogiannis, Nick Papanikolaou, Yongheng Yang

Steady State Simulations of a Hybrid HVAC/HVDC Network using OS Based ARM Devices 2266
Ioan Catalin Damian, Mircea Eremia

Experimental Comparison of FPGA-Implemented Model Predictive Voltage Control to Cascaded
Proportional Resonant Control for a Three-Phase Four-Wire Three-Level Grid-Forming Inverter of
250 kVA 2276
Jarren Lange, Dominik Schmies, Karl Stephan Stille, Joachim Böcker, Oliver Wallscheid

Experimental Study of Interleaved Y-Inverter Performance 2285
Yusuke Endo, Masataka Minami, Hamzeh J. Jaber, Alberto Castellazzi

Design of a GaN-Based Reconfigurable Resonant Converter for High Frequency On-Board
Charger of Battery Electric Vehicles 2293
*Manh Tuan Tran, Haaris Rasool, Dai Duong Tran, Mohamed El Baghdadi, Philippe Lataire,
Omar Hegazy*

Transient Liquid Phase Bond Reliability Evaluation of Die-Attach for Power Module Packaging 2301
Laxma R. Billa, Yangang Wang, Thomas Grant, Xiang Li, Harley Neal, Muhammad Morshed

Experimental Evaluation on Observer-Based Delay-Compensating Active Damping for LC-Filters 2308
Michael Schütt, Hans-Günter Eckel

Influence of Static Rotor Imbalance on the Roller Bearing Damage Due to Inverter-Induced
Bearing Currents 2316
Martin Weicker, Omid Safdarzadeh, Andreas Binder

Novel Current Balancing Method for HF Interleaved Converters with Reduced Control Effort 2327
Christian Beckemeier, Jens Friebe

dV/dt-Based Filter Design for Motor Inverters with Continuous Output Voltage 2334
Sabrina Ulmer, Stevan Bugarski, Gernot Schullerus, Ertugrul Sönmez

Evaluation of Core Losses in Transformers for Three-Phase Multi-Level DAB Converters 2344
Babak Khanzadeh, Yuriy Serdyuk, Torbjörn Thiringer

A Quasi-Offline Condition Monitoring Method of DC-Link Capacitor Banks in Accelerator Power
Converters 2355
*Timm Felix Baumann, Konstantinos Papastergiou, Raul Murillo Garcia, Dimosthenis
Peftitsis*

Minimizing Voltage Stress in Auxiliary Resonant Commutated Pole Inverters using Saturable
Inductors 2366
Markus Zocher, Norbert Grass, Ralph Kennel

Adaptive Dead-Time Control in a Resonant Wireless Power Transfer System 2375
Tim Krigar, Martin Pfost

Multilevel Battery Converter with Cascaded H-Bridges on Cell Level-Battery Management System
Or a Renewed Attempt for Power Electronic Building Blocks? 2383
*Max Rothenburger, Markus Horn, Xiao Yu, Gerold Schulze, Koenraad Muyllaert, Peter
Zacharias, Ludwig Brabetz, Hartmut Hillmer*

Design and Potential of EMI cm Chokes with Integrated DM Inductance 2392
Mohammad Ali, Rehnuma Bushra, Jens Friebe, Axel Mertens

Implementation Options of a Fully SiC Buck-CSI for Advanced Motor Drive Application.......................... 2402
Yonghwa Lee, Alberto Castellazzi

Optimized Control Scheme to Achieve ZVS for the Complete Pre-Charging Phase of
Supercapacitors with a 500 kHz SiC- And GaN-Based Dual Active Bridge 2413
Patrick Lenzen, Martin Pfost

Fault Blocking Capability in the DC-MMC with Reduced Number of Sub-Modules.................................... 2422
J. D. Páez, F. Morel, S. Bacha, P. Dworakowski

An Open-Source FEM Magnetic Toolbox for Calculating Electric and Thermal Behavior of Power
Electronic Magnetic Components 2432
Nikolas Förster, Jonas Hölscher, Till Piepenbrock, Philipp Rehlaender, Oliver Wallscheid,
Frank Schafmeister, Joachim Böcker

Comparison of Dual-Active-Bridge-Based Topologies for Single-Phase Single-Stage EV On-Board
Chargers 2441
Daniel Gaona, Denis Pauls, Eduardo Facanha De Oliveira

Design Concepts for Medium Voltage DC Networks Supplying the Future Circular Collider (FCC)........... 2451
Manuel Colmenero, Francisco R. Blanquez, Ramon Blasco-Gimenez

A Novel Dual CC-CV Output Wireless EV Charger with Minimal Dependency on Both Coil
Coupling and Load Variation 2462
Subhranil Barman, Kishore Chatterjee

A High-Performance EMI Filter Based on Laminated Ferrite Ring Cores 2470
Marcin Kacki, Marek S. Rylko, John G. Hayes, Charles R. Sullivan

Investigation of the Static Performance and Avalanche Reliability of High Voltage 4H-SiC
Merged-PiN-Schottky Diodes 2477
Chengjun Shen, Saeed Jahdi, Phil Mellor, Juefei Yang, Erfan Bashar, Jose Ortiz-Gonzalez,
Olayiwola Alatise

On Chain-Link Based Multi-Port Converters Able to Connect HVDC and MVDC to AC
Transmission Network.. 2486
Daniele Falchi, Oriol Gomis-Bellmunt, Eduardo Prieto-Araujo, Olivier Despouys

Voltage Control Scheme for Multilevel Interfacing PV Application: Real-Time MRAC-Based
Approach 2496
Mohammad Sadegh Orfi Yeganeh, Mehdi Rahmani, Nenad Mijatovic, Tomislav Dragicevic,
Frede Blaabjerg, Pooya Davari

Control Principles for Island Operation and Black Start by Offshore Wind Farms Integrating Grid-
Forming Converters.. 2504
Daniela Pagnani, Lukasz Kocewiak, Jesper Hjerrild, Frede Blaabjerg, Claus Leth Bak

Experimental Study of the Reduction and Removal of Turn-On Snubber for IGCT Based MMC
Submodule using Fast Silicon Diodes... 2515
Arthur Boutry, Cyril Buttay, Besar Asllani, Bruno Lefebvre, Eric Vagnon, Dong Dong

Characterisation of a Ferrite-Polymer Based Magnetic Material 2526
Johan Le Leslé, Guillaume Lefevre, Julien Morand, Rémi Perrin, Pierre-Yves Pichon,
Guillaume Regnat

Model Predictive-Based Control Technique for Fault Ride-Through Capability of VSG-Based Grid-Forming Converter.. 2537
    Mobina Pouresmaeil, Amir Sepehr, Basit Ali Khan, Jafar Adabi, Edris Pouresmaeil

Grounding Points in HV/MV Hybrid Transformer Auxiliary Converters.. 2544
    Adrian Wiemer, Jürgen Biela

Non-Parasitic Induced Transient Overvoltage in ANPC Topology Due to Critical Switching Sequences 2554
    Michael Geiss, Robert Kragl, Jürgen Thoma, Benjamin Volzer

Open-Delta SBC: A New Converter Topology with Low Number of Sub-Modules for MV Applications.. 2564
    D. Lanzarotto, P. B Steckler, K. Vershinin, F. Morel

Characterising the Effect of an Inverter on the Regulation of the AC Voltage using a Frequency Response Identification Technique 2574
    Mohamed Aldarmon, Joan Marc Rodriguez, Adria Junyent-Ferre

Artificial-Intelligence Based DC-DC Converter Efficiency Modelling and Parameters Optimization 2581
    Fanghao Tian, Diego Bernal Cobaleda, Wilmar Martinez

Analysis of the Loss Distribution of a 6 kW Two Stage Power Supply for 600 V DC Applications............ 2588
    Lukas Fräger, Sascha Langfermann, Michael Owzareck, Dennis Kampen, Jens Friebe

Study on the Gate Loop Design and Its Impact on Switching Characteristics of GaN Transistors............... 2596
    Xiaomeng Geng, Carsten Kuring, Oliver Hilt, Mihaela Wolf, Joachim Würfl, Sibylle Dieckerhoff

Analysis of Current Sharing in the Parallel Connection of GaN Transistors 2607
    Frederik Stalleicken, Sibylle Dieckerhoff, Karsten Handt, Sebastian Nielebock

Verification of GaN-HEMT Spice Models using an S-Parameters Approach 2618
    Alonso Gutierrez, Nasri Said, Emmanuel Marcault, Mathieu Gavelle

Power Loss Modelling of GaN HEMT-Based 3L-ANPC Three-Phase Inverter for Different PWM Techniques.. 2628
    Salvatore Mita, Arjun Sujeeth, Giuseppe Aiello, Dario Patti, Francesco Gennaro, Giacomo Scelba, Mario Cacciato

Generalized Core and Winding Area Ratio - Trends for Inductors and Transformers in Power Electronics with High Switching Frequencies.. 2638
    Siqi Lin, Leon Fauth, Wilmar Martnez, Jens Friebe

Active Substrate Termination of Discrete and Monolithic Bidirectional GaN HEMTs in a T-Type Inverter 2644
    Carsten Kuring, Yannic Lange, Xiaomeng Geng, Oliver Hilt, Mihaela Wolf, Joachim Würfl, Sibylle Dieckerhoff

Transformer Design Optimization and Comparison for a DC-DC Converter Used in PV Micro-Inverters... 2655
    Tobias Manthey, Meriem Khader, Jens Friebe

Automated Gate Impedance Network Design for SiC MOSFETs using SPICE Solver Interfaced with MATLAB Environment 2661
    Pawel Piotr Kubulus, Szymon Michal Beczkowski, Stig Munk-Nielsen, Asger Bjørn Jørgensen

An Improved Multi-Loop Resonant and Plug-In Repetitive Control Schemes for Three-Phase
Stand-Alone PWM Inverter Supplying Non-Linear Loads 2670
Ahmad Ali Nazeri, Peter Zacharias

High Switching Frequency Operation of a Single-Phase Five-Level Hybrid Active Neutral Point
Clamped Inverter with a Model Predictive Control Approach 2682
Mohammad Najjar, Mahdi Shahparasti, Rasool Heydari, Morten Nymand

Design of Planar Coupled Inductor Applied to Zero-Current Switching Clamped Current Converter 2689
Vinicius Freire Bezerra, Tobias Manthey, Montiê Alves Vitorino, Jens Friebe

Characterization of Online Junction Temperature of the SiC Power MOSFET by Combination of
Four TSEPs using Neural Network 2698
Kanuj Sharma, Simon Kamm, Kevin Muñoz Barón, Ingmar Kallfass

Novel Extended Robust Disturbance Observer for Improved Cogging Force Compensation in
Permanent Magnet Linear Motors 2706
Franz Luckert, Axel Mertens

Improvement of a Self-Powered Gate Driver Power Supply 2715
*Mariana Raya, Oriol Aviñó, Sergio Busquets-Monge, Xavier Perpiñá, Miquel Vellvehi, Xavier
Jordà*

Optimization and Scaling of a Compact High-Power IGCT Capacitor Charger Based on Simulation
and Measurements with a 300 kW/3.3 kV Demonstrator 2726
Felix Haag, Fabian Albrecht, Volker Brommer, Oliver Liebfried, Klaus F. Hoffmann

Multilayer Busbars for Medium Voltage ANPC Converter Dedicated to Battery Energy Storage
Systems 2736
Mamadou Lamine Beye, Luc Bimmel, Anthony Bier, Jérémy Martin

A Simulation Model for SiC MOSFET Switching Transients Controlled by an Adaptive Gate
Driver with the Capability of Reducing Switching Losses and EMI Across the Full Operating
Range 2744
Zheming Li, Robert W. Maier, Mark-M. Bakran, Franz-J. Niedernostheide, Daniel Domes

Phase-Shift Modulation for Flying-Capacitor DC-DC Converters 2754
Philipp Rehlaender, Frank Schafmeister, Joachim Böcker

An EV Integrated Isolated DC Charger using a Six-Phase Synchronous Machine 2763
Sukhjit S Ghumman, Mehanathan Pathmanathan, Peter W Lehn

Configurable ISOP-IPOP DC-DC Converter for Universal Solid-State Transformer 2773
Pramod Apte, Jens Friebe, Lukas Fräger

Using System-On-Chip Boards for the Deployment of Controller for Verification and Prototyping 2780
Adeel Jamal, Gerd Griepentrog

Utilizing the Reactive Current Control Capability of an MMC-Fed AC/DC Converter for Volt-
Second Balancing in Medium Frequency Transformers 2788
*Kaveh Pouresmaeil, Maurice Roes, Jorge Duarte, Korneel Wijnands, Nico Baars, George
Papafotiou*

Cost Comparison for Different PV-Battery System Architectures Including Power Converter
Reliability 2795
*Martijn Deckers, Leander Van Cappellen, Glenn Emmers, Fereshteh Poormohammadi, Johan
Driesen*

Insulation Design and Analysis of a Medium Voltage Planar PCB-Based Power Bus Considering Interconnects and Ancillary Circuit Integration 2806
Joshua Stewart, Rolando Burgos, Dushan Boroyevich

Modular Multilevel Converter Control with using a General Space Vector PWM Method in Medium Voltage Hydro Power Application 2813
Chengjun Tang, Torbjörn Thiringer

A Technical Overview of Single-Stage Three-Port DC-DC-AC Converters 2824
Sebastian Neira, Zoe Blatsi, Michael M. C. Merlin, Javier Pereda

Common-Mode EMI Noise Modeling of Three-Level T-Type Inverter for Adjustable Speed Drive Systems 2835
Vefa Karakasli, Abdelmoumin Allioua, Gerd Griepentrog

A Condition Monitoring Scheme for Semiconductor Devices in Modular Multilevel Converters with Cascaded H-Bridge Submodules 2843
Mohsen Asoodar, Mehrdad Nahalparvari, Christer Danielsson, Hans-Peter Nee

Particular Requirements on Drive Inverters for Safe and Robust Operation on an Open Industrial DC Grid 2852
Simon Puls, Jan-Niklas Koch, Martin Ehlich, Holger Borcherding

Investigation About Operation and Performance of Gate Drivers for Power Electronics Converters for Cryogenic Temperatures 2860
Mustafeez-Ul-Hassan, Yuxuan Wu, Vyacheslav Solovyov, Fang Luo

Synchronization Angle Determination in DVCSFO of DFIM Naval Propulsion 2869
Youssef Drimizi, Maria Pietrzak-David, Pascal Maussion

Power Control of LCR-DAB Converter with Phase Shift in Fixed Switching Frequency 2877
Seung-Hyuk Baek, Jaehong Lee, Seung-Hwan Lee, Sungmin Kim

A Simplified Braking Method for Direct Matrix Converter-Fed PMSM Drives with Consideration of Avoiding Regenerative Energy 2885
Jun Xie, Dustin Henneberg, Martin Suberski, Thomas Ellinger, Uwe Radel, Jürgen Petzoldt

Inverter-Machine Parametric Co-Design for Energy Efficient Electric Drives 2893
Jaedon Kwak, Alberto Castellazzi

Bidirectional Cuk Converter in Partial-Power Architecture with Current Mode Control for Battery Energy Storage System in Electric Vehicles 2903
J. S. Artal-Sevil, J. Anzola, V. Ballestín-Bernad, I. Aizpuru

Design Space Exploration for a Capacitive 36V, 4A, 4:1 DCDC Converter with GaN Switches using a Performance-Cost-Matrix Including Uncommon Topologies 2912
Adrian Gehl, Malte Kempchen, Simon Disselkamp, Markus Olbrich, Bernhard Wicht

A Fast Control for a Three-Switch Multi-Input DC-DC Converter 2919
Simone Cosso, Andrea Formentini, Mario Marchesoni, Massimiliano Passalacqua, Luis Vaccaro

Impact on the Torque and on the Copper Losses Under Fault-Tolerant Control of 5-Phase PMSG 2930
A. Dieng

Weighting Factor Design for FS-MPC in VSCs: A Brain Emotional Learning-Based Approach 2939
Mohammad Sadegh Orfi Yeganeh, Arman Oshnoei, Saeed Peyghami, Nenad Mijatovic, Tomislav Dragicevic, Frede Blaabjerg

A Strategy for Smooth Microgrid Transitions Without Phase Misalignment and Voltage Mismatch 2948
Gabriel Silva Rocha, Amiron Wolff Dos Santos Serra, Cesar Augusto Santana Castelo Branco, Hercules Araujo Oliveira, Jose Gomes De Matos, Luiz Antonio De Souza Ribeiro

Subtle Design and Performance Comparison of WF-FSM and DC-VRM for Large-Scale Direct-Drive Wind Power Generation 2958
Udochukwu B. Akuru, Maarten J. Kamper, Zi-Qiang Zhu

Analysis and Implementation of Different Non-Isolated Partial-Power Processing Architectures Based on the Cuk Converter 2967
J. S. Artal-Sevil, J. Anzola, V. Ballestín-Bernad, J. L. Bernal-Agustín

GaN HEMT and SiC Diode Commutation Cell Based Dual-Buck Single-Phase Inverter with Premagnetized Inductors and Negative Gate Driver Turn-Off Voltage 2977
Tobias Brinker, Hendrik Gräber, Jens Friebe

Determination of Optimal Associated Discrete Circuit Switch Model Parameters for Real-Time Simulation of Dual-Active Bridge Converters 2985
Marija Stevic, Ravinder Venugopal

Integrated Motor Drive: A Multidisciplinary Approach 2996
Betty Lemaire-Semail, Nadir Idir, Eric Semail, Souad Harmand

Hardware in the Loop Test of an Electric Aircraft Powertrain 3005
Sebastian Mönninghoff, Moritz Scholjegerdes, Kay Hameyer

A Multi-Port Smart Transformer for Green Airport Electrification 3014
Giampaolo Buticchi, Giovanni De Carne, Thiago Pereira, Kangan Wang, Xiang Gao, Jiajun Yang, Youngjong Ko, Zhixiang Zou, Marco Liserre

Improvement of EMI Filter Attenuation using Shielding 3022
Mohammad Ali, Rehnuma Bushra, Jens Friebe, Axel Mertens

Implementation of Onsite Junction Temperature Estimation for a SiC MOSFET Module for Condition Monitoring 3031
Farzad Hosseinabadi, Shahid Jaman, Sachin Kumar Bhoi, Md. Mahamudul Hasan, Sajib Chakraborty, Mohamed El Baghdadi, Omar Hegazy

Energy Storage Systems for Airborne Wind Generators 3037
Bakr Bagaber, Axel Mertens

Design Interactions of AC- And DC-Side Filters for Traction Drives with SiC Inverters 3048
Hedieh Movagharnejad, Benjamin Knebusch, Axel Mertens, Bernd Ponick

Investigation of an Interleaved Current-Fed Single Active Bridge DC-DC Converter for PV Applications 3059
Lucas Vinícius De Araújo Gomes, Tobias Manthey, Montiê Alves Vitorino, Jens Friebe

Real-Time Thermal Characterization of Power Semiconductors using a PSO-Based Digital Twin Approach 3067
Johannes Kuprat, Yoann Pascal, Marco Liserre

Self-Sensing Design and Control for an Induction Machine with an Additional Short-Circuited Rotor Coil 3075
Stefan Luecke, Axel Mertens

Calculating the Tractive Power and Power Conversion Efficiency of Battery Electric Vehicles using a Global Navigation Satellite System and a Road Elevation Database 3084
Shinichi Domae, Alberto Castellazzi, Hamzeh J. Jaber, Tenghui Dong, Taketsune Nakamura

PCB Layer Optimization of Planar Medium Frequency Transformer for On-Board EV Chargers 3092
Fabian Groon, Hamzeh Beiranvand, Thiago Pereira, Görkem Can, Marco Liserre

Fault Current Capability Assessment of Low-Voltage Side Inverters in Smart-Transformers 3101
Thiago Pereira, Luis Camurca, Francisco Santos, Marco Liserre

Adaptive Resonant-Valley Switching for a GaN HEMT Direct AC-AC Auxiliary Resonant Commutated Pole Converter 3112
Kyle Steyn, Johan Beukes

The Variation of Core Loss in High-Frequency Transformers Under Different Load Conditions 3120
Navid Rasekh, Jun Wang, Xibo Yuan

A Complete PFC Inductor Design for Lighting Equipment Applications 3130
Wai Keung Mo, Kasper M. Paasch, Thomas Ebel

Automatic Generation Control-Based Charging/Discharging Strategy for EV Fleets to Enhance the Stability of a Vehicle-To-Weak Grid System 3140
Majid Mehrasa, Mehrdad Gholami, Reza Razi, Khaled Hajar, Antoine Labonne, Ahmad Hably, Seddik Bacha

Model-Based Converter Control for the Emulation of a Wind Turbine Drive Train 3149
Alexander Ernst, Wilfried Holzke, Dawid Koczy, Nando Kaminski, Bernd Orlik

A Novel Grid-Demanded Power Point Tracking (GPPT) Control Method for Wind Turbines to Preserve Grid Stability with High Wind Energy Penetration 3159
David Matthies, Alexander Ernst, Henning Sauerland, René Reimann, Wilfried Holzke, Bernd Orlik

Extension and Implementation of a Model-Based Lifetime Monitoring System with Parallel Calculation of Multiple Power Semiconductors 3169
Steffen Menzel, Wilfried Holzke, Michael Hanf, Holger Groke, Bernd Orlik, Nando Kaminski

Smart Charging Strategy for Electric Vehicles using an Optimized Fuzzy Logic System 3179
M. Gholami, M. Mehrasa, R. Razi, K. Hajar, A. Hably, S. Bacha, A. Labonne

Analysis and Discussion of a Concept for an Adjustable Inductance Based on an Impact of an Orthogonal Magnetic Field 3188
Guido Schierle, Michael Meissner, Klaus F. Hoffmann

A Field Programmable and Dynamic Configurable Power Electronic Converter Concept 3198
Bjarte Hoff

DAB Converter Discrete ADRC Control into Real-Time CHIL Simulation of a MVDC/LVDC Power Grid 3206
Alessio Clerici, Riccardo Chiumeo, Diego Raggini, Alessandro Veroni

SNNFT: Sequential Neural Network-Fuzzy Thermal Early Warning System for Lithium-Ion Batteries 3215
   Marui Li, Chaoyu Dong, Yunfei Mu, Qian Xiao, Jingming Cao, Hongjie Jia

Fine-Grained Dynamics Representation and Stability Analysis for MMC-Based Hybrid AC/DC Power Systems 3225
   Jingming Cao, Chaoyu Dong, Qian Xiao, Marui Li, Xiaodan Yu, Hongjie Jia

Adaptive Pontryagin's Minimum Principle-Inspired Supervised-Learning-Based Energy Management for Hybrid Trains Powered by Fuel Cells and Batteries 3235
   Hujun Peng, Feifei Li, Zhu Chen, Kai Deng, Sebina Jeschke, Kay Hameyer

A Case Study of Pole-Phase Changing Induction Machine Performance 3246
   Konstantina Bitsi, Sjoerd G. Bosga

New Topology of Superconducting Fault Current Limiter with Bypass Resistor 3254
   D. Baimel, Eli Barbi, S. Bronstein, N. Baimel, A. Kuperman

A Pre- And Discharge Unit for Capacitive DC-Links Based on a Dual-Switch Bidirectional Flyback Converter 3262
   Madlen Hoffmann, Martin März

Control and Integration of a Multiphase Brushless Wounded Synchronous Motor Drive 3272
   Remi Perrin, Guilherme Bueno-Mariani

A Way Forward to Achieve Interoperability in Multi-Vendor HVDC Systems 3282
   Adil Abdalrahman, Ying-Jiang Häfner, Philippe Maibach, Christoph Haederli

Model Predicitve Position Control of Electrical Drives on an Industrial PC 3292
   Fabian Karau, Michael Leuer

Bidirectional Active EMC Filter for Industrial Power Converters 3301
   Bernhard Wunsch, Stanislav Skibin, Ville Forsstrom

A General Method to Measure Parasitic Capacitance of Transformer using Guarding Technique 3309
   Shaokang Luan, Stig Munk-Nielsen, Bruce Wakelin, Magnus Hortans, Jan Schupp, Hongbo Zhao

Inductance Analysis of Electric Machines by Classical and Numerical Methods 3318
   J. J. Germishuizen, T. J. E. Miller

Dynamic Wireless Power Transfer DWPT Time Domain Model: Xyz Position and Speed Coupling Effect 3327
   Iosu Aizpuru, Eneko Agirrezabala, Mikel Mazuela, Unai Iraola, Estanis Oyarbide, Carlos Bernal

Dynamic Average Small Signal Model of the SAB Converter 3336
   Alexis A. Gómez, Alberto Rodríguez, Marta M. Hernando, Diego G. Lamar, Javier Sebastián, Ibán Ayarzaguena, Jose Manuel Bermejo, Igor Larrazabal, David Ortega, Francisco Vázquez

Algorithm for Optimal Selection of Drive Motor Transmission Combination 3344
   Santiago Ramos Garces, Dries Jacques, Stijn Derammelaere, Simon Houwen, Nick Van Oosterwyck, Bart Vanwalleghem

Evaluation of Drain-Source Voltage in Switch Transient Time Intervals as Gate Oxide Degradation Precursor of SiC Power MOSFETs 3353
   Javad Naghibi, Sadegh Mohsenzade, Kamyar Mehran, Martin P. Foster

Active Output LLC Converter Topology 3362
Hannes Börngen, Eyke Liegmann, Sriram Jagannath, Ralph Kennel

Short Circuit Type II and III Behavior of 1.2 kV Power SiC-MOSFETs................................. 3373
Xing Liu, Xupeng Li, Thomas Basler

Analog MPPT Comparison for Interplanetary Small Satellites Missions 3382
C. Torres, A. Garrigós, J. M. Blanes, P. Casado, D. Marroquí, C. Orts

Feasibility Assessment of Variable-Speed Generator Set Concepts with Focus on Rating of Power
Electronic Equipment 3391
Hendrik Fehr, Albrecht Gensior, Andreas Möckel, Frank Atzler, Tilo Roß, Carsten Reincke-Collon

Bus Voltage Regulation using Sequentially Switched ZVZCS Converters for Spacecraft Power
Systems.. 3401
A. Garrigós, C. Orts, D. Marroquí, J. M. Blanes, C. Torres, P. Casado

A Standardized and Modular Power Electronics Platform for Academic Research on Advanced
Grid-Connected Converter Control and Microgrids.. 3411
Frank S. R., Schulz D., Stefanski L., Schwendemann R., Hiller M.

Gate Input Capacitance Characterization for Power MOSFETs using Turn-On and Turn-Off
Switching Waveforms 3420
Yota Nishitani, Michiko Inoue, Takashi Sato, Michihiro Shintani

AC Battery: Modular Layout with Cell-Level Degradation Control 3429
Claudio Burgos-Mellado, Marcos Orchard, Diego Muñoz-Carpintero, Tomislav Dragicevic, Lorenzo Reyes-Chamorro, Jacqueline Llanos

Analysis of Test Methods for Measurement of Leakage and Magnetising Inductances in Integrated
Transformers 3440
Sajad A. Ansari, Jonathan N. Davidson, Martin P. Foster, David A. Stone

A Topology-Morphing Series Resonant Converter for Photovoltaic Module Applications........... 3450
Grigorios Sergentanis, Liliana De Lillo, Lee Empringham, C. Mark Johnson

A Novel Parameter for the Evaluation of Protective Circuits for IGBT Explosion Protection in
Submodules of MMC 3460
Christoph Junghans, Hans-Guenter Eckel

Sub-Modules Switching Algorithms for Dual Active Bridge Modular Multilevel Converters to
Optimize Capacitor Voltage Deviation Versus Power Efficiency.............................. 3470
Peizhou Xia, Chuantong Hao, Stephen Finney, Michael Merlin

Systematic Adaptive Robust State Feedback Control for Active Front-End Rectifiers 3480
Aidar Zhetessov, Giri Venkataramanan

An Optimized Compensation Strategy of Direct Matrix Converter-Fed PMSM Drives with Field
Weakening Under Unbalanced Supply Conditions 3491
Jun Xie, Dustin Henneberg, Martin Suberski, Manuel Kusebauch, Uwe Rädel, Jürgen Petzoldt

Double Inverter Concept for High-Speed Drives Without Motor Filters 3501
Henning Kasten, Stephan Beineke, Matthias Bachmann

A Universal Single Stage Current-Fed Bidirectional Converter with Both AC and DC Input Power Source Compatibility ... 3511
Manish Kumar, Sumit Pramanick, Bijaya Ketan Panigrahi

Optimization of Electric Vehicle Charge Scheduling with Consideration of Battery Degradation ... 3518
Raka Jovanovic, Sertac Bayhan, Islam Safak Bayram

Onboard ESU Sizing and Dynamic IPT Charging Scenarios for a Tramway Application ... 3529
Endika Bilbao Muruaga, Irma Villar, Florian Legay, Pierre Prenleloup, Jean-François Reynaud

Investigations on the Active Reduction of Common Mode Noise with Opposing Noise Sources ... 3536
Philipp Marx, Felix Seybold, Philipp Ziegler, David Hirning, Jörg Roth-Stielow

Knowledge Based Grey Box Modeling of Inaccessible Circuits for System EMC-Simulation in Time Domain ... 3545
Jan-Philipp Roche, Jens Friebe, Oliver Niggemann

Novel Quasi-Direct Rotor Position Estimator for Permanent Magnet Synchronous Machines Based on the Back-Electromotive Force using Current Oversampling ... 3555
Georg Lindemann, Viktor Willich, Axel Mertens

Design Considerations for Fast On-State Voltage Measurement Circuits ... 3565
Mathias C. J. Weiser, Manuel Rueß, Ingmar Kallfass

Analytical, FEM and Experimental Study of the Influence of the Airgap Size in Different Types of Ferrite Cores ... 3574
Asier Arruti, Francisco Jose Perez-Cebolla, Jon Anzola, Iosu Aizpuru, Mikel Mazuela

Design Method of a High Frequency GaN-Based Half-Bridge with Bottom-Side Cooled Transistors using Multi-PCB Assembly ... 3582
Loris Pace, Florian Chevalier, Thierry Duquesne, Nadir Idir

A 30 kW Dynamic Wireless Inductive Charging System for EVs ... 3590
Zariff Meira Gomes, José Renes Pinheiro, Gilney Damm, Karim Kadem, Hassan Moussa

Dynamic Control of the Switching Behavior of SiC MOSFETs in Converter Operation ... 3599
Jochen Henn, Laurids Schmitz, Rik W. De Doncker

A Series Resonant Balancing Converter for Bipolar DC Grids on Ships ... 3607
Sachin Yadav, Zian Qin, Pavol Bauer

A V2G-Enabled Seven-Level Buck PFC Rectifier for EV Charging Application ... 3615
Anekant Jain, Ritika Agarwal, Krishna Kumar Gupta, Sanjay K. Jain

Experimental Demonstration of a 2.2kW Active-Clamp Converter for High-Current Wide-Voltage-Transfer Ratio Applications ... 3625
Philipp Rehlaender, Bastian Korthauer, Frank Schafmeister, Joachim Böcker

A Simplified Model for the Battery Ageing Potential Under Highly Rippled Load ... 3636
Tomáš Kacetl, Jan Kacetl, Nima Tashakor, Stefan Goetz

System Modeling and Design of a Hybrid Renewable Energy System for a Cable Network Head-End Station in Rural Area ... 3646
Tobias Schillinger, Thomas Schuhmann, Martin Eckart

Comparison of System-Level Availability in Industrial Grids 3655
G. Emmers, J. Driesen

Ageing Mitigation and Loss Control in Reconfigurable Batteries in Series-Level Setups 3665
Tomáš Kacetl, Jan Kacetl, Nima Tashakor, Stefan Goetz

Characterization of Conventional and Advanced Current Measurement Techniques Suitable for
WBG Semiconductor Devices 3676
Severin Klever, André Thönnessen, Rik W. De Doncker

Zero-Sequence Voltage Reduces DC-Link Capacitor Demand in Cascaded H-Bridge Converters for
Large-Scale Electrolyzers by 40% 3686
Roland Unruh, Frank Schafmeister, Joachim Böcker

Thermal Behavior Impact on the Electric Motor Shape Multi-Objective Optimization 3696
Aissam Riad Meddour, Anthony Babin, Nassim Rizoug, Christopher Vagg, Richard Burke,
Laid Degaa

Modelling Approaches of Power Systems Considering Grid-Connected Converters and Renewable
Generation Dynamics 3704
Jaume Girona-Badia, Vinícius Albernaz Lacerda, Eduardo Prieto-Araujo, Oriol Gomis-
Bellmunt, Stephan Kusche, Florian Pöschke, Horst Schulte

Efficiency and Lifetime Analysis of Several Airborne Wind Energy Electrical Drive Concepts 3711
Bakr Bagaber, Daniel Heide, Bernd Ponick, Axel Mertens

Design and Performance Analysis of Single-Phase Axial Flux Permanent Magnet Motor for
Coaxial Cascade 3722
Chu Wang, Xiaowei Hu, Xiaoya Wang, Weiwei Geng, Qiang Li, Jingning Hou

Comparison of Pulse Current Capability of Different Switches for Modular Multilevel Converter-
Based Arbitrary Wave Shape Generator Used for Dielectric Testing of High Voltage Grid Assets 3729
Dhanashree Ashok Ganeshpure, Ajeeth Phrassanna Soundararajan, Thiago Batista Soeiro,
Mohamad Ghaffarian Niasar, Peter Vaessen, Pavol Bauer

Accurate Modeling of IGBT-Based Converters in PLECS 3740
Anne Von Hoegen, Philipp Tillmann, Tetsuya Kojima, Rik W. De Doncker

Novel Analytical Method for Estimating the Junction-To-Top Thermal Resistance of Power
MOSFETs 3750
José Miguel Sanz-Alcaine, Francisco Jose Perez-Cebolla, Carlos Bernal-Ruiz, Asier Arruti,
Iosu Aizpuru

DC-Side Impedance for Handling Interoperability of Multi-Vendor Multi-Terminal HVDC
Systems 3757
Ashkan Nami, Adil Abdalrahman, Ying-Jiang Häfner, Malaya Kumar Sahu, Khirod Kumar
Nayak

Utilizing the Electroluminescence of SiC MOSFETs as Degradation Sensitive Optical Parameter 3766
Lukas A. Ruppert, Michael Laumen, Rik W. De Doncker

Characterization of GaN-On-AlN/SiC Transistors Towards Monolithic Integrability 3775
Nick Wieczorek, Xiaomeng Geng, Carsten Kuring, Oliver Hilt, Frank Brunner, Mihaela Wolf,
Joachim Würfl, Sibylle Dieckerhoff

Optimal Frequency for Dynamic Wireless Power Transfer 3786
Mincui Liang, Khalil El Khamlichi Drissi, Christophe Pasquier

A Wide-Input-Voltage-Range 50W Series-Capacitor Buck Converter with Ancillary Voltage Bus for Fast Transient Response in 48V PoL Applications...3796
Nameer Khan, James Xu, Gerard Villar Piqué, John Pigott, Henk Jan Bergveld, Alaa El Sherif, Olivier Trescases

Four-Level Boost Inverter Based on ANPC Topology with Switched-Capacitor Branch............................3804
Robert Stala, Adam Penczek, Stanislaw Piróg, Aleksander Skala, Andrzej Mondzik, Zbigniew Waradzyn, Krishna Kumar Gupta, Pallavee Bhatnagar, Sanjay K. Jain, Kasinath Jena

Comparative Evaluation of Partially-Rated Energy Storage Integration Topologies for High Voltage Modular Multilevel Converters...3813
Zoe Blatsi, Sebastian Neira, Stephen Finney, Michael M. C. Merlin

Influence of Current Collapse Due to V_{ds} Bias Effect on GaN-HEMTs I_d-V_{ds} Characteristics in Saturation Region...3822
Xuyang Lu, Arnaud Videt, Ke Li, Soroush Faramehr, Petar Igic, Nadir Idir

Deep-Learning Fault Detection and Classification on a UAV Propulsion System 3831
Pierre-Yves Brulin, Fouad Khenfri, Nassim Rizoug

A Compact Solid State Transformer for Replacing Conventional Medium Power Transformer in Weight-Critical Applications...3838
Leon Fauth, Felix Willer, Jens Friebe

Comparative Study of Single-Phase and Three-Phase DAB for EV Charging Application..........................3846
Nicola Blasuttigh, Hamzeh Beiranvand, Thiago Pereira, Marco Liserre

Dynamic Load Emulation for Automotive Power IC Robustness Validation 3855
Alexander Ulbing, Daniel Kostynski, Markus Sievers

DAB Frequency Decoupling Control with Current Minimization 3862
Simon Uicich, Jean-Yves Gauthier, Xuefang Lin-Shi, Bruno Allard, Arnaud Plat

Design and Performance Analysis of a Modified Proportional Multi-Resonant (PMR) Controller for Three-Phase Voltage-Source Inverters 3871
Ahmad Ali Nazeri, Mahmoud Saeidi, Peter Zacharias

Proposition and Comparison of Several Solutions for High Induced Voltage Across Inactive Transmitting Coils in a Series-Series Compensation DIPT System 3883
Wassim Kabbara, Tanguy Phulpin, Mohamed Bensetti, Antoine Caillierez, Serge Loudot, Daniel Sadarnac

Modeling and Measuring the Bearing Capacitance of Radially Loaded Bearings 3893
Stefan Quabeck, Daniel C. Rodriguez, Rik W. De Doncker

Comprehensive Control of Matrix Converters in On-Board Electric Drive Applications............................3903
Galina Mirzaeva

Power System Simulation Tool for Quick Benchmarking of Innovative MVDC Grids in E-Mobility Applications..3910
Daniel Siemaszko, Philippe Noisette

An Artificial Intelligence Pipeline for Critical Equipment Thermal Conditioning System Design 3920
Raik Orbay, Athanasios Tzanakis, Inko Marcaide, Jonas Löfgren, Torbjörn Thiringer, Thomas Bernichon

Aspects of Stability Issues of HVAC/HVDC Coupled Grids.. 3928
  Gianni Bakhos, Kosei Shinoda, Juan-Carlos Gonzalez-Torres, Abdelkrim Benchaib, Luigi Vanfretti, Seddik Bacha

Measurement of Coss-V Characteristic of the 1.7kV/900A SiC Power Module and Estimation of the Channel Current... 3938
  Jacek Rabkowski, Fernando Gonzalez-Hernando, Mariusz Zdanowski, Irma Villar, Uxue Larrañaga

In-Slot Cooling of Electrical Machines using Traditional Techniques and Additive Manufacturing 3947
  Ahmed Hembel, Gokhan Cakal, Bulent Sarlioglu

Comparison of High-Power 2-Level and 3-Level Converters in Terms of Power Density, Costs and Performance.. 3957
  Ludwig Schlegel, Wilfried Hofmann

Autonomous Characterization of Lithium-Ion Battery Model Parameters Utilizing a Mathematical Optimization Methodology 3966
  Hamzeh Beiranvand, Helge Krüger, Sandra Hansen, Marco Liserre, Christian Werlig, Andreas Würsig

SOC Governed Algorithm for an EV Cascaded H-Bridge Connected to a DC Charger 3975
  Giulia Tresca, Andrea Formentini, Filippo Gemma, Federico Lusardi, Riccardo Leuzzi, Pericle Zanchetta

Shaping the Transition from Si-Based Power Devices to SiC MOSFETs and GaN HEMTs 3984
  Gerald Deboy

Reinventing Batteries Through Nanotechnology 3986
  Yi Cui

Advancing GaN Power ICs: Efficiency, Reliability & Autonomy.. 3987
  Dan Kinzer

Electrification Strategy of Volkswagen Group.. 3989
  Alexander Krick

Make it Fly — the Future of Sustainable Aviation.. 3991
  Tanja Neuland

The Instrumental but Extremely Challenging Role of Hydrogen Towards a Decarbonized Society 3992
  Stefan Linder

Short Circuit Behavior of Dual Three-Phase Permanent Magnet Synchronous Motors with Different Mutual Inductance in Electric Propulsion Application 3993
  Yinghui Yang, Georg Möhlenkamp

Hybrid Silicon-SiC Inverter – Combining the Best of Both Worlds 4003
  Hans-Günter Eckel, Felix Kayser, Pham Ha Trieu To

Robustness of SiC Trench MOSFETs 4004
  Christian Felgemacher

3D Predictive Fatigue Modeling of Power Modules 4005
  Ben Samples, Brandon Passmore

Heterogeneous Integration of Power Conversion using Power Supply on Chip and Power Supply in Package...... 4006
Cian Ó Mathúna, Seamus O'Driscoll

Driving Innovations for Power Electronics with Integratable and Sustainable Magnetics...... 4008
Matt Wilkowski

Impact of Package Technology on the Switching Behavior of High-Voltage GaN FETs...... 4011
Sebastian Klötzer

Impact of Power Electronics on Battery Operation 4012
Dirk Uwe Sauer

Trends in Power Electronics and Batteries for Electrified Vehicle Infrastructure...... 4013
Torsten Leifert

Impact of High Frequency Current Pulses on Battery Ageing 4014
Julia Kowal

Aircraft Electrification – System-Level Potentials for Aviation Decarbonization 4015
Kathrin Ebner, Antoine Habersetzer, Arne Seitz

About Power Electronics Challenges in Aviation 4016
Marco Bohllaender

Development of Electric Motors for Aircraft Applications...... 4017
Simon Wolfstädter

Powertrain Trends in Electric Trucks...... 4018
Luciana C. Afonso

Modulation Strategy Impact of BEV Inverters on the Voltage Ripple and the High-Voltage Traction System Stability 4019
Cornelius Rettner

Zero Emission Trucks & Bodies 4020
Martin Glaser

Integrating Offshore Wind & Hydrogen - An Operator's View 4021
Florian Gremme

Status Quo and Future Prospects of Power Electronic Solutions for Electrolysis Plants 4022
Sven Schumann

Modular Power Supply System for Large Scale Water Electrolyzers...... 4023
Ralf Juchem, Klaus Rigbers

Properties of a Lithium-Ion Battery as a Partner of Power Electronics...... 4025
Alexander Blömeke, Katharina Lilith Quade, Dominik Jöst, Weihan Li, Florian Ringbeck, Dirk Uwe Sauer

Author Index

A Theoretical Comparison of Different Virtual Synchronous Generator Implementations on Inverters

Patrick Körner, Andrea Reindl, Hans Meier, Michael Niemetz
Faculty of Electrical Engineering and Information Technology
Ostbayerische Technische Hochschule Regensburg
Seybothstraße 2, 93053 Regensburg, Germany
Email: patrick1.koerner@st.oth-regensburg.de, andreareindl@ieee.org,
{hans.meier, michael.niemetz}@oth-regensburg.de

Keywords

≪Virtual Synchronous Generator (VSG)≫, ≪Virtual impedance≫, ≪Distributed Generation≫, ≪Voltage Source Inverter (VSI)≫, ≪Grid-connected inverter≫

Abstract

The goal to overcome the global climate crisis leads to a rising demand for the usage of Renewable Energy Sources (RES). Decentralized control strategies are needed to allow the integration of RES into the grid. The Virtual Synchronous Generator (VSG) is proposed as a method to add virtual inertia to the grid by emulating the rotating mass of a Synchronous Generator (SG) on the control algorithm of an inverter. This paper presents the VSG control structure as well as the mathematical description in a unified form. Due to the fact that classical droop control can be seen as a special form of the VSG, their correlation is highlighted by evaluating the steady state output characteristics of the inverter. Furthermore, a theoretical comparison between different VSG topologies, including the VISMA-Method 2 and the synchronverter, is given. In order to achieve better voltage stability, principles to add virtual impedance to the inverter's output are described.

Introduction

The rising penetration of renewable energy sources in the grid leads to the need for decentralized and communication-less control strategies. Furthermore, microgrids consisting of many Distributed Generators (DGs) are emerging. Power electronic devices, called inverters, are connecting these DGs to the grid. Inverters are usually supplied from a DC-line or a DC-stage [1]. The two categories, Current Source Inverter (CSI) and Voltage Source Inverter (VSI), can be classified into grid-forming, grid-feeding and grid-supporting converters. CSIs hardly have the ability to regulate the grid voltage. Due to this, it can't operate in islanded mode without having a grid-forming VSI in parallel. [2]

Large synchronous generators in centralized power plants can operate as grid-forming voltage sources that have good load sharing characteristics and provide large inertia to the grid [3,4]. On the other hand, the increasing number of inverters that do not have rotating masses and therefore stored kinetic energy, decreases the overall inertia, which results in a decreasing reliability and supply quality of the grid [5–8].

Droop control is an established method to allow a parallel operation of DGs. This control technique emulates the steady-state droop characteristics of the governor and the Automatic Voltage Regulator (AVR) of a SG, allowing participation in frequency and voltage regulation [4]. This primary control method has a lack of inertia [7,8]. As for, this results in a higher Rate of Change of Frequency (RoCoF) during transient load peaks or grid faults, than with a SG.

The VSG is a proposed solution for this problem. By implementing the droop characteristics (P-ω-droop and Q-V-droop) and the small-signal model of a SG on the control algorithm of an inverter, virtual inertia can be added to the grid. The key is the emulation of the swing equation of a SG and a suitable Energy Storage System (ESS).

It has to be taken into account that there are current-controlled and voltage-controlled VSGs. The Virtual Synchronous Machine VISMA-Method 1 [9] is a current-controlled VSG, while the VISMA-Method 2 [10], synchronverters [11–13] and VSGs only based on the electromechanical swing equation [14–19] are voltage-controlled ones. Since the CSI's control strategies rely on a Phase-Locked Loop (PLL) to obtain the angular frequency from the grid, it can not operate in islanded mode [20]. Current research mostly focuses on enhanced control strategies for VSG implementations on VSIs, to achieve better transient characteristics [16]. In the reference [17] a new damping strategy for active power oscillation damping was developed and verified by experimental measurements.

This paper focuses on VSGs for VSIs in order to allow an islanded operation and to draw a comparison to classical droop control [21]. Therefore a brief description of the control algorithm is given in the next Section by using a unified control structure. The following Section presents a method used in high order models, to add virtual impedance to the system, while the last Section closes with a comparison between the VISMA-Method 2 and the synchronverter.

Unified structure of a Virtual Synchronous Generator

The basic control structure of a VSG, consisting of a virtual inertia emulating block and a virtual AVR, is shown in Fig. 1. The three-phase output of the inverter is filtered through a LC-filter (L_f, C_f). A cur-

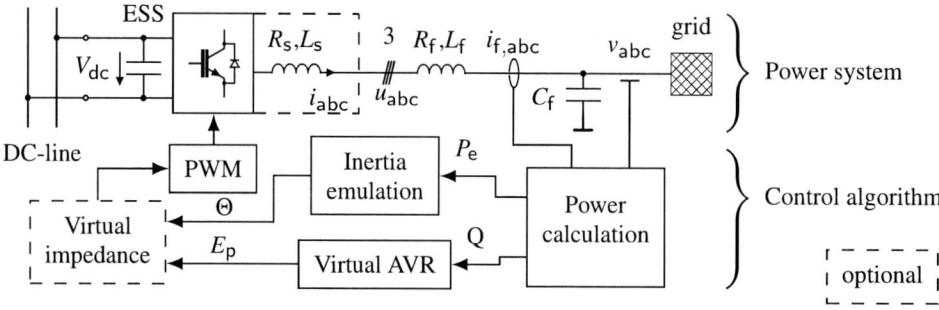

Fig. 1: Basic control structure of the VSG algorithm on a VSI for inertia emulation in a distributed generation system that is connected to the grid [22, 23] and adds virtual impedance (R_s, L_s) to the output. Combining the control algorithm with an suitable ESS adds virtual inertia to the grid by emulating the mechanical swing equation of a SG in the inertia emulating block.

rent and voltage measurement is used to obtain the corresponding output characteristics. The literature describes various approaches for applying the control structure in different frame representations. Either the control is based on the natural *abc*-frame [24], the stationary αβ-frame [25] or the synchronous rotating *dq*-frame [1, 17, 19, 26, 27]. This implies the usage of Clarke- and/or Park-transformation. Reference [28] describes a method called "double decoupled synchronous reference frame", which can be used to control a single-phase or an unbalanced three-phase VSG by using a combination of all three frames [23].

Active power control

The active power control on a VSG is done by emulating a virtual governor and the mechanical swing equation, the latter is the key to add virtual inertia to the system. The obtained active power P_e is fed into the VSG model. As an output, the angular position $\Theta = \int \omega dt$ of the rotor field to the reference axis is calculated. Whereas ω is the angular frequency. [9]

The swing equation

$$J \cdot \ddot{\Theta} = T_m - T_e - D_m \cdot \dot{\Theta} \tag{1}$$

describes the mechanical behavior of the rotor [29]. Where $\ddot{\Theta}$ is the angular acceleration, J is the moment of inertia, T_m is the mechanical torque of the prime mover, T_e is the electrical torque and D_m is the damping coefficient due to mechanical friction [21]. All VSG implementations (except VISMA) use a modified version of (1) since it is more accurate to model the damping effect with the term $D_p \cdot (\dot{\Theta} - \dot{\Theta}_0)$, which represents the much larger electrical damping effects [21]. This damping is caused by induction currents in the amortisseur windings due to a relative motion ($\Delta\dot{\Theta} = \dot{\Theta} - \dot{\Theta}_0$) between the rotor and the stator magnetic field [21]. As follows, the swing equation should be rewritten as:

$$J \cdot \ddot{\Theta} = T_m - T_e - D_p \cdot (\dot{\Theta} - \dot{\Theta}_0) \tag{2}$$

Since the relative difference between $\dot{\Theta}$ and $\dot{\Theta}_0$ is negligible, (2) can be multiplied with the nominal angular frequency $\dot{\Theta}_0 = \omega_0$ on both sides [15, 21], which leads to:

$$J \cdot \omega_0 \cdot \frac{d\omega}{dt} = P_m - P_e - D \cdot (\omega - \omega_0) \tag{3}$$

where P_m and P_e are the mechanical and electrical powers and $J \cdot \omega_0$ is called the inertia constant of the synchronous generator.

The virtual governor acts like a deviating secondary frequency regulator and can be written as [1]:

$$P_m = P_0 - k_\omega \cdot (\omega - \omega_0) \tag{4}$$

P_0 is the nominal active power and k_ω represents the governor droop factor. This factor usually is a proportional controller or a PI controller and can be realized to only act, when the deviation $(\omega - \omega_0)$ exceeds certain values [21]. The mechanical governor of a SG usually has a large delay, therefore an exact mechanical analogon would include a first order lag unit with parameter T_d [8].

When using the Laplace transformation on (3) and (4), the result can be combined and simplified to

$$P_0 - P_e - k_p \cdot (\omega - \omega_0) = J \cdot \omega_0 \cdot s \cdot \omega \tag{5}$$

where $k_p = D_p + k_\omega$ is used as droop factor. Furthermore the equation for the angular frequency can be rewritten as:

$$\omega = \frac{1}{J \cdot \omega_0 \cdot s + k_p} \cdot (P_0 - P_e + k_p \cdot \omega_0) \tag{6}$$

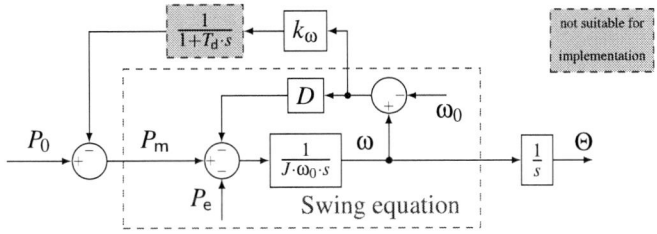

Fig. 2: Unification of the active power control block for the VSG, consisting of the swing equation with a damping term D, the governor [21] and the modeling of the mechanical governor delay T_d [8]. The governor is realized as a proportional droop controller with droop factor k_ω. As an output, the angular position Θ is obtained.

The analysis of the system's state space eigenvalues in the references [7, 8] proves, that the in Fig. 2 shown model with and without the optional governor delay is stable. It can be observed, that increasing values for the inertia constant $J \cdot \omega_0$ leads to a convergence of the eigenvalues onto the origin. Thus, $J \cdot \omega_0$ affects the oscillation frequency of the active power during transient events while its damping effect can be neglected. As for an increasing T_d, the eigenvalue analysis and the simulations results from [7] show, that the system gets more oscillatory and has higher RoCoF. Therefore, the mechanical governor delay should not be emulated in the VSG. [7, 8]

Virtual inertia can only be emulated, when the ESS (cf. Fig. 1) is able to supply the VSI with the needed electrical energy [3, 8].

Reactive power control

The virtual automatic voltage regulator is realized in the reactive power control (Fig. 3), same as with a SG [25]. An internal voltage reference V_{ref} is controlled by drooping the reactive power deviation between the measured reactive power Q and the setpoint Q_0 with the factor D_q [14]. Where V_0 is the voltage setpoint. The low pass filter represents the exciter system whereas for Q-V-droop, T_f should be set as the time constant of the LC power filter (Fig. 1) in order to achieve better reactive power sharing among all VSIs [26]. E_p is the amplitude of the back Electromotive Force (EMF), which is used to calculate the output voltage u_{abc} of the VSI [25].

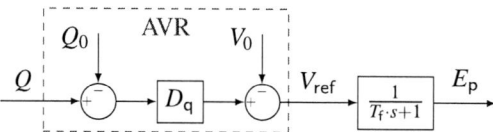

Fig. 3: Reactive power control by implementing the AVR as a proportional Q-V-droop controller with an additional low pass filter as representation for the exciter system [26]. The output E_p is the amplitude of the back EMF.

The AVR can be realized differently on each VSG and therefore it is not possible to present a unified control structure. For further interest, reference [26] investigates different AVRs for VSGs.

Steady state output characteristics

The VSI can take an active role in frequency- and voltage-control, by applying the mentioned drooping techniques. These steady state output characteristics (Fig. 4) allow parallelization of multiple VSGs.

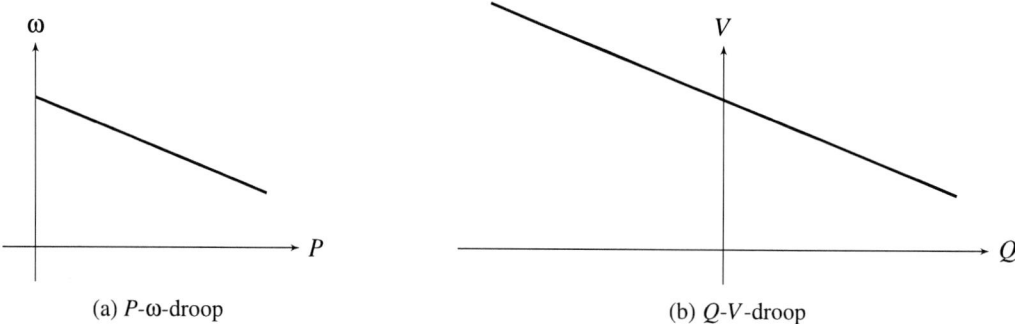

(a) P-ω-droop   (b) Q-V-droop

Fig. 4: Steady state output characteristics of a VSG for active and reactive power [30].

To draw a comparison to classical droop control, the virtual moment of inertia J in (6) can be removed by setting it to 0, which leads to the equation for classical droop control [8]:

$$\omega = \frac{1}{k_p} \cdot (P_0 - P_e) + \omega_0 \tag{7}$$

Due to the missing first order lag unit, which inserts a frequency drooping delay, (7) has worse transient characteristics than (6). This results in smaller values for the RoCoF. As a consequence, existing virtual inertia improves the transient abilities and enhances the frequency support of the inverter [4] as experimental results [8] and simulations [31] show.

Virtual Winding for High Order VSG models

Virtual impedance (R_s, L_s in Fig. 1) is added, because the stator impedance of a SG typically is in the range of $1.5\,\text{p.u.} - 2.0\,\text{p.u.}$, while the output filter impedance of a VSI is significantly lower ($0.05\,\text{p.u.} - 0.2\,\text{p.u.}$) [13, 32, 33]. Furthermore, the LC-filter values are fixed due to the switching frequency of the VSI and it is more energy efficient to rely on changeable virtual resistances, rather than fixed, non-virtual ones [21]. A larger output impedance leads to more robustness against voltage fluctuations [13] and enhances the voltage profile [34]. There are different approaches to add virtual impedance to the output of a VSI. One is referred as virtual winding of the stator and is only used in high order VSG models, that try to mimic the geometry of the stator inductances [26]. For the interested reader, a brief mathematical description of the SG geometry is given in [11, 35].

Assuming a constant excitation current i_f for the rotor field and $\dot{\Theta} \approx \text{const.}$, the induced EMF $e = \begin{bmatrix} e_a & e_b & e_c \end{bmatrix}^T$ can be written as an expression of the mutual inductance M_f between the stator and rotor windings:

$$e = \dot{\Theta} \cdot M_f \cdot i_f \cdot \widetilde{\sin}(\Theta) \approx E_p \cdot \widetilde{\sin}(\Theta) \tag{8}$$

$$\widetilde{\sin}(\Theta) = \begin{bmatrix} \sin(\Theta) & \sin\left(\Theta - \frac{2\pi}{3}\right) & \sin\left(\Theta + \frac{2\pi}{3}\right) \end{bmatrix}^T \tag{9}$$

The expression $\widetilde{\sin}(\Theta)$ is given due to the geometry of the stator windings. [11] Following, the output voltage vector $u_{abc} = \begin{bmatrix} u_a & u_b & u_c \end{bmatrix}^T$ can be defined via a voltage drop due to the output current $i_{abc} = \begin{bmatrix} i_a & i_b & i_c \end{bmatrix}^T$:

$$u_{abc} = e - R_s \cdot i_{abc} - L_s \cdot \frac{di_{abc}}{dt} \tag{10}$$

When implementing (8) and (10) in the control algorithm, a virtual impedance can be added to the VSI's output.

Experimental results, that prove the concept of virtual impedance implementation, can be obtained from [13]. Reference [32] compares an inverter, where virtual impedance is implemented, to one, where none is added. The overall benefits are verified via simulation and measurements [32, 36].

Different Virtual Synchronous Generator Implementations

In this Section, a discussion of different VSG topologies, including the VISMA-Method 2 and the synchronverter, is given. VSG implementations only based on the mechanical swing equation are not shown. Therefore the authors refer to the comprehensive review in reference [20] and states that the control structure in Fig. 2 is a unified one.

VISMA

The voltage to current model VISMA-Method 1 was invented by Beck and Hesse in 2007 and is the most accurate and highest order model of a SG [37]. Their improved version called VISMA-Method 2 is a current to voltage model which is more suitable for implementation, due to the more common Pulse-Width Modulation (PWM) inverters. Since it is considered as a 7th order model, there are only a few simplifications that are made in comparison to a real SG. These include the neglection of the winding coupling, which would result in additional mutual inductances, as well as the effect of damping in the windings. [20]

Fig. 5 shows the implementation of VISMA-Method 2, which includes the virtual winding, the virtual governor and the swing equation. The corresponding literature does not state what exact expression is used for $f(s)$ in the damping term of the swing equation [21]. Furthermore, Fig. 5 does not include the virtual AVR, which is used to calculate E_p for the back EMF (Fig. 3).

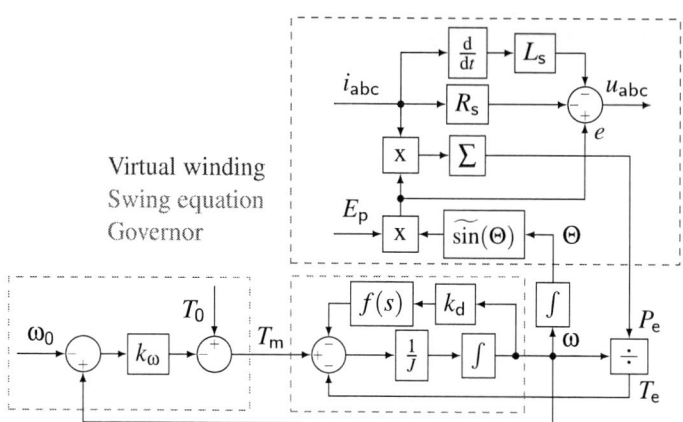

Fig. 5: Control structure of the VISMA-Method 2 [10], which includes the modeling of the virtual winding and the mechanical swing equation in form of torque, as well as the virtual governor.

The VISMA-Method 2 can operate without further hardware configurations as a grid supporting inverter in grid-connected and islanded mode. However, it has a poor dynamic response time when changing from grid-connected to islanded mode. When a change in grid frequency happens (in a poor grid for example), it shows a large RoCoF response. Advantages of the VISMA are a small frequency deviation and a small RoCoF response during load disturbances. The same applies to the dynamic response during power steps. [20]

Due to VISMA's complexity, a lot of computation power is required to run the algorithm on a VSI. Therefore, the VISMA is hardly used in real power systems. [20]

Synchronverter

The synchronverter is a 2nd order model of the SG and was firstly introduced in 2009 [20, 29]. Both authors presented further enhancements, that increases the virtual impedance and that allows the synchronverter to self-synchronize itself before connecting to the grid [11, 13, 38].

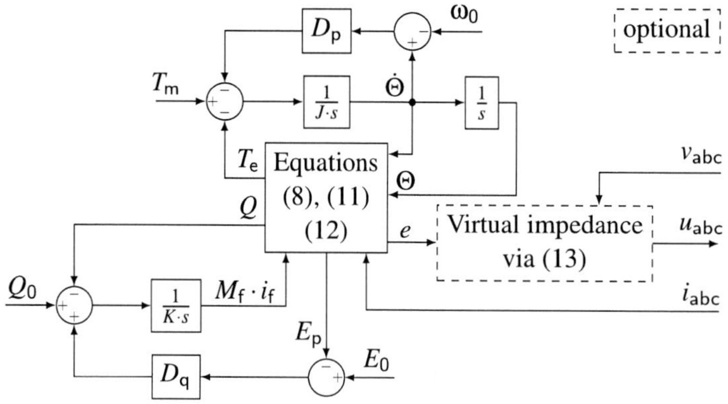

Fig. 6: Control structure of the synchronverter [29], based on the electromagnetic equations, which includes an optional virtual impedance emulating block [36].

Same as the VISMA, the synchronverter realizes the swing equation and therefore the active power control in form of torque. A simplification is made by not emulating the virtual winding and therefore the VSI only has an output impedance that is determined by the output filter (R_f, L_f, C_f in Fig. 1). [20] Synchronverter's control structure is shown in Fig. 6, whereas the equations used in the model are the following [11, 13]:

$$T_e = M_f \cdot i_f \cdot \langle i_{abc}, \widetilde{\sin(\Theta)} \rangle \tag{11}$$

$$Q = \dot{\Theta} \cdot M_f \cdot i_f \cdot \langle i_{abc}, \widetilde{\sin(\Theta - \frac{\pi}{2})} \rangle \tag{12}$$

$$u_{abc} = \frac{(n-1) \cdot v_{abc} + e}{n} \tag{13}$$

$$\Leftrightarrow e = v_{abc} + n \cdot (u_{abc} - v_{abc}) = v_{abc} + n \cdot \Delta u \tag{14}$$

Although the virtual winding technique is missing, reference [13] stated a simple method to virtually increase the output impedance of the LC-filter with the factor n by implementing (13) in the algorithm. This imposes u_{abc} as the output voltage of the inverter, although v_{abc} (cf. Fig. 1) is used in the calculation. It follows that $R_s = (n-1) \cdot R_f$ and $L_s = (n-1) \cdot L_f$. The resulting one phase model of the synchronverter (Fig. 7) can be obtained from (14). Using this method results in a more robust design and positively influences the current peak values during asymmetric faults [13, 39, 40].

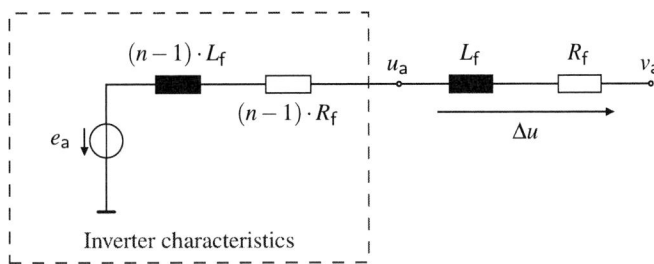

Fig. 7: One phase model of a synchronverter that adds virtual impedance (the ones multiplied by $(n-1)$) to the inverter's output. As a result, the overall output impedance of the inverter is multiplied by the factor n. [13]

Further enhancements for an active power oscillation suppression can be added to the swing equation of the synchronverter in form of an additional virtual damping term [36].

In comparison to the VISMA-Method 2, the implementation of the synchronverter is far simpler to realize and it has better dynamic characteristics when changes from grid-connected to islanded mode occur. On the other hand, it shows a larger frequency deviation response when changes in the grid frequency happen and it has worse dynamic properties during power steps or load disturbances than the VISMA. [20]

Synchronverter is a suitable solution for HVDC-transmission, where it can work as a synchronous motor at the sending-end and as a SG at the receiving-end [12].

Conclusion

This paper stated the importance of VSGs for future renewable energy sources dominated grid compositions, including HVDC-transmission and microgrids. Their ability to decrease the RoCoF and damp power oscillations was shown by the mathematical investigation of a unified VSG control structure. The emulation of the swing equation in compound with a ESS adds virtual inertia to the grid. The steady state output characteristics of a VSG were shown in order to draw a comparison to classical droop control. The VISMA-Method 2 and the synchronverter implementations were presented while stating the difference between both of them. For both models, a comparison of their behavior during load and frequency changes and changes from grid-connected to islanded mode was given. Since the synchronverter is one of the most promising solutions and can be implemented with reasonable computation effort, further enhancements were listed and referenced. As one of the most important design enhancements, the method to implement virtual impedance was shown for both the VISMA-Method 2 and the synchronverter. The verification of the obtained theoretical statements was done by referencing corresponding simulation results and experimental measurements from the literature. An investigation of other VSG topologies only based on the swing equation is left for future work.

References

[1] X. Meng, J. Liu, and Z. Liu, "A Generalized Droop Control for Grid-Supporting Inverter Based on Comparison Between Traditional Droop Control and Virtual Synchronous Generator Control," *IEEE Trans. on Power Electronics*, pp. 5416–5438, 2019.

[2] J. Rocabert, A. Luna, F. Blaabjerg, and P. Rodríguez, "Control of Power Converters in AC Microgrids," *IEEE Trans. on Power Electronics*, pp. 4734–4749, 2012.

[3] P. Unruh, M. Nuschke, P. Strauß, and F. Welck, "Overview on Grid-Forming Inverter Control Methods," *Energies*, p. 2589, 2020.

[4] M. Chen, D. Zhou, and F. Blaabjerg, "Impact of Synchronous Generator Replacement with VSG on Power System Stability," in *IEEE 21st Workshop on Control and Modeling for Power Electronics*, 2020, pp. 1–7.

[5] J. Driesen and K. Visscher, "Virtual Synchronous Generators," in *IEEE Power and Energy Society General Meeting - Conversion and Delivery of Electrical Energy in the 21st Century*, 2008, pp. 1–3.

[6] M. H. J. Bollen, Y. Yang, and F. Hassan, "Integration of Distributed Generation in the Power System - A Power Quality Approach," in *13th International Conference on Harmonics and Quality of Power*, 2008, pp. 1–8.

[7] J. Liu, Y. Miura, and T. Ise, "Dynamic Characteristics and Stability Comparisons between Virtual Synchronous Generator and Droop Control in Inverter-Based Distributed Generators," in *International Power Electronics Conference (IPEC-Hiroshima 2014 - ECCE ASIA)*, 2014, pp. 1536–1543.

[8] J. Liu, Y. Miura, and T. Ise, "Comparison of Dynamic Characteristics Between Virtual Synchronous Generator and Droop Control in Inverter-Based Distributed Generators," *IEEE Trans. on Power Electronics*, pp. 3600–3611, 2016.

[9] Y. Chen, R. Hesse, D. Turschner, and H.-P. Beck, "Improving the Grid Power Quality Using Virtual Synchronous Machines," in *International Conference on Power Engineering, Energy and Electrical Drives*, 2011, pp. 1–6.

[10] Y. Chen, R. Hesse, D. Turschner, and H.-P. Beck, "Comparison of methods for implementing virtual synchronous machine on inverters," *Renewable Energy and Power Quality Journal*, pp. 734–739, 2012.

[11] Q.-C. Zhong and G. Weiss, "Synchronverters: Inverters That Mimic Synchronous Generators," *IEEE Trans. on Industrial Electronics*, pp. 1259–1267, 2011.

[12] R. Aouini, B. Marinescu, K. Ben Kilani, and M. Elleuch, "Synchronverter-Based Emulation and Control of HVDC Transmission," *IEEE Trans. on Power Systems*, pp. 278–286, 2016.

[13] V. Natarajan and G. Weiss, "Synchronverters With Better Stability Due to Virtual Inductors, Virtual Capacitors, and Anti-Windup," *IEEE Trans. on Industrial Electronics*, pp. 5994–6004, 2017.

[14] Y. Du, J. M. Guerrero, L. Chang, J. Su, and M. Mao, "Modeling, Analysis, and Design of a Frequency-Droop-Based Virtual Synchronous Generator for Microgrid Applications," in *IEEE ECCE Asia Downunder*, 2013, pp. 643–649.

[15] M. Guan, W. Pan, J. Zhang, Q. Hao, J. Cheng, and X. Zheng, "Synchronous Generator Emulation Control Strategy for Voltage Source Converter (VSC) Stations," *IEEE Trans. on Power Systems*, pp. 3093–3101, 2015.

[16] J. Liu, Y. Miura, H. Bevrani, and T. Ise, "Enhanced Virtual Synchronous Generator Control for Parallel Inverters in Microgrids," *IEEE Trans. on Smart Grid*, pp. 2268–2277, 2017.

[17] M. Chen, D. Zhou, and F. Blaabjerg, "Active Power Oscillation Damping Based on Acceleration Control in Paralleled Virtual Synchronous Generators System," *IEEE Trans. on Power Electronics*, pp. 9501–9510, 2021.

[18] X. Xiong, C. Wu, and F. Blaabjerg, "An Improved Synchronization Stability Method of Virtual Synchronous Generators Based on Frequency Feedforward on Reactive Power Control Loop," *IEEE Trans. on Power Electronics*, pp. 9136–9148, 2021.

[19] X. Xiong, C. Wu, P. Cheng, and F. Blaabjerg, "An Optimal Damping Design of Virtual Synchronous Generators for Transient Stability Enhancement," *IEEE Trans. on Power Electronics*, pp. 11 026–11 030, 2021.

[20] M. Chen, D. Zhou, and F. Blaabjerg, "Modelling, Implementation, and Assessment of Virtual Synchronous Generator in Power Systems," *Journal of Modern Power Systems and Clean Energy*, pp. 399–411, 2020.

[21] X. Meng, Z. Liu, J. Liu, T. Wu, S. Wang, and B. Liu, "Comparison between Virtual Synchronous Generator and Droop Controlled Inverter," in *IEEE 2nd Annual Southern Power Electronics Conference (SPEC)*. Piscataway, NJ: IEEE, 2016, pp. 1–6.

[22] T. Shintai, Y. Miura, and T. Ise, "Oscillation Damping of a Distributed Generator Using a Virtual Synchronous Generator," *IEEE Trans. on Power Delivery*, pp. 668–676, 2014.

[23] J. Liu, M. Yushi, T. Ise, J. Yoshizawa, and K. Watanabe, "Parallel Operation of a Synchronous Generator and a Virtual Synchronous Generator under Unbalanced Loading Condition in Microgrids," in *IEEE 8th International Power Electronics and Motion Control Conference (IPEMC-ECCE Asia)*, 2016, pp. 3741–3748.

[24] L. Xia and L. Hai, "Comparison of Dynamic Power Sharing Characteristics between Virtual Synchronous Generator and Droop Control in Inverter-Based Microgrid," in *IEEE 3rd International Future Energy Electronics Conference and ECCE Asia (IFEEC 2017 - ECCE Asia)*. IEEE, 2017, pp. 1548–1552.

[25] O. Mo, S. D'Arco, and J. A. Suul, "Evaluation of Virtual Synchronous Machines With Dynamic or Quasi-Stationary Machine Models," *IEEE Trans. on Industrial Electronics*, pp. 5952–5962, 2017.

[26] M. Chen, D. Zhou, and F. Blaabjerg, "Voltage Control Impact on Performance of Virtual Synchronous Generator," in *International Power Electronics Conference (IPEC-Hiroshima - ECCE ASIA)*, 2014, pp. 1981–1986.

[27] C. Zhang, Q. Zhong, J. Meng, X. Chen, Q. Huang, S. Chen, and Z. Lv, "An Improved Synchronverter Model and its Dynamic Behaviour Comparison with Synchronous Generator," in *2nd IET Renewable Power Generation Conference (RPG 2013)*, 2013, pp. 1–4.

[28] Y. Hirase, O. Noro, E. Yoshimura, H. Nakagawa, K. Sakimoto, and Y. Shindo, "Virtual Synchronous Generator Control with Double Decoupled Synchronous Reference Frame for Single-Phase Inverter," in *International Power Electronics Conference (IPEC-Hiroshima 2014 - ECCE ASIA)*, 2014, pp. 1552–1559.

[29] Q. Zhong and G. Weiss, "Static Synchronous Generators for Distributed Generation and Renewable Energy," in *IEEE/PES Power Systems Conference and Exposition*, 2009, pp. 1–6.

[30] J. Roldan-Perez, A. Rodriguez-Cabero, and M. Prodanovic, "Harmonic Virtual Impedance Design for Parallel-Connected Grid-Tied Synchronverters," *IEEE Journal of Emerging and Selected Topics in Power Electronics*, pp. 493–503, 2019.

[31] G. Júnior, T. Nascimento, and L. S. Barros, "Comparison of Virtual Synchronous Generator Strategies for Control of Distributed Energy Sources and Power System Stability Improvement," *Simpósio Brasileiro de Sistemas Elétricos - SBSE*, 2020.

[32] R. Rosso, S. Engelken, and M. Liserre, "Robust Stability Analysis of Synchronverters Operating in Parallel," *IEEE Trans. on Power Electronics*, pp. 11 309–11 319, 2019.

[33] L. Zhang, L. Harnefors, and H.-P. Nee, "Power-Synchronization Control of Grid-Connected Voltage-Source Converters," *IEEE Trans. on Power Systems*, pp. 809–820, 2010.

[34] A. Tarrasó, J. I. Candela, J. Rocabert, and P. Rodriguez, "Grid Voltage Harmonic Damping Method for SPC based Power Converters with Multiple Virtual Admittance Control," in *IEEE Energy Conversion Congress and Exposition (ECCE)*, 2017, pp. 64–68.

[35] E. Brown and G. Weiss, "A study of the use of synchronverters for grid stabilization using simulations in SimPower," Master's thesis, Tel Aviv University, 2015.

[36] Z. Shuai, W. Huang, Z. J. Shen, A. Luo, and Z. Tian, "Active Power Oscillation and Suppression Techniques Between Two Parallel Synchronverters During Load Fluctuations," *IEEE Trans. on Power Electronics*, pp. 4127–4142, 2020.

[37] H.-P. Beck and R. Hesse, "Virtual Synchronous Machine," in *9th International Conference on Electrical Power Quality and Utilisation*, 2007, pp. 1–6.

[38] Q.-C. Zhong, P.-L. Nguyen, Z. Ma, and W. Sheng, "Self-Synchronized Synchronverters: Inverters Without a Dedicated Synchronization Unit," *IEEE Trans. on Power Electronics*, pp. 617–630, 2014.

[39] R. Rosso, J. Cassoli, G. Buticchi, S. Engelken, and M. Liserre, "Robust Stability Analysis of LCL Filter Based Synchronverter Under Different Grid Conditions," *IEEE Trans. on Power Electronics*, pp. 5842–5853, 2019.

[40] L. He, Z. Shuai, X. Zhang, X. Liu, Z. Li, and Z. J. Shen, "Transient Characteristics of Synchronverters Subjected to Asymmetric Faults," *IEEE Trans. on Power Delivery*, pp. 1171–1183, 2019.

Linear Flux-Switching Machine Design - A Multiobjective Optimization

Hendrik Marks, Henning Schillingmann, Sridhar Balasubramanian, Markus Henke
INSTITUTE FOR ELECTRICAL MACHINES, TRACTION AND DRIVES (IMAB)
TU Braunschweig
Braunschweig, Germany
Email: h.marks@tu-braunschweig.de
URL: https://www.tu-braunschweig.de/imab

Keywords

≪Linear Machine≫, ≪Flux-Switching Machine≫, ≪Electromagnetic Model≫, ≪Optimization≫, ≪NSGA-II≫

Abstract

In this paper, a linear electric machine is designed for the application of a free-piston engine to generate electrical energy. Due to the high thermal load on the mover, only machine topologies with a passive mover are considered. It is shown that flux-switching machines (FSM) are capable of fulfilling the requirements.

A preceding thermal analysis provides the temperature distribution within the machine. The selection of the magnetic materials is determined based on the results. Subsequently an electromagnetic model for a 12 slot/14 pole FSM is created and optimized. Due to the application, the use of three design targets is appropriate: The efficiency η, the mass of the mover m_M and the permanent magnet (PM) mass m_{PM}. To take all the objectives into account, a multicriterial optimization using the NSGA-II algorithm is performed.

I Introduction

Free-piston engines with a linear generator, as shown in fig. 1, are used to generate electric energy out of combustion. They consist of a combustion chamber (1), a linear generator (2) and a rebound device (3), all linked by a piston (4). The piston is moved periodically back and forth by the force of the combustion and the rebound device. The electric generator converts the mechanical energy into electrical energy. Free-piston engines can be used as a range extender for electric vehicles [1] or as combined heat and power plants for decentralized energy supply [2].

Due to the linear movement, a crankshaft is not needed. This has the advantage of a variable compression and thus a variable fuel like petrol, methane or hydrogen. Another benefit is the lower number of mechanical components, which decrease the mechanical losses.

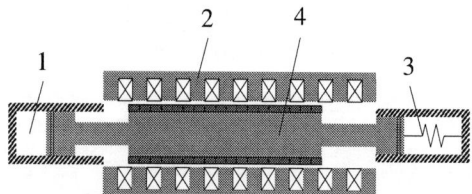

Fig. 1: Free-piston engine components:
1 - combustion chamber, 2 - linear electric generator,
3 - rebound device, 4 - moving piston

A major challenge in operating these engines is the thermal load due to the combustion process. Because of the low thermal resistance between the combustion chamber and the mover of the electrical machine the cooling requirements for the electric machine and in particular its mover are high.

The maximum temperatures are determined by the permissible temperature rise in the winding and the permanent magnets. In addition, due to the higher temperature of the mover, PM with expensive and environmentally harmful additives, such as dysprosium, have to be used. Electric machines with passive movers on the other hand either have no PM or the magnets are located in the stator, which is thermally better connected to the cooling medium. The passive structure of the mover consequently increases its maximum permissible temperature.

Possible linear machine topologies with a passive mover structure that could be used for the free-piston machine are those without any magnets, such as switched or synchronous reluctance machines [3] and topologies with PM placed in the stator, like doubly-salient, flux-reversal or flux-switching machines [4, 5]. A literature study and an initial design study using finite-element analysis (FEA) show that a tubular flux-switching linear machine (FSM) with 12 stator teeth and 14 mover teeth (12/14-FSM) is the most suitable for the application, as it fulfills all force requirements with the lightest mover.

Mechanical requirements

Piston dynamic simulation dictates the mechanical requirements that the linear generator has to meet. A maximum stroke of $s_{max} = 80$ mm is specified by the piston. Thereby, the generator must apply an effective value of the braking force of $F_{rms} = 1.2$ kN and a maximum braking force of $F_{max} = 2$ kN, as shown in Tab. I. The machine is designed for the rms value of the force. The maximum value is obtained by a higher current density for a short time.

Table I: Requirements from piston dynamic simulation

Parameter	Value
Maximum stroke s_{max}	80 mm
Force (rms) of electric generator F_{rms}	1.2 kN
Force (max) of electric generator F_{max}	2 kN
Airgap width δ_g	1 mm

II Investigation of thermal effects

A thermal analysis is carried out for the 12/14-FSM by means of a 2d-FEA. For this purpose, a cooling concept is first developed and then the thermal parameters required for the simulation are determined. The thermal simulation is performed before the optimization to estimate whether the electric machine is sufficiently cooled and also to choose a material for the permanent magnets. Although the subsequent optimization changes the boundary conditions, it is expected to improve the efficiency and thus reduce the power dissipation and the temperature rise.

Water cooling with axial channels is chosen as the cooling concept. For the given geometry, laminar flow results in a heat transfer coefficient from the stator to the coolant of $h = 440 \, \text{W m}^{-2} \text{K}^{-1}$. The coolant inlet temperature is $60 \,°\text{C}$.

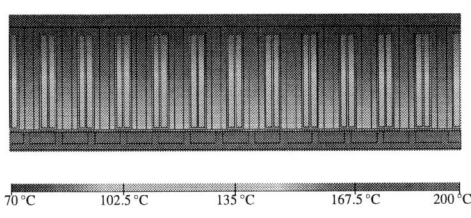

70 °C   102.5 °C   135 °C   167.5 °C   200 °C

Fig. 2: Results of the thermal simulation

The temperature distribution of the electric machine is shown in fig. 2. The maximum temperature can be found in the mover as a heat input from the combustion process results in an assumed boundary condition of $200 \,°\text{C}$ in the mover. The temperature of the stator decreases continuously in the radial direction. Here, the stator iron and the permanent magnets have approximately the same temperature axially. The slot shows a slightly higher temperature due to the ohmic losses.

The maximum temperature of the permanent magnets is $97 \,°\text{C}$. This relatively low temperature means that magnet materials can be used that have a minimum proportion of dysprosium and thus have a higher remanence flux density. VACODYM 247 [6] with coercive field strength of $730 \, \text{kA m}^{-1}$ at $100 \,°\text{C}$ is chosen as PM material. This material is free of dysprosium according to the datasheet. Due to the cylindrical geometry, a 3d magnetic flux is present. Therefore the Somaloy 700 3P 800 MPa from the Höganäs [7] is selected as soft magnetic material to reduce the eddy-current losses.

In summary, the thermal analysis shows that the cooling concept with a water cooling for the free-piston engine is sufficient. The low temperatures of the permanent magnets mean that the use of dysprosium can be avoided. It is expected that machines with modified geometry will be sufficiently cooled too, since there is a high difference between prevailing temperatures and permissible temperatures.

III Electromagnetic optimization of the FSM

The optimal design of an electric machine usually represents a compromise between multiple contradicting objectives. For the application illustrated in this paper, a high efficiency is crucial. Furthermore, the reciprocating mover mass must be kept as low as possible, since the power is inversely proportional to the square root of the moving mass. Moreover the mass of the permanent magnets should be considered, since it accounts for the majority of the material costs.

This chapter explains the applied optimization algorithm. Afterwards the design space of the optimization and the consideration of the iron saturation are discussed. Finally the results are presented.

Multiobjective optimization

The nature of multiobjective optimization is that there are several independent design targets. In most of the cases, there is no solution where all objective variables are at the optimum at the same time. In many applications, they even contradict each other [8].

A possible approach to this problem is defining the individual variables as partial objectives and converting them into a common objective function by means of weighting factors. The multicriterial optimization is then reduced to a single overall criterion [9]. However, a decisive disadvantage of this method is the necessary definition of the optimization premises prior to the optimization. The design space is only analyzed according to these assumptions. The position of the found solution relative to the unknown pareto front cannot be determined. Thus, no conclusion about the goodness of the found solution according to the pareto optimality is possible.

NSGA-II

A method more suitable is a multicriterial optimization, where different objective functions are minimized simultaneously. In the case with contradicting objectives, there is no unique optimal solution, but rather the goal is to identify the set of pareto optimal solutions.

A common optimization algorithm is the *Non-dominated Sorting Genetic Algorithm*, NSGA-II, which was described by *Deb et al.* in 2002 [10]. It is a variation of the Genetic Algorithm for multicriterial optimization in which individual designs are ranked using a non-dominated sorting process. The NSGA-II quickly became popular for multiobjective optimization problems as this algorithm is less subject to errors, fast and robust, and therefore superior to other optimization methods.

It has also frequently been used for optimizing electric machines, like a switched reluctance machine [11], a permanent magnet synchronous machine [13], induction machines [12], or a flux-switching machine [14].

The process of the algorithm is illustrated in fig. 3. After defining the number of decision variables x, the number of objective functions $f(x)$, the maximum number of generations i_{max} and the population size n_{pop}, a random population of chromosomes (i.e. individual machine designs) is initialized. This population is evaluated by magnetostatic FEA simulation. All individuals are then sorted according to the non-dominated sorting principle, checking which one is superior to another with respect to the target variables (forming the pareto front). Chromosomes with the same rank are furthermore sorted by crowding distance. This ensures diversity among the individuals.

In order to produce a new generation that differs from the starting population, a parent population is defined. The choice of the parent generation is based on binary tournament selection: Two random pairs

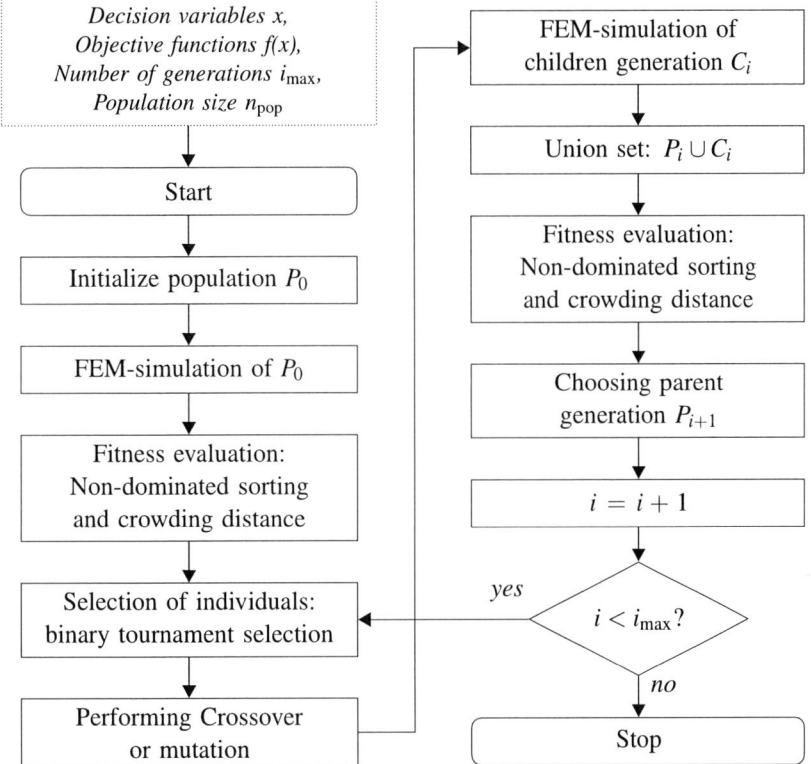

Fig. 3: NSGA-II optimization procedure

of individuals are compared and the superior ones (lower rank or higher crowding distance) are selected as parents. The following generation of individuals is generated by crossover (with a probability of 80 %) or by mutation of a certain gene (with a probability of 20 %).

After a subsequent FEM-simulation of this children generation, the parent and children generation are merged and the previously described sorting process is performed again. The n_{pop} fittest individuals are then selected as the parent generation for the next iteration.

Design space for the optimization

The parametrization of the FSM and the chosen design space is explained. Fig. 4 presents all the relevant geometric parameters for designing the electric machine.

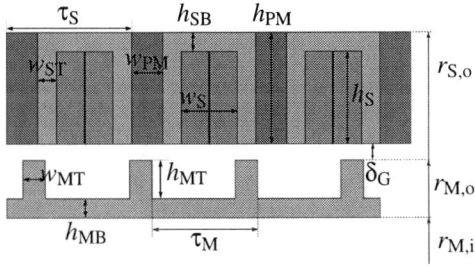

Fig. 4: Geometry parameters of the FSM

For the optimization of the FSM eight input parameters are considered. These are stator outer radius ($r_{S,o}$), the mover length, the mover outer radius ($r_{M,o}$), width of stator slots and teeth (w_S, w_{ST}), width and height of the mover teeth (w_{MT}, h_{MT}), inner radius of the mover ($r_{M,i}$) and thus the height of the mover back iron ($h_{M,b}$).

For the genetic optimization it is recommended to use relative parameters to describe the geometry. One advantage over absolute parameters is that unfeasible designs like a negative width of the permanent magnets are avoided. Another benefit is that the combination of two machines in the process of genetic evolution is always a feasible solution. This is not guaranteed with absolute parameters. Therefore the following relative parameters are defined in Tab. II.

Table II: Relative design parameters

Parameter	Range	Parameter	Range
$k_1 = \dfrac{r_{S,o}}{r_{S,o,max}}$	$\in [0.3, 1.0]$	$k_5 = \dfrac{w_{ST}}{\tau_S - w_S}$	$\in [0.2, 0.4]$
$k_2 = \dfrac{r_{S,o}}{l_M}$	$\in [0.2, 0.5]$	$k_6 = \dfrac{w_{MT}}{\tau_M}$	$\in [0.1, 0.6]$
$k_3 = \dfrac{r_{M,o}}{r_{S,o}}$	$\in [0.3, 0.8]$	$k_7 = \dfrac{r_{M,i}}{r_{M,o}}$	$\in [0.4, 0.9]$
$k_4 = \dfrac{w_S}{\tau_S}$	$\in [0.2, 0.6]$	$k_8 = \dfrac{h_{MT}}{r_{M,o} - r_{M,i}}$	$\in [0.2, 0.8]$

The maximum radius of the stator is set to $r_{S,o,max} = 140\,\text{mm}$ by a preceding parameter study. In the same way the boundaries for the coefficients k_x are chosen. The correct choice of the boundaries is necessary for proper results. Too big intervals lead to a poorly filled design space and too small intervals result in suboptimal designs. The distribution of the final design parameters of all optimal machines shows that no boundary condition is set too narrow, as there is no accumulation near the boundaries.

The three objectives are the negative efficiency $f(1) = -\eta$, the mass of the mover $f(2) = m_M$ and the mass of the permanent magnets $f(3) = m_{PM}$. The used algorithm leads to a minimization of each objective. For maximization of the efficiency, a negative sign is used.

Saturation of the iron

Each design is initially simulated with a current density of $6\,\text{A\,mm}^{-2}$ and the generated force is evaluated. Due to the requirement of a low mover mass, there can be designs with highly saturated iron in the mover. Therefore the dependency between current density and force is nonlinear. To calculate the correct current density the Newton-Raphson algorithm is used.

Based on the initial current density and a step width, the gradient of the force with the current density in this point is calculated. Then the new linear approximation for the current density is calculated. This process is repeated until the simulated force is equal to the required force.

This method considers the saturation of the iron precisely. However it increases the simulation time drastically. The average number of iterations per individual is 3. In each iteration the calculation of a gradient is done, which needs two simulations. Thus the total computation time increases by a factor of six.

Reference machine

The mechanical modeling is not part of this paper. Instead, the velocity distribution from [2] is assumed for a mover with a mass of $m_{M,ref} = 3\,\text{kg}$. The mass consists of the mass of the mover of the electrical machine and a constant mass of $1.5\,\text{kg}$ which considers the additional moving parts like the piston. Therefore a root mean square value of $v_{rms} = 4\,\text{m\,s}^{-1}$ is assumed. Together with the rms force the following mechanical power is calculated for the reference machine:

$$P_{mech,ref} = F_{rms} \cdot v_{rms,ref} = 4.8\,\text{kW} \tag{1}$$

It can be shown that for a constant force the mechanical power is reduced by the square root of the mover mass caused by the lower velocity. The dependency of the mechanical power of a design in relation to the power of the reference is:

$$P_{mech}(m_M) = P_{mech,ref} \cdot \sqrt{\frac{m_{M,ref}}{m_M}} \tag{2}$$

This equation is used for the calculation of the efficiency together with the ohmic losses. Iron losses are not calculated during the optimization, because this would need the calculation of the magnetic flux density for several time steps, which would increase the calculation time drastically.

One problem with a multiobjective optimization like NSGA-II is that even designs with low efficiency are possible optimal designs. Because the total number of designs is limited, a boundary for the lowest efficiency of $\eta_{min} = 60\%$ is used. Any individual with a lower value is ranked worse than one with a suitable value. In this way the result is a more dense filled parameter space. [10]

Optimization results

For the multiobjective optimization 50 generations with 1500 individuals each with a total number of designs of 75000 are calculated. Out of these designs, there are some points in which the improvement of one objective requires the decrease of another objective. All these points are optimal designs and form the pareto front which is shown in figure 5. The pareto front is a surface in the three-dimensional space due to the three objectives. Additionally the two-dimensional pareto fronts of two objectives each are shown. In the figures only the relevant designs are depicted. The one with a higher mass of the mover or the permanent magnets are neglected.

In the two-dimensional plots there are points which have a higher value of both objectives. For these points the third objective has to be taken into account which decreases. Therefore these are the optimal designs.

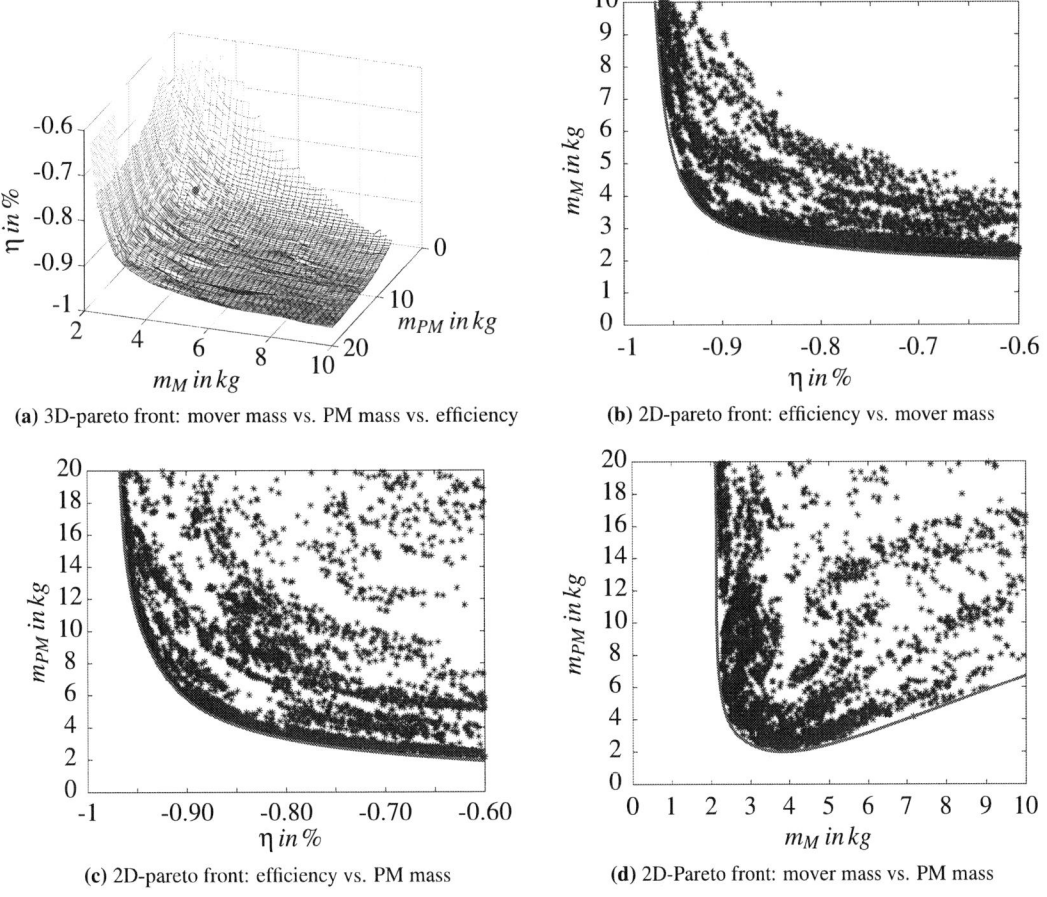

(a) 3D-pareto front: mover mass vs. PM mass vs. efficiency

(b) 2D-pareto front: efficiency vs. mover mass

(c) 2D-pareto front: efficiency vs. PM mass

(d) 2D-Pareto front: mover mass vs. PM mass

Fig. 5: Pareto front of the FSM

The next step is to choose one of the optimal designs which is suitable for this specific application. One approach is to specific a normal vector which determines exactly one point on the pareto front. In this case the point on the surface with the highest curvature in figure 5a is chosen. This point is marked red. Other points on the surface are also suitable. But for example for points with a high mass of the permanent magnets like 15 kg a rise in mass of the permanent magnets leads to a small change in efficiency. It would be an unbalanced optimum.

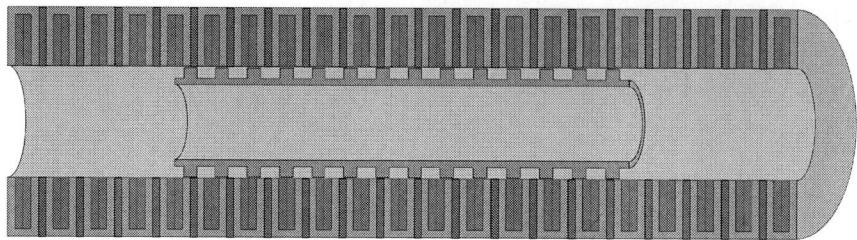

Fig. 6: Final flux-switching machine design

The chosen design has a mass of the mover of 3.77 kg, a mass of the permanent magnets of 7.33 kg and an efficiency of 85.88 %. The final machine design is shown in Fig. 6, table III presents the geometric parameters.

Due to the mover length of 280 mm, the stator length of 486 mm and the maximum stroke of 80 mm, it is possible to reduce the number of stator teeth to 16. This increases the efficiency to 89.63 % and reduce the mass of the permanent magnets to 5.5 kg. Thereby all three objectives are sufficient for the application in a free-piston machine.

The maximum braking force of 2 kN is obtained with a current density of 9.67 A mm^{-2}. This shows the non-linearity between current and force as the required current density is about 7 % higher than expected from a linear correlation.

Concerning demagnetization in the final design, it can be observed that even at maximum current, the field strength in the magnets does not exceed the coercivity H_C of 730 kA m^{-1}.

It can be seen that the three dimensional pareto front is convex. This is especially useful for the optimization. Deterministic algorithms converge without the trouble of being trapped into local optima. In this case the fully deterministic Hooke-Jeeves-algorithm is also used to find an optimum. The found optimum is a part of the pareto front. A drawback of the Hooke-Jeeves-algorithm is the strong dependency on the chosen parameter for the one dimensional objective function.

Table III: Final design parameters

Parameter	Value	Parameter	Value
Number of stator teeth (total) $N_{S,total}$	21	k_1	0.49
Number of stator teeth (active) N_S	12	k_2	0.25
Number of mover teeth N_M	14	k_3	0.47
Copper fill factor k_{Cu}	0.5	k_4	0.42
Air gap width δ_g	1 mm	k_5	0.34
Radius of the stator $r_{S,o}$	69 mm	k_6	0.41
Radius of the mover $r_{M,o}$	333 mm	k_7	0.69
Length of the mover l_M	280 mm	k_8	0.62

Calculation time

About 15 generations are needed until an approximation of the pareto front is recognizable. Nevertheless up to the 50th generation there are improvements in the pareto front. The front is more densely populated and there are some better designs due to crossover and mutation.

For the optimization 50 generations are calculated in total. The optimization is done in parallel, using 2d finite element simulation with the open-source software FEMM. Each model is composed of approximately 60 000 mesh elements. Using 24 parallel calculations, one generation needs a computation time of about 2 h. So the total time is abut 100 h. The processor used for the optimization is an Intel Xeon Platinum 8268 with 24 cores and a base clock speed of 2.90 GHz.

IV Conclusion

This paper deals with the design of a linear generator for use in a free-piston machine. It must meet the requirements in terms of braking force and have the highest possible force density and efficiency. Due to the thermal load on the mover, only machines with a passive mover were considered. The Flux-Switching Machine turned out to be most appropriate and was then further investigated. A cooling concept was analyzed to assess the thermal situation.

Afterwards the electromagnetic design was optimized via the optimization algorithm NSGA-II. The chosen design space and the consideration of the iron saturation were discussed. The result is a three-dimensional pareto front. The chosen design is a compromise with respect to the three objectives, which are the mover mass, the mass of the permanent magnets and the efficiency. Furthermore the machine meets the force requirements.

References

[1] M. Hanipah Razali, R. Mikalsen and A. P. Roskilly, "Recent commercial free-piston engine developments for automotive applications", Applied Thermal Engineering 75, pp. 493-503, 2017, doi: 10.1016/j.applthermaleng.2014.09.039

[2] H. Schillingmann, Q. Maurus and M. Henke, "Linear Generator Design for a Free-Piston Engine with high Force Density", 2019 12th International Symposium on Linear Drives for Industry Applications (LDIA), pp. 1-6, 2019, doi: 10.1109/LDIA.2019.8770996

[3] U. S. Deshpande, J. J. Cathey and E. Richter, "High-force density linear switched reluctance machine", in IEEE Transactions on Industry Applications, vol. 31, no. 2, pp. 345-352, 1995, doi: 10.1109/28.370283

[4] J. Wang, W. Wang, R. Clark, K. Atallah, and D. Howe, "A tubular flux-switching permanent magnet machine", Journal of Applied Physics, 103(7):07F105, 2008, doi: 10.1063/1.2830541

[5] M. Cheng, W. Hua, J. Zhang and W. Zhao, "Overview of Stator-Permanent Magnet Brushless Machines", in IEEE Transactions on Industrial Electronics, vol. 58, no. 11, pp. 5087-5101, 2011, doi: 10.1109/TIE.2011.2123853

[6] VACUUMSCHMELZE GmbH & Co. KG: VACODYM material data. `https://vacuumschmelze.com/products/Permanent-Magnets/NdFeB-Magnets---VACODYM` [Online, 24.05.2022]

[7] Höganäs AB: Somaloy 3p material data. `https://www.hoganas.com/en/powder-technologies/products/somaloy/somaloy-3p` [Online, 24.05.2022]

[8] G. Bramerdorfer, J. A. Tapia, J. J. Pyrhönen and A. Cavagnino, "Modern Electrical Machine Design Optimization: Techniques, Trends, and Best Practices", in IEEE Transactions on Industrial Electronics, vol. 65, no. 10, pp. 7672-7684, 2018, doi: 10.1109/TIE.2018.2801805

[9] X.-S. Yang, "Multi-Objective Optimization" in Nature-Inspired Optimization Algorithms, Elsevier, ch. 14, 2014, p. 197-211, ISBN: 10.1109/TIE.2018.2801805

[10] K. Deb, A. Pratap, S. Agarwal and T. Meyarivan, "A fast and elitist multiobjective genetic algorithm: NSGA-II", in IEEE Transactions on Evolutionary Computation, vol. 6, no. 2, pp. 182-197, 2002, doi: 10.1109/4235.996017

[11] M. El-Nemr, M. Afifi, H. Rezk and M. Ibrahim, "Finite element based overall optimization of switched reluctance motor using multi-objective genetic algorithm (NSGA-II)", in Mathematics 2021, 9(5)-576, 2021, doi: 10.3390/math9050576

[12] I. M. Alsofyani, N. R. N. Idris, M. Jannati, S. A. Anbaran and Y. A. Alamri, "Using NSGA II multiobjective genetic algorithm for EKF-based estimation of speed and electrical torque in AC induction machines", 2014 IEEE 8th International Power Engineering and Optimization Conference (PEOCO2014), pp. 396-401, 2014, doi: 10.1109/PEOCO.2014.6814461

[13] Y. Ma, T. W. Ching, W. N. Fu and S. Niu, "Multi-Objective Optimization of a Direct-Drive Dual-Structure Permanent Magnet Machine", in IEEE Transactions on Magnetics, vol. 55, no. 10, pp. 1-4, 2019, doi: 10.1109/TMAG.2019.2922475

[14] F. Mahmouditabar, A. Vahedi, M. R. Mosavi and M. H. B. Bafghi, "Sensitivity analysis and multiobjective design optimization of flux switching permanent magnet motor using MLP-ANN modeling and NSGA-II algorithm", in IEEE International Transactions on Electrical Energy Systems, vol. 30, no. 9, 2020, doi: 10.1002/2050-7038.12511

Single-Arm MMC-based Converter for Transformerless Rail Interties

Simon Beck, Simon Fuchs and Jürgen Biela
LABORATORY FOR HIGH POWER ELECTRONIC SYSTEMS (HPE) / ETH ZÜRICH
beck@hpe.ee.ethz.ch / https://hpe.ee.ethz.ch

Keywords

≪AC-AC converter≫, ≪Solid-State Transformer≫, ≪Modular Multilevel Converters (MMC)≫, ≪Traction Application≫

Abstract

MMC-based converters are suitable to efficiently intertie the 50 Hz three-phase grid with the 16.7 Hz single-phase railway grid without a 16.7 Hz transformer. In this paper, a new single-arm MMC-based converter topology with a multi-winding 50 Hz transformer is presented and simulated. The findings are compared to the direct and indirect MMC variants. The comparison shows a substantial reduction of the required number of modules, switches and installed semiconductor power with the proposed topology, while the system reliability is improved at the same time.

1 Introduction

In many railway systems in Europe, a single-phase medium voltage alternating current (MVAC) system at 15 kV or 25 kV and 16.7 Hz is used for supplying the trains. Initially, the supply voltage was

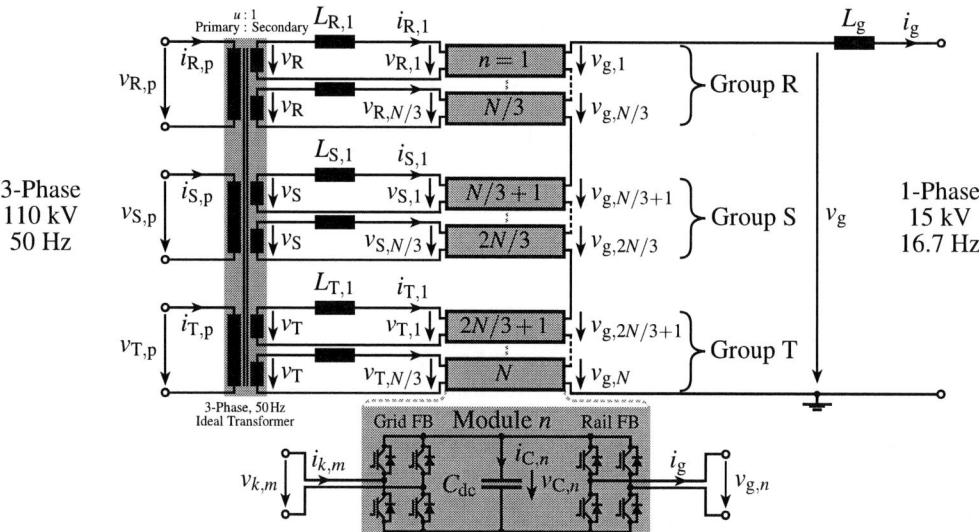

Fig. 1: Basic structure of the proposed single-arm converter with N modules consisting of back-to-back full bridges connected to a common module capacitor C_{dc} (equal for all $n \in \{1, 2, \dots, N\}$). The modules are arranged in three groups, that correspond to the grid phases $k \in \{R, S, T\}$. The PWM carrier for generating $v_{k,m}$ is phase-shifted depending on the module's position $m \in \{1, 2, \dots, N/3\}$ within each group. Thus, the current ripples of $i_{k,m}$ flowing through the combined transformer leakage and module inductance $L_{k,m}$ are cancelled out in the primary phase currents $i_{k,p} = \frac{N/3}{u} i_k = \frac{1}{u} \sum_{m=1}^{N/3} i_{k,m}$. On the 16.7 Hz side, all N modules produce the catenary voltage $v_g = \sum_{n=1}^{N} v_{g,n}$ using a distributed PWM algorithm.

generated by single-phase generators or from the three-phase 50 Hz national grid by rotary converters at interie stations [1]. With the advent of power electronics, 2L or 3L power electronic converters connected in between a 50 Hz transformer at the grid's side and a 16.7 Hz transformer on the railway's side were introduced. While these converters themselves feature a relatively high efficiency, the required transformers and filters reduce the overall efficiency significantly. To eliminate the 16.7 Hz transformer and filters, various systems based on modular multilevel converters (MMCs) have been proposed [2–6]. These converters have been denoted as "transformerless" interties, static frequency converters (SFCs), modular multilevel frequency converters (MMFCs), power electronic transformers (PETs), or solid-state transformers (SSTs). The indirect and direct MMCs are two interesting PET concepts for grid-to-rail interties as discussed in [3]. While the indirect MMC needs to store more energy in total than the direct MMC, the direct MMC requires more switches. In general, only relatively few publications about MMC systems designed specifically for 16.7 Hz rail interties are available.

Striving for cost effective and more compact solutions, this paper presents a new modular single-arm converter, which connects directly to the 16.7 Hz railway supply line. The new converter minimizes the number of installed semiconductors compared to other "transformerless" solutions by using multiple windings per phase in the three-phase 50 Hz transformer.

In the following, the basic converter structure and control principles are introduced in section 2. Then, the required number of modules is derived and transformer design considerations are presented in section 3.1. Thereafter, the required number of semiconductors and the required installed power of semiconductors for the proposed converter are derived in section 3.2. The required module capacitance as well as the required amount of energy stored within the converter are calculated in section 3.3. In section 3.4, semiconductor losses and reliability are investigated. Finally in section 4, key converter characteristics are compared to the direct and indirect MMC variants as discussed in [3].

2 Basic converter structure and control principles

The basic converter structure is shown in Fig. 1. A single module arm consisting of N modules is inserted between the rail catenary and ground, i.e. the train rails. Each module consists of two back-to-back full bridges sharing a common module capacitor C_{dc} charged to the nominal module voltage V_C on average. The back-to-back full bridges enable a bidirectional power transfer between the grid and the railway sides. Possible bypasses to short malfunctioning modules on the railway side are not shown for the sake of brevity. The full bridge connected to the 50 Hz grid side is denoted *Grid FB*. The full bridge on the 16.7 Hz rail grid side is denoted *Rail FB*. Their switching frequencies are typically in the range of a few hundred Hertz with the *Grid FBs* typically operating at a higher switching frequency as will be explained in more detail in subsections 2.1, 2.2 and 3.4. The *Rail FBs* are used to connect all N modules in series on the railway side. The catenary current i_g flows through all modules and is assumed to be known in the following.

On the three-phase 50 Hz side, the modules are split into three groups of $N/3$ modules each. Hence, N is a multiple of three. Each group of modules corresponds to one of the three 50 Hz phases $k \in \{R, S, T\}$. The index $m \in \{1, 2, \ldots, N/3\}$ denotes the position of each module within the respective group. Every module is connected to its own secondary transformer winding via the combined transformer leakage and module inductance $L_{k,m}$. These inductances are assumed to be equal for all modules in this paper. The transformer turns ratio $u = v_{k,p}/v_k$ is chosen such that the three primary-side grid voltages $v_{k,p}$ are transformed to the three secondary-side phase voltages v_k at $N/3$ secondary-side windings each. Consequently, the secondary-side currents $i_{k,m}$ flowing through the modules and their associated inductances $L_{k,m}$ translate to primary-side transformer currents:

$$i_{k,p} = \frac{N/3}{u} i_k = \frac{1}{u} \sum_{m=1}^{N/3} i_{k,m}, \qquad k \in \{R, S, T\} \tag{1}$$

In steady-state, the module voltages fluctuate due to the currents flowing through the module capacitors. On the railway side, all N modules in series must be able to generate the catenary voltage v_g at all times.

Hence, the minimal module voltage is defined by

$$V_{C,\min} = (1 - r)V_C = \hat{v}_{g,\mathrm{ov},n} = \frac{\hat{v}_{g,\mathrm{ov}}}{N} \tag{2}$$

Thereby, a nominal module voltage V_C, a maximum module voltage fluctuation $\pm r V_C$, and a perfect PWM without overmodulation of the module output voltages are assumed. Peak values are in general denoted by the hat symbol. The highest expected overvoltage $\hat{v}_{g,\mathrm{ov}}$ must be considered instead of the nominal peak catenary voltage $\hat{v}_g$. This overvoltage is typically specified to be 20% higher than the nominal peak catenary voltage [3].

With the nominal (and minimal) module capacitor voltage defined, the transformer turns ratio u is calculated such that the required peak module output voltage $\hat{v}_{k,m}$ on the 50 Hz side does not exceed the minimal module capacitor voltage. If the current ripples in the inductor currents are neglected, i.e. $i_{k,m} = i_k = \pm \frac{2P}{N\hat{v}_k}\cos(\omega t - \varphi_k)$, the maximum phase voltage amplitude $\hat{v}_k$ is given considering the module inductor voltages $v_{L,k,m}$ as

$$v_{k,m} = v_k - v_{L,k,m} = v_k - L_{k,m}\frac{\mathrm{d}}{\mathrm{dt}}i_{k,m} = \hat{v}_k\cos(\omega t - \varphi_k) \pm \omega L_{k,m}\frac{2P}{N\hat{v}_k}\sin(\omega t - \varphi_k) \leq V_{C,\min} \tag{3}$$

$$\Longleftrightarrow \quad \hat{v}_{k,m} = \sqrt{\hat{v}_k^2 + 4\omega^2 L_{k,m}^2 P^2/(N^2\hat{v}_k^2)} \leq V_{C,\min} \tag{4}$$

$$\Longleftrightarrow \quad \hat{v}_k^2 \leq \frac{1}{2}\left(V_{C,\min}^2 + \sqrt{V_{C,\min}^4 - 16\omega^2 L_{k,m}^2 P^2/N^2}\right) \quad \text{and} \quad L_{k,m} \leq \frac{V_{C,\min}^2}{4\omega P/N} \tag{5}$$

Here, $\varphi_k \in \{0, 2\pi/3, 4\pi/3\}$ indicates the phase shift between the phases $k \in \{R, S, T\}$ and only a purely active power transfer is considered in this paper, i.e. $S = P$. However, reactive energy transfer, consumption, and/or generation is generally possible. With the purely active power transfer, v_k and i_k are in phase or 180° shifted if the power flow is inverted. As on the railway side, $\hat{v}_k$ must be chosen below its theoretical maximum to account for the specified overvoltages $\hat{v}_{k,\mathrm{ov}}$ in the 50 Hz grid. It can be seen in the first inequality in eq. (5) that with a bigger inductance $L_{k,m}$, the maximum phase voltage $\hat{v}_k$ decreases. From the second inequality in eq. (5), it is also evident that the inductance is constrained by the chosen minimal capacitor voltage and the transferred power per module.

Since the catenary current i_g flows through all modules on the railway side and equal module output voltages $v_{g,n} = v_g/N$ are generated on average on the railway side, the averaged transferred power is shared equally among all modules, i.e. (using RMS values)

$$P_{\mathrm{mod,rail}} = V_{g,n}I_g = \frac{V_g I_g}{N} = \frac{P}{N} \tag{6}$$

Consequently, each module group transfers one third of the total power, which is expected since each group corresponds to one of the three grid phases. With the control system presented in the following sections, this is still the case when the grid voltages become unbalanced. Considering further that the averaged energy (or capacitor voltage) within each module must remain constant over time, the power balance on the grid and railway side of each module is evaluated (neglecting current ripples):

$$\frac{P}{N} = P_{\mathrm{mod,rail}} = V_{g,n}I_g = P_{\mathrm{mod,grid}} = V_k I_k \tag{7}$$

In this paper, $V_k = V_{g,n} = v_g/N$ is chosen, which consequently also implies $I_k = I_g$. Using these assumptions and further assuming a perfect modulation, the current flowing through the n-th module capacitor is

$$i_{C,n} = |i_{k,m}| - |i_g| = |i_k| - |i_g| = \hat{i}_g\left(|\cos\omega t - \varphi_k| - |\cos\omega_g t|\right), \qquad |i_{C,n}| \leq \hat{i}_g \tag{8}$$

The current $i_{C,n}$ consists of two rectified sinusoidal components with the respective frequencies of the grid and the railway side. Thus, the lowest frequency component of this current (and hence of the module energy fluctuation) is twice the frequency of the unrectified current component with the lowest frequency, i.e. $2f_{\mathrm{g}} = 33.4\,\mathrm{Hz}$. Unlike in a traditional MMC, no circulating currents are needed to balance the module voltages because each module can control its grid side current $i_{k,m}$ individually.

The module capacitors act as a kind of modular DC-link. Hence, the grid and railway sides are decoupled and can be controlled independently as will be shown in the following subsections.

2.1 Control principles for the railway side

On the railway side, all N modules are used in series to generate the bipolar catenary voltage with $2N+1$ voltage levels:

$$v_{\mathrm{g}} = \sum_{n=1}^{N} v_{\mathrm{g},n} \approx v_{\mathrm{g}}^{*} = \hat{v}_{\mathrm{g}} \cos\left(\omega_{\mathrm{g}} t\right), \qquad v_{\mathrm{g},n} \in \left\{ -v_{\mathrm{C},n}, 0\,\mathrm{V}, +v_{\mathrm{C},n} \right\} \tag{9}$$

To do so, a standard distributed PWM with a module sorting and balancing algorithm can be used, where all module voltages are assumed equal and constant during one PWM cycle [2,7,8]. Alternatively, also more advanced distributed PWM methods such as [9], where unequal but constant voltages are assumed, or an adapted version of [10] where unequal and varying module voltages are considered, can be used. These modulation schemes ensure that all modules are evenly charged or discharged (depending on the power flow direction) by the catenary current, if the converter is operated within its operating range. The catenary voltage v_{g} is generated based on the catenary current i_{g}, which is defined by the load and is assumed to be known. Since only active power is transferred, v_{g} is chosen to be either in phase (power transfer from grid to rail) or 180° out of phase (power transfer from rail to grid) with i_{g}. The catenary voltage amplitude is consequently given by the desired transferred power P using the relation $\hat{v}_{\mathrm{g}} = 2P/\hat{i}_{\mathrm{g}}$. Thus, the converter acts as a voltage source. Fig. 2a illustrates simulated catenary voltage and current waveforms. Of course, power must also be exchanged with the 50 Hz grid to not accumulate energy within the converter. This is discussed in the next section.

2.2 Control principles for the grid side

This paper's control system for the 50 Hz grid side is based on a cascaded PI-controller structure to keep the module voltages close to their nominal value V_{C} and the secondary-side transformer currents $i_{k,m}$ in phase (or 180° out of phase depending on the power flow direction) with the grid phase voltages v_k to achieve purely active power transfer. However, other suitable control systems are also conceivable in the future. The chosen controller structure is shown in Fig. 3. The controller structure is implemented in each module. The locally unavailable current and voltage measurements are transferred from a central entity to each module using a suitable communication protocol as described e.g. in [11]. The inner control loop consists of a relatively fast PI-controller, which is shown in red in Fig. 3. It utilizes the transformer's

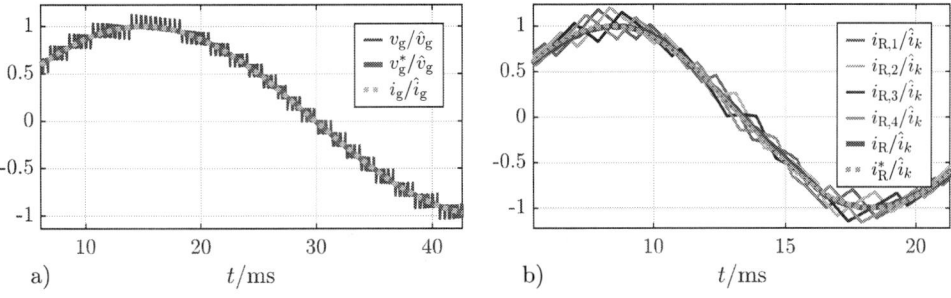

Fig. 2: Simulated voltage and current waveforms (normalized) using the system parameters given in Table I and Table II. Reference waveforms are marked by an asterisk. a) Catenary voltage and current utilizing a distributed PWM scheme that assumes equal and constant module voltages. b) Module currents and the resulting grid current (referred to the secondary transformer side) on the 50 Hz side for phase R.

leakage inductance and/or dedicated module inductors (jointly denoted as $L_{k,m}$) to control the current $i_{k,m}$ flowing through this inductance by generating a suitable voltage reference $v_{k,m}^* = v_k - v_{L,\text{PI}}$ for the *Grid FB*'s voltage modulator. The PWM introduces a delay of $T_{\text{PWM}}/2$ in the generated module output voltage $v_{k,m}$. To compensate this delay, a PLL is used to measure the grid voltage angle $\varphi = \omega t$. Thereafter, $\omega \cdot T_{\text{PWM}}/2 = \omega/2f_{\text{PWM}}$ is added to the PLL's measured grid angle to compensate the PWM delay of $T_{\text{PWM}}/2$. Hence, a time-shifted (into the future) version of the secondary-side phase voltage v_k is fed forward as stated below.

$$v_{k,m}^*(t) = v_k(t + T_{\text{PWM}}/2) - v_{L,\text{PI}}(t) = \hat{v}_k \cos(\omega t + \omega/2f_{\text{PWM}} - \varphi_k) - v_{L,\text{PI}}(t) \tag{10}$$

Since a transformer leakage inductance is always present, the primary-side grid voltages $v_{k,\text{p}} = u v_k$ are measured and transformed to the corresponding secondary-side voltages v_k using the known turns ratio u. The resulting reference value $v_{k,m}^*$ is generated using a standard 3L modulator with PWM period $T_{\text{PWM}} = 1/f_{\text{PWM}}$. A low PWM frequency is desirable to keep the switching losses low. However, with a constant and finite inductance $L_{k,m}$, the current ripple in $i_{k,m}$ increases as the PWM frequency decreases. To mitigate this issue, the structure of the proposed topology can be advantageously used because only the sum of one phase's secondary-side currents appears on the primary side, cf. eq. (1). Hence, the module current ripples are largely cancelled on the primary transformer side if the PWM carriers within a given group's modules are interleaved by one $N/3$-th of the PWM period. Fig. 2b illustrates this effect. Therefore, the proposed topology allows to have relatively large module current ripples. Consequently, also smaller module inductances $L_{k,m}$ and/or a relatively low PWM frequency are possible, while the primary-side currents $i_{k,\text{p}} = N/3u \cdot i_k$ are almost ripple-free. On the other hand, large current ripples increase the current RMS values on the secondary-side and therefore also the conduction losses. Thus, a suitable trade-off between conduction and switching losses must be chosen. Large current ripples also make the current control more difficult since the current tracking error is not supposed to consider the current ripple. Therefore, the measured current must typically be low-pass filtered, which introduces a delay and deteriorates the control performance. Again, using the proposed topology advantageously, the primary-side transformer currents may be measured, which represent, due to the cancelled current ripples, a kind of "low-pass filtered" version of the secondary-side currents but without any filter delay. This enables a relatively fast tracking of the current reference. However, the actual module currents $i_{k,m}$ must still be measured and controlled since only controlling the primary-side currents may lead to diverging module currents, which still sum to the desired primary-side currents, cf. eq. (1). Hence, using the weighting factor w, a combination of both transformer sides' currents is used for the measured

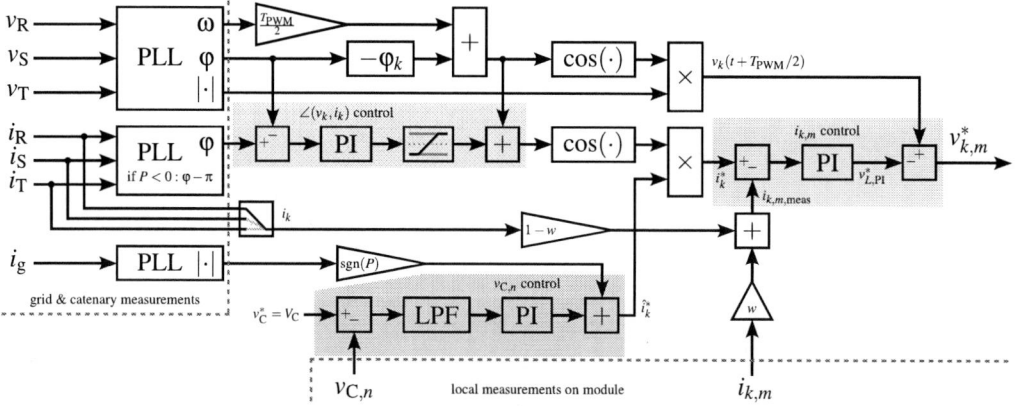

Fig. 3: Block diagram of the considered cascaded controller structure for the 50 Hz-side module currents $i_{k,m}$ and the module voltages $v_{C,n}$ to be implemented on each module. The power flow direction (i.e. $\text{sgn}(P)$) is assumed to be known. The catenary current i_g could also be measured locally on each module.

module current value, i.e.

$$i_{k,m,\text{meas}} = (1-w)i_k + wi_{k,m} \tag{11}$$

This calculated current is then used to track the module current reference i_k^*, which is defined by two outer control loops: One for the current amplitude reference $\hat{i}_k^*$ and one for the current angle reference as explained below.

The current amplitude reference's control loop consists of a relatively slow PI controller, which is used to track the module voltage reference set point $v_C^* = V_C$. This controller is not supposed to track the (unavoidable) module voltage fluctuation due to the catenary and grid currents. Instead, it is only supposed to track the average module voltage deviation from V_C over multiple grid periods. To accomplish this, the measured module voltage error is low-pass filtered (LPF) with a cross-over frequency below $2f_g = 33.4\,\text{Hz}$, cf. eq. (8). Lastly, the desired module current amplitude set point $\hat{i}_{k,\text{ff}}^* = \text{sgn}(P)\hat{i}_g$, is fed forward and added to the PI-controller's output to derive the current amplitude reference as shown in blue in Fig. 3.

Despite the compensation of the PWM delay by adjusting the measured grid angle as explained above, the inductor current will still slightly lag its reference due to the finite control performance. To counteract this in steady-state, another PLL measures the primary transformer side current angle and an additional outer PI-control loop is used to bring the voltage and current angles in phase (or 180° out of phase depending on the power flow direction) by altering the reference current's angle as shown in green in Fig. 3.

3 System design

In this section, key system characteristics are discussed and the considered simulated system for demonstrating the performance of the proposed system is specified.

3.1 Required number of modules and transformer design

The required number of modules N is primarily determined by the maximum catenary overvoltage amplitude $\hat{v}_{g,\text{ov}}$, which the series-connected converter arm must be able to block in worst case. Furthermore, N must be a multiple of three due to the arrangement in three module groups within the arm. Therefore, N can be calculated based on the system parameters given in Table I. The result is given in Table III.

To minimize the number of required secondary windings of the 50 Hz transformer, a low number of modules and a high module voltage is desirable. Just as in the MMC systems presented in [3], 4.5 kV IGBTs with a useable maximum module voltage of approximately 2.9 kV are chosen for the setup in this paper to simplify the comparison. This results in $N = 12$ modules and hence also 12 secondary-side transformer windings. Since the last module ($n = N$) is grounded at one catenary terminal, the corresponding required transformer winding insulation against ground can be calculated as $V_{\text{insul,N}} = \hat{v}_C + \hat{v}_{k,\text{ov}}$, if the voltage across the module inductor is neglected. For the remaining modules, the winding insulation requirements increase step by step by one module's maximum voltage per winding, i.e.

$$V_{\text{insul},n} = (N+1-n)\hat{v}_C + \hat{v}_{k,\text{ov}}. \tag{12}$$

However, for symmetry reasons and to achieve equal transformer leakage inductances, all windings are likely insulated for the maximum voltage $V_{\text{insul}} = V_{\text{insul,1}} = N\hat{v}_C + \hat{v}_{k,\text{ov}}$. With the power balance in eq. (7) follows that each winding must be designed to transfer one N-th of the total transferred power. Despite the required multi-winding design, the transformer can be built using a standard 3-legged core (as indicated in Fig. 1).

3.2 Semiconductor requirements

Every module consists of eight switches, hence the total number of switches is given by $N_{\text{sc}} = 8N$. All switches within a module are connected to the same DC capacitor C_{dc}. Therefore, the required voltage rating is equal to the maximum module capacitor voltage, $V_{\text{sc}} = \hat{v}_C = (1+r)V_C$. Note that the switches' "usable" blocking or operating voltage is considered here, which is typically approximately 60% of the rated voltage. Depending on the chosen voltage safety margins on the catenary and on the grid side,

Table I: System parameters adapted from [3].

Variable	Meaning	Value(s)
$S = P$	Transferred power	$15\,\text{MW}$
$\hat{v}_\text{g}$	Rated catenary voltage amplitude	$\sqrt{2} \cdot 15\,\text{kV}$
$\hat{v}_{\text{g,ov}}$	Maximum catenary voltage amplitude	$\sqrt{2} \cdot 18\,\text{kV}$
$\hat{v}_{\text{g,}n} = \hat{v}_\text{g}/N$	Rated catenary-side module voltage amplitude	$\sqrt{2} \cdot 1.25\,\text{kV}$
$\hat{v}_k$	Secondary-side phase voltage amplitude	$\sqrt{2} \cdot 1.25\,\text{kV}$
$\hat{i}_\text{g}$	Rated catenary current amplitude	$\sqrt{2} \cdot 1\,\text{kA}$
$\hat{i}_k$	Rated secondary-side current amplitude	$\sqrt{2} \cdot 1\,\text{kA}$
V_C	Nominal module capacitor voltage	$1\ldots 5\,\text{kV}$
$\pm r$	Relative module voltage fluctuation	$\pm 10\%$
f	Grid/transformer frequency	$50\,\text{Hz}$
f_g	Rail/catenary frequency	$16.7\,\text{Hz}$
$\hat{v}_\text{g}/\sqrt{3}$	3-ϕ-side phase voltage amplitude (direct & indirect MMC)	$\sqrt{2/3} \cdot 15\,\text{kV}$
$\hat{i}_\text{g}/\sqrt{3}$	3-ϕ-side phase current amplitude (direct & indirect MMC)	$\sqrt{2} \cdot 575\,\text{A}$

Table II: Simulated system parameters. If not stated otherwise, parameters given Table I are used.

Variable	Meaning	Value
N	Number of modules	12
V_C	Nominal module capacitor voltage	$2.6\,\text{kV}$
C_dc	Module capacitance	$15.7\,\text{mF}$
$L_{k,m}$	Transformer leakage and module inductance	$3.2\,\text{mH}$
u	Transformer turns ratios	50.8
w	Current weighting factor	0.2
f_PWM	Grid-side PWM frequency per module	$400\,\text{Hz}$
-	Catenary-side PWM frequency per module	$200\,\text{Hz}$
$f_\text{g,PWM}$	Resulting catenary-side PWM frequency	$2.4\,\text{kHz}$
-	IGBT model (4.5 kV, 1200 A)	Infineon FZ1200R45HL3

Table III: Comparison of intertie concepts using the parameters given in Table I. The module capacitance and the averaged stored energy for the indirect MMC are omitted, since there are various schemes to optimize these values for the three-phase side, e.g. [12–14].

Characteristic	Variable	Single-arm MMC	Direct MMC	Indirect MMC
Number of arms	-	1	6	10
Modules per arm	N	$3\left\lceil \frac{\hat{v}_\text{g,ov}/3}{V_\text{C,min}} \right\rceil$	$\left\lceil \frac{\hat{v}_\text{g,ov}}{V_\text{C,min}} \right\rceil$	$\left\lceil \frac{\hat{v}_\text{g,ov}}{V_\text{C,min}} \right\rceil$
Number of modules	N_total	N	$6N$	$10N$
Number of switches	N_sc	$8N$	$24N$	$20N$
Catenary voltage levels	-	$2N+1$	$2N+1$	$2N+1$
Installed blocking voltage	$V_\text{sc,total}$	$8N\hat{v}_\text{C}$	$24N\hat{v}_\text{C}$	$20N\hat{v}_\text{C}$
Semicond. current rating	I_sc	$\hat{i}_\text{g}$	$\frac{2+\sqrt{3}}{6}\hat{i}_\text{g}$	$\frac{1+\sqrt{3}}{6}\hat{i}_\text{g}$ (3-ϕ-side) $^3/_4 \cdot \hat{i}_\text{g}$ (1-ϕ-side)
Installed semicond. power	P_sc	$8N\hat{v}_\text{C}\hat{i}_\text{g}$	$4(2+\sqrt{3})N\hat{v}_\text{C}\hat{i}_\text{g}$	$2(4+\sqrt{3})N\hat{v}_\text{C}\hat{i}_\text{g}$
Module capacitance	C_dc	$\approx 0.28\frac{\hat{i}_\text{g}}{\omega_\text{g}rV_\text{C}}$	$\approx 0.28\frac{P}{2\omega_\text{g}rNV_\text{C}^2}$	-
Nominally stored energy	E_total	$\approx 0.14N\frac{V_\text{C}\hat{i}_\text{g}}{\omega_\text{g}r}$	$\approx 0.14\frac{P}{2\omega_\text{g}r}$	-

the switches' current ratings may vary for the two full bridges. As mentioned above, equal current amplitudes on the rail and grid side of each module are assumed in this paper. The current rating for all switches is in this case $I_{sc} = \hat{i}_g$, if additional current ripples caused by modulation are neglected. This means that all semiconductors can be of the same type and they are equally utilized in terms of blocked voltage and equally utilized in terms of conducted current. The installed semiconductor power can then be derived and it is given in Table III. Also, the chosen IGBT type is given in Table II.

3.3 Module capacitor and energy storage requirements

The required module capacitance C_{dc} is defined by the chosen module capacitor voltage level V_C, its allowable relative fluctuation r and the currents flowing through the module capacitors $i_{C,n}$. Numerically integrating the nominal current waveforms and assuming $f \approx 3f_g$ yields a good approximation for the required module capacitance and, based on the number of modules, the nominally stored energy within the converter, which are both again given in Table III.

3.4 Semiconductor losses and reliability

As mentioned above, the proposed topology utilizes the voltage and current capabilities of every installed semiconductor well. However, the *Grid* and *Rail FBs* must be operated at different frequencies, which is explained in the following.

On the railway side, the PWM frequency per module can be kept low due to the distributed PWM scheme, which results in an effective PWM frequency that is N times higher as stated in Table II.

On the three-phase grid side, a higher PWM frequency is necessary since each module must control its grid-side current $i_{k,m}$ individually. Because relatively high current ripples of $i_{k,m}$ cancel out on the primary transformer side, the PWM frequency can still be kept relatively low. In this paper, $f_{PWM} = 400\,\text{Hz}$ is chosen. Nevertheless, simulations have shown that PWM frequencies as low as $300\,\text{Hz}$ and lower are also feasible in steady-state with only minor effects on the quality of the grid currents.

To estimate the semiconductor loss factor of the IGBTs and the built-in diodes, a PLECS Blockset and Simulink simulation is used with thermal models for the IGBTs and the diodes provided by Infineon. Depending on the selected module inductance $L_{k,m}$, the voltage safety margin on the $50\,\text{Hz}$ grid side, the PWM frequency and the controller tuning, a semiconductor loss factor in the range of 1.15%-1.4% can be achieved. The semiconductor loss factor is defined as in [3] as the ratio of the thermally lost power within the semiconductors and the total transferred power: $P_{sc,loss}/P$.

MMCs and also the proposed converter are modular systems in which all modules must be operational in order for the converter to work (neglecting backup/failover modules). If a given switch (incl. driver/control/supply circuits) has an expected mean time to failure (MTTF) of $T_{sc,mttf}$ and the converter system consists of N_{sc} identical switches, the expected MTTF for the entire system follows as

$$T_{mttf} = T_{sc,mttf}/N_{sc} \tag{13}$$

It is evident that a small number of switches is desirable for the reliability of the system as the MTTF of the overall system scales with $1/N_{sc}$. Thus, the proposed converter's small number of required switches has a positive impact on the expected reliability of the system. Furthermore, the proposed converter requires only three failover modules (one per module group) to provide $N-1$-redundancy.

4 Comparison to existing MMC-based rail intertie concepts

In this section, the proposed converter is compared to the direct and the indirect MMC variants discussed in [3]. The comparison is conducted using the system parameters given in Table I and the derived system characteristics given in Table III. Resulting key characteristics are shown in Fig. 4 for the considered systems according to Table I and Table II. Fig. 5 illustrates these characteristics as functions of a variable module voltage level V_C.

The proposed converter requires only about 17%-21% of the number of modules of the direct MMC and only about 10%-12% of the indirect MMC. Also, only about 33%-43% of the number of switches of the direct MMC and only 40%-51% of the indirect MMC are required. This substantially reduces the number of active hardware components, such as switches, gate drives, auxiliary supplies or measurement circuits. According to eq. (13), the mean time to failure is expected to be doubled or even tripled compared to the

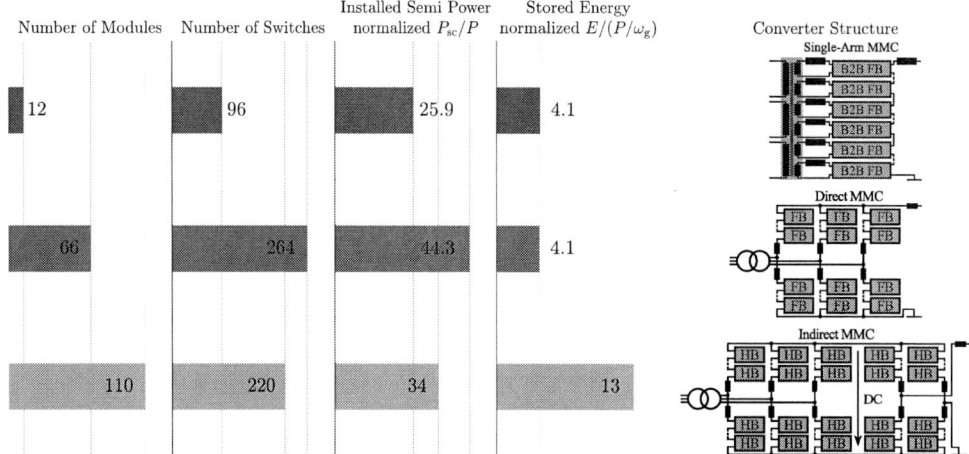

Fig. 4: Key characteristics of this paper's investigated single-arm MMC and the direct and indirect MMC concepts as discussed in [3]. Parameters are given in Table I and Table II with $V_C = 2.6\,\text{kV}$. Full bridge modules are denoted "FB", while half bridge modules are denoted "HB" and back-to-back full bridge modules are denoted "B2B FB".

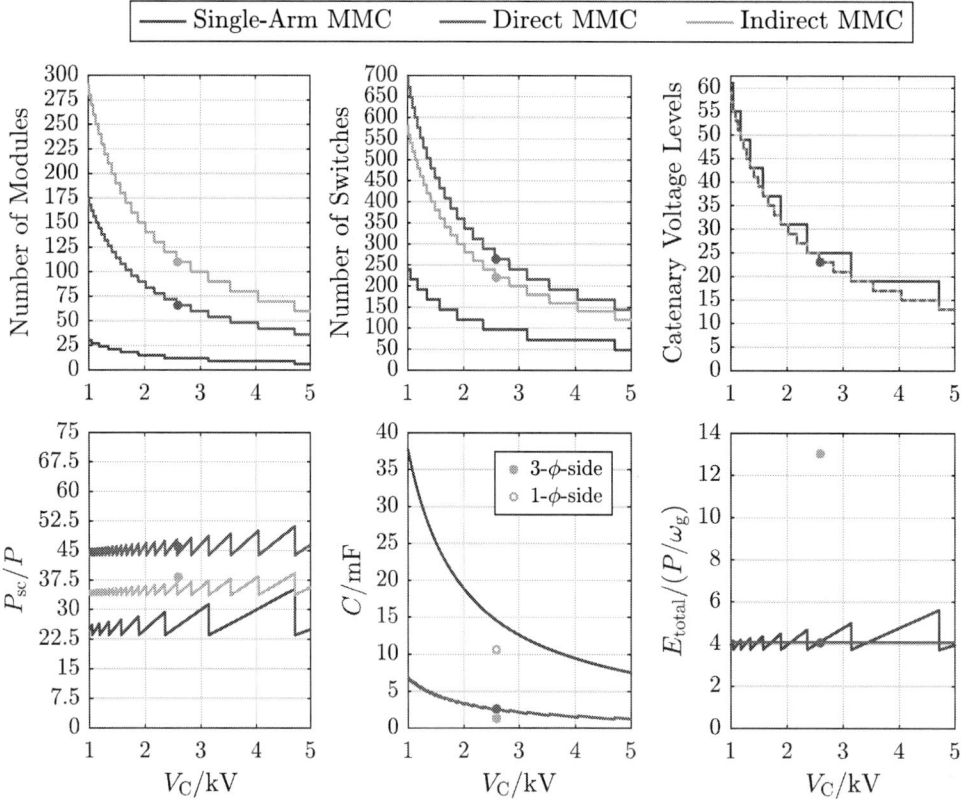

Fig. 5: Comparison of key characteristics of the direct and the indirect MMC concepts presented in [3] with the proposed converter as functions of the chosen module voltage V_C. The parameters given in table Table I are considered. Parameter values of the studied systems in [3] are marked by dots.

indirect and direct MMCs thanks to the reduced number of switches. The number of available catenary voltage levels of the proposed converter are the same or more due to the fact that the number of modules must be a multiple of three.

In the proposed converter, switches with a higher current rating are required, which is about 38% higher compared to the direct MMC and 54% higher compared to the indirect MMC on the three-phase side and 25% higher than on the indirect MMC's single-phase side. This is due to the fact that the total required semiconductor voltage blocking capability (installed blocking voltage) of the proposed system is much lower, while the same power is transferred. This is reflected in the resulting installed semiconductor power requirement, which is up to 46% lower compared to the direct MMC and up to 30% lower compared to the the indirect MMC.

Regarding the energy storage requirements, substantially less energy needs to be stored within the converter compared to the indirect MMC in [3], while the energy storage requirements are comparable to the direct MMC. This is mainly due to the fact that the indirect MMC has to buffer energy fluctuations caused by relatively low frequent 16.7 Hz arm currents and voltages on the single-phase side. In the proposed converter and the direct MMC, a combination of catenary and 50 Hz grid currents and voltages cause less energy fluctuation in the modules. Since the proposed single-arm topology requires only about ⅕ the modules compared to the direct MMC but it must store a comparable amount of energy, the module capacitances need to be about 5 times as big at equal module voltage levels.

In [3], the semiconductor loss factor of both the direct and indirect MMC systems have been calculated to be 1.1%, which is comparable to the 1.15%-1.4% derived from a comprehensive PLECS Blockset and Simulink simulation of the proposed converter using thermal IGBT and diode models by Infineon.

Since the proposed converter's modules require an insulated 50 Hz grid side voltage, a more complex transformer design is needed with $N \geq 3$ insulated secondary windings instead of only three. The power rating per winding is however only one N-th of the total transferred power instead of one third.

5 Conclusion

In this paper, a single-arm MMC-based converter is proposed as a "transformerless" solution for interconnecting the 50 Hz three-phase grid with the 16.7 Hz single-phase railway grid. It is shown that, compared to two MMC variants found in literature, only about 10%-21% of the number of modules are required. The expected mean time to failure is at least doubled thanks to 49%-72% less switches. The total installed semiconductor power is reduced by up to 46%. A disadvantage is the required more sophisticated 50 Hz transformer design. The proposed converter's feasibility is verified by simulation and thereby the semiconductor loss factor is determined to be within 1.15%-1.4% of the total transferred power.

References

[1] Steimel A.: Power-electronic grid supply of AC railway systems, 13th International Conference on Optimization of Electrical and Electronic Equipment (OPTIM 2012), pp. 16–25

[2] Lesnicar A. and Marquardt R.: An innovative modular multilevel converter topology suitable for a wide power range, 2003 IEEE Bologna Power Tech Conference Proceedings Vol. 3, pp. 6–

[3] Winkelnkemper M., Korn A., and Steimer P.: A modular direct converter for transformerless rail interties, IEEE 2010 International Symposium on Industrial Electronics, pp. 562–567

[4] Ängquist L., Haider A., Nee H.-P., and Jiang H.: Open-loop approach to control a modular multilevel frequency converter, Proceedings of the 2011 14th European Conference on Power Electronics and Applications, pp. 1–10

[5] Vasiladiotis M., Cherix N., and Rufer A.: Single-to-three-phase direct AC/AC modular multilevel converters with integrated split battery energy storage for railway interties, 17th European Conference on Power Electronics and Applications 2015 (EPE'15 ECCE-Europe), pp. 1–7

[6] Nami A., Jiang H., and Subramanian S.: A modular multilevel converter for use in a high voltage traction system, Patent WO2 018 091 065A1, Nov. 15, 2016

[7] Li Z., Gao F., Xu F., Ma X., Chu Z., Wang P., Gou R., and Li Y.: Power module capacitor voltage balancing method for a ±350-kV/1000-MW modular multilevel converter, IEEE Transactions on Power Electronics Vol. 31, pp. 3977–3984

[8] Rohner S., Bernet S., Hiller M., and Sommer R.: Modulation, losses, and semiconductor requirements of modular multilevel converters, IEEE Transactions on Industrial Electronics Vol. 57 no 8, pp. 2633–2642

[9] Fehr H., Gensior A., and Bernet S.: Experimental evaluation of PWM-methods for modular multilevel converters, 18th European Conference on Power Electronics and Applications 2016 (EPE'16 ECCE Europe), pp. 1–10

[10] Fuchs S., Beck S., and Biela J.: High output voltage precision pwm for modular multilevel converters, 19th European Conference on Power Electronics and Applications 2017 (EPE'17 ECCE Europe), pp. 1–10

[11] Rietmann S., Fuchs S., Hillers A., and Biela J.: Field bus for data exchange and control of modular power electronic systems with high synchronisation accuracy, 2018 International Power Electronics Conference (IPEC-Niigata 2018 -ECCE Asia), pp. 2301–2308

[12] Korn A. J., Winkelnkemper M., and Steimer P.: Low output frequency operation of the modular multi-level converter, 2010 IEEE Energy Conversion Congress and Exposition, pp. 3993–3997

[13] Engel S. P. and Doncker R. W. D.: Control of the modular multi-level converter for minimized cell capacitance, Proceedings of the 2011 14th European Conference on Power Electronics and Applications, pp. 1–10

[14] Ilves K., Antonopoulos A., Harnefors L., Norrga S., Ängquist L., and Nee H.-P.: Capacitor voltage ripple shaping in modular multilevel converters allowing for operating region extension, IECON 2011 - 37th Annual Conference of the IEEE Industrial Electronics Society, pp. 4403–4408

Medium Voltage Diode Rectifier Design for High Step-Up DC-DC Converter

Pierre Le Métayer[1,2], Cyril Buttay[2], Drazen Dujic[3], Piotr Dworakowski[1]

[1]SUPERGRID INSTITUTE
23 Rue Cyprian
69100 Villeurbanne, France
+33 7 63 66 19 15
pierre.lemetayer@supergrid-institute.com
https://www.supergrid-institute.com

[2]Univ Lyon, CNRS, INSA Lyon, Université Claude Bernard Lyon 1, Ecole Centrale de Lyon,
Ampère, UMR 5005,
69621 Villeurbanne, France
http://www.ampere-lab.fr

[3]Power Electronics Laboratory (PEL)
École Polytechnique Fédérale de Lausanne
1015 Lausanne, Switzerland
https://www.epfl.ch/labs/pel/

Acknowledgements

This work was supported by a grant overseen by the French National Research Agency (ANR) as part of the "Investissements d'Avenir" Program (ANE-ITE-002-01)

Keywords

«Diode», « Medium voltage converter », « Silicon Carbide », « DC-DC power converter », «Photovoltaic »

Abstract

A medium voltage rectifier, relying on series connection of SiC Schottky diodes, is designed and tested within a unidirectional isolated DC-DC converter. The design is realized considering efficiency, static and dynamic balancing and failure mode considerations. The model of the rectifier circuit is improved compared to the previous literature enabling to precisely represent the voltage oscillations due to switching, facilitating the design of the snubbers. Experimental results are presented to validate the design. The developed model results show good fit with the measurements.

Introduction

Medium voltage direct current (MVdc) networks are considered in numerous works for the integration of renewable energy sources [1]–[4]. In particular, the photovoltaic (PV) application is discussed in [3], [4]. Step-up DC-DC converters are required to interface low voltage PV clusters to an MVdc collection network. The MVdc collection network is expected to be in the order of ±10 kV, the actual value depending on various design parameters, specific to particular application [3].

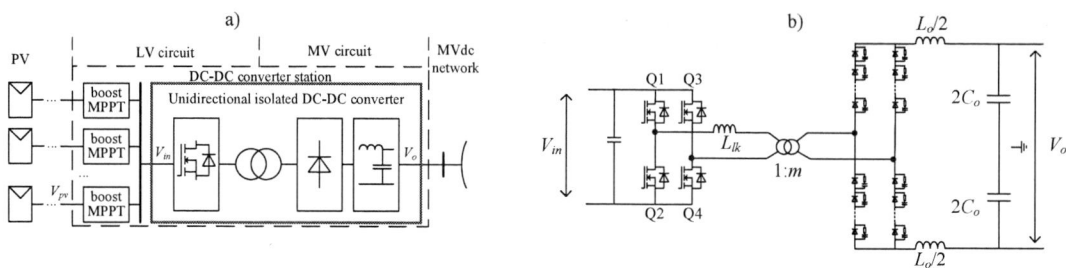

Fig. 1: a) Unidirectional isolated DC-DC converter in MVdc PV application, b) PSFB with a medium voltage rectifier using series connected diodes.

Unidirectional power flow is considered in PV application, as the power is only exported towards the network. Multiple maximum power point tracking (MPPT) boost DC-DC converters are used to interface PV strings with a Phase Shifted Full Bridge (PSFB) converter, selected as the unidirectional isolated step-up DC-DC converter [5] (this later converter having significantly higher power ratings than individual MPPT boosts). The PSFB circuit (Fig. 1b) uses a diode rectifier on the MV side (as compared to transistors for, e.g. the dual active bridge), making it an attractive solution from the control and cost point of view. The rectifier bridge components have a current rating of a few tens of amperes [5]. Diodes of such rating are most commonly found in discrete component packages with voltage ratings up to 3.3 kV [6]. A series connection of diodes is addressed in this paper, and presented in details below, focusing on the detailed design of the snubber components to limit switching overvoltage.

The *LC* output filter of the PSFB was shown in [7] to be advantageous compared with a simple output capacitor when the converter experiences faults on the MV side. However, the presence of an inductor between the rectifier and the output capacitor means that the rectifier bridge voltage is not clamped to the output voltage. Voltage oscillations occur at the diode turn-off, due to resonances between the transformer leakage inductance and parasitic capacitances of transformer and diodes, as described in [8], [9]. It is desirable to keep this voltage ringing low, so as to limit the peak voltage experienced by the diodes, and so as to limit electro-magnetic interferences (EMI). The *RC* snubber design for the series connection of SiC diodes takes into account the dynamic voltage balancing as well as the voltage ringing in the PSFB converter.

This paper presents the design of a medium voltage (4 kVdc) medium frequency rectifier, intended to operate with non-clamped voltage, based on series-connected SiC Schottky diodes. The choices of diode rating and number of series connected diodes are discussed considering efficiency and reliability. The voltage oscillations are modelled based on [9], adapted to series-connected assemblies of diodes, with the inclusion of the effects of transformer primary-secondary capacitance and limited switching speeds. This simplified model enables rapid snubber design. Tests are performed on the designed rectifier and measured waveforms are compared with the results expected from derived models.

Design of the rectifier

Diode assemblies are commercially available for medium voltage applications [10]. However, these assemblies usually use silicon bipolar diodes. Thus, reverse recovery happens at diode turn-off [11], yielding power losses. In [3] it is shown that step-up DC-DC converters for MVdc collection network in PV applications are expected to reach efficiencies similar to that of state of the art PV inverter cascaded with transformer for MVac collection network. Thus, very high efficiency is needed for all components of the converter. SiC Schottky diodes exhibit little to no reverse recovery current [11] and they are thus adapted for medium frequency, high efficiency applications. As a consequence, this article investigates the use of SiC diodes in the output rectifier of PSFB converter.

The rectifier presented in this paper is designed for a voltage of 4 kVdc, and maximum currents I_{mean} = 6 A, I_{rms} = 9 A, I_{peak}= 20 A as in the converter design presented in [5]. The switching frequency is 20 kHz [5].

Commercially available diodes with ratings in the range of the specified currents are mostly found in discrete packages, with voltage ratings up to 1.7 kV (3.3 kV diodes are less common and have less available current ratings). In order to limit the number of series connected diodes only 1.2 kV and 1.7 kV diodes are considered. One diode of each voltage rating and of similar current ratings is picked from the same manufacturer for comparison.

The choice of the number of series connected diodes is motivated by the presence of overvoltages at switching instant and failure mode considerations. Fig.2 shows a simplified rectifier voltage waveform with an overvoltage at switching instant, considered to be limited by the snubber design to 1.25 times the steady state voltage. The number of diodes n_{diodes} is chosen so as n_{diodes}-1 diodes rated blocking voltage is sufficient to withstand the switching instant overvoltage. Indeed, it is shown in [12] that for series connected diodes, an event of cosmic radiation interaction does not lead to single event burnout, as the blocking voltage is transferred to the other diodes of the series connection. However, this is only possible if the remaining diodes have enough blocking capability. If this is the case, the impacted diode does not fail and its blocking capability is recovered after a period of time dependent on the applied voltage [12]. Number of series connected diodes and resulting conduction losses (at nominal current) for both voltage ratings are presented in Table I. The junction temperature is considered to be 125°C, as a worst-case scenario. Losses weighted with European efficiency coefficients (standardized efficiency measure considering the power distribution for a PV system operating under the European climate) [13] are also presented as a way to evaluate the impact of losses at lower powers on the comparison. It must be noted that these values have no real physical meaning but they are relevant for comparisons.

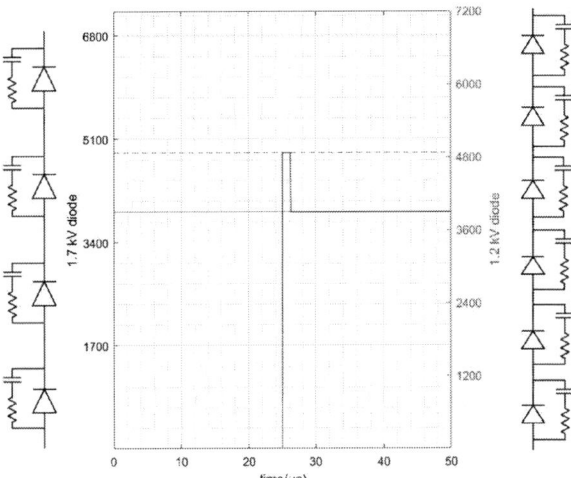

Fig. 2: Simplified rectifier waveform (blue) with overvoltage at switching instant, considered to be limited by the snubber design to 1.25 times the steady state voltage. Implementations with 6x1.2 kV or 4x1.7 kV series-connected diodes are represented. No static balancing resistors are represented as the leakage current of used diodes is considered sufficiently low.

Table I: Comparison of rectifier bridge performance for 2 selected SiC Schottky diodes

Diode reference	Voltage rating	Number of diodes	Individual diode conduction losses at nominal current	Full bridge rectifier conduction losses at nominal current	Full bridge rectifier losses with EU efficiency coefficients
Genesic GB10MPS17-247	1700 V	4	10.6 W	170 W	68.2 W
Genesic GD15MPS12-247	1200 V	6	9.4 W	225 W	91.7 W

One can note that the 1.2 kV diode has inferior individual losses compared to the 1.7 kV model. However, it is seen that 1.7 kV diodes choice leads to lower rectifier losses thanks to lower number of diodes, and this rating is thus selected. The losses weighted with EU efficiency coefficients confirm this choice, with proportionally an even slightly larger losses reduction compared to the 1.2 kV option. Thermal management of diode losses is met by individual heatsinks [14] at floating electrical potential, in natural convection condition. Each heatsink is referenced to the cathode of the diode, thus its potential varies according to ground in rectifier operation. This is beneficial for diode balancing as the parasitic capacitances to the ground are expected much lower than in a one common grounded heatsink scenario (such capacitances degrade voltage balancing in the series string [15]). It must be noted that the presented rectifier prototype is designed to validate electrical properties and the thermal aspects are not presented further in this paper. The cooling of the rectifier of a converter such as in Fig. 1 would beneficiate from being designed together with the medium frequency transformer and the output filter. Snubbers are placed in parallel to each diode of the series connection, in order to ensure dynamic voltage distribution during commutations and to keep overvoltages lower than 1.25 times the steady state voltage (voltage after oscillating transient). The snubber is composed of a capacitor $C_{d1diode}$ and a resistor $R_{d1diode}$ in series. It is proposed to select the snubber values based on following equations, adapted from [16]:

$$C_{d_{1diode}} = k_{sc} \cdot C_{j_{1diode}} \qquad R_{d_{1diode}} = k_{sr} \cdot \sqrt{\frac{m \cdot L_{lk}}{C_{d_{1diode}}}} \qquad (1)$$

With $C_{j1diode}$ the junction capacitance of a single diode of the series connection, m the transformer ratio and L_{lk} the transformer leakage inductance (see Table II). The value of $C_{j1diode}$ is dependent on the reverse voltage applied to the diode. The $C_{j1diode}$ capacitance value at 50% of the operating voltage (80 pF for the selected diode) is selected in order to simplify the calculations. Parameters k_{sc} and k_{sr} are the design parameter that can be adjusted in order to reduce oscillating transient overvoltages at switching instants. It is proposed to initially set $k_{sc} = 3$ and $k_{sr} = 1$ following typical design rules of generic RC snubbers [16] and tune them for overvoltages concerns using the model presented in the following section. The final values selected for this prototype are $C_{d1diode} = 270$ pF and $R_{d1diode} = 650$ Ω.

High value resistor in parallel of each diode can be added for static voltage balancing required due to diode leakage current deviations. However, considering the low value of leakage current of the selected diode (9 µA at 175 °C of junction temperature) they are not implemented in this design.

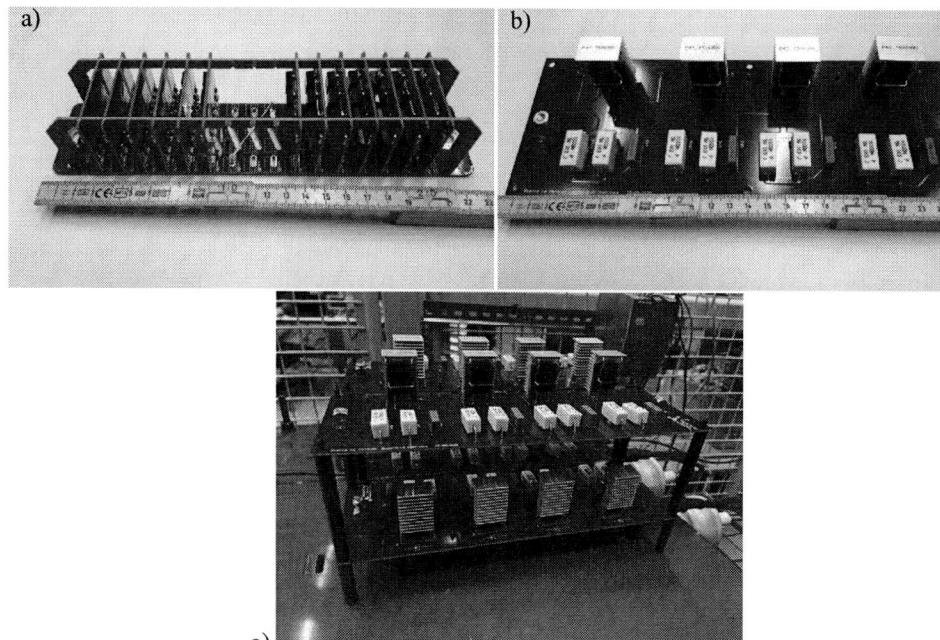

Fig. 3: a) An example of commercially available high voltage diode assembly, intended for forced air or oil cooling; b) Diode assembly composed of four series connected diodes, designed for natural convection cooling c) Full-bridge rectifier composed of 4 diode assemblies.

Model and simulations

In order to analyze the voltage oscillations occurring at diode turn-off, the rectifier and transformer are modeled taking into account their parasitic elements. The model is derived according to [9], adapted to a full bridge rectifier and including the primary-secondary parasitic capacitance of the medium frequency transformer (MFT) modeled with lumped elements. The values of the parasitic elements of the MFT used in the test set-up can be found in Table II. The circuit model is presented in Fig. 4. The voltage source is driven by rate-limited step function representing the limited switching speed of the input bridge.

Table II: MFT parameters

Transformation ratio	Leakage inductance	Primary winding capacitance	Secondary winding capacitance	Primary to secondary winding capacitance
$m = 2.33$	$L_{lk} = 34\ \mu H$	$C_p = 16\ pF$	$C_s = 100\ pF$	$C_{ps} = 30\ pF$

The diodes in Fig. 4 represent the series connection of n_{diodes}. Thus, the junction capacitance C_j, snubber capacitor C_d, and snubber resistor R_d are defined as:

$$C_j = \frac{C_{j1diode}}{n_{diodes}} \qquad C_d = \frac{C_{d1diode}}{n_{diodes}} \qquad R_d = R_{d1diode} \cdot n_{diodes} \tag{2}$$

In Fig. 4b, the simplified model is represented from the primary side of the MFT; secondary side components are moved to the primary following equations (3). At switching instant, the parasitic components of the pair of diodes in blocking state are in parallel to each other in the equivalent circuit of the converter. This results in the values of the aggregated components of the simplified model given in Fig. 4b.

$$C_s' = C_s \cdot m^2 \quad C_j' = C_j \cdot m^2 \quad C_d' = C_d \cdot m^2 \quad R_d' = \frac{R_d}{m^2} \tag{3}$$

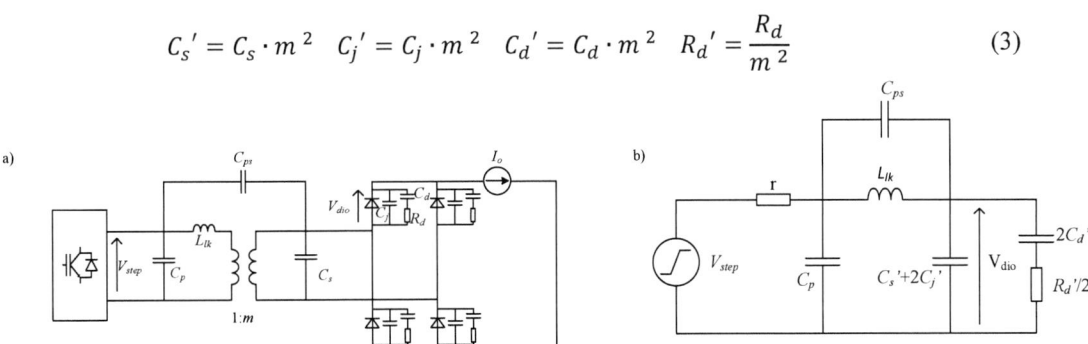

Fig. 4: a) Model of the PSFB converter including parasitic elements, b) Simplified PSFB model at diode turn-off used to study the voltage ringing phenomenon.

The quantity of interest is the voltage across each diode, V_{dio}. The effect of the switching is shown in Fig. 5a where the circuit is driven by different rate-limited step functions. The effect of variation of snubber parameters is also shown in Fig. 5b.

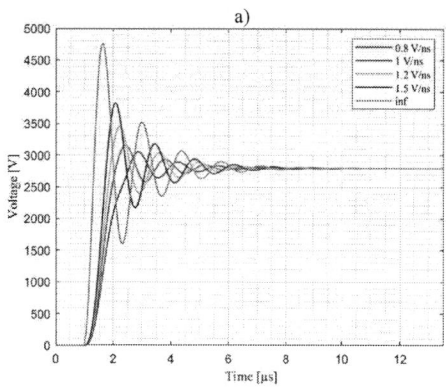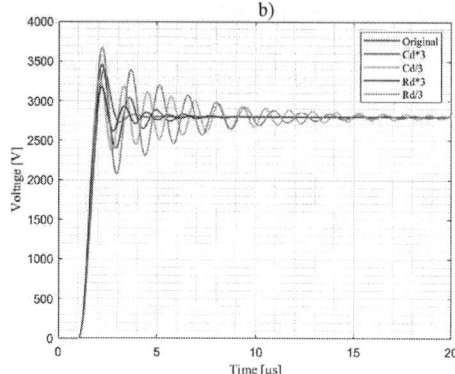

Fig. 5: Resulting $V_{dio}*m$ (voltage across a diode string) from Fig. 4b model, with a) different switching speeds of V_{step} (base case 1.2 V/ns), b) snubber components parameter variations.

It is observed that oscillations are more severe for high switching speed, with important overvoltages. Thus, designing the snubber to attenuate the overvoltages associated with a perfect step input would lead to oversizing compared with a more realistic case.

In order to validate the simplified PSFB model, a detailed circuit model is implemented using the Simulink Simscape Electrical library. Each diode is modeled with its junction capacitance (with voltage dependency taken into account via a lookup table) and RC snubber. The input inverter bridge is also modeled. Rectifier voltage oscillations obtained with the detailed model (Fig. 6) are similar to those presented in Fig. 5a (simplified model) for similar rates of change in voltage. This detailed simulation with each individual diode also enables to verify the voltage distribution across the series-connected diodes. Results are shown in Fig. 6.

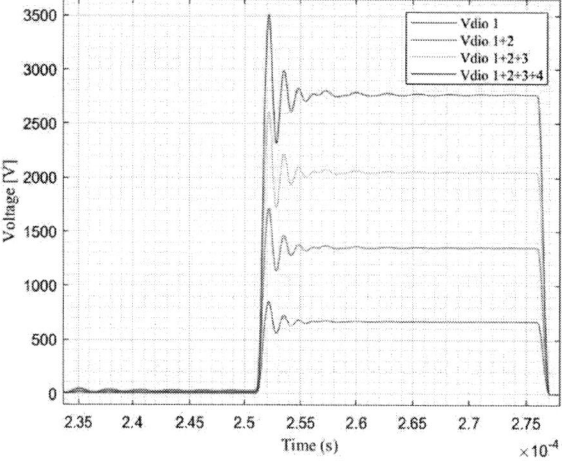

Fig. 6: Detailed simulation of a diode series connection, voltage distribution between the 4 diodes of the series connection. Voltages are stacked from the anode-side diode to the cathode-side diode.

The effect of value deviation (caused by manufacturing tolerances) of the snubber components on the voltage balancing can be seen (Figs. 7a and b). Snubber capacitor value deviation has a larger influence on balancing than that of snubber resistance.

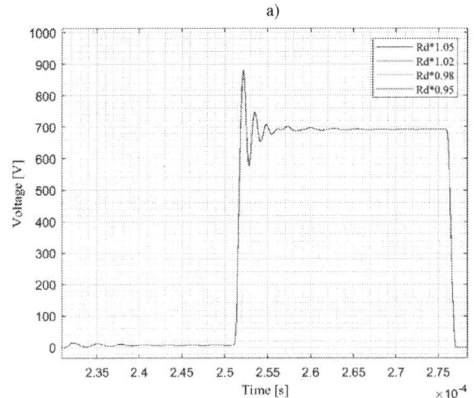 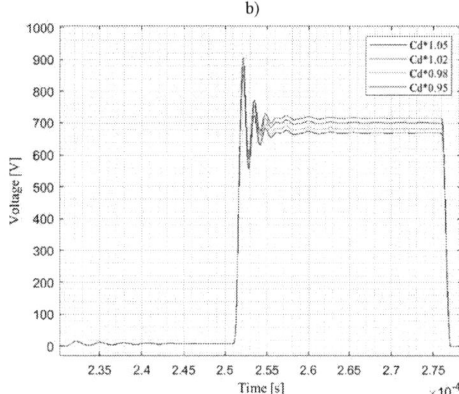

Fig. 7: Detailed simulation of a diode series connection, a) snubber resistor value deviation effect on voltage distribution, b) snubber capacitor value deviation effect on voltage distribution.

One should note that although complete simulations of the switching circuit enables to design the snubber precisely, the addition of so many parasitic components slows down the simulation that is already using a very small time step (typically inferior to 0.1 µs). Previous model presented in Fig. 4b is thus preferred when iterating on snubber components selection with regards to overvoltage limitation.

Experimental results

The designed rectifier is tested within the DC-DC converter test setup shown in Fig. 8. The inverter is operated at 20 kHz, with a full wave modulation. The MFT parasitic elements shown in Table II were characterized following the procedure given in [17].

Fig. 8: Test setup: a 1200 Vdc power supply feeds an inverter controlled in open loop to produce square waves at 20 kHz. The inverter is connected to the low voltage side of the MFT of ratio 1:m. The high voltage side of the MFT is connected to the studied rectifier, made of 4 assemblies shown in Fig. 3b.

The rectifier is tested in a non-clamped voltage configuration (inductive output) with an input voltage of V_{in} = 1200 V. The voltage distribution between the 4 diodes of an assembly is shown in Fig. 9. The maximum measured voltage imbalance is around 10%. The measured waveforms are similar to simulation results.

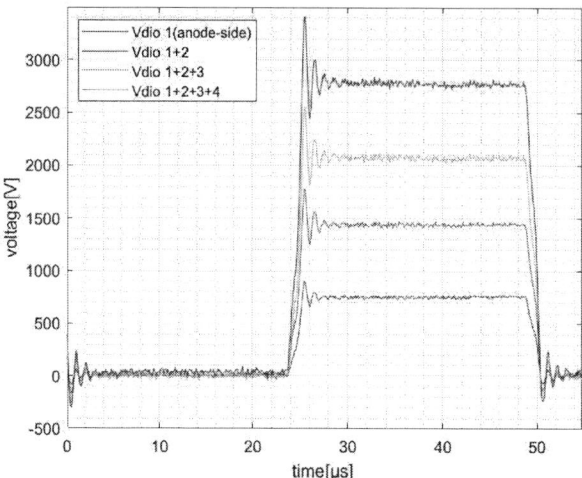

Fig. 9: Voltage distribution between the 4 diodes of the series connection. Voltages are stacked from the anode-side diode to the cathode-side diode.

The simplified model presented in the previous section (Fig. 4b) is ran with the same switching speed of the input bridge as observed in the measured waveforms (1.2 V/ns), the snubber components values from the design section, and the parasitic elements from Table II. The measured voltage across one diode string is shown in Fig. 10, superimposed with the model result.

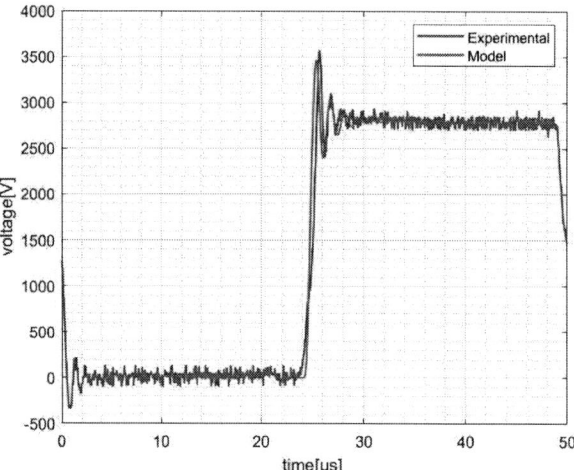

Fig. 10: Comparison between simulation results using the simplified model and experimental measurements of a diode assembly at turn-off.

The model result is close to the measured waveform. We can observe only small differences in terms of peak voltage, frequency of the oscillation and speed of the voltage step. These are considered to be due to inaccuracies in the transformer parasitic capacitances, as such small values (a few pF or tens of pF) are complicated to characterize with precision. However, the results still show the relevance of the simplified model for designing the snubber components in order to limit the overvoltage.

Conclusion

A rectifier for medium voltage medium frequency DC-DC converter has been designed and tested. The design choice in terms of number of series connected diodes is motivated by efficiency and reliability. The particular operation of the rectifier with an *LC* output filter ("unclamped operation") is studied and a simplified model is improved based on the literature. The model allows fast simulations of turn-off voltage oscillations, enabling snubber design taking into account DC-DC converter parasitic elements. Measurements confirm the proposed design and model.

References

[1] M. De Prada Gil, J. L. Domínguez-García, F. Díaz-González, M. Aragüés-Peñalba, and O. Gomis-Bellmunt, 'Feasibility analysis of offshore wind power plants with DC collection grid', *Renewable Energy*, vol. 78, pp. 467–477, Jun. 2015.

[2] CIGRE WG C6.31, 'Medium voltage direct current (MVDC) grid feasibility study', Feb. 2020.

[3] P. Le Métayer *et al.*, 'Break-even distance for MVDC electricity networks according to power loss criteria', in *2021 23rd European Conference on Power Electronics and Applications (EPE'21 ECCE Europe)*, 2021, pp. 1–9.

[4] H. A. B. Siddique, S. M. Ali, and R. W. De Doncker, 'DC collector grid configurations for large photovoltaic parks', in *2013 15th European Conference on Power Electronics and Applications (EPE)*, 2013, pp. 1–10.

[5] P. Le Métayer, Q. Loeuillet, F. Wallart, C. Buttay, D. Dujic, and P. Dworakowski, 'Unidirectional Isolated dc-dc converter for Photovoltaic MVdc Power Collection Networks', submitted to *IEEE Journal of Emerging and Selected Topics in Power Electronics*, 2022.

[6] 'Silicon Carbide (SiC) Schottky Diode - GeneSiC Semiconductor', *GeneSiC Semiconductor, Inc.* [Online]. Available: https://www.genesicsemi.com/sic-schottky-mps/. [Accessed: 06-Dec-2021].

[7] P. Dworakowski, P. Le Métayer, C. Buttay, and D. Dujic, 'Unidirectional step-up isolated DC-DC converter for MVDC electrical networks', accepted for CIGRE Session 2022, 2022.

[8] I. Ferencz, D. Petreus, and T. Pătărău, 'Comparative Study of Three Snubber Circuits for a Phase-Shift Converter', in *2020 International Symposium on Power Electronics, Electrical Drives, Automation and Motion (SPEEDAM)*, 2020, pp. 763–768.

[9] G. R. Zhu, D. H. Zhang, W. Chen, and F. Luo, 'Modeling and analysis of a rectifier voltage stress mechanism in PSFB converter', in *2012 Twenty-Seventh Annual IEEE Applied Power Electronics Conference and Exposition (APEC)*, 2012, pp. 857–862.

[10] 'High Voltage Rectifier Unit Series'. [Online]. Available: https://www.hv-semi.com/gyzldyxl. [Accessed: 06-Dec-2021].

[11] Z. Zeng *et al.*, 'Performance Comparison of FRD and SiC Schottky Diode in Si/SiC Hybrid Switch Power Module', in *2020 IEEE 9th International Power Electronics and Motion Control Conference (IPEMC2020-ECCE Asia)*, 2020, pp. 1890–1893.

[12] X. Liao, Y. Liu, J. Li, J. Cheng, and Y. Yang, 'A possible single event burnout hardening technique for SiC Schottky barrier diodes', *Superlattices and Microstructures*, vol. 160, p. 107087, Dec. 2021.

[13] Z. Salam and A. Rahman, 'Efficiency for photovoltaic inverter: A technological review', presented at the 2014 IEEE Conference on Energy Conversion, CENCON 2014, 2014, pp. 175–180.

[14] 'C247-025-1VE - Ohmite Mfg Co'. [Online]. Available: https://www.ohmite.com/catalog/c-series-heatsink/C247-025-1VE. [Accessed: 30-Mar-2022].

[15] Y. He and D. J. Perreault, 'Diode Evaluation and Series Diode Balancing for High-Voltage High-Frequency Power Converters', *IEEE Transactions on Power Electronics*, vol. 35, no. 6, pp. 6301–6314, Jun. 2020.

[16] CDE Cornell Dubilier, 'Application guide, Designing RC Snubber Networks'. .

[17] C. Liu, L. Qi, X. Cui, and X. Wei, 'Experimental Extraction of Parasitic Capacitances for High-Frequency Transformers', *IEEE Transactions on Power Electronics*, vol. 32, no. 6, pp. 4157–4167, Jun. 2017.

Fast Switching Planar Inductance Current Source ZETA Converter with Integrated Common Mode Filter

Benjamin H. Zacher, Member, IEEE, Christian Schumann, Senior Member, IEEE
P3E - Power Electronics, Electronics and EMC
Kaiserslautern University of Applied Sciences
Schoenstr. 11
Kaiserslautern, Germany
Phone: +49 (0) 631-3724-2242
Email: benjamin.zacher@hs-kl.de, christian.schumann@hs-kl.de
URL: http://www.hs-kl.de/p3e

Acknowledgments

This research has been funded by the German Federal Ministry for Education and Research (project number 13FH541KX9). We thank the SEW-EURODRIVE GmbH & Co KG in Bruchsal, Germany for supporting this project.

Keywords

≪Planar magnetics≫, ≪Wide bandgap≫, ≪DC-DC power converter≫, ≪Current source≫, ≪Coupled inductor≫.

Abstract

The ZETA topology provides unique opportunities as a source for current-fed inverter systems. This topology enables coupling of input and output inductor to reduce the current ripple. In this paper, design aspects for a coupled inductor with an integrated common mode filter are described and the advantages to provide a current source characteristic using a ZETA converter are discussed. Relevant parasitic inductances and capacitances are calculated using finite element analysis and analytical models. The behavior of the multiphysics system consisting of electrical and magnetical domain is considered using a numerical system simulation including parasitic elements. This simulation shows good results and is compared to experiments. The experimental setup is using gallium-nitride eHEMT transistors with switching frequencies up to 1 MHz.

Introduction

Due to the increasing widespread use and high availability of fast switching wide-bandgap devices, the current source inverter topology is becoming more and more a focus also for the use at small power ranges [1][2][3][4]. When using fast switching devices the design of the dc link inductor is still challenging. Especially when dealing with low inductance loads, the current source inverter performance decreases if the dc link current ripple is increased. Therefore a minimal current ripple is beneficial. When using ac input to feed a current source inverter system the common solution is an active rectifier which is used to control the dc link current [5]. When using dc input (e.g. a battery) the variety of different possible topologies to realize current source characteristic for the dc link is huge. To choose and optimize the best possible circuit and control it is necessary to focus on the advantages and features while using synergetic effects. A basic example is the buck-boost synergetic control strategy described in [6] which shows that it is possible to use a buck converter inductance as dc link and combining the inverter boost capability with the basic buck converter topology. The key in the design process is to use existing components to create new topological features.

In this paper the well-known ZETA topology [7][8] has been investigated. When removing the output capacitor the ZETA converter is suitable for a current source characteristic. It is also a common practice to use coupled inductors to decrease the output current ripple drastically [8]. Adding another inductance to the coupled inductor enables a new feature by integrating an output common mode filter.

The proposed topology is shown in fig. 1. On the one hand, in order to achieve a high power density, it is necessary to use high switching frequencies to minimize the inductor volume [11]. On the other hand high switching frequencies lead to increased EMI problems. With the aim of low manufacturing costs it is shown advantageous to use planar inductors based on PCB winding systems. Therefore a coupled inductor based on a PCB winding and standard E-I core was designed.

ZETA Converter

The basic ZETA topology provides a positive output voltage from a variable positive input voltage and can be operated in both step-up and step-down mode. Using two active switches it can be operated in bidirectional power mode. With inverted power flow the ZETA converter then operates as a SEPIC converter. Thus the converter is operating in continuous conduction mode and in order to achieve a current source characteristic, the output capacitor has to be removed. The output inductor then becomes

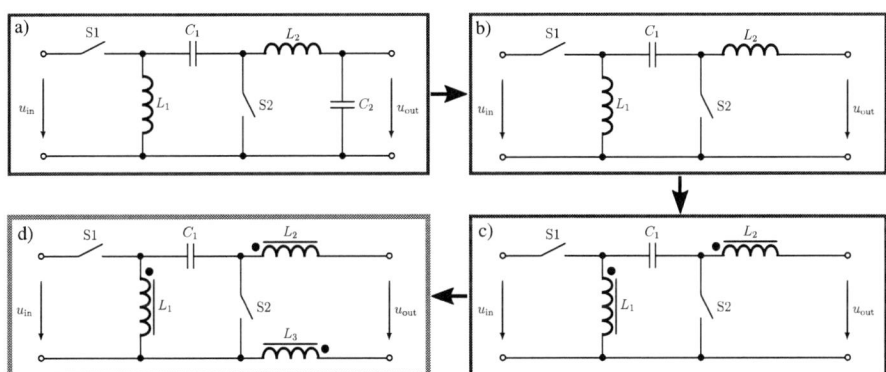

Fig. 1: a) basic bidirectional ZETA topology, b) removed output capacitor, c) added a coupling of the inductors, d) splitting the output inductor

the dc link inductor for use in current source inverter systems. To reduce the common mode noise the dc link inductor is often designed as a common mode choke [5] using coupled inductors. It is possible to further integrate this setup in the ZETA converter as shown in fig.1.

Proposed Topology and Application Ratings

Consequently the proposed topology is shown in fig. 2. The target application is a current source inverter

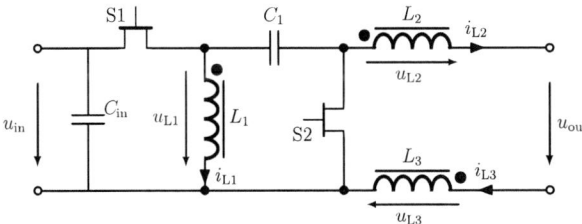

Fig. 2: Enhanced ZETA topology with integrated common mode filter and eHEMT transistors

Table I: Nominal Converter Ratings

input voltage	U_{in}	50 V
output current	I_{out}	5 A
output voltage	$U_{out,max}$	100 V
output power	$P_{out,max}$	500 W
switching frequency	f_{sw}	1 MHz

for a low power electric motor. The desired nominal converter ratings are shown in table I.

Coupled Inductor Design

The coupled inductor must meet multiple requirements. On the one hand the coupling should reduce common mode noise. On the other hand the inductor stores energy for the power conversion and reduces the output current ripple. Additionally the inductor should be reduced in size and use of material. The design process is similar to the design process of a dual-mode choke. A dual-mode choke decreases both common mode and differential mode noises and can be constructed in a single core geometry [12]. However, unlike a standard filter design the ZETA topology has to be considered.

Proposed Inductor Topology

The geometrical setup of the coupled inductor is shown in fig. 3. The inductor winding L_1 is the primary

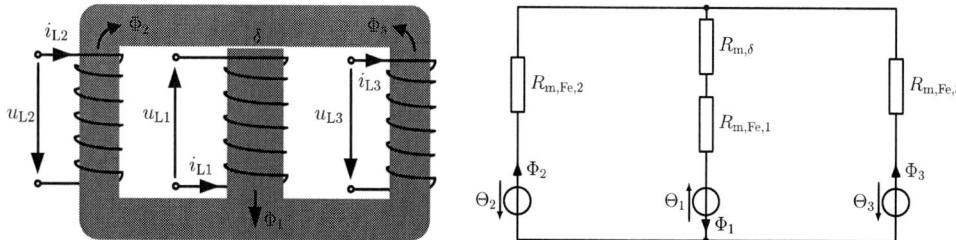

Fig. 3: Core geometry, winding setup and magnetic equivalent circuit

inductor of the ZETA converter and the flux, due to the current waveform, is inducing an additional negative voltage to the corresponding L_2 and L_3 winding terminals. This causes a decreased terminal voltage and thus a decreased current slew rate. The windings L_2 and L_3 are symmetrically coupled with winding L_1. An air gap δ is necessary for energy storage but also prevents saturation of the magnetic core material due to the high dc current in the windings. Thus resulting in a lower volume of the magnetic material. When applying a common mode current to L_2 and L_3 the magnetic flux components add to an opposing field which blocks the common mode noise. Because of the low magnetic resistance path the coupling between L_2 and L_3 is high. The magnetic equivalent circuit as shown in fig. 3 describes the mathematical relations of the coupled winding system. The following analytical terminal voltage model of the coupled inductor also considers the winding resistances R_{L1}, R_{L2} and R_{L3}.

$$\begin{pmatrix} u_{L1} \\ u_{L2} \\ u_{L3} \end{pmatrix} = \begin{pmatrix} R_{L1} & 0 & 0 \\ 0 & R_{L2} & 0 \\ 0 & 0 & R_{L3} \end{pmatrix} \begin{pmatrix} i_{L1} \\ i_{L2} \\ i_{L3} \end{pmatrix} + \begin{pmatrix} L_{11} & L_{12} & L_{13} \\ L_{21} & L_{22} & -L_{23} \\ L_{31} & -L_{32} & L_{33} \end{pmatrix} \cdot \frac{\mathrm{d}}{\mathrm{d}t} \begin{pmatrix} i_{L1} \\ i_{L2} \\ i_{L3} \end{pmatrix} \tag{1}$$

With the described magnetic circuit the inductance matrix can be analytically calculated and is defined by the flux distribution ignoring the leakage fluxes. The inductance values depend on the permeability of the used magnetic material, the geometrical dimension (length, cross-section and air gap) and number of turns of each winding. To consider leakage and nonlinear behavior of the used materials finite element calculations are necessary.

Finite Element Analysis (FEA)

To perform the finite element calculations a 3D model of the inductor is created. The used planar E-I core is a standard catalog part and therefore highly available and cost efficient. Each winding has a turn number of 6 and is distributed on two PCB layers as shown in fig. 4.

Inductance Matrix and Inductive Coupling Coefficients

The FEA yields the inductance matrix and the inductive coupling coefficients of the winding terminals. The calculated values are shown in table II. The coupling coefficient between L_2 and L_3 is limited by the core permeability. The calculated values are used to extend the model described in equation 1 which is then used in the numerical simulation.

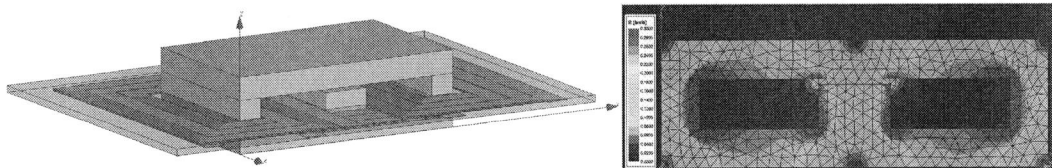

Fig. 4: 3D model of the planar coupled inductance, flux density at maximum power rating

Table II: Calculated inductance values and coupling coefficients

$L_{i,j}$	$i = 1$	$i = 2$	$i = 3$
$j = 1$	22.358 µH	10.975 µH	10.975 µH
$j = 2$	10.975 µH	81.389 µH	69.308 µH
$j = 3$	10.975 µH	69.308 µH	81.389 µH

$k_{i,j}$	$i = 1$	$i = 2$	$i = 3$
$j = 1$	1	0.257	0.257
$j = 2$	0.257	1	0.852
$j = 3$	0.257	0.852	1

Parasitic Capacitances

Because of the expected high voltage gradient du/dt parasitic capacitances can not be ignored. Many different equivalent circuits for parasitic capacitances in transformers and coupled inductors has been presented [13][14][15]. A common way is to differ between the winding self capacitance and the coupled capacitances between windings as shown in fig. 5. To determine the self capacitance values an analytical

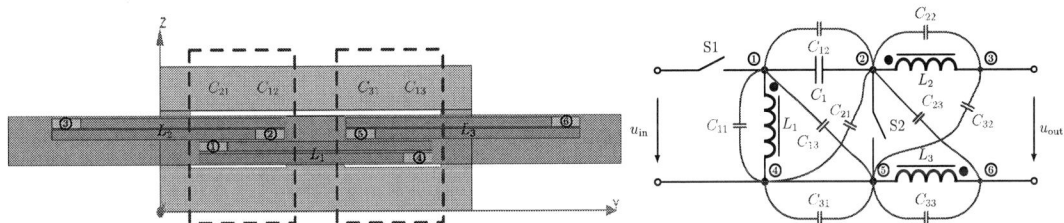

Fig. 5: Schematic layer structure and parasitic capacitances

calculation method for planar inductors is shown in [15], which is proven by FEA and experimental results. The winding coupling capacitances are much harder to determine but the parasitic effect can be avoided or minimized when changing the capacitive coupling of winding systems. In fig. 5 it is shown that due to the special core and winding arrangement the coupling of windings is minimized. Implementing the coupled inductor into the ZETA topology shows that some parasitic capacitances are not relevant (e.g. shorted) or can be used parallel to existing capacitances. The biggest parasitic effect have the self capacitances of each winding, which has to be considered in the simulation and experimental design. Therefore the capacitances have been minimized by using wide distances and a low number of turns. The capacitance matrix of the setup has been calculated by FEA. The relevant coupling capacitance and the self capacitances have been calculated yielding $C_{21} = 90$ pF and $C_{11} = C_{22} = C_{33} = 70$ pF.

Simulation Setup and Results

In order to determine the functionality a full parametric numerical system simulation is performed using matlab, simulink and the simscape library. The simulation contains the power electronic system for the ZETA topology, the load model and additionally the magnetic system for the coupled inductor model. Both domains are linked within the conservative simulation environment as displayed in the simulation schematic in fig. 6. Relevant parasitic components as described in the previous section are also consid-

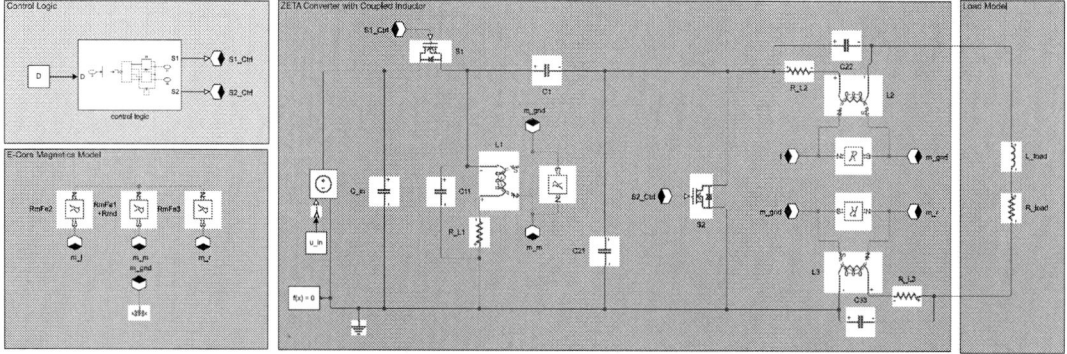

Fig. 6: Multi-domain numerical system simulation schematic in simulink

ered in the simulation. Due to the full parametric approach the influence of the switching frequency can easily be highlighted. Fig. 7 shows the resulting current ripple at different switching frequencies. Also

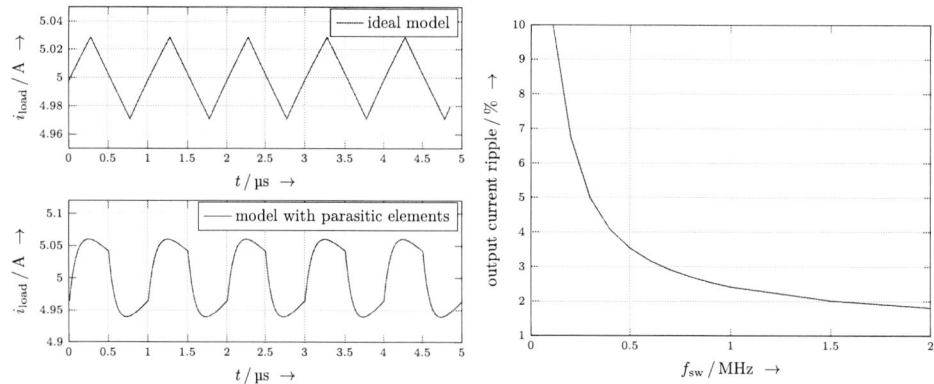

Fig. 7: Simulation results of an ideal converter and with parasitic elements at 1 MHz switching frequency (left). Switching frequency and corresponding current ripple (right).

the parasitic effects are seen in the results in fig. 7. Unfortunately it was not possible to include a dynamic transistor model because the existing models are not compatible with the used simulation environment. To get even better simulation results this can be done in the future. The simulation is using a fixed duty cycle so that the maximum power rating is achieved. Further a small deadtime which is also necessary in the experiment has been inserted.

Experimental Setup and Results

For the experimental evaluation a schematic and PCB layout has been created which is shown in fig. 8. To minimize the current density in the inductor windings parallel layers as well as a stacked PCB are used. The stacked PCB design also further reduces the parasitic capacitances because of the increased distance between winding layers. The used standard E-I core is modified so that the middle path has a 400 μm air gap and is then pressed on the PCB as shown in fig. 8. A standard connector ensures the

Fig. 8: 3D model of the experiment (left) and realized PCB (right)

safe connection to the controller board. The experimental setup uses GaN Systems eHEMT to realize 1 MHz switching frequency. Further to provide the necessary gate configuration an Infineon Technologies EiceDRIVER is used. The driver provides an isolated gate and supports switching frequencies up to 3 MHz. The driver is typically used to drive ohmic gate contact GaN transistors. Therefore the gate circuit had to be modified for the used schottky gate contact GaN Systems transistor. The gate voltage and the resulting drain-source voltages are shown in fig. 9. The gate voltage is set to negative at the off-switching phase to ensure safe off operation. Also in fig. 9 the drain source voltage of both switches show a ringing. This could not be optimized further because of the used gate circuit and is still a remaining problem. Because of the hard switching application the switching operation defines the switching losses and heat generation. Further a small deadtime is inserted which is improving the switching oper-

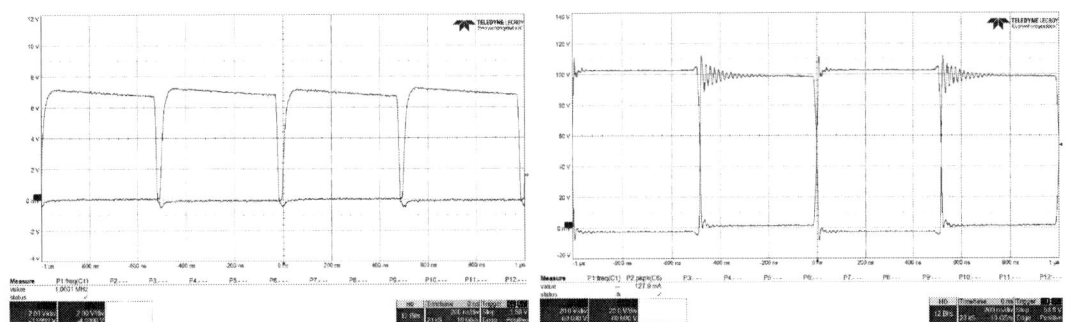

Fig. 9: Gate voltage (left) and drain source voltage (right) for both switches S1 (blue) and S2 (red)

ation. The deadtime has to be compensated in the control algorithm due to the nonlinear transfer function.

In order to measure the ouput current ripple a low inductance resistor was used. The inductance of the load including cables was determined below 8 µH. Because of this parasitic inductance the measurement results can not directly be compared to the simulation results but still the general functionality can be determined. Fig. 10 shows the measured output current at 100 kHz, 500kHz and 1 MHz switching frequency. At low switching frequencies the current ripple is defined by the current slew rate due to the duty cycle of the converter operation. Fig. 10 also shows that the ringing in the switching operation is affecting the resulting current ripple. At high switching frequencies the ringing which is caused by the switching operation is dominating the output current ripple so it remains unchanged at 500 kHz and 1 MHz respectively.

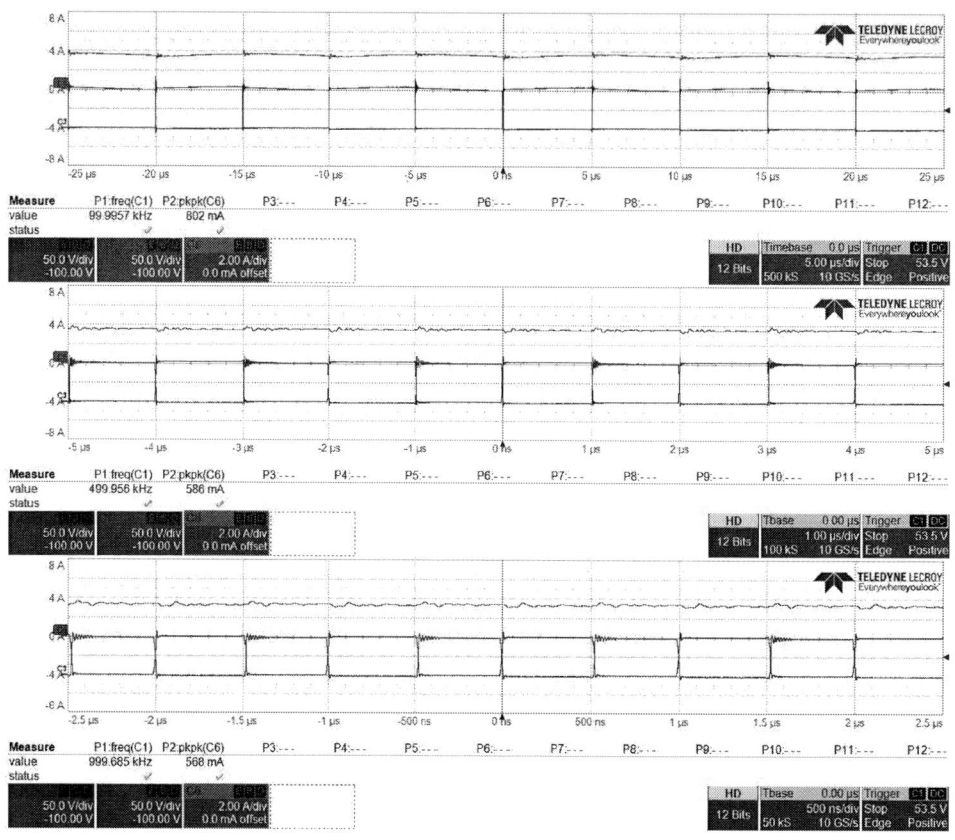

Fig. 10: Measured output current at different switching frequencies. Output current (green). Drain source voltage of S1 (blue) and S2 (red)

Conclusion

In this paper, the ZETA topology has been investigated, which is a good candidate for feeding battery powered current source inverter systems due to its unique possibility for coupled inductors. The further integration of an additional coupled winding system enables the feature of reducing common mode noise. The design of the coupled inductor is challenging and needs to consider parasitic elements. The parasitic inductances and capacitances have been calculated with FEA and are used to enhance the numerical simulation models. The numerical simulations show very promising results. An experiment was performed and it was shown that the basic converter function is working. It was not possible to match and verify the simulation results because the switching operations could not be sufficiently optimized. This research enables a deeper understanding for using PCB winding inductors for converters with high switching frequencies and high current densities. Further it opens up a field for multi-objective optimization of this system. When optimizing the layout and the switching operation the proposed system is a good candidate for current-fed inverters.

References

[1] H. Dai and T. M. Jahns, "Comparative investigation of PWM current-source inverters for future machine drives using high-frequency wide-bandgap power switches," 2018 IEEE Applied Power Electronics Conference and Exposition (APEC), 2018, pp. 2601-2608, doi: 10.1109/APEC.2018.8341384.

[2] H. Dai, T. M. Jahns, R. A. Torres, D. Han and B. Sarlioglu, "Comparative Evaluation of Conducted Common-Mode EMI in Voltage-Source and Current-Source Inverters using Wide-Bandgap Switches," 2018 IEEE Transportation Electrification Conference and Expo (ITEC), 2018, pp. 788-794, doi: 10.1109/ITEC.2018.8450157.

[3] R. A. Torres, H. Dai, W. Lee, T. M. Jahns and B. Sarlioglu, "Current-Source Inverters for Integrated Motor Drives using Wide-Bandgap Power Switches," 2018 IEEE Transportation Electrification Conference and Expo (ITEC), 2018, pp. 1002-1008, doi: 10.1109/ITEC.2018.8450127.

[4] R. A. Torres, H. Dai, W. Lee, T. M. Jahns and B. Sarlioglu, "Development of Current-Source-Inverter-based Integrated Motor Drives using Wide-Bandgap Power Switches," 2019 IEEE 15th Brazilian Power Electronics Conference and 5th IEEE Southern Power Electronics Conference (COBEP/SPEC), 2019, pp. 1-6, doi: 10.1109/COBEP/SPEC44138.2019.9065675.

[5] T. Friedli, S. D. Round, D. Hassler and J. W. Kolar, "Design and Performance of a 200-kHz All-SiC JFET Current DC-Link Back-to-Back Converter," in IEEE Transactions on Industry Applications, vol. 45, no. 5, pp. 1868-1878, Sept.-oct. 2009, doi: 10.1109/TIA.2009.2027538.

[6] M. Guacci, M. Tatic, D. Bortis, J. W. Kolar, Y. Kinoshita and H. Ishida, "Novel Three-Phase Two-Third-Modulated Buck-Boost Current Source Inverter System Employing Dual-Gate Monolithic Bidirectional GaN e-FETs," 2019 IEEE 10th International Symposium on Power Electronics for Distributed Generation Systems (PEDG), 2019, pp. 674-683, doi: 10.1109/PEDG.2019.8807580.

[7] S. Shringi, S. K. Sharma and K. S. Rathode, "Comparative Study of Buck-Boost, Cuk and Zeta Converter for Maximum output Power Using P&O technique with solar," 2019 2nd International Conference on Power Energy, Environment and Intelligent Control (PEEIC), 2019, pp. 538-542, doi: 10.1109/PEEIC47157.2019.8976534.

[8] Dimna Denny C and Shahin M, "Analysis of bidirectional SEPIC/Zeta converter with coupled inductor," 2015 International Conference on Technological Advancements in Power and Energy (TAP Energy), 2015, pp. 103-108, doi: 10.1109/TAPENERGY.2015.7229600.

[9] J. C. Schroeder and F. W. Fuchs, "Detailed characterization of coupled inductors in interleaved converters regarding the demand for additional filtering," 2012 IEEE Energy Conversion Congress and Exposition (ECCE), 2012, pp. 759-766, doi: 10.1109/ECCE.2012.6342743.

[10] C. Østergaard, C. S. Kjeldsen and M. Nymand, "Calculation of Planar Transformer Capacitance Based on the Applied Terminal Voltages," 2020 IEEE 21st Workshop on Control and Modeling for Power Electronics (COMPEL), 2020, pp. 1-7, doi: 10.1109/COMPEL49091.2020.9265797.

[11] Z. Ouyang, O. C. Thomsen and M. A. E. Andersen, "Optimal Design and Tradeoff Analysis of Planar Transformer in High-Power DC–DC Converters," in IEEE Transactions on Industrial Electronics, vol. 59, no. 7, pp. 2800-2810, July 2012, doi: 10.1109/TIE.2010.2046005.

[12] Y. Shiraki, S. Yoneda, K. Omae and T. Nagao, "Inductance Analysis for Compact Dual-Mode Choke Considering Magnetic Saturation," 2018 International Symposium on Electromagnetic Compatibility (EMC EUROPE), 2018, pp. 630-635, doi: 10.1109/EMCEurope.2018.8485094.

[13] C. Hebedean, C. Munteanu, A. Racasan, C. Pacurar and D. Augustin, "The influence of parameters on the parasitic capacitance values in a planar transformer," 2015 9th International Symposium on Advanced Topics in Electrical Engineering (ATEE), 2015, pp. 838-843, doi: 10.1109/ATEE.2015.7133942.

[14] V. K. N., S. Satpathy and L. N., "Analysis and design methodology for Planar Transformer with low self-capacitance used in high voltage flyback charging circuit," 2016 IEEE International Conference on Power Electronics, Drives and Energy Systems (PEDES), 2016, pp. 1-5, doi: 10.1109/PEDES.2016.7914510.

[15] L. Dalessandro, F. da Silveira Cavalcante and J. W. Kolar, "Self-Capacitance of High-Voltage Transformers," in IEEE Transactions on Power Electronics, vol. 22, no. 5, pp. 2081-2092, Sept. 2007, doi: 10.1109/TPEL.2007.904252.

System Level Simulation of Moisture Propagation and Effects in Wind Power Converters

WENZEL, Johannes C. and MERTENS, Axel
LEIBNIZ UNIVERSITY HANNOVER – Institute for Drive Systems and Power Electronics
Welfengarten 1, 30167 Hannover, Germany
Phone: +49 511 762 - 188 35
Fax: +49 511 762 - 30 40
Email: johannes.wenzel@ial.uni-hannover.de
URL: http://www.ial.uni-hannover.de

July 11, 2022

Acknowledgement

This work is part of the project ReCoWind and was funded by the Federal Ministry for Economic Affairs and Climate Action on the basis of a decision by the German Bundestag. Funding number: 0324336E.

Keywords

≪Reliability≫, ≪Modelling≫, ≪Environment≫, ≪Mission profile≫, ≪AC/AC converter≫

Abstract

The aim of this work is to create a better understanding of the propagation of moisture and its effects in power electronic converters with the help of computer-aided simulations. Due to the interaction of humidity and temperature, a combination of multiple simulation models is required. Suitable measurement methods for parameter identification and validation will also be presented.

Introduction

The increasing use of power electronics even under extreme climatic conditions (e.g., e-mobility or offshore applications) can cause moisture-induced degradation mechanisms in electrical components [1, 2]. A more detailed understanding of moisture diffusion acquired by using computer-aided modelling at a system level should help to prevent critical degradation in the future.

The goal of such modelling and simulation is to identify critical combinations of load and climate, for example, with the aim to adjusting the variables that can be influenced (load) via the operational management. This includes preheating the power modules of the converter when the power electronic system is restarted after longer downtimes, during which moisture may have penetrated the power module. When the moisture has made its way to the chips, they can suffer from various degradation mechanisms when re-applying the operating voltage [3, 4].

Humidity Modelling under Dynamic Temperature Cycles

For an accurate simulation of moisture propagation in such complex systems as power electronic converters, the first step is to create a simulation environment that can reproduce the diffusion processes of water molecules. There are already many commercially available options for simulating diffusion processes, such as those of heat. However, these cannot be applied equivalently to diffusion and transfer of mass due to moisture concentration gradients across material boundaries, which are strongly dependent on the solubility of water in the materials [5].

The first publications in the field of moisture modelling in power electronics were realised with the help of electrical analogies [6, 7]. Just as resistors and capacitors can be used for thermal modelling, R-C networks were used to represent the diffusion and storage behaviour in so-called hygrothermal models. However, if thermodynamic effects are correctly represented, the use of electrical analogies quickly reaches its limits and becomes unnecessarily convoluted.

In comparison to modelling using electrical analogies, another modelling approach is based on Fick's second diffusion law, as proposed one dimensionally in [8, 9]

$$\frac{\partial c}{\partial t} = D(T) \cdot \frac{\partial^2 c}{\partial x^2} \qquad \text{with} \qquad D(T) = D_0 \cdot \mathrm{e}^{-\frac{Q}{k \cdot T}} \tag{1}$$

where c is the concentration of the particles, which can also be seen as the absolute humidity h_{abs}. D_0 is the diffusion constant of the respective material, Q is the activation energy of water, and k is the Boltzmann constant. This exists in analogy to the thermal behaviour based on the heat transfer equation

$$\frac{\partial T}{\partial t} = \frac{\lambda}{\rho \cdot c_{\mathrm{p}}} \cdot \frac{\partial^2 T}{\partial x^2} \tag{2}$$

where Lambda is the coefficient of thermal conductivity, ρ is the material-dependent density, and c_{p} its specific heat capacity related to the mass.

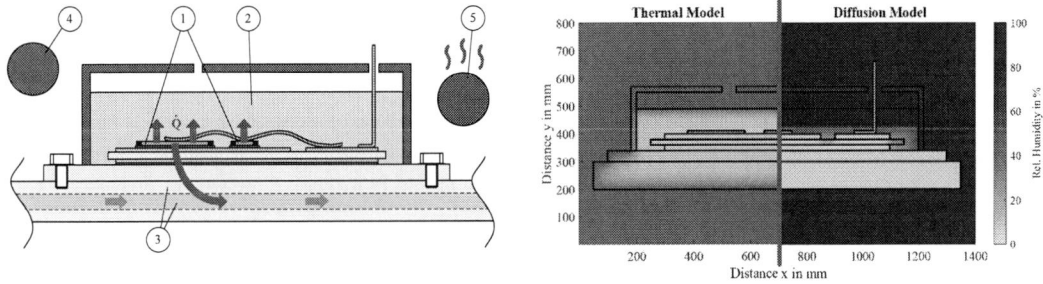

Fig. 1: Left: Cross-section of an IGBT module with power semiconductors (1) encapsulated in silicone (2) on a heat sink (3) with simplified additional heat sources, e.g., chokes (5) and moisture reservoirs, e.g., plastic components (4). Right: Example of the qualitative module cross-section shown in (a) viewed in the two-dimensional hygrothermal model - thermal on the left and moisture diffusion on the right.

The modelling approach in this work based on Fick's diffusion laws is to be performed by a one-, two- or three-dimensional mapping of temperature and absolute humidity for the materials and topologies of the systems considered. An example is shown in Fig. 1 for a two-dimensional module cross-section, where the aim is to identify and predict critical scenarios with the help of the moisture distribution shown on the right. Here, the power semiconductor losses and the heat sink are not represented. Coupling with thermal models is always necessary, as visualised in Fig. 2, since the interrelations considered are strongly temperature dependent and linked via the diffusion coefficient $D(T)$. A schematic representation is given in Fig. 2, where the ambient temperature ϑ_{env} and the power dissipation at the chip are first passed to the thermal part of the model as input variables and the ambient humidity $h_{\mathrm{rel,env}}$ is added afterwards in the diffusion model. It should be noted that the thermal model is not the same as in conventional thermo-electric modelling of power converters, since the temperature distribution especially in the opposite direction to the heat sink (through the electrically insulating silicone) is needed. The output goal of the model is to obtain the relative humidity $h_{\mathrm{rel,comp}}$ at a certain place of interest in the system. This value can be used from experience of critical moisture levels in power modules to make predictions on the degradation state at these locations. This makes it possible to define location-specific moisture thresholds.

Thus, in addition to the ambient humidity and temperature, the load profile of the converter or drive system must be considered, leading to strong temperature fluctuations and therefore to fluctuations in the diffusion behaviour within the materials. It is necessary to identify heat sources within the converter cabinet whose dynamic cycles depend on the load profile. This can be achieved with the help of a higher-level system model (loss model), as presented in [10], to simulate load cycles and losses on the basis of realistic load profiles. The resulting power losses can be entered into the thermal part of the hygrothermal diffusion model (here in one dimension) via:

$$\frac{P_{\text{loss}}}{A} = \dot{q} = -\lambda \cdot \frac{\mathrm{d}T}{\mathrm{d}x} \tag{3}$$

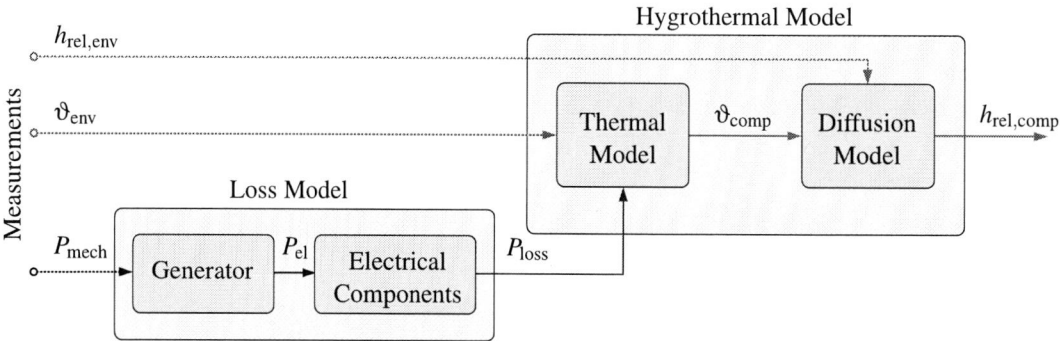

Fig. 2: Internal operation of the loss and hygrothermal model with weak coupling through thermal paths

Since the time constants of the material diffusion are significantly higher than the thermal ones, even at high temperatures, it is sufficient to use a quasi-stationary loss model using averaged losses, for instance over a period of the fundamental frequency, particularly when significantly longer time series of up to several months are considered.

In air, in addition to the mass flow J_{D} of diffusion, mass transport of water molecules occurs significantly faster via convection – forced by the fan or caused naturally by heat sources. For now, due to the long time constant associated with diffusion, this can be handled by making the assumption that the air has a uniform humidity. Computational fluid dynamics (CFD) simulation tools are suitable for more detailed consideration. CFD simulations are able to evaluate pressure differences of the ambient air volume so that a vector field of the air flow is created, which could be added to the diffusion mass flow J_{D} in future work.

Parameter Identification and Validation Methods

A theoretical analysis of such complex processes as diffusion and storage of moisture is made considerably more difficult by strongly material-dependent aspects. Parameters from literature might be hardly transferable due to the diversity of material compositions and must be verified with the help of test specimens using the materials investigated. Furthermore, an abstraction of the overall system is required due to the large number of different components used in a converter system. In the following, two pre-developed test concepts will be presented, which can be used for the parameterisation of hygrothermal models. Several types of materials are especially important due to their ability to store moisture: insolation materials like polypropylene (PP) or polyvinylchloride (PVC); epoxy as used for PCBs or insolating construction material (FR4); and of course silicone gel, as used in the power semiconductor modules.

Gravimetric Measurements

Critical parameters, such as the diffusivity or solubility of the plastic materials, can be determined, for example, via gravimetric measurements under controlled climatic conditions. The DIN EN ISO 12572 standard presents a method which uses a dehumidifying substance in an airtight cup which is covered and sealed by a thin specimen of the material to be tested. Here, the diffusion constant for a given temperature can be determined gravimetrically via

$$J_D = \frac{\Delta \dot{m}}{A_1} = -D \cdot \frac{dc}{dx} \tag{4}$$

where A_1 is the cross-sectional area of the material and $\Delta \dot{m}$ is the weight increase per unit time. In this work, a simpler method is used. The mass of the tested material specimen in this paper is measured (without any enclosure or cup) at regular time intervals on a high-precision scale. All specimens have a nearly identical volume with a surface-to-volume ratio as large as possible. The tested materials (PP as used in component housings, FR4 with and without coating for PCBs, and PVC as used in cable insulation) need to be dried over an extended period of time (here at 10% relative humidity and 85°C for one month) [11]. The weights shown in Fig. 3 are normalised relative to the minimum weight in the dry state: $m = m_{\text{meas}}/m_{\text{dry}} \cdot 100\% - 100\%$. It will not be possible to achieve a completely dry condition in a finite time due to the ambient humidity which remains for technical reasons. After the drying, a humidity step is applied, also at a defined environmental temperature, which leads to an exponential approach of the weight of the materials to the steady state.

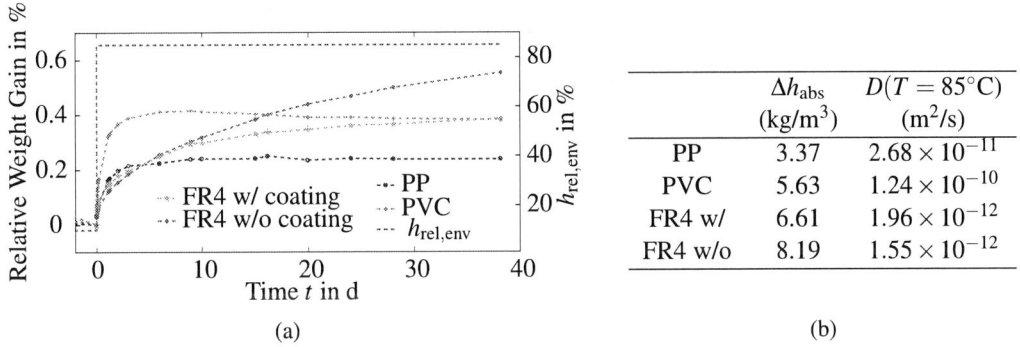

	Δh_{abs} (kg/m^3)	$D(T = 85°C)$ (m^2/s)
PP	3.37	2.68×10^{-11}
PVC	5.63	1.24×10^{-10}
FR4 w/	6.61	1.96×10^{-12}
FR4 w/o	8.19	1.55×10^{-12}

(a)   (b)

Fig. 3: (a) Gravimetric measurement of different materials at a temperature of 85°C; (b) Listed absolute water absorption and calculated diffusion constants

The different diffusion constants are clearly visible. Based on Fick's diffusion behaviour, the ratio of $m_{0.5}$, the mass at the time when 50% of the total moisture has been absorbed, to the total mass increase m_∞ can be calculated as a function of the square root of time. This can be rearranged to give D for the one-dimensional case of a thin test specimen with a thickness l [12]:

$$\frac{m_{0.5}}{m_\infty} = 4 \cdot \sqrt{\frac{D \cdot t_{0.5}}{\pi \cdot l^2}} \qquad \Leftrightarrow \qquad D = \frac{\pi}{16} \cdot \left(\frac{m_{0.5}}{m_\infty}\right)^2 \cdot \frac{l^2}{t_{0.5}} \tag{5}$$

It should be noted that PP has already reached its maximum weight after one week, while this has not yet been reached in the FR4 samples even after more than one month. Also with some materials, such as PVC, the weight decreases after reaching a maximum, which could be due to the fact that the material itself reacts with the water molecules and the reaction products are outgassed due to the relatively high environmental temperature. The easiest way to circumvent this effect is to precondition the samples in advance with high ambient humidity.

Diffusion Time Behaviour using Moulded Humidity Sensors in Silicone Gel

Due to its physical contact with the power semiconductor, the insulating silicone gel inside the modules plays a key role in the accumulation of water on the chip surface. Therefore, a method is presented here for recording dynamic humidity curves and their temperature dependency with the help of commercially available sensors moulded into the silicone. This method can be used to derive the coefficients of the specific silicone by matching, or to validate the simulation with parameters from other tests. For this purpose, a stainless steel cup is filled with silicone as shown in Fig. 4 (a), in which the sensors of type Sensirion SHT31-ARP on a tiny PCB are mounted and connected via the bottom of the cup. The only diffusion exchange possible with the environment is through the upper surface. Based on [8], different stationary temperatures are set, while a dynamic humidity step is specified in both directions in order to measure the temperature diffusion dependency. In addition, it can be seen in Fig. 4 (b) that the sensor (in this case, only the measurement profiles of the sensors 1 and 4 are represented) show increasingly low-pass filtered behaviour compared to the environment with increasing depth of the sensors.

The dashed lines show the corresponding simulated temperatures and relative humidities from a 2D version of the silicone encapsulated test specimen. The identical ambient conditions were given to the simulation model.

The large deviation from the measurement between days 2 and 3 is discussed in the following paragraph. In addition to the temperature-dependent diffusion, another effect can be observed in Fig. 4: between

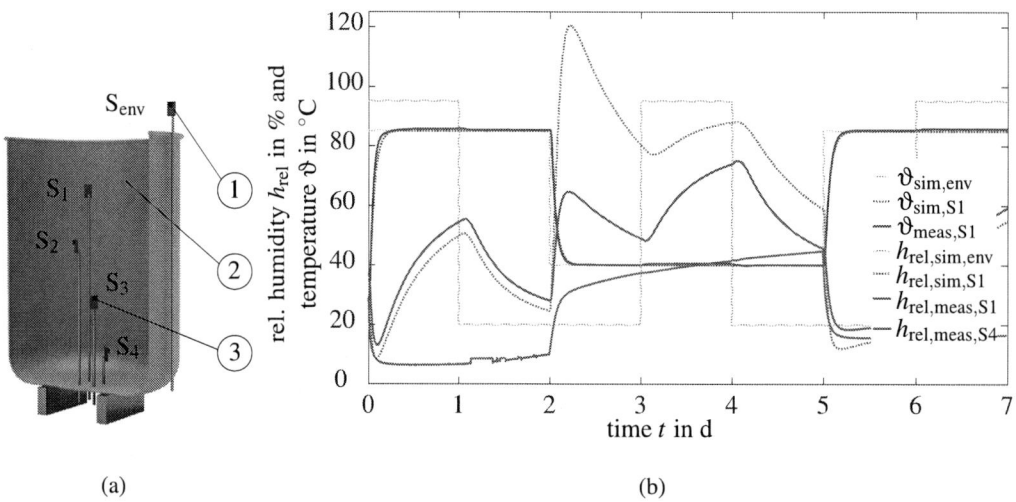

(a)   (b)

Fig. 4: (a) DUT: (1) environmental sensor, (2) silicone cast in stainless steel cup (diameter: 10 cm, height: 13.5 cm) and (3) moulded sensors at 30 mm spacing; (b) Dynamic moisture diffusion behaviour within the silicone at different ambient temperatures and relative humidities using data from the measuring setup

days 2 and 5, the humidity at sensor S_4 increases continuously, although the ambient humidity at sensor S_{env} fluctuates significantly. This can be attributed to the storage effect of the silicone layers lying over sensor S_4. It can thus be confirmed that moisture storage effects within the materials can lead to strongly deviating humidities in systems compared to their environment. This measuring method, however, is only useful in potting compounds such as silicone. For plastics like the ones shown earlier, a gravimetric measuring method is still recommended.

Another observation results when the measured relative humidity in the silicone is converted into an absolute humidity using the known dew point curve for air, which indicates the ratio of absolute, maximum humidity to temperature (Fig. 5 (a)). This approach is important to consider because the sensors are calibrated for relative humidity in air and not in silicone.

The difference between simulation and measurement of the relative humidity between days 2 and 3 in Fig. 4 (b) can be explained by taking a look at the calculated absolute humidity in Fig. 5 (b), where the data from sensor S_1 shows a drop within the measured absolute humidity. This is an effect that the simulation model based on Fick's diffusion (correctly) does not represent. A drop of this magnitude can only be explained by the antiproportional dependence on the increasing temperature at this point. Here it can be seen that the use of the dew point curve for air in silicone is not valid. As a further investigation, described in the following section, a dew point curve equivalent to that in Fig. 5 (a) must be determined for silicone.

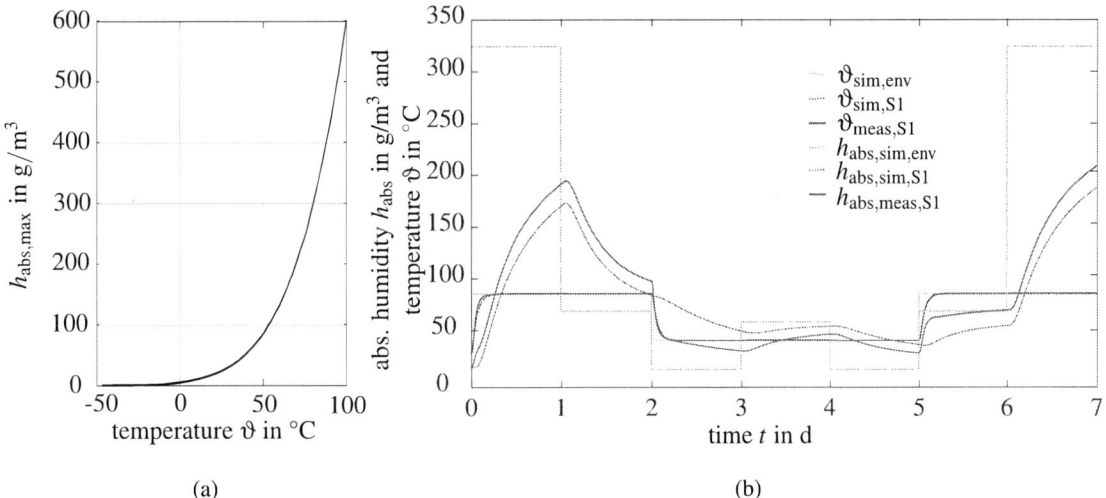

(a)   (b)

Fig. 5: (a) Dew point curve of water vapour in air; (b) absolute humidity in the silicone calculated with the dew point curve from (a) based on the measurements and simulations from Fig. 4

Optical Measuring Method for Silicone Dew Point Detection

Due to the fact that commercially available sensors like the Sensirion SHT31-ARP used in this work are calibrated for the relative humidity in air, the dew point behaviour of the silicone gel must also be determined. It is assumed that the so-called fogging of the gel occurs when the dew point is reached and small water droplets appear inside the gel. This leads to an intransparency of the material which can be detected. The basic procedure of the measurement is as follows: first, a specific relative humidity (for example 85%) is set in the environment of the test specimen at a high temperature like 85°C for faster diffusion. As soon as a steady state of relative humidity is reached inside the silicone, the ambient temperature can be lowered as quickly as the climatic chamber allows, until an optical measuring sensor reveals a clearly perceptible cloudiness in the otherwise transparent silicone. The temperature existing in the silicone at this point corresponds to the dew point temperature.

To determine the absolute humidity in the silicone associated with the previously measured dew point temperature, a gravimetric measurement of the silicone must also be carried out. This is accomplished under the same initial environmental conditions as the optical measurements. Here, too, the same procedure can be used as for the gravimetric measurements performed on the plastics described above [13, 14]. This procedure is to be repeated for different relative ambient humidities, resulting in a dew point curve for the silicone. With the help of this curve, it is possible to obtain information about the relative humidity in the silicone using commercial sensors encapsulated in the silicone, which is of special interest when investigating the degradation mechanisms affecting the power semiconductors [15].

Conclusion

For a better understanding of moisture penetration within complex systems – which are exposed not only to ambient conditions but also to thermal fluctuations due to load cycles – it is necessary to develop hygrothermal simulation models which, above all, accurately represent the interaction between temperature and humidity. For this purpose, a number of different parameter identification and validation procedures need to be carried out, including those presented here. Furthermore, next to the analysis of the relative humidity as a prediction of condensation risks, a look at the absolute humidity is always relevant as a verification of whether the scenarios considered (for example in sealing compounds) are valid. In addition, a coupling with further simulation models, which addresses both the convection overlaying diffusion and the realistic calculation of power losses as thermal input for these models, is the object of current analyses.

References

[1] Conseil, Helene, et al. "Humidity build-up in a typical electronic enclosure exposed to cycling conditions and effect on corrosion reliability." IEEE Transactions on Components, Packaging and Manufacturing Technology 6.9 (2016): 1379-1388.

[2] Kremp, Sebastian, Oliver Schilling, and Verena Mueller. "Empirical study on humidity conditions inside of power modules under varying external conditions." CIPS 2016; 9th International Conference on Integrated Power Electronics Systems. VDE, 2016.

[3] Zorn, Christian, and Nando Kaminski. "Temperature humidity bias (THB) testing on IGBT modules at high bias levels." CIPS 2014; 8th International Conference on Integrated Power Electronics Systems. VDE, 2014.

[4] Kostka, Benedikt Rafael, et al. "A Concept for Detection of Humidity Driven Degradation of IGBT Modules." IEEE Transactions on Power Electronics (2021).

[5] Comyn, John, ed. Polymer permeability. Springer Science and Business Media, 2012.

[6] Bayerer, Reinhold, Matthias Lassmann, and Sebastian Kremp. "Transient hygrothermal-response of power modules in inverters—The basis for mission profiling under climate and power loading." IEEE Transactions on Power Electronics 31.1 (2015): 613-620.

[7] Staliulionis, Ž., S. Mohanty, and J. H. Hattel. "Resistor-Capacitor Approach for Modelling of Temperature and Humidity Response Inside Electronic Enclosures." 2019 20th International Conference on Thermal, Mechanical and Multi-Physics Simulation and Experiments in Microelectronics and Microsystems (EuroSimE). IEEE, 2019.

[8] Quast, Fabian, and Andreas Nagel. "Humidity in Traction Converters." PCIM Europe 2018; International Exhibition and Conference for Power Electronics, Intelligent Motion, Renewable Energy and Energy Management. VDE, 2018.

[9] Schuster, Oskar, Andreas Nagel, and Bernd Laska. "Observation and simulation of dynamic humidity in power converters for railway applications due to moisture diffusion in plastics." 2021 23rd European Conference on Power Electronics and Applications (EPE'21 ECCE Europe). IEEE, 2021.

[10] Morisse, Marcel, et al. "Dependency of the lifetime estimation of power modules in fully rated wind turbine converters on the modelling depth of the overall system." 2016 18th European Conference on Power Electronics and Applications (EPE'16 ECCE Europe). IEEE, 2016.

[11] Qiu, Zhijie, et al. "Study on Moisture Absorption Characteristics of Power Plastic Packaging Devices." 2018 21st International Conference on Electrical Machines and Systems (ICEMS). IEEE, 2018.

[12] Dermitzaki, E., et al. "Structure Property Correlation of epoxy resins under the influence of moisture and temperature; and comparison of Diffusion coefficient with MD-simulations." 2008 2nd Electronics System-Integration Technology Conference. IEEE, 2008.

[13] Hatori, Kenji, et al. "Humidity Absorption Behavior of Silicone Gel in HVIGBT Modules." 2021 23rd European Conference on Power Electronics and Applications (EPE'21 ECCE Europe). IEEE, 2021.

[14] Harley, Stephen J., Elizabeth A. Glascoe, and Robert S. Maxwell. "Thermodynamic study on dynamic water vapor sorption in Sylgard-184." The Journal of Physical Chemistry B 116.48 (2012): 14183-14190.

[15] Nakamura, Keiichi, et al. "The test method to confirm robustness against condensation." 2019 21st European Conference on Power Electronics and Applications (EPE'19 ECCE Europe). IEEE, 2019.

PWM-Based Optimization-Free Active Voltage-Balancing Control of 7-Level Active Neutral-Point-Clamped Flying-Capacitor Multicell Inverters

Vahid Dargahi
University of Washington
1900 Commerce St
Tacoma, WA, USA
Phone: +1 (253) 692-5812
Email: vdargahi@uw.edu

Keywords

≪Active balancing≫, ≪A-NPC topology≫, ≪Cost-function elimination≫, ≪Flying-capacitor≫, ≪Logic-equations≫, ≪Optimization-free method≫, ≪Pulse-width modulation≫, ≪Voltage control≫

Abstract

This paper introduces a new pulse-width-modulation (PWM) -based control technique for 7-level active neutral-point-clamped (A-NPC) flying-capacitor multicell (FCM) inverters. It regulates inverter's output voltage, current, and balances FCs at reference values. It is realized on basis of active control approach which is achieved using instantaneous measurements taken from inverter, and then being interpreted into the logic quantities. Contrary to space-vector modulation (SVM) and model predictive control (MPC) methods, the proposed solution is deemed a direct approach that excludes cost/objective functions and time/software-consuming optimization processes to generate the switching states. It utilizes the measured values of the FC voltages and load current to derive the logic variables and equations that directly determine the inverter's switching states according to the reference voltage-levels. As its advantage, any commonly used modulators such as phase-shifted-carrier (PSC) or phase-disposition (PD) strategy could be readily embedded into control unit to generate inverter's reference voltage-levels. The experimental results are provided to validate the suggested active balancing technique, its effectiveness and fast execution-time.

1 Introduction

Multilevel converters provide high-efficiency performance for a myriad of industrial applications ranging from low-voltages to medium/high-voltages [1]. The flying-capacitor (FC)-based topologies including FC-multilcell (FCM) as well as active neutral-point-clamped (A-NPC) converters are gaining increased popularity, specifically for the data center power supplies and electric aircraft applications owing to the prominent feature of high power-density that is originated from the use of multiple switching cells offering reduced voltage rating for switches and FCs [2].

A 10-level GaN-based FCM dc-dc boost converter has been reported in [3] for spacecraft applications where it achieves a power density of 24 kW/kg with an efficiency of 99.1%. Similarly, a 7-level GaN-based FCM inverter has been implemented in [4] which features a power density of 216 W/in^3 with a peak efficiency of 97.6%. The design and implementation of a 1 kV to 380 V inverter based on a 9-level FCM topology which is realized using the 200 V GaN-FET switches has been reported in [5]. This dual-interleaved converter achieves a peak efficiency of 98.6% at 6 kW power level with density of 15 kW/kg.

The 3-level A-NPC converter is derived from 3-level diode-clamped topology by replacing the passive clamping-power-diodes with active switches [6]. This new additions alleviates the thermal issues by bringing down the junction temperature of the semiconductors which is resulted from an improved power loss distribution among the devices [7,8].

Despite the better thermal characteristics that are gained from the aforementioned modifications, the resultant A-NPC topology can generate only 3-level staircase voltages within its each phase. The insertion of the FC-based switching-cells in a tandem arrangement is an innovative solution that marries the 3-level A-NPC converter with FCM topology. This hybrid configuration realizes the extension of the voltage-levels beyond 3, and more importantly, paves the way for reaching high-voltage and high-power ranges. For instance, PCS 8000 drive series manufactured by ABB uses this configuration to provide power in the range of 6 MVA up to 100 MVA [9].

It is worth bringing the system's stability issue to the fore since the use of the FC-based multicells imposes a significant complexity on converter, and puts the overall hybrid ANPC-FCM topology into the verge of instability. This inherent phenomenon originates from the fact that the voltage of the FCs within switching power cells tends to drift away and to deviate from their reference values given that an appropriate control strategy is not adopted or not provided. This might result in the failure of the switches, and consequently, it deteriorates the power quality of the generated voltage and current waveforms drastically [10].

Thus, a correct voltage stabilization of the FCs is indispensable for the safe functioning of converter system which consequently averts the excessive voltage stress on the switches. The FCs' voltage drift and imbalance issues have been addressed in the previous work [11–13].

Space-vector modulation (SVM) and model predictive control (MPC) techniques are the most commonly adopted methods for this purpose [14–17]. They use the deviation of the FC voltages from their reference values in order to construct the objective functions. The optimal switching state must be found in the real time by iteratively minimizing the cost functions which leads to a least-squares-based optimization problem. The solution of these non-linear equations requires optimization techniques, such as gradient-search, which in turn, introduces extra mathematical computational burden and complexity. This could potentially impede the industrial applications of the inverters and/or hinders converters from achieving high switching-frequency ranges.

This paper proposes a PWM-based control technique for 7-level ANPC-FCM inverters. It regulates the inverter's generated voltage, current, and ensures that the FCs are balanced at reference voltages. It employs an active voltage control approach which is achieved by using instantaneous measurements taken from the inverter. Unlike the SVM and MPC methods, the introduced solution is a direct approach that excludes cost/objective functions and time/software-consuming optimization processes. It utilizes the measured values of the FC voltages in conjunction with the load current to derive logical variables and equations that directly determine inverter's switching states according to the staircase reference voltage-levels. The experimentally taken results confirm the proposed active voltage balancing technique, its effectiveness, and fast execution-time.

2 Inverter Topology and Active Balancing

The phase-leg of a 7-level ANPC-FCM inverter is shown in Fig. 1. It is realized by four line-frequency (LF) power switches like the insulated-gate bipolar transistors (IGBTs) of S and S', as shown in Fig. 1, on the dc-link side of the converter, and a four-level FCM stage which is placed on the ac side, and is comprised of three switching cells. This multicell segment is made up of high-frequency (HF) switches like IGBTs of $T_3, T_3', T_2, T_2', T_1, T_1'$ and two FCs of FC_2, FC_1 with voltage ratings of 2 p.u. and 1 p.u., respectively, assuming that the total dc-link voltage is 6 p.u.. The two LF switches of S are turned-on only during the positive half-cycle of the output voltage, and consequently they connect the dc-link of V_{dc} with 3 p.u. voltage to the 4-level FCM inverter section. This configuration generates the PWM voltage-levels of +3, +2, +1, and 0 at the output stage. The LF switches of S' are gated-on only when a negative voltage is required. Hence, the dc-link of V_{dc} with -3 p.u. voltage feeds the FCM inverter during negative half-cycle.

This topology synthesizes the PWM voltage-levels of -3, -2, -1, and 0. All the switching states are listed in Table. I.

The following logical functions are defined accordingly as the determining variables in the active

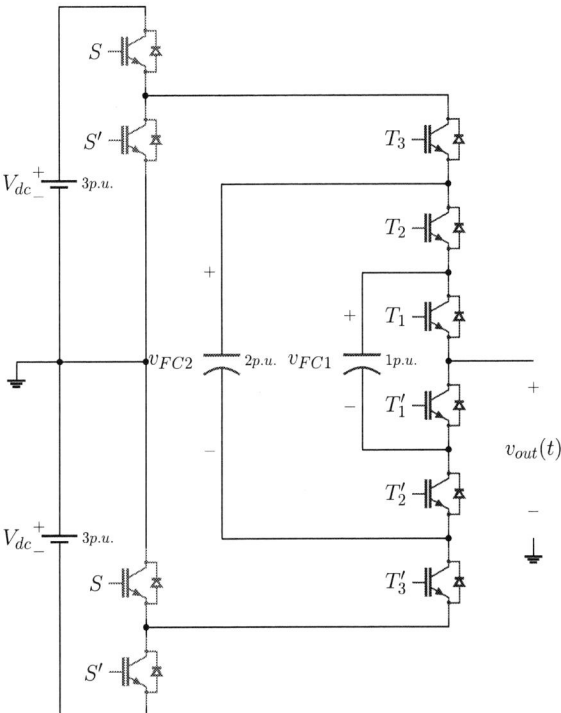

Fig. 1: Single-phase 7-level ANPC-FCM inverter topology.

voltage balancing of the 7-level ANPC-FCM inverter [10].

$$\mu(t) = \begin{cases} 1 & i_{out}(t) \geq 0 \\ 0 & i_{out}(t) < 0 \end{cases} \tag{1}$$

$$\delta_1(t) = \begin{cases} 1 & v_{FC1}(t) \geq \dfrac{V_{dc}}{3} \\ 0 & v_{FC1}(t) < \dfrac{V_{dc}}{3} \end{cases} \tag{2}$$

$$\delta_2(t) = \begin{cases} 1 & v_{FC2}(t) \geq \dfrac{2 \times V_{dc}}{3} \\ 0 & v_{FC2}(t) < \dfrac{2 \times V_{dc}}{3} \end{cases} \tag{3}$$

$$\Delta_1(t) = 1 - \dfrac{3 \times v_{FC1}(t)}{V_{dc}} \tag{4}$$

$$\Delta_2(t) = 1 - \dfrac{3 \times v_{FC2}(t)}{2 \times V_{dc}} \tag{5}$$

Table I: Switching & redundant states in 7-level ANPC-FCM inverter.

PWM Voltage Level	Switching States of the IGBTs (S,S')	(T_3,T_2,T_1)	Δv_{FC2}	Δv_{FC1}	i_{out}	Redundant States
$\lambda_{P3}:+3$	$(1,0)$	$(1,1,1)$	0	0	$\pm$	1
$\lambda_{P2}:+2$	$(1,0)$	$(1,1,0)$	0	$+$	$+$	3
			0	$-$	$-$	
	$(1,0)$	$(1,0,1)$	$+$	$-$	$+$	
			$-$	$+$	$-$	
	$(1,0)$	$(0,1,1)$	$-$	0	$+$	
			$+$	0	$-$	
$\lambda_{P1}:+1$	$(1,0)$	$(1,0,0)$	$+$	0	$+$	3
			$-$	0	$-$	
	$(1,0)$	$(0,0,1)$	0	$-$	$+$	
			0	$+$	$-$	
	$(1,0)$	$(0,1,0)$	$-$	$+$	$+$	
			$+$	$-$	$-$	
$\lambda_0:0$	$(1,0)$	$(0,0,0)$	0	0	$\pm$	2
	$(0,1)$	$(1,1,1)$	0	0	$\pm$	
$\lambda_{N1}:-1$	$(0,1)$	$(1,1,0)$	0	$+$	$+$	3
			0	$-$	$-$	
	$(0,1)$	$(1,0,1)$	$+$	$-$	$+$	
			$-$	$+$	$-$	
	$(0,1)$	$(0,1,1)$	$-$	0	$+$	
			$+$	0	$-$	
$\lambda_{N2}:-2$	$(0,1)$	$(1,0,0)$	$+$	0	$+$	3
			$-$	0	$-$	
	$(0,1)$	$(0,0,1)$	0	$-$	$+$	
			0	$+$	$-$	
	$(0,1)$	$(0,1,0)$	$-$	$+$	$+$	
			$+$	$-$	$-$	
$\lambda_{N3}:-3$	$(0,1)$	$(0,0,0)$	0	0	$\pm$	1

$$\Pi_1(t) = \begin{cases} 1 & |\Delta_1(t)| \geq |\Delta_2(t)| \\ 0 & |\Delta_1(t)| < |\Delta_2(t)| \end{cases} \tag{6}$$

$$\Pi_2(t) = \begin{cases} 1 & |\Delta_2(t)| \geq |\Delta_1(t)| \\ 0 & |\Delta_2(t)| < |\Delta_1(t)| \end{cases} \tag{7}$$

where $v_{FC,\iota}(t)$ and $\dfrac{\iota \times V_{dc}}{3}$ are the measured and the reference voltages of the ιth-FC within the 3-cell FCM, and $\delta_\iota(t)$ is its voltage deviation value ($\iota = 1, 2$). The variable $\Pi_\iota(t)$ is the priority of the ιth-FC. As $\Pi_\iota(t)$ equals 1, it means that the pertinent ι^{th}-FC possesses the largest voltage deviation value among the FCs within the FCM section, and it needs to be charged or discharged based on the output current direction. This is implemented through the logical functions of $\gamma_\iota(t)$ and $\chi_\iota(t)$ which are defined in 8-11 where $+, \cdot, \bar{\ }, \oplus$ are the logical OR, AND, NOT, and XOR operators, respectively.

$$\gamma_1(t) = \Pi_1(t) \cdot \left(\mu(t) \cdot \overline{\delta_1(t)} + \overline{\mu(t)} \cdot \delta_1(t) \right) = \Pi_1(t) \cdot \left(\mu(t) \oplus \delta_1(t) \right) \tag{8}$$

$$\gamma_2(t) = \Pi_2(t) \cdot \left(\mu(t) \cdot \overline{\delta_2(t)} + \overline{\mu(t)} \cdot \delta_2(t) \right) = \Pi_2(t) \cdot \left(\mu(t) \oplus \delta_2(t) \right) \tag{9}$$

$$\chi_1(t) = \Pi_1(t) \cdot \left(\overline{\mu(t) \cdot \overline{\delta_1(t)} + \overline{\mu(t)} \cdot \delta_1(t)} \right) = \Pi_1(t) \cdot \left(\overline{\mu(t) \oplus \delta_1(t)} \right) \tag{10}$$

$$\chi_2(t) = \Pi_2(t) \cdot \left(\overline{\mu(t) \cdot \overline{\delta_2(t)} + \overline{\mu(t)} \cdot \delta_2(t)} \right) = \Pi_2(t) \cdot \left(\overline{\mu(t) \oplus \delta_2(t)} \right) \tag{11}$$

After taking the similar derivation steps for each voltage-level, as explained in [10], the ultimate switching pulses for (T_3, T_2, T_1) can be derived based on the redundancies listed in Table. I and the logical functions in 8-11, as follows:

$$T_3(t) = \lambda_{P3}(t) + \lambda_{P2}(t) \cdot \overline{\chi_2(t)} + \lambda_{P1}(t) \cdot \gamma_2(t) + \lambda_{N2}(t) \cdot \overline{\chi_2(t)} + \lambda_{N1}(t) \cdot \gamma_2(t) \tag{12}$$

$$\begin{aligned} T_2(t) &= \lambda_{P3}(t) + \lambda_{P2}(t) \cdot \left(\overline{\chi_1(t) + \gamma_2(t)} \right) + \lambda_{P1}(t) \cdot \left(\gamma_1(t) + \chi_2(t) \right) \\ &+ \lambda_{N2}(t) \cdot \left(\overline{\chi_1(t) + \gamma_2(t)} \right) + \lambda_{N1}(t) \cdot \left(\gamma_1(t) + \chi_2(t) \right) \end{aligned} \tag{13}$$

$$T_1(t) = \lambda_{P3}(t) + \lambda_{P2}(t) \cdot \overline{\gamma_1(t)} + \lambda_{P1}(t) \cdot \chi_1(t) + \lambda_{N2}(t) \cdot \overline{\gamma_1(t)} + \lambda_{N1}(t) \cdot \chi_1(t) \tag{14}$$

3 Experimental Results

To validate the proposed logic-equation-based active voltage balancing method, the prototype of 7-level ANPC-FCM was built in the laboratory, as shown in Fig. 2. The FF150R12ME3G half-bridge modules consisting of two 1.2-kV 200-A IGBT/diode dies were used for the realization of power switches and the inverter's phase-leg. The LA 55-P and LV 25-P Hall-effect current and voltage transducers were employed for measuring the output current and the FC voltages. The NI-9684 GPIC board in conjunction with sbRIO-9607 chassis and Zynq-7020 FPGA developed by the National Instruments were used to implement the active voltage balancing control method. An RL load of 48.7 Ω-12 mH was used in experiments. The ANPC-FCM inverter was fed by a ±300 V dc-voltage supply ($2\times V_{dc} = 2\times300$ V = 600 V), and was controlled by the PSC-PWM modulator with modulation depth of 0.9. The fundamental frequency of the output voltage was 60 Hz. The switching frequency (carrier's frequency) was set at 20-kH, and hence the voltage harmonic clusters occur at 60-kHz, 120-kHz, 180-kHz, $\cdots$. The phase-shift between the adjacent carriers was $\dfrac{2\pi}{3}$ radians.

The channels of the oscilloscope are designated as follows unless otherwise is specified: **CH4** with 100 $volts/div$ illustrates the total dc-link voltage of ±300 V ($2\times V_{dc} = 2\times300$ V = 600 V); **CH5** with 100 $volts/div$ shows the voltage of FC_2; **CH6** with 100 $volts/div$ illustrates the voltage of FC_1; **CH7** with 100 $volts/div$ shows the generated voltage of the 7-level ANPC-FCM inverter ($v_{out}(t)$); and **CH8** with 5 $amperes/div$ illustrates the output current. The steady state, harmonic spectrum of output voltage, and transient dynamics are illustrated in Figs. 3-14. Figs. 3-7 show the steady state of the voltages as well as output current along with harmonic spectrum of v_{out}. As it is illustrated, the proposed logic-equations balance the voltage of FC_2, FC_1 at 200 V and 100 V, respectively, and generate a 7-level voltage of ±300 V with steps of 100 V at the output stage according to the PSC-WPM pattern.

The transient dynamics of the inverter with a changing dc-voltage source are illustrated in Figs. 8-10. As shown in these results, the voltage of FC_1 and FC_2 increases from 50 V and

PWM-Based Optimization-Less Active Voltage-Balancing Control of 7-Level Active
Neutral-Point-Clamped Flying-Capacitor Inverters

DARGAHI Vahid

Fig. 2: Implemented single-phase 7-level ANPC-FCM inverter.

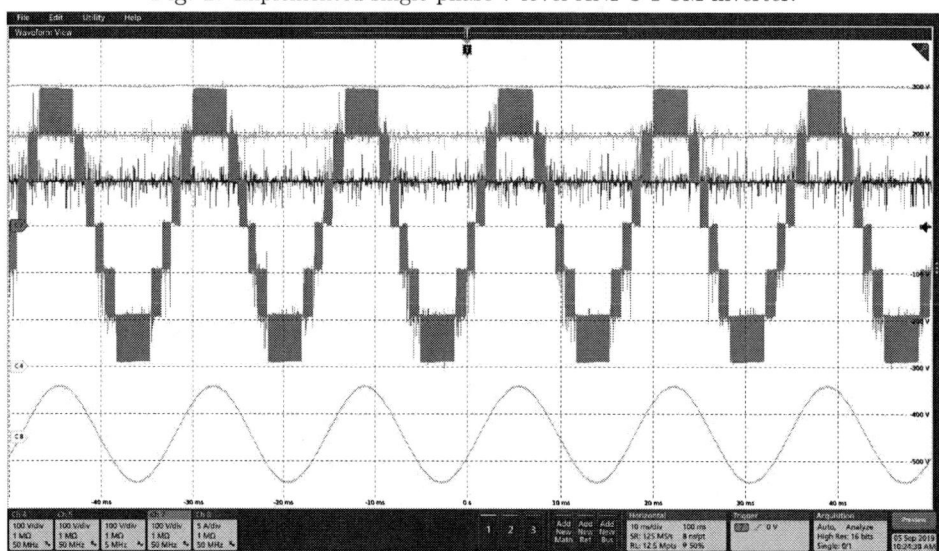

Fig. 3: Steady-state dynamics: $2 \times V_{dc} = 600$ V, $m_a = 0.9$, $f_{sw} = 20$ kHz, $f_o = 60$ Hz.

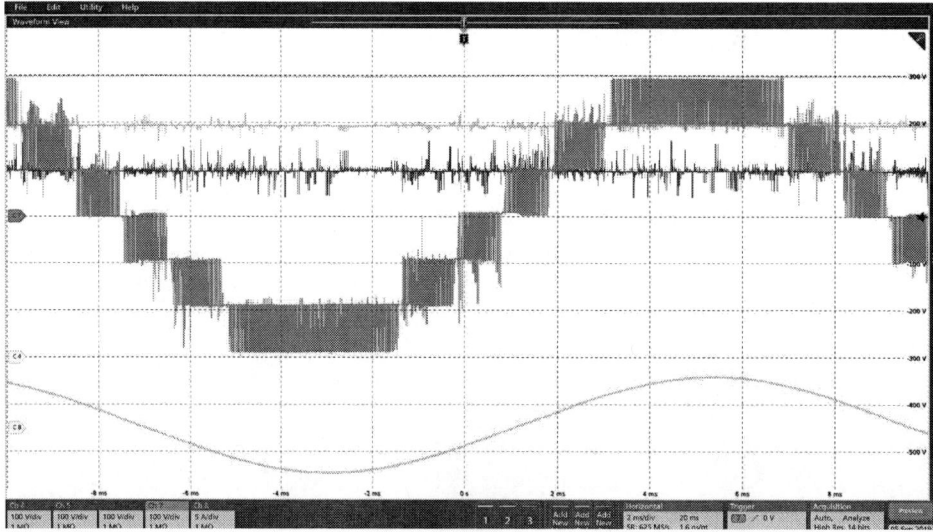

EPE'22 ECCE Europe

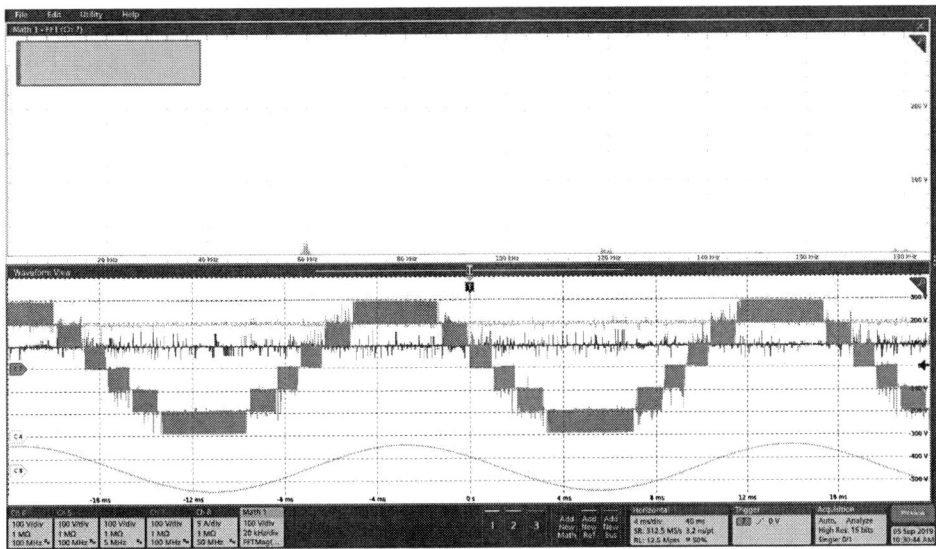

Fig. 5: Harmonic clusters under steady-state dynamics: $2 \times V_{dc} = 600\text{V}$, $m_a = 0.9$, $f_{sw} = 20\text{kHz}$, $f_o = 60\text{Hz}$.

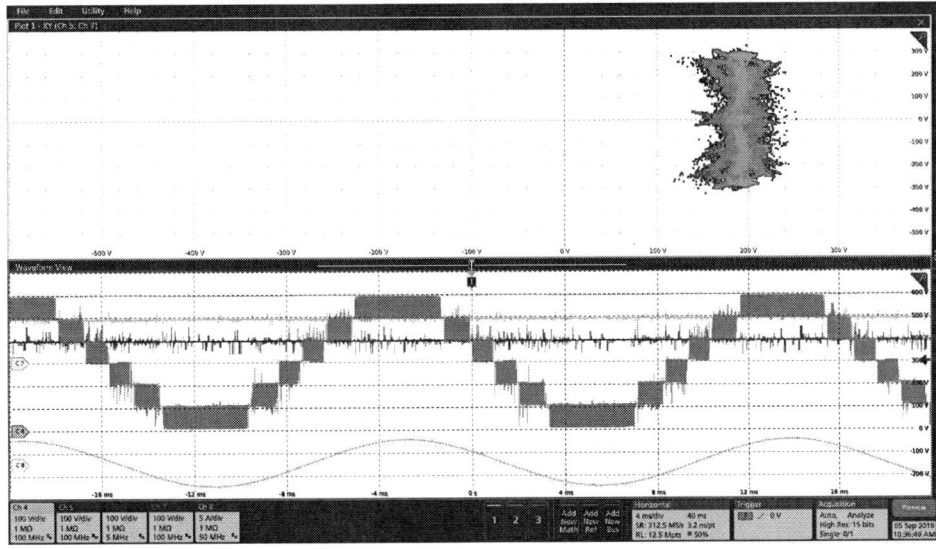

Fig. 6: v_{out} versus v_{FC2} under steady-state dynamics: $2 \times V_{dc} = 600\text{V}$, $m_a = 0.9$, $f_{sw} = 20\text{kHz}$, $f_o = 60\text{Hz}$.

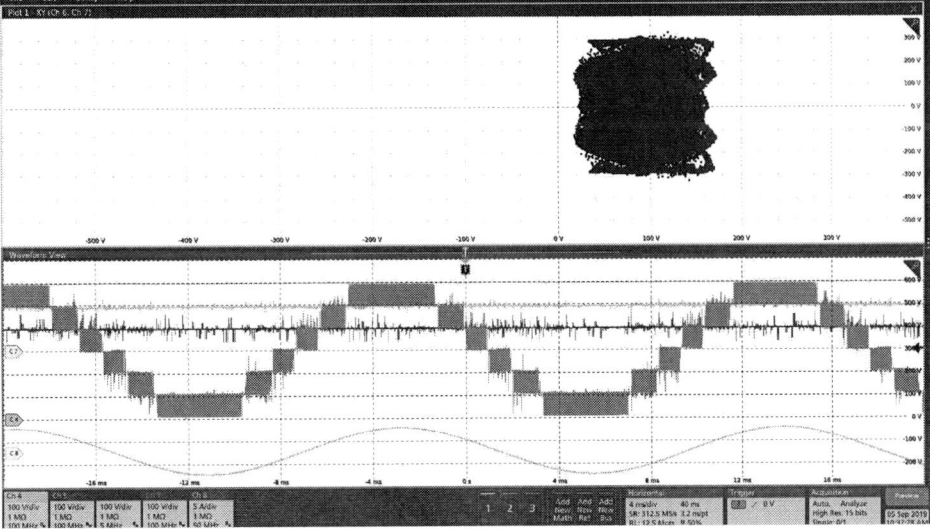

PWM-Based Optimization-Less Active Voltage-Balancing Control of 7-Level Active
Neutral-Point-Clamped Flying-Capacitor Inverters

DARGAHI Vahid

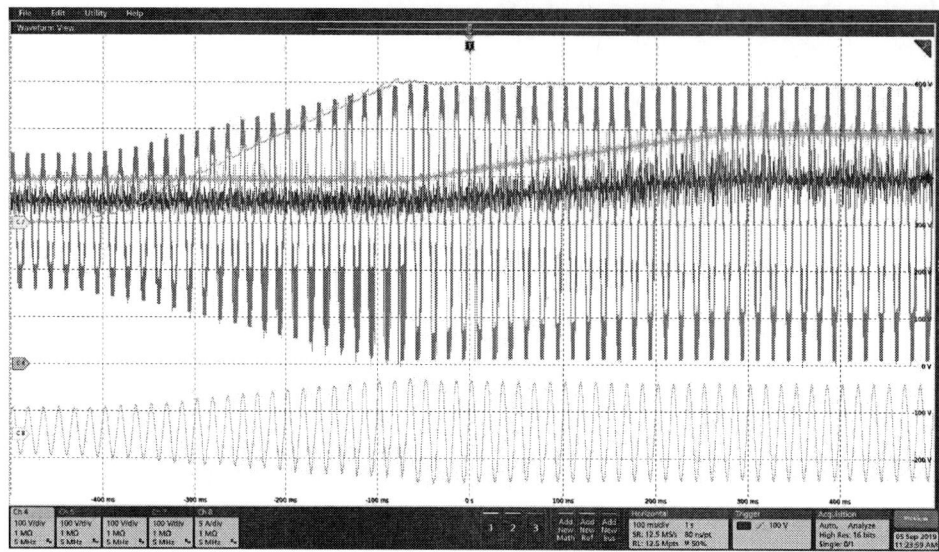

Fig. 8: Transient dynamics as $2 \times V_{dc}$ increases from 300V to 600V, $m_a = 0.9$, $f_{sw} = 20$ kHz, $f_o = 60$ Hz.

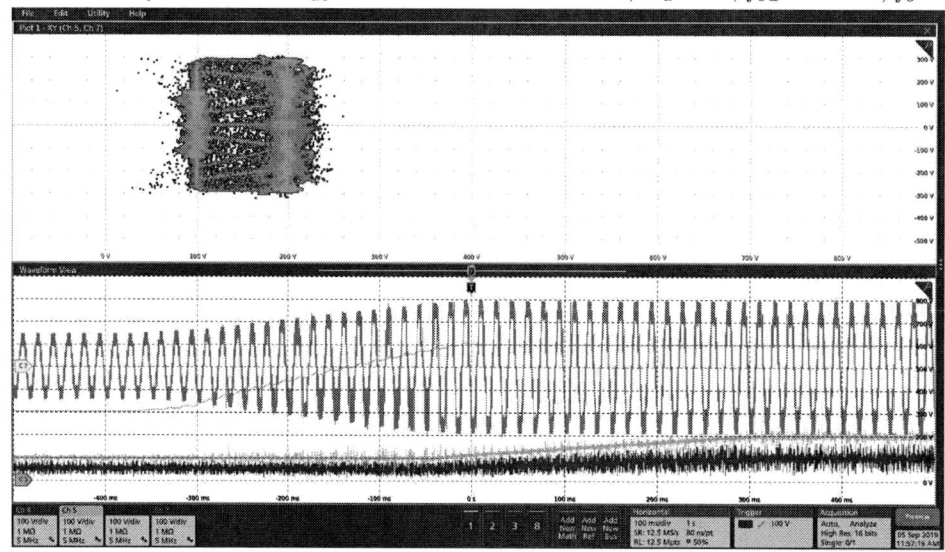

Fig. 9: v_{out} versus v_{FC2} as $2 \times V_{dc}$ increases from 300 V to 600 V, $m_a = 0.9$, $f_{sw} = 20$ kHz, $f_o = 60$ Hz.

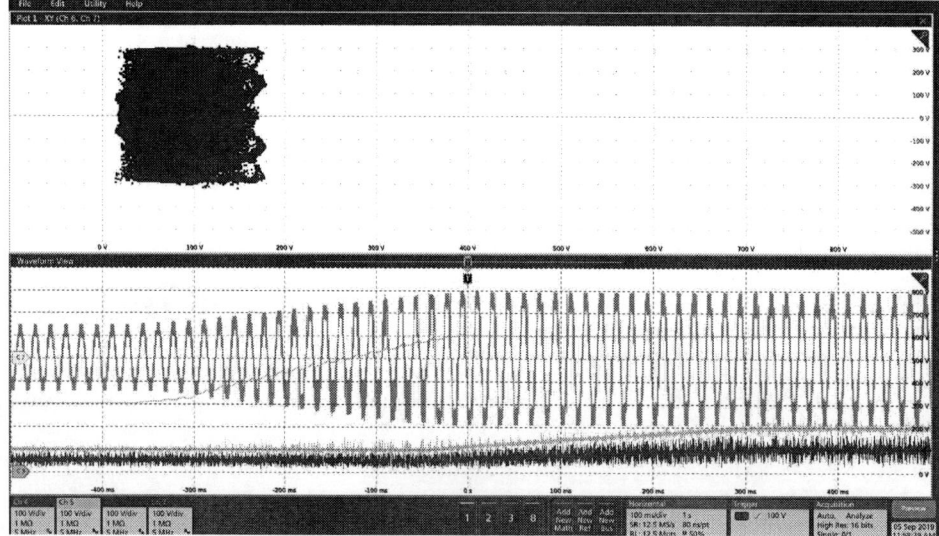

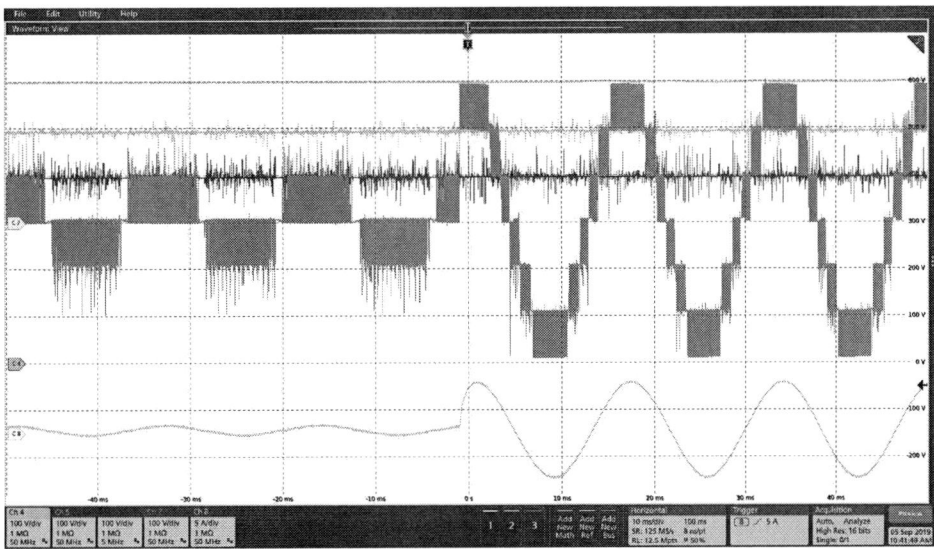

Fig. 11: Transient dynamics when m_a changes from 0.1 to 0.9: $2 \times V_{dc} = 600\text{V}$, $f_{sw} = 20\text{kHz}$, $f_o = 60\text{Hz}$.

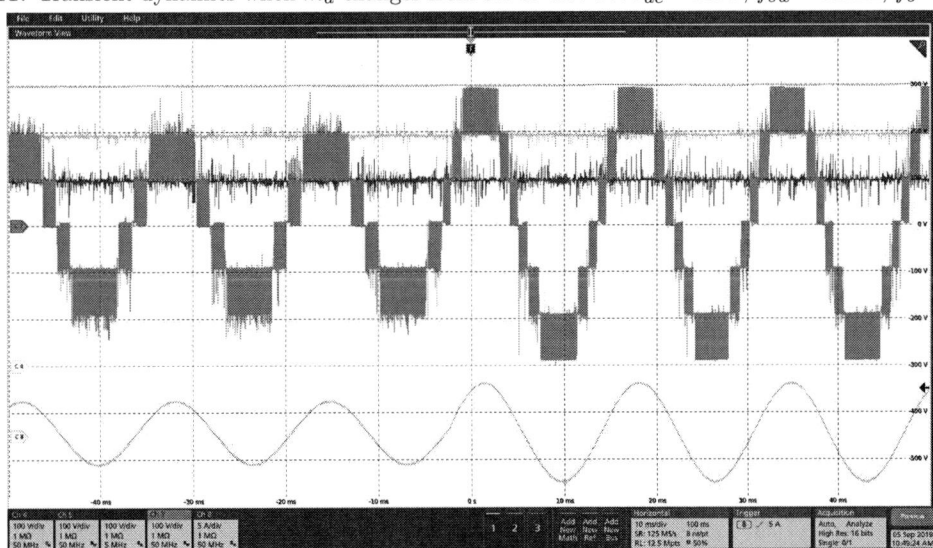

Fig. 12: Transient dynamics when m_a changes from 0.6 to 0.925: $2 \times V_{dc} = 600\text{V}$, $f_{sw} = 20\text{kHz}$, $f_o = 60\text{Hz}$.

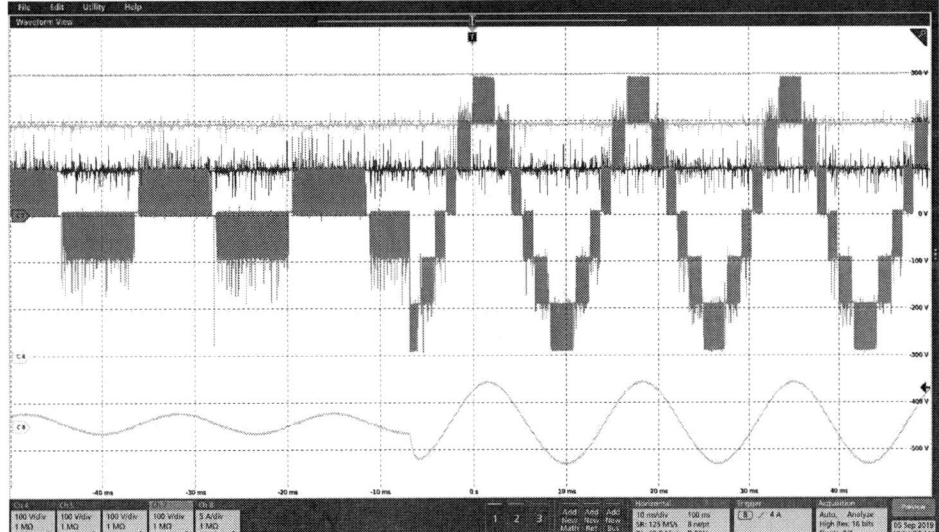

100 V, respectively, to 100 V and 200 V when the total dc-link voltage steps up from 300 V to 600 V. The transient dynamics under the varying modulation indexes are demonstrated in Figs. 11-14. These results show that the inverter exhibits a full range of modulation index for its operation without losing its stability and/or suffering the voltage drift issues even under the low modulation index values. The provided experimental results in Figs. 3-14 demonstrate that the proposed active voltage balancing technique, which is based on a set of simple and effortless logic-equations, stabilized the FC voltages at their reference values, generated the required voltage-level commanded by PSC-PWM, and retained the converter stable during disturbances including the variations of m_a and V_{dc}.

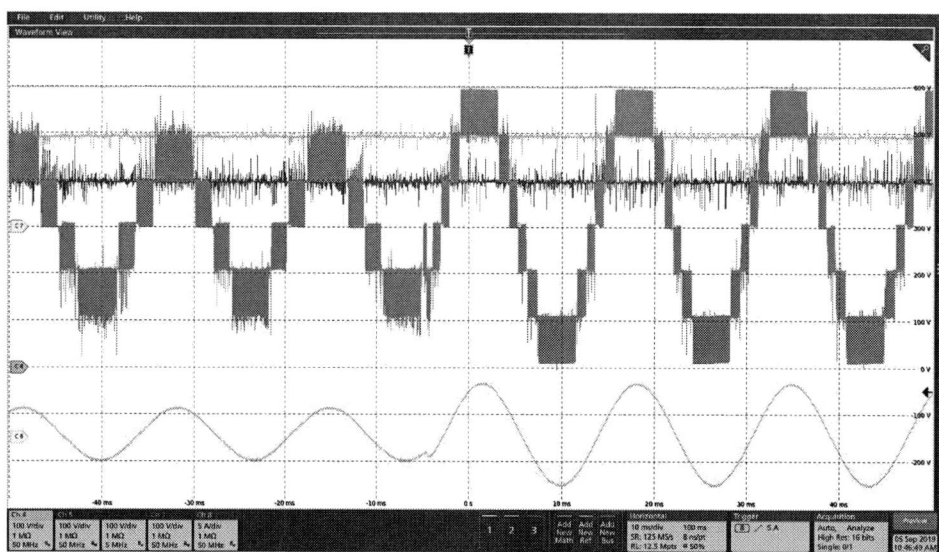

Fig. 14: Transient dynamics when m_a changes from 0.5 to 0.95: $2 \times V_{dc} = 600$V, $f_{sw} = 20$kHz, $f_o = 60$Hz.

4   Conclusion

The provided experimental results demonstrated that proposed logic-equations were capable of regulating the flying-capacitor voltages at their reference values along with generating the PWM-based output voltage-level with fast dynamics. Furthermore, the inverter remains stable during transient disturbances including varying m_a and V_{dc} which is critical for drive applications.

References

[1] N. R. Zargari, Z. Cheng, and R. Paes, "A guide to matching medium-voltage drive topology to petrochemical applications," *IEEE Transactions on Industry Applications*, vol. 54, no. 2, pp. 1912–1920, March 2018.

[2] N. Pallo, T. Foulkes, T. Modeer, S. Coday, and R. Pilawa-Podgurski, "Power-dense multilevel inverter module using interleaved gan-based phases for electric aircraft propulsion," in *2018 IEEE Applied Power Electronics Conference and Exposition (APEC)*, 2018, pp. 1656–1661.

[3] S. Coday, A. Barchowsky, and R. C. Pilawa-Podgurski, "A 10-level gan-based flying capacitor multilevel boost converter for radiation-hardened operation in space applications," in *2021 IEEE Applied Power Electronics Conference and Exposition (APEC)*, 2021, pp. 2798–2803.

[4] Y. Lei, C. Barth, S. Qin, W.-c. Liu, I. Moon, A. Stillwell, D. Chou, T. Foulkes, Z. Ye, Z. Liao, and R. C. Pilawa-Podgurski, "A 2 kw, single-phase, 7-level, gan inverter with an

active energy buffer achieving 216 w/in3 power density and 97.6% peak efficiency," in *2016 IEEE Applied Power Electronics Conference and Exposition (APEC)*, 2016, pp. 1512–1519.

[5] T. Modeer, N. Pallo, T. Foulkes, C. B. Barth, and R. C. N. Pilawa-Podgurski, "Design of a gan-based interleaved nine-level flying capacitor multilevel inverter for electric aircraft applications," *IEEE Transactions on Power Electronics*, vol. 35, no. 11, pp. 12153–12165, 2020.

[6] l. Mayor, M. Rizo, A. Rodríguez Monter, and E. J. Bueno, "Commutation behavior analysis of a dual 3l-anpc-vsc phase-leg pebb using 4.5-kv and 1.5-ka hv-igbt modules," *IEEE Transactions on Power Electronics*, vol. 34, no. 2, pp. 1125–1141, 2019.

[7] V. Dargahi, K. Corzine, and A. K. Sadigh, "A pyramid-type (pt) multilevel converter topology," in *2020 IEEE Energy Conversion Congress and Exposition (ECCE)*, 2020, pp. 3926–3933.

[8] R. R. Khorasani, S. Farzamkia, F. Wu, A. Khoshkbar-Sadigh, M. T. Brady, and V. Dargahi, "Power loss modeling and thermal comparison of sic-mosfet-based 2-level inverter and 3-level flying capacitor multicell inverter," in *2021 IEEE Applied Power Electronics Conference and Exposition (APEC)*, 2021, pp. 2607–2612.

[9] J. A. Anderson, E. J. Hanak, L. Schrittwieser, M. Guacci, J. W. Kolar, and G. Deboy, "All-silicon 99.35% efficient three-phase seven-level hybrid neutral point clamped/flying capacitor inverter," *CPSS Transactions on Power Electronics and Applications*, vol. 4, no. 1, pp. 50–61, March 2019.

[10] V. Dargahi, A. K. Sadigh, R. R. Khorasani, and J. Rodriguez, "Active voltage balancing control of a seven-level hybrid multilevel converter topology," *IEEE Transactions on Industrial Electronics*, vol. 69, no. 1, pp. 74–89, 2022.

[11] C. Li, R. Lu, W. Li, and Y. Wang, "Space vector modulation for sic si hybrid active neutral point clamped converter," in *2018 IEEE International Power Electronics and Application Conference and Exposition (PEAC)*, Nov 2018, pp. 1–6.

[12] P. Acuna, R. P. Aguilera, B. McGrath, P. Lezana, A. Ghias, and J. Pou, "Sequential phase-shifted model predictive control for a single-phase five-level h-bridge flying capacitor converter," in *2018 Asian Conference on Energy, Power and Transportation Electrification (ACEPT)*, Oct 2018, pp. 1–7.

[13] A. Mora, R. P. Aguilera, R. Cárdenas, P. Lezana, and D. D. C. Lu, "Phase-shifted model predictive control of a three-level active-npc converter," in *2018 IEEE 27th International Symposium on Industrial Electronics (ISIE)*, June 2018, pp. 270–276.

[14] F. Salinas, M. A. González, M. F. Escalante, and J. de León Morales, "Control design strategy for flying capacitor multilevel converters based on petri nets," *IEEE Transactions on Industrial Electronics*, vol. 63, no. 3, pp. 1728–1736, March 2016.

[15] J. Druant, T. Vyncke, F. D. Belie, P. Sergeant, and J. Melkebeek, "Adding inverter fault detection to model-based predictive control for flying-capacitor inverters," *IEEE Transactions on Industrial Electronics*, vol. 62, no. 4, pp. 2054–2063, April 2015.

[16] F. Salinas, M. A. González, and M. F. Escalante, "Finite control set-model predictive control of a flying capacitor multilevel chopper using petri nets," *IEEE Transactions on Industrial Electronics*, vol. 63, no. 9, pp. 5891–5899, Sep. 2016.

[17] F. Salinas, M. Ghanes, J. P. Barbot, M. F. Escalante, and B. Amghar, "Modeling and control design based on petri nets for serial multicellular choppers," *IEEE Transactions on Control Systems Technology*, vol. 23, no. 1, pp. 91–100, Jan 2015.

Model Predictive Power Sharing Algorithm for Fuel Cell Integration in a Dual Inverter Electric Vehicle Drivetrain

Mehanathan Pathmanathan, Caniggia Viana, Sukhjit Singh, Peter W. Lehn
Department of Electrical and Computer Engineering
University of Toronto, Canada

Acknowledgement

This work was supported by the Natural Sciences and Engineering Research Council of Canada (NSERC) under Grant CRDPJ 513206-17.

Keywords

≪Fuel Cell≫, ≪Power Sharing≫, ≪electric vehicle≫, ≪drivetrain≫, ≪modeling≫.

Abstract

Fuel Cell hybrid electric vehicles require power electronic converter topologies and control algorithms which allow the slow dynamics and unidirectional nature of fuel cells to be respected throughout a drive cycle. This paper introduces a model predictive control algorithm which allows a dual inverter topology to achieve the required constraints for fuel cell integration in electric vehicles.

Introduction

The global push towards decarbonization can be facilitated by transportation electrification. Fuel cell hybrid electric vehicles (FCVs) which use a proton exchange membrane (PEM) hydrogen fuel cell stack as the primary EV energy source are one avenue to transportation electrification [1].

To avoid fuel cell damage due to the fuel starvation phenomenon, the rate of change of fuel cell power needs to be limited. Furthermore, fuel cells are unidirectional sources, meaning that they cannot absorb energy. These constraints make it impossible for a fuel cell to be the sole energy source of an EV, since an EVs drivetrain must be capable of achieving fast power dynamics to accelerate quickly, and must be able to absorb power during regenerative braking transients [2].

Accordingly, a secondary energy source is always present on FCVs. Commercial vehicles such as the Toyota Mirai use a battery pack to provide the instantaneous power needed for fast vehicle dynamics, and to absorb the power generated during regenerative braking transients [3]. Traditionally, a unidirectional DC/DC converter is used to connect the FC stack to the battery pack [4]. This DC/DC converter has a dual role of stepping up the fuel cell voltage, and coordinating the power flow of the fuel cell to ensure that the fuel cell power follows a slowly changing reference.

The presence of a dedicated DC/DC converter for FC integration adds weight and volume to the FCV, especially due to the magnetic energy storage stage which is needed. A method of achieving FC integration without the need of a dedicated DC/DC converter (and associated magnetics) is to use a dual inverter drivetrain (Fig 1a) [5, 6]. In this topology, an open-winding traction motor is used, with a three-phase, two-level inverter on each side of the motor. The FC is used as the DC energy source on one inverter, while the battery pack is used as the DC energy source of the other inverter. The ability to use lower-voltage traction modules with lower switching energy also provides the opportunity to reduce power electronic losses in the system [7].

An algorithm for control and modulation of the dual inverter is needed to ensure that the unidirectional, slowly changing reference of the fuel cell can be achieved. This paper introduces an approach utilizing finite control set (FCS) model predictive control to achieve these goals as an alternative to the vector-based power sharing introduced in [6]. It is forseen that the ability of MPC to combine the dual objectives of motor torque production and FC power reference tracking into a single cost function could reduce the complexity of the control implementation relative to the approach of [6].

System Architecture

The fuel cell studied in the simulation section of this work is the Ballard FCmove [8]. The VI curve of this fuel cell is plotted using an exponential approximation in Fig. 1b, along with a linear approximation. The linear approximation allows the fuel cell to be modelled as an internal voltage source of 520 V and a series resistance of 0.85 Ω (E_{FC} and R_{FC} in Fig. 1a.)

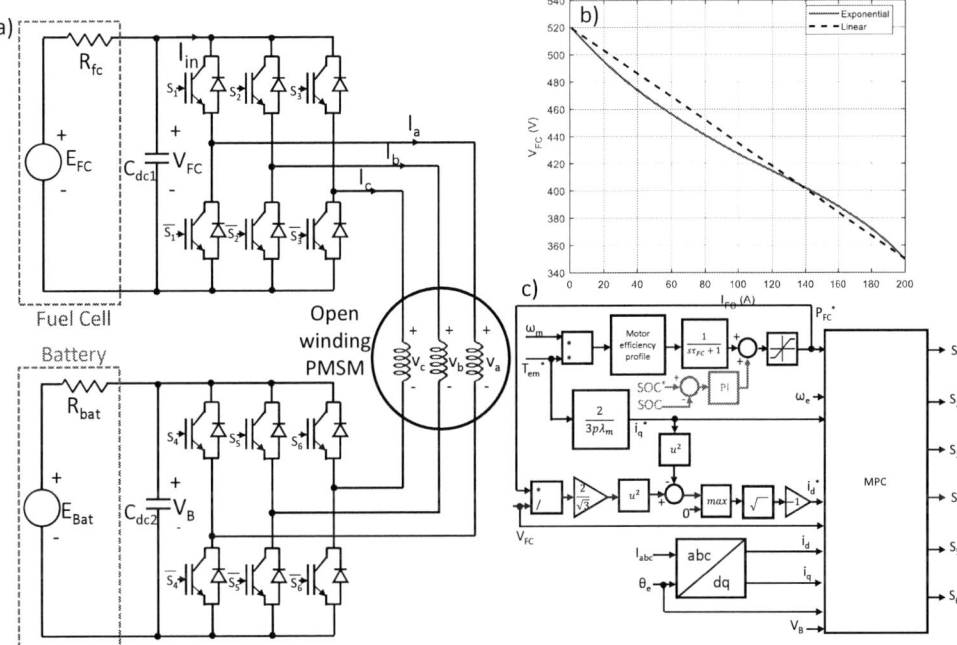

Fig. 1: Topology of proposed system (a), and Exponential model of Ballard FCMove V-I curve and linear model using E_{FC} and R_{FC} (b), and control diagram of the MPC based FC/battery dual inverter system (c)

Control Algorithm

The control diagram used in this study is shown in Fig. 1c. The predictive controller acts by selecting optimal switching states of the dual inverter in order to achieve regulation of the motor torque (via tracking reference dq currents) and fuel cell power. The following sections will describe the operation of the control algorithm.

Reference Generation

The fuel cell power reference (P_{FC}^*) is generated from low-pass filtering the motor mechanical power, as shown in Fig. 1c [2]. A battery SOC controller (shown in grey) can be used to increase or decrease the power reference. This controller is not implemented in this work, but is included in Fig. 1c for explanatory purposes. The saturation block in Fig. 1c) is used to limit the fuel cell power reference between its maximum and minimum values. A minimum power reference greater than zero is required to ensure that the fuel cell is not shut down while the FCV is idling, or the vehicle is performing regenerative braking. In the simulation section of this work, a minimum P_{FC}^* of 5 kW is used and a maximum P_{FC}^* of 50 kW are used.

The motor q-axis current reference is given by (using non-salient PMSM assumption):

$$i_q^* = \frac{2T_{em}^*}{3p\lambda_m} \tag{1}$$

p is the motor pole pairs, and T_{em}^* is the torque reference. For a non-salient PMSM not operating in the field weakening region, the d-axis current reference should normally be set to zero. In this paper, an exception to this rule exists during control events where a sharp reduction in torque reference magnitude is requested. From (1), it is apparent that i_q^* scales linearly with T_{em}^*. However, since P_{FC}^* is a low-pass filtered value of the drive mechanical power, it is apparent that P_{FC}^* can be larger than the mechanical power when a reduction in torque is requested.

The maximum power which can be extracted by from the fuel cell at a given torque reference is:

$$P_{FCM} = \frac{3}{2}|\vec{V_{FCM}}||\vec{I}| \tag{2}$$

$|\vec{V_{FCM}}|$ is the maximum possible value of the FC inverter voltage vector (considered to be $\frac{V_{FC}}{\sqrt{3}}$ in this work), while $|\vec{I}|$ is the magnitude of the motor current vector. The minimum current vector magnitude needed to ensure a fuel cell power reference can be delivered is:

$$I_{PS} = \frac{2P_{FC}^*}{\sqrt{3}V_{FC}} \tag{3}$$

In the above equation, V_{FC} is the FC terminal voltage. If the calculated i_q^* value from (1) is smaller than I_{PS}, a non-zero i_d^* is needed to ensure that the requested fuel cell power can be delivered. This i_d^* lengthens the current vector, increasing $|\vec{I}|$ so that it is equal to I_{PS} The required value of i_d^* is given by:

$$i_d^* = \begin{cases} 0 & |i_q^*| > |I_{PS}| \\ -\sqrt{(I_{PS})^2 - (i_q^*)^2)} & |i_q^*| < |I_{PS}| \end{cases} \tag{4}$$

The second row of (4) only uses the negative solution of i_d^* so the injected reactive current acts as a field weakening current for the motor.

Switching Vector Selection

A dual inverter has 64 (2^6) possible switching vectors. In order to minimize the computation time required by the predictive control algorithm, it is beneficial to reduce the number of switching states. The switching states shown in the Appendix (Table II) are used, which eliminate all redundancies and also non-redundant states which are not essential for operation of the predictive control algorithm. In this table the terms $S_{aFC}, S_{bFC}, S_{cFC}$ are the gating signals of the upper switches in the FC inverter and S_{aB}, S_{bB}, S_{cB} are the corresponding signals of the battery inverter.

Fig. 2a illustrates the voltage vectors in the $\alpha\beta$ frame for the motor voltage ($V_{\alpha\beta}$), as well as the FC and battery inverter voltage vectors individually ($V_{\alpha\beta FC}$) and $V_{\alpha\beta B}$) when the full set of 64 switching vectors is used where $V_B = 1pu$ and $V_{FC} = 0.7pu$. In contrast, Fig 2b shows the corresponding vectors achieved when the reduced set of vectors described in the Appendix are utilized. In the experimental implementation, operation with this reduced set of 30 switching vectors was found to result in a computation time of 46.8 μs, compared to 90.6 μs with the full set of 64 switching vectors, when a 200 MHz Texas Instruments F28379D digital signal processor was used as the control platform.

Motor Current Prediction

The motor current prediction algorithm evaluates the dq current values which will be generated by each of the switching states shown in the Appendix. As a starting point, a rotational transformation is applied to the motor $\alpha\beta$ components from each of the available switching vectors to obtain the dq voltages, using

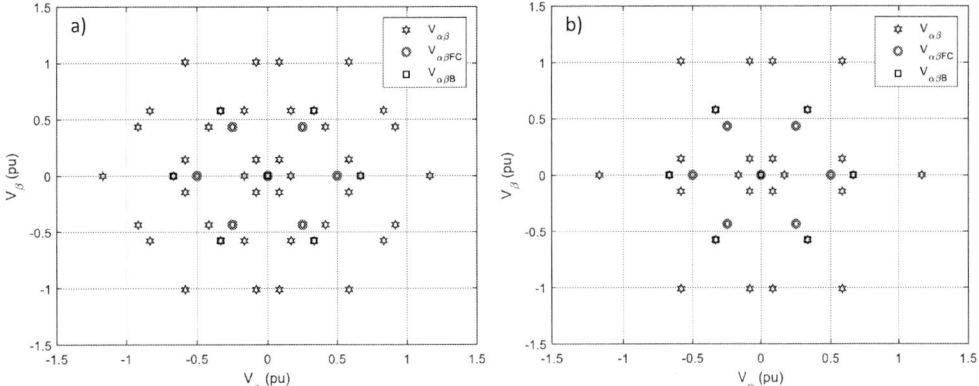

Fig. 2: Full set of available voltage vectors (a), reduced set utilized in this study (b) for a case where $V_{FC} = 0.75\,pu$, $V_B = 1\,pu$

the measured rotor electrical position θ_e:

$$
\begin{bmatrix} v_d^{k+1} \\ v_q^{k+1} \end{bmatrix} = \begin{bmatrix} \cos\theta & \sin(\theta_e) \\ -\sin\theta & \cos(\theta_e) \end{bmatrix} \begin{bmatrix} v_\alpha^{k+1} \\ v_\beta^{k+1} \end{bmatrix}
\tag{5}
$$

The predicted motor d and q currents can be obtained by performing a forward Euler discretization on the differential equations describing the two-axis model of the PMSM:

$$
i_d^{k+2} = (1 - \frac{R_s T_s}{L_d}) i_d^{k+1} + \frac{T_s}{L_d} v_d^{k+1} + \frac{\omega_e L_q T_s}{L_d} i_q^{k+1}
\tag{6}
$$

$$
i_q^{k+2} = (1 - \frac{R_s T_s}{L_q}) i_q^{k+1} + \frac{T_s}{L_q} v_q^{k+1} + \frac{\omega_e T_s}{L_q} (L_d i_d^{k+1} + \lambda_m)
\tag{7}
$$

Please note that $k+2$ prediction is used in this study, meaning that the predicted currents will be two steps ahead from the measured current. The predicted one-step ahead d and q axes currents (i_d^{k+1} and i_q^{k+1}) are calculated using the optimal voltage vectors from the previous time step and measured dq current values in place of i_d^{k+1} and i_q^{k+1}. [9].

Fuel cell power prediction

The FC inverter DC current (I_{in} in Fig. 1a) is a function of the motor phase currents and fuel cell inverter switching functions. Its one step ahead value is given by:

$$
I_{in}^{k+1} = S_{aFC} i_a^{k+1} + S_{bFC} i_b^{k+1} + S_{cFC} i_c^{k+1}
\tag{8}
$$

As shown in Fig. 1a, the voltage across the DC link capacitor is equal to the fuel cell terminal voltage. Thus, the predicted fuel cell voltage can be obtained by using a forward Euler approximation to discretize the first-order differential equation describing the DC-link capacitor voltage:

$$
V_{FC}^{k+2} = \frac{T_s}{C_{dc}} (I_{FC}^{k+1} - I_{in}^{k+1}) + V_{FC}^{k+1}
\tag{9}
$$

In this equation, I_{FC}^{k+1} is the one-step ahead fuel cell current, given by:

$$
I_{FC}^{k+1} = \frac{E_{FC} - V_{FC}^{k+1}}{R_{FC}}
\tag{10}
$$

Additionally, the $k+2$ fuell cell current can be calculated using V_{FC}^{k+2} in (10). Finally the two-step-ahead predicted FC power is obtained by $P_{FC}^{k+2} = I_{FC}^{k+2} V_{FC}^{k+2}$.

Cost function

The objective of the MPC algorithm is to select the switching state which minimized the error between control parameters (motor dq currents, P_{FC}) and their reference values. These errors can be expressed in terms of a cost function (where the current variables are in Amperes, and the power variables are in Watts):

$$g = (i_d^* - i_d^{k+2})^2 + (i_q^* - i_q^{k+2})^2 + k(P_{FC}^* - P_{FC}^{k+2})^2 \tag{11}$$

The variable k is the weighting function applied to the fuel cell power error (0.02 in this work). The optimal switching state is ultimately the switching state with the minimum value of g.

Simulation Results

A simulation model in PLECS blockset and Simulink was constructed to verify the performance of the proposed algorithm. The parameters of the proposed system are shown in Table I. A torque reference of 180 Nm was originally specified for the PMSM. This torque was achieved by requesting a i_q value of 190 A, with $i_d = 0$ A as can be seen from Fig. 2a. This torque causes the PMSM to accelerate from t = 0 s to t = 2.5 s, since the magnitude of the motor electromagnetic torque (T_{em}) is greater than that of the load torque (T_L, set to -100 Nm). At t = 2.5s, the torque reference is changed to 20 Nm, which causes i_q^* to become 25 A. At the same time, the load torque profile (T_L) is changed to be -20 Nm, which causes the motor speed to be constant. This reduction of T_{em} is achieved by a step reduction in i_q, as seen in Fig. 3. A negative i_d^* is injected at the time of the i_q^* reduction, to ensure the minimum current vector magnitude $|I_{PS}|$ (3) for attaining the FC power reference is maintained.

At t = 4 s, the motor torque reference is changed from 20 Nm to -180 Nm. At this time, the motor begins to brake, meaning that it is generating power. The fuel cell power drops to its minimum value of 5 kW, and maintains this value despite the power which is generated by the braking action of the motor. At this time, the battery power (P_{bat}) has a step decrease towards a minimum value of -40 kW, as the braking power of the generator and the FC power are absorbed by the battery.

Table I: Simulation parameters

Parameter	Description	Value
E_{FC}	Fuel cell internal voltage	520 V
R_{FC}	Fuel cell ESR from linear approx.	0.85 Ω
E_{Bat}	Battery internal voltage	400 V
C_{dc1}, C_{dc2}	DC link capacitance	4.5 mF
R_{bat}	Battery ESR	0.1 Ω
τ_{FC}	FC controller time constant	1 s
T_s	Sampling time	50 μs
$L_d = L_q$	Motor d and q inductances	800 μH
R_s	Motor phase resistance	45 mΩ
λ_m	Motor flux linkage	0.127 Wb

Experimental Results

A reduced scale experimental setup was built to verify the performance of the MPC algorithm (Fig. 3). The open-wound PMSM was fed by a dual inverter controlled by a Texas Instruments F28379D DSP. The two energy sources on the DC-links of the dual inverter were implemented using EA PSB750 power supplies. The R_{FC} and E_{FC} FC model elements depicted in Fig. 1a were implemented via the EA PSB750 internal resistance operating mode. In the experimental setup, E_{FC} was set to 150 V, R_{FC} was

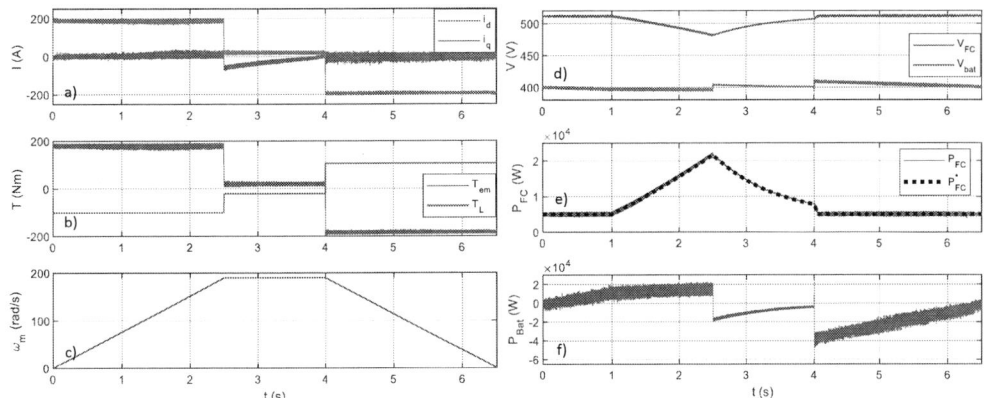

Fig. 3: Motor dq currents (a), torque (b), speed (c), FC and battery voltages (d), FC power reference and power (e), and battery power (f)

set to 1.667 Ω. The battery was modelled as a constant voltage 150 V source. Additionally, the prototype dual inverter was constructed with busbars designed to accommodate 500 μF DC-link capacitors. As such, the experiments were conducted with this value of capacitance.

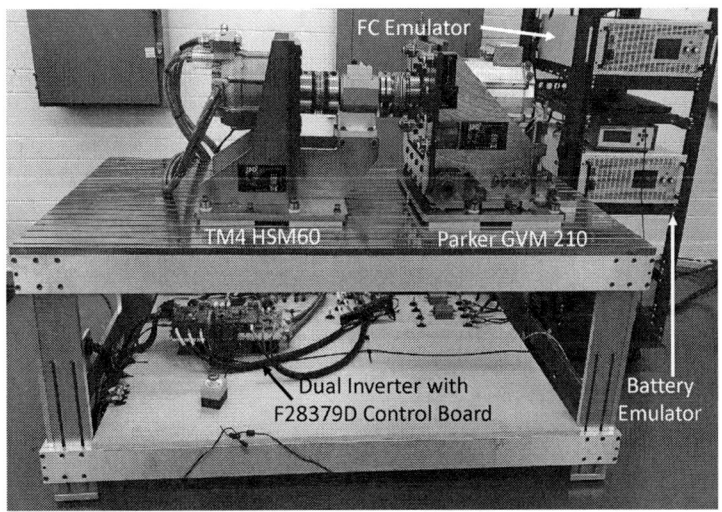

Fig. 4: Experimental setup used to verify predictive control algorithm

In Fig. 5a, results are shown for a regenerative transient in motor torque at a constant speed of 500 rpm. Prior to the transient, the motor was operating with a reference torque of 30 Nm. At the transient time, the reference was changed to -15 Nm. From CH4 in this figure, it is apparent that the torque changes rapidly at this time. CH1 and CH2 show the FC voltage and current respectively. A slow transition in FC voltage and current is noted, due to the influence of the low pass filter with time constant τ_{FC} which is used to generate the fuel cell power reference.

The motor phase 'A' current is provided as CH3. The magnitude of this current waveform does not drop rapidly when the regenerative transient is enacted. This is due to the injection of reactive current, which allows the fuel cell power to slowly ramp down even when mechanical power is reduced drastically, as described in by (4).

A second set of experimental results are shown in Fig. 5b, where the motor torque reference is changed from 10 Nm to 30 Nm at a speed of 500 rpm. In this case, the FC current and voltage once again

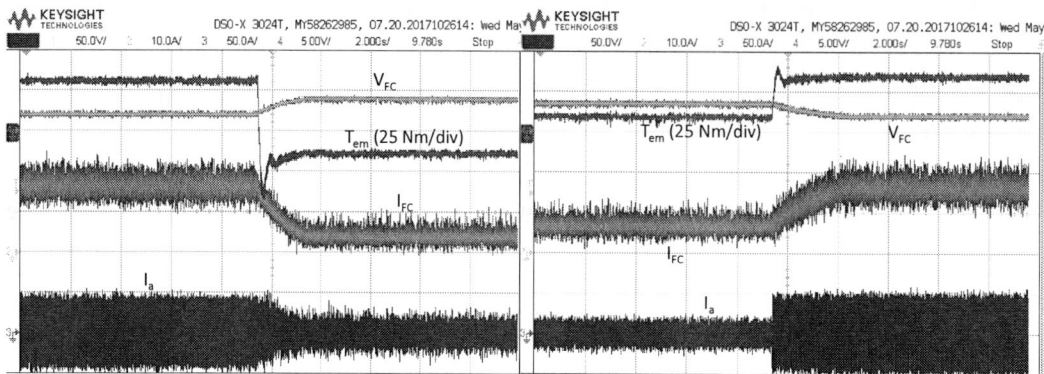

Fig. 5: Experimental results for +30 Nm to -15 Nm regenerative braking transient (a) and 10 Nm to 30 Nm torque increase transient at 500 rpm

change slowly due to the influence of the slowly-changing fuel cell power reference. In contrast with the previous experimental result, the magnitude of the phase 'A' current increases rapidly at the time of the torque transient. This is because no reactive current injection is needed to ensure that the fuel cell power can track its reference, during increases in current magnitude.

Conclusion

A finite control set model predictive control algorithm for a dual inverter drivetrain in a FCV is introduced. This algorithm allows fuel cell power reference tracking and motor torque reference tracking to be achieved simultaneously. In particular, this approach ensures that the fuel cell power tracks a slowly changing reference value, even when rapid torque (and hence power) dynamics are requested from the traction motor. Additionally, the algorithm ensures that the fuel cell power remains positive even when the motor performs regenerative braking.

Simulation results are presented showcasing the ability of the proposed algorithm in meeting these twin goals of fuel cell power tracking and motor torque control. Finally, experimental results are shown validating the practical performance of the proposed algorithm.

References

[1] U.R Prasanna, P. Xuewei, A.K Rathore, and K. Rajashekara, "Propulsion system architecture and power conditioning topologies for fuel cell vehicles," *IEEE Trans. on Ind. Elec.*, vol. 51, no. 1, pp. 640-650.

[2] J. Bauman, and M. Kazerani, "A comparative study of fuel-cell-battery, fuel-cell-ultracapacitor, and fuel-cell-battery-ultracapacitor vehicles," *IEEE Trans. on Vehicular Technology*, vol. 57, no. 2, pp. 760-769.

[3] Y. Hasuka, H. Sekine, K. Katano, and Y. Nonobe, "Development of boost converter for mirai," *SAE Techinical Paper*, pp. 1–6, 2015.

[4] N. Elsayad, H. Moradisizkoohi, and O. A. Mohammed, "A new single-switch structure of a dc-dc converter with wide conversion ratio for fuel cell vehicles: analysis and development," *IEEE Journal of Emerging and Selected Topics in Power Electronics*, vol. 8, no. 3, pp. 2785–2800, 2020.

[5] R. Shi, S. Semsar, and P.W. Lehn, "Single-stage hybrid energy storage integration in electric vehicles using vector controlled power sharing," *IEEE Trans. on Ind. Elec.*, vol. 68, no. 11, pp. 10623–10633, 2021.

[6] M. Pathmanathan, S. Semsar, C. Viana and P.W. Lehn, "Power Sharing Control Algorithm for Direct Integration of Fuel Cells in a Dual-Inverter Electric Vehicle Drivetrain," in *IEEE Transactions on Transportation Electrification*, vol. 8, no. 2, pp. 2490-2500, June 2022

[7] Y. Wang, M. Pathmanathan and P.W. Lehn, "Loss Comparison of Electric Vehicle Fuel Cell Integration Methods," International Symposium on Industrial Electronics (ISIE) 2022

[8] Ballard, "FCmoveHD Product Data Sheet,"[Online]. Available: https://www.ballard.com /docs/default-source/motive-modules-documents/fcmovetm.pdf?sfvrsn=6a83c3806

[9] J. Rodriguez, and P. Cortes, "Predictive Control of Power Converters and Electrical Drives," John Wiley and Sons, 2012.

[10] K.A. Corzine, S.D. Sudhoff, and C.A. Whitcomb, "Performance characteristics of a cascaded two-level converter," *IEEE Trans. on Energy Conv.*, vol. 14, no. 3, pp. 433–439, 1999.

Appendix

The dual inverter switching states which are utilized by MPC algorithm are shown in Table II below. S_{aFC}, S_{bFC} and S_{cFC} are the gating signals of the upper IGBTs in legs a, b and c of the FC inverter. Likewise, S_{aB}, S_{bB} and S_{cB} are the corresponding gating signals of the battery inverter. $v_{\alpha FC}$ and $v_{\beta FC}$ are the stationary reference frame voltages produced by the FC inverter (in terms of the FC voltage, V_{FC}). The stationary reference frame representation of the motor voltages (represented in terms of V_{FC} and the battery voltage, V_B are provided as v_α and v_β.

Table II: Switching states utilized in the fuel cell/battery dual inverter predictive controller

State No.	S_{aFC}	S_{bFC}	S_{cFC}	S_{aB}	S_{bB}	S_{cB}	$v_{\alpha FC}$	$v_{\beta FC}$	v_α	v_β
0	0	0	0	0	0	0	0	0	0	0
1	1	0	0	0	0	0	$\frac{2V_{FC}}{3}$	0	$\frac{2V_{FC}}{3}$	0
2	0	1	0	0	0	0	$-\frac{V_{FC}}{3}$	$\frac{V_{FC}}{\sqrt{3}}$	$-\frac{V_{FC}}{3}$	$\frac{V_{FC}}{\sqrt{3}}$
3	1	1	0	0	0	0	$\frac{V_{FC}}{3}$	$\frac{V_{FC}}{\sqrt{3}}$	$\frac{V_{FC}}{3}$	$\frac{V_{FC}}{\sqrt{3}}$
4	0	0	1	0	0	0	$-\frac{V_{FC}}{3}$	$-\frac{V_{FC}}{\sqrt{3}}$	$-\frac{V_{FC}}{3}$	$-\frac{V_{FC}}{\sqrt{3}}$
5	1	0	1	0	0	0	$\frac{V_{FC}}{3}$	$-\frac{V_{FC}}{\sqrt{3}}$	$\frac{V_{FC}}{3}$	$-\frac{V_{FC}}{\sqrt{3}}$
6	0	1	1	0	0	0	$-\frac{2V_{FC}}{3}$	0	$-\frac{2V_{FC}}{3}$	0
8	0	0	0	1	0	0	0	0	$-\frac{2V_B}{3}$	0
9	1	0	0	1	0	0	$\frac{2V_{FC}}{3}$	0	$\frac{2V_{FC}-2V_B}{3}$	0
14	0	1	1	1	0	0	$-\frac{2V_{FC}}{3}$	0	$\frac{-2V_{FC}-2V_B}{3}$	0
18	0	1	1	1	0	0	$-\frac{V_{FC}}{3}$	$\frac{V_{FC}}{\sqrt{3}}$	$\frac{V_B-V_{FC}}{3}$	$\frac{V_B-V_{FC}}{\sqrt{3}}$
19	1	1	0	0	1	0	$\frac{V_{FC}}{3}$	$\frac{V_{FC}}{\sqrt{3}}$	$\frac{V_{FC}+V_B}{3}$	$\frac{V_{FC}-V_B}{\sqrt{3}}$
20	0	0	1	1	0	0	$-\frac{V_{FC}}{3}$	$-\frac{V_{FC}}{\sqrt{3}}$	$\frac{V_B-V_{FC}}{3}$	$\frac{-V_{FC}-V_B}{\sqrt{3}}$
21	1	0	1	0	1	0	$\frac{V_{FC}}{3}$	$-\frac{V_{FC}}{\sqrt{3}}$	$\frac{V_{FC}+V_B}{3}$	$\frac{-V_{FC}-V_B}{\sqrt{3}}$
23	1	1	1	0	1	0	0	0	$\frac{V_B}{3}$	$-\frac{V_B}{\sqrt{3}}$
26	0	1	0	1	1	0	$-\frac{V_{FC}}{3}$	$\frac{V_{FC}}{\sqrt{3}}$	$\frac{-V_{FC}-V_B}{3}$	$\frac{V_{FC}-V_B}{\sqrt{3}}$
27	1	1	0	1	1	0	$\frac{V_{FC}}{3}$	$\frac{V_{FC}}{\sqrt{3}}$	$\frac{V_{FC}-V_B}{3}$	$\frac{V_{FC}-V_B}{\sqrt{3}}$
28	0	0	1	1	1	0	$-\frac{V_{FC}}{3}$	$-\frac{V_{FC}}{\sqrt{3}}$	$\frac{-V_{FC}-V_B}{3}$	$\frac{-V_{FC}-V_B}{\sqrt{3}}$
29	1	0	1	1	1	0	$\frac{V_{FC}}{3}$	$-\frac{V_{FC}}{\sqrt{3}}$	$\frac{V_{FC}-V_B}{3}$	$\frac{-V_{FC}-V_B}{\sqrt{3}}$
31	1	1	1	1	1	0	0	0	$\frac{-V_{FC}}{3}$	$\frac{-V_B}{\sqrt{3}}$
34	0	1	0	0	0	1	$\frac{-V_{FC}}{3}$	$\frac{V_{FC}}{\sqrt{3}}$	$\frac{V_B-V_{FC}}{3}$	$\frac{V_{FC}+V_B}{\sqrt{3}}$
35	1	1	0	0	0	1	$\frac{V_{FC}}{3}$	$\frac{V_{FC}}{\sqrt{3}}$	$\frac{V_B+V_{FC}}{3}$	$\frac{V_{FC}+V_B}{\sqrt{3}}$
36	0	0	1	0	0	1	$-\frac{V_{FC}}{3}$	$-\frac{V_{FC}}{\sqrt{3}}$	$\frac{V_B-V_{FC}}{3}$	$\frac{V_B-V_{FC}}{\sqrt{3}}$
37	1	0	1	0	0	1	$\frac{V_{FC}}{3}$	$-\frac{V_{FC}}{\sqrt{3}}$	$\frac{V_B+V_{FC}}{3}$	$\frac{V_B-V_{FC}}{\sqrt{3}}$
39	1	1	1	0	0	1	0	0	$\frac{V_B}{3}$	$\frac{V_B}{\sqrt{3}}$
40	0	0	0	1	0	1	0	0	$-\frac{V_B}{3}$	$\frac{V_B}{\sqrt{3}}$
42	0	1	0	1	0	1	$-\frac{V_{FC}}{3}$	$\frac{V_{FC}}{\sqrt{3}}$	$\frac{-V_B-V_{FC}}{3}$	$\frac{V_B+V_{FC}}{\sqrt{3}}$
43	1	1	0	1	0	1	$\frac{V_{FC}}{3}$	$\frac{V_{FC}}{\sqrt{3}}$	$\frac{V_{FC}-V_B}{3}$	$\frac{V_B+V_{FC}}{\sqrt{3}}$
44	0	0	1	1	0	1	$-\frac{V_{FC}}{3}$	$-\frac{V_{FC}}{\sqrt{3}}$	$\frac{-V_B-V_{FC}}{3}$	$\frac{V_B-V_{FC}}{\sqrt{3}}$
45	1	0	1	1	0	1	$\frac{V_{FC}}{3}$	$-\frac{V_{FC}}{\sqrt{3}}$	$\frac{V_B-V_{FC}}{3}$	$\frac{V_B-V_{FC}}{\sqrt{3}}$
47	1	1	1	1	0	1	0	0	$\frac{-V_B}{3}$	$\frac{V_B}{\sqrt{3}}$
48	0	0	0	0	1	1	0	0	$\frac{2V_B}{3}$	0
49	1	0	0	0	1	1	$\frac{2V_{FC}}{3}$	0	$\frac{2V_{FC}+2V_B}{3}$	0
54	0	1	1	0	1	1	$-\frac{2V_{FC}}{3}$	0	$\frac{2V_B-2V_{FC}}{3}$	0

Comparative evaluation of the 5-phase Vienna and the 5-phase PWM rectifiers under DC voltage control

A. Dieng

LER- Université Cheikh Anta Diop de Dakar, IREENA – UNIVERSITY OF NANTES

Tel : 00 221 776665138

abdoulayendaw.dieng@ucad.edu.sn

Keywords

PWM rectifier, Vienna rectifier, Permanent Magnet Synchronous Generator, Voltage Control

Abstract

Here a comparative evaluation of the 5-phase Vienna and the 5-phase PWM rectifiers is done. A double closed control loops strategy with an internal current loop and an external voltage loop is investigated. An itemized analysis is performed then the simulations and experimental results are done.

Introduction

In the context of renewable energy source exploitation, the fault-tolerant energy conversion chains is an attractive solution. Compared to the classic association of a 3-phase PWM rectifier with conventional three-phase machine, the polyphase electrical drives offers many advantages [1]-[4]. Many research works investigate the DC bus control strategies of the Vienna rectifier and the PWM rectifier as in [5]-[7]. Compared to others rectifiers, VIENNA rectifier presents more advantages. It is the favorite choice for its advantages such as less numbers of the switches, simple structure, high power density and ability to realize unity power factor with appropriate control strategy. The drawback of the VIENNA rectifier is the neutral-point (NP) voltage unbalance and oscillation [8]-[9]. Here a comparative evaluation of the 5-phase Vienna and the 5-phase PWM rectifier, In the context of renewable energy source exploitation, is done. This converters are supplied by the 5-phase PMSG non-sinusoidal EMF. The 5-phase Vienna rectifier compared to the 5-phase PWM rectifier has several advantages. Its particularity is that it has 5 active power components combined with diodes so it is deemed to be robust. In [10] the comparison to the Vienna rectifier with the conventional PWM rectifier show that the switching losses are reduced by a factor of 6 for frequencies below to 50 kHz. Another advantage of the Vienna rectifier, in the context of renewable energy source exploitation, is its non-reversibility. Fig. 1 et Fig. 2 show respectively the control scheme of the 5-phase PWM rectifier supplied by the 5-phase PMSG and the control scheme of the 5-phase Vienna rectifier supplied by the 5-phase PMSG.

This work focuses on the DC bus voltage control performances of the two rectifiers and the current control performances of the 5-phase PMSG. The operation in V-f mode is considered. The generator speed is considered constant. The inverter, located after the DC bus regulates the voltage across the load. The converter, generator side, regulates the DC bus voltage and the power transfer. The power delivered by the generator is imposed by the load. Here the inverter – load set can be replaced by a single-phase load purely resistive. the generator speed being considered constant, the DC bus voltage regulator generates the torque reference which permits to calculate the current references. Many current control strategy of 5-phase permanent magnet synchronous machine have already been proposed in [11]-[15]. Here, all harmoninics of the EMF is exploited.

The DC bus voltage is controlled on the generator side by controlling the electromagnetic torque of the 5-phase PMSG non sinusoidal EMF. The controller for the regulation the DC bus voltage is a conventional PI regulator. In the abcde frame, the current references is not constant. Then a robust and accurate AC current controller which also controls the switching frequency of the power switches and is slightly sensitive to the system's electrical parameters is necessary as proposed in [15]-[20]. For a good comparison, the machine neutral is connected to the midpoint of the DC bus for both rectifers. An itemized analysis is done. Simulation and experimental results are presented.

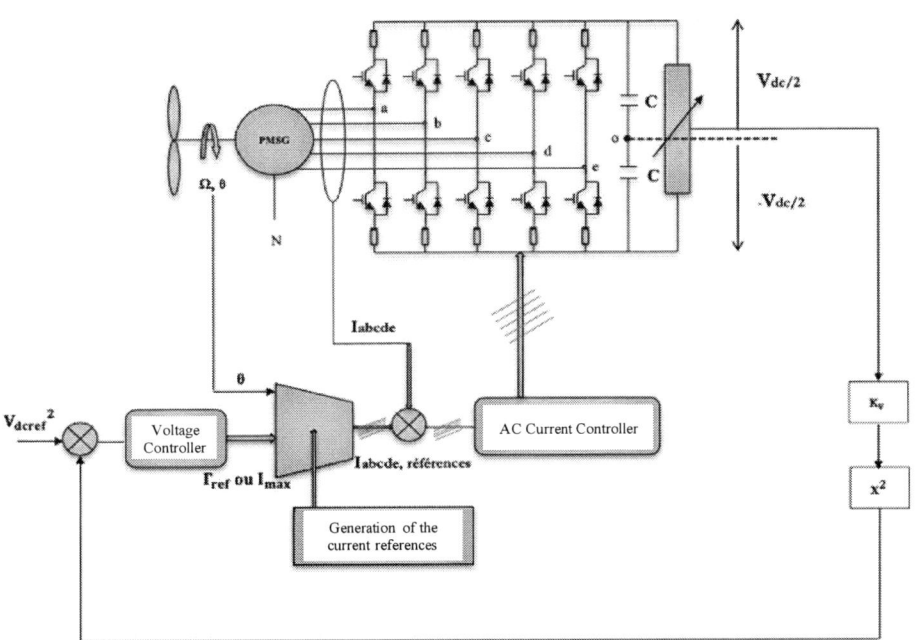

Fig. 1: DC voltage Control scheme of the 5-phase PWM rectifier supplied by the 5-phase PMSG non sinusoidal EMF

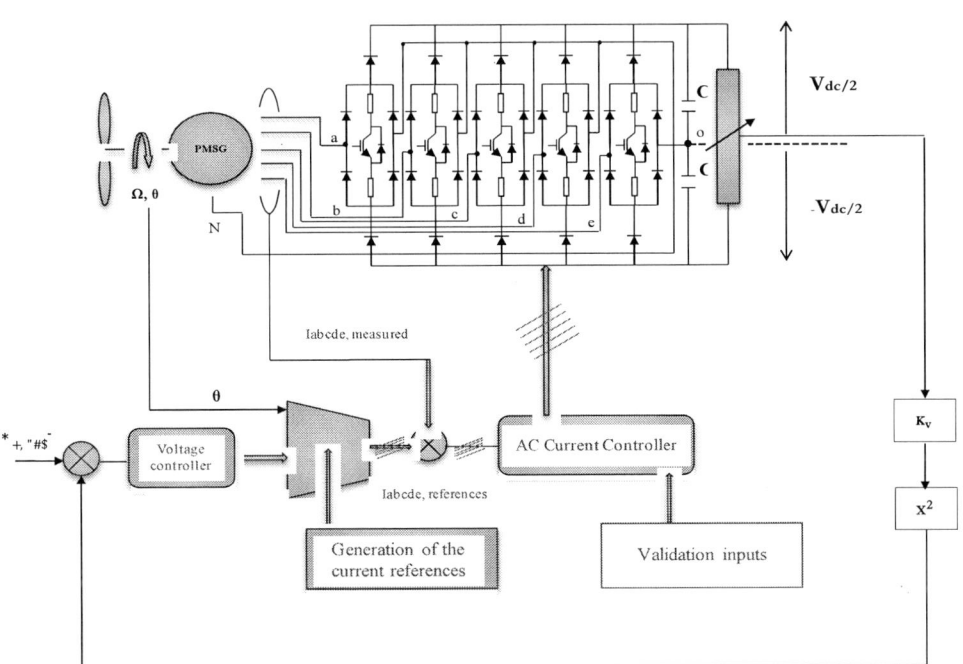

Fig. 2: DC voltage Control scheme of the 5-phase Vienna rectifier supplied by the 5-phase PMSG non sinusoidal EMF

Electrical model of the Two converters and the 5-phase PMSG

Electrical model of the 5-phase PMSG – PWM rectifier SET

The electrical model of the 5-phase PMSG – PWM rectifier is given by:

$$[E] = [R][I] + [L]\frac{d}{dt}[I] + [T_T][S_T]\frac{V_{dc}}{2} + \frac{1}{5}E_t[I_5] \tag{1}$$

where $[S_T] = \begin{bmatrix} S_a \\ S_b \\ S_c \\ S_d \\ S_e \end{bmatrix}$ $[I] = \begin{bmatrix} I_a \\ I_b \\ I_c \\ I_d \\ I_e \end{bmatrix}$ $[E] = \begin{bmatrix} E_a \\ E_b \\ E_c \\ E_d \\ E_e \end{bmatrix}$

$$[T_T] = \frac{1}{5}\begin{bmatrix} 4 & -1 & -1 & -1 & -1 \\ -1 & 4 & -1 & -1 & -1 \\ -1 & -1 & 4 & -1 & -1 \\ -1 & -1 & -1 & 4 & -1 \\ -1 & -1 & -1 & -1 & 4 \end{bmatrix} \quad [R] = \begin{bmatrix} r & 0 & 0 & 0 & 0 \\ 0 & r & 0 & 0 & 0 \\ 0 & 0 & r & 0 & 0 \\ 0 & 0 & 0 & r & 0 \\ 0 & 0 & 0 & 0 & r \end{bmatrix} \quad [L] = \begin{bmatrix} L_1 & L_2 & L_3 & L_3 & L_2 \\ L_2 & L_1 & L_2 & L_3 & L_3 \\ L_3 & L_2 & L_1 & L_2 & L_3 \\ L_3 & L_3 & L_2 & L_1 & L_2 \\ L_2 & L_3 & L_3 & L_2 & L_1 \end{bmatrix}$$

$$[I_5] = \begin{bmatrix} 1 \\ 1 \\ 1 \\ 1 \\ 1 \end{bmatrix} \quad \text{and} \quad E_t = E_a + E_b + E_c + E_d + E_e$$

S_z represents the conduction status of the branch supplying the z-phase (1 or -1 depending on the switches status), z = a, b, c, d, e.

The dynamic model of the 5-phase PMSG in abcde frames is developed in [2][15].

Electrical model of the 5-phase PMSG – Vienna rectifier SET

In contrary to the classic PWM rectifier, the equivalent model of the 5-phase Vienna rectifier is a complex. Establishing the equivalent model of the Vienna rectifier requires to know the conduction states of each diode. The control strategy adopted under normal operation requires that the current and FEM vectors are collinear. In this case, it is possible to not consider the conduction status of each diode of the Vienna rectifier. The electrical model of the 5-phase PMSG – Vienna rectifier is simple. If the neutral of the machine is not connected, the relation which links the generator simple voltages to the Vienna output voltages can be written in the form:

$$\begin{bmatrix} V_a \\ V_b \\ V_c \\ V_d \\ V_e \end{bmatrix} = \frac{1}{5}\frac{V_{dc}}{2}\begin{bmatrix} 4 & -1 & -1 & -1 & -1 \\ -1 & 4 & -1 & -1 & -1 \\ -1 & -1 & 4 & -1 & -1 \\ -1 & -1 & -1 & 4 & -1 \\ -1 & -1 & -1 & -1 & 4 \end{bmatrix}\begin{bmatrix} (1-S_a)\text{sign}(i_a) \\ (1-S_b)\text{sign}(i_b) \\ (1-S_c)\text{sign}(i_c) \\ (1-S_d)\text{sign}(i_d) \\ (1-S_e)\text{sign}(i_e) \end{bmatrix} + \frac{1}{5}E_t[I_5] \tag{2}$$

where $[I_5] = \begin{bmatrix} 1 \\ 1 \\ 1 \\ 1 \\ 1 \end{bmatrix}$ and $E_t = E_a + E_b + E_c + E_d + E_e$

S_z represents the conduction status of the branch supplying the z-phase (1 or -1 depending on the switches status), z = a, b, c, d, e.

Therefore, the electrical model of the 5-phase PMSG – Vienna rectifier is given by:

$$[E] = [R][I] + [L]\frac{d}{dt}[I] + [T_T][S_{TV}]\frac{V_{dc}}{2} + \frac{1}{5}E_t[I_5] \tag{3}$$

where $[S_{TV}] = \begin{bmatrix} (1-S_a)\text{sign}(i_a) \\ (1-S_b)\text{sign}(i_b) \\ (1-S_c)\text{sign}(i_c) \\ (1-S_d)\text{sign}(i_d) \\ (1-S_e)\text{sign}(i_e) \end{bmatrix}$ $[I] = \begin{bmatrix} I_a \\ I_b \\ I_c \\ I_d \\ I_e \end{bmatrix}$ $[E] = \begin{bmatrix} E_a \\ E_b \\ E_c \\ E_d \\ E_e \end{bmatrix}$

$$[T_T] = \frac{1}{5}\begin{bmatrix} 4 & -1 & -1 & -1 & -1 \\ -1 & 4 & -1 & -1 & -1 \\ -1 & -1 & 4 & -1 & -1 \\ -1 & -1 & -1 & 4 & -1 \\ -1 & -1 & -1 & -1 & 4 \end{bmatrix} [R] = \begin{bmatrix} r & 0 & 0 & 0 & 0 \\ 0 & r & 0 & 0 & 0 \\ 0 & 0 & r & 0 & 0 \\ 0 & 0 & 0 & r & 0 \\ 0 & 0 & 0 & 0 & r \end{bmatrix}$$

$$[L] = \begin{bmatrix} L_1 & L_2 & L_3 & L_3 & L_2 \\ L_2 & L_1 & L_2 & L_3 & L_3 \\ L_3 & L_2 & L_1 & L_2 & L_3 \\ L_3 & L_3 & L_2 & L_1 & L_2 \\ L_2 & L_3 & L_3 & L_2 & L_1 \end{bmatrix}$$

If the neutral of the machine is connected to the DC bus midpoint, equation (2) becomes :

$$\begin{bmatrix} V_a \\ V_b \\ V_c \\ V_d \\ V_e \end{bmatrix} = \frac{V_{dc}}{2}\begin{bmatrix} (1-S_a)\text{sign}(i_a) \\ (1-S_b)\text{sign}(i_b) \\ (1-S_c)\text{sign}(i_c) \\ (1-S_d)\text{sign}(i_d) \\ (1-S_e)\text{sign}(i_e) \end{bmatrix} \tag{4}$$

Thus the electrical model of the 5-phase PMSG – Vienna rectifier can be written:

$$[E] = [R][I] + [L]\frac{d}{dt}[I] + [S_{TV}]\frac{V_{dc}}{2} \tag{5}$$

DC Voltage Control strategy

This part is developed in [19].

The Fourier analysis of the EMF of the machine under consideration is summarized in Table 1. Table 1 summarizes the normalized magnitude of each harmonic of the considered machine.

Table 1. Fourier analysis of the EMF profile

EMF Harmonic	1	3	7	9
Magnitude/Fundamental %	100%	30%	0.2%	0.7%

The Fourier analysis of the EMF shows that the ninth harmonic and the seven harmonic are very low and can be neglected. Then the fundamental and the third harmonic of EMF are only considered. In this part a control strategy to keep the DC bus voltage constant is developed. neglecting all losses and applying the principle of power conservation, we can write:

$$P \approx P_{dc} + P_{ch} \approx V_{dc}I_c + P_{ch} \tag{6}$$

P : the generator electromagnetic power

P_{ch} : the power available across the load

V_{dc} : the DC voltage across the load

I_c : the current flowing through the equivalent capacitor

P_{dc} : the power received by the equivalent capacitor

C_{eq} : the equivalent capacitor

By replacing the expression of the generator electromagnetic power, the following equations are deduced:

$$C_{eq}V_{dc}\frac{dV_{dc}}{dt} = \Gamma\Omega - \Gamma_0\Omega \tag{7}$$

$$V_{dc}{}^2(s) = \frac{2\Omega}{C_{eq}s}(\Gamma - \Gamma_0) \tag{8}$$

The seven harmonic is very low and is neglected. The spectrum of the current references is reduced to the fundamental and to the third harmonic. We show that the expression of the electromagnetic power is given by [15]:

$$P = [E]^t[I] = \frac{5}{2}(E_mI_m + E_sI_s) \tag{9}$$

Where

E_m and E_s are respectively the maximum magnitude of the EMF fundamental and the EMF third harmonic

I_m and I_s are respectively the maximum magnitude of the current fundamental and the current third harmonic
So, Equation (8) can be rewritten in the form:

$$V_{dc}{}^2(s) = \frac{5E_m(1+x^2)}{C_{eq}s}(I_m - I_0) \tag{10}$$

where $x = \frac{I_s}{I_m} = \frac{E_s}{E_m}$ (the average torque is maximized and the copper losses is minimized [15])

Generation of the optimal current references

This part is developed in [15]. The power transfer is optimal when the losses are minimal. In this case an optimal control strategy that minimizes the reactive power and the copper losses is proposed. The fundamental and the third harmonic EMF are only considered.
In this case the E_a, E_b, E_c, E_d, E_e expressions are written:

$$[E] = \omega\Phi_m[G(\theta)] + 3\omega\Phi_s[G(3\theta)] \tag{11}$$

Where :

$$[G(\theta)] = \begin{bmatrix} A(\theta) \\ B(\theta) \\ C(\theta) \\ D(\theta) \\ E(\theta) \end{bmatrix} = \begin{bmatrix} \sin(\theta) \\ \sin\left(\theta - \frac{2\pi}{5}\right) \\ \sin\left(\theta - \frac{4\pi}{5}\right) \\ \sin\left(\theta - \frac{6\pi}{5}\right) \\ \sin\left(\theta - \frac{8\pi}{5}\right) \end{bmatrix} \quad [G(3\theta)] = \begin{bmatrix} A(3\theta) \\ B(3\theta) \\ C(3\theta) \\ D(3\theta) \\ E(3\theta) \end{bmatrix} \begin{bmatrix} \sin(3\theta) \\ \sin3\left(\theta - \frac{2\pi}{5}\right) \\ \sin3\left(\theta - \frac{4\pi}{5}\right) \\ \sin3\left(\theta - \frac{6\pi}{5}\right) \\ \sin3\left(\theta - \frac{8\pi}{5}\right) \end{bmatrix}$$

Where Φ_m, Φ_s are respectively the maximum magnitudes of the flux in the main and in the secondary machine, $\omega = p\Omega$ is the electrical angular speed and p is the number of pair poles and it is equal to 3 for the considered machine.
The total optimal current reference for each phase is given by [15] :

$$I_{zref} = \frac{E_z}{\sum_{z=a}^{e}E_z{}^2}\Gamma_{ref}\Omega \tag{12}$$

The self-oscillating phase-shifted modulator and current controller [20]

The next figure shows a basic scheme where the chosen current controller has been used in DC/AC converters [17][18]. It runs in sliding mode and offers an accurate current control. At high and low frequencies, it operates in different ways, i.e. at high frequency it operates for frequency switching control and at low frequency it operates for current control.

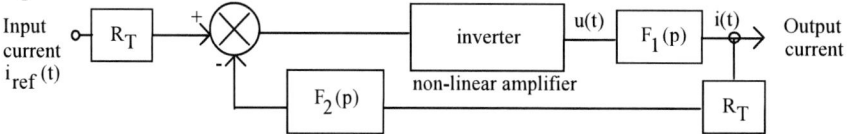

Figure 3: Scheme of the inverter current control loop

The control loop is used to create a current to voltage feedback. The inverter feeds the voltage signal u(t) to the load. Then, the transfer functions F1(p), F2(p) and RT respectively depict the load voltage to current transfer function, a second order low-pass filter transfer function and a current sensor transfer one.

$$F_1(p) = \frac{I(p)}{U(p)} = \frac{1}{R_L + L_L p} \tag{13}$$

$$F_2(p) = \frac{1}{1 + 2\xi\dfrac{p}{\omega_0} + \dfrac{p^2}{\omega_0^2}} \tag{14}$$

Thanks to the third order low-pass transfer function, the linear part gives birth to a self-oscillating mode. Then, the maximum switching frequency of the power switches depends on the f_{osc} oscillation frequency given by [17][18].

$$\frac{f_{osc}}{f_0} = \frac{\omega_{osc}}{\omega_0} = \sqrt{1 + \frac{2\,\xi}{\omega_0\,\tau_1}} = \sqrt{1 + 2\,\xi.\frac{f_{c1}}{f_0}} \tag{15}$$

It is the peculiarity of this current controller which thus limits the maximum switching frequency to a known value. Moreover, an equivalent gain of the non-linear stage, i.e. the inverter, can be defined. It is done in [17].

Simulations and Experimental results

This work focuses on the DC bus voltage control performances of the Two rectifier and the current control performances of the 5-phase PMSG. The generator speed is constant. Simulations are carried out using Matlab Simulink. The experimental prototype used to validate the proposed control strategy involves a 3 kW DC machine, 3kW 5-phase PMSG, 5-phase PWM rectifier, a 4 quadrant inverter which controls the DC machine in order to keep the generator speed constant and a dspace 1103.

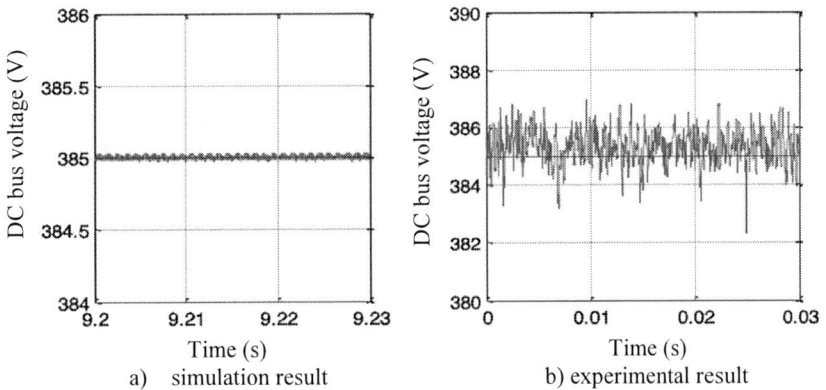

a)   simulation result   b) experimental result

Fig. 4: PWM rectifier, DC bus Voltage

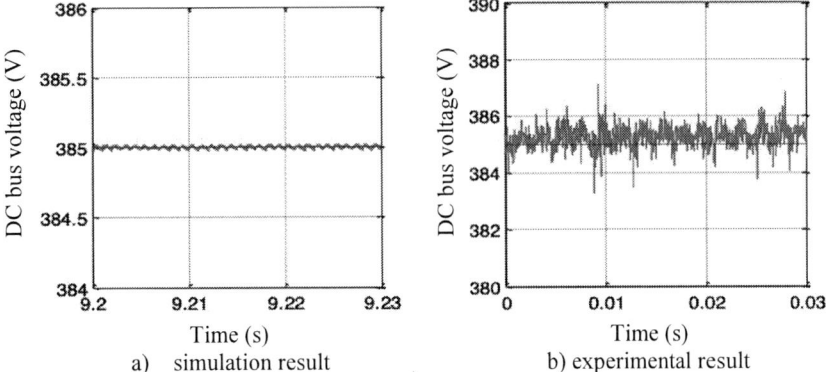

a)   simulation result   b) experimental result

Fig. 5: Vienna rectifier, DC bus Voltage

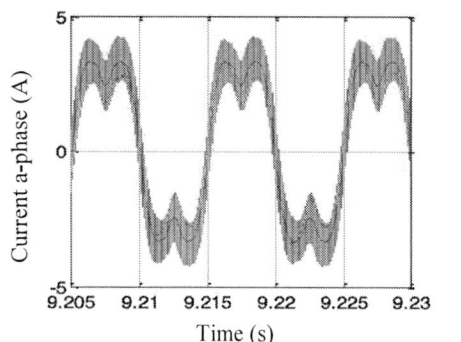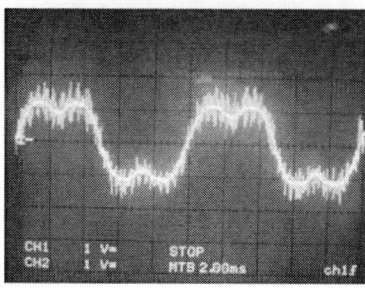

Fig. 6: PWM rectifier, Current in the a-phase

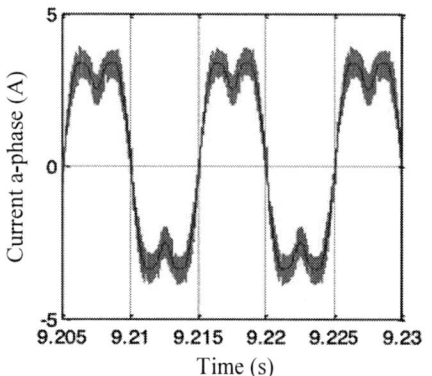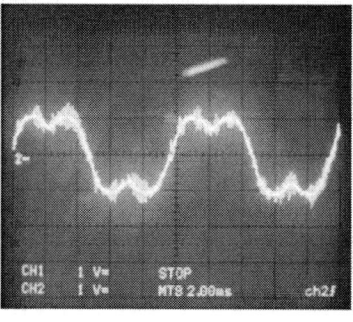

Fig. 7: Vienna rectifier, Current in the a-phase

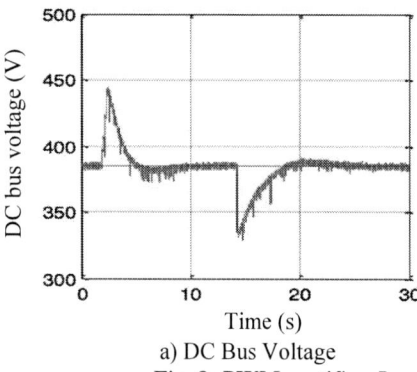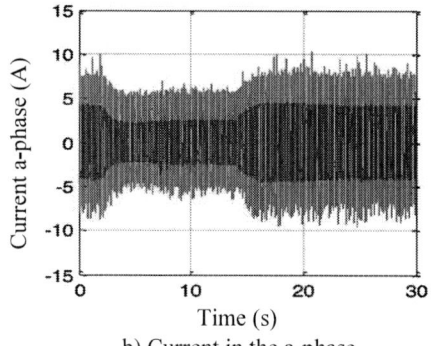

a) DC Bus Voltage   b) Current in the a-phase

Fig. 8: PWM rectifier, Load impact 50 %, experimental result

Fig. 4, Fig. 5, Fig. 6 and Fig. 7 show the simulation and experimental results in the case of the closed loop voltage control. The seventh harmonic of the FEM is exploited. The developed control strategy made it possible to regulate the DC bus voltage (Fig. 4 and Fig. 5). Even with a 50% load impact, the DC bus voltage stabilizes at his reference value (Fig. 8). Without the measurement noises and the switching effect, no disturbance has been observed. Fig. 6 and Fig. 7 show a good tracking performance of AC current controller used to control the imposed sinusoidal current in abcde frame.

Now, the performances of the 5-phase Vienna rectifier and the 5-phase PWM rectifier are compared. All simulated results are in accordance with the experimental results. By viewing Fig. 6 and Fig. 7, current ripples are very high in the case of the 5-phase PWM rectifier. The ripple amplitude of the DC output voltages is lower when using the 5-phase Vienna rectifier in comparison with the 5-phase PWM rectifier as shown in Fig. 4 and Fig. 5.

Conclusion

A comparative evaluation of the 5-phase Vienna and the 5-phase PWM rectifier has been done. Simulation and experimental results prove the effectiveness of the developed DC bus voltage control strategy. All simulated results are in accordance with the experimental results. In conclusion, the analysis has shown that the current ripples and the ripple amplitude of the DC output voltages is lower when using the 5-phase Vienna rectifier.

References

[1] A Dieng, Impact of the seventh harmonic magnitude for an optimal control of five-phase PMSM, 2020 International Conference on Advanced Electrical and Energy Systems, IOP Conf. Series: Earth and Environmental Science 582 (2020) 012008 doi:10.1088/1755-1315/582/1/012008, pp. 1-9, 2020

[2] A Dieng, J C Le Claire, A B Mboup, M F Benkhoris and M Ait-Ahmed, Analysis of five-phase permanent magnet synchronous motor, Revue Roumaine Sciences Techniques – Électrotechnique et Énergétique. Vol. 61, 2, pp. 116–120, Bucarest, 2016

[3] Kestelyn X, Semail E and Hautier J P, Vectorial Multi-machine modeling for a five-phase machine, International Congress on Electrical Machines, ICEM, Belgium, 2002, CD-ROM

[4] Toliyat H. A, Analysis and simulation of five-phase variable-speed induction motor drives under asymmetrical connections, IEEE Transactions on Power Electronics, Vol. 13, No. 4, pp. 748-756, 1998

[5] Debranjan M and Debaprasad K, Voltage Sensorless Control of VIENNA Rectifier in the Input Current Oriented Reference Frame, IEEE Transactions on Power Electronics, Vol. 34, Issue: 8, Aug. 2019

[6] Hongyan Z , Trillion Q Z , Yan L , Jifei D and Pu S, Control and Analysis of Vienna Rectifier Used as the Generator-Side Converter of PMSG-based Wind Power Generation Systems, Journal of Power Electronics, Vol. 17, No. 1, pp. 212-221, January 2017

[7] Hui W and Hui Q, Study of Control Strategies for Voltage-Source PWM Rectifier, Proceedings of the 2nd International Conference on Computer Science and Electronics Engineering (ICCSEE 2013), Vol. 34, pp. 1268-1271, ISBN 978-90-78677-61-1

[8] Qi E, Luo Z, Chen H and Zhu G, Modelling and Control of Single Phase VIENNA Rectifier, 2016 International Conference on Industrial Informatics - Computing Technology, Intelligent Technology, Industrial Information Integration (ICIICII), 3-4 Dec. 2016, Wuhan, China

[9] Lai R, Wang F, Burgos R, Boroyevich D, Jiang D and Zhang D, Average modeling and control design for VIENNA-type rectifiers considering the dc-link voltage balance, IEEE Transactions on Power Electronics, vol. 24, no. 11, pp. 2509-2522, Nov. 2009

[10] Pathak D, Locher R E, 3-Phase Power Factor Correction Using Vienna Rectifier Approach and Modular Construction for Improved overall Performance, Efficiency and Reliability, IXYS CORP. Santa Clara, CA USA.

[11] Zong Z L, Wang K, Zhang J. Y, Control strategy of five-phase PMSM utilizing third harmonic current to improve output torque, 2017 Chinese Automation Congress (CAC), 20-22 Oct. 2017, Jinan, China

[12] Baudart F, Dehez B, Matagne E, Telteu-Nedelcu D, Alexandre P and Labrique F, Torque control strategy of polyphase permanent-magnet synchronous machines with minimal controller reconfiguration under open-circuit fault of one phase, IEEE Transactions on Industrial Electronics, Vol. 59, Iss.6, pp. 2632-2644, 2012

[13] Baudart F, Matagne E, Dehez B and Labrique F, Optimal current waveforms for permanent magnet synchronous machines with any number of phases in open circuit, Mathematics and Computers in Simulation, Elsevier, Vol. 90, pp.1-14, 2013

[14] Kestelyn X and Semail E, A vectorial approach for generation of optimal current references for multiphase permanent-magnet synchronous machines in real time, IEEE Transactions on Industrial Electronics, Vol. 58, No. 11, pp. 5057–5065, 2011

[15] A. Dieng, J.C. Le Claire, A. B. Mboup, M.F. Benkhoris and M. Ait-Ahmed, "An improved torque control strategy of five-phase PMSG-PWM rectifier set for marine current turbine applications" 2019 IEEE 13th International Conference on Compatibility, Power Electronics and Power Engineering, CPE-POWERENG 2019

[16] J C Le Claire, S Siala, J Saillard and R Le Doeuff, Method and device for controlling switches in a control system with variable structure, with controllable frequency, US patent n° 6.376.935 B1, April 23, 2002

[17] J C Le Claire, Power Electronic Converters – PWM Strategies and Current Control Techniques», Chapter 14: Current Control using Self-Oscillating Currents Controllers, ISTE, London, United Kingdom, 2011, WILEY, Hoboken, USA, 2011, pp 417- 447

[18] J C Le Claire, S Siala, J Saillard and R Le Doeuff, A new Pulse Modulation for Voltage Supply Inverter's Current Control, 8th European Conference on Power Electronics and Applications, Lausanne, Switzerland, September 1999, CD-ROM ref. ISBN 90-75815-04-2

[19] A Dieng and J C Le Claire, Performances Assessment of five-phase Vienna rectifier - PMSG SET: Experimental validation of DC bus voltage control, Proceedings of the 15th IEEE Conference on Industrial Electronics and Applications, ICIEA 2020, pp. 291–295, 9248244, 2020

[20] A Dieng, M F Benkhoris, M Ait-Ahmed and J C Le Claire, Modeling and Optimal Current Control of Five-Phase PMSG - PWM Rectifier SET Non-Sinusoidal EMF under Open-Circuit Faults, 21st European Conference on Power Electronics and Applications, EPE 2019 ECCE Europe, 8915578, pp. 1-9, 2019

Modelling and Control of a 50kW SiC-based Isolated DAB Converter for Off-Board Chargers of Electric Vehicles

Haaris Rasool[1,2], Manh Tuan Tran[1,2], Sajib Chakraborty[1,2], (Member, IEEE), Joeri Van Mierlo[1,2], (Senior Member, IEEE)
Thomas Geury[1,2], (Member, IEEE), Mohamed El Baghdadi[1,2], (Member, IEEE) and Omar Hegazy[1,2], (Senior Member, IEEE)

[1]MOBI-EPOWERS Research Group, ETEC Department, Vrije Universiteit Brussel (VUB), Pleinlaan 2,1050
Brussels, Belgium
[2]Flanders Make, 3001 Heverlee, Belgium

*Corresponding author: Omar Hegazy (omar.hegazy@vub.be).

Keywords

«Bi-directional», «Charging infrastructure for EVs », «Wide bandgap», «Power flow control»,

Abstract

The paper proposes the design of a 50kW isolated DC-DC Dual Active Bridge (DAB) converter for a high-power off-board charger for Electric vehicles (EVs) applications. The detailed electro-thermal simulation of the wide band gap (WBG) (i.e., SiC MOSFETs)-based bidirectional DAB is performed to determine the performance of the system in terms of efficiency at high-power operation. A linear model based on system identification has been created to design the control approach accurately. Dual-loop phase shift constant-current (CC) and constant-voltage (CV) control strategy is implemented on the dynamic simulation model at a higher switching frequency to validate the stability of the designed controller. The proposed DAB operates with an acceptable ripple and dual-loop voltage-current control that comprehensively tracks reference commands, while the maximum efficiency achieved is approximately 97.5% at rated power.

1. Introduction

With increasing awareness of climate change, research efforts are focusing on decarbonizing vehicle emissions. The automotive and energy sector are focusing on battery electric vehicles (BEVs) and renewable energy resources (RERs). Therefore, this also involves new challenges such as an increasing energy demand by charging BEVs and grid stability issues because of the intermittent nature of RERs. However, today it is necessary to develop appropriate BEV chargers that allow bidirectional power flow and intelligent strategies to manage the charging process [1].

There are two types of charging systems for EVs: conductive charging and inductive charging. Conductive charging systems are better established than inductive charging, which is widely accepted in electrified transport. In conductive charging, the vehicle stays in direct contact with the supply through the socket to transfer power. Conductive charging is further classified into on-board and off-board charging systems [2]. The Society of Automotive Engineers (SAE) and Electric Power Research Institute (EPRI) have categorized EV charging levels as level-1, level-2, level-3, and next-generation ultra-fast charging. The level-1 and level-2 chargers are considered on-board chargers to inject power in batteries. However, a level-3 charger typically works as an external converter and can effectively manage high power flow. Fast DC charger is associated with level-3 and next-generation charging [3], [4]. The off-board charger reduces on-board circuitry to reduce the overall weight of the vehicle [5]. During inductive charging, there is no physical contact with the supply to transfer power to the vehicle. The wireless charging technique utilizes an electromagnetic field to transfer power.

The low-frequency or high-frequency transformer provides galvanic isolation between the EV and the grid. The filter is used to eliminate unwanted harmonic currents. Power Electronic Converters (PECs) convert the three-phase AC power from the grid into DC power used to charge the EV battery. The control unit operates the switches of PEC to adjust the voltage and current level, which the BEV can accept. Since the charger is characterized by high voltages and currents, galvanic isolation between the electrical grid and the BEV is required to ensure a safe operation. This can be realized by two transformer topologies: a low-frequency transformer (LFT) or a high-frequency transformer (HFT). Typically, the topology with an isolated DC-DC converter with HFT is considered for high power Off-board charger design (i.e., 50 kW and beyond). It comprises converting incoming AC power to a fixed DC output using an Active Front End (AFE) converter, which is then converted to the demanded voltage of the EV using an isolated DC-DC converter. This topology contains an LCL filter, an AC/DC bidirectional converter, a DC filter, an isolated DC-DC bidirectional converter with a high frequency (HF) transformer and a DC link filter [6], as shown in Fig. 1. The bidirectional off-board charger connects to the battery pack of EVs, which later supplies the inverter and drives the electric motor. The battery pack is also connected with a buck DC-DC converter inside the vehicle, which provides the auxiliary power supply of EVs.

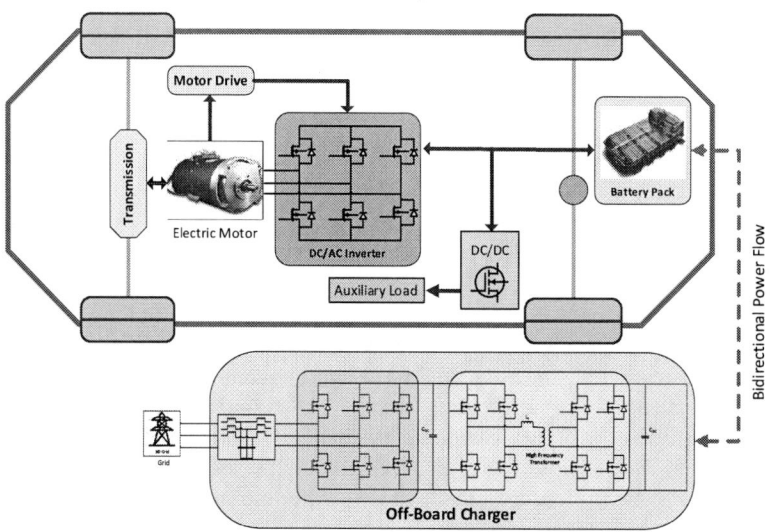

Fig 1. Off-Board charging arrangement for fast DC charging of BEVs.

The different topologies of isolated DC-DC converters are discussed in the literature, such as full-bridge LLC converter (FBLLC), phase-shifted full-bridge converter (PSFB) and dual active bridge (DAB) converter. The DAB converter is best suited for high voltage applications for bidirectional power flow thanks to its higher efficiency, soft switching commutation capabilities and a reduced number of devices [7]. The modularity and synchronization structure in DAB topologies, allows converters to achieve higher output power and facilitates the two-way operation of power flow for battery charging (G2V) and discharging (V2G) applications. Therefore, DAB topology is beneficial where the following factors are required: galvanic isolation, power density, high voltage conversion ratio and reliability [7], [8]. These factors make an ideal charging station and energy storage system. Thus, this paper proposes a DAB DC-DC converter to design a 50 kW off-board charger for BEVs. Furthermore, the stepwise design methodology and control design of the DAB converter is analyzed in this paper. The aim of this study is to integrate the WBG SiC semiconductor and high-switching frequency control approach for the emerging application of off-board chargers. At the same time, this paper depicts the design of an isolated bidirectional DC-DC converter for an off-board fast charger. The converter provides isolation between the grid and the battery of the EV. The objective of designing an isolated DC-DC converter is to get higher efficiency and high specific power by eliminating the classical low-frequency grid-side transformer. The conventional battery charging technique is known as constant current-constant voltage (CC-CV) mode. The idea is that the battery is charged with maximum constant current according to the battery cell capacity up to cut-off voltage and then charged at constant voltage until the drawn current decreases to C/10 or less, where C states the charge or discharge rate of the battery over one hour [4]-[5]. The proposed control approach allows batteries to be charged in different modes based on different voltages and current levels to keep the battery life maximal [9].

This paper is arranged into different sections. Section 2 of this paper describes the system and control architecture. Section 3 illustrates the design methodology of the DAB converter. In Section 4, the detailed loss modelling of DAB converter in a dynamic simulation model is explained. Section 5 is dedicated to control design, whereas Section 6 describes the simulation results. Finally, the conclusions are provided in Section 7.

2. System and control architecture

In this paper, an isolated DAB DC-DC converter is designed and simulated. It consists of a primary side capacitor, coupling inductor, high-frequency transformer, and secondary side capacitor. The SiC-based power modules are utilized to design and operate at a high switching frequency. A controller is required to drive eight switches of the DAB converter and regulate the power according to the requirement of the battery capacity. The topology and bi-directional control architecture of the DAB converter is illustrated in Fig. 2.

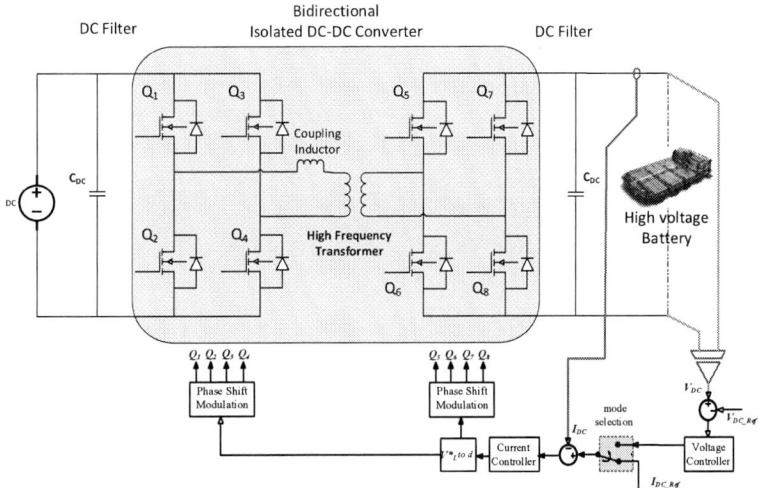

Fig 2. Isolated DAB DC-DC converter and control architecture.

3. DAB Phase Shift and Inductor Selection

The phase shift of the converter is dependent on the value of the inductor L of primary side of the DAB converter. In this paper, the turns ratio of the transformer considered, n, is 1 and the switching frequency f_s is 40kHz. The required total leakage inductance is calculated from Eq. 1, 2, and Fig. 3 (a) [7]-[8]. It is approximately 36 µH at a phase shift of 60° for the 50kW power. The planar transformer's primary side leakage inductance measured in the lab is 4µH. The coupling inductor of 32 µH is introduced on the primary side of the DAB converter with a combination of the leakage inductor of the transformer. Therefore, a total inductor of 36µH is required to achieve the desired power transfer with phase shift control.

$$\phi = \frac{\pi}{2}\left(1 - \sqrt{1 - 8\frac{nf_s LP_{out}}{V_{in}V_{out}}}\right) \tag{1}$$

where,
$$n = \frac{V_{out}}{V_{in}} \tag{2}$$

Another analysis is performed between the output power and phase shift ϕ, to analyze the maximum allowable power of DAB DC-DC converter with a phase shift. In this investigation, all other parameters are considered constant in Eq. 3, such as turns ratio n is 1, inductance L is 36µH, and the switching frequency f_s is 40kHz. The plot of Fig. 3 (b) shows that the peak power is transferred when the phase shift is 90°. This analysis is very important to check the power safety margin while designing a converter at a given power rating. This analysis shows that, maximum rated power can be delivered without any interruption at phase shift of 60°. The sinusoidal curve of the power versus phase shift illustrates the flow of power in bidirectional mode such as V2G and G2V for the off-board charger. The limit in phase shift will be applied for the closed loop control implementation in order to limit the power of the converter.

$$P_{out} = \frac{V_{in}V_{out}\phi\left(1 - \left|\frac{\phi}{\pi}\right|\right)}{2\pi f_s nL} \tag{3}$$

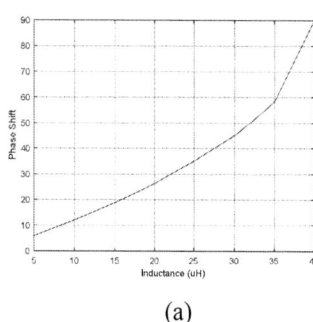 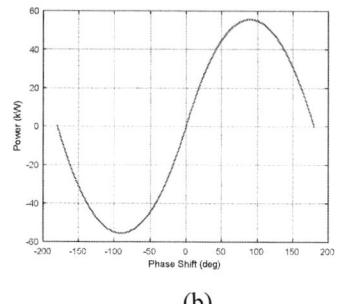

(a)   (b)

Fig 3. Inductor selection: (a) Phase shift vs inductance, (b) power versus phase shift.

4. DAB Power Loss Modelling

In this paper, the power loss model of the DAB converter is developed in simulation to predict the overall efficiency. The power loss model includes half-bridge (CAS300M17BM2) module loss, high-frequency transformer loss and passive components loss.

4.1 Converter Loss Model

A. Conduction Power Losses

The conduction power losses of the power electronic converter depend on the voltage-current characteristic of the transistor and the diode. It can be calculated using Eqs. (4-6) [10]–[12]. V_{ds}[V] and i_d [A] of Eq. 4, represent the forward saturation voltage and drain current of the MOSFET. V_{ds} is obtained as a function of i_d and the junction temperature. It can be estimated by using an interpolation technique on the datasheet characteristics of the half-bridge module. Integrating the instantaneous power losses over a switching cycle T_{sw} (sec) gives an average value of the MOSFET conduction losses as in Eq. (4).

$$P_{CM} = \frac{1}{T_{sw}} \int_0^{T_{sw}} P_{cond_{loss}}(t)dt = \frac{1}{T_{sw}} \int_0^{T_{sw}} V_{ds}(t)i_d(t)\, dt \tag{4}$$

The conduction loss of the body diode is calculated using Eq. (5).
V_f (V) and i_f (A) represent the forward saturation voltage and current of the diode, respectively. V_f is a function of the junction temperature and the current i_f.

$$P_{CD} = \frac{1}{T_{sw}} \int_0^{T_{sw}} P_{D\,cond_{loss}}(t)dt = \frac{1}{T_{sw}} \int_0^{T_{sw}} V_f(t)i_f(t)dt \tag{5}$$

The total conduction loss is calculated by adding up the average power losses of the MOSFET and the diode, as represented in Eq. (6).

$$P_{total_C} = P_{CM} + P_{CD} \tag{6}$$

B. Switching Power Losses:

The estimation of the switching losses in the device is done by using datasheet characteristics. A complex strategy with a small step size is needed to estimate the losses that occur during the device's turn-on and turn-off transients. However, a lookup table method is more feasible and easier to implement for estimating the switching power losses. The switching energies E_{swon} (mJ) and E_{swoff} (mJ) are used to estimate the switching power loss as expressed in Eq. (7). The switching energies of MOSFETs are defined as functions of drain current, junction temperature and drain voltage [13]–[16].

$$E_{total} = E_{sw_{on}} + E_{sw_{off}} \tag{7}$$

Hence, the switching power losses can be estimated from the energy losses expressed in Eq. (8).

$$P_{Msw_{loss}} = \left(E_{sw_{on}} + E_{sw_{off}}\right).f_S \tag{8}$$

where, f_S is the operating switching frequency.
The switching losses for a diode are defined by the reverse recovery characteristics but can also be calculated from the reverse recovery charge and reverse recovery time. However, for SiC technology-based MOFET, the switching losses of the diode are negligible. It can be calculated using the relation expressed in Eq. (9), which is dependent on E_{rr}.

$$P_{Dsw_{loss}} = E_{rr}f_S \tag{9}$$

The total average power losses of the MOSFET and body diode can be computed with Eq. (10).

$$P_{total_{sw}} = P_{Dsw_{loss}} + P_{Msw_{loss}} \tag{10}$$

4.2 Transformer Loss Model

Transformers and inductors impact the size and weight of isolated DC-DC converter designs. Increasing the operating frequency reduces the requirement for passive filters. However, increasing the switching frequency beyond a certain value increases the power loss and reduces the efficiency, because of the skin effect and proximity effect. More interleaving can be achieved with a planar transformer to reduce the effect of proximity. The planar transformers have some advantages over conventional transformers and are therefore used in this research. The power density of planar magnets is high. It allows more interleaving, further reducing the conductor losses. Tight control is possible through planar magnets. The compact size of the transformer can support the integration of an additional leakage or coupling inductor with the transformer without the need for a separate component on the board [7][8]. The planar transformer is chosen for the isolated 50kW DAB DC-DC converter design. The manufacturer specification of the high-frequency transformer is depicted in table-I. These parameters are used in simulation design and power loss modelling of a transformer.

Table I: High-frequency transformer specification.

Parameters	Values	Parameters	Values
Power (kW)	50	Turns Ratio Primary to Secondary	10:10
Operating Frequency (kHz)	10-70	Cooling Type	Liquid cold plate
Primary Inductance (mH)	2.3	Estimated Power Loss Core (W)	101 ($R_{ac} = 0.0125\Omega$)
Leakage inductance (µH)	4	Estimated Power Loss Winding (W)	108 ($R_{dc} = 0.0133\Omega$)
Input Voltage (V)	550-900	Estimated Maximum Temperature Rise (°C)	80
Output Voltage (V)	550-900	Length x Width x Height (mm)	293 x 204.5 x 75

4.3 Passive Component Loss Model

The passive filters power loss significantly contributes to the converter design's efficiency. Therefore, the power loss of the inductor and capacitor are calculated using mathematical equations [10].

A. Inductor Losses Calculation

The inductor's total losses depend on the windings and core power losses. It can be determined by Eq. (11).

$$P_{ind_{Total\ loss}} = P_{ind_{loss}} + P_{ind_{core}} \tag{11}$$

The power losses in the inductor windings are represented in Eq. (12).

$$P_{ind_{loss}} = I_L^2 R_{LDC} + \Delta I_L^2 R_{LAC} \tag{12}$$

where I_L (A) is the inductor current, ΔI_L (A) is the inductor current ripple, R_{LDC} (mΩ) is the inductor resistance and R_{LAC} (mΩ) is the AC winding resistance due to the skin effect. R_{LDC} is quantified by Pouillet's law as shown in Eq. (13).

$$R_{LDC} = \frac{\rho N l_T}{A_w} \tag{13}$$

where, ρ (mm.Ω) is the wire resistivity, N is the number of turns, l_T [mm] is the length of the turn and the cross-sectional area of the wire is denoted by A_w (mm²). The wire for the inductor core can be selected, according to A_w and by the current passing through the inductor.

R_{LAC} can be calculated based on the skin depth S_d [mm] and F_R is the winding resistance ratio [10], as expressed in Eq. (14). Skin depth can also be used for the selection of Litz wire with strand size not bigger than three times the skin depth.

$$R_{LAC} = R_{LDC} F_R \tag{14}$$

Another factor on which the total inductor losses are the core losses, represented by Eq. (15).

$$P_{ind_{core}} = k f_s \left(\frac{2}{\pi^2} 4 f_S\right)^{\alpha-1} B_{pk}^\beta V_{core} \tag{15}$$

where k, α, β are the Steinmetz parameters and V_{core} (L) is the volume of the core, which can be extracted from the datasheet of the inductor core material. B_{pk} is the peak magnetic flux density.

B. Capacitor Losses Calculation:

The equivalent series resistance (ESR (mΩ)) of the capacitor is an important parameter to estimate the power loss. The ESR is normally mentioned in the datasheet of the capacitor. The power loss of the capacitor is calculated using Eq. (16) [17].

$$P_C = I_{CRMS}^2 ESR\ (f_S) + I_{leak} V_C \tag{16}$$

where, I_{CRMS} (A) is the RMS current passing through the capacitor, V_C (V) is the average capacitor voltage and I_{leak} (A) is the leakage current passing through the capacitor.

4.4 Efficiency Estimation

System efficiency is a very important variant for the observation of the overall performance of the system. In this paper, the efficiency of an isolated DAB DC-DC converter is estimated using the complete power losses, such as power losses of the SiC MOSFETs, losses in the passive filters and high-frequency transformer loss. The loss of a high-frequency transformer is mentioned in the manufacturer's specifications. The efficiency in percentage is calculated using the DC output power and the total average power losses of the converter according to Eq. (17).

$$\eta = \frac{P_{DC}}{P_{DC} + P_{loss\_MOSFET} + P_{loss_{psve}} + P_{loss\ TR}}\ 100\ (\%) \tag{17}$$

5. DAB Converter Control Design

This paper proposes dual-loop control for the DAB converter. The PI controllers are tuned to obtain the desired response in terms of overshoot, rise time, and settling time, and tested in a dynamic simulation. All sensors' delays are included in the control system design process. The design and tuning of the controllers have been carried out using gain margin (*gM*) at gain margin frequency (*ωcg*) and phase margin (PM) at crossover frequency (*ωcp*). The block diagram of dual-loop voltage and current control of the DAB DC-DC converter are shown in Fig. 4. It consists of an outer loop constant voltage control, regulating the output voltage in CV mode, and an internal control loop is constant current control, regulating the output current in CC mode. The output of the voltage controller is the reference signal of current controller. The output control signal of current controller is the phase delay, it can be transformed into time delay (i.e., 60° phase delay = $\frac{60^o}{360^o f_S}$ sec time delay). Since the charge control begins with constant current mode, the delay should be limited to 60° to avoid overcharging current. The time delay impacts the delay of the triangular carrier wave for the generation of 50% duty cycle PWM digital signals.

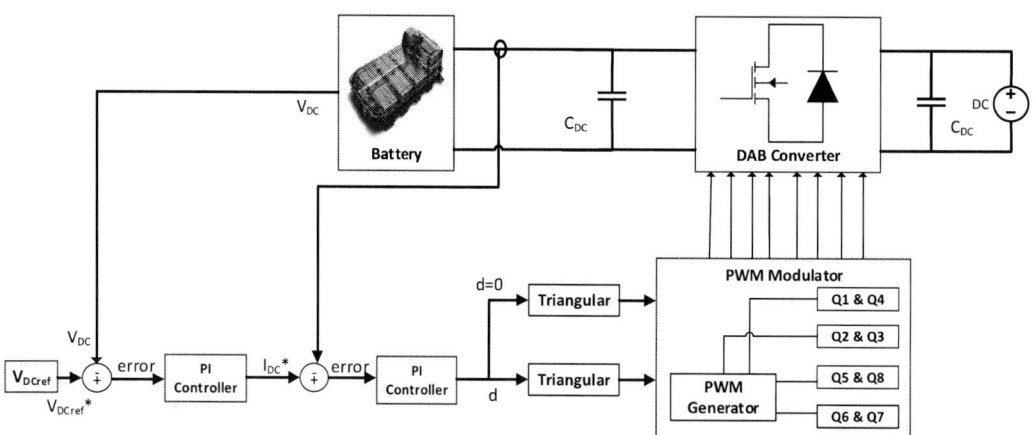

Fig. 4: Dual loop control block diagram of DAB converter.

The closed loop constant current control is designed with the linear model of current $G_{CC_{plant}}$. The transfer function of load current is given in Eq. 18, the input is phase delay and output is load DC current. This second-order transfer function of load current is identified using MATLAB system identification tool (ident). For the system identification, input and output data are logged from the dynamic open-loop simulation model of DAB DC-DC converter. The input and output is further applied for the identification of transfer function using MATLAB tool. The best match is achieved at 99% with the second-order transfer function $G_{CC_{plant}}$.

$$G_{CC_{plant}} = \frac{I_{DC}}{delay} = \frac{647.7s + 2.098 \times 10^6}{s^2 + 4011s + 2.533 \times 10^6} \quad (18)$$

The constant current control loop is designed at a crossover frequency of 4kHz with sufficient phase margin, overshoot, rise time, and settling time to ensure fast dynamics in the charging process. The dual loop constant voltage control loop is shown in Fig. 5. The voltage control is designed using the transfer function of DC converter current, output capacitance and maximum load. The transfer function of load DC voltage $G_{VV_{plant}}$ is given in Eq. 19, the input is DC reference voltage and output is load DC voltage.

$$G_{VV_{plant}} = \frac{V_{DC}}{I_{DC_{Ref}}} = \frac{\left(k_p + \frac{k_i}{s}\right)G_{CC_{plant}}}{\left(k_p + \frac{k_i}{s}\right)G_{CC_{plant}} + 1} \times \frac{\frac{1}{sC_{DC}}}{\frac{1}{sC_{DC}} + \frac{1}{R_o}} \quad (19)$$

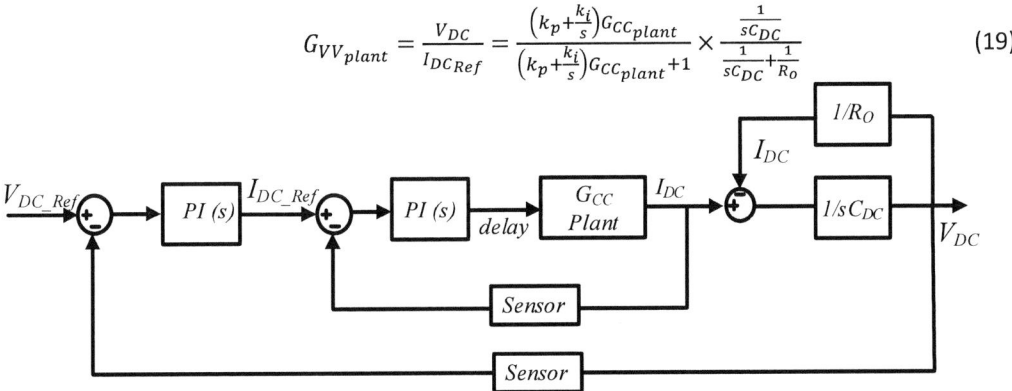

Fig. 5: Voltage loop control of DAB converter.

The closed-loop control design includes the sensors and PWM delays, i.e., 10µs for voltage sensor (LEM DVC 1000-P) delay, 3µs for current sensor (ISB-425-A-802) and 0.5µs for PWM. After close-loop analysis, the PI controllers are tuned according to margins and frequencies. The outer constant voltage control is designed at 50Hz. The inner loop current PI controllers are designed at <1/10 of the switching frequency, which is <4kHz. The control parameters, close loop overshoot, gM, ωcg, PM and ωcp of voltage and current control are displayed in Table II. The discrete controller will be implemented in the real-time FPGA platform of dSPACE MicrolabBox.

Table II: Controller design parameters

Controller	Control Parameters	Overshoot (%)	GM (dB)	PM (deg)	Wcp (rad/sec)	Wcg (rad/sec)
Current Control	K_p=37.1, K_i=91453.4	4.6	39.7	86	2.41×10^4	8.14×10^4
Voltage Control	K_p=0.22, K_i=27.4	6	22.1	62	326	1.8×10^3

6. Simulation Results

A dynamics simulation model of the DAB DC-DC converter is established in MATLAB/Simulink. The dual loop voltage-current control strategy is designed and implemented in simulation using specifications at rated power.

6.1 Specification

The specification of DAB DC-DC converter is shown in Table-III. These parameters are used for simulation design and performance validation.

Table III: Simulation design parameters

Parameters	Values
Power (kW)	50
Operating frequency (kHz)	40

Input voltage (V)	700-800
Output voltage (V)	700-800
Half-bridge module	CAS300M17BM2
Primary to secondary transformer ratio	10:10
Total inductance (µH)	36
DC link capacitance (µF)	200

6.2 Simulation Results

The voltage-current controller of the DAB converter has been implemented in simulation at 40kHz switching frequency. The DC voltage reference tracking has been tested by changing the reference command from 550 to 750V, its efficiency and primary-secondary MOSFETs gate signals, primary-secondary transformer voltages, and inductor current are shown in Fig. 6 (a), (b), and (c) respectively. The DC current reference tracking has been tested by changing the reference command from 20 to 60A, its efficiency and primary-secondary MOSFETs gate signals, primary-secondary transformer voltages, and inductor current are shown in Fig. 7 (a), (b), and (c) respectively. Both controllers i.e., constant voltage and constant control, are practically observed to track the reference signal successfully.

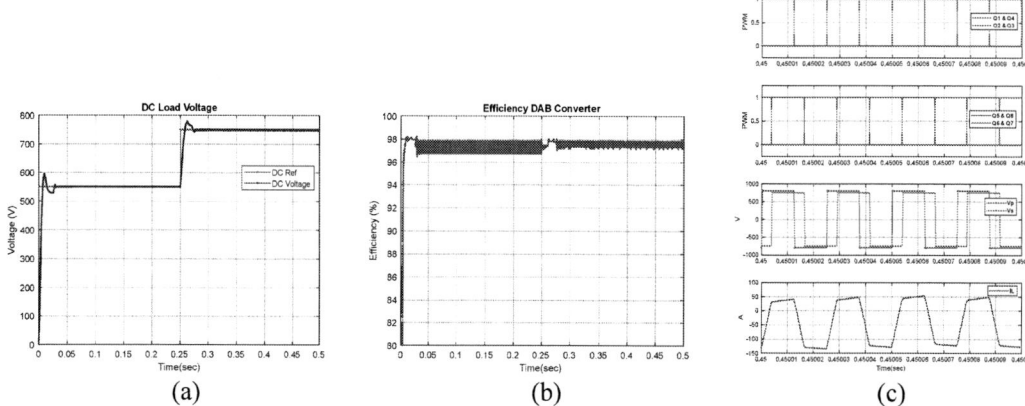

Fig 6. Constant voltage control: (a) DC voltage tracking, (b) efficiency of DAB converter, c) PWM Signals, transformer primary-secondary voltages and inductor current.

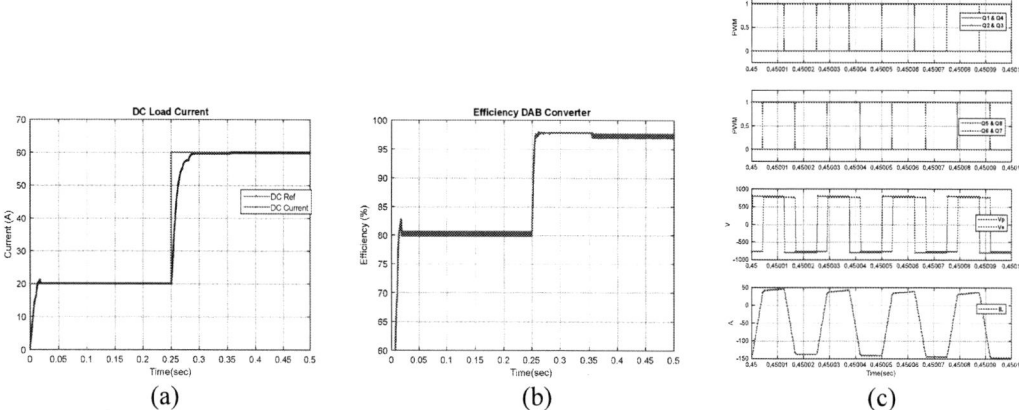

Fig 7. Constant current control: (a) DC current tracking, (b) efficiency of DAB converter, (c) PWM Signals, transformer primary-secondary voltages and inductor current.

7. Conclusions

In this paper, a 50kW SiC technology-based isolated DAB converter is designed for off-board charging stations. The work is carried out to develop a dynamic simulation model and loss model to estimate the power loss and efficiency of the DAB converter. MATLAB System Identification tool is used to create linear models of DAB converters to perform static stability, controllability analysis and offline control system design. The proposed CC-CV control strategy is implemented in simulation at a 40kHz switching frequency to charge the batteries at different modes. The closed loop dynamic simulation model is executed to estimate power losses and efficiency accurately. The efficiency of the 50kW three-phase low-frequency transformer (DTF50) is 97%, and its weight is 230kg, which is mentioned in the datasheet. While the total efficiency of the DAB converter is 97.5%, and the estimated weight is about 20kg. It includes power electronics modules, a high-frequency transformer, liquid cooling plate, passive components, gate drivers, PCBs, etc. The estimated parameters depict that the isolated DAB converter has an advantage over low-frequency transformers because of its higher efficiency and less weight. The provided results of controllers prove a correct operation of the entire control system and thus enable an isolated DAB converter for BEVs used in electric vehicle chargers.

Acknowledgements

The authors are grateful to VLAIO (ex. IWT) and Flux50, national funding schemes in Belgium, for the support to the current work, performed within the BELLA project (project ID: HBC.2021.0800). We also acknowledge Flanders Make for the support to our research group.

References

[1] N. S. Pearre and H. Ribberink, "Review of research on V2X technologies, strategies, and operations," *Renewable and Sustainable Energy Reviews*, vol. 105. Elsevier Ltd, pp. 61–70, May 2019, doi: 10.1016/j.rser.2019.01.047.

[2] S. Schey, "Electric vehicle charging infrastructure deployment guidelines British Columbia," 2009.

[3] E. Langer, "Liquid Cooling for EV Charging—What to Know to Keep Electric Vehicles on the Go."

[4] S. Habib, M. M. Khan, F. Abbas, and H. Tang, "Assessment of electric vehicles concerning impacts, charging infrastructure with unidirectional and bidirectional chargers, and power flow comparisons," *Int. J. Energy Res.*, vol. 42, no. 11, pp. 3416–3441, 2018.

[5] T. Braunl, "EV charging standards," *Univ. West. Aust. Perth, Aust.*, pp. 1–5, 2012.

[6] M. Brenna, F. Foiadelli, C. Leone, and M. Longo, "Electric Vehicles Charging Technology Review and Optimal Size Estimation," *J. Electr. Eng. Technol.*, pp. 1–14, 2020.

[7] H. Ramakrishnan, "Bi-Directional dual active bridge reference design for level 3 electric vehicle charging stations," *Syst. Eng. Texas Instruments, India*, 2019.

[8] A. Kulkarni, "Design of Gallium Nitride Transistor Based Dual Active Bridge DC-DC Converter," 2021.

[9] V. Monteiro *et al.*, "Assessment of a battery charger for electric vehicles with reactive power control," in *IECON 2012-38th Annual Conference on IEEE Industrial Electronics Society*, 2012, pp. 5142–5147.

[10] H. Rasool *et al.*, "Design Optimization and Electro-thermal Modelling of an Off-Board Charging System for Electric Bus Applications," *IEEE Access*, 2021.

[11] H. Rasool, A. Zhaksylyk, S. Chakraborty, M. El Baghdadi, and O. Hegazy, "Optimal Design Strategy and Electro-Thermal Modelling of a High-Power Off-Board Charger for Electric Vehicle Applications," in *2020 Fifteenth International Conference on Ecological Vehicles and Renewable Energies (EVER)*, 2020, pp. 1–8.

[12] H. Rasool, M. El Baghdadi, A. M. Rauf, A. Zhaksylyk, and O. Hegazy, "A Rapid Non-Linear Computation Model of Power Loss and Electro Thermal Behaviour of Three-Phase Inverters in EV Drivetrains," in *2020 International Symposium on Power Electronics, Electrical Drives, Automation and Motion (SPEEDAM)*, 2020, pp. 317–323.

[13] D. Graovac, M. Purschel, and A. Kiep, "MOSFET power losses calculation using the data-sheet parameters," *Infineon Appl. note*, vol. 1, pp. 1–23, 2006.

[14] U. Nicolai and A. Wintrich, "Determining switching losses of SEMIKRON IGBT modules," *SEMIKRON Appl. Note, AN*, vol. 1403, 2014.

[15] P. Semiconductors, "Application Manual Power Semiconductors."

[16] J. Guo, "Modeling and design of inverters using novel power loss calculation and dc-link current/voltage ripple estimation methods and bus bar analysis." 2017.

[17] R. W. Erickson and D. Maksimovic, *Fundamentals of power electronics*. Springer Science & Business Media, 2007.

Impact of Cyber Attacks on Cost Oriented Power Routing Schemes in Microgrids

Kirti Gupta[1], Subham Sahoo[2], Bijaya Ketan Panigrahi[1], and Frede Blaabjerg[2]

[1]Department of Electrical Engineering, Indian Institute of Technology, Delhi, 110016, India
Email: {Kirti.Gupta, Bijaya.Ketan.Panigrahi}@ee.iitd.ac.in
[2]Department of Energy, Aalborg University, Aalborg, 9220, Denmark
Email: {sssa, fbl}@energy.aau.dk

Keywords

≪Cooperative energy management (CEM)≫, ≪economic dispatch (ED)≫, ≪real- time (RT) simulation≫, ≪microgrid (MG)≫, ≪cyber attack≫.

Abstract

The distributed economic dispatch (ED) algorithm carried out in an AC microgrid (MG) is a promising solution which guarantees flexibility, scalability and reliability over single point failure as compared to the centralized approach. Not to mention, the integration of communication infrastructure for information exchanges on one hand adds feasibility for the distributed operation but at the same time, is a threat to the smart grid. The attackers can penetrate in the communication links and inject malicious data in order to gain economical benefits, disrupt the proper functioning of the system etc. Hence, the investigation of the effect of cyber attacks on cooperative energy management (CEM) is an important concern both theoretically and practically. This paper analyses the impact of cyber attacks on a CEM, optimizing ED to a sub-optimal value in an islanded AC MG. The response of the system over false data injection (FDI) and hijacking attacks is further demonstrated on a real-time (RT) co-simulation platform.

Introduction

With the increase in environmental pollution and energy crisis globally, the shift towards cleaner and greener energy solutions have been escalating. Such a shift leads to the development of a smart grid, integrated with advanced control technologies, communication infrastructures, etc. MG plays a major role in integrating these future energy sources which are environmental-friendly. MGs with the hierarchical control structure consists of primary, secondary and tertiary control layers. The later being slowest of all with a time-scale of operation seconds to minutes [1] is responsible for cost-oriented power routing schemes i.e, ED and unit commitment. However, in future with large integration of inverter based distributed generations (DGs), with intermittent nature would degrade the economic efficiency if ED is a part of tertiary control. In order to reduce the time gap of operation and increase the economic efficiency, ED is integrated in the secondary control layer with a time-scale of operation between 100 ms and 1 s.

Further, with integration of information and communication technologies (ICT), distributed control and optimization is preferred to enhance the flexibility, scalability and reliability with respect to its traditional counterpart. Hence, distributed algorithms are promising solutions for ED problem in a MG. The CEM allocates multiple DG units to meet the demand to minimize the total generation cost in a distributed manner [2]. The RT cooperative control plays a vital role in achieving the objectives of frequency restoration, proportional reactive power sharing and ED at the same time [3]. Moreover, the communication cost is also reduced as the information regarding loads are not required.

The information exchanges makes the system vulnerable to cyber attacks. The cyber attacks can disrupt the data confidentiality, integrity and availability [4]. with the increase in integration of DGs and hence the communication links between them has deepened the problem further. An attacker can penetrate in the communication link and malicious attack the information being exchanged. This can cause a chain reaction and affect the normal operation of the system. Hence, this work is dedicated to investigate the impact of cyber attacks on CEM system in an AC MG. Among data integrity and data availability attacks [5, 6], this paper primarily focuses on the later one [7, 8, 9], which can be either FDI and hijacking attacks [10, 11, 12]. These attacks may lead to instability, uneconomic operation, or even shut down of the system. Hence, exploring the impact of various cyber attacks is of a practical value. The paper investigates the impact of cyber attacks on ED optimization problem, affecting the generation cost of a DG to settle at a sub-optimal value. Further, a RT simulation has been performed on a co-simulation testbed. The analysis has been carried out on a four bus test MG system integrated with DGs.

The main contributions of this paper are summarized as follows:
- investigating the impact of cyber attacks on information exchanged, optimizing the generation cost of a DG to a sub-optimal value;
- demonstrating FDI and hijacking attacks and its effect on cost of generation, and the objectives of frequency restoration, proportional reactive power sharing and ED;
- RT testing of a CEM of an islanded AC MG on a co-simulation platform.

The remainder of the paper is organized as: preliminaries on graph theory, control of MG in islanded mode of operation and CEM is presented at first. Various cyber vulnerabilities like FDI, hijacking attacks and variation of cost parameters are discussed further followed by the experimental results. Finally, the work is concluded.

Preliminaries

This section presents some useful preliminaries required for analysis. Various notations used in the further sections are illustrated in this section.

Graph Theory

Let us consider an islanded MG with 'N' DGs connected via communication links. The communication topology of the system can be expressed as a graph with nodes (V) being the DGs and the edges (E) representing the communication links. The graph can be expressed as $G = (V, E, A)$, where $V = 1, 2, ..., N$; $E \subset V \times V$; adjacency matrix, $A = (a_{ij})_{N \times N}$ where (i,j $\in$ V). The graph considered in this work is bidirectional. Each entry $(a_{ij}$ of the adjacency matrix (A) represents the communication weight. The weight $a_{ij} > 0$ if $(i, j) \in E$, otherwise $a_{ij} = 0$. Further, N_i is denoted as a set of neighbouring DGs to i^{th} DG expressed as, $N_i = \{j \in V | (j, i) \in E\}$. The laplacian matrix (L) can be expressed in terms of A and in-degree matrix (D) as $L = D - A$. Here, D can be expressed as $diag(d_1, d_2, ...d_N) \in R^{N \times N}$ where, $d_i = \sum_{j \in N_i} a_{ij}$ is known as weighted in-degree of node i [13].

Primary and Secondary Control of Microgrid

The basic control structure of an islanded MG consisting of primary and secondary controllers is presented in the Fig. 1a. The primary control comprises of three control loops namely, droop control, voltage control and current control. As primary control is not sufficient to drive the system to zero steady state error, hence secondary control is integrated to generate frequency and voltage correction terms to achieve this objective.

The droop control for i^{th} DG can be expressed as:

$$\omega^i = \omega_{ref} - m_p^i . P^i \tag{1}$$

$$V_d^i = V_{ref} - n_q^i . Q^i \tag{2}$$

$$V_q^i = 0 \tag{3}$$

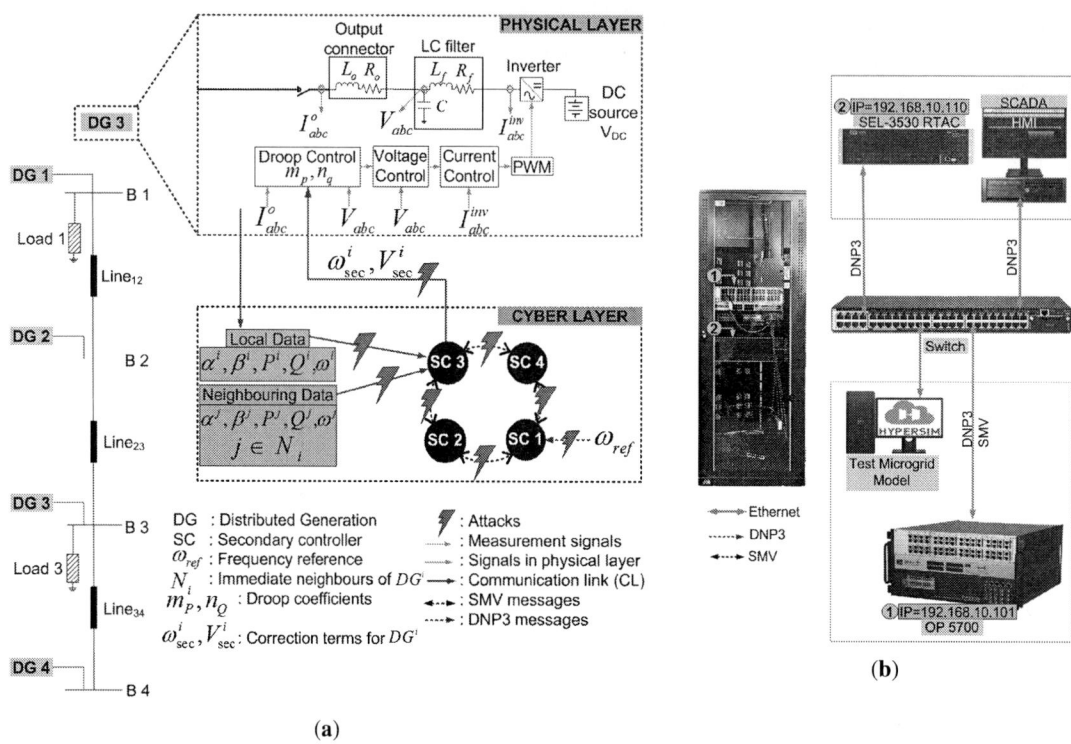

(a)

(b)

Fig. 1: RT co-simulation platform (a) Test MG model with four DGs, (b) Testbed setup.

where, ω_{ref} and V_{ref} are the reference frequency and voltage of the system. Further, m_p and n_q are the constant active and reactive power droop coefficients; P and Q are the active and reactive power; V_d and V_q are the d-axis and q-axis voltages; ω is the frequency. Moreover, superscript 'i' denotes the corresponding quantities of an ith DG.

The secondary control is integrated to compensate for the frequency and voltage deviations caused by the primary control. Conventionally, centralized framework was incorporated but with the advancement of ICT, distributed control architecture is preferred. It is scalable, flexible, relieves computational burden, support plug and play functionalities [14]. The overall equation consisting of both primary and secondary control can be expressed as:

$$\omega^i = \omega_{ref} - m_p^i.P^i + \omega_{sec}^i \tag{4}$$

$$V_d^i = V_{ref} - n_q^i.Q^i + V_{sec}^i \tag{5}$$

where, ω_{sec} and V_{sec} are the frequency and voltage correction terms generated by the secondary controller. Such a distributed secondary control architecture helps to restore frequency to the reference values, to proportionally share active/reactive power among the DGs.

Cooperative Energy Management

The term ED refers to allocating the resources (say DGs) to meet the demand in a most economic way. To estimate the total cost of the output power provided by the DGs, cost function is used. Assuming the cost of generation for i^{th} DG (C^i) be expressed as a quadratic function [15, 16], represented by (6)

$$C^i(P^i) = \alpha^i(P^i)^2 + \beta^i P^i + \gamma^i \tag{6}$$

where, P^i denotes the power generated by i^{th} DG; and α^i, β^i and γ^i are the cost coefficients. In order to minimize the total generation cost while maintaining the balance of supply and demand, it is mandatory to equalize the incremental cost of each DG. Upon differentiating (6) with respect to P^i, we obtain incremental cost function expressed by (7)

$$\eta^i(P^i) = 2\alpha^i P^i + \beta^i \tag{7}$$

The overall system can be regarded as a multiagent system, with each DG as an agent. The agents exchange the information related to active power, incremental cost to achieve the objective of economic dispatch and at the same time balancing the generation and load demand through distributed algorithm. Furthermore, incorporating this objective in the secondary controller of an islanded MG, the objectives of frequency restoration, ED and proportional reactive power sharing are achieved at the same time.

The overall frequency control equation comprising of primary and secondary controllers can be expressed by (8), and the frequency correction term generated by is represented by (9), with $K_{p\omega}^i$ and $K_{i\omega}^i$ being the proportional and integral gains [17]. The error term is further represented by (10). Similarly, the equations corresponding to cooperative voltage controller can be derived.

$$\omega^i = \omega_{ref} - \eta^i(P^i) + \omega_{sec}^i \tag{8}$$

$$\omega_{sec}^i = K_{p\omega}^i \dot{e}_\omega^i + K_{i\omega}^i e_\omega^i \tag{9}$$

$$\dot{e}_\omega^i = -\sum_{j\in N_i} a_{ij}\left(\eta^i(P^i) - \eta^j(P^j)\right) - \sum_{j\in N_i} a_{ij}\left(\omega^i - \omega^j\right) - \sum_{j\in N_i} g_i\left(\omega^i - \omega_{ref}\right) \tag{10}$$

The objectives of this control architecture can be mathematically expressed as:

1. To restore the frequency of each DG (ω^i) to the reference value (ω_{ref}):

$$\lim_{t\to\infty} \omega_i(t) = \omega_{ref} \tag{11}$$

2. To achieve the optimal active power sharing:

$$\lim_{t\to\infty}[\eta^j(P^j) - \eta^i(P^i)] = 0 \tag{12}$$

3. To realize the proportional reactive power sharing:

$$\lim_{t\to\infty}[n_q^j Q^j(t) - n_q^i Q^(t)] = 0 \tag{13}$$

where j$\in N_i$. i.e., all the immediate neighbors of ith DG.

Cyber Vulnerabilities

The adversaries can target either the nodes of communication links in the cyber-physical model of MG [18]. Depending on the target, attack can be classified in two categories which can be described as:

- The attacks targetted by adversaries on the communication links can be grouped under false data injection (FDI) attacks;

- The attack targetting the controllers to generate the unfair commands can be agrregated as hijacking attacks.

Based on these definitions, the attacked entities and the attack equations can be formulated as described in the section further.

False-Data Injection Attacks:

The well-crafted FDI attack can hide its presence, commonly termed as *deception (or stealth)* attacks. The attacker can later on inject unfair attack value, commonly termed as *destablization* attacks, to disrupt the control functionality leading to system instability. Assuming the information exchange vector be $x^i(t)=[\alpha^i(t), \beta^i(t), P^i(t), Q^i(t), \omega^i(t)]$ and constant attack element be x^{iA}, then FDI attack is expressed as:

$$x^{iF}(t) = x^i(t) + x^{iA}(t) \tag{14}$$

Hijacking Attacks:

The hijacking attacks can be represented by:

$$x^{iH}(t) = (1 - \varphi^H)x^i(t) + \varphi^H x^{iA}(t) \tag{15}$$

where, φ^H is a binary number, indicating hijacking attack if 1, otherwise no attack. It poses difficulty in detecting the attacked agent as such an attack causes all the agents to behave in an abnormal way.

The characteristic feature of FDIA is that, the attack value is added to the existing signal, this may although allow to reach a consensus but the converged value may be incorrect. On the contrary, the hijacking attacks are carried out by completely replacing the existing signal. This disrupts the update process of consensus algorithm. To address the formulation of such attacks and response of CEM system under such attacks, a four bus islanded AC MG is studied on a RT co-simulation testbed.

Variation in Cost of Generation:

As discussed, the third-party adversary can attack on any of the variables being exchanged, deviating the optimal solution to a sub-optimal value. Substituting (6), (7), (9) and (10) in (8), we get

$$\eta^i(P^i) = \left[1 + K_{p\omega}^i \sum_{j\in N_i} a_{ij}\right]^{-1} \left[\begin{array}{c} \omega_{ref} - \omega^i + K_{p\omega}^i \sum_{j\in N_i} a_{ij}\eta^j - K_{p\omega}^i \omega^i \sum_{j\in N_i} a_{ij} + K_{p\omega}^i \sum_{j\in N_i} a_{ij}\omega^j - \\ K_{p\omega}^i \omega^i \sum_{j\in N_i} g_i + K_{p\omega}^i \omega_{ref} \sum_{j\in N_i} g_i + K_{i\omega}^i e_\omega^i \end{array}\right] \tag{16}$$

From (16), (7) and (6), we can further find the deviation of cost of generation of a DG in terms of the attacked variables. The variation of cost of generation is further analyzed with attack on cost parameter (α^1, α^2) and active power (P^1) in the next section.

Experimental Results

The co-simulation testbed with a dedicated Ethernet-based network integrated via switch is shown in the Fig.1b. The testbed comprises of OP-5700, which is a RT simulator to emulate the MG test system; SEL-3530 Real-Time Automation Controller (RTAC) hardware integrated with ACSELERATOR RTAC SEL-5033 software for monitoring application. The cyber-physical layer of MG comprising of the primary and secondary control layers are modelled in HYPERSIM software and integrated with OP-5700. A human machine interface (HMI) is developed in ACSELERATOR Diagram Builder SEL-5035 software monitoring and controlling functionalities (locally/remotely). The cyber layer of the MG is linked through various communication protocols such as, sampled message values (SMV) and distributed network protocol (DNP3). The information of frequency, active/reactive power, cost parameters are exchanged via SMV to achieve the objectives of frequency restoration, proportional active/reactive power, ED through CEM architecture. Further, DNP3 protocol is used to monitor the network parameters [9].

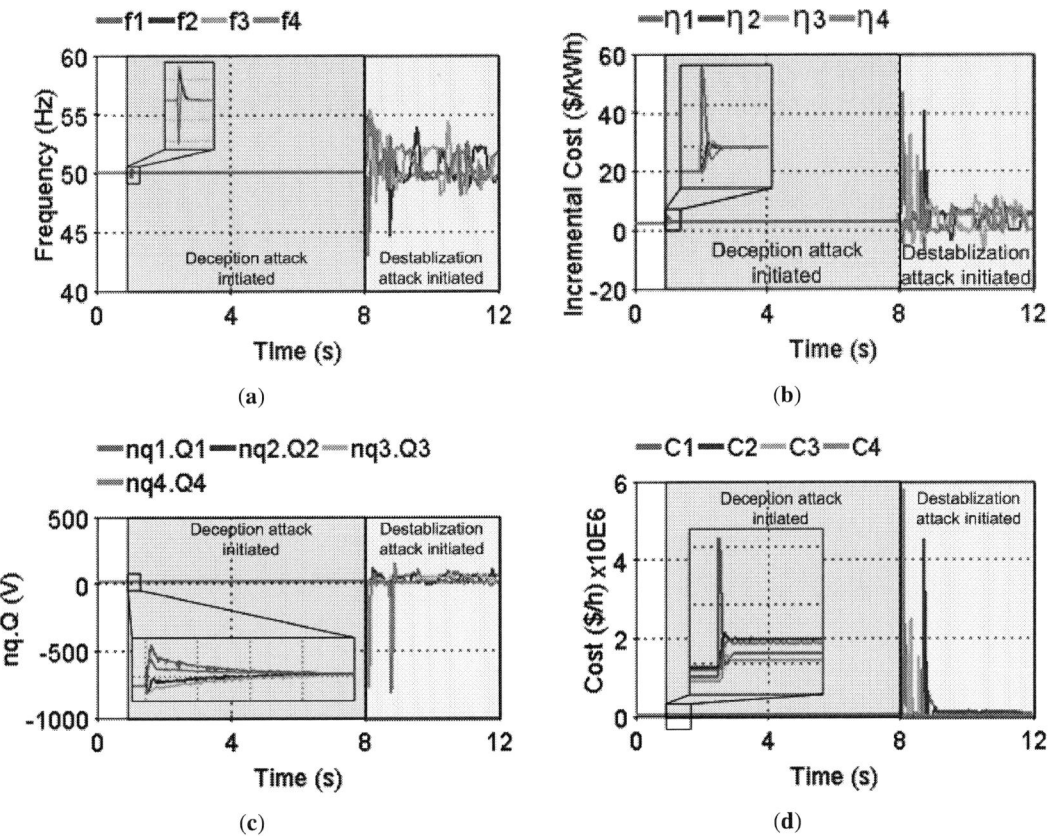

Fig. 2: RT simulation results on HYPERSIM with FDIA initiated at 1 s with small value and increased to unfair value at 8 s (**a**) frequency restoration; (**b**) ED; (**c**) proportional reactive power sharing; and (**d**) generation cost.

In Fig.1a, a 400 V/50 Hz islanded AC MG is considered composed of four DGs each of 40 kVA rating. The DG parameters are given in Table I. The various vulnerable points are also highlighted. The attacker can target these vulnerable entities to launch cyber attacks. The system response to FDI and hijacking attacks on DG 1 are further demonstrated. in this section.

The variations of system objectives and cost of generation on launching FDI attacks are first presented. Each of the simulation results are divided in two shaded portions. The green shaded section represents the system under stealthy FDI attack and the orange section represents the system under destablization attack. It can be observed in Fig. 2 that prior to FDI attack, system was operating normally satisfying all the objectives. Later deception attack was initiated at 1 s (appearing similar to a load change) and destablization attacks was further initiated at 8 s. Although the system objectives were satisfied during deception attack but it became unstable on initiating destablization attack.

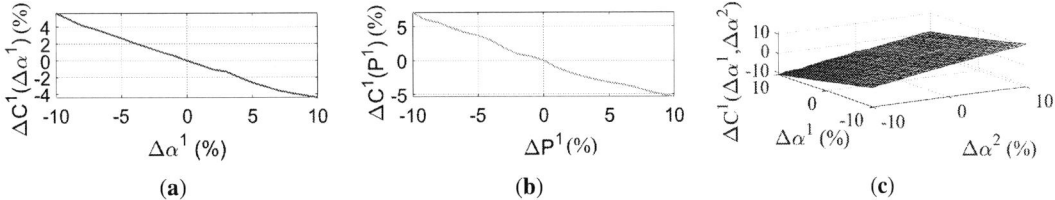

Fig. 3: Variation of cost of generation (**a**) α^1 attacked; (**b**) P^1 attacked; (**c**) α^1 and α^2 attacked.

Table I: DG PARAMETERS

Active power droop coefficient	m_p	9.4 x 10-5 rad/Ws	
Reactive power droop coefficient	n_q	1.3 x 10-3 V/VAr	
Voltage controller proportional gain	K_{pv}	0.2	
Voltage controller integral gain	K_{iv}	1	
Current controller proportional gain	K_{pc}	5	
Current controller integral gain	K_{ic}	100	
Line parameters	$Line_{12}$	$R_{12}= 0.23\ \Omega$	$L_{12}= 318\ \mu H$
	$Line_{23}$	$R_{23}= 0.35\ \Omega$	$L_{23}= 1847 \mu H$
	$Line_{34}$	$R_{34}= 0.23\ \Omega$	$L_{34}= 318\ \mu H$
Load parameters	Load 1	P1= 36 kW	Q1= 36 kVAr
	Load3	P3=45.9 kW	Q3=22.8 kVAr
Cost parameters	$\alpha^1= 0.094\ \$/kW^2 h$	$\beta^1= 1.22\ \$/kWh$	$\gamma^1= 51\ \$/h$
	$\alpha^2= 0.078\ \$/kW^2 h$	$\beta^2= 3.41\ \$/kWh$	$\gamma^2= 31\ \$/h$
	$\alpha^3 =0.105\ \$/kW^2 h$	$\beta^3= 2.53\ \$/kWh$	$\gamma^3= 78\ \$/h$
	$\alpha^4 =0.082\ \$/kW^2 h$	$\beta^4= 4.02\ \$/kWh$	$\gamma^4= 42\ \$/h$

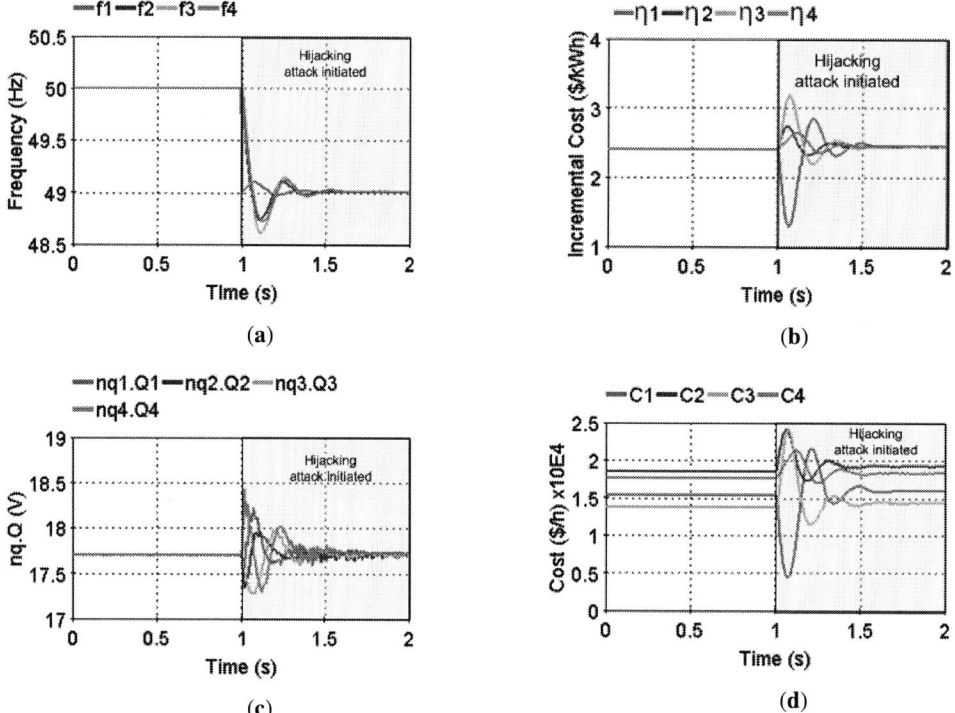

Fig. 4: RT simulation results on HYPERSIM with hijacking attack initiated at 1 s with a small value (**a**) frequency restoration; (**b**) ED; (**c**) proportional reactive power sharing; and (**d**) generation cost.

Assuming $\Delta C^i(x) = C^{Ai}(x) - C^i(x)$ for DG^i, where $\Delta C^i(x)$ is deviation in cost of generation when $C^i(x)$ is attacked by x to get the new value of $C^{Ai}(x)$. Fig. 3 reflects the variation in cost of generation for DG^1, with individual attack on α^1 and P^1 and combined attack by α^1 and α^2. Each attack vector $(\alpha^1, \alpha^2, P^1)$ are deviated in range of $\pm 10\%$. The maximum deviation in cost, $\Delta C^1(\Delta \alpha^1) = 5.66\%$ and $\Delta C^1(\Delta P^1) = 7.10\%$ was observed for $\Delta \alpha^1 = -10\%$, $\Delta P^1 = -10\%$, as shown in Fig. 3a and Fig. 3b, respectively. It was further observed that a larger deviation in cost, $\Delta C^1 = 11.34\%$ was obtained when attack was done in a combined manner $(\Delta \alpha^1 = -10\%, \Delta \alpha^1 = 10\%)$ as shown in Fig. 3c.

Further, the variations of system objectives and cost of generation on launching hijack attack at 1 s are presented in the Fig. 4. The system objectives converges to a new attacked value presented in the blue shaded portion. Similar to FDI attacks, the effect of variations of different attack elements can be plotted in case of hijack attacks as well.

Conclusion

This paper analyzes the impact of cyber risks on generation cost in a CEM system. It has been observed that combined attack on the cost parameters (α^1, α^2) affected the generation cost to a larger extent than the individually, for the same deviation in the attack parameters. The influence of the attack on P^1 was further investigated. RT testing of the CEM under FDI and hijacking attacks has also been demonstrated.

References

[1] Z. Li, Z. Cheng, J. Liang, J. Si, L. Dong and S. Li, "Distributed Event-Triggered Secondary Control for Economic Dispatch and Frequency Restoration Control of Droop-Controlled AC Microgrids," in *IEEE Trans. Sustain. Energy*, vol. 11, no. 3, pp. 1938-1950, July 2020, doi: 10.1109/TSTE.2019.2946740.

[2] Y. Han, K. Zhang, H. Li, E. A. A. Coelho and J. M. Guerrero, "MAS-Based Distributed Coordinated Control and Optimization in Microgrid and Microgrid Clusters: A Comprehensive Overview," *IEEE Trans. Power Electron.*, vol. 33, no. 8, pp. 6488-6508, Aug. 2018, doi: 10.1109/TPEL.2017.2761438.

[3] S. Sahoo, T. Dragičević and F. Blaabjerg, "Resilient Operation of Heterogeneous Sources in Cooperative DC Microgrids," *IEEE Trans. Power Electron.*, vol. 35, no. 12, pp. 12601-12605, Dec. 2020, doi: 10.1109/TPEL.2020.2991055.

[4] X. Fu, G. Chen and D. Yang, "Local False Data Injection Attack Theory Considering Isolation Physical-Protection in Power Systems," *IEEE Access*, vol. 8, pp. 103285-103290, 2020, doi: 10.1109/ACCESS.2020.2999585.

[5] P. Li, Y. Liu, H. Xin and X. Jiang, "A Robust Distributed Economic Dispatch Strategy of Virtual Power Plant Under Cyber-Attacks," *IEEE Trans. Ind. Informat.*, vol. 14, no. 10, pp. 4343-4352, Oct. 2018, doi: 10.1109/TII.2017.2788868.

[6] G. Chen and Z. Guo, "Distributed Secondary and Optimal Active Power Sharing Control for Islanded Microgrids With Communication Delays," *IEEE Trans. Smart Grid*, vol. 10, no. 2, pp. 2002-2014, March 2019, doi: 10.1109/TSG.2017.2785811.

[7] J. Duan and M. Y. Chow, "A Novel Data Integrity Attack on Consensus-Based Distributed Energy Management Algorithm Using Local Information," *IEEE Trans. Ind. Informat.*, vol. 15, no. 3, pp. 1544-1553, March 2019, doi: 10.1109/TII.2018.2851248.

[8] H. Pourbabak, J. Luo, T. Chen and W. Su, "A Novel Consensus-Based Distributed Algorithm for Economic Dispatch Based on Local Estimation of Power Mismatch," *IEEE Trans. Smart Grid*, vol. 9, no. 6, pp. 5930-5942, Nov. 2018, doi: 10.1109/TSG.2017.2699084.

[9] G. D. Torre and T. Yucelen, "Adaptive architectures for resilient control of networked multiagent systems in the presence of misbehaving agents," *Int. J. Control*, vol. 91, no. 3, pp. 495-507, 2018.

[10] K. Gupta, S. Sahoo, R. Mohanty, B. K. Panigrahi and F. Blaabjerg, "Decentralized Anomaly Characterization Certificates in Cyber-Physical Power Electronics Based Power Systems," *2021 IEEE 22nd Workshop on Control and Modelling of Power Electronics (COMPEL)*, 2021, pp. 1-6, doi: 10.1109/COMPEL52922.2021.9645984.

[11] S. Sahoo and J. C. -H. Peng, "A Localized Event-Driven Resilient Mechanism for Cooperative Microgrid Against Data Integrity Attacks," *IEEE Trans. Cybernetics*, vol. 51, no. 7, pp. 3687-3698, July 2021, doi: 10.1109/TCYB.2020.2989225.

[12] S. Sahoo, J. C. -H. Peng, S. Mishra and T. Dragičević, "Distributed Screening of Hijacking Attacks in DC Microgrids," *IEEE Trans. Power Electron.*, vol. 35, no. 7, pp. 7574-7582, July 2020, doi: 10.1109/TPEL.2019.2957071.

[13] B. Huang, Y. Li, F. Zhan, Q. Sun and H. Zhang, "A Distributed Robust Economic Dispatch Strategy for Integrated Energy System Considering Cyber-Attacks," *IEEE Trans. Ind. Informat.*, vol. 18, no. 2, pp. 880-890, Feb. 2022, doi: 10.1109/TII.2021.3077509.

[14] K. Gupta, S. Sahoo, B. K. Panigrahi, F. Blaabjerg, and P. Popovski, "On the Assessment of Cyber Risks and Attack Surfaces in a Real-Time Co-Simulation Cybersecurity Testbed for Inverter-Based Microgrids," *Energies*, vol. 14, no. 16, p. 4941, 2021. [Online]. Available: https://www.mdpi.com/1996-1073/14/16/4941.

[15] W. Zeng, Y. Zhang and M. -Y. Chow, "Resilient Distributed Energy Management Subject to Unexpected Misbehaving Generation Units," *IEEE Trans. Ind. Informat.*, vol. 13, no. 1, pp. 208-216, Feb. 2017, doi: 10.1109/TII.2015.2496228.

[16] Z. Cheng and M. -Y. Chow, "Resilient Collaborative Distributed Energy Management System Framework for Cyber-Physical DC Microgrids," *IEEE Trans. Smart Grid*, vol. 11, no. 6, pp. 4637-4649, Nov. 2020, doi: 10.1109/TSG.2020.3001059.

[17] M. S. Sadabadi, S. Sahoo and F. Blaabjerg, "A Fully Resilient Cyber-Secure Synchronization Strategy for AC Microgrids," *IEEE Trans. Power Electron.*, vol. 36, no. 12, pp. 13372-13378, Dec. 2021, doi: 10.1109/TPEL.2021.3091587.

[18] S. K. Mazumder et al., "A Review of Current Research Trends in Power-Electronic Innovations in Cyber–Physical Systems," *IEEE J. Emerg. Sel. Topics Power Electron.*, vol. 9, no. 5, pp. 5146-5163, Oct. 2021, doi: 10.1109/JESTPE.2021.3051876.

Response of IGBT chip characteristics due to critical stress

Kohei Yamauchi[1] and Rik W. De Doncker[2]

[1] FUJI ELECTRIC CO., LTD
Matsumoto Tsukama 4 - 18 - 1
Nagano, Japan
Tel.: +81(0) 263 - 277405.
E-Mail: yamauchi-
kouhei@fujielectric.com
URL: https://www.fujielectric.com/

[2] INSTITUTE FOR POWER ELECTRONIC
AND ELECTRICAL DRIVES
RWTH AACHEN UNIVERSITY
Jaegerstr. 17-19
Aachen, Germany
Tel.: +49 (0) 241-8096992
URL: http://www.isea.rwth-aachen.de

Abstract

Focusing on the change in power semiconductor chip characteristics in response to mechanical stress applied to the chip, we confirmed the change of IGBT (Insulated Gate Bipolar Transistor) characteristics by bending test using a PCB (Printed Circuit Board) substrate to which an IGBT chip was mounted. The critical stress under which IGBT chip would fail was calculated by mechanical stress simulation, and the corresponding bending deformation was then applied to the PCB board and the characteristics of the IGBT chip has been investigated. Specifically gate capacitance has decreased after critical tensile stress loaded. By considering the stress direction of the each bending test and the structure of the IGBT chip, some hypothesis that can explain the mechanism of the characteristics change was suggested. As the next step, this hypothesis will be tested and the research regarding this chip characteristics change will be continued.

Keywords

«Condition Monitoring», «IGBT», «Lifetime», «Power semiconductor device», «Reliability».

Introduction

In an effort to realize a sustainable society, the electrification of automobiles is progressing based on the CO_2 reduction target. Especially the demand of power modules for traction inverter used in automobiles are growing at a phenomenal rate. As the size of power modules continue to shrink and become more multifunctional, the technologies of handling chips efficiently, lifetime diagnosis, and lifetime control are becoming more important [1-9]. Conventionally, it has been considered that the chip characteristics do not change if a local and direct mechanical load is applied to a chip itself [10-11]. However, since the situation has changed from the past in a trend that the chip become thinner and more multifunctional, the thin Si chip was investigated whether its characteristics would change with respect to the failure limit bending stress. Although IGBT chip is mounted in module for survey purpose, in this investigation IGBT chips is mounted on PCB board.

1. Experimental methods and modeling

1-1. Experimental model

Fig. 1 shows an IGBT mounted PCB board used for the experimental test. The PCB board is made of a high heat-resistant resin material that can cope with the soldering temperature, and its thickness is

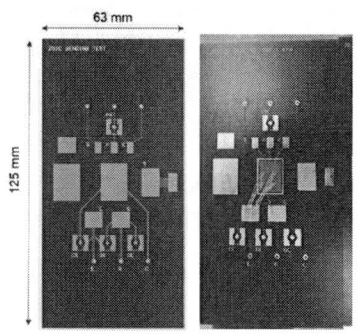

Fig. 1 PCB model and experimental sample.

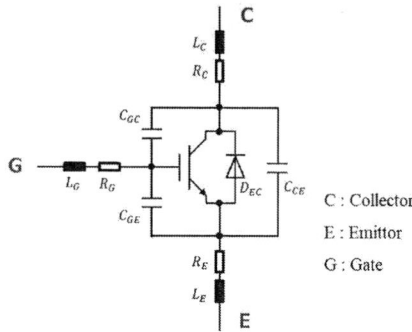

C : Collector

E : Emittor

G : Gate

Fig. 2 Electrical model of RC-IGBT.

optimized according to the bending test deformation. The IGBT chips used in this samples are thin chips made of Si material and are soldered to the mounting board using lead-free high strength solder. The IGBT chip is in the center of the board. The gate, sense, and emitter wires are connected to the electrode pads on the PCB substrate by wire bonding. Metal pins are soldered to through holes in the substrate to provide external connections.

1-2. Chip characteristics analysis method

In this study, impedances have taken into account as a composite parameter to represent the characteristic changes in chip. Fig. 2 shows the electrical circuit model of the chip applied in this study. Since RC-IGBT (Reverse Conducting IGBT) chip, which is FWD is built in IGBT is used, a diode is built into the electrical circuit. The gate (G), collector (C) and emitter (E) are connected respectively on the impedance measurement and totally 6 impedances are measured with one module (G-E, E-G, G-C, C-G, C-E, E-C). After the bending test, impedances (analysis frequency from 100 Hz to 100 MHz) were measured using the impedance analyzer (Keysight 4294A) as shown in Fig. 3 and Fig. 4. The absolute and phase values of the impedance at each frequency are obtained by frequency analysis.

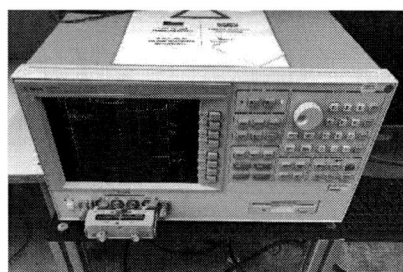

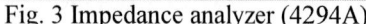

Fig. 3 Impedance analyzer (4294A).

Fig. 4 Impedance measuring of experimental sample.

1-3. Chip characteristics comparing method

To analyze the chip characteristics in detail, various resistance and capacitance values inside the chip were extracted and each values were compared before and after the bending test. Since the impedance acquired by the impedance analyzer is calculated as a composite impedance value including the RLC values present in the electrical circuit, it is difficult to identify which RLC values have changed. Therefore, extraction calculations were carried out according to the flow shown in Fig. 5 and all the RLC values shown in Fig. 2 were extracted. The extraction process involves reworking the assumed electrical circuit into an equation showing the impedance for each connection, substituting appropriate values for each RLC values, comparing the calculated composite impedance with the measured impedance and performing an optimization calculation to make the difference smaller, thereby obtaining appropriate RLC values. As shown in Fig. 2, a total of 13 RLCs were used as variables (outputs), and six types of composite impedance (GE, EG, GC, CG, CE and EC) and 2 types of impedance data (absolute and phase value) were used as 12 input data to represent in terms of entire impedance values. 12 impedance data were set as inputs and the impedance equations for each connection were set up by combining theoretical and experimental values. Some of the equations are shown in Equations (1) and (2), and the results of the optimization calculations are shown in Fig. 6. The dotted line shows the measured data and the solid line shows the data calculated by the optimization calculations. It was confirmed that the measured and optimized data matched well and that the circuit was set up appropriately.

$$Z_{GE} = \frac{1}{2\pi f j (C_{GE} + \frac{1}{\frac{1}{C_{GC}} + \frac{1}{C_{CE}}})} + (R_G + R_E) + 2\pi f j (L_G + L_E) \ \dots \ (1)$$

$$Z_{EG} = \frac{1}{2\pi f j (C_{EG} + \frac{1}{\frac{1}{C_{CG}} + \frac{1}{C_{EC} + 1/(2\pi f j * D_{EC})}})} + (R_G + R_E) + 2\pi f j (L_G + L_E) \ \dots \ (2)$$

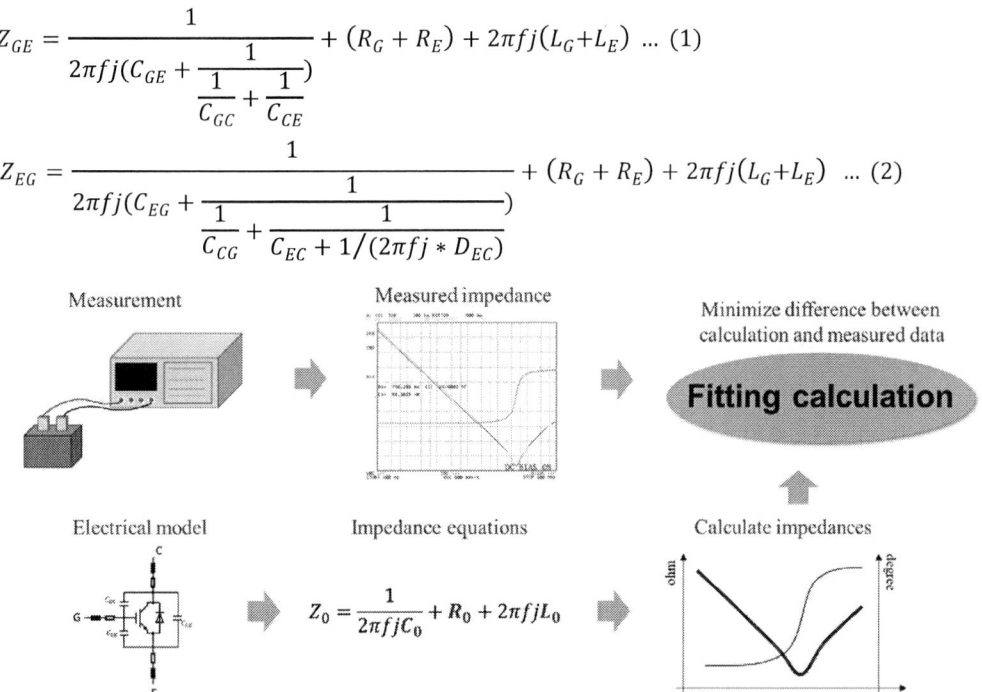

Fig. 5 RLC values extracting calculation flow. Comparing measured impedance data and calculated impedance based on the electrical model and equations.

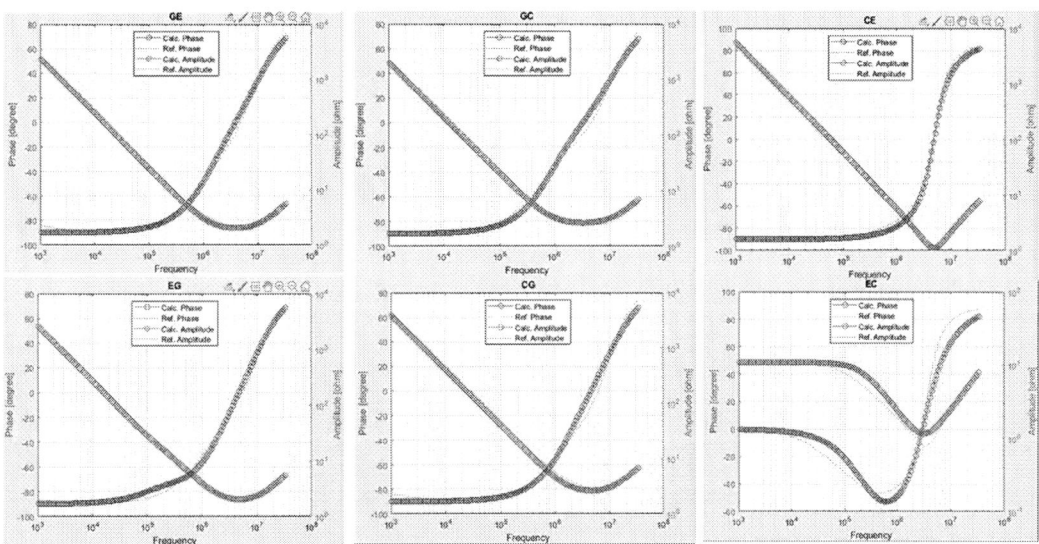

Fig. 6 Comparing measured impedance data and calculated impedance data by optimization calculation using GA with MATLAB.

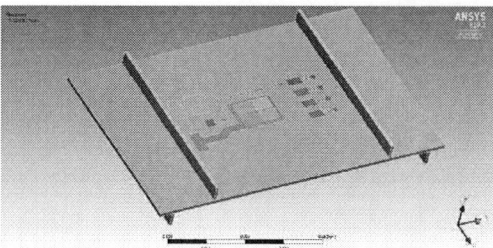

Fig. 7 Bending stress simulation model with detailed chip model.

Fig. 8 Bending test equipment (Zwick/Roell Z150).

Fig. 9 A photograph taken during four point bending test.

1-4. Mechanical calculation model

Fig. 7 shows the model used for the bending stress simulation. Basically, the model reproduces the same shape as the experimental test sample shown in Fig. 1. One of the features of the model is that the electrodes and insulation structure of the chip are taken into consideration so that detailed stress changes inside the IGBT chip can be confirmed in response to bending stresses.

1-5. Bending test method

Fig. 2 shows the bending test equipment used for the bending tests. The compressive stress test by four point bending test is shown in Fig. 4, and the tensile stress test by three point bending test is performed with same equipment. After the bending test, impedances (analysis frequency from 100 Hz to 100 MHz) were measured using the impedance analyzer (Keysight 4294A).

2. Bending stress simulation result

Bending stress was calculated to obtain the correlation between the bending stress applied to the IGBT chip and the bending deformation of the entire PCB substrate. Fig. 10 shows the results of compressive stress simulation by four point bending test. There are four fulcrum points, which are in contact with the PCB substrate but they are not adhesive, in order to simulate the actual bending test by applying an appropriate amount of friction. The results of the three-point bending stress simulation are shown in Fig. 11. The required amount of deformation was obtained by increasing the amount of bending deformation until it reached the reference value of the chip failure. In the case of the three point bending test, the same simulation method was used to obtain the required amount of applied bending. Fig. 12 shows stress distribution of IGBT device and the other materials. This stress distribution confirmed that the Si layer of the IGBT chip can reaches to failure critical state.

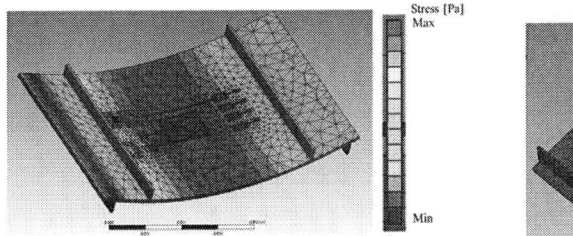

Fig. 10 Compressive stress simulation result of four point bending test.

Fig. 11 Tensile stress simulation result of three point bending test.

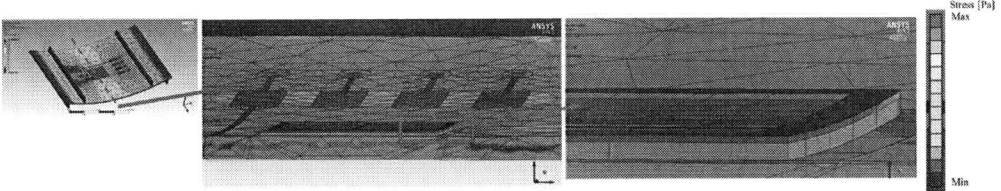

Fig. 12 Stress distribution in IGBT device under bending test deformation.

3. Experimental bending test result

3-1. Bending test result

Bending tests were carried out with the electromechanical tensile and compression testing machine (Zwick/Roell Z150), the specific attachment shown in Fig. 13. The bending test in progress is shown in Fig. 14. In order to verify that the applied bending deformation obtained from the bending stress simulation was appropriate, the simulated bending deformation which can provide critical stress against IGBT chip was loaded to the experimental sample and it was confirmed that the test sample has broken as shown in Fig. 15.

Fig. 13 Specific attachment for bending test.

Fig. 14 State under four point bending test.

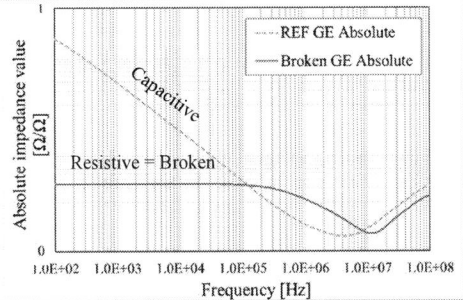

(a) Absolute impedance value comparison.

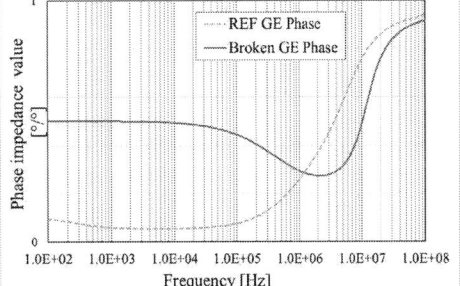

(b) Phase impedance value comparison.

Fig. 15 Impedance value comparison normal status and broken status by bending test.

3-2. Measured impedance and RLC values comparison

Fig. 16 shows the absolute value and phase values of the frequency dependence impedance measured before and after the tensile bending test. Analysis samples are subjected to stress at the fracture limit and evaluated the status prior to complete fracture. Impedance was measured at 6 patterns according to electrical circuit diagram of Fig. 2. GE impedance measurement results show an increase in absolute impedance in the low-frequency range. The increase in impedance is expected to indicate a high possibility that the synthetic capacitance value is decreasing. No significant change was observed in other measured impedance results. Since this comparison does not show quantitatively how much the impedance changed, the resistance and capacitance values are extracted from the impedance curve and compared in Fig. 17 based on the RLC extraction calculation as explained at 1-3. Focusing on the gate resistance, it showed a slight increasing trend after compressive stress was applied, and a decreasing trend after tensile stress was applied. As for the gate capacitance value, it was almost unchanged after compressive stress was applied, but showed a decreasing trend after tensile stress was applied. From the above, it was confirmed that the chip characteristics changed after loaded critical stress.

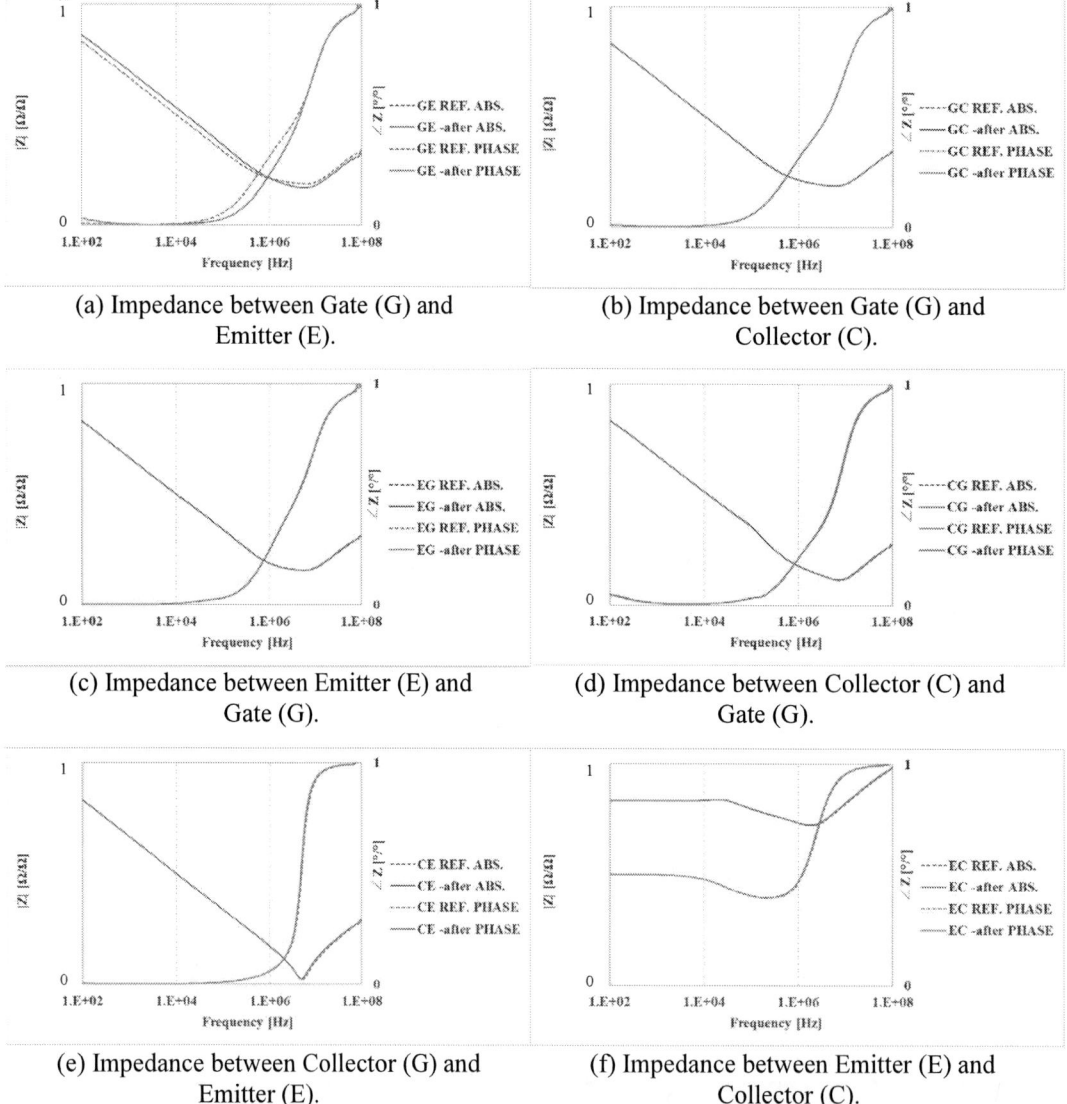

(a) Impedance between Gate (G) and Emitter (E).

(b) Impedance between Gate (G) and Collector (C).

(c) Impedance between Emitter (E) and Gate (G).

(d) Impedance between Collector (C) and Gate (G).

(e) Impedance between Collector (G) and Emitter (E).

(f) Impedance between Emitter (E) and Collector (C).

Fig. 16 Impedance values comparison before and after tensile stress loading at each connection of Gate, Collector and Emitter.

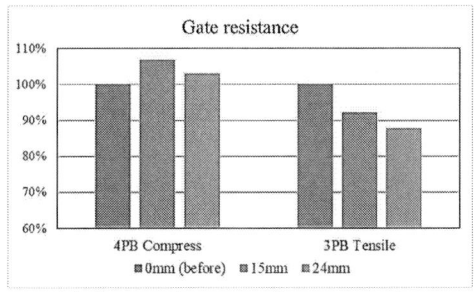
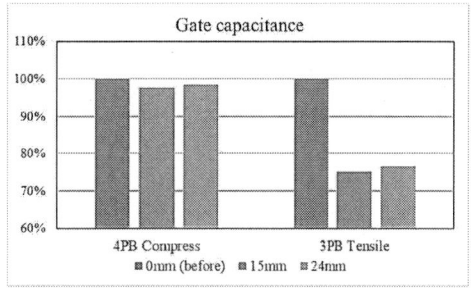

(a) Gate resistance value comparison.   (b) Gate capacitance value comparison.

Fig. 17 Gate resistance value comparison among bending deformation 0 mm (start), 15mm and 24 mm.

4. Discussion

In order to verify if this change was caused by something other than the chip, such as solder deformation or PCB deformation, the amount of warpage of the entire PCB board after the bending test was measured and verified. Fig. 18 shows a total 3D position of entire PCB and the center zone shows the area where IGBT chip is positioned, and measure deformation through the bending test. Few μm was measured as entire deformation but if the center area was focused, there was almost none deformation.

In addition to considering PCB deformation, ion concentration effect has investigated. Tested samples have passed through over 250 °C thermal process to dissipate ions inside of IGBT chip and impedance has measured and RC values have compared as shown in Fig 19. From the above discussion it was unlikely that the characteristics changed due to the influence of the other components.

The results of the change in resistance and capacitance values in response to compressive and tensile stresses are discussed in terms of the relationship between the structure of the IGBT chip and the physical properties. Fig. 13 shows the status of each stress loaded and the cross-sectional scheme of the IGBT with stress vectors. The formula for calculating the capacitance value is shown in Equation (3).

$$C = \varepsilon \frac{A}{d} . \quad (3)$$

ε indicates the $\varepsilon_0 \varepsilon_r$ (relative permittivity) and d is the thickness of the insulate layer which also means the distance between electrodes, and A is the surface area. The situation of the IGBT trench while the compressive stress loading with the four point bending test is depicted using the cross-sectional view of the IGBT trench structure. When compressive stress is applied, each part is expected to shrink as it tends to compress. Therefore, the distance d tends to shorten and the surface area A tends to decrease, and the permittivity ε is expected to increase due to compression. On the other hand, when tensile stress is applied by the three point bending test, d and A tend to expand, and the permittivity is expected to decrease.

In addition, since the bending stress which is very critical for breakdown in the bending test was applied, it can be inferred that the permittivity tends to decrease more prominently. This is because slight degraded but almost broken parts can appear in the insulating layer before the complete failure. Furthermore the thickness and the surface area also can change if the IGBT structure has deformed. However, the cross section investigation of after bending test sample confirmed that the IGBT structure didn't change significantly.

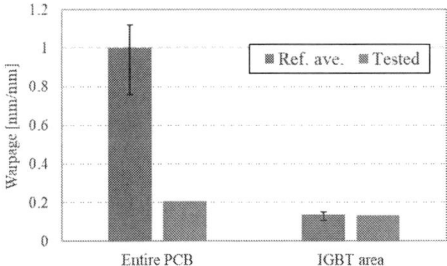
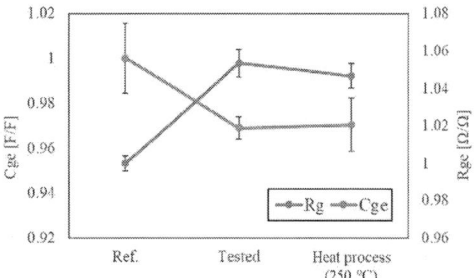

Fig. 18 Deformation of test samples before and after bending test.

Fig. 19 Rg values trend through bending test and thermal process (250 °C).

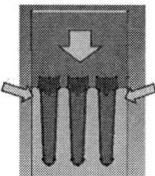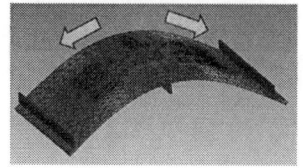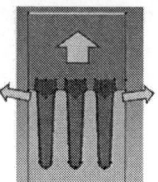

(a) Compression stress loaded status.   (b) Tensile stress loaded status.

Fig. 20 Compression and tensile stress loading state by four and three point bending test with corresponding cross section of IGBT structure.

Conclusion

The chip characteristics under the state just before electrical breakdown were evaluated by applying critical mechanical stress, which is calculated by the bending stress simulation, to the experimental sample. No significant change was observed in the chip characteristics after compressive stress was applied by four point bending test, but the gate capacitance decreased after tensile stress was applied by three point bending test. Under tensile stress, capacitance decrease is considered as a result of relative permittivity decrease, since the IGBT chip has a possibility to introduce micro cracks in insulation layer under this critical stress. Based on the current findings in this paper, in order to develop a method to predict module lifetime, we will investigate characteristic changes under electrical stress during long-term reliability test at the next step.

Acknowledgements

The authors wish to acknowledge the people who supported us at the Corporate R&D Headquarters and the Semiconductors Business Group of Fuji Electric Co., Ltd. during the research stay at ISEA, RWTH.

References

[1] Gu B.: Condition monitoring of press-pack IGBT devices using Deformation Detection Approach, PEMD 2020

[2] Colla E. L.: Characterisation of the fatigued state of ferroelectric PZT thin-fdm capacitors, Microelectronic Engineering 29, 1995, pp. 145-148

[3] Timothy A.: Designing Power Modules for Degradation Sensing, ECCE 2019

[4] van der Broeck C. H.: Monitoring 3-D Temperature Distributions and Device Losses in Power Electronics Modules, IEEE transactions on Power Electronics vol. 34, no. 8, pp. 7983-7995

[5] Stippich A.: Significance of Thermal Cross-Coupling Effects in Power Semiconductor Modules, SPEC 2016

[6] Baker N.: IGBT Junction Temperature Measurement via Peak Gate Current, IEEE transactions on power electronics, vol. 31, no. 5, pp. 3784-3793

[7] Ye X.: Online Condition Monitoring of Power MOSFET Gate Oxide Degradation Based on Miller Platform Voltage, IEEE transactions on power electronics, vol. 32, no. 6, pp. 4776-4784

[8] van der Broeck C. H.: In-situ Thermal Impedance Spectroscopy of Power Electronic Modules for Localized Degradation Identification, PCIM 2019

[9] Kalker S.: Reviewing Thermal Monitoring Techniques for Smart Power Modules, IEEE Journal of Emerging and Selected Topics in Power Electronics

[10] Tepper T.: Correlation between microstructure particle size dielectric constant and electrical resistivity of nano-size amorphous SiO2 powder, NanoStructured Materials 1999, vol. 11, vo. 8, pp. 1081–1089

[11] Kahn H.: Mechanical fatigue of polysilicon: Effects of mean stress and stress amplitude, Acta Materialia 2006, vol. 54, pp. 667-678

Mega-hertz High-power WPT system with Parallel-connected inverters using current balance circuit

Masamichi Yamaguchi, Keisuke Kusaka and Jun-ichi Itoh
NAGAOKA UNIVERSITY OF TECHNOLOGY
1603-1, Kamitomioka-chou, Nagaoka, Niigata, Japan
Tel.: +81 / (258) − 47.9533.
Fax: +81 / (258) − 47.9533.
E-Mail: m_yamaguchi@stn.nagaokaut.ac.jp, Kusaka@vos.nagaokaut.ac.jp,
Itoh@vos.nagaokaut.ac.jp
URL: http://itohserver01.nagaokaut.ac.jp/itohlab/en/index.html

Acknowledgements

This work was supported by Council for Science, Technology and Innovation(CSTI), Cross-ministerial Strategic Innovation Promotion Program (SIP), "Energy systems of an Internet of Energy (IoE) society" (Funding agency : JST).

Keywords

«Current balancing», «Transformer», «Gallium Nitride (GaN)», «Wireless power transmission», «High frequency power converter».

Abstract

This paper proposes a new current balancer for the megahertz WPT system. The proposed balancer achieves a current balance between the parallel connection inverter on the transmission side of the WPT system. The balancer consists of two transformers without any additional control to achieve a current balance. The output side of a conventional balancer is not isolated from the input side of the balancer. The proposed balancer isolates the output side and the input side of the balancer. A turn ratio of two transformers is derived by magnet flux under the current balancing condition. Furthermore, the proposed balancer also has the capability of the impedance matching. The leakage inductance of two transformers is applied to achieve an impedance matching with an additional capacitor. The experimental results reveal that the current balance is achieved even if the phase of the output voltage of each inverter has a difference of approximately 8 ns.

Introduction

In recent years, battery chargers have been developed rapidly by increasing the interest in small mobility, for example, drones. Notably, wireless power transfer (WPT) systems for battery chargers have been actively studied for safety and convenience [1–2]. Then, the rapid charging is required to reduce the charging time. Although, the WPT system charger takes a long time to charge the onboard batteries. Thus, increasing the output power of the WPT system to at least ten kilo watts is urgent.

On the other hand, the WPT systems with a transmission frequency of 85 kHz have been commonly used because the WPT system for EVs charger is normalized with 85 kHz. Although, over ten kilowatt system with 85 kHz is bulky and heavy as the charging system of the small mobility. One of the leading causes is a transfer coil. The transfer coil in 85 kHz has a ferrite core at a transmission coil and a receiver coil. Then, the cursing distance of the mobility is affected directly by the weight of the receiver coil that is mounted to the mobility.

Then, applying megahertz band frequency is one of the solutions to decrease the weight and the size of the transfer coil because the air-core coil is able to be applied as the transfer coil by using megahertz frequency. Thus, the megahertz operation in the WPT system has been actively studied [3–8]. The WPT system employs a GaN device [9–10] to achieve the megahertz operation in a kilowatt order system.

Thus, the increase of the output power of the WPT system in megahertz operation is necessary for the small mobility. A parallel connection of the GaN devices or the inverter circuits [11–12] has been studied to achieve a kilowatt order system. Then, the current unbalance occurs between paralleled circuits or devices to delay the switching timing. The current unbalance causes thermal unbalance, which destroys the devices. Thus, the current balancing method is necessary for paralleled circuits or devices.

However, it is difficult to apply an accurate control to synchronize switching timings of each device or circuit still in the megahertz operation because the control period should be shorter than a few nanosecond. Thus, the current balancing method should be composed of the passive components without additional current control. So far, a current balancer circuit for the paralleled inverters in megahertz operation has been studied [13]. The balancer consists of two transformers without any additional control. Although, the wire connection at the balancer circuit is complicate especially of the over three parallel configurations because the primary side is non-isolated from and the secondary side of the two transformers.

This paper proposes the new current balancer, including impedance matching capability, to increase the output power with the parallel connection inverters. The advantage of the proposed balancer is that the configuration of the circuit is simple, and it is easy to expand into the three paralleled configurations. Moreover, the proposed balancer has the capability of an impedance transform to match the impedance of the load. Thus, Zero-Voltage-Switching(ZVS) is achieved in the parallel-connected inverters in any load condition. First, the balance circuit is theoretically analyzed to derive the circuit parameters in the balance condition. Then, the operation of the proposed balance circuit and ZVS is demonstrated by the simulation and the fundamental experiments in the megahertz operation in the several conditions.

Conventional balancer circuit

Figure 1 shows the configuration of the parallel-connected inverter using a non-isolated balancer-[13]. The balancer consists of two transformers. The primary side of each transformer is connected to the output of each inverter. The secondary side of each transformer is connected to a transfer coil in series. Two transformers have the same design parameters and have no magnetic coupling with each other.

On the other hand, two transformers have a leakage inductance. The leakage inductance appears between the inverters and the transfer coil. Thus, the resonant capacitance should be modified to keep the resonant condition. When the leakage inductance has a large value, the voltage of the transfer coil is decreased even if the resonant condition is achieved by adjusting the resonant capacitance. Thus, the transfer power is regulated by the leakage inductance of the balancer circuit.

Proposed balancer circuit

Figure 2 shows the proposed configuration of the balancer circuit. The primary side and the secondary side of each transformer are isolated. The primary side of each transformer is also connected to the output of each inverter.

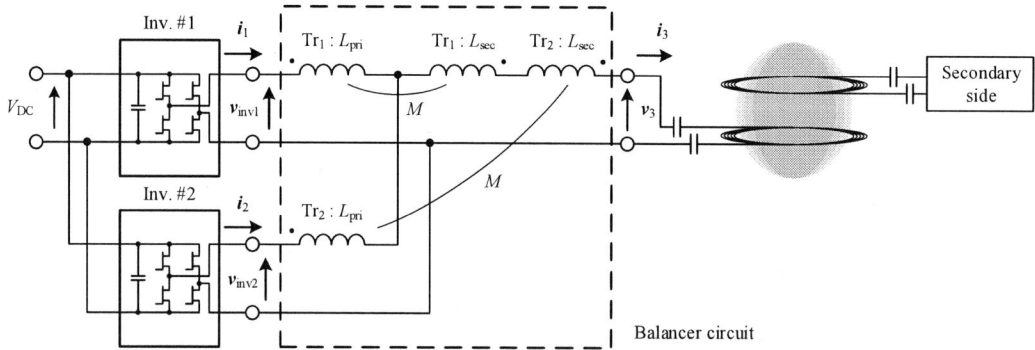

Fig. 1. Configuration of the conventional balancer circuit in parallel-connected inverter system.

Parameters of balancer circuit

Figure 3 shows the detail of the balancer circuit. In the balancer circuit, the transformer currents i_1, i_2, and i_3 are sinusoidal because the transmission coil and the capacitor in the WPT system consider the resonant load. Each voltage v_1, v_2, and v_3 are expressed as

$$v_1 = N_1 \frac{d\Phi}{dt},$$

$$= N_1 \frac{d}{dt}(\Phi_1 - k\Phi_2), \tag{1}$$

$$v_2 = N_1 \frac{d}{dt}(\Phi_3 - k\Phi_4), \tag{2}$$

$$v_3 = N_2 \frac{d}{dt}(\Phi_2 - k\Phi_1) + N_2 \frac{d}{dt}(\Phi_4 - k\Phi_3), \tag{3}$$

$$v_3 = \frac{1}{2}(v_1 + v_2), \tag{4}$$

$$i_1 = i_2 = \frac{i_3}{2}, \tag{5}$$

where Φ_1 is the magnet flux produced by the primary side of the Tr_1, Φ_2 is the magnet flux produced by the secondary side of the Tr_1, Φ_3 is the magnet flux produced by the primary side of the Tr_2, Φ_4 is the magnet flux produced by the secondary side of the Tr_2, N_1 is the number of turns in the primary side, and N_2 is the number of turns in the secondary side. Each magnetic flux under the current condition (5) is considered as

$$\Phi_1 = \Phi_3, \tag{6}$$

$$\Phi_2 = \Phi_4, \tag{7}$$

thus, the voltage equation is expressed by (1), (2), (3), (4), (6), and (7) as

$$2N_2 \frac{d}{dt}\{2\Phi_2 - 2k\Phi_1\} = 2N_1 \frac{d}{dt}(\Phi_1 - k\Phi_2), \tag{8}$$

then, the magnetic flux Φ_1 and Φ_2 are considered as the same value when the coupling factor k is unity.

$$k = 1, \tag{9}$$

$$\Phi_1 = \Phi_2, \tag{10}$$

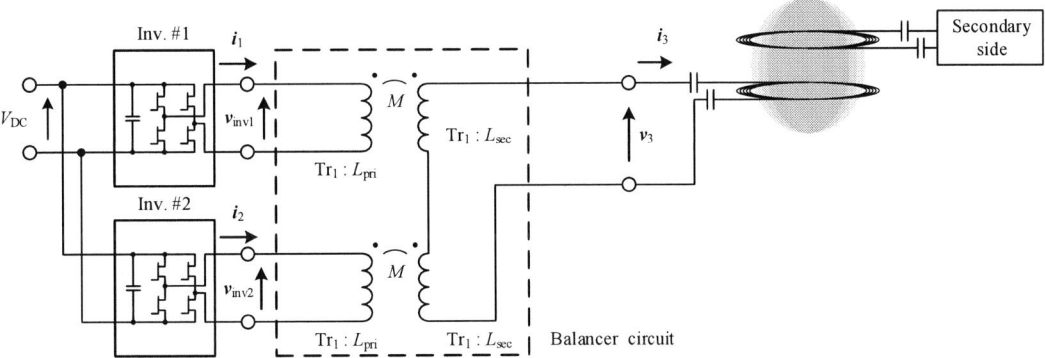

Fig. 2. Configuration of the proposed balancer circuit in parallel-connected inverter system.

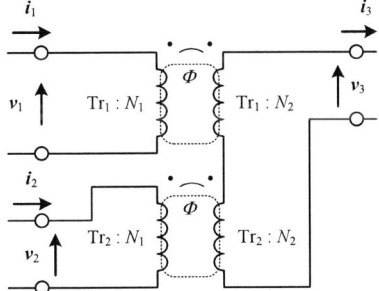

Fig. 3. Configuration of proposed balancer circuit.

The turn ratio of the transformer to achieve the current balance is expressed by using (8), (9), and (10) as

$$\frac{N_1}{N_2} = 2 \,, \tag{11}$$

the turn ratio for the current balance (11) is the same as for the non-isolated balancer[13].

Impedance matching using leakage inductance

Two transformers in the proposed balancer also have the leakage inductance between the inverters and the transfer coil. The leakage inductance of the proposed balancer is the same as that of the non-isolated balancer[13] because the turn ratio is also the same. The leakage inductance of the balancer L_{leakage}, which is composed of two transformers, is derived in [13] as

$$L_{\text{leakage}} = 4L_{\text{sec}}(1-k) \,, \tag{12}$$

where L_{sec} is the self-inductance of the secondary side of each transformer. Then, the leakage inductance appears in series on the secondary side of the balancer[13].

Although the balancer has the leakage inductance, the inductance has the capability of the impedance matching using the additional capacitor. Figure 4 (a) shows the connection of the additional capacitor, and (b) shows the equivalent circuit for the impedance matching capability in the proposed balancer. The leakage inductance of the balancer L_{leakage} and the additional capacitor C_m compose the impedance matching circuit. The load considers the resistor R_L because the power factor of the WPT system is unity under operation in resonant frequency.

The additional capacitor C_m is determined to match the impedance of the load and the inverter. Thus, the modification of the resonant capacitance is not needed to achieve the resonant condition. Moreover, the ZVS operation of two inverters becomes easy because the impedance matching circuit adjusts the impedance to the ZVS condition at the input terminal of the matching circuit. Then, the input impedance $\boldsymbol{Z}_{\text{in}}$ is expressed as

$$\boldsymbol{Z}_{\text{in}} = jX_L - \frac{jR_L X_C}{R_L - jX_C} \,. \tag{13}$$

Then, the input impedance $\boldsymbol{Z}_{\text{in}}$ must be satisfied the following condition to achieve the impedance matching.

$$R_{\text{out}} = \text{Re}\left[\boldsymbol{Z}_{\text{in}}\right] \,, \tag{14}$$

$$X_{\text{out}} = \text{Im}\left[\boldsymbol{Z}_{\text{in}}\right] \,, \tag{15}$$

where R_{out} is the resistance of the output impedance, X_{out} is the reactance of the output impedance at the inverter circuit. The capacitance C_m and the inductance L_{leakage} under the impedance matching is expressed as

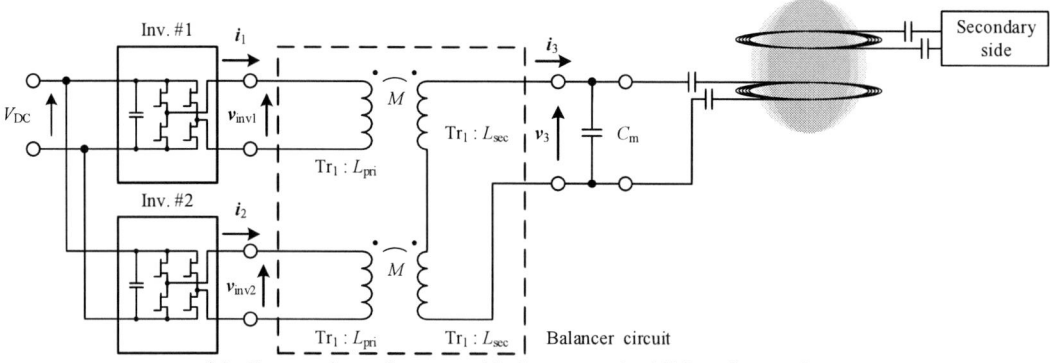

(a) Connection of proposed balancer and additional capacitor.

(b) Equivalent circuit

Fig. 4. Impedance matching capability in proposed configuration.

$$C_{\mathrm{m}} = \frac{1}{\omega R_{\mathrm{L}}}\sqrt{\frac{R_{\mathrm{L}}}{R_{\mathrm{out}}} - 1}, \quad (16)$$

$$L_{\mathrm{leakage}} = \frac{X_{\mathrm{out}}}{\omega} + \frac{C_{\mathrm{m}}R_{\mathrm{L}}^{2}}{1 + \omega^{2}C_{\mathrm{m}}^{2}R_{\mathrm{L}}^{2}}. \quad (17)$$

Thus, the self-inductance of the transformer by using (12) and (17) is determined as

$$L_{\mathrm{sec}} = \frac{1}{4(1-k)}\left\{\frac{X_{\mathrm{out}}}{\omega} + \frac{C_{\mathrm{m}}R_{\mathrm{L}}^{2}}{1 + \omega^{2}C_{\mathrm{m}}^{2}R_{\mathrm{L}}^{2}}\right\}. \quad (18)$$

Simulation results

Table I shows the simulation parameters. The transfer coil and the secondary side of the WPT system consider the resonant load. The capacitor of the GaN-FET between drain and source is considered as C_{ds} in the simulation to accurately simulate the output voltage of the inverter. Then, the load resistance R_{L} is selected to 50 Ω to check the impedance matching, although the required impedance is around 15 Ω to achieve the ZVS operation. The simulation was implemented by PLECS*(Plexim Inc.)*.

Table I. Simulation parameters.

Main circuit		
DC link voltage	V_{DC}	300 V
Switching frequency	F_{s}	6.78 MHz
Duty	d	40%
Parasitic capasitance at drain-source of GaN-FET	C_{ds}	127 pF
Balancer		
Output resistance of inverter	R_{out}	10.0 Ω
Output reactance of inverter	X_{out}	10.0 Ω
Impedance matching capacitor	C_{m}	939 pF
Leakage inductance of balancer	L_{leakage}	0.70 μH
Coupling factor	k	0.9
Inductance of secondary side	L_{sec}	1.76 μH
Inductance of primary side	L_{pri}	7.0 μH
Load		
Resonant inductance	L_{r}	5.8 μH
Resonant capacitance	C_{r}	95.0 pF
Load resistance	R_{L}	50.0 Ω

Current balance working

Figure 5 shows the each inverter's output voltage and current without the amplitude difference and the phase difference of the output voltage. Each output voltage is the same amplitude and phase. The coupling factor k is 0.9. The output current of each inverter is balanced.

Figure 6 shows the output voltage and the current of each inverter with the phase difference of the output voltage. The phase of the inverter #2 is set to a 5.0% delay. Each current becomes the same amplitude with the phase difference of the output voltage. The output voltage of the each inverter achieves ZVS operation.

Figure 7 shows the output voltage and current of each inverter when each output voltage has a different phase and amplitude. The voltage amplitude of the inverter #2 is decreased to half the value of the inverter #1. The phase of the inverter #2 is set to a 10.0% delay. Each current is also made balanced, and the ZVS operation is almost achieved.

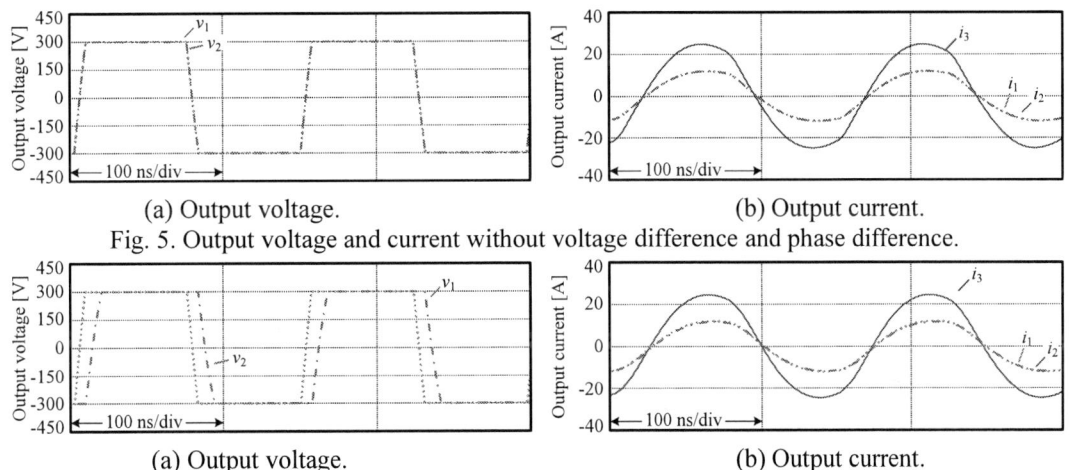

(a) Output voltage.   (b) Output current.

Fig. 5. Output voltage and current without voltage difference and phase difference.

(a) Output voltage.   (b) Output current.

Fig. 6. Output voltage and current with phase difference.

Impedance matching

Figure 8 shows the output voltage and the current of each inverter with different load resistance. The load resistance R_L is selected to 30 Ω and 100 Ω to check the effect of the impedance matching. Then, the capacitor C_m and the leakage inductance $L_{leakage}$ is modified to match the impedance. Table II shows the matching parameters with the R_L changing. The phase of the inverter #2 is set to a 5.0% delay. Each current i_1 and i_2, and the total output current i_3 have no difference with load difference between 30 Ω and 100 Ω. Moreover, both inverters also achieve ZVS operation with load difference. This result indicates that the input impedance of the balancer kept constant by the additional capacitor C_m and the leakage inductance $L_{leakage}$.

Experimental results

Figure 9 shows the prototype balancer and the parallel connection inverters in the experiment. The prototype balancer consists of two air-core transformers. The output of the balancer connects with the resonant load as the transfer coil and the secondary side of the WPT system. Table III shows the each parameter of the experimental verification. The GaN-transistors (PGA26E07BA: 600 V, 26 A, Panasonic) are used for the mega-hertz switching in each inverter.

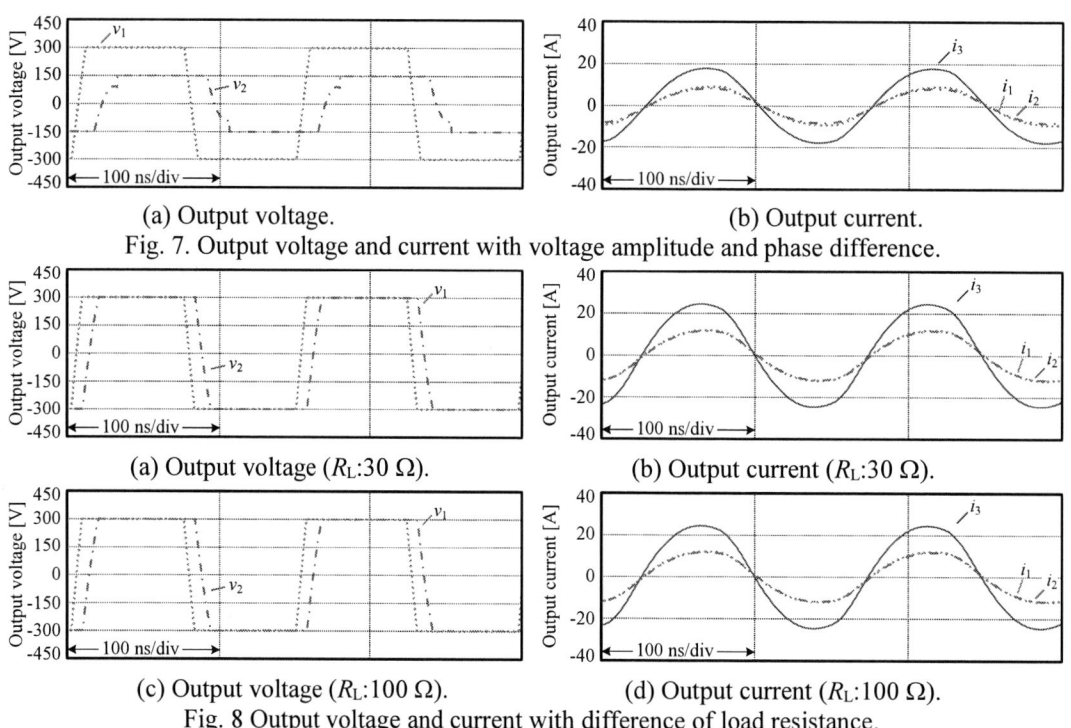

(a) Output voltage. (b) Output current.

Fig. 7. Output voltage and current with voltage amplitude and phase difference.

(a) Output voltage (R_L:30 Ω). (b) Output current (R_L:30 Ω).

(c) Output voltage (R_L:100 Ω). (d) Output current (R_L:100 Ω).

Fig. 8 Output voltage and current with difference of load resistance.

Table II. Balancer parameter with load difference.

Load resistance R_L = 30.0 Ω

Impedance matching capacitor	C_m	1107 pF
Leakage inductance of balancer	$L_{leakage}$	0.57 µH
Inductance of secondary side	L_{sec}	1.42 µH

Load resistance R_L = 100 Ω

Impedance matching capacitor	C_m	704 pF
Leakage inductance of balancer	$L_{leakage}$	0.94 µH
Inductance of secondary side	L_{sec}	2.35 µH

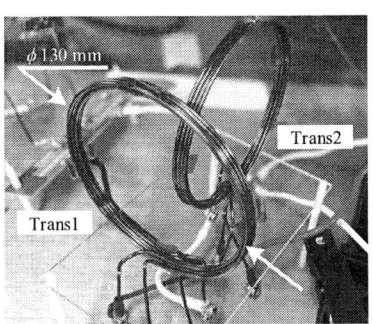
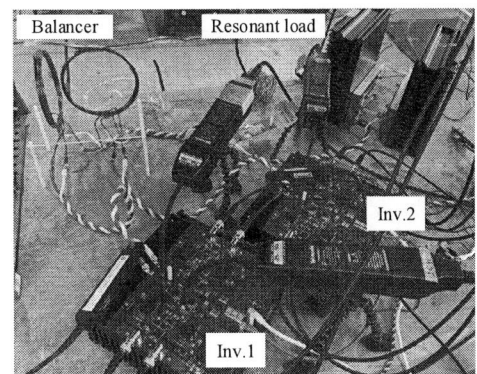

(a) Prototype balancer   (b) Overview of experimental circuit.

Fig.9 Experimental circuit.

Table III. Balancer parameters.

Main circuit		
DC link voltage	V_{DC}	200 V
Switching frequency	F_s	6.78 MHz
Duty	d	20%
Balancer		
Output resistance of inverter	R_{out}	10.0 Ω
Output reactance of inverter	X_{out}	10.0 Ω
Impedance matching capacitor	C_m	940 pF
Inductance of primary side	L_{pri}	4.7 µH
Inductance of secondary side	L_{sec}	1.3 µH
Coupling factor	k	0.84
Load		
Resonant inductance	L_r	5.3 µH
Resonant capacitance	C_r	105 pF
Load resistance	R_L	50.0 Ω

Current balance working

Figure 10 shows the output voltage and the current of the paralleled inverters in the experiment using the prototype balancer. Note that the impedance matching capacitor C_m is not implemented to confirm the fundamental balance operation. Thus, the load resistance is 20.0 Ω to achieve the ZVS. The operation frequency is 5.25 MHz because the resonant frequency of the resonant load is decreased by the leakage inductance in the balancer. The DC-link voltage is 150V. The output current i_{inv1} and i_{inv2} is balanced even if the phase of the voltage has a difference of approximately 5 ns in figure 10(b). The difference between the maximum value of the current is 5.6%. Therefore, the balancer working is achieved fundamentally without any additional control in megahertz operation.

Impedance matching

Figure 11 shows the output voltage and the current of the paralleled inverters with impedance matching capacitor C_m. The load resistance is 50.0 Ω to verify the impedance matching. The operation frequency is 6.78 MHz. The phase difference between the output voltage v_{inv1} and v_{inv2} is approximately 8 ns in figure 11(b). The output current i_{inv1} and i_{inv2} is balanced even if the output voltage has the phase difference. The output voltage v_{inv2} achieves the ZVS operation completely. The v_{inv1} does not achieve ZVS operation because of the phase of the v_{inv1} is delayed from the v_{inv2}. The total output is 620 W at the load resistor RL in figure 11(a).

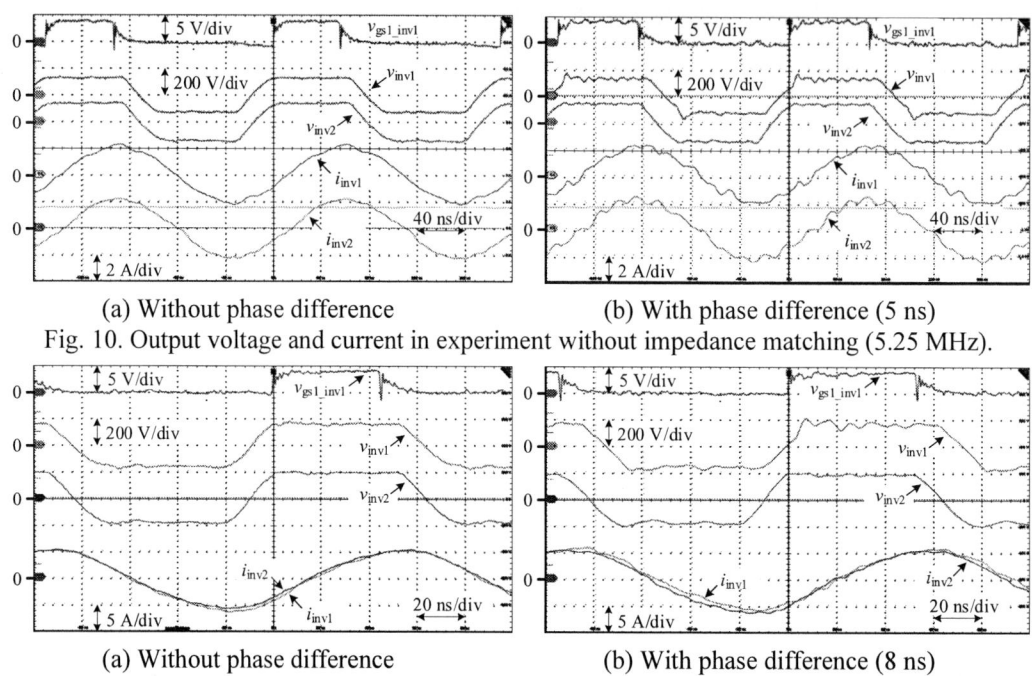

(a) Without phase difference   (b) With phase difference (5 ns)

Fig. 10. Output voltage and current in experiment without impedance matching (5.25 MHz).

(a) Without phase difference   (b) With phase difference (8 ns)

Fig. 11. Output voltage and current in experiment with impedance matching (6.78 MHz).

Conclusion

In this paper, the new current balancer, including impedance matching capability for megahertz operation, was proposed. The proposed balancer consists of two transformers without any additional control to achieve the current balance. The turn ratio of two transformers to achieve the current balancing was derived theoretically based on the magnet flux under the current balancing condition. The impedance matching condition is also derived by the leakage inductance of the balancer and the additional capacitor. The simulation results verified the operation of the proposed balancer under the amplitude and the phase difference condition. The input impedance of the balancer kept constant with the impedance matching capability. The experimental result with proposed balancer achieved current balance with a phase difference of the output voltage approximately 8 ns. In addition, the ZVS operation is almost achieved in the each inverter by impedance conversion using additional capacitor.

References

[1] J. Li and D. Costinett, "Comprehensive Design for 6.78 MHz Wireless Power Transfer Systems", 2018 IEEE Energy Conversion Congress and Exposition (ECCE), Portland, OR, 2018, pp. 906-913.

[2] N. K. Trung, T. Ogata, S. Tanaka and K. Akatsu, "Attenuate Influence of Parasitic Elements in 13.56-MHz Inverter for Wireless Power Transfer Systems," in IEEE Transactions on Power Electronics, vol. 33, no. 4, pp. 3218-3231, April 2018.

[3] N. K. Trung, T. Ogata, S. Tanaka, K. Akatsu, "Analysis and PCB Design of Class D Inverter for Wireless Power Transfer Systems Operating at 13.56MHz", IEEJ Journal of Industry Applications, 2015, vol 4, no.6, pp. 703-713.

[4] S. Suzuki, T. Shimizu, "A Study on Efficiency Improvement of High-frequency Current Output Inverter Based on Immittance Conversion Element", IEEJ Journal of Industry Applications, 2015, vol 4, no.3, pp. 220-226.

[5] M. Fu, Z. Tang and C. Ma, "Analysis and Optimized Design of Compensation Capacitors for a Megahertz WPT System Using Full-Bridge Rectifier," in IEEE Transactions on Industrial Informatics, vol. 15, no. 1, pp. 95-104, Jan. 2019, doi: 10.1109/TII.2018.2833209.

[6] M. Liu, M. Fu, Y. Wang and C. Ma, "Battery Cell Equalization via Megahertz Multiple-Receiver Wireless Power Transfer," in IEEE Transactions on Power Electronics, vol. 33, no. 5, pp. 4135-4144, May 2018, doi: 10.1109/TPEL.2017.2713407.

[7] J. Song, M. Liu and C. Ma, "Efficiency optimization and power distribution design of a megahertz multi-receiver wireless power transfer system," 2017 IEEE PELS Workshop on Emerging Technologies: Wireless Power Transfer (WoW), Chongqing, China, 2017, pp. 54-58, doi: 10.1109/WoW.2017.7959364.

[8] M. Fu, H. Yin and C. Ma, "Megahertz Multiple-Receiver Wireless Power Transfer Systems With Power Flow Management and Maximum Efficiency Point Tracking," in IEEE Transactions on Microwave Theory and Techniques, vol. 65, no. 11, pp. 4285-4293, Nov. 2017, doi: 10.1109/TMTT.2017.2689747.

[9] L. Jiang and D. Costinett, "A single-stage 6.78 MHz transmitter with the improved light load efficiency for wireless power transfer applications," 2018 IEEE Applied Power Electronics Conference and Exposition (APEC), San Antonio, TX, USA, 2018, pp. 3160-3166, doi: 10.1109/APEC.2018.8341553.

[10] L. Jiang and D. Costinett, "A High-Efficiency GaN-Based Single-Stage 6.78 MHz Transmitter for Wireless Power Transfer Applications," in IEEE Transactions on Power Electronics, vol. 34, no. 8, pp. 7677-7692, Aug. 2019, doi: 10.1109/TPEL.2018.2879958.

[11] J. Shi, L. Zhou and X. He, "Common-Duty-Ratio Control of Input-Parallel Output-Parallel (IPOP) Connected DC–DC Converter Modules With Automatic Sharing of Currents," in IEEE Transactions on Power Electronics, vol. 27, no. 7, pp. 3277-3291, July 2012, doi: 10.1109/TPEL.2011.2180541.

[12] N. K. Trung and K. Akatsu, "Design challenges for 13.56MHz 10 kW resonant inverter for wireless power transfer systems", 2019 10th International Conference on Power Electronics and ECCE Asia (ICPE 2019 - ECCE Asia), Busan, Korea (South), 2019, pp. 1-7.

[13] M. Yamaguchi, K. Kusaka and J. Itoh, "Current Balancing Method in Parallel Connected Inverter Circuit for Megahertz WPT System," 2021 IEEE 30th International Symposium on Industrial Electronics (ISIE), 2021, pp. 1-6, doi: 10.1109/ISIE45552.2021.9576365.

Investigation and Mitigation of Common-mode Voltage in Four-level NPC Converters Modulated by Redundant Level Modulation

Jun Wang*, Wei Xu, Xibo Yuan and Lihong Xie
Department of Electrical and Electronic Engineering, University of Bristol, Bristol, UK
*Email: jun.wang@bristol.ac.uk

Keywords

«DC-AC converter», «Multi-level inverters», «Modulation strategy», «Power quality», «Capacitor voltage balancing»

Abstract

Redundant level modulation (RLM) has emerged as a powerful modulation-based voltage balancing method for a range of multilevel converters. For four-level neutral point clamped (4L-NPC) topologies, utilizing redundant voltage levels is the mandatory solution in principle to enable the single-end 4L-NPC converters to keep a voltage balance under all operating conditions, without requiring auxiliary circuits. However, the side effects of RLM caused by the extra switching actions have not been fully studied, with the common-mode voltage in particular. Hence, this work investigates the worsened common-mode voltage problem induced by RLM through theoretical analysis and simulations, followed by a simple carrier-based mitigation method that is verified in experiments.

Introduction

Multilevel converters, e.g. neutral-point-clamped (NPC) converters, have been intensively researched as promising alternatives to conventional two-level converters, because they in general offer lower switching loss, lower blocking voltage requirement and lower output harmonics/EMI. For example, the three-level NPC topologies, such as diode-clamped [1], T-type [2] and active-clamped [3] NPC topologies, have been widely implemented in low/medium-voltage power conversion systems to improve the system efficiency and power density, such as in [2], [4]. As the derivation of three-level topologies, four-level topologies, such as π-type NPC [5], active-clamped NPC [6] and hybrid clamped [7] variations, offer further reduced switching voltage and more output voltage steps, for which a typical system and operation principle are illustrated in Fig. 1 and Fig. 2. However, due to the inherent lack of capacitor-self-balancing capability, the application of four-level NPC converters has been significantly hindered [8]. For example, the four-level converter implemented by Schneider Electric [9] requires auxiliary 'dc bus balancers' circuits to maintain the capacitor voltages, which introduces extra loss and additional hardware.

Fortunately, the recently proposed redundant level modulation (RLM) [10] has solved the voltage balancing issue completely, which is a modulation-based approach that relies on more switching events rather than requiring additional hardware. RLM shares the same principles as other existing approaches, such as the overlapped carrier [11] and virtual vector [12] approaches, in terms of utilizing the redundant voltage levels. RLM has been proven effective and implemented in [13], [14] with its most significant side effect, the extra switching loss, well evaluated and understood. However, there have been concerns raised about the common-mode (CM) voltage in RLM-modulated four-level converters since the additionally switching events can very likely lead to worsened CM voltage, which is reported to cause issues such as the reduced bearing lifetime, increased insulation stress and

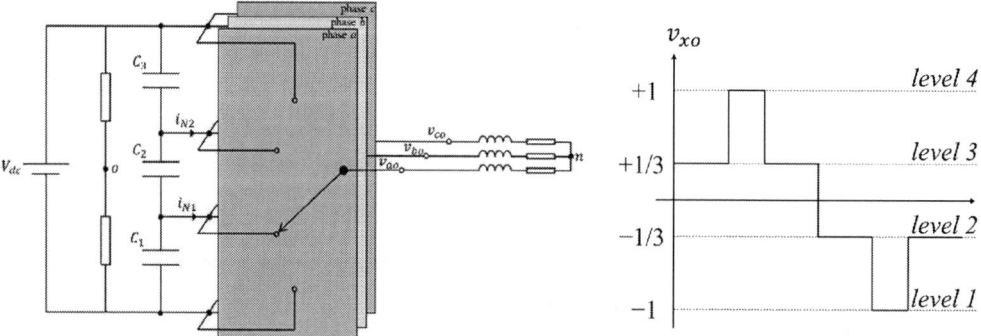

Fig. 1. A four-level NPC three-phase inverter-load system   Fig. 2. Illustration of the four-level phase output voltage

electromagnetic interference (EMI) problems. While the RLM is deemed mandatory for four-level NPC converters to maintain their basic functionality without auxiliary circuits, this potential CM voltage issue need to be investigated and mitigated. Although there are previous studies conducted on the CM voltage of four-level NPC converters and the mitigation methods, e.g. [15], [16], they are all based on the assumption of having auxiliary voltage balancing circuits.

Hence, as the contribution, this work investigates the CM voltage in RLM-modulated four-level NPC converters for the first time and subsequently proposes a modulation-based simple solution to mitigate it. The investigation starts from the theoretical analysis of the generation principles of CM voltage comparing the regular modulation and RLM cases. This issue is then evaluated in simulation with a novel mitigation method implemented, which is later verified in an experimental setup.

Overview of Redundant Level Modulation in 4L-NPC converters

To start with, the principle of the RLM is explained as follows. In contrast to the regular modulation, RLM introduces one output voltage level in each carrier cycle as illustrated in Fig. 3 to gain extra control of capacitor voltages, while the volt-second product of this carrier cycle remains the same. This can be understood as altering the high-frequency behavior of the converter output by adding more switching actions. In four-level NPC converters, RLM can be easily realized through splitting and altering the modulating waves as demonstrated in [10], with regular level-shifted carriers in place as illustrated in Fig. 4 and the output states defined in Table I.

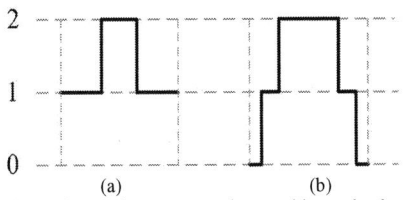

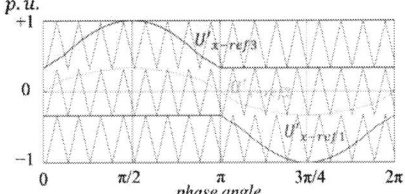

Fig. 3. Illustration of phase output voltages with equal volt-second product (a) regular modulation (b) redundant modulation

Fig. 4. Implementation of RLM with level-shifted carriers [10].

Table I Output states of a four-level NPC phase leg

Level l	Output states per unit	Output voltage v_{xo}
4	$+1$	$+V_{dc}/2$
3	$+1/3$	$+V_{dc}/6$
2	$-1/3$	$-V_{dc}/6$
1	-1	$-V_{dc}/2$

Fig. 5 further illustrates the principle of this carrier-based implementation of RLM in 4L-NPC converters. It can be seen that, with an offset value added to space away two adjacent split modulating waves, one extra voltage level can be added to the output voltage, which brings a new degree of freedom to manipulate the duty ratio of certain voltage levels to control the capacitor voltages. The closed-loop control based on this concept can be found in [10]. Note that the RLM pattern showing in Fig. 5 means that one phase leg compulsorily outputs three voltage levels in one carrier cycle, either *level 4/3/2* (when $V_{x\text{-ref}} \geq 0$) or *level 3/2/1* (when $V_{x\text{-ref}} < 0$).

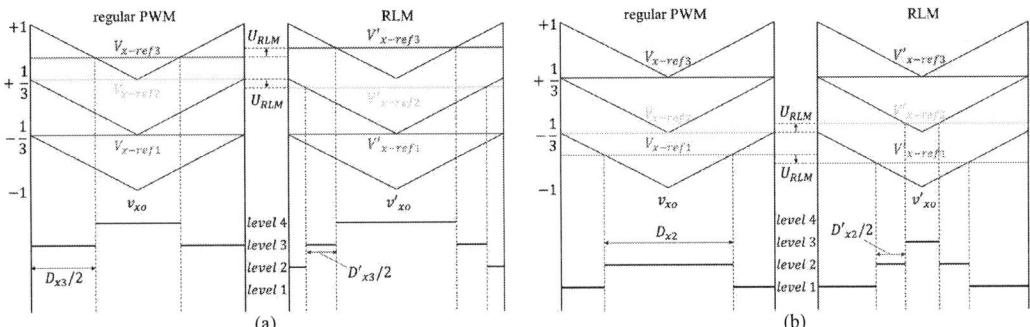

Fig. 5. Illustration of one carrier cycle under regular PWM and RLM when (a) $V_{x\text{-ref}} \geq 0$ (b) $V_{x\text{-ref}} < 0$

The drawback of RLM is the extra switching events, e.g. four switching transitions in one carrier cycle instead of two transitions in regular modulation. This drawback leads to extra switching losses as analyzed in [10]. Meanwhile, it also has impacts on the common-mode voltage which will be shown in the next section.

Common-mode voltage in RLM-modulated 4L-NPC converters

The common-mode (CM) voltage in a three-phase inverter-load system is expressed as

$$v_{no} = \frac{1}{3}(v_{ao} + v_{bo} + v_{co}) \tag{1}$$

Where n denotes the load neutral point; o is the dc-link neutral point (normally grounded); a, b, c represents the three phases. To investigate the CM voltage impacted by RLM, a simulation model has been built based on the following specifications in a 1200V, three-phase, four-level ANPC [13], [14] configuration. The switching frequency f_{sw} is lowered to 2 kHz to show the phenomenon more clearly.

TABLE II SYSTEM SPECIFICATIONS

DC-link voltage V_{dc}	1200 V
Nominal capacitor voltage E	400 V
Carrier frequency f_{sw}	2 kHz
Fundamental frequency f_0	50 Hz
C1 = C2 = C3	450 µF
Modulation index M	0.92
R	4.57 Ω per phase
L	3 mH per phase

Firstly, to theoretically analyze the CM voltage, the three phase output voltages (V_a, V_b, V_c) in one switching cycle are re-classified into a dominant voltage V_1 and two possible cancelling voltages V_2 and V_3. Starting from a four-level converter modulated by regular Sinusoidal Pulse Width Modulation (SPWM), the possible common mode voltages are listed in Table III, referring to the combinations of V_1, V_2 and V_3. All the cases are illustrated in Fig. 6 throughout one fundamental cycle. In the majority of the cases, V_2 and V_3 fully cancel each other, which leaves V_1 as the only term left in (1), such as Case A2, A3, A4, A5 in Table III, and yields possible CM voltage amplitude of 1/6 of V_{dc} and 1/12 of V_{dc}. In some cases, V_2 and V_3 can only partly cancel each other (e.g. $+1$ and $-1/3$), which yields a higher CM voltage amplitude of 5/18 of V_{dc}, such as Case A1 and A6 in Table III. A typical switching window under regular SPWM is shown in Fig. 7(a), where the peak CM voltage is 333.3V (5/18 of 1200V).

Table III Common-mode voltage levels in 4L-NPC with regular SPWM

Case	Sum of voltage level l	V_1	V_2	V_3	V_{CM}
A1	10	$+1$	$+1$	$-1/3$	$5/18 \cdot V_{dc}$
A2	9	$+1$	$+1/3 \ (+1)$	$-1/3 \ (-1)$	$1/6 \cdot V_{dc}$
A3	8	$+1/3$	$+1$	-1	$1/12 \cdot V_{dc}$
A4	7	$-1/3$	$+1$	-1	$-1/12 \cdot V_{dc}$
A5	6	-1	$+1/3(+1)$	$-1/3(-1)$	$-1/6 \cdot V_{dc}$
A6	5	-1	-1	$+1/3$	$-5/18 \cdot V_{dc}$

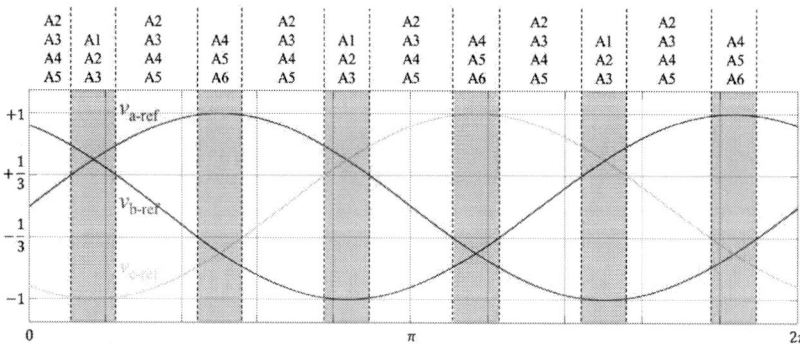

Fig. 6. Illustration of CM voltage case regions over one fundamental cycle (*modulation index M = 1*).

In the case of RLM, the converter phase leg mandatorily outputs three voltage levels in one switching cycle, rather than two levels, which leads to two more additional combinations of voltages as listed in Table IV. As can be seen, in Case B1 and Case B8, V_2 and V_3 have the same sign and therefore they reinforce rather than cancel each other. These two extra cases leads to a higher CM voltage amplitude of 7/18 of V_{dc} to appear in RLM four-level converters. This is illustrated in Fig. 7(b) where a peak CM voltage of 466V (7/18 of 1200V) can be observed. It can also be seen that the output phase voltage V_{xo} shows the RLM pattern as intended in Fig. 5. The cause of this is that the highest voltage levels of three phases all locate at the central of a switching cycle, showing a 'hill' shape.

Table IV Common-mode voltage levels in 4L-NPC with RLM

Case	Sum of voltage level l	V_1	V_2	V_3	V_{CM}
B1	11	+1	+1	+1/3	$7/18 \cdot V_{dc}$
B2	10	+1	+1	−1/3	$5/18 \cdot V_{dc}$
B3	9	+1	+1/3 (+1)	−1/3 (−1)	$1/6 \cdot V_{dc}$
B4	8	+1/3	+1	−1	$1/12 \cdot V_{dc}$
B5	7	−1/3	+1	−1	$-1/12 \cdot V_{dc}$
B6	6	−1	+1/3(+1)	−1/3(−1)	$-1/6 \cdot V_{dc}$
B7	5	−1	−1	+1/3	$-5/18 \cdot V_{dc}$
B8	4	−1	−1	−1/3	$-7/18 \cdot V_{dc}$

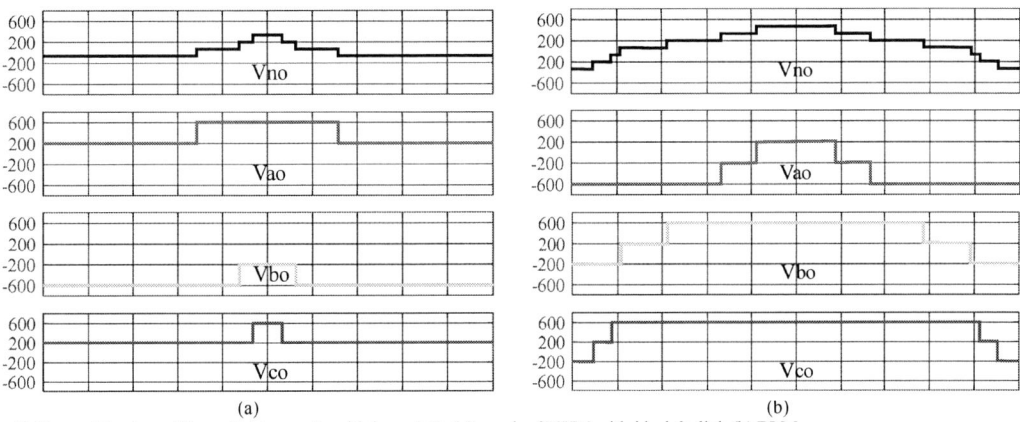

Fig. 7. Zoomed in view of the switching cycle with largest V_{no} (a) regular SPWM with ideal dc-link (b) RLM

Fig. 9 shows a comparison of common-mode voltage extracted from simulation between regular modulation and RLM cases. Note that the regular SPWM case assumes the dc-link is idea, i.e. the three capacitor voltages in Fig. 1 constantly stay at the nominal value - so there is no voltage balancing problem. This ideal case is not realistic without any external voltage balancing circuits or three isolated dc sources connected in series. Therefore, the regular modulation case here is just for the reference purpose.

It can be seen that the common-mode voltage in the RLM case is worse than the regular modulation case. In the regular modulation, the V_{no} shows six steps as analyzed in Table III, in which the peak value is ±333 V ($5/18 \cdot V_{dc}$). In contrast, in the RLM case, the V_{no} shows two more steps (one extra step towards each polarity), which brings the peak value to ± 466 V ($7/18 \cdot V_{dc}$ as analyzed in Table IV). The FFT analysis of the CM voltage is plotted in Fig. 9, which shows the amplitude of CM voltage harmonics has a spike of 300 V at the switching frequency. Note there is also a low-frequency CM voltage component at the 3rd order as can be observed in Fig. 9, which is caused by the zero-sequence voltage added for voltage balancing purpose [10]. This finding establishes that the RLM does worsen the CM voltage problem in a 4L-NPC converter.

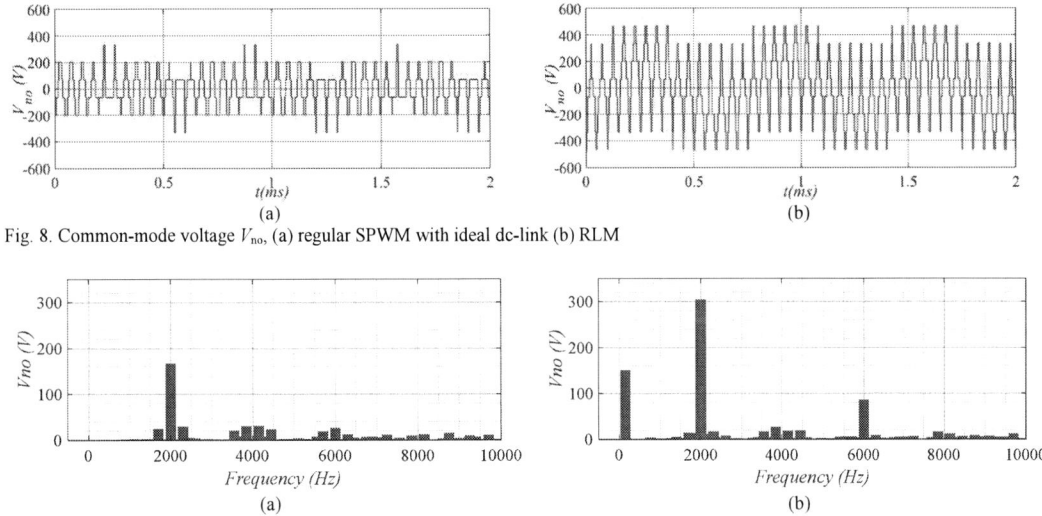

Fig. 8. Common-mode voltage V_{no}, (a) regular SPWM with ideal dc-link (b) RLM

Fig. 9. FFT of common-mode voltage V_{no} (a) regular SPWM with ideal dc-link (b) RLM

A simple carrier-based mitigation method of common-mode voltage

To mitigate the CM voltage, this work proposes a simple carried-based approach as follows. To counter the worst-case scenario discovered in the last section, the proposed solution is to apply modified carriers with 180° phase shift (flipped vertically) when $V_{\text{ref}} \geq 0$, as shown in Fig. 10. When $V_{\text{ref}} < 0$, the carriers stay the same. This approach leads to a time-domain redistribution of voltage levels in one carrier cycle. As can be seen, this redistribution results in the output *level 4* pushed to the two sides of one carrier cycle when $V_{\text{ref}} \geq 0$, instead of the middle, which result in a 'valley' shape instead of a 'hill' shape. In this case, when $V_{\text{ref}} \geq 0$, the middle part of one switching cycle will be *level 2* instead of *level 4*, which can avoid the CM voltage to fall on the *level 4 + level 4 + level 2* case in Fig. 7. By applying approach, the output voltages v_{xo} (when $V_{\text{ref}} \geq 0$) and v_{xo} (when $V_{\text{ref}} < 0$) show a symmetric pattern with reference to the dc-link neutral point voltage v_o, in contrast to the regular case in Fig. 5. This approach is referred as the 'flipped carrier'.

Given sinusoidal reference voltages, there is always one phase having the opposite signum to the other two as illustrated in Fig. 6. In this case, the 'valley' shaped voltage with flipped carriers would show in at least one phase , which leads to V_2 and V_3 fully/partly cancelling each other at the center of one switching cycle, instead of reinforcing each other as in the case of Fig. 7(b). Therefore, the worst cases B1 and B8 are mitigated.

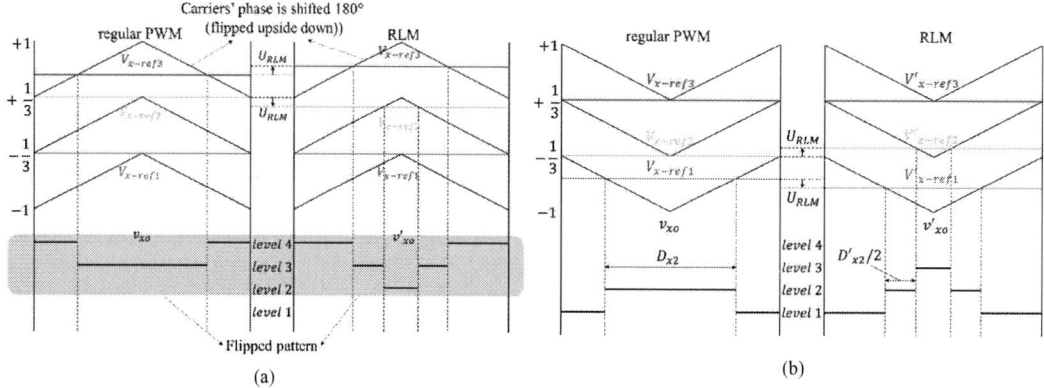

Fig. 10. Illustration of one carrier cycle under regular PWM and RLM (a)$V_{\text{x-ref}} \geq 0$ **with flipped carriers** (b)$V_{\text{x-ref}} < 0$ (same as Fig. 5)

The proposed flipped carrier approach is evaluated in simulation under the same conditions. Fig. 11 shows the CM voltages in this case, where a clear improvement can be observed in comparison to Fig. 8. In the RLM case, the largest amplitude is mitigated from 466V to 333V, which only appears four brief times in one fundamental cycle under this operating point. For most of the time, the CM voltage is only 1/6 or 1/12 of the dc-link voltage. The FFT analysis of the CM voltage is shown in Fig. 12(b). As can be seen, the amplitude of CM voltage at around the switching frequency is damped from 300V to <100V, which is a considerable improvement compared to Fig. 9(b). Note the flipped carrier is also effective for the SPWM case. The peak CM voltage is also improved to 1/6 of V_{dc}

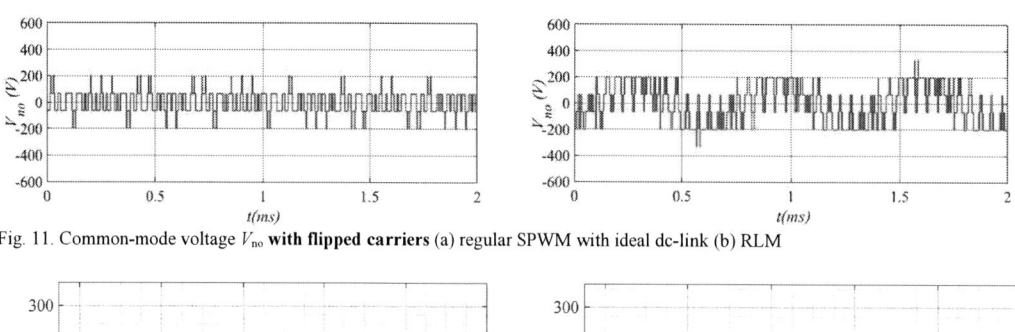

Fig. 11. Common-mode voltage V_{no} **with flipped carriers** (a) regular SPWM with ideal dc-link (b) RLM

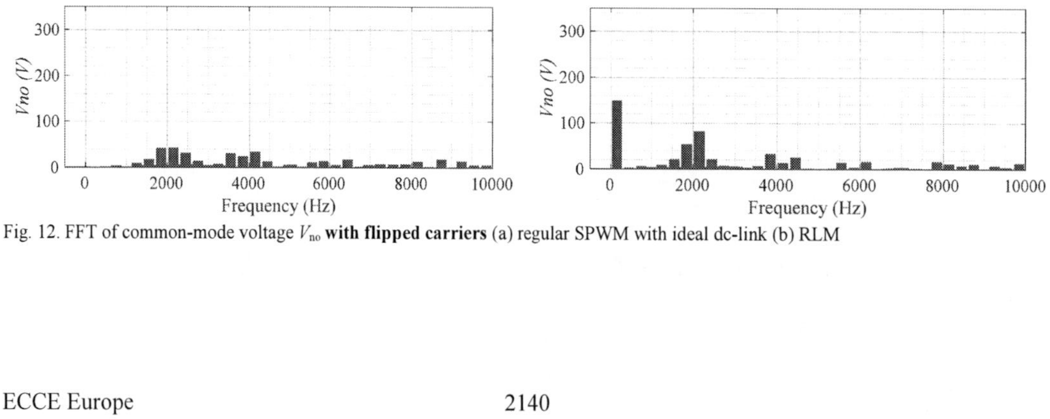

Fig. 12. FFT of common-mode voltage V_{no} **with flipped carriers** (a) regular SPWM with ideal dc-link (b) RLM

As plotted in Fig. 13(b), the worst-case center part in one switching cycle becomes *level 2 + level 4 + level 4*, which has an amplitude of 333V (5/18 V_{dc}) and shows much less frequently in contrast to Fig. 7. Note that the V_2 and V_3 (the top levels of V_{ao} and V_{bo} at the center in Fig. 13(b)) does not always completely cancel each other since the width of them does not always equal to each other. The widths of these levels depend on the voltage balancing algorithm which is corelated to the load current level, switching frequency and the capacitances as demonstrated in [10]. Therefore, there are still possible high CM voltage at 7/18 of V_{dc} that 'leaks out', i.e. the brief spikes reaching 466 V observed in Fig. 13(b). These spikes can also occur at the boundaries between switching cycles.

In other words, because this flipped carrier method is not an "active" CM voltage mitigation method (e.g. by actively selecting the appropriate voltage vectors as in [17]), it cannot completely and consistently suppress the peak CM voltage throughout fundamental cycles depending on the operating point, while it can effectively damp the CM voltage harmonics as demonstrated. Note that, with regular SPWM assuming auxiliary voltage balancers, the proposed flipped carrier also shows effective CM voltage suppressing effect comparing Fig. 8(a)/Fig. 9(a) against Fig. 11(a)/Fig. 12(a).

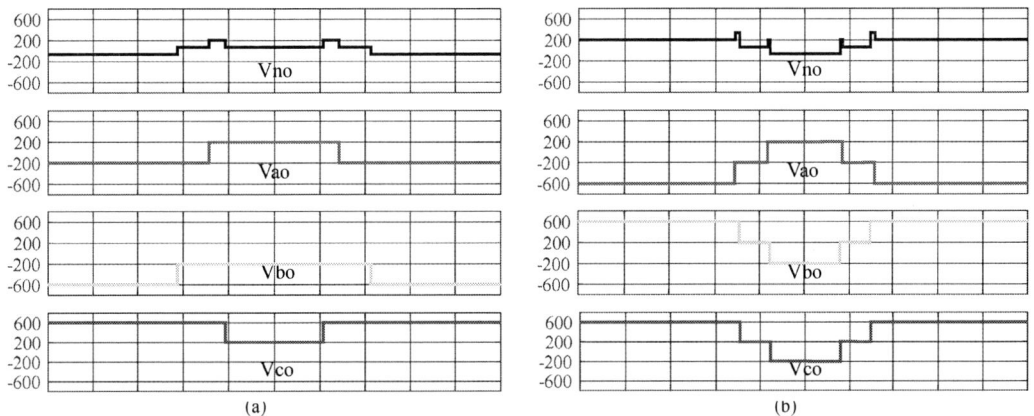

Fig. 13. Zoomed in view of the switching cycle showing largest V_{no} **with flipped carriers** (a) regular SPWM with ideal dc-link (b) RLM

This flipped carrier approach can be straightforwardly implemented through the process illustrated in Fig. 14, with the polarity of the reference voltage $V_{x\text{-ref}}$ (before introducing RLM) as the criteria to determine the phase angle of the carriers in each switch period T_{sw}. This approach can also be referred as the symmetrical modulation since the output voltage pattern in the positive and negative cycle of $V_{x\text{-ref}}$ is symmetrical with respect to the x axis.

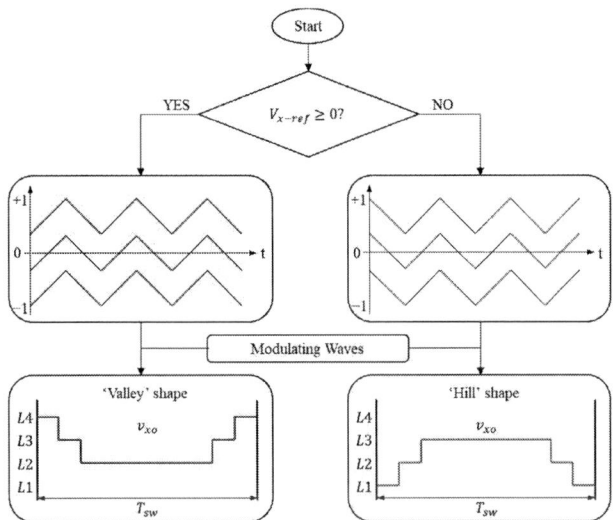

Fig. 14. Flow chart of the proposed 'flipped carrier' approach

However, the flipped carrier approach is found to trade off the quality of differential-mode (DM) line-to-line voltage while mitigating the common-mode voltage, which is a common trade-off as reported in [17]. This trade-off is reflected in Fig. 15, where it can be seen that the THD and harmonic at f_{sw} is worse in the case of the flipped carriers. Meanwhile, it does not impact the phase voltage quality as noticeably. Therefore, this trade-off between CM voltage and DM line voltage quality needs to be considered when the flipped carrier approach is applied.

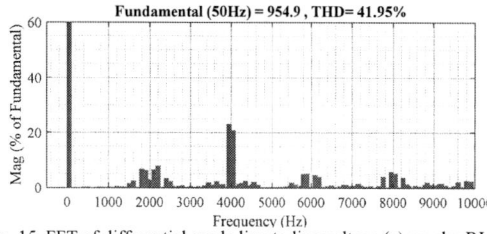 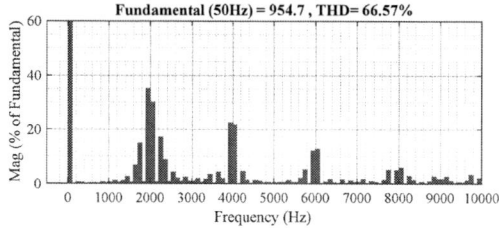

Fig. 15. FFT of differential-mode line-to-line voltage (a) regular RLM (b) RLM + flipped carriers

Experimental Evaluation

The presented CM voltage analysis and the proposed mitigation method are evaluated in an experimental prototype consisting of Silicon Carbide (SiC) modules and four-level active NPC topology as presented in [14], of which a photo is shown in Fig. 16. The functionality of the original RLM has been proven as shown in Fig. 17 in a single-end inverter operating at a high modulation index and near-unity power factor, which shows well-balanced capacitor voltages given the four visible straight-line output voltage levels in the phase voltage. The converter operates at a dc-link voltage V_{dc} of 1000V, a switching frequency f_{sw} of 5 kHz, a modulation index M of 0.9 and a power factor $cos\,\varphi$ of 0.99 in this case as a proof of concept.

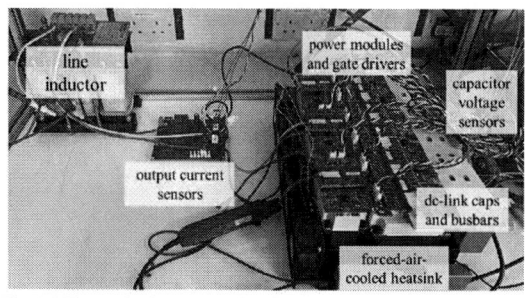 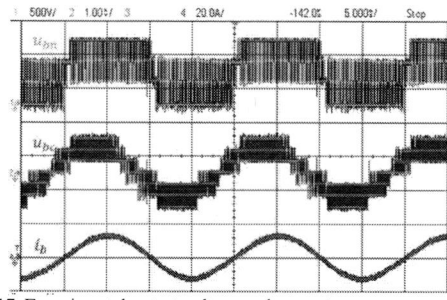

Fig. 16. Photo of a three-phase four-level inverter prototype

Fig. 17. Experimental output voltage and current

With the dc-link voltage reduced to 300 V, the waveforms are captured and shown below, with FFT performed on the measured CM voltage. The RLM-based voltage balancing algorithm and the flipped carrier approach are programmed in a DSP (TMS320F28335) based controller.

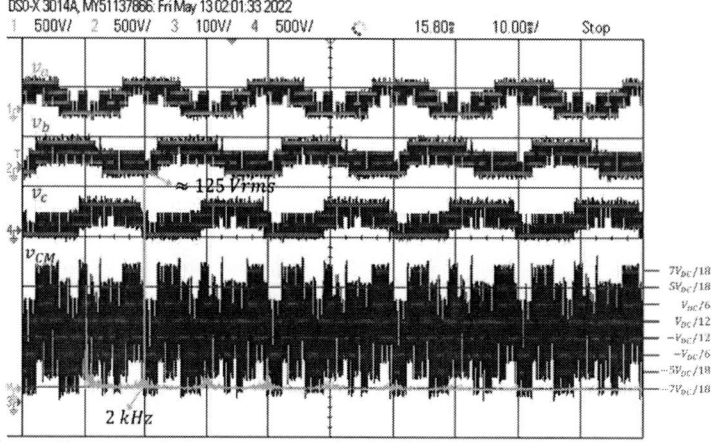

Fig. 18. Experimental waveforms with RLM and regular carriers (f_{sw} = 2 kHz, M = 0.8, V_{DC} = 300 V)

In the case with RLM and regular carriers in Fig. 18, it can be seen that the CM voltage often reaches the $7V_{dc}/18$ level, and the largest harmonic around the switching frequency reaches around 125 V. With the flipped carriers implemented, the improved waveforms are shown in Fig. 19. Although there are still spikes reaching $7V_{DC}/18$, it is visible that the CM voltage is improved overall. This is proved by the FFT results showing that the largest harmonic around the switching frequency is significantly reduced to 33 V from 125 V. In summary, this comparison verifies the effectiveness of the proposed flipper carrier as a CM voltage mitigation method.

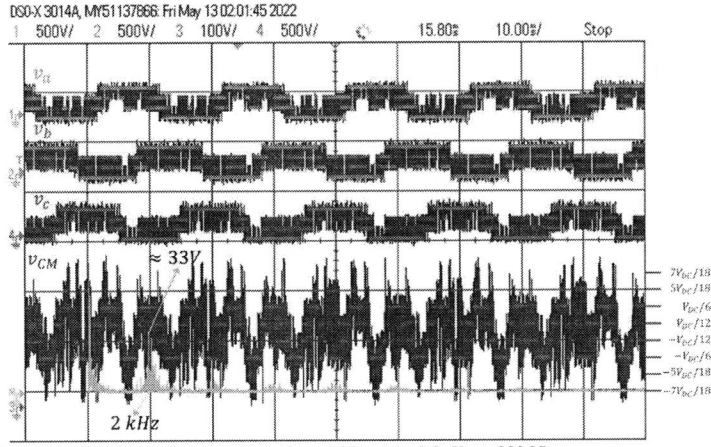

Fig. 19. Experimental waveforms with RLM and flipped carriers (f_{SW} = 2 kHz, M = 0.8, V_{DC} = 300 V)

Fig. 20 shows a zoom-in view of the waveforms. In the highlighted switching cycle, it can be seen that v_b and v_c show a 'valley' shape while v_a shows a 'hill' shape as intended in Fig. 10. This results in the CM voltage mostly ranging between $+V_{dc}/6$ and $-5V_{dc}/18$, with a narrow spike reaching $-7V_{dc}/18$, which is explained by Fig. 13(b).

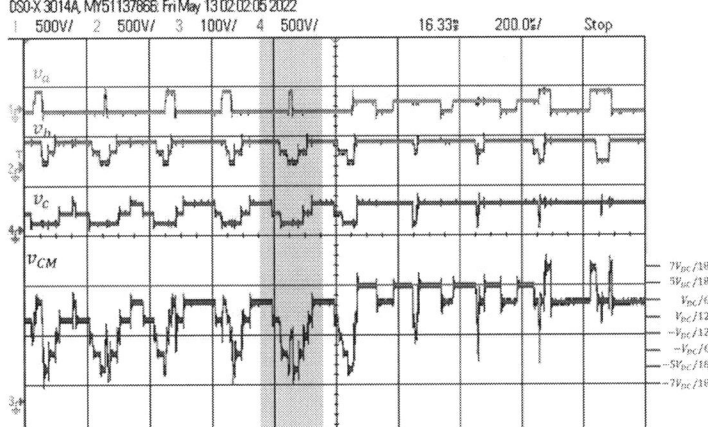

Fig. 20. Zoomed-in experimental waveforms with RLM and flipped carriers (f_{SW} = 2 kHz, M = 0.8, V_{DC} = 300 V)

Conclusion

This paper studied the common-mode voltage issue in the RLM-modulated four-level NPC converters. While the RLM is mandatory for the voltage balancing purpose, its negative impact on the CM voltage has been confirmed by theoretical analysis and simulation, which shows a considerable increase of switching-frequency harmonics and peak amplitude (from $5V_{DC}/18$ to $7V_{DC}/18$). To address this issue, this work proposes a simple carrier-based method with flipped carriers to mitigate the CM voltage, which has proven to be effective in the simulation and experiments.

References

[1] A. Nabae, I. Takahashi, and H. Akagi, "A new neutral-point-clamped PWM inverter," *IEEE Trans. Ind. Appl.*, vol. 1A-17, no. 5, pp. 518–523, 1981.

[2] M. Schweizer and J. W. Kolar, "Design and implementation of a highly efficient three-level T-type converter for low-voltage applications," *IEEE Trans. Power Electron.*, vol. 28, no. 2, pp. 899–907, 2013.

[3] P. Barbosa, P. Steimer, J. Steinke, M. Winkelnkemper, and N. Celanovic, "Active-Neutral-Point-Clamped (ANPC) multilevel converter technology," in *European Conference on Power Electronics and Applications*, 2005.

[4] D. Zhang, J. He, and D. Pan, "A megawatt-scale medium-voltage high-efficiency high power density ' SiC + Si ' hybrid three-level propulsion systems," *IEEE Trans. Ind. Appl.*, vol. 55, no. 6, pp. 5971–5980, 2019.

[5] X. Yuan, "A four-level π-type converter for low-voltage applications," in *IEEE Proc. European Conference on Power Electronics and Applications*, 2015.

[6] B. Wang, "Four-level neutral point clamped converter with reduced switch count," in *IEEE Annual Power Electronics Specialists Conference*, 2008, pp. 2626–2632.

[7] K. Wang, Z. Zheng, L. Xu, and Y. Li, "A four-level hybrid-clamped converter with natural capacitor voltage balancing ability," *IEEE Trans. Power Electron.*, vol. 29, no. 3, pp. 1152–1162, Mar. 2014.

[8] J. Pou, R. Pindado, and D. Boroyevich, "Voltage-balance limits in four-level diode-clamped converters with passive front ends," *IEEE Trans. Ind. Electron.*, vol. 52, no. 1, pp. 190–196, Feb. 2005.

[9] H. R. Nielsen, "3-phase high-power UPS," US8385091B2, 2013.

[10] J. Wang, X. Yuan, and B. Jin, "Carrier-based closed-loop dc-link voltage balancing algorithm for four level NPC converters based on redundant level modulation," *IEEE Trans. Ind. Electron.*, vol. 68, no. 12, pp. 11707–11718, 2021.

[11] K. Wang, Z. Zheng, and Y. Li, "A novel carrier-overlapped PWM method for four-level neutral-point clamped converters," *IEEE Trans. Power Electron.*, vol. 34, no. 1, pp. 7–11, Jan. 2019.

[12] S. Busquets-Monge, J. Bordonau, and J. Rocabert, "A virtual-vector pulsewidth modulation for the four-level diode-clamped DC–AC converter," *IEEE Trans. Power Electron.*, vol. 23, no. 4, pp. 1964–1972, Jul. 2008.

[13] X. Yuan, J. Wang, I. Laird, and W. Zhou, "Wide-bandgap device enabled multilevel converters with simplified structures and capacitor voltage balancing capability," *IEEE Open J. Power Electron.*, vol. 2, no. May, pp. 401–410, 2021.

[14] J. Wang, I. Laird, X. Yuan, and W. Zhou, "A 1.2 kV 100 kW four-level ANPC inverter with SiC power modules and capacitor voltage balance for EV traction applications," in *IEEE Proc. Energy Conversion Congress and Exposition (ECCE)*, 2021.

[15] J. Pribadi, B. Park, and D. C. Lee, "Operation of four-Level ANPC inverter based on space-vector modulation for common-mode voltage reduction," in *IEEE Proc. International Symposium on Electrical and Electronics Engineering (ISEE)*, 2019, pp. 281–285.

[16] J. Pribadi, D. C. Lee, and B. M. Han, "Suppression of common-mode voltage based on space-vector modulation for four-level hybrid inverters," in *IEEE Proc. International Future Energy Electronics Conference (IFEEC)*, 2019.

[17] X. Yuan, Y. Li, and C. Wang, "Objective optimisation for multilevel neutral-point-clamped converters with zero-sequence signal control," *IET Power Electron.*, vol. 3, no. 5, pp. 755–763, 2010.

Ferrite optimization for a three-phase wireless power transfer system for electric vehicles

Shuang Nie, Mehanathan Pathmanathan, Peter W. Lehn
Department of Electrical Computer Engineering, Faculty of Applied Science and Engineering
University of Toronto, Toronto, Canada
shuang.nie@mail.utoronto.ca

Acknowledgement

This work was supported by the Natural Sciences and Engineering Research Council of Canada (NSERC) under Grant CRDPJ 513206-17.

Keywords

≪Wireless power transmission≫, ≪Electric vehicle≫, ≪Charging Infrastructure for EV´s≫, ≪Passive component≫, ≪Modelling≫

Abstract

This paper proposes a new design for placement of the ferrite in high power three-phase EV wireless power transfer systems. The paper analyzes ferrite designs with variations in orientation, size and position. The proposed design reduces ferrite usage, lowers required excitation voltage and improves DC-DC efficiency.

Introduction

Over the last decade, electric vehicles (EVs) have become a viable option for transportation electrification where pollution from transportation emission can be massively reduced. Developing EV charging facilities is critical for resolving the drive range concern which is the largest barrier for EV market expansion. Wireless power transfer (WPT) is a contactless charging technique for EV which can minimize the electrocution risk and ensure user safety during the charging process.

Three-phase coil designs in wireless power transfer have been investigated in recent years as alternative to conventional wireless power transfer prototypes such as the circular pad (CP), bipolar pad DD pad and DDQ pad [1]. Pathmanathan et al. [2] proposed a method of designing a three-phase WPT transmitter analogous to the stator windings of a three-phase, two-pole electrical machine. Pries et al. [3] developed a three-phase inductive WPT systems with bipolar phase windings to improve power density and specific power of wireless charging systems for high-power applications.

Most existing commercial WPT three-phase pads for stationary charging use large quantities of the magnetic material ferrite [3][4]. Ferrite can significantly improve the magnetic performance and power transfer, but has the disadvantage of being brittle and can increase the system cost [5]. Mohammad et al. proposed a design and optimization method of ferrite cores for a single coil to single coil wireless charging system to improve its misalignment tolerance and minimize the core loss [6]. This paper proposes a ferrite optimization design of the three-phase stationary wireless charging system. The proposed design achieved higher DC-DC efficiency at receiver misaligned case after reducing ferrite by around 40% and lowered the required driving DC bus voltage by 62% at max in the experimental measurement.

Ferrite optimization for a three-phase transmitter

Three-phase transmitter modeling

The coil structure for a typical three-phase wireless power transfer system is presented in Fig. 1. The power is delivered by injecting AC currents into three transmitter coils (named as coil A, coil B and coil C) to generate a compound magnetic flux waveform at the receiver coil. The receiver coil picks up the magnetic flux and the resulting induced voltage can provide power to a resistor or battery load. The SAE J2954 standard states that the leakage flux outside the charging zone has to be less than 27 uT at 100 kHz to ensure user safety. As a result, shielding material such as ferrite and aluminum are added on both transmitter and receiver sides to reshape the magnetic field into an enclosed loop contained inside the charging zone.

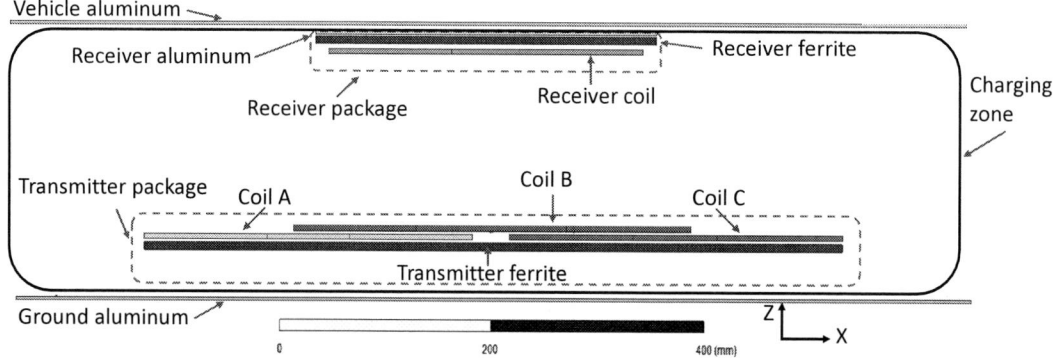

Fig. 1: Three-phase coil structure with ferrite and aluminum

The impact of ferrite on the system appears as variation in the inductance matrix. For a three-phase charging system, a 4×4 inductance matrix is constructed: (all value are in rms):

$$
\begin{bmatrix} \mathbf{V_a} \\ \mathbf{V_b} \\ \mathbf{V_c} \\ 0 \end{bmatrix} = \begin{bmatrix} Z_a & j\omega M_{ab} & j\omega M_{ac} & j\omega M_{ar} \\ j\omega M_{ba} & Z_b & j\omega M_{bc} & j\omega M_{br} \\ j\omega M_{ca} & j\omega M_{cb} & Z_c & j\omega M_{cr} \\ j\omega M_{ra} & j\omega M_{rb} & j\omega M_{rc} & Z_r + R_{L,eq} \end{bmatrix} \begin{bmatrix} \mathbf{I_a} \\ \mathbf{I_b} \\ \mathbf{I_c} \\ \mathbf{I_r} \end{bmatrix}, \quad Z_n = j\omega L_n - \frac{j}{\omega C_n} + R_n \tag{1}
$$

where R_n is the effective coil total resistance, $R_{L,eq}$ is the equivalent resistance of load, L_n is the self-inductance, M_{nm} is the mutual inductance, ω is the angular frequency of the excitation current, and C_n is the series compensation capacitance. The effective coil total resistance of each transmitter phase R_n is a sum of litz wire AC resistance at ω, additional effective resistance caused by aluminum sheet eddy current loss and additional effective resistance caused by ferrite core loss. For coil n, the coil is self-compensated where $C_n = \frac{1}{\omega^2 L_n}$. The matrix is symmetric with $M_{nm} = M_{mn}$.

Table I: Characteristics of each entries of the inductance matrix

Description	Parameter	Main system effect	Correlation
T-T mutual inductance	M_{ab}, M_{ac}, M_{bc}	excitation voltage	positive
T-R mutual inductance	M_{ar}, M_{br}, M_{cr}	system coil efficiency	positive
Self inductance	L_a, L_b, L_c	capacitor voltage stress	positive
Effective coil total resistance	R_a, R_b, R_c	system coil efficiency	negative

T-T: transmitter-transmitter, T-R: transmitter-receiver

Based on past WPT research, the characteristic of each matrix entry is summarized in the Table I. For example, from row 1, a positive correlation indicates that higher transmitter-transmitter mutual inductance requires a high excitation voltage, thus a high DC bus voltage. All the parameters in the Table I can be affected by the placement of ferrite and aluminum. As shown in the Fig. 1, the size and position of the vehicle aluminum sheet and ground aluminum sheet are fixed to contain the entire charging zone. The receiver ferrite is commonly the same size as the receiver coil with a small ferrite-coil distance. In this paper, the transmitter ferrite is optimized to determine its orientation, size and location.

Transmitter-transmitter mutual inductances and DC bus voltage

For a practical three-phase wireless power transfer application, a low excitation and DC bus voltage is required to ensure the interoperability of the three-phase wireless system and available DC power supply equipment. The DC bus voltage is mainly impacted by the transmitter-transmitter mutual inductances. By reformatting each row of (1) and substituting $C_n = \frac{1}{\omega^2 L_n}$, the transmitter voltage for each phase can be derived as:

$$\mathbf{V}_a = \mathbf{I}_a R_a + j\omega \left(M_{ar}\mathbf{I}_r + M_{ab}\mathbf{I}_b + M_{ac}\mathbf{I}_c \right) \tag{2}$$

$$\mathbf{V}_b = \mathbf{I}_b R_b + j\omega \left(M_{br}\mathbf{I}_r + M_{ba}\mathbf{I}_a + M_{bc}\mathbf{I}_c \right) \tag{3}$$

$$\mathbf{V}_c = \mathbf{I}_c R_c + j\omega \left(M_{cr}\mathbf{I}_r + M_{ca}\mathbf{I}_a + M_{cb}\mathbf{I}_b \right) \tag{4}$$

where,

$$\mathbf{V}_a = V_{a,rms}\underline{/\theta_a}, \mathbf{V}_b = V_{b,rms}\underline{/\theta_b}, \mathbf{V}_c = V_{c,rms}\underline{/\theta_c} \tag{5}$$

$$V_{dc} = \max\{V_{a,rms}, V_{b,rms}, V_{c,rms}\}\pi/\sqrt{2} \tag{6}$$

From (2)-(4), the reactive voltage of each phase is determined by the product of transmitter phase current and transmitter-transmitter mutual inductance. As transmitter-transmitter mutual inductance increases, the magnitude of each phase voltage increases. The DC bus voltage required for exciting the three-phase wireless system is determined by the maximum phase voltage among all three phases as shown in (6). Overall, large transmitter-transmitter mutual inductances lead to high phase voltages thus a high DC bus voltage.

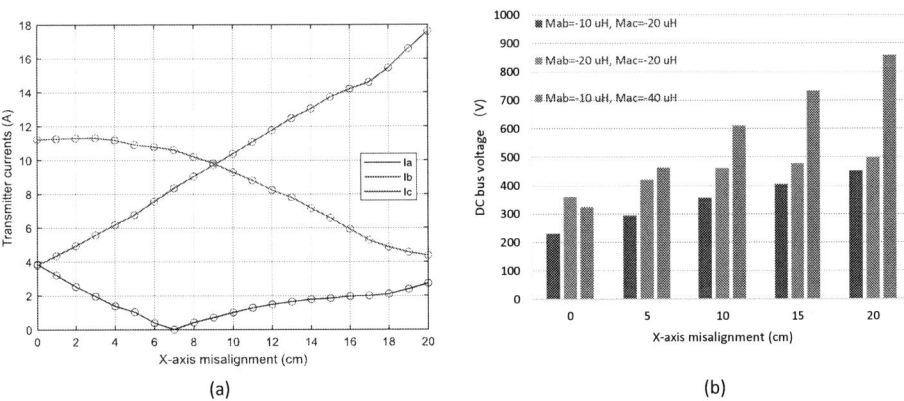

(a)   (b)

Fig. 2: a) Three-phase system transmitter current magnitude b) required DC bus voltage for different transmitter-transmitter mutual inductances under receiver X-axis misalignment

Fig. 2 (a) presents the transmitter current distribution for a typical three-phase wireless system under different receiver X-axis misalignment [7]. The transmitter current distribution is optimized to achieve the highest coil efficiency for different receiver X-axis misalignment. Fig. 2 (b) displays the required DC bus voltage for a 1 kW three-phase wireless power transfer under different receiver X-axis misalignment. The DC bus voltages are calculated based on the transmitter currents in Fig. 2 (a) and different combinations of transmitter-transmitter mutual inductances M_{ab}, M_{bc} and M_{ac}. From Fig. 1, M_{ab} equals M_{bc} due to the symmetry of three-phase coil structure. $M_{ab} = -10\,\mu H, M_{ac} = -20\,\mu H$ is selected as the base case since it is difficult to construct a perfectly decoupled transmitter pad in practice. The magnitude of M_{ab} is typically lower than the magnitude of M_{ac} since the overlapping distance among adjacent coils can be adjusted more easily whereas the side coil A,C distance is limited by the total transmitter pad size. Fig. 2 (b) presents how M_{ab} or M_{ac} impacts the DC bus voltage. By comparing the orange bar with the blue bar, it is seen that M_{ab} has a larger impact on the DC bus voltage as when the receiver is perfectly aligned. This is apparent as the DC bus voltage for the orange bar is 56% greater than the blue bar at X=0 cm but only 10% greater at X=20 cm. By comparing the green bar with the blue bar, it is clear that M_{ac} has a larger impact on the DC bus voltage when the receiver is misaligned. This can be seen from the DC bus voltage of the green bar being 90% higher than the blue bar at X=20 cm but only 40% higher at X=0 cm. It is also important to highlight that the DC bus voltage increases as receiver misalignment increases for all transmitter-transmitter mutual inductance values. The height of the blue bar is almost doubled from X=0 cm to X=20 cm receiver misalignments. Thus, in order to reduce the required DC bus voltage, both M_{ab} and M_{ac} need to be minimized by the ferrite placement optimization. In particular, M_{ac} needs to be minimized to reduce the maximum DC bus voltage which occurs at receiver extreme misalignment.

Ferrite orientation

Many WPT designs place a large ferrite pad beneath the transmitter pad to reduce the eddy current loss in the aluminum shield and increase the transmitter-receiver mutual inductances for efficiency improvement. However, a larger ferrite pad can generate high core loss in the ferrite itself due to the eddy current conducting along the internal loop. A typical method to reduce the core loss is to replace one integral ferrite pad with many small ferrite blocks. Air gap are fixed between each ferrite block which prevents the eddy current flowing and reduces the core loss.

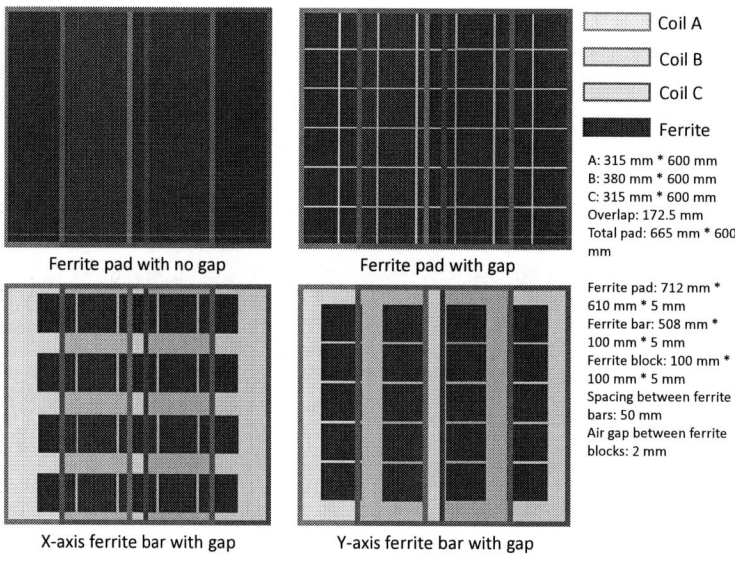

Fig. 3: Transmitter ferrite placement orientation

To optimize the ferrite placement, the first step is to determine the ferrite orientation. Fig. 3 shows the view from the top of the three-phase transmitter pad where three types of ferrite orientation are compared. The dimensions of each coil and different ferrite placements are given. The X-axis direction is defined in Fig. 1 which is the door-door direction of a vehicle while the Y-axis is the in vehicle driving direction. Typically the X-axis misalignment of the receiver is a challenge as the Y-axis misalignment can be resolved by driving the vehicle forward and backward. This paper adopts PC95 power ferrite from TDK which has a high saturation flux density 0.5 T.

In Tabel II, the effective coil total resistance at 85 kHz for each orientation with and without air gap is simulated through ANSYS Maxwell loss models. By comparing the top two rows of Table II, it is obvious that the air gap can reduce the core loss significantly. The effective coil total resistance values of the three orientations are similar since the core loss difference is minimized by the air gap.

Table II: Simulated effective coil total resistance value of different orientation

Orientation	$R_a(m\Omega)$	$R_b(m\Omega)$	$R_c(m\Omega)$
Ferrite pad (no gap)	604	950	604
Ferrite pad (with gap)	233	370	233
X-axis ferrite bar (with gap)	214	374	214
Y-axis ferrite bar (with gap)	212	344	212

The inductance matrices are derived from ANSYS Maxwell for all four types of orientations.

$$\mathbf{L}_{pad,no\,gap} = \begin{bmatrix} 478.1 & -6.9 & -62.3 & 3.5 \\ -6.9 & 539.2 & -8.1 & 26.9 \\ -62.3 & -8.1 & 472.5 & 3.5 \\ 3.5 & 26.9 & 3.5 & 183.2 \end{bmatrix} \mu H \quad \mathbf{L}_{pad,gap} = \begin{bmatrix} 425.4 & -7.2 & -37.2 & 3.1 \\ -7.2 & 475.8 & -7.5 & 23.5 \\ -37.2 & -7.5 & 422.8 & 3.1 \\ 3.1 & 23.5 & 3.1 & 182.8 \end{bmatrix} \mu H$$

$$\mathbf{L}_{Xbar,gap} = \begin{bmatrix} 329.6 & -33.5 & -46.5 & 0.4 \\ -33.5 & 421.0 & -33.1 & 20.7 \\ -46.5 & -33.1 & 328.2 & 0.4 \\ 0.4 & 20.7 & 0.4 & 185.0 \end{bmatrix} \mu H \quad \mathbf{L}_{Ybar,gap} = \begin{bmatrix} 297.8 & -23.1 & -19.8 & 1.6 \\ -23.1 & 310.0 & -23.1 & 16.5 \\ -19.8 & -23.1 & 297.8 & 1.6 \\ 1.6 & 16.5 & 1.6 & 184 \end{bmatrix} \mu H$$

By comparing the inductance matrices, it is obvious that the Y-axis ferrite bar with gap case has the advantages of reducing the transmitter-transmitter mutual inductance M_{ac}. As discussed in the previous section, the low transmitter-transmitter mutual inductance can reduce the required input DC voltage. However, the Y-axis ferrite bar with gap case decreases the transmitter-receiver mutual inductance compared with the ferrite pad with gap case. The transmitter-transmitter mutual inductance between adjacent coils, M_{ab}, M_{bc} increase due to the two center ferrite bars placed at the intersection region of the adjacent coils as shown in Fig. 3. Both the transmitter-receiver mutual inductance and the adjacent transmitter-transmitter mutual inductance can be improved by further optimization of the Y-axis ferrite bar size and location.

Ferrite orientation experimental measurement

Fig. 4 shows the three-phase transmitter pad constructed using NELD 1100/40SNSN type 2 litz wire (AWG 10) from New England Wire Technologies and the three types of ferrite orientation with gap. A single continuous piece of ferrite with sufficiently large dimensions to construct the ungapped ferrite pad is not available (as the PC95 power ferrite was only available in maximum dimensions of 100 mm * 100 mm * 5 mm). The later analysis in this manuscript is based on ferrite placements all with air gap considered. Table III displays the measured effective coil total resistances for the three types of orientation. The measured resistance values have small discrepancy with the simulated resistance values in the previous section.

Table III: Measured effective coil total resistance value of different orientation

Orientation	$R_a(m\Omega)$	$R_b(m\Omega)$	$R_c(m\Omega)$
Ferrite pad (with gap)	286	440	280
X-axis ferrite bar (with gap)	257	433	255
Y-axis ferrite bar (with gap)	273	419	263

The inductance matrices for all three types of orientation are measured by the Hioki impedance analyzer IM3570 (matrices from left to right, ferrite pad with gap, X-axis ferrite bar with gap, Y-axis ferrite bar with gap. The unit is in μH).

$$\begin{bmatrix} 436.4 & -5.4 & -41.3 & 3.5 \\ -5.4 & 492.1 & -6.1 & 24.2 \\ -41.3 & -6.1 & 435.3 & 3.3 \\ 3.5 & 24.2 & 3.3 & 171.4 \end{bmatrix} \begin{bmatrix} 336.1 & -40.2 & -53.7 & 1.2 \\ -40.2 & 430.3 & -36.8 & 21.0 \\ -53.7 & -36.8 & 335.9 & 1.1 \\ 1.2 & 22.1 & 1.1 & 170.8 \end{bmatrix} \begin{bmatrix} 297.2 & -19.8 & -20.3 & 1.9 \\ -19.8 & 318.9 & -26.0 & 17.1 \\ -20.3 & -26.0 & 299.0 & 1.7 \\ 1.9 & 17.1 & 1.7 & 171.2 \end{bmatrix}$$

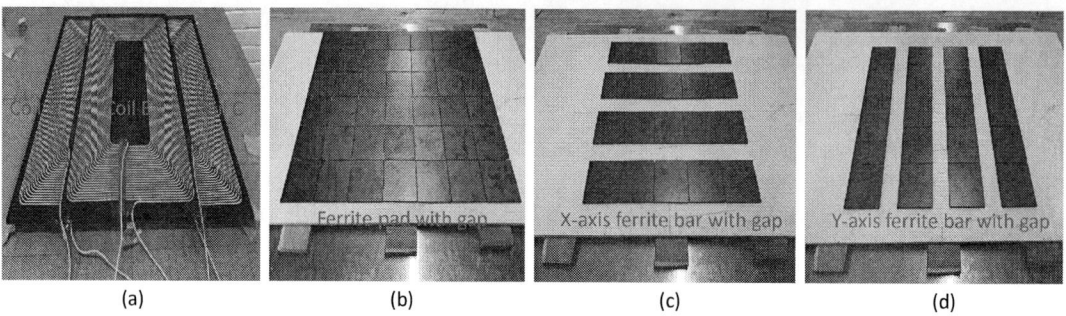

Fig. 4: (a) Three-phase transmitter coils, (b) Ferrite pad with gap, (c) X-axis ferrite bar with gap, (d) Y-axis ferrite bar with gap

The measured inductance matrices have small discrepancy with the simulated inductance matrices in the previous section. Based on the measured inductance matrices , the Y-axis ferrite bar with gap is selected as the ferrite orientation due to its ability to decrease the transmitter-transmitter mutual inductance M_{ac}.

Ferrite size and position

With the ferrite bar orientation determined to be along the Y-axis, the ferrite size and position are optimized to increase the transmitter-receiver mutual inductance and further reduce the transmitter-transmitter mutual inductances. From Fig. 2 (b), it is known that the maximum DC bus voltage required occurs at receiver extreme misalignment. Therefore, the ferrite optimization process starts by assuming receiver centered above the transmitter, and ends with receiver at extreme misalignment to ensure that the final optimized DC bus voltage does not exceed the power electronics limit. To simplify the optimization process, one ferrite bar is chosen to enhance the coupling of each coil. A 25 mm (width) * 508 mm (length) * 5 mm (thickness) thin bar is placed beneath the center of each coil, as per Fig. 5 (a). The thin bar adopt five 25 mm * 100 mm * 5 mm ferrite blocks with a 2 mm air gap between each two blocks. The ferrite width and the ferrite-coil Z distance are selected as design variables since these two variables have highest impact on the transmitter-transmitter mutual inductances M_{ab} and M_{ac}.

Six critical parameters M_{ac}, M_{ab}, V_{dc}, R_b, M_{br} and η_{coil} are plotted in Fig. 5 (b), (c), (d), (e), (f) and (g). For each combination of ferrite width and ferrite-coil Z distance, the inductance matrix and effective coil total resistances of the three-phase coil system can be derived from ANSYS Maxwell simulation. Based on the inductance matrix and effective coil total resistances, the current distribution of each phase can be obtained through the optimal excitation method in [7]. The DC bus voltage V_{dc} can be derived from the current distribution and the transmitter-transmitter mutual inductances following (2)-(6). The red plane in Fig. 5 (d) indicates the boundary of V_{dc} for which the points below the plane are considered as preferred solution range. The highest DC bus voltage preferred due to the power electronic limit is 400 V. From Fig. 2 (b), the ratio between the DC bus voltage at receiver extreme misalignment (X=20 cm) and the DC bus voltage at receiver perfect alignment (X=0 cm) varies from 1.4 to 2.6. The median 2 is selected to be the ratio between DC bus voltages at X=20 cm and X=0 cm. Therefore, to ensure that the DC bus voltage is below 400 V at X=20 cm, the DC bus voltage limit (red plane in (d)) for the X=0 cm case is set as 400/2=200 V.

Based on the derived current distribution and the simulated effective coil total resistances, the coil efficiency can be calculated as:

$$\eta_{coil} = \frac{P_{out}}{P_{out} + (I_a^2 * R_a + I_b^2 * R_b + I_c^2 * R_c + I_r^2 * R_r)} \tag{7}$$

Observing Fig.5 (d) and Fig.5 (g) together, the optimized ferrite width and ferrite-coil Z distance for the receiver centered case are determined from the point (red star in (g)) with the highest coil efficiency within the feasible range determined from figure (d). The ferrite width is selected to be 100 mm while the ferrite-coil Z distance is selected to be 10 mm.

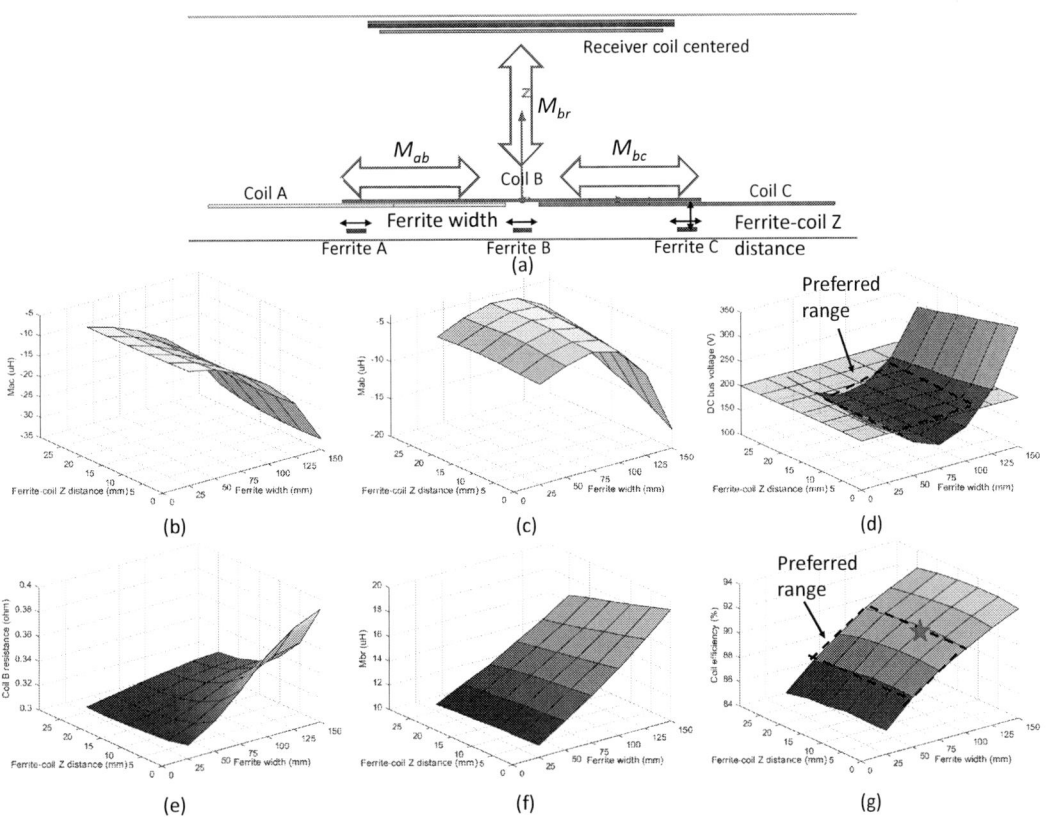

Fig. 5: (a) Three-phase coil structure as receiver coil centered, (b) M_{ac}, (c) M_{ab}, (d) V_{dc}, (e) R_b, (f) M_{br}, (g) η_{coil}

From the above design, the optimal ferrite width for all ferrites (A, B and C) and ferrite-coil Z distance are determined under receiver perfect alignment case. For the receiver extreme misalignment case (X=20 cm), the three-phase system performs similar to the single coil system where the side coil (A or C) is conducting the highest current as shown in Fig. 2 (a). Therefore, ferrite A and C ferrite can be re-optimized to enhance the performance of side coil when receiver misaligned as shown in Fig.6 (a). Width of ferrite A,C and ferrite-ferrite center distances are selected as design variables. To ensure symmetry of the three-phase coil system, the width of ferrite A is equal to the width of ferrite C and the ferrite A-B center distance is equal to the ferrite B-C center distance. Based on the previous optimization result, width of ferrite B remains constant as 100 mm. Ferrite-coil Z distance for all ferrites remain constant as 10 mm.

Side ferrite A and C are optimized by plotting M_{ac}, M_{ab}, V_{dc}, R_a, M_{ar} and η_{coil} with respect to ferrite A,C width and ferrite A-B(B-C) center distance in Fig. 6. The red plane in Fig. 6 (d) indicates the boundary of V_{dc} for which the points below the plane are considered as preferred solution range. As mentioned above, the maximum DC bus voltage due to power electronics limit is 400 V. Therefore, the red plane is set as 400 V to filter out the combinations of ferrite A,C width and ferrite A-B(B-C) center distance which results in a high DC bus voltage. Fig. 6 (g) presents the coil efficiency at receiver misaligned case derived from (7). From the feasible range determined in Fig. 6 (d), the same range can be plotted in Fig. 6 (g). The red star with the highest coil efficiency in the preferred range is selected to be the optimal point.

The final ferrite width are 200 mm for ferrite A, 100 mm for ferrite B and 200 mm for ferrite C. The final ferrite-ferrite center distance is 200 mm for A-B and 200 mm for B-C. The final ferrite-coil Z distance is 10 mm for all three ferrite bars.

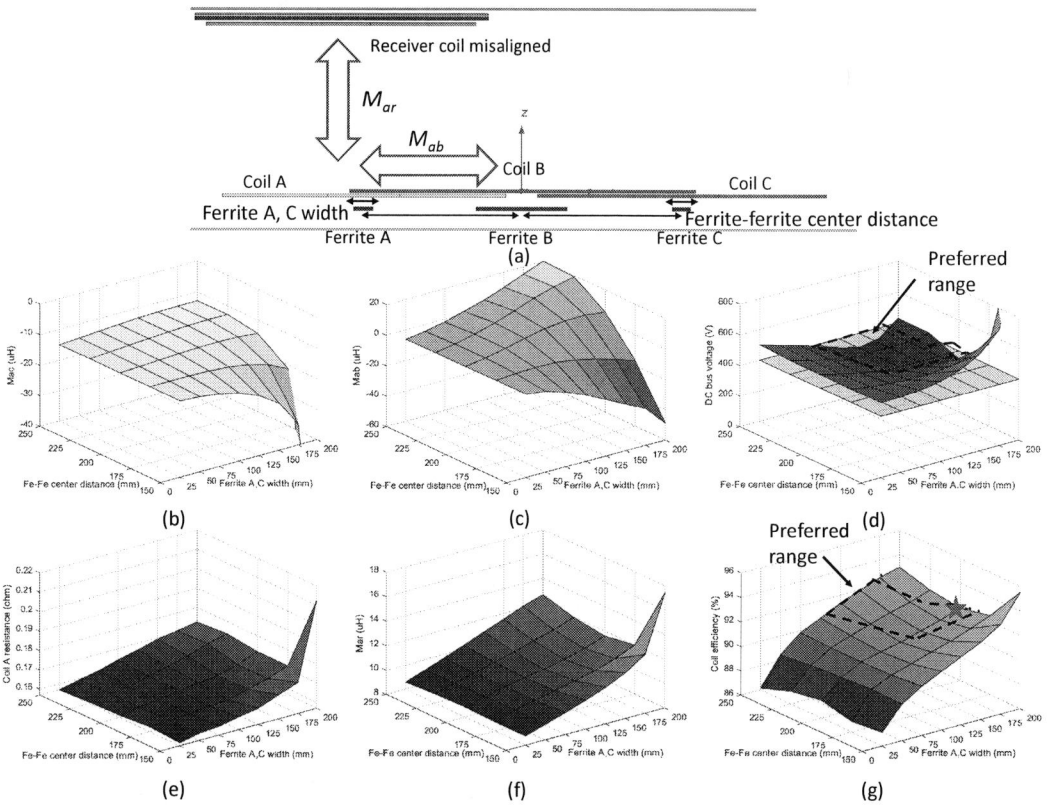

Fig. 6: (a) Three-phase coil structure as receiver coil misaligned to X=20 cm, (b) M_{ac}, (c) M_{ab}, (d) V_{dc}, (e) R_a, (f) M_{ar}, (g) η_{coil}

Ferrite size and position experimental measurement

Five different pairs of ferrite width and ferrite-coil Z distance are selected to validate the simulation result shown in Fig. 5. Table IV displays the measured inductance matrix values and effective coil total resistances for the selected pairs. The comparison among pairs P1, P2 and P3 indicates how the inductance matrix values and effective coil total resistances vary with ferrite width when the ferrite-coil Z distance remain constant. The comparison among P2, P4 and P5 indicates how the inductance matrix values and effective coil total resistances vary with the ferrite-coil Z distance when the ferrite width remain constant. The measured inductance and resistance values matches the simulated result in Fig. 5 within an acceptable error. The last two columns of Table IV display the estimated DC bus voltage and coil efficiency derived from the measured inductance matrix and effective coil total resistances in Table IV. Among all the pairs P1-P5, P3 is disqualified due to the large V_{dc} values which exceeded the preferred solution range $V_{dc} \leq 200V$. By comparing remaining pairs, P5 is selected to be the optimal pair since it has the highest coil efficiency. Therefore, the optimal ferrite width is 100 mm while the optimal ferrite-coil Z distance is 10 mm.

Based from the result in the last two columns of Table IV, the width of ferrite B is set to 100 mm and the ferrite-coil Z distance is set to 10 mm. Five different pairs of ferrite A,C width and ferrite-ferrite center distance are selected to validate the simulation result shown in Fig. 6. Table V displays the measured inductance matrix values and effective coil total resistances for the considered pairs. The comparison among P6, P7 and P8 indicates how the inductance matrix values and effective coil total resistances vary with ferrite A,C width when the ferrite-ferrite center distance remain constant. The comparison among P8, P9 and P10 indicates how the inductance matrix values vary with the ferrite-ferrite center distance when the ferrite A,C width remain constant. The measured values matches the simulated result

Table IV: Experimental measurement for ferrite B optimization

Pair	FW (mm)	FCZD (mm)	$M_{ab}(\mu h)$	$M_{ac}(\mu h)$	$M_{br}(\mu h)$	$R_b(mohm)$	$V_{dc}(V)$	$\eta_{coil}(\%)$
P1	50	0	-4.87	-9.03	11.57	342	172	88.27
P2	100	0	-5.06	-16.19	15.26	389	173	90.21
P3	150	0	-17.75	-39.77	18.45	441	392	91.83
P4	100	25	-4.61	-15.65	13.76	362	186	89.09
P5	100	10	-4.90	-16.42	14.63	336	178	90.68

FW: ferrite width, FCZD: ferrite-coil Z distance

Table V: Experimental measurement for ferrite A,C optimization

Pair	FW2 (mm)	FCD (mm)	$M_{ab}(\mu h)$	$M_{ac}(\mu h)$	$M_{ar}(\mu h)$	$R_a(mohm)$	$V_{dc}(V)$	$\eta_{coil}(\%)$
P6	100	175	-5.21	-19.51	10.69	162	542	91.27
P7	150	175	-12.80	-20.08	11.92	170	565	91.08
P8	200	175	-16.78	-24.56	13.05	175	616	93.29
P9	200	200	4.23	-16.78	12.76	174	382	93.01
P10	200	225	22.02	-15.64	13.09	177	456	92.94

FW2: ferrite A,C width, FCD: ferrite-ferrite center distance

displayed in Fig. 6 within an acceptable error. The last two columns of Table V display the estimated DC bus voltage and coil efficiency derived from the measured inductance matrix and effective coil total resistances in Table V. Among all the pairs P6-P10, only P9 is qualified due to its low DC bus voltage which is within the preferred solution range $V_{dc} \leq 400V$. Therefore, the optimal ferrite A,C width is 200 mm while the optimal ferrite-ferrite center distance is 200 mm.

Three-phase wireless system performance with optimized ferrite

In this section, the experimental measurements are completed for the ferrite pad with gap case and the optimized Y-axis ferrite bar with gap case under 1 kW output power, 300 V battery voltage and 85 kHz operating frequency. The ferrite pad with gap case is selected as benchmark to evaluate the performance of the optimized Y-axis ferrite bar with gap case regarding DC bus voltage and system efficiency. In particular, both the coil efficiency without power electronic losses and the DC-DC efficiency considering power electronic losses are measured to demonstrate the impact of ferrite placement towards power electronic losses. Fig. 7 shows the power electronic topology and the experiment setup for the three-phase wireless power transfer system.

Fig. 8 shows the waveform of the transmitter voltage and current of the three phase system using a full ferrite pad with gap at 5 cm receiver misalignment. Fig. 9 shows the waveform of the transmitter voltage and current of the three phase system using the optimized Y-axis ferrite bar with gap at 5 cm receiver misalignment. The yellow square voltage waveform in Fig. 8 and Fig. 9 is the voltage of phase B. By comparing the height of the yellow square voltage, it is obvious that the DC bus voltage required by the ferrite pad with gap case is higher than the optimized Y-axis ferrite bar with gap case. It is also important to notice that the optimized Y-axis ferrite bar with gap case can achieve soft switching while the ferrite bar with gap case is hard switching.

The required DC bus voltage for the 1 kW output power for the ferrite pad with gap case and optimized Y-axis ferrite bar with gap case are plotted in Fig. 10 left figure as a function of receiver misalignment. Due to the limitation of accessible DC bus voltage, the experiment validation is completed for receiver misalignment at 0 cm, 5 cm and 10 cm. The peak DC bus voltage difference is around 360 V at receiver 10 cm misalignment. The simulated, measured coil efficiencies and measured DC-DC efficiencies for the ferrite pad with gap case and optimized Y-axis ferrite bar with gap case are plotted in Fig. 10 right figure as a function of receiver misalignment. For coil efficiency, the ferrite pad with gap case is around 2.8% higher than the optimized Y-axis ferrite bar with gap case at 0 cm receiver misalignment since

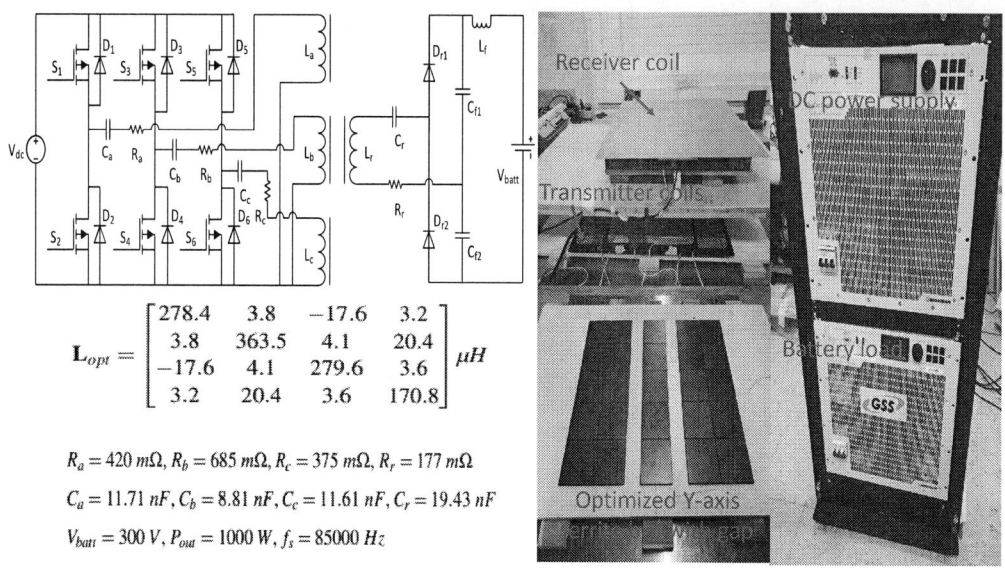

Fig. 7: Power electronic topology and experiment setup for the three-phase wireless system

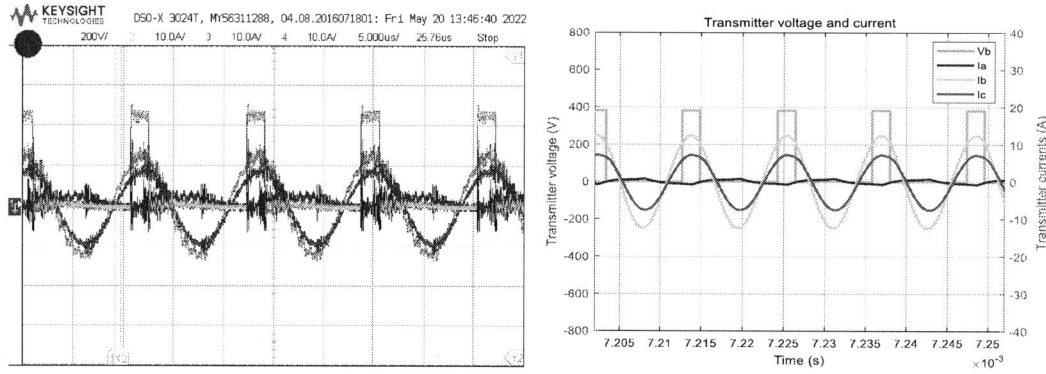

Fig. 8: Transmitter voltage and current for ferrite pad with gap at 5 cm receiver misalignment, oscilloscope measurement (left), simulation (right).

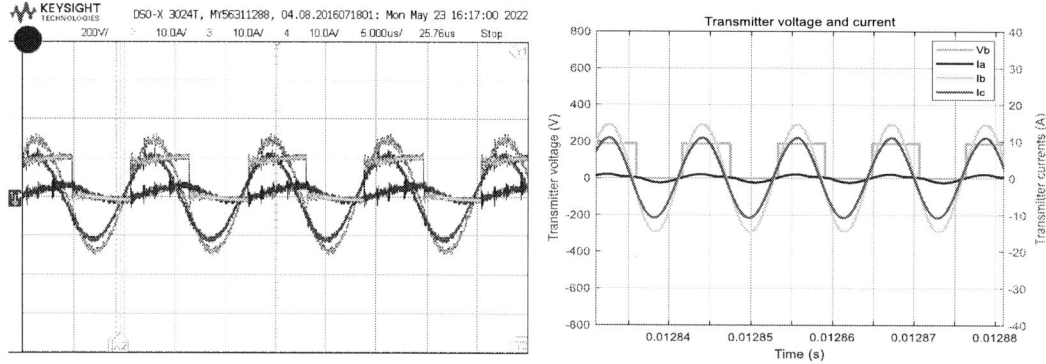

Fig. 9: Transmitter voltage and current for optimized Y-axis ferrite bar with gap at 5 cm receiver misalignment, oscilloscope measurement (left), simulation (right).

it has larger transmitter-receiver mutual inductances. However, if taking the power electronic loss into account, the DC-DC efficiency of the ferrite pad with gap case drops dramatically as receiver misaligned

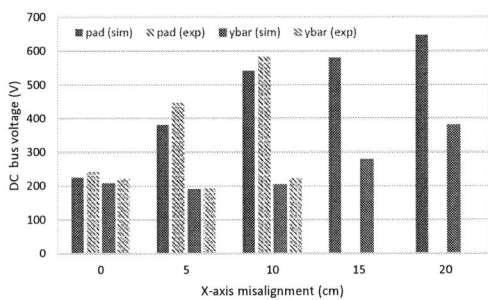 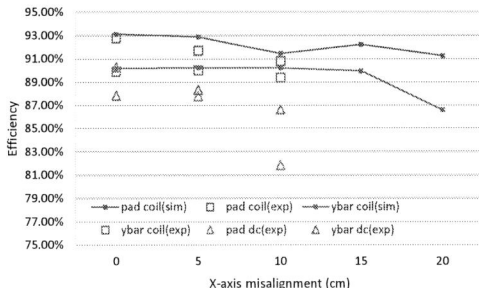

Fig. 10: Left figure: DC bus voltage for ferrite pad with gap and optimized Y-axis ferrite bar with gap at different receiver X-axis misalignment, right figure: efficiency for ferrite pad with gap and optimized Y-axis ferrite bar with gap at different receiver X-axis misalignment .

due to hard switching as shown in Fig. 8. The DC-DC efficiency of ferrite pad with gap case is around 4.8 % lower than the optimized Y-axis ferrite bar with gap case at 10 cm receiver misalignment.

Conclusion

This paper compares the ferrite placement in the aspects of orientation, size and position based on their impact to the inductance matrix. The proposed ferrite design reduces the ferrite usage by 40% but is capable of achieving higher DC-DC efficiency at receiver misalignment case and lowers the DC bus voltage by 62% at max.

References

[1] H. Feng, R. Tavakoli, O. C. Onar and Z. Pantic, "Advances in High-Power Wireless Charging Systems: Overview and Design Considerations," in IEEE Transactions on Transportation Electrification, vol. 6, no. 3, pp. 886-919, Sept. 2020

[2] M. Pathmanathan, S. Nie, N. Yakop and P. W. Lehn, "Field-Oriented Control of a Three-Phase Wireless Power Transfer System Transmitter," in IEEE Transactions on Transportation Electrification, vol. 5, no. 4, pp. 1015-1026, Dec. 2019

[3] J. Pries, V. P. N. Galigekere, O. C. Onar and G. Su, "A 50-kW Three-Phase Wireless Power Transfer System Using Bipolar Windings and Series Resonant Networks for Rotating Magnetic Fields," in IEEE Transactions on Power Electronics, vol. 35, no. 5, pp. 4500-4517, May 2020

[4] A. U. Ibrahim, W. Zhong and M. D. Xu, "A 50-kW Three-Channel Wireless Power Transfer System With Low Stray Magnetic Field," in IEEE Transactions on Power Electronics, vol. 36, no. 9, pp. 9941-9954, Sept. 2021

[5] M. G. S. Pearce, G. A. Covic and J. T. Boys, "Reduced Ferrite Double D Pad for Roadway IPT Applications," in IEEE Transactions on Power Electronics, vol. 36, no. 5, pp. 5055-5068, May 2021

[6] M. Mohammad, S. Choi, Z. Islam, S. Kwak and J. Baek, "Core Design and Optimization for Better Misalignment Tolerance and Higher Range of Wireless Charging of PHEV," in IEEE Transactions on Transportation Electrification, vol. 3, no. 2, pp. 445-453, June 2017

[7] S. Nie, M. Pathmanathan, N. Yakop, Z. Luo, and P. W. Lehn, "Field orientation based three-coil decoupled wireless transmitter for electric vehicle charging with large lateral receiver misalignment tolerance," IET Power Electronics, vol. 14, no. 5, pp. 946–957.

Frequency and Modulation Index Related Effects in Continuous and Discontinuous Modulated Y-Inverter for Motor-Drive Applications

Hamzeh J. Jaber, and Alberto Castellazzi
Kyoto University of Advanced Science
Kyoto, Japan
hamzeh.j.jaber@kuas.ac.jp

Acknowledgments

The authors gratefully acknowledge the support received from Grant 20H02138 of the Japanese Society for the Promotion of Science (JSPS) under the Kakenhi-Kiban-B scheme.

Keywords

≪DC-AC converters ≫, ≪Discontinuous pulse-width modulation≫, ≪GaN transistors≫.

Abstract

The effect of the pulse-width modulation scheme on the efficiency of a three-phase Y-inverter with a wide range of output voltages and currents is investigated experimentally in this paper. The efficiency measurements obtained while conducting experiments on a GaN-based Y-inverter prototype indicate that no pulse-width modulation scheme results in higher efficiency of the Y-inverter over the entire range of motor operation. In other words, depending on the operating conditions, a discontinuous PWM results in a higher or lower efficiency than a continuous sinusoidal PWM. As a result, there is a chance that using a hybrid modulation strategy will improve the efficiency of the Y-inverter. Furthermore, the distortion effect caused by inappropriate inductor and capacitor value selection is highlighted.

Introduction

Higher rotation speed machines with higher pole numbers and smaller size for the same torque ratings will be required to advance the state-of-the-art in electric motor drives. Such requirements translate primarily into the need for higher switching frequency solutions in the electrical power converter, but clearly without sacrificing efficiency. Wide-band-gap (WBG) semiconductors are viewed as key enabling technologies of improvement from this perspective. However, it is now widely accepted that simply replacing silicon (Si) transistors with silicon carbide (SiC) or gallium nitride (GaN) transistors yields incremental benefits but does not fully justify the higher cost of the technology. The following are bottlenecks to fully utilizing WBG semiconductors in conventional power conversion solutions:

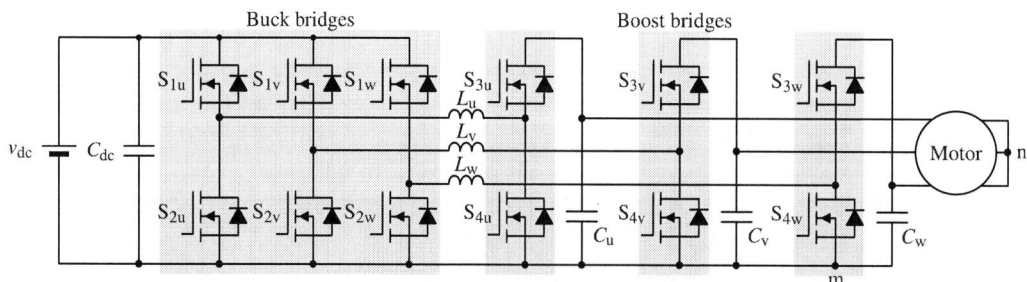

Fig. 1: Circuit configuration of a three-phase buck-boost Y-inverter.

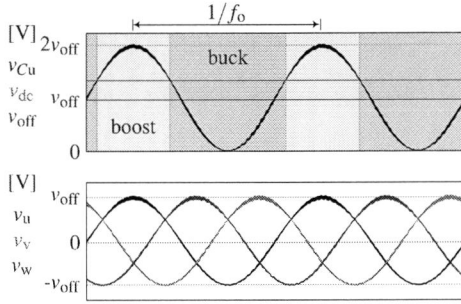
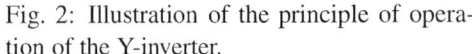
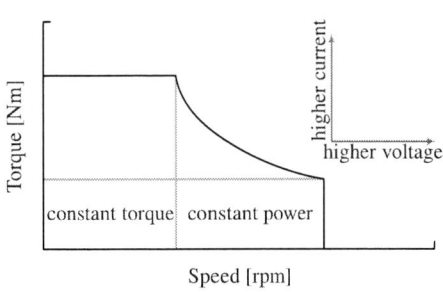

Fig. 2: Illustration of the principle of operation of the Y-inverter.

Fig. 3: Torque-speed curve of a permanent magnet synchronous motor.

1. High dv/dt: When going from Si to SiC or GaN, switching transitions can easily be faster by a factor of 10. The much higher dv/dt values pose critical limitations to both functionality (e.g., common mode noise in parasitic capacitance, voltage overshoots, and wave reflections in cables and connectors), and reliability (e.g., degradation of motor winding insulation, and corrosion of motor bearings due to parasitic current flow),

2. Electromagnetic emission: The increase in the switching frequency that can be achieved in established solutions without sacrificing efficiency is closer to a factor of 4 to 5, which means that the spectrum of electromagnetic emission is shifted by the same amount towards higher frequencies. This makes meeting typical emission requirements, which typically impose lower frequency limits, extremely difficult.

To overcome the limitations mentioned above, multilevel inversion has been proposed as an evolutionary solution. Indeed, 3 or 5 level inversion can help to improve voltage inversion performance to some extent. However, the large number of components of the multilevel inverters makes it difficult for such approaches to gain widespread acceptance for use with WBG devices.

Recently, new topologies of two-level inverters have been proposed to exploit the high switching frequency capability of the WBG devices. Specifically, two approaches have been pursued; current source inversion [1, 2] and Y-inversion [3, 4]. The latter is derived from a well-known concept of creating an inverter from 3 Y-connected dc-dc converters, but has recently gained significant momentum, in particular by its implementation as a non-inverting buck-boost converter (4-switch cell). Both topologies do away with the above-mentioned issues, allowing the semiconductor devices to operate at significantly higher switching frequencies. Moreover, both offer built-in voltage boosting capability, so that no boost chopper or dc-link stage are required. They are thus ideal for high-performance motor-drive systems. This paper focuses on the three-phase buck-boost Y-inverter proposed in [3]. In particular, this paper presents an experimental investigation of the impact of the pulse-width modulation (PWM) strategy on the efficiency of a three-phase Y-inverter with a wide range of output voltages and currents.

Three-phase buck-boost Y-inverter

Circuit configuration

Fig. 1 shows the circuit configuration of the three-phase buck-boost Y-inverter (hereinafter referred to as the Y-inverter). Here, v_{dc} represents the dc voltage source, and C_{dc} is the input-side filter capacitor. S_{1u} and S_{2u} are the switching devices of the buck bridge of phase u, S_{1v} and S_{2v} are the switching devices of the buck bridge of phase v, S_{1w} and S_{2w} are the switching devices of the buck bridge of phase w, S_{3u} and S_{4u} are the switching devices of the boost bridge of phase u, S_{3v} and S_{4v} are the switching devices of the boost bridge of phase v, and S_{3w} and S_{4w} are the switching devices of the boost bridge of phase w. L_u, L_v, and L_w are the inductors connected between the middle points of the buck bridges and the boost bridges of phase u, v, and w, respectively. C_u, C_v, and C_w are the output capacitors of phase u, v, and w, respectively. m is the ground, and n is the neutral point of the Y-connected load.

Table I: Experimental circuit parameters used for efficiency measurements.

Parameter	Symbol	Value
Input dc voltage	v_{dc}	140 V
Inductance	L	13.6 µH
Capacitance	C	6 µF
Switching frequency	f_{sw}	100 kHz

Table II: Experimental circuit parameters that corresponds to Fig. 14.

Parameter	Symbol	Value
Input dc voltage	v_{dc}	60 V
Inductance	L	20 µH
Capacitance	C	4.7 µF
Switching frequency	f_{sw}	150 kHz

Principle of operation

The Y-inverter consists of three non-inverting buck-boost converters (NIBB). Each NIBB produces a strictly positive voltage, where the duty ratios of the buck and the boost bridges can be appropriately adjusted to produce a sine-shaped voltage as shown in Fig. 2. Here, v_{Cu} is the voltage across C_u which has a sinusoidal shape with a frequency of f_o, an offset voltage of v_{off}, and a peak-to-peak voltage of $2v_{off}$. The NIBB of phase u operates in the buck mode when $v_{dc} > v_{Cu}$, and in boost mode when $v_{dc} < v_{Cu}$. Similarly, the NIBBs of phase v and phase w generates v_{Cv} and v_{Cw}, respectively, where v_{Cu}, v_{Cv}, and v_{Cw} are $2\pi/3$ radians apart. v_{off} is eliminated from the phase-to-neutral voltages of the Y-connected load.

Experimental investigation of the effect of modulation scheme on efficiency and voltage quality

Fig. 3 shows a typical torque-speed curve of permanent magnet synchronous motor (PMSM). For a wide range of operations, the motor voltages and currents can vary significantly in which a relatively high voltage (corresponds to a relatively low current) and a relatively low voltage (corresponds to a relatively high current) are required for constant-power operation. Therefore, the Y-inverter has to operate under these conditions, where it may be required to produce a peak-to-peak load voltage below or above the dc source voltage.

In [3], sinusoidal PWM (SPWM) and discontinuous PWM (DPWM) schemes were employed for the Y-inverter. It is shown in [3] that the DPWM results in lower switching losses. However, a more careful investigation should be conducted to evaluate the effect of the modulation scheme on the efficiency. For this purpose, experiments were performed on a 3-kW GaN-based Y-inverter prototype to measure the efficiency under various operating conditions.

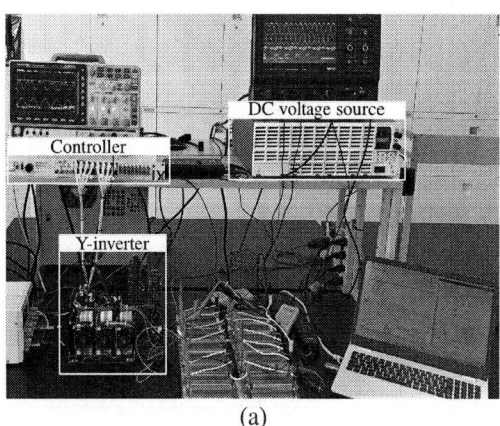
(a)

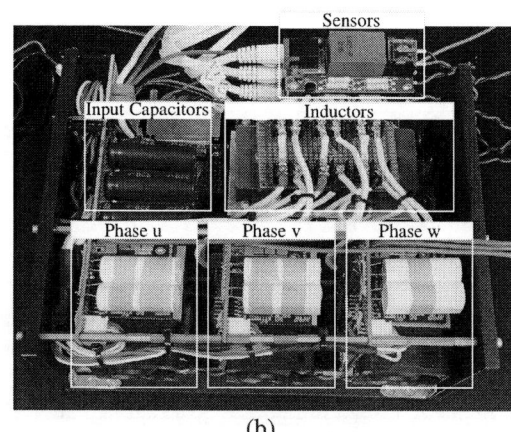
(b)

Fig. 4: Photograph of (a) the experimental setup and (b) the Y-inverter prototype.

Table III: Measured efficiency of the Y-inverter prototype under various operating conditions.

Load RMS voltage	Load voltage frequency	Output power	Modulation scheme	Efficiency
31.8 V	500 Hz	599 W	SPWM	93.97%
			DPWM	92.54%
		1184 W	SPWM	97.35%
			DPWM	94.41%
49.5 V	1.0 kHz	585 W	SPWM	95.01%
			DPWM	93.27%
		1081 W	SPWM	99.11%
			DPWM	98.73%
74.2 V	1.5 kHz	601 W	SPWM	87.74%
			DPWM	90.13%
		1316 W	SPWM	93.83%
			DPWM	94.39%

Experimental conditions

Table I summarizes the experimental circuit parameters used for efficiency measurements, where the experimental setup is shown in Fig. 4 (a), and the Y-inverter prototype is shown in Fig. 4 (b) in which 650-V, 25 mΩ GaN transistors (GS66516B) from GaN Systems are used. Each experiment corresponds to a different set of output capacitor peak voltage, $\hat{v}_C$, frequency, f_o, load output power, P_o, and modulation scheme. The following experiments were carried out.

1. $\hat{v}_C = 90$ V, $f_o = 500$ Hz, $P_o = 599$ W, and both SPWM and DPWM were used.
2. $\hat{v}_C = 90$ V, $f_o = 500$ Hz, $P_o = 1184$ W, and both SPWM and DPWM were used.
3. $\hat{v}_C = 140$ V, $f_o = 1.0$ kHz, $P_o = 585$ W, and both SPWM and DPWM were used.
4. $\hat{v}_C = 140$ V, $f_o = 1.0$ kHz, $P_o = 1081$ W, and both SPWM and DPWM were used.
5. $\hat{v}_C = 210$ V, $f_o = 1.5$ kHz, $P_o = 601$ W, and both SPWM and DPWM were used.
6. $\hat{v}_C = 210$ V, $f_o = 1.5$ kHz, $P_o = 1316$ W, and both SPWM and DPWM were used.

In Fig 5 to Fig 12, the waveforms of d_{u1}, d_{u2}, v_{Cu}, and v_{dc}, are captured by GW Instek MDO-2204EX oscilloscope, and the waveforms of v_u, v_v, v_w, and i_{Lw}, are captured by WaveSurfer 3104z Teledyne LeCroy oscilloscope. where d_{u1} and d_{u2} are the duty ratios of S_{1u} and S_{3u}, respectively, and v_u, v_v, v_w, are the phase-to-neutral voltages of phase u, phase v, and phase w, respectively.

Micsig DP10013 voltage probe was used to measure v_u, Ivytech P5205A voltage probes were used to measure v_v and v_w, Tektronix TRCP0300 current probe was used to measure i_{Lw}, and GW Instek GTP-200B-4 voltage probes were used to measure v_{Cu}, and v_{dc}. Imperix B-Box RCP 3.0 was used for control and PWM generation.

Experimental results

The operating conditions of the experiments performed on the Y-inverter prototype are summarized in Table III along with the corresponding measured efficiency values. Fig.5, Fig.6, Fig.7, and Fig.8 show the experimental waveforms when the Y-inverter operated in the buck-only mode, whereas Fig.9, Fig.10, Fig.11, and Fig.12 show the experimental waveforms when the Y-inverter operated in the buck-boost mode. Fig. 13 shows the measured efficiency values for different load RMS voltages where the data points in Fig. 13 (a) and Fig. 13 (b) are grouped based on the output power values, which are relatively close to each other in each group.

The boundary between the buck-only and the buck-boost modes corresponds to operating conditions where $v_{dc} = \hat{v}_C$. These operating conditions are depicted in the Table III, and correspond to the data where the load RMS voltage is 49.5 V ($v_{dc} = \hat{v}_C = 140$ V). The data in Table III and Fig. 13 shows that the SPWM results in higher efficiency when the Y-inverter operates in the buck-only operation, while DPWM results in higher efficiency when the Y-inverter operates in the buck-boost operation.

Frequency and modulation index related effects in continuous and discontinuous modulated Y-Inverter for motor-drive applications

JABER Hamzeh J.

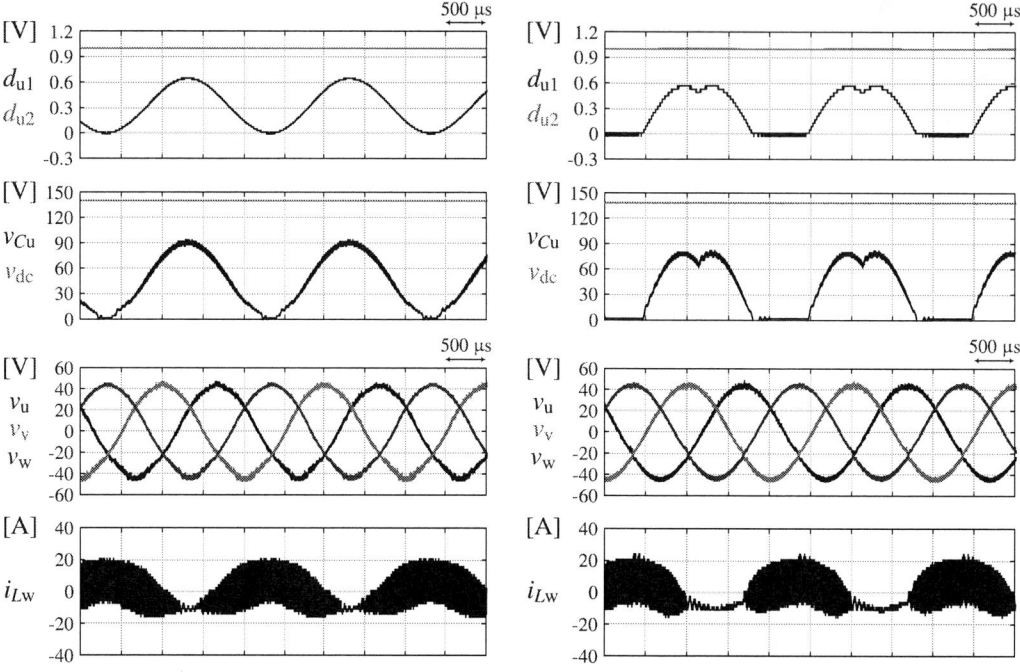

Fig. 5: Experimental waveforms when $\hat{v}_C = 90\,\text{V}$, $f_o = 500\,\text{Hz}$, $P_o = 599\,\text{W}$, and SPWM is used.

Fig. 6: Experimental waveforms when $\hat{v}_C = 90\,\text{V}$, $f_o = 500\,\text{Hz}$, $P_o = 599\,\text{W}$, and DPWM is used.

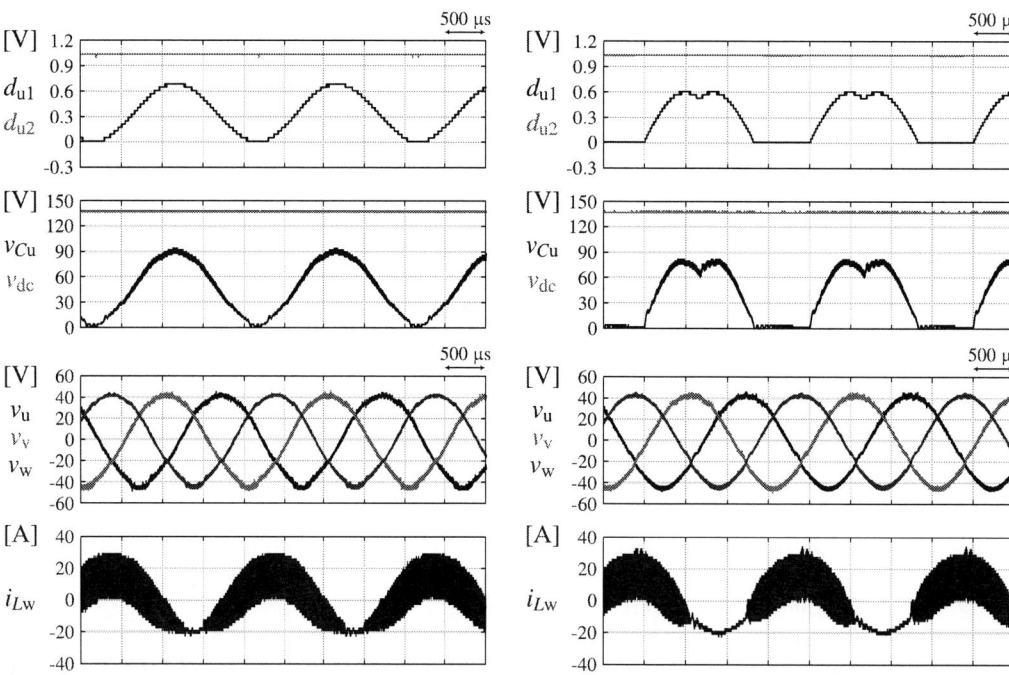

Fig. 7: Experimental waveforms when $\hat{v}_C = 90\,\text{V}$, $f_o = 500\,\text{Hz}$, $P_o = 1184\,\text{W}$, and SPWM is used.

Fig. 8: Experimental waveforms when $\hat{v}_C = 90\,\text{V}$, $f_o = 500\,\text{Hz}$, $P_o = 1184\,\text{W}$, and DPWM is used.

EPE'22 ECCE Europe

Frequency and modulation index related effects in continuous and discontinuous modulated Y-Inverter for motor-drive applications

JABER Hamzeh J.

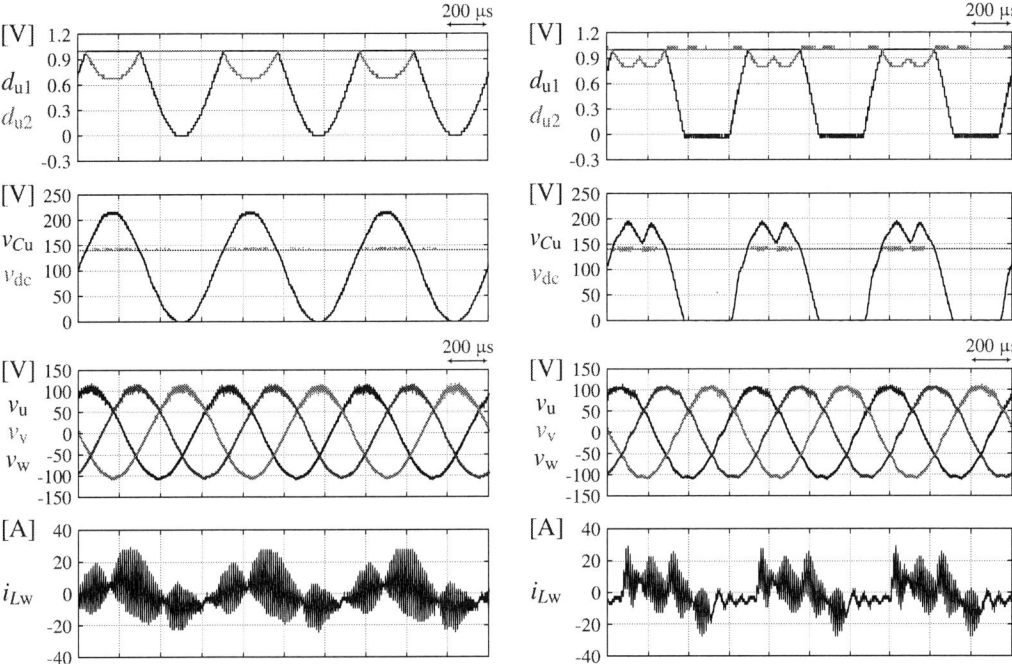

Fig. 9: Experimental waveforms when $\hat{v}_C = 210\,\text{V}$, $f_\text{o} = 1.5\,\text{kHz}$, $P_\text{o} = 601\,\text{W}$, and SPWM is used.

Fig. 10: Experimental waveforms when $\hat{v}_C = 210\,\text{V}$, $f_\text{o} = 1.5\,\text{kHz}$, $P_\text{o} = 601\,\text{W}$, and DPWM is used.

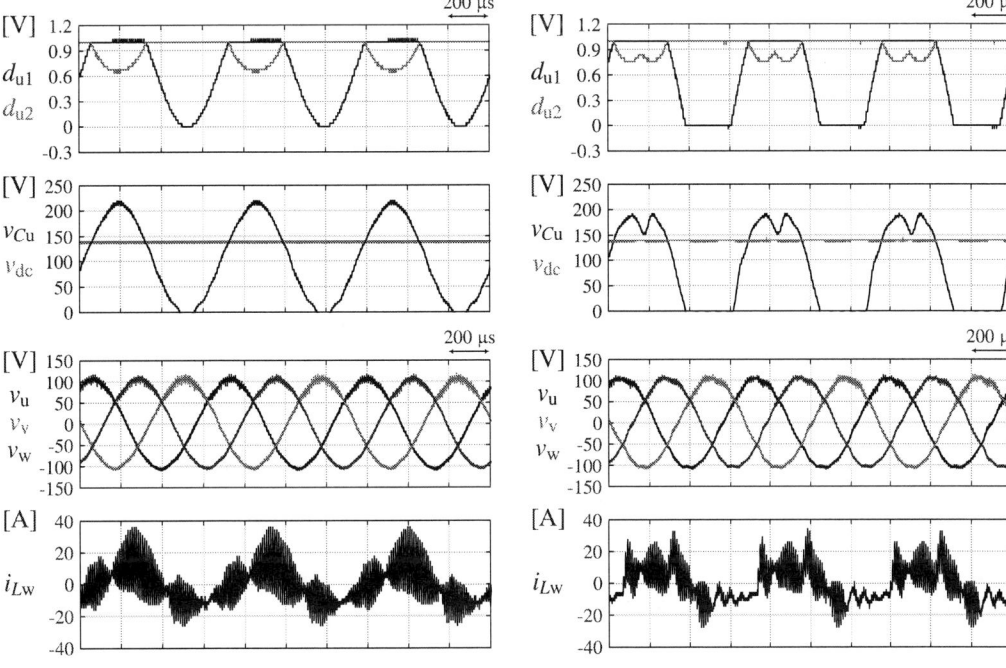

Fig. 11: Experimental waveforms when $\hat{v}_C = 210\,\text{V}$, $f_\text{o} = 1.5\,\text{kHz}$, $P_\text{o} = 1316\,\text{W}$, and SPWM is used.

Fig. 12: Experimental waveforms when $\hat{v}_C = 210\,\text{V}$, $f_\text{o} = 1.5\,\text{kHz}$, $P_\text{o} = 1316\,\text{W}$, and DPWM is used.

EPE'22 ECCE Europe

Frequency and modulation index related effects in continuous and discontinuous modulated Y-Inverter for motor-drive applications   JABER Hamzeh J.

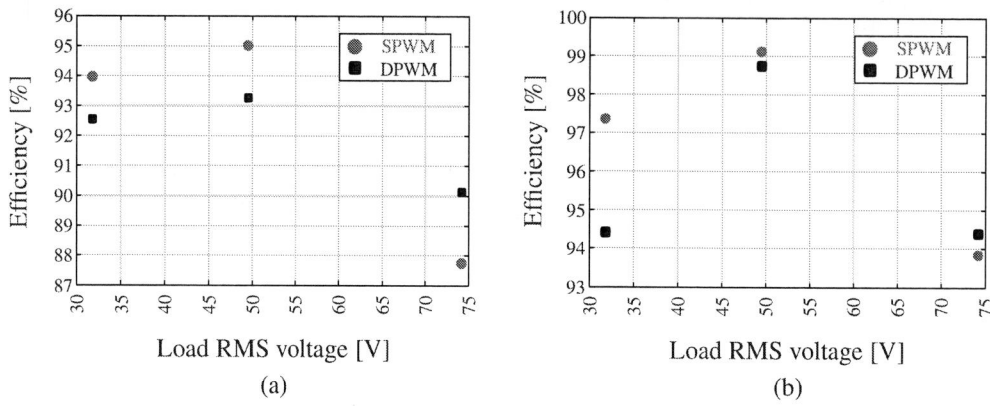

(a)   (b)

Fig. 13: Visualization of Table III data where (a) corresponds to lower power conditions, and (b) corresponds to higher power conditions.

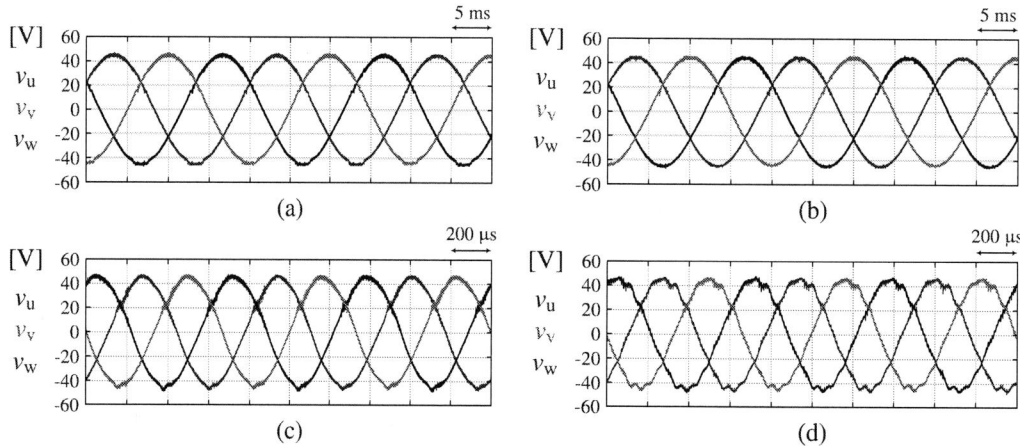

(a)   (b)

(c)   (d)

Fig. 14: Experimental phase-to-neutral voltages (a) when $f_o = 50\,\text{Hz}$, and SPWM is used, (b) when $f_o = 50\,\text{Hz}$, and DPWM is used, (c) when $f_o = 1.5\,\text{kHz}$, and SPWM is used, and (d) when $f_o = 1.5\,\text{kHz}$, and DPWM is used.

Although the DPWM reduces the switching losses, other losses appear to outweigh the reduction of the switching losses during the buck-only operation. These results show that there is a potential for the development of a hybrid PWM scheme in which SPWM is used during the buck-only operation and DPWM is used during the buck-boost operation. A detailed theoretical analysis of the losses should be done and it is left for future work.

Besides the effect of the modulation scheme on efficiency, the effect on the voltage quality has been investigated. Carefully looking at v_{Cu} in Fig. 10 and Fig. 12, a distortion in f_o voltage component can be observed. To further investigate the effect of f_o and the modulation scheme on the load voltage quality, the phase-to-neutral voltages were measured for relatively low and relatively high values of f_o and for both DPWM and SPWM.

Fig. 14 shows the phase-to-neutral voltages, where the corresponding circuit parameters are shown in Table II. In Fig. 14 (a) and Fig. 14 (c) SPWM is used, while in Fig. 14 (b) and Fig. 14 (d) DPWM is used. In Fig. 14 (a) and Fig. 14 (b), $f_o = 50\,\text{Hz}$, while in Fig. 14 (c) and Fig. 14 (d), $f_o = 1.5\,\text{kHz}$. As shown in Fig. 14, as the f_o increases, more pronounced distortion of the output waveforms is detected. Clearly, the value of f_o where distortion becomes non-negligible depends on both the switching frequency and the values of L and C used in the design of the power cell. These findings have been confirmed by simulations

EPE'22 ECCE Europe   2162

as well, although not shown in this paper. Taking the aforementioned findings into consideration, it is possible to conclude that the switching frequency cannot simply be increased further, and that an appropriate selection of the L and C values should be considered

Conclusion

This paper has highlighted differences in the use of SPWM and DPWM schemes at the boundary of the operational regimes of the Y-Inverter in motor-drive applications. The results indicate that inverter performance optimization should include the possibility of hybrid modulation schemes (i.e., applying DPWM when the motor operates at low-torque high-speed conditions and SPWM at low-speed high-torque conditions). The development of an algorithm that enables automatic selection of the modulation strategy depending on the instantaneous operational condition is under study.

References

[1] R. Amorim Torres, H. Dai, W. Lee, B. Sarlioglu and T. Jahns, "Current-Source Inverter Integrated Motor Drives Using Dual-Gate Four-Quadrant Wide-Bandgap Power Switches," in IEEE Transactions on Industry Applications, vol. 57, no. 5, pp. 5183-5198, Sept.-Oct. 2021, doi: 10.1109/TIA.2021.3096179.

[2] R. A. Torres, H. Dai, W. Lee, T. M. Jahns and B. Sarlioglu, "Current-Source Inverters for Integrated Motor Drives using Wide-Bandgap Power Switches," 2018 IEEE Transportation Electrification Conference and Expo (ITEC), 2018, pp. 1002-1008, doi: 10.1109/ITEC.2018.8450127.

[3] M. Antivachis, D. Bortis, L. Schrittwieser and J. W. Kolar, "Three-phase buck-boost Y-inverter with wide DC input voltage range," 2018 IEEE Applied Power Electronics Conference and Exposition (APEC), 2018, pp. 1492-1499, doi: 10.1109/APEC.2018.8341214.

[4] M. Antivachis, N. Kleynhans and J. W. Kolar, "Three-Phase Sinusoidal Output Buck-Boost GaN Y-Inverter for Advanced Variable Speed AC Drives," in IEEE Journal of Emerging and Selected Topics in Power Electronics, vol. 10, no. 3, pp. 3459-3476, June 2022, doi: 10.1109/JESTPE.2020.3026742.

Performance Evaluation of Sinusoidal-Flux Reluctance Machine for Improving Power Density with Reduced Torque and Input-Current Ripples

Kiwa Nagayasu, Masaki Iida, Kazuhiro Umetani, Mastaka Ishihara, Eiji Hiraki
OKAYAMA UNIVERSITY / GRADUATE SCHOOL OF NATURAL SCIENCE AND
TECHNOLOGY
3-1-1 Tsushimanaka, Kita-ku
Okayama, Japan
Tel.: +81 / (86) − 251.8121.
Fax: +81 / (86) − 251.8115.
E-Mail: p4pg68f5@s.okayama-u.ac.jp
URL: https://www.cc.okayama-u.ac.jp/~eng_epc/

Acknowledgements

This study is supported by Nagamori Foundation Research Grant 2022.

Keywords

«Switched reluctance drive», «Electrical machine», «Ripple minimization», «Synchronous Reluctance Machine (SynRM)»

Abstract

Reluctance machines are attractive for vehicle propulsion for being free from the permanent magnets, although conventional reluctance machines, such as the synchronous reluctance machine (SynRM) and the switched reluctance machine (SRM), suffer from low power density or large input-current and torque ripples. To solve these problems, a recent study has proposed the sinusoidal-flux reluctance machine, which is operated with the sinusoidal phase flux waveform. This preceding study has confirmed the operating principle of this machine, although little information has been provided on the performance compared to the existing reluctance machines. The purpose of this study is to elucidate the benefits of the sinusoidal-flux reluctance machine compared to SynRM and SRM. This study experimentally tested the performance of the sinusoidal-flux reluctance machine, SynRM, and SRM, designed under the conditions of the same stator core and the same rotor outer diameter. The experiment revealed that the sinusoidal-flux reluctance machine can reduce the peak flux compared to the SynRM with smaller toque and input-current ripples than the SRM, suggesting that the sinusoidal-flux reluctance machine is promising for vehicle propulsion.

Introduction

Propelled by the recent concern about global warming, electrified vehicles, such as electric vehicles and hybrid vehicles, are attracting the researchers' attention for reducing the carbon dioxide emission. These vehicles are propelled by the electric machines installed in the body of the vehicle. As the vehicle propulsion needs to cover a wide range of driving conditions, these electric machines are required to be operated under a wide range of torque and rotation speed. Therefore, the majority of these electrified vehicles are currently adopting the permanent magnet synchronous machines (PSMS) [1]–[3] for propulsion systems, because they can offer large output torque at a comparatively small rotation speed, which is a difficult operation condition for many electric machines. However, these machines need permanent magnets, which suffer from expensive material costs and unstable material supply. Besides, the permanent magnets are mechanically fragile and thermally degradable, which requires the delicate mechanical design of the electric machines and a strong cooling system. These problems of the PMSMs may hinder the electrification of the low-cost vehicles, which are prevailing in number, particularly in developing countries, and therefore tend to have a great effect on carbon dioxide emission.

A possible approach to solve these problems is to adopt the reluctance machines for the propulsion of electrified vehicles. The reluctance machines do not need permanent magnets for their torque generating mechanism. Therefore, these machines can have simple but robust mechanical construction, high thermal tolerance, and strong cost-effectiveness, all of which are promising for installation in the vehicle. Therefore, many studies have investigated the application of the reluctance machines for vehicle propulsion [4]–[11]. The reluctance machines have a wide variety of mechanical structures and control schemes. However, the majority of these studies have focused on the two typical reluctance machines: The switched reluctance machine (SRM) [6], [7], [9] and the synchronous reluctance machine (SynRM) [8], [10], [11].

From the viewpoint of electric machine drive, the major difference between these two reluctance machines lies in the connection of the phase winding, the phase inductance (or reluctance) profile, and the phase current

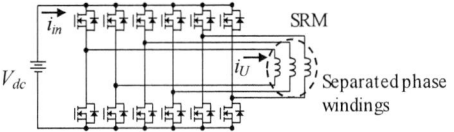

(a) Switched reluctance machine (SRM)

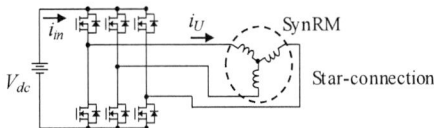

(b) Synchronous reluctance machine (SynRM)

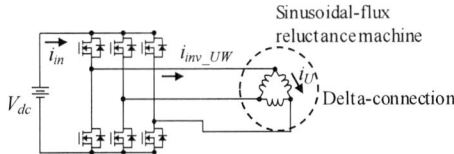

(c) Sinusoidal-flux reluctance machine

Fig. 1: Phase winding connection and inverter circuit of 3 reluctance machines

waveform to drive these machines. The SRM is designed to separate each phase winding from the others, as illustrated in Fig. 1(a). As shown later, the phase inductance was designed to have a triangular profile; the inverter should supply the phase current in the square waveform. The prominent merit of the SRM is the ability of comparatively larger torque output with smaller copper loss than SynRM, particularly at a low rotation speed. Contrarily, the SRM has severe drawbacks of the large input-current and torque ripples. The input-current ripple can deteriorate the battery lifespan due to the increase in the high-order harmonics in the battery current; the torque ripple can decrease the driving comfort due to the increase in the noise vibration. Therefore, the large input-current ripple and the large output torque ripple of the SynRM are severe drawbacks for vehicle propulsion. Another drawback of the SRM is that the normal three-phase inverters cannot be utilized to drive the SRM as the sum of the phase current is not zero [12]. Some recent studies [13][14] have certainly developed the control technique to greatly reduce both of the input-current and output torque ripples. However, even with these new technologies, the SRM does not accept the normal three-phase inverter, which entails a cost-up for the motor driving system.

On the other hand, the SynRM is designed to have the delta connection of the phase winding, as shown in Fig. 1(b), and the sinusoidal phase inductance profile. The inverter that drives this machine should supply the phase current in the sinusoidal waveform [15]. Theoretically, the SynRM can be operated without the input-current ripple and the output torque ripple. However, the SynRM tends to have low power density because the SynRM is susceptible to magnetic saturation due to the third harmonics contained in the phase flux density, as discussed in the subsequent section. Therefore, the size of the SynRM tends to be large for vehicle propulsion, hindering the installation in the limited space of vehicle.

As reviewed above, both of these two reluctance machines still have drawbacks for vehicular applications. To solve this difficulty, a novel reluctance machine concept has been proposed by recent studies [16][17]. Unlike the SynRM and the SRM, this machine has the phase windings connected in the delta connection, as shown in Fig. 1(c), the sinusoidal reluctance profile, and is driven with the sinusoidal phase flux waveform. Hereafter, this paper refers to this machine as the sinusoidal-flux reluctance machine. This machine is expected to operate without generating the input-current ripple and the output torque. Furthermore, this machine can be driven with the normal three-phase inverter. In addition to these attractive features, this machine can be operated with smaller peak phase flux and therefore can have greater power density than the SynRM.

The basic operating principles of the sinusoidal-flux reluctance machine have been experimentally and analytically confirmed in these preceding studies. However, the performance comparison with the SRM and SynRM was not performed in these studies because the prototype of the sinusoidal-flux reluctance machine was not optimally designed for fair comparison with the SynRM and SRM. The purpose of this paper is to report the performance comparison results of the sinusoidal-flux reluctance machine with the SynRM and the SRM by fairly designing this reluctance machine under the same restrictions are SynRM and SRM. For observing the performance difference only by the basic operation principles, these three machines were designed only by changing the rotor outer periphery, winding connection, and number of the winding turns.

The remainder of this paper is divided into three sections. Section II briefly reviews the sinusoidal-flux reluctance machine in comparison with the SRM and the SynRM. Sections III and IV perform the simulation and the experiment, respectively, to confirm the performance of this reluctance machine in comparison with the SRM and the SynRM. Finally, section V gives the conclusions.

Review of Sinusoidal-Flux Machine in Comparison with SRM and SynRM

This section briefly reviews the operating principles of the sinusoidal-flux machine. For this purpose, the operating principles of SRM and SynRM are reviewed in advance. Hereafter, these reluctance machines are supposed to have the 3 phases, namely phases U, V, and W, as is common for these machines. The magnetic saturation is neglected in this section for simplifying the discussion.

In the reluctance machines, the phase inductance profile plays an essential role for electric energy conversion to the torque output. The phase inductance profile has the wavenumber of 2 when plotted as a function of the electric angle. The instantaneous torque τ and the input current i_{in} can be formulated as a function of the phase current i as,

$$\tau = \sum_{k=U,V,W} \frac{P}{2} \frac{dL_k}{d\theta} i_k^2, \quad i_{in} = \frac{1}{V_{dc}} \sum_{k=U,V,W} v_k i_k = \frac{1}{V_{dc}} \sum_{k=U,V,W} \frac{dL_k i_k}{dt} i_k = \frac{\Omega}{V_{dc}} \sum_{k=U,V,W} \left(2\tau + \frac{PL_k}{2} \frac{di_k^2}{d\theta} \right). \tag{1}$$

where P is the number of the rotor pole pairs; L is the phase inductance; θ is the electric angle; i is the phase current; V_{dc} is the voltage of the DC power supply to the inverter; v is the phase voltage; Ω is the angular velocity of the rotor, and subscription U, V, W are the indicator of the phase. On the other hand, the phase magnetic flux ϕ generated in a phase winding can be formulated as follows, if the number of turns of the phase winding denotes N:

$$\phi_k = L_k i_k / N. \tag{2}$$

The SRM has the triangular phase inductance profile. The phase windings are supplied with the square-shaped current with the same wavenumber as the inductance profile, as depicted in Fig. 2. The phase current flows for 60 degrees alternatingly in each phase during the increase of the phase inductance at a constant rate. Therefore, the constant torque and the constant input current are expected according to (1). However, the existence of the phase inductance prohibits the sharp drop of the phase current at the magnetizing and

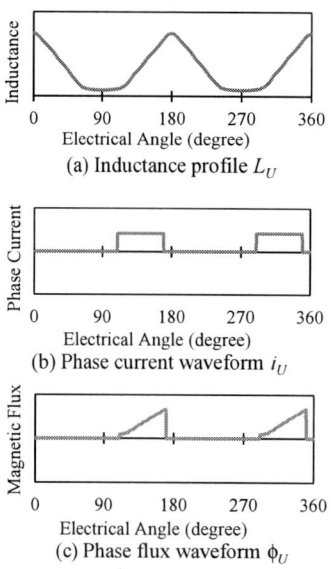

(a) Inductance profile L_U

(b) Phase current waveform i_U

(c) Phase flux waveform ϕ_U

Fig. 2: Waveforms of SRM

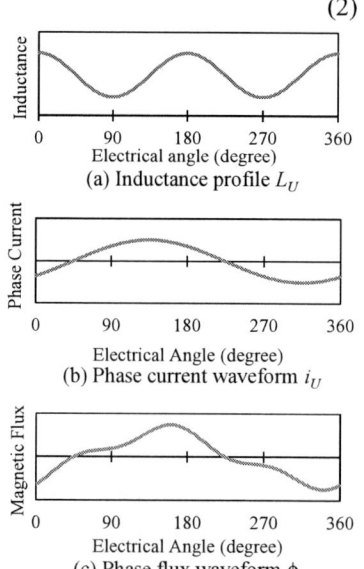

(a) Inductance profile L_U

(b) Phase current waveform i_U

(c) Phase flux waveform ϕ_U

Fig. 3: Waveforms of SynRM

demagnetizing transients. Therefore, the actual phase current deviates from the ideal waveform, particularly at high rotating speed, causing large input-current and torque ripples.

On the other hand, the SynRM has the sinusoidal phase inductance profile as depicted in Fig. 3. The phase windings are supplied with the sinusoidal current waveform with the half wavenumber as the inductance profile. According to (1), this also results in the constant torque and the constant input current. Unlike the SRM, the phase current is not required to vary suddenly. Therefore, the SynRM can exhibit small input-current and torque ripples in a wide range of practical operations. However, according to (2), the magnetic flux waveform is not purely sinusoidal but contains the third harmonics. This harmonic increases the peak magnetic flux, although this harmonic does not contribute to the torque output. Therefore, the SynRM is susceptible to magnetic saturation, which restricts the maximum torque output and reduces the power density.

Unlike the SRM and the SynRM, the sinusoidal-flux reluctance machine has the sinusoidal phase reluctance profile, as depicted in Fig. 4, with the wavenumber of 2. Because the phase windings are connected in the delta-connection, each phase winding is applied with the sinusoidal voltage output from the inverter. Therefore, the phase magnetic flux has the sinusoidal waveform with the half wavenumber as the reluctance profile.

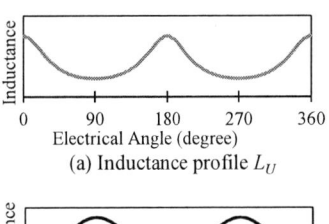

(a) Inductance profile L_U

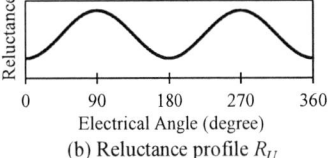

(b) Reluctance profile R_U

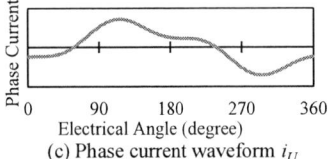

(c) Phase current waveform i_U

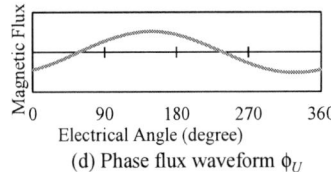

(d) Phase flux waveform ϕ_U

Fig. 4: Waveforms of sinusoidal-flux reluctance motor

Note that the phase reluctance R is formulated as $R=N^2/L$. Therefore, (1) can be rewritten as

$$\tau = -\sum_{k=U,V,W} \frac{P}{2}\frac{dR_k}{d\theta}\phi_k^2, \quad i_{in} = \frac{\Omega}{V_{dc}}\sum_{k=U,V,W} \frac{PR_k}{2}\frac{d\phi_k^2}{d\theta}. \tag{3}$$

Therefore, this machine also achieves the constant input current and the constant output torque, suggesting that this machine is beneficial in smaller input-current and torque ripples than SRM, similar to the SynRM. Furthermore, the phase magnetic flux waveform is purely sinusoidal without harmonics. Therefore, this machine is expected to have a smaller peak magnetic flux than the SynRM, which will result in an improvement in the maximum output torque and therefore the power density.

Despite the aforementioned attractive features, the phase current waveform of the sinusoidal-flux reluctance machine contains the third harmonics, which circulates in the delta-connected windings. Therefore, the inverter output current is sinusoidal without harmonics, similar to the SynRM, thus enabling the normal three-phase inverter to operate this machine. However, this circulating current can increase the copper loss compared to the SynRM, although the absence of the third harmonics in the phase magnetic flux will reduce the iron loss.

It is worth noticing that the phase flux waveform can have the phase shift from that depicted in Fig. 4, similar to the SynRM, which also accepts the phase shift of the phase current waveform from Fig. 3. The phase shift of the phase flux waveform does not increase the input-current nor torque ripples. Therefore, in practical design, the optimal phase shift can be designed under various design considerations. For example, the optimal phase shift can be determined so that the copper loss is minimized under the same torque output, which was adopted in the design of the sinusoidal-flux reluctance motor and the SynRM in the next section. Consequently, the sinusoidal-flux reluctance machine can avoid the drawbacks of the SynRM and the SRM.

Simulation-Based Design of Sinusoidal-Flux Reluctance Machine

For performance comparison of the three reluctance machines, i.e. the proposed machine, the SynRM, and the SRM, these machines were designed by utilizing the electromagnetic simulator JMAG19.1 (JSOL Corp.). The three reluctance machine models were constructed based on the commercially available SRM, which is used as the SRM. The specifications of the SRM are listed in Table I. All machine models have the same stator with concentrated windings. However, the number of turns was different depending on the connection of the phase winding: The number of turns was set

Table I: Specifications of Commercial SRM

Model number	RB165SR-96CSRM (Motion System Tech Inc.)
Rated value	1.2 kW, 96V, 6000 r/min
Pole number	Stator: 12 poles, Rotor: 8 poles
Number of turns	14 turns/pole
Min. gap b/w stator and rotor	0.3 mm
Stack length	40 mm

at 14 for the SynRM and the SRM, whereas the number of turns was set at 24 for the sinusoidal-flux reluctance machine because the voltage applied to the phase winding is $\sqrt{3}$ times greater than that of the SynRM due to the delta-connection. Beside of the number of turns, the only difference lies in the rotor shape. Commonly SynRMs have a rotor shape with multiple flux barriers. However, for simplifying the difference among the three machines, we designed the rotors of the proposed reluctance machine and the SynRM by only modifying the rotor's outer periphery shape of the commercial SRM without implementing the flux barriers.

Design of Rotor Geometry

The rotor geometry of the sinusoidal-flux reluctance machine and the SynRM was determined by approximating the rotor geometry by a 288-gon and optimizing the gap length between each vertex and the inner diameter of the stator. This section hereafter describes the rotor geometry determination process of the sinusoidal-flux reluctance machine, although a similar procedure was also taken to determine the rotor geometry for the SynRM.

According to the preceding study [16], the design of the sinusoidal-flux reluctance machine should consider the following two points to have good efficiency: 1. The maximum phase inductance, i.e. the inductance at the aligned position, should have as great value as possible; 2. The ratio of the maximum phase reluctance to the least phase reluctance should be around 3. Therefore, this study designed the rotor shape according to these instructions.

As the sinusoidal-flux reluctance machine should be designed to have the sinusoidal phase reluctance profile, a straightforward method for designing the rotor geometry is to determine the gap length $l_p(\theta)$ from the rotor outer periphery to the stator inner diameter by the sinusoidal wave with the wavenumber of 2. Hence, $l_p(\theta)$ is determined by the following equation, where l_{p0} and l_{p1} are the positive values determined by the design.

$$l_p(\theta) = l_{p0} - l_{p1} \cos 2\theta. \tag{4}$$

According to the instruction the gap length at the aligned position, i.e. $l_{p0} - l_{p1}$ should be set at the minimum possible gap length accepted by the mechanical restrictions; l_{p1} should be set to have the appropriate value of the ratio between the maximum and minimum reluctance. However, this straightforward design resulted in the reduction of the maximum inductance because the gap $l_p(\theta)$ starts to increase at any small deviation from the aligned position, i.e. $\theta = 0°, 180°$. Therefore, for maximizing the inductance at the aligned position, the increase of the gap length should be suppressed near the aligned position. For this purpose, this paper determined $l_p(\theta)$ according to the following equation where l_{p2} is an additional value to be determined by design:

$$l_p(\theta) = l_{p0} - l_{p1} \cos 2\theta + l_{p2} \cos 4\theta. \tag{5}$$

As can be seen in Fig. 5, this function can suppress the gap length increase near the aligned position to increase the maximum inductance. Nonetheless, adding the third right-hand term causes deviation of the phase reluctance profile from the sinusoidal waveform, increasing the input-current and torque ripples. Therefore, l_{p0}, l_{p1}, and l_{p2} should be determined to increase the maximum phase inductance at the aligned position while considering the acceptable ripples. This paper adopted the input-current ripple ratio of 200% and the torque ripple ratio

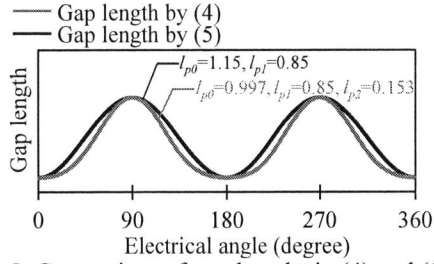

Fig.5 : Comparison of gap lengths in (4) and (5) in the sinusoidal-flux reluctance machine

of 80% as the acceptable ripple rations, where the ripple ratio is defined as the difference between the maximum and minimum values normalized by the average value. These acceptable ripple ratios were set to be below the commercial SRM. This design may be far from an optimal design. Better design for the sinusoidal-flux reluctance machine should be sought in future studies.

The gap length at the aligned position, i.e. $l_{p0}-l_{p1}+l_{p2}$, was set at 0.3mm, which is the minimum gap length of the commercial SRM. Under this restriction, the optimal values for three parameters l_{p0}, l_{p1}, and l_{p2} were searched according to the flowchart shown in Fig. 6. This flowchart indicates the trial-and-error approach for designing these parameters. Initial parameters were determined by setting l_{p2} at zero: $L_{p0} = 0.8$, $L_{p1} = -0.5$, and $L_{p2} = 0.0$. Firstly, the rotor shape was designed for l_{p0}, l_{p1}, and l_{p2} using 3D CAD design software (Fusion360). Based on this rotor shape, the inductance profile $L_U(\theta)$ was calculated using the electromagnetic field simulator (JMAG-Designer). Based on this result, the optimal phase shift angle of the phase flux was analytically calculated so that the output torque or 1Nm was output with the smallest copper loss. Then the input-current and torque ripples were calculated to check whether these ripples are within the acceptable ripple ratios. By repeating this process while gradually increasing l_{p2}, the optimal design of the rotor shape was determined as $L_{p0} = 0.997$, $L_{p1} = -0.85$, and $L_{p2} = 0.153$, which has the minimum copper loss within the acceptable ripple ratios.

Simulation Results

To confirm the design of the sinusoidal-flux reluctance machine, the electromagnetic field simulation was carried out using the simulator JMAG (JSOL Corp.). This simulation tested the design of the three reluctance machines, i.e. the sinusoidal-flux reluctance machine, the SynRM, and the SRM. Figure 7 shows the simulation models of these three machines; Table 2 lists the materials adopted for the simulation models, although the magnetic saturation was not considered in this simulation.

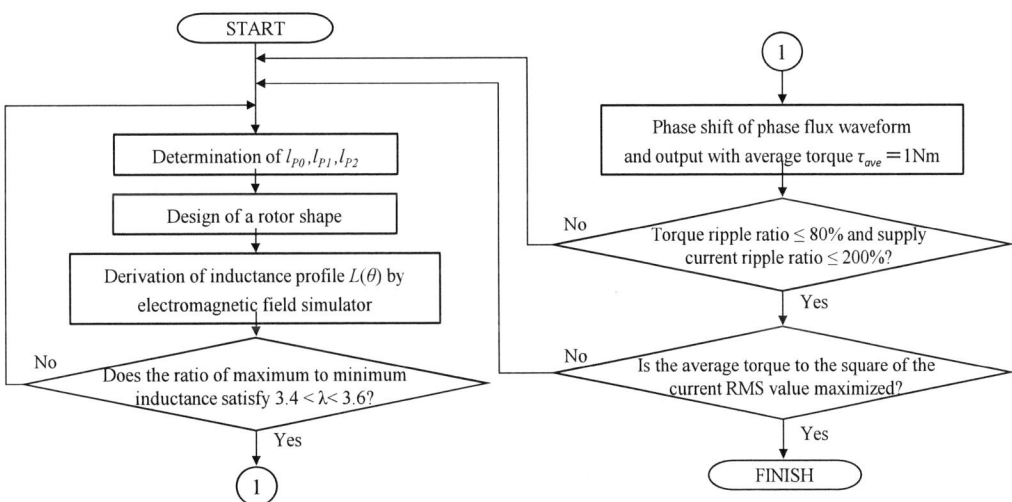

Fig.6 : Design algorithm for the rotor geometry of the sinusoidal-flux reluctance machine

Table II: Materials used in FEM model

Component	Material or relative magnetic permeability
Stator	Material name: 35H300 (Nippon Steel Corp.)
Rotor	Material name: 35H300 (Nippon Steel Corp.)
Winding	Relative magnetic permeability: 1
Shaft	Relative magnetic permeability: 1

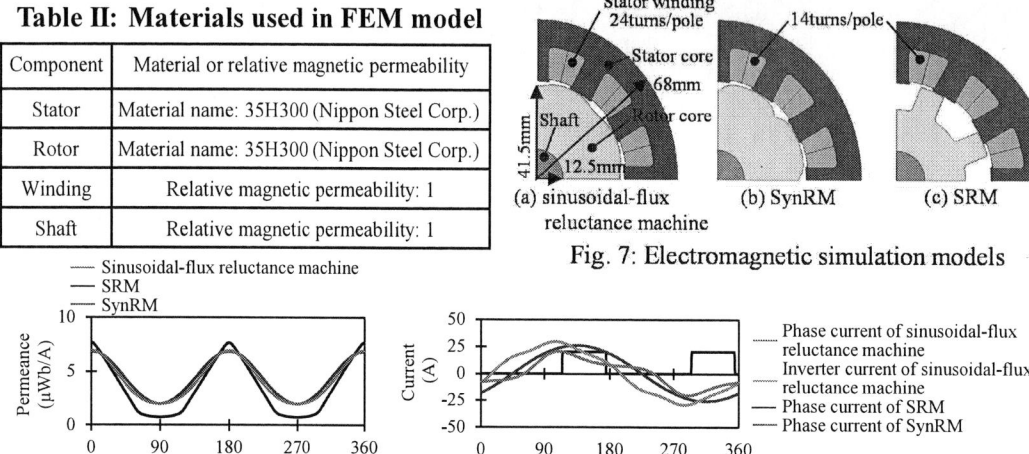

Fig. 7: Electromagnetic simulation models

Fig. 8: Permeance profile for simulation   Fig. 9 Current waveform for outputting 1Nm at 2000 r/min

Figure 8 shows the permeance of the three reluctance machines. This figure plots the permeance profile instead of the inductance profile because the number of turns of the phase windings is different among the three machines due to the difference in the phase winding connection. The results revealed that the sinusoidal-flux reluctance machine exhibited maximum permeance similar to the SynRM but slightly smaller than the SRM. The ratio between the maximum and minimum inductance was 3.5 in the sinusoidal-flux reluctance machine, which was within the target of the design.

Based on the permeance profile calculated in Fig. 8, a motor behavior model was constructed to calculate the input-current and torque ripples by utilizing the inverter circuit simulator. For this purpose, the behavior models of the three reluctance machines were constructed according to [18] and operated using the driving systems shown in Fig. 1, which were modeled in the model space of the circuit simulator PSIM2021a (Myway Corp.). The inverters of the driving systems were supplied with the DC power of the voltage Vdc=96 V. This simulation did not contain any losses and consider the magnetic saturation.

The inverter output current was controlled by the hysteresis control with the hysteresis width of 0.5 A to follow the current command values. Figure 9 shows the command values for 1 Nm at 2000 r/min.

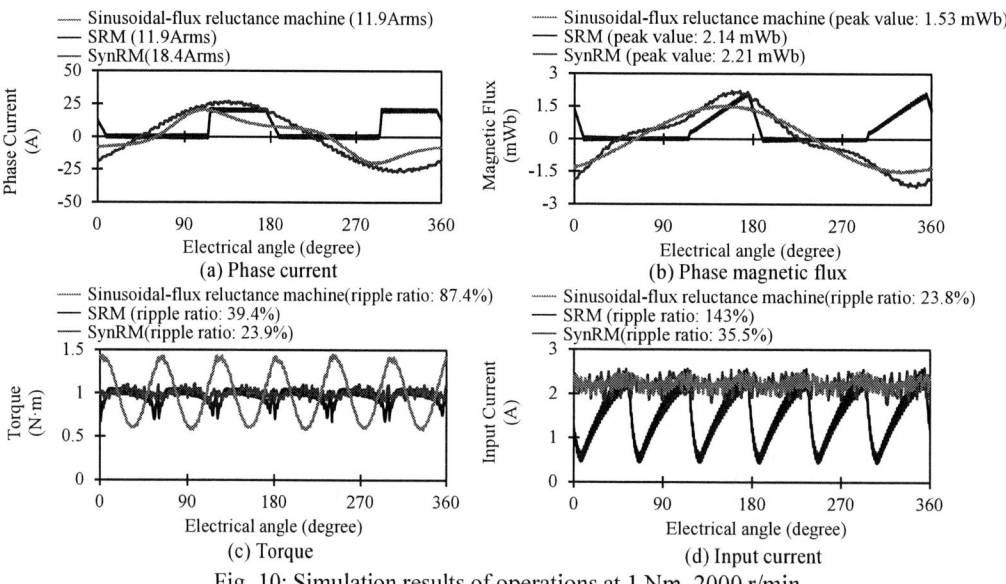

Fig. 10: Simulation results of operations at 1 Nm, 2000 r/min

Figure 10 shows the simulation results at the output torque of 1 Nm and the rotation speed of 2000 r/min. As can be seen in the figure, the sinusoidal-flux reluctance machine exhibited the smallest peak magnetic flux, which is a promising feature to mitigate the magnetic saturation and improve the power density. As for the phase current waveforms, the sinusoidal-flux reluctance machine exhibited the smallest rms values. However, this is because this machine has much greater number of turns, i.e. 24 turns, compared to the SynRM and the SRM, which has the 14 turns. If the resistance of the phase winding is simply approximated to be proportional to the square of the number of turns, the copper loss of the sinusoidal-flux reluctance machine is 1.22 times greater than the SynRM and 2.94 times greater than the SRM.

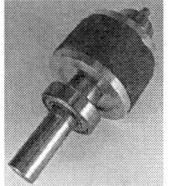 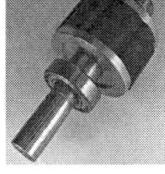 

(a) Sinusoidal-flux reluctance machine (b) SynRM (c) SRM

Fig. 11: Photographs of rotors of experimental reluctance machines

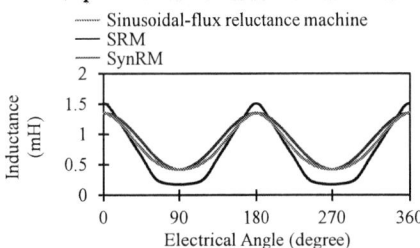

Fig. 12: Measurement result of the inductance profile

This figure also shows the simulation results of the input-current and torque ripples. The sinusoidal-flux reluctance machine exhibited a significant torque ripple. This may have been caused by the insufficient optimization of the rotor geometry, which should be improved in future study. However, the sinusoidal-flux reluctance machine exhibited a small input-current ripple similar to the SynRM, which is far smaller than the SRM. Therefore, the simulation results imply the potential power density improvement by the sinusoidal-flux reluctance machine in comparison to the SynRM with a increase in the copper loss, although the further design optimization method of the rotor geometry should be investigated to reduce the torque ripple to the similar value as the SynRM.

Experiment

The experiment was carried out to test the performance of the sinusoidal-flux reluctance machine in comparison with the SRM and the SynRM. The SRM and the SynRM incorporated the same stator with the phase windings of 14 turns, which is the stator of the commercial SRM specified in Table I. The sinusoidal-flux reluctance machine also adopts the same stator except that the number of the phase winding is changed to 24. The rotors of these machines were fabricated according to the design obtained in the previous section. Figure 11 shows the photographs of the rotors. These electric machines exhibited inductance profiles similar to the simulation result of the previous section, as shown in Fig. 12. The current command values for the inverters were determined based on Fig. 12.

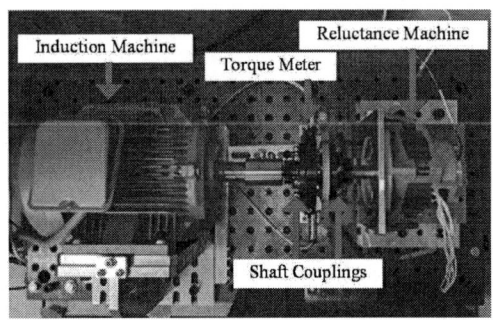

Fig. 13: Machine test bench

Table III : List of Motor Bench

Used equipment	Model No,(manufacturer)
Motor for measurement	RB165SR-96CSRM (Motion System Tech Inc.)
Torque meter	T40B (HBM Co.)
Induction Machine	TFO-K(Hitachi, Ltd.)

Table IV : List of used equipment

Used equipment	Model No.(Manufacturer)
Oscilloscope	HDO4034A (Teledyne LeCroy)
Current Probe Amplifier	TCPA300 (Tektronix, Inc)
Current Probe(i_{in})	TCP305A (Tektronix, Inc)
Current Probe(i_{U})	TCP303 (Tektronix, Inc)
High Voltage Differential Probe(v_{U})	P5200A (Tektronix, Inc)
DC power supply(V_{dc})	GP0110-50R (Takasago, Ltd)
Digital multimeter(Measurement of V_{dc})	34461A (Keysight Technologies)

In this experiment, the operating waveforms of the sinusoidal-flux machine, SynRM, and SRM were evaluated using the reluctance machine test bench shown in Fig. 13. The specifications of the test bench are listed in Table III. In this test bench, the reluctance machine was mechanically connected to the induction machine, which serves as the power load, via an instantaneous torque meter and coupling. The reluctance machine was supplied with ac power via an inverter according to the circuit diagram shown in Fig. 1. The inverter was supplied with the dc voltage of 96 V. The inverter was controlled to output the ac current according to the current command value using the same hysteresis control as described in the previous section. The hysteresis width was adjusted to operate these experimental electric machines approximately at the inverter switching frequency of 30kHz.

Firstly, the operating waveforms were measured under two driving conditions corresponding to the unsaturated magnetization and the saturated magnetization of the sinusoidal-flux reluctance machine. Specifically, the former condition outputs the torque of 1 Nm, whereas the latter condition outputs the torque of 4 Nm. Similar to the simulation, the experiment evaluated the following four waveforms: 1. Phase current, 2. Phase voltage, 3. Input current, and 4. Instantaneous torque. Table IV lists the instrument employed for the measurement. The phase current, phase voltage, and input current were measured at the rotation speed of 2000 r/min. However, the instantaneous torque was measured at the rotation speed of 100 r/min for the torque ripple to be within the frequency range of the instantaneous torque meter and not to induce the mechanical resonance of the motor test bench.

Figure 14 shows the operating waveforms of the three reluctance machines, measured at the output torque of 1 Nm. The peak phase flux of the sinusoidal-flux reluctance machine was found to be the smallest among the three electric machines. Specifically, the maximum phase flux of the sinusoidal-flux reluctance machine was 20% smaller than that of the SynRM, indicating that the sinusoidal-flux machine is less susceptible to magnetic saturation. The sinusoidal-flux reluctance machine exhibited the effective reduction of the input current ripple compared to the SRM, as is similar to the SynRM. Meanwhile, the sinusoidal-flux reluctance machine did not show an effective reduction of the torque ripple. As discussed in the simulation, this insufficient reduction effect of the torque ripple was caused by the insufficient design optimization of the sinusoidal-flux reluctance machine, which will be investigated in the future study.

Figure 15 shows the operating waveforms at the output torque of 4 Nm. In this condition, the sinusoidal-flux reluctance machine again exhibited the smallest peak phase flux among the three machines. Furthermore, this machine also exhibited the smallest input-current ripple and a similar torque ripple as the SynRM. Therefore, the sinusoidal-flux reluctance machine kept the input-current and torque ripples

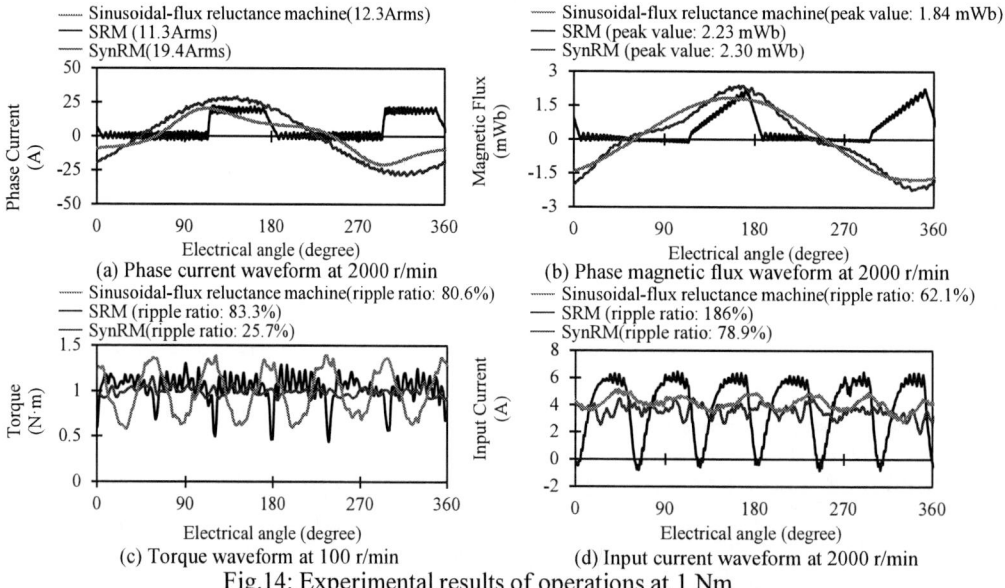

Fig.14: Experimental results of operations at 1 Nm

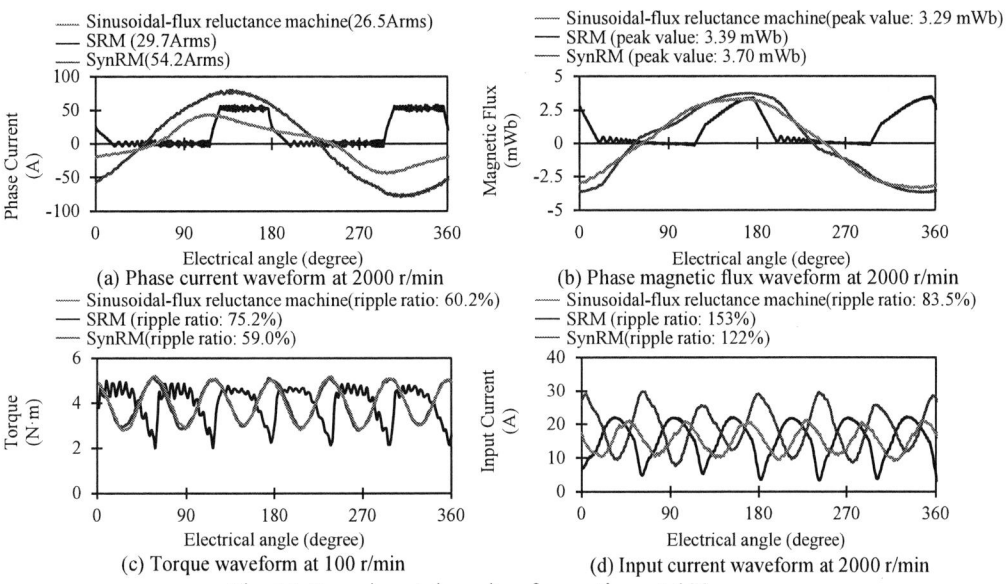

(a) Phase current waveform at 2000 r/min
(b) Phase magnetic flux waveform at 2000 r/min
(c) Torque waveform at 100 r/min
(d) Input current waveform at 2000 r/min

Fig. 15: Experimental results of operations at 4 Nm

small even under high torque output, whereas the SynRM increased rapidly the input-current and torque ripples as the output torque increases. The reason for this difference lies in maximum phase flux. Because the SynRM generates a large phase flux than the sinusoidal-flux reluctance machine, the SynRM tends to increase the ripples at lower output torque than the sinusoidal-flux reluctance machine. Therefore, these results support that the sinusoidal-flux reluctance machine can improve the power density compared to the SynRM.

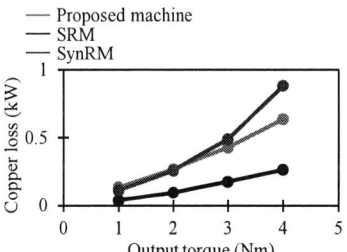

Fig. 16: Comparison of copper loss at various output torque

The simulation results reported in the previous section pointed out that the major drawback of the sinusoidal-flux reluctance machine is large copper loss. Therefore, the rms values of the phase current were evaluated at various output torque to estimate the copper loss of the phase windings and compare the result among the three reluctance machines. Figure 16 presents the results. Certainly, the sinusoidal-flux reluctance machine exhibited far greater copper loss than the SRM, similarly to the SynRM. However, if compared to the SynRM, the sinusoidal-flux reluctance machine exhibited less copper loss at high output torque operation. This is also caused by the reduction in the peak phase flux. Because the SynRM more profoundly saturates than the sinusoidal-flux reluctance machine, the SynRM tends to generate large copper loss at high output torque operations. Therefore, this result also supports that the sinusoidal-flux reluctance machine can improve the power density compared to the SynRM.

Conclusions

Reluctance machines are attractive for vehicular propulsion, although conventional reluctance machines as the SynRM and the SRM suffers from low power density or large input-current and torque ripples. To overcome these difficulties, the novel reluctance machine with sinusoidal phase flux has been recently proposed. This paper constructed this reluctance machine and tested the performance in comparison with the SynRM and the SRM using the simulation and the experiment. The sinusoidal-flux reluctance machine and the SynRM were designed from the commercially available SRM, which was used as the SRM, by only changing the rotor outer periphery geometry and the number of turns of the phase windings, in order to examine the features originated from the fundamental concept of these three reluctance machines. The simulation and experiment indicated that the sinusoidal-flux reluctance machine can generate a far smaller input-current ripple than the SRM. Furthermore, the sinusoidal-flux reluctance machine was found to reduce the peak phase flux compared to the SynRM. The experiment also supported that the sinusoidal-flux reluctance machine generates similar torque ripple and smaller input-current ripple with smaller copper loss compared to the SynRM at high output torque operation, which suggests the power density improvement by the sinusoidal-flux reluctance machine. At the same time, however, the sinusoidal-flux reluctance machine exhibited a large torque ripple similar to the SRM, which is unexpected from the theory. This discrepancy from the theory is attributed to the insufficient rotor design of the sinusoidal-flux reluctance machine. Therefore, better rotor shape design should be investigated in a future study to reduce the torque ripple.

References

[1] Du J. Wang X., Lu H.: Optimization of magnet shape based on efficiency map of IPSMS for EVs, IEEE Trans. Appl. Supercond, vol. 26, no. 7, pp. 1-7, Oct. 2016, Art no. 0609807.

[2] Jung. H., Park G., Kim D., Jung S.: Optimal design and validation of IPMSM for maximum efficiency distribution compatible to energy consumption areas of HD-EV, IEEE Trans. Magn., vol. 53, no. 6, pp. 1-4, Jun. 2017, Art no. 8201904.

[3] Hwang Y., Lee J.: HEV motor comparison of IPMSM with Nd sintered magnet and heavy rare-earth free injection magnet in the same size, IEEE Trans. Appl. Supercond., vol. 28, no. 3, pp. 1-5, Apr. 2018, Art no. 5206405.

[4] Zhu Z. Q., Chan C. C.: Electrical machine topologies and technologies for electric, hybrid, and fuel cell vehicles, Proc. IEEE Vehicle Power Propulsion Conf., Harbin, China, pp. 1–6, Sept. 2008.

[5] Raminosoa T. Torrey D. A., El-Refaie A. M., Grace K., Pan D., Grubic S., Bodla K., Huh K.-K.: Sinusoidal reluctance machine with dc winding: an attractive non-permanent-magnet option, IEEE Trans. Ind. Appl., vol. 52, no. 3, pp. 2129-2137, May/Jun. 2016.

[6] Uddin W., Husain T., Sozer Y., Husain I.: Design methodology of a switched reluctance machine for off-road vehicle applications, IEEE Trans. Ind. Appl., vol. 52, no. 3, pp. 2138-2147, May/Jun. 2016.

[7] Martin R., Widmer J. D., Mecrow B. C., Kimiabeigi M., Mebarki A., Brown N. L.: Electromagnetic considerations for a six-phase switched reluctance motor driven by a three-phase inverter, vol. 52, no. 5, pp. 3787-3795, Sept./Oct. 2016.

[8] Bianchi N., Bolognani S., Carraro E., Castiello M.: Electric vehicle traction based on synchronous reluctance motors, IEEE Trans. Ind. Appl., vol. 52, no. 6, pp. 4762-4769, Nov./Dec. 2016.

[9] Zhu J., Cheng K. W. E., Xue X., Zou Y.: Design of a new enhanced torque in-wheel switched reluctance motor with divided teeth for electic vehicles, IEEE Trans. Magn., vol. 53, no. 11, pp. 1-4, Nov. 2017, Art no. 2501504.

[10] Kumar G. V., Chuang C.-H., Lu M.-Z., Liaw C.-M.: Development of an electric vehicle synchronous reluctance motor drive, IEEE Trans. Vehicular Tech., vol. 69, no. 5, pp. 5012-5025, May 2020.

[11] Credo A., Fabri G., Villani M., Popescu M.: Adopting the topology optimization in the design of high-speed synchronous reluctance motors for electric vehicles, IEEE Trans. Ind. Appl., vol. 56, no. 5, pp. 5429-5438, Sept./Oct. 2020.

[12] Suppharangsan W., Wang J.: Experimental validation of a new switching technique for DC-link capacitor minimization in switched reluctance machine drives, Proc. IEEE Int. Electric Machines Drives Conf., Chicago, USA, pp. 1031–1036, May 2013.

[13] Kusumi T., Hara T., Umetani K., Hiraki E.: Phase-current waveform for switched reluctance motors to eliminate input-current ripple and torque ripple in low-power propulsion below magnetic saturation, IET Power Electronics, vol. 13, no. 15, pp. 3351–3359, Nov. 2020.

[14] Kusumi T., Hara T., Umetani K., Hiraki E.: Simultaneous tuning of rotor shape and phase current of switched reluctance motors for eliminating input current and torque ripples with reduced copper loss, IEEE Trans. Ind. Appl., vol. 56, no. 6, pp. 6384–6398, Nov. 2020.

[15] Payza O., Demir Y., Aydin M.: Investigation of losses for a concentrated winding high-speed permanent magnet-assisted synchronous reluctance motor for washing machine application, IEEE Trans. Magn., vol. 54, no.11, 8207606, 2018.

[16] Iida M., Kusumi T., Umetani K., Hiraki E.: Feasibility of sinusoidal flux drive design of switched reluctance motor for reducing torque and input current ripples with three-leg inverter, Proc. IEEE Intl. Power Electron. Motion Ctrl. Conf. (PEMC2020), Apr.2020, Gliwice, Poland, pp. 439–446.

[17] Iida M., Umetani K., Kusumi T., Ishihara M., Hiraki E.: Sinusoidal-flux reluctance machine driven with three-phase inverter for improving power density with reduced torque and input current ripples, Proc. IEEE European Conf. Power Electron. Appl.(EPE2021), Ghent, Belgium, Sept. 2021, pp., 1–10.

[18] Hara T., Kusumi T., Umetani K., Hiraki E.: A simple behavior model for switched reluctance motors based on magnetic energy, Proc. Intl. Power Electron. Motion Control Conf. (IPEMC2016), Hefei, China, May. 2016.

Power Hardware-In-the-Loop test of low-voltage battery for a plug-in hybrid electric vehicle

Ronan GERMAN[1], Florian TOURNEZ[1], Alain BOUSCAYROL[1],

Aurélien LIEVRE[2], Betty LEMAIRE-SEMAIL[1],

[1] Univ. Lille, Arts et Metiers Paris Tech, Centrale Lille, JUNIA-Hauts-de-France, EA 2697- L2EP, F-59000 Lille, France

[2] Valeo Equipements Electriques Moteur, 2 rue André Boulle, Créteil, F-94017 Cedex, France

Keywords

Battery testing, electric vehicle, hardware-in-the-loop, energetic macroscopic representation.

Abstract

A methodology developed for a quicker validation of various sub-systems is applied in this paper. It is based on two steps, i.e. the simulation validation and the power Hardware-in-the-Loop (HiL) testing. A new plug-in hybrid electric vehicle with low-voltage battery has been validated on a demo car. A new battery with higher power density is proposed to minimize the energy losses. Before its integration in car, simulation and hardware-in-the-loop tests are achieved. For the HiL part, a real-time simulation of the powertrain is coupled to the battery to test various driving conditions. A dedicated emulation interface is developed. The tests in this paper demonstrate the ability of the battery to operate in a safe condition for the vehicle using a WLTC driving cycle.

Introduction

The development of electrified vehicles is a on the road to reduce the greenhouse gases of the transport sector [1]-[3]. The European Commission is managing several research projects to speed up the development of battery electric vehicles, hybrid electric vehicles and fuel cell vehicles [4]-[6]. As example, the H2020 PANDA European project [7] deals with a unified modelling method for virtual and real testing of electrified vehicles and their components. In that aim, the Energetic Macroscopic Representation (EMR) formalism [8] is used to organize the simulation models of these electrified vehicles, for different development tasks. Siemens has integrated an EMR library in its well-known Simcenter AMESIM simulation package [9]. Multi-level models have thus been developed for batteries and electric drives of different electrified vehicles in a cloud to be used for virtual tests of new components but also for real tests (i.e. hardware-in-the-loop testing).

Low-voltage batteries are developed in hybrid electric vehicles to avoid a too high DC bus and associated safety constraints [10][11]. Valeo is developing a low-voltage hybrid powertrain that has already been implemented in a demo-car [12]. A gasoline car has been thus retrofitted using a 48V battery. This plug-in hybrid electric vehicle has also been simulated using the PANDA method and the actual battery leads to a small gain in terms of energy consumption [12]. A new battery with a higher power density is now considered to increase the energy saving. Indeed, high power batteries are designed to have low series resistance. As a consequence, the joule losses are minimized. The new battery must be tested before its implementation in the demo-car. The HIL (Hardware-In-the-Loop) method [13][14] is used for this test. This paper deals with the HIL test of this battery using the real-time model of the hybrid traction subsystem of the studied plug-in hybrid electric vehicle (P-HEV).

Studied vehicle and new battery

A. Low-voltage plug-in hybrid vehicle with original battery

The studied vehicle is a 308 Peugeot SW car that has been retrofitted by Valeo to test an innovative traction subsystem based on a 48V battery and low-voltage electrical machines [12].

The traction subsystem is composed of a thermal engine, 2 electrical drives and a low-voltage (48 V) battery (Fig. 1). For the front part of the vehicle, the 94 kW internal combustion engine and the 4 kW front electrical machine are coupled through a belt. For the rear part, a 25 kW electrical machine drives the rear wheels. Both permanent magnet synchronous machines are supplied by voltage-source-inverters and a 48V battery. This low-voltage hybridization avoids strong modification without specific safety subsystems. The 48V set is enough to improve significantly the fuel consumption [12].

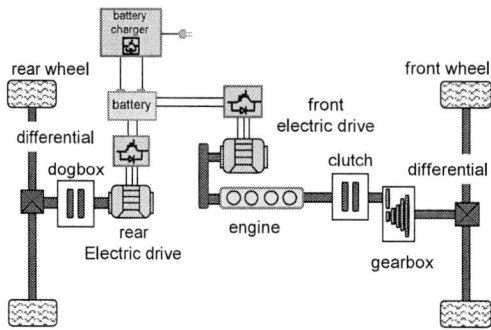

Fig. 1. Hybrid traction subsystems for the studied vehicle

B. New battery to be integrated

An initial self-made Li-ion battery (5 kWh, 111 Ah) has been embedded in the demo-car. The maximal discharge current is 600A (5.4 C). The rated voltage is 44.4 V. As the total battery weight is 50 kg, the energy density is 100 Wh/kg and the power one is 266 W/kg. This energy density seems to be low but it includes the water-cooling system.

A new Li-ion NMC battery is manufactured by the Bluways company. It has a pretty high-power capability (1.92 kW/kg). This amount of power could allow to use it without a cooling system. Thus, in the next sections, the power tests will be achieved without any cooling system to validate this. For a module (Fig. 2), the maximal C-rate is 15 C, the maximal discharge current is 1200 A, the maximal charge current is 240 A. A module is composed of 30 cells that lead to a 142 Wh/kg energy density.

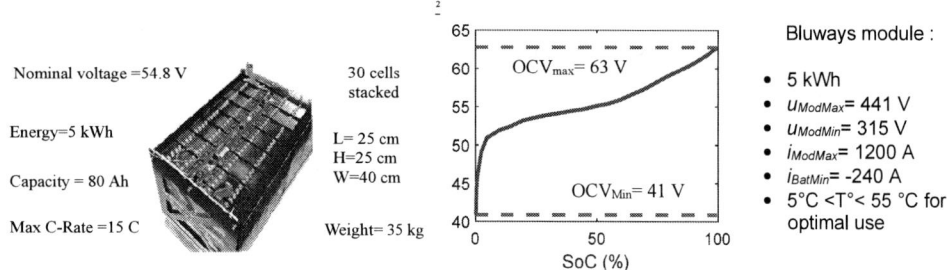

Fig. 2. Characteristic of the new battery module

In order to get the vehicle performances, two battery modules are connected in parallel (Fig. 3) to obtain a 10 kWh energy source that double the capacity for only 70 kg (40% of supplementary mass). The nominal voltage is slightly higher (56 V) that lead to reduces the current peak for the same traction power.

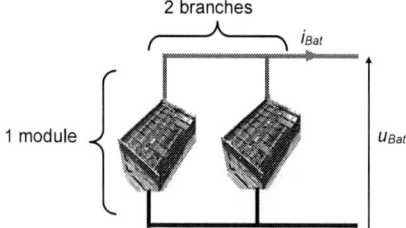

Fig. 3. Battery pack for the studied P-HEV

C. Simulation of the studied P-HEV

First, the complete hybrid traction subsystem is modelled using EMR (Fig. 4) according to the PANDA project. The energy sources are depicted by green oval pictograms (e.g. the battery). The energy conversion elements are depicted by orange pictograms (e.g. the electric drive). The input of the simulation and the power test is the velocity as a function of the time (driving cycle). As a consequence, the velocity should be controlled to generate the right amount of power to the tested battery. The control scheme (light blue pictograms) is directly deduced by an inversion of the EMR [8]. Moreover, an Energy Management Strategy (EMS, dark-blue pictogram) distributes the energy within the subsystem in order to reduce the energy consumption. Equations and details on the control are given in [12].

The hybrid powertrain is then implemented in the Simcenter AMESIM software using the EMR library developed by Siemens within the PANDA project [9]. The simulation model is available in the PANDA cloud.

It can be noted that a static map of efficiency is used for the electric drive in this simulation. However, the PANDA cloud has multi-level models of the electric drive that can be seamlessly interchanged according to the EMR organization, for more in-depth studies [12]. The use of a static map is also a way to preserve the confidentiality of this innovative low-voltage electrical machine.

The simulation study results are given from *Fig. 5 to Fig. 9*. For a WLTC class3 reference cycle (*Fig. 5*) that corresponds to a total cumulated distance of 23.3 km under various conditions (i.e. urban to highway). Simulation results on battery current (*Fig. 7*), battery voltage (*Fig. 8*) show that the electrical limits (Fig. 2, Fig. 3) are not crossed. The evolution of the SoC gives an idea of the driving range in hybrid mode ($\approx$180km). This simulation is used as a pre-validation step. The power test can be achieved in the next section.

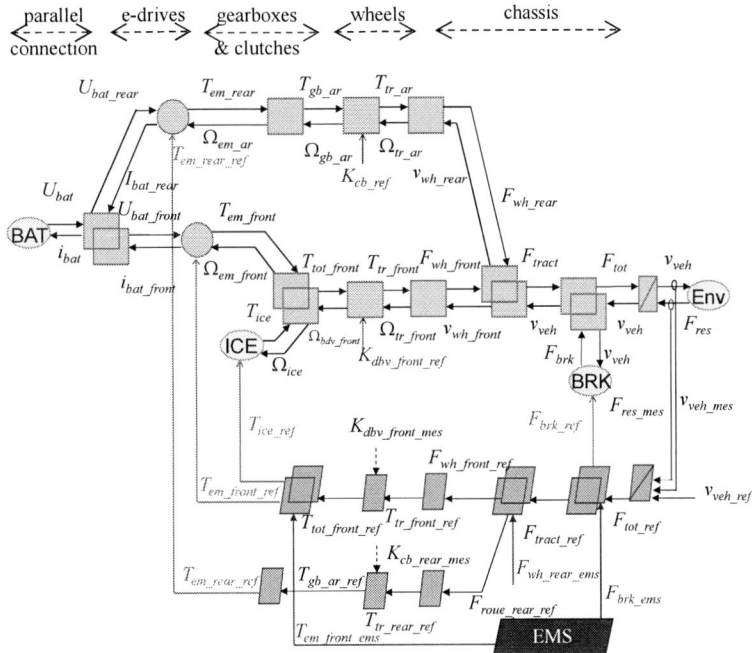

Fig. 4. EMR and control for the studied P-HEV

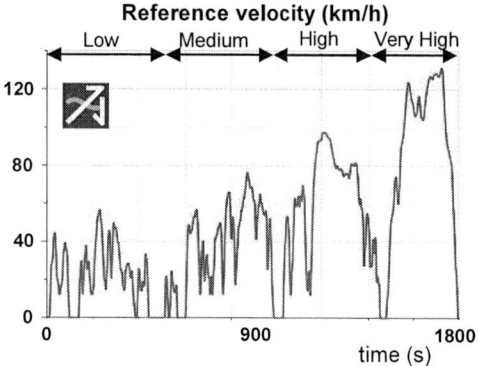

Fig. 5 WLTC reference velocity cycle

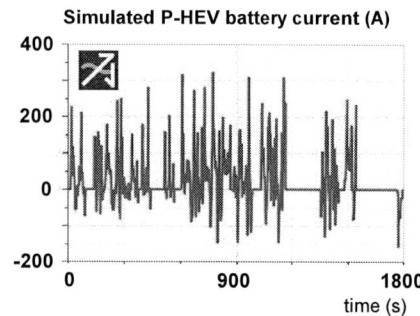

Fig. 7 Simulated battery current

Fig. 6 Cumulated distance

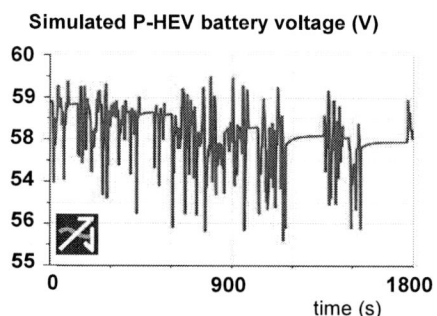

Fig. 8 Simulated battery voltage

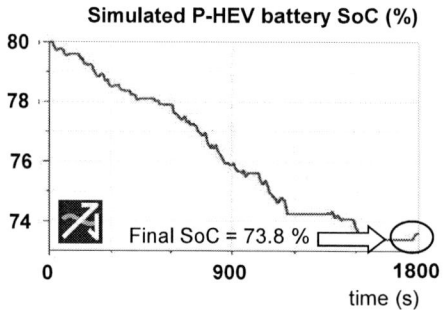

Fig. 9 simulated battery SoC

Power Hardware-in-the-Loop testing

A. Power HIL principle

As the simulation study has given results compatible with the tested battery, the power test is achieved. In order to test the battery, the battery model is replaced by a real battery in the simulation loop (Fig. 10). The hybrid traction subsystem is simulated in real-time in a dedicated ECU (Electronic Control Unit). As the simulation only delivers signal variables, a power amplifier is needed to generate the load current to the tested battery. This power amplifier is an interface between the real device under test and the simulation part.

The power amplifier is a current source and is directly connected to the battery to test. The measured experimental battery voltage is used as an input for the real-time simulation while the current runs through the battery. Thus, there is a real time interaction between the two parts.

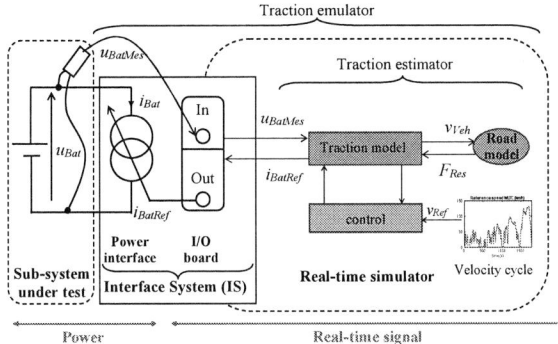

Fig. 10. HIL organization

B. HIL set-up

A Typhoon ECU is in charge of the real-time simulation of the hybrid traction subsystems (Fig. 11). A dedicated compiler has been developed to translate the Simcenter AMESIM model from the cloud to the ECU code. This ECU is connected to a Cinergia power amplifier to generate the load current for the battery. A unique Blueways module is considered in this test and the traction current is divided by two in order to test the module as in the real vehicle (where 2 modules are connected in parallel).

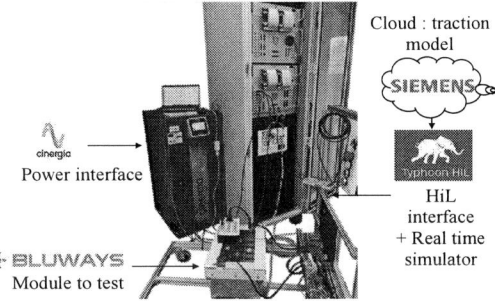

Fig. 11. HIL experimental set-up

C. Experimental results

The same driving cycle is imposed (Fig. 12, Fig. 13) for the power test. One module is tested instead of the whole battery (Fig. 3). The current is measured (Fig. 14). The module voltage measured from the BMS and the voltage imposed to the power amplifier are very close (Fig. 15). The State-of-Charge (SoC) has the same evolution (Fig. 16) than in the pure simulation (Fig. 7). Experimental test allows to get more information than simulation. Here, the experimental test bench is instrumented with thermocouples. Thus, the evolution of the module temperature can be recorded. During the test, the temperature has variation within the battery limits (Fig. 17). This confirms that a cooling system is of no interest with the new battery.

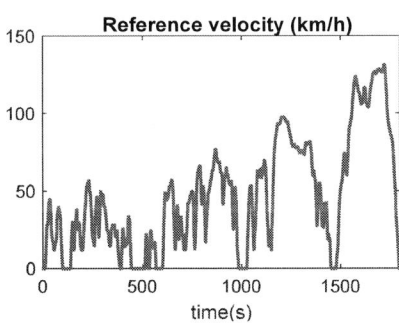

Fig. 12 WLTC reference velocity cycle

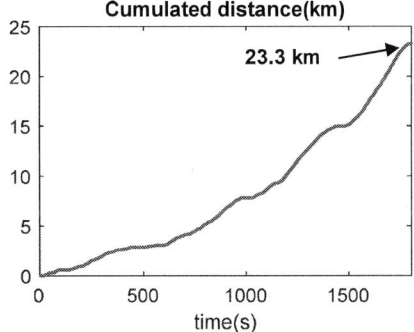

Fig. 13 Cumulated distance

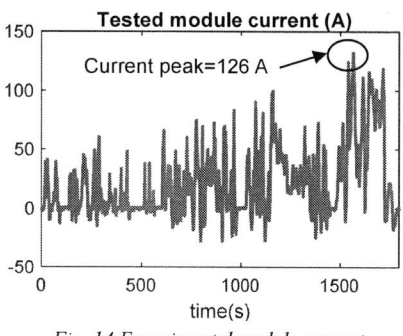

Fig. 14 Experimental module current

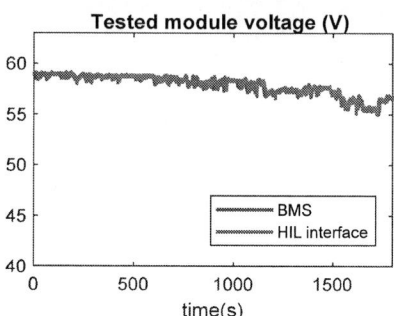

Fig. 15 Experimental module voltage

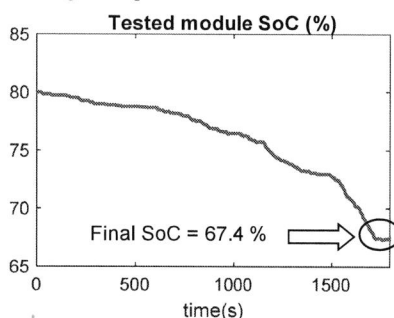

Fig. 16 Experimental module SoC

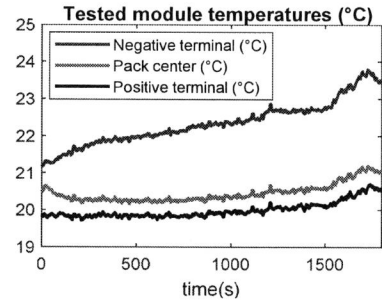

Fig. 17 Experimental temperature

Conclusion

A new battery has been tested before its integration on a demo hybrid electric vehicle. A real-time simulation model has been used for the power hardware-in-the-loop testing of the battery. The new battery can achieve the performances requested by the vehicle. This kind of test also gives additional indications. As a matter of fact the negligible experimental self-heating of the new battery indicates that a cooling system is optional.

Acknowledgment

This paper has been realized within the framework of the PANDA project which has received funding from the European Union's Horizon 2020 research and innovation program under grant agreement no. 824256 (PANDA).

References

[1] "Global EV outlook 2018, towards cross-modal electrification", International Energy Agency report, 2018.

[2] C. C. Chan, 'Overview of electric, hybrid and fuel cell vehicles', in Encyclopedia of Automotive Engineering. Hoboken, ch. 5, NJ, USA: Wiley, 2015

[3] M. Ehsani, Y. Gao, S. E. Gay, A. Emadi, 'Modern Electric, Hybrid Electric, and Fuel Cell Vehicles'. Elec. Eng., CRC Press, ISBN 9781138330498, February 2018.

[4] G. Falchetta , N. Noussan, "Electric vehicle charging network in Europe: An accessibility and deployment trends analysis" Transportation Research Part D: Transport and Environment, Vol. 94, May 2021, #102813

[5] K. Osieczkoa, D. Zimona, E. Płaczekb, I. Prokopiukc, "Factors that influence the expansion of electric delivery vehicles and trucks in EU countries, Journal of Environmental Management, Vol. 296, October 2021, ref. 113177.

[6] "Virtual product development and production of all types of electrified vehicles and components", Horizon 2020 programme of the European Comission, Online: https://cordis.europa.eu/programme/id/H2020_LC-GV-02-2018/fr , Consulted in December 2021

[7] A. Bouscayrol, A. Lepoutre, C. Irimia, C. Husar, J. Jaguemont, A. Lièvre, C. Martis, D. Zuber, V. Blandow, F. Gao, W. Van Dorp, G. Sirbu, J. Lecoutere, "Power Advanced N-level Digital Architecture for models of electrified vehicles and their components", *Transport Research Arena 2020*, Helsinki.

[8] A. Bouscayrol, J. P. Hautier, B. Lemaire Semail, "Systemic design methodologies for electrical energy systems Analysis, Synthesis and Management", Chapter 3: Graphic formalisms for the control of multi-physical energetic system: COG and EMR, *ISTE and Wiley*, ISBN 978-1-84821-3888-3, 2012

[9] C. Husar, M. Grovu, C. Irimia, A. Desreveaux, A. Bouscayrol, M. Ponchant, P. Magnin, "Comparison of Energetic Macroscopic Representation and structural representation on EV simulation under Simcenter Amesim", IEEE-VPPC'19, Hanoi (Vietnam), October 2019.

[10] S.D Lee, J. Cherry, M. Safoutin, J. McDonald, M. Olechiw, "Modeling and Validation of 48V Mild Hybrid Lithium-Ion Battery Pack", *SAE International Journal of Alternative Powertrains*, Vol. 7, no. 3, pp. 273-288, April 2018.

[11] J Yuan, L Yang, "Predictive energy management strategy for connected 48V hybrid electric vehicles", *Energy*, 2019, Vol. 187, November 2019, ref. 115952.

[12] F. Tournez, R. Vincent, W. Lhomme, S. Roquet, A. Bouscayrol, M. Ahmed, B. Lemaire-Semail, A. Lièvre, "Difference between average efficiency and efficiency map of the electric drive on fuel saving estimation for P-HEV", *IEEE-VPPC'21*, Gijon (Spain), October 2021.

[13] A. Bouscayrol, "Hardware-In-the-Loop simulation" in Industrial Electronics Handbook, Chicago:CRC Press, Taylor & Francis group, pp. 33-1-33-15, March 2011.

[14] A. Sharma, R. Nusrat, M.A. Buhuiya, M.Z. Youssef, "Hardware-in-the-Loop Validation of Different Power Train Topologies' Models in Electric Vehicles: A Plug-and-Play Capability", IEEE open journal of vehicular technology, vol. 2, 2021, pp.365-376.

Stability Analysis of DFIG System connected with High-Frequency Capacitive Grid based on Closed-Loop Current Control and Direct Power Control

Bin Hu[1], Heng Nian[1], Subham Sahoo[2], Frede Blaabjerg[2], Yaqian Zhang[3], Zixiao Xu[4]

[1]College of Electrical Engineering
Zhejiang University
Hangzhou, China
11810031@zju.edu.cn
nianheng@zju.edu.cn

[3]College of Electrical Engineering
Southeast University
Nanjing, China
yaqianzhang83@seu.edu.cn

[2]Department of Energy Technology
Aalborg University
Aalborg, Denmark
sssa@energy.aau.dk
fbl@energy.aau.dk

[4]Department of Electrical Engineering
Northwestern Polytechnical University
Xi'an, China
xuzixiao_9602@mail.nwpu.edu.cn

Acknowledgements

This work was supported by the National Natural Science Foundation of China under Grant 51977194.

Keywords

«Impedance analysis», «Coupling characteristics», «Direct power control», «Doubly-Fed Induction Generator», «Stability analysis»

Abstract

This paper analyzes the stability of doubly-fed induction generator (DFIG) system connected with high-frequency capacitive grid based on closed-loop current control (CCC) and direct power control (DPC). There will be some high-frequency resonance issues when employs DPC. Consequently, the reason of this high-frequency resonance and frequency coupling characteristic are studied.

Introduction

Doubly-fed induction generator (DFIG) has the advantages of low cost, small converter rating, competitive durability and flexible power adjustment capability, which is popular for the commercial wind power generation system around the globe [1], [2]. Nowadays, the grid network has been more complex due to the diversified equipment, such as parallel compensation, long transmission cable, high voltage DC (HVDC), and STATCOM [3]. It is widely adopted that the interconnected grid with DFIG will change from inductive grid to high-frequency capacitive grid [3]. An impedance sweeping test has been employed in the AC network of the Guangxi-side of Luxi district in China. It can be found that the AC grid behaves as a capacitance at high frequency, and the cut-off frequency is 866 Hz [4].

Closed-loop current control (CCC) and direct power control (DPC) are two common control strategies for DFIG system [5]. Compared with CCC, the DPC can directly control the active power and reactive power, which can enhance the power dynamic response [6]. Furthermore, Ohnishi proposed a simple structure of DPC combining the instantaneous power theory and direct torque control [7]. This DPC is based on the PWM modulation without the selection of single voltage vector and the application of hysteresis controller [8], [9].

This paper obtains an interesting phenomenon, where DFIG system based on CCC is always stable under the high-frequency capacitive grid, while the DFIG system based on DPC has some resonance issues. And there will be high-frequency coupling characteristics during the high-frequency oscillation of DFIG system. However, the DFIG system based on CCC may behave as a passivity inductance at high frequency, which is often misinterpreted as the origin of high-frequency resonance and high-frequency coupling characteristics.

Topology of DFIG system connected with high-frequency capacitive grid

The investigated configuration diagram of DFIG system is depicted in Fig. 1. The control signals of DFIG system are generated from the rotor side converter (RSC) and grid side converter (GSC). Normally, the GSC can keep the stable dc voltage for DFIG system, and the RSC can achieve the maximum power tracking (MPPT) and provide the maximum expected active power.

Except the inductive weak grid, there are two common configurations of weak networks to satisfy the requirement of power quality compensation and long-distance transmission: 1) parallel compensated network with grid inductance L_g and compensated capacitance C_{com}; 2) long transmission cable modelled as the series connection of several Π units with the cable resistor R_1, the cable inductor L_1 and the cable shunt capacitor $C_1/2$.

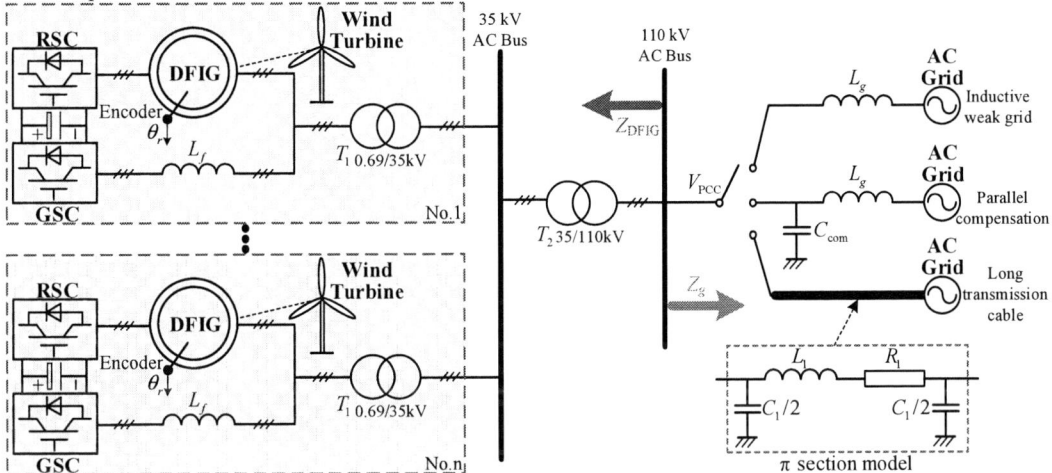

Fig. 1: Investigated configuration diagram of DFIG system

Compared with the inductive weak grid, the parallel compensated network and long transmission cable contain the parallel capacitance, which always present capacitive at high frequency. An impedance sweeping test has been employed in the AC network of the Guangxi-side of Luxi district in China. It can be found that the AC grid can be equivalent to the high-frequency capacitive grid in Fig. 2 with the cut-off frequency of 866 Hz [4]. To be honest, it is more reasonable to model the long transmission cable with multiple Π units. However, this paper pays more attention to the stability issues of different control strategies when there is a parallel capacitor in the weak grid, so the long transmission cable is also regarded as a high-frequency capacitive grid for simplify.

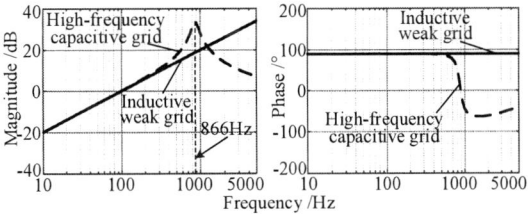

Fig. 2: Bode diagram of inductive weak grid and high-frequency capacitive grid

Fig. 3 depicts the simplified topology of DFIG system connected with high-frequency capacitive grid. According to Fig. 3, the DFIG system is connected to a high-frequency capacitive grid, such that the additional resistor R_g and capacitance C_g will be parallel with the grid inductance L_g. The other parameters of DFIG system is shown in Table I.

Table I: Parameters of DFIG system

Symb	Parameter	Value
U_s	Rated voltage	690 V
P_s	Rated power	1.5 MW
f_1	Fundamental frequency	50 Hz
f_r	Rotor frequency	35 Hz
n_p	Pole pairs	2
V_{dc}	Dc-link voltage	1050 V
L_{ls}	Stator leakage	0.06 mH
L_{lr}	Rotor leakage	0.083 mH
L_m	Mutual inductance	4.43 mH
R_s	Stator resistance	2.4 mΩ
R_r	Rotor resistance	2 mΩ
K_e	Turns ratio	0.33
f_s	Switching frequency	5 kHz

The GSC is ignored in this paper and the dc-link voltage V_{dc} is assumed to be constant, the reason is as follow: 1) The overall impedance of DFIG system can be equal to the parallel connection of DFIG+RSC and GSC, but the one with larger magnitude will play a leading role; 2) the magnitude of impedance is related to the output power under the same port voltage, while the GSC can only output the limited slip power; 3) there is a L or LCL type filter between the point of common coupling (PCC) and the GSC, which further increase the magnitude of GSC.

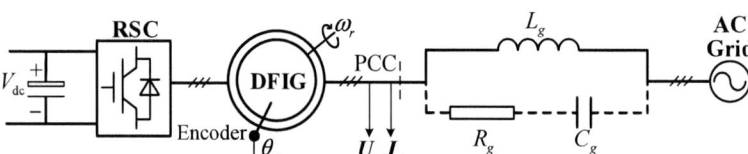

Fig. 3: Simplified topology of DFIG system connected with high-frequency capacitive grid

High-frequency resonance issues when switching from CCC to DPC

The resonance issues of DFIG system based on DPC connected with high-frequency capacitive grid is shown in Fig. 4. When switched from DPC to CCC, the resonance issues can be suppressed. Fig. 5 is the FFT analysis, it can be found that this high-frequency resonance issues have a frequency coupling characteristics, indicating that there will be two frequencies differing by 100 Hz in the spectrum together. However, the frequency coupling always exist in an asymmetrical way in the controller. It is worth notifying that the influence of controller has been ignored due to the limited bandwidth. In addition, when the proportional integral (PI) parameter of DPC controller is increased, there will be multiple coupling frequencies.

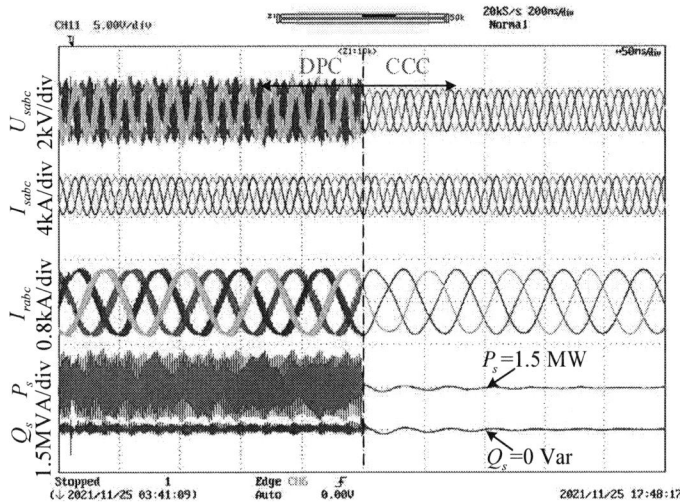

Fig. 4: Resonance issues of DFIG system based on DPC connected with high-frequency capacitive grid

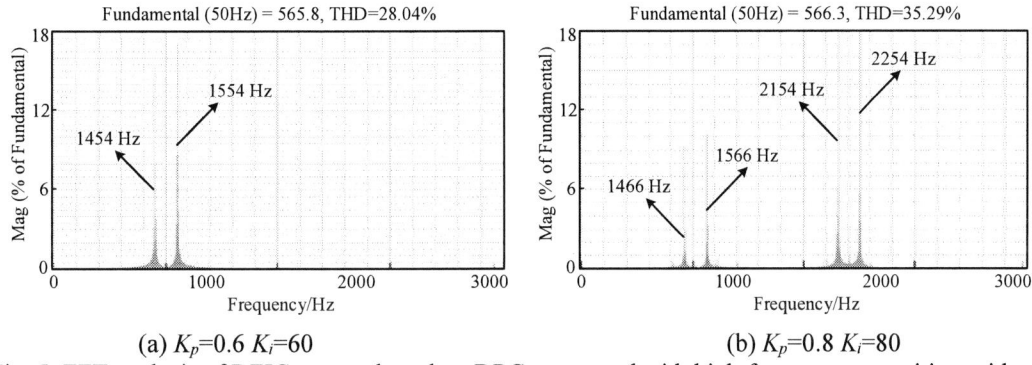

(a) K_p=0.6 K_i=60

(b) K_p=0.8 K_i=80

Fig. 5: FFT analysis of DFIG system based on DPC connected with high-frequency capacitive grid

Impedance modeling and analysis of Nyquist diagram

This paper establishes the impedance model of DFIG system based on CCC and DPC as shown in Fig. 6. And Fig. 7 depicts the Nyquist diagram for the developed impedance model.

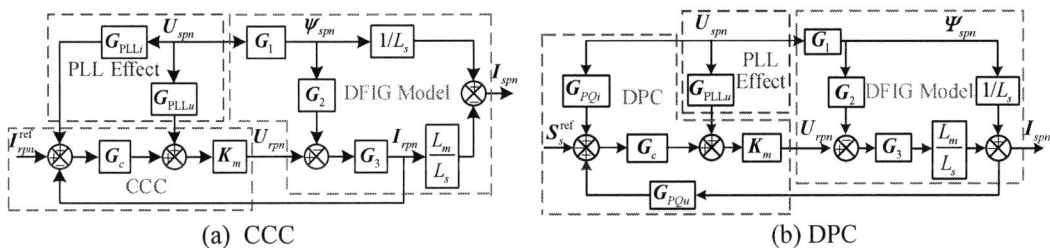

(a) CCC

(b) DPC

Fig. 6: Impedance model of DFIG system based on CCC and DPC

$$\boldsymbol{G}_1 = \begin{bmatrix} \dfrac{1}{s} & 0 \\ 0 & \dfrac{1}{s - 2\mathrm{j}\omega_g} \end{bmatrix} \quad \boldsymbol{G}_2 = \dfrac{L_m}{L_s} \begin{bmatrix} s - \mathrm{j}\omega_r & 0 \\ 0 & s + \mathrm{j}\omega_r - 2\mathrm{j}\omega_g \end{bmatrix}$$

$$\boldsymbol{G}_3 = \begin{bmatrix} \dfrac{1}{R_r + \left(s - \mathrm{j}\omega_r\right)L_r\sigma} & 0 \\ 0 & \dfrac{1}{R_r + \left(s + \mathrm{j}\omega_r - 2\mathrm{j}\omega_g\right)L_r\sigma} \end{bmatrix} \tag{1}$$

$$\boldsymbol{G}_c = \begin{bmatrix} \dfrac{K_{pc}\left(s - \mathrm{j}\omega_g\right) + K_{ic}}{s - \mathrm{j}\omega_g} & 0 \\ 0 & \dfrac{K_{pc}\left(s - \mathrm{j}\omega_g\right) + K_{ic}}{s - \mathrm{j}\omega_g} \end{bmatrix} \tag{2}$$

$$\boldsymbol{K}_m = \begin{bmatrix} e^{-1.5s/f_s} & 0 \\ 0 & e^{-1.5(s - 2\mathrm{j}\omega_g)/f_s} \end{bmatrix} \tag{3}$$

$$\boldsymbol{G}_{\mathrm{PLL}i} = \frac{1}{2}\begin{bmatrix} -\boldsymbol{I}_{rdq0}H_{\mathrm{PLL}}(s) & \boldsymbol{I}_{rdq0}H_{\mathrm{PLL}}(s) \\ \boldsymbol{I}^*_{rdq0}H_{\mathrm{PLL}}(s) & -\boldsymbol{I}^*_{rdq0}H_{\mathrm{PLL}}(s) \end{bmatrix} \qquad \boldsymbol{G}_{\mathrm{PLL}u} = \frac{1}{2}\begin{bmatrix} \boldsymbol{U}_{rdq0}H_{\mathrm{PLL}}(s) & -\boldsymbol{U}_{rdq0}H_{\mathrm{PLL}}(s) \\ -\boldsymbol{U}^*_{rdq0}H_{\mathrm{PLL}}(s) & \boldsymbol{U}^*_{rdq0}H_{\mathrm{PLL}}(s) \end{bmatrix} \tag{4}$$

$$\boldsymbol{G}_{PQu} = \begin{bmatrix} \boldsymbol{U}_{sdq0} & 0 \\ 0 & \boldsymbol{U}^*_{sdq0} \end{bmatrix} \quad \boldsymbol{G}_{PQi} = \begin{bmatrix} 0 & \boldsymbol{I}_{sdq0} \\ \boldsymbol{I}^*_{sdq0} & 0 \end{bmatrix} \tag{5}$$

where the $\boldsymbol{G}_1$, $\boldsymbol{G}_2$ and $\boldsymbol{G}_3$ are the DFIG parameter matrices. $\boldsymbol{G}_c$ is the current controller matrix, that K_{pc} and K_{ic} are the proportional gain and integral gain of current controller. $\boldsymbol{K}_m$ is the system delay matrix. $\boldsymbol{G}_{\mathrm{PLL}i}$ and $\boldsymbol{G}_{\mathrm{PLL}u}$ are the PLL matrices related to the rotor current coordinate transformation and rotor voltage coordinate transformation. $\boldsymbol{G}_{\mathrm{PLL}i}$ and $\boldsymbol{G}_{\mathrm{PLL}u}$ are the power calculation matrices. The subscripts p and n denote the positive and negative sequence components. The superscript ref is the reference value. * represents the conjugate operator. The subscript 0 denotes the steady-state component. $H_{\mathrm{PLL}}(s) = H_P(s)/[U_{sd0}H_P(s) + s - \mathrm{j}\omega_g]$. The PI controller of the PLL is denoted as $H_P(s) = K_{pp} + K_{ip}/(s - \mathrm{j}\omega_g)$, in which K_{pp} and K_{ip} are the proportional gain and integral gain of PLL. K_{pc} and K_{ic} are the proportional gain and integral gain of current or power controller. ω_g is the grid angular frequency, and ω_r is the rotor angular frequency. $\sigma = 1 - L_m^2/(L_s \cdot L_r)$.

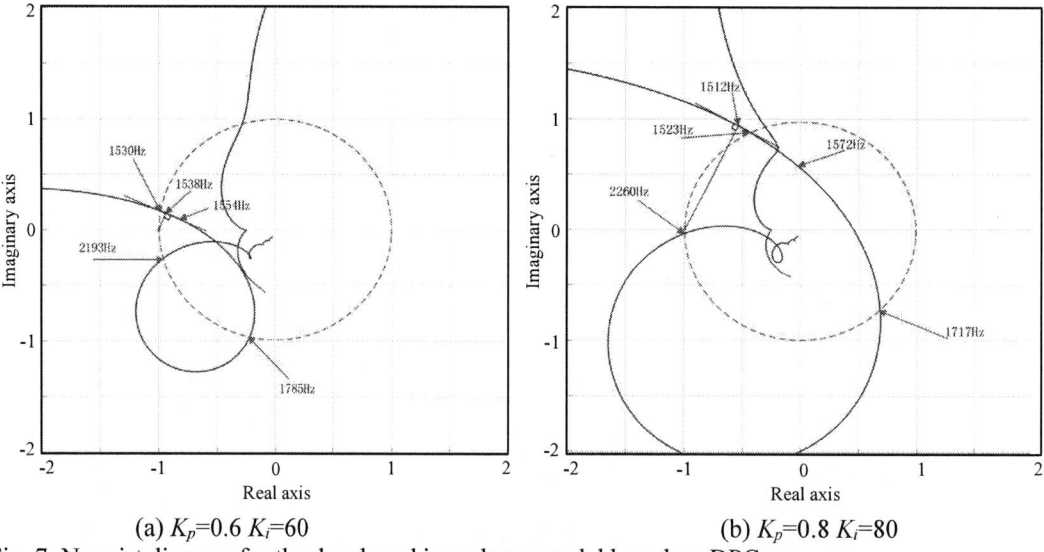

(a) K_p=0.6 K_i=60   (b) K_p=0.8 K_i=80

Fig. 7: Nyquist diagram for the developed impedance model based on DPC

According to the Nyquist diagram in Fig. 7, it can be found that the Nyquist trajectory will pass through the unit circle several times, which is the reason of insufficient phase margin. Apart from that, there will be some errors between FFT results in Fig. 5 and Nyquist diagram in Fig.7. It is caused by the output rotor voltage limitation or PWM limitation. We can modify the impedance model by some describing function when considering these limitations, but these limitations will not influence the cause analysis of resonance issues.

Frequency coupling analysis for DPC and CCC

The simplified model assumes $L_s \approx L_r \approx L_m$ to ignore leakage inductance, and assume $R_g \approx 0$. The influence of PLL can be ignored at high frequency due to limited PLL bandwidth. The $j\omega_g$ and $j\omega_r$ are much smaller than s at high frequency, then the impedance model in Fig. 6 (a) and (b) can be simplified in (6) and (7).

$$
\begin{cases}
Z_{CCC11} = Z_{CCC22} \overset{high\,f}{\approx} \dfrac{s^2 L_m^2 \sigma + s L_m K_{pc} e^{-1.5 s/f_s}}{s L_m + K_{pc} e^{-1.5 s/f_s}} \overset{s\to\infty}{\approx} s L_m \sigma + K_{pc} e^{-1.5 s/f_s} \\
\\
Z_{CCC12} = Z_{CCC21} \overset{high\,f}{\approx} 0
\end{cases}
\tag{6}
$$

$$
\begin{cases}
Z_{DPC11} = Z_{DPC22} \overset{high\,f}{\approx} \dfrac{s L_m \sigma + U_{sdq0} K_{pc} e^{-1.5 s/f_s}}{\sigma + 1} \overset{s\to\infty}{\approx} s L_m \sigma + K_{pc} e^{-1.5 s/f_s} \\
\\
Z_{DPC12} = Z_{DPC21} \overset{high\,f}{\approx} \dfrac{s L_m \sigma + U_{sdq0} K_{pc} e^{-1.5 s/f_s}}{I_{sdq0} K_{pc} e^{-1.5 s/f_s}} \overset{s\to\infty}{\approx} \dfrac{s L_m \sigma}{I_{sdq0} K_{pc} e^{-1.5 s/f_s}}
\end{cases}
\tag{7}
$$

It can be found that there is no frequency coupling element for CCC, but the DPC has four elements at high frequency. When the s approaches infinity, and consider $\sigma \approx 0$ and $U_{sdq0}=1$, the Z_{CCC11} and Z_{CCC22} have the same form with Z_{DPC11} and Z_{DPC22}. So the most significant difference is that Z_{DPC12} and Z_{DPC21} will not approach to 0 at high frequency, which cause the frequency coupling. The effect of time delay does not decay in frequency coupling element at high frequency, thus the Nyquist trajectory will pass through the unit circle several times in Fig. 7.

In addition, as for inner-current loop and outer-power loop, the time delay will also affect the frequency coupling element when the bandwidth of outer-power loop is large enough.

Conclusion

This paper compares the impedance characteristic of DFIG system between CCC and DPC. From the analysis, it can be found some high-frequency resonance issues under high-frequency capacitive grid when employing DPC. This high-frequency resonance has some interesting phenomenon, such as multiple coupling frequency at high frequency. And this paper analyzes the cause of frequency coupling from DPC, that the time delay in frequency coupling element will deteriorate the stability. In brief, it is not recommended to employ the DPC connected to high-frequency capacitive grid without any impedance shaping methods.

References

[1] B. Hu, H. Nian, M. Li, Y. Xu, Y. Liao and J. Yang, "Impedance-based analysis and stability improvement of DFIG system within PLL bandwidth," *IEEE Trans. Ind. Electron.*, vol. 69, no. 6, pp. 5803-5814, Jun. 2022.

[2] H. Nian, B. Hu, Y. Xu, C. Wu, L. Chen and F. Blaabjerg, "Analysis and reshaping on impedance characteristic of DFIG system based on symmetrical PLL," *IEEE Trans. Power Electron.*, vol. 35, no. 11, pp. 11720-11730, Nov. 2020.

[3] R. Wang, Q. Sun, W. Hu, Y. Li, D. Ma and P. Wang, "SoC-based droop coefficients stability region analysis of the battery for stand-alone supply systems with constant power loads," *IEEE Trans. Power Electron.*, vol. 36, no. 7, pp. 7866-7879, Jul. 2021.

[4] C. Zou et al., "Analysis of resonance between a VSC-HVDC converter and the AC Grid," *IEEE Trans. Power Electron.*, vol. 33, no. 12, pp. 10157-10168, Dec. 2018.

[5] B. Hu, H. Nian, M. Li and Y. Xu, "Impedance characteristic analysis and reshaping method of DFIG system based on DPC without PLL," *IEEE Trans. Power Electron.*, vol. 68, no. 10, pp. 9767-9777, Oct. 2021.

[6] B. Hu, H. Nian, J. Yang, M. Li and Y. Xu, "High-frequency resonance analysis and reshaping control strategy of DFIG system based on DPC," *IEEE Trans. Power Electron.*, vol. 36, no. 7, pp. 7810-7819, Jul. 2021.

[7] T. Ohnishi, "Three phase PWM converter/inverter by means of instantaneous active and reactive power control," *in Proc. Int. Conf. on Ind. Electron.*, Nov. 1991, pp. 819–824.

[8] Z. Xie, W. Wu, Y. Chen and W. Gong, "Admittance-based stability comparative analysis of Grid-connected inverters with direct power control and closed-loop current control," *IEEE Trans. Ind. Electron.*, vol. 68, no. 9, pp. 8333-8344, Sept. 2021.

[9] B. Hu, H. Nian, M. Li, Y. Liao, J. Yang and H. Tong, "Impedance characteristic analysis and stability improvement method for DFIG system within PLL bandwidth based on different reference frames," *IEEE Trans. Ind. Electron.*, to be published. doi: 10.1109/TIE.2022.3150092.

Full-Bridge Modular Multilevel Converter for the Four-Quadrant Supply of High Power Magnets in Particle Accelerators

Manuel Colmenero*, Ricardo Vidal-Albalate[†], Francisco R. Blanquez*, Ramon Blasco-Gimenez[+]

*CERN - European Organization for Nuclear Research

[†] Universitat Jaume I de Castelló

[+]Universitat Politècnica de València

*1 Espl. des Particules, Meyrin, Switzerland

[†]Av. de Vicent Sos Baynat, s/n, Castellon, Spain

[+]Camino de Vera, s/n, Valencia, Spain

Email: mcolmene@cern.ch

URL: http://www.cern.ch

Keywords

≪Modular Multilevel Converters (MMC)≫, ≪Medium voltage converter≫, ≪Converter Control≫,≪Particle accelerators≫.

Abstract

Many particle accelerators require to supply chains of magnets with high quality, high magnitude, cycling currents. To do this, the power converters need to provide high output voltages, reaching in some cases tens of kilovolts. Additionally, converters are required to store the magnet energy during de-magnitasion cycles. For such application, Full-bridge Modular Multilevel Converters (FB-MMC) could be used given their capacity to store energy, their inherent reliability and their good harmonic performance. This paper studies how this converter topology could be used for this application, proposing a method to recover and store the energy of the magnet using the converter submodules.

Introduction

In particle accelerators, it is desired to precisely regulate the magnetic field generated by a set of magnets to control particle trajectories. However, the magnetic field is not always constant: it is small at the beginning of the cycle, when particles are injected into the machine, it increases during acceleration, remains constant during extraction and is brought back to the initial value at the end of the cycle. In some machines, this cycle happens every few seconds, leading to fast variations of the magnetic field and thus, of the current required to generate it.

In some machines, high currents, in the order of kA, are required for bending the trajectory of particles. This is the case of the CERN's Proton Synchrotron accelerator, where switch-mode power converters are used to supply the main dipole magnets [1]. This system generates high-quality pulses of current, up to 6 kA, every few seconds. To this avail, the converter must generate a voltage across the magnet of ± 10 kV. In total, the system is dimensioned to handle an output power of 60 MW.

The design of such a powering system had to cope with several technical issues. The first concerns the rating of the semiconductors: the high current demanded by the application requires to connect several switches in parallel to form a basic commutation cell, which is then connected in series with others to meet the high voltage requirement. The second concerns the high power demand. Although the converters are connected to a strong distribution network, the large power fluctuations are enough to cause power quality issues. To avoid these quality issues, large DC capacitors banks are installed to store the magnetic energy and shave the peak power absorbed from the network.

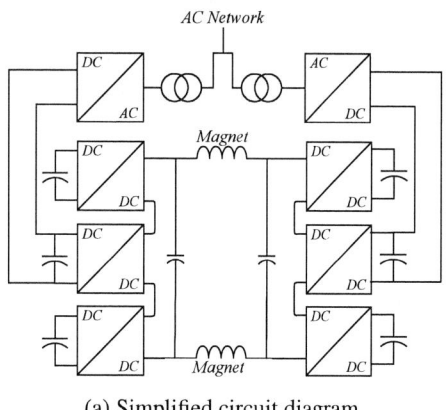

(a) Simplified circuit diagram.

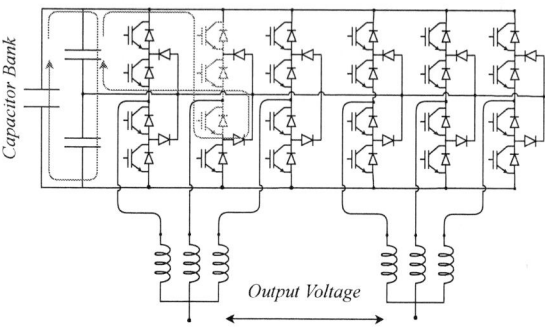

(b) A semiconductor failure can cause the discharge of the capacitor bank (red).

Fig. 1: Architecture of the powering system for the Proton Synchrotron at CERN [1].

The system described in [1], has been under operation since 2011, and constists of three-phase, three-level Neutral-Point Clamped converters as the basic switching cell. Two of these cells form a DC/DC converter, and there are six DC/DC converters connected in series with the magnets. To store the magnetic energy, each of the DC/DC converters is connected to a large capacitor bank. The losses on the converter and on the magnets are provided by two basic switching cells used as Active Front Ends, which are connected to two of the capacitor banks. To balance the rest of capacitors, a energy balancing strategy that distributes the energy is implemented. Fig. 1a shows the simplified circuit diagram of the system whereas Fig. 1b shows the diagram of one of the DC/DC converters.

One of the main disadvantages of this system is the lack of modularity. In particular, if a device fails, the affected DC/DC converter and, thus, the whole system needs to be stopped for repairing or moved to a degraded operation mode. Moreover, if the device fails into a short-circuit, then there is risk that a significant part of the stored energy discharges through the faulty converter leg (see Fig. 1b). Considering the large size of the capacitor banks, such a failure could cause considerable damage to the equipment concerned, leading to a long unavailability period. Another disadvantage concerns the high ratings required for the semiconductors. Since the commercial availability of high-voltage, high-current components is limited, there is little room for the optimization of the switching cells.

Therefore, in the framework of the studies for the construction of a new accelerator at CERN [2], this paper proposes to use Modular Multilevel Converters based on Full-Bridge submodules [3] (FB-MMC) as an alternative to the existing technology when high output voltages and energy storage are required. A system based on the FB-MMC can achieve high output voltages employing conventional low voltage semiconductors, allowing a better optimization of the converter. Besides, reliability can be significantly improved since the high modularity of the MMC allows to bypass a cell in case of failure without significantly impacting the operation of the converter [4]. Additionally, by using the submodules to store the energy from the magnets, the safety of the system can be increased by reducing the energy that can be potentially discharged in the case of a fault, which would be only a fraction of the total, and whose damage would be limited to a single cell, easily replaceable. Besides these features, there are other well-known advantages of using MMCs, as their high efficiency [5] and their low harmonic distortion [6], which would allow to reduce the size and cost of the filters.

Accordingly, the following sections describe the implementation of a particle accelerator powering system using MMCs. It explains how the submodules need to be dimensioned to store the magnet energy and how the control needs to be modified to incorporate this new feature. Then, the concepts developed are validated by simulation.

Magnet Powering System using Full-Bridge MMCs

The architecture of a magnet powering system based on Full-bridge MMCs is shown in Fig. 2. The system is composed of two conventional MMC AC/DC converters in back-to-back configuration, with the magnet chain, which is split in two, connected between the positive and negative terminals. To control the current, each one of the MMCs produces an output voltage across its terminals, which is similar to the one produced by the other converter but with opposite polarity. By doing this, the voltage that the stations need to produce is only half the voltage required by the magnets and the maximum voltage to ground the magnets see is significantly reduced (to one quarter of the total voltage in this case).

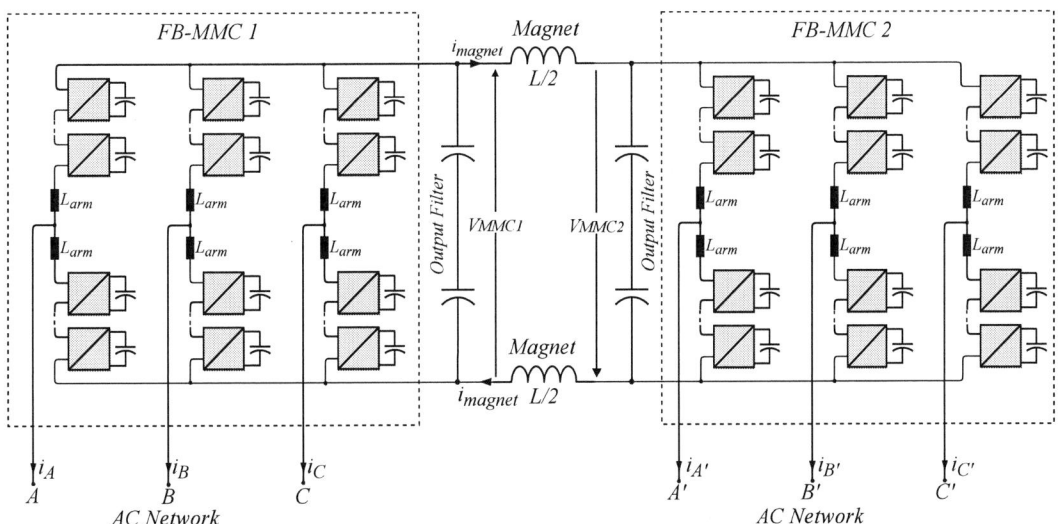

Fig. 2: Diagram of the proposed magnet power supply system based on Full-Bridge Modular Multilevel Converters.

A significant difference with the system described in [1] is that there are not dedicated converters working as Active Front Ends. Instead, the MMCs allow to manage directly the connection to the AC network while keeping the control of the output current and voltage [7]. On the other hand, contrary to conventional back-to-back systems, where power flows from one converter's network to the other [8], this application requires that both converters absorb power from the AC network. Absorbed power, which is a fraction of the total, is used to cover only the system losses. The energy required to create the magnetic field in the magnets comes exclusively from the MMC submodule capacitors. Therefore, and contrary to what happens in typical MMC applications, the voltages of the submodules are subjected to large voltage swings during the ramp-up and ramp-down of the magnets.

For this application, the large capacitor banks are replaced by a distributed storage using the MMC submodules. Converter design for this application requires a compromise between the number of submodules, their maximum and minimum voltages and the selection of the capacitors. Given the amount of energy to store, the converters must be able to generate the peak output voltage at the end of the ramp-up, where the charge, and thus the voltage of the submodule is minimal. On the other hand, a compromise between modularity and submodule voltage needs to be found: a high submodule voltage allows to reduce the number of semiconductors but requires higher ratings for IGBTs and would result in a poorer harmonic performance. Conversely, a high number of submodules would add extra complexity to the system and would result in excessive losses.

For the system under consideration, the converters must be able to generate an output voltage of 5 kV at the end of the ramp-up period. A rated submodule voltage of 1.2 kV offers a good compromise between number of cells and component availability. In fact, many commercially available MMCs employ similar voltages [9]. Finally, the total energy to be stored is 14 MJ, split between the two converters. With this

Table I: Main System Parameters

Parameter	Value
Rated DC voltage	± 5 kV
Rated AC voltage	2.6 kV
Number of Submodules	8 FB
Submodule Capacitance	0.29 F
Rated submodule voltage	1.2 kV
Minimum Submodule Voltage	0.625 kV

data, eight submodules per arm are selected, which results in a minimum submodule voltage of 625 V at the end of the ramp-up transient, enough to produce the 5 kV required by the magnets. A submodule capacitance of 0.29 F guarantees that the voltage does not drop below this value. Using these values, the rest of the converter parameters can be determined using some of the methods found in literature. Table I summarizes the main parameters of the converters.

Converter Control

The final objective of the MMCs is the accurate control of the magnet current while managing the energy balance between the magnets and the internal storage. To control the converter, several cascaded loops are required. At the highest level, there is a current regulator, where the measured current is compared with the reference given by the accelerator operators to generate the voltage references for the converters. The required magnet voltage, $V_{magnet}*$, is divided by two and sent to the converters with opposite signs. The individual references, ($V_{MMC1}*$ $V_{MMC2}*$), are passed directly to the low level control to generate the DC components of the arm voltages.

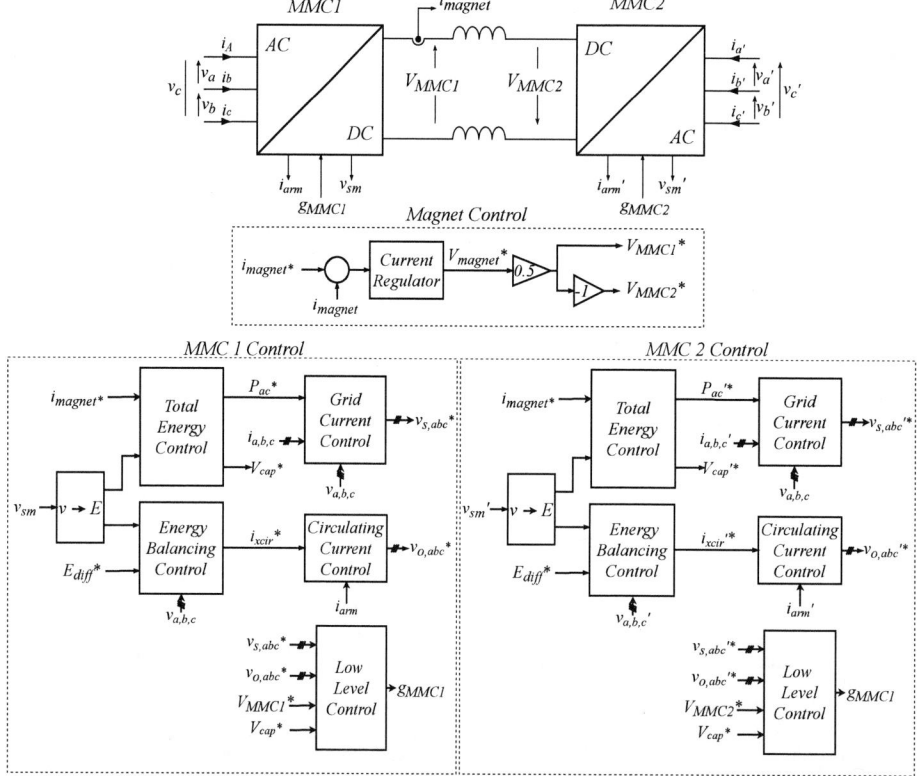

Fig. 3: Simplified diagram of the control of the system.

As previously explained, the control must ensure that the converter only absorbs from the AC grid the active power required to cover the converter and magnet losses. Hence, the magnetic energy needs to be provided exclusively from the submodule capacitors and has to be recovered at the end of each acceleration cycle. The relation between the converter energy (E), the AC (P_{AC}) and DC (P_{DC}) powers is:

$$\frac{dE}{dt} = P_{\text{in}} - P_{\text{out}} = P_{\text{AC}} - P_{\text{DC}} \tag{1}$$

where P_{out} is the power demanded by the magnets and P_{in} the power required form the AC grid to keep the MMC energy at its reference value. For the MMC energy control, P_{DC} can be considered as a disturbance and the MMC energy is controlled by means of P_{AC}. Neglecting losses, if the power imported from the ac grid is zero, the magnet power have to be obtained from the MMC (cell capacitors) at the expense of reducing their stored energy.

$$\frac{dE}{dt} = -P_{\text{DC}} \tag{2}$$

Therefore, in order to fed the magnets using the MMCs energy, the stored energy needs to be modify according to the magnet energization and de-energization cycle, that is, the cell capacitor voltages have to be reduced as the magnets current increases and vice versa. For this purpose, the following relation between the magnets and the MMC energy can be written:

$$\frac{1}{2} \cdot L \cdot i_{\text{magnet}}(t)^2 = \frac{1}{2} \cdot 6 \cdot C \cdot N \cdot (V_{\text{o,cap}}^2 - V_{\text{cap}}(t)^2) \tag{3}$$

where L is the magnet inductance, C is the submodule capacitance, N is the number of cells per arm, $V_{o,cap}$ is the initial capacitor voltage, $i_{magnet}(t)$ is the instantaneous magnet current and $V_{cap}(t)$ is the instantaneous cell capacitor voltage. Given that the magnets inductance and current waveform are known, the following cell voltage reference can be defined:

$$V_{\text{cap}}(t) = \sqrt{V_{\text{o,cap}}^2 - \frac{L}{6 \cdot C \cdot N} \cdot i_{\text{magnet}}(t)^2} \tag{4}$$

Using (4), the desired trajectory for the total energy of the MMC is constructed. Since the capacitor voltage measurements are available, it is possible to obtain the instantaneous energy stored by the converter and to compare it with the desired trajectory. The error is passed to the total energy controller, which regulates the power that needs to be absorbed by setting the d-axis AC current reference.

Horizontal and vertical balancing of leg and arm energies is achieved by controlling the circulating currents flowing through the arms. The horizontal balancing, or the difference between the legs energies, is controlled by setting a DC component on the circulating current reference. On the other hand, vertical balancing, or the difference between the upper and lower arms energy, is controlled by injecting AC components to the circulating currents. These components are of positive sequence to balance the common energy differences between upper an lower arms and of negative sequence to balance the differential energy differences between them. The current references generated by the three energy balancing controllers are passed to the circulating current control, which ensures their proper tracking by means of proportional-resonant regulators.

Finally, for the generation of the switching signals, a phase-shifted PWM modulation strategy is used considering the low number of submodules employed. To guarantee the individual balancing of the submodules, additional components are added to the modulation signal sent to each submodule to modify the connection/disconnection times and, subsequently, the charge of the capacitor. Additionally, the phase-shifts between the legs and arms PWMs are selected to maximize the harmonic performance of the converter. The simplified diagram of the control is shown in Fig. 3.

Simulation Results

A simulation model of the system has been developed in MATLAB/Simulink to verify the proper behavior of the proposed strategy. The model consists of two Full-Bridge MMC converters connected as proposed across a magnet chain of inductance L=0.9 H and resistance R=0.16 Ohms.The models for the converters include a detailed representation of the submodules, which allows to verify the proper behaviour of the balancing loops and the low-level controller. The modulation strategy used is the phase-shifted PWM, with a carrier frequency of Fc=250 Hz. To filter the output voltage ripple, second order output filter formed by the arm reactors and an output capacitor is added to the output terminals of the two stations. On the AC side, the converters are supplied by two wye/delta transformers with the delta on the secondary to allow the injection of third order harmonics. The supply voltage is 2.6 kV. The rest of parameters are collected in Table I.

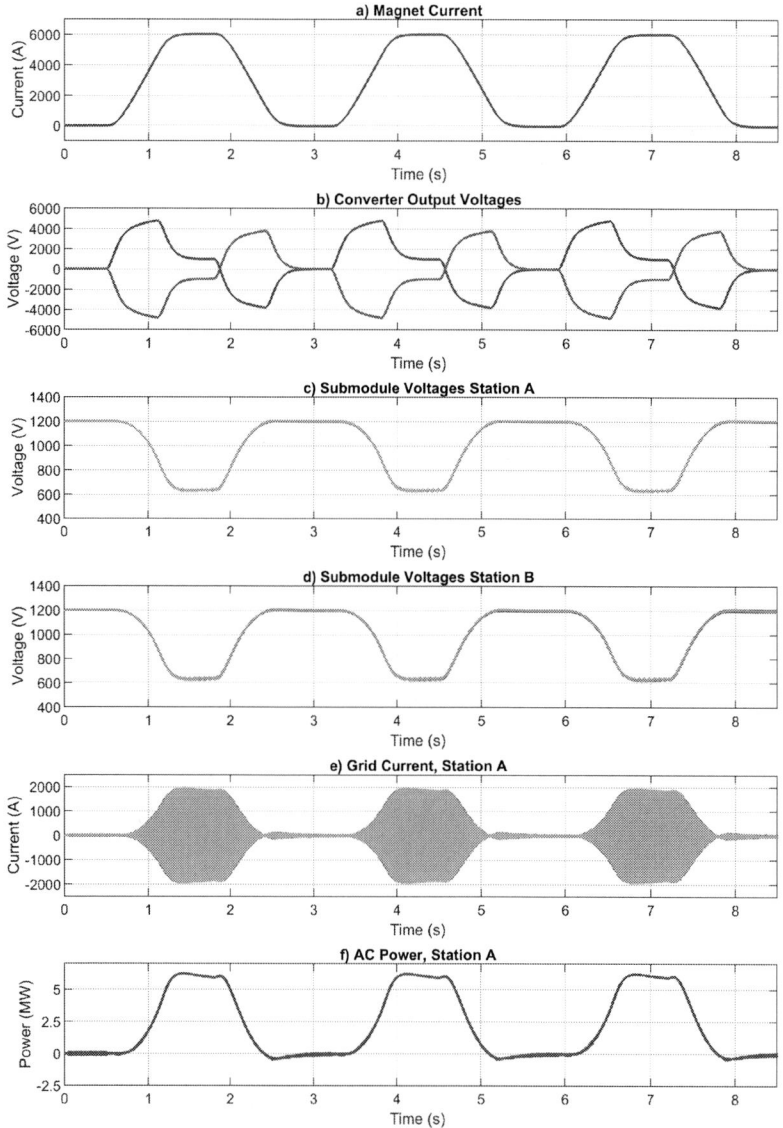

Fig. 4: Main operation curves of the power converter. a) Magnet current. b) Converter output voltages. c) Submodule Voltages of station 1. d) Submodule Voltages of station 2. e) AC current absorbed by the converter 1. d) AC power absorbed by converter 1.

Fig. 4 shows the simulation results of the system supplying the magnets with trapezoidal currents. Fig 4a shows the current flowing through the magnets, which reaches 6000 Amps at the end of the acceleration cycle. To achieve the desired current increasing and decreasing rates, the converters apply voltages with opposite polarity (Fig 4b). The peak voltage is produced at the end of the ramp-up, where the drop across the magnet resistance and the voltage caused by the $L \cdot di/dt$ add up, reaching the expected peak value of 5 kV on both converter outputs. The discharge and charge of the submodule capacitors is shown in Fig 4c and Fig 4d, where a selection of several submodule capacitor voltages has been done for simplicity. The proper dimensioning of the submodule capacitors together with a suitable control action guarantees that these do not discharge below the 625 V, allowing the generation of the required output voltage at the end of the ramp-up. Since the submodule capacitance is high, the voltage fluctuations due to the connection/bypass of the cells are small, playing a little role for the dimensioning of the cells, which, as explained, is based mainly on the voltage generation capabilities of the converter. Additionally, these figures also show the proper behaviour of the individual capacitor balancing algorithm: no deviations between the submodule voltages are observed.

On the other hand, Fig 4e shows the current absorbed by one of the stations whereas Fig 4f shows the resulting power. As explained, the total energy balancing loop requires to absorb only the power dissipated by resistive elements. The proper behavior of the current loop allows to follow the current reference given by the total energy controller. As it can be seen, the burden on the AC network is significantly reduced: the power supplied to the magnets can reach values as high as 60 MW whereas the power absorbed from the AC network, summing the contribution of the two stations, remains below 13 MW.

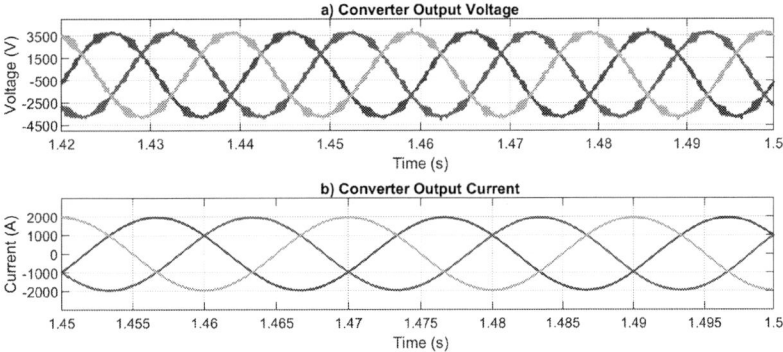

Fig. 5: Output MMC AC voltages and currents during current flat-top.

Finally, Fig. 5, shows the MMC AC voltages. During the flat-top, where the current is at its maximum level, the MMCs can synthesize high quality output waveforms with eight voltage levels. In this sense, the harmonic performance of the MMCs is better than that of the 3-Level converters used in [1]. Note that, during the phase of the cycle where the submodule capacitors are fully-charged, and thus fewer levels are used, the currents absorbed are small. Therefore, the impact on AC network quality is minimal.

Discussion and Conclusion

This paper demonstrates the feasibility of using Full-Bridge MMCs for providing the pulsed currents required by the magnet chains of a particle accelerator. By using this converter topology, it is possible to shave the peak power absorbed from the AC grid by allowing the exchange of energy between magnets and the submodule capacitors. Compared with the existing solution, using MMCs would improve the reliability of the system by distributing the energy storage among a large number of cells. Besides, the high modularity of the solution would allow to bypass a faulty module, improving the availability of the machine. On the other hand, the lower voltage ratings required for the semiconductors would allow a better optimization of the cells and, thus, of the converter performance. In this sense, it has been showed that the MMC can obtain the required output current using a low switching frequency. However, the reduction of the switching frequency might not result in a reduction of the losses compared with the existing solution considering that more semiconductors are needed.

Acknowledgements

R. Vidal-Albalate and R. Blasco-Gimenez would like to thank the Spanish Agency of Research (AEI) for its partial support of this work . Work supported by research grant PID2020-112943RB-I00 funded by MCIN/AEI/10.13039/501100011033.

References

[1] Boattini, F., Burnet, J. & Skawinski, G. POPS: The 60MW power converter for the PS accelerator: Control strategy and performances. *2015 17th European Conference On Power Electronics And Applications (EPE'15 ECCE-Europe)*. pp. 1-10 (2015)

[2] Benedikt, M., Blondel, A., Brunner, O., Capeans Garrido, M., Cerutti, F., Gutleber, J., Janot, P., Jimenez, J., Mertens, V., Milanese, A., Oide, K., Osborne, J., Otto, T., Papaphilippou, Y., Poole, J., Tavian, L. & Zimmermann, F. FCC-ee: The Lepton Collider: Future Circular Collider Conceptual Design Report Volume 2. Future Circular Collider. (CERN,2018,12), https://cds.cern.ch/record/2651299

[3] Lin, W., Jovcic, D., Nguefeu, S. & Saad, H. Full-Bridge MMC Converter Optimal Design to HVDC Operational Requirements. *IEEE Transactions On Power Delivery*. 31, 1342-1350 (2016)

[4] Alharbi, M., Isik, S. & Bhattacharya, S. Reliability Comparison and Evaluation of MMC Based HVDC Systems. *2018 IEEE Electronic Power Grid (eGrid)*. pp. 1-5 (2018)

[5] Rohner, S., Bernet, S., Hiller, M. & Sommer, R. Modulation, Losses, and Semiconductor Requirements of Modular Multilevel Converters. *IEEE Transactions On Industrial Electronics*. 57, 2633-2642 (2010)

[6] Mishra, P. & Bhesaniya, M. Comparison of Total Harmonic Distortion of Modular Multilevel Converter and Parallel Hybrid Modular Multilevel Converter. *2018 2nd International Conference On Trends In Electronics And Informatics (ICOEI)*. pp. 890-894 (2018)

[7] Cui, S. & Sul, S. A Comprehensive DC Short-Circuit Fault Ride Through Strategy of Hybrid Modular Multilevel Converters (MMCs) for Overhead Line Transmission. *IEEE Transactions On Power Electronics*. 31, 7780-7796 (2016)

[8] Colmenero, M., Blanquez, F. & Kahle, K. Transient Voltage Dip Mitigation System Based On Hybrid Modular Multilevel Converters. *2020 22nd European Conference On Power Electronics And Applications (EPE'20 ECCE Europe)*. pp. P.1-P.10 (2020)

[9] Deng, F., Lü, Y., Liu, C., Heng, Q., Yu, Q. & Zhao, J. Overview on submodule topologies, modeling, modulation, control schemes, fault diagnosis, and tolerant control strategies of modular multilevel converters. *Chinese Journal Of Electrical Engineering*. 6, 1-21 (2020)

Deep Neural Network for Magnetic Core Loss Estimation using the MagNet Experimental Database

Xiaobing Shen, Hans Wouters, Wilmar Martinez
KU Leuven – EnergyVille
Department of Electrical Engineering (ESAT)
Leuven - Genk, Belgium
Email: xiaobing.shen, hans.wouters, wilmar.martinez@kuleuven.be

Keywords

≪Magnetic Device≫, ≪Machine Learning≫, ≪Core Loss Modelling≫, ≪Neural Network≫.

Abstract

Magnetic components play a critical role in power electronics systems and their evolution towards higher power density and efficiency. Nevertheless, accurately modelling magnetic core losses is not a trivial task, requiring extensive measurements. In the context of the general advances of Machine Learning technologies in power electronics applications, this paper presents a Deep Neural Network (DNN) approach to core loss estimations. Various internal parameters of the DNN are tested and compared, to identify the optimal DNN structure for the core loss estimation, including the number of hidden layers, number of neurons, data transformation, and different activation functions. The training data-set comprises the MagNet database for N87 toroid magnetic cores, based on an experimental data acquisition system capable of automatically measuring various magnetic cores under arbitrary excitation signals. The results of the DNN models indicate that a DNN with suitable parameters can robustly and accurately model the core losses. The attainable accuracy is well within the required range for magnetic core losses. The optimal structure proposed in this paper consists of 10 hidden layers with sigmoid activation functions, 10 neurons in each layer, integrating a log-transformation and data normalization. The model is validated with extensive experimental tests similar to the MagNet measurement system. Furthermore, tests at higher switching frequencies up to 1MHz indicate that the model can predict losses for parameters outside the range of its training data. With the achieved performance, the DNN can benefit various power electronics engineering challenges such as loss estimation for inductor design.

Introduction

Core loss estimation is not a trivial task and due to its complexity, it is not an entirely solved matter neither. Although magnetic characterization has been studied extensively for over a century, the distinctive conditions in power electronics have introduced many new challenges. This imposes an important area of research, as magnetic components play a critical role in most power electronic applications. These new challenges for the characterization, modelling and measurements of magnetic components arise from various developments. Where traditionally, the excitation of such magnetic materials has dominantly been sinusoidal, various modulation control methods such as pulse width modulation (PWM) in power electronics lead to a diversity of excitation waveforms [1]. Because of this modulation, they generally contain a significant amount of harmonic content [2]. Furthermore, advances in semiconductor technology towards wide bandgap devices have led to an increase in switching frequencies into the MHz range in order to realize compact power converters [3].

Different models and equations have historically been used to characterize magnetic materials. Many of these parameters can be traced back to the relationship between the magnetic flux density in the material and the applied magnetic field strength. The magnetic permeability of the material relates these quantities in the BH curve, often referred to as the hysteresis loop. An established model to characterize the hysteresis loop is the Steinmetz equation, first introduced in 1892 [4]. Many variations of this equation are developed, to broaden the range in which the equation applies and to take various additional parameters into account such as temperature, frequency, and the type of excitation [5], [6]. One of the most recent advances is the Steinmetz Premagnetization Graph presented in [7], based on the improved Generalized Steinmetz Equation (iGSE) [8]. This considers non-sinusoidal excitation, as well as the effect of DC magnetic flux bias. These models, however, are computationally expensive, require specific material properties, and often require extensive measurements.

Recently, novel methodologies based on Machine Learning (ML) are researched for characterizing the core losses [1], [9]-[11], as these algorithms are proven to be effective in solving nonlinear problems [12]. The concept of Deep Neural Networks (DNN), first proposed by G.E. Hinton et al. [13], refers to a machine learning process that obtains a multi-level deep network structure, based on sample data [14]. DNNs, as a new research field of ML, gains breakthroughs in applications such as speech recognition and computer vision. This paper presents a Deep Neural Network based approach and investigates the effect of the many parameters that can be fine-tuned in the DNN, to provide a robust and accurate model.

Data-set for DNN Model

Machine learning algorithms are to a large extent data based. With larger data-set, better performances in the training process of a DNN will be achieved. This paper presents a DNN based on the MagNet database [12]. Analogous to the ImageNet database [15], which is widely used in computer vision and general deep learning research, MagNet provides data for various magnetic research purposes. It is based on a data acquisition system capable of automatically generating large data-sets. The measurement setup is shown in Fig. 1 The power stage of this setup can generate arbitrary excitation waveforms. The main components consist of a power amplifier (V1) to generate sinusoidal excitations, a single-phase bridge (Qs) to generate the PWM excitations, and AO4444 MOSFETs for all power switches. Various magnetic components can be connected as devices under testing (DUT), including varying geometries, materials, and sizes. Measurements are made of the current through the primary winding and voltage over the secondary winding. Shunt resistors are used for the current measurements, voltage dividers for the voltage measurements.

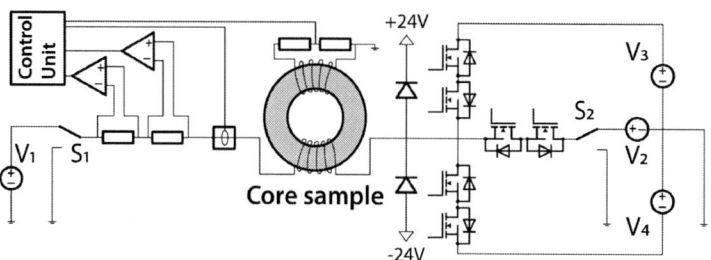

Fig. 1: MagNet magnetic core loss measurement setup

In this paper, the MagNet database is limited to data on toroid cores N87 material. A hypothesis of [12] is that the physics that govern the core loss for a material, are analogous for similar materials. Therefore, knowledge gained of magnetic characterization using a certain waveform or material, can be used as input for characterizing similar materials and/or with different waveforms. This hypothesis will be tested against the constructed DNN. This paper uses the data-set acquired by applying a sinusoidal excitation on the N87 toroid as DUT, as this is popularly used in power electronics. The frequency range of the data-set is 10kHz to 500kHz. The duty ratio is fixed, at a constant 0.50 value. The data-set provides the flux density [mT] and power loss density [kW/m³] as a function of the frequency [Hz], totaling over 10,000 data-points.

DNN Model Structure and Optimization

As illustrated in Fig. 2a, the starting point of the algorithm is a function with unknown parameters. The equations governing the magnetic losses, however, are complicated and contain many parameters characterizing the material properties, geometry, excitation waveform, etc. This large number of unknown parameters is defined as , alongside a loss function $L(\theta)$ for each of them. This loss indicates the approximation of the estimate values from their true value. Then, the gradient decent method is used, a linear iterative optimization algorithm which is established in both linear and nonlinear problems due to its robust convergence properties. With it, the minima of the loss function are located, providing the optimal solutions for minimum losses and their parameters θ. These are the unknown parameters required for the model to accurately compute the core losses for new inputs.

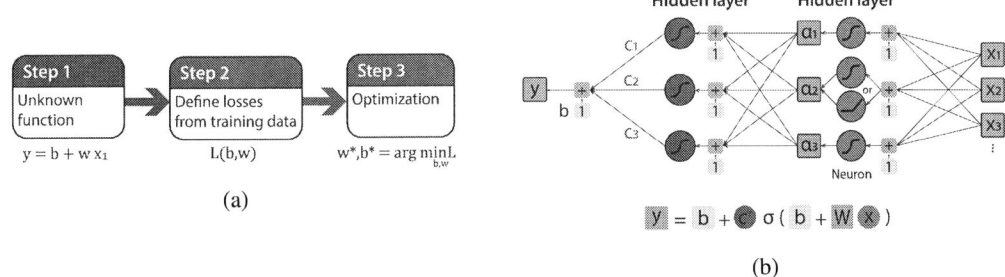

(a)

(b)

Fig. 2: (a) General training workflow in machine learning, (b) General DNN structure

A DNN structure is proposed based on the workflow of machine learning. Fig. 2b shows the typical structure of a DNN, containing multiple inputs and outputs. Different neurons and layers are connected, resulting in a neural network. The last layer has a fixed number of neurons, representing the number of outputs. The rest of the layers are the hyperparameters of the structure, also named hidden layers. With the expansion of the number of neurons and layers, the DNN strengthens its processing power for more complex data. The required computation time, however, also increases along with its complexity, resulting in a trade-off.

Other important hyperparameters are the batch and epoch number. The batch size stands for the number of samples processed per training iteration. Each sample is one measurement of the magnetic losses and flux density as a function of the excitation frequency. An epoch is comprised of multiple batches and represents the number of times the DNN iterates over all samples.

Furthermore, the DNN training process involves splitting the data-set of the magnetic core losses into the training and validation set. In this paper, 10-fold cross validation is employed, which implies a random partition and classification of the original core loss data-set. After initializing the weights and biases of the DNN and the hyperparameters, the mean square error (MSE) is used to evaluate the accuracy of the DNN and update its parameters in an iterative manner.

DNN Analysis Parameters

The DNN structure is physical-free as discussed above. To prevail over alternative data processing and optimization algorithms, a robust, flexible, and accurate DNN is required, especially for power electronics. Structural parameters of the DNN can significantly influence its behavior, including: (1) the structure of DNN, i.e., choices of activation function, neurons, and hidden layers; (2) data scaling techniques such as input and output data normalization and transformation; (3) training processing, including the splitting method and optimization algorithm on the data. This paper evaluates five different DNN structures, as described in Table I.

Table I: Overview of the DNN analysis parameters

Model ID	Hidden layers	Neurons	Data transformation	Data normalisation	Activation function
M1	1	1	None	None	None
M2	2	5+1	None	None	Sigmoid
M3	2	5+1	Log	Yes	Sigmoid
M4	5	5+1	Log	Yes	Sigmoid
M5	10	5+1	Log	Yes	Sigmoid

Results of Selected DNN Structures

According to the wide range of parameters presented in Table II, the proposed DNN models are investigated in MATLAB, making intensive use of its NN-toolbox. All DNNs detailed above are validated by a 10-fold cross-validation. The results are displayed in Fig. 3 and summarized in Table II, providing the difference in the resulting errors and the epoch number at which the best validation performance (BVP) is reached.

Table II: Overview of the DNN analysis parameters

Model ID	Max. error	RMS error	BVP epoch no.
M1	/	/	7
M2	25.10%	13.70%	914
M3	0.53%	0.04%	108
M4	0.28%	0.00%	21
M5	0.18%	0.00%	42

Model 1, with no data transformation or normalization, nor any layers and neurons, is a simple and well-known linear regression model. it provides fast convergence but low performance, as shown in Fig. 3a. The MSE of M2 is shown in Fig. 3b. Compared to model 1, it consists of 2 hidden layers with sigmoid activation functions, 5 neurons for the hidden layer, and 1 neutron for the output layer. As visible in the graph, this drastically decreases the MSE. The epoch number, however, increases significantly as well.

Considering that the 2-dimensions of inputs have a wide range of magnitude, which may result in numerical problems for a model, data normalization and log-transformation for both input data and output data are implored. This scales the data-set, improving the accuracy of the training process while also increasing the training speed. Hereby, model 3 with the same number of neurons and layers as model 2 performs significantly better. This is illustrated in Fig. 3c.

Furthermore, the MSE of models 4 and 5 are shown in Fig. 3d and Fig. 3e respectively. Compared with the third model, M4 and M5 go deeper with 5 and 10 hidden layers. The results show that M5 with data normalization, log-transformation and 10 layers shows the best performance. Thus, one can conclude that by going deeper while using data transformation, higher performance can be obtained.

In summary, a deeper structure and variable processing methods (variable normalization and transformation) contribute to a more robust and accurate DNN. More specifically, with variable processing methods, the performance improves dramatically. However, there is a limit to how deep the model should go to obtain acceptable accuracy. In our specification, 5 layers with data processing techniques, as introduced above, are acceptable to build a fully connected DNN for magnetic core loss estimation, based on the MagNet database. Moreover, to get a deeper structure as opposed to a 'fatter' DNN, a 'skinny' DNN is maintained by fixing the number of neurons in each layer.

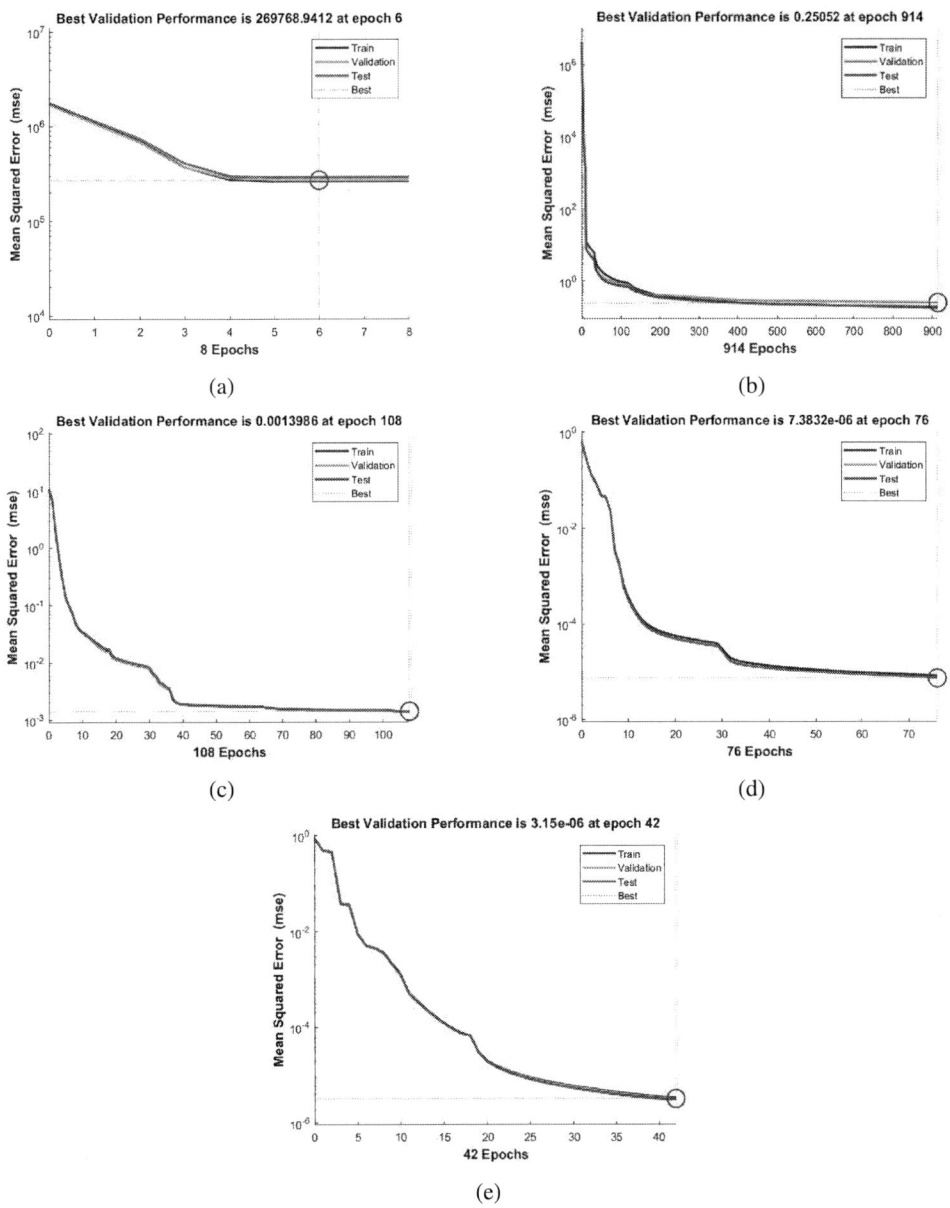

Fig. 3: MSE of the different DNNs, (a) model 1, (b) model 2, (c) model 3, (d) model 4, (e) model 5

Experimental Comparison

To validate both the MagNet measurement data and the DNN model, extensive experimental validation is conducted. The DUT in these experiments is a N87 toroid core. The R25 size is selected and 32 windings are used, as shown in Fig. 4c. A high frequency full-bridge GaN inverter produces the excitation signals with the same frequency and duty cycle as used in the MagNet database. This measurement setup is shown in Fig. 4a. Both the current through the DUT and the induced voltage over a secondary winding are measured, which excludes the effect of the DUT's resistance from the voltage measurement. The data is then processed to construct the BH-curve, and subsequently calculate the core losses. As such, both the training data and the DNN output can be validated. Furthermore, this is also used to test the ability of the DNN for accurately predicting core losses outside the range of its training data. Therefore, measurements are done at switching frequencies of 500kHz to 1MHz as well.

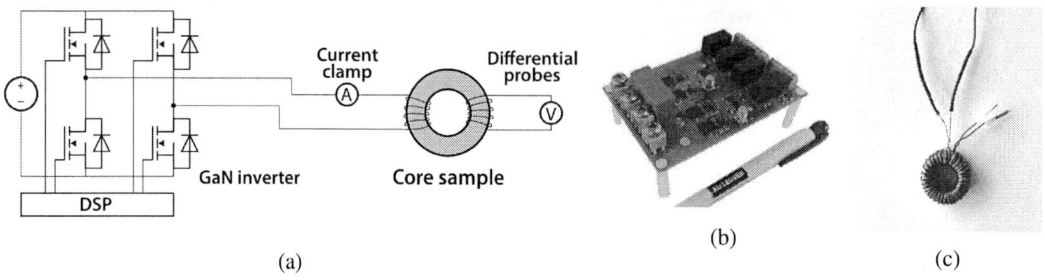

Fig. 4: (a) Measurement setup schematic, (b) GaN inverter and control circuit, (c) Toroid N87 R25 DUT

To perform the measurements, the GaN inverters are controlled by a 150MHz DSP with a unipolar modulation, shown in Fig. 4b. The inverter works within a frequency range of 10 kHz to 1MHz. The current through the primary winding and the voltage over the secondary are measured by an oscilloscope capable of collecting 10M points per measurement with a maximum sampling frequency of 2.5 GS/s real time. Fig. 3a shows the measured currents and voltages at a 50Hz fundamental frequency with varying carrier frequencies and a fixed magnetic flux density of 200mT in the DUT.

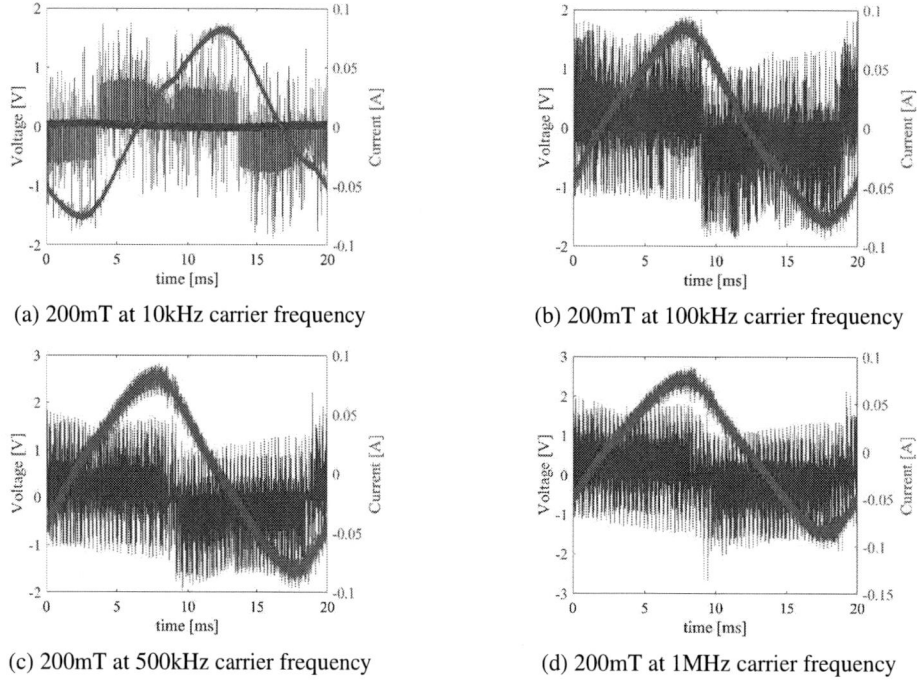

(a) 200mT at 10kHz carrier frequency

(b) 200mT at 100kHz carrier frequency

(c) 200mT at 500kHz carrier frequency

(d) 200mT at 1MHz carrier frequency

Fig. 5: Measured current and voltage waveforms of the DUT at 50 Hz fundamental frequency

Based upon the collected data, the flux density and core losses within the frequency range of 10kHz-1MHz are determined. Hereby, the BH-curves at different frequencies are derived as well, as shown in Fig. 6a. The flux density can be obtained from the induced voltage over the secondary winding using Faraday's law, provided in equation (1). In addition, the magnetic field strength in the inductor can be derived from the current through the primary winding based on Ampère's law, see equation (2).

$$B(t) = \frac{1}{N_2 A_e} \int v_2(t) dt \tag{1}$$

$$H(t) = \frac{N_1 i_1(t)}{l_e} \tag{2}$$

Here, N_1 and N_2 are the number of turns of the primary and secondary winding respectively, A_e is the effective area of the core, and l_e is the length of the main flux path. These can be combined to calculate the magnetic core loss density in equation (3), based on the detailed mathematical procedure in [16]:

$$P_{Fe} = \frac{N_1 f}{N_2 A_e l_e \rho} \int_0^1 i_1(t) v_2 dt \tag{3}$$

where is the material mass density, which for N87 is equal to 4850 kg/m³. Due to the sinusoidal voltage, the induced magnetic flux density will be of sinusoidal shape as well. This leads to the BH-curves as shown in Fig. 6a and the magnetic losses of Fig. 6b as a function of the switching frequency.

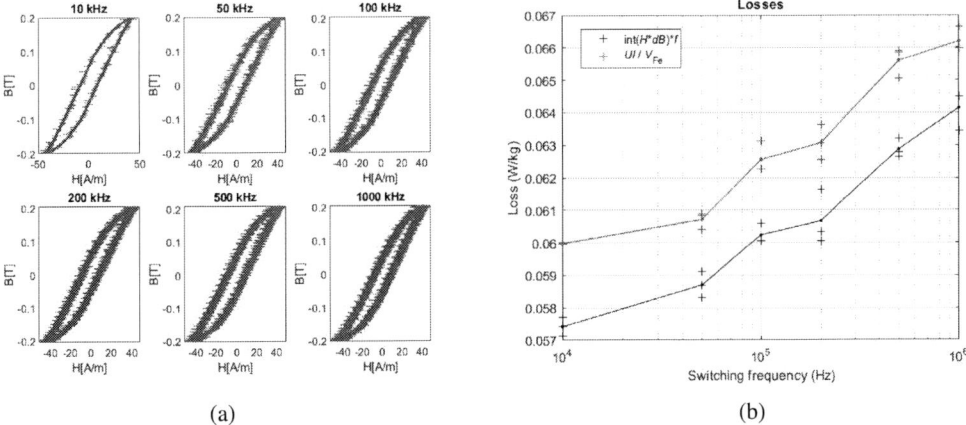

(a)   (b)

Fig. 6: Measurement (a) BH-curves and (b) Magnetic losses at $B_{max} = 200mT$ for the N87 R25 toroid

As such, at a certain switching frequency such as 1MHz, the magnetic losses can be predicted for an entire range of flux densities, as illustrated in Fig. 7. This shows a comparison between the measured results and the DNN model for different flux densities at a fixed switching frequency of 1MHz. As such, it provides the error distribution of the DNN model at different flux densities.

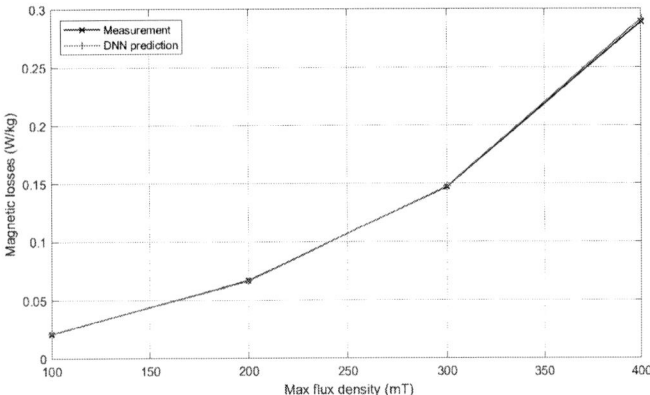

Fig. 7: Comparison of the measured and predicted losses for a switching frequency of 1MHz

It can be observed that the error is relatively small, with an error of only 0.2% at 1MHz and 300mT. The maximum error is 0.90%, at the maximal flux density of 400mT. This larger error at higher flux densities is an expected outcome, due to the nonlinear behavior of the core when saturating. Furthermore, these accurate results at 1MHz indicate the possibility of using the DNN model in high frequency applications such as magnetic design for high frequency power converters and LC filters. As such, the fast model can be used to analyze and reduce the losses in magnetic components. The trained DNN model could work as a surrogate model for the characterization of these magnetic losses.

Conclusion

This paper explores the potential of Deep Neural Networks to solve the complex challenges arising from the calculation of magnetic core losses. By comparing different DNN structures, a new DNN model based on experimental data is presented and validated. Developmental procedures of the DNN model are discussed, such as data processing, cross-validation, and deep layers with fixed neurons. The results indicate that the resulting DNN-based model is capable of producing accurate results compared to the experimental magnetic core measurements. The accuracy obtained is well within the acceptable range for magnetic core loss estimation. Its use in high frequency applications is experimentally validated at up to 1MHz, as well as the use of the DNN model for switching frequencies outside of the range of its training data.

Future work could be done on improving the DNN model to capture more scenarios, such as different excitation waveforms, a wider range of input frequencies, and different core materials, shapes and sizes. With more measurement data, the model can be further enhanced and optimized. As for magnetic design, it could function as a surrogate model, which yields the same behaviors and performance as existing magnetic loss models while reducing the number of required measurements and setups. This way, the computational and testing efforts can be reduced. The proposed structure is envisioned to be implemented for magnet loss estimation for inductor and transformer design and optimization, as it can provide a well-trained, robust, and accurate model.

References

[1] Z. Zhao et al., "Modeling Magnetic Hysteresis Under DC-Biased Magnetization Using the Neural Network," IEEE Trans. Magn., vol. 45, no. 10, pp. 3958–3961, Oct. 2009, doi: 10.1109/TMAG.2009.2023070

[2] W. Martinez and C. Suarez, "Total Harmonic Distortion Analysis in Magnetic Characterization using High Frequency GaN Inverter in the MHz Order," in EPE '19 ECCE Europe, Sep. 2019, p. P.1-P.9. doi: 10.23919/EPE.2019.8915546

[3] M. Parvez et al., "Wide Bandgap DC–DC Converter Topologies for Power Applications," Proc. IEEE, vol. 109, no. 7, pp. 1253–1275, Jul. 2021, doi: 10.1109/JPROC.2021.3072170

[4]] Chas. P. Steinmetz, "On the Law of Hysteresis," Trans. Am. Inst. Electr. Eng., vol. IX, no. 1, pp. 1–64, Jan. 1892, doi: 10.1109/T-AIEE.1892.5570437

[5] J. Reinert, A. Brockmeyer, and R. De Doncker, "Calculation of losses in ferro- and ferrimagnetic materials based on the modified Steinmetz equation," IEEE Trans. Ind. Appl., vol. 37, no. 4, pp. 1055–1061, Jul. 2001, doi: 10.1109/28.936396

[6] J. Li, T. Abdallah, and C. R. Sullivan, "Improved calculation of core loss with nonsinusoidal waveforms," in IEEE 36th IAS Annual Meeting, Sep. 2001, vol. 4, pp. 2203–2210 vol.4. doi: 10.1109/IAS.2001.955931

[7] J. Mu¨hlethaler et al., "Core losses under DC bias condition based on Steinmetz parameters," in The 2010 ECCE ASIA -, Jun. 2010, pp. 2430–2437. doi: 10.1109/IPEC.2010.5542385

[8] K. Venkatachalam et al., "Accurate prediction of ferrite core loss with nonsinusoidal waveforms using only Steinmetz parameters," CIPE, 2002. Proceedings., Jun. 2002, pp. 36–41. doi: 10.1109/CIPE.2002.1196712

[9]] I. Kucuk, "Prediction of hysteresis loop in magnetic cores using neural network and genetic algorithm," J. Magn. Magn. Mater., vol. 305, no. 2, pp. 423–427, Oct. 2006, doi: 10.1016/j.jmmm.2006.01.137

[10] X. Zhao and Y. Tan, "Modeling Hysteresis and Its Inverse Model Using Neural Networks Based on Expanded Input Space Method," TCST., vol. 16, no. 3, pp. 484–490, 2008, doi: 10.1109/TCST.2007.906274

[11] S. Shimokawa et al., "Fast 3-D Optimization of Magnetic Cores for Loss and Volume Reduction," IEEE Trans. Magn., vol. 54, no. 11, pp. 1–4, Nov. 2018, doi: 10.1109/TMAG.2018.2841364

[12] H. Li et al., "MagNet: A Machine Learning Framework for Magnetic Core Loss Modeling," in 2020 IEEE 21st COMPEL, Nov. 2020, pp. 1–8. doi: 10.1109/COMPEL49091.2020.9265869

[13] G. E. Hinton, S. Osindero, and Y.-W. Teh, "A fast learning algorithm for deep belief nets," Neural Comput., vol. 18, no. 7, pp. 1527–1554, Jul. 2006, doi: 10.1162/neco.2006.18.7.1527

[14] Y. Bengio, "Learning Deep Architectures for AI," Found Trends Mach. Learn, 2007, doi: 10.1561/2200000006

[15] J. Deng, W. Dong et al., "ImageNet: A large-scale hierarchical image database," in 2009 IEEE Conference on Computer Vision and Pattern Recognition, Jun. 2009, pp. 248–255. doi: 10.1109/CVPR.2009.5206848

[16] P. Rasilo et al., "Simulink Model for PWM-Supplied Laminated Magnetic Cores Including Hysteresis, Eddy-Current, and Excess Losses," IEEE Trans. Power Electron., vol. 34, no. 2, pp. 1683–1695, Feb. 2019

Hybrid circuit board structure for power electronics

Gerrit Braun, Deniz-Heinz Moldenhauer
SMA Solar Technology AG
Sonnenallee 1, 34266 Niestetal, Germany
Tel.: +49-5619522423507.
E-Mail: Gerrit.Braun@sma.de
URL: http://www.sma.de

Acknowledgements

Parts of the results presented in this paper were obtained within the joint research projects Netprosum2030 (FKZ 0350021B) & Voyager-PV (FKZ 03EE1057B) supported by funding of the German Federal Ministry for Economic Affairs and Climate Action, on the basis of a decision of the German Bundestag.

Keywords

«Gallium Nitride (GaN)», «Half-bridge», «Hybrid», «Wide bandgap devices», «Thermal design»

Abstract

The objective of this paper is to show a novel approach for integrating GaN semiconductors as a discrete SMD component using an additional printed circuit board (PCB) for the power electronics of a PV inverter. This additional PCB is designed as an integrated metal substrate (IMS) circuit board to conduct the dissipated heat from the semiconductor into the heat sink. The IMS board is soldered with a BGA connection to a larger mother board. Therefore, usual production steps of electronics production are used to make the structure as simple as possible.

Introduction

Current power electronics in PV inverters largely use discrete (through-hole technology / THT or surface-mount technology / SMT) semiconductor packages or so-called power modules with solder or press-fit pins. The use of power modules allows a higher circuit complexity per area and better cooling. Thereby, the conventional power modules have quite high parasitic inductances, due to the construction technology with pins, leads and bond wires. Those parasitic inductances have a negative effect on the electrical performance of fast switching power electronics with wide band gap (WBG) semiconductors such as voltage overshoot [1]. Semiconductors for the surface-mounting technology (SMT) manufacturing process packaged as surface mount device (SMD) have smaller parasitic inductances [2]. Many GaN semiconductors are available in a SMD package. If such discrete packages are used, the PCB design is essential for electrical and thermal performance.

In this paper a novel hybrid construction approach with IMS boards will be presented to integrate GaN semiconductors as a SMD package on a larger power electronics assembly. Ball Grid Arrays (BGA) are used to connect the IMS board with a larger FR4 motherboard. GaN semiconductors on IMS boards have already shown their potential [4]. With the presented approach it is possible to integrate GaN semiconductors into PV and storage inverters to benefit on system level.

The boundary conditions for the chosen assembly are to maintain electrical insulation of the electronics to the heatsink. The power electronic should bridge the gap between the motherboard and the heatsink to maintain a simple heatsink design with a flat surface without a pedestal. As thermal interface material a thermal grease shall be used. Therefore, a construction approach with a castellated board or a PCB with thermal vias (like they are proposed in application notes) together with an insulating foil as thermal interface material could not be used.

Insulated metal substrate boards

The simplest structure of an IMS circuit board consists of a metal substrate, an insulating prepreg layer and the copper layer with the desired layout. IMS circuit boards are available with a copper or aluminum substrate. The thickness of the available substrate can vary. Standard thicknesses of 1.0, 1.5 or 2.0 mm are available or can be ordered on request. Usual thicknesses for the insulating layer are 50 to 100 μm. Typical values for the thermal conductivity of such prepregs are 3 to 4 W/(m·K) but values up to 12 W/(m·K) are available, too [5], [6]. The typical thermal conductivity of standard FR4 is between 0.3 to 0.4 W/(m·K). In comparison thermal enhanced prepregs for IMS boards have roughly 10 times better thermal conductivity than standard FR4 material. Multilayer IMS boards with more than one layer and up to six layers are also available. Fig. 1 shows a schematic cross section of a two-layer IMS PCB. The two layers can be connected with microvias. Those vias can be designed as blind vias or as copper filled vias. Copper filled vias can be used to improve the thermal conductivity of the FR4 core between the two copper layers. However, the aspect ratio, the ration between drill diameter and hole depth, of the PCB manufacturer must always be taken into account.

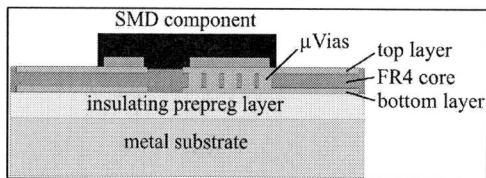

Fig. 1: Schematic cross section of a two-layer IMS board

Construction of a demo board

The structure of the power electronics demo board consists of two circuit boards. The IMS circuit board with the power switches and a motherboard which, in addition to the IMS circuit boards, is holding other components like DC link capacitors, filter components, galvanic isolated driver supply, digital isolator and additional circuits for evaluation.

The connection between the IMS board and the motherboard is realized through a so-called Ball Grid Array. Normally BGAs are used to connect packages with integrated circuits with a variety of solder joints with a PCB. In this approach the BGAs are used to connect the half bridge with the DC+, DC-, bridge-out potential and the driver IOs from the driver integrated power switch to the motherboard. Fig. 2 shows a schematic cross-section of the IMS connected to the motherboard by BGAs. Because of the dimensions of the components mounted on top of the IMS it is necessary to add a milling into the motherboard. Through this milling the components are visible, and it is possible to evaluate the semiconductors with a thermal camera. The IMS is soldered on the bottom side of the motherboard. This board is screwed to a heatsink.

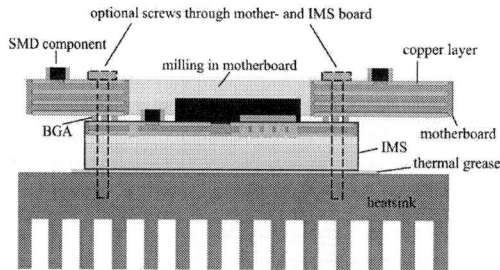

Fig. 2: Schematic cross-section of a hybrid PCB-IMS construction with SMD components and the BGA solder joints between the motherboard and IMS board

The circuit on the IMS board contains a half bridge with samples of the LMG3422R050 from Texas Instruments. This device includes a 600V GaN chip with an integrated driver, temperature reporting and protection features [7]. For the driver to function properly it is necessary to place some components close to the device on the IMS. Therefore, the IMS board can be divided into a power part and a driver part with the additional components and solder joints. Fig. 3 shows the designed IMS board.

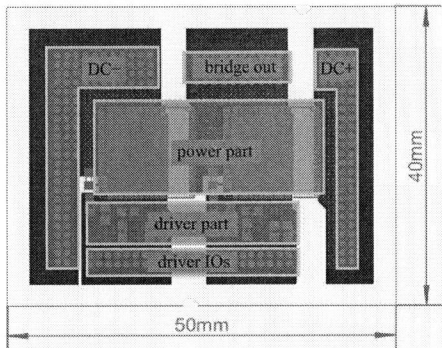

Fig. 3: Designed IMS board with marked areas for the BGA solder joints and components

The designed IMS board is 40 mm by 50 mm and features 224 solder pads for the BGA solder joints. In the datasheet of the LMG3422R050 a four-layer stack-up is recommended but for a better thermal performance a two-layer design with copper filled microvias was chosen. Most of the vias are placed under the exposed thermal pad of the semiconductor to create a better thermal conductivity from layer one to layer two. With this design approach, it was not possible to place a DC link capacitor between 'DC+' and 'DC-' directly onto the IMS and minimizing the influence of the DC link inductance. Instead, the DC link capacitor was placed onto the motherboard. As metal substrate 1.0 mm thick copper was used. The insulating prepreg layer has a thermal conductivity of 3 W/(m·K) and is about 100µm thick. In the driver section of the IMS board the vias are placed to connect the ground pins with a ground plane from the driver part.

The production process was carried out with the typical steps of an electronics manufacturing and consists of the following steps:

1. IMS board is squeegeed with solder paste in a stencil printer with a stencil
2. Components are placed on IMS board by a pick-and-place machine
3. Soldering through a reflow oven of the IMS
4. IMS boards are placed inside a tray or reel for the pick-and-place machine
5. Bottom side of the motherboard is squeegeed with solder paste
6. Components and IMS board(s) are placed on bottom side of the motherboard
7. Soldering through a reflow oven of the motherboard
8. If needed, repeat last three steps for top side of motherboard

X-Ray Analysis

After the production process had been completed, the construction of an IMS board soldered onto a motherboard was examined using an x-ray device. Fig. 4 shows the top view of the soldered PCBs and the soldered balls are visible as darker circles. The balls have a much higher contrast in comparison to the copper layer around. That is an indication of a correct connected solder joint between the two circuit boards.

Fig. 4: X-Ray of IMS board soldered to the motherboard

In comparison Fig. 5 shows an x-ray of a test run in which only one circuit board was prepared with solder paste. Therefore, there is less solder paste available to connect the two pads and form a correct solder joint. The solder pads with higher contrast are outlined in green and the solder pads with less contrast is outlined in red. It is visible that the ball pads in the red marker have less contrast in comparison to the others. This is an indication that the pads of the motherboard and the IMS board have no contact to each other. The assumption is as follows: When enough solder paste has been applied to bridge the gap between the two pads, the ball forms like a pillar between them. If there is not enough solder paste, it forms some sort of hill due to the surface tension on the pad. In an x-ray image thicker material or materials with higher density appear darker. Fig. 6 illustrates this assumption.

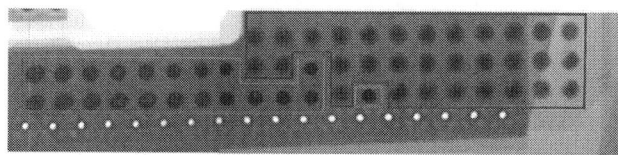

Fig. 5: X-ray of a BGA solder joint connection between a standard PCB and an IMS board; Green: BGA with high contrast; Red: BGA with less contrast

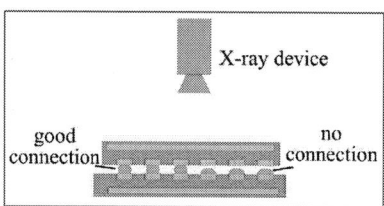

Fig. 6: Schematic x-ray of a BGA with good connections and no connection between two PCBs

This assumption could be checked with a continuity test on some driver IO connections in the test run with less solder paste. Some of those connections appeared darker and a visible difference in contrast was evident and only two balls are connected in parallel. The continuity test with a multimeter confirmed the assumption.

In Fig. 4 some balls show little bright spots. These spots are an indication of solder voids. Various effects concerning the lifetime and reliability of BGA solder joints were already investigated in [8] and [9]. Further investigations similar to [3] must show if those voids can lead to a failure of the soldered balls in combination with the heat coming from the power semiconductors and the presence of the contact pressure from the screw connection to the heatsink.

Thermal Performance

The thermal performance was investigated by determining the thermal resistance between the device junction and the heatsink base. A reverse DC current through the device was applied. The current and the voltage across the device were measured with a Yokogawa Power Analyzer. The temperature was measured with the infrared camera FLIR T365. A thermo couple was placed at a blind hole in the heatsink base underneath the device to measure the heatsink temperature underneath but as close as possible to the bottom side of the IMS.

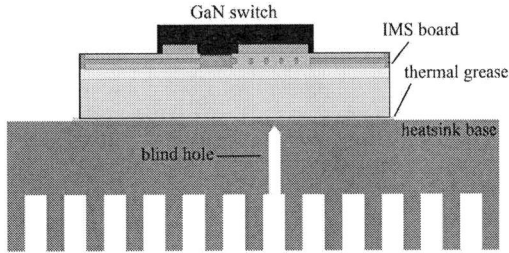

Fig. 7: Schematic cross-section of the test setup for determining the temperatures with a blind hole in the heatsink base

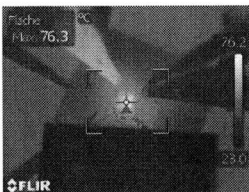

Fig. 8: Thermal image of the DUT with an applied Power P_{loss} of 20,0W

The thermal impedance $R_{th,j-h}$ can be calculated with the measured temperature values:

$$R_{th,j-h} = \frac{T_{Junction} - T_{Heatsink\,base}}{P_{loss}} = \frac{76,3°C - 34,4°C}{20,0W} = 2,095\ \frac{K}{W} \tag{1}$$

The thermal impedance $R_{th,j-h}$ is around 2,1 K/W even though the x-ray image showed a larger number of voids in the solder joint of the device to the IMS board.

Switching Performance

Switching performance had been evaluated with an oscilloscope with 1 GHz bandwidth and 20 GS/s and differential probes with a bandwidth of 400 MHz. With the proposed converter the maximum achieved switching speed was 40 V/ns with a DC link voltage of 400V. The efficiency in buck converter mode was around 98% at a power level of 2 kW. However, the max. voltage peak during switching transition was 738V resulting in an overvoltage of 338V and significant ringing.

From these results it can be concluded that the DC link inductance in the layout is too high. It seems to be necessary to arrange the switches in side by side with additional ceramic capacitor to achieve a significant lower commutation loop.

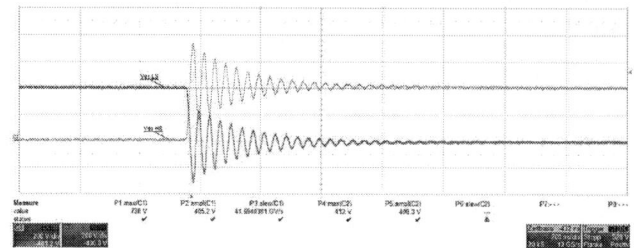

Fig. 9: Switching waveforms of the high-side turn-off switching transition

Conclusion

This paper presents the design of a power electronics stage with a hybrid PCB construction. The construction is based on a copper based IMS board and the surface mount technology (SMT) production process in a typical electronics manufacturing. BGAs were successfully used to connect the two boards. Based on x-ray pictures, the solder joint connection between a standard PCB and an IMS board were investigated and except for a number of small voids in the solder joint no failure due to production process were detectable. First tests of the IMS board were successfully performed with an inverter motherboard.

References

[1] A. Anthon, J. C. Hernandez, Z. Zhang and M. A. E. Andersen, "Switching investigations on a SiC MOSFET in a TO-247 package," IECON 2014 - 40th Annual Conference of the IEEE Industrial Electronics Society, 2014, pp. 1854-1860

[2] A. Lemmon, S. Banerjee, K. Matocha and L. Gant, "Analysis of Packaging Impedance on Performance of SiC MOSFETs," PCIM Europe 2016; International Exhibition and Conference for Power Electronics, Intelligent Motion, Renewable Energy and Energy Management, 2016, pp. 1-8.

[3] C. Schwabe, N. Thönelt, T. Basler, "Reliability investigation of SiC MOSFETs under switching operation in various packages," CIPS 2022 - 12th International Conference on Integrated Power Electronics Systems, 2022

[4] J. L. Lu, D. Chen, L. Yushyna, "A high power-density and high efficiency insulated metal substrate based GaN HEMT power module," 2017 IEEE Energy Conversion Congress and Exposition (ECCE), 2017, pp. 3654-3658

[5] LeitOn, Design Rules for Aluminum IMS, online available:https://www.leiton.de/technology-alu-printed-circuits.html

[6] LeitOn, Design Rules for Copper IMS, online available: https://www.leiton.de/technology-copper-printed-circuits.html

[7] Texas Instruments, Datasheet, LMG342xR050 600-V 50-mΩ GaN FET with Integrated Driver, Protection, and Temperature Reporting datasheet (Rev. A), online available: https://www.ti.com/product/LMG3422R050

[8] Qiang Yu, T. Shibutani, Y. Kobayashi and M. Shiratori, "The Effect of Voids on Thermal Reliability of BGA Lead Free Solder Joint and Reliability Detecting Standard," Thermal and Thermomechanical Proceedings 10th Intersociety Conference on Phenomena in Electronics Systems, 2006. ITHERM 2006., 2006, pp. 1024-1030, doi: 10.1109/ITHERM.2006.1645457.

[9] R. Yano and Q. Yu, "The life cycle impact assessment that the variabilities of BGA solder connection makes," 2017 IEEE 19th Electronics Packaging Technology Conference (EPTC), 2017, pp. 1-6, doi: 10.1109/EPTC.2017.8277488.

Active control of gear mesh vibration using a permanent-magnet synchronous motor and simultaneous equation method

Dominik Reitmeier, M.Sc.
Leibniz University Hannover - Institute for Drive Systems and Power Electronics
Welfengarten 1
30167 Hannover, Germany
Phone: +49 (0)511 762 4231
Fax: +49 (0)511 762 3040
Email: dominik.reitmeier@ial.uni-hannover.de
URL: https://www.ial.uni-hannover.de

Keywords

≪Acoustic noise≫, ≪Active damping≫, ≪Adaptive control≫, ≪Mechatronics≫, ≪Control of drive≫

Abstract

This paper presents an active vibration control (AVC) for gear mesh vibrations. The vibrations are reduced by an additional motor torque. To determine the amplitude and phase of the torque without prior identification of the controlled system parameters, a narrowband AVC using the simultaneous equation method is presented.

Introduction

Many drivetrains use gearboxes or transmissions to convert speed and torque from one rotating power source to another. During power transmission, internal excitation mechanisms generate periodic vibrations. These vibrations propagate through the shafts and bearings to the housing, where they are emitted as an acoustic noise. The spectrum of the noise contains individual multiples of the speed and is audible as a whining sound. Gear whine is a problem particularly in electric vehicles, since the combustion engine no longer masks this noise. This vibration and monotonous noise can also be disturbing in other applications.

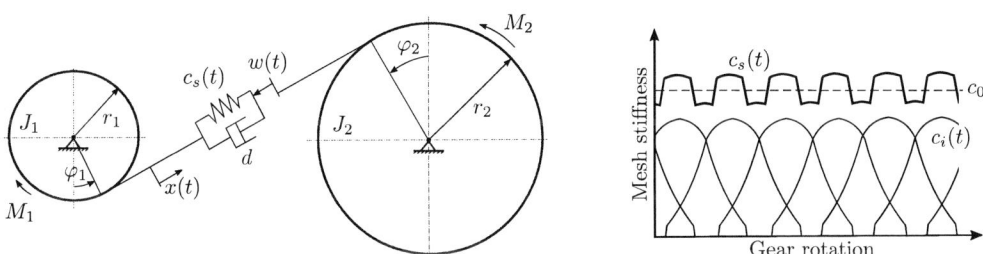

Fig. 1: Dynamic model of a one stage gear system [1]

Several sources contribute to the generation of noise in gearboxes [2]. Some of the excitations are caused by manufacturing-related deviations $w(t)$, such as gear manufacturing errors, profile errors or shaft misalignments. Under load, a major excitation mechanism is the non-constant mesh stiffness $c_s(t)$. The mesh stiffness varies during the rotation because there are not always the same number of teeth in mesh and because the stiffness of a single tooth $c_i(t)$ depends also on its position. Mathematically, this parametric oscillation can be described with the gear model in Figure 1 and the equations given in (1) (2). The equations can be further simplified to a single spring-mass system (6) [1].

$$J_1\ddot{\varphi}_1 = -c_s(t)\,r_1\,(r_1\varphi_1 - r_2\varphi_2 - w(t)) - dr_1\,(r_1\dot{\varphi}_1 - r_2\dot{\varphi}_2 - \dot{w}(t)) + M_1 \tag{1}$$

$$J_2\ddot{\varphi}_2 = c_s(t)\,r_2\,(r_1\varphi_1 - r_2\varphi_2 - w(t)) + dr_2\,(r_1\dot{\varphi}_1 - r_2\dot{\varphi}_2 - \dot{w}(t)) + M_2 \tag{2}$$

$$x = r_1\varphi_1 - r_2\varphi_2 - w(t) \quad (3) \qquad M = \frac{M_1 r_1}{J_1} - \frac{M_2 r_2}{J_2} \quad (4) \qquad \frac{1}{m} = \frac{r_1^2}{J_1} + \frac{r_2^2}{J_2} \tag{5}$$

$$\ddot{x} + \frac{d}{m}\dot{x} + \frac{c_s(t)}{m}x + \ddot{w}(t) = M \tag{6}$$

Methods for noise and vibration reduction

In addition to a high manufacturing quality, noise reduction methods can be divided into constructive, passive and active methods. Constructive methods try to achieve a low variation in the gear stiffness $c_s(t)$ by changing the tooth design. Passive methods use damping materials to reduce the transmission of vibration and sound. Active vibration control (AVC) systems measure the vibrations with an accelerometer and try to generate a vibration with the same amplitude and opposite phase that is superimposed on the undesired vibration. This theoretically cancels the vibration and the resulting noise. AVC methods differ in the number, placement, and choice of actuator used to generate the compensating vibration (Figure 2).

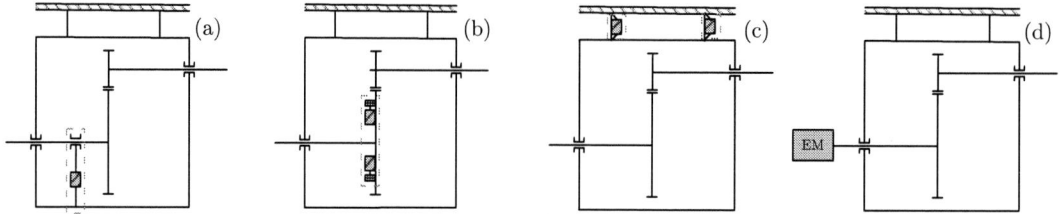

Fig. 2: Overview of methods for active vibration control of gearboxes [3]

Concept (a) presented by Montague et. al. [4] is the first research in which active vibration control has been applied to transmissions. A piezoelectric actuator was used which transmits a compensation force to the shaft via an additional bearing. At a gear meshing frequency of 4500 Hz, vibrations could be damped by 70%. Chen and Brennan [5] suggested a control system in which three inertial mass actuators mounted directly on the gear generate centrifugal forces to suppress vibrations (b). Experimental results showed a reduction of about 7 dB in acceleration at gear meshing frequencies between 150 Hz and 350 Hz. Concepts based on figure (c) employ active controlled struts to minimize vibrations in the recording structure [6]. The concept (d) published by Benzel [7] [8] achieves noise reduction by a harmonic motor torque M_c. The idea is based on converting the parameter oscillation from Equation (6) into a forced harmonic oscillation (7) and the compensation of the applied force term with M_c.

$$\ddot{x} + \frac{d}{m}\dot{x} + \frac{c_{s,0} + \sum_{n=1}^{N} c_{s,i}\sin(n\omega t + \varphi_n)}{m}x = M \;\Rightarrow\; \ddot{x} + \frac{d}{m}\dot{x} + \frac{c_{s,0}}{m}x = M - \frac{\sum_{n=1}^{N} c_{s,i}\sin(n\omega t + \varphi_n)}{m} + M_c \tag{7}$$

Experimental studies on a drive train with a power of 0.8 kW and a two-stage gearbox show a reduction in acceleration of 14 dB. One advantage of this concept is that no additional actuator is required. A disadvantage is that an experimental analysis is required in advance of how the drive torque affects the error signal, known as secondary path identification. An inaccurate identification of the secondary path, or a change during operation, can cause the algorithm to become unstable. The limited bandwidth of the current controller can furthermore exclude the application range for high frequencies. Therefore, this paper presents a method to determine the required compensation torque online and without prior experimental analysis of the secondary path. In addition, a control structure for controlling the high-frequency currents is presented.

Narrowband active vibration control

The basic configuration of the drive system consists of a spur gear, a permanent-magnet synchronous machine and an inverter (Figure 3). The current control is designed as a field-oriented control and the rotor position is measured with an incremental encoder. For active vibration control, an accelerometer is attached to the gearbox to measure the vibrations. The AVC algorithm controls the phase and amplitude of the required compensation torque respectively the compensation current. The current is then set by a controller located in parallel with the dq current control.

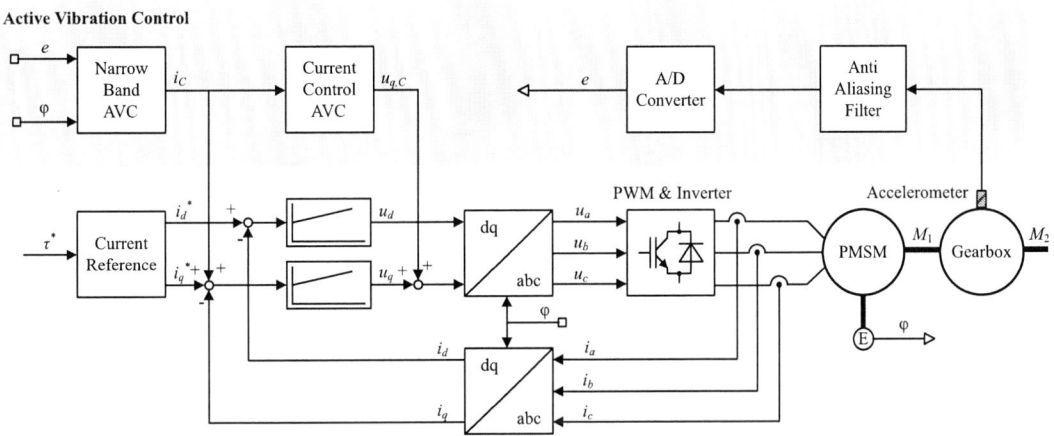

Fig. 3: Fundamental structure of the control system

The active vibration control is a narrowband implementation as shown in Figure 4. Narrowband AVC algorithms offer the advantage that individual frequency components can be selectively damped [9]. The individual frequency components operate in parallel. The system controls the required output signal by adaptive filtering of a periodic reference signal. The reference signal can be generated synthetically or measured by a measurement system. Since the gear mesh vibrations are multiples of the mechanical rotor frequency, the measured rotor position is used as the reference signal. To determine the frequency component of the gear meshing frequency, the rotor position is multiplied by the number of teeth of the pinion and the respective harmonic z_n. A sine and cosine component is formed from the reference signal, and the amplitude and phase of the output signal are adjusted by adjusting the weights of the adaptive filter (see equation (8),(9)). The transfer function $S(z)$ describes the secondary path, which is the path between the output of the adaptive filter and the input of the error sensor. In this application, the secondary path consists of the transfer behavior of the current controller, the PWM, the inverter, the electric machine, the gearbox, the accelerometer and the anti-aliasing filter.

$$\underline{w} = w_a + i w_b \qquad (8) \qquad \underline{i}_{C,n}(k) = \underline{w} \cdot e^{i z_n \varphi(k)} \qquad (9) \qquad \underline{E} = e_{a,n} + i e_{b,n} \qquad (10)$$

$$\underline{e}_n(k) = \underline{E} \cdot e^{i z_n \varphi(k)} = \underline{P} \cdot e^{i z_n \varphi(k)} + \underline{S} \cdot \underline{w} \cdot e^{i z_n \varphi(k)} \qquad (11)$$

The weights are adjusted by the simultaneous equation method [10]. The basic idea of the this method is to measure the output behavior of the filter weights and to calculate the new filter weights based on the change in the error signal. Compared to other methods such as the FxLMS algorithm [11], no accurate model of the secondary path is required. The simultaneous equation algorithm works blockwise. The initial weights of the algorithm are adjusted only every N time steps and are kept constant during that time. The algorithm assumes that the complex error signal of the past time period $\underline{E}_0$, the current time

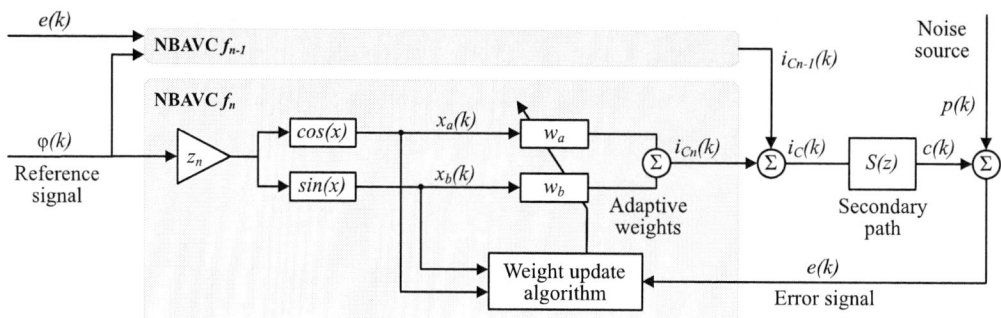

Fig. 4: Narrow band active vibration control

period $\underline{E}_1$ and the future time period $\underline{E}_2$ for one frequency component can be described with the following equations.

$$\underline{E}_0 = \underline{P}_0 + \underline{S} \cdot \underline{w}_0 \quad (12) \qquad \underline{E}_1 = \underline{P}_1 + \underline{S} \cdot \underline{w}_1 \quad (13) \qquad 0 = \underline{P}_2 + \underline{S} \cdot \underline{w}_{opt} \quad (14)$$

This system of equations can be used to determine the optimal weights w_{opt}:

$$\underline{w}_{opt} = \frac{\underline{E}_0 \cdot \underline{w}_1 - \underline{E}_1 \cdot \underline{w}_0}{\underline{E}_0 - \underline{E}_1} \tag{15}$$

Since the error signal consists of several frequencies, the complex Fourier coefficients for the frequency component must be calculated.

$$e_{a,n} = \frac{1}{N}\sum_{k=0}^{N-1} e(k) \cdot \cos\left(\frac{2\pi z_n k}{N}\right) = \frac{1}{N}\sum_{k=0}^{N-1} e(k) \cdot \cos\left(z_n \varphi(k)\right) = \frac{1}{N}\sum_{k=0}^{N-1} e(k) \cdot x_a(k) \tag{16}$$

$$e_{b,n} = \frac{1}{N}\sum_{k=0}^{N-1} e(k) \cdot \sin\left(\frac{2\pi z_n k}{N}\right) = \frac{1}{N}\sum_{k=0}^{N-1} e(k) \cdot \sin\left(z_n \varphi(k)\right) = \frac{1}{N}\sum_{k=0}^{N-1} e(k) \cdot x_b(k) \tag{17}$$

An abrupt change of the filter weights excites the mechanical system to oscillate. In addition, the error signal is provided with measurement uncertainties. For these reasons, the update of the filter weights is done stepwise (18). At a step size of $\mu = 1$, the algorithm converges fast. Smaller step sizes result in slower but more stable convergence behavior.

$$\underline{w}_2 = (1 - \mu) \cdot \underline{w}_1 + \mu \cdot \underline{w}_{opt} \tag{18}$$

Control of the harmonic currents

In addition to determining the required filter weights and the required compensation current, the current must also be controlled for higher frequencies. The existing dq current control might not have the required bandwidth. Also, the compensation current can be in the range of the measurement noise of the current sensor. For these reasons, compensation current control is implemented as feedforward control in parallel with the dq current control (Figure 3). The control is performed by a PDT1 controller (Figure 5) whose output signal is added to the q-component of the desired voltage.

The feedforward controller (equation 19) is parameterized so that the numerator polynomial of the controller compensates the denominator polynomial of the controlled system (equation 20). For this ap-

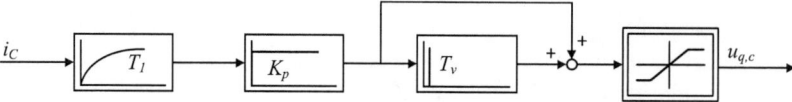

Fig. 5: Current controller for active vibration control

plication case of a permanent-magnet synchronous machine controlled via a field-oriented control, T_s describes the time constant of the motor winding. The derivative time T_v of the differentiator is set to the time constant T_s. T_σ is the equivalent time constant of the actuator, in this case the inverter including the PWM. The parameter T_1 can be used to set the cut-off frequency and thus the usable range of the control. The parameter T_1 was set equal to the time constant T_σ. In addition, the output of the feedforward control was limited.

$$G_s = \frac{K_s}{1 + s \cdot T_s} \cdot \frac{1}{1 + s \cdot T_\sigma} \qquad (19)$$

$$G_c = K_p \cdot \frac{1 + s \cdot T_v}{1 + s \cdot T_1} \qquad (20)$$

$$K_p = \frac{1}{K_s} = R_s \quad (21) \qquad T_v = T_s = \frac{L_q}{R_s} \quad (22) \qquad T_\sigma \leq T_1 \leq T_s \quad (23)$$

Test bench and experimental results

The experimental tests were carried out on the test bench shown in Figure 6. The drive train consists of a three-stage spur gear and a two-stage planetary gear (3), which are connected via a differential. The planetary gear is driven by an induction machine (2) with a power of 60kW. A permanent magnet synchronous machine (PMSM) (1) with a rated power of 75kW is mounted on the spur gear. Both electrical machines are powered by a three-phase inverter with a switching frequency of 36kHz. Active vibration control was tested on the first stage, the high speed stage, of the spur gearbox. The PMSM is used as actuator. The first gear stage consists of helical gearing with a ratio of i = 4. The pinion has 19 teeth, resulting in a gear mesh frequency of $f_{z1} = \frac{19 \cdot n}{60}$ for the first harmonic and $f_{z2} = \frac{38 \cdot n}{60}$ for the second harmonic.

Fig. 6: Test bench for experimental evaluation

Measurement sensors are 4 accelerometers (4) mounted on the gearbox and a microphone (5) as well as a measurement of the DC and AC current of the inverter (6). The accelerometer for the control is mounted directly on the bearing of the first stage and measures the radial acceleration. Besides there is another accelerometer (sensor 1) which is used for evaluation. Another accelerometer (sensor 2) measures the axial movement of the gearbox and is also located near the bearing. Sensor 3 is mounted on the gearbox housing near the permanent magnet synchronous machine. The microphone is placed at a distance of approximately 1 m from the drive train.

The results for the steady-state operating point with the speed of $n = 7000 \text{min}^{-1}$ and a load of $M = 15\text{Nm}$ are shown in Fig. 7. Shown are the accelerations of sensors 1 and 2 and the sound pressure level of the microphone. The blue line is the signal without active vibration control. In orange is drawn the spectrum of the algorithm using the simultaneous equation method. The algorithm updates its output weights for the first gear mesh frequency every 5 rotations and calculates the fourier analysis of the error signal over the same period. The value of the step size μ is 0.1. For comparison, the results of the filtered-x-LMS (FxLMS) algorithm were also shown. For the FxLMS algorithm, the secondary path was determined

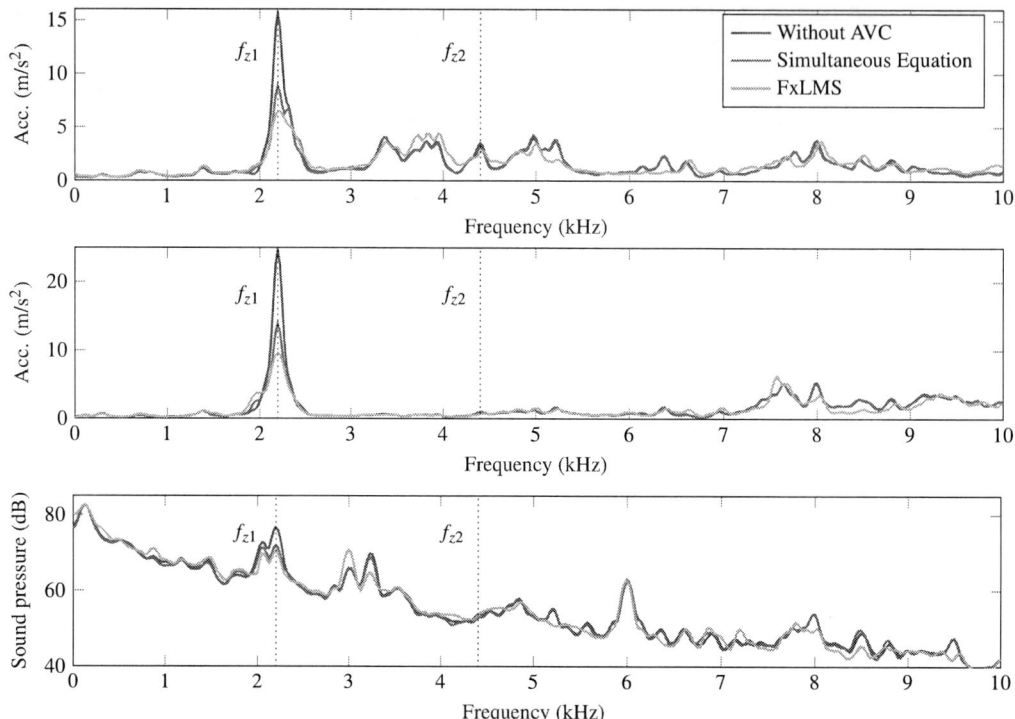

Fig. 7: Operating point 7000min^{-1} 15Nm; sensor 1 (top), sensor 2 (middle), microphone (bottom)

using a random noise signal. The step size μ is also set to 0.1, and the algorithm updates its output weights at a frequency of 100Hz.

When comparing the signals, it can be seen that both methods can reduce the vibrations in radial and axial direction. With the microphone, a reduction in noise of about 5 dB can also be measured for the first gear mesh frequency. The second gear mesh frequency causes only minor radial forces and is negligible for this operating point. The frequency spectrum also shows further acceleration components, which can be assigned to the other gear stages and the electrical machines. Furthermore, it can be seen that the vibrations are damped slightly better with the FxLMS algorithm. With both algorithms, the amplitude of the compensation current I_c is limited to 25A, which is about 10% of the rated current. By increasing this limit, the vibrations can be further damped.

The power consumption measured at the input of the inverter is 11.27kW without compensation and 11.68kW with compensation (SE method). In the DC and AC current of the inverter, the components of the active vibration control are clearly noticeable (Figure 8).

Table I shows the acceleration of the first mesh frequency for other operating points. The acceleration with compensation here refers to the simultaneous equation method. When comparing the operating

Speed	Torque	Sensor 1		Sensor 2		Sensor 3		Microphone	
		Without	With	Without	With	Without	With	Without	With
5000min^{-1}	5Nm	3.6m/s^2	1.8m/s^2	3.2m/s^2	0.9m/s^2	0.4m/s^2	0.4m/s^2	70dB	68dB
5000min^{-1}	15Nm	2.7m/s^2	1.4m/s^2	3.3m/s^2	1.6m/s^2	0.4m/s^2	0.4m/s^2	70dB	68dB
7000min^{-1}	5Nm	15m/s^2	9.1m/s^2	20m/s^2	13m/s^2	4.2m/s^2	2.7m/s^2	74dB	70dB
7000min^{-1}	15Nm	15m/s^2	8.7m/s^2	24m/s^2	14m/s^2	7.0m/s^2	4.1m/s^2	77dB	72dB

Table I: Accelerations without and with active noise control

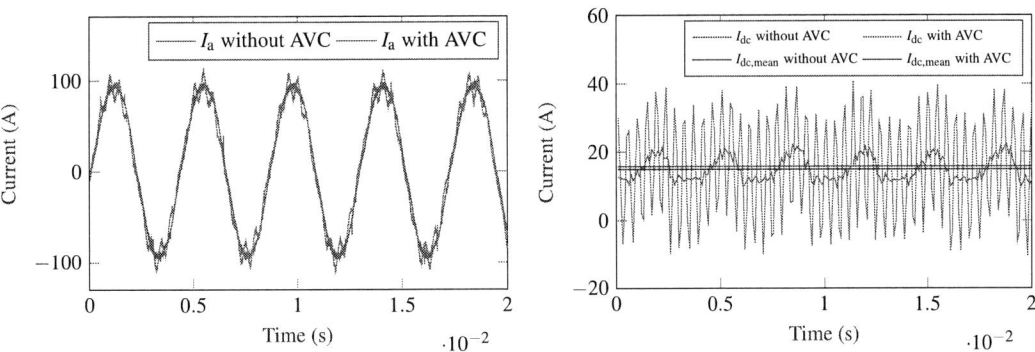

Fig. 8: Measured currents of the inverter, alternating current (left) and direct current (right)

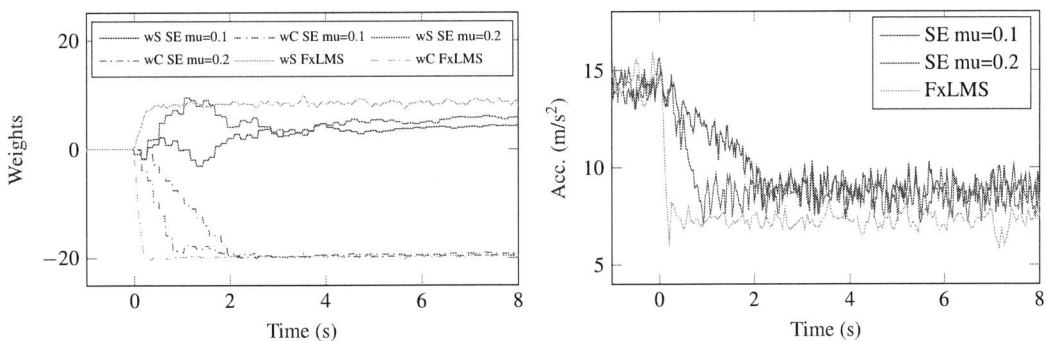

Fig. 9: Evolution of filter weights (left) and acceleration at first mesh frequency (right)

points, it can be seen that the acceleration increases from 5000min^{-1} to 7000min^{-1}. The achievable vibration and noise reduction is greater for the 7000min^{-1} speed.

The convergence behavior of the algorithm for the steady-state operating point at a speed of 7000min^{-1} and a torque of $M = 15\text{Nm}$ is shown in Figure 9. The left figure shows the two output weights and the right figure shows the amplitude of the acceleration of the first mesh frequency $|E_{fz1}|$. The algorithm is activated at time $t = 0$. Increasing the step size μ from 0.1 to 0.2 results in faster convergence. The variance of the output weights for the steady state value is similar for these two step sizes. The FxLMS algorithm converges even faster due to its higher update rate. It also achieves a slightly different final value for the sinusoidal output weight and achieves a slightly better reduction.

Figure 10 shows the behavior of the algorithm with a change in speed. On the left side are again the filter

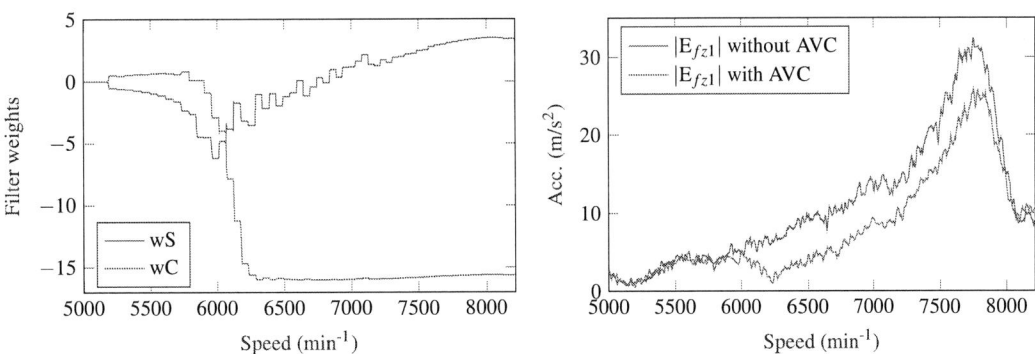

Fig. 10: Behavior under speed change: filter weights (left) and acceleration (right)

weights and on the right side the acceleration of the first gear mesh frequency. The algorithm is activated at a speed of 5200min^{-1} and then also converges to the values in Figure 9. As the filter weights reach this value, a reduction in acceleration is measurable. The acceleration increases up to a speed of 7700min^{-1} and then decreases again. At this speed or excitation frequency, there is a resonance of the mechanical structure. In general, it can be seen that the algorithm remains stable despite changes in speed. The reduction remains the same over the speed range.

Conclusion

The non-constant gear stiffness generates torque pulsations during power transmission in gearboxes. This leads to unwanted vibrations and noise. An approach was presented that reduces the torque pulsations with the electric machine and actively damps the resulting vibrations and noise. For this purpose, a method was investigated with which the amplitude and the phase of the compensation current can be calculated without a model of the secondary path. In addition, a control structure was presented with which the high-frequency current can be controlled. Experimental tests of the method show that active damping of vibration and noise is possible. A comparison with the simultaneous equation method versus the FxLMS algorithm shows that the FxLMS algorithm achieves a slightly better reduction. If a determination of the secondary path is not possible, the simultaneous equation method can be used. In the other case the use of the FxLMS algorithm is preferable.

References

[1] Sanzenbacher S.: Reduction of transmission noises by mitigation of structure-borne sound, Stuttgart: Institut für Maschinenelemente, 2016, ISBN 978-3-936100-66-2

[2] Smith J. D.:Gear noise and vibration, CRC press, 2nd edition, 2013, ISBN 978-0824741297

[3] Zech, P.: Aktive Reduktion modulierter Zahneingriffsvibrationen von Planetengetrieben. *Shaker Verlag*, 2019

[4] Montague, G. T., Kascak, A. F., Palazzolo, A., Manchala, D.: Feedforward control of gear mesh vibration using piezoelectric actuators. *Shock and Vibration*, vol. 1, no. 5, pp. 473-484, 1994

[5] Chen, M. H., Brennan, M. J.: Active control of gear vibration using specially configured sensors and actuators. *Smart Materials and Structures*, vol. 9, no. 3, pp. 342, 2000

[6] Sutton, T. J., Elliott, S. J., Brennan, M. J., Heron, K. H., Jessop, D. A. C.: Active isolation of multiple structural waves on a helicopter gearbox support strut. *Journal of Sound and Vibration*, vol. 205, no. 1, pp. 81-101, 1997

[7] Benzel, T., Möckel, H. A.: Active control of gear pair vibration with an electronically commutated motor as actuator. In *4th International Electric Drives Production Conference (EDPC)*, pp. 1-6, IEEE, 2014

[8] Benzel, T., Möckel, H. A.: Active gear pair vibration control during non-static load and speed with an electronically commutated motor as actuator. In *10. ETG/GMM-Symposium Innovative small Drives and Micro-Motor Systems (IKMT)*, pp. 1-6, VDE, 2015

[9] Kuo, S. M., Morgan, D. R.: Active noise control: a tutorial review. *Proceedings of the IEEE*, 87.6, pp. 943-973, 1999

[10] Zech, P., Plöger, D. F., Rinderknecht, S.: Active control of planetary gearbox vibration using phase-exact and narrowband simultaneous equations adaptation without explicitly identified secondary path models. *Mechanical Systems and Signal Processing*, vol. 120, pp. 234-251, 2019

[11] Burgess, John C.: Active adaptive sound control in a duct: A computer simulation. *The Journal of the Acoustical Society of America*, vol. 70, no. 3, pp. 715-726, 1981

Research Laboratory for Testing Grid Connected Devices under Grid Voltage / Grid Impedance Variations and Microgrid Conditions

Swen Bosch, Jochen Staiger, Heinrich Steinhart
University of Applied Sciences Aalen
Laboratory of Power Electronics and Electrical Drives
Beethovenstraße 1
73430 Aalen, Germany
Tel.: +49 7361 576 4233
E-Mail: swen.bosch@hs-aalen.de
URL: https://www.hs-aalen.de/de/facilities/84

Acknowledgements

The German Federal Ministry for Economic Affairs and Climate Action has supported this work based on a decision by the German Bundestag under the Project 0350029D.

Keywords

«Fault ride-through», «Grid-connected converter», «Impedance analysis», «Test bench», «Transmission of electrical energy»

Abstract

Distributed (renewable) energy sources and a growing amount of loads are connected to the grid via (switched) power electronics. To investigate the grid-integration of these devices, a test bench was built up, which offers the possibility to vary the impedance and the voltage (in amplitude, waveform and frequency) at the PCC during operation. For this purpose, the test bench includes a transformer-based *Grid Voltage Emulator*, a *Grid Impedance Emulator* as well as a *4Q-AC-Source*. Due to the increasing use of power electronics, the grid tends to behave capacitively in some load cases. To cover this topic as well, the test stand also includes a binary graded capacitor bank to emulate volatile capacitive loads. The design, realization and operation of the respective components is described in this paper. In addition, various experimental results regarding the grid voltage and impedance emulator are presented focusing e.g. on the current commutation of highly inductive loads. Furthermore, measurements of different devices under test, like an impedance analyzer or an active resonance damper, are shown.

Introduction

The electric power supply in many countries is changing from a centralized energy generation to a distributed and decentralized renewable energy generation [1]. Many of these renewable energy sources are connected to the grid via power electronics. Due to the decentralized energy generation, the power distribution network is getting less stable and harder to operate, why only centralized power plants like coal-fired or nuclear power plants cannot realize the grid control. In addition, the small and medium decentralized power plants have to take part in the grid control. Therefore, new control strategies for grid-connected inverters like the grid-forming control [2,3] instead of the grid-following control are needed. Compared to centralized power plants, the decentralized energy generation and grid-feeding via power electronics is offering a way higher dynamic behavior, but also a higher volatility regarding the amount of the provided energy, what must also be taken into consideration.

The grid integration of these decentralized power plants can be investigated in different ways: simulations, field tests and laboratory tests. Simulative analysis is a fast approach, but often simplifications have to be done, why the real and detailed behavior and interaction of the grid-connected devices cannot be figured out. Practical field tests and measurements with the real hardware setup are the most realistic way of investigation, but they are very expensive and time consuming. Furthermore, the parameters like e.g. the grid impedance cannot be influenced, why investigations have to be done at different places to cover a broad range of scenarios. As the previous explanations reveal, laboratory tests are offering a cost and time saving option to investigate the grid-integration of power electronic devices. Additionally, the investigations can be carried out in a safe environment, what is advantageous especially when new hardware or control setups are investigated. To emulate the relevant grid characteristics, appropriate devices had to be designed and developed for the laboratory test bench.

Beside the frequency, the amplitude and the waveform of the voltage as well as the impedance at the (internal) point of common coupling ((I)PCC) are the main characteristics of a grid. Due to e.g. load changes, voltage sags/swells as well as interruptions, caused by failures, can occur. The short-circuit power and with this the impedance of the grid is also varying, especially when the grid is fed by a relevant amount of renewable energy sources, since the energy fed to the grid is depending on the time of day and the season.

This paper is dealing with the design and the development of a *Grid Voltage Emulator* and a *Grid Impedance Emulator*. Additionally, a (commercial) four-quadrant grid-simulator (*4Q-AC-Source*) can be integrated to vary not only the amplitude of the voltage, but also the waveform and the frequency to realize e.g. distorted voltages or frequency variations at the PCC. Furthermore, a *Resonance Generator* consisting of binary graded capacitors can be used to emulate a capacitive load / a capacitive grid. This allows the creation of a resonance point, whose resonance frequency depends on the capacitance of the *Resonance Generator* and the inductance of the *Grid Impedance Emulator*.

One main target of the development was to vary the grid parameters online, that means while the devices under test (DUTs) are connected to the grid and are operating according to their respective purpose. Hereby, not only the steady state, but also the transient behavior of the DUTs can be investigated. Exemplary steady state investigations are the behavior while long-term under-/overvoltage or the operation under distorted/unsymmetrical PCC voltages. Transient investigations are e.g. fault ride through tests (symmetrical or unsymmetrical) or the emulation of switching operations in the grid, which can result in a change of the short-circuit power of the grid and with this in a change of the impedance at the PCC. For the online variation of the grid parameters, IGBTs instead of e.g. thyristors or electromechanical switches are applied to change between voltage or impedance levels. Different issues as switching of highly inductive loads or short-circuit protection of the transformers secondary windings when switching between the voltage levels were considered and will be discussed. Experimental tests and measurements of the emulators itself as well as of selected DUTs, e.g. of the impedance analyzer, will be presented.

For the medium voltage-level, several publications can be found regarding a grid emulation, e. g. [4,5]. Some groups of researchers were already dealing with the topic of voltage and impedance emulation or micro grid at the low-voltage level, e.g. [6,7] in a laboratory scale (up to 30 kVA) or [8] (several 100 kVA, depending on the configuration). Also, several members of the *European Distributed Energy Resources Laboratories* (*DERlab*, ref. [9]) are focusing on this topic.

Nevertheless, only a few articles regarding voltage and impedance emulation can be found in the literature, although the topic is relevant and actual in the context of the transformation of the electric power supply. Therefore, this article would like to offer a contribution to this topic regarding the design, realization and operation of such a test bench.

The article is structured as follows: First, the setup of the components of the test bench will be described. Subsequently, the entire test stand and possible components are discussed, what will be followed by the experimental results. The article is concluded by a summary and an outlook.

Setup of the Grid Voltage Emulator, the Grid Impedance Emulator and the Resonance Generator

The single-phase setups of the *Grid Voltage Emulator* and the *Grid Impedance Emulator* are shown in Figs. 1a and 1b, respectively. The main component of the *Grid Voltage Emulator* is a three-phase four-wire transformer (primary voltage 400 V, rated power 50 kVA, 4 % short-circuit voltage) with eight taps for different secondary voltage levels (20%, 40%, 60%, 70%, 80%, 90%, 100%, 110%). The voltage levels can be preselected by manual switches. Switching within operation is realized via the IGBT switches, which will be discussed in detail afterwards.

Beside the IGBTs for switching during operation, the *Grid Impedance Emulator* consists of inductors, which are connected in series. A preselection can be done by the manual switches. Additionally, it is possible to short-circuit the inductors to vary the inductance of the preselection. For this purpose, the taps of the inductors are wired to connectors at the front of the cabinet (see Fig. 6, the first connectors from the bottom).

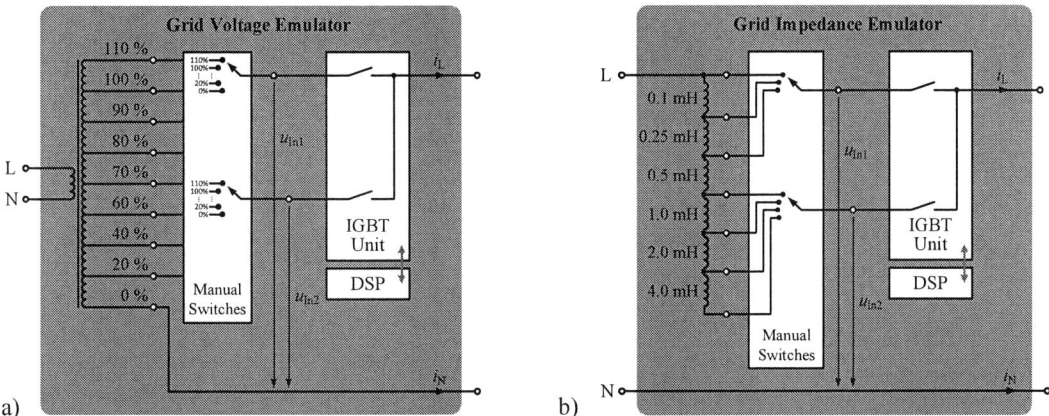

Fig. 1: Single phase setup of a) the Grid Voltage Emulator and b) the Grid Impedance Emulator

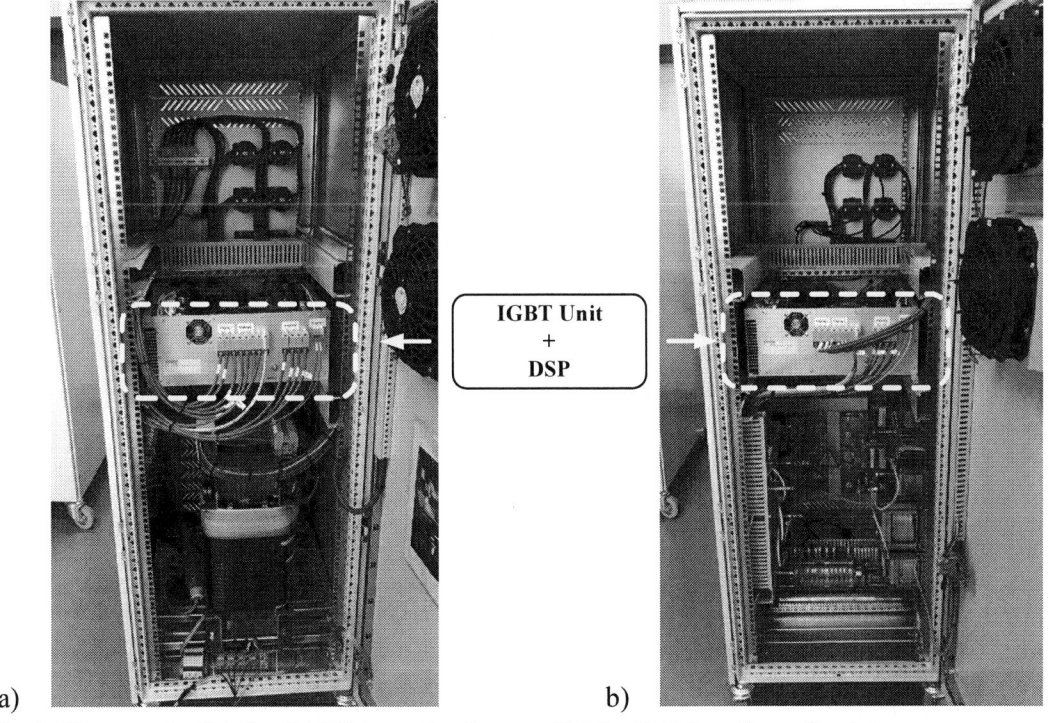

Fig. 2: Photograph of a) the Grid Voltage Emulator and b) the Grid Impedance Emulator

At both emulators, the IGBT switches are controlled by a Texas Instruments TMS320F28377D digital signal processor (DSP). Both input voltages u_{In1} and u_{In2} as well as the three phase currents i_L and the neutral conductor current i_N are measured. By this, an overvoltage and an overcurrent protection can be realized. In addition, the phase angle of the voltages can be derived by the voltage measurement and with this the phase angle of the switching operation can be chosen (e.g. in the zero crossing or the maximum of the voltage, see experimental results). This can be advantageous for some applications like the suppression of inrush currents at capacitive loads. The single-phase schematic of the IGBT unit is shown in Fig. 3a.

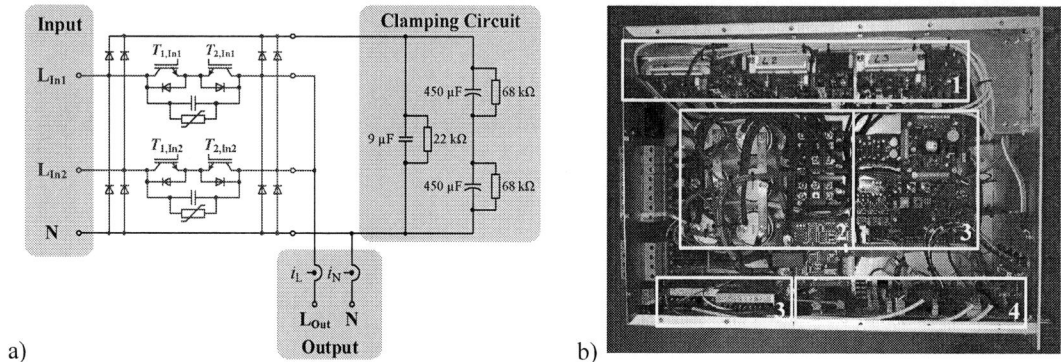

a) b)

Fig. 3: IGBT unit: a) Single-phase setup b) photograph (IGBT modules (1), electrolytic capacitors and diodes of the clamping circuit (2), power supply (3) and DSP and analog signal processing (4))

Per input branch, two antiserial connected IGBTs (Semikron SKM400GM17E4, V_{CES} = 1700 V, I_{Cnom} = 400 A, ref. [10]) are used. The current carrying capacity of the IGBTs is overrated intentionally far above the maximum currents (I_{max} = 60 A_{RMS}) of the emulators in order to take e.g. inrush currents of capacitive loads or grid-connected induction machines into account. To protect the IGBTs from high du/dt stress and overvoltage, a snubber capacitor and a varistor are connected directly on the screw terminals of the IGBT modules. For the switching of highly inductive loads, an additional clamping circuit was integrated into the IGBT units, consisting of a low-inductive film capacitor and electrolytic capacitors, also including discharging resistors. The clamping is realized by diode rectifier modules (International Rectifier 90MT160K, V_{RRM} = 1600 V, I_{FSM}(50Hz) = 770 A). In Fig. 3b, the afore mentioned components can be seen in the photograph of the IGBT unit.

The design of the clamping circuit was carried out on the basis of a single-phase worst-case scenario:

- the maximum inductance of the *Grid Impedance Emulator* is about 8 mH. Taking the inductances of the *Grid Voltage Emulator* and a possible inductive load into account, the maximum series inductance is assumed to be about L_{max} = 12 mH,
- the maximum load current is 60 A_{RMS},
- the load gets completely disconnected from the IPCC,
- the load is disconnected at the time, the load current has reached its peak value.

With these worst-case assumptions, the maximum energy stored in the inductances is

$$E_{L,max} = \frac{1}{2} \cdot L_{max} \cdot \left(\sqrt{2} \cdot I_{max} \right)^2 = \frac{1}{2} \cdot 12 \, \text{mH} \cdot \left(\sqrt{2} \cdot 60 \, \text{A} \right)^2 = 43{,}2 \, \text{Ws}. \tag{1}$$

Neglecting any ohmic losses, this energy has to be stored in the capacitors in addition to the initial energy of the capacitor. The voltage depending energy difference stored in a capacitor is

$$\Delta E_C = \frac{1}{2} \cdot C \cdot \left(U_2^2 - U_1^2 \right), \tag{2}$$

where the initial voltage U_1 is the peak input voltage of 325 V und U_2 is the final voltage. With this, the final voltage U_2 at the capacitor can be calculated:

$$U_2 = \sqrt{\frac{E_{L,max}}{\frac{1}{2} \cdot C} + U_1^2} = \sqrt{\frac{36 \, \text{Ws}}{\frac{1}{2} \cdot 234 \, \mu\text{F}} + \left(325 \, \text{V} \right)^2} \approx 689 \, \text{V}. \tag{3}$$

This voltage is within the permissible voltage range, which is limited to 800 V due to the clamping circuit capacitors rated voltage.

Another component of the test bench, the *Resonance Generator*, is used to emulate a variable capacitive load and with this a frequency-variable resonance point in combination with the inductance of the *Grid*

Impedance Emulator. The single-phase setup of the *Resonance Generator* is shown by Fig. 4a and consists of three binary graded capacitors, which can be switched by anti-parallel thyristors. The series inductors are offering an additional degree of freedom regarding possible resonance frequencies, which are in the range of about 200 Hz to 3 kHz. A more detailed view on the *Resonance Generator* is given by [11].

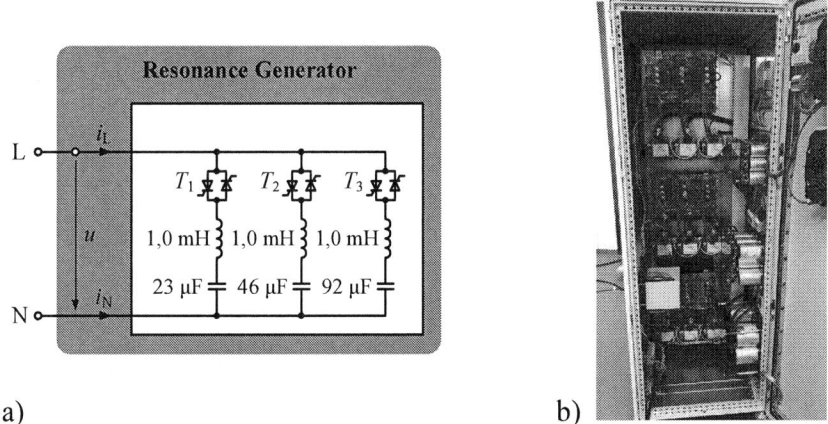

a)   b)

Fig. 4: Resonance generator: a) single phase setup and b) hardware setup

Overall Setup of the Research Laboratory

The overall setup of the research laboratory including an exemplary selection of grid-connected devices is given by Fig. 5.

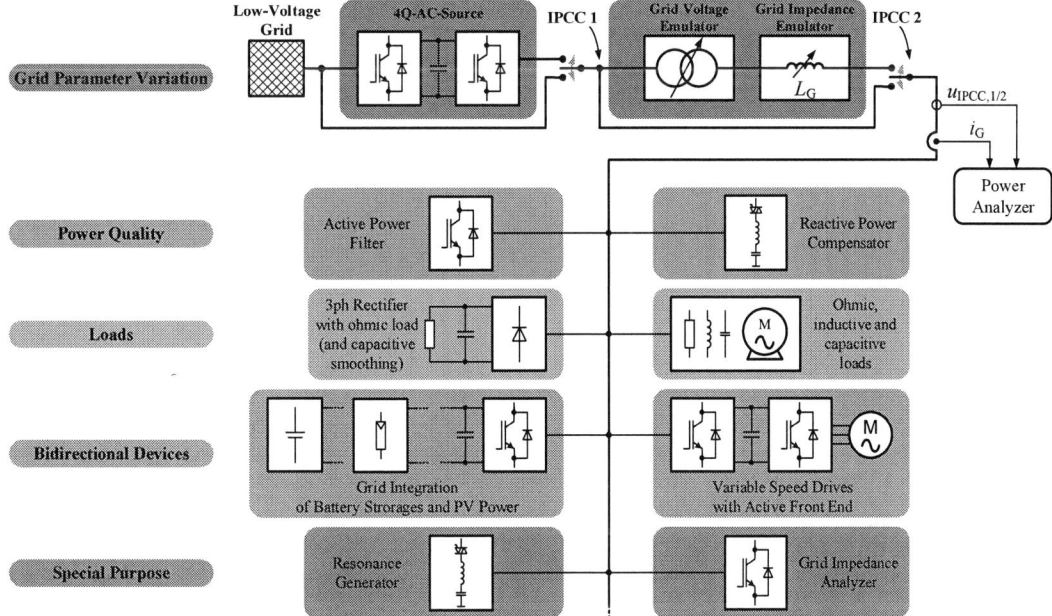

Fig. 5: Overall setup of the research laboratory with an exemplary selection of grid connected devices

The setup can be divided in the following groups:

- Grid Parameter Variation: Including the previously mentioned *Grid Voltage Emulator* and *Grid Impedance Emulator* as well as the *4Q-AC-Source*.
- Power Quality: Compensation of fundamental reactive and distortion power by an *Active Power Filter* (*APF*, with voltage controlled resonance damping capability) and a *Reactive Power Compensator*.

- Loads: Depending on the objective of the investigation, e.g. linear loads like passive devices or induction machines as well as non-linear loads like rectifiers can be integrated into the test bench.
- Bidirectional Devices: Grid feed of renewable energy or regenerative inverters.
- Special Purpose: The previously mentioned *Resonance Generator* to emulate a variable capacitive load and with this a frequency-variable resonance point in combination with the inductance of the *Grid Impedance Emulator*. The *Grid Impedance Analyzer* (also self-developed) is able to determine the impedance at the (I)PCC by means of a monofrequent current injection.

The photograph depicted in Fig. 6 is showing the *4Q-AC-Source*, the *Grid Voltage Emulator*, the *Grid Impedance Emulator* and the *Power Analyzer* (see also Fig. 5, blue highlighted and white, respectively).

Fig. 6: *4Q-AC-Source, Grid Voltage Emulator, Grid Impedance Emulator* and Dewetron DEWE2600 *Power Analyzer* (from left to right)

Experimental Results

In the following, an exemplary selection of tests and measurements of the emulators itself as well as of selected DUTs will be presented. The components used and a description of the test are presented in the form of a list, which is supplemented by the corresponding diagrams of the measurements. Components that are not mentioned are not used in the respective measurement.

Online Switching Between Voltage Levels

- *Grid Voltage Emulator*:
  - Input voltage: 230 V_{RMS} phase-neutral.
- Load:
  - $R = 10\ \Omega$, $L = 2.1$ mH.
- Description:
  - Grid voltage emulator in stand-alone mode.
  - Switching at t = 0 s (the time of the maximum of the load current) from 60% to 100% output voltage.
  - The output voltage is not interrupted, so the load is connected to the voltage source continuously.
  - After the switching operation, small oscillations can be seen in the output voltage
  - After the step in the output voltage, the output current is increasing according to the time constant of the load, which is about 0.2 ms.
  - Since all components are freely programmable, voltage sags and swells can be realized by switching between different voltage levels.
- Diagrams:

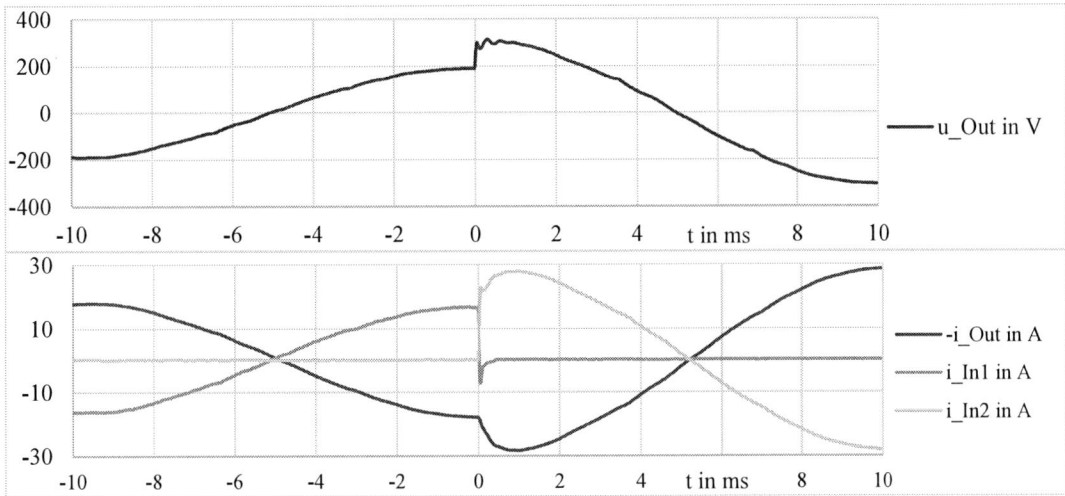

Fig. 7: Measured quantities: Online switching between voltage levels

Three-Phase Voltage Start up

- *Grid Voltage Emulator*:
  - Input voltage: 230 V_{RMS} phase-neutral.
  - Output voltage: 70% of the input voltage.
- *Grid Impedance Emulator*:
  - Stationary inductance: 1 mH.
- Load:
  - 5.5 kW induction machine in star connection with neutral conductor.
- Description:
  - Both emulators are connected in series.
  - The output voltages are switched on at their respective zero crossing.
  - Due to the series inductance, the high start-up currents are causing a reduction of the output voltages.
  - The output voltages are reaching their nominal level after the start-up currents decayed.
- Diagrams:

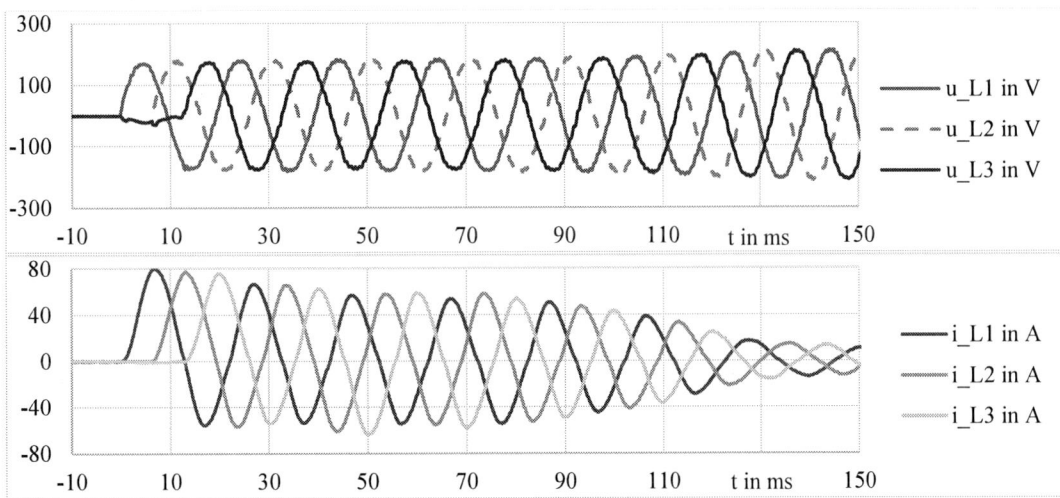

Fig. 8: Measured quantities: Three-phase voltage start up

Online Impedance Variation and Analyzation

- *Grid Impedance Emulator*
- *Impedance Analyzer*:
  - Monofrequent current injection, 15 A_{RMS}, 75 Hz.
- Description:
  - The grid impedance emulator switches from $L = 1.5$ mH / $R = 0.08$ Ω (values determined by dc measurement) to $L = 0$ mH / $R = 0$ Ω at t = 0 s.
  - At t = 1 s, the impedance analyzer has stationary tracked the change in the impedance ($\Delta L = 1.62$ mH, $\Delta R = 0.071$ Ω).
  - Compared to the dc measurement, the error is about 8% (inductance) and 11% (resistance), and with this in an acceptable range.
- Diagrams:

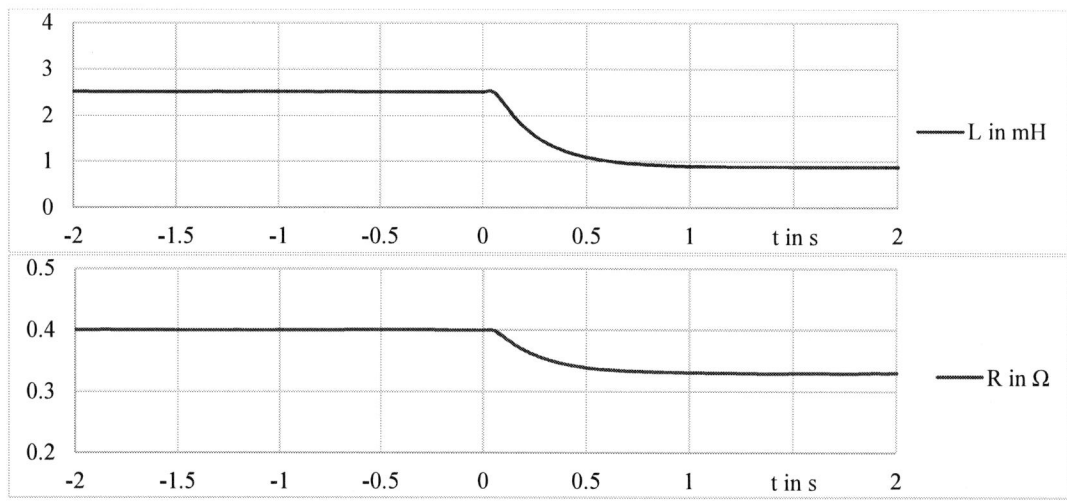

Fig. 9: Estimated quantities: Online Impedance Variation and Analyzation

Resonance Generation

- *4Q-AC-Source*:
  - Output voltage U_{IPCC2} (ref. Fig. 5): 140 V_{RMS} phase-neutral.
  - Distortion by a 5th harmonic with $U_{RMS,5} = 20$ V.
- *Grid Impedance Emulator*:
  - $L = 1.0$ mH.
- *Resonance Generator*:
  - $L = 1.0$ mH, $C = 92$ μF, resonance point at about 290 Hz.
- *Active Power Filter*:
  - Connected to the grid, but no resonance damping activated.
- Description:
  - A resonance current can be clearly seen in the current of the Resonance Generator i_{RG}.
  - This resonance current is causing an additional distortion in u_{IPCC2}, resulting in a 5th harmonic of 60 V_{RMS}, which is three times the source voltage of 20 V_{RMS}.
  - Due to the limited bandwidth of the *APF*s fundamental current control and the massive distortion in the u_{IPCC2} voltage, the *APF* is also drawing a 5th harmonic current from the grid.
- Diagrams:

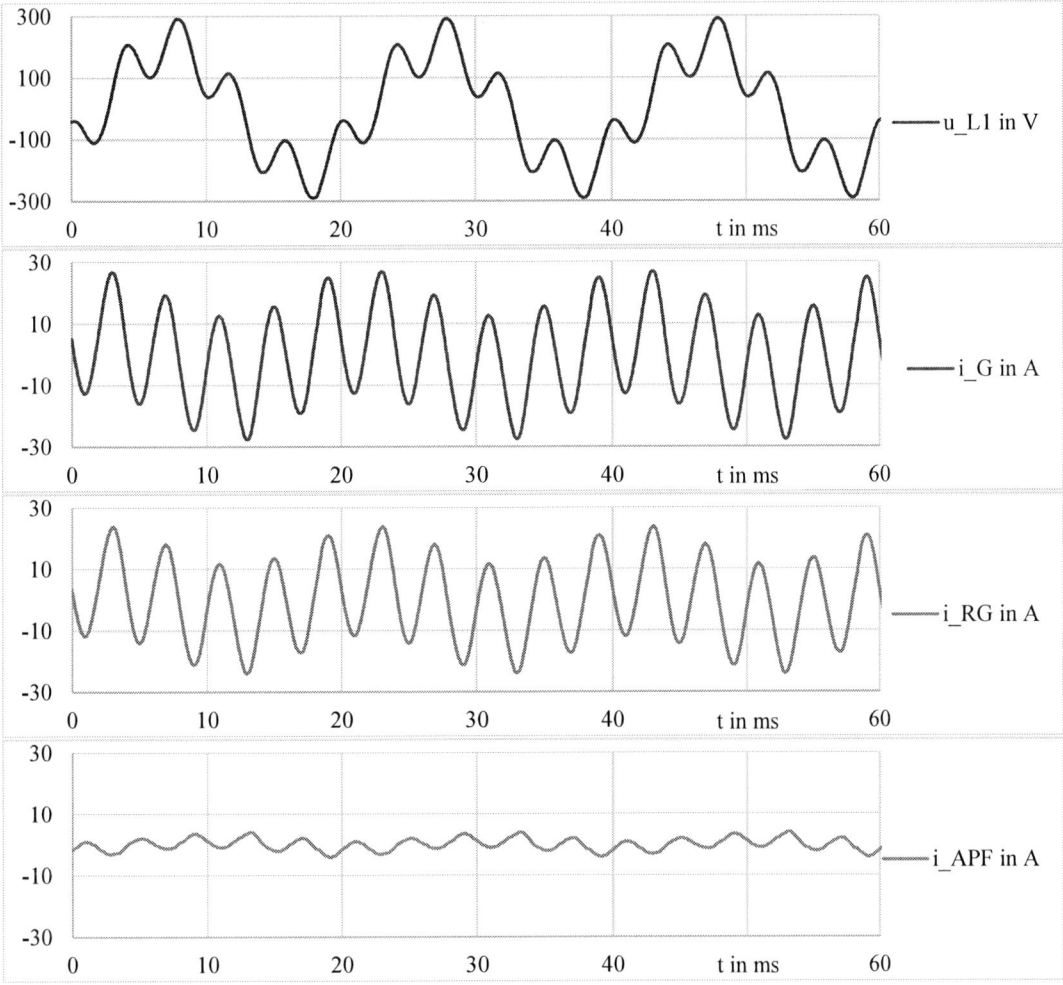

Fig. 10: Measured quantities: Resonance Generation

Resonance Generation and Damping

- Same setup of the *4Q-AC-Source*, the *Grid Impedance Emulator* and the *Resonance Generator* as described for the measurement "Resonance Generation".
- Active Power Filter:
  - Connected to the grid, resonance damping activated.
- Description:
  - The resonance damping of the *APF* is activated. Since the resonance damping is voltage controlled, the control objective is to compensate the distortion at the 5th harmonic in the voltage.
  - Due to the control objective, the *APF* is injecting a 5th harmonic current, which is eliminating the voltage distortion at the 5th harmonic.
  - With this, the excitation of the resonance circuit gets eliminated and the resonance current disappears. The current i_{RG} of the *Resonance Generator* consists mainly of the fundamental current.
- Diagrams:

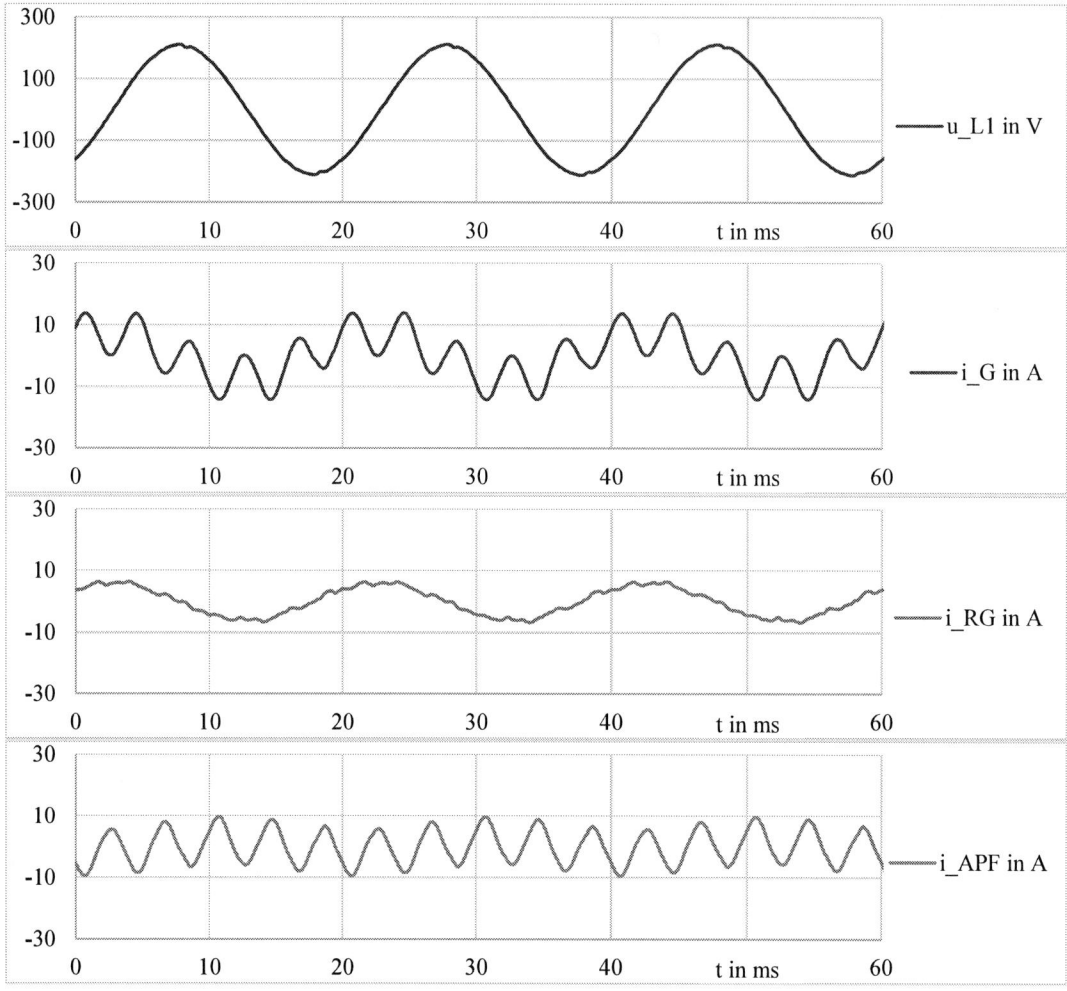

Fig. 11: Measured quantities: Resonance Generation and Damping

Conclusion and Outlook

More and more renewable energy sources and consumers are connected to the grid via (switched) power electronics. Since the grid parameters like voltage, frequency and impedance can vary, it is important to investigate the grid-integration and the interaction of these devices in a proper way and a wide range of parameters to ensure a safe and reliable operation.

For this purpose, a test bench was designed, developed and built up, which offers the possibility to carry out investigations in a laboratory environment and with this in a time and cost saving way. With the *Grid Voltage Emulator* and the *Grid Impedance Emulator* in combination with the *4Q-AC-Source* it is possible to influence all the mentioned grid parameters. In addition, the *Resonance Generator* offers the possibility of emulating capacitive loads and thus, in combination with the *Grid Impedance Emulator*, generating a defined resonance point in the grid. Tests and measurements show the operation and the performance of the components of the test bench, regarding the test bench itself as well as the devices under test.

For future work, it is planned to use the test bench regarding to its basic functionality for online voltage and impedance variation tests. Also, a major topic will be further investigations on resonance damping. Moreover, upcoming control schemes for grid-connected inverters like the grid-forming control will be a main focus of research.

References

[1] German Federal Ministry for Economic Affairs and Climate Action (Deutsches Bundesministerium für Wirtschaft und Energie): Renewable Energies in Numbers (Erneuerbare Energien in Zahlen), October 2020, pp. 1-88

[2] Y. Lin *et al.*, "Pathways to the Next-Generation Power System With Inverter-Based Resources: Challenges and recommendations," in *IEEE Electrification Magazine*, vol. 10, no. 1, pp. 10-21, March 2022, doi: 10.1109/MELE.2021.3139132.

[3] A. Tuckey and S. Round, "Grid-Forming Inverters for Grid-Connected Microgrids: Developing "good citizens" to ensure the continued flow of stable, reliable power," in *IEEE Electrification Magazine*, vol. 10, no. 1, pp. 39-51, March 2022, doi: 10.1109/MELE.2021.3139172.

[4] K. Biligiri, S. Harpool, A. von Jouanne, E. Amon and T. Brekken, "Grid emulator for compliance testing of Wave Energy Converters," *2014 IEEE Conference on Technologies for Sustainability (SusTech)*, 2014, pp. 30-34, doi: 10.1109/SusTech.2014.7046213.

[5] Gevorgian, V. "Grid Simulator for Testing a Wind Turbine on Offshore Floating Platform." (Feb. 2012). Available: http://www.nrel.gov/docs/fy12osti/53813.pdf

[6] F. W. Fuchs *et al.*, "Research laboratory for grid integration of distributed renewable energy resources - design and realization -," *2012 IEEE Energy Conversion Congress and Exposition (ECCE)*, 2012, pp. 1974-1981, doi: 10.1109/ECCE.2012.6342570.

[7] F. W. Fuchs et al., "Research laboratory for grid-integration of distributed renewable energy resources - Integration analysis of DERs -," 2012 15th International Power Electronics and Motion Control Conference (EPE/PEMC), 2012, pp. LS7a.4-1-LS7a.4-8, doi: 10.1109/EPEPEMC.2012.6397504.

[8] E. Nasr-Azadani *et al.*, "The Canadian Renewable Energy Laboratory: A testbed for microgrids," in *IEEE Electrification Magazine*, vol. 8, no. 1, pp. 49-60, March 2020, doi: 10.1109/MELE.2019.2962889.

[9] European Distributed Energy Resources Laboratories (DERlab) e. V.. Link: https://der-lab.net/

[10] Semikron SKM400GM17E4 IGBT Module Datasheet. Link: https://www.semikron.com/dl/service-support/downloads/download/semikron-datasheet-skm400gm17e4-22895160/

[11] S. Bosch, J. Staiger and H. Steinhart, "Test Bench for the Investigation of Resonances in Low-Voltage Grids," *PCIM Europe 2022; International Exhibition and Conference for Power Electronics, Intelligent Motion, Renewable Energy and Energy Management,* Nuremberg, Germany, 2022, pp. 1-7. DOI:10.30420/565822264, ISBN 978-3-8007-5822-7

Reducing the Impact of Skin Effect Induced Measurement Errors in M-Shunts by Deliberate Field Coupling

Hauke Lutzen[1], Jonas Müller[1], Vladimir Polezhaev[2],
Till Huesgen[2], Nando Kaminski[1]

[1] University of Bremen, Institute for Electrical Drives, Power Electronics and Devices (IALB), Bremen, Germany
[2] University of Applied Science Kempten, Electronics Integration Lab, Kempten Germany

E-Mail: hauke.lutzen@uni-bremen.de

Acknowledgements

This project has been supported in the frame of the ECPE Joint Research Program.

Keywords

« Device Characterisation » « Sensor » « Measurement » « Noise » « Pulsed Power »

Abstract

The ever-increasing switching speed of semiconductor devices requires a precise measurement of steep current transients. The M-shunt concept offers high signal fidelity, good cooling, and simple manufacturing. Depending on the resistive material used, temperature as well as skin and proximity effects impede static and dynamic measurements to a different degree. A step forward has been derived from the ideal coaxial shunt, so far, a purely theoretical concept, which is hardly producible due to its sophisticated structure. By transferring this concept to the M-shunt structure with its improved PCB manufacturing technologies it can now be realised in practice. Nevertheless, the calibration and the correct degree of delay compensation remain challenging and are investigated more closely within this paper. Furthermore, it will be discussed why the conventional method of bandwidth determination doesn't work for the M-shunt structure. In addition to the low inductance introduced into the load circuit, the high bandwidth of the shunts could be demonstrated, as well as the possibility to extend this by design rules. Supplemented by the advantages of the lower load inductance, the M-shunt will become the tool of choice for characterising switching transients at least up to 200 MHz required bandwidth eventually. Although it is obviously difficult to improve the 3 dB bandwidth with suitable design rules, the range of nearly entirely unaffected measurement frequencies (e.g. < 1 dB) can be significantly extended by limited coupling. For even higher frequencies, measurements of the current M-shunt models, as well as for the coaxial shunts used as reference, should be corrected by post processing to get precise measurement results.

Introduction

Modern power electronics utilise wide band-gap (WBG) semiconductor devices made from SiC or GaN. These devices have properties superior to those of silicon and conquer a growing market share in power applications like automotive, aerospace, microgrid, renewable energy, and many more. One of the devices' main characteristics is their extremely fast switching with switching times down to the nanosecond range. On the one hand, this enables low switching losses and high-frequency operation, but on the other hand, it is a challenge for dielectrics and may cause severe EMI, especially when interacting with parasitics. In other words, a detailed and high-fidelity characterisation of the switching transients is required to avoid SOA or EMI problems. While voltage measurements are basic craftmanship, it is challenging to introduce a current sensor into the power loop without introducing additional parasitics, thereby making measurement errors due to the high dv/dt, di/dt, EMI, or cross-talk, which come in addition to the usual measurement inaccuracies. For high fidelity measurements, the coaxial shunt concept is state of the art and the M-shunt concept was proposed as an alternative, offering good signal fidelity, better cooling, lower inductance introduced into the power loop, and lower cost due to simple manufacturing [1]. Furthermore, the M-shunt allows for a simple and reproducible, though only partial compensation of the skin and proximity effect [2]. The concept was theoretically proposed for the coaxial shunt [3], but has just not been feasible in (mass-) production. However, the choice of resistive materials and the temperature impact significantly influence the actual measurement performance with conflicting optimisation targets.

This paper shows the quantitative extinction of coupled fields and the influences of the skin effect. In particular, the use of temperature-compensated resistive materials is influencing the reduction of inductance in the measuring circuit, such that a non-compensated material with post-processing for temperature compensation could be a better option. Furthermore, a delay-free measurement requires different design rules for the M-shunt and offers additional degrees of freedom, which provide another advantage over the coaxial shunt. [2] The investigation was carried out through repetitive double pulse tests (DPT), comparing the fidelity of the current measurements. Moreover, transfer characteristics of the manufactured shunts were recorded with a network analyser to obtain a better knowledge of the shunt's measurement behaviour and precision.

II.   Double pulse measurement and parasitics

The measurements for this investigation were carried out with a double pulse test. Figure 4 shows the DPT board with its integrated PCB-based M-shunt, which facilitates the lowest additional stray inductance due to the shortest connecting PCB traces and due to adhering strictly to the principle of coplanar conductors. Furthermore, particular attention was paid to reducing parasitics, which would otherwise result in oscillations, and current or voltage overshoot during the fast switching of the WBG devices [4]. Such parasitic inductances and capacitances have an adverse influence on the measurement and lead to deviation from the ideal switching waveforms [5]. Figure 6 shows the schematic of the DPT set-up including parasitics. Especially the parasitic capacitance C_L of the load inductor, the parasitics capacitance C_{FWD} of the free-wheeling diode, and the overall stray inductance of the power loop cause significant overshoot and ringing [6]. Because slowing down the switching by increasing R_G or C_{GS} is not an option, the reduction of the overall parasitic inductance in the DPT board, also including the measuring system, was the main design objective.

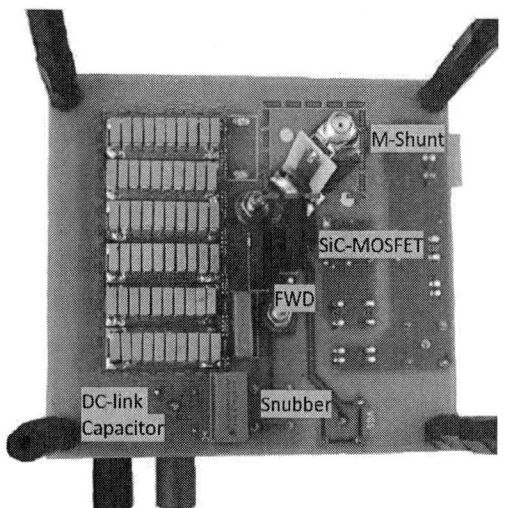

Figure 1 M-shunt integrated into the measurement board of the DPT set-up (red box top right)

Figure 2 Bottom side of the measurement board shown in Figure 1 with a coaxial shunt measurement in series (red box bottom right)

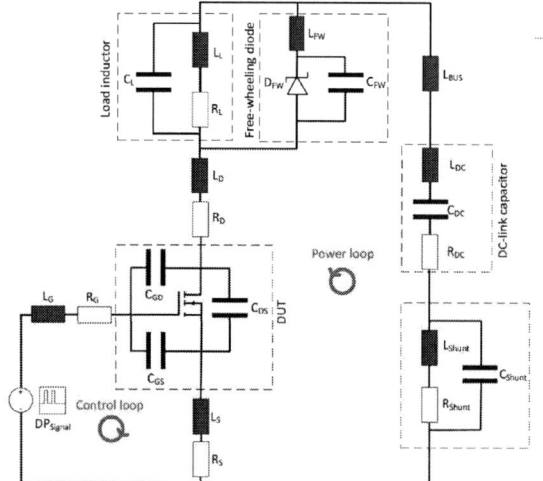

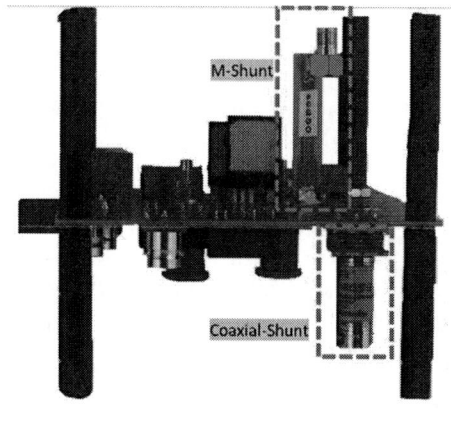

Figure 3 Circuit diagram of the DPT set-up including parasitics

Figure 4 Side view of the measurement board for shunt comparison with the M-shunt sticking out upwards and the coaxial shunt sticking out downwards

III. Current measurement methods and their limitations

There are several well-established current measurement techniques based on shunt resistors, current transformers, Rogowski coils, the Hall effect with open and closed-loop sensing, and fluxgate [7]. All of these techniques are well-suited for low frequency or slow transient measurements. In contrast, fast switching applications utilising WBG semiconductors are still a challenge due to the high di/dt and dv/dt rates. On the one hand, the bandwidth has to be very high to precisely measure the steep current slopes, but on the other hand, inductance inserted into the power loop must be negligibly small to avoid disturbance of the circuitry to be measured. Here a trade-off between bandwidth, accuracy, power limitation, interference with the circuitry, and of course cost occurs. Commercial Rogowski coils are not suited for steep transients, though, Rogowski coils with a higher bandwidth have already been reported [8] [9]. However, these are still limited due to inductive or capacitive couplings from the power loop, inserted resistance, and saturation or filter effects. State-of-the-art for fast switching devices are shunt resistors, especially the coaxial shunt with claimed bandwidths up to 2 GHz [10]. However, the coaxial shunt shows important limitations because it introduces a high additional inductance into the power circuits, e.g. 2 to 8 nH for the 100 mΩ version [11], and is difficult to cool, thus, it is quite limited in the energy of a single measurement pulse and power dissipation during continuous operation. An alternative is the M-shunt, which introduces only little inductance into the power circuit. Mostly less than 1 nH for the versions used. [11] [2] Additionally, it is much easier to cool. For comparison: The 25 mΩ coaxial shunt models are specified with a maximum energy of 3 J. [10] Depending on the specific model, the M-shunts used here were classified with maximum energies ranging from 100 J to significantly more than 200 J, respectively, while not requiring a larger installation space on the circuit board. [12] A comparison of the pros and cons of the main current measurement methods is listed in Table 1.

Table 1 Pros and Cons of selected current measurement methods, c.f. [1].

	Rogowski coil	Coaxial shunt	M-shunt
Advantages	• Galvanic insulation • Almost no additional loop inductance • High current measurement	• High bandwidth (claimed up to 2 GHz) • High accuracy • No auxiliary power required	• High bandwidth • High accuracy • Easy integration • Easy cooling • No auxiliary power required • Low additional inductance
Disadvantages	• Low bandwidth • Low accuracy, dependent on coil position and dv/dt • Not suited for DC measurement • Auxiliary power required • Signal delay caused by integrator	• No galvanic insulation • Additional loop inductance • Difficult cooling • Pulse energy is quite limited • Large mechanical size	• No galvanic insulation • Additional loop inductance

Within the scope of this paper, the M-shunts manufactured, will be compared to commercial coaxial shunts as the reference. The reference chosen is one of the most common versions. For the DPT the 25 mΩ was used for proper comparability. However, the overall inductance in the loop will be increased by two current sensors in series leading to overshoots due to the introduced inductance. The comparison will show differences in the output of the measurements and allows a dedicated comparison of the shunt models.

IV. Shunt Measurement of high fidelity

The general cross-sections of the coaxial shunt and the M-shunt are shown in Figure 5 and Figure 6, respectively. The most important advantage of the coaxial shunt is the ohmic current measurement in a field-free region, which is ensured by the radial symmetry of the resistive cylinder and which allows for direct measurement without the necessity of a compensation circuitry. However, parasitic inductance is indeed introduced into the circuit to be measured. This inductance can be reduced by large radii (leading to low H–Field), and small radius differences (resulting in small flux), which makes the design more complicated and the device larger. The M-shunt is a planar construction with the same cross-section. It can approach the field-free condition if designed according to the usual rules of coplanar conductors, i.e. mainly the dielectric between the conductors needs to be as thin as possible to avoid stray inductance. A pivotal point of the M-shunt design is the current splitting part "current distributor", which was already introduced in [2] and is shown in the box marked red on the left side of Figure 7 and which ensures an even current flow through the top and bottom traces of the shunt. Another challenge of the M-shunt design is avoiding edge effects – or turning them into a feature. The M-shunts characterised within this paper are based on the manufacturing and design concepts of [12], although some of the newer variants are equipped with SMA sockets, as can be seen in Figure 8.

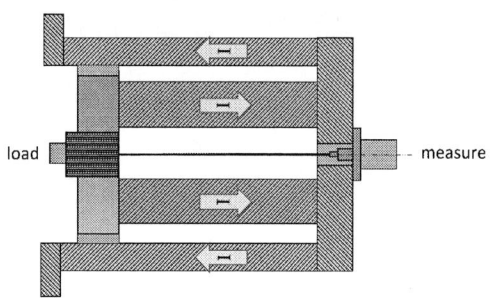

Figure 5: Schematic cross-section of the rotational symmetric coaxial shunt [3]

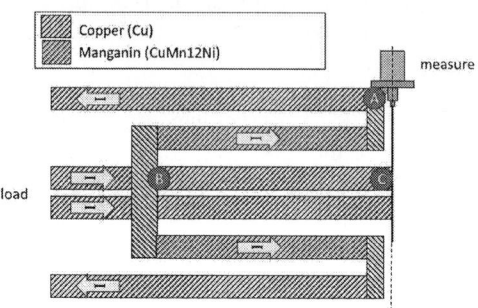

Figure 6: Schematic cross-section of planar 6-layer PCB M-shunt [1]

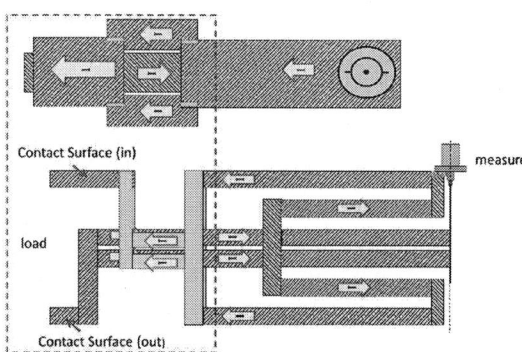

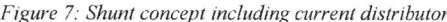

Figure 7: Shunt concept including current distributor

Figure 8: Picture of an M-shunt reduced to 25% size with an SMA socket for the measurement (left) and for load connection (right)

Fundamentally, there is a severe frequency limit for the coaxial- and M-shunt if the thickness of the resistive layer is larger than the skin depth. The current density and ohmic field will not reach the inner region of the resistive layer. For coaxial shunts, this is a known phenomenon [3] and maybe best understood using the concept of "field diffusion". As shown in Figure 9, the skin and proximity effects cause a current displacement towards the inner edge of the individual top and bottom power traces that make up the M-shunt, or to the centre of the tubes of the coaxial shunt, respectively. This uneven current distribution across the conductors leads to a delay in the measurement signal picked up by the taps at the outer edges. The delay of the measured signal concerning the actual current signal is shown in Figure 10 and comes on top of the inaccuracy caused by the higher effective resistance, i.e. measurement signal is larger than it should be. Even worse, the temperature-dependent increase of the resistivity in the shunt would cause an elusive change of skin depth. This additional portion is difficult to determine and could be avoided if the resistive layer was made from a temperature compensated material like Manganin ® (registered trademark of Isabellenhütte Heusler GmbH & Co. KG). The higher resistivity of Manganin (copper-mangenese-nickel alloy), compared with pure copper generates a larger skin depth per se and accordingly less delay at steep transients. To achieve the identical resistance value at room temperature, the Manganin must be about 25 times thicker than copper, which substantially increases the influence of the skin effect, as shown in Figure 10. At the same time, the skin depth improves only by a factor of less than 16. Depending on the ringing frequency and the overall layer thickness, the influence of the skin effect can increase significantly.

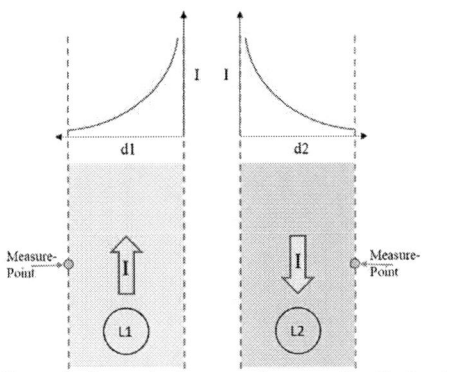

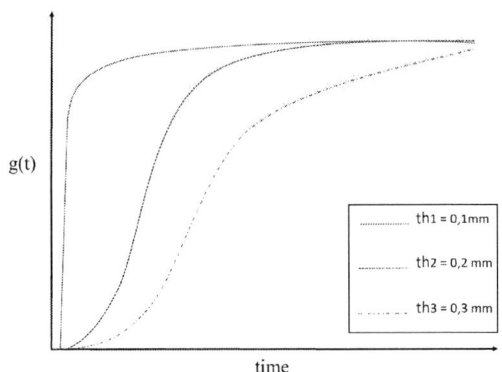

Figure 9: Principle diagram of current distribution in the conductor (L1) and resistor (L2) in coaxial- and M-shunts [2]

Figure 10: Sketch of the step response of coaxial shunts at high frequencies depending on the wall thickness (th_1=0.1 mm; th_2 = 0.2 mm th_3 = 0.3 mm) [3]

V.   On the Difficulties of Bandwidth

Bandwidth determination is straightforward for the coaxial reference shunt. In most cases the +3 dB limit is determined for the shunt's transfer function. For the coaxial reference the magnitude decreases steadily and would run out of the range, shown in Figure 11, at its +3 dB limit. For the 50 mΩ version characterised within this work this point lies around 400 MHz. This determination method is based on the equivalent circuit model of a resistive current sensor connected to an oscilloscope, shown in Figure 12, while the resistance of the oscilloscope Z_l is set to 50 Ω for all measurements within this paper.

This two-port model consists out of the total parasitic inductance L_{ins} and the mutual inductance L_m caused by the magnetic coupling between load and measuring circuit. [13] The L_s seen in the Figure 12 represents the inductance component introduced exclusively into the load circuit, consisting essentially of the current distributor, but not the resistor, so the route from mark A to B in Figure 6 is excluded from the total inserted inductance L_{ins}.

In this case the transfer function is basically described by:

$$G = \frac{V}{I} = R + j\omega L$$

The situation is more difficult in the case of the determination of the M-shunt's bandwidth. Considering the curves for shunts of the basic design, only changed in their overall size, a completely different behaviour is obtained. The additional design freedom, which will be used for the dedicated coupling described, works here as a capacitive component, so the circuit diagram has to be changed and consists of further capacitive and inductive components but is not fully explained so far.

To determine the limiting frequency, the +3 dB gain limit seems to be method of choice and is given for the reference model used at 400 MHz by the datasheet [10] and measured as 45 MHz, shown in Figure 11. Figure 13 shows a picture of the versions compared, which are all designed as 25mΩ resistances, but changed in the overall area, through simultaneous reducing of the length and the width to 75 %, 50 % and 25 % of original 100 % length and width, respectively.

This option, tagged as f1 in Table 1, seems to be too optimistic for the M-shunt because the negative trend upfront already exceeds comparable deviations, but in the opposite direction (damping). To transfer the 3dB limit in the negative direction on the other hand seems to be too pessimistic because the amplification factor of 0.7 (-3 dB) in other terms would be much lower than the 1.41 of the +3dB boundary. Permitting the same deviation from the 0 dB mark, the resulting frequency would be even higher but still arbitrary. In any case, it can be ensured that the most pessimistic value (±3 dB) allows equal or smaller deviations than the positive gain limit of the coaxial shunt. Based on this pessimistic variant, a useful bandwidth of values between 141.6 and 214.74 MHz results for the considered shunts. In any case, this is still lower than the values given in the datasheet for the coaxial reference (400 MHz at 25 mΩ), whereby the reference shunts used in this study, as in other studies, fall behind the datasheet promise by far. [13].

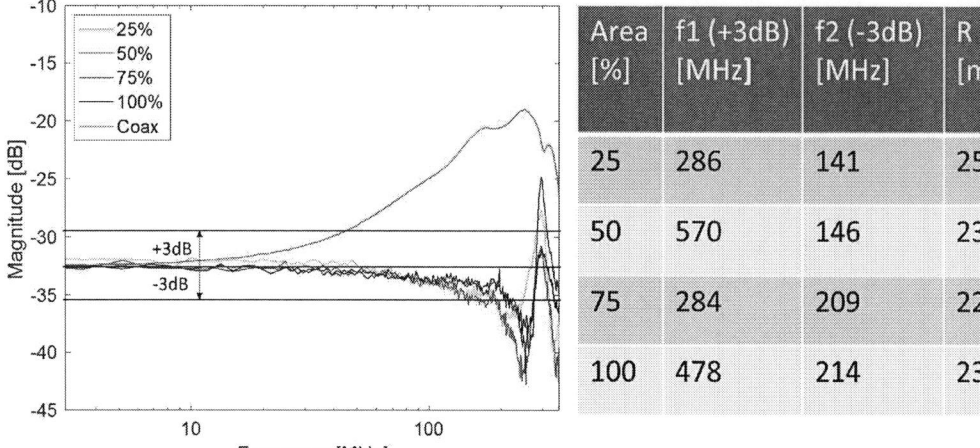

Area [%]	f1 (+3dB) [MHz]	f2 (-3dB) [MHz]	R [mΩ]
25	286	141	25.3
50	570	146	23.6
75	284	209	22.6
100	478	214	23.4

Figure 11 S-Parameter measurement for M-shunts of different size but with nearly same resistance (also shown in Figure 13) and reference coaxial shunt (-1dB Offset for better comparative view)

Table 1 Summary of the positive and negative 3 dB limits for shunts of different area

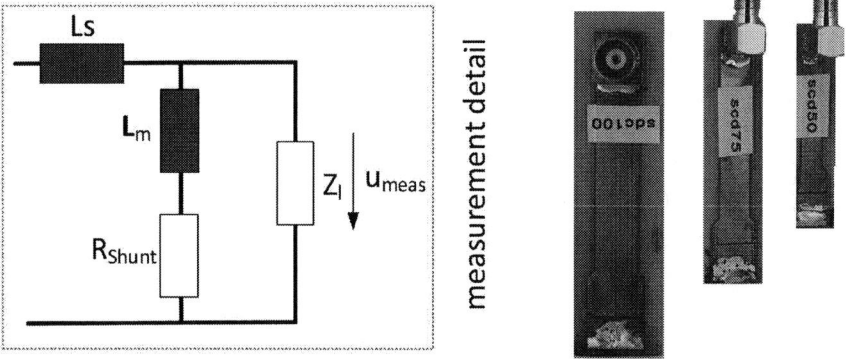

Figure 12 Equivalent circuit for resistive current measurement with termination resistor [13]

Figure 13 Picture of the shunts of different area

VI.   Concept of adjustable field coupling for delay compensation

The basic idea of the designs analysed pursues influencing design-related inconsistencies and measurement errors through targeted field coupling as introduced for the ideal coax-shunt. (Figure 14) [3] For this purpose, the basic design, presented in [12] was modified in three different ways. Figure 15 shows the basic concept of the delayed measurement signal, and the intended field coupling. The central goal is to achieve the ideal step response from the superposition of the delayed measurement signal and the signal components generated by coupling [2], at least in the range of the usual measuring frequencies.

In the first step, only the outer lead of the M-shunt's cross-section (conduction layer) was widened, while the measurement tap and resistor layer remained the same width. Without changes at the measurement tap nearly no changes could be detected for the measured signal (design "CL"). In the second step, in addition to the outer lead, also the measurement taps were widened equally (Figure 16 of design "CLML"). The number written in the identification of the "CL" or "CLML" curves is representing the proportional width extension, compared to the width of the resistance layer, as illustrated in Figure 16. In this case, a clear influence on the measured output signal can be noticed (Figure 18). A wider overhanging surface (tested up to 50 % width) results in a steeper transient and a larger overshoot. By comparing to coaxial shunt measurements performed in series with the M-shunt measurements shown, it can be concluded, that the coupling is (as expected) only part of the measurement and is not influencing the load circuit itself. (Figure 19) The reference measurements of all curves, performed with the coaxial shunts in series, show almost a perfect match. Therefore, one may conclude that the M-shunt model is suitable to transfer the concept of the ideal coaxial shunt into practice to remain capable of measuring at high frequencies. In principle, these improved frequency responses should be recognisable through an improved transfer function, which will assist to determine the right amount of overhang in order not to overcompensate for the delay.

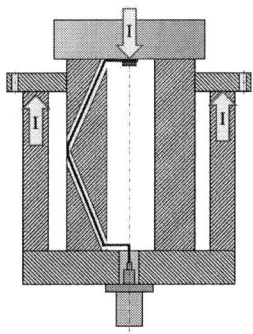

Figure 14 Cross section of a coaxial shunt with improved measuring tap [3]

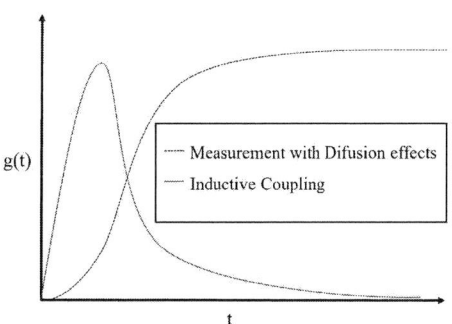

Figure 15 Visualisation of the signal components [2]

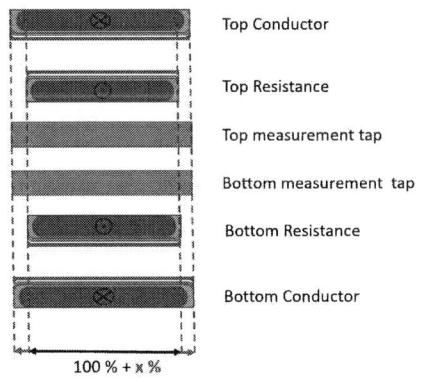

Figure 16 Design Cross section drawing for shunts with dedicated overhanging coupling areas in conductor only (CL) and conductor plus measurement Layer (CLML) [2]

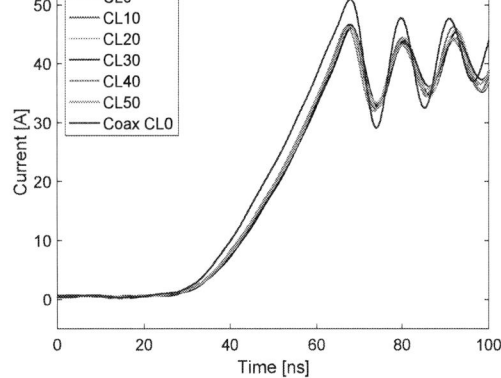

Figure 17 Comparison of turn-on behaviour measured by Shunts with wider conduction layer compared to their coaxial reference shunt measurement in series

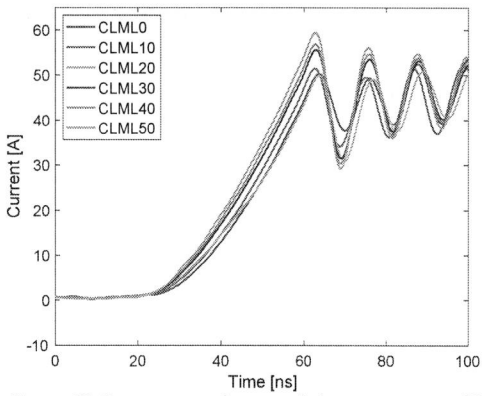

Figure 18 Comparison of turn-on behaviour measured by Shunts with wider conduction and measurement layer from 0 to 50 % overhang (CLML)

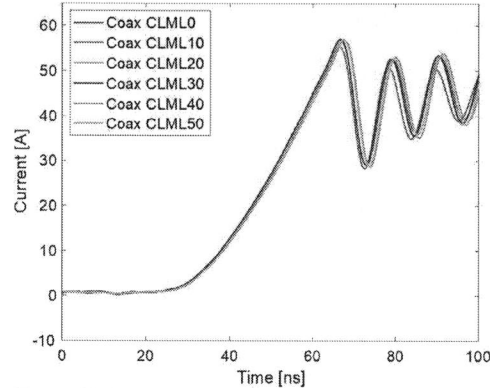

Figure 19 Comparison of coaxial shunt measurements of turn-on behaviour measured in series of the CLML designs. The differences seen in Figure 18 are only measured and not traceable in the circuit itself

VII. Verification of the Transfer Function

In case of the coaxial shunt, the following method, i.e. the calculation of the transfer function from the measurement signals, is used to verify the bandwidth [13]. For the M-shunt, it is rather used to correctly determine the transfer function as a whole, because the bandwidth determination has its difficulties as mentioned. Figure 20 shows the measurement result recorded with the network analyser of the already introduced CLML designs. The change of the gain over the frequency response as a function of the exceeding coupling surface is evident and emphasises the observations described in section VI. Shunts with an overhang of more than 10% obviously overcompensate the measurement delay already as was also shown by the curves of the double pulse test. The longest-lasting bandwidth with almost no deviation (<1dB) seems to be in the range of a coupling area around 5 % overhang (between 0 % and 10 %). Table 2 shows the decreasing bandwidth via the coupling surface set at the +3dB limit. As mentioned the 0% coupling version has the highest bandwidth per definition. On the other hand, due to its previous damping (-3 dB limit), it would basically get a rather bad value, as described in section VI. The calculated measuring inductance is therefore only a mathematical factor and should be treated with caution in a physical context due to the above-mentioned difficulties of the equivalent circuit diagram for the M-shunt.

In the previous section Figure 18 shows the comparison of the measurement curves of the CLML variants. Figure 21 in this section shows the same measurement curves if adjusted by their transfer function to the 'real current' considering [7]:

$$i_{meas}(j\omega) = i_{real}(j\omega) \cdot \frac{G(j\omega)}{R}$$

respectively:

$$i_{real}(t) = \frac{1}{R} \cdot \mathcal{F}^{-1}\left\{\frac{\mathcal{F}\{i_{meas}(t)\}}{G(j\omega)}\right\}$$

To verify the concept of the postprocessing Figure 22 shows the measured signals of the "CL0" M-shunt compared to its coaxial reference, like it was already shown in Figure 17. In addition, it is shown how the signal looks after postprocessing of the 44 MHz signal shown. Considering the measurements performed combined by the aid of these calculations, it can be stated that the coaxial shunt clearly falls behind its promise and has a comparable bandwidth to the M-shunt. However, within their 3 dB limits respectively for low frequencies below 150 MHz, the average deviation, i.e. the measurement error at low frequencies, is significantly larger for the coaxial shunt than for the M-shunt. Nevertheless, the comparison shows that not all measurement inaccuracies can be fully prevented by the post-processing, which is the reason for the differences between the final signals.

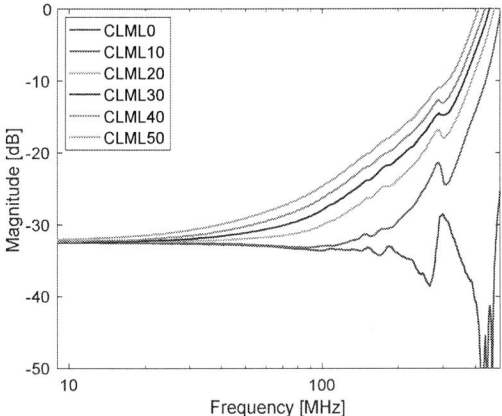

Figure 20 S-Parameter measurement for shunts with wider conduction and measurement layer from 0 to 50 % overhang (CLML)

Table 2 Overview on the ±3dB bandwidth for the CLML M-shunts with different amount of overhang.

Area [%]	R [mΩ]	f_BW [MHz]	L [pH]
0	24,16	290	13,22
10	24,26	202	19,09
20	24,39	113	34,23
30	24,43	82	46,97
40	24,47	62	62,67
50	24,48	47	83,72

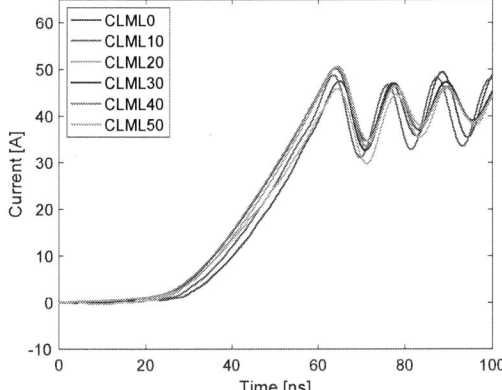

Figure 21 Comparison of turn-on behaviour measured by Shunts with wider conduction and measurement layer from 0 to 50 % overhang (CLML) after post-processing

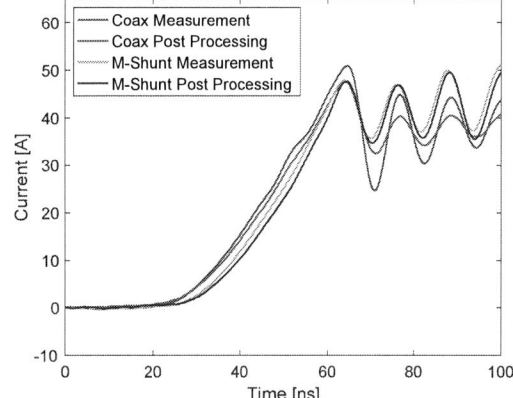

Figure 22 Comparison of turn on characterisation measured with M-shunt and coaxial shunt. Furthermore same measurement results after postprocessing

Conclusion

In summary, the series of experiments presented here prove the possibility to shape the output signal of the M-shunt according to the principle of the ideal coaxial shunt. However, the theoretical idea of a step response is naturally subject to delays and limits the validity of a clean calibration by means of controlled coupling. Nevertheless, the "almost" instantaneous step responses of GaN switching can certainly be used for iterative approximation towards the "ideal M-shunt", although it should be noted that there is always a narrow line between glossing over results and genuine calibration or correction of errors. It must be mentioned that the regular bandwidth definition is not directly transferable for the M-shunt, due to its capacitive components. Nevertheless, at least at frequencies up to 200 MHz, the M-shunt is an equivalent measuring device to the coaxial shunt without any post processing or compensation. Supplemented by the advantages of the lower introduced inductance [11], the M-shunt will become the tool of choice for the characterisation of switching transients at least up to 200 MHz required bandwidth, eventually. Although it is obviously difficult to improve the 3dB bandwidth with suitable design rules, the range of almost entirely unaffected measurement frequencies (e.g. < 1dB) can be significantly extended through defined coupling areas. For even higher frequencies, measurements of the current M-shunt models, as well as for the coaxial shunts, should be corrected by post processing to get precise measurement results.

References

[1] **Bödeker, C., Adelmund,M., Kaminski, N.** The M-Shunts Structure Applied to Printed Circuit Boards. *CIPS 2018; 10th International Conference on Integrated Power Electronics Systems.* 2018.

[2] **Lutzen, H., Mitsui, K., Silber, D., Wada, K., Kaminski, N.** Optimisation and Proof of Concept Studies for the M-Shunt applied to Printed Circuit Boards. Berlin, Germany : CIPS, 2020.

[3] **Malewski, R., Nguyen, C., Feser, K., Hylten-Cavallius, N.** Elimination of the skin effect error in heavy current shunts. s.l. : IEEE Transactions on Power Apparatus and Systems, April 1981. S. 1333-1340.

[4] **Oladele, O K, et al.** Optimizing Switching Performance of Cascode-Light Sic JFET Bidirectional Switch for Matrix Converter. *IEEE International Power Electronics and Application Conference and Exposition (PEAC).* 2018.

[5] **Zhang, Z., et al.** Methodology for switching characterization evaluation of wide band-gap devices in a phase-leg configuration. Forth Worth, TX, : IEEE APEC, 2014.

[6] **Sellers, A. J.** Effects of parasitic inductance on the performance of 600-V GaN devices. IEEE Electric Ship Technologies Symposium (ESTS) : s.n., 2017.

[7] **Ziegler, S., et al.** Current Sensing Techniques: A Review. *IEEE Sensors Journal.* 9, 2009, Bd. 4.

[8] **Hain S., Bakran M.** New Rogowski coil Design with a High dV/dt Immunity and High Bandwidth. *2013 15th European Conference on Power Electronics and Applications (EPE).* 2013.

[9] **Wang, J.,Hedayati, M., Liu, D., Adami, S-E., Dymond, H., Dalton, J., Stark, B.** Infinity Sensor: Temperature Sensing in GaN Power Devices using Peak di/dt. *2018 IEEE Energy Conversion Congress and Exposition (ECCE).* 2018. S. 884-890.

[10] **Shunts, T&M.** Coaxial Shunt - general information. *IB Billmann.* [Online] 2020. www.ib-billmann.de/koax_e.php.

[11] **Wilhelmi, F., Schmid, A., Lindemann, A.** Assessment of State-of-the-Art Current Sensors for Fast Switching. 2022.

[12] **Lutzen, H., Polezhaev, V., Bahadur Rawal, K., Ahmmed, K., Huesgen, T., Kaminski, N.** Temperature Compensated M-Shunts for Fast Transient and Low Inductive Current Measurements. s.l. : CIPS, 2022.

[13] **Zhang, Wen, Zhang, Zheyu und Wang, Fred.** Review and Bandwidth Measurement of Coaxial Shunt Resistors for Wide-Bandgap Devices Dynamic Characterization. Baltimore, MD, USA : IEEE Energy Conversion Congress and Exposition (ECCE), 2019.

Grid Forming Control for HVDC Systems: Opportunities and Challenges

Adil Abdalrahman[*], Ying-Jiang Häfner, Malaya Kumar Sahu,
Khirod Kumar Nayak and Ashkan Nami
Hitachi Energy - HVDC, Ludvika, Sweden
[*]Email: adil.abdalrahman@hitachienergy.com

Keywords

≪Grid-forming control≫, ≪HVDC≫, ≪TSO≫

Abstract

The aim of this paper is to highlight the opportunities and challenges of Grid Forming Control (GFC) for High Voltage Direct Current (HVDC) applications. A special focus will be on contradictory GFC performance requirements from European Transmission System Operators (TSOs). The impacts of these requirements on AC grid, HVDC performance and hardware are discussed.

Introduction

Due to the increasing integration of renewable energy sources (RES) and a corresponding reduction of conventional generating units, there is a demand from the power-electronic converters to provide grid-forming behavior through proper control of the converter systems nowadays. There are different ways of implementing GFC. Reference [1] gives an overview on GFC methods. The most interesting methods are: Voltage Source Control [2], Droop-based Grid Forming Control [3], Power Synchronization Control [4], Virtual Synchronous Machine Control [5] and [6], and PLL-Based Modified Current-Control methods [7]. The common feature of GFC is that the control should be equipped with functionalities that enable inertia and frequency support, islanding operation, black-start, and synchronization capabilities. Many papers have been published in GFC area, but GFC for HVDC systems is scarcely addressed. Nevertheless, the so-called "GFC" has been successfully applied in the HVDC installations where the GFC is necessary, for example, HVDC converters connected to weak or extremely weak AC grids or islanded network (offshore wind) with zero or very limited inertia [8] and [9]. In fact, according to [10], GFC may not be appropriate when the AC grid is very strong and it may become unstable under such conditions.

The aim of this paper is to highlight the opportunities and challenges of GFC for HVDC applications. A special focus will be on contradictory GFC performance requirements from European Transmission System Operators (TSOs). The impacts of these requirements on AC grid, HVDC performance and hardware are discussed.

The paper is organized as follows: a brief description about the operating principle of HVDC is given. Following that, requirements on Grid-forming behavior according to existing Grid codes/standards are summarized. Next, some of requirements which deserve further efforts from both TSOs and vendors are highlighted in order to reach a common goal of using HVDC to support a greener AC grid in the best way. Finally, some conclusions based on observations from simulation results are presented.

Principle of VSC-HVDC operation

Fig. 1 shows a schematic diagram for a VSC-HVDC system based on Modular Multilevel Converter (MMC) with Half Bridge cells. The two stations can be connected to the same or different AC grids. If the two stations are connected to the same grid the HVDC link is referred to as "embedded HVDC link". Ignoring the active power losses in the transformer and phase reactor, the phasor diagram for the point of common coupling (PCC) voltage (U_{pcc}), the converter valve voltage (U_v) and the current at valve side of transformer (I_v) is shown in Fig. 2. The active and reactive power at PCC can be expressed as:

$$P = U_{pcc} I_v \cos\varphi \approx \frac{U_v U_{pcc}}{X} \sin\delta \tag{1}$$

$$Q = U_{pcc} I_v \sin\varphi \approx \frac{U_{pcc}(U_v \cos\delta - U_{pcc})}{X} \tag{2}$$

where $X = \dfrac{X_r}{2} + X_t$ i.e., X is the total equivalent reactance of the transformer short circuit reactance (X_t) and arm reactor (X_r).

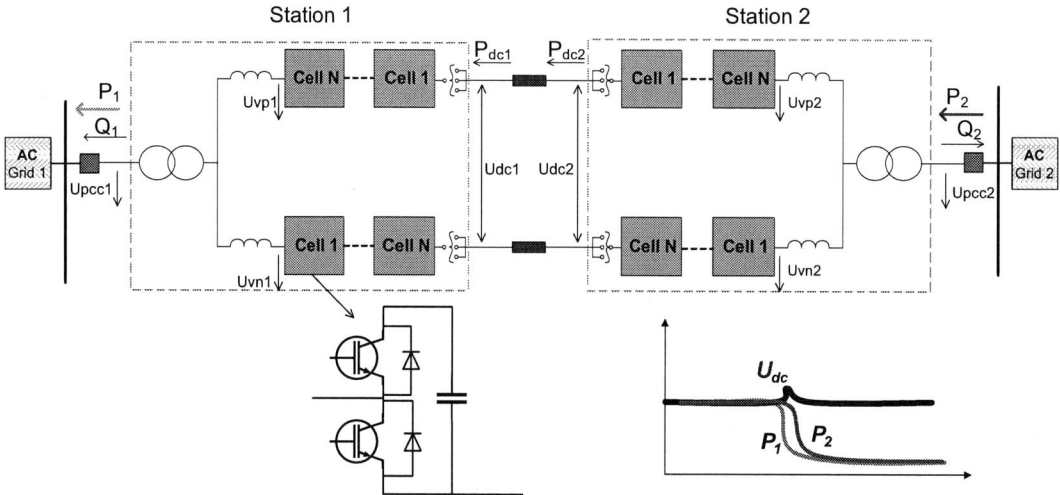

Fig. 1: A schematic diagram of a point-to-point VSC-HVDC system.

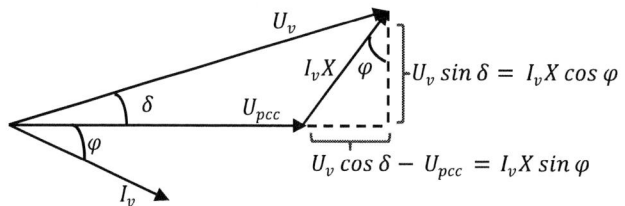

Fig. 2: Phasor diagram of voltages and current (quantities are in p.u.).

The active power is controlled by controlling the angle difference between U_v and U_{pcc}, whereas the reactive power is controlled by amplitude of U_v. Active power should be controlled in a coordinated way. One station controls the active power (station 1 in this case), and other station controls the DC voltage (station 2 in this case) such that the active power on the DC side is balanced. If the losses in the HVDC system is ignored, the active power in the DC voltage control (U_{dc}-control) station P_2 follows the active power in the active power control (P-control) station P_1 except for transient as shown in Fig. 1 supposing that the active power and DC voltage are controlled in a coordinated way. This means that HVDC system (within purple dash-line) is a transmission medium.

However, differing from AC transmission mediums, each station can control AC voltage or reactive power independently which can support the respective AC grid voltage, and it can also vary the active power (manually or automatically) to support AC grid frequency or facilitate power trading. It is also possible to switch the P-control mode to Frequency & Voltage control which is used during grid restoration (black start) or for forming grids of an islanded network e.g., offshore wind farms [2]. In that case the other station should operate in U_{dc}-control mode.

Various benefits can be achieved with HVDC if it's operated stably a coordinated manner. However, if the performance requirements are not coordinated e.g., if great deal of emphasis is put on certain requirements while overlooking other requirements or impacts, there is a risk of jeopardizing the HVDC system stability and/or having negative impact on the AC system stability. Examples of such requirements (which are associated with grid-forming behavior) in existing grid codes/technical specifications are discussed through simulation results presented in this paper.

Grid-forming behavior – requirements from Grid codes/standards

Requirements for Grid Forming behavior are already in place in German Grid Code [11] and [12]. In [12] the expected behavior is elaborated with illustrative plots for different test cases such phase jump, frequency ramp AC faults, Islanding etc. The plots can be used as references cases when verifying GFC behavior. It's however unclear if the tests and reference plots in [12] are all applicable for both P-controlling and U_{dc}-controlling stations in an HVDC link. Furthermore, the assumptions used for converter energy storage (i.e., finite or infinite) when the reference plots were created are also unclear. Requirements for Grid Forming Capability has recently been added to the National Grid (UK) grid code [13], where new terms such as Phase Jump Power, Inertia Power and Control-based Power have been added. For each type of power there is a certain requirement. What's more pronounced in the requirements in [13] is that the use of virtual impedance in emulating voltage source behavior is not permitted. From the control structure presented in [14], it seems it's desired to strictly emulate synchronous machine behavior. There is also a requirement to withstand a voltage phase jump as high as 60 degrees. The ENTSO-E technical report [15] sets the following Grid Forming requirements:

- Creating system voltage
- Contributing to fault level
- To act as sink for harmonics
- To act as sink for unbalance
- Contribution to inertia
- System survival to allow effective operation of Low Frequency Demand Disconnection (LFDD)
- Preventing adverse control interactions

IEEE has recently published the standard P2800 [16] which, when it comes to HVDC, is only applicable at point of connection of a VSC-HVDC connecting Isolated inverter-based resources (IBR). There are no clear requirements for other types of HVDC such HVDC connecting two different AC networks (interconnector) or an HVDC connected to two busbars within a meshed synchronous AC system (referred to as an Embedded HVDC [17]).

The grid codes/Technical reports share one common thing i.e., the converter should behave as a voltage source behind impedance to provide inherent support for frequency and voltage support in response to grid disturbances such as phase jump and voltage dip/swell. The current limitation philosophy during transients is hardly discussed and it's left as an area of research for industry/academia.

Apart from [14], no other document discusses the energy storage limitation in HVDC systems. In [14] the limited energy storage available in HVDC system is briefly discussed along with the measures to be taken if grid-forming behavior such phase jump power or inertia power is expected from the onshore station in an HVDC system connecting offshore wind farms. The conclusion in [14] is that there is a need for additional energy storage to be installed in the onshore station if HVDC and wind turbines are to be operated in the traditional way as i.e., HVDC offshore station in Frequency & Voltage control mode, HVDC onshore station in U_{dc}-control mode, and wind turbines in grid-following mode. It's also unclear from [11] and [13] whether it's expected or not that both HVDC stations operate simultaneously in Grid Forming mode.

Grid-forming behavior - unharmonized requirements

Inertial response versus active power response

In recent European grid codes [11], [12] and [13], there are new requirement of providing inertial support with a certain active power contribution during transients, such as phase jump or load rejection. Fig. 3 shows the active power responses (left side figure) for a transient event of +10 degrees phase jump in the network voltage. The simulated phase jump is in the grid connected to the station in P-control mode. The considered AC network short circuit ratios (SCRs) for P-control and U_{dc}-control stations are 6 and 10 (based on the system parameters listed in Table I), respectively. During this transient, the power response is driven primarily by the phase jump [18]:

$$\Delta P_\delta = \frac{U_v U_{pcc}}{X} \sin(\Delta\delta) \tag{3}$$

where $\Delta\delta$ is the change in the angle between U_v and U_{pcc}.

In addition, there are also the contributions from inertia power (proportional to RoCoF); frequency droop dependent power order change (proportional to frequency deviation) and damping power (due to emulating damping in synchronous machine). The figure shows that the higher the inertia, the more the power variation is. The figure on the right side shows the corresponding power from U_{dc}-control station where the DC voltage is maintained within ±6%. It shows that the source of energy for inertial response provided by the P-control station is actually the AC network connected to U_{dc}-control station.

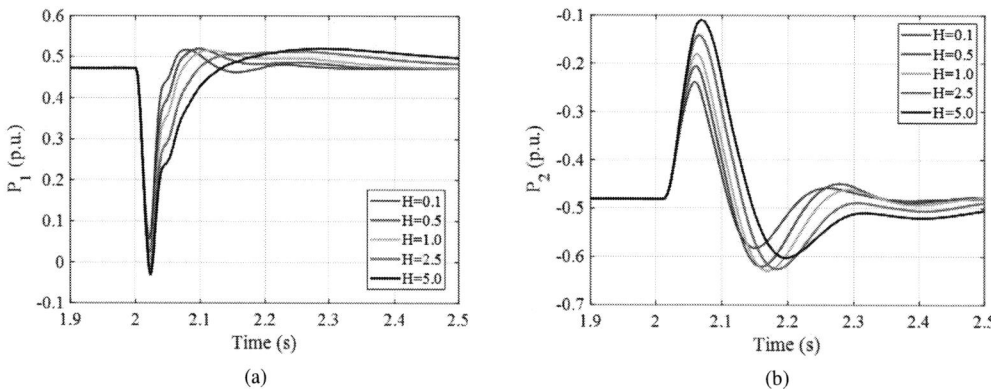

Fig. 3: Active power responses to a +10 degrees phase jump in AC network connected to the P-control station while different inertia constants (H) are considered (for the P-control station): **(a)** from P-control station; **(b)** from U_{dc}-control station.

It's to be noted that in this paper in all simulations made in the converter the convention for active power is positive toward the grid. Moreover, the simulation is made based on a typical HVDC link with realistic main circuit parameters, control & protection.

While introducing new requirement related to inertial response, the traditional requirements are still maintained as they have been e.g., HVDC system to recover its active power faster post AC faults and to change its active power rapidly either during a step change in active power order or during Emergency Power Control (EPC). Typically, the traditional requirement for post fault active power recovery is to recover to 90% of pre-fault power within 150 - 200 ms. Regarding EPC, different speeds of active power change may be required, but in some cases reduction from full power to zero power is required to take place within 5 – 10 ms. It should be realized that the requirement on the inertia will have impact on the speed of P-control. Fig. 4 shows the active power a 10% step change in active power simulated for different inertia constants (H). It can be seen from the figure that the step response time increases as H is increased.

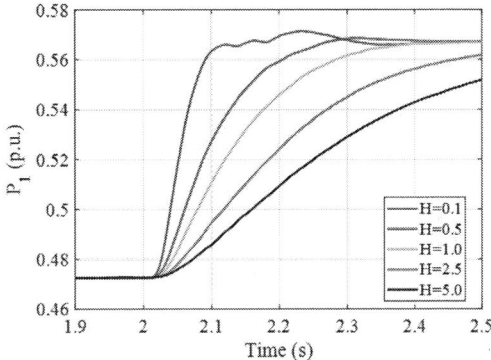

Fig. 4: Active power under a 10% step change in active power reference for different inertia constants.

It is clear from Fig. 4 that increasing inertia will slow down the active power step response. Thus, the requirement on inertial response can also affect the performance of a fast-acting function like EPC and also power oscillation damping (POD) controller. Similarly, increasing inertia on the U_{dc}-control station will slow down the DC voltage step response.

Requirements of high inertia versus limited energy in HVDC system

In UK there is a plan to specify how much inertia that each grid-forming converter should provide [13]. The range could be 2 - 25 MWs/MVA, as per [14]. As mentioned previously, HVDC is just a transmission medium and the energy available in HVDC system (MMC submodules & DC cables/overhead lines, etc.) is limited typically 100 - 200 ms which is defined by full discharge at rated power. However, availability of more than 90% of this energy is a prerequisite for producing the necessary level of AC voltage which allows the converter to operate properly. Thus, the energy for grid support needs to come from AC grids connected to remote station. As shown in Fig. 3, it is possible for the P-control station to provide a large inertia as long as the U_{dc}-control is sufficiently fast to maintain the DC voltage stability and the AC network connected to U_{dc}-control station can tolerant transient power variation. However, if one of the two mentioned preconditions is not fulfilled e.g., the U_{dc}-control is not fast enough for other design requirement reason, the large inertia on the P-control station could affect the DC voltage, and the DC voltage in turn can negatively affect the AC voltage stability which will be further addressed in the last subsection). Furthermore, there is a risk of trip if inertia is high when there is a large grid disturbance (e.g. a solid 3-phase fault) in close vicinity of U_{dc}-control station when operates as an inverter.

Large phase jump versus limited impedance

In [12], there is a requirement to withstand a phase jump up to 60 degrees and it's not permitted to use a virtual impedance in GFC. Equation (3) shows that the transient power depends on the AC grid voltage phase change and reactance X. The equivalent reactance X of HVDC converter (typically less than 0.3 p.u.) is lower than synchronous machine reactance. Suppose that the voltage is 1 p.u., then a phase jump of 60 degrees would lead to more than 1 p.u. power change. This implies that the converter is forced to operate outside its power capability, and the protection would act, which can eventually trip the converter. This trip can be mitigated by increasing X in (3) via using an additional virtual impedance, i.e., X can be increased without physically modifying any main circuit equipment by introducing an additional virtual impedance in the control. Fig. 5 shows the simulation of an event of 60 degrees phase jump in the grid connected to station 1 (see Fig. 1) while the considered AC network SCRs for both stations are 40. In Fig. 5, the blue curves are the results from a GFC with a proper virtual impedance included whereas the red curves are the results from a GFC without any virtual impedance. The figure shows that with a proper virtual impedance it is possible for the converter to limit the transient currents under a sudden large phase jump in AC grid, which in turn limits the power into/from the converter. In this way, the trip of HVDC converter can be avoided allowing the HVDC to remain in service to support the AC grid.

As stated earlier, the internal impedance of HVDC converter station is typically smaller than a synchronous machine impedance in per-unit value. Another key point to mention is that the transient current capability of converter is also smaller than a synchronous generator. On the other hand, HVDC converter has the advantage of being flexible in control. Therefore, it is important to utilize its flexibility in control, for instance, to use a virtual impedance to limit the current without significantly affecting its behavior as a voltage source behind an impedance.

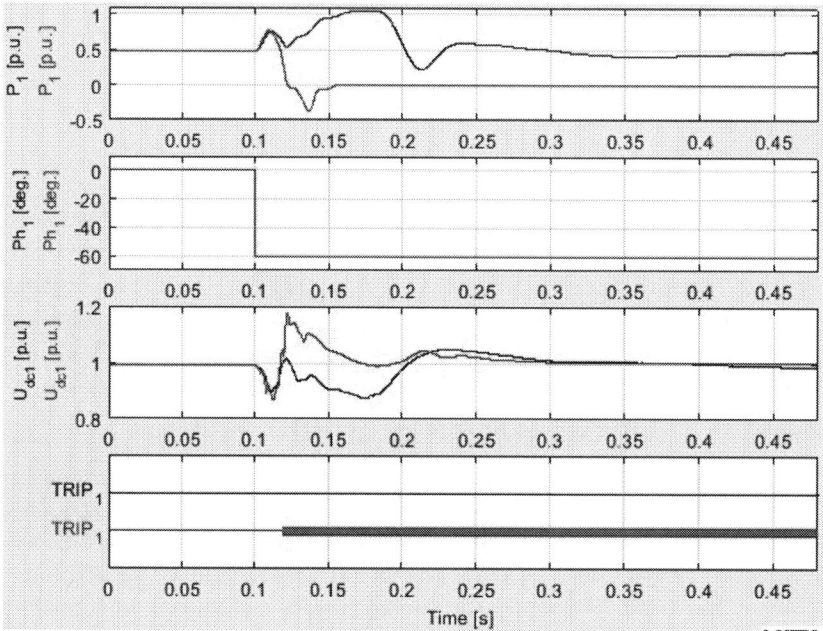

Fig. 5: Simulation of an event of 60 degrees phase jump in the grid connected to station 1. Blue curves: with virtual impedance; red curves: without virtual impedance. Subplot 1: active power at PCC of P-control station, subplot 2: voltage phase of AC grid connected to P-control station, subplot 3: dc voltage of P-control station, subplot 4: trip indication for P-control station.

Inertia support versus stability of DC voltage

Providing grid forming behavior has become mandatory requirement [11], [12] and [13]. Furthermore, some TSO is planning to specify how much inertia that each grid-forming converter should provide [13]. As discussed earlier, these requirements may be fulfilled for the P-control station under two important pre-conditions. One of the pre-conditions is the fast U_{dc}-control. However, increasing inertia in the U_{dc}-control station will slow down the DC voltage response (in a similar way as it does for P-control station, see Fig. 4). Obviously, the U_{dc}-control station can not be designed to provide large inertia for the purpose of accommodating the P-control station providing sufficient inertia by maintaining the DC voltage stability. It's important to note that large transient disturbances in P-control station as well as in AC grid (e.g. a 3-phase faults near converter station) demand a very fast response from U_{dc}-control station to avoid any un-acceptable DC voltage excursion outside the normal voltage range. The DC voltage may become too high or too low depending on if the P-control station is in inverter or rectifier operation prior to disturbances. If the DC voltage is too high, it could hit the limitation of main circuit equipment and protection would trip converters. If the DC voltage is too low, it would lead to converter being not able to generate the desired AC voltage which either leads to un-controlled current or negative impact on the AC side.

Fig. 6 shows as an example of an HVDC interconnector under a 3-phase fault near the P-control station. The connected AC grids on both stations are relatively weak (SCR=4). The U_{dc}-control station with grid forming capability and two different inertia constants are considered: blue curves are obtained from the maximum inertia which the HVDC system is permitted (H_{DCN}), whereas the red curves are obtained with the inertia increased by a factor of 2 (from H_{DCN}). Fig. 6 shows that it is not possible for the DC link to recover after this disturbance if the U_{dc}-control station has higher inertia than permitted, as the U_{dc}-control cannot react fast enough to bring down the DC voltage below the trip level. The permitted inertia (H_{DCN}) is typically below 0.3 s.

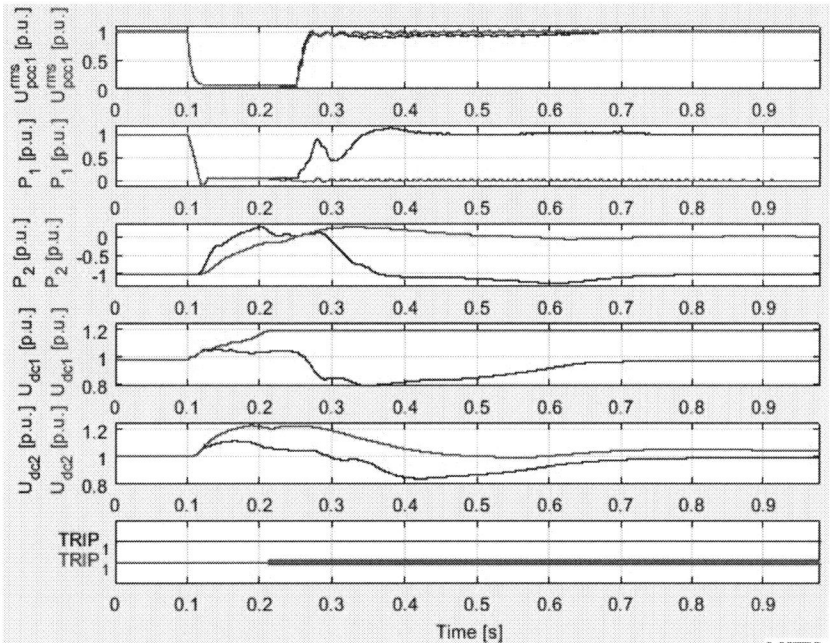

Fig. 6: 3-phase to ground fault near the P-control station. Subplot 1: PCC voltage, subplot 2: active power at PCC of P-control station, subplot 3: active power at PCC of P-control station, subplot 4: dc voltage of P-control station, subplot 5: dc voltage of U_{dc}-control station, subplot 6: trip indication for P-control station.

In the VSC-HVDC application of connecting offshore wind farm, inertia support for onshore grid from the onshore converter will be quite limited even if the onshore converter control is designed with grid forming capability. Suppose that the wind turbines operate in the traditional way i.e., grid-following mode. In that case the offshore converter of HVDC controls the voltage and frequency of the offshore grid and the onshore converter control the DC link voltage. With this control, the wind turbine generators in grid-following mode with maximum power point tracking (MPPT) and the active power generated will be immediately transmitted to onshore grids. This means that HVDC can only deliver the power generated by wind farms, or in another word offshore converter can be considered as a constant power source that does not contribute to U_{dc}-control. Thus, the task of maintaining the DC system voltage stability (or keeping the active power in balance) can only be achieved by the onshore station. On the other hand, grid forming behavior demands the onshore converter to supply/absorb the active power required by the AC grid depending on the type of transient events e.g., voltage angle jump due to trip of AC lines, generators, or loads. Obviously, there will be a conflict between "maintaining the DC voltage stability" and "grid forming behavior". The possible consequences are that the DC voltage stability is jeopardized while providing high inertia support or maintaining the DC voltage stability and the inertia support gets diminished by the U_{dc}-control action. Evidence of that is provided in Fig. 7 where the considered SCR for the onshore grid is 20.

It can be seen from the figure above an additional active power is immediately injected in the grid by HVDC when the phase angle changes from 0 to -10 degree at t=0.1 sec. The active power infeed from offshore converter remains constant (3rd subplot) throughout the disturbance period. This means

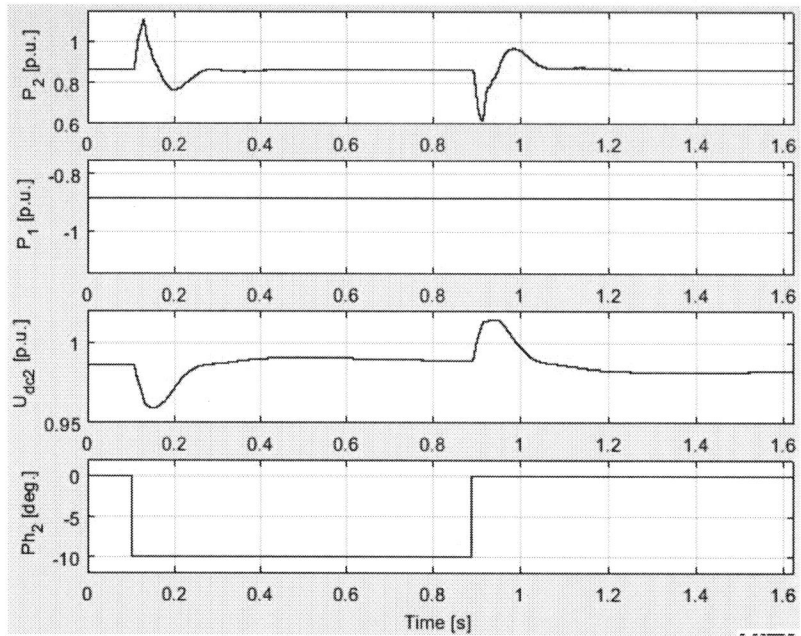

Fig. 7: Grid voltage phase jump of -10 degrees – onshore grid of a VSC-HVDC connecting offshore wind farm. Subplot 1: Active power at PCC of onshore converter, subplot 2: Active power at PCC of offshore converter, subplot 3: dc voltage of onshore converter and subplot 4: grid voltage phase (onshore grid represented as a Thevenin source).

that the energy is taken from DC link (or converter submodule capacitance and cable capacitance) and the DC voltage starts to fall quickly. Soon after that, the U_{dc}-control reacts to restore the DC voltage by taking the energy from the onshore grid. Eventually, the average power from HVDC onshore converter is maintained the same (as the pre-disturbance value) which means that the active power support diminishes to zero. Had the DC control not reacted faster, the DC voltage would have gone so low risking the stability of HVDC link.

It's to be noted that for VSC-HVDC connecting of offshore wind farms, a possible way to provide inertia/frequency support for a longer period without jeopardizing the DC link stability is to install an additional energy storage in the onshore converter on DC side. The other option is to have grid forming capability in wind turbines and operate them below the MPPT with a headroom (impact on cost needs be evaluated). In that case the offshore HVDC converter can operate in U_{dc}-control mode (in grid following mode) and onshore converter operates in P-control mode (in grid forming mode). The energy required for onshore grid support can be extracted from wind turbines if wind is available.

Conclusion

A brief review on the requirements related to grid-forming behaviors in existing Grid codes/standards has been made. Challenges from some of the requirements are discussed. Via simulation examples, it is shown that:

- A very limited inertia support can be provided by the converter in U_{dc}-control mode for HVDC inter-connection applications.
- Diminished inertia support from onshore converter in U_{dc}-control mode for HVDC offshore wind power integration.
- Inertia support can be shaped according to requirement by the converter in P-control mode under the pre-conditions of sufficiently fast U_{dc}-control in other station and the AC network connected to U_{dc}-control station can tolerant transient power variation.

- High inertia by the converter in P-control mode may sacrifice traditional functions such as EPC, POD, etc.

Therefore, there is a need for further joint efforts from both TSOs and vendors to reach a common understanding so that unharmonized requirements are removed to avoid unnecessary problems or risk of worsening performance in comparison with the existing HVDC design.

Appendix

The simulation throughout the paper has been implemented in PSCAD/EMTDC using the system parameters detailed in Table I.

Table I: System parameters

Parameter		Value
Point-to-point HVDC link (asymmetrical monopole)	$U_{ac,rated}$	400 kV
	P_{rated}	1000 MW
	$U_{dc,rated}$	525 kV
	f_0	50 Hz
HVDC connected OWF (symmetrical monopole)	$U_{ac,rated}$	400 kV (onshore)
	P_{rated}	900 MW
	$U_{dc,rated}$	± 320 kV
	f_0	50 Hz

References

[1] Unruh, Peter, et al. "Overview on grid-forming inverter control methods." Energies 13.10 (2020): 2589.

[2] Y. Jiang-Hafner "HVDC system and method to control a voltage source converter in a HVDC system," U.S. Patent 8760888, 14 June 2007.

[3] Torres, L.A.B. et al. "Power supply synchronization without communication," In Proceedings of the 2012 IEEE Power & Energy Society General Meeting, San Diego, CA, USA, 22–26 July 2012.

[4] L. Harnefors et al., "Robust Analytic Design of Power-Synchronization Control," in IEEE Transactions on Industrial Electronics, vol. 66, no. 8, pp. 5810-5819, Aug. 2019.

[5] D'Arco et al., "Virtual synchronous machines—Classification of implementations and analysis of equivalence to droop controllers for microgrids,". In Proceedings of the IEEE Grenoble PowerTech, Grenoble, France, 16–20 June 2013; IEEE: Piscataway, NJ, USA, 2013.

[6] P. Rodriguez et al., "Control of grid-connected power converters based on a virtual admittance control loop," 15th European Conference on Power Electronics and Applications (EPE), 2013, pp. 1-10.

[7] E. Rokrok et al. "Effect of Using PLL-Based Grid-Forming Control on Active Power Dynamics Under Various SCR," In Proceedings of the IECON 2019—45th Annual Conference of the IEEE Industrial Electronics Society, Lisbon, Portugal, 14–17 October 2019; IEEE: Piscataway, NJ, USA, 2019; pp. 4799–4804.

[8] Y. Häfner et al., "Stability Enhancement and Blackout Prevention by VSC Based HVDC,". Paper 14-02, CIGRE 2011, Bologna, Italy.

[9] Y. Häfner et al., "HVDC with Voltage Source Converters – A Powerful Standby Black Start Facility,". IEEE PES T&D conference, Chicago, USA, April 21-24, 2008.

[10] Y. Li et al., "Revisiting Grid-Forming and Grid-Following Inverters - A Duality Theory", May 2021. Available online at: https://arxiv.org/abs/2105.13094.

[11] German Grid Code: VDE-AR-N 4131.

[12] FNN Guideline: Grid forming behaviour of HVDC systems and DC-connected PPMs. Supplement to VDE-AR-N 4131 for dynamic frequency/active power behaviour and dynamic voltage control without reactive current specification.

[13] The Grid Code, issue 6, revision 12, 09 March 2022. Available at: https://www.nationalgrideso.com/electricity-transmission/document/162271/download.

[14] GC0137: Minimum Specification Required for Provision of GB Grid Forming (GBGF) Capability (formerly Virtual Synchronous Machine). Available online at: https://www.nationalgrideso.com/industry-information/codes/grid-code-old/modifications/gc0137-minimum-specification-required.

[15] ENTSO-E Technical Report "High Penetration of Power Electronic Interfaced Power Sources and the Potential Contribution of Grid Forming Converters," Available at: https://euagenda.eu/upload/publications/untitled-292051-ea.pdf.

[16] IEEE standard 2800 – 2022 "IEEE Standard for Interconnection and Interoperability of Inverter-Based Resources (IBRs) Interconnecting with Associated Transmission Electric Power Systems," Available at: https://standards.ieee.org/ieee/2800/10453.

[17] Technical Brochure "Influence of Embedded HVDC Transmission on System Security and AC Network Performance", Cire WG C4/B4/C1.604, 2013.

[18] Annex 8 - Workgroup Meeting 3 Presentation .pdf. Available online at: https://www.nationalgrideso.com/industry-information/codes/grid-code-old/modifications/gc0137-minimum-specification-required.

A Highly Integrated and Modular High Speed Electric Drive for Lightweight Electric Mountain Bikes

Matthias Hofer, Mario Nikowitz, Manfred Schrödl
Technische Universität Wien
Institute of Energy Systems and Electrical Drives
Gußhausstraße 25-27
A-1040 Vienna, Austria
Phone: +43 1 58801-370230
Email: matthias.hofer@tuwien.ac.at
URL: http://www.tuwien.ac.at

Keywords

≪Electrical Drive≫, ≪Permanent Magnet Motor≫, ≪Electric Bicycle≫, ≪Pedelec≫, ≪Gear Box≫

Abstract

In this work the application of a high-speed electric machine combined with a high gear ratio is investigated for a lightweight electric mountain bike. The proposed drive utilizes the high power density of high rotational speeds of electric machines to reach low system mass. The system integration and their components are described and compared to commercially available products. First experimental results of the proposed electric drive are presented.

Introduction

In the last years, the high annual sales figures of electric powered assisted cycles (EPAC), also known as pedelecs (max. speed 25kph, 250W continuous power, 600W peak power), confirm the increasing demand on emission free light electric vehicles (LEV), e.g. 1.2 million e-bikes were sold during the first half-year 2021 in Germany, which is nearly a +10% annual increase [1]. Several reasons like the environmental friendliness, higher urban mobility attractiveness by avoiding traffic jams, higher driving range, increased personal health awareness or even semiprofessional sports activities are well known.

Depending on the scope of the e-bike (city bike, trekking bike, all terrain bike, race bike, mountain bike) the requirements of an electric drivetrain differ in a very wide range. Electric drives can be located at several places at the bicycle frame and have different motor topologies [2]. Direct drive in-wheel motor topologies are often used for conversion kits of conventional bikes and use an outer rotor machine topology [1], but due to a high motor mass and especially a high unsprung mass at the wheels such topologies are not applicable for sportive bike applications. At mountain bikes the electric drivetrain has to fulfill very strong requirements defined by the sportive application. The electric drive shall be fully integrated into the bike frame for handling reasons and protection in the terrain, very powerful for long distance uphill driving, very compact, have a low mass and good bike balance. These targets are exactly the main focus of the electric drivetrain proposed in this paper.

Today, bike manufacturers often use a family of electric drive units which mainly differ in power, size and driving range to serve the complete range of commercial pedelecs. Therefore, in most cases hub motor topologies arranged at the bike treadle are used. Such concepts are modular and useable in a wide range but the bike integration capability is limited. Although frame integrated electric drive solutions are available (e.g. from supplier Fazua) they have limited performance. The proposed e-mountain bike shall feel and look like a conventional mountain bike by using a high integration of the electric drive. Finally, the overall e-bike performance depends on an interdisciplinary design approach with respect to system design, component design, control and testing [3],[4].

In Table I an overview of widespread available e-bike motors in 2021 is presented [5]. The mass of the e-bike motors (including treadle, gear box, control unit, but without battery) varies from 1,92 – 3,9kg and provide a peak power from 340W to 580W. Although these values differ in a wide range, the specific torque is approximately 30Nm/kg for each e-bike motor. It is well known that the size of electric machines is directly linked to the machine torque. This characteristic is specified by the so called Esson number C [6]. The torque density is mainly related to the machine type and the machine cooling which is similar at all e-bike motors. Therefore, the application of a high gear ratio combined with high rotational electric machine speeds is a possibility to reduce size and weight of electric drivetrains. According to Table I the commercial e-bike motors use a total gear ratio in the range of approximately 12 to 44. The Yamaha PW-X2 motor has the highest gear ratio of 44,1 although the overall specific torque is the lowest in Table I. Thus, the fundamental structural benefit of high gear ratios is not realized in existing products. In this paper the long-term goal is to reduce the weight of the electric drivetrain for e-mountain bikes by application of high-speed electric machines combined with a high gear ratio. In the first step, the competitiveness shall be shown and finally, by optimization a significant improvement shall be realized.

Table I: Available e-bike motors in 2021 [5]

Supplier	Type	Peak-Torque in Nm	Peak-Power in W	Motor-Weight in kg	Total Gear Ratio	Torque density in Nm/kg
Bosch	Perfomance Line CX	85	580	2,79	12,2	30,5
Brose	Drive S Mag	90	340	2,98	25,6	30,2
Fazua	Evation Ride 50	60	400	1,92 *)	24,7	31,25
Shimano	EP 8	85	unknown	2,57	40	33
TQ	HPR 120 S	120	unknown	3,9	37	30,7
Yamaha	PW-X2	80	480	3,06	44,1	26,1

*) only motor unit without treadle axle and treadle sensor (+1,3kg)

The Proposed Electric Drive System

In this work a modular and lightweight electric drive system for a full integration in the mountain bike frame is investigated. Due to legal restrictions a peak power of 600W is selected and limited by software. The first prototype concept to evaluate technological aspects has the target requirements of peak torque of 80 Nm, peak power of 600W and motor weight of max. 2,5kg. The electric motor unit consists of an inverter, an electric machine, a planetary gear stage and a bevel gear stage including treadle with torque and speed sensor. The complete drive pack as well as the battery is integrated in the bike frame. The drive pack can be inserted from the bottom of the frame, as shown in Fig. 1. The proposed drive concept is completely located in the lower tube of the bike frame and the components use a common outer diameter of 72mm as shown in Fig. 2. The drive pack is designed modular and scalable by length adjustment of the electric machine. The battery pack including the battery management system (BMS) is modular as well and fully integrated in the bike frame. Thus, several power ratings combined with variable driving ranges by battery variation can be realized to serve different bike applications.

The bike's chain drive is setup with one fixed gear on the treadle and a conventional switchgear on the rear wheel. The electric drive support works up to a treadle cadence of 122rpm, which is equal to an electric machine rotational speed of 11.000rpm combined with a total gear ratio of 1:90. This high ratio is reached with a planetary stage and a bevel gear topology. This concept shows a significantly higher gear ratio and significantly higher electric machine speeds compared to existing commercial e-bike motors. By a high geometric integration of the gear stages, the electric machine and the inverter a compact drive pack can be realized which is inserted into the frame according to Fig. 1. Finally, after system optimization in a subsequent project step a higher torque and power density compared to current available e-bike motors is proposed. The target is to reach a specific torque of 40Nm/kg after optimization.

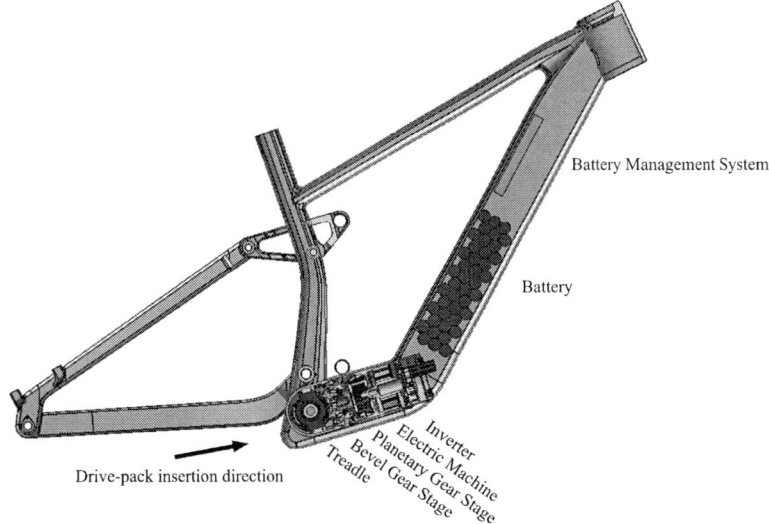

Fig. 1: Proposed frame setup including electric drive components

The e-Bike Components

The e-bike drive pack is designed modular with common interfaces and consists of several sub-components. The mass break down of the propsed design is presented in Tab. II. The total mass of the electric drive pack without battery is designed with 2,48kg with a peak torque of >81Nm.

Table II: Mass break down of the proposed E-bike drive

	Volume in cm^3	Mass in g
Bevel Gear with Treadle and Freewheel	392	995
Planetary Gear Stage	193	568
Electric Machine	211	736
Inverter	143	175
Battery (without BMS)	950	2200

The Inverter

The inverter is located directly on the electric machine and currently consists of three boards, a power board, a controller board and an interface board as shown in Fig. 3. To communicate with a display and the battery management system a CAN-interface and a wireless Ant+ interface is available. A classic three phase voltage source inverter with MOSFET power transistors is implemented. The hardware is prepared for sensorless field oriented control of the electric machine, which allows the operation without any rotor angular position sensor. In contrast to the work [7] a sensorless control method without additional injection is proposed [8]. A full integration of the 3 boards into a single inverter and a common housing is planned in the next project step.

The Gear Stages

The gear stage consists of two planetary gear stages combined with a bevel gear stage to turn the rotational axis to the treadle axle. Further, a torque sensor and a freewheel, which separates the gearbox and the motor from the treadle shaft in thrust mode, are integrated. The gearbox has a total gear ratio of 1:90 while the planetary stages reach a transmission ratio of 1:60. This high transmission ratio in a small installation space with a high output torque is made possible by an unconventional kinematic coupling of the degrees of freedom of the two planetary stages. This would enable transmission ratios of up to 1:600 in the same installation space.

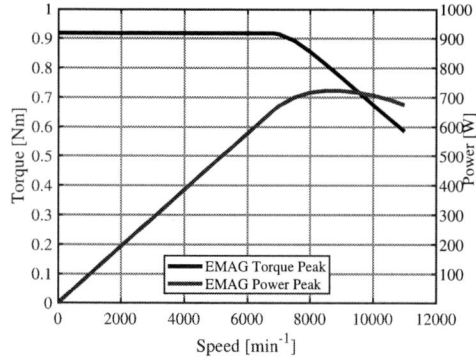

Fig. 3: Inverter concept

Fig. 2: CAD model of the treadle axle with gearbox, electric machine and inverter

The Electric Machine

For a high efficient electric drivetrain a permanent magnet synchronous machine is proposed. The electric machine is designed for high rotational speeds up to 11.000rpm at compact physical dimensions. The machine provides an inductance difference in L_d and L_q to ensure sufficient sensorless control at standstill and low speeds. For simple production a fractional slot winding with concentrated coils (Fig. 4) was chosen. The proposed stator design allows the implementation of flat-wire windings for reaching higher copper fill factors and a higher machine efficiency. The simulated electromagnetic peak performance (>0.9Nm and 700W) at a phase current of 17Arms is presented in Fig. 5. For a sufficient high speed operation with low harmonics the rotor is skewed by discrete skewing. The machines active electrical components are currently designed to a simple and low cost production even at small production volumes using common materials.

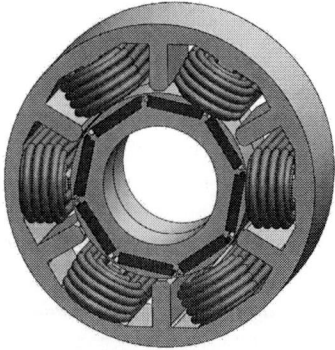

Fig. 4: CAD model of the active electrical machine components

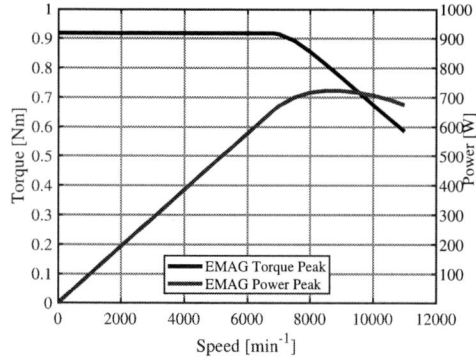

Fig. 5: Simulated electric machine characteristics

The Battery

In contrast to conventional battery packs with a nominal voltage of 36V this e-bike proposes a 48V nominal voltage. In general, the battery space within the bike frame can be equipped with several battery configurations. The proposed work considers a battery, which provides 500Wh usable energy at an operating voltage range from 46,2 to 58,8V. Thus, 42 Li-Ion round cells, type 18650 (14 in series and 3 parallel) and a battery management system (BMS) are installed, which communicates with the inverter during operation.

Experimental Results

For experimental verification of the proposed electric drive concept the electric machine and inverter prototypes are built according the design, see Fig. 6. The power board is directly mounted on the electric machine, the controller board and the interface board can be connected by board-to-board connectors.

Fig. 6: Prototypes of the electric machine and the inverter's power board

A field oriented control with maximum torque per ampere (MTPA) at lower speeds and maximum torque per volt (MTPV) for flux weakening is implemented in the inverter control. The rated direct axis and quadrature axis inductances of the machine are measured (Fig. 7). The inductances show a sufficient difference to apply sensorless control methods. Further, the electromagnetic performance of the electric machine prototype at 25°C rotor temperature for several voltage levels is depicted in Fig. 8. As expected, for higher voltage levels a peak power up to 800W can be reached. By these tests the target performance of the proposed electric machine and the inverter is confirmed.

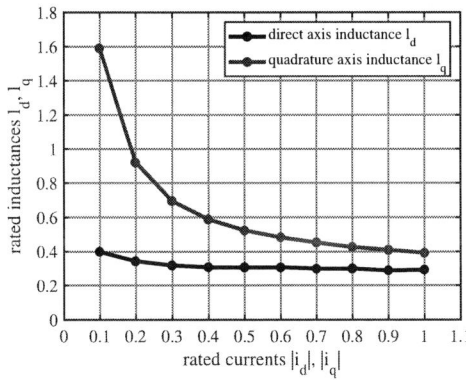

Fig. 7: Measured inductance characteristic of the electric machine

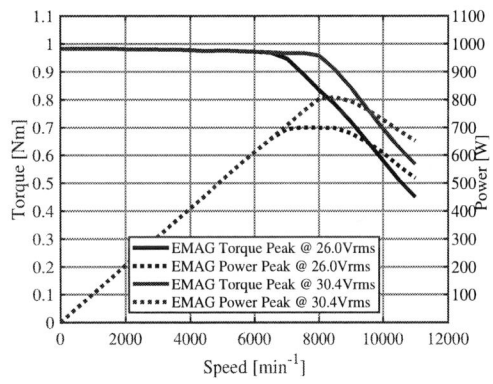

Fig. 8: Measured power and torque characteristic of the electric machine

Further Optimization

In the current project phase the complete proposed electric drive system according to Fig. 1 is investigated on a system test rig for a final confirmation of the competitiveness. After an experimental performance analysis of the complete drivetrain an optimization phase will be continued. The overall performance will be optimized by an interdisciplinary approach using several system parameters to reach a high efficiency at a low mass. First, the total gear ratio can be adjusted in a wide range without changing the installation space. The electric machine can be adjusted by changing electrical steel and permanent magnet materials. Thus, a mass and size reduction of the electric machine up to 10% at same torque and power is expected by optimization. Based on the inverter prototype design the inverter will be redesigned in collaboration with an industrial electronic manufacturer to reach a compact design and an effective high volume production capability.

Conclusion

In this paper an electric bicycle drivetrain with a high-speed electric machine and a gearbox with a very high gear ratio is investigated. By an innovative and compact gearbox design and an integrated inverter, the electric drivetrain is able to be fully integrated in the bike frame. Further, a modular concept is reached to adjust the drivetrain to several bike configurations and types. The prototype concept realizes significantly higher gear ratios as currently available e-bike motors. First components of the proposed e-bike drivetrain were successfully tested. The experimental results confirm the competitiveness to currently offered e-bike drivetrains. The proposed design represents the basis for further system improvements to point out the structural benefit of high power densities at high-speed electrical drives.

References

[1] https://www.ziv-zweirad.de/presse-medien/pressemitteilungen, Marktdaten 1. Halbjahr 2021[Online]
[2] W. Chlebosz, G. Ombach, J. Junak: Comparison of permanent magnet brushless motor with outer and inner rotor used in e-bike, Proceedings of the XIX International Conference on Electrical Machines - ICEM 2010, doi: 10.1109/ICELMACH.2010.5608000
[3] M. Schmitt, S. Decker, M. Doppelbauer: Measuring and Characterization of a Pedal Electric Cycle (Pedelec) on a Full System Test-Bench with Full Range Emulation of a Cyclist, Proceedings o the 21st European Conference on Power Electronics and Applications (EPE '19 ECCE Europe), 2019, doi:10.23919/EPE.2019.8915567
[4] G. Thejasree, R. Maniyeri, P. Kulkami: Modeling and Simulation of a Pedelec Proceedings of the 2019 Innovations in Power and Advanced Computing Technologies (i-PACT), 2019, doi:10.1109/i-PACT44901.2019.8960086
[5] https://ebike-mtb.com/der-beste-emtb-motor-test/ [Online]
[6] A. Binder: Elektrische Maschinen und Antriebe, Grundlagen, Betriebsverhalten, Springer Verlag, 2012, doi: 10.1007/978-3-540-71850-5
[7] S.R.Filho, L. Sun, T. Lambert, M. Ikhlas, Y. Yang, A. Emadi: Low-Speed Sensorless Control of a Surface Mounted Permanent Magnet Motor in an e-Bike Application, Proceedings of the 2021 IEEE Transportation Electrification Conference and Expo (ITEC), 2021, doi: 10.1109/ITEC51675.2021.9490158
[8] M.Hofer, M.Nikowitz, M.Schroedl: Sensorless control of a reluctance synchronous machine in the whole speed range without voltage pulse injections, The IEEE 3rd International Future Energy Electronics Conference and ECCE Asia, 2017, doi: 10.1109/IFEEC.2017.7992211 Vanderkeyn Ralf W.: Example of fast switching component, EPE Journal Vol 20 no 5, pp. 48- 56

Performance Enhancement of Power Conditioning Systems in More Electric Aircrafts

Nick Rigogiannis and Nick Papanikolaou
DEMOCRITUS UNIVERSITY OF
THRACE
Department of Electrical and Computer
Engineering, Electrical Machines
Laboratory, 67132
Xanthi, Greece
Tel.: +30 – 25410.79921
E-Mail: nrigogia@ee.duth.gr,
npapanik@ee.duth.gr

Yongheng Yang
ZHEJIANG UNIVERSITY
College of Electrical Engineering, 310027
Hangzhou, China
Tel.: +86 – 571.8795.2980
E-Mail: yang_yh@zju.edu.cn

Acknowledgements

The research work of Nick Rigogiannis was supported by the Hellenic Foundation for Research and Innovation (HFRI) under the 3rd Call for HFRI PhD Fellowships (Fellowship Number: 5547).

Keywords

«More Electric Aircraft», «Aerospace», «Microgrid», «Power Quality», «DC-DC Power Converter», «Digital Control», «Supercapacitors».

Abstract

Power quality improvement constitutes a critical issue in the DC distribution networks of modern and future More Electric Aircrafts (MEAs). In this paper, an improved control strategy for power conditioning systems (i.e., aiming to mitigate DC bus voltage transients, mainly in feeders with critical loads) in MEA microgrids is proposed, to enhance the system dynamics. The studied power conditioner is based on the widely proposed bidirectional buck-boost DC-DC converter configuration, along with a supercapacitor bank. The proposed controller design enhances the system performance, employing an effective control loop (based on a current control method), taking into consideration the detailed converter average model, so as to enhance the performance of conventional power conditioning systems during transient operation. The proposed control strategy is validated by experimental tests, presented on a SiC-based scaled-down hardware prototype.

Introduction

Over the last decade the aviation industry is moving towards the More Electric Aircraft (MEA), as it is considered one of the most promising solutions for more efficient air transportations with minimum environmental footprint [1], [2]. The electrification level of commercial aircrafts is constantly increasing, mainly thanks to the notable development of power electronics-dominated power systems (as the on-board distribution network of the MEA), whilst according to *Airbus'* predictions, the power rating of civil aircraft microgrids should finally reach 20 MW in the next few decades [3]. In this context, new types of power sources, hybrid energy storage units and more electronically controlled loads with various characteristics (e.g., constant power, pulsed power, intermittent operation, etc.) are emerging, imposing stringent requirements, in terms of stability, robustness, flexibility and dynamic performance, for the on-board microgrid [4].

As for the MEA microgrid architecture, currently several hybrid (both DC and AC) configurations have been employed (e.g., *Airbus* A380 and *Boeing* B787), whereas purely DC configurations have been investigated and proposed for future MEAs, within various research frameworks, such as Clean Sky JU

[5] and MOET EU [6], coming as a result of the emerging technology of DC microgrids [7]. In particular, in the MEA distribution network, Constant Power Loads (CPLs) i.e., tightly-regulated electronic loads such as inverter driven electric motors, and Pulsed Power Loads (PPLs) are mainly jeopardizing the DC microgrid stability and power quality, imposing new challenges for performance enhancement, as well as for advanced and sophisticated control schemes for the power conditioning converters [8]. In order to overcome these power quality issues, stabilize CPLs and effectively accommodate PPLs, the incorporation of enhanced (in terms of robustness, dynamic performance, stability and flexibility) power conditioning systems, especially in feeders with priority loads, becomes imperative.

The aforementioned PPLs are very common critical loads in on-board microgrids (e.g., wing de-icing systems, radars, sonars, etc.), demanding a large amount of energy within very short time intervals, featuring so a pulsed behavior with high power characteristics. In details, according to [8] and [9] in an MEA microgrid, the peak power may last for approximately 20-200 ms, whereas the peak-to-average power ratio may be more than 5-to-1, across a time scale of 50-500 ms. Hence, voltage transients may occur in the microgrid DC buses, leading to poor power quality and evoking critical conditions for various loads, when the DC voltage exceeds its allowed limits. In many cases, large PPLs require excess power source capacity or dedicated power generation to protect other aircraft equipment.

Generally, in DC microgrids, the term "power quality" denotes constant DC voltage waveforms throughout microgrid buses (or within clearly specified voltage levels), oscillations-free power supply in steady-state operation and fast transient response during dynamic conditions [10]-[12]. Moreover, the aforementioned transient phenomena may occur either due to normal disturbances, such as electric load steps (pulsed / intermittent operation) and engine speed changes or due to abnormal disturbances, such as momentary power interruptions and fault clearings. According to the *MIL-STD-704F* and *ISO 1540:2006* (standards related to aircraft electric power characteristics), transients are classified into three categories, namely lesser, normal and abnormal transients [13], [14]. Lesser transients are those that do not exceed the steady state limits; normal transients may exceed the steady state limits but they remain within the specified normal transient region; finally, transients that exceed normal transient limits (as a result of an abnormal disturbance) and eventually return to steady state limits are defined as abnormal transients.

In order to maintain the bus voltage within the specified limits (imposed by the relevant standards), power conditioning units are employed, comprising bidirectional DC-DC converters and Energy Storage Systems (ESSs) [15], [16], usually connected as parallel active filters (PAF); typically supercapacitor (SC) banks are utilized (as they feature high power density and fast dynamics) to counterbalance the capacity limit of the main energy sources, which are usually designed to supply the average loads, rather than meet the peak power demands (short duration transients), due to techno-economic considerations [8]. These systems may operate, either exclusively during transients, or under a cooperative energy management strategy, e.g., in a hybrid ESS, comprising both batteries (high energy density / slow dynamics) and SCs (high power density / fast dynamics), or in a multi-source DC MG with slow ramp rate energy sources. Such systems have been proposed and investigated in the literature over the last few years [15]-[18].

In light of the above, this work studies a power conditioning system with enhanced performance, by means of an improved control scheme, so as to effectively compensate for DC bus voltage transients (e.g., sags or swells) by sourcing or sinking power to / from the DC bus, respectively. The converter mathematical model is developed, taking into account the power losses; this leads to the proposal of a simple current controller with improved characteristics, minimizing the error between the reference and the real value during transients. Experimental tests are carried out on a scaled-down hardware prototype, indicating the functionality and improved performance of the proposed controller.

System Description

The studied power conditioning system has been proposed in [15] and it comprises a bidirectional DC-DC converter that interfaces a supercapacitor bank into the microgrid main DC bus. The incorporation of the studied system in the MEA DC microgrid, as well as a detailed schematic diagram are depicted in Fig. 1. The well-established bidirectional buck-boost DC-DC converter topology has been selected, which is in favor for various two-quadrant applications (such as energy recovery and active power filtering) as it constitutes a compact configuration with minimum semiconductors and passive components count [15]-[20].

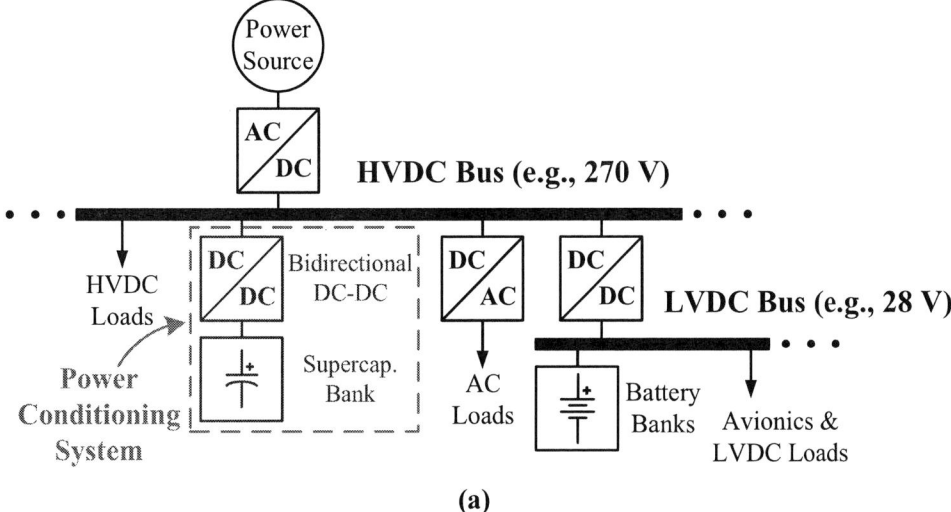

(a)

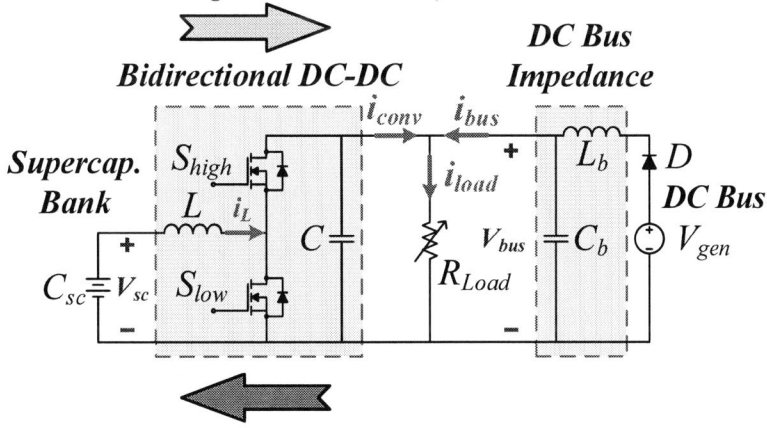

(b)

Fig. 1: A simplified block diagram of the MEA system under study: (a) incorporation of the proposed power conditioning system into the MEA DC microgrid and (b) detailed schematic diagram.

In this scheme, supercapacitors provide ultra-high energy storage capability and high power density, but low rated voltages [20]. On the other hand, the growing on-board power demand of MEAs is inevitably accompanied by increasing DC bus voltage levels of the distribution network, from the actual 540 V [15] to the 3-kV of the *E-Fan X* prototype, (i.e., a hybrid-electric demonstrator, developed by *Siemens*, *Airbus* and *Rolls Royce*) [3], and reasonably further. Therefore, the accommodation of high-voltage

ESSs in future MEAs is hindered, mainly because of ineffective, bulky and costly configurations. For this reason, the converter high voltage side is connected to the main DC bus, whereas its low voltage side is connected to the supercapacitor bank, where the inductor smoothes the current ripple, in order to extend the supercapacitors lifetime.

Owing to their aforementioned characteristics, supercapacitors are able to compensate for steep transients and effectively support the DC bus. When voltage sags occur, the bidirectional converter operates in the step-up mode, providing power to the DC bus, discharging the supercapacitor bank, as it is depicted in Fig. 1(b). On the other hand, in a voltage swell case the converter operates in the step-down mode, sinking power from the DC bus, charging the supercapacitor bank, as Fig. 1(a) depicts. In such a way, transient phenomena are effectively mitigated. As already discussed, the converter operates solely during transient intervals.

It is noted that the proposed concept can be also applied at any other DC network (either with single-bus or with multi-bus configuration), the voltage level of which needs to be preserved within strictly defined limits, ensuring high power quality. However, despite the functionality, robustness and compactness of the bidirectional buck-boost DC-DC topology, the controller design remains challenging, especially in the MEA microgrid application, where transients occur in the time scale of milliseconds; hence, rapid dynamic response is imperative [8], [9]. In this regard, the inductor current reference derivation by the aid of Proportional-Integral (PI) controllers is ineffective, due to their inherent phase lagging. In order to design an effective controller, the alternative of utilizing the converter average model, incorporating its power losses, is studied. As it will be shown in next Sections, the proposed model is of efficacy in steady-state and "slow" transients' conditions. However, the terms "slow" and "fast" transients may refer to various time constants and are defined according to the MEA demand profile requirements (e.g., specific DC bus feeder, load type, generation unit capacity, etc.) [16]-[18].

Controller Design

According to [15], the control scheme is based on the load current monitoring (assuming that the power conditioning system is connected to a DC bus feeder with critical loads) and the decoupling of its AC component via a digitally implemented low-pass filter, so as the converter to operate solely during voltage transients. The necessary internal current control loop is based on the popular Peak Current Control (PCC) method [19]. This work focuses on the derivation of the inductor (i.e., the supercapacitor bank) current reference from the DC bus current reference, as it is presented in Fig. 2. It is assumed that the inductor peak and average current values are very close (continuous conduction mode with small current ripple), in order for the PCC method to operate effectively. Similarly, the converter average model for step-down operation can be derived.

In addition, according to the schematic diagram of Fig. 3, the average model of the bidirectional buck-boost converter (in step-up mode) under steady-state or "slow" transients' operation, can be developed, so as to derive the current transfer function. Apparently, the use of the average model is rational, as long as the converter switching period is quite smaller than the time constant of the occurred transient phenomenon, in order for the inductor volt-second balance (during a switching period) to be considered valid. Hence, in the current work the term "slow" transients refers to time constants at the millisecond-scale or above (assuming switching frequencies above 10 kHz). In this way, the proposed power conditioning scheme is suitable for the above-mentioned normal transients that occur in MEA power systems.

Finally, based on the assumption that the converter operates solely during transients providing its full power (i.e., pulsed current), no significant thermal rise occurs; thus, changes in parasitic elements (such as supercapacitors Equivalent Series Resistance – ESR, inductor DC resistance or MOSFETs on-resistance) can be neglected. Last but not least, it is worth noting that for the considered bidirectional DC-DC converter both high-voltage and low-voltage sides can become input or output, although for the following mathematical analysis the high-voltage side current is referred as "output current" (i.e., i_{conv}), since step-up operation (sourcing power to the DC bus) is considered.

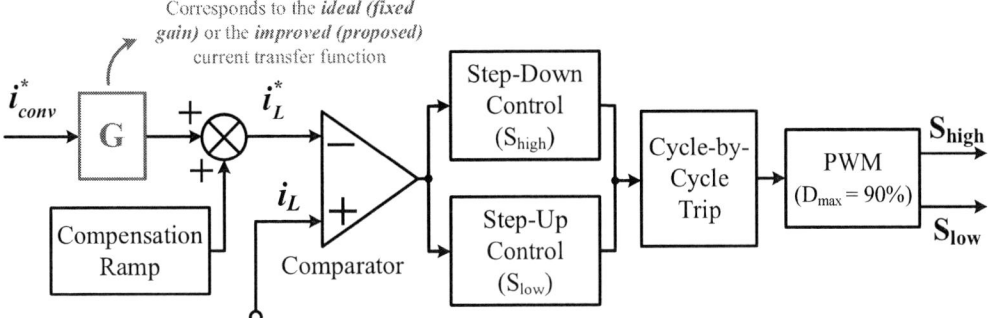

Fig. 2: Control concept, based on the PCC method (derivation of the inductor current reference from the DC bus current reference).

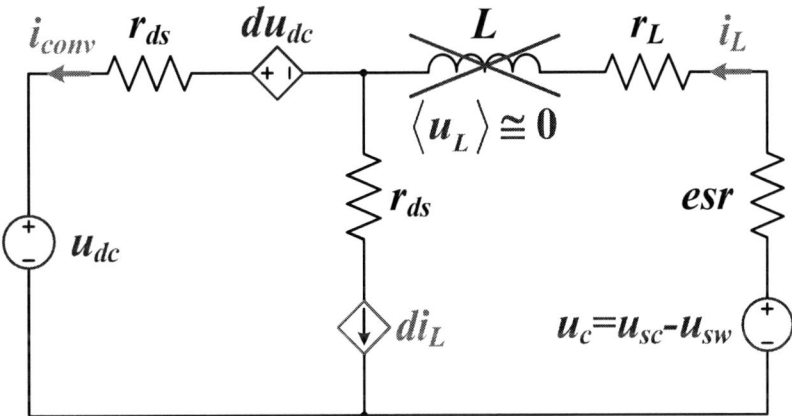

Fig. 3: Equivalent circuit for the average model development, considering step-up operation (voltage sag – supercapacitors discharging case).

According to the equivalent circuit of Fig. 3 and by applying the Kirchoff's current law (KCL) and voltage law (KVL), it is acquired:

$$KCL : i_L = di_L + i_{conv} \Rightarrow i_L \cdot (1-d) = i_{conv} \Rightarrow d = 1 - \frac{i_{conv}}{i_L} \tag{1}$$

$$KVL : u_c - u_{dc} = -du_{dc} + i_L \cdot (esr + r_L) + i_{conv} r_{ds} \tag{2}$$

where i_L and i_{conv} are the inductor (supercapacitors) and the converter output currents, respectively, d stands for the converter duty cycle, u_{dc} is the DC bus voltage (considered constant), u_{sc} is the supercapacitor bank voltage (considered constant), r_L is the inductor internal resistance, esr is the supercapacitors ESR value, r_{ds} is the power MOSFETs on resistance, and u_{sw} is a voltage drop which corresponds to the converter switching losses; it can be extracted by the total (turn-on and turn-off) energy losses, according to the semiconductor switches datasheet (given a specific switching frequency). After mathematical manipulations, the following quadratic equation is derived:

$$(esr + r_L) \cdot i_L^2 + (i_{conv} r_{ds} - u_c) \cdot i_L + i_{conv} u_{dc} = 0 \tag{3}$$

$$\Delta = (i_{conv} r_{ds} - u_c)^2 - 4u_{dc} \cdot i_{conv} \cdot (esr + r_L) \tag{4}$$

To ensure that the discriminant is greater or equal to zero (real roots), it should be:

$$(i_{conv} r_{ds} - u_c)^2 - 4u_{dc} \cdot i_{conv} \cdot (esr + r_L) \geq 0 \Rightarrow (i_{conv} r_{ds} - u_c)^2 \geq 4u_{dc} \cdot i_{conv} \cdot (esr + r_L) \tag{5}$$

By using the quadratic formula, it is obtained that:

$$i_{L_{1,2}} = \frac{\left(u_c - i_{conv}r_{ds}\right) \pm \sqrt{\left(i_{conv}r_{ds} - u_c\right)^2 - 4u_{dc}i_{conv}\left(esr + r_L\right)}}{2\left(esr + r_L\right)} \tag{6}$$

Both roots in (6) may be positive; however, the first one leads to an unacceptably high current ratio (i_L/i_{conv}), and thus to an unacceptable duty cycle value and poor efficiency. Hence, only the second root is valid (practical), as it leads to an acceptable current ratio.

$$i_L = \frac{\left(u_c - i_{conv}r_{ds}\right) - \sqrt{\left(i_{conv}r_{ds} - u_c\right)^2 - 4u_{dc}i_{conv}\left(esr + r_L\right)}}{2\left(esr + r_L\right)} \tag{7}$$

Eq. (7) is used for the proposed improved controller design, whereas for comparative assessment, the following ideal current transfer function (fixed gain) is also considered:

$$\left(\frac{i_L}{i_{conv}}\right)_{ideal} = \frac{u_o}{u_c} \Rightarrow i_L = \frac{u_o}{u_c} \cdot i_{conv} \tag{8}$$

Experimental Validation

In order to assess the performance of the proposed controller, a laboratory-scale SiC-based bidirectional buck-boost DC-DC converter is designed and constructed. The controller is digitally implemented by the aid of a TMS320F28027 microcontroller unit, which is dedicated for current control applications as it accommodates analog comparators that can be connected to its on-chip PWM (Pulse Width Modulation) module, whereas the slope compensation unit (ramp generator) is on-chip [19]; it is noted that the addition of a compensation ramp is imperative in current control applications, whilst the procedure for the design of a compensation ramp with properly calculated slope in the studied converter is presented in [15]. The foremost electronic components and parameters of the constructed power conditioning system are summarized in Table I.

In parallel, since the aim of this work is to study and evaluate the proposed controller effectiveness, various output current reference scenarios are considered, for both steady-state and transient operation. Hence, experimental results focus on the deviation between the real (measured) output current and the reference value. The latter one is arbitrary and thus no specific DC bus voltage waveforms are given.

Table I: Main components and parameters of the experimental test bench.

Description	Symbol - Product Code	Value [Unit]
DC bus voltage (ct.)	u_{dc} *(electronic load in constant-voltage mode / EA-EL 9760-25 2400W 800R)*	45 [V]
supercapacitors voltage (ct.)	u_{sc} *(power supply emulated / SM 70-AR-24)*	30 [V]
converter output (DC bus) capacitance	*C*	470 [μF]
SiC Power MOSFETs	NVHL020N120SC1 [21]	-
MOSFETs on-resistance	r_{ds} *(typical @ 25°C)*	20 [mΩ]
converter switching frequency	f_s	100 [kHz]
inductance value	*L (custom / comprising two E55/28/21 ferrite cores)*	188 [μH]
DC inductor resistance	r_L	30 [mΩ]
supercapacitors ESR	*esr (power supply internal set / SM 70-AR-24)*	70 [mΩ]
microcontroller unit	TMS320F28027 [22] *(C2000 Piccolo MCU F28027 LaunchPad evaluation board)*	-
current sensor	LTS 25-NP [23]	-

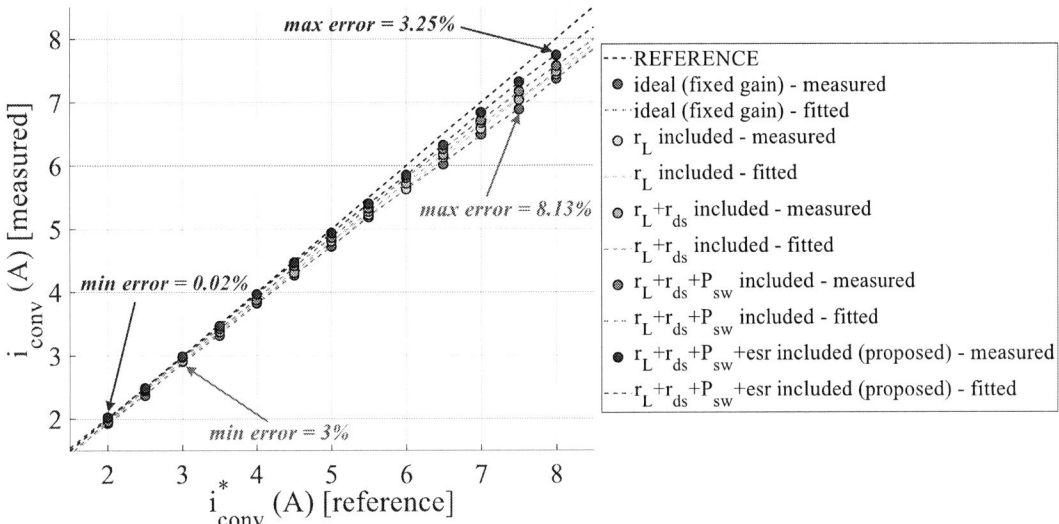

Fig. 4: Experimental results for steady-state operation.

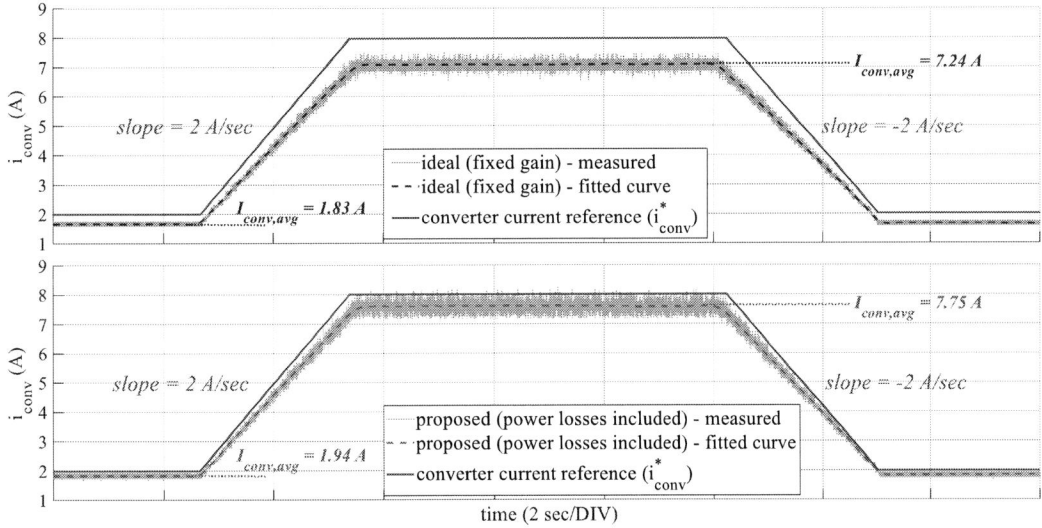

Fig. 5: Experimental results for "slow" transients' operation (2 A → 8 A → 2A) with a slope value of 2 A/sec.

At first, the proposed controller performance is investigated, in steady-state converter operation. In Fig. 4, the converter output current (provided to the DC bus) is illustrated and compared to the converter current reference. The proposed controller (blue line) which takes into consideration the overall power losses excels the performance of the fixed gain ideal transfer function (red line), whereas three additional cases are given, for the inclusion of only a specific fraction of the overall losses. The real output current measurements indicate the enhanced controller performance, keeping the error among the reference and the measured values below 3.3% for a wide current reference range. On the other hand, this deviation for the fixed gain controller (which takes into account the ideal converter transfer function) may reach 8.2 %, exhibiting poor transient performance and thus ineffective load compensation.

Next, in Fig. 5, a slow transient occurs, driving the converter current reference (i_{conv}^*) from 2 A to 8 A and vice versa, with a slope of 2 A/sec. The improved performance that can be achieved with the proposed controller, compared to the ideal (fixed gain) transfer function is highlighted. Finally, in Fig. 6 steeper transients with positive slope (i.e., 4 A/sec and 6 A/sec) are depicted, indicating the fact that the proposed controller operates effectively for relatively "slow" transients (in the sec-time scale), minimizing successfully the error between the real (measured) and the reference current value, injected to the DC bus.

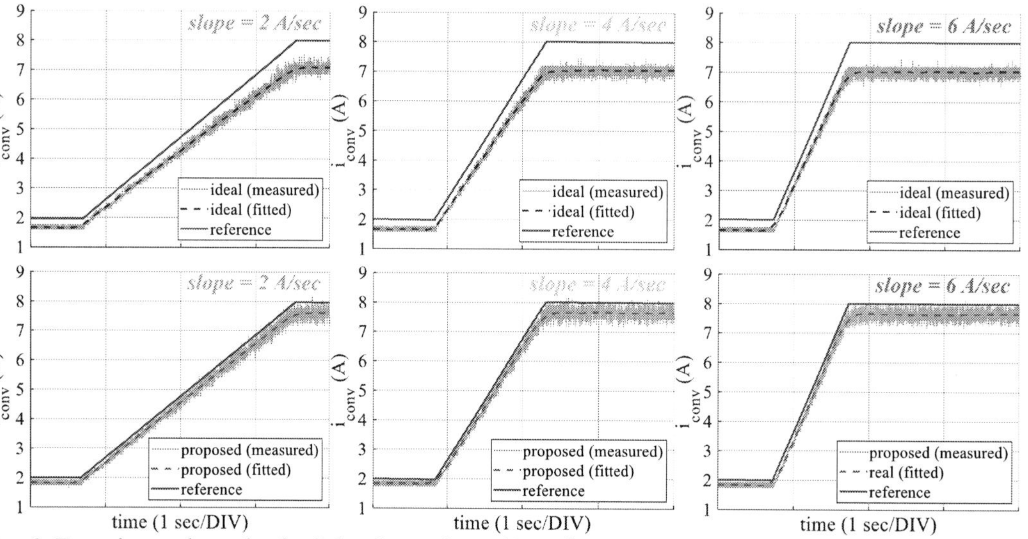

Fig. 6: Experimental results for "slow" transients (2 A → 8 A) with various slope values (2 A/sec, 4 A/sec and 6 A/sec).

Conclusion

This paper proposes an improved controller design for power conditioning systems, applicable to MEA microgrids, taking into account the detailed bidirectional DC-DC converter average model (including power losses). The developed control scheme is able to significantly enhance the converter transient behavior, ensuring high power quality, whereas it is of extremely low computational burden. The presented experimental results on an all-SiC scaled-down hardware prototype highlight the functionality of the proposed (digitally implemented) control strategy, as well as the enhanced converter dynamic performance.

References

[1] B. Sarlioglu, C. T. Morris, "More Electric Aircraft: Review, Challenges, and Opportunities for Commercial Transport Aircraft," *IEEE Trans. Transp. Electrif.*, vol. 1, no. 1, pp. 54-64, June 2015.

[2] P. Wheeler, S. Bozhko, "The More Electric Aircraft: Technology and challenges.," *IEEE Electrif. Mag.*, vol. 2, no. 4, pp. 6-12, Dec. 2014.

[3] M. Guacci, D. Bortis, J. W. Kolar, "High-Efficiency Weight-Optimized Fault-Tolerant Modular Multi-Cell Three-Phase GaN Inverter for Next Generation Aerospace Applications," *2018 IEEE Energy Conversion Congress and Exposition (ECCE)*, 2018, pp. 1334-1341.

[4] A. Barzkar, M. Ghassemi, "Components of Electrical Power Systems in More and All-Electric Aircraft: A Review," *IEEE Trans. Transp. Electrif.*, (Early Access), [Online]. Available: doi: 10.1109/TTE.2022.3174362.

[5] M. A. A. Mohamed, M. Rashed, X. Lang, J. Atkin, S. Yeoh, S. Bozhko, "Droop control design to minimize losses in DC microgrid for more electric aircraft, *Electric Power Systems Research*, vol. 199, p. 107452, Oct. 2021.

[6] T. Yang, S. Bozhko, G. Asher, "Application of Dynamic Phasor Concept in Modeling Aircraft Electrical Power Systems," *SAE Int. J. Aerosp.* vol. 6, no. 1, pp. 38-48, Sept. 2013.

[7] D. Baros, D. Voglitsis, N. P. Papanikolaou, A. Kyritsis, N. Rigogiannis, "Wireless Power Transfer for Distributed Energy Sources Exploitation in DC Microgrids," *IEEE Trans. Sust. Energy*, vol. 10, no. 4, pp. 2039-2049, Oct. 2019.

[8] Q. Xu, N. Vafamand, L. Chen, T. Dragicevic, L. Xie, F. Blaabjerg, "Review on Advanced Control Technologies for Bidirectional DC/DC Converters in DC Microgrids," *IEEE J. of Emer. and Sel. Topics in Power Electron.*, vol. 9, no. 2, pp. 1205-1221, April 2021.

[9] X. Lang, T. Yang, G. Bai, S. Bozhko, P. Wheeler, "Active Disturbance Rejection Control of DC-Bus Voltages within a High-Speed Aircraft Electric Starter/Generator System," *IEEE Trans. Transp. Electrif.*, (Early Access), [Online]. Available: doi: 10.1109/TTE.2022.3164351.

[10] International Standard, IEC 61000-4-30:2015, "Electromagnetic compatibility (EMC) - Part 4-30: Testing and measurement techniques - Power quality measurement methods," Feb. 2015.

[11] International Standard, IEEE 1159-2019, "IEEE Recommended Practice for Monitoring Electric Power Quality," Aug. 2019.

[12] International Standard, EN 50160:2010, "Voltage characteristics of electricity supplied by public electricity networks," July 2010.

[13] International Standard, MIL-STD-704F, "Aircraft Electric Power Characteristics," Mar. 2004.

[14] International Standard, ISO 1540:2006, "Aerospace - Characteristics of aircraft electrical systems," Feb. 2006.

[15] N. Rigogiannis, D. Voglitsis, T. Jappe, N. Papanikolaou, "Voltage Transients Mitigation in the DC Distribution Network of More/All Electric Aircrafts," *Energies*, vol. 13, no. 16, p. 4123, Aug. 2020.

[16] G. C. Christidis, I. C. Karatzaferis, M. Sautreuil, E. C. Tatakis, N. P. Papanikolaou, "Modeling and analysis of an innovative waste heat recovery system for helicopters," *2013 15th European Conference on Power Electronics and Applications (EPE)*, 2013, pp. 1-10.

[17] Chen, Q. Song, S. Yin, J. Chen, "On the Decentralized Energy Management Strategy for the All-Electric APU of Future More Electric Aircraft Composed of Multiple Fuel Cells and Supercapacitors," *IEEE Trans. Ind. Electron.*, vol. 67, no. 8, pp. 6183-6194, Aug. 2020.

[18] J. Chen, Q. Song, "A Decentralized Energy Management Strategy for a Fuel Cell/Supercapacitor-Based Auxiliary Power Unit of a More Electric Aircraft," *IEEE Trans. Ind. Electron.*, vol. 66, no. 7, pp. 5736-5747, July 2019.

[19] N. Rigogiannis, D. Voglitsis, N. Papanikolaou, "Microcontroller Based Implementation of Peak Current Control Method in a Bidirectional Buck-Boost DC-DC Converter," *2018 20th International Symposium on Electrical Apparatus and Technologies (SIELA)*, 2018, pp. 1-4.

[20] I. Karatzaferis, E. C. Tatakis, N. Papanikolaou, "Investigation of Energy Savings on Industrial Motor Drives Using Bidirectional Converters," *IEEE Access*, vol. 5, pp. 17952-17961, 2017.

[21] Datasheet, "Silicon Carbide (SiC) MOSFET – 20 mohm, 1200V, M1, TO-247-3L NVHL020N120SC1," *onsemi*, [Online]. Available: https://www.onsemi.com/pdf/datasheet/nvhl020n120sc1-d.pdf.

[22] Datasheet, "TMS320F2802x Microcontrollers," *Texas Instruments*, [Online]. Available: https://www.ti.com/lit/ds/symlink/tms320f28027.pdf?ts=1655898109371&ref_url=https%253A%252F%252Fwww.ti.com%252Fproduct%252FTMS320F28027.

[23] Datasheet, "Current Transducer LTS 25-NP," *LEM*, [Online]. Available: https://www.lem.com/sites/default/files/products_datasheets/lts_25-np.pdf.

Steady State Simulations of a Hybrid HVAC/HVDC Network Using OS Based ARM Devices

Ioan Catalin Damian, Mircea Eremia
UNIVERSITY POLITEHNICA OF BUCHAREST
Splaiul Independenței 313, Sector 6
Bucharest, Romania
E-Mail: ioan_catalin.damian@upb.ro, eremia1@yahoo.com
URL: http://www.dsee.upb.ro

Keywords

«AC-DC converter», «Multi-terminal HVDC», «Hybrid simulation».

Abstract

For steady state calculations, all approaches make use of established computing devices (mostly PCs with x86-64 or x86-32 architecture) and environments that offer excellent performance but at a high cost. This paper presents an innovative alternative that is based on emerging technologies which have an ARM architecture, are still able to run powerful software, yet meet cost requirements that are less than 60 USD.

Introduction

The past 20 years have revealed a growing need to transmit large amounts of power over longer distances (greater than 500 km). This has made high voltage DC (HVDC) networks extremely enticing, considering numerous advantages over high voltage AC (HVAC) transmission, such as: lower losses, greater controllability, lower overhead/underground line footprint and possibility to use existing HVAC pillars. While development of HVDC started with point-to-point links, it continued with multiterminal networks (e.g. Nan'ao project in 2013, Zhoushan in 2014, ZhangBei in 2020 etc.). This has prompted the need to grow algorithms that can be used to perform steady-state calculations for systems that have both HVAC and HVDC. Presently, there are two major categories of steady-state algorithms: a unified method and a sequential method [1,2,3].

The unified approach evaluates the complete system model and performs all calculations in one sweep. The sequential method alternates the calculation process between the AC region and the DC region, with the use of established iterative methods such as Seidel-Gauss, Newton-Raphson and other advanced methods [2]. Because in the sequential method the modified zone is treated only in the perspective of admittance matrix alterations and does not imply algorithm changes in the remaining zones, this approach will be exploited in the following sections.

The computing tools that enable streamlined simulations for steady-state analysis are mostly based on x86-64 or x86-32 PC architectures and have greatly over time. Moreover, computing performance has become so great that it takes only tens of seconds to simulate highly complex networks. However, these computing instruments cost at least 500 USD, without taking into consideration licensing fees related to advanced software packages. The Advanced RISC Machines (ARM) architecture was first proposed in the late 1980s, but it gained substantial traction after 2000, propelled by the expanding market of smartphones. However, in 2012, the Raspberry Pi Foundation launched the first single-board computer based on the ARM architecture. This device reached consumer acclaim for its versatility and low cost and paved the way for an accelerated development of automation projects such as: robot controllers, smart home hubs, factory controller etc. [4]. The current iteration of this device (version 4B) is sufficiently powerful to run even demanding operating systems such as Windows 10. However, no one has attempted to perform complex HVAC/HVDC network steady state simulations on such a device. This paper approaches this topic, using mathematical models that are detailed in the following paragraphs and are implemented using the C# programming language on a PC and on a Raspberry Pi 4, both of them running Windows 10 natively.

Steady State Model of a Hybrid HVAC/HVDC Network

A steady state model of a hybrid network consists of three zones of interest. One zone corresponds to the AC network, one relates to the DC network and lastly a zone that reflects the interface between the AC and DC regions. PCC represents the point of common coupling, between the AC system and the HVDC converter station.

Considering that a sequential method is used to perform steady state calculations, this leads to an alternation between the AC zone and the DC zone.

AC System Calculations

The AC region calculations can be performed using well developed techniques, such as Newton-Raphson. In this method, the goal is to iteratively adjust parameters using a Jacobian matrix, until a convergence condition is reached [2,3,5].

For each PQ node, the following powers are calculated:

$$P_i^{imp} = P_{g,i} - P_{c,i}$$
$$Q_i^{imp} = Q_{g,i} - Q_{c,i}$$

(1)

$P_{g,i}$ and $Q_{g,i}$ are the active and reactive generated powers (at node i), $P_{c,i}$ and $Q_{c,i}$ are the active and reactive consumed powers (at node i).

In the first iteration, node voltages ($U_i^{(0)}$) are initialized using 1 p.u. value and the phase angles ($\theta_i^{(0)}$) are set at 0.

Looking at the PU nodes, the following powers are calculated:

$$P_i^{imp} = P_{g,i} - P_{c,i}$$
$$Q_i^{min} = Q_{g,i}^{min} - Q_{c,i}$$
$$Q_i^{max} = Q_{g,i}^{max} - Q_{c,i}$$

(2)

Q_i^{min} and Q_i^{max} are the maximum and minimum reactive powers (at node i).

The slack bus nodes are regarded only for voltage calculation, while the angle remains 0:

$$U_i = U_i^{imp}; \; \theta_i = 0$$

In the following step, it is necessary to calculate node powers, as shown in (3):

$$P_i^{(p)} = real\left(\underline{S}_i^{(p)}\right) = \sum_{k=1}^{n} U_i^{(p)} U_k^{(p)} \left[G_{ik} \cos\left(\theta_i^{(p)} - \theta_k^{(p)}\right) + B_{ik} \sin\left(\theta_i^{(p)} - \theta_k^{(p)}\right)\right]$$
$$Q_i^{(p)} = imag\left(\underline{S}_i^{(p)}\right) = \sum_{k=1}^{n} U_i^{(p)} U_k^{(p)} \left[G_{ik} \sin\left(\theta_i^{(p)} - \theta_k^{(p)}\right) - B_{ik} \cos\left(\theta_i^{(p)} - \theta_k^{(p)}\right)\right]$$

(3)

If the iteration number p is greater than 2, for all PU nodes, the reactive power confinement limits must be verified:

- If $Q_i^{(p)} < Q_i^{min}$, then the node becomes PQ, having $Q_i^{imp} = Q_i^{min}$;

- If $Q_i^{(p)} > Q_i^{min}$, then the node becomes PQ, having $Q_i^{imp} = Q_i^{max}$.

Next, for PU and QP nodes, power corrections are evaluated:

- If the node is PU, then:

$$\Delta P_i^{(p)} = P_i^{imp} - P_i^{(p)} \tag{4.a}$$

- If the node is PQ, then:

$$\Delta P_i^{(p)} = P_i^{imp} - P_i^{(p)} \tag{4.b}$$

$$\Delta Q_i^{(p)} = Q_i^{imp} - Q_i^{(p)} \tag{4.c}$$

The convergence test is performed using (5.a) and (5.b).

$$\max_i \left\{ \left| \Delta P_i^{(p)} \right| \right\} \le \varepsilon \tag{5.a}$$

$$\max_i \left\{ \left| \Delta Q_i^{(p)} \right| \right\} \le \varepsilon \tag{5.b}$$

If the convergence check shows that the condition is valid, the Newton-Raphson procedure can end with grid power-flow calculations and DC network calculations can commence. However, if the established threshold is bigger than the power corrections, the algorithm must continue with the calculation of the Jacobian matrix ($J^{(p)}$) as in (6), using the terms defined in (7.a) and (7.b).

$$\mathbf{J}^{(p)} = \begin{array}{|c|c|} \hline \mathbf{H}^{(p)} & \mathbf{K}^{(p)} \\ \hline \mathbf{M}^{(p)} & \mathbf{L}^{(p)} \\ \hline \end{array} \tag{6}$$

$$H = \begin{bmatrix} \dfrac{\partial P_1}{\partial \theta_1} & \cdots & \dfrac{\partial P_1}{\partial \theta_N} \\ \vdots & \ddots & \vdots \\ \dfrac{\partial P_N}{\partial \theta_1} & \cdots & \dfrac{\partial P_N}{\partial \theta_N} \end{bmatrix} \qquad K = \begin{bmatrix} \dfrac{\partial P_1}{\partial U_1} U_1 & \cdots & \dfrac{\partial P_1}{\partial U_{n_c}} U_{n_c} \\ \vdots & \ddots & \vdots \\ \dfrac{\partial P_N}{\partial U_1} U_1 & \cdots & \dfrac{\partial P_N}{\partial U_{n_c}} U_{n_c} \end{bmatrix} \tag{7.a}$$

$$M = \begin{bmatrix} \dfrac{\partial Q_1}{\partial \theta_1} & \cdots & \dfrac{\partial Q_1}{\partial \theta_N} \\ \vdots & \ddots & \vdots \\ \dfrac{\partial Q_{n_c}}{\partial \theta_1} & \cdots & \dfrac{\partial Q_{n_c}}{\partial \theta_N} \end{bmatrix} \qquad L = \begin{bmatrix} \dfrac{\partial Q_1}{\partial U_1} U_1 & \cdots & \dfrac{\partial Q_1}{\partial U_{n_c}} U_{n_c} \\ \vdots & \ddots & \vdots \\ \dfrac{\partial Q_{n_c}}{\partial U_1} U_1 & \cdots & \dfrac{\partial Q_{n_c}}{\partial U_{n_c}} U_{n_c} \end{bmatrix} \tag{7.b}$$

The corrections for the phase angle $\Delta\boldsymbol{\theta}^{(p)}$ and the relative voltage $\left(\dfrac{\Delta \mathbf{U}}{\mathbf{U}}\right)^{(p)}$ can be calculated using (8):

$$\begin{array}{|c|c|} \hline \mathbf{H}^{(p)} & \mathbf{K}^{(p)} \\ \hline \mathbf{M}^{(p)} & \mathbf{L}^{(p)} \\ \hline \end{array} \begin{array}{|c|} \hline \Delta\boldsymbol{\theta}^{(p)} \\ \hline \left(\dfrac{\Delta \mathbf{U}}{\mathbf{U}}\right)^{(p)} \\ \hline \end{array} = \begin{array}{|c|} \hline \Delta P^{(p)} \\ \hline \Delta Q^{(p)} \\ \hline \end{array} \tag{8}$$

With the previous corrections, it is now possible to determine the values of the phase angle and the node voltage for the following iteration:

$$\theta_i^{(p+1)} = \theta_i^{(p)} + \Delta\theta_i^{(p)} \tag{9}$$

- If the node is PQ then:

$$U_i^{(p+1)} = U_i^{(p)} \left(1 + \frac{\Delta U_i^{(p)}}{U_i^{(p)}} \right) \tag{10}$$

- If the node was initially of type PU but is now a PQ node:

- If $Q_i^{imp} = Q_i^{min}$ and $U_i^{(p+1)} < U_i^{imp}$ then $U_i^{(p+1)} = U_i^{imp}$ and the node returns to PU type

- Also, if $Q_i^{imp} = Q_i^{max}$ and $U_i^{(p+1)} > U_i^{imp}$ then $U_i^{(p+1)} = U_i^{imp}$ and the node returns to PU type

Next, the computation procedure continues with the next iteration, by following the same algorithm shown from (3) onwards.

DC System and Converter Interface Calculations

The DC system, with converter interface, can be approached using the Newton-Raphson iterative method [3,6,7,8,9].

In order to determine steady-state values inside the DC network (as well as for converter interface), the values previously determined for the AC region can now be used to initialize the computation process of the DC region. In the DC zone, as well as for the converter region (Figure 1), the Newton-Raphson method can again be used, with the help of several main equations. These are:

- The active and reactive power flow between the PCC and the AC system:

$$f_1 = P_s - \sum_{k=1}^{n} U_s U_k [G_{sk} \cos(\theta_s - \theta_k) + B_{sk} \sin(\theta_s - \theta_k)] \tag{11}$$

$$f_2 = Q_s - \sum_{k=1}^{n} U_s U_k [G_{sk} \sin(\theta_s - \theta_k) - B_{sk} \cos(\theta_s - \theta_k)] \tag{12}$$

where (s) is the PCC node and (k) is the AC node.

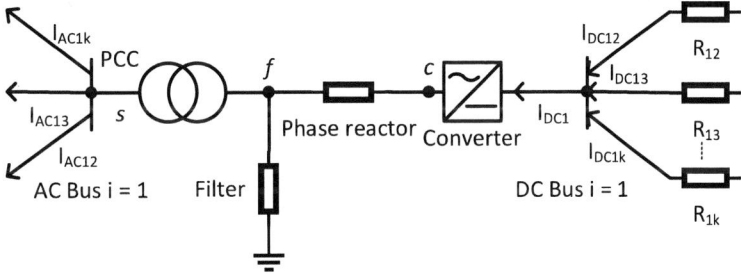

Fig. 1. AC/DC interface

- The power flow between the PCC node (s) and the converter (c):

$$f_3 = P_s - U_s \{-U_s G_{cs} + U_c [G_{cs} \cos(\theta_s - \theta_c) + B_{cs} \sin(\theta_s - \theta_c)]\} \tag{13}$$

$$f_4 = Q_s - U_s \{-U_s G_{cs} + U_c [G_{cs} \sin(\theta_s - \theta_c) - B_{cs} \cos(\theta_s - \theta_c)]\} \tag{14}$$

- The power balance between the AC side and the DC side of the converter (losses are neglected), leads to a 5th function:

$$f_5 = U_{DC} I_{DC} - U_c \{U_c G_{cs} - U_s [G_{cs} \cos(\theta_c - \theta_s) + B_{cs} \sin(\theta_c - \theta_s)]\} \tag{15}$$

- The current injection at the DC side,

$$f_6 = I_{DC,i} - \sum_{\substack{k=1 \\ k \neq i}}^{n} G_{ik} U_{DC,k} \tag{16}$$

where i is any DC side node.

As in the AC zone calculation procedure, it is necessary to define the Jacobian matrix. Afterwards, using power and voltage corrections, new values are calculated until the convergence threshold is reached.

Depending on the type of operational mode the converter is set, it is possible to specify additional functions:

1. Constant $P - Q$:

$$P_s = P_s^{imp} => f_7 = P_s - P_s^{imp} \tag{17.a}$$

$$Q_s = Q_s^{imp} => f_8 = Q_s - Q_s^{imp} \tag{18.a}$$

2. Constant $V_{DC} - Q$:

$$V_{DC} = V_{DC}^{imp} => f_7 = V_{DC} - V_{DC}^{imp} \tag{17.b}$$

$$Q_s = Q_s^{imp} => f_8 = Q_s - Q_s^{imp} \tag{18.b}$$

3. Constant $P - V_{AC}$:

$$P_s = P_s^{imp} => f_7 = P_s - P_s^{imp} \tag{17.c}$$

$$V_s = V_s^{imp} => f_8 = V_s - V_s^{imp} \tag{18.c}$$

4. Constant $V_{DC} - V_{AC}$:

$$V_{DC} = V_{DC}^{imp} => f_7 = V_{DC} - V_{DC}^{imp} \tag{17.d}$$

$$V_s = V_s^{imp} => f_8 = V_s - V_s^{imp} \tag{18.d}$$

The unknown variables, which can be determined using the Newton-Raphson method are identifiable from functions $f_1, f_2, ... f_8$: P_s, U_s, θ_s for f_1; Q_s, U_s, θ_s for f_2; $P_s, U_s, U_c, \theta_s, \theta_c$ for f_3; $Q_s, U_s, U_c, \theta_s, \theta_c$ for f_4; $U_{DC}, I_{DC}, U_c, U_s, \theta_s, \theta_c$ for f_5; U_{DC}, I_{DC} for f_6; while f_7 and f_8 depend on the control strategy. The system is solvable since there are eight unknown variables and eight equations. These equations form the Jacobin matrix J. The corrections vector, ΔX, which has the structure in (9) can now be calculated.

$$\Delta X = [\Delta P_s, \Delta Q_s, \Delta U_s, \Delta \theta_s, \Delta U_c, \Delta \theta_c, \Delta U_{DC}, \Delta I_{DC}]^T \tag{19}$$

The correction vector adjusts the values of various electrical parameters, and the calculation process either continues with a new iteration or finishes, depending on the convergence criterion ε. If the DC network steady-state calculations reach the finish criterion, then all resulting values can be used to reinitialize the calculation process for the AC zone. The process repeats until a global convergence criterion is met.

Case Study

The aim of the case study is to demonstrate the validity of the concept that, using OS based ARM devices, it is possible to perform steady-state calculations for complex hybrid HVAC/HVDC networks, in an efficient and fast manner, that is comparable to the performance of expensive and highly advanced PCs.

Description of the Computing Setup

Simulation is performed on 2 types of devices. The first type consists of a PC with an Intel Core i7 1065G7, with 4 cores (8 logical processors), that has a maximum boost frequency of 3.9 GHz. The second device is a Raspberry Pi 4, that has a quad core Cortex A72 ARM CPU, running at 1.5 GHz. Both devices run Windows 10 natively.

The steady-state calculation program is developed using the C# programming language, in the Universal Windows Platform (UWP). Furthermore, there are three versions of the same program: one which targets x86-64, one which targets 32bit emulation on Raspberry Pi 4 and the other one which targets native ARM64 on Raspberry Pi 4.

The concept of running Windows 10 apps on a Raspberry Pi 4 is extremely new, and it is based on the "Windows on Raspberry Project" [9], in which open-source drivers are developed in order to facilitate the interface between the hardware layer and the operating system layer. With this approach, it is possible to run apps in two manners: deploying native apps that are compiled for 64bit ARM or using an emulation sandbox, that accompanies the operating system and enables compilation and deployment of x86-32 apps.

The steady-state software is designed to offer flexibility by allowing the user to change the global error margin as well as the number of consecutive runs. Additionally, the program is configured to calculate the steady-state values using an error margin of 0.00003 s (AC zone error, DC zone error and global error). This value is chosen in order to increase computation effort and to test the limits of the computing equipment. Furthermore, to evaluate performance on the two different devices while also considering equipment-specific thermal limits, 10 consecutive runs are implemented.

Description of the Analyzed Hybrid Network

A complex HVAC/HVDC network is brought forward in order to test the performance of an ARM device. It consists of two onshore AC systems (system *A*, with nodes *A0* and *A1*, and system *B*, with nodes *B0*, *B1*, *B2* and *B3*), 4 offshore AC systems (system *C*, with *C1* and *C2*, system *D*, with *D1*, system *E*, with *E1* and system *F*, with *F1)*, 2 DC nodes which have no connection to the AC networks (*B4* and *B5*) and 3 DC systems (*DCS1*, with *A1* and *C1*, *DCS2*, with *B2*, *B3*, *B5*, *F1* and *E1*, and *DCS3*, with *A1*, *C2*, *D1*, *E1*, *B1*, *B4* and *B2*).

The entire hybrid network is shown in Figure 2, and it is based on a benchmark network [11] by *Cigre Working Group B4.57*, in which an additional line between nodes *Ba-A1* and *Ba-B1* has been added.

In the DC zone, there are symmetric monopole busses (identified with notation *Bm*) and bipole busses (*Bb*). There are also 2 DC-DC converters (*Cd-B1* and *Cd-E1*). The *DCS1* and *DCS2* systems have a voltage rating of ± 200 kV, while *DCS3* has a rated voltage of ± 400 kV. The AC onshore zones operate at 380 kV and the offshore zones operate at 145 kV.

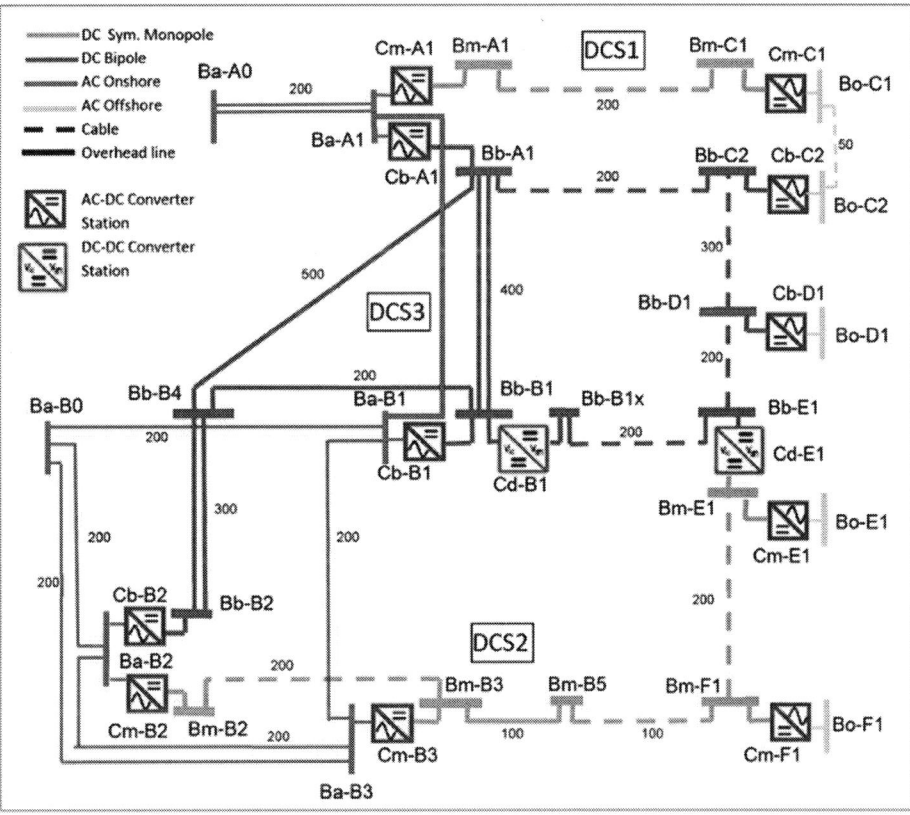

Fig. 2. Hybrid HVAC/HVDC network, adapted from [11]

Table I shows HVAC-HVDC network parameters: generation and consumption data, line parameters and converter data (with operational setpoints).

Table I. HVAC-HVDC network parameters [11]

AC bus data											
Bus	A0	A1	B0	B1	B2	B3	C1	C2	D1	E1	F1
Bus type	Slack	PQ	Slack	PQ	PQ	PQ	PQ	PQ	PQ	PQ	PQ
Generation [MW]	-	2000	-	1000	1000	1000	500	500	1000	0	500
Load [MW]	-	1000	-	2200	2300	1900	0	0	0	100	0

AC-DC converter data								
AC-DC conv. station	Power rating [MVA]	Converter arm reactance [mH]	Transf. leakage reactance [mH]	Transf. resistance [Ω]	Transf. primary voltage [kV]	Transf. secondary voltage [kV]	Operation mode setpoints	
A1	800	29	35	0.363	380	220	Q = 0	V_{DC} = 1 pu
C1	800	58	69	0.726	145	220	V/f control	
B2	800	29	35	0.363	380	220	Q = 0	V_{DC} = 0.99 pu
B3	1200	19	23	0.242	380	220	V_{AC} = 1 pu	P = 800 MW
E1	200	116	58	29	29	19	V/f control	
F1	800	29	35	0.363	145	220	V/f control	
A1	2*1200	19	23	0.242	380	220	V_{AC} = 1 pu	V_{DC} = 1.01 pu
B1	2*1200	19	23	0.242	380	220	V_{AC} = 1 pu	P = 1900 MW
B2	2*1200	29	35	0.363	380	220	V_{AC} = 1 pu	P = 1700 MW
C2	2*400	58	69	0.726	145	220	V_{AC} = 1 pu	P = −600 MW
D1	2*800	29	35	0.363	145	220	V/f control	

AC and DC line data					
Line data	r_0 [Ω/km]	l [mH/km]	c [μF/km]	g_0 [μS/km]	Max. current [A]
DC OHL ± 400 kV	0.0114	0.9356	0.0123	-	3500
DC OHL ± 200 kV	0.0133	0.8273	0.0139	-	3000
DC cable ± 400 kV	0.011	2.615	0.1908	0.048	2265
DC cable ± 200 kV	0.011	2.615	0.2185	0.055	1962
AC cable 145 kV	0.0843	0.2526	0.1837	0.041	715
AC OHL 380 kV	0.02	0.8532	0.0135	-	3555

Analysis of the Steady-State Results

Results from the steady-state calculation are visually presented in Figure 3, where the active power, reactive power, DC power, AC and DC voltages as well as AC voltage angle are brought forward.

In the DC region, the lines with the highest loading are *B3-F1* (691.9 MW), *D1-E1* (911.31 MW) and *B4-A1* (983.4 MW). Furthermore, all DC node voltages retain a voltage value that is close to 1 p.u., within the range of 0.983 p.u. and 1.015 p.u. Power-flow in the DC network is effectively regulated with the help of DC/DC converters, that ensure no lines are overloaded.

In the AC region, because most of the converters operate in constant voltage control mode, the AC voltages remain close to 1 p.u. Furthermore, because there is a high consumption in the north-western zone of the AC onshore grid, most of the power required in *A1* node is brought through AC lines from the slack bus, which leads to a voltage angle in node *A1* of -4.3° (which is the highest value considering the entire AC onshore grid. However, the highest loading can be observed in the south-western AC network (approx. 3400 MW), and it is covered by slack bus *Ba-Bo*, and with contribution from north-western AC zone and the off-shore AC zone (through the DC network).

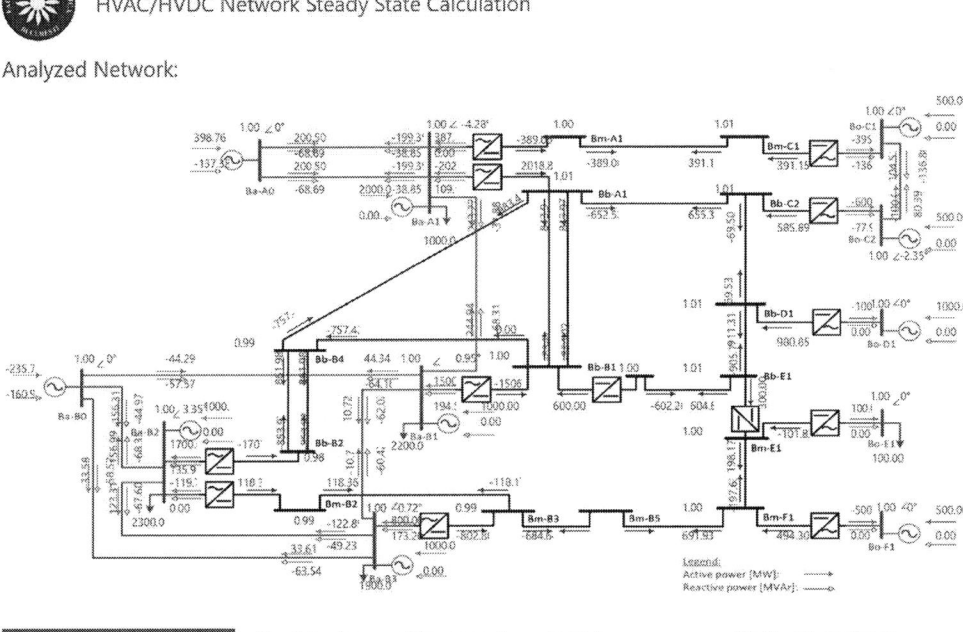

Fig. 3. Software interface with steady-state results

Evaluation of ARM Device Performance

The steady-state calculation, performed on both types of devices shows that for the specified error, there are a total of 2 global iterations. This demonstrates the effectiveness of the Newton-Raphson method, even applied for a hybrid HVAC/HVDC network. The number of iterations which are necessary for the AC zone and for the DC zone are summarized in Figure 4.*a*. One global iteration consists of several iterations for the AC zone and a number of iterations for the DC zone. In the second global iteration, the last values computed for the DC zone (with AC interface) are used as initialization parameters for the AC zone.

Next, an analysis is performed regarding the time necessary to perform 10 consecutive runs (it is done on both devices). The number of milliseconds for each run is shown in Figure 4.*b*. One simulation amounts to a complete run of the steady-state calculations, for both the AC and the DC zones (two global iterations). Figure 4.*b* also shows the linear trendlines for all scenarios, without including the first simulation run to eliminate possible memory allocation delays.

It can be observed that the best performing platform is the x86-64 PC, followed by the native ARM64 Raspberry Pi 4. The difference between the two solutions amounts to approximately 22 ms. Furthermore, the least performing platform is the 32bit emulation on the Raspberry Pi 4, which computes the same calculations in additional 55 ms compared to x86-64.

Another observation relates to the fact that there is a higher duration for the first run, for the native and emulation approach on Raspberry Pi 4. More specifically, the first simulation shows doubled duration for the ARM64 solution, and a 70% increase in the time necessary to run the first run on the 32bit emulation.

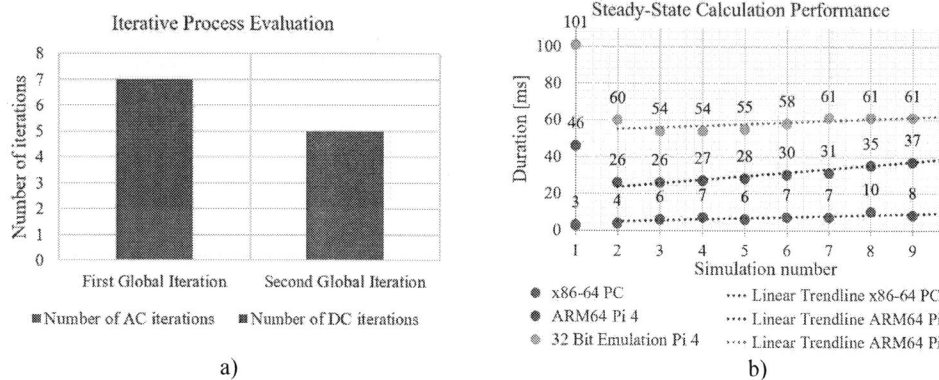

a) b)

Fig. 4. a) Overview of the iterative process; b) Simulation time for three use-case scenarios

Additionally, as the number of simulations increase, so does duration. This is an expected behavior, because after each run, processor temperature increases. Therefore, in order to remain within thermal limits, the computing unit decreases frequency. This is visible for all solutions.

The case-study analysis shows that there are differences regarding simulation time, between the three approaches. Furthermore, the most efficient manner in which steady-state simulations should be performed on an ARM device is to compile a program for the native environment (which is ARM64). However, time benefits for the classical x86-64 over ARM64 are less than 43 ms, which can be considered negligible. Therefore, taking into consideration that an ARM device, such as the Raspberry Pi 4, costs less than 60 USD and is able to compute complex steady-state simulations in the order of milliseconds, it is clearly a feasible option.

Even though for longer-term computations thermal limits are stricter in this form-factor (PCs have extensive active and passive cooling capabilities, while the Raspberry Pi 4 only relies on natural convection), good steady-state computation performance on a Raspberry Pi 4 shows that there is potential regarding the possibility to deploy dynamic simulations on this device,

Another relevant aspect concerns the fact that Raspberry Pi 4 provides 40 general-purpose input/output (GPIO) pins, which can be used to communicate with other equipment. As such, by combining the steady-state capabilities with the possibility to interact with other devices in a supervisory control and data acquisition (SCADA) system, it is possible to envision scenarios in which an OS based Raspberry Pi 4 can be converted into a network state estimator.

Finally, even though the proposed solution from the case study was applied on hybrid HVAC/HVDC transmission networks, it can also be extended to hybrid distribution networks.

Conclusion

The scope of this paper was to evaluate the feasibility regarding the usage of an OS based ARM device for steady-state calculations of complex HVAC/HVDC networks, considering the expansion of such devices beyond traditional IoT applications.

The first part of this article consisted of a review of methods which enable steady-state calculations for hybrid networks, consisting of multiple HVAC and HVDC zones. It was concluded that one flexible approach is the usage of a sequential algorithm, which alternates between the AC zones and the DC zones and relies on the iterative Newton-Raphson method.

In the second part, focus was directed towards the actual analysis of the computing performance of an OS based Raspberry Pi 4 for steady-state calculations. This original concept was steadily analyzed by implementing a software that was built on the C# programming language, in the Universal Windows Platform. It was demonstrated that, even though classical approaches are faster compared to the ARM solution, the differences are negligible (in the order of tens of milliseconds).

Summing up, the study performed in this article validated the feasibility of using an OS based ARM device for steady-state calculation using the sequential method and Newton-Raphson.

References

[1]   Beerten J., Cole S. and Belmans R.: Generalized Steady-State VSC MTDC Model for Sequential AC/DC Power Flow Algorithms, IEEE Transactions on Power Systems, vol. 27, no 2, May 2012

[2]   Damian I. C.: Supply of Large Cities Using Modular Multilevel High Voltage Direct Current Converters. PhD Thesis, University Politehnica of Bucharest, 2020

[3]   Eremia M., Liu C. C. and Edris A. A. (Eds.): Advanced Solutions in Power Systems - HVDC, FACTS and Artificial Intelligence, Hoboken, New Jersey: IEEE and Wiley Publishing Press, 2016

[4]   Ghael H. D., Solanki L., Sahu G.: A Review Paper on Raspberry Pi and its Applications, International Journal of Advances in Engineering and Management, Vol. 2, Issue 12, pp. 225-227, 2020

[5]   Eremia M. (Ed.): Electric Power Systems: Electric Networks, Bucharest, Romanian Academy Publishing, 2006

[6]   Chaudhuri N. and Chaudhuri B.: Multi-terminal Direct-Current Grids, Wiley & Sons, 2014

[7]   Jovcic D.: High Voltage Direct Current Transmission - Converters, Systems and DC Grids, Second Edition, Wiley & Sons, 2019

[8]   Hertem D. V., Gomis-Bellmunt O. and Liang J.: HVDC Grids for Offshore and Supergrid of the Future, Hoboken, New Jersey: John Wiley & Sons, 2016

[9]   Kim C. K., Moon S. I., Hur K., Kim J. M. and Jang G.: HVDC Transmission - VSC HVDC Based MMC Topology in Power Systems, World Scientific, 2021

[10]   Windows on Raspberry Project. [Online]. Available: https://www.worproject.ml/ (accessed on December 2020).

[11]   Wachal R., Jindal A., Dennetiere S., Saad H., Rui O., Cole S., Barnes M., Zhang L., Song Z., Jardini J., Garcia J. C., Mosallat F., Suriyaarachich H., Le-Huy P., Totterdell A., Zeni L., Kodsi S., Deepak T., Thepparat P., Beddard T., Velasquez J., D'Arco S., Morales A., Kono Y., Vrana T. K. and Yanh Y.: Guide for the Development of Models for HVDC Converters in a HVDC Grid, Working Group B4.57, Cigre, December 2014

Experimental Comparison of FPGA-Implemented Model Predictive Voltage Control to Cascaded Proportional Resonant Control for a Three-Phase Four-Wire Three-Level Grid-Forming Inverter of 250 kVA

Jarren Lange, Dominik Schmies, Karl Stephan Stille, Joachim Böcker, Oliver Wallscheid
Paderborn University, Department of Power Electronics and Electrical Drives
Warburger Str. 100
33098 Paderborn, Germany
Phone: +49 (5251) 60-2189
Email: {lange, schmies, stille, boecker, wallscheid}@lea.upb.de
URL: http://lea.upb.de

Keywords

≪Controller benchmark≫, ≪Converter control≫, ≪MPC (Model-based Predictive Control)≫, ≪Multi-level inverters≫,≪Proportional Resonant Control≫, ≪Voltage control≫.

Abstract

Modern microgrid systems require inverters capable of forming an acceptable grid voltage, during islanded operation, both supplying and absorbing power, through unbalanced load conditions and transients. In order to address this concern, the steady state and transient performance of two control methods are compared which are both capable to supply unbalanced three-phase loads. The first method is a cascaded proportional resonant (PR) control, which is a state-of-the-art controller for single-phase-capable voltage-source inverters but suffers from lower performance in transient conditions. The second method is a new field-programmable gate array (FPGA)-implemented finite control set model-based predictive control (FCS-MPC) which shows high performance for transients as well as black-start capability. Experimental results show the model-based predictive controller (MPC) to have better steady-state performance, with an average total-harmonic-distortion (THD) over the entire steady-state operation range of 1.4 % vs. 2.2 % for the PR controller, and a voltage regulation error of 0.71 % for the MPC vs. 1.1 % for the PR controller over the entire operating range. The MPC also shows advantages over the PR controller during transient response conditions, enabling settling times within 600 µs vs. 100 ms of the PR controller.

1 Background

In the expansion of renewable electrical power systems with inverter-based generation systems, the ratio of inertia-based generation is decreasing. This places greater responsibility on the performance of both grid-supporting and grid-forming inverter control systems, especially in islanded conditions. Under islanded conditions, inverters are required to respond faster to maintain acceptable voltage conditions despite the presence of unbalanced loads or transient events.

As the technology of power electronic devices advances, the switching frequencies of inverters increases. This results in shorter control cycles for grid-connected inverters and, therefore, allowing better controller performance in grid-related applications. However, devices of power rating in the 100 kW range and above are still operating at switching frequencies of about 1 kHz or even at some 100 Hz. Hence, the control for these devices is limited in bandwidth [1]. This presents challenges for the modern power system, where increasing controller bandwidth is important to avoid controller-controller interactions. As computational performance increases, it does not necessarily benefit these larger power devices.

At the moment, a state-of-the-art industrial controller for the control of voltage sourcing inverters is the PR controller. This single-phase-capable controller allows for the control of voltage-source inverters in the presence of imbalances and other non-ideal conditions. However, the nature of PR control, to operate on the average-voltage model, means that the transient performance of control is limited. In practical applications the control can only act upon the analog-to-digital converter (ADC) sample representing the average value (say in the middle of sampling interval, which may be inconsistent with multi-level converters) and must calculate the appropriate action before the next pulse width modulation (PWM) update.

The MPC, presented in [2], however, is not dependent on any controller states and hence acts only upon the most recent sample of measured values and produces a reaction immediately. This places a greater emphasis on the processing requirements, described in [3], as a valid control decision should be computed as quickly as possible. The implementation of the finite control set model-based predictive control (FCS-MPC), however, lends itself to the parallel operation of modern FPGA processors.

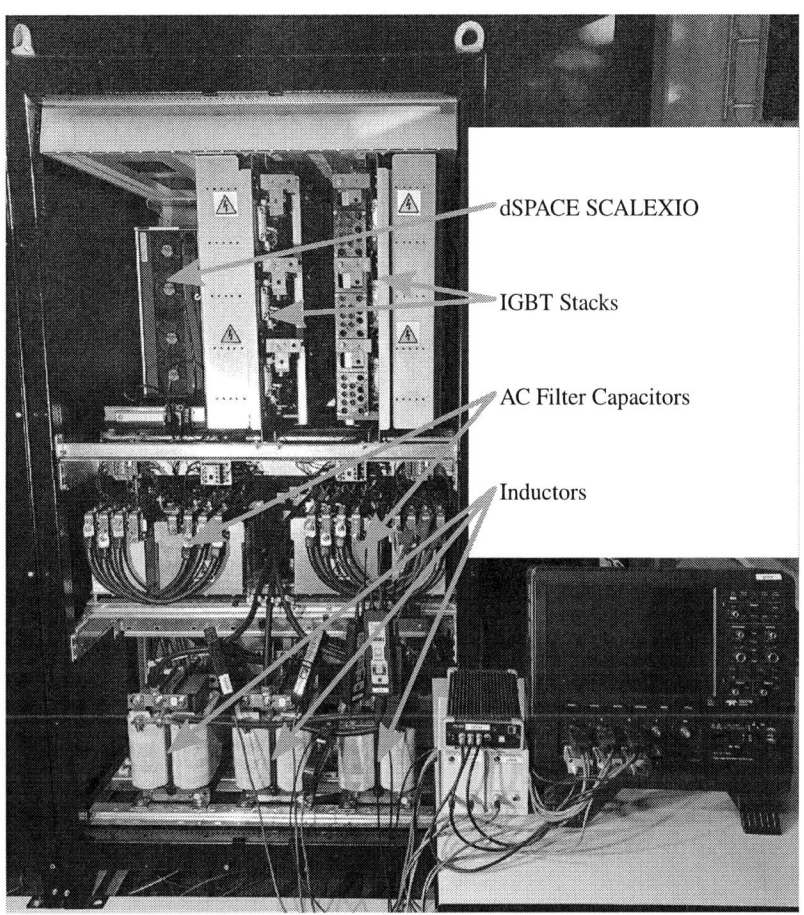

Fig. 1: Test hardware: two back-to-back 250 kVA inverters

Max. power	250 kVA per device
Topology	3-level NPC
Filter inductance	70 µH
Filter capacitance	250 µF
Control hardware	dSPACE DS6001, DS6601

Table I: Technical specifications of test hardware

This investigation focuses on the control of a single custom 250 kVA voltage-source inverter shown in Fig. 1. It is controlled via a combination of FPGA and CPU dSPACE SCALEXIO rapid control prototyping system. One rack system consists of two back-to-back 250 kVA 3-level neutral-point-clamped (NPC) insulated gate bipolar transistor (IGBT) converters, where one is operated as the voltage forming inverter under test and one is operated as an AC-load. Each converter is equipped with a configurable filter, where the IGBT network output of the grid forming source connects to a LC filter and the load side to a L filter. An overview of the experimental setup is shown in Fig. 2.

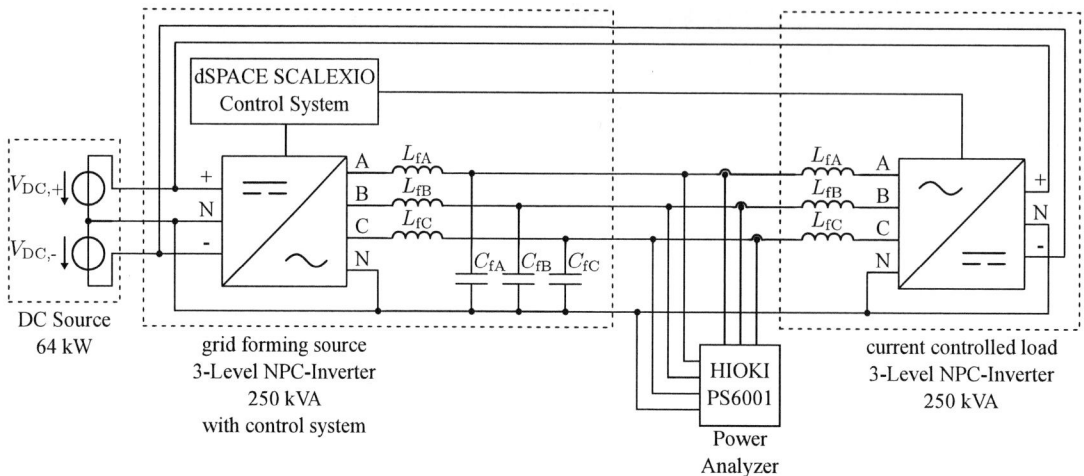

Fig. 2: Experimental setup used for the performance evaluation

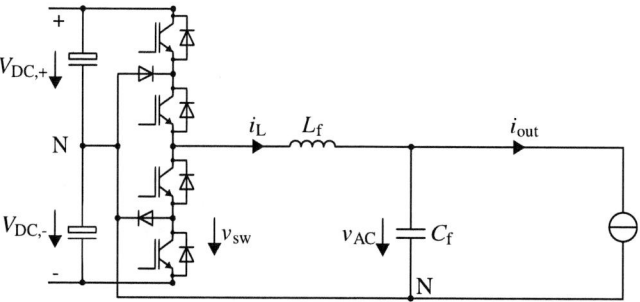

Fig. 3: Single-phase circuit diagram, indicating the used measurements and states

2 Plant model and control

The control objective for both considered controllers is the correct regulation of the symmetrical three phase, four wire (three phases and neutral) voltages, with minimal distortion, under a set of steady-state and transient conditions. Further constraints on the control is the limitation of the IGBT currents to protect the system, and avoid exceeding any protection limits. Since the LC filter is connected via the external neutral line, the analysis is performed in the per-phase representation. The LC network is shown in Fig. 3, where i_L is the inductor current and v_{AC} the capacitor voltage of the output filter. v_{SW} describes the output voltage of the IGBT network and i_{out} is the load current. The dynamics of the LC network are described as follows:

$$
\begin{aligned}
\frac{\mathrm{d}}{\mathrm{d}t}\begin{pmatrix} i_L \\ v_{AC} \end{pmatrix} &= \begin{pmatrix} 0 & -\frac{1}{L_f} \\ \frac{1}{C_f} & 0 \end{pmatrix}\begin{pmatrix} i_L \\ v_{AC} \end{pmatrix} + \begin{pmatrix} \frac{1}{L_f} \\ 0 \end{pmatrix}(v_{SW}) + \begin{pmatrix} 0 \\ -\frac{1}{C_f} \end{pmatrix}(i_{out}), \\
\frac{\mathrm{d}}{\mathrm{d}t}x &= Ax + Bu + Ew, \\
y &= I_2 x = Cx,
\end{aligned}
\tag{1}
$$

where I_2 is a 2x2 identity matrix.

2.1 Proportional resonant control

In this approach, a cascaded control modelling (Fig. 4) was implemented, with a voltage controller G_{PR}^v and a subordinated current controller G_{PR}^i. In comparison of control strategies for grid-connected converters, linear proportional-integral (PI) controllers in the stationary frame have the disadvantage of a remaining steady-state error. As the control has to cope not only with symmetrical three-phase loads, the common control in a rotating dq-frame is also not suitable. Instead, three single-phase controls are implemented, each with a resonant controller action in the voltage controller as well as the current controller with the base frequency ω_0 in order to be able to track a sinusoidal reference without a steady-state error [4].

	Current controller G_{PR}^i	Voltage controller G_{PR}^v
K	$0.001\ \frac{1}{A}$	$0.452\ \frac{A}{V}$
K_{R0}	500000	500000
ω_0	$314.159\ \frac{rad}{s}$	$314.159\ \frac{rad}{s}$
d_0	$5\cdot 10^{-6}$	$1\cdot 10^{-6}$
K_{R1}	10000	15000
ω_1	$942.478\ \frac{rad}{s}$	$942.478\ \frac{rad}{s}$
d_1	$10\cdot 10^{-6}$	$2\cdot 10^{-6}$
ω_I	$62.832\ \frac{rad}{s}$	0

Table II: Proportional resonant controller parameters

The measurements of the average inductor currents of the inverter by Regular Sampling approach delivered unconvincing results. The necessary interlock between the semiconductors combined with rather complex commutation processes of a 3-level neutral-point-clamped topology and a relatively small inductor led to high measurement errors. Instead, the currents were sampled 500 times per switching period and from the results, the average value of the previous switching period was calculated. That caused an additional delay in the current control loop which had to be compensated by a discrete-time prediction model based on (1) using the matrix-exponential method, in order to avoid stability problems.

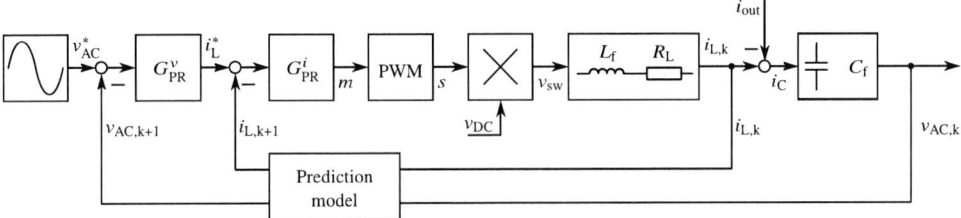

Fig. 4: Cascaded proportional resonant control modelling

It is shown in [5] that the PR controller in the stationary frame can be derived by transforming two separate PI-controllers from the positive and negative sequence frames. The infinite gain at the base frequency ω_0 can lead to stability problems in the practical implementation, because small control errors could cause saturation of the controller output. Therefore it is helpful to use a non-ideal integrator with the damping factor d. In a four-wire-system, 3rd harmonics can easily show up, therefore a second resonant term with three times the base frequency was added. Furthermore, to achieve better accuracy for controlling DC-offsets, an integral term was added to the controller, which is only active in the current controller. The plant transfer function of the current control loop does not show integrating behaviour over the complete frequency range. The corner frequency resulting from the gain factor K and the integral part, functionally similar to a PI-controller, was set to the corner frequency of the inductor with its series resistance to cancel out the pole of the low-pass and achieve setpoint tracking for low frequencies. The plant transfer function of the voltage control loop shows integrating behaviour for the full frequency range, hence an additional integrator is not needed. The resulting transfer function for the proposed proportional resonant controller is

$$G_{PR}(s) = K\left(1 + \frac{2K_{R1}d_0\omega_0 s}{s^2 + 2d_0\omega_0 s + \omega_0^2} + \frac{2K_{R2}d_1\omega_1 s}{s^2 + 2d_1\omega_1 s + \omega_1^2} + \frac{\omega_I}{s}\right) \tag{2}$$

under the condition $d_i \ll 1$ and combined with a gain factor K_{Ri}.

The discrete-time implementation of (2) similar to the described solution in [6] is

$$G_{PR}(z) = K\left(1 + \frac{b_{2,0}z^2 + b_{1,0}z + b_{0,0}}{a_{2,0}z^2 + a_{1,0}z + a_{0,0}} + \frac{b_{2,1}z^2 + b_{1,1}z + b_{0,1}}{a_{2,1}z^2 + a_{1,1}z + a_{0,1}} + \frac{\omega_I T_s z}{z - 1}\right) \tag{3}$$

with the coefficients of Tab. III.

	First resonance term	Second resonance term
	$\omega_0 = 2\pi \cdot 50\,\frac{\text{rad}}{\text{s}}$	$\omega_1 = 2\pi \cdot 150\,\frac{\text{rad}}{\text{s}}$
a_0	$T_s^2\omega_0^2 - 4\omega_0 d_0 T_s + 4$	$T_s^2\omega_1^2 - 4\omega_1 d_1 T_s + 4$
a_1	$2T_s^2\omega_0^2 - 8$	$2T_s^2\omega_1^2 - 8$
a_2	$T_s^2\omega_0^2 + 4\omega_0 d_0 T_s + 4$	$T_s^2\omega_1^2 + 4\omega_1 d_1 T_s + 4$
b_0	$-4K_{R0}T_s\omega_0 d_0$	$-4K_{R1}T_s\omega_1 d_1$
b_1	0	0
b_2	$-b_0$	$-b_0$

Table III: Proportional resonant controller coefficients

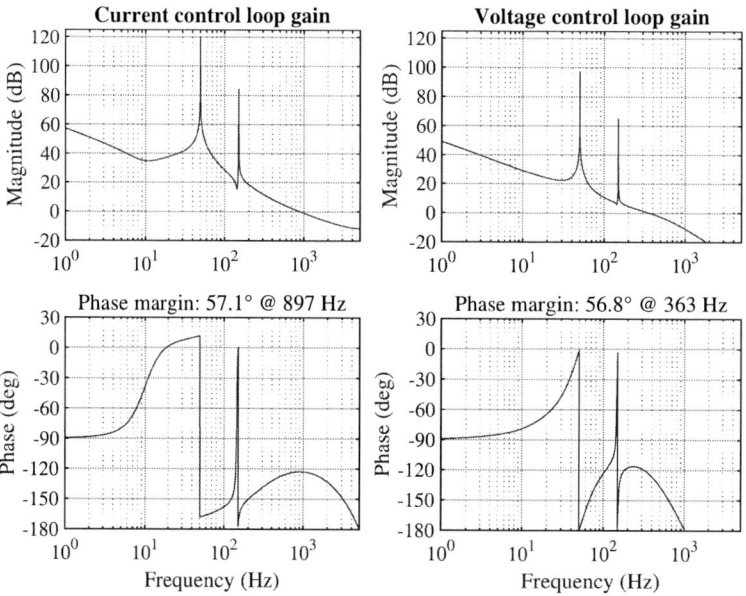

Fig. 5: Open-loop gain of the current- and voltage control loops for the PR approach

The Bode plots of the open-loop transfer functions for the discrete-time models are shown in Fig. 5 with the parameters listed in Tab. II. The control design objective is a gain for the first resonance term as high as possible with a phase margin of at least 50° for both voltage- and current-controllers. The specific controller parameters from Tab. II have been optimized by manual tuning.

2.2 MPC approach

The MPC used is an adaptation of the MPC introduced in [2], illustrated in Fig. 6. Here, the implemented MPC is geared towards the control of the capacitor voltage while limiting the IGBT current. The MPC uses the per-phase model of the LC filter to determine the predicted states. The model is discretized using the matrix-exponential method, and v_{sw} is evaluated for all 3 possible switch states $[V_{\text{DC}+}; 0; -V_{\text{DC}-}]$, providing 3 predictions per phase. The prediction is then repeated to obtain the x_{k+2} prediction, resulting in 9 predictions per phase. The error function is:

$$J(x_{k+1}) = \left(v_c^* - v_{c,k+1}\right)^2 + f_{\text{oc}}\left(i_{\text{l},k+1}, i_{\text{lim}}\right)$$

where,

$$f_{\text{oc}}\left(i_{\text{l},k+1}, i_{\text{lim}}\right) = \begin{cases} k_{\text{i,lim}}\left(i_{\text{l},k+1} - i_{\text{lim}}\right)^2 & \text{if } \left|i_{\text{l},k+1}\right| > i_{\text{lim}} \\ 0 & \text{otherwise.} \end{cases}$$

(4)

and i_{lim} is the maximum allowed instantaneous current value. The weighting term for the over-current term, $k_{\text{i,lim}}$ was arbitrarily set to a value of 10A^{-2}. The current limit is implemented as a soft-constraint, to enable continuous operation during over-current conditions, allowing the controller to choose an output state that minimises the

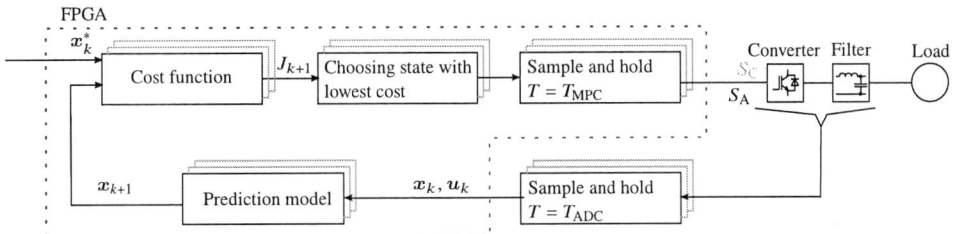

Fig. 6: Principle of FPGA implemented MPC operation

over-current condition.

The switch selection, which produces the lowest error function result, is then selected every T_{MPC}. This switch state is then applied until the next evaluation, T_{MPC} later. For an average effective switching frequency of approximately 10 kHz, T_{MPC} was set to 21 µs. This is significantly slower than the ADC sample time of $T_{\mathrm{ADC}} = 200$ ns.

The entirety of the control logic was implemented in the FPGA (dSPACE DS6601) interfaced with multi-I/O modules (dSPACE DS6651). The controller current limit, i_{lim}, was set to 600 A, to limit operation outside of the nominal operating region.

3 Experimental comparison results

The two controllers are compared for a limited number of steady-state and transient performance conditions. There are several similarities and differences in the operation principle, as well as the implementation of the two controller approaches shown in Tab. IV.

3.1 Metrics

The RMS and THD calculations are performed independantly of the control systems, by a HIOKI power analyzer. The RMS calculation is

$$V_{\mathrm{RMS}} = \sqrt{\frac{1}{M} \sum_{k=0}^{M-1} (v(k))^2} \qquad (5)$$

where M is the number of samples ($v(k)$) taken within a period of the voltage. Absolute voltage regulation error is calculated as

$$V_{\mathrm{RMS,Err}} = \left| \frac{V_{\mathrm{RMS}} - V_{\mathrm{nom}}}{V_{\mathrm{nom}}} \right| \qquad (6)$$

where V_{nom} is the nominal voltage setpoint. The THD is calculated

$$THD_{v_{\mathrm{AC}}} = \frac{\sqrt{\sum_{k=2}^{K} (u_k)^2}}{u_1} * 100\%. \qquad (7)$$

Here, K is the number of harmonics (u_k) used for the harmonic calculation. Harmonic orders are referenced from the fundamental component (u_1). For the measurements performed $K = 50$, with harmonic measurements complying to the IEC 61000-4-7:2002 standard.

3.2 Steady-state performance

One comparison of the two controller structures is their behaviour across a range of load conditions. To demonstrate this, the controllers were operated with constant set-point voltages (three-phase, 230 V RMS per phase) and loaded by a balanced current source across the operating range of the device (0 kVA to 250 kVA) for a full rotation of the phase angle. The results shown in Fig. 7 uses the data obtained from the HIOKI PW6001 power analyzer, connected as shown in Fig. 2.

Both controllers show acceptable steady-state performance, seen in Fig. 7 over the entire test region, with a maximum THD of 2.5% with a maximum voltage regulation error of 3%. This harmonic content performance

	PR control	FCS-MPC
Computing hardware	CPU	FPGA
Switching frequency	10 kHz (fixed)	Appr. 10 kHz (variable)
Control cycle time	100 µs	21 µs
Computation time	12 µs	320 ns
Sampling concept	Oversampling with averaging	Single measurement of most recent sample
Manipulated variable	Modulation index for PWM	Finite switching vector

Table IV: Comparison of the FCS-MPC and the PR control approach

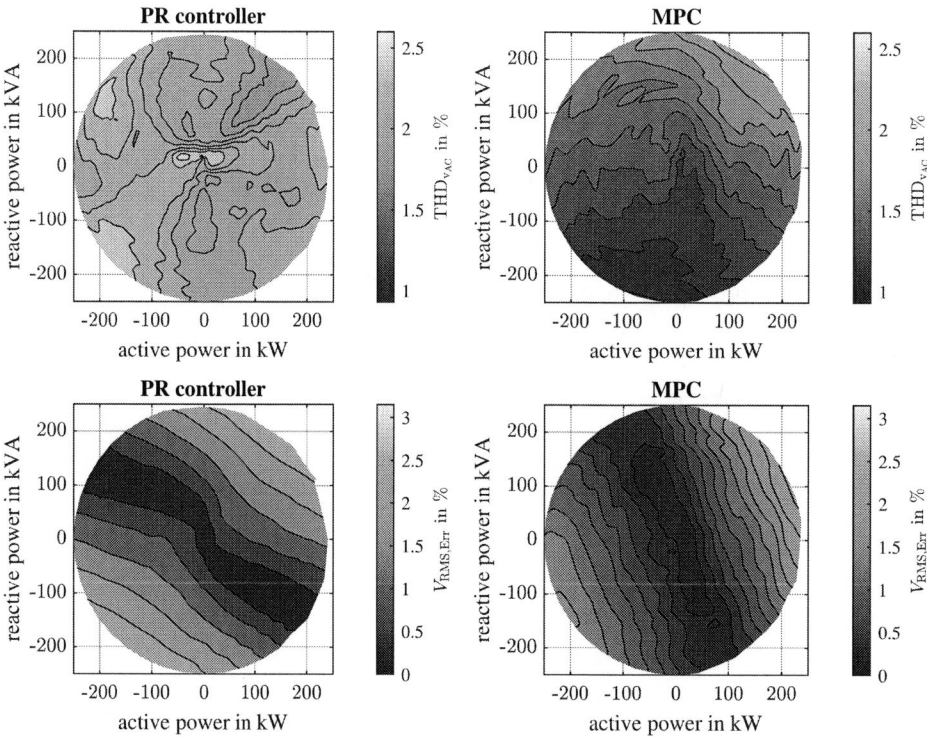

Fig. 7: Three-phase load voltage total harmonic distortion and absolute RMS voltage error

is well below the IEEE STD 519-2014 requirements of a maximum 5% THD. Voltage regulation is also for both controllers within the ±10% requirements for most voltage source applications (IEC 64020-3 for example). The voltage regulation of both controllers is heavily dependant on the model parameters used (both controllers used the ideal parameters of Tab. I). However, when the average performance over the entire range is averaged and compared in Tab. V, it is clear, by the metrics provided that the FCS-MPC performs better under steady-state conditions.

3.3 Dynamic response

In order to compare the dynamic response of both controllers their performance is demonstrated through two different high dynamic test conditions; a black start condition (the establishment of the three-phase voltages after all three-phase voltages have reached 0 V) and a load step. These represent some of the more demanding test cases for a voltage forming inverter in the field. The results shown are from the measurements using the sensors and processing units on the device.

	PR control	FCS-MPC
mean $(\text{THD}_{v_{AC}})$	2.2%	1.4%
mean $(V_{\text{RMS,Err}})$	1.1%	0.71%

Table V: Steady-state performance result comparison

3.3.1 Black-start

In order to reduce the initial currents into the filter capacitors, the voltage per phase reference to the controllers was only initiated from the internal voltage reference zero crossing point. From there on the voltage reference, for both controllers, was set to the nominal 230 V at 50 Hz. The results of the test, both voltage and currents are shown in Fig. 8.

The results show, as expected the MPC has stabilized the voltage to the setpoint within a few microseconds of the setpoint being non-zero. The PR controller, due to the resonant term of the controller and the delays of the cascaded controller design, requires approximately 80 ms before settling to the steady-state condition. It is noted that the more rapid control of the voltage does result in higher inductor current ripple for the MPC approach, resulting in higher filter capacitor currents (approx. 30 % higher compared to the PR approach).

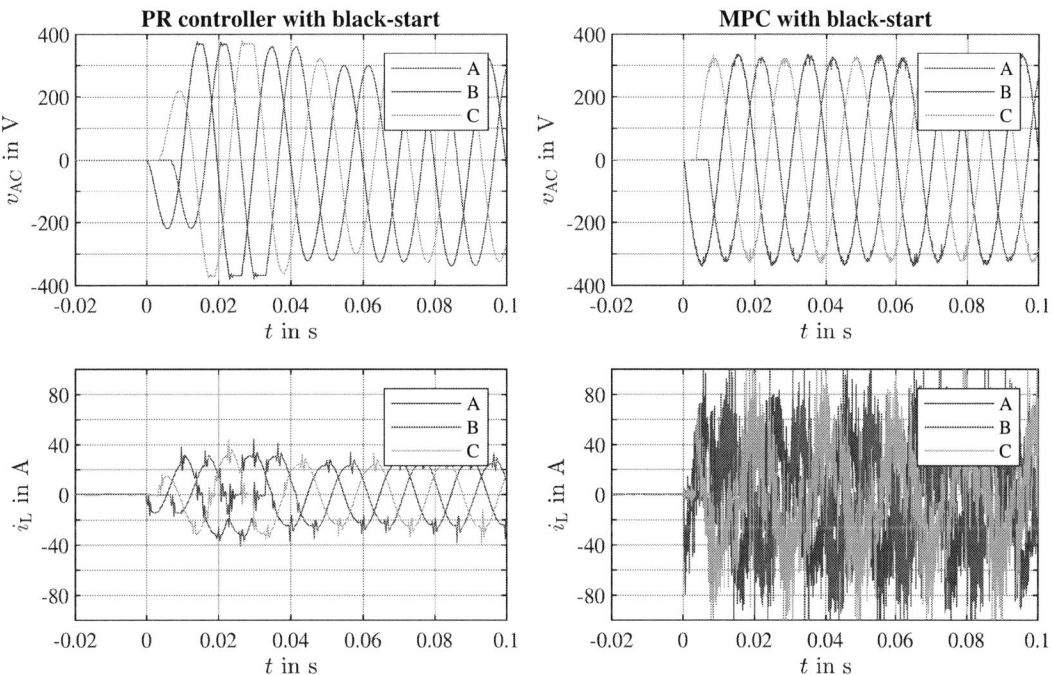

Fig. 8: Capacitor voltage v_{AC} and inductor currents i_L during black-start with no load connected

3.3.2 Load step

In order to demonstrate a load step, a load with a power factor of 1.0 was changed from a nominal load of 25 % (63 kW) to 75 % (188 kW). The load was provided with the phase angle information from the voltage controllers and approximates an ideal current source. The results of the load step test are shown in Fig. 9.

During the load step, the PR controller recovers from the load step within approximately 100 ms, including an overshoot of the output voltage, which was limited by the DC bus voltage of the converter during the experiment. In comparison the MPC shows a minor dip in the output voltage, lasting 600 μs. The duration of this dip was extended due to the inductor current limit (600 A). It is again noted that the MPC technique does result in higher inductor current ripple.

4 Conclusion and outlook

Experimental tests were conducted on an experimental three-phase four-wire three-level 250 kVA inverter to compare the behaviour of a voltage forming PR controller to a voltage forming FCS-MPC. The PR controller was designed and implemented with oversampling and averaging using fixed time windows of 100 μs fitting the PR controller cycle time to control a standard PWM modulation index. The FCS-MPC, however, utilized the FPGA to implement the algorithm which held a single switch state for a fixed period of time. The steady-state performance of the two controllers is similar with both controllers meeting standard performance metrics. However, the MPC provided a lower voltage THD and better voltage regulation across the entire tested operational range at the expense of higher capacitor currents. During transient tests, a black-start test and a load-step, the MPC reached

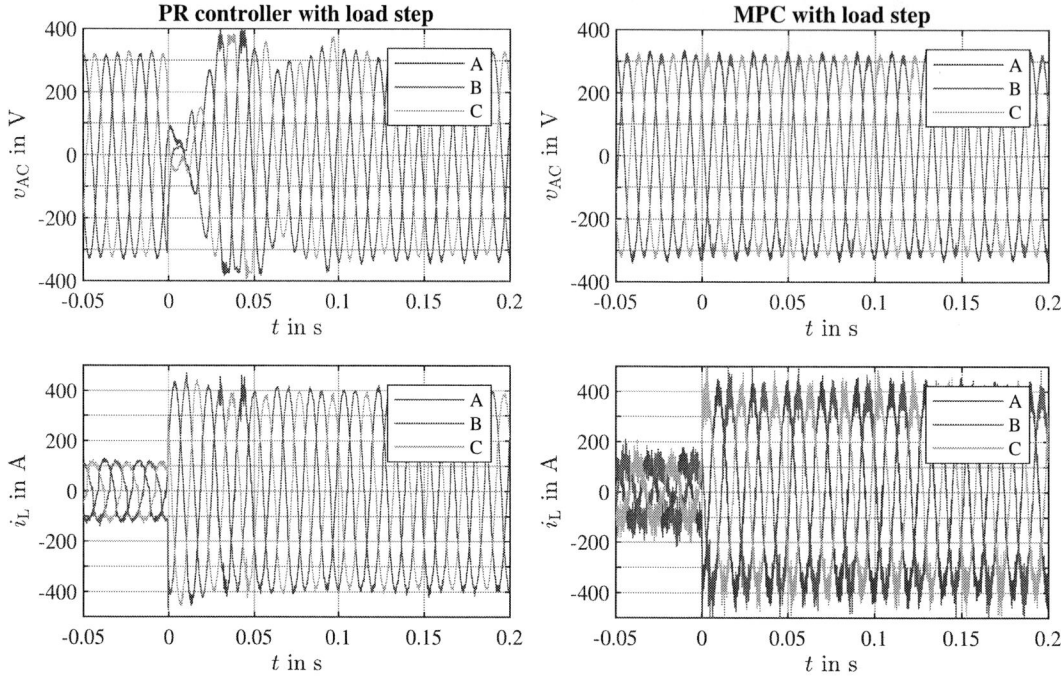

Fig. 9: Capacitor voltage v_{AC} and inductor currents i_L during load step of both controllers from 25 % to 75 % (of rated 360 A capacity, with a power factor of 1)

a stable steady-state condition within 600 µs compared to the 100 ms response time needed by the PR to stabilize. Further investigations into the MPC controller include removing the time gating logic (T_{MPC}) to utilize the available FPGA architecture with the intention of improving dynamic performance. Comparisons against other controller architectures are also required.

References

[1] J. Böcker, S. Beineke, and A. Bähr, "On the control bandwidth of servo drives," in *2009 13th European Conference on Power Electronics and Applications*, 2009.

[2] S. Kouro, P. Cortes, R. Vargas, U. Ammann, and J. Rodriguez, "Model predictive control, a simple and powerful method to control power converters," *IEEE Transactions on Industrial Electronics*, vol. 56, no. 6, pp. 1826–1838, 2009.

[3] J. Böcker, B. Freudenberg, A. The, and S. Dieckerhoff, "Experimental comparison of model predictive control and cascaded control of the modular multilevel converter," *IEEE Transactions on Power Electronics*, vol. 30, no. 1, pp. 422–430, 2015.

[4] B. A. Francis and W. M. Wonham, "The internal model principle of control theory," *Automatica*, vol. 12, no. 5, pp. 457–465, 1976.

[5] R. Teodorescu, F. Blaabjerg, M. Liserre, and P. C. Loh, "Proportional-resonant controllers and filters for grid-connected voltage-source converters," *IEE Proceedings on Electric Power Applications*, vol. 153, no. 5, 2006.

[6] T. D. C. Busarello, J. A. Pomilio, and M. G. Simoes, "Design procedure for a digital proportional-resonant current controller in a grid connected inverter," in *IEEE 4th Southern Power Electronics Conference (SPEC)*, 2018.

Experimental study of interleaved Y-Inverter performance

Yusuke Endo, Masataka Minami
KOBE CITY COLLEGE OF TECHNOLOGY
8-3, Gakuenhigashi-machi, Nishi, Kobe,
Hyogo 651-2194, Japan
Tel.: +81-78-795-3232
E-Mail: kcct-minami@g.kobe-kosen.ac.jp
URL: http://www.kobe-kosen.ac.jp/~minami/ja/index.html

Hamzeh J. Jaber, Alberto Castellazzi
KYOTO UNIVERSITY OF ADVANCED SCIENCE
18 Yamanouchi Gotanda-cho, Ukyo-ku,
Kyoto 615-8577, JAPAN
Tel.: +81-75-496-6504
E-Mail: hamzeh.j.jaber@kuas.ac.jp; alberto.castellazzi@kuas.ac.jp
URL: https://www.kuas.ac.jp/en/faculty-of-engineering/faculty-and-research/faculty

Acknowledgements

The authors gratefully acknowledge the support received from Grant 20H02138 of the Japanese Society for the Promotion of Science (JSPS) under the Kakenhi-Kiban-B scheme.

Keywords

«DC-AC converter», «Inverter design», «Interleaved inverters», «Wide Bandgap devices», « High frequency power converter», « High power density systems».

Abstract

The Y-Inverter is rapidly gaining popularity as an alternative to the "dc-boost plus inverter" configuration in motor drive applications. Recently, advantages of designing with multiple interleaved cells per phase, as opposed to simply paralleling devices, have been pointed out and solutions using both coupled and non-coupled inductors have been proposed. This paper presents a detailed experimental benchmark study of the inverter performance when using non-interleaved and interleaved cells; for the interleaved case, both non-coupled and coupled inductors are considered. The results clearly show that interleaving enables an important reduction of the current stress on the output and input filter capacitors, as well as the possibility to design the magnetic components with higher energy density and higher efficiency. Interleaving clearly appears as a superior alternative to the straightforward paralleling of semiconductor devices, when the current ratings are such as to require multiple transistors anyway. Specific differences in the circuit performance optimization potential exist between the case of coupled and non-coupled inductors. Only the interleaved non-coupled case is presented here in detail, with some comparison results. All cases will be dealt with in detail in the final paper.

Introduction

The Y-Inverter is a sinusoidally modulated inverter topology introduced fairly recently and which has been shown to feature a number of superior properties over conventional Voltage Source Inverters (VSI), particularly to the aim of fully exploiting the characteristics of modern wide-band-gap (WBG) semiconductor devices [1-3]. Fig. 1 illustrates the single-phase circuit schematic (a) and the modulation signals for the buck and boost cells (b). With this modulation, the single cell produces an output voltage which is an offset sinusoidal waveform,

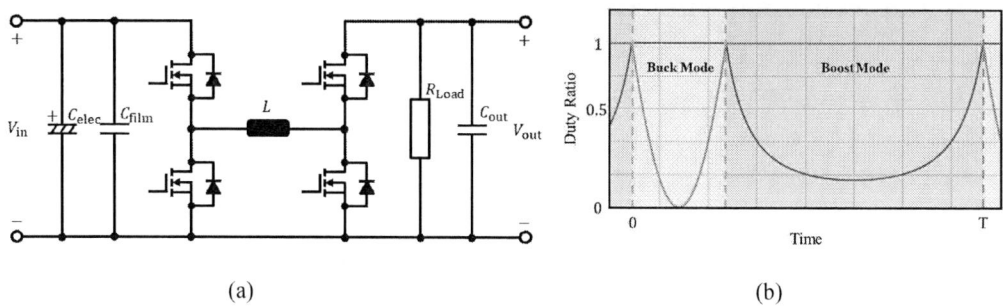

(a)   (b)

Fig. 1: Circuit schematic of a single cell of buck-boost Y-Inverter power cell (a); duty and inverse-duty cycles (i.e., high side transistor) for the buck and boost semi-cells, respectively (b).

$$V_{out} = \frac{V_{peak}}{2} \cdot [1 + \sin(\omega t)] \tag{1}$$

Fig.2 which shows the 3-phase system, where all cells share the common ground terminal connection (Y-configuration). In this configuration, the dc-offset gets cancelled and the phase-to-neutral voltage has amplitude equal to Vpeak/2. Fig.3 shows the phase-to neutral voltages and one phase inductor current for the 3-phase load powered by the Y-Inverter operated with continuous sinusoidal S-PWM [1]. Here, the circuit parameters and test conditions are: Vin=60 V; $Vpeak$=90 V; f_{LOAD}=50 Hz; fs=200 kHz.

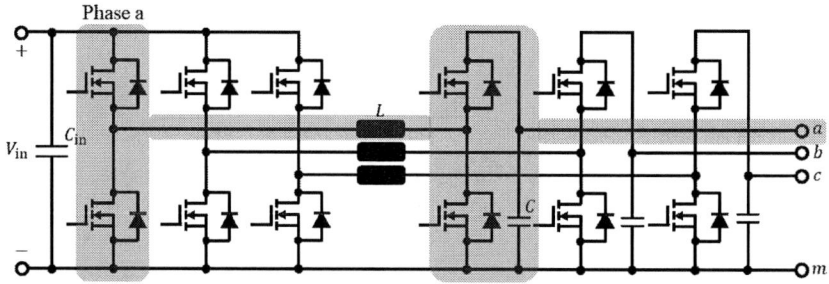

Fig. 2: Three-phase non-inverting buck-boost Y-Inverter.

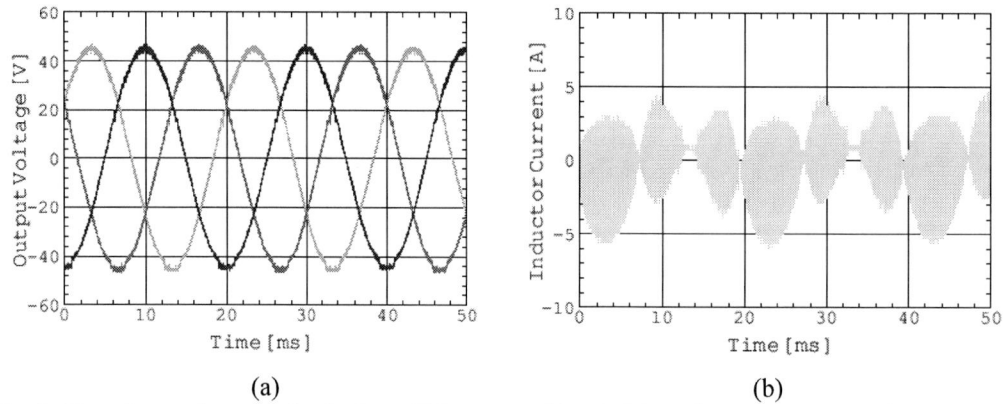

(a)   (b)

Fig. 3: Experimental results for 3-phase Y-Inverter with S-PWM: phase-to-neutral voltages 3(a); one phase inductor current 3(b). The low-level distortion (flat portion) in the voltage waveforms and corresponding zero current are due to an imperfection in the modulation signals during test.

Important innovative aspects of this topology over standard VSI solutions include:

- sinusoidal output voltage synthesis (as opposed to square wave modulated), which: a) eliminates dV/dt related noise and stress, allowing to fully exploit the switching speed of WBG transistors; b) reduces harmonics to essentially the load fundamental and the switching frequency components; c) reduces, on average, the device output-side switching losses, since transitions take place at sinusoidally decreasing drain-source voltage, enabling higher switching frequencies;
- no high-voltage DC-link capacitor: this is oftentimes one of the bulkiest and reliability critical components in the system and, in applications where electrolytic capacitor technology is not admitted (e.g., avionics), a rather costly one, too; indeed, the Y-Inverter requires good film technology capacitors on the output, due to relatively high current levels, but their value can be typically contained within few µF`s; removal of the large DC-link capacitor is an important asset towards higher integration levels.
- integrated voltage boost capability, allowing for the use of devices with different technology and different voltage ratings on the input (i.e., buck cell) and output side cells (i.e., boost cell). Also, the switching frequencies for the two cells can be chosen to be different, so as to optimize the converter design and performance [4].

On the other hand, one inherent challenge of the Y-Inverter is that the inductor current scales in some proportion to the output voltage boost ratio. This has motivated designs using interleaved cells with or without inductor coupling [4, 5], to reduce the magnetic flux density in the inductor and contain its losses and volume, for the same number of semiconductor switches, which is simply dictated by considerations about the current rating: that is, the interleaved configuration is proposed as an alternative to the straightforward paralleling of multiple semiconductor dies.

Experimental prototype test results

Specifically, here two cases are considered: 1) interleaving with 180 degrees phase-shift of the PWM carrier and no inductor coupling; 2) interleaving with 180 degrees phase-shift of the PWM carrier and cross-coupled inductors. The corresponding circuit schematic is detailed ahead for each case. The transistors are all normally-off GaN HEMTs with 650V rating. The value of the electrolytic capacitor on the input side does not correspond to any specific performance target, but rather to the capability of testing in the presence of relatively long cables to the power supply. For all test conditions and results, the following data apply: Vin=60 V; $Vpeak$=140V; f_{Load}=50 Hz and fs=200 kHz; R_{Load}=50 Ω. The test setup includes a Kikusui PWR400L DC power supply, Headspring HECS-B/A Controller, Fig. 4 (a), a Teledyne LeCroy WaveSurfer3024z 200 MHz 4 GS/s oscilloscope, Fig, 4 (b), and PP011 differential voltage probe, CP30 50 MHz and CP150 10 MHz current probes.

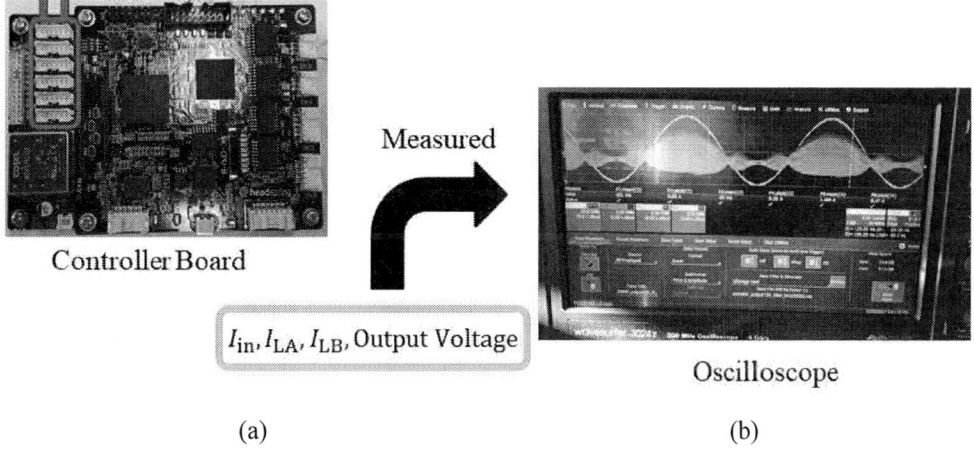

Controller Board

Measured

$I_{in}, I_{LA}, I_{LB},$ Output Voltage

Oscilloscope

(a) (b)

Fig. 4: (a) headspring controller; (b) oscilloscope with capture of experimental current and voltage waveforms.

Single-phase interleaved cells with non-coupled inductors

Fig. 5 shows the circuit schematic for the implemented prototype. With this configuration, interleaving yields a reduction of the current ripple at the output capacitor, with some benefit for the capacitor current stress and voltage ripple. Fig. 6 (a) shows the resulting output capacitor voltage, while Fig. 6 (b) a detail of the inductor currents and their ripple. The overall inductor current waveforms are shown in Fig. 7. Each inductor has a value of 120 μH (more details are included in the next Section). In this configuration, the measured cell efficiency was 97.3%.

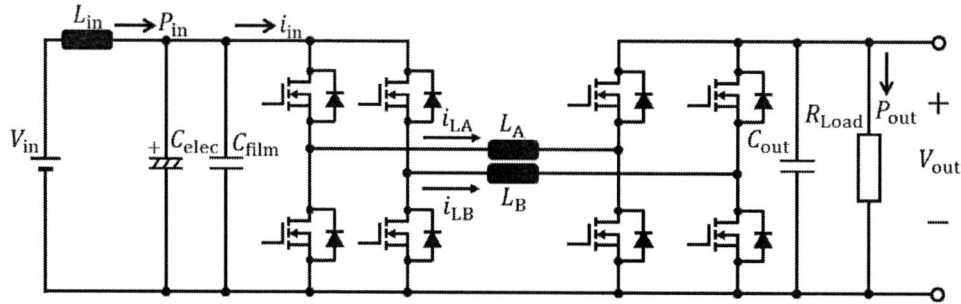

Fig. 5: Single cell of interleaved non-inverting buck-boost Y-Inverter with non-coupled inductors.

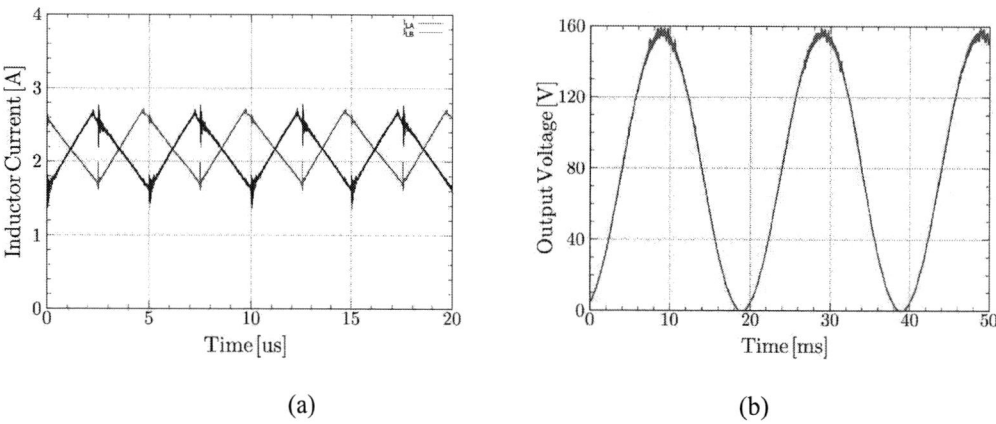

(a)   (b)

Fig. 6: (a) output voltage; (b) detail of inductor current ripple.

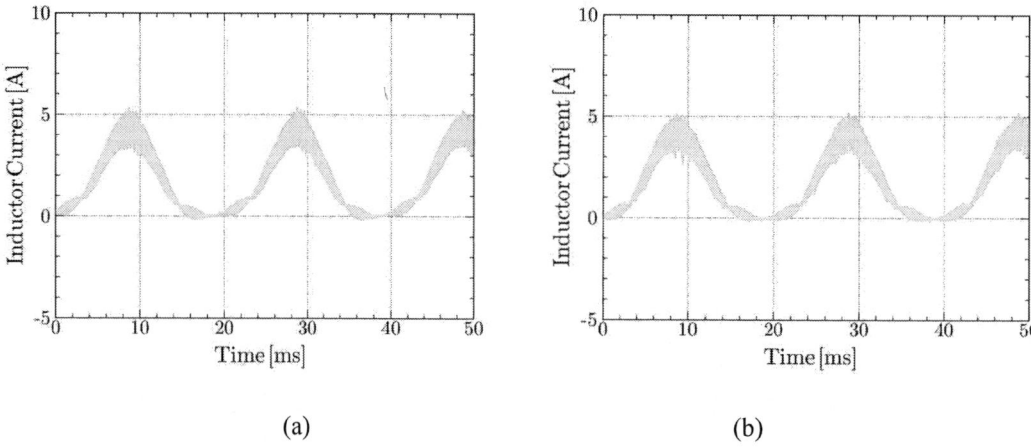

(a)   (b)

Fig. 7: (a) current in inductor L_A; (b) current in inductor L_B.

Single-phase interleaved cells with cross-coupled inductors

For this case, the corresponding topology is illustrated in Fig. 8. With this configuration, on top of the reduction of the current ripple at the output capacitor, an effective reduction of the magnetizing inductor current can be achieved, which allows a more compact inductor design.

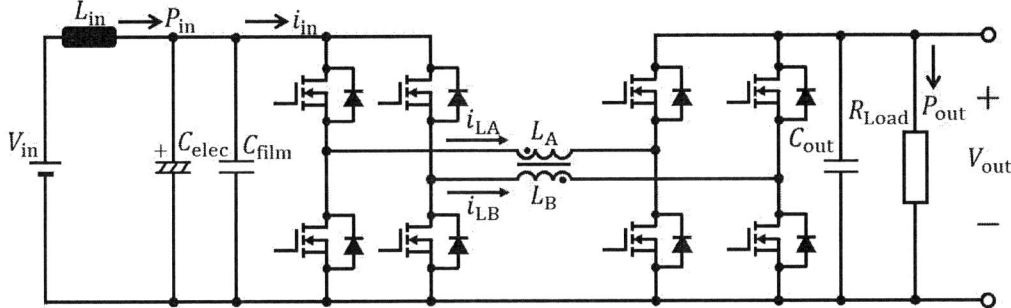

Fig. 8: Single cell of interleaved non-inverting buck-boost Y-Inverter with cross-coupled inductors.

The equivalent inductor model is shown in Fig. 9 and assumes identical values for L_1 and L_2; $M=kL$ is the mutual inductance and $L+M=(1+k)L$ the parasitic inductance of each strand. In our case, the actual inductor was fabricated on a E-shaped core using core material PC40EC70X69X16 manufactured by TDK-Lambda Corporation (Mn-Zn, $A_L = 4845 \pm 25\%$ nH/turns²), with a nominal gap of 0.2 mm (varied to trim k) and using high frequency Litz wire. For a target self inductance value of 120 µH, 12 turns were wound. It should be noted that this is also the inductor which was used for the tests with separate inductors, in which two separate cores were used, each with a single winding.

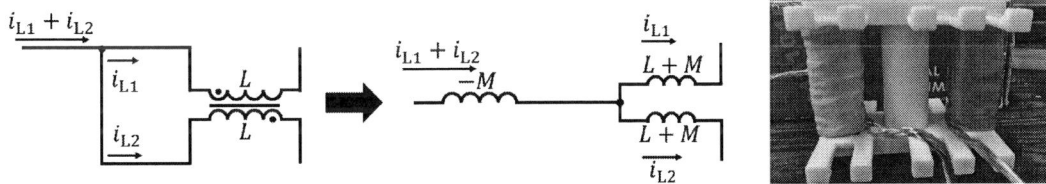

Fig. 9: (a) Cross-coupled inductor equivalent model and (b) inductor prototype.

However, here, it becomes very important to trim the value of the coupling coefficient k to achieve high efficiency. The inductor prototype is shown in Fig. 9 (b), while Fig. 10 shows extensive characterisation results. Specifically: Fig.10 (a) and (b) show the frequency characteristics of inductance and resistance on the secondary side, respectively, when the primary side is open; Fig. 10 (c) and (d) show the frequency characteristics of inductance and resistance on the primary side, respectively, when the secondary side is open. Fig. 10 (e) and (f) show the frequency characteristics of inductance and resistance on the secondary side, respectively, when the primary side is shorted; finally, Fig. 10 (g) and (h) show the frequency characteristics of inductance and resistance on the primary side, respectively, when the secondary side is shorted. At th switching frequency value, 200 kHz, the corresponding values can be summarized as in Table I. From these results, the value of k can be estimated at about 0.58, which yielded best results, with an efficiency of 97%.

Table II: Measured inductor parameters at 200 kHz.

Parameter	Inductance	Resistance
$L_{1\text{open}}$	120.2588 µH	494.048 mΩ
$L_{2\text{open}}$	120.6908 µH	491.577 mΩ
$L_{1\text{short}}$	99.67244 µH	353.06 mΩ
$L_{2\text{short}}$	100.3371 µH	359.287 mΩ

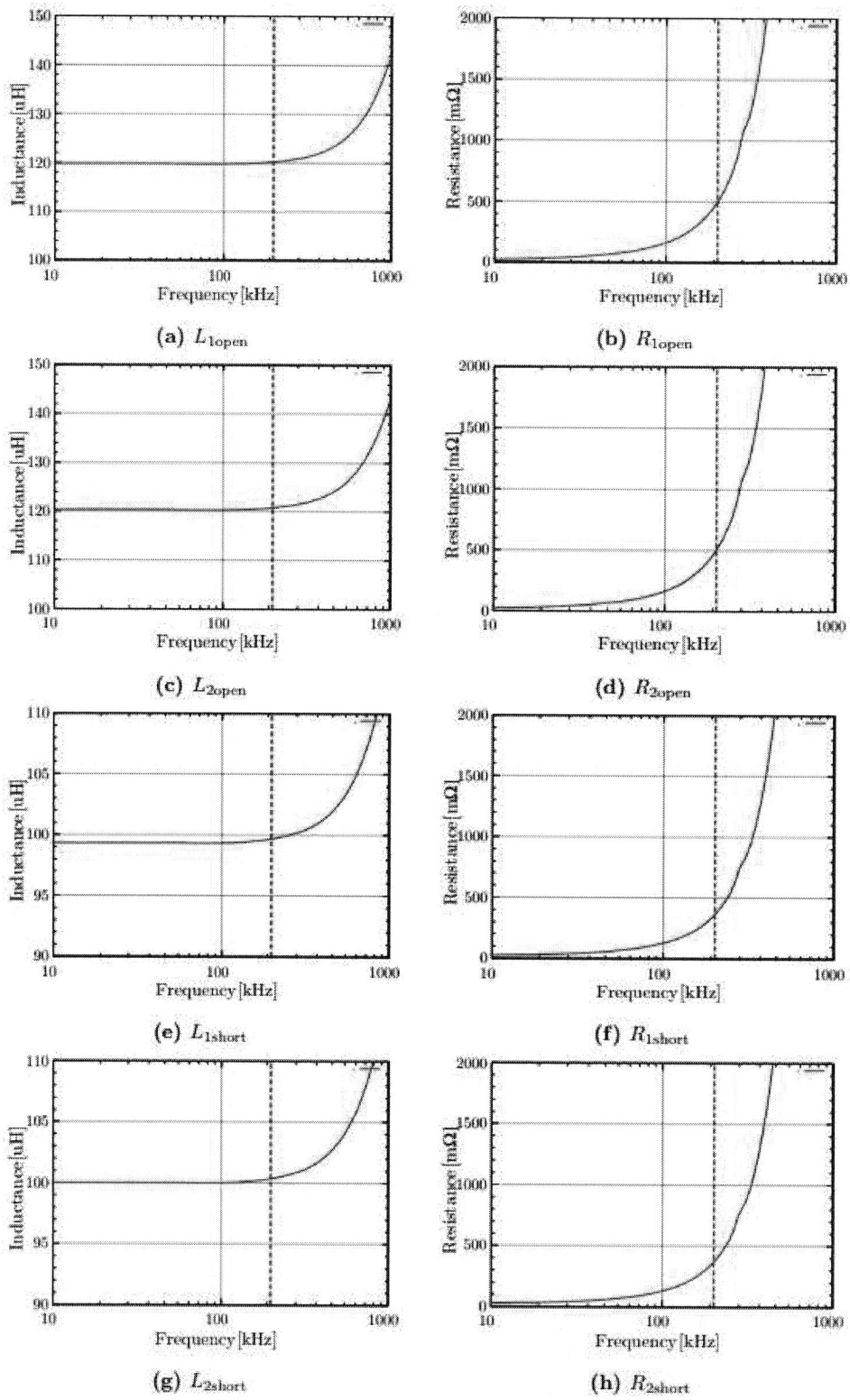

Fig. 10: Frequency characteristics of the coupled inductor; (a),(b): the primary side is open, (c),(d): the secondary side is open, (e),(f): the primary side is short, (g),(h): the secondary side is shorted.

Representative voltage and current waveforms for his case are shown in Figs. 11 and 12. Fig. 11 (a) shows the output capacitor voltage: as can be seen, the ripple is identical to the case of Fig. 6. For direct comparison, Fig. 11 (b) shows the same waveform when the phase-shift in the PWM carrier signals of the two interleaved cells is zeroed.

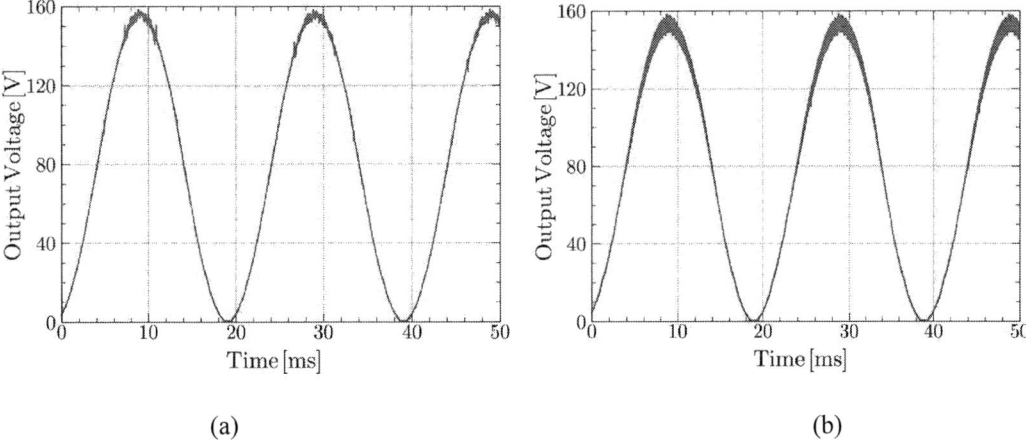

(a)   (b)

Fig. 11: (a) output capacitor voltage with PWM carriers phase-shift of 180 degrees ; (b) output capacitor voltage with PWM carriers phase-shift of 0 degrees.

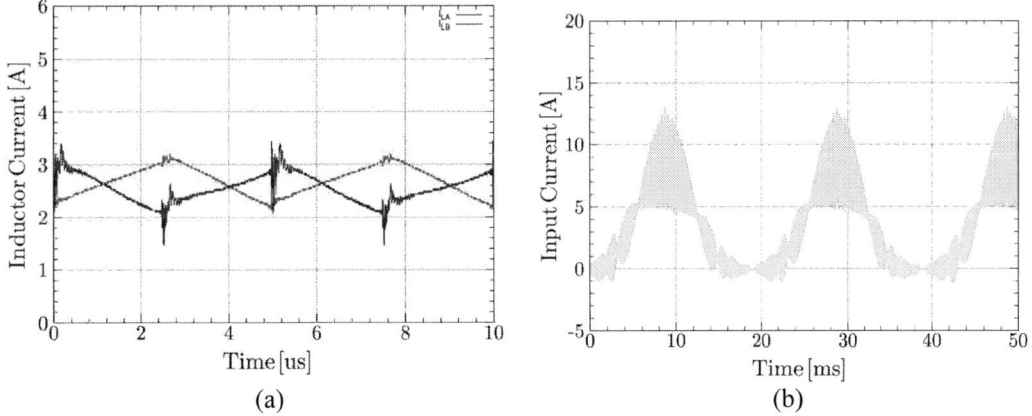

(a)   (b)

Fig. 12: (a) detail of current ripple in the two windings of the inductor; (b) total inductor current.

Fig. 12 shows results for the inductor current: in (a), details of the ripple in each winding of the inductor are given, whereas in (b) the total inductor current is shown.

Conclusion

The Y-Inverter is receiving increasing interest for the development of advanced motor-drives. In particular, its operating characteristics enable the full exploitation of the superior features of wide-band-gap semiconductors to yield disruptive improvements in efficiency and power density. The inverter features not-only benefit the power electronics, but also the machine, which can be designed for much higher efficiency and power density target values. Focusing on the interleaved implementation of the Y-Inverter, this paper showed that such approach can yield important advantages over straightforward device paralleling for meeting current rating requirements. The use of cross-coupled or non-coupled inductors when applying interleaving is left as an option to optimize either the size and efficiency of the magnetics (e.g., in relatively high current applications) or to reduce the stress in the capacitors (e.g., in the case of relatively high-voltage applications). The achievable reduction of core-size with interleaving using cross-coupled inductors has not been quantified here, but it is deemed considerable for high current applications.

References

[1] M. Antivachis, D. Bortis, L. Schrittwieser, J.W. Kolar, *Three-Phase Buck-Boost Y-Inverter with Wide DC Input Voltage Range*, in Proc. APEC 2018, San Antonio, TX, USA, March 4-8, 2018.

[2] J. W. Kolar, J. Huber, *Next-Generation SiC/GaN Three-Phase Variable-Speed Drive Inverter Concepts*, in Proc. PCIM2021, Nuremberg, Germany, May 2021.

[3] M. Antivachis, N. Kleynhans, J. W. Kolar, *Three-Phase Sinusoidal Output Buck-Boost GaN Y-Inverter for Advanced Variable Speed AC Drives,* IEEE Journal of Emerging and Selected Topics in Power Electronics - Special Issue on Electric Machine Drives and Converters for Automotive Applications (Early Access), 2020.

[4] H. N. Tran, A. Castellazzi, S. Domae, T. Dong, T., Nakamura, *Low-voltage-battery powered hybrid-GaN/SiC Y-Inverter dual-rotor Halbach motor drive for light electric vehicles*, in Proc. ICEMS2021, Busan, South Korea, Oct. 30th – Nov. 2nd, 2021.

[5] M. Minami, Y. Endo, A. Castellazzi, *A Numerical Comparison of Current Reduction of Single-Phase Buck-Boost Y-inverter via Interleaved Technique, SPC-21-110/MD-21-097, IEEJ, 17th September, 2021.*

Design of a GaN-Based Reconfigurable Resonant Converter for High Frequency On-Board Charger of Battery Electric Vehicles

Manh Tuan Tran[1,2], Haaris Rasool[1,2], Dai Duong Tran[1,2], Mohamed El Baghdadi[1,2] Philippe Lataire[1,2] and Omar Hegazy[1,2*], (Member, IEEE).

[1] Vrije Universiteit Brussel (VUB) & MOBI Research Group, ETEC Department, Pleinlaan 2, 1050 Brussels, Belgium
[2] Flanders Make, 3001 Heverlee, Belgium
*Corresponding author: Omar Hegazy (omar.hegazy@vub.be).

Keywords

«Electric vehicle On-Board charger», «CC-CV (constant current constant voltage) charging», «Reconfigurable resonant network».

Abstract

In this paper, a reconfigurable resonant converter for On-board Chargers (OBCs) and its' design consideration is proposed. This approach allows high frequency operation with maximum efficiency and Zero-Voltage Switching (ZVS) for all primary switches over a wide range of load conditions. The proposed converter is a combination of a Full Bridge LLC (FB-LLC) and an LCL compensator to form a high order resonant network. Constant Current (CC) and Constant Voltage (CV) charging profiles can be implemented automatically by reconfiguring the resonant network on the secondary side. In this design, GaN-FETs and a high frequency planar transformer are employed to further take advantage of high frequency operation. A 300kHz-3.3kW hardware prototype was built. Simulation and experiment results are provided with a maximum efficiency of 97%.

I. Introduction

On-board charger (OBC) is one of the key components in plug-in hybrid electric vehicles (PHEVs) and Battery electric vehicles (BEVs). The automobile industry has recently seen a surge in demand for high power density and efficiency. As a result, increasing switching frequency is required to reduce the size of an OBC system. However, switching losses, conduction losses, and passive component losses, which are proportional to switching frequency, dramatically reduce efficiency. Among several isolated DC-DC topologies for EVs (Electric Vehicles) charging application, LLC resonant emerged as the most promising candidate due to its advantages such as very wide range of zero voltage switching turn-on (ZVS) and Zero current switching (ZCS) turn-off in primary and secondary devices [1][2]. The high efficiency operation maintained around resonant frequency thanks to soft switching achieving and low circulating current. However, in EVs charging applications, a wide range of frequency regulation is required for Constant Current (CC) and Constant Voltage (CV) implementation [2]. This leads to the high ratio of resonant inductor over magnetizing inductor to extend regulation range[3][4]. The efficiency would be degraded due to the high RMS current value and large magnetizing current in resonant tank. In previous work [5], the author introduced a multi-MHz wireless power charger system with a combination of a LCL network [6] and a constant voltage topology, namely S-LCC, to perform a CC-CV charging profile at a fixed resonant frequency. In this study, the proposed approach will be further investigated in On-board charger applications. This paper presents a reconfiguration resonant DC-DC. In which the constant voltage can be easily achieved by a LLC-DCX [7]. For the CC mode, a T-Type LCL is cascaded with an LLC converter on the secondary side of the transformer to convert the constant voltage source into a constant current source. The major characteristics of the proposed converter and contributions of this paper can be summarized as follows:

1) Very simple control strategy for CC-CV implementation without frequency or phase shift modulation. Hence, a larger magnetizing inductance can be designed to reduce the circulating current and make voltage and current gain function less sensitive with a larger tolerance of resonant components.
2) This approach allows more freedom in designing charging specifications for EV applications thanks to its high order resonant network.
3) Optimal magnetic parameters ensure soft-switching ZVS for all primary switches and nearly ZCS for secondary diodes across the entire load range, allowing high frequency and high efficiency operation.
4) Two additional bidirectional switches can be used to minimize conduction losses without any switching losses.
5. A design procedure based on the Multi-Objective Optimazation-based GA (MOGA) framework is developed for proposed coverter.

II. Proposed Reconfigurable Full-Bridge Resonant Converter

The detailed architecture of the proposed on-board charger consists of a front end totem pole Power Factor Correction (PFC) and a back-end reconfigurable resonant converter as illustrated in Fig. 1. The scope of the paper mainly studied the proposed DC-DC converter. To meet the wide range of output voltage requirements as in Fig.3 (a), proposed hybrid resonant topology is a combination of a voltage source converter, which can be LLC or CLLC topology, and a T-type LCL network. The LCL resonant network can be on the primary side or secondary side as depicted in Fig.2 (a), (b), and (c). For high voltage EV battery charger applications, the LCL circuit can be inserted on either the primary or secondary side.

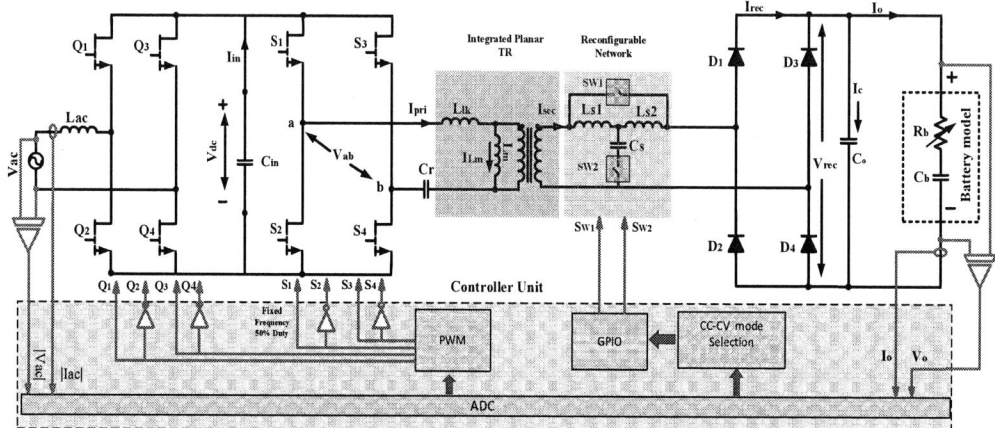

Fig. 1: Configuration of the proposed DC-DC converter system

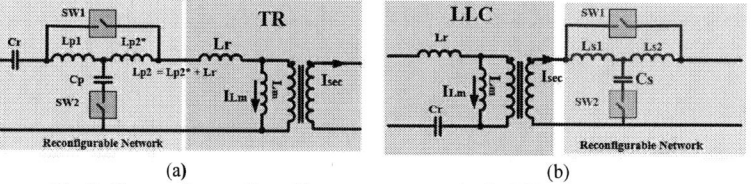

(a)   (b)

Fig. 2. Proposed reconfigurable resonant networks for EVs charger system:
(a) configurable network in primary side, (b) configurable network in secondary side.

Compared to the conventional control method for LLC converter, the proposed dc-dc converter with a reconfigurable resonant network allows fulfilling CC-CV operation at a fixed frequency as illustraled in Fig.3 (b). As a result, DC-DC converter stage can operate as a DC transformer (DCX) with a constant gain. Thus, the overall efficiency is maximized. During the CC mode, the bidirectional switch SW_1 is turn OFF and SW_2 is turn ON, the high order resonant network LLC + LCL forms a current source. As battery voltage increase to maximum voltage, SW_1 is turn ON and SW_2 is turn OFF, the converter is configured as voltage source. Thanks to soft-switching operation over the entire load range at the high switching frequency, the GaN (Gallium nitride) Mosfets and Planar Transformer can be used to achieve a high-power density system.

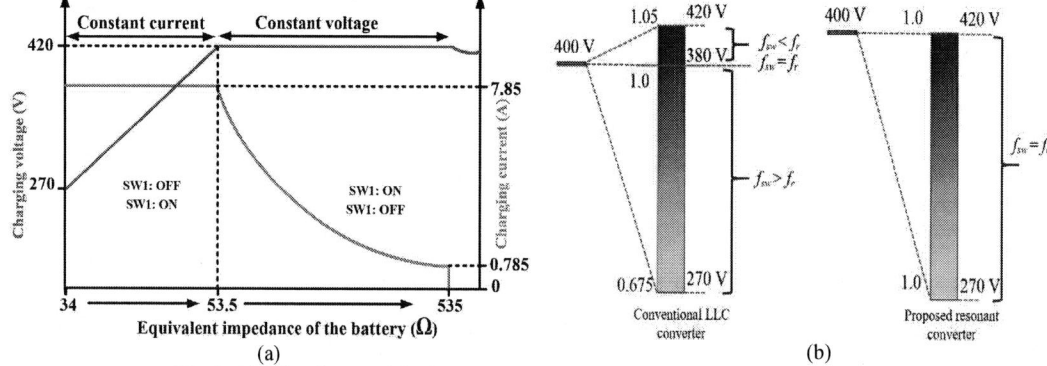

(a)   (b)

Fig. 3. (a) Charging specifications , (b) Required voltage range of DC-DC conveter

III. Steady state analysis of proposed DC-DC converter in CC and CV charging mode

The first harmonic approximation (FHA) method is adopted for deriving the steady-state gain function of the proposed converter. The equivalent AC circuits in CC and CV are described as in Fig. 4 and Fig.5. AC input voltage V_{in} is supplied by a full-bridge circuit. Integrated leakage inductor L_r, Magnetizing inductor L_m, C_r, and additional components L_{s1}, L_{s2}, C_2 form a high order resonant network on the primary and secondary sides, respectively. The fundamental components of input voltage V_{in}, output voltage V_{O_ac} and output current I_{O_ac} are derived from (1), (2) and (3). In which, $n = Np/Ns$ is transformer turn ratio.

$$
\begin{cases}
V_{in} = \dfrac{4V_{dc}}{\pi} \sum_{k=1,3,5}^{\infty} \dfrac{1}{k}\sin(2\pi k f_{sw}t) & (1) \\[3mm]
V_{O\_ac} = \dfrac{4nV_O}{\pi} \sum_{k=1,3,5}^{\infty} \dfrac{1}{k}\sin(2\pi k f_{sw}t - \psi) & (2) \\[3mm]
I_{O\_ac} = \dfrac{\pi I_O}{2n}\sin(2\pi k f_{sw}t - \psi) & (3)
\end{cases}
$$

The rms AC equivalent values are calculated as in eq. (4), (5) and (6). The AC equivalent load resistor R_{O_ac} reflected from secondary side can be obtained as in (7).

$$
\begin{cases}
V_{in_{rms}} = \dfrac{2\sqrt{2}V_{dc}}{\pi} & (4) \\[3mm]
V_{O_{rms}} = \dfrac{2\sqrt{2}nV_O}{\pi} & (5) \\[3mm]
I_{O_{rms}} = \dfrac{\pi I_O}{2\sqrt{2}n} & (6) \\[3mm]
R_{O\_ac} = \dfrac{8n^2 R_O}{\pi^2} & (7)
\end{cases}
$$

A. Converter configuration for CC charging mode.

As depicted in Fig.4(a), the equivalent AC circuit of CC mode charging is configured by turning ON SW_2, turning OFF SW_1. Where L_r is integrated resonant inductor, L_m is magnetizing inductor of Planar transformer. Additional resonant LCL parameters are reflected into the primary side and denoted as $n^2 L_{s1}, n^2 L_{s2}, C_2/n^2$.

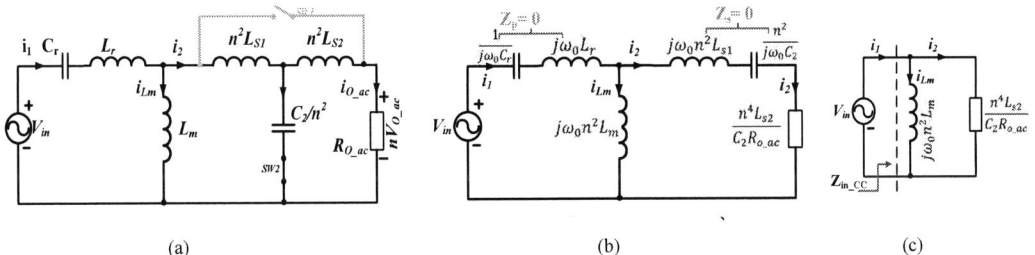

(a)   (b)   (c)

Fig. 4. Configuration of DC-DC converter in CC mode

$$
Z_{eq} = \frac{\dfrac{n^2}{j\omega_0 C_2}(n^2 j\omega_0 L_{s2} + R_{o\_ac})}{\dfrac{n^2}{j\omega_0 C_2} + n^2 j\omega_0 L_{s2} + R_{o\_ac}} = \frac{n^4 L_{s2}}{C_2 R_{o\_ac}} + \frac{n^2}{j\omega_0 C_2} \tag{8}
$$

The equivalent impedance Z_{eq} is combined of $n^2 L_{s2}$, C_2/n^2 and R_{o_ac} as calculated in (8). When the conditions are satisfied as in (9), (10). Fig.4 (b) and Fig.4 (c) are deduced to derive output current as in (11). It can be observed from (11) that the output current is independent to the load resistor value. As a result, the constant current is performed at the fixed switching frequency.

$$
j\omega_0 L_r + \frac{1}{j\omega_0 C_r} = 0 \tag{9}
$$

$$j\omega_0 n^2 L_{s1} + \frac{n^2}{j\omega_0 C_2} = j\omega_0 n^2 L_{s2} + \frac{n^2}{j\omega_0 C_2} = 0 \tag{10}$$

$$\begin{cases} I_2 = \frac{V_{in} C_2 R_{o\_ac}}{n^4 j\omega_0 L_{s2}} & (11) \\ \\ I_{O\_ac} = \frac{I_2 Z_{eq}}{n^2 L_{s2} + R_{o\_ac}} = \frac{V_{in}}{n^2 j\omega_0 L_{s2}} & (12) \\ \\ I_O = \frac{8V_{dc}}{\pi^2 n j\omega_0 L_{s2}} & (13) \end{cases}$$

As seen from (13), The DC output current value can be designed based on selecting inductor value L_{s2}. It can be noticed that the higher designed switching frequency will lead to the smaller inductor value.

B. Converter configuration for CV charging mode.

The constant voltage operation can be realized by closing SW_1, and opening SW_2. In this mode, the AC equivalent circuit is configured as conventional LLC topology.

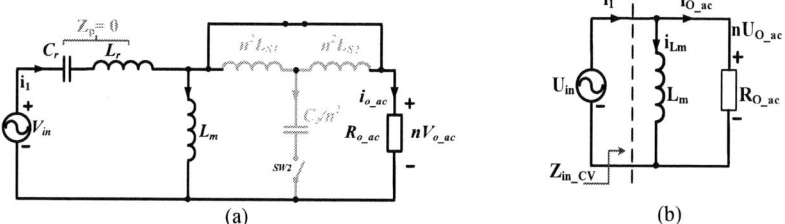

(a)   (b)

Fig. 5. Configuration of DC-DC converter in CV mode

Using the fundamental harmonic approximation method for equivalent circuit in Fig.5(a), the input-to-output voltage gain is expressed in (14). And the unity gain can be obtained at resonant frequency ω_0 fulfilled (15).

$$\begin{cases} Gv = \left| \frac{\frac{2\sqrt{2}nV_O}{\pi}}{\frac{2\sqrt{2}V_{dc}}{\pi}} \right| = \left| \frac{(j\omega_0 L_m)//R_{O\_ac}}{j\omega_0 L_r + (j\omega_0 C_r)^{-1} + (j\omega_0 L_{m]})//R_{O\_ac}} \right| = \frac{nV_o}{V_{in}} = 1 & (14) \end{cases}$$

$$Z_p = j\omega_0 L_r + \frac{1}{j\omega_0 C_r} = 0 \tag{15}$$

$$Z_{in\_cv} = \frac{j\omega_0 L_m R_{O\_ac}}{j\omega_0 L_m + R_{O\_ac}} \tag{16}$$

Since the input impedance Z_{in_cv} is a function of load R_{O_ac} and L_m as in (16), the phase of Z_{in_cv} is determined by $arg[Im\{Z_{in\_cv}\}/Re\{Z_{in\_cv}\}] = arg[-R_{o\_ac}/Z_m]$. Therefore, the most important aspect of the design is to maximize the magnetizing impedance Z_m to minimize the reactive power but still guarantee the soft switching operation. The value of magnetizing inductance is limited as (17) to guarantee the ZVS of primary GAN switches.

$$L_m{}^* \le \frac{T_d}{8C_{oss}f_O} \tag{17}$$

The value of L_m is selected by Optimization method as described in the section III.C to ensure the required current fully discharging ouput capacitance C_{oss} of the switches during deadtime interval T_d.

C. Design procedure and consideration of proposed converter

The designed procedure of the proposed converter is illustrated in Fig.6. The input specifications are defined by the required CC-CV charging profile in Fig.3 (a). Resonant components significantly affect efficiency and power density for the proposed EVs converter design method. Therefore, the Multi-Objectives optimization framework based Matlab environment is built to select the optimal design parameters for resonant components.

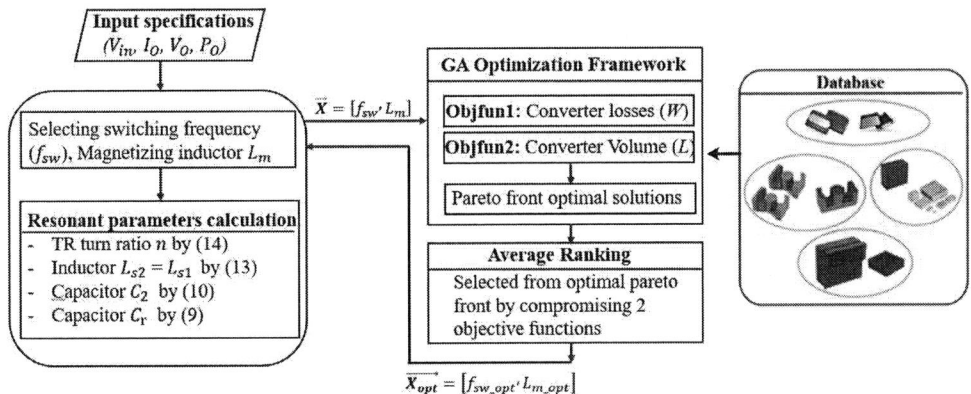

Fig. 6. Design framework of proposed converter system

The design procedure begins by defining input specifications such as input voltage, output voltage, and power rating via created user interface. The optimization algorithm implementation is briefly explained as follows:

Step1. Firstly, the objective functions are defined as (18)

$$Objectives:\ Min \begin{cases} ObjFunc1\ (\vec{X}) = P_{loss}(\vec{X}) \\ ObjFunc2\ (\vec{X}) = Vol(\vec{X}) \end{cases} \tag{18}$$
$$Where:\ \vec{X} = [f_{sw}, L_m]$$

The objective functions of power losses and volume are formulated based on the works in [8][9].

Step2. The constraints and boundary values are selected to generate rational results as given in (19)

$$Subject\ to: \begin{cases} 100\ kHz \leq fsw \leq 500\ kHz \\ L_m \leq L_m{}^* \end{cases} \tag{19}$$

Where the boundary value of magnetizing inductance of transformer is defined as in (17)

Step3. Multi Objective Genetic Algorithm (MOGA) is applied to solve the non-convex and multiple objectives. The multi- solutions are presented in a set of Pareto-optimal solutions, as shown in Fig.7

Step4. The average ranking method is utilized to select the most desire solution from the set of Pareto-fronts. A single objective function can be created from two objectives functions by using a weighting factor δ as in (20). For On-board charger system, the power density is critical. In this design, δ coefficient is 0.7, meaning power density has higher priority over power losses. The optimal solution is marked as a red star in Fig. 7.

$$f_{composite} = \delta ObjFunc2\ (\vec{X}) + (1 - \delta)ObjFunc1\ (\vec{X})$$
$$Where:\ 0 \leq \delta \leq 1 \tag{20}$$

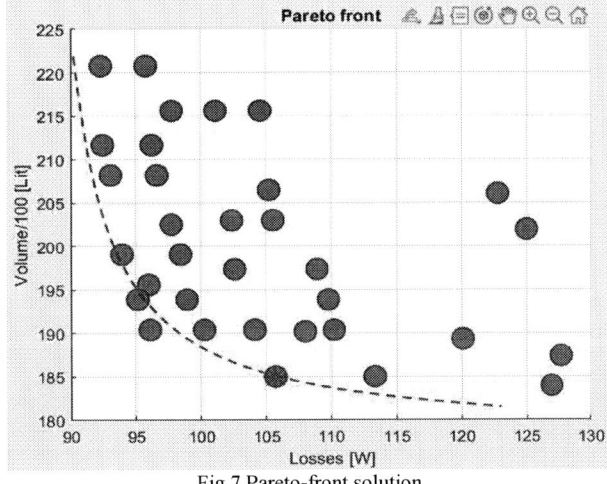

Fig.7 Pareto-front solution

Step5. After launching GA based optimization solver, the optimal parameters of switching frequency f_{sw_opt} and L_{m_opt} is obtained. According to the (13), (12), (10) and (9), the resonant parameters are calculated as (21)

$$\begin{cases} L_{s1} = L_{s2} = \dfrac{8V_{dc}}{\pi^2 nj\omega_{opt}I_O} \\ C_2 = \dfrac{1}{\left(\omega_{opt}\right)^2 L_{s2}} \\ C_r = \dfrac{1}{\left(\omega_{opt}\right)^2 L_r} \end{cases} \quad (21)$$

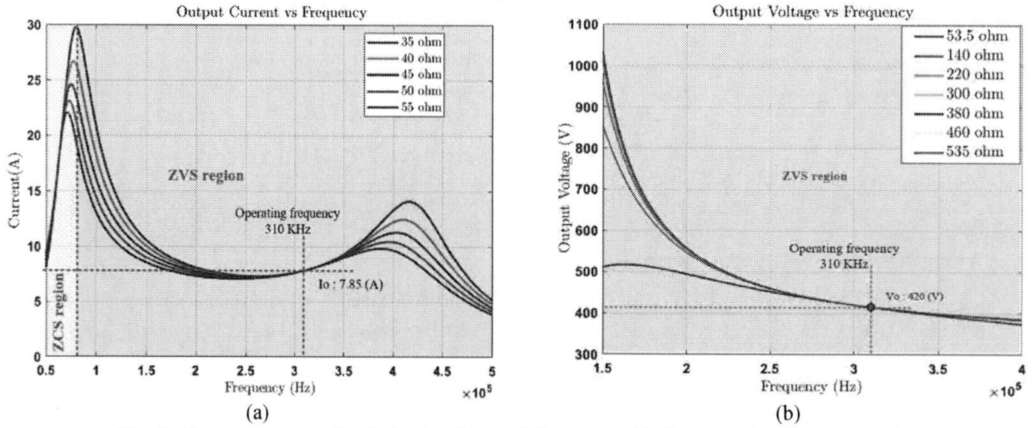

Fig. 8. Gain curve according to load resistor and frequency: (a) Current gain, (b) voltage gain

To validate selected parameters, the output voltage and the output current curves versus the operating switching frequency with various loads are plotted in Fig. 8. The charging values of the converter are maintained constant at resonant frequency regardless of load conditions. Hence, the proposed design method is suitable to implement CC-CV charging for battery charring applications.

IV. Simulattion and Experiment Results

In this section, simulation model is built by PSIM to validate the key waveforms of proposed converter as illustrated in Fig.9. The experiment results are then povided by a 300kHz- 3.3kW reconfigurable resonant converter prototype as shown in Fig.10. Design parameters of resonant components are listed in Table I. There is a small deviation between simulated and practical parameters. Therefore, the output value is correctified by slightly adjusting the switching frequency to guarantee operating at a resonant point.

Table I: Specification and components of proposed converter

Parameters	Symbols	Values
Power	P_O	3.3 kW
Input voltge	V_{in}	400 V
Output Voltage	V_{out}	270V-420V
Transformer's turn ratio	N_P/N_S	1.12
Switching frequency	f_{sw}	300 kHz
Primary inductor	L_r	2.8 uH
Magnetizing Inductor	L_m	50uH
Additional secondary inductors	L_{s1}, L_{s2}	22 uH
Primary capacitor	C_r	94 nF
Additional Secondary capacitor	C_2	14.2 nF
Components	Symbols	Part number
Pimary GAN switches	$S_1 - S_4$	GS66508T
Secondary rectifier Schottky diodes	$D_1 - D_4$	SCS230AE2

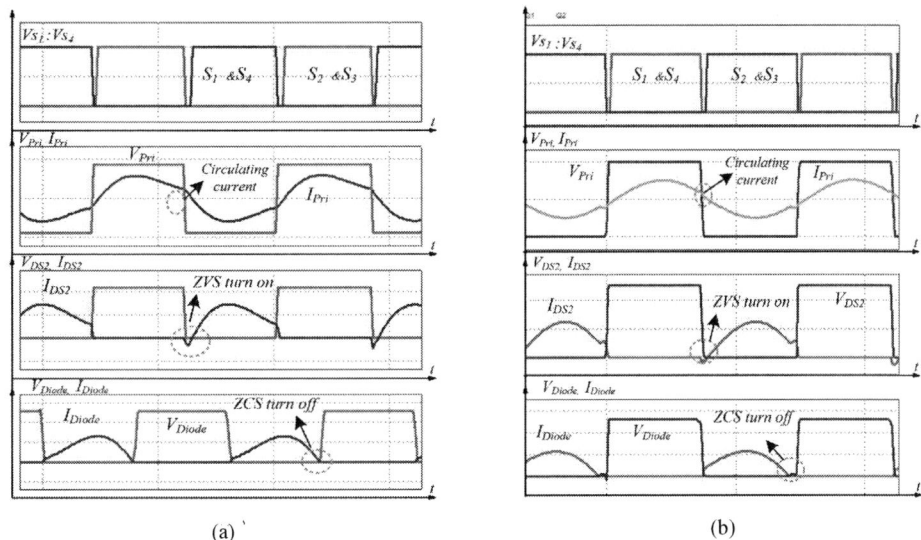

Fig. 9. Key waveforms of the proposed converter in CC mode (a), and CV mode (b)

The soft switching characteristics of the proposed converter are satisified in both CC mode and CV mode as shown in Fig.9. Where ZVS for all primary devices and ZCS for secondary diodes are clearly demonstrated.

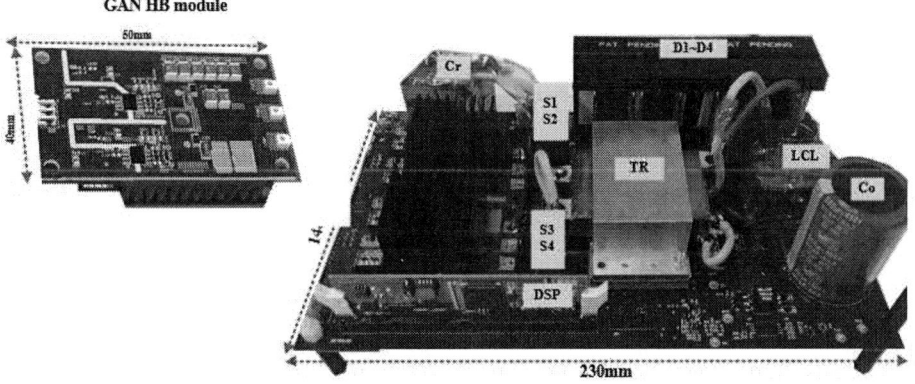

Fig. 10.   Photograph of the prototype DC-DC converter

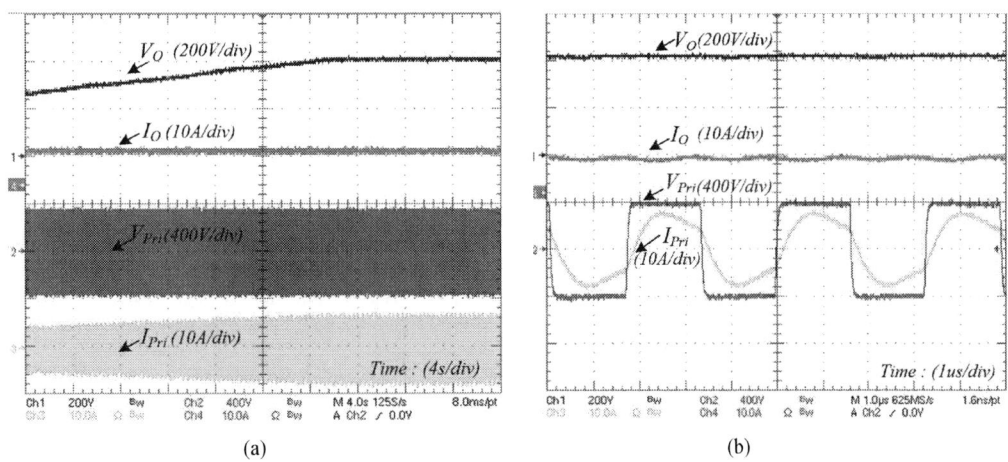

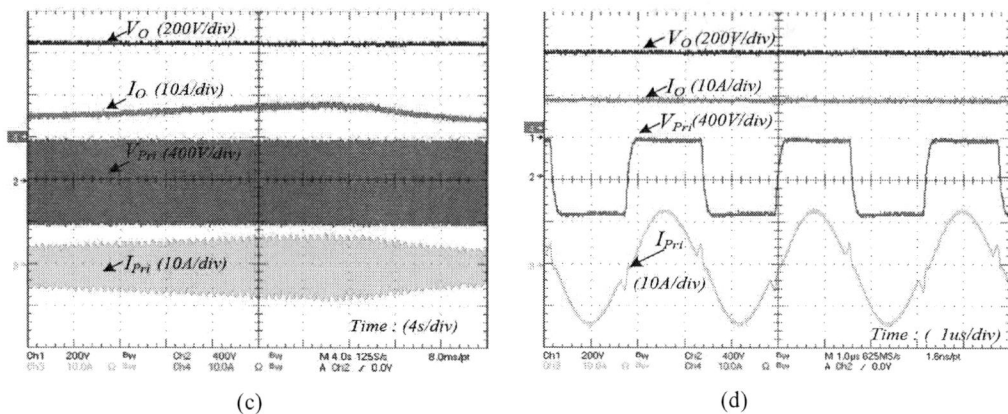

(c)   (d)

Fig.11. Experiment results 3 kW with variable resistor load: (a) Constant current performance, (b) Key waveforms in CC mode, (c) Constant voltage performance, (d) Key waveforms in CV mode

As shown in Fig.11, the constant current and constant voltage charging can be performed at a fixed frequency of 310 kHz. The output current of 7.8 A is kept constant during CC mode regardless of load variation as in Fig. 11(a). Fig.11 (b) shows that the primary current leads primary voltage, and the ZVS condition is achieved. While Fig.11 (c) shows the constant voltage of 420 V with key waveforms zoomed in Fig.11 (d).

V. Conclusion

In this paper, a GaN-Based reconfigurable DC-DC converter has been proposed for On-board charger applications. The CC/CV charging profile is performed without a dedicated controller design. The converter operates at the fixed-high switching frequency and optimal condition to maximum efficiency thank to proposed reconfigurable network. The optimization design framework is developed for sizing passive components, which contributed a substantial portion of the overall volume of the converter. The design methodology of the 3.3kW DC-DC converter is validated at switching frequency of 300 kHz hardware prototype with a peak efficiency 97.5%.

ACKNOWLEDGEMENT

This research has received funding from the European Union's Horizon 2020 research and innovation programme under grant agreement No 101006943, under the title of URBANIZED (https://urbanized.eu/). We also acknowledge Flanders Make for the support to our research group.
.

References

[1] D. Junjun, L. Siqi, H. Sideng, C. C. Mi, and M. Ruiqing, "Design methodology of LLC resonant converters for electric vehicle battery chargers," IEEE Trans. Veh. Technol., vol. 63, no. 4, pp. 1581–1592, May 2014.

[2] F. Musavi, M. Craciun, D. S. Gautam and W. Eberle, "Control Strategies for Wide Output Voltage Range LLC Resonant DC–DC Converters in Battery Chargers," in *IEEE Transactions on Vehicular Technology*, vol. 63, no. 3, pp. 1117-1125, March 2014, doi: 10.1109/TVT.2013.2283158.

[3] R. Beiranvand, B. Rashidian, M. R. Zolghadri and S. M. Hossein Alavi, "A Design Procedure for Optimizing the LLC Resonant Converter as a Wide Output Range Voltage Source," in *IEEE Transactions on Power Electronics*, vol. 27, no. 8, pp. 3749-3763, Aug. 2012, doi: 10.1109/TPEL.2012.2187801.

[4] R. Beiranvand, B. Rashidian, M. R. Zolghadri and S. M. H. Alavi, "Using LLC Resonant Converter for Designing Wide-Range Voltage Source," in *IEEE Transactions on Industrial Electronics*, vol. 58, no. 5, pp. 1746-1756, May 2011, doi: 10.1109/TIE.2010.2052537.

[5] M. T. Tran and W. Choi, "Design and Implementation of a Constant Current and Constant Voltage Wireless Charger Operating at 6.78 MHz," in *IEEE Access*, vol. 7, pp. 184254-184265, 2019, doi: 10.1109/ACCESS.2019.2959981.

[6] M. Borage, K. V. Nagesh, M. S. Bhatia and S. Tiwari, "Resonant Immittance Converter Topologies," in *IEEE Transactions on Industrial Electronics*, vol. 58, no. 3, pp. 971-978, March 2011, doi: 10.1109/TIE.2010.2047835.

[7] T. M. Tuan and W. Choi, "A Novel Two-Stage Power Conversion Method suitable for LDCs of the Electric Vehicles," *2019 10th International Conference on Power Electronics and ECCE Asia (ICPE 2019 - ECCE Asia)*, 2019, pp. 1-7, doi: 10.23919/ICPE2019-ECCEAsia42246.2019.8797020.

[8] Z. Li *et al.*, "A High-efficiency DC/DC Converter with SiC Devices and LLC topology for Charging Electric Vehicles," *2018 Asian Conference on Energy, Power and Transportation Electrification (ACEPT)*, 2018, pp. 1-7, doi: 10.1109/ACEPT.2018.8610743.

[9] H. -N. Vu, M. Abdel-Monem, M. El Baghdadi, J. Van Mierlo and O. Hegazy, "Multi-Objective Optimization of On-Board Chargers Based on State-of-the-Art 650V GaN Power Transistors for the Application of Electric Vehicles," *2019 IEEE Vehicle Power and Propulsion Conference (VPPC)*, 2019, pp. 1-6, doi: 10.1109/VPPC46532.2019.8952196.

Transient Liquid Phase Bond Reliability Evaluation of Die-attach for Power Module Packaging

Laxma R. Billa, Yangang Wang, Thomas Grant, Xiang Li, Harley Neal, Muhammad Morshed
Dynex Semiconductor Ltd.
Doddington Road, Lincoln, UK, LN6 3LF
E-Mail: muhammad.morshed@dynexsemi.com

Keywords

Transient liquid phase bond, Sn-Cu, Power semiconductors, Module packaging

Abstract

This paper presents the low temperature and pressure-less Sn-Cu solder technology process by means of Transient Liquid Phase (TLP) diffusion phenomena for high temperature power module packaging. The Sn-Cu diffusion process is developed for bonding a large area dies and its reliability are evaluated by the bond strength and accelerated active and passive thermal cycling tests.

1. Introduction

Strong demand of high melting die-attach materials has been driven by rapid progress in power electronics for medium and high-power modules. High lead (Pb) solder alloys with melting temperatures above 280 °C are not allowed to be use for some application sectors such as automotive and aerospace due to the RoHS (Restrictions of Hazardous Substances) regulations of the European Union. Popular Pb-free solder alloys, such as Sn-Ag-Cu and Sn-Sb, exhibit very poor thermo-mechanical reliability [1]. However, the sinter materials such as silver and copper are good candidates for high temperature applications because of their better electrical and thermal conductive properties, but the material cost is high and the process is complex [2,3]. An alternating solution to the Pb-free die attach process is Transient Liquid Phase (TLP) bonding, also known as solid-liquid interdiffusion (SLID) boding [4-6]. It involves the incorporation of a low melting temperature metal into another high melting temperature metal matrix to create an intermetallic compound (IMC) with an intermediate melting temperature. It enables the use of a low process temperature for creating a high re-melting temperature (>400°C) joint and shorter process time as the kinetics of liquid-solid diffusion is much faster than solid-state diffusion. In addition, the IMC joints with a high re-melting temperature improve creep resistance at elevated temperatures compared to solder joints [4,7]. Sn-based alloy, specifically Sn-Cu metal matrix, is one of the attractive alloys in the electronics industry as Sn reacts to the Cu quickly and creates intermetallic compounds (IMCs) in the entire bond.

The process development for the large-area die attach is one of the key technologies to enable highly-reliable next-generation standardized power module manufacturing for railway traction inverter applications [8]. The high melting temperature joining technology could be implemented with high-power density, high switching frequencies with a smaller size of wide band gap (WBG) devices such as silicon carbide (SiC) and gallium nitride (GaN) for high temperature operation capability and improved reliability.

This paper describes the large area die attach (area > 100mm^2, FRD and IGBT) by TLP diffusion soldering process on which reliability analysis has been performed by thermal shock and active power cycling test methods. The TLP diffusion soldering process has been optimized by die shear strength using smaller die size whereas bend tests have been carried out for the larger size of dies. Static electrical characteristic tests have been performed before and after the thermal shock reliability test. Scanning acoustic microscopy (SAM) was carried out to detect any die bond degradation after reliability tests whereas SEM-EDS analysis was carried out to investigate the complete formation of the Cu-Sn IMC structure.

2. Experimental works

The design choices of the experiments prepared with Ag back metallization dies and Si_3N_4 ceramic DBC substrates were made to produce binary high melting temperature Cu-Sn IMC. A conventional reflow oven was used for the pressureless TLP soldering process. The Cu-Sn paste was applied onto the substrate via a 100μm thick stencil. The diffusion soldering process was optimized by considering different process temperatures (255°C -290°C) and isothermal holding times under a N_2 environment. The diffusion process conditions were optimized by achieving a full conversion of the Cu-Sn matrix into the IMCs in the entire bond and minimizing the void level. Table I shows the process parameters considered for TLP soldering of die size 3x3mm².

Table I: Process conditions for each run.

Run No.	Temperature (°C)	Holding time (minute)	Process environment
1	255	3	N_2
2	290	3	N_2
3	255	10	N_2
4	290	10	N_2

Die shear strength is one of the test methods normally used to check the bond strength for optimizing the soldering process. The best characterization method is the hot die shear test, but this is not available in-house. A similar method of combining isothermal ageing and die shear test was used instead. The thermal ageing was carried out at 175°C for 1000 hours at atmospheric pressure. The die shear strength of thermally aged samples was analyzed, and the best performing process parameters for large area die soldering were selected. SEM-EDS cross-sectional analysis was carried out to investigate the full conversion of IMC matrix which are Cu_3Sn and Cu_6Sn_5.

After optimization of the TLP soldering process for smaller die size, the large size of Si FRD and IGBT (>100mm²) were considered for TLPS soldering with a lower percentage of voids. The die bond adhesion for large size die was examined with the aid of a qualitative bend test where the mandrel diameter used for the bend test was 13mm. The reliability of large area die bond was carried out by thermal shock and power cycling tests. The test condition for the thermal shock test is $\Delta T=225°C$ (-50°C to 175°C) and dwell time 30 minutes. The power cycling test parameters are $\Delta T=120°C$ ($T_{jmin}= 30°C$ and $T_{jmax}=150°C$), $I_c=120A$ to 150A, coolant temperature 15°C and $t_{on}/t_{off}=0.5s/3s$. Here T_{jmax}, $T_{jmin,}$ and I_c left variable to keep ΔT constant. Figure 1 shows a typical power cycling test sample. The reliability test samples were inspected optically and with SAM (Scanning acoustic microscopy) to identify any cracks or delamination initiated in the die bond area. The static electrical test was performed at room temperature and at elevated temperature levels on thermal shock samples after 1000 cycles.

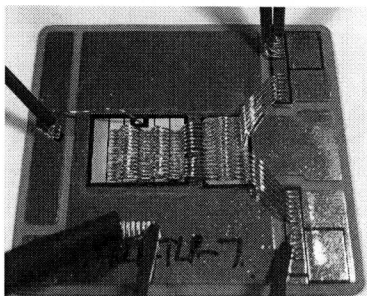

Fig. 1: A typical power cycling test sample of TLP soldered IGBT and FRD (1200V/200A).

3. Results and discussion
3.1. Shear strength analysis

Die shear test was carried out with samples consisting of 3mm×3mm die with varying process conditions (see table I). Die shear test was also carried out of thermal aging samples which were TLP soldered with similar process parameters. Figure 2 shows the average shear strength of as-soldered and after isothermal aging samples. The shear strength drops in Run 2 to Run 4 but increases in Run 1 after thermal aging. This indicates the negative correlation between the post-aging bond strength and process temperature. According to Tatsumi et al., at high temperature ageing, Cu_6Sn_5 IMC structure will transform to Cu_3Sn structure with two possible reactions [9].

$$9Cu + Cu_6Sn_5 \rightarrow 5Cu_3Sn \tag{1}$$

$$Cu_6Sn_5 \rightarrow 2Cu_3Sn + 3Sn \tag{2}$$

The reaction of equation (1) will be the main pathway according to Gibbs free energy change which is a negative value of ~89kJ/mol whereas equation (2) is a positive value of ~10kJ/mol [10]. The hypothesis for such correlation is that higher temperature produces more Cu_3Sn species, which in turn develops Kirkendall voids during thermal aging. For this reason, a lower process temperature or time is preferred in order to minimize the Cu_3Sn phase at the die bonding stage. In Figure 2, a low temperature and shorter holding time (Run1) might have improved the shear strength of the TLP joint up to a 175°C ageing temperature by a transformation from Cu_6Sn_5 to Cu_3Sn. The TLP process of Run 1 was considered for large area die attach on DBC substrate.

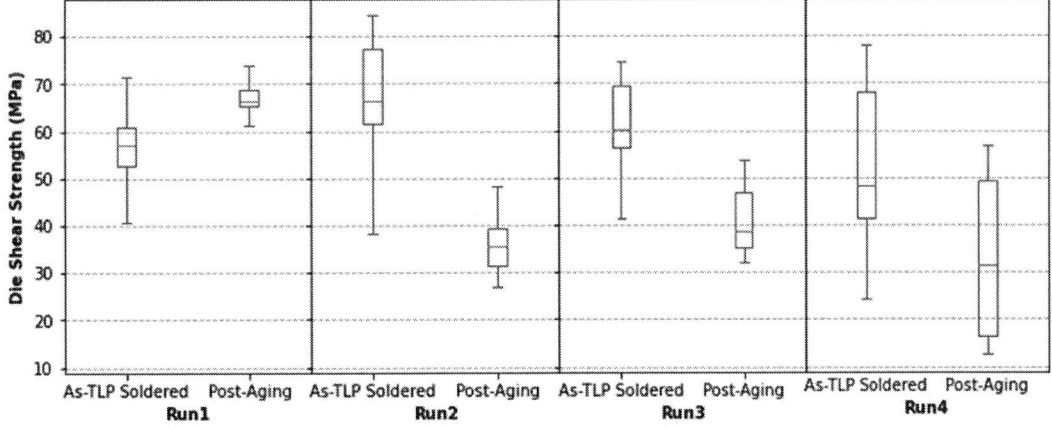

Fig. 2: TLP die bond shear strength, as-soldered and post-ageing in different process conditions.

3.2. TLP Bond Characterization

The metallographic cross-sectional analysis was carried out by SEM-EDX (Scanning Electron Microscopy-Energy Dispersive Spectroscopy) for microstructure evaluation of Sn-Cu matrix's TLP diffusion. Figure 3 shows the full conversion of Sn into intermetallic phases with Cu particles in a bulk structure and at the interface between TLP bond and substrate/die metallization, respectively. Figure 4 also shows the SEM-EDX analysis of the TLP bond. It is confirmed that all bulk Sn is consumed into IMC structure where the atomic ratio of Cu and Sn for Cu_3Sn phase is about 3:1 and for Cu_6Sn_5 phase is about 1.2:1. In addition, a continuous layer of IMC was observed at the die and substrate interface, confirming good adhesion on both interfaces. EDX analysis also confirmed the proportion of Cu_3Sn phase in the bond increases with increasing process temperature and holding time. This is expected as higher temperature causes diffusion growth of Cu_3Sn phase by transforming the Cu_6Sn_5 phase.

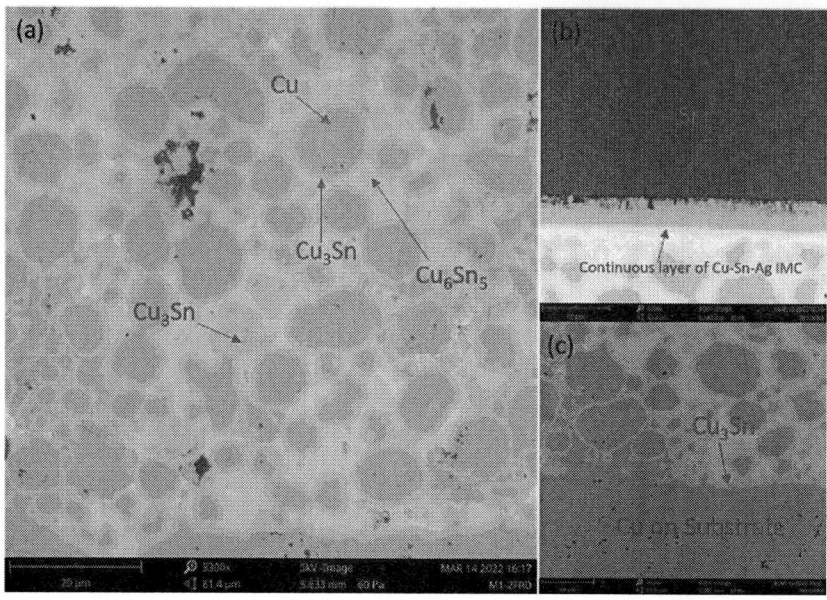

Fig. 3: SEM-EDX analysis on: (a) TLP bond region, (b) the interface between Si and TLP bond, and (c) the interface between TLP bond and substrate.

Figure 4 shows an X-ray image of IGBTs and FRDs bonded on a DBC substrate which has been soldered by the Run1 process. The maximum percentage of voids was observed in a large die area (IGBT) and measured at about 7%. At least three samples were prepared in each reflow process, and the results found the process is repeatable with low voids.

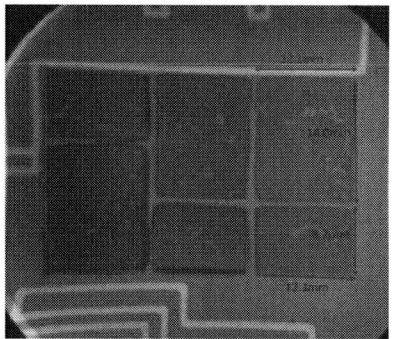

Fig. 4: X-ray image of TLP large area dies (IGBTs and FRDs) attached by Run 1 process.

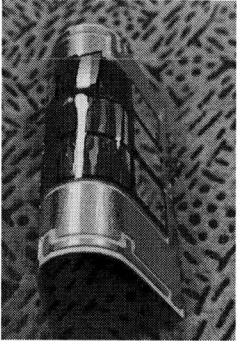

Fig. 5: A typical bend test using a 13mm mandrel diameter for a large area die size.

3.3. Bend Test

Quantitate bend test was carried out for large die samples. The TLP die bond adhesion was also examined with the aid of the mandrel bend test. Here mandrel diameter used was 13mm for introducing a higher bend. Figure 5 shows a typical bend test result where die cracks occur due to cohesive failure indicating good adhesion of die bonding.

3.4. Reliability Tests

A total of 1000 cycles with $\Delta T=215°C$ thermal shock test was carried out. SAM scan analysis showed (as shown in Figure 6) there was no die TLP solder degradation or die cracks visible after the test indicating the proposed TLP process is good life under high thermomechanical stress. The electrical functionality of the die-attach samples did not deteriorate after the 1000 cycles of the thermal shock test. Figure 7 shows static electrical functionality tests carried out in different cycles of thermal shock. The FRD dies to functional at the room temperature, and elevated temperature levels (125ºC and 150ºC) without a significant change of their I-V characteristic confirm the TLP die bonds could perform well at high-temperature operation. It is believed that the TLP process could also be implemented in high temperature power module packaging applications such as SiC.

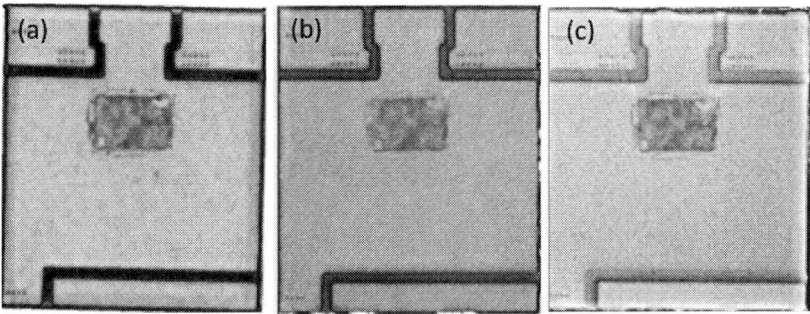

Fig. 6: SAM images: (a) Pre-thermal shock test, (b) after 600 cycles, and (c) after 1000 cycles.

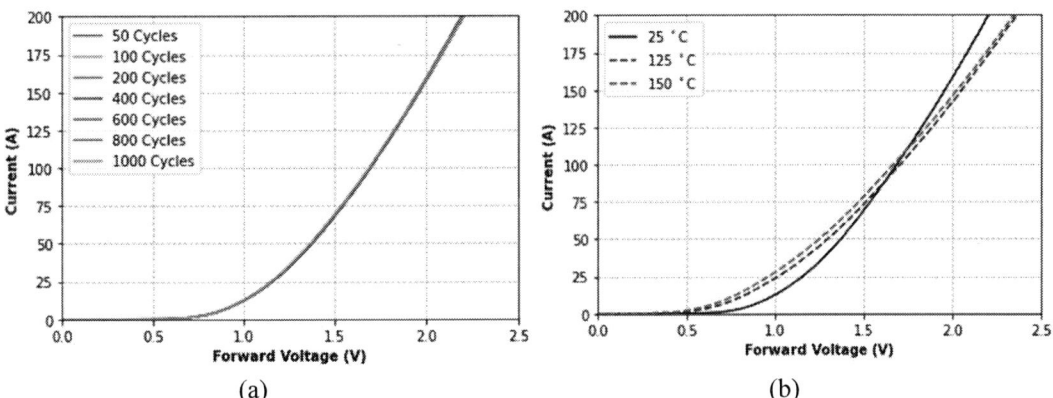

Fig. 7: Electrical test was done on FRDs: (a) comparing the I-V characteristics of FRD after thermal shock test with the number of different cycles, and (b) comparing the I-V characteristics of FRD electrical testing done at different temperature levels after 1000 cycles of thermal shock test.

Figure 8(a) shows the power cycling test results on IGBT, where the failure criteria was a 5% increase of Vce(on). A total of 53000 cycles were recorded at which the wirebonds started to deteriorate and lift off. An overshoot of Vec(on) was observed at around 63000 cycles due to die short-circuit, which was caused by the localized overheating on the surface of the chip after which wirebonds lift-off. Figure 8(b) also shows the wire bond lifted off confirmed by optical microscopy inspection. The wire bonds lift-off could have been caused by the degradation of the TLP solder material, which increased the thermal resistance. Figure 9 shows the structural function diagram (R_{th} vs. C_{th}) where an increase of the thermal resistance by the degradation of TLP die solder material after the power cycling test can be seen. Higher thermal resistance inside the TLP die solder area could increase the localized chip surface temperature and accelerate wirebond lift-off. Figure 9 shows the degradation of TLP solder material where thermal resistance increased after the power cycling test, confirming that the wire bond lifted off due to localized chip temperature increase.

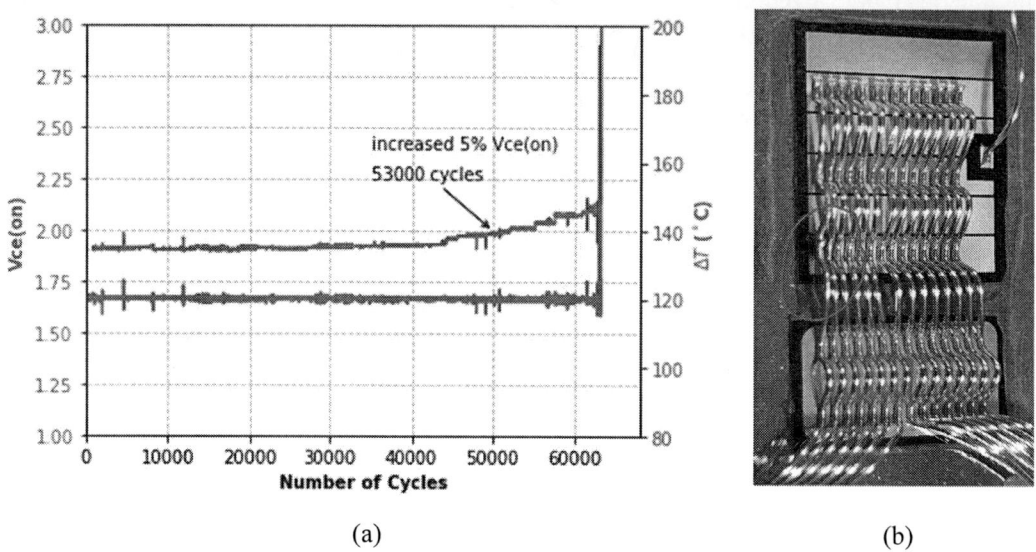

(a)   (b)

Fig. 8: (a) Number of cycles to failure with respect to increasing in Vce(on), and (b) Critical failure location of wire bonds during Power cycling test.

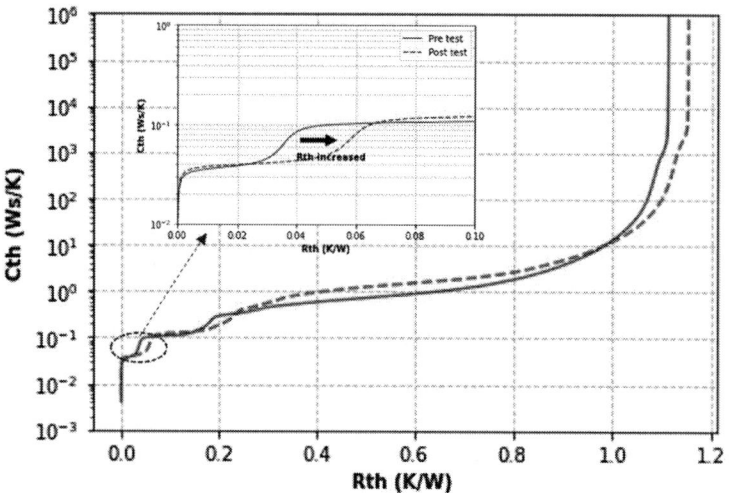

Fig. 9: Rth measurement of TLP soldered die sample using the structural diagram.

4. Conclusion

The TLP bonding for die-attach applications in power module packaging could replace the use of high Pb solder alloys and is a cost-effective solution over Cu/Ag sinter bonds. The TLP diffusion soldering process has been successfully optimized for large area die attach with a fully transformed high melting temperature intermetallic structures with a low percentage of soldered voids. The optimized TLP process improved the die shear strength to an average of 55 MPa, which helped to use SLID soldering of large size die on DBC substrate. The thermomechanical reliability and active power cycling tests confirmed that TLP diffusion soldering for large area die attach can apply to high temperature power module packaging applications.

References

[1] Zeng, Guang, Stuart McDonald, and Kazuhiro Nogita."Development of high-temperature solders." *Microelectronics Reliability* 52.7, 2012: pp.1306-1322.

[2] B. Schellscheidt, J. Richter, O. Lochthofen and T. Licht, "Throughput Optimization of a Sintering Die Attach Process," *2021 23rd European Microelectronics and Packaging Conference & Exhibition (EMPC)*, 2021, pp. 1-4.

[3] Calabretta, Michele, et al. "Silver Sintering for Silicon Carbide Die Attach: Process Optimization and Structural Modeling." *Applied Sciences* 11.15, 2021, pp. 7012.

[4] Kang, Hyejun, Ashutosh Sharma, and Jae Pil Jung. "Recent Progress in Transient Liquid Phase and Wire Bonding Technologies for Power Electronics." *Metals* 10.7, 2020, pp. 934.

[5] Shao, Huakai, et al. "Interfacial reaction and mechanical properties for Cu/Sn/Ag system low temperature transient liquid phase bonding." *Journal of Materials Science: Materials in Electronics* 27.5, 2016, pp. 4839-4848.

[6] Guth, Karsten, Dirk Siepe, Jens Görlich, Holger Torwesten, Roman Roth, Frank Hille, and Frank Umbach. "New assembly and interconnects beyond sintering methods." In *Proc. PCIM*, 2010. pp. 232-237.

[7] Attari, Vahid, Supriyo Ghosh, Thien Duong, and Raymundo Arroyave. "On the interfacial phase growth and vacancy evolution during accelerated electromigration in Cu/Sn/Cu microjoints." *Acta Materialia* 160, 2018, pp.185-198.

[8] Li, Xiang, Daohui Li, Guiqin Chang, Wei Gong, Matthew Packwood, Daniel Pottage, Yangang Wang, Haihui Luo, and Guoyou Liu. "High-Voltage Hybrid IGBT Power Modules for Miniaturization of Rolling Stock Traction Inverters." *IEEE Transactions on Industrial Electronics* 69, no. 2, 2021, pp. 1266-1275.

[9] Tatsumi Hiroaki, Adrian Lis, Hiroshi Yamaguchi, Tomoki Matsuda, Tomokazu Sano, Yoshihiro Kashiba, and Akio Hirose. "Evolution of transient liquid-phase sintered Cu–Sn skeleton microstructure during thermal aging." *Applied Sciences* 9, no. 1, 2019, pp. 157.

[10] Bao, Y.; Wu, A.; Shao, H.; Zhao, Y.; Liu, L.; Zou, G. Microstructural evolution and mechanical reliability of transient liquid phase sintered joint during thermal aging. J. Mater. Sci., 54, 2018, pp. 765–776.

Experimental Evaluation on Observer-Based Delay-Compensating Active Damping for LC-Filters

Michael Schütt, Hans-Günter Eckel
UNIVERSITY OF ROSTOCK
Albert-Einstein-Str. 2
D-18059 Rostock, Germany
Tel.: +49 (0) 381 – 498 7116
Fax: +49 (0) 381 – 498 7102
E-Mail: michael.schuett@uni-rostock.de
URL: http://www.iee.uni-rostock.de

Acknowledgments

This paper was made within the framework of the research project *Netz-Stabil* and financed by the European Social Fund (ESF/14-BM-A55-0015/16). This paper is part of the qualification program *Promotion of Young Scientists in Excellent Research Associations - Excellence Research Programme of the State of Mecklenburg-Western Pomerania.*

Keywords

«Active Damping», «Current observer», «Discrete-time», «Digital control», «Frequency-Domain Analysis»

Abstract

This work presents experimental results of a discrete-time observer-based delay-compensating active damping technique for LC filters. This method is state-feedback-based and consequently very robust regarding the system parameter such as the grid impedance. Nonetheless, the limits of this strategy are discussed. The experimental setup is a low voltage representation of a 5 MW wind converter. Much know-how was invested in transferring the electrical conditions from high to low power. Particular mention should be made of the adaptation concerning the naturally slower switches in the high voltage and the compensation of the higher damping of the passive components in the low voltage class.

Introduction

[1] presented an active damping technique for LC filters using observers. The paper outlines the premise that state-feedback-based solutions for active damping (such as [1–3]) are inherently more robust regarding the grid impedance compared to forward-path-based strategies (like [4]). Further, [1] explains that merely using state-feedback directly for fast signals such as LC-resonances for active damping can yield inferior results due to delays in the system, such as the computational delay of the controller itself. That paper, consequently, provides an observer-based solution to compensate for the delays in the system. These observer structures can also replace the necessity of a second current sensor stage. The presented outcomes utilize only inverter current and capacitor voltage measurements. Using another current measurement (grid or capacitor current) would decrease the complexity and improve the robustness of the proposed strategy.

This paper presents the experimental results of this observer-based active damping technique and outlines its limits.

Control Technique

Figure 1 illustrates the implemented current control scheme with active damping using a cascaded observer structure as proposed in [1].

Since the grid current is just an estimated disturbance and not an observed state, the observer's estimate-phase lags depending on the observer's controller bandwidth [1]. [1] further argues that the dead-beat observer implementation ensures that the current information lags by one sample step. Therefore, three stages of the cascaded observer structure (for $k + 2$ estimations) have to be implemented to establish a proper phase of the estimated grid-current, as shown in Figure 1. Note that a second current sensor stage significantly reduces the observer cascade's effort, and the grid current becomes an observed state instead of an estimated disturbance, yielding much better dynamic properties regarding phase lags and robustness.

However, the third stage of the cascaded observer structure lacks information on the inverter voltage of the sample instance $k + 2$. There might be the argument that the future state information and an estimate of the future current reference could also be used to estimate that voltage reference information.

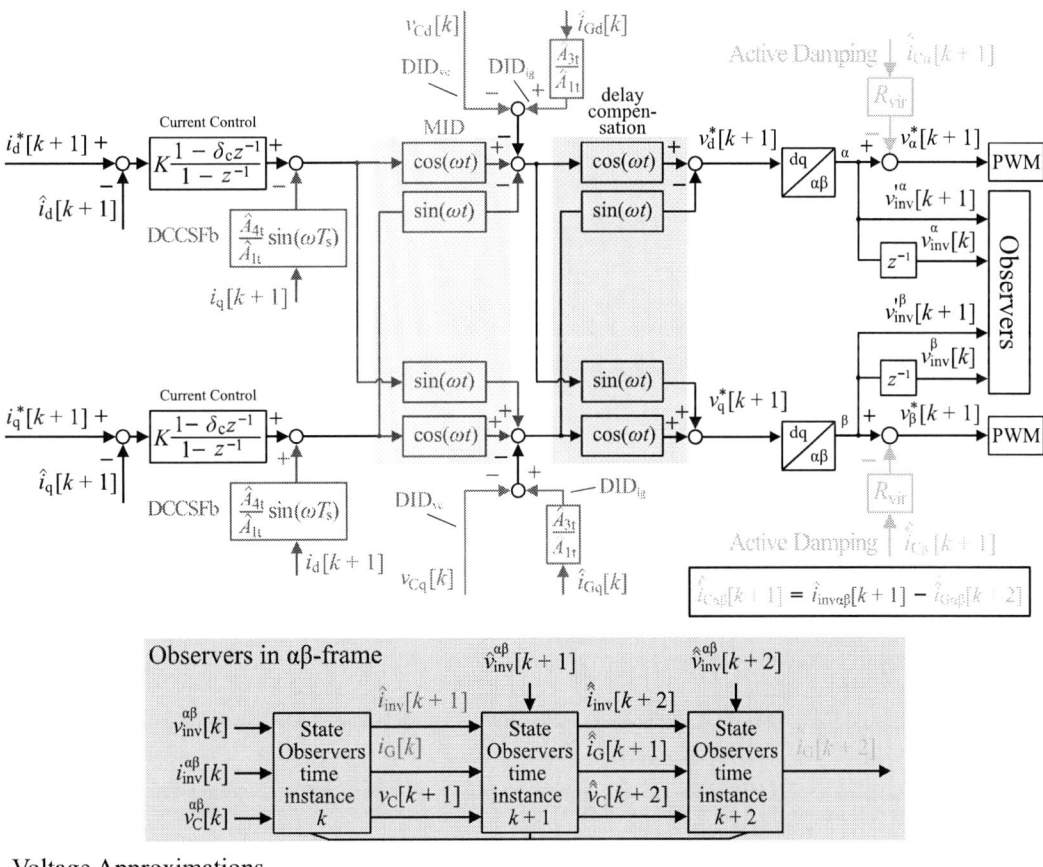

Figure 1: Block diagram of the current control structure with delay compensation via a cascaded Luenberger-style observer structure with active damping implementation and decoupling techniques: DCCSFb – Decoupling Cross-Coupling State Feedback, DIC – Disturbance Input Coupling; DID – Disturbance Input Decoupling (index: vc – capacitor voltage, ig – grid-current). DID – Disturbance Input Decoupling (for: vc-capacitor voltage, ig-grid current), MIC – Manipulated Input Coupling (from dq-transformation), MID – Manipulated Input Decoupling (inverse of MIC)

However, using the reference voltage of the previous sample instance does yield very close results already. Investigating the estimation of the voltage reference for the sample instance $k + 2$ could be a subject of future research nonetheless.

Simulation Results

The simulation results of this section are based on a 5 MW wind turbine model (parameter in Table 1). This model includes most of the dynamic behavior of the real system down to the switch level – such as computational delay and PWM, yielding an overall dead time of 2 μs.

Figure 2 shows the resulting frequency response estimation plots for varying parameter estimation errors up to 30 % with the current control structure of Figure 1 using active damping, disturbance input decoupling, decoupling cross-coupling state feedback based on the present, and future state information estimated from the proposed cascaded discrete Luenberger-style observer structure. It should be noted that no stable operation was found for operation without active damping for 3 kHz sampling (single-sampling) or active damping without delay compensation. The proposed delay compensation structure, however, achieves a very robust and well-damped system.

Experimental Setup

The experimental setup was designed with great care regarding the timing and resistivity of the introduced low voltage components. Very low resistive MOSFETs were chosen to represent the original IGBT switches and slowed down (via gate resistors) to match the switching and conduction behavior.

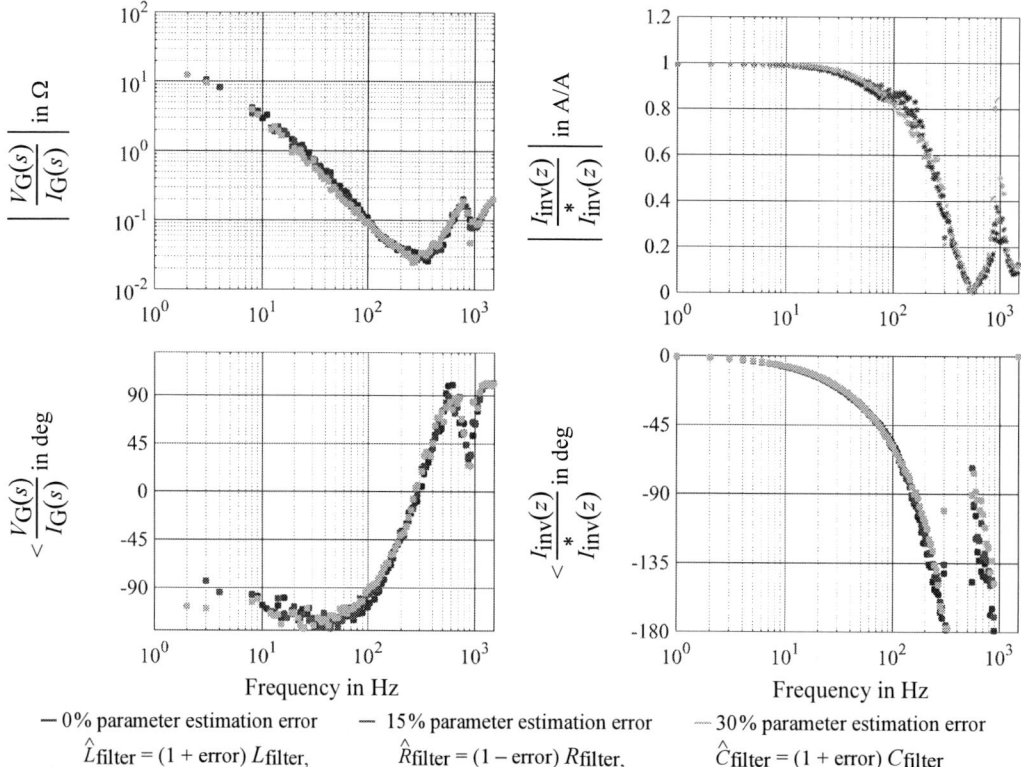

— 0 % parameter estimation error　　— 15 % parameter estimation error　　— 30 % parameter estimation error

$$\hat{L}_{\text{filter}} = (1 + \text{error})\, L_{\text{filter}}, \qquad \hat{R}_{\text{filter}} = (1 - \text{error})\, R_{\text{filter}}, \qquad \hat{C}_{\text{filter}} = (1 + \text{error})\, C_{\text{filter}}$$

Figure 2: *Simulation*: Dynamic analysis for different parameter estimation errors ($\hat{L}_{\text{filter}}$, $\hat{C}_{\text{filter}}$, $\hat{R}_{\text{filter}}$) of the discrete current controller for LC filters (w/ computational delay and 3 kHz PWM and 3 kHz single-sampling implementation) with active damping (Rvir = 30 mΩ) using a deadbeat-based cascaded discrete Luenberger-style observer structure for estimation of present and future state information for the grid-current, inverter current, and capacitor voltage. *Left*: dynamic stiffness (impedance seen from the grid), *Right*: command tracking (bode-plot)

Table 1 Parameters of the original 5 MW Wind Turbine and the scaled test bench representation

Parameter		5 MW Wind Turbine	25 V Lab test bench
L_f	– filter inductance	25 μH	379 μH (400 μH intended)
R_f	– filter resistance	0.3 mΩ	18 mΩ (5 mΩ intended)
C_f	– filter capacitance	4 mF	240 μF
V_{DC}	– DC-link voltage	1100 V	40 V
V_G	– grid voltage	690 V	25 V
I_N	– base current	4500 A	10 A
C_{DC}	– DC-link capacitance	92 mF	5.52 mF
f_{sw}	– switching frequency	3 kHz	3 kHz
t_{dead}	– dead time (half bridge switching)	appr. 2 μs	appr. 2 μs
Switches		1.7 kV Trench/Fieldstop IGBT 4	100 V HEXFET Power MOSFET

Further, inductive components scale very poorly into low-power applications. This meant that most components off the shelf had too high resistive attributes compared to the original 5 MW reference and thus would dampen the system too much to show any resonant properties. Therefore, the 25V components were overdesigned for their respective currents (factor of 10-100).

With single-sampling, the control was too oscillatory without active damping to find any working operating point. This instability was already present in the simulations.

Figure 3 shows the command tracking plot for different damping coefficients for single-sampling (3 kHz switching and 3 kHz sampling). The grid impedance of the laboratory setup is unknown. Thus, the simulation plots are not exact replicas, but the experiments yield similar attributes and trends. Future work includes both simulation and experimental work to estimate the full-state model of the system more accurately to reproduce the experimental results more closely.

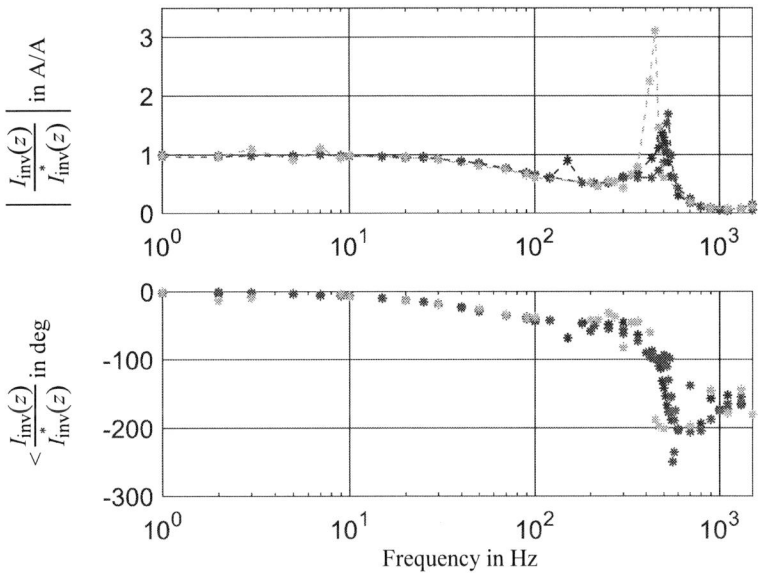

—*— optimal damping: R_{vir}=150 mΩ,   —*— over-damped: R_{vir}=220 mΩ,   —*— under-damped: R_{vir}=70 mΩ

Figure 3: *Measurement*: Command tracking FRF for different active damping coefficients of the discrete current controller for LC filters on the test bench setup (w/ computational delay, 3 kHz sampling) using a deadbeat-based cascaded discrete Luenberger-style observer structure for estimation of present and future state-information for the grid-current, inverter-current, and capacitor-voltage.

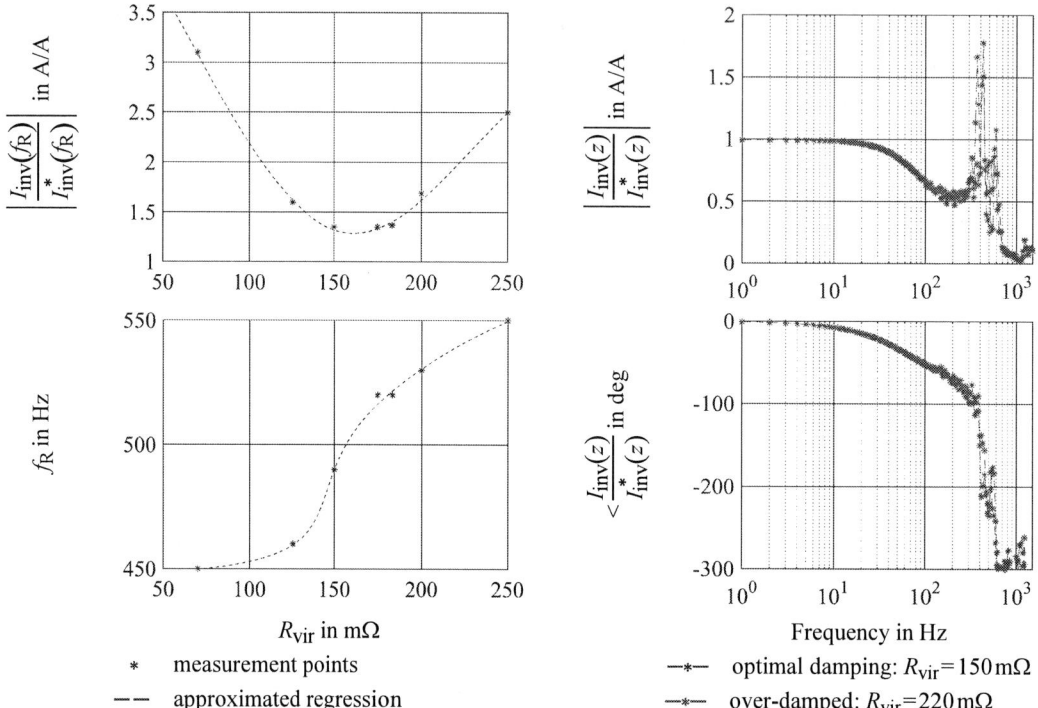

Figure 4: *Left*: *Measurement*: *Top*: Oscillatory peak in the command tracking FRF (*Frequency Response Function*) over the damping coefficient R_{vir}, *Bottom*: Frequency of the oscillatory peak in the command tracking FRF over the damping coefficient — *Right*: *Simulation*: Command tracking FRF with reduced grid-impedance by one order of magnitude – similar behavior to measurements achieved.

In a first attempt, the lab's grid connection is estimated to be an order of magnitude stronger than the original wind turbine application. The anti-resonance is not visible within the Nyquist frequency, which indicates a very low grid-impedance.

This strong grid example seems unrealistic for the given application. However, the effect was investigated further. Figure 4 (right) shows similar attributes in simulations with much smaller grid

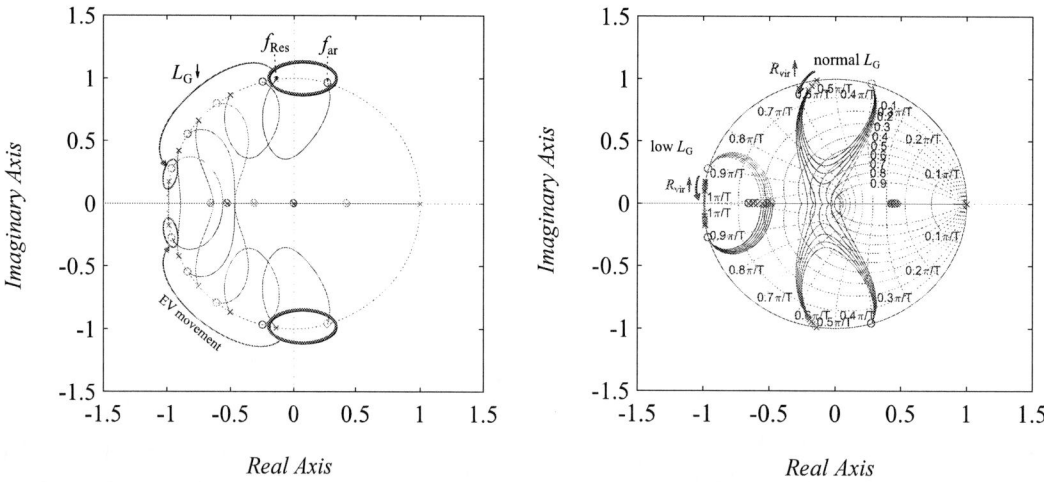

Figure 5: *Left*: eigenvalue movement of the LC-plant for decreasing grid impedance (L_G); *Right*: eigenvalue movement for increasing R_{vir} for two cases of grid impedances (low and normal).

impedances. The location of the EVs depends on the grid impedance. These grid impedances can change over time, but most applications should not drastically change in the order of magnitude.

Figure 5 (left) shows the eigenvalue movement for the variation in grid impedances. Further, Figure 5 (right) shows that lower damping coefficients yield stronger relative eigenvalue movement for a low grid impedance. In this case, over-damping and instability happen much sooner than in the weaker grid example. Figure 4 (left) documents the iterative tuning process of the damping coefficient for the experimental setup.

To summarize, the order of magnitude of the grid impedance dictates the order of magnitude for effective damping coefficients. Therefore, in the case of grid-impedances varying in large regions, adaptive tuning of the damping coefficients is necessary to guarantee stability and well-behaved dynamic attributes for the proposed control scheme.

It should be noted that double sampling (6 kHz sampling for 3 kHz switching) solves most of the discussed issues. The current control was stable even without active damping in that case. However, the produced harmonic impedance of the inverter seen from the grid does not feature the well-damped resonant properties without the active damping (Figure 2 – *left*). A high bandwidth disturbance source must be installed to show these impedance attributes in the test bench case. This hardware implementation is part of the upcoming work.

Figure 6 shows the time domain plots of the α and β inverter currents (grid currents are almost perfect sinusoidal in comparison). The prominent harmonics of the single-sampling version are not caused by the resonance but are in the 150 Hz region (in dq 150 Hz, in αβ in the 100 Hz and 200 Hz region). These issues are caused mainly by the single-sampling PWM scheme, which is discussed in much detail in [5].

Further, measuring currents of inductive components during the maxima and minima of the PWM carrier works very well in filtering out the fundamental switching harmonics. However, using the same sampling instance for the voltage measurement at the capacitors does cause further harmonic issues, which also get attenuated with the double-sampling scheme.

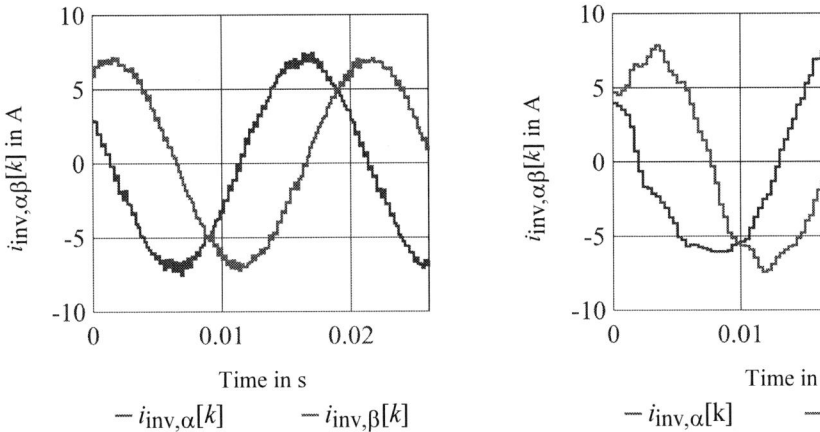

Figure 6: Measurements: Time domain plot of the α,β inverter currents with the proposed current control scheme for 6 kHz double-sampling on the *left* and 3 kHz single-sampling on the *right*.

1 — FPGA Xilinx Spartan

2 — DSP TMS320

3 — Sensor inputs (OP-amps)

4 — PWM-output signals

5 — Line driver

6 — RS232: programming & reading

Figure 7: Photographs of the control board of the 25 V test bench − DSP/FPGA combination and peripherals.

Conclusion

This work demonstrates an observer-based algorithm for active damping implemented on a discrete DSP/FPGA-based system (Figure 7 and Figure 8). Furthermore, the design choices are closely linked to the impedance seen from the grid (dynamic stiffness), as this metric provides the most significant insight as to whether actual damping is provided to the overall system. The control structure only measures the inverter current and capacitor voltage. Introducing another current measurement stage would vastly decrease computational effort and dramatically increase overall robustness.

The simulation and experimental results illustrate that the proposed active damping scheme provides vital and dynamically well-behaved damping attributes.

The simulation and experimental results display effective damping of the LC resonance without information on the grid-impedance. While the proposed method does not need the information on the grid impedance, a very strong grid (an order of magnitude smaller than the original setup) still causes issues for the system's stability.

1 — PWM-input signals (from FPGA)

2 — Isolation chips

3 — DC-link

4 — Half-bridge drivers

5 — Low-side phase switches

6 — High-side phase switches

7 — Chopper

8 — Placeholder for alternative microcontroller

Figure 8: Photographs of the inverter board of the 25V laboratory test bench − phase switches, drivers, and peripherals.

References

[1] M. Schütt and H. Eckel, *Discrete active damping control design for LCL-resonances using observers*, 2019 21st European Conference on Power Electronics and Applications (EPE '19 ECCE Europe), 2019, pp. P.1-P.10, doi: 10.23919/EPE.2019.8914981.

[2] B. Abdeldjabar, Xu Dianguo, X. Wang, and F. Blaabjerg, *Robust active damping control of LCL filtered grid connected converter based active disturbance rejection control*, in: 2016 IEEE 8th International Power Electronics and Motion Control Conference (IPEMC-ECCE Asia), 2016, pp. 2661–2666.

[3] H. Ge, Y. Zhen, Y. Wang, and D. Wang, *Research on LCL filter active damping strategy in active power filter system*, in: 2017 9th International Conference on Modelling, Identification and Control (ICMIC), 2017 9th International Conference on Modelling, Identification and Control (ICMIC), 2017, pp. 476–481.

[4] J. Dannehl, C. Wessels, and F.W. Fuchs, *Limitations of Voltage-Oriented PI Current Control of Grid-Connected PWM Rectifiers With LCL Filters*, IEEE Transactions on Industrial Electronics 56 (2009), pp. 380–388

[5] C. M. Wolf, M. W. Degner, and F. Briz, *Analysis of current sampling errors in PWM*, VSI drives, in: 2013 IEEE Energy Conversion Congress and Exposition, 2013, pp. 1770–1777.

[6] B. Abdeldjabar, Xu Dianguo, X. Wang, and F. Blaabjerg, Robust active damping control of LCL filtered grid connected converter based active disturbance rejection control, in: 2016 IEEE 8th International Power Electronics and Motion Control Conference (IPEMC-ECCE Asia), 2016, pp. 2661–2666.

[7] H. Ge, Y. Zhen, Y. Wang, and D. Wang, Research on LCL filter active damping strategy in active power filter system, in: 2017 9th International Conference on Modelling, Identification and Control (ICMIC), 2017 9th International Conference on Modelling, Identification and Control (ICMIC), 2017, pp. 476–481.

[8] H. Yuan and X. Jiang, A simple active damping method for Active Power Filters, in: 2016 IEEE Applied Power Electronics Conference and Exposition (APEC), 2016 IEEE

[9] Hu Wei, Zhou Hui, Sun Jian-jun, Jiang Yi-ming, and Zha Xiao-ming, Resonance analysis and suppression of system with multiple grid-connected inverters, in: 2015 IEEE 2nd International Future Energy Electronics Conference (IFEEC), 2015 IEEE 2nd International Future Energy Electronics Conference (IFEEC), 2015, pp. 1–6.

Influence of static rotor imbalance on the roller bearing damage due to inverter-induced bearing currents

Martin Weicker, Omid Safdarzadeh, Andreas Binder
Technische Universität Darmstadt
Institute for Electrical Energy Conversion
Landgraf-Georg-Straße 4
64283 Darmstadt, Germany
Tel.: +49 / (0) 6151 16 24191
E-Mail: mweicker@ew.tu-darmstadt.de
URL: http://www.ew.tu-darmstadt.de

Acknowledgements

This IGF Project IGF-Nr. 21488 N (FVA Nr. 650 III) of the Research Association for Drive Technology (FVA) was supported via AiF within the program for promoting the Industrial Collective Research (IGF) of the Federal Ministry of Economic Affairs and Climate Action (BMWK), based on a resolution of the German Parliament. We acknowledge also the support of our industrial partner *SEW Eurodrive.*

Keywords

«bearing currents», «voltage source inverter», «induction motor», «static rotor imbalance»

Abstract

The influence of static rotor imbalance on the generation of bearing fluting due to bearing currents is experimentally investigated. For that six inverter-fed induction machines with the rated power of 1.5 kW and three with the rated power of 11 kW were tested with especially prepared bearings. The machines were operated with three different static rotor imbalances (U0, U1, U2) to test the imbalance influence on the fluting effect due to discharge and rotor-to-ground bearing currents. The damaging effect of inverter-induced bearing currents is influenced by the parameters "rotor speed", "bearing temperature", "bearing grease" and "grounding conditions" of the inverter-motor-system. The impact of a static rotor imbalance on the damaging process caused by the parasitic bearing currents is so far unknown. Therefore it shall be clarified, if the combination of static rotor imbalance and bearing currents may damage the bearing raceway surfaces much faster than without imbalance. As a concluding result our test procedures showed no acceleration of bearing deterioration due to static rotor imbalances.

Introduction

High frequency bearing currents may occur due to the common-mode inverter output voltage in inverter-fed AC machines [3]. If the apparent bearing current density surpasses a limit of 0.1 ... 0.3 A/mm² the bearing currents may cause damage and thus increased wear to the mechanical roller bearings [3]. This might lead to unexpected early bearing failures, increasing the machine downtime and reducing the bearing lifespan. The four main types of these inverter-induced bearing currents are circulating, electrical discharge machining (EDM), capacitive and rotor-to-ground bearing currents [3]. Parameters that influence the type and magnitude of these bearing currents are rotor speed, radial and axial bearing forces, bearing temperature, viscosity and conductivity of the bearing grease, the resulting electrical impedance of the bearing and the grounding conditions of the inverter-motor-system [3], [4], [6].

Early research on the mechanism of bearing damage due to electrical bearing currents was already done in 1943 [5], focusing on the fluting damage. The initiation mechanism of fluting from starting with craters on the raceway due to "sparking" in the lubrication film between raceway and roller elements, leading to vibrations and the raceway fluting pattern, is still not understood in detail. So also recent publications deal with the propagation of fluting at the bearing raceway [2]. The process of generation of the fluting pattern on the bearing raceway surface due to bearing currents is still under research. We investigated, if an additional mechanical vibration may accelerate the fluting process and may damage the bearing surface faster. For that, we investigated the influence of different static rotor imbalances at TEFC standard cage-induction machines with a rated power of 1.5 kW and 11 kW, respectively, in combination with EDM and rotor-to-ground bearing currents, when being operated at voltage source PWM IGBT inverters.

Test set-up and measurement system

Figure 1 a) shows for the rated motor power of 1.5 kW the three investigated four-pole ($2p = 4$) induction machines with a squirrel cage rotor. Figure 3 shows one of the investigated four-pole induction machines with rated power of 11 kW with its additional rotor imbalance. The machine stator windings are fed by IGBT-two-level-voltage source inverters with $U_d = 560$ V DC link voltage, type FC-302, Company *Danfoss*. At all measurements the switching frequency of the inverters is kept constant at 10 kHz. The induction machines are modified for bearing current measurements by insulated bearing seats. This bearing seat insulation is bridged by a short copper cable to measure the by-passing bearing currents as i_m (Fig. 7) via a current probe, IWATSU SS-250, instead of the real bearing current i_b. The schematic cross section of the machine in Figure 1 b) exhibits also the sliding carbon brush contact to measure the bearing voltage drop u_b.

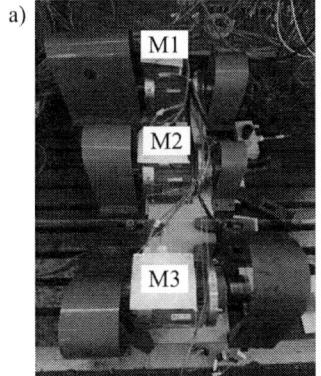 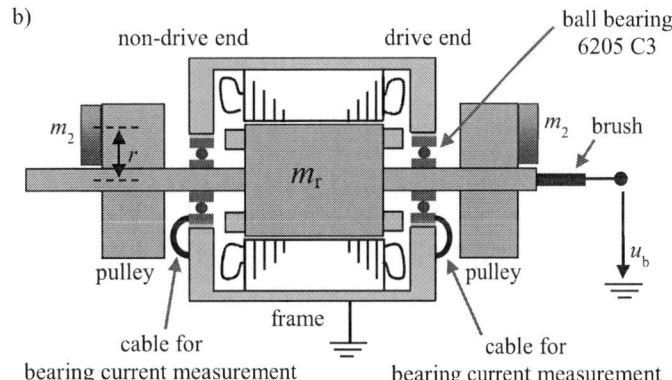

Fig. 1: a) Three TEFC standard cage induction machines M1, M2, M3 with rated power 1.5 kW at 400 V AC, are fed by three IGBT-two-level-voltage source inverters with $U_d = 560$ V DC link voltage, type FC-302, Company *Danfoss*. b) Schematic cross section of one cage induction machine (m_r: rotor mass) with an additional imbalance rotor mass m_2, mounted on the pulley at the drive end and non-drive end side. Both ball bearings, type 6205 C3, are electrically insulated towards the bearing seat. This bearing seat insulation is bridged via a short copper cable to measure the by-passing bearing currents via a current probe.

The motors were equipped at NDE and DE on the shaft with pulleys to increase the radial forces via the belt tension. Here the belt is removed to study only the influence of radial bearing forces, caused by rotor mass gravity, by the single sided magnetic pull due to a parasitic rotor eccentricity and by the rotor static imbalance forces. Figure 2 shows one of the three test motors of the test bench (Fig. 1 a) with a rated power of 1.5 kW with additional rotor masses attached on both pulleys (with removed belt) on drive end and non-drive end side, pointing in the same radial direction to cause an additional static rotor imbalance. Without the masses only the natural individual rotor imbalance U0 is present. With two different masses, equal at NDE and DE, two different bigger imbalances U1 and U2 are created. The imbalance U1 is simply realized by the cylindrical mass m at the radial distance r of the rotor shaft center (Fig. 2b), yielding $\Delta U1 = 2 \cdot m \cdot r$, with masses m on both NDE and DE. With (1) the additional imbalance $\Delta U2$ of the two semicircular "U-shaped" masses (Fig. 2 c) is calculated by inner and outer radii r_1 and r_2 as well as the axial length l and the mass density γ of steel with 7.9 g/cm³. The fixing screw is assumed to fill completely the drilled hole.

$$\Delta U2 = \tfrac{2}{3} \cdot l \cdot \gamma \cdot (r_2^3 - r_1^3) \cdot 2 \tag{1}$$

The resulting radial bearing force F_r consists of the rotor gravity force $F_g = m_r \cdot g$, the force $F_U = U \cdot (2\pi \cdot n)^2$ caused by static rotor imbalance and the inevitable single-sided magnetic pull F_m of the induction motor. With horizontal shaft, the rotor gravity force F_g acts at all time vertically downwards. The static rotor imbalance F_U rotates with the rotor speed n. The single-sided magnetic pull F_m acts in the direction of the smallest air gap δ at an assumed residual rotor eccentricity [1], which is at least one half of the radial internal bearing clearance of e.g. $P_d/2 = 10.25\ \mu m$ at the ball bearing type 6205 C3. The single-sided magnetic pull F_m is calculated in Table I by (2), [1], considering the stator inner diameter d_{si}, the axial iron core length l_{Fe}, the radial flux density amplitude in the air gap $\hat{B}_{\delta s1} = 1\ T$ and the relative eccentricity ε, which is the eccentricity e related to the radial air gap width δ according to (3). The value $e = P_d/2$, but might be increased by a) static and b) dynamic rotor eccentricity due to a) manufacturing influence and b) due to rotor shaft bending [1]. Here is $e = P_d/2$, $\varepsilon = 4\%$.

$$F_m = \tfrac{\pi}{4 \cdot \mu_0} \cdot d_{si} \cdot l_{Fe} \cdot \varepsilon \cdot \hat{B}_{\delta s1}^2 \qquad\qquad 2p \geq 4 \tag{2}$$

$$\varepsilon = \tfrac{e}{\delta} \tag{3}$$

Table I gives an overview of the calculated mechanical and magnetic forces F_g, F_U and F_m at the two test benches with the 1.5 kW- and 11 kW-induction motors, where $e = P_d/2$ is assumed. The static rotor imbalance force is calculated with the highest investigated imbalance level $\Delta U2$ at a rotor speed of 1000 min^{-1}.

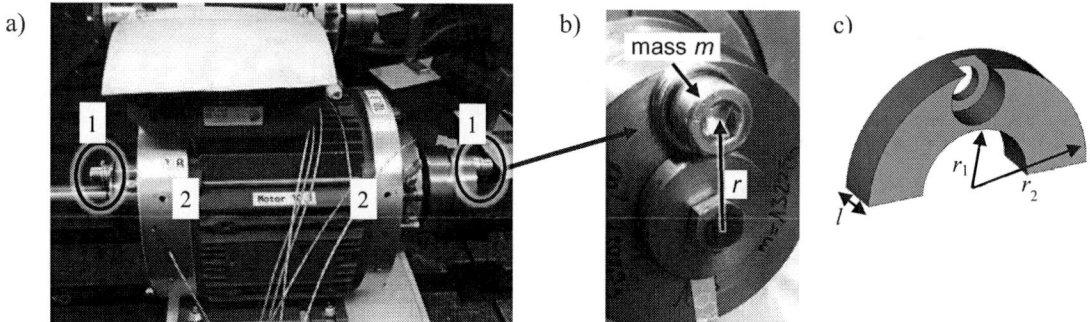

Fig. 2: a) Four-pole squirrel cage induction machine 1.5 kW, 400 V Y, prepared for bearing current measurements, with the additional static rotor imbalance $\Delta U1$ (1), according to Fig. 2 b). At both bearing end shields are mounting holes (2) in radial direction for the fixation of the vibration sensor.
b) Static rotor imbalance $\Delta U1 = 1500$ g·mm by the two additional masses m, here shown at the drive end side of the machine, at a distance r from the center of the rotor shaft. c) Sketch of one of the two additional "U-shaped" masses for the increased static rotor imbalance $\Delta U2 = 6264$ g·mm.

Fig. 3: a) One of the three four-pole TEFC squirrel cage standard induction machines with the rated power of 11 kW, prepared for bearing current measurements, with an additional static rotor imbalance $\Delta U2$ (1). b) Additional "U-shaped" mass m to increase the static rotor imbalance to $\Delta U2 = 40031$ g·mm at the drive end and non-drive end side of the machine.

Table I: Calculated rotor gravity forces, static rotor imbalance forces and magnetic forces of the two investigated induction machine power levels

P_N / kW	m_r / kg	F_g / N	$\Delta U2$ / (g·mm)	ΔF_{U2} / N ($n = 1000$ min^{-1})	F_m / N
1.5	7.5	74	6264	68.7	99.9
11	39.6	388	40031	439.0	241.0

The residual imbalance due to the rotor manufacturing process could be measured on the balancing rotor test bench. Here for simplicity it was assumed that the maximum residual imbalance G6.3 according to DIN ISO 1940 Part 1 occurred at rated speed $n_N = 1500$ min^{-1}. The corresponding residual distance e_S of the rotor center of gravity from the rotor rotational axis is $e_S = G/(2\pi \cdot n_N)$. Hence the residual rotor imbalance force is $F_{U,res} = m_r \cdot e_S \cdot (2\pi \cdot n)^2$ at the rotor speed n, yielding U0 $= m_r \cdot e_S$. The resulting increased rotor imbalances are U1 $= \Delta$U1 + U0 and U2 $= \Delta$U2 + U0.

Figure 4 shows the calculated orbit of the resulting radial bearing force F_r during motor operation at the investigated 1.5 kW-induction machines at a rotor speed of $n = 1000$ min^{-1} with residual imbalance (Fig 4. a) and increased imbalance (Fig 4. b), always with additional single sided magnetic pull F_m. Only between three and five of the nine balls of the bearing 6205 C3 are radially loaded during motor operation at the lower half of the bearing, labeled as bearing load zone.

If the rotor speed n rises strongly or the static rotor imbalance U increases, the resulting radial bearing force is increased in that way, that the bearing load zone rotates in the bearing at the whole bearing circumference raceway.

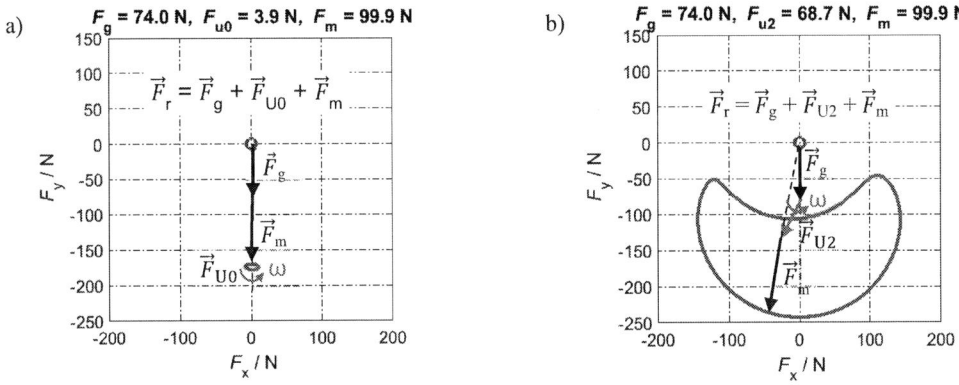

Fig. 4: Calculated orbit of the radial bearing force vector $\vec{F}_r$ during motor operation of the 1.5 kW-induction machine at a rotor speed $n = 1000$ min^{-1} ($\omega = 2\pi \cdot n$). The single-sided magnetic pull F_m acts in direction of the smallest air gap.
a) Considering rotor mass gravity force F_g, magnetic pull F_m and rotational force due to residual imbalance $F_{U,res} = F_{U0}$.
b) Like a), but with increased static rotor imbalance force F_{U2}.

Bearing Current Measurements with static rotor imbalance

EDM bearing currents at 1.5 kW-induction motor

Figure 5 shows the measured peak-to-peak EDM bearing current amplitudes immediately after lubrication film breakdown of induction motor M3 (rated power 1.5 kW) at three different static rotor imbalances U0, U1, U2 (Table II). The shown values are the average of 50 EDM events (z) along with the 1σ-deviation band ($z \pm \sigma$), measured in steps of $\Delta n = 150$ min^{-1}. In case of Fig. 5 c) without additional static rotor imbalance only the residual imbalance U0 is active. Here the EDM bearing currents occur in a speed range of 300 … 1900 min^{-1} with an average bearing current peak-to-peak value of $I_{m,av} = 0.65$ A. Below 300 min^{-1} the lubrication film is too thin, so metallic contact may occur, prohibiting a film breakdown. Above 1900 min^{-1} the film thickness is big enough, so the electric field strength $E_b = \widehat{U}_b/h$ is smaller than the breakdown field strength E_D. The voltage $\widehat{U}_b$ is the bearing voltage shortly before the EDM breakdown occurs (Fig. 5d). The higher the rotor imbalance is, the smaller is the rotor speed range, where EDM bearing currents occur as well as the higher is the average bearing current peak-to-peak value up to $I_{m,av} = 1.10$ A. In the grey marked rotor speed range, there was no measurement possible due to high mechanical motor vibrations, caused by the additional rotor imbalance. An explanation for the reduced speed range of EDM currents at higher imbalance is missing until now. Obviously the lubrication film thickness is increased.

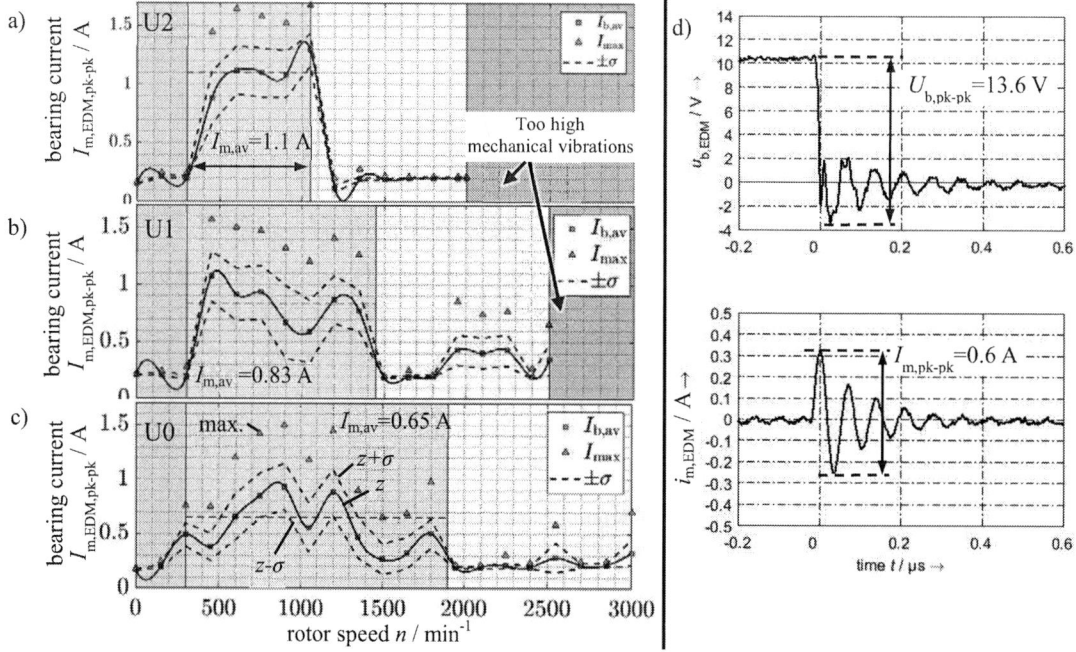

Fig. 5: Measured peak-to-peak values of EDM bearing currents over rotor speed (0 ... 3000 min^{-1}) at NDE side of inverter-fed motor M3 for a) a high static rotor imbalance U2, b) small static rotor imbalance U1, and c) with residual rotor imbalance U0 due to manufacturing process. d) Measured typical EDM bearing voltage and current at NDE side of the inverter-fed motor M3 at rotor speed $n = 750$ min^{-1} and with rotor imbalance U0.

Rotor-to-ground bearing currents at 1.5 kW-induction motor

The rotor of the 1.5 kW-test motor M3 (Fig. 1a) is grounded via a rotor-sliding carbon brush connected to a grounding cable. To force the bearing current through the test bearing at the DE side, the NDE side bearing is electrically insulated towards the bearing end shield. The resulting equivalent capacitive circuit of the electric machine is shown in Fig. 7. Figure 6 shows the measured peak-to-peak rotor-to-ground bearing currents $I_{m,DE}$ of the 1.5 kW-induction motor M3 at the three different static rotor imbalances U0, U1, U2 (Table II). The rotor-to-ground bearing current i_b occurs at each change of the inverter output common-mode voltage u_{CM} as a common-mode effect and is limited by the impedance of the whole grounding path from stator winding potential to ground potential. So the influence of the bearing impedance is only a small part of that impedance and is due to $R_{b,DE} \gg 1/(\omega_{HF} \cdot C_{b,DE})$ mainly an ohmic bearing current. This current is obviously only slightly increasing with speed and with imbalance, so it is mainly independent of additional rotor imbalance forces. The i_m-occurrence rate is $6 \cdot 10$ kHz = 60 kHz.

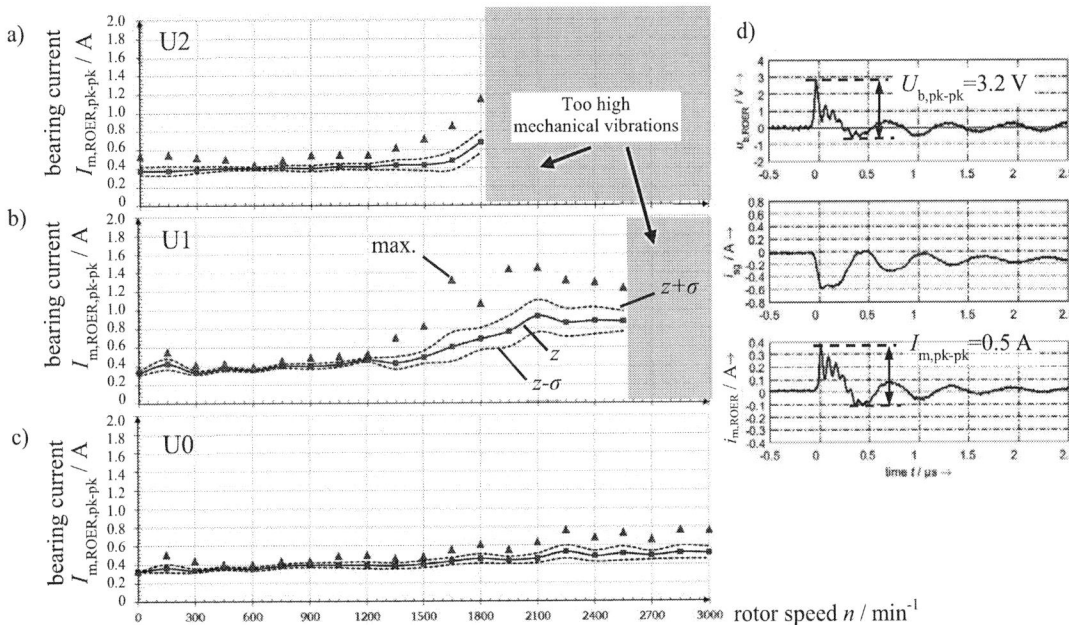

Fig. 6: Like Fig. 5, but with rotor-to-ground bearing currents i_{rg} (abbreviation: ROER) via the rotor grounding. The average of 50 current events (z) is shown along with the 1σ-deviation and the maximum amplitude (max.) out of 50 events. d) Measured typical rotor-to-ground bearing voltage, stator-to-ground current i_{sg} and bearing current at DE side of inverter-fed 1.5 kW motor M3 at rotor speed $n = 1500$ min^{-1} and with rotor imbalance U1.

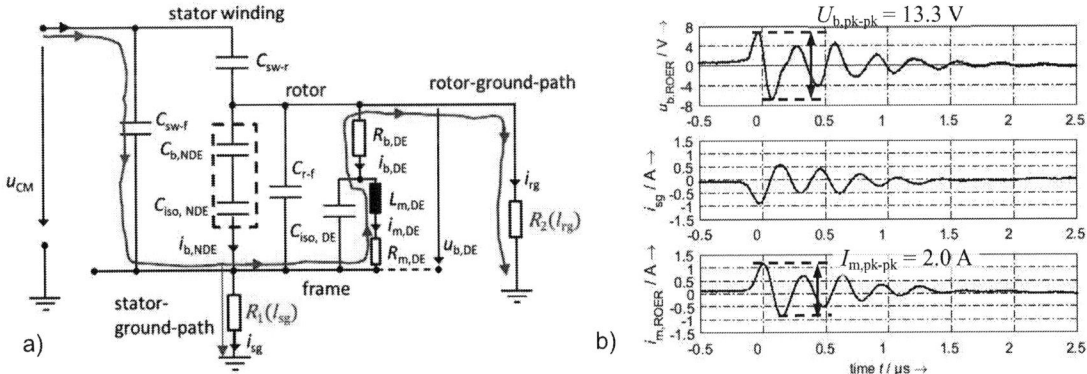

Fig. 7: a) Equivalent capacitive circuit for cage induction-motor without considering the stator winding inductance, for HF AC common-mode currents and a rotor-to-ground bearing current i_{rg}. The test bearing is at DE. The NDE bearing is electrically insulated towards the bearing end shield.
b) Measured typical rotor-to-ground bearing voltage, stator-to-ground current i_{sg} and bearing current at DE side of inverter-fed 11 kW-induction motor of long-term test of Fig. 13 a) at 1022 h.

Table II: Calculated static rotor imbalances of the two investigated induction machines

P_N / kW	Maximum permissible residual imbalance U0 at balance quality G6.3 / (g·mm)	ΔU1 / (g·mm)	ΔU2 / (g·mm)
1.5	352	1498	6322
11	1702	not tested	40379

Long-term test with EDM and rotor-to-ground bearing currents at 1.5 kW-induction motor

The change in the bearing raceway surface is analyzed after 1000 h-long-term tests with EDM (Fig. 8) and rotor-to-ground bearing currents (Fig. 9) at increased imbalances ΔU1 and ΔU2. In case of EDM bearing currents and a static rotor imbalance ΔU1 only a grey racetrack surface (grey "frosting") at the bearing surface was detected after 1000 h of operation, as the EDM bearing currents occur only with roughly 1/1000 of the rotor-to-ground occurrence rate. So the craters on the racetrack due to sparking are smoothened by the by-passing roller elements between two EDM current events.

With rotor-to-ground bearing currents and a static rotor imbalance U2 the bearing raceway surface of the lower half of the outer bearing ring shows fluting after 1000 h of operation, with the bearing current amplitude is much smaller than with EDM currents. Due to the 60 kHz-occurrence rate there is no time for smoothing the raceway by the roller elements between two bearing current events. So the fluting process may be started.

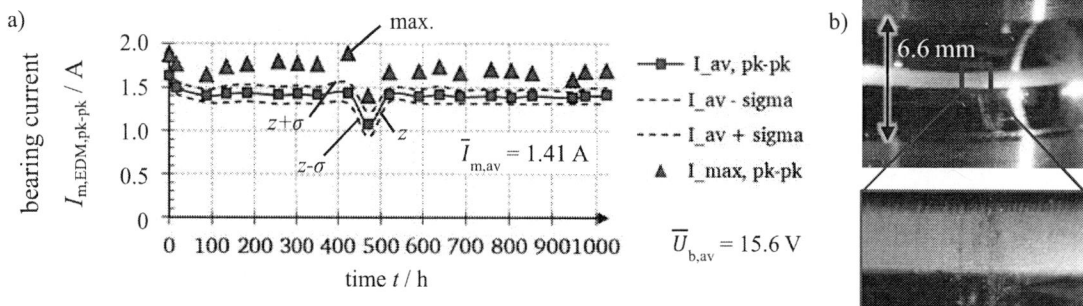

Fig. 8 a) Static rotor imbalance ΔU1: Measured peak-to-peak values of EDM bearing currents at the 1.5 kW-induction motor M3 at NDE side during operating time of 1024 h at rotor speed $n = 1000$ min^{-1} and average bearing temperature $\vartheta_b = 50\ °C$.
b) Grey racetrack on outer bearing raceway after 1024 hours of operation.

With an average bearing current peak-to-peak value of $I_{m,pk\text{-}pk} = 0.37$ A and an analytically calculated *Hertz*'ian area of $A_{Hertz} = 0.434$ mm², the apparent bearing current density [3] is calculated for rotor-to-ground bearing currents as $J_b \approx (2/3) \cdot I_{m,pk\text{-}pk}/A_{Hertz} = 0.57$ A/mm², but $J_b \approx 2.17$ A/mm² for EDM currents. With an apparent bearing current density limit of 0.1 A/mm² for DC and low frequency AC bearing currents, due to 0.57 A/mm² > 0.1 A/mm² fluting is predicted [3]. The 1/1000-times lower occurrence rate of EDM bearing currents allows the surface smoothening effect, so the much higher value $J_b \approx 2.17$ A/mm² > 0.1 A/mm² does not lead to fluting.

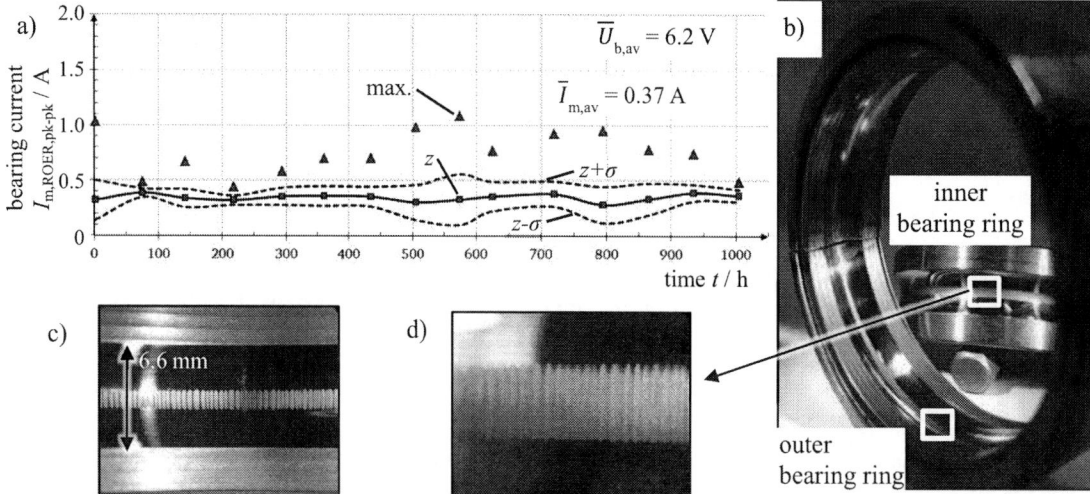

Fig. 9 a) Static rotor imbalance ΔU2: Measured peak-to-peak values of rotor-to-ground bearing currents at 1.5 kW-induction motor M3 at DE side during operating time of 1024 h at a rotor speed $n = 1000$ min^{-1} and at an average bearing temperature $\vartheta_b = 50\ °C$.
b) Damaged bearing with fluting at the racetrack surface of the lower half part (load zone) of the outer bearing ring, type 6205 C3.
c) Fluting at the surface only in the load zone of the outer bearing ring.
d) Fluting at the whole surface of the inner bearing ring, due to continuous ring rotation.

Figure 10 shows the measured mechanical radial acceleration at the DE side of the bearing end shield versus frequency during the operating time t. The mechanical acceleration at the characteristic frequency of 2.4 kHz exceeds at 700 h the limit of 2 m/s² for safe bearings, according to [9]. The fluting effect for rotor-to-ground bearing currents was not accelerated by increased imbalances $\Delta U1$, $\Delta U2$ in comparison to U0, whereas with EDM bearing currents no fluting occurred, independent of U0, $\Delta U1$, $\Delta U2$.

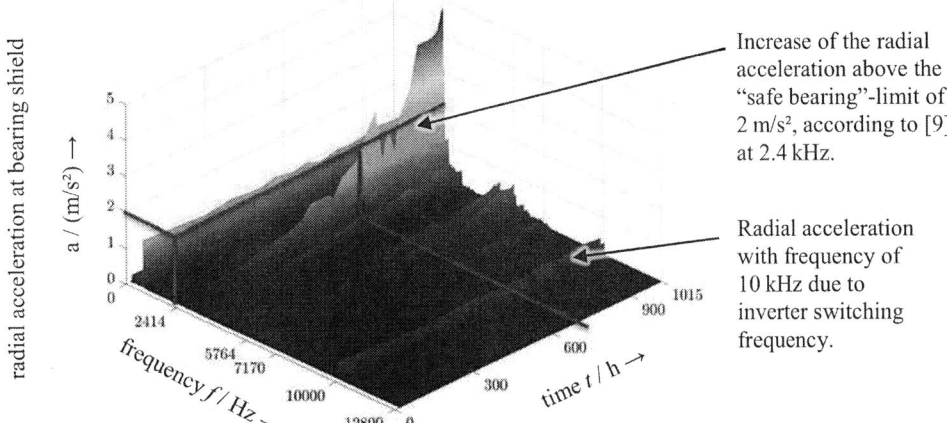

Increase of the radial acceleration above the "safe bearing"-limit of 2 m/s², according to [9] at 2.4 kHz.

Radial acceleration with frequency of 10 kHz due to inverter switching frequency.

Fig. 10 a) Effect of the static rotor imbalance $\Delta U2$: Measured radial mechanical acceleration (rms-value) with *Smart Balance 2* from Company *Schenck* at the DE bearing end shield of the 1.5 kW-induction motor M3 during operating time of 1024 h. Test with rotor-to-ground bearing currents, rotor speed $n = 1000$ min^{-1}, average bearing temperature $\vartheta_b = 50\ °C$, inverter switching frequency $f_{IGBT} = 10$ kHz.

Long-term test with EDM and rotor-to-ground bearing currents with 11 kW-induction motors

The three investigated four-pole 11 kW-induction machines (Fig. 3) were operated in parallel for 1040 hours with increased static rotor imbalance $\Delta U2$ (Table II) with the occurrence of EDM bearing currents, caused by the sparking of the lubrication film due to the common-mode bearing voltage exceeding frequently the discharge threshold. The measured peak-to-peak EDM bearing currents at the DE side of the 11 kW-induction motor M1 in dependence of the operating time are shown in Fig. 11.

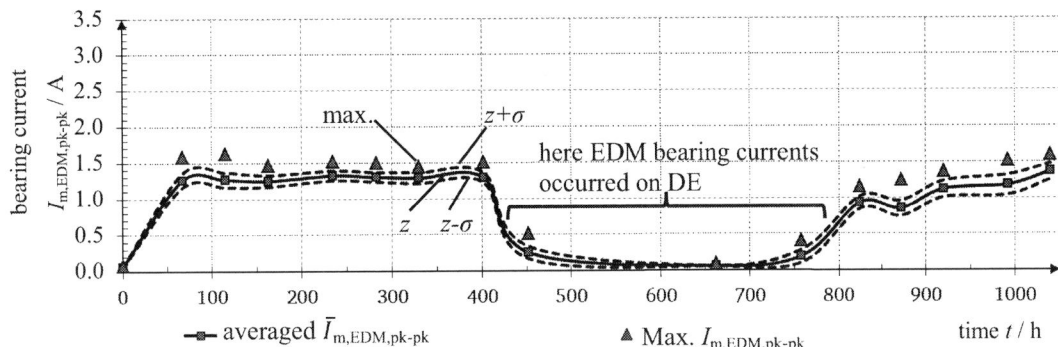

Fig. 11: Static rotor imbalance $\Delta U2$: Measured peak-to-peak EDM bearing currents at the 11 kW-induction motor M1 at NDE side during operating time of 1040 h at rotor speed $n = 1000$ min^{-1} and average bearing temperature $\vartheta_b = 60\ °C$.

EDM bearing currents occurred, which are statistically distributed between the bearing DE and NDE side. Table III shows the results of the long-term test with all three 11 kW-induction machines. The maximum apparent bearing current density occurs with 0.41 A/mm² at non-drive end side of motor M2. A grey raceway (Fig. 12a) occur on the investigated ball bearings, although the maximum apparent bearing current density was rather big 0.41 A/mm² > 0.1 A/mm², due to the low occurrence rate of EDM events. If $J_b < 0.1$ A/mm², even no "grey frosting" occurs, but the raceway surface is unchanged Fig. 12b).

 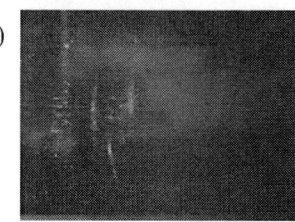

Fig. 12: According to Fig. 11: a) Inner bearing ring racetrack of the DE side bearing with a grey raceway. b) Inner bearing ring racetrack of the NDE side, showing no change of raceway surface.

Table III: Static rotor imbalance $\Delta U2$, 1000 min^{-1}, measured EDM bearing currents: Long-term test with three 11 kW-induction machines M1, M2, M3.

Drive end side	motor M1	motor M2	motor M3
av. bearing voltage $U_{b,av}$	10.66 V	12.46 V	8.47 V
confidence interval $\sigma_{u,av}$	1.23 V	1.27 V	1.48 V
av. bearing current $I_{m,EDM,pk-pk,av}$	0.94 A	0.12 A	0.47 A
confidence interval $\sigma_{i,av}$	0.08 A	0.01 A	0.08 A
apparent bearing current density J_b	**0.39 A/mm² > 0.1**	**0.05 A/mm² < 0.1**	**0.20 A/mm² > 0.1**
bearing surface of inner ring	grey raceway	no surface change	grey raceway
Non-drive end side			
av. bearing voltage $U_{b,av}$	10.56 V	13.44 V	4.46 V
confidence interval $\sigma_{u,av}$	1.34 V	1.27 V	1.18 V
av. bearing current $I_{m,EDM,pk-pk,av}$	0.08 A	0.98 A	0.05 A
confidence interval $\sigma_{i,av}$	0.01 A	0.09 A	0.01 A
apparent bearing current density J_b	**0.03 A/mm² < 0.1**	**0.41 A/mm² > 0.1**	**0.02 A/mm² < 0.1**
bearing surface of inner ring	no surface change	grey raceway	no surface change

In comparison with the same tests at residual static imbalance U0, no influence of the increased rotor imbalance $\Delta U2$ on the bearing surface was detected. Similar to the long-term test of Fig. 11, a long-term test with rotor-to-ground bearing currents was investigated with the static rotor imbalance $\Delta U2$ (Fig. 13). The three investigated four-pole 11 kW-induction machines (Fig. 3) were operated over 1023 hours with static rotor imbalance $\Delta U2$ and rotor-to-ground bearing currents at DE side. The measured rotor-to-ground bearing currents at DE side of 11 kW-induction motor M1 over operating time are shown in Fig. 13. Table IV summarizes the results of the long-term test with rotor-to-ground bearing currents at 11 kW-induction machines over 1023 hours. Again no influence of the increased static rotor imbalance $\Delta U2$ on the bearing raceway damage was found.

In a similar way to Fig. 13, a long-term test with 11 kW-induction machines was performed at imbalance $\Delta U2$, but at grid operation 50 Hz, 400 V, $n = 1500$ min^{-1}. No bearing currents occurred during operating time of 1031 h, and no change of the raceway surfaces of the bearings were detected.

Table IV: Static rotor imbalance $\Delta U2$, 1000 min^{-1}, measured rotor-to-ground bearing currents: Long-term test at three 11 kW-induction machines M1, M2, M3.

Drive end side	motor M1	motor M2	motor M3
av. bearing voltage $U_{b,av}$	14.30 V	16.78 V	12.56 V
confidence interval $\sigma_{u,av}$	1.31 V	0.98 V	0.73 V
av. bearing current $I_{m,ROER,pk-to-pk,av}$	2.13 A	5.50 A	5.08 A
confidence interval $\sigma_{i,av}$	0.18 A	0.27 A	0.18 A
apparent bearing current density J_b	**0.89 A/mm^2 > 0.1**	**2.29 A/mm^2 > 0.1**	**2.12 A/mm^2 > 0.1**
bearing surface of inner ring	grey raceway	starting fluting	starting fluting

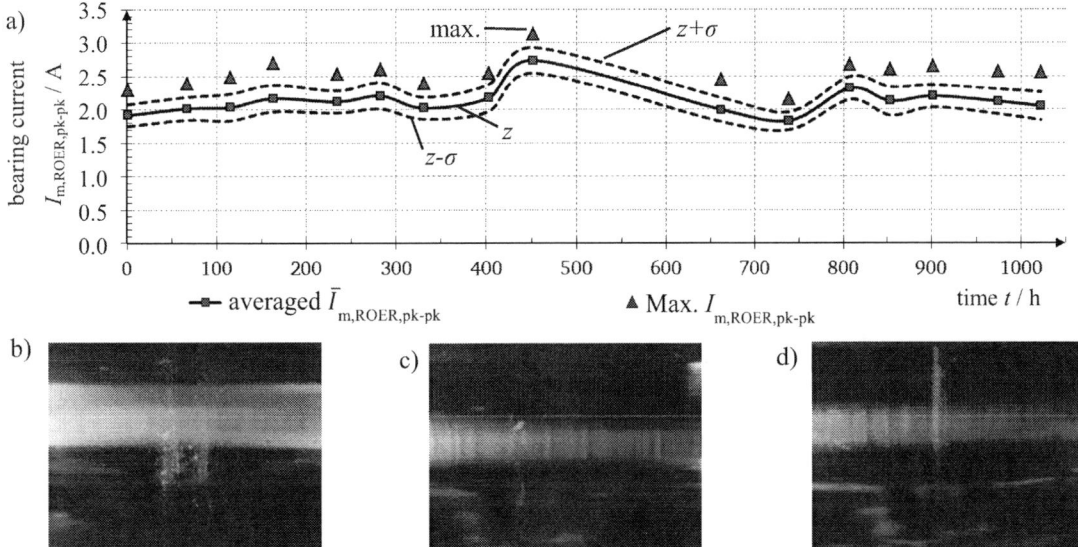

Fig. 13 a) Static rotor imbalance $\Delta U2$: Measured peak-to-peak rotor-to-ground bearing currents at the 11 kW-induction motor M1 at the DE side during an operating time of 1023 h at a rotor speed $n = 1000$ min^{-1} and an average bearing temperature $\vartheta_b = 60$ °C.
b) Inner bearing ring of the DE side bearing of motor M1 with a grey raceway.
c) Inner bearing ring of the DE side bearing of motor M2 with beginning of fluting.
d) Inner bearing ring of the DE side bearing of motor M3 with beginning of fluting.

Conclusion

With three 1.5 kW- and three 11 kW-4-pole cage induction standard TEFC motors (frame sizes 90 mm and 160 mm) short-term and 1000 h long-term tests were performed to find out, whether an additional static rotor imbalance accelerates the bearing damage under an inverter-induced bearing current flow. Two types of inverter-induced bearing currents were investigated: EDM bearing currents due to electrical lubrication film breakdown, caused by the common-mode bearing voltage, and the common-mode rotor-to-ground bearing currents in case of grounded rotor, e.g. via a gear or conductive belt. The latter have the same occurrence rate of 6-times inverter switching frequency and a similar time signal shape as the circular inverter-induced bearing currents, which occur in bigger AC machines typically above frame size 200 mm [3]. The static rotor imbalance was enlarged by additional rotor masses by a factor of 4 and 16 as $\Delta U1$ and $\Delta U2$ for the 1.5 kW-induction motors with respect to the residual imbalance limit U0 according to G6.3 of DIN ISO 1940 Part 1. For the 11 kW-induction motors the imbalance was enlarged with $\Delta U2$ by a factor 16. The occurring n-frequent vertical and horizontal bearing vibration could accelerate the bearing fluting damage, which is believed according to [5] to be generated by bearing vibrations, caused by the small bearing surface craters of the sparking bearing current flow. Although these additional vibrations occurred, these additional imbalances have no influence on an accelerated electrical bearing current damage. The EDM bearing currents have, like without additional rotor imbalance, a roughly 1000-times smaller occurrence rate than the rotor-to-ground bearing currents, which flow at each inverter-switching instant, thus with 6-times the switching frequency, so here 60 kHz. Hence, the well-known limit of 0.1 A/mm² of apparent bearing current density to avoid fluting, which was derived from DC and LF AC bearing currents, did only apply to the rotor-to-ground bearing currents. For EDM currents even apparent bearing current densities up to 2 A/mm² did not cause any fluting, but resulted only in a "grey frosted" bearing racetrack, which still allows stable operation, so the influence of increased static rotor imbalance on the electric bearing damage may be disregarded at least within the scope of the presented experimental data, whereas the type of bearing current, either EDM or rotor-to-ground bearing currents is decisive for the electrical bearing damage.

References

[1] Werner U., Binder A.: "Rotor dynamic analysis of asynchronous machines including the finite-element-method for engineering low vibration motors," *Proc. of international Symposium on Power Electronics, Electrical Drives, Automation and Motion* (SPEEDAM), 23. - 26.05.2006, Taormina, Italy, pp. 88-96.

[2] Sunahara K. et al., "Effect of lubrication regime on washboard/fluting pattern formation due to electrical pitting", poster presentation at 43rd Leeds-Lyon Symposium on Tribology, 2016.

[3] Muetze A., Binder A., Vogel H., Hering J., "Experimental evaluation of the endangerment of ball bearings due to inverter-induced bearing currents," *Proc. of the 2004 IEEE Industry Applications Conf.*, Seattle, WA, USA, 03. - 07. October, 2004, vol. 3, pp. 1989 - 1995.

[4] Binder A., Muetze A., "Scaling effects of inverter-induced bearing currents in AC machines," *IEEE Transactions on Industry Applications*, vol. 44, no. 3, pp. 769 - 776, June 2008.

[5] Kohaut A.: "Riffelbildung in Wälzlagern infolge elektrischer Korrosion, (English: Fluting in bearings due to electrical corrosion)," Julius-Maximilians-Universität, Würzburg, Germany, habilitation thesis, 1943.

[6] Wittek E., Kriese M., Tischmacher H., Gattermann S., Ponick B., Poll G., "Capacitances and lubricant film thicknesses of motor bearings under different operating conditions," *Proc. of 2010 XIX International Conference on Electrical Machines*, 6 pages, CD ROM, Rome, Italy, 06. - 08. September, 2010.

[8] VDI-Guideline 3832, Measurement of structure-borne sound of rolling element bearings in machines and plants for evaluation of condition, April 2013, Düsseldorf, Germany.

[9] IEC 60034-14:2018, Rotating electrical machines - Part 14: Mechanical vibration of certain machines with shaft heights 56 mm and higher – Measurement, evaluation and limits of vibration severity (IEC 60034-14:2018), Geneva, Switzerland.

[10] DIN 620-4:2004-06, Rolling bearings - Rolling bearing tolerances - Part 4: Radial internal clearance (DIN 620-4:2004), Beuth, Berlin, Germany.

Novel current balancing method for HF interleaved converters with reduced control effort

Christian Beckemeier, Jens Friebe
Institute for Drive Systems and Power Electronics
Welfengarten 1
Hannover, Germany
Phone: +49 511 762-12228
Fax: ++49 511 762-3040
Email: christian.beckemeier@ial.uni-hannover.de
URL: https://www.ial.uni-hannover.de/

Acknowledgments

This work was supported by the Deutsche Forschungsgemeinschaft (German Research Foundation, DFG) through the Germany's Excellence Strategy - EXC 2163/1 - Sustainable and Energy Efficient Aviation - Project-ID 390881007.

Keywords

≪Interleaved converter≫, ≪Current sharing≫, ≪High frequency power converter≫, ≪Microcontrollers≫, ≪Multi-level inverters≫

Abstract

The Active Neutral Point Clamped (ANPC) topology as a three level inverter is particularly well suited for subdivision into fundamental switching semiconductors and modulating fast switching semiconductors. This allows each semiconductor to be selected for its best characteristics. One part of a hybrid ANPC inverter can be realized with high frequency interleaved stages based on wide bandgap power semiconductors. A challenge is the calculation effort of the duty cycle for this high switching frequency, especially due to the interleaved based stages. It is often considered that the reduction of the calculation effort has a negative impact on the current accounting. This paper proposes a simplified method for the balancing of the current in the interleaved stages with significantly reduced control frequency. An active current symmetry should be implemented, which would be easier due to the reduced control effort. In this paper, this method is experimentally validated without active current symmetry control.

Introduction

Multi-level inverter allows to use power semiconductor with less blocking voltage as the DC-Link voltage. This allows to use 650 V wide bandgap devices in a inverter with 800 V DC-Link. Wide bandgap devices allow a high frequency stage to reduce the size of the inverter. To increase the output power of the inverter, several HF stages can be connected in parallel. This also has the advantage that the interleaving concept can be used to reduce the output current ripple. The disadvantage of parallel stages is a higher control effort because of the necessity to implement a current balancing [1, 2, 3, 4]. In this paper an interleaved DC/AC converter with a reduced control afford is investigated. In the beginning, the hybrid ANPC topology is presented. Then the challenges of interleaved inverters are discussed and a method to significantly reduce the control effort of the inverter is presented. This method is validated with a measurement on the introduced inverter. The reduction in control effort should enable microcontrollers to be used in HF interleaved converters, avoiding the necessity for e.g. an FPGA.

1 Hybrid ANPC Topology

The ANPC Topology is especially well qualified for an interleaved DC/AC Converter and is shown in Fig. 1. It consists of a low frequency section (S1 - S4) which is switched with a fundamental frequency of $f_0 = 50\,\mathrm{Hz}$. For the positive halfwave the switches S_1 and S_3 are activated and for the negative halfwave the switches S_2 and S_4. The switches have almost only conduction losses because they switch with a low frequency and for restive load at zero current. With the switches S_5, S_6 and inductance L the output voltage will be modulated with a high switching frequency f_s. A additional buffer capacitor C_c is closely placed parallel to the high frequency switches to provied a low impedance path for HF current. Due to the separation into low frequency and high frequency section, the best semiconductor type for the respective parameters can be selected. The ANPC Topology is particularly benefitial for this case, because only four low frequency switches are needed. In the inverter, six HF bridges with GaN-HEMTs are used parallel with a switching frequency of $f_s = 200\,\mathrm{kHz}$ and four Si-IGBTs at the LF section [5].

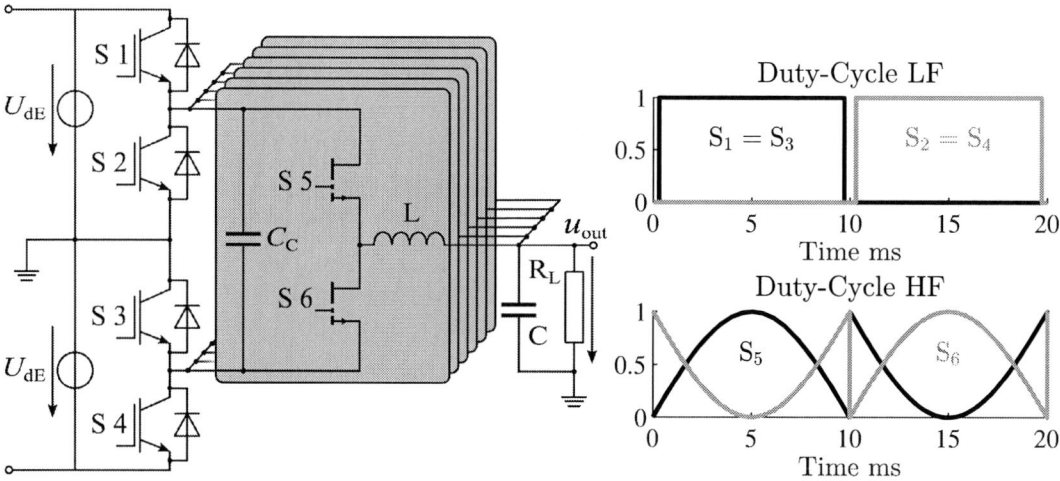

Fig. 1: The ANPC Topologie (left) and the associated control signals (right)

The increased output power and the lower filter effort are the advantages of the six parallel stages. The disadvantages are higher control effort and a possible current deviation between the stages. This arises due to small differences between the stages, for example a different switching speed, mismatched PWM-frequency or unequal inductance values. A current deviation leads to an unequal load distribution, so that an overload of the stage with the highest output power can result. To balance the stages, each current $i_{L,j}$ has to be measured to compensate the current error with an appropriate control strategy.

A current deviation between the stages has multiple reasons. The largest influence on the deviation is due to varying inductor parameters such as inductance and resistance value. Also unequal control signals like dead time, gate voltage, etc. have an influence, therefore a central duty cycle generation is [6].

Control requirements

To generate interleaved PWM signals that are as identical as possible, an FPGA or a microcontroller with independent PWM modules can be used. An FPGA has a higher calculation speed and can perform parallel tasks, which makes it suitable for this application. In this paper it is shown that even with limited calculation power it is possible to control this ANPC, using a microcontroller. Especially for grid application the normally high calculation effort has no advantage. To control the ANPC with the microcontroller, a high generation speed of the duty cycle is required, which corresponds to a compact C code. In the left part of Fig. 2 the generation of the duty cycle with optional current deviation is shown.

The duty cycle will be generated in a separate CLA task (Control Law Accelerator). Optionally a current deviation can be compensated. In this paper, the CLA frequency f_{cla} is the control frequency of the

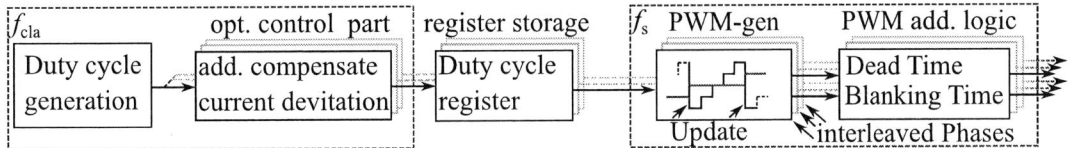

Fig. 2: Creating interleaved PWM signals with different control f_{cla} and switching frequency f_{s}

inverter. The duty cycle from each stage will be written in the duty cycle register. The generation of the PWM signals is shown in the right part of Fig 2. The duty cycles are refreshed in each switching period of the HF stage. In this example, the PWM signals are generated by a comparison between a triangular signal (up-counter) and the duty cycle which will be refreshed at the triangle counter zero. Therefore, the duty cycle signal is updated when the PWM signal is set high.

Each interleaved PWM module of the microcontroller updates from the Duty Cycle register once per switching period. With 6 stages this results in a control frequency of $f_{\text{cla}} = 6 \cdot f_{\text{s}} = 1.2\,\text{MHz}$. This is shown in Fig 3 (middle), where colored values correspond to a refreshed duty cycle. But this high control frequency cannot be realized with normal microprocessors. A control with 6 predefined logic tables transmitting coordinated values from the DMA to the PWM modules will achieve this speed, but this makes current compensation or grid synchronization practically impossible and is therefore not attempted.

Reduction of control effort

This high control effort is not necessary for grid applications and will be reduced in this paper. To ensure a reduction in control frequency does not have a strong effect on current symmetry, although PWM module updated with old duty cycle values. With a reduction of the control frequency f_{cla} to the switching frequency f_{s}, always the same stage gets the updated duty cycle value. This method is shown in Fig. 3 (top) where stage one gets the updated value, visible by a colored number. The new duty cycle is always taken at the rising edge of the corresponding stage, and stages two to six (number marked in black) are given the last calculated delayed duty cycle from the stage with the color from the arrow.

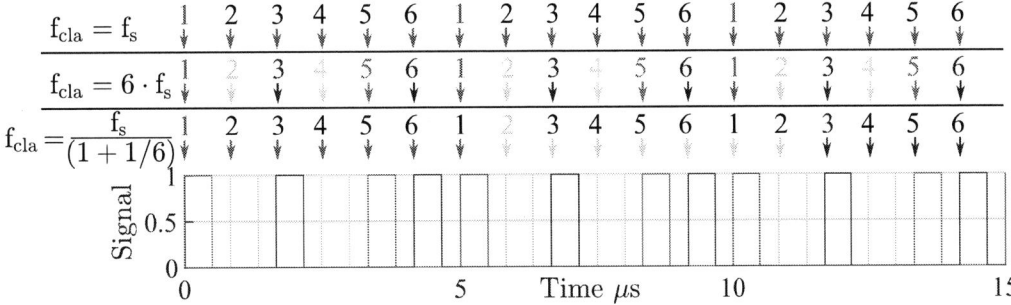

Fig. 3: Different control frequencys for a six stage interleaved converter. Each stage updates with the rising edge, colored stage numbers with a new duty cycle, black stage numbers with the last calculated

With $f_{\text{cla}} = f_{\text{s}}$ an asymmetrical refresh of all stages results, which leads to another reason for deviation from the target currents. By prioritizing the individual stages, it can be assumed that stage 1 has a larger current when the voltage slope is positive and a smaller current when the voltage slope is negative than the other stages. The last stage (in this case 6) has the opposite behavior to stage 1, since it experiences the increase/decrease of the duty cycle only with a delay. As a result, the output potential is already changed accordingly, whereby a smaller voltage time surface is present at the inductance 6. The simulated deviations for otherwise ideal components are shown in Fig. 4 (top). There all average inductor currents of the individual stages $\bar{i}_{\text{L},s}$ are shown and the deviation between Stage 1 and Stage 6 is up to $1.2\,\text{A}$. The series of deviations corresponds to the delayed stage. The simulated output power of the converter is $16.6\,\text{kW}$.

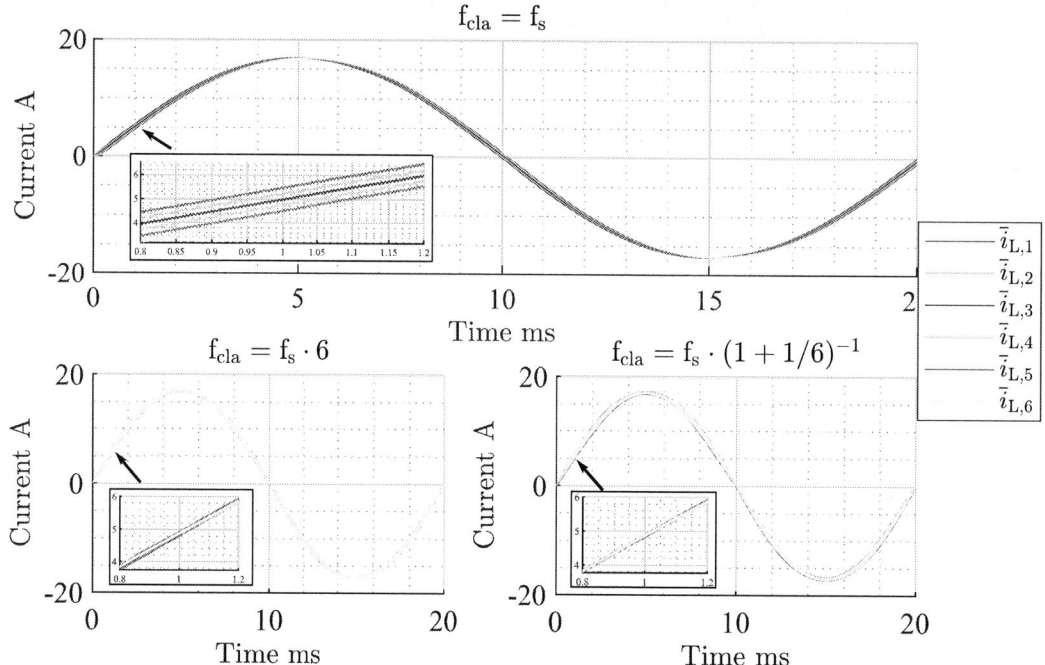

Fig. 4: Illustration of the simulated influence of the described control frequencies. The control frequency of $f_{cla} = f_s$ leads to a strong deviation of a symmetrical current load (top). The proposed method $f_{cla} = f_s \cdot (1+1/6)^{-1}$ (bottom right) leads to a symmetrical current load as well as a reduced control effort with a similar result as with a strongly increased control frequency $f_{cla} = 6 \cdot f_s$ (bottom left)

The asymmetry can be eliminated by increasing the control frequency according to the number of stages n. With 6 stages this results in a control frequency of $f_{cla} = 6 \cdot f_s = 1.2\,\text{MHz}$, so that each stage gets an updated duty cycle. This results in a symmetrical current in each stage, but also significantly increases the control effort. This method is shown in Fig 4 at the bottom, left side.

In this paper, it is proposed to set the control frequency of the inverter according to $f_{cla} = f_s \cdot (k \pm 1/n)^{-1}$. Here, n the number of stages and $k \in \mathbb{R}^+$ is a natural Number to reduce more the control frequency. With a control frequency of $f_{cla} = f_s \cdot (1+1/6)^{-1}$ the control effort is again lower than $f_{cla} = f_s$. Similar to method $f_{cla} = f_s$, one stage per switching frequency is updated with the refreshed duty cycle, but in the following period the next stage gets the refreshed duty cycle. So that each stage is handled equally and this is shown in Fig. 3 (bottom). With this approach, the simulation of the ANPC also leads to an quasi ideal superposition of the individual average currents of the inductor stages $\bar{i}_{L,s}$ shown in Fig. 4 at the bottom, right side.

Further possibilities of a more symmetrical control at a lower control frequency would be separately calculated duty cycle values for each stage. With the help of a linearization at the operating point, each stage could receive an approximate duty cycle. This method is not considered due to the additional calculation effort.

Measurement

In this section, different control frequencies f_{cla} are tested. All of the measurements were performed with the presented inverter in [5]. The measurements were carried out with the switching frequency of $f_s = 200\,\text{kHz}$ and an output voltage of $u_{out} = 230\,\text{V}$ at 50 Hz. The reasons for possible current deviations are minimised by a very symmetrical structure. A microcontroller generates all PWM signals leading to simular blanking and dead time for each stage. The gate loops are low inductive, low gate resistors and equal PCB design for each stage are realized.

Measurement of the Inductance

The deviations of each inductance of the HF stages are measured with a Bode 100. The inductances have a difference up to 4 µH. All the values of the inductors are shown in table I. In [2] the influence of a deviation in resistance of the inductance and semiconductors are analysed. A deviating in inductance resistance leads mainly to a deviation in the DC current. In the AC converter, the deviation are shown in the peaks of the sinusoidal output voltage. The influence of an inductance is reflected in the response time. A lower inductance leads to a faster change in current while the duty cycle changes. This deviation mainly happens near the zero crossing.

HF-Stage	inductivity	resistance	phase offset
1	28.788 µH	415 mΩ	0 °
2	26.834 µH	304 mΩ	60 °
3	29.621 µH	317 mΩ	120 °
4	30.626 µH	319 mΩ	180 °
5	28.596 µH	1623 mΩ	240 °
6	29.703 µH	317 mΩ	300 °

Table I: Measurement of the inductors of each HF stage by the switching frequency $f_s = 200\,\mathrm{kHz}$. Phase offset is the PWM modul offset of each interleaved stage

Current Measurement results

In the following section a part of the measurement results is presented. To get a better comparison the error current of each stage with respect to the average current is used. The error current $\bar{e}_{c,k}$ is the deviation from each stage current to the targeted (average) stage current $\bar{e}_{c,k} = \bar{i}_{L,k} - \frac{\sum_{j=1}^{n} \bar{i}_{L,j}}{n}$. The output power of the inverter is 4.8 kW, so each stage should have a symmetric peak output current of 5 A by $u_{\mathrm{out}} = 230\,\mathrm{V}$.

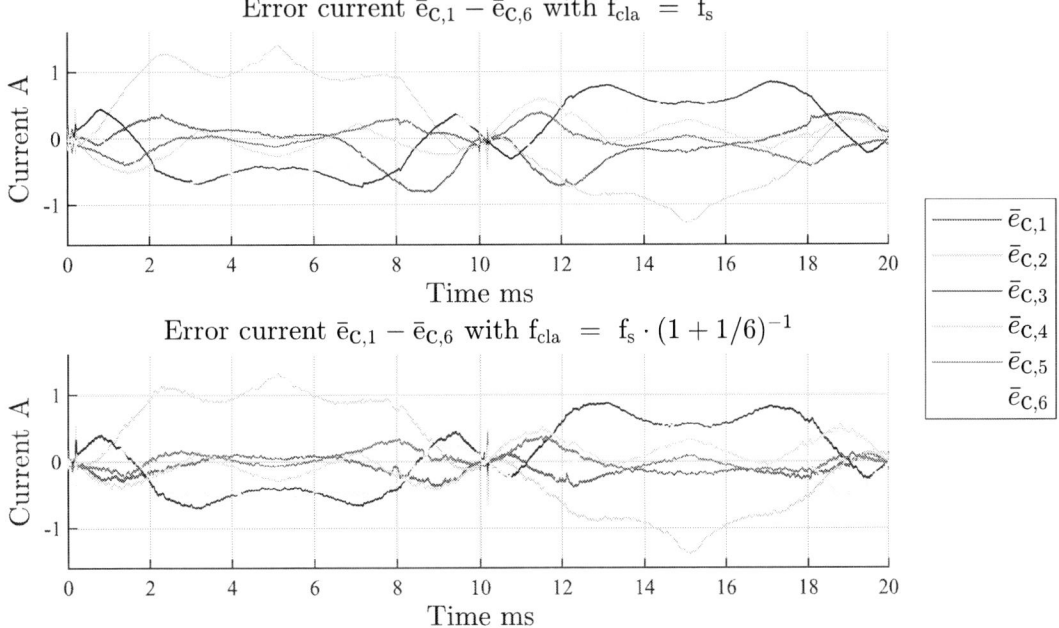

Fig. 5: Error current by a control frequency $f_{\mathrm{cla}} = f_s$ (top) and $f_{\mathrm{cla}} = f_s \cdot (1 + 1/6)^{-1}$ (bottom). The asymmetrical components have been strongly minimized by $f_{\mathrm{cla}} = f_s \cdot (1 + 1/6)^{-1}$. No current feedback control is implemented

In Fig. 5 (top) the error current with $f_{cla} = f_s$ is shown and can be splitted into symmetric and asymmetric parts. In the Fig. 5 a strong asymmetry deviation at the current can be seen. Especially in the case of error current $\bar{e}_{c,1}$ and $\bar{e}_{c,6}$, a significant deviation can be seen in both half waves. With the proposed method $f_{cla} = f_s \cdot (1 + 1/6)^{-1}$ Fig. 5 (bottom), the asymmetric deviation could be greatly minimized. A slight widening of the average current of each stage can be seen. This can be explained by the partial delay of the duty cycle and by the resulting superposition of an oscillation with $f_{cla} = f_s \cdot (1 + 1/6)^{-1}$ frequency. This has no significant negative impact on the current symmetry itself. A comparison between Fig. 4 and Fig. 5 is difficult because other parasitic effects in the measurement dominate. In particular, the different values of the inductors (shown in Table I) make a direct comparison impossible. A more suitable comparison is performed in the following section.

Measurement comparison

In order to evaluate the symmetry of a sinusoidal wave, the mathematical symmetry condition equation 1 (left) be used with $\Delta\varphi = 5\,\text{ms}$ for the positive half wave and $\Delta\varphi = 15\,\text{ms}$ for the negative half wave.

$$f(t + \Delta\varphi) = f(-t + \Delta\varphi) \quad \Rightarrow \quad \Delta\bar{e}_{c,k}(t) = \bar{e}_{c,k}(t + \Delta\varphi) - \bar{e}_{c,k}(-t + \Delta\varphi) \tag{1}$$

The deviation between a symmetrical sinus and the measured stage current are calculated with 1 (right) and is shown in Fig. 6 on the left side for the positive half wave (pos HW) and for the negative in the right side (neg HW). A negative delta error at positive half wave describes a larger current contribution before the maximum of the sine. A positive delta error at negative half wave describes a larger current contribution before the minimum of the sine.

With $f_{cla} = f_s$ the current deviation is similar to the simulated deviation in Fig. 4. At the positive half wave the current of the first stage $\bar{e}_{c,1}$ has the largest asymmetric current percentage at the rising sine and the smallest at the falling sine. This can also be seen in Fig. 6 by the largest negative delta error $\Delta\bar{e}_{c,1}$. The order of the stages can also be seen, as the error current increases with the corresponding stage. Considering the negative current in the negative half wave, the measurement results also agree with the simulation results.

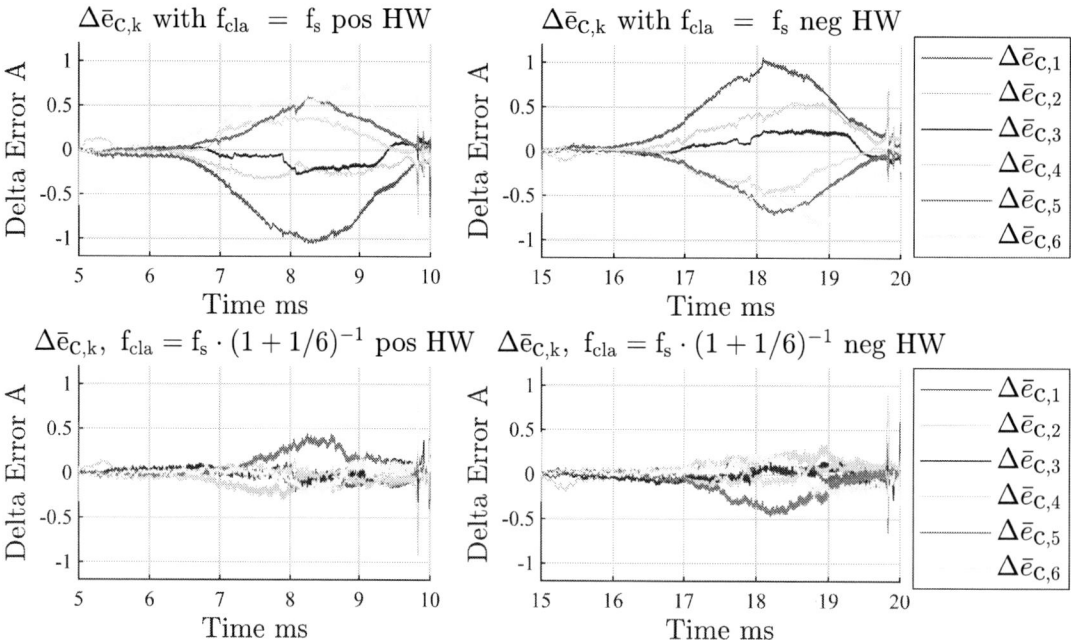

Fig. 6: Comparison of the asymmetric error current reduction with the control frequencies $f_{cla} = f_s$ and $f_{cla} = f_s \cdot (1 + 1/6)^{-1}$. The asymmetric current can be reduced with less control effort

The delta error with the lower control frequency $f_{cla} = f_s \cdot (1 + 1/6)^{-1}$ in both half waves can be significant reduced. Only the fifth stage has a current deviation of about 0.5 A, which is due to an higher resistance of inductor 5.

Conclusion

In this paper, a method has been presented to reduce the control effort without significantly affecting the current balance in interleaved converters. This was achieved by systematically adjusting the control frequency according to the presented equation. The improvements are shown in Fig. 6. This method was successfully tested with an interleaved converter with 6 stages and a switching frequency of $f_s = 200\,\text{kHz}$.

References

[1] Xunwei Zhou, Peng Xu and F. C. Lee, "A novel current-sharing control technique for low-voltage high-current voltage regulator module applications," in IEEE Transactions on Power Electronics, vol. 15, no. 6, pp. 1153-1162, Nov. 2000, doi: 10.1109/63.892830.

[2] J. A. Abu Qahouq, L. Huang and D. Huard, "Sensorless Current Sharing Analysis and Scheme For Multi-phase Converters," in IEEE Transactions on Power Electronics, vol. 23, no. 5, pp. 2237-2247, Sept. 2008, doi: 10.1109/TPEL.2008.2001897.

[3] J. A. Abu-Qahouq, "Analysis and Design of N-Phase Current-Sharing Autotuning Controller," in IEEE Transactions on Power Electronics, vol. 25, no. 6, pp. 1641-1651, June 2010, doi: 10.1109/TPEL.2009.2037892.

[4] H. Kim, M. Falahi, T. M. Jahns and M. W. Degner, "Inductor Current Measurement and Regulation Using a Single DC Link Current Sensor for Interleaved DC–DC Converters," in IEEE Transactions on Power Electronics, vol. 26, no. 5, pp. 1503-1510, May 2011, doi: 10.1109/TPEL.2010.2084108.

[5] L. Fauth, C. Beckemeier and J. Friebe, "A Hybrid Active Neutral Point Clamped Converter consisting of Si IGBTs and GaN HEMTs for Auxiliary Systems of Electric Aircraft," 2021 IEEE Energy Conversion Congress and Exposition (ECCE), 2021, pp. 1917-1923, doi: 10.1109/ECCE47101.2021.9595512.

[6] J. A. Abu Qahouq, L. Huang and D. Huard, "Sensorless Current Sharing Analysis and Scheme For Multi-phase Converters," in IEEE Transactions on Power Electronics, vol. 23, no. 5, pp. 2237-2247, Sept. 2008, doi: 10.1109/TPEL.2008.2001897.

[7] P. Antoszczuk, R. G. Retegui, M. Funes and D. Carrica, "Optimized Implementation of a Current Control Algorithm for Multiphase Interleaved Power Converters," in IEEE Transactions on Industrial Informatics, vol. 10, no. 4, pp. 2224-2232, Nov. 2014, doi: 10.1109/TII.2014.2362071.

dV/dt-Based Filter Design for Motor Inverters with Continuous Output Voltage

Sabrina Ulmer, Stevan Bugarski, Gernot Schullerus, Ertugrul Sönmez

Reutlingen University
Electronics and Drives
Oferdingerstr. 50
72768 Reutlingen, Germany
+49 (7121) 271-7080

{sabrina.ulmer, stevan.bugarski, gernot.schullerus, ertugrul.soenmez}
@reutlingen-university.de
www.electronics-and-drives.de

Acknowledgments

This work is supported by the German Federal Ministry of Education and Research.

Keywords

≪DC-AC Converter≫, ≪Inverter-output filter≫, ≪Passive filters≫, ≪Gallium Nitride (GaN)≫, ≪dV/dt≫

Abstract

Wide bandgap semiconductors enable high switching frequencies and thus the integration of filters into the inverter. The current paper addresses the design problem for the filter components such that a given dV/dt at the inverter terminals is obtained.

Introduction

The application of wide bandgap (WBG) semiconductors opens the option for higher switching frequencies in a range well above 500 kHz. Due to these switching frequencies, a new inverter concept becomes feasible, where filters are integrated into the inverter printed circuit board (PCB) to produce quasi-continuous output voltages. In [1] such a concept was presented for a three-phase inverter for fractional power synchronous machines.

The research activities conducted so far, illustrated, that in spite of the high switching frequencies the passive components are still significantly determining the size of this converter type. As one of the main benefits of this converter type is given by the reduction of the dV/dt values at the motor terminals, a minimum size of the filter components can be obtained by specifying a maximum dV/dt and deriving the filter inductor and capacitor size such, that this requirement is satisfied with minimum passive component sizes. This question is addressed in [2] for a LRC filter, that is, a filter with a passive damping, based on the analysis of the filter step response or in [3] based on transfer function considerations for an LC filter. An additional aspect is discussed in [4], where the effect of motor cables is considered as well.

In [2], as well as in other publications, the need for damping of the LC resonant structure is emphasized and a passive resistor based damping is applied. In the concept discussed in [1], an active damping based on inductor current feedback is introduced. The current publication discusses the LC filter design for the concept from [1]. A new design procedure is given, that determines the filter parameter such that a maximum dV/dt is respected. To preserve the active damping capabilities, the influence of the design on the inductor current feedback based damping is considered in the design process, as well.

The paper is organized as follows: First, the power electronics module with the output filter for quasi-continuous output voltage is described. Then, the relationship between the filter parameters and the dV/dt of the output voltage is derived, followed by design considerations for the LC filter parameter. Next, some simulations and measurements are illustrated. Finally, conclusions are given.

Inverter system

The inverter for three-phase motor control with quasi-continuous output voltage from [1], considered in this paper, is illustrated in Fig. 1. Each phase consists of a half-bridge (HB) with DC voltage V_B, where the modulation is achieved by a asynchronous delta-sigma ($\Delta\Sigma$) modulator. The input signals of each modulator are the reference input signal v_{ORef}, the voltage v_{HB} at the HB switching node and the filter inductor current i_{LF} for the respective phase. This is illustrated in more detail for one phase in Fig. 2.

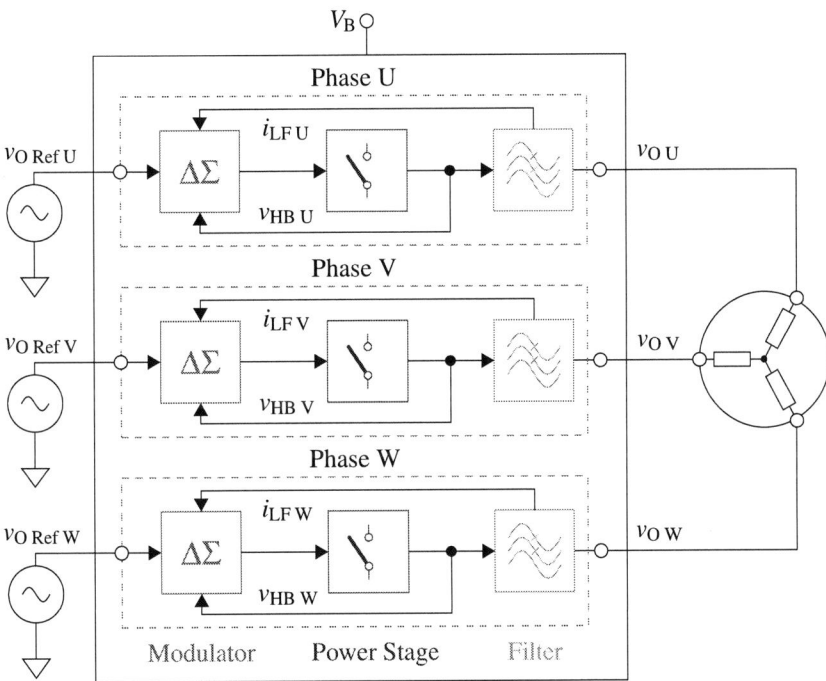

Fig. 1: GaN-based motor inverter with quasi-continuous output voltage

The output voltage v_O for each phase results from filtering the pulsed voltage v_{HB} with an LC filter. For active damping as in [1] an inductor current feedback with the transfer characteristic G_{FB} [5] is applied. A motor winding is connected to the filter output terminals, where the back-emf is not illustrated for simplicity. In the following section, design considerations for the parameters L_F and C_F of the filter are given.

Filter design

The design rule calculation approach depends on the relationship of the filter resonance frequency and the switching frequency [2]. If the filter resonance is higher than the switching frequency, then the dV/dt analysis can be done based on the step response of the filter as e.g. in [2]. For the discussion given below the motor winding will not be considered yet, that is, the phase module will be operated with open terminals. The influence of the motor winding on the results, obtained below, will be investigated in simulations.

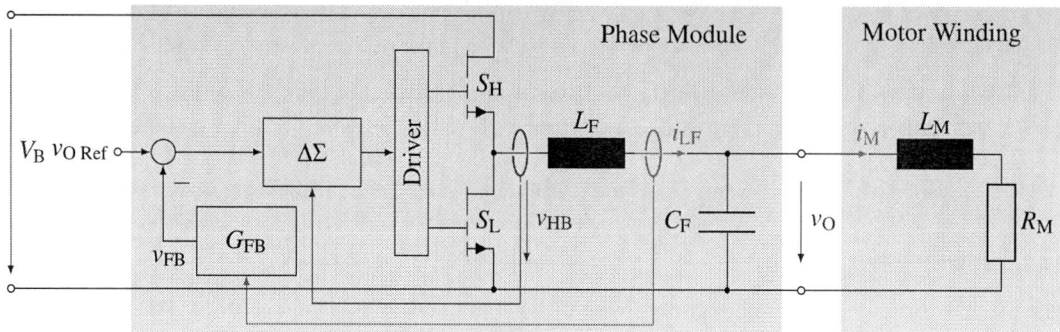

Fig. 2: Phase module with motor winding

We denote with

$$G_{VO} = \frac{1}{L_F C_F s^2 + 2\sqrt{L_F C_F}\, s + 1} \tag{1}$$

the transfer function with v_{HB} as input and v_O as output variable using the feedback rule

$$G_{FB} = 2\sqrt{\frac{L_F}{C_F}}$$

as in [1]. The step response $h_{VO}(t)$ of this transfer function is then calculated using (1) and the inverse Laplace transform. Differentiating the step response $h_{VO}(t)$ yields

$$\frac{dh_{VO}}{dt} = \frac{t}{L_F C_F} e^{-\frac{t}{\sqrt{L_F C_F}}}$$

with the maximum at

$$t = \sqrt{L_F C_F} \qquad \text{as} \qquad \left.\frac{dh_{VO}}{dt}\right|_{t=\sqrt{L_F C_F}} = \frac{1}{e\sqrt{L_F C_F}},$$

corresponding to the maximum dV/dt, given an input voltage step of 1 V. Note, that this value will slightly differ from the value in [2], where dV/dt is determined based on two selected points in the step response.

In the case, when the filter resonance frequency is between the signal frequency and the switching frequency, the approach given above is not valid. Here, a new approach for determining a relationship between the dV/dt, that is, the slew rate, values and the filter parameters are developed. It is based on the idea that the filter structure from Fig. 2 is the output filter of a buck converter. In the sequel, we assume a triangular current shape during one switching interval of duration T_S as illustrated in the left-hand side of Fig. 3.

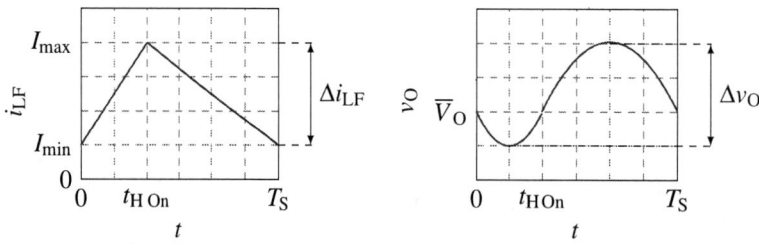

Fig. 3: Inductor current i_{LF} and output voltage v_O

In addition, we denote by t_{HOn} the on-time duration of the upper switch S_{H} of the half-bridge, by $\overline{V}_{\mathrm{O}}$ the output voltage at the beginning of the switching interval in steady state, Δi_{LF} the inductor current ripple and $\Delta v_{\mathrm{O}}(t) = v_{\mathrm{O}}(t) - \overline{V}_{\mathrm{O}}$. Then, we obtain after some calculations for $t \in [0 \quad t_{\mathrm{HOn}}]$

$$\Delta v_{\mathrm{O}}(t) = \frac{\Delta i_{\mathrm{LF}}}{2\,C_{\mathrm{F}}}\left(\frac{t^2}{t_{\mathrm{HOn}}} - t\right) \qquad \Rightarrow \qquad \frac{\mathrm{d}\Delta v_{\mathrm{O}}}{\mathrm{d}t} = \frac{\mathrm{d}v_{\mathrm{O}}}{\mathrm{d}t} = \frac{\Delta i_{\mathrm{LF}}}{2\,C_{\mathrm{F}}}\left(\frac{2\,t}{t_{\mathrm{HOn}}} - 1\right) .$$

The maximum slew rate is obtained for $t = t_{\mathrm{HOn}}$

$$\left.\frac{\mathrm{d}v_{\mathrm{O}}}{\mathrm{d}t}\right|_{t=t_{\mathrm{HOn}}} = \frac{\Delta i_{\mathrm{LF}}}{2\,C_{\mathrm{F}}} \quad \text{where} \quad \Delta i_{\mathrm{LF}} = \frac{V_{\mathrm{B}}(1-d)}{L_{\mathrm{F}}}dT_{\mathrm{S}} \quad \text{as} \quad \left.\frac{\mathrm{d}v_{\mathrm{O}}}{\mathrm{d}t}\right|_{t=t_{\mathrm{HOn}}} = \frac{V_{\mathrm{B}}(1-d)}{2\,C_{\mathrm{F}}L_{\mathrm{F}}}dT_{\mathrm{S}} , \quad (2)$$

where

$$d = \frac{t_{\mathrm{HOn}}}{T_{\mathrm{S}}} \qquad\qquad (3)$$

is the duty cycle. For a more detailed analysis, the modulator structure and the corresponding signals are illustrated in Fig. 4.

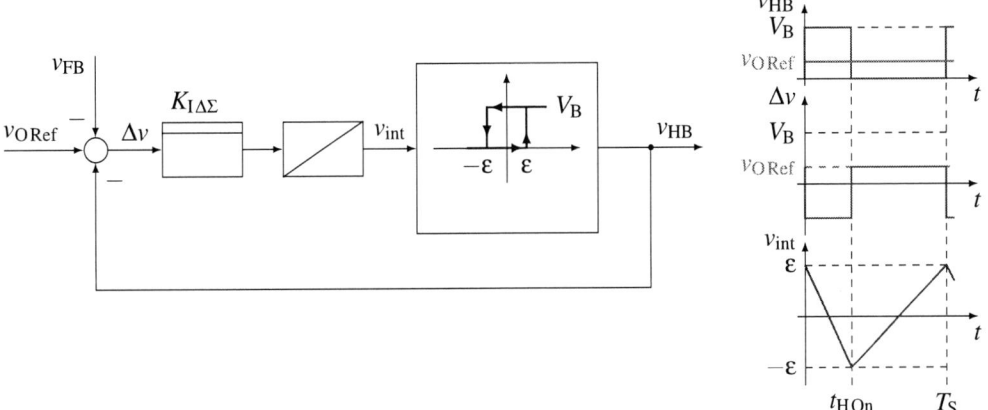

Fig. 4: Modulator structure and modulator signals

First, we do not consider the current feeeback, that is, $v_{\mathrm{FB}} = 0$. From Fig. 4 one obtains

$$t_{\mathrm{HOn}} = -\frac{2\,\varepsilon}{K_{\mathrm{I}\Delta\Sigma}(v_{\mathrm{ORef}} - V_{\mathrm{B}})} ,$$

$$T_{\mathrm{S}} - t_{\mathrm{HOn}} = \frac{2\,\varepsilon}{K_{\mathrm{I}\Delta\Sigma}\,v_{\mathrm{ORef}}} \qquad \Rightarrow \qquad T_{\mathrm{S}} = \frac{2\,\varepsilon}{K_{\mathrm{I}\Delta\Sigma}\,v_{\mathrm{ORef}}}\frac{V_{\mathrm{B}}}{V_{\mathrm{B}} - v_{\mathrm{ORef}}} . \qquad (4)$$

From (4) one obtains after some calculations

$$(1-d)dT_{\mathrm{S}} = \frac{K}{V_{\mathrm{B}}} \qquad\qquad \text{with} \qquad\qquad K = \frac{2\,\varepsilon}{K_{\mathrm{I}\Delta\Sigma}} , \qquad (5)$$

where K is proportional to the modulator hysteresis width $2\,\varepsilon$ and to the inverse of the integrator gain $K_{\mathrm{I}\Delta\Sigma}$. Introducing the result from (5) into (2) we obtain

$$\Delta i_{\mathrm{LF}} = \frac{K}{L_{\mathrm{F}}} \qquad\qquad (6)$$

and

$$\frac{dv_O}{dt}\bigg|_{t=t_{HOn}} = \frac{K}{2\,C_F L_F}\,.\tag{7}$$

This result illustrates that, assuming the filter resonance frequency below the switching frequency, the dV/dt value can be influenced by both output filter component values, as well as the constant K, depending on the modulator design parameters ε and $K_{I\Delta\Sigma}$. From (4) it is obvious, that K is related to the switching period T_S or equivalently the switching frequency f_S, such that this relationship is reasonable, as well.

Besides the slew rate of the output voltage, the inductor current ripple Δi_{LF} is relevant for the system design. A high current ripple increases the stress in the switches. It increases the inductor losses and influences the inductor design as the core size will typically increase in order to avoid saturation [6]. In addition, a high current ripple influences the modulator via the voltage v_{FB} (see Fig. 4). A detailed analytical evaluation of this effect is beyond the scope of this paper. However, the influence will be evaluated in a simulation.

The design steps are basically the same as in [2] or [3]. Note however, that the design equations are different.

1. Specify the maximum admissible current ripple $\Delta I_{LF\,max}$ and the maximum dV/dt value,
2. Calculate L_F from (6) given K,
3. Calculate C_F from (2) given K and L_F from step 2,
4. Check whether the filter resonance frequency is below the minimum switching frequency and adapt design requirements and/or K if necessary.

If the design results in a filter resonance frequency that is too high, the switching frequency can be increased by modifying K. Alternatively, the filter parameters can be modified to reduce the filter resonance frequency. This will either reduce the inductor current ripple or the dV/dt value or both.

These considerations were developed based on the buck converter concept, thus implicitly assuming a constant output voltage v_O. For sinusoidal output voltages, these results will give a reasonable approximation, if the frequency of the sinusoidal signal is below the resonance frequency by a reasonable factor. This will be illustrated in simulations.

Simulations

For simulation, the system illustrated in Fig. 2 was implemented in a Simulink-Simscape environment with the parameters given in Table I. To illustrate the influence of the filter design on the output voltage, three different values were used for the filter inductance L_F and the output capacitor C_F, respectively.

V_B	ε	$K_{I\Delta\Sigma}$	G_{FB}		L_F	C_F	f_0
48 V	0.1	2×10^4	$\dfrac{T_F\,k_{IL}}{T_F\,s+1}$	$k_{IL}=2\sqrt{\dfrac{L_F}{C_F}}$	4.7 µH	330 nF	127.8 kHz
				$T_F=1\,\mu\mathrm{s}$	15 µH	680 nF	49.8 kHz
					33 µH	1.36 µF	23.8 kHz

Table I: System parameters

Fig. 5 illustrates the trajectories of the inductor current i_{LF}, the output voltage v_O and the corresponding estimated value for dv_O/dt estimated from v_O using the filter

$$\frac{s}{1\times10^{-7}s+1}\tag{8}$$

in the time interval marked in Fig. 6 by the red rectangle for the output voltage v_O for the variations of L_F and C_F from Table I.

Fig. 6 displays the output voltage v_O for a input reference signal $v_{ORef} = (24 + 15 \sin(2\pi f\, t))\text{V}$ where f is $1\,\text{kHz}$. Note, that the offset in v_{ORef} results from the HB configuration as shown in Fig. 2 producing a DC offset of $\frac{1}{2}V_B = 24\,\text{V}$. The simulation data in Fig. 5, 6 and 8 was produced with a load resistance of $R_L = 1\,\text{k}\Omega$ at the output terminals of the phase module, thus simulating a situation close to the open terminal case.

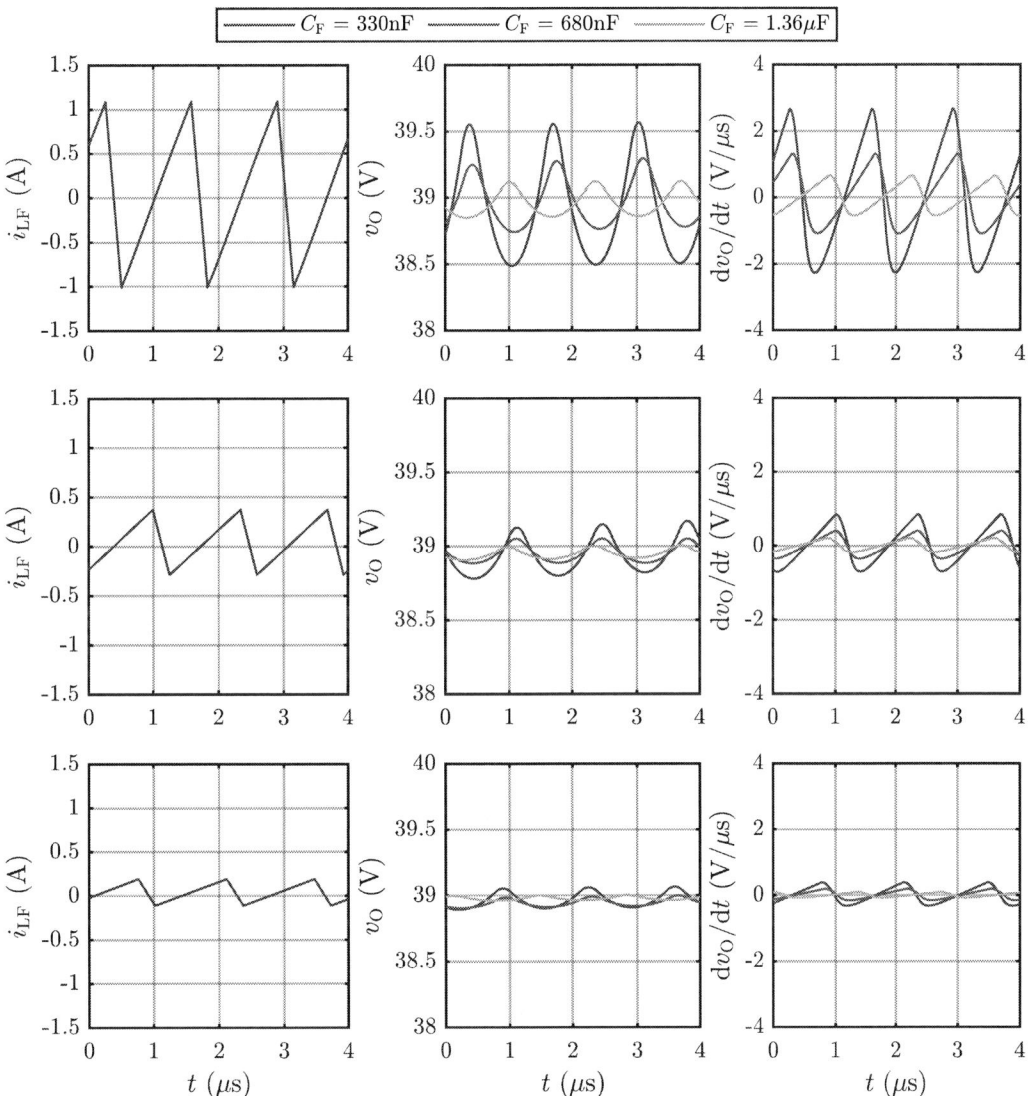

Fig. 5: Detailed view of simulation results for $R_L = 1\,\text{k}\Omega$ with variation of L_F, C_F: Upper row: $L_F = 4.7\,\mu\text{H}$, middle row: $L_F = 15\,\mu\text{H}$, lower row: $L_F = 33\,\mu\text{H}$

In Table II the inductor current ripple values Δi_{LF} and the slew rate dv_O/dt for all combinations of L_F and C_F given in Table I are calculated using (6) and (7), respectively, for comparison with the trajectories in Fig. 5. Note, that according to (6) the current ripple Δi_{LF} does not depend on the filter capacitor C_F, such that it is the same for all three capacitor values. The slew rates given in Table II correspond to the maximum values in the right-hand side plots in Fig. 5.

In Fig. 7, the effect of filter parameter variation is illustrated in simulations, where a motor winding with

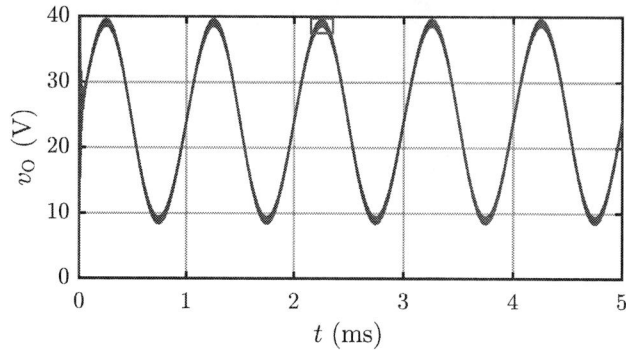

Fig. 6: Simulation result for the output voltage v_O with $L_F = 4.7\,\mu\text{H}$, $C_F = 330\,\text{nF}$

L_F	Δi_{LF}	C_F	330 nF	680 nF	1.36 µF
4.7 µH	2.13 A		3.22 V/µs	1.56 V/µs	0.78 V/µs
15 µH	670 mA	$\mathrm{d}v_O/\mathrm{d}t$	1.01 V/µs	0.49 V/µs	0.24 V/µs
33 µH	300 mA		0.46 V/µs	0.22 V/µs	0.11 V/µs

Table II: Values calculated with (6) and (7)

$R_M = 0.4\,\Omega$ and $L_M = 1\,\text{mH}$ is connected to the phase module terminals as illustrated in Fig. 2. The results and the comparison with Fig. 5 illustrate, that the effect of the motor winding on the slew rate of the phase module output voltage is small, as expected due to the large difference between the filter inductance L_F and the motor inductance L_M.

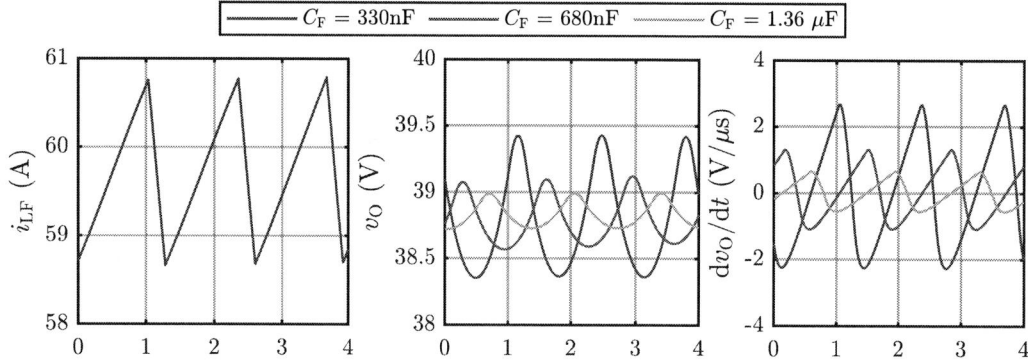

Fig. 7: Detailed view of simulation results with motor winding, $L_F = 4.7\,\mu\text{H}$ and variation of C_F

Finally, Fig. 8 displays the effect of the current ripple due to a low filter inductance value on the modulation. It is observed, that for $L_F = 0.47\,\mu\text{H}$ the switching frequency increases with respect to the case with $L_F = 15\,\mu\text{H}$ due to the higher values of Δv. From the given filter parameter one obtains a filter resonance frequency in the range of the switching frequency such that this design parameter set is out of the range given by the above mentioned design procedure.

Measurements

The hardware setup is given in Fig. 9. Fig. 9b illustrates the laboratory measurement setup with phase module, oscilloscope, power supply and signal generator for generating the input reference signal $v_{O\text{Ref}}$, whereas Fig. 9a displays the phase module PCB, where the GaN HB is marked by a red rectangle. The output voltage is measured with open terminals of the phase module, using a voltage probe with a bandwidth of 500 MHz connected to an oscilloscope with a bandwidth of 500 MHz. The inductor current trajectory is not displayed in the measurements as the bandwidth of the current sensor integrated into the phase module for current feedback is given by 80 kHz. Although this is sufficient for the feedback illustrated in Fig. 2 it is not sufficient for displaying the current ripple.

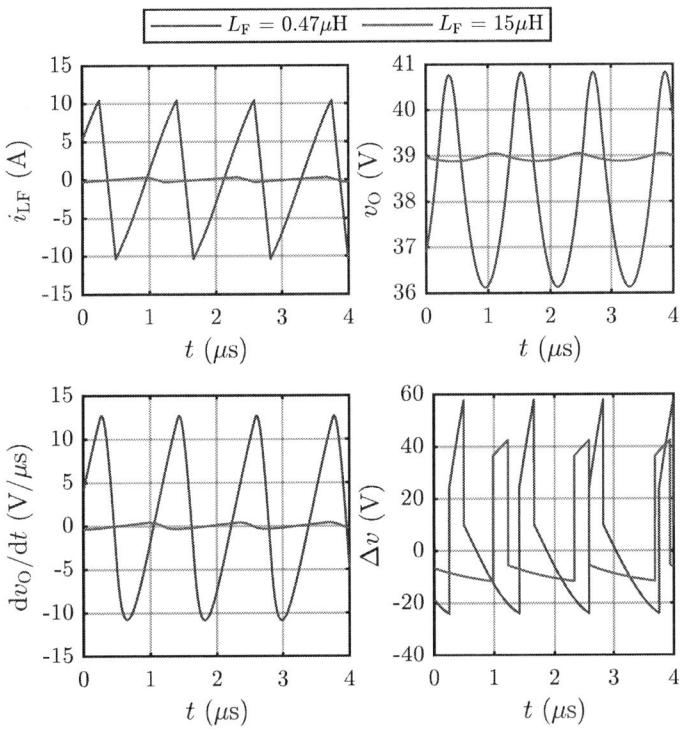

Fig. 8: Simulation results with influence of current ripple on the modulator behavior for $C_F = 680\,\text{nF}$

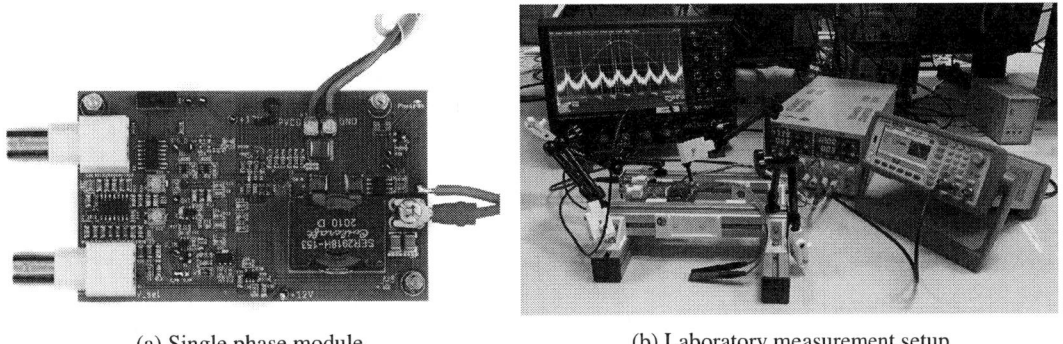

(a) Single phase module   (b) Laboratory measurement setup

Fig. 9: Hardware setup

Fig. 10 shows the output voltage v_O for a signal period based on $v_{O\text{Ref}}$ used in simulations with the parameters $L_F = 15\,\mu\text{H}$ and $C_F = 1.36\,\mu\text{F}$. Fig. 11 displays a detailed view of the measurements, marked in Fig. 10 by a red rectangle, for the parameters given in Table I. Although some deviations between simulations and measurements are observed, which can be attributed to device tolerances, the measurements support the simulation results and illustrate the theoretical discussion.

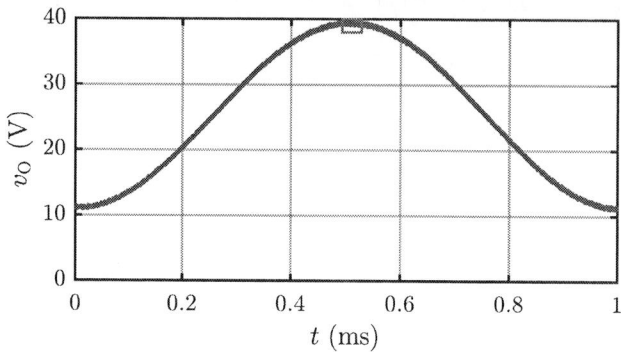

Fig. 10: Measurement result of the output voltage for one period

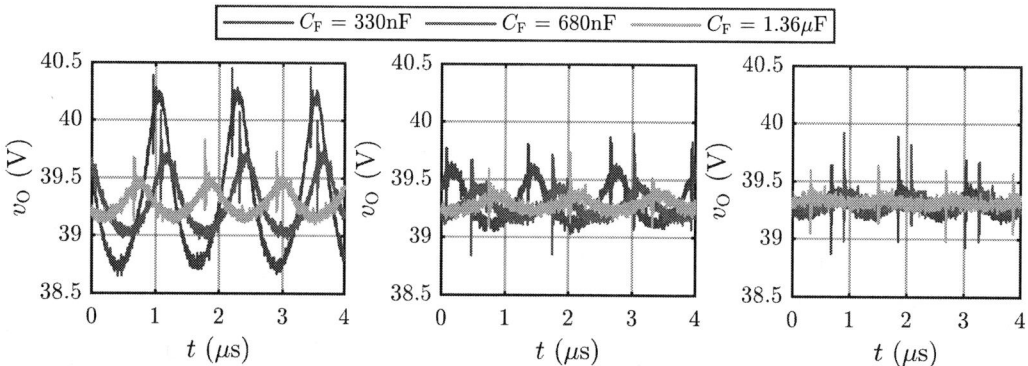

Fig. 11: Detailed view of measurement result of the output voltage with variation of L_F, C_F: Left: $L_F = 4.7\,\mu H$, middle: $L_F = 15\,\mu H$, right: $L_F = 33\,\mu H$

Conclusion

This contribution presents a new filter design approach for designing the output LC filter of a GaN based inverter such that a required dV/dt at the power electronic module is obtained with minimum filter component parameters. The design method is based on the idea, that the output filter can be considered as the output filter of a buck converter. A relation between the output filter parameters and the output voltage slew rate was developed and the effect of different filter values on the output voltage slew rate was investigated in simulations and measurements.

References

[1] Ulmer S., Walz-Lange A., Maatz A., Schullerus G., Sönmez E. and Hennig E.: Active Filter Damping for a GaN-Based Three Phase Power Stage with Continuous Output Voltage, 23rd European Conference on Power Electronics and Applications (EPE ECCE Europe), 2021

[2] Kim H., Anurag A., Acharya S. and Bhattacharya S.: Analytical Study of SiC MOSFET Based Inverter Output dv/dt Mitigation and Loss Comparison With a Passive dv/dt Filter for High Frequency Motor Drive Applications, IEEE Access, Vol. 9, 2021

[3] Vadstrup C., Wang X. and Blaabjerg F.: LC Filter Design for Wide Band Gap Device Based Adjustable Speed Drives, International Power Electronics and Application Conference and Exposition (PEAC), 2014

[4] Chen X., Xu D., Liu F. and Zhang J.: A Novel Inverter-Output Passive Filter for Reducing Both Differential- and Common-Mode dv/dt at the Motor Terminals in PWM Drive Systems, IEEE Transactions on Industrial Electronics, Vol. 54, Iss. 1, 2007

[5] Ulmer S., Schullerus G. and Sönmez E.: High Pass Design in Active Filter Damping, International Exhibition and Conference for Power Electronics, Intelligent Motion, Renewable Energy and Energy Management (PCIM Europe), 2022

[6] Hurley W. G. and Wölfle W. H.: Transformers and Inductors for Power Electronics. Theory, Design and Applications, Wiley, 2014

Evaluation of Core Losses in Transformers for Three-phase Multi-level DAB Converters

Babak Khanzadeh[*], Yuriy Serdyuk[†], and Torbjörn Thiringer[‡]
Chalmers University of Technology
Department of Electrical Engineering, SE-412 96
Gothenburg, Sweden
Emails: [*]ababak@chalmers.se, [†]yuriy.serdyuk@chalmers.se,
and [‡]torbjorn.thiringer@chalmers.se
URL: https://www.chalmers.se/

Acknowledgments

The authors gratefully acknowledge the financial support from the Swedish Energy Agency (Energimyndigheten). The authors also would like to thank Dr. Alexandre Giraud for his valuable suggestions.

Keywords

≪Dual active bridge (DAB) dc-dc converter≫, ≪Multi-level converters≫, ≪Solid-state transformer (SST)≫, ≪Core loss modelling≫, ≪Three-phase system≫.

Abstract

This paper provides closed-form formulas for estimating the core losses of three-phase transformers excited with multi-level converters for dual active bridge (DAB) applications. The formulas are derived by applying the improved generalized Steinmetz equation with approximated flux waveforms generated from one of the windings. The effect of different winding configurations on the estimated losses is studied. The results are validated with MATLAB simulations, and their applicability for three-phase multi-level DAB converters is investigated.

It is shown that for the YY and the ΔΔ configurations, an estimation error of less than 7.5% and 11.5%, respectively, can be achieved for most phase shifts between 0 and 20 degrees. It is highest when the distribution of the leakage inductance is more uniform between the primary and the secondary sides. Moreover, effect of the transition times of the bridges on the estimation error is also studied. It is established that estimating the losses from the winding, whose transition time is the largest, results in lower estimation error. It is demonstrated that using the parameters of the Δ side results in less estimation error for the YΔ and the ΔY configurations and most of the studied designs. The maximum estimation error for these two winding configurations is below 8.5% for phase shifts below 20 degrees.

Introduction

The DAB converters have become a hot topic for research in recent years. This is due to their capability to provide bidirectional power flow, wide soft-switching operation range, and last but not least, ensured galvanic isolation between the dc networks [1]. For medium- and high-voltage applications, DAB converters constructed with multi-level converter topologies can eliminate the need for a series connection of semiconductor devices [2]. For these converters, it has been shown that the quasi-two-level (Q2L) operation can reduce the required capacitor storage of the converter, thus improving the power density [3,4].

Aside from the power density, evaluating the overall efficiency of the converter is one of the main design steps. In this regard, evaluating the losses of the transformer is crucial in the design process of the

DAB converter. One of the main components of transformer losses is iron losses. To estimate these losses, different methods can be used [5,6]. If research focuses on the transformer design, finite element methods (FEM) can be used for detailed and accurate iron loss analysis. However, for research with a focus on the power electronics aspects, FEM analysis is time inefficient. Nonetheless, the impact of power electronics characteristics (e.g., any modulation technique or control strategy) on the core losses should be considered. A practical solution for the estimation of these losses is empirical methods [7]. One such empirical method is the so-called improved generalized Steinmetz equation (IGSE) [8]. According to IGSE, the time-average power loss per unit volume of a core, $\overline{P}_V$, can be calculated as

$$\overline{P}_V = \frac{1}{T}\int_0^T k_i \left|\frac{dB}{dt}\right|^\alpha (\Delta B)^{\beta-\alpha}\, dt, \qquad k_i = \frac{k}{(2\pi)^{\alpha-1}\int_0^{2\pi}|\cos(\theta)|^\alpha 2^{\beta-\alpha}\, d\theta} \tag{1}$$

where α, β, and k are the Steinmetz coefficients of the core material; ΔB is the peak-to-peak magnetic flux density; T is the fundamental period of the flux waveform. It has been shown that IGSE tends to be a good compromise between the accuracy of the model and the number of parameters required to model the losses [7]. The three required Steinmetz coefficients can be extracted from the material datasheet or from experiments, and the flux waveforms can be used to estimate the losses using (1).

A few papers have studied different aspects of three-phase multi-level DAB converters with Q2L modulation [3,4,9,10]. However, the effects of the modulation strategy on the transformer core losses are not investigated. An option for this study is IGSE. Nonetheless, direct implementation of (1) in design optimization of the power electronics is time inefficient. This paper provides simple closed-form formulas to tackle this issue. The formulas are derived using IGSE to link the core losses and the modulation parameters. In addition, mathematical expressions are provided to couple the maximum flux density and the required core cross-section with the specifications of the converters. The derived formulas can be applied to core-type single-phase transformers connected in three-phase configuration or E-core three-phase transformers.

Converter Topology and Waveforms

Fig. 1a depicts an overall schematic of a DAB converter. The power flow is controlled by introducing a phase shift between the transformer's primary and the secondary phase voltages [1]. Different two- or multi-level converter topologies can be used based on the application and the required dc-links voltages [2]. In this paper, the analysis is performed for a general case of a $(2M+1)$-level converter. Additionally, the converter is assumed to be modulated with the quasi-two-level (Q2L) modulation [9]. The phase-to-ground voltages of a $(2M+1)$-level converter operating with the Q2L modulation are shown in Fig. 1b. The time spent on each voltage level is defined as dwell time ($\theta_k/\omega\ \forall k\in[0,M-1]$ where $\omega = 2\pi f_{sw}$ and f_{sw} is the switching frequency of the converter). Moreover, the total time to alter the polarity of the voltage from its maximum positive value to its minimum negative value (or vice versa) is defined as transition time, τ. Here, it is assumed that all dwell angles are equal (i.e., $\psi \triangleq 2\theta_0 = \omega\tau/(2M-1)$) and $0 \le \omega\tau \le \pi/3$.

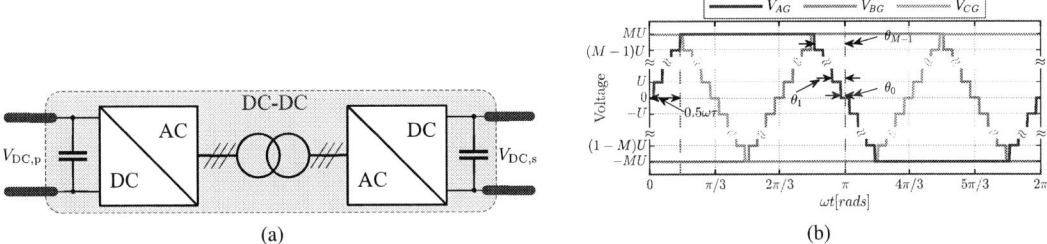

Fig. 1: A $(2M+1)$-level 3ϕ DAB converter (a) The topology. (b) The phase-to-ground voltages.

Depending on the value of M and θ_k, Fig. 1b can represent waveforms of different converters. As an example, with $M = 1$ and $\theta_0 = 0$, the typical waveforms of a conventional two-level converter are obtained. In turn, $M = 1$ and $\theta_0 \in [0, \pi/2]$ can describe three-level NPC or T-type converters. Waveforms of converters like the modular multi-level converter (MMC) or controlled transition bridge (CTB) can be achieved for higher values of M. In these converters, the submodule capacitors are inserted and bypassed sequentially to form the Q2L waveform shown in Fig. 1b. The modulation methods of these converters are out of the scope of this paper. A valuable source for modulation techniques of these converters is [10].

Similar to the converter topologies, different winding configurations of the transformer can be suggested based on the application [11]. In this paper, only Y and Δ connections of the windings will be considered, resulting in four different combinations for a two-port DAB converter. The phase-to-phase and phase-to-neutral voltages are applied to the windings for the Δ- and Y-connected windings, respectively. By considering Fig. 1b as a reference, the phase voltages of the Y- and Δ-connected windings, $v_{\phi Y}$ and $v_{\phi\Delta}$, can be expressed as

$$v_{\phi\Delta}(\omega t) = \sum_{H=1}^{\infty} \sum_{k=0}^{M_\Delta-1} \left(\frac{8U_\Delta}{H\pi} \cos(H\theta_{k,\Delta}) \cos\left(H\omega t - \frac{H\pi}{3}\right) \sin\left(\frac{H\pi}{3}\right) \right) \tag{2a}$$

$$v_{\phi Y}(\omega t) = \sum_{H=1}^{\infty} \sum_{k=0}^{M_Y-1} \left(\frac{4U_Y}{H\pi} \cos(H\theta_{k,Y}) \sin(H\omega t) \right) \tag{2b}$$

where $H \in \{2h - 1 | h \in \mathbb{N}, \; 3 \nmid 2h - 1\}$ is the harmonic order, and U is the magnitude of each step in the phase-to-ground voltage as shown in Fig. 1b. Hereafter, it is assumed that the primary side is the supply side, and the secondary side is the load side (i.e., the active power flows from the primary dc-link to the secondary dc-link). Moreover, subscripts **p** and **s** denote the primary-side and the secondary-side parameters, respectively. Similarly, subscripts Δ and Y are used to distinguish the parameters of the Δ- and Y-connected windings.

Flux and Core Loss Modeling

This study is focused on core-type single-phase transformers connected in three-phase configuration and E-core three-phase transformers. The total instantaneous flux linkage in the core is a complex function of voltage waveforms of both windings, leakage fluxes, winding resistances, and the core geometry. For simplicity, the flux is assumed to be distributed homogeneously across the core cross-section and also remain homogeneous along the core. Furthermore, the resistance of the windings usually is small. Thus, the voltage drop across them can be neglected. With these assumptions, the total instantaneous flux in the core, $\Phi^{k_\sigma}(\omega t)$, can be calculated as

$$\dot{\Phi}^{k_\sigma}(\omega t) = \frac{d\Phi^{k_\sigma}(\omega t)}{dt} = \frac{k_\sigma}{N_p} v_{\phi p}(\omega t) + \frac{1 - k_\sigma}{N_s} v_{\phi s}(\omega t), \quad \text{and} \quad k_\sigma \triangleq \frac{L_{\sigma s}}{L_{\sigma,s}^{\text{tot}}} \tag{3}$$

where N_p and N_s are the numbers of turns of the primary and the secondary windings, respectively; $L_{\sigma s}$ is the leakage inductance of the secondary winding; $L_{\sigma,s}^{\text{tot}}$ is the secondary-side-referred total leakage inductance (i.e., $L_{\sigma,s}^{\text{tot}} = L_{\sigma s} + L_{\sigma p}'$); $v_{\phi p}$ is the primary-side phase voltage; $v_{\phi s}$ is the secondary-side phase voltage. By using (2) and (3), the waveform of the flux for the YY configuration can be derived as

$$\begin{aligned}
\Phi_{YY}^{k_\sigma}(\omega t) = &-\frac{k_\sigma}{N_p} \sum_{H=1}^{\infty} \sum_{k=0}^{M_p-1} \left(\frac{4U_p}{H^2\pi\omega} \cos(H\theta_{k,p}) \cos(H\omega t) \right) \\
&-\frac{1 - k_\sigma}{N_s} \sum_{H=1}^{\infty} \sum_{k=0}^{M_s-1} \left(\frac{4U_s}{H^2\pi\omega} \cos(H\theta_{k,s}) \cos(H(\omega t - \varphi)) \right)
\end{aligned} \tag{4}$$

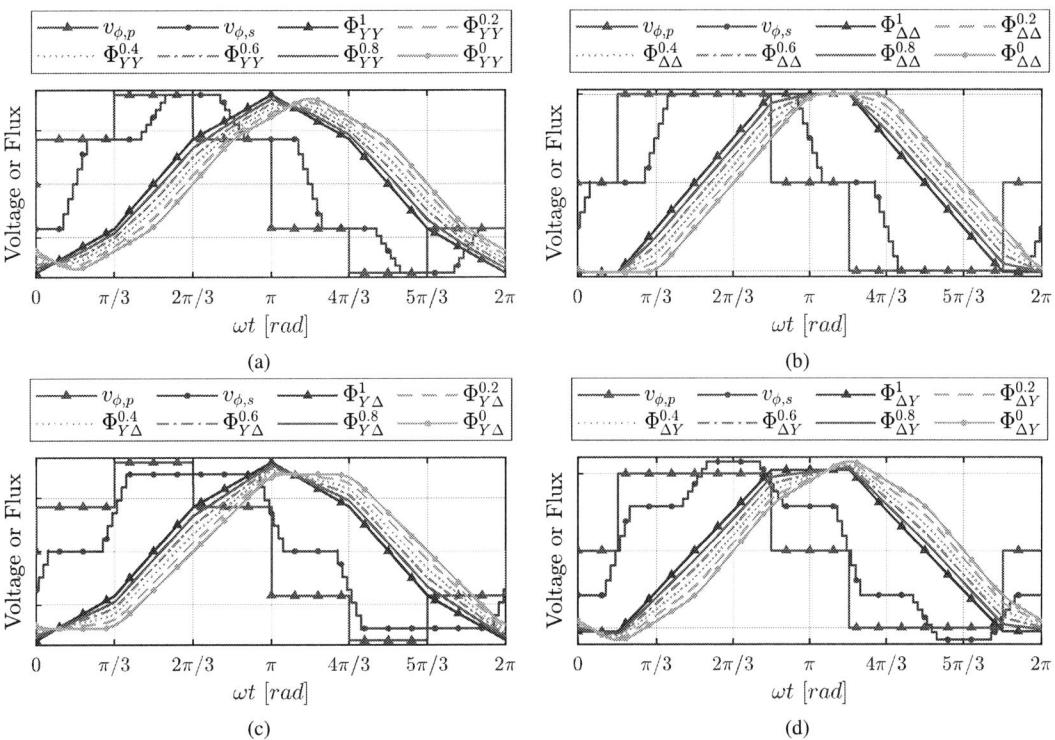

Fig. 2: Phase voltages and flux waveforms, Φ^{k_σ}, of a multi-level DAB converter at $\varphi = \pi/6$ for different values of k_σ. (a) YY configuration. (b) $\Delta\Delta$ configuration. (c) YΔ configuration. (d) ΔY configuration.

where $\varphi \in [0, \pi/2]$ is the phase shift between the primary and the secondary phase voltages, similarly, the flux waveform for the $\Delta\Delta$, the YΔ and the ΔY configurations can be calculated as

$$
\begin{aligned}
\Phi_{\Delta\Delta}^{k_\sigma}(\omega t) = &+ \frac{k_\sigma}{N_p} \sum_{H=1}^{\infty} \sum_{k=0}^{M_p-1} \left(\frac{8U_p}{H^2\pi\omega} \cos(H\theta_{k,p}) \sin\left(H\left(\omega t - \frac{\pi}{3}\right)\right) \sin\left(\frac{H\pi}{3}\right) \right) \\
&+ \frac{1-k_\sigma}{N_s} \sum_{H=1}^{\infty} \sum_{k=0}^{M_s-1} \left(\frac{8U_s}{H^2\pi\omega} \cos(H\theta_{k,s}) \sin\left(H\left(\omega t - \varphi - \frac{\pi}{3}\right)\right) \sin\left(\frac{H\pi}{3}\right) \right)
\end{aligned}
\tag{5a}
$$

$$
\begin{aligned}
\Phi_{Y\Delta}^{k_\sigma}(\omega t) = &- \frac{k_\sigma}{N_p} \sum_{H=1}^{\infty} \sum_{k=0}^{M_p-1} \left(\frac{4U_p}{H^2\pi\omega} \cos(H\theta_{k,p}) \cos(H\omega t) \right) \\
&+ \frac{1-k_\sigma}{N_s} \sum_{H=1}^{\infty} \sum_{k=0}^{M_s-1} \left(\frac{8U_s}{H^2\pi\omega} \cos(H\theta_{k,s}) \sin\left(H\left(\omega t - \varphi - \frac{\pi}{2}\right)\right) \sin\left(\frac{H\pi}{3}\right) \right)
\end{aligned}
\tag{5b}
$$

$$
\begin{aligned}
\Phi_{\Delta Y}^{k_\sigma}(\omega t) = &+ \frac{k_\sigma}{N_p} \sum_{H=1}^{\infty} \sum_{k=0}^{M_p-1} \left(\frac{8U_p}{H^2\pi\omega} \cos(H\theta_{k,p}) \sin\left(H\left(\omega t - \frac{\pi}{3}\right)\right) \sin\left(\frac{H\pi}{3}\right) \right) \\
&- \frac{1-k_\sigma}{N_s} \sum_{H=1}^{\infty} \sum_{k=0}^{M_s-1} \left(\frac{4U_s}{H^2\pi\omega} \cos(H\theta_{k,s}) \cos\left(H(\omega t - \varphi + \frac{\pi}{6})\right) \right).
\end{aligned}
\tag{5c}
$$

Fig. 2 shows the phase voltages and flux waveforms of a 3ϕ multi-level DAB converter for different winding configurations and $\varphi = \pi/6$. The flux inside the core is depicted for different values of k_σ. When $k_\sigma = 0$, there is no leakage flux from the secondary winding, or one can say that only the secondary winding is excited. Thus, the flux waveform inside the core depends just on the secondary phase voltage. Similarly, if $k_\sigma = 1$, there is no leakage flux from the primary winding (or one can say that only the primary winding is excited). In this case, the flux waveform inside the core depends only on the primary phase voltage. In reality, there will be leakage flux from both windings, and $k_\sigma \in \{0,1\}$ are two extreme cases. Therefore, the flux waveform will depend on the primary and the secondary voltages and the phase shift between them. As can be seen from Fig. 2, the higher the k_σ, the closer the flux waveform to the case where only the primary winding is excited. Similarly, the lower the k_σ, the closer the flux waveform to the case where only the secondary winding is excited.

Eventually, to calculate the core losses, one should apply (1) on the waveforms of (4) or (5). However, calculating the losses using (4) and (5) is complicated and time demanding. To further simplify the loss calculation, one can use the extreme cases of $k_\sigma \in \{0,1\}$ and assume that only one of the windings is excited. Then the voltage applied to the terminals of the energized winding determines the flux shape inside the core. This is an impactful assumption, and its effects will be discussed later. Considering this assumption, the total instantaneous flux linkage in the core, λ, can be calculated using (2) and the relationship between MMF and flux linkage (i.e., $v = d\lambda/dt$) as

$$\lambda_\Delta = \sum_{H=1}^{\infty} \sum_{k=0}^{M_\Delta-1} \left(\frac{8U_\Delta}{H^2\pi\omega} \sin\left(H\omega t - \frac{H\pi}{3}\right) \sin\left(\frac{H\pi}{3}\right) \cos(H\theta_{k,\Delta}) \right) \tag{6a}$$

$$\lambda_Y = \sum_{H=1}^{\infty} \sum_{k=0}^{M_Y-1} \left(\frac{-4U_Y}{H^2\pi\omega} \cos(H\omega t) \cos(H\theta_{k,Y}) \right). \tag{6b}$$

The maximum value of the flux linkage can be calculated assuming that the flux is distributed evenly in a cross-section of the core as

$$\widehat{\lambda}_\Delta = N_\Delta B_{\max,\Delta} A_\Delta = \frac{V_{dc,\Delta}}{6 f_{sw,\Delta}} \tag{7a}$$

$$\widehat{\lambda}_Y = N_Y B_{\max,Y} A_Y = \frac{V_{dc,Y}}{9 f_{sw,Y}} \left(1 - \frac{3M_Y\psi_Y}{4\pi} \right) \tag{7b}$$

where N is the number of turns of the winding, A is the cross-section of the core, V_{dc} is the dc-link voltage, and $B_{\max}$ is the maximum flux density in the core. As can be seen, the maximum flux linkage of the Δ-type winding is independent of the dwell time and the number of levels. In contrast, the maximum flux linkage reduces by increased waveform levels and higher dwell time for the Y-type winding. This is because by increasing $\omega\tau$ up to 60 degrees the area under phase voltage stays the same for the Δ-type winding while it reduces for the Y-type winding.

Since the flux is calculated from one of the windings, it is no longer dependent on the phase shift between the bridges or the configuration of the windings. This substantially simplifies the loss calculation. By applying (1) on (6), the time-average power loss per unit volume of the core can be calculated as

$$\overline{P}_{V,\Delta} = k B_{\max,\Delta}^\beta f_{sw,\Delta}^\alpha C_\Delta \tag{8a}$$

$$\overline{P}_{V,Y} = k B_{\max,Y}^\beta f_{sw,Y}^\alpha C_Y \tag{8b}$$

where C_Δ and C_Y are design factors dependent on the parameters of the converter as

$$C_\Delta = 4 \times \frac{\frac{\pi}{3} - (M_\Delta - 0.5)\psi_\Delta + \psi_\Delta \sum_{k=1}^{2M_\Delta-1} \left(\frac{k}{2M_\Delta}\right)^\alpha}{\left(\frac{\pi}{3}\right)^\alpha \int_0^{2\pi} |\cos(\theta)|^\alpha d\theta} \tag{9a}$$

$$C_Y = 4 \times \frac{(1+2^{\alpha-1})\left(\frac{\pi}{3} - (2M_Y-1)\psi_Y\right) + \psi_Y \left(\sum_{k=1}^{M_Y-1} \left(\frac{k}{M_Y}\right)^\alpha + \sum_{k=1}^{2M_Y-1} \left(1+\frac{k}{2M_Y}\right)^\alpha\right)}{\left(\frac{2\pi}{3} - \frac{M_Y\psi_Y}{2}\right)^\alpha \int_0^{2\pi} |\cos(\theta)|^\alpha d\theta}. \tag{9b}$$

Fig. 3 visualizes C_Y and C_Δ as a function of the number of levels and dwell angle for three values of $\alpha \in \{1.5, 1.75, 2\}$. The value C_Y reduces by increasing α for a given M_Y and ψ_Y. On the contrary, larger α results in higher C_Δ. In Fig. 3c and Fig. 3f, the isolines of $\omega\tau$ are also plotted. The higher the $\omega\tau_Y$, the higher the value of C_Y. Consequently, the loss density in the core increases if $B_{\max,Y}$ is kept constant. However, the increase is marginal because a decrease in the core cross-section compensates for the reduction in $\frac{d\phi_Y}{dt}$ to keep $B_{\max,Y}$ constant. On the other hand, C_Δ reduces considerably by increasing $\omega\tau_\Delta$ as shown in Fig. 3f. This is because $N_\Delta B_{\max,\Delta} A_\Delta$ is independent of $\omega\tau_\Delta$. Therefore, increasing $\omega\tau_\Delta$ only reduces $\frac{dB_\Delta}{dt}$, resulting in lower loss density.

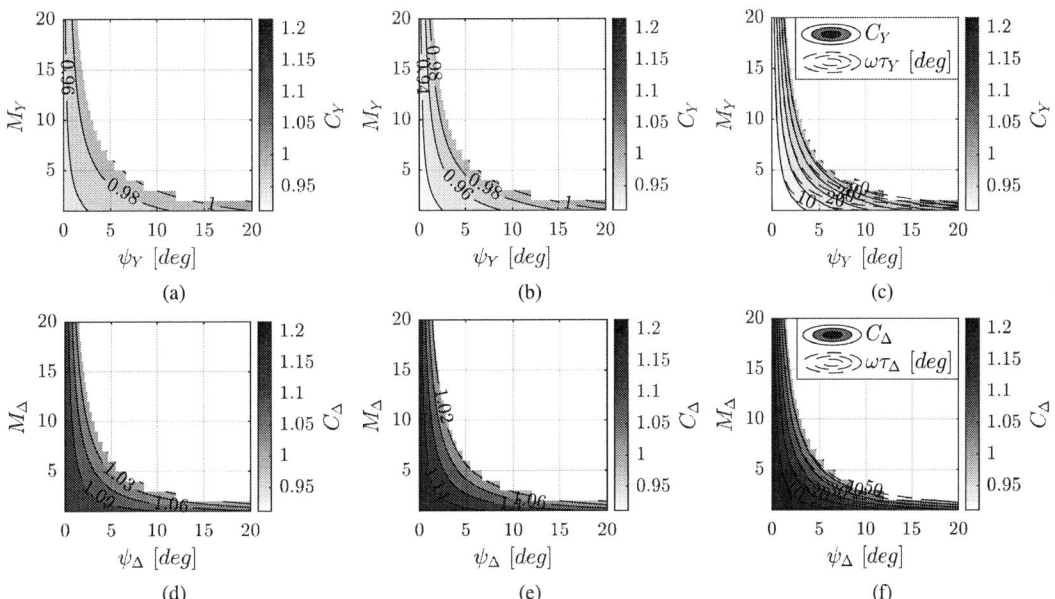

Fig. 3: C_Y and C_Δ as a function of M and ψ. The equivalent $\omega\tau$ is overlayed for $\alpha = 2$. (a) C_Y @ $\alpha = 1.5$. (b) C_Y @ $\alpha = 1.75$. (c) C_Y @ $\alpha = 2$. (d) C_Δ @ $\alpha = 1.5$. (e) C_Δ @ $\alpha = 1.75$. (f) C_Δ @ $\alpha = 2$.

Simulation and Verification

For the derivation of (8), it is assumed that $k_\sigma \in \{0, 1\}$. However, in practical applications, there is a leakage flux from both sides ($k_\sigma \notin \{0, 1\}$), and the losses depend on the distribution of the leakage inductance. Therefore, a model of a DAB converter is developed in PLECS to verify the derived formulas. The specification of the simulated converters is presented in Table I. The primary bridge is assumed to be a two-level converter ($M_p = 1$, $\theta_{0,p} = 0$), and the secondary bridge is an MMC with six sub-modules per arm ($M_s = 3$). The primary side switches are modulated with a 50% duty cycle, and the secondary bridge is modulated with the Q2L modulation technique. The primary and the secondary dc-links nominal voltage are selected to be equal for simplicity ($V_{dc,p,nom} = V_{dc,s,nom} = 5$ kV). The converter has a nominal power of $P_{nom} = 2$ MW and a switching frequency of $f_{sw} = 5$ kHz. A controller regulates the active power flow by adjusting the phase shift between the two bridges.

Table I: Specifications of the simulated dc-dc converters

Parameter	Value	Parameter	Value	Parameter	Value	Parameter	Value
$V_{dc,p,nom}$	5 kV	φ_{nom}	20°	$N_{p,YY} : N_{s,YY}$	11 : 11	$L_{\sigma,YY,p}^{tot}$	84.2 μH
$V_{dc,s,nom}$	5 kV	θ_{0s}	1.8°	$N_{p,\Delta\Delta} : N_{s,\Delta\Delta}$	19 : 19	$L_{\sigma,\Delta\Delta,p}^{tot}$	252.4 μH
P_{nom}	2 MW	M_s	3	$N_{p,Y\Delta} : N_{s,Y\Delta}$	11 : 19	$L_{\sigma,Y\Delta,p}^{tot}$	80.2 μH
f_{sw}	5 kHz	$R_p = R_s'$	10 mΩ	$N_{p,\Delta Y} : N_{s,\Delta Y}$	19 : 11	$L_{\sigma,\Delta Y,p}^{tot}$	240.6 μH

It is assumed that the transformer has a three-phase E-core structure. The core is modeled in the magnetic domain of PLECS with linear magnetic permeances. Different winding configurations are realized by electrical connections in the electrical domain of PLECS. It is assumed that the primary winding has a resistance value of $R_p = 10$ mΩ. The primary-side-referred secondary winding resistance is also assumed to be $R'_s = 10$ mΩ. A nanocrystalline magnetic material, VITROPERM 500 F [12], from Vacuumschmelze is selected as the core material for this study. The Steinmetz coefficients of the material are extracted from the typical loss curve provided in the material datasheet as $\alpha \approx 1.8$, $\beta \approx 2.09$, $k \approx 0.0093$. The core cross-section is assumed to be constant throughout the study and is selected to be $A = 100$ cm^2.

The value of the leakage inductance for each winding configuration is selected such that the nominal power of the converter is transferred at $\varphi_{nom} = 20°$. It is important to note that designing the converter with high values of φ is undesirable due to high reactive power flow. The value of total leakage inductance, L_σ^{tot}, is kept constant while its distribution between the primary and secondary sides is altered to validate the assumptions. This is done by tuning the leakage permeances of the primary and the secondary windings. The values of the total leakage inductances and the turn ratios of the windings for different winding configurations are provided in Table I.

The waveforms of the fluxes in different parts of the core are captured in the steady-state and are saved for post-processing. Eventually, (1) is applied to the obtained waveforms from the simulation to calculate the loss density. The loss density is also estimated for two extreme cases of $k_\sigma \in \{0, 1\}$ using (7) and (8). As mentioned earlier, when $k_\sigma = 0$, there is no flux leakage from the secondary side winding. Therefore, the secondary bridge waveforms should be used to estimate the losses. Similarly, the primary bridge waveforms should be used to estimate the losses when $k_\sigma = 1$. The estimation of the core loss density, $\Delta \overline{P}_{V,x}$ (where x $\in \{p,s\}$), is defined as

$$\Delta \overline{P}_{V,x} = \frac{\overline{P}_V^{sim} - \overline{P}_{V,x}^{calc}}{\overline{P}_V^{sim}} \times 100 \tag{10}$$

where $\overline{P}_V^{sim}$ is the loss density obtained from applying (1) on the flux waveforms of PLECS, and $\overline{P}_{V,x}^{calc}$ (where x $\in \{p,s\}$) is the loss density calculated from (8) using the waveforms of either the primary (i.e., $\overline{P}_{V,p}^{calc}$) or the secondary side ($\overline{P}_{V,s}^{calc}$).

Fig. 4 depicts $\Delta \overline{P}_{V,x}$ at different load levels, P_{load}, for YY and ΔΔ winding configurations. The primary bridge parameters are used to calculate the losses for Fig. 4a and Fig. 4c, whereas for Fig. 4b and Fig. 4d, the parameters of the secondary bridge are used. It can be seen that the estimation error is zero if there is a leakage flux only from one of the windings (i.e., $k_\sigma \in \{0, 1\}$). This is because (8) is derived under these conditions, and it is the most accurate when $k_\sigma \approx \{0, 1\}$. Using the primary bridge parameters for both winding connections and $k_\sigma \lesssim 0.7$ overestimates the losses. This is because of higher $\frac{dB}{dt}$ and (or) higher B_{max} when $k_\sigma \approx 1$, as shown in Fig. 2a and Fig. 2b. For the YY configuration, increasing load (increasing φ) reduces B_{max} (see Fig. 2a). On the other hand, for the ΔΔ configuration, B_{max} remains constant while $\frac{dB}{dt}$ reduces as load increases (see Fig. 2b). Therefore, the higher the load (or φ), the higher the estimation error for both winding configurations. This effect is more pronounced when $k_\sigma \approx 0.5$, as shown in Fig. 4.

The value of $|\Delta \overline{P}_{V,p}|$ is larger than $|\Delta \overline{P}_{V,s}|$ for most of the points in the plots (excluding the cases where k_σ is close to 1). To highlight this further, the primary bridge is changed to an MMC with six submodules per arm. The dwell angles of both bridges are swept between 0 and 12 degrees for three cases where $(k_\sigma, P_{load}) \in \{(0.4, P_{nom}), (0.5, P_{nom}), (0.6, P_{nom})\}$. Fig. 5 depicts $|\Delta \overline{P}_{V,min}| \triangleq \min(|\Delta \overline{P}_{V,p}|, |\Delta \overline{P}_{V,s}|)$ for both winding configurations under the conditions above. The points where $|\Delta \overline{P}_{V,p}| < |\Delta \overline{P}_{V,s}|$ are marked with a checkerboard pattern. For both winding configurations and all three conditions, the boundary $|\Delta \overline{P}_{V,p}| = |\Delta \overline{P}_{V,s}|$ is identified as $\omega \tau_p = \omega \tau_s$. The values of the Steinmetz coefficients are altered to see whether they impact the boundary $\omega \tau_p = \omega \tau_s$ or not. Simulations results established that the boundary is independent of Steinmetz coefficients value for the YY and the ΔΔ winding configurations.

As seen from Fig. 5, using the primary bridge parameters results in less estimation error when $\omega \tau_p > \omega \tau_s$. Looking at Fig. 2a and Fig. 2b, one can see that the flux waveform obtained from the bridge with higher

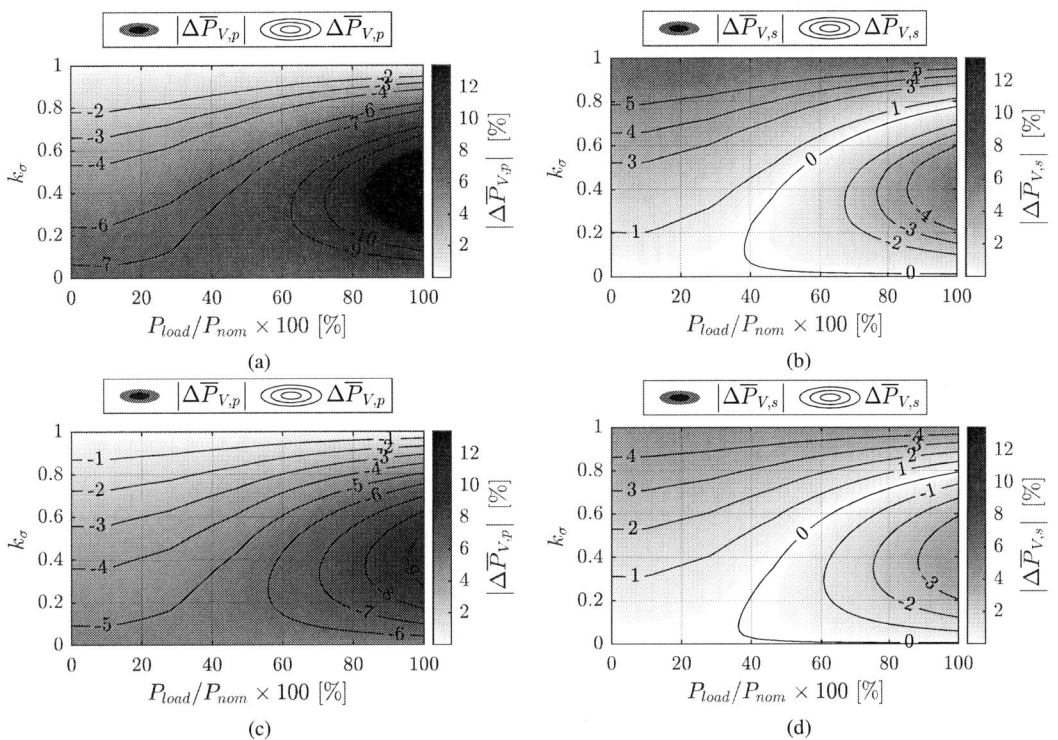

Fig. 4: The estimation error of core loss density, $\Delta\overline{P}_V$. (a) $\Delta\overline{P}_{V,p}$ for YY configuration. (b) $\Delta\overline{P}_{V,s}$ for YY configuration. (c) $\Delta\overline{P}_{V,p}$ for $\Delta\Delta$ configuration. (d) $\Delta\overline{P}_{V,s}$ for $\Delta\Delta$ configuration.

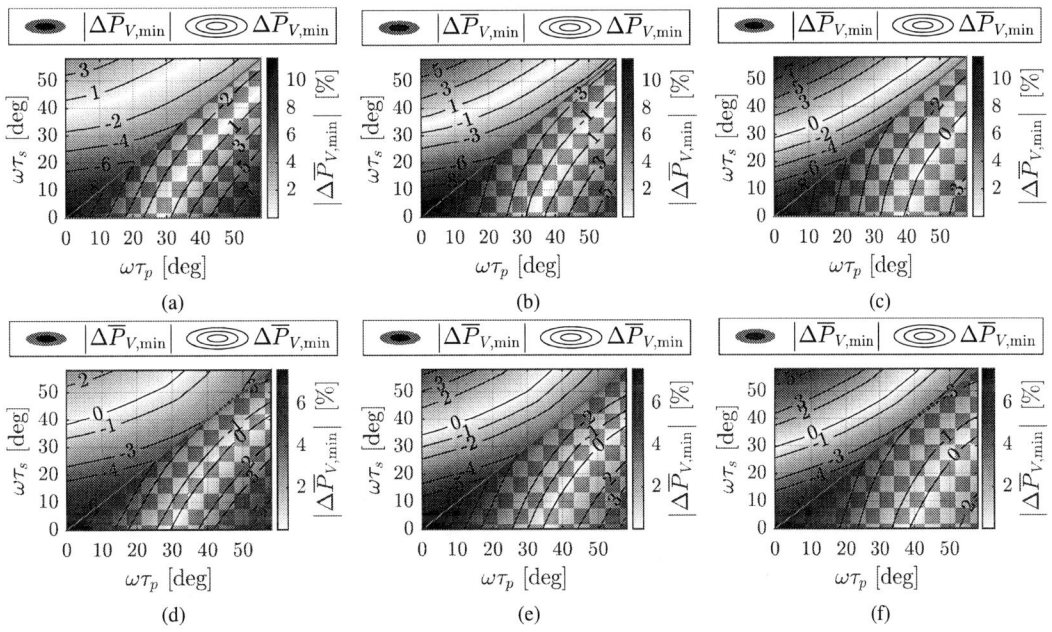

Fig. 5: The value of $|\Delta\overline{P}_{V,\min}|$ at $P_{\text{load}} = P_{\text{nom}}$. (a) For YY and $k_\sigma = 0.4$. (b) For YY and $k_\sigma = 0.5$. (c) For YY and $k_\sigma = 0.6$. (d) For $\Delta\Delta$ and $k_\sigma = 0.4$. (e) For $\Delta\Delta$ and $k_\sigma = 0.5$. (f) For $\Delta\Delta$ and $k_\sigma = 0.6$.

$\omega\tau$ is a smoother and better representative of the actual flux in the core. Therefore, the estimation error is less when the parameters of the bridge with the highest $\omega\tau$ are used for the loss density estimation. As seen from Fig. 5, the maximum estimation error is below 7.5% for most parameter combinations and the $\Delta\Delta$ winding configuration. This value is 11.5% for the YY configuration. The higher estimation error of the YY configuration is due to pronounced flux peaks, especially at low $\omega\tau$.

Fig. 6 shows $|\Delta\overline{P}_V|$ for ΔY and YΔ winding configurations. Similar to the YY and the $\Delta\Delta$ configurations, the estimation error is lower when the value k_σ is close to 1 or 0 and the waveforms of the winding with the lowest leakage flux is used. For the YΔ winding configuration and the cases where k_σ diverges from 0 or 1, an estimation using the Δ-side waveforms gives a considerably lower error. However, for the ΔY configuration, $|\Delta\overline{P}_{V,p}| \approx |\Delta\overline{P}_{V,s}|$ when k_σ diverges from 0 or 1. To investigate this further, the topology of the primary bridge is changed to an MMC with six submodules per arm. The dwell angles of both bridges are swept between 0 and 12 degrees for three cases of $(k_\sigma, P_{\text{load}}) \in \{(0.4, P_{\text{nom}}), (0.5, P_{\text{nom}}), (0.6, P_{\text{nom}})\}$.

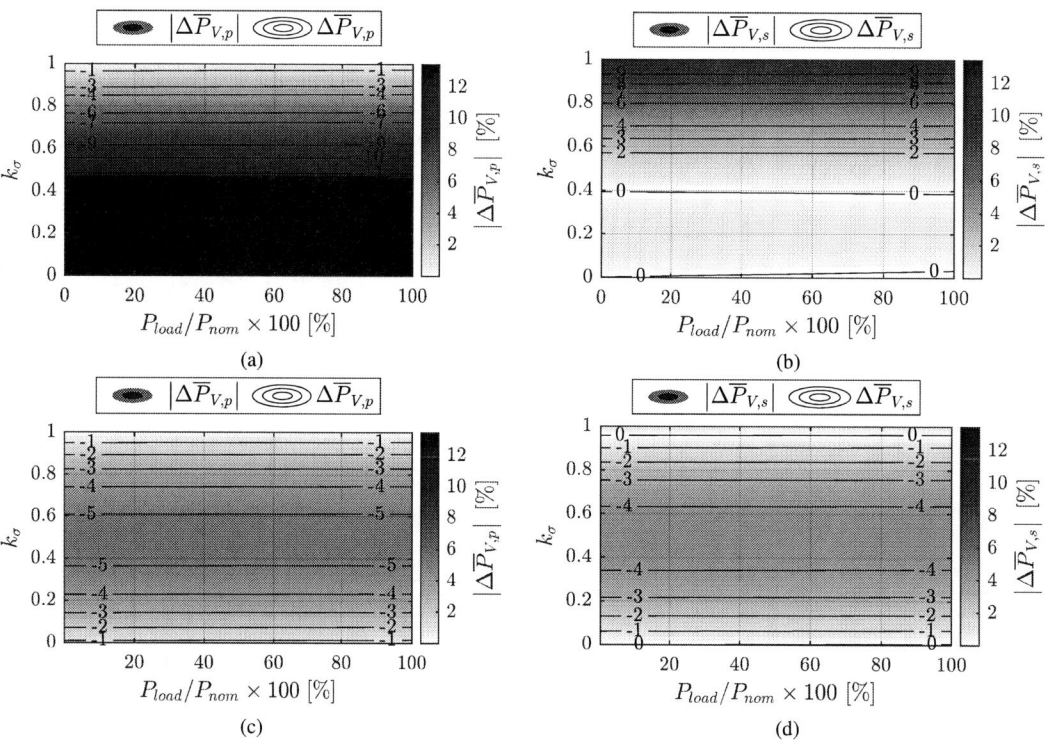

Fig. 6: The estimation error of core loss density, $\Delta\overline{P}_V$. (a) $\Delta\overline{P}_{V,p}$ for YΔ configuration. (b) $\Delta\overline{P}_{V,s}$ for YΔ configuration. (c) $\Delta\overline{P}_{V,p}$ for ΔY configuration. (d) $\Delta\overline{P}_{V,s}$ for ΔY configuration.

Fig. 7 shows $\Delta\overline{P}_{V,\min} \triangleq \min(|\Delta\overline{P}_{V,p}|, |\Delta\overline{P}_{V,s}|)$ for YΔ and ΔY winding configurations for the three cases. The domain where $|\Delta\overline{P}_{V,p}| < |\Delta\overline{P}_{V,s}|$ is marked with a checkerboard pattern. The boundary where $|\Delta\overline{P}_{V,p}| = |\Delta\overline{P}_{V,s}|$ is identified as $\omega\tau_\Delta \approx 1.4\omega\tau_Y - 22.4°$. This boundary remains unchanged for phase shifts below 20°. Using the waveforms of the Δ-side results in lower estimation error for both winding configurations and $\omega\tau_\Delta \geq 1.4\omega\tau_Y - 22.4°$. Moreover, the maximum estimation error is below 8.5% for both winding configurations. For these two winding configurations, the boundary where $|\Delta\overline{P}_{V,p}| = |\Delta\overline{P}_{V,s}|$ is dependent on the values of the Steinmetz coefficients. The larger the α, the smaller the domain where $|\Delta\overline{P}_{V\Delta}| < |\Delta\overline{P}_{V,Y}|$. This domain also reduces with a decrease in β. Nonetheless, using the waveforms of the Δ-side still results in lower estimation error for most of the points shown in Fig. 7.

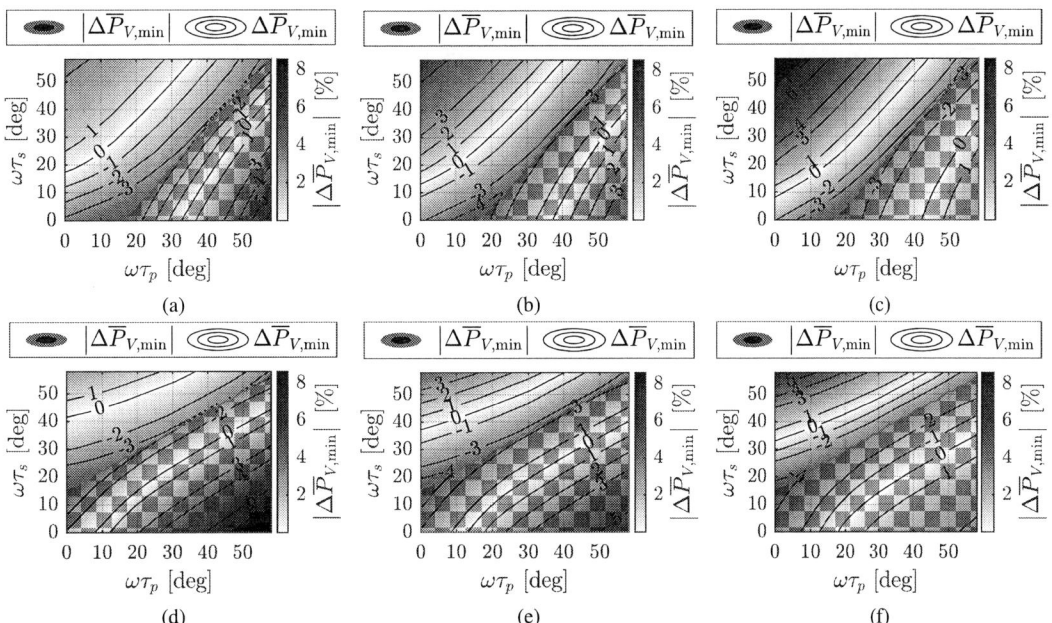

Fig. 7: The value of $|\Delta \overline{P}_{V,\min}|$ at $P_{\text{load}} = P_{\text{nom}}$. (a) For YΔ and $k_\sigma = 0.4$. (b) For YΔ and $k_\sigma = 0.5$. (c) For YΔ and $k_\sigma = 0.6$. (d) For ΔY and $k_\sigma = 0.4$. (e) For ΔY and $k_\sigma = 0.5$. (f) For ΔY and $k_\sigma = 0.6$.

Conclusion

Mathematical formulas have been derived to couple the maximum flux density and the required core cross-section with the specifications of three-phase multi-level DAB converters. It is shown that the maximum flux linkage of the Δ-type winding is independent of the transition time and the number of converter levels. In contrast, the maximum flux linkage reduces by increased levels and higher transition time for the Y-type winding.

In addition, simple closed-form formulas have been derived using IGSE to link the core losses to the modulation parameters. The obtained formulas are applicable to three-phase E-core transformers and three single-phase transformers with three-phase electrical connections. Moreover, effect of the transition times of the bridges on the estimation error is also studied. It is shown that using the parameters of the bridge with the longest transition time for the YY and ΔΔ configurations results in less estimation error. Following this guideline for the phase shifts below 20°, estimation errors lower than 7.5% and 11.5% can be achieved for the ΔΔ and the YY configurations, respectively. It is demonstrated that using the parameters of the Δ side results in less estimation error for the YΔ and the ΔY configurations and most of the studied designs. The maximum estimation error for these two winding configurations is below 8.5%.

References

[1] R. De Doncker, D. Divan, and M. Kheraluwala, "A three-phase soft-switched high-power-density dc/dc converter for high-power applications," *IEEE Transactions on Industry Applications*, vol. 27, no. 1, pp. 63–73, 1991.

[2] G. P. Adam, I. A. Gowaid, S. J. Finney, D. Holliday, and B. W. Williams, "Review of dc–dc converters for multi-terminal hvdc transmission networks," *IET Power Electronics*, vol. 9, no. 2, pp. 281–296, 2016.

[3] B. Khanzadeh, Y. Okazaki, and T. Thiringer, "Capacitor and switch size comparisons on high-power medium-voltage dc–dc converters with three-phase medium-frequency transformer," *IEEE Journal of Emerging and Selected Topics in Power Electronics*, vol. 9, no. 3, pp. 3331–3338, 2021.

[4] B. Khanzadeh, T. Thiringer, and Y. Okazaki, "Capacitor size comparison on high-power dc-dc converters with different transformer winding configurations on the ac-link," in *2020 22nd European Conference on Power Electronics and Applications (EPE'20 ECCE Europe)*, pp. P.1–P.7, 2020.

[5] M. A. Bahmani, E. Agheb, T. Thiringer, H. Høidalen, and Y. Serdyuk, "Core loss behavior in high frequency high power transformers—i: Effect of core topology," *Journal of Renewable and Sustainable Energy*, vol. 4, no. 3, p. 033112, 2012.

[6] E. Agheb, M. A. Bahmani, H. Høidalen, and T. Thiringer, "Core loss behavior in high frequency high power transformers—ii: Arbitrary excitation," *Journal of Renewable and Sustainable Energy*, vol. 4, no. 3, p. 033113, 2012.

[7] A. Krings and J. Soulard, "Overview and comparison of iron loss models for electrical machines," *Journal of Electrical Engineering*, vol. 10, no. 3, pp. 8–8, 2010.

[8] K. Venkatachalam, C. R. Sullivan, T. Abdallah, and H. Tacca, "Accurate prediction of ferrite core loss with nonsinusoidal waveforms using only steinmetz parameters," in *2002 IEEE Workshop on Computers in Power Electronics, 2002. Proceedings.*, pp. 36–41, IEEE, 2002.

[9] I. Gowaid, G. Adam, A. M. Massoud, S. Ahmed, D. Holliday, and B. Williams, "Quasi two-level operation of modular multilevel converter for use in a high-power dc transformer with dc fault isolation capability," *IEEE Transactions on Power Electronics*, vol. 30, no. 1, pp. 108–123, 2014.

[10] I. A. Gowaid, G. P. Adam, S. Ahmed, D. Holliday, and B. W. Williams, "Analysis and design of a modular multilevel converter with trapezoidal modulation for medium and high voltage dc-dc transformers," *IEEE Transactions on Power Electronics*, vol. 30, no. 10, pp. 5439–5457, 2015.

[11] N. H. Baars, J. Everts, C. G. Wijnands, and E. A. Lomonova, "Performance evaluation of a three-phase dual active bridge dc–dc converter with different transformer winding configurations," *IEEE Transactions on Power Electronics*, vol. 31, no. 10, pp. 6814–6823, 2015.

[12] "VITROPERM 500 F: iron-based nanocrystalline material with soft-magnetic properties." https://vacuumschmelze.com/Nanocrystalline-Material.

A quasi-offline condition monitoring method of DC-link capacitor banks in accelerator power converters

Timm Felix Baumann[1,2], Konstantinos Papastergiou[2], Raul Murillo Garcia[2], Dimosthenis Peftitsis[1]

[1]DEPARTMENT OF ELECTRIC POWER ENGINEERING NTNU - NORWEGIAN UNIVERSITY OF SCIENCE AND TECHNOLOGY

[2]CERN - EUROPEAN ORGANIZATION FOR NUCLEAR RESEARCH

[1]Gløshaugen, Trondheim, Norway

[2]Esplanade des Particules 1, Geneva, Switzerland

E-mail: timm.felix.baumann@cern.ch

Keywords

« DC-link », « Capacitors », « Condition Monitoring », « Data analysis », « Accelerators »

Abstract

This paper proposes a condition monitoring scheme for DC-link electrolytic capacitors used in regenerative medium power converters. The monitoring scheme is based on capacitor bank voltage measurements in quasi-offline mode. The aim of this work is to identify the voltage measurement precision and sampling requirements to enable a reproducible and reliable capacitance calculation. The proposed method is analysed theoretically and supported by simulation results and validated experimentally.

1. Introduction

In large-scale power converter farms a key objective is the reliable operation that is achieved with an optimum maintenance schedule. A common way to enhance reliability and expand their lifespan is by derating the converters' operation at a lower power in order to reduce the anticipated electrothermal stress for the components. One of the converter's components that is sensitive to aging is the DC-link capacitor [1]. Capacitors exhibit degradation effects during normal usage, which are usually caused by the degradation of the dielectric material. Degradation of the dielectric is pronounced with capacitors in cycling applications due to the internal heating of terminals and the higher RMS current. This results in lower capacitance, higher leakage current and an increased series impedance. Considering the operating constraints of the converter, these degradation effects might be tolerated within an acceptable range. However, a significant degree of degradation deteriorates the performance of the converter, in terms of voltage and current ripple and increased power losses.

Therefore, a condition monitoring (CM) scheme is valuable for assessing the health condition of the capacitors and for determining maintenance and replacement intervals before degradation reaches a critical level or failure of the component.

The operation of condition monitoring systems is based on data acquisition and offline data processing. There are different methods to acquire these data. The most straightforward way is to measure these data using sensing circuits. However, such circuits are cost intensive and can be a further source of failure, whereas they can -in the worst case- decrease the reliability of the application. A way to overcome this challenge, is to utilize data which are already measured for control purposes of the power converter, or to just model and simulate the degradation under the given load profile. Nevertheless, the second approach necessitates the development of accurate electrothermal models by also considering the accurate modelling of various materials comprising the physical components. Such models are based on a large amount of experimental data acquired under a large variety of operating conditions that usually require long testing procedures, unless field-data on failures are available [2].

There are three different data acquisition methods for CM as shown in Fig. 1. These methods are classified based on whether they require online or offline measurements. Online measurements refer to data acquired during the normal operation of the power converter, and while the measurement circuit does not interfere with the converter's operation. Online measurements usually suffer from lower accuracy and must also be designed and executed in a way to have the lowest possible impact on the converter's operation. On the other hand, offline measurements require the converter or the components to be disconnected from the applications before performing CM measurements. Offline measurements are more accurate than the online method, but they are cost-intensive and require an interruption of the electrical energy conversion process. In addition to these methods, it is possible to perform CM measurements when the converter is not in operation but still connected with the source and load. This type of measurement is termed *quasi-offline* and they are utilized in the presented work [3].

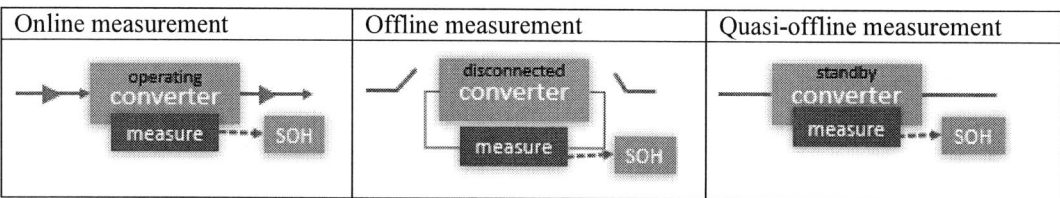

Online measurement	Offline measurement	Quasi-offline measurement

Fig. 1: Classification of CM methods.

Each of these four basic CM principles exhibits advantages, which depend on the specific application [3] , [4]. This study focuses on CM by using the existing sensors already incorporated on the power converters, making the measurements of type quasi-offline.

The performance indicators for electrolytic capacitors are mainly the values of the capacitance and the equivalent series resistance (ESR). This combination can also be described as a loss angle or dissipation factor and utilised as a CM parameter [5]. It is also possible to analyse the frequency dependence of the series impedance [6]. This can be done by assessing the switching harmonics during normal operation or by injecting signals superimposed with the modulation to get a specific frequency response [7]. However, for this application injecting superimposed signals is not tolerable because of the high precision output current control. Most of such online measurements require accurate voltage and current measurements in combination with a high sampling rate. Moreover, especially for existing converters, the remining processing and network capacity during operation is quite limited for doing online CM. Another possible method to determine the capacitance and ESR is to perform a discharge with a network of different switchable resistors, which can be realized as quasi-offline CM [8]. However, it should work without hardware changes.

This paper proposes a quasi-offline CM scheme for DC-link capacitors that can be implemented without the need for disconnecting the capacitors from the power converters. Moreover, the proposed scheme eliminates the need for external sensing circuits for monitoring the state of health (SOH) of capacitors, and it makes use of standard sensing circuits already implemented on the converters for their control.

More specifically, the capacitor's voltage decay is recorded each time the power converter is switched off by the accelerator operators. A low-tolerance and low-thermal drift discharge resistor, that is already connected in the converter's DC-link, is used for dissipating the energy. The voltage is recorded by the DC-link voltage measurement sensor and post-processed to identify the capacitors' SOH.

The paper is organized as follows. Section 2 introduces the design and operating principles of the accelerator converters at CERN. Then, Section 3 analyses the prospered CM system, including the mathematically background for optimal sampling intervals. The experimental validation of the CM scheme is presented in Section 4. Finally, the conclusions are summarized in Section 5.

2. Design and operating principles of power converters at CERN

Particle accelerators facilities employ hundreds of power converters for the steering of particle beams towards a particle storage ring, such as the Large Hadron Collider, or towards targets that create secondary particle beams for experiments, such as the CLOUD experiment. The power converters supply powerful electromagnets often in a cycling mode operating 24 hours a day and 7 days a week.

To control and monitor more than 5000 power converters at the European Laboratory of Particle Physics, their control hardware is connected to a technical network that gives access to significant diagnostic information. This includes state of various critical components as well as real time measurements from sensors inside the power converters. Eventually, the magnet current is recorded with a 10-kHz sampling rate by sensors offering a measurement accuracy in the order of 1 part per million (ppm).

One of the recent developments is the SIRIUS [9] power converter, that employs energy storage in electrolytic capacitors for recovering the magnet energy after each cycle. A simplified schematic of SIRIUS converters is depicted in Fig. 2.

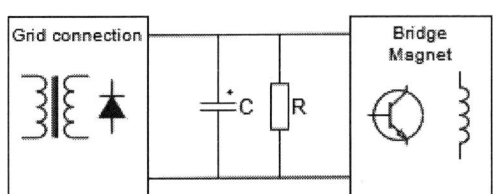

Table I: Data of SIRUIS converter (one brick, basic configuration) [9]

SIRIUS key data	
Grid connection	3ph/400V/32A$_{rms}$
Output voltage	450V
Output current peak	450A
Output current RMS	200A

Fig. 2: Schematic diagram of the fundamental circuit of a SIRIUS accelerator converter [9]

Electrolytic capacitors in SIRIUS are subject to more than 15 million cycles per year and, hence, a systematic monitoring is required as part of the preventive maintenance plan.

3. Proposed condition monitoring scheme for DC-link capacitors

In power converters a number of measurements such as the DC-link voltage, input and output currents are used for control and regulation while other signals such as the state of switchgear and thermal switches are used for the safety of the equipment [10]. In the majority of the present systems, failures of individual components will shut down the system immediately. On the contrary, a CM can predict the SOH of the components allowing the converter operators to schedule the replacement or maintenance of the components before a failure occurs.

Usually, discharging resistors are connected to DC-link capacitors enabling a safe way to dissipate their stored energy when a converter's shut-down is requested. It is important that these capacitors are discharged for safety and maintenance reasons within a certain time interval. In the simplest case there is a parallel-connected resistor that enables the discharging process within a few minutes. The capacitors are connected in series (e.g., C_1, C_2 in Fig. 3) to reach higher operating voltages. Due to manufacturing tolerances (i.e., capacitance and leakage current variations), balancing resistors (R_{B1}, R_{B2} in Fig. 3) are required across each series-connected capacitor to balance the steady-state voltages and avoid voltage drifts. These balancing resistors have the same effect as the discharging resistor. Fig. 3 shows an example of a capacitor bank with the already analysed discharging ohmic paths. The leakage current of the capacitors is modelled as parallel-connected resistors, R_{L1} and R_{L2}.

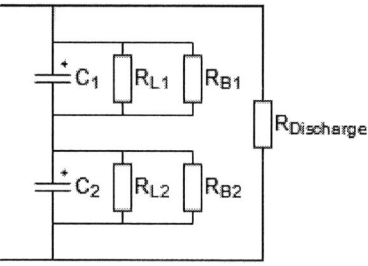

Fig. 3: Schematic diagram of a typical capacitor bank with the discharging resistors.

Using the discharging process of the DC-link capacitors after shutting down the converter is an ideal condition for conducting measurements. Since the power semiconductor devices are not switching, such measurements will not be impacted by electromagnetic noise. In this case the capacitor bank is only discharged through the discharging resistor, balancing resistors, and leakage current paths.

The discharging process follows the well-known Equation 1 for resistive discharging of a capacitor:

$$V_c(t) = V_0 \cdot e^{\frac{-t}{\tau}} \; ; \; \tau = R \cdot C \tag{1}$$

This equation expresses the capacitor voltage V_C as a function of the initial voltage V_0 at $t=0$ and the time constant τ. V_0 refers to the voltage of the capacitor bank when the converter shuts down. The time constant τ is a function of the discharging resistance R and capacitance C, where R is the resulting resistance of all ohmic paths (discharge, balance, and leakage current modelled as resistor) of the bank and C is the resulting capacitance of all capacitors.

The values of the discharging and balancing resistors do not change significantly due to degradation, but the leakage current modelled as further ohmic path will do. The decreasing capacitance and increasing leakage current will both cause a quicker discharge resulting in a shorter time constant τ. Thus, it makes sense to use the change of this time constant as a degradation indicator. The series impedance of the capacitor will increase due to degradation, but it will still be negligible in this slow discharging process, which usually takes several minutes.

The input to the proposed CM scheme is the recorded capacitor bank voltage. In principle, it is sufficient to measure the voltage at two different time instants during the discharging process and calculate the time constant τ by rearranging Equation 1 as shown in Equation 2, where (V_1, t_1) refers to the first sample and (V_2, t_2) refers to the second sample.

$$\tau = -\frac{t_2 - t_1}{ln(\frac{V_2}{V_1})} \tag{2}$$

Due to measurement inaccuracies and remining noise, it is is crucial to choose the time instants to acquire the voltage samples in a sophisticated way . Therefore, Equation 2 is analysed to study the impact of time and voltage accuracy. In order to find a general solution, it is possible to normalize the discharge in the voltage and time domain. This is done in time domain by defining $x = t/\tau$ and in amplitude domain by defining $y = V_c(t)/V_0$. By doing this, a general discharge according to Equation 3 can be found. Such a discharge is plotted in Fig. 4. This plot corresponds basically to $\tau=1s$ and $V_0=1V$.

$$y(x) = e^{-x}, \; y = \frac{V_c}{V_0}, x = \frac{t}{\tau} \tag{3}$$

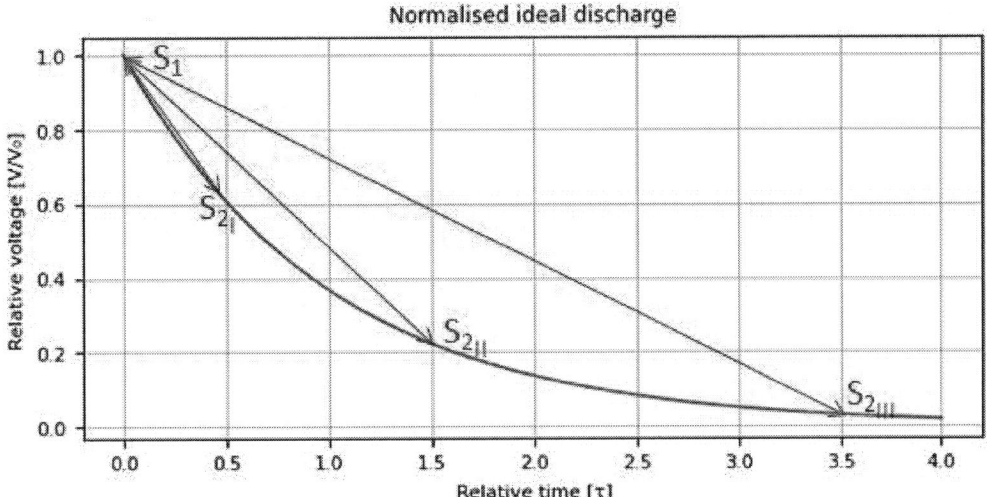

Fig. 4: Normalized discharge curve with sample pair suggestion.

Each sample has a timestamp, $t_{timestamp}$, with a limited time accuracy. Equation 4 shows the two different errors that might be contained in the timestamps of the sample point ($t_{samplepoint}$).

$$t_{timestamp} = (t_{samplepoint} + t_{shiftConstant} + t_{jitter}) \tag{4}$$

This additional time errors can be constant ($t_{shiftConstant}$) or random (t_{jitter}). Examples for such additional time errors are network latency, impact of pre-emptive multitasking systems and analogue digital conversion.

For the additive errors, they must be strictly distinguished between the constant part and the random part. Because of the time difference calculation, constant time shifts ($t_{shiftConstant}$) have no influence on accuracy of τ calculation. From Equation 5, it is shown that $t_{shiftConstant}$, which is equal for both sample points, disappears in the time difference calculation.

$$
\begin{aligned}
\Delta t &= t_2 - t_1 \\
&= (t_{2samplepoint} + t_{shiftConstant} + t_{2jitter}) - (t_{1samplepoint} + t_{shiftConstant} + t_{1jitter}) \\
&= t_{2samplepoint} - t_{1samplepoint} + t_{2jitter} + t_{1jitter} \\
&= t_{2samplepoint} - t_{1samplepoint} + t_{jitterResulting}
\end{aligned}
\tag{5}
$$

Therefore time shifts of the system clock and constant dead-times in the processing chain are not relevant. The jitter of both sample times can be modulated as a resulting jitter. By taking a higher time difference, the influence of time jitter becomes less relevant.

The voltage measurement can be influenced by two error types. This is an additional error (V_{Err}) and a linear error ($Err_{Amplify}$) according to Equation 6. Additional errors are noise and offset from amplifiers in the circuit. Linear errors are due to resistor tolerances in voltage dividers or incorrect amplification factors.

$$V_{measured} = (V_{real} + V_{Err}) \cdot (1 - Err_{Amplify}) \tag{6}$$

For the τ calculation, only the relative relation of the two samples is relevant, because of the division of the two voltage samples (see Equation 6).

$$\frac{V_2}{V_1} = \frac{(V_{2real} + V_{2Err}) \cdot (1 - Err_{Amplify})}{(V_{1real} + V_{1Err}) \cdot (1 - Err_{Amplify})} = \frac{V_{2real} + V_{2Err}}{V_{1real} + V_{1Err}} \tag{7}$$

This means that only a linear representation of the voltage is required, and the linear error ($Err_{Amplify}$) has no impact on the operation of the CM scheme.

In Equation 8 the τ calculation formula with all relevant errors is presented and a relative error of τ can be calculated using Equation 9.

$$\tau_{withErr} = -\frac{(t_2 - t_1 + t_{jitter})}{ln(\frac{V_2 + V_{2Err}}{V_1 + V_{1Err}})} \tag{8}$$

$$Err_{\tau relative} = abs(\tau - \tau_{withErr}) \tag{9}$$

By filling in the errors in Equation 8 and calculate an absolute error according to Equation 9, an optimal time between two samples can be found in the normalised time and voltage domain (see Equation 3). The relative error on τ is defined according to Equation 10. To minimise the effect of noise on the voltage (V_1), the first sample is taken at the beginning of the discharging curve ($x_1=0$). With these simplifications (see Equation 11), the error function (Equation 12) is defined.

$$Err_{\tau relative} = abs\left(1 - \frac{-(x_2 - x_1 + x_{jitter})}{ln(\frac{V_{RelativIdeal}(x_2) + V_{2ErrRelaiv}}{V_{RelativIdeal}(x_1) + V_{1ErrRelativ}})}\right) \tag{10}$$

$$V_{RelativIdeal}(x) = e^{-x} , x_1 = 0 \tag{11}$$

$$Err_{\tau relative}(x_2) = abs(1 + \frac{(x_2 + x_{jitter})}{ln(\frac{e^{-x_2} + V_{2ErrRelaive}}{1 + V_{1ErrRelative}})}) \tag{12}$$

Equation 12 is analysed numerically to assess the impact of the measurement errors and optimization potential. An example demonstrating the impact of the measurement errors by setting each error contribution to 1% as shown in Table II has been considered. The results of this analysis are plotted in Fig. 5.

Table II: Error calculating basic

$V_{1ErrRelaive}$	$V_{2ErrRelative}$	x_{ijtter}
-0.01	0.01	0.01

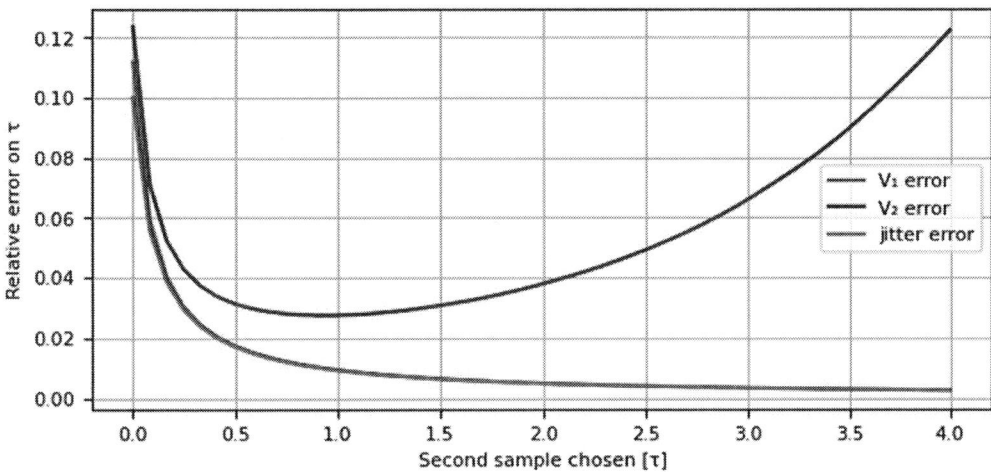

Fig. 5: Impact of the choice of the second time stamp on different errors

As shown in Fig. 5, the inaccuracy of V_1 and time jitter becomes less relevant for a delayed choice of sample 2. Most interesting is the impact of V_2 error. For this error the optimal sampling point is at τ. It should be noted that both voltage samples are sensed from the same sensor which means that they are impacted by the same noise level. In general, the superposition of the different errors cannot be calculated in a simple way. According to this calculation, it is found that $t_2 = \tau$ where the error on V_2 has a minimum impact and the other error contributions are still quite low. Of course, the exact τ value is unknown, but it can be approximated based on analytical calculations or by plotting a discharge record.

For further reduction of noise, it is possible to filter the signal and to consider the average values of different sampling pairs. To be robust against outliners, the median principle can improve monitoring. It is noted that the amount of noise is just an indicator for the quality of the measurement but has no impact on the assessment of the capacitors bank's SOH. However, low noise levels are required for detecting the slight changes of the time constant, which will indicate capacitors degradation.

4. Experimental validation

This section analyses experimentally the impact of optimal voltage sampling on the validation of the capacitance measurement. Fig. 6 shows photos of the experimental setup of the SIRIUS converter and the capacitor banks at CERN laboratory, which were used for performing the experimental evaluation of the proposed CM scheme. Two types of DC-link capacitor banks having slightly different characteristics have been used. To observe the difference in the time constant, measurements with two different capacitor banks (Bank "A" and Bank "B") were conducted. The circuit of the capacitor bank remains the same, but different capacitor types are used. Despite the same nominal data, the type "B" has a slightly higher typical capacitance values compared to type "A" counterparts.

SIRIUS converter

Capacitor Bank "A"

Capacitor Bank "B"

Fig. 6: Photos of the laboratory prototypes of SIRIUS converter and the two different capacitor banks

Optimal voltage sampling

When a converter is shut down or trips due to an internal or external fault, the data acquisition process starts. A voltage sample is collected every 30 s until the voltage reaches 20 V. Usually, the first sample has a lower voltage than the nominal direct voltage of 900 V, because the initialization of the recording is a bit delayed.

In optimal conditions, the value of τ should always be the same irrespective of the samples' choice. Due to different types of noise and measurement errors, there are differences between the outcomes. As analysed in the previous section, the accuracy can increase by choosing the optimal time between two samples. For assessing this further, experimental data of a discharging process has been recorded and plotted in see Fig. 7. To study the impact of sample pair choices, two methods are compared. Method 1 calculates τ just by taking into account two adjacent samples. On the other hand, Method 2 takes the second sample with the optimal time distance. This optimal time distance is approximately 1τ later, which corresponds to 11 samples later in the analysed discharge. In Fig. 7 an example pair for Method 1 is marked in with a yellow arrow and for Method 2 with a green arrow.

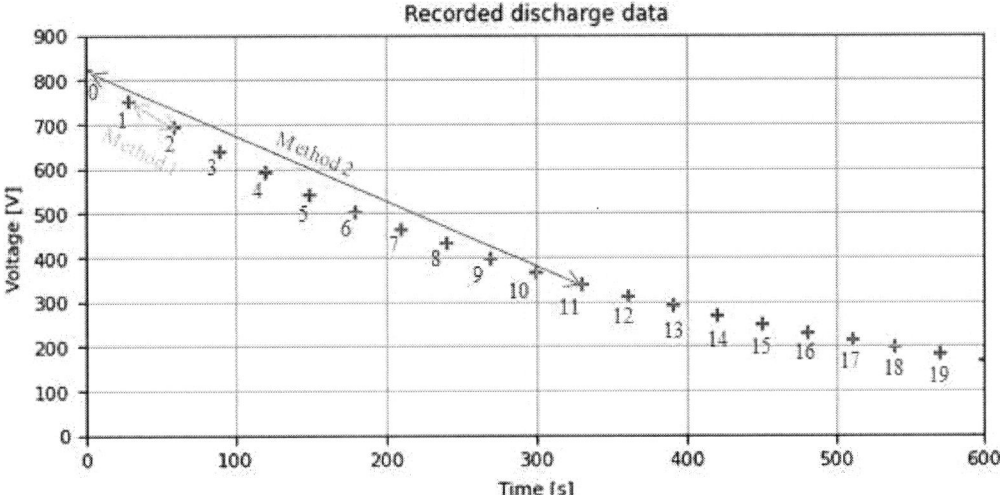

Fig. 7: Example of a recorded discharge data points. The yellow arrow shows a sample pair to calculate τ according to Method 1 and the green arrow according to Method 2.

To enable a statistical analysis of the data, τ is estimated from several sample pairs using both methods. The calculated outcome is indexed with the index of the first sample (S_1). In Table III the mean and Standard Deviation (SD) values are calculated for τ using both sampling methods. These two statistical indicators bases on the first five calculated τ, and again for the following next five calculations, and so on. This to show the impact of early and late samples.

Table III: Differences of sampling methods

Index of S_1	Method 1		Method 2	
	mean [s]	SD	mean [s]	SD
0 ... 4	328.0	9.38	339.7	2.18
5 ... 9	335.7	5.00	339.4	1.41
10 ... 15	340.3	5.66	342.2	2.87
15 ... 20	338.7	29.11	343.7	5.56

A good indicator of the robustness of the calculation is the SD. Table III shows that with Method 2 the SD is lower compared to Method 1 for each sampling case. Furthermore, the SD is lower for earlier samples, even when in the given data the SD of the τ of the first five samples indicates a higher value. It seems to be an effect of an outliner in the samples.

Therefore, potentially the best practise is to calculate τ from the first sample and the sample one τ later. Because of the remaining noise also on these two samples, it is wiser to use a higher number of τ calculations based on different sampling points and estimate an average value of τ. In the current implementation, τ is calculated with Method 2 from the first five sample pairs. From these values, the median is calculated, as the result for the discharge analysis. With the median principle, the τ calculation is more stable against outliners when compared to the case when only the mean is considered.

Verification of measurement principle

First an analysis of all the capacitors and resistors of the SIRIUS DC-link is done, to calculate the expected τ. The nominal value of the DC-link capacitance is 58.1 mF with a parallel resistance of 6.04 kΩ. Therefore, the nominal expected τ is 351.4 s.

For the two configurations with Bank "A" and Bank "B" the time constant is calculated analytically by doing network analysis to get reference values for the measurements. This is done by calculating the resulting capacitance and the resulting resistance for the DC-link. These values are listed for both configurations in Table IV.

Table IV: Test configuration

Test configuration	Bank "A"	Bank "B"
DC-Link voltage	900V	
Calculated total capacitance	54.7mF	61.3mF
Calculated total resistance	6.04kΩ	
Calculated τ	330.4s	370.4s

Discharging tests were performed for both types of capacitor banks. For each bank, the test was performed three times, to validate the reproducibility. On each occasion, the converter was turned-on and when the DC-link reached 900V, it was turned-off to record the capacitor discharge. In Fig. 8 the recorded data points from the first measurement of the capacitor banks "A" and "B" are plotted.

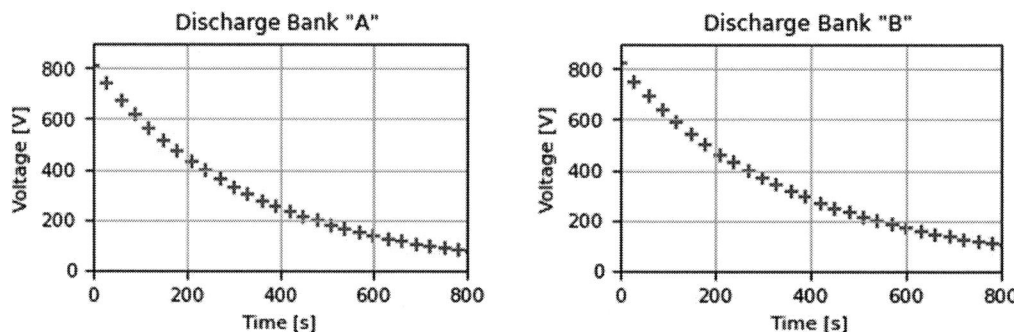

Fig. 8: Experimental discharging plots for capacitor Bank "A" and "B" (measurement 1).

From this data points, τ is calculated and the results for the three repeated measurements with both banks are summarized in Table V.

Table V: Estimated τ values based on experimental data

Test results	Bank "A"	Bank "B"
Calculated τ	330.4s	370.4s
τ measurement 1	336.3s	381.2s
τ measurement 2	336.7s	379.9s
τ measurement 3	336.3s	381.2s

The repetition of the experiments has shown a good reproducibility of the measurements, and the extracted τ values are close to the calculated ones. All the measurements are conducted in sequence to minimize the effect of temperature changes.

Measuring the time constant in the proposed way can be used as an indicator for the capacitor bank CM. The change of the capacitor type from "A" to "B" has increased the capacitance by 12%, which is a quite significant change to validate the scheme. The proposed method makes a one-point overall measurement of all capacitors connected to the DC link. This gives a good overview of the condition of the connected banks.

However, the anticipated limitation using this method is the weakness of assessing the health status of each single capacitor, especially in the cases when several capacitor banks were connected to the converter. In the current investigations thermal dependencies were neglected, which also influence the value of the capacitance and the resistors. An increase of temperature of the discharging resistor increases of 5% yields an increase of τ from 330.4 s to 340.2 s for Bank "A" of capacitors. Therefore, such aspects must be taken into the calculation to accurately estimate τ.

With the proposed method the focus is on the ageing mechanism of the capacitance decrease and leakage current increase. For many applications, a low series impedance is essential. An increase of this impedance will not be detected by the proposed method.

5. Conclusion

This work proposes the use of standard precision (~0.1-1%) measurements using the DC-link voltage sensor to acquire a capacitive voltage decay. Using an experimental setup with two different types of capacitor banks (54.7mF and 61.3mF) it was demonstrated that it is possible to detect up to 1mF of capacitance difference. Noise and measurement inaccuracy has lowest impact when the time difference between two voltage samples is 1τ. This system is, therefore, able to detect a capacitance decrease of 10% due to degradation which is the degradation alert threshold for the application under study.

It is possible to perform the voltage measurement with the existing sensors as used for control. This helps to reduce cost and keeps the hardware complexity low. Therefore, this system can even be introduced for existing converters without any hardware changes. The use of the regular shut-down periods of the converters for performing quasi-offline measurements enable automatized measurements with reduced noise impact. Moreover, computing power of the converter controller and network capacity for CM will be available after the operational turn-off. The voltage sampling can be performed in an optimal way to reduce the impact of noise and the limited measurement accuracy.

References

[1] S. Peyghami, Z. Wang and F. Blaabjerg, "A Guideline for Reliability Prediction in Power Electronic Converters," *IEEE TRANSACTIONS ON POWER ELECTRONICS,* vol. 35, no. 10, pp. 10958-10968, 2020.

[2] S. Zhao and H. Wang, "Enabling Data-Driven Condition Monitoring of Power Electronic Systems With Artificial Intelligence: Concepts, Tools, and Developments," *IEEE POWER ELECTRONICS MAGAZINE,* pp. 18-27, 23 March 2021.

[3] Z. Zhao, P. Davari, W. Lu, H. Wang and F. Blaabjerg, "An Overview of Condition Monitoring Techniques for Capacitors in DC-Link Applications," *IEEE TRANSACTIONS ON POWER ELECTRONICS,* vol. 36, no. 4, p. 3692, 2021.

[4] P. Sundararajan, M. Sathik, F. Sasongko, C. Seng Tan, J. Pou, F. Blaabjerg and A. K. Gupta, "Condition Monitoring of DC-Link Capacitors Using Goertzel Algorithm for Failure Precursor Parameter Goertzel Algorithm for Failure Precursor Parameter," *IEEE TRANSACTIONS ON POWER ELECTRONICS,* vol. 35, no. 6, pp. 6384 - 6396, 2020.

[5] M. Ghadrdan, S. Peyghami, H. Mokhtari, H. Wang and F. Blaabjerg, "Dissipation Factor as a Degradation Indicator for Electrolytic Capacitors," *Emerging and Selected Topics in Power Electronics.*

[6] M. Asoodar, M. Nahalparvari, C. Danielsson, R. Söderström and H.-P. Nee, "Online Health Monitoring of DC-Link Capacitors in Modular Multilevel Converters for FACTS and HVDC Applications," *IEEE TRANSACTIONS ON POWER ELECTRONICS,* vol. 36, no. 12, pp. 13489-13503, 2021.

[7] T. Li, J. Chen, P. Cong, X. Dai, R. Qiu and Z. Liu, "Online Condition Monitoring of DC-Link Capacitor for AC/DC/AC PWM Converter," *IEEE TRANSACTIONS ON POWER ELECTRONICS,* vol. 37, no. 1, pp. 865-878, 2022.

[8] W. Yu and X. Du, "A VEN Condition Monitoring Method of DC-Link Capacitors for Power Converters," *IEEE TRANSACTIONS ON INDUSTRIAL ELECTRONICS,* vol. 66, no. 2, pp. 1296-1306, 2019.

[9] "SIRIUS," CERN Medium Power Converters Section, 20 June 2022. [Online]. Available: https://section-mpc.web.cern.ch/content/sirius. [Accessed 20 June 2022].

[10] "Documentation SIRIUS," CERN Medium Power Converters Section, 19 December 2021. [Online]. Available: https://section-mpc.web.cern.ch/content/documentation#Sensors%20and%20interfaces. [Accessed 19 December 2021].

Minimizing voltage stress in Auxiliary Resonant Commutated Pole Inverters Using saturable Inductors

Markus Zocher
ELSYS, Technische
Hochschule Nürnberg
Nuremberg, Germany
Tel. +49 (911) 5880 - 1814
E-Mail: markus.zocher@th-nuernberg.de
URL: http://www.th-nuernberg.de/elsys

Norbert Grass
ELSYS, Technische
Hochschule Nürnberg
Nuremberg, Germany
Tel. +49 (911) 5880 - 1814
E-Mail: norbert.grass@th-nuernberg.de
URL: http://www.th-nuernberg.de/elsys

Ralph Kennel
EAL, Technical University
of Munich
Munich, Germany
Tel. +49 (89) 289 - 28358
E-Mail:
ralph.kennel@tum.de
URL:
http://www.epe.ed.tum.de

Keywords

Resonant converter, Resonant Peak Damping Strategy, Reverse Recovery, Voltage Source Inverter, Zero Voltage Switching

Abstract

This paper demonstrates a method to reduce the voltage stress of the auxiliary transistors inside an ARCP inverter. A saturable inductance softens the diode reverse recovery process. The proposed method has been successfully tested on a prototype converter over a wide range of currents, DC link voltages and switching frequencies.

Introduction

Switching losses of semiconductors used in either grid-connected or drive-side inverters are the main limitation to the switching frequency. To allow higher switching frequencies several resonant topologies are known, that can reduce the switching stress of the power semiconductors and allow soft-switching. In contrast to intrinsic resonant topologies like a LLC converter, the Auxiliary Resonant Commutated Pole (ARCP) structure, which has been investigated since the 1990s [1]-[5], adds an auxiliary circuit to a standard two-level-inverter. This circuit provides a resonant phase, which occurs while the main power devices are switched. The standard ARCP structure adds an auxiliary leg to each half-bridge as shown in Fig. 1. In recent years also different ARCP topologies have been published [6][7]. Recently, the company Pre-Switch Inc. presented an ARCP converter based on silicon carbide (SiC) MOSFETs which is controlled by artificial intelligence [9]. Most of the investigations assume a linear or almost linear resonant tank. However, this assumption usually leads to large resonant elements L_r and C_r, having a low resonant frequency and therefore limiting the viable switching frequency. This paper shows the challenges of small resonant tanks and the benefits of using a nonlinear partly saturable resonant inductance L_r.

Principle of operation

Behavior of the ideal circuit

A current commutation from diode D_2 to the IGBT S_1 with ideal semiconductors is displayed in Fig. 2. When the auxiliary transistor S_{A1} is switched on, the resonant inductance L_R takes over the load current I_{Ph}. As soon as $i_R > I_{Ph}$, the bridge output voltage rises and the collector-emitter-voltage of S_1 decreases. When the collector-emitter-voltage of the main IGBT S_1 reaches zero, S_1 is switched on at t_3 without significant turn on-losses.

EPE'22 ECCE Europe

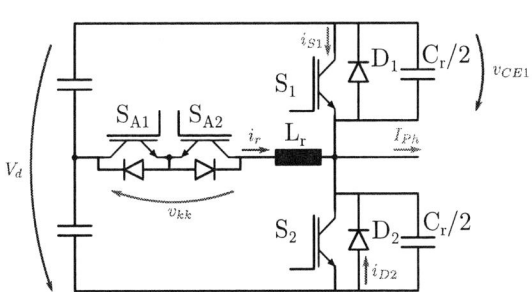

Fig. 1. Schematic half bridge with ARCP

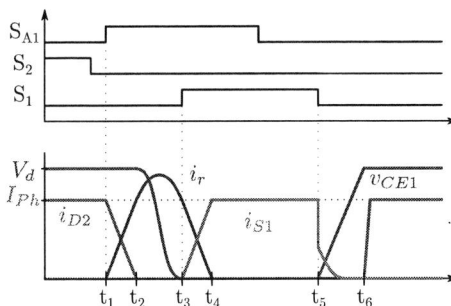

Fig. 2. ARCP idealized waveforms

When the IGBT S_1 is turned off, the load current charges the resonant capacitance C_r, the rise-time of the collector-emitter-voltage is approximately

$$t_r = t_6 - t_5 = C_r \cdot V_D / I_{Ph} \qquad (1)$$

Since the rise-time of the voltage is increased due to the resonant capacitance, the IGBT's turnoff-losses are lower when compared with a hard switching topology.

However, when the IGBT is turned off at low currents, the rise time $(t_6 - t_5)$ increases substantially like Fig. 3 shows. To fulfil dead-time limitations and avoid a hard turn-on of the main IGBT S_2 it might be necessary to speed up the voltage rise. This is done using the auxiliary IGBT S_{A2} (Fig. 4).

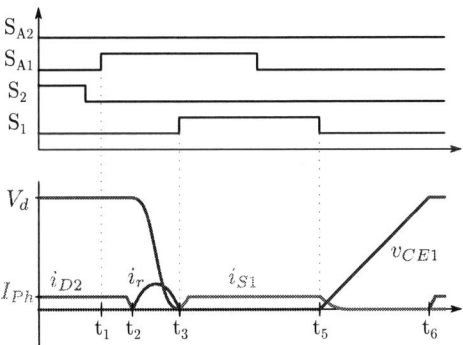

Fig. 3. Waveforms at low current without S_{A2}

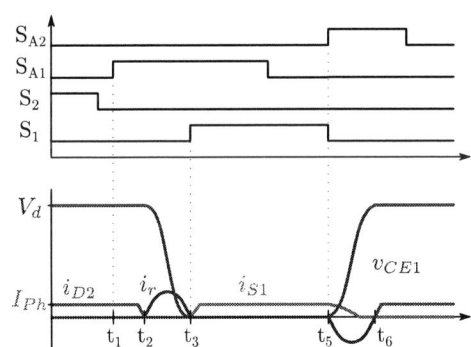

Fig. 4. Waveforms at low current with S_{A2}

The turn-off time can be decreased even further at the cost of an increase in switching losses when IGBT S_{A2} is turned on before IGBT S_1 is switched off. This operation introducing a "boost" current is discussed in [1].

Reverse Recovery of auxiliary diodes

The reduction of the ratings of the resonant elements L_r and C_r of an ARCP lead to a faster commutation time, with a lower amount of stored energy, resulting in lower losses in the resonant tank. However, when the inductor L_r is designed with a small inductance value, the reverse recovery behavior of the auxiliary diodes must be considered. During the reverse recovery (starting from t_4) the transistors S_1 and S_{A1} are constantly switched on and the auxiliary diode of S_{A2} behaves like a nonlinear capacitance. To investigate the reverse recovery behavior, the full circuit can be simplified according to Fig. 5. The capacitance C_{A2} is the sum of the diffusion capacitance containing the charge Q_a and the junction capacitance, which is charged up with Q_b.

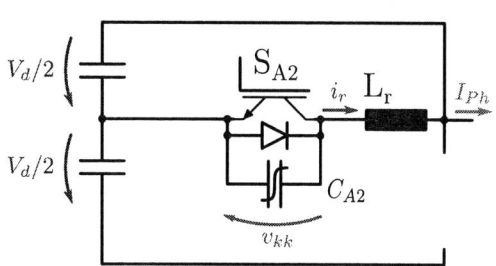

Fig. 5. Schematic during reverse recovery

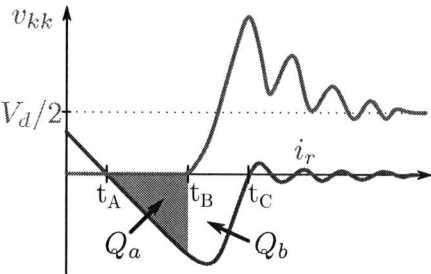

Fig. 6. Waveforms during reverse recovery

Because of the reverse recovery effect, the resonant current i_r can become negative and magnetic energy is stored again in the resonant inductor L_r. In Fig. 6 the point t_4 of the ideal circuit is expanded to the time range $t_A \dots t_C$.

After removing the charge Q_a the diode is building up the blocking voltage. The magnetic energy is transferred to the resulting output capacitance C_{A2} of the auxiliary IGBT and diode.

The value of the resonant current at t_B can be calculated using the approach

$$\int_{t_A}^{t_B} i_r(t)\,dt = -\int_{t_A}^{t_B} \frac{V_D}{2 \cdot L_r}\,dt \stackrel{!}{=} -Q_a \qquad (2)$$

Solved to the current value

$$i_r(t_B) \approx -\sqrt{\frac{V_D \cdot Q_a}{L_r}} \qquad (3)$$

To minimize the stored energy in L_r during reverse recovery, $|i_r(t_B)|$ should be minimized. This can be achieved by choosing a fast recovery diode with small recovery charge or using a large inductance L_r.

The voltage and current of the resonant circuit shape between t_B and t_C cannot be calculated straight forward, but the nonlinear capacitance C_{A2} requires the solution of a system of nonlinear differential equations.

Reduction of diode overvoltage

To keep the conduction losses of the auxiliary elements low and to find fast devices with low Q_a, it is desirable to use auxiliary transistors and diodes S_{A1} and S_{A2} with a low blocking voltage. In the best case, the maximum voltage occurring at the auxiliary semiconductor might be lower than the DC link voltage. Therefore, it is important to minimize the overvoltage of v_{KK}.

To decrease the overvoltage, the reverse current must be minimized and auxiliary transistors and diodes with a low output capacitance and a low reverse recovery charge should be preferred. However, experimental measurements with an ARCP inverter in the section "Experimental Results" will show that even modern discrete IGBT5 in parallel with fast recovery diodes cause a high overvoltage.

A further solution to lower the overvoltage of v_{kk} would benefit from increased damping of the parasitic resonant tank consisting of L_r and C_{A2}, using RC-snubbers and TVS-diodes in parallel to the auxiliary diode. However, adding these elements will add extra losses to the circuit. Moreover, the negative resonant current and consequently the energy stored in L_r will increase, leading to even higher voltage spikes. As adding RC-snubbers and TVS-diodes will create new disadvantages, this approach is not discussed further.

The presented approach starts from the fact, that the reverse recovery current will decrease, when the negative slope of the resonant current i_r is decreased. At a given DC link voltage this can be achieved using a larger resonant inductance L_r. Since this large inductance value is only needed for resonant currents close to zero, a combination of a linear and a saturable inductance is promising to decrease the reverse current and maintain the benefits of a small inductance. With this combination, the benefits of using a small resonant tank are maintained and the reverse recovery process is optimized.

The saturable part of the inductance can be built using a core with high permeability e.g., the standard ferrite N87 from TDK [12], the linear part using low permeability materials like iron powder.

As the flux density inside the ferrite core is expected to change fast, the eddy current losses inside the ferrite core provide extra damping while the ferrite core is not saturated. Therefore, the saturable inductor additionally works as a magnetic snubber in the critical moment. A similar technique is already known for resonant DC-DC-converters [11].

Temperature dependency of saturable core

For practical experiments ferrite cores with the material N87 from the manfacturer EPCOS TDK were used. Fig. 7 which was extracted from [12] shows, that the core saturates more quickly at higher temperatures. Moreover, the loss density decreases at high temperatures according to Fig. 8 . A small loss density is equivalent to a lower imaginary part of the complex permeability and thus, providing less damping. Therefore, the ferrite core should be chosen regarding satisfactory behavior at high temperatures.

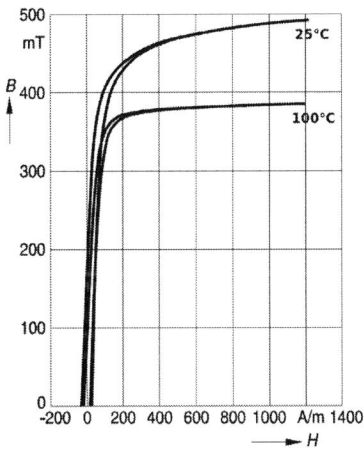

Fig. 7. N87 Saturation [12]

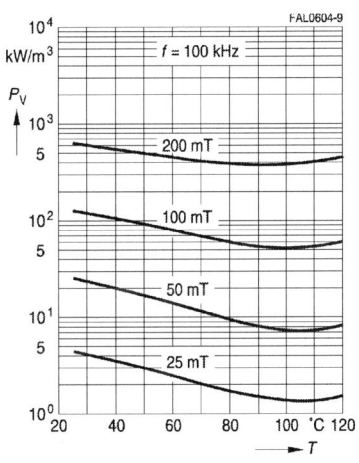

Fig. 8. N87 Loss density [12]

Experimental Results

For practical verification, a three-phase two-level inverter with ARCP extension has been built.

Fig. 9. Converter Prototype

Fig. 10. Auxiliary Leg without saturable core

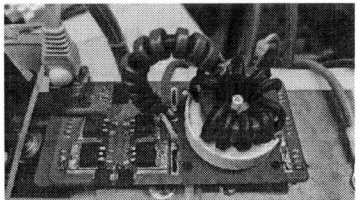

Fig. 11. Auxiliary Leg with saturable core

Fig. 10 and Fig. 11 show the auxiliary legs that are compared. The linear core is identical, the saturable inductor contains 10 extra N87 ferrite cores. Small cores are used as the ratio between surface area and core volume is high. Therefore, the dissipation of the heat resulting from the core losses is more effective. Simulating the behavior of the complete circuit is difficult, since a detailed nonlinear model of the N87 ferrite core and a detailed diode model containing the reverse recovery behavior is needed. Therefore, a more practical approach was preferred for selecting the ferrite cores.

The number of ferrite cores was selected in a way that the resulting inductance without saturation is at least 10 times as high as in the saturated case. The following table gives an overview of the used components.

TABLE I.   Components Prototype Converter

Component	Type	Max. Voltage	Nom. Current
Main IGBTs	Semikron SKM100GB12T4	1200V	100A
Auxiliary IGBTs	Infineon IGB50N65S5	650V	50A
Auxiliary Diodes	Vishay ETH3006S	600V	30A
DC link	Kemet C44UM: 1100uF	1200V	114A
Resonant Inductance (linear)	Carbonyl Iron Powder Core, Amidon (T130-2): ≈ 650 nH		
Resonant Inductance (nonlinear)	10x Epcos N87 Core (T10): ≈ 9.0 uH		
Resonant Capacitance $(C_r/2)$	Two 1206 2.2nF (NP0) in series	2x630V	

Since the timing of the auxiliary leg cannot be calculated straight forward the method proposed in [10] is used. For double pulse measurements an inductive load was connected between DC+ and the bridge output of one half bridge, the measurements with periodic signals were done in a full-bridge configuration.

Double Pulse Measurements

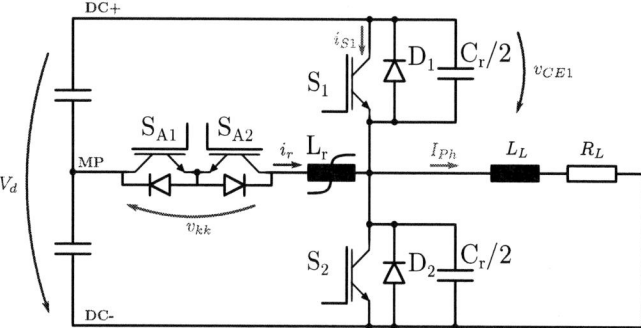

Fig. 12. Double pulse setup

The behavior of the linear and the saturable inductors are compared using double-pulse measurements. Fig. 13 shows the collector-emitter-voltages and the collector currents of the main transistor S_1 which only subtly differ.

As it can be seen from Fig. 14, the linear inductor causes a large overvoltage at the resonant transistors. At a DC link voltage of 500V the maximum overvoltage of v_{kk} reaches 574V (green waveform of Fig. 14) which is close to 600V - the maximum blocking voltage of the auxiliary diode. With the extra saturable inductor, the maximum voltages could be reduced down to 273V (blue waveform of Fig. 14) which is only slightly higher than half of the DC link voltage. Since the ferrite cores saturate quickly, the overall commutation time is not increased significantly.

As inverters with 1200V IGBTs are usually used up to a DC link voltage of about 800V, the operation of an ARCPI with a linear inductance that provides a small commutation time and auxiliary transistors of the 600V/650V class is not feasible. However, with the saturable core the DC link voltage of 800V is achievable.

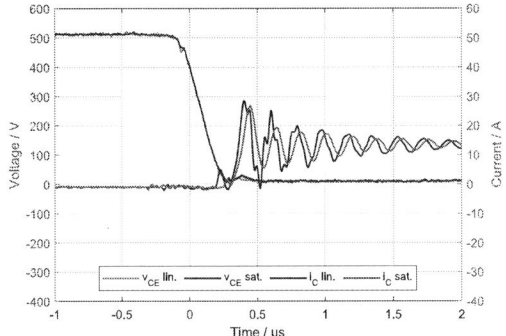

Fig. 13. Double Pulse Measurements,
$V_D = 500\ V$, $I_{Ph} = 15\ A$, main circuit

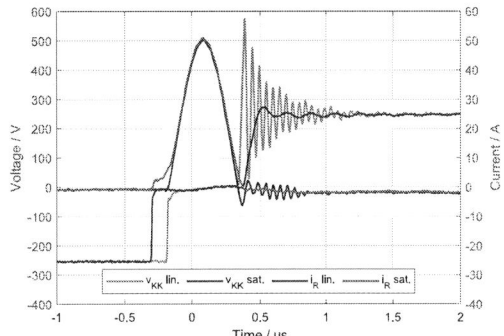

Fig. 14. Double Pulse Measurement,
$V_D = 500\ V$, $I_{Ph} = 15\ A$ resonant circuit

Loss Estimation

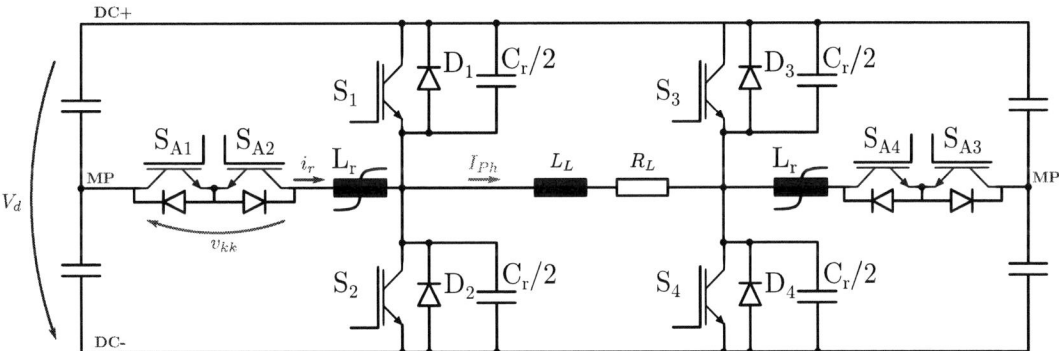

Fig. 15. Periodic operation setup

The improvement in the system behavior comes at the expense of losses inside the ferrite core. To estimate the losses, the ARCP converter is switched periodically in full bridge configuration with an inductive load ($L_L \approx 1mH$) between two phase legs and a 50Hz sinusoidal current with the amplitude I_L.

$$I_{Ph}(t) = I_L \cdot \sin(2\pi \cdot 50\text{Hz} \cdot t) \tag{4}$$

A direct loss measurement in the electrical domain using voltage and current measurements was found to be too inaccurate, therefore a calorimetric measurement is preferred. During operation the temperature is measured with an infrared camera. The following thermal properties were obtained for a single toroid core (R10x6x4) using the cool down behavior from [13]. The temperature was measured with an infrared camera and an additional K-type thermocouple. A single ferrite core has the following thermal properties

$$C_{th} \approx 0.72\frac{\text{J}}{\text{K}}, \tau_{th} \approx 80s \rightarrow R_{th} = \frac{\tau_{th}}{C_{th}} \approx 110\frac{\text{K}}{\text{W}} \tag{5}$$

At an ambient temperature of 22°C a temperature of 100°C of the hottest core equals a loss of 700mW. In this case, the sum of the losses of all 10 ferrite cores is less or equal to 7.0W per phase. However, it was already shown [10] with the same prototype that the turn on losses are reduced from 15mJ to less than 500μJ at the nominal operating point ($V_d = 600V$ and $I_C = 100A$).

Therefore, the extra losses generated from the saturable core are neglectable compared to the savings in switching losses in a hard-switched inverter.

Parameter Sweeps

Fig. 16 illustrates that the temperature of the ferrite core increases linearly with the switching frequency. This behavior is like expected since the core losses are proportional to the number of saturation cycles.

A higher DC link voltage leads to a faster saturation and higher eddy currents inside the ferrite. The measurements in Fig. 17 show, that the core temperature increases almost linearly with the DC link voltage.

The load current dependency on the temperature displayed in Fig. 18 might seem unintuitive at first glance. At higher output current amplitudes, the ferrite core stays cooler. This is caused by the switching actions when the current is commutated from an IGBT to the complimentary diode. At low currents the turn off must be accelerated using the auxiliary circuit like displayed in Fig. 4. When the amplitude of the sinusoidal output current is high, this speed-up is only needed during the zero crossings of the output current and not at each switching action.

In summary, the highest losses in the saturable core occur at the highest DC link voltage, the highest switching frequency and low output current amplitudes, which is the most critical point of operation.

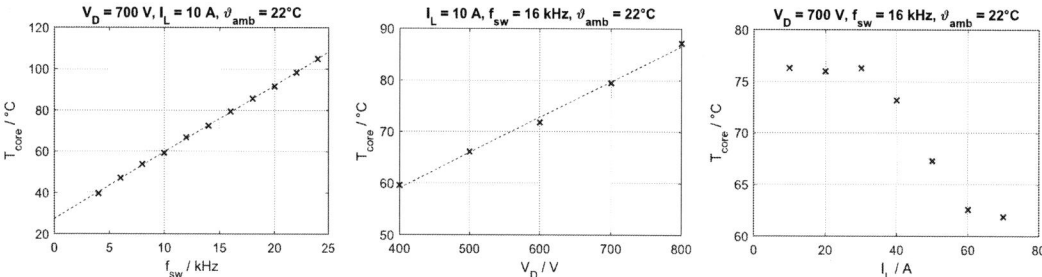

Fig. 16. f_{sw} dependency on T_{Core} Fig. 17. V_d dependency on T_{Core} Fig. 18. I_L dependency on T_{Core}

Temperature dependency of ferrite

Like explained in the theoretical part of this paper, it is important to consider the temperature dependency of the ferrite core itself. The saturation flux density as well as the core losses of N87 material decrease when the temperature increases from 25°C to 100°C [12]. These two effects lead to a larger overvoltage at higher core temperature. A periodic operation at 700V, 22kHz and a low output current amplitude of 15A leads to a core temperature of approximately 100°C like the thermal measurement in Fig. 19 shows. The according current and voltage signals at this point is shown in Fig. 20 and compared to a measurement at 25°C. The maximum voltage increases from 402V to 510V. Compared to the maximum blocking voltage of 600V, this voltage spike is still acceptable.

Therefore, it must be ensured, that the auxiliary voltage is below the maximum blocking voltage at the largest ambient temperature in combination with the most critical point of operation.

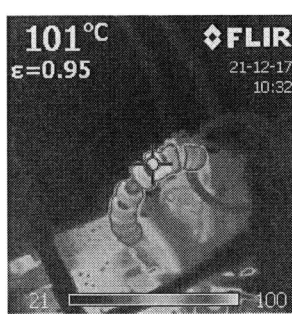

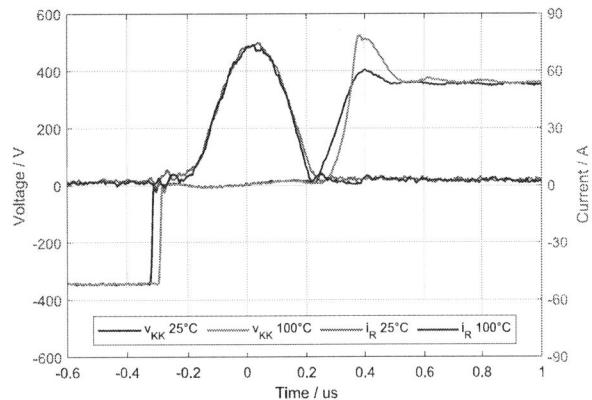

Fig. 19. Thermal camera, $V_D = 700\,V$, $f_{sw} = 22kHz$, $I_{Ph} = 15\,A$

Fig. 20. $V_D = 700\,V$, $I_{Ph} = 15\,A$, comparison temperature behaviour of nonlinear resonant circuit

Conclusion

This paper showed that a saturable core, working as a magnetic snubber, significantly reduces the overvoltage, which results from the reverse recovery of the auxiliary diodes, inside the auxiliary leg of a ARCP inverter. This allows ARCP designs with smaller resonant tanks and higher switching frequencies, where the reverse recovery behavior of the ARCP diodes is critical. The resulting core losses are usually small, compared to the rated power and the savings in switching losses of the inverter, but the temperature must be considered for the saturation and damping behavior of the saturable core. The most critical operation point is at the highest switching frequency and DC link voltage and low output currents. This operation point can be used to investigate and optimize the design of the partial saturable inductor.

References

[1] R.W. De Doncker and J.P. Lyons: "The Auxiliary Resonant Commutated Pole Converter".Proceedings IEEE-IAS, Seattle, pp. 1228-1235, 1990

[2] H.-J. Pfisterer and H. Spath, "Switching behaviour of an auxiliary resonant commutated pole (ARCP) converter," *7th IEEE International Power Electronics Congress. Technical Proceedings. CIEP 2000 (Cat. No.00TH8529)*, 2000, pp. 359-364, doi: 10.1109/CIEP.2000.891440.

[3] R. Teichmann and H. Gueldner: "Analysis of transfer ratio limitations in auxiliary resonant commutated pole converters," *7th IEEE International Power Electronics Congress. Technical Proceedings. CIEP 2000 (Cat. No.00TH8529)*, 2000, pp. 15-20, doi: 10.1109/CIEP.2000.891385.

[4] R. Teichmann, "Control parameter selection in auxiliary resonant commutated pole converters," *IECON'01. 27th Annual Conference of the IEEE Industrial Electronics Society (Cat. No.37243)*, 2001, pp. 862-869 vol.2, doi: 10.1109/IECON.2001.975870.

[5] F. Hinrichsen and G. Tareilus: "Simple Design and Control of ARCP-Inverter for Universal Power Range", SPEEDAM, Ravello, 2002

[6] H. Morii, M. Yamamoto and S. Funabiki, "Capacitor-Less Auxiliary Resonant Commutated Pole (ARCP) voltage source soft switching inverter suitable for EV," *2009 13th European Conference on Power Electronics and Applications*, 2009, pp. 1-8.

[7] E. Chu, L. Huang and Z. Fu, "Research on an active double auxiliary resonant commutated pole soft-switching inverter," 2014 IEEE 23rd International Symposium on Industrial Electronics (ISIE), 2014, pp. 637-642, doi: 10.1109/ISIE.2014.6864686.

[8] A. Charalambous, X. Yuan, N. McNeill, S. Walder, Q. Yan and C. Frederickson, "Controlling the output voltage frequency response of the auxiliary commutated pole inverter," *IECON 2016 -*

42nd Annual Conference of the IEEE Industrial Electronics Society, 2016, pp. 3305-3310, doi: 10.1109/IECON.2016.7793828.

[9] Pre-Switch Inc.: "Pre-Switch demonstrates efficacy of AI-based soft switching using 200kVA inverter reference", https://www.pre-switch.com, 2021

[10] M. Zocher, N. Grass and R. Kennel, "Auxiliary Resonant Commutated Pole Inverter (ARCPI) Operation Using online voltage measurements," 2022 IEEE Applied Power Electronics Conference and Exposition (APEC), 2022, pp. 1592-1597, doi: 10.1109/APEC43599.2022.9773560.

[11] K. Harada, Y. Ishihara and Toshiyukitodaka, "An improved magnetic snubber circuit for the diode reverse recovery in DC-to-DC converters," PESC 98 Record. 29th Annual IEEE Power Electronics Specialists Conference (Cat. No.98CH36196), 1998, pp. 701-706 vol.1, doi: 10.1109/PESC.1998.701975.

[12] TDK Epcos, Datasheet N87 Ferrite, https://www.tdk-electronics.tdk.com/en/529404/products/product-catalog/ferrites-and-accessories/epcos-ferrites-and-accessories/ferrite-materials

[13] P. Papamanolis, T. Guillod, F. Krismer and J. W. Kolar, "Transient Calorimetric Measurement of Ferrite Core Losses up to 50 MHz," in IEEE Transactions on Power Electronics, vol. 36, no. 3, pp. 2548-2563, March 2021, doi: 10.1109/TPEL.2020.3017043.

Adaptive Dead-Time Control in a Resonant Wireless Power Transfer System

Tim Krigar, Martin Pfost
TU Dortmund University
Martin-Schmeisser-Weg 4
44227 Dortmund, Germany
Phone: +49 231-7556731
Email: tim.krigar@tu-dortmund.de
URL: https://ewa.etit.tu-dortmund.de/

Acknowledgments

This work was supported by the German Federal Ministry of Education and Research under grant 16ES1023. The authors are responsible for the content of this publication.

Keywords

≪Wireless power transmission≫, ≪Contactless Power Supply≫, ≪Parasitics≫, ≪Zero-voltage switching≫, ≪Dead-time≫

Abstract

The influence of parasitic capacitances and of the output load on the ZVS behavior of a resonant wireless power transfer system is analyzed in this work. For this, the charging process of the output capacitors of a GaN-based inverting half bridge is described in detail. Furthermore, the influence of the parasitic capacitances and of the load on the resonant current is discussed.

The analyzed system has a peak efficiency of 92.9 % at 500 W with an operating frequency of 2 MHz. Based on the ZVS analysis, an adaptive dead-time control is presented to raise the average efficiency. For example, the efficiency can be raised by 0.5 % while transferring 250 W.

Introduction

To avoid bulky cables or abrasive brushes, wireless power transfer (WPT) systems are attractive for modern industrial applications to supply moving actuators. The most common WPT systems are inductive WPT (iWPT) systems with capacitors to compensate the large stray inductances of the weakly coupled transmitting and receiving coil. (Other approaches exist that use capacitive WPT, but iWPT has the advantage that the transmission ratio can be changed by the turn number ratio.)

In compact iWPT systems, high switching frequencies in the MHz range are needed, which asks for soft switching capability of the half bridge stage of the primary side. To realize zero voltage switching (ZVS), a resonant LLC topology can be used as shown in [1, 2]. In MHz range resonant converter systems, the dead-time has to be set particularly accurate to achieve ZVS and high efficiency. The ZVS capability is heavily influenced by the load, by the switching frequency, and by parasitic capacitances, see [3–6].

This paper presents an analysis of the ZVS capability of a 2 MHz 500 W resonant WPT system with step-down characteristic from 400 V to 48 V. Based on this analysis, an adaptive dead-time control of the WPT system without any communication with the receiving side is presented. This control scheme improves the efficiency over a wide range and is very attractive for WPT systems with a passive receiving side.

System Overview

Fig. 1 shows an overview of the proposed system. The rectangular voltage v_{sw} is generated by a half bridge consisting of two GaN HEMTs (GS66508T), the operating frequency is 2 MHz. The use of GaN HEMTs enables fast switching due to their small parasitic capacitances and because they do not exhibit a reverse-recovery effect. Especially in resonant systems with ZVS and operating frequencies in the MHz range, the output capacitances of the half bridge switches must be small to enable a fast charging and discharging process of the switching node.

The half bridge supplies the resonant network formed by the resonant capacitor C_r, the transmitting coil L_1, and the center-taped receiving coil L_{2a}/L_{2b}, cf. Fig. 1. The system uses an LLC topology to compensate the stray inductances. For this, the all-primary-referred equivalent circuit of the transformer is used, see Fig. 2. The stray inductances are aggregated in L_r. L_m is the mutual inductance of the transformer. This topology achieves a compensation with only one capacitor on the primary side of the iWPT system. The input voltage of 400 V is stepped down to an output voltage of 48 V. With only two rectifying diodes, the receiving side is kept passive and simple.

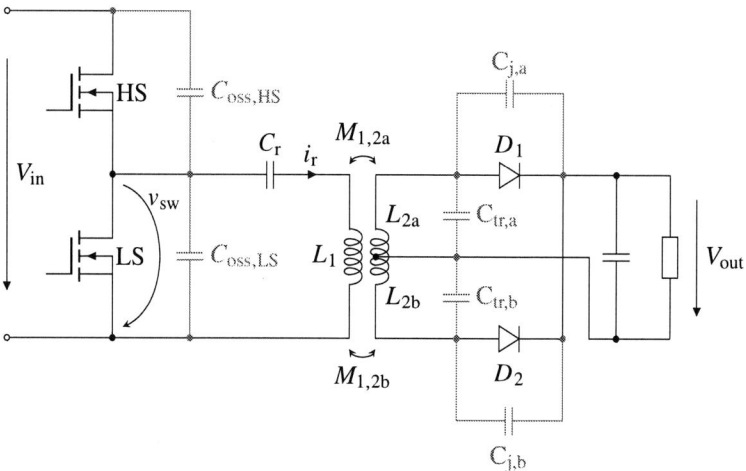

Fig. 1: Schematic of components are shown in black with the parasitic capacitances in red.

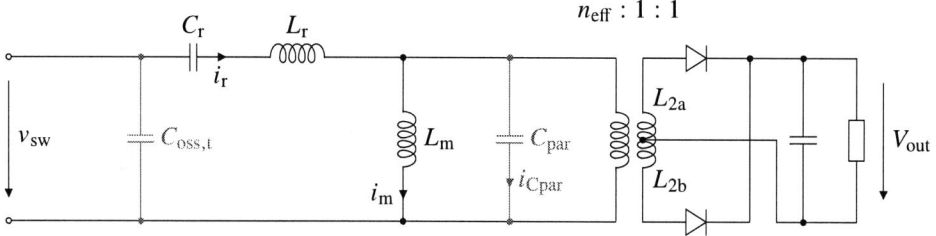

Fig. 2: All-primary-referred equivalent circuit. The parasitic capacitances which influence the ZVS behavior are marked red.

$C_{oss,HS}$ and $C_{oss,LS}$ are the effective output capacitances of the GaN HEMTs. They are summarized in the total output capacitance in Fig. 2 as

$$C_{oss,t} = C_{oss,HS} + C_{oss,LS}. \tag{1}$$

The junction capacitors of the rectifying diodes are represented by $C_{j,a}$ and $C_{j,b}$. $C_{tr,a}$ and $C_{tr,b}$ are the parasitic capacitances of the receiving coils L_{2a} and L_{2b}. Note that parasitic capacitances between the receiving and the transmitting coil can be neglected due to the large air gap of 5 mm between the coils of the WPT system. For circuit analysis, the receiving side parasitic capacitances are summarized and transferred to the primary side in Fig. 2 as

$$C_{\text{par}} = \frac{1}{n_{\text{eff}}^2} (C_{\text{j,a}} + C_{\text{tr,a}} + C_{\text{j,b}} + C_{\text{tr,b}}).$$

(2)

Dead-Time Analysis

During the dead-time, $C_{\text{oss,t}}$ is charged by the resonant current $i_{\text{r}}(t)$, cf. Figs. 1 and 2, to provide ZVS capability. This process is influenced by several factors, especially by C_{par} and by the output power P_{out}. To analyze this behavior of the iWPT system, circuit simulation is used. The values of the components and of the parasitic capacitances which are used in the simulation are listed in Tab. I. The results of the simulation can be seen in Fig. 3.

Note that only a switch-on event is shown, because a switch-off event behaves the same way, only negated. The figure includes three columns, where the first column is a reference column. It shows the behavior of the analyzed system with a summarized parasitic capacitance $C_{\text{par}} = 50\,\text{pF}$ while transferring 500 W. The other two columns show the system behavior with varying C_{par} (middle column) and P_{out} (right column).

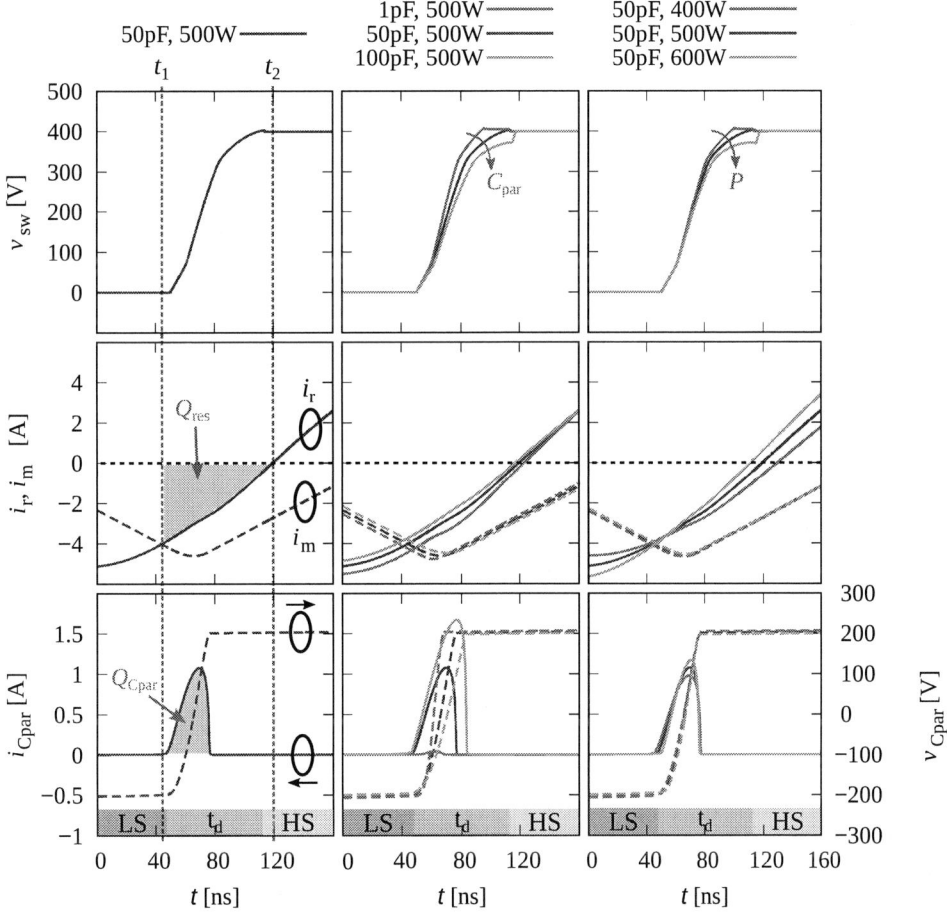

Fig. 3: ZVS behavior dependencies of the presented system with $t_{\text{d}} = 65\,\text{ns}$. Left column: proposed system with $C_{\text{par}} = 50\,\text{pF}$ and $P_{\text{out}} = 500\,\text{W}$. Center column: influence of the parasitic capacitance C_{par} from 1 pF to 100 pF. Right column: influence of output power for P_{out} from 400 W to 600 W.

Table I: Simulation parameters

C_r	690 pF
L_r	8.8 μH
L_m	5.26 μH
$C_{oss,t}$	200 pF
C_{par}	50 pF
n_{eff}	4.11
V_{in}	400 V

There are two time points t_1 and t_2 that have to be considered. At t_1, the resonant and magnetizing current intersect (middle row in Fig. 3). The resonant current $i_r(t)$ then starts to charge the parasitic capacitances, which results in an increase in voltage V_m across the mutual inductance of the transformer (bottom row in Fig. 3). The required charge must be considered in addition to the charge which is needed to charge $C_{oss,t}$ to 400 V, which is why the parasitic capacitances influence the ZVS behavior of the system. The charging process of $C_{oss,t}$ begins together with the dead-time. At t_2, the resonant current crosses zero. After that, $i_r(t)$ discharges $C_{oss,t}$. Thus, the dead-time should not exceed t_2. The maximum available charge Q_{res} of the resonant current can be calculated as

$$Q_{res} = \int_{t_1}^{t_2} i_r(t)\, dt. \tag{3}$$

$i_r(t)$ is determined using the fundamental harmonic approximation as described in [7]:

$$i_r(t) = \frac{v_{1,sw}(t)}{Z_{eq}} = \frac{v_{1,sw}(t)}{i\omega L_r + \dfrac{1}{i\omega L_m - \dfrac{i}{\omega L_m} + \dfrac{1}{R_{ac}}} - \dfrac{i}{\omega C_r}} \quad \text{with} \quad R_{ac} = n_{eff}^2 \frac{8}{\pi^2} R_L. \tag{4}$$

This complex expression contains α, the phase shift between $v_{1,sw}(t)$ and $i_r(t)$, and $|i_r(t)|$. Both are defined as

$$\alpha = \angle i_r(t) = \text{atan}\left[R_{ac}\left(\left(\omega C_{par} - \frac{1}{\omega L_m}\right)^2 + \frac{1}{R_{ac}^2} \right) \left(-\omega L_r - \frac{-\omega C_{par} + \dfrac{1}{\omega L_m}}{\left(\omega C_{par} - \dfrac{1}{\omega L_m}\right)^2 + \dfrac{1}{R_{ac}^2}} + \frac{1}{\omega C_r} \right) \right] \tag{5}$$

and

$$|i_r(t)| = \frac{\hat{V}_{1,sw}}{\sqrt{\left(\omega L_r + \dfrac{-\omega C_{par} + \dfrac{1}{\omega L_m}}{\left(\omega C_{par} - \dfrac{1}{\omega L_m}\right)^2 + \dfrac{1}{R_{ac}^2}} - \dfrac{1}{\omega C_r} \right)^2 + \dfrac{1}{R_{ac}^2\left(\left(\omega C_{par} - \dfrac{1}{\omega L_m}\right)^2 + \dfrac{1}{R_{ac}^2}\right)^2}}}. \tag{6}$$

The influence of the parasitic capacitances and of the load on the resonant current can be seen in Fig. 4 (a) and (b) respectively. With this, the ZVS behavior can be explained as followed. A system with higher C_{par} results in a lower amplitude and a smaller phase shift. A rise of the output load increases the amplitude but lowers the phase shift. These two modifications lower Q_{res} down to a value where Q_{res} is not sufficient to fully load $C_{oss,t}$. This leads to incomplete ZVS.

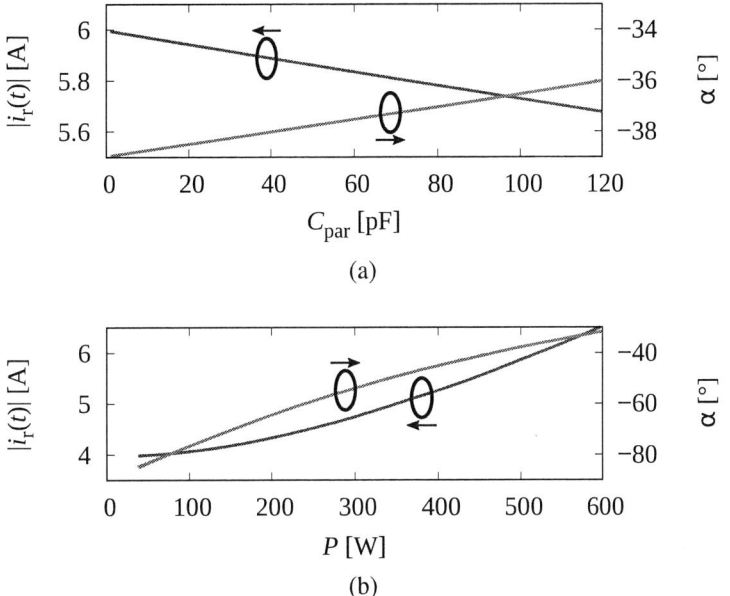

(a)

(b)

Fig. 4: Influence of (a) the parasitic capacitance C_{par} ($P = 500\,\text{W}$) and of (b) the load P ($C_{par} = 50\,\text{pF}$) on the resonant current $i_r(t)$. The influence on the magnitude of $i_r(t)$ and on α, the phase shift between $i_r(t)$ and $v_{1,sw}(t)$ is shown.

Experimental Results

The described ZVS behavior is confirmed by measurement results, see Fig. 5. While transmitting 300 W, ZVS is achieved with both dead-times of 45 ns and 65 ns. The system works more efficiently with a dead-time of 45 ns due to the longer lossy reverse conduction mode that comes with a dead-time of 65 ns. (Since the GaN HEMTs are driven with on- and off-state gate voltages of 6 V and −4 V to prevent parasitic turn-on, there is a large reverse conduction voltage of approx. 7 V, which is why the reverse conduction mode should be as brief as possible.) While transferring 500 W, the system is more efficient with a dead-time of 65 ns because a dead-time of 45 ns is too short to fully charge $C_{oss,t}$, which leads to incomplete ZVS.

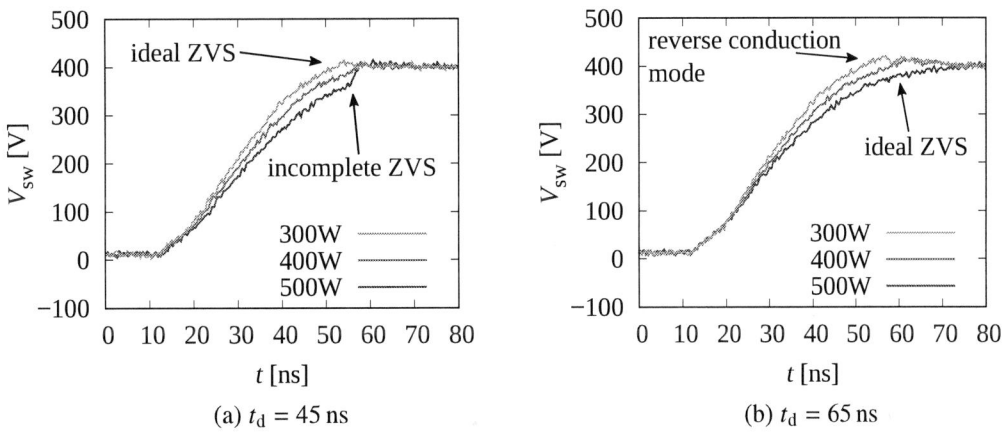

(a) $t_d = 45\,\text{ns}$ (b) $t_d = 65\,\text{ns}$

Fig. 5: Measurement results of a switch-on event with different output power and dead-time t_d of (a) 45 ns and (b) 65 ns

In systems similar to the presented one, where the values of L_m and L_r are in the same range, C_{par} influences the ZVS behavior more than in conventional LLC converters due their comparatively smaller magnetizing current caused by higher values of L_m. Because of this, rectifying diodes with a low junction capacitance and a transformer design with a low parasitic capacitance are mandatory.

Adaptive Dead-Time Control

As described before, the ZVS behavior of the system is load-dependent, which asks for a dead-time control to achieve higher efficiency. In the presented WPT system the switching frequency is set to a fixed value of 2 MHz. The receiving side is passive and does not communicate with the transmitting side, meaning that the output load is unknown to the transmitting side. Therefore, an adaptive control is realized in which the input current is measured and the dead-time is set accordingly. In Fig. 6, the experimental setup is shown schematically.

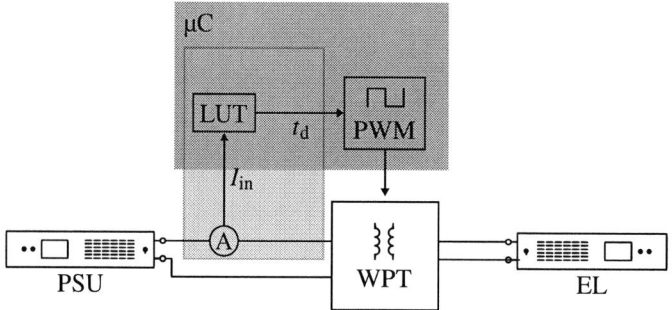

Fig. 6: Schematic of the experimental setup. The WPT system is placed between the power supply unit (PSU) and the electronic load (EL). The efficiency is measured with a precision power analyzer (not shown). The green area marks the dead-time adaption. There, the input current is measured with a hall sensor (A). Based on this value, the dead-time is adapted with a look-up table (LUT).

The input current is measured with a hall sensor (TMCS1101) which is directly connected to the μC. There, a predetermined dead-time is selected with a look-up table (LUT) based on the measured input current I_{in}. To set the dead-time accurately, a μC (STM32F334R8T6) with high-resolution PWM (HRPWM) is used, allowing to adjust the dead-time during operation with an accuracy of 217 ps. To fill the LUT, the input-current-dependent efficiency is measured for different dead-times as shown in Fig. 7. As described before, the optimal switching moment is when the charging of $C_{oss,t}$ is just completed.

The peak efficiency of 92.9 % at $I_{in} = 1.32$ A is achieved with a dead-time of 65 ns. Between an input current of 0.99 A and 1.28 A, a dead-time of 55 ns is more efficient. Below 0.99 A, the efficiency can be increased with $t_d = 45$ ns. The enveloping curve in Fig. 7 shows that a finer resolution of the LUT is not needed. Furthermore, it can be seen that there is no further rise of the efficiency with a dead-time of 75 ns due to incomplete ZVS at higher values of I_{in} and a longer duration of the dead-time than the zero crossing of $i_r(t)$. Thus, $C_{oss,t}$ is slightly discharged again before the switching moment.

With the determined boundaries, an adaptive control of the dead-time can be realized. The system efficiency with dead-time control (green line) and with a fixed dead-time of 65 ns (dashed purple line) is shown in Fig. 8. With the dead-time control, the average efficiency can be raised. For example, the efficiency can be raised from 90.6 % to 91.1 % and from 91.6 % to 92 % while transferring 250 W and 300 W respectively, see Fig. 8.

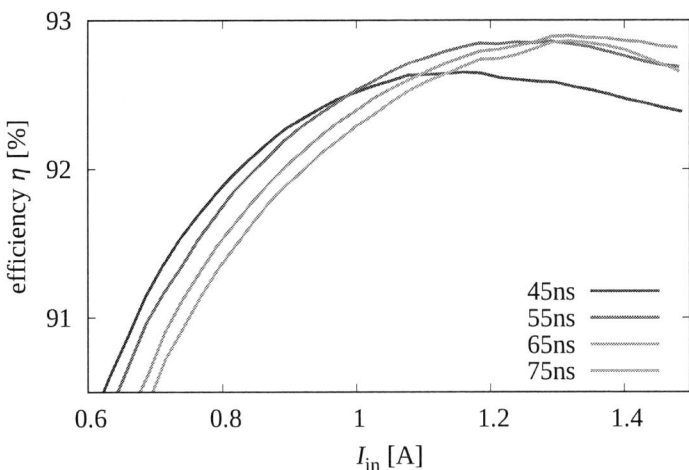

Fig. 7: Efficiency of the system with different dead-times depending on I_{in}.

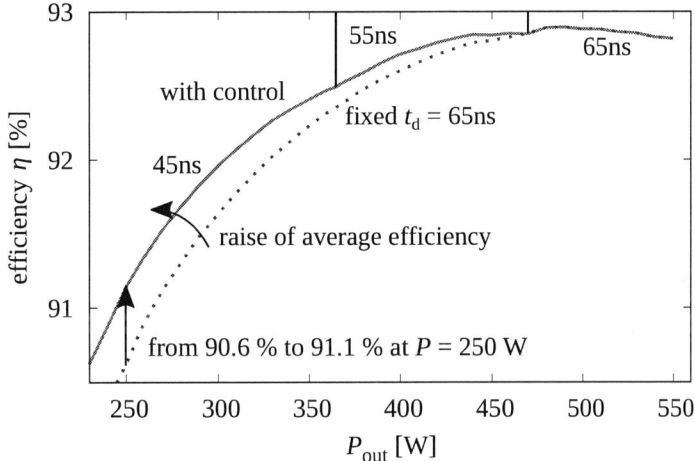

Fig. 8: System efficiency with and without adaptive dead-time control

Conclusion

This paper presents a detailed analysis of the ZVS behavior of a resonant inductive WPT system operating in the MHz range based on the LLC topology. The recharging process of the output capacitance of the half bridge and how it is influenced by parasitic capacitances and by load changes is investigated. Based on this, an adaptive dead-time control is realized by tuning the dead-time depending on the input current. This improves the efficiency over a wide range. This dead-time control is easily implemented and attractive for inductive resonant WPT systems where the receiving side does not communicate with the transmitting side.

References

[1] L. Gu, G. Zulauf, A. Stein, P. A. Kyaw, T. Chen, and J. M. R. Davila, "6.78-MHz wireless power transfer with self-resonant coils at 95% DCDC efficiency," *IEEE Transactions on Power Electronics*, vol. 36, no. 3, pp. 2456–2460, Mar. 2021.

[2] T. Krigar and M. Pfost, "2-MHz compact wireless power transfer system with voltage conversion from 400 V to 48 V," in *2021 IEEE Wireless Power Transfer Conference (WPTC)*, 2021.

[3] H. Wen, J. Gong, C.-S. Yeh, Y. Han, and J. Lai, "An investigation on fully zero-voltage-switching condition for high-frequency GaN based LLC converter in solid-state-transformer application," in *2019 IEEE Applied Power Electronics Conference and Exposition (APEC)*, 2019, pp. 797–801.

[4] H. Wen, Y. Liu, and J.-s. Lai, "Analysis on the effect of secondary side devices for the operation of GaN based LLC resonant converter," in *2020 IEEE Applied Power Electronics Conference and Exposition (APEC)*, 2020, pp. 2214–2218.

[5] H. Chen and X. Wu, "Analysis on the influence of the secondary parasitic capacitance to ZVS transient in LLC resonant converter," in *2014 IEEE Energy Conversion Congress and Exposition (ECCE)*, 2014, pp. 4755–4760.

[6] W. Qin, L. Zhang, and X. Wu, "Re-examination of ZVS condition for MHz LLC converter operating at resonant frequency," in *2018 IEEE International Power Electronics and Application Conference and Exposition (PEAC)*, 2018, pp. 1–4.

[7] C. Oeder and T. Duerbaum, "ZVS investigation of LLC converters based on FHA assumptions," in *2013 Twenty-Eighth Annual IEEE Applied Power Electronics Conference and Exposition (APEC)*, 2013, pp. 2643–2648.

Multilevel battery converter with cascaded H-bridges on cell level – battery management system or a renewed attempt for Power Electronic Building Blocks?

Max Rothenburger[1,2], Markus Horn[1], Xiao Yu[1], Gerold Schulze[2], Koenraad Muyllaert[2], Peter Zacharias[1], Ludwig Brabetz[1], Hartmut Hillmer[1]

[1] UNIVERSITY OF KASSEL
Wilhelmshoeher Allee 71
D-34121 Kassel, Germany
Tel.: +49 / 561 804 – 4103
max.rothenburger@uni-kassel.de
https://www.uni-kassel.de

[2] p&e power&energy GmbH
Universitaetsplatz 12
D-34127 Kassel, Germany
Tel.: +49 / 561 95379 – 721
gerold.schulze@p-and-e.com
https://www.p-and-e.com/

Acknowledgements

Parts of the presented results have been sponsored by the BMBF (German Federal Ministry of Education and Research). Founding reference number 16ME0143.

GEFÖRDERT VOM

Bundesministerium
für Bildung
und Forschung

Keywords

«Multi-level converters», «Pulsed current», «Battery», «Battery Management Systems (BMS)», «Cascaded H-Bridge

Abstract

The combination of battery management system and power electronics, using multi-level topologies in general and cascaded H-bridges in particular, offers advantages over the state of the art. Their evaluation must consider possible effects on battery life. Experimental studies of battery cells show that these effects are small compared to typical loads.

Multi-level topologies and the advantage in battery storage applications with Lithium-based battery cells

To achieve the goal of functional and monetary improvements in battery-electric storage systems, the merger of previously separate functions, like battery management and power control, has great potential.

There are a variety of power electronic topologies for converting electrical power – long known solutions, that have so far been used in specialised areas only, are multi-level topologies with more than three levels [1].

Two main motivators drive the use of multi-level topologies, especially cascaded H-bridges. On the one hand, the insufficient dielectric strength of the components, e.g., in HVDC applications. On the other hand, the better efficiency/higher resulting switching frequency with lower blocking voltage of the components with the consequence of lower expenses in the filter technology.

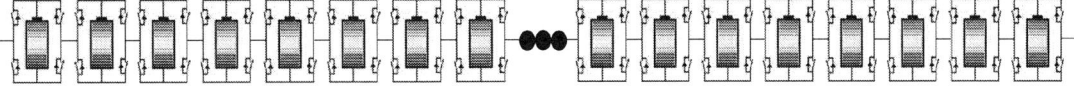

Fig. 1

A disadvantage of most multi-level topologies is the higher complexity of control. Sophisticated control strategies are required especially for voltage balancing of the necessary energy storage devices – in most cases capacitor intermediate circuits.

This is a major advantage of using cascaded H-bridges in battery storage systems. Lithium-based battery cells store more energy per volume compared to capacitors. This allows charge or energy equalization between stages in seconds or minutes rather than milliseconds, several orders of magnitude slower than using a capacitor as a storage device.

The voltage and, if possible, also the temperature have to be monitored for each individual battery cell in the serial connection. Charge equalisation, known as balancing, should be possible between serially connected cells. These functions are usually fulfilled by a Battery Management System (BMS) [2].

This leads to the basic idea – each battery cell gets an H-bridge that brings together the power electronics and the BMS functionality (see Fig. 1) – to borrow Einstein's idea – make things as simple as possible, but not simpler.

There are proposed topologies to include battery cells in a cascaded H-bridge converter, but they are using battery-modules with multiple serial connected cells on each converter-level [3][4]. The use of battery modules with serially connected battery cells requires further cell balancing procedures within the module [3] as well as an additional BMS to monitor the battery cells.

With a battery cell voltage between 2.9 V and 3.6 V (LFP/graphite), at least 120 cascaded stages are required for transformerless grid coupling (230 V_{rms}). Since each battery cell level is equipped with its own H-bridge, the current can be conducted through the cell in positive or negative direction, as well as bypassing the cell. Active cell balancing is possible with operating current up to the nominal current. The small voltage steps between the stages allow operation without high switching-frequency of the H-bridges. Both, DC and AC, voltage can be output.

The proposed topology offers further advantages in terms of safety, availability, repairability and applicability for different applications.

- In the high-impedance state of the bridges, no voltage is present at the output terminals of the battery pack, thus preventing short circuits at the terminals of the battery module during transport and installation without further measures.
- The current can be bypassed at conspicuous, low-power or, in some fault cases, even defective battery cells.
- The failure of one cell does not necessarily lead to the failure of the entire system.
- Due to the active full-power balancing, battery cells with large capacity differences can be used in a system without restrictions. The proposed topology is also suitable for the use of aged, second use cells and grade B cells, as well as for the use of different cell manufacturers or cell types.
- Cascading a simple structure like the H-bridge into a complete system offers a lower market entry barrier for all, who want to do the power electronics integration into their system, e.g. battery cell manufacturers, battery pack manufacturers, battery system suppliers etc.
- Generalised, the presented design offers a competitive solution for battery systems from 12 V to 1500 V output voltage (DC or AC) and for cell capacities from 20 Ah to 500 Ah

With the advantages described, the topology is most suitable for areas of application in which higher battery cell capacities are required, such as electromobility (buses, trains, ships, etc.), stationary energy storage, construction site supply, and so on. This assessment is linked to the attempt to establish a Power Electronic Building Block (PEEB), which has already been carried out several times in power electronics [5–7]. In this case, a PEBB for the power electronics of battery storage systems.

In order to reap the benefits mentioned above, some challenges need to be overcome. Let's name the ones for which there are already promising solutions, but which are still waiting for experimental and economic proof:

- There is a suitable electromechanical connection between the battery cell and PCB-based power electronics that offers low contact resistance and compensation for mechanical tolerances.
- A detachable power connection between battery cell and PCB-based power electronics offers improved repair possibilities.
- The proportional specific costs (€/W) for the proposed power electronics should be lower than those of conventional photovoltaic converters.
- The total costs of the topology are lower compared to a conventional inverter with external BMS due to the integrated BMS functionality and the advantages mentioned (active balancing, availability, etc.) (€/W).
- Despite the higher number of components, there is a lower or equal probability of failure compared to conventional approaches (first considerations below).

The old challenges for new design approaches – temperature and switching overvoltage

For single-phase, transformer less coupling to the 230 V_{rms} AC grid, the sum of the battery cell voltages must exceed the grid peak voltage. Depending on the discharge voltage limit of the battery cells, a series connection of 120 to 140 stages is required. These stages could be combined to cell packs with multiple stages on one PCB (see Fig. 2). Due to the number of stages needed, a large number of MOSFETs is required in the system. Can a low utilization of the components increase their lifetime to such an extent that their number cancels out?

Fig. 2: Top view of a PCB, designed for a nominal current of 150 A, with H-Bridges for a 4s cell pack of 280 Ah cells (72 mm x 173 mm x 208 mm), available from different manufacturers like CATL, EVE, Lishen and others

A simple way to increase the reliability of components is to operate them well below their rated specifications. This method is known as derating. For example, if a component designed to operate reliably at +125°C is only operated at +55°C, lifetime will increase. As a rule of thumb, the life of a component doubles for every 10 K drop in temperature. This relationship between temperature and service life is based on the theoretical framework of the Arrhenius equation, which establishes a relationship between temperature and ageing acceleration. This works for electronic components as well as for the battery cells [1, 8].

Concerning temperature, the limiting part in a battery energy storage system (BESS) is the battery cell itself, which will degrade the least at temperature around 20 to 30°C. With a limited ambient temperature range of 5 to 35°C for the whole system, this leaves enough space for natural cooling of the power semiconductors, which normally work with base temperatures above 80°C.

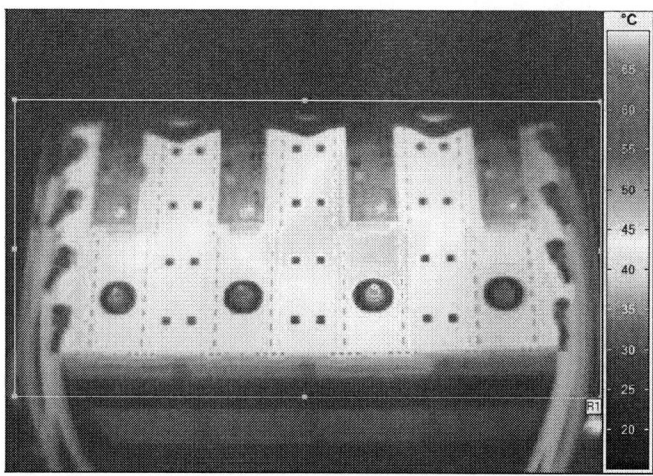

Fig. 3: Infrared image of the PCB shown in Fig. 2 at a current of 150 A, hotspot temperature 50°C

Additionally, investment in silicon MOSFETs also leads to peak efficiencies of over 99 % and thus low additional heat losses. As shown in Fig. 3, at a current of 150 A (around 0.5C) the temperature increases only about 25 K, which leaves the semiconductors at a cosy temperature of 50°C at an ambient temperature of 25°C.

In addition to the low temperature, the low voltage utilization of the MOSFETs also contributes to an increase in lifetime. Stationary the drain-source voltage is the cell voltage and with max. 3.6 V, for LFP cells, is far below the permissible voltage of a 25 to 40 V MOSFET. But what is about switching overvoltage?

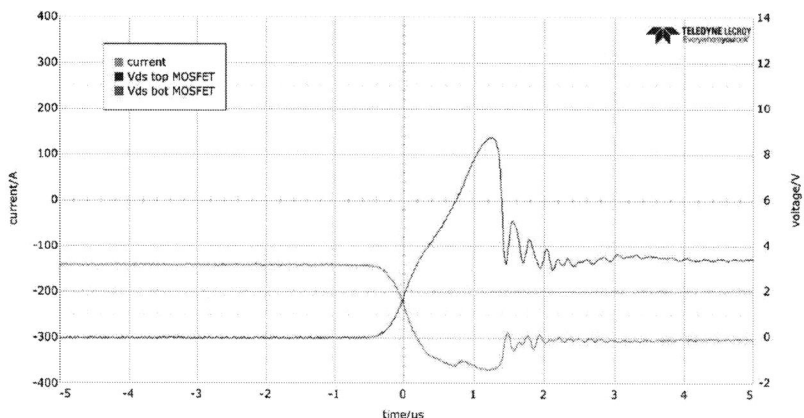

Fig. 4: Switching from positive to bypass state at 200 A discharging current (yellow), drain-source-voltage at bottom MOSFET (red), drain-source-voltage at top MOSFET (blue)

The voltage curve of a commutation event at 200 A discharge current shown in Fig. 4 with 8.8 V_{peak} is safely within the permissible voltage range of the MOSFET. In the proposed topology the battery cell is located in the commutation circuit (see Fig. 7). To avoid an additional DC link capacitor, the cell inductance plays an important role for the design of the commutation circuit.

Determination of the battery inductance

Overvoltage across transistors can be caused by the commutation inductance L_c, due to the rapid current reduction during commutation transient. Together with the battery cell voltage, this overvoltage can endanger the MOSFETs used in the battery converter (see Fig. 7). The absolute maximal drain-source voltage limitation alone is not enough to guarantee a safe operation of the transistor. Fig. 5 specifies the maximum allowable pulse currents that a transistor can withstand when it is in on-state without exceeding the voltage limitation of 25V.

Another approach is to use the transient thermal resistance of the transistor to estimate the maximal allowable power P_{max} generated during the switching transient:

$$P_{max}(t_p) = \frac{T_{jmax} - T_{amb}}{Z_{th}(t_p)} \tag{1}$$

The switching power P_{sw} can be approximated as the power averaged over the period t_p:

$$P_{sw} \approx \frac{1}{t_p} \int_0^{t_p} \left(-\frac{U_d}{t_p} \cdot t + U_d\right) \cdot \left(\frac{I_d}{t_p} \cdot t\right) \cdot dt = \frac{U_d I_d}{6} \tag{2}$$

For those one-time events, such as overcurrent, which lead to the shutdown of the transistors, the maximal permissible current can be estimated for a given pulse duration t_p.

$$\frac{U_d I_d}{6} = P_{sw} < P_{max}(t_p), \quad I_{d,max} \approx \frac{6 \cdot P_{max}(t_p)}{U_d} \tag{3}$$

Therefore, a modeling of the commutation inductance is reasonable for the overvoltage and the maximal switchable current investigations during the switching transients. To figure out the inductance introduced by the battery cell, single pulse tests on a battery cell have been implemented, whereby the behavior of the battery cell under different test conditions, such as different states of charge (SoC), temperatures, current directions, and current intensities, has been considered. It is assumed that the pulse length is short and has no influence on the SoC during the test.

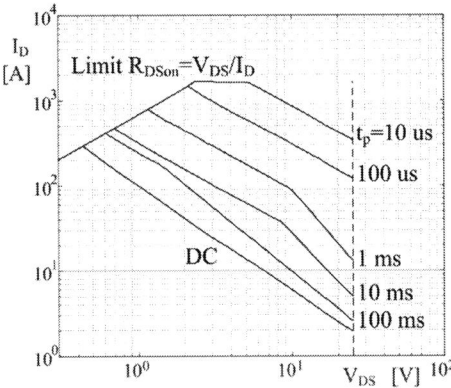
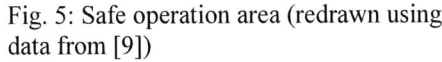

Fig. 5: Safe operation area (redrawn using data from [9])

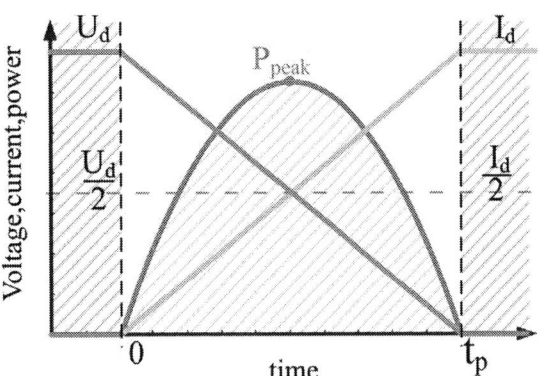

Fig. 6: Simplified voltage (blue), current (yellow) and power (green) profiles during the switching transient

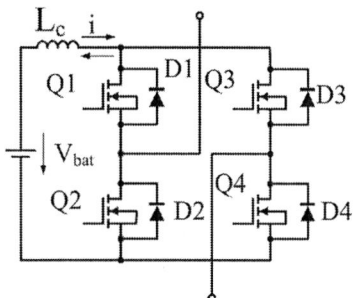

Fig. 7: Commutation inductance

Table I: Test Conditions

Peak Current [A]	136	272	408	544
Current Direction	Positive	Negative		
SoC	20%	50%	80%	
Battery Temperature [°C] (Heating or cooling for at least 5 hours)	5	20	50	

All test conditions are listed in Table I and 72 combinations of test conditions were tested. The object under test is a battery cell (Lishen Battery) with a nominal capacity of 272 Ah. It was tested up to a current peak of 544 A (2C). If the current flows into the battery, it is defined as positive current, and in reverse, it is considered as negative (see Fig. 8). The measurement setup is illustrated in Fig. 9 which is familiar with the widely used double pulse test configuration (DPT) for characterizing of semiconductors. The only difference is that 10 diodes are connected in series in the commutation loop, since the threshold voltage of the lower Schottky diode alone is not enough to block the battery nominal voltage of 3.2 V.

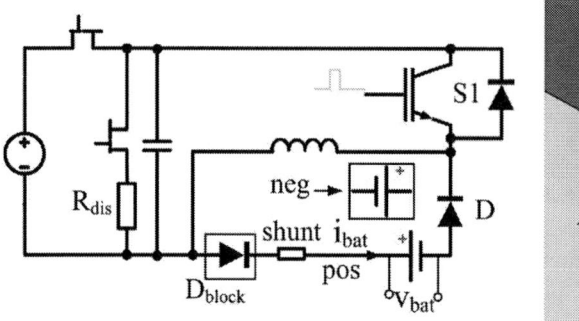

Fig. 8: Schematic of the pulse test

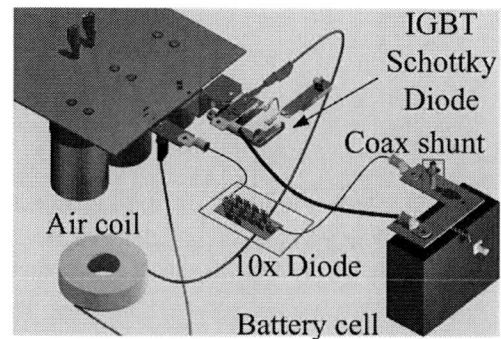

Fig. 9: Measurement setup

For measuring the battery current, the coax shunt SDN-414-10 from T&M research products was connected in series with the battery cell, while the battery voltage v_{bat} was directly measured between two battery terminals. To approximate the effective inductance, the measured currents were filtered, and their first-order derivatives were calculated during the data post-processing. The inductance can be estimated as the ratio between the absolute maximum of the measured battery voltage and the absolute maximum of the current derivative using $L_{bat} \approx V_{bat,max}/(di_{bat}/dt|_{max})$. Fig. 10 shows two of the measurement results under different measuring conditions and the corresponding approximated inductance values. All derived inductance values fall in the range of 4 to 7 nH.

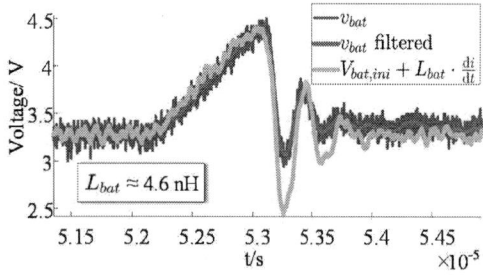

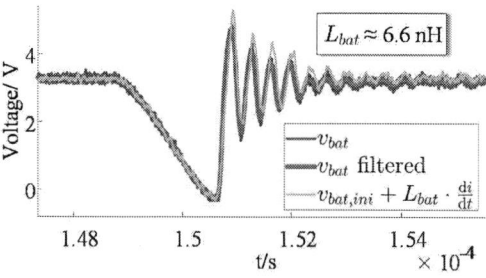

Fig. 10: Approximation of inductance introduced by battery cell: 20 % SoC, positive, 0.5C, 5°C (left); 80 % SoC, negative, 2C, 50°C (right)

What does the battery say about a pulsating load?

A configuration of battery cells with the semiconductor full bridges can provide (almost) any output voltages and frequencies. In mains parallel operation, the voltage is controlled depending on the present voltage value by switching the half-bridges to high and low impedance. Depending on the charging and discharging of the electrical storage system, the cells can be charged and discharged with sections of the mains current.

The influences of the expected load, such as calendar and cyclical ageing, as well as the change in electrical parameters associated with ageing, such as internal resistance and double-layer capacitances, have hardly been researched at present. Two publications on the subject consider on the one hand the load with pure square-wave current and on the other hand the current with a high-frequency AC component superimposed on the charging and discharging current [10, 11].

In addition to the development of the multilevel inverter, cell ageing and the change in electrical parameters during the cycle tests were further investigated.

A test bench designed for the study is cycling lithium iron phosphate cells from two different manufacturers since March 2021. In total, the setup contains 16 cells, 8 cells EVE LF50K and 8 cells Benergy BXL-LFP-50AHP with 50 Ah each (nominal value).

The cell array is divided into two groups, of which one control group is charged and discharged with DC current. The other group is charged and discharged with the pulse sequence shown in Fig. 11. The individual pulses are each a sine half-wave cut in and out at different angles. The pulse sequence thus corresponds in sections to the magnitude of the mains current and thus represents a practical cell current.

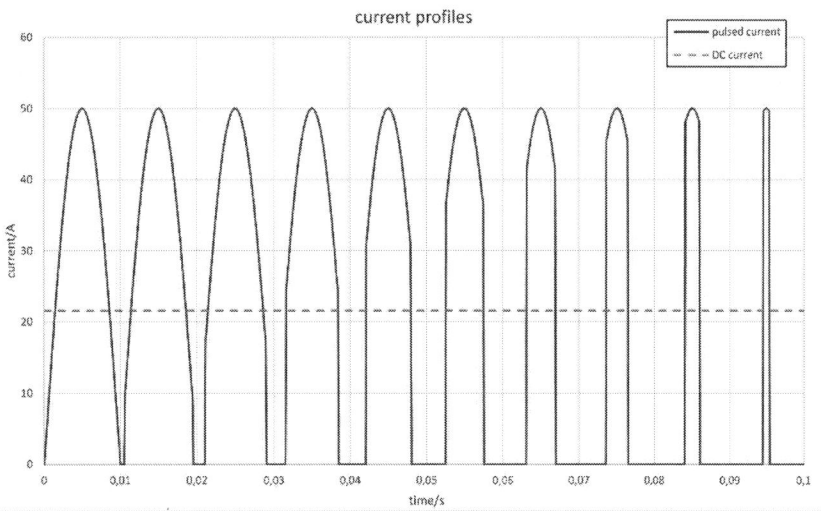

Fig. 11: Imprinted current profiles for DC- and pulsed current with same C-rate

The cell test bench shown in Fig. 12 performs load jumps and capacitance measurements at regular intervals for both load profiles in addition to cycling with DC and the current profile shown. The load steps allow the determination of the electrical parameters mentioned. With the help of the capacitance measurements, it is possible to derive the State of Health.

Fig. 12: Test bench for single cell test

A National Instruments PXI takes control of the test bench and records the measured quantities of voltage, current and temperature for each cell in the setup. Electronic loads imprint the current profile and load step shown in Fig. 11 for the electrical parameters sampled at regular intervals.

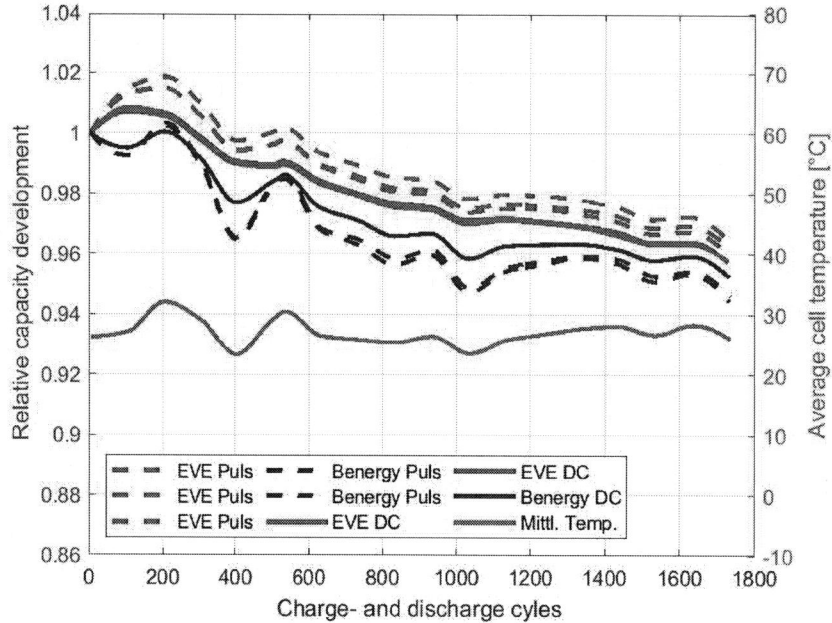

Fig. 13: Relative capacity degradation with different current profiles

So far, a selection of cells has completed almost 1800 cycles between 2.8 V and 3.4 V. Fig. 13 shows the capacity curve of the batteries that have been cycling since March 2021. So far, the pulsating current load appears to cause one of the battery types to age less. The other battery type seems to cycle age less under DC load.

Since, the test bench does not have air conditioning, the temperature has a strong influence on the capacity measurement and is therefore shown in the graph.

Conclusion

The proposed use of cascaded H-bridges for individual battery cells offers a number of advantages. In addition to full coverage of the typical BMS functions for the battery cells, active cell balancing with nominal current on cell level and highest efficiencies is possible. The topology allows arbitrary AC or DC output voltages, so no additional converters are needed. A three-phase AC-system can be built by combining three of the proposed systems.

With regard to lifetime, negative or positive effects of the operating mode on the battery cells cannot be clearly determined at present for unfiltered pulsed currents. What can be clearly stated is that the observed differences are less than 1%.

The investigated topology has no voltage on the output terminals when switched off, thus preventing short circuits. In addition, the presented approach improves repairability and allows the free use of different cell manufacturers – if the battery cell has two poles.

References

[1] J. Rodriguez, J.-S. Lai, and F. Z. Peng, Eds., *Multilevel inverters: a survey of topologies, controls, and applications*, 2002.

[2] *Secondary cells and batteries containing alkaline or other non-acid electrolytes - safety requirements for secondary lithium cells and batteries for use in industrial applications*, IEC 62619, Internationale Elektrotechnische Kommission, Geneva, 2017.

[3] M. Chen, B. Zhang, Y. Li, G. Qi, and J. Liu, "Design of a multi-level battery management system for a Cascade H-bridge energy storage system," in *2014 IEEE PES Asia-Pacific Power and Energy Engineering Conference (APPEEC 2014): Kowloon, Hong Kong, 7 - 10 December 2014*, Hong Kong, 2014, pp. 1–5. Accessed: May 18 2022.

[4] Z. Ling, Z. Zhang, Z. Li, and Y. Li, "State-of-charge balancing control of battery energy storage system based on cascaded H-bridge multilevel inverter," in *2016 IEEE 8th International Power Electronics and Motion Control Conference (IPEMC-ECCE Asia) took place 22-26 May 2016 in Hefei, P.R. China*, Hefei, China, 2016, pp. 2310–2314. Accessed: Mar. 2 2022.

[5] T. Ericsen and A. Tucker, "Power Electronics Building Blocks and potential power modulator applications," in *Conference Record of the Twenty-Third International Power Modulator Symposium (Cat. No. 98CH36133)*, 1998, pp. 12–15.

[6] F.C.Y. Lee and D. Peng, *Power electronics building block and system integration*, 2000.

[7] T. Ericsen, N. Hingorani, and Y. Khersonsky, *PEBB - Power Electronics Building Blocks from Concept to Reality*, 2006.

[8] D. Sauer, F. Ringbeck, M. Kuipers, and M. Faber, *Bedarf, Funktion und Konzepte für thermische Managementsysteme von Batteriesystemen in Elektrofahrzeugen*, 2018.

[9] Nexperia B.V., "PSMN0R9-25YLD: Product data sheet," 2016.

[10] A. Ghassemi, P. C. Banerjee, Z. Zhang, A. Hollenkamp, and B. Bahrani, "Aging Effects of Twice Line Frequency Ripple on Lithium Iron Phosphate (LiFePO 4) Batteries," in *2019 21st European Conference on Power Electronics and Applications (EPE '19 ECCE Europe)*, Genova, Italy, 2019, P.1-P.9. Accessed: Apr. 25 2021.

[11] A. Bessman, R. Soares, O. Wallmark, P. Svens, and G. Lindbergh, "Aging effects of AC harmonics on lithium-ion cells," *Journal of Energy Storage*, vol. 21, pp. 741–749, 2019, doi: 10.1016/j.est.2018.12.016.

Design and Potential of EMI CM Chokes with Integrated DM Inductance

Mohammad Ali, Rehnuma Bushra, Jens Friebe, Axel Mertens
LEIBNIZ UNIVERSITY HANNOVER
Institute for Drive Systems and Power Electronics
Welfengarten 1
30167 Hannover, Germany
Tel: +49 / (0) - 511 762 3778
Fax: +49 / (0) - 511 762 3040
E-mail: mohammad.ali@ial.uni-hannover.de
URL: www.ial.uni-hannover.de

Acknowledgments

This work was supported by Forschungsvereinigung Antriebstechnik e.V. (FVA) within Project FVA 637V.

Keywords

≪Choke≫, ≪3-D Electromagnetic Modeling≫, ≪DM Inductance≫,≪Electromagnetic Interference (EMI) Filters≫, ≪Mutual Couplings≫, ≪Filter Optimization≫.

Abstract

A common-mode (CM) choke is one of the major filtering components used in electromagnetic interference (EMI) filters that impede the flow of common-mode current in the system. Besides, a common-mode choke also offers a finite differential-mode (DM) inductance in the form of leakage inductance. In general, the leakage inductance is small, which necessitates extra DM inductors and leads to a large filter size. In this paper, a single-phase CM choke is designed with an integrated DM inductance for use in EMI filters, based on a flux bypass inside the toroidal core. The DM inductance of these chokes can be altered by redesigning the windings, while the CM inductance remains identical to a conventional CM choke. The design and effectiveness of them are further supported by theoretical, simulated, and measurement results and core saturation analysis. Finally, an extension to three-phase CM chokes with integrated three-phase DM inductance is discussed.

I. Introduction

The introduction of wide-bandgap power semiconductors has made EMI filter design quite challenging when aiming to meet the EMC regulations according to IEC and CISPR standards. EMI filters have to attenuate high emission levels, especially at higher frequencies in the MHz range generated from the steep $\frac{\mathrm{d}v}{\mathrm{d}t}$ switching rate of the wide-bandgap power semiconductor devices. An EMI filter usually attenuates both common-mode (CM) and differential-mode (DM) noise. A common-mode (CM) choke is an indispensable part of an EMI filter that helps to attenuate the CM noise between the phase and neutral lines with reference to the ground. The CM choke typically consists of a soft magnetic toroidal core with windings wound around the core. It usually provides a large CM inductance. For fixed core dimensions, the CM inductance is directly proportional to the square of the number of turns in each winding. However, as 100 % coupling does not occur between the windings of the CM choke, leakage inductance exists in the operation of a conventional CM choke. This leakage inductance acts like the DM inductance and suppresses the DM noise between the phase and neutral lines. In most cases, however, the leakage inductance of a choke is not sufficient to suppress the DM noise [1], [2], [3], [4]. To improve the DM performance, a single-phase common-mode choke usually uses two additional external inductors, which result in a larger filter. Therefore, a CM choke with an integrated DM inductor can be the solution when trying to reduce the size of the inductor of an EMI filter and improve the DM attenuation [5], [6]. According to [5], an integrated DM inductor with a conventional CM choke may reduce an EMI filter's volume up to 30 %.

However, most of the literature on common-mode chokes does not address how to improve the DM inductance. Only limited literature especially analyzes the improvement of the DM inductance of a common-mode choke [5], [7], [8]. In [9], [10], an asymmetric winding structure is proposed to attenuate DM noise along with the CM noise. However, an asymmetric winding structure may create radiated emissions at higher frequencies and requires careful analysis with respect to core saturation. In [5], a new core structure consisting of two cores, a toroidal core, and a solenoid core, is proposed. However, the effect of different parameters on the performance of the new choke is not clearly discussed. This paper shows the effect of different parameters (number of turns on the block core, air-gap effect, relative permeability, and core material effect) on the performance of a conventional CM choke with a block core and its attenuation in an EMI filter. In addition, a new choke structure with a Y-core to improve the DM inductance of a three-phase CM choke is also proposed in this paper.

II. Winding Structures to Improve DM Inductance

A. Structure of a Single-Phase CM Choke with Block Core

Fig. 1 shows a single-phase CM choke with a block core and its simulation model. After inserting the block core, the DM inductance of the respective choke can be further improved by introducing a number of turns on the block core. There are six independent copper windings on the cores: four on the toroidal core ($T_1\prime$, $T_1\prime\prime$, $T_2\prime$, $T_2\prime\prime$), and two on the block core (B_1, B_2), as indicated in Figs. 1e and 1f [5]. The coils $T_1\prime$, B_1, $T_1\prime\prime$ and $T_2\prime$, B_2, $T_2\prime\prime$ each comprise two separate windings of this single-phase CM choke. The four windings on the toroidal core have the same number of turns, and the two windings on the block core also have an identical number of turns to ensure a symmetrical and homogeneous structure.

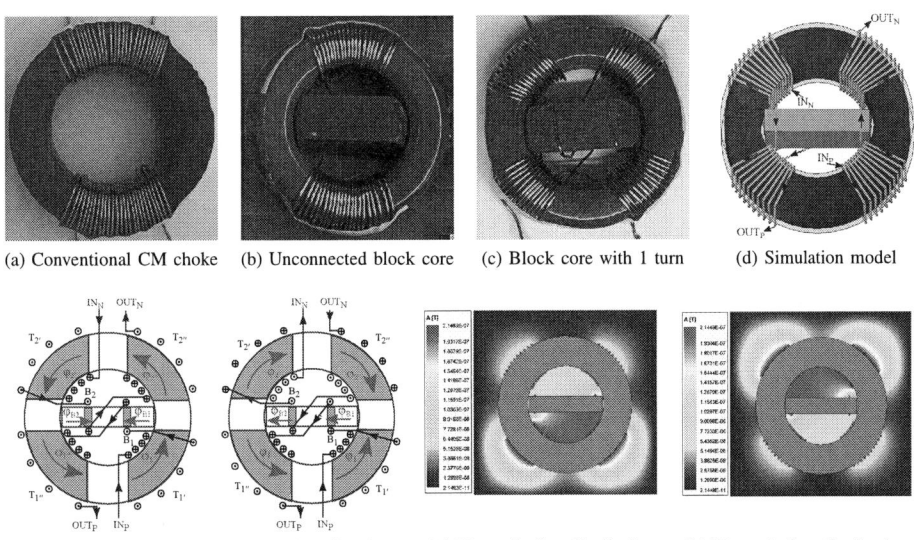

(a) Conventional CM choke   (b) Unconnected block core   (c) Block core with 1 turn   (d) Simulation model

(e) CM flux direction   (f) DM flux direction   (g) Magnetic flux distribution   (h) Magnetic flux distribution

Fig. 1. Single-phase CM choke: (a) conventional CM choke; (b) CM choke with unconnected block core; (c) CM choke with 1 turn in block core; (d) Q3D simulation model; (e) CM flux direction; (f) DM flux direction; (g) Magnetic flux distribution (winding 1); (h) Magnetic flux distribution (winding 2)

The flux direction in this new choke is shown in Figs. 1e and 1f. The red arrows in the figure show the magnetic flux direction, and the light blue and dark blue arrows indicate the current direction in the choke. Current entering the plane of the diagram is indicated by the circled cross and current exiting the plane is indicated by the circled dot. In determining the direction of the flux, it is assumed that the current flows from the cross to the dot. As shown in Fig. 1e, the flux induced by the CM current is oriented anticlockwise, and for the two windings on the toroidal, the fluxes add together to provide high CM inductance, as in a conventional CM choke. However, because of the opposite induced flux in the block core, the two fluxes oppose each other without significantly affecting the CM inductance of the choke. On the one hand, the directions of the magnetic fluxes created by the DM current in the toroidal core exactly oppose each other, as shown in Fig. 1f. In addition, the windings on the block core create two fluxes with the same direction and produce more magnetic flux inside the core, thus resulting in more leakage flux in the air gap between the toroidal and block core. For the magnetic flux density analysis, a simulation model was developed in ANSYS Q3D, as shown in Fig. 1d. The distribution of magnetic flux in the CM choke with the block core is shown in Figs. 1g and 1h. These figures show higher magnetic flux distribution in the inner portion of the toroidal core due to the block core and increased leakage inductance along the magnetic path created by the block core. As a result, a high DM inductance can be obtained from this configuration.

B. Magnetic and Electrical Circuits of a Single-Phase CM Choke with a Block Core

Magnetic circuits are primarily concerned with the behavior of magnetic fields within a single component or a group of components. Therefore, it is quite easy to obtain the electrical circuit of an electromagnetic component, e.g., an inductor, from the magnetic circuit. In the magnetic circuit, the magnetomotive force (MMF) is represented by NI, where N denotes the number of turns in the winding, and I represents the current flowing through it. The polarity of the MMF is determined by the flux direction. In the magnetic circuit, $\mathfrak{R}_T$, $\mathfrak{R}_B$ denote the reluctance of the toroidal core and block core, respectively. The reluctance of the air gap between the block core and the toroidal core is denoted by $\mathfrak{R}_G$. When there are no windings on the block core, or when the winding on the block core is not connected, no MMF is generated in the block core. However, due to the existence of the magnetic path, the reluctance of the block core and air gap exists in the equivalent magnetic circuit of a CM choke with an unconnected block core.

An equivalent circuit of the CM choke with different numbers of turns on the block core can be easily obtained from the equivalent magnetic circuit shown in Fig. 2. In the equivalent electrical circuit, the reluctances of the windings $T_1\prime$, $T_1\prime\prime$ are combined and represented by an inductance L_{T1} for simplification, while an inductance $L_{B2}\prime$ symbolizes the reluctance of the block core $\mathfrak{R}_{B2}\prime$. Therefore, L_{T1} and $L_{B2}\prime$ together denote the first winding of the CM choke. Similarly, L_{T3} and L_{B2} characterize the second winding. The CM and DM configurations of the CM choke with a block core are shown in Figs. 2b and 2d, respectively. The inductance matrix is completely populated, with all non-zero mutual inductances; in Figs. 2b and 2d, couplings are shown only between neighboured windings to reduce complexity.

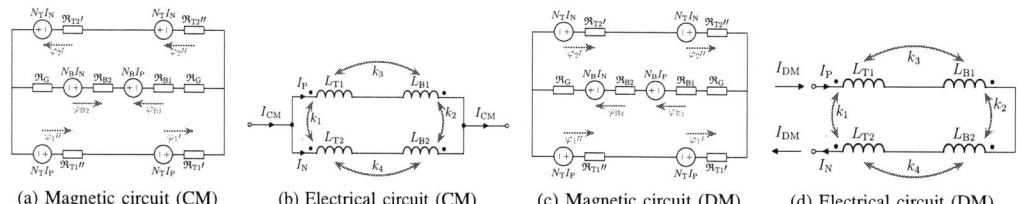

| (a) Magnetic circuit (CM) | (b) Electrical circuit (CM) | (c) Magnetic circuit (DM) | (d) Electrical circuit (DM) |

Fig. 2. Magnetic and electrical circuits of single-phase CM choke with block core [5]

C. Common-Mode Inductance of the CM Chokes with a Block Core

The inductance of an inductor can be expressed in terms of the ratio of induced voltage to the rate of change of current.

$$L = \frac{V}{\frac{dI}{dt}} \tag{1}$$

In a single-phase system, the CM current flows in the same direction through the phase and neutral conductors and returns through the ground path. These equal currents create an equal voltage drop across both the phase conductor winding and the neutral conductor winding of the CM choke. The currents in the phase conductor and neutral conductor and the voltage in CM configurations can be expressed as follows:

$$I_{P\_Conv(CM)} = I_{N\_Conv(CM)} = \frac{I_{CM}}{2} \tag{2a}$$

$$V_{P\_Conv(CM)} = V_{N\_Conv(CM)} = V_{CM} \tag{2b}$$

The voltage drops across the phase and neutral windings can be written in the form of a matrix [5]

$$\begin{bmatrix} V_{P\_Conv} \\ V_{N\_Conv} \end{bmatrix} = \begin{bmatrix} L_{leakage\_1} + \frac{N^2}{4\mathfrak{R}_T} & \frac{N^2}{4\mathfrak{R}_T} \\ \frac{N^2}{4\mathfrak{R}_T} & L_{leakage\_2} + \frac{N^2}{4\mathfrak{R}_T} \end{bmatrix} \begin{bmatrix} \frac{dI_{P\_Conv}}{dt} \\ \frac{dI_{N\_Conv}}{dt} \end{bmatrix} \tag{3}$$

In this equation, $L_{leakage\_i}$ denotes the leakage inductance of the ith winding, N is the number of turns on a toroidal core in a conventional CM choke, and $\mathfrak{R}_T$ refers to the reluctance of a quadrant of the toroidal core and can be expressed using the following equation.

$$\mathfrak{R}_T = \frac{\frac{l_{e\_T}}{4}}{\mu_{r\_T}\mu_0 A_{e\_T}} \tag{4}$$

Here, l_{e_T} and A_{e_T} indicate the mean length and effective cross-section of the toroidal core, respectively, and can be expressed as follows:

$$A_{e\_T} = (r_{out} - r_{in}) \cdot h \tag{5}$$

$$l_{e\_T} = \frac{2\pi \cdot \ln\left(\frac{r_{out}}{r_{in}}\right)}{\frac{1}{r_{in}} - \frac{1}{r_{out}}} \tag{6}$$

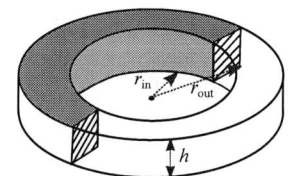

Fig. 3. Toroidal core

Here, r_{in}, r_{out} denote the inner and outer radii of the core, respectively, and h is the height of the toroidal core.

Using Equations (1), (2) and (3), the CM inductance of a conventional CM choke can be expressed as follows:

$$L_{CM\_Conv} = \frac{L_{leakage\_1}}{2} + \frac{N^2}{4\mathfrak{R}_T} \cong \frac{N^2}{4\mathfrak{R}_T} \tag{7}$$

In the equivalent equation for the CM inductance written above, the leakage inductance part was ignored because the leakage inductance does not affect the CM inductance significantly. Using the CM inductance of a conventional CM choke from Equation (7), the CM inductance of a single-phase CM choke with a block core can be expressed as follows [5]:

$$L_{\text{CM\_New}} \cong \frac{L_{\text{leakage\_1}}\prime + L_{\text{leakage\_1}}\prime\prime + L_{\text{leakage\_B1}}}{2} + \frac{N^2}{\mathfrak{R}_{\text{T}}} \cong \frac{N_{\text{T}}^2}{\mathfrak{R}_{\text{T}}} \tag{8}$$

Here, $L_{\text{leakage\_1}}\prime$, $L_{\text{leakage\_1}}\prime\prime$ and $L_{\text{leakage\_B1}}$ are together considered as the leakage inductance of a winding, since $T_1\prime$, $T_1\prime\prime$ and B_1 comprise a single winding of this choke. When calculating the CM inductance, these leakage parts are ignored because the fluxes in the block core cancel each other out. Comparing Equations (8) and (7), it can be seen that $L_{\text{CM\_Conv}} = L_{\text{CM\_New}}$ when $N_{\text{T}} = \frac{N}{2}$, where N is the number of turns in each winding of the conventional CM choke and N_{T} denotes the number of turns in each winding of the new-design choke.

D. Differential-Mode Inductance of the CM Choke with a Block Core

Since the DM current flows through the phase conductor and the neutral conductor in opposite directions in a single-phase system, the phase and neutral currents and voltage can be expressed as follows:

$$I_{\text{P\_Conv(DM)}} = -I_{\text{N\_Conv(DM)}} = I_{\text{DM}} \tag{9a}$$

$$V_{\text{P\_Conv(DM)}} = -V_{\text{N\_Conv(DM)}} = V_{\text{DM}} \tag{9b}$$

The DM inductance of a conventional single-phase CM choke can be expressed using Equations (1), (9) and (3)

$$L_{\text{DM\_Conv}} \cong L_{\text{leakage\_1}} \tag{10}$$

In this equation, $\frac{N^2}{4\mathfrak{R}_{\text{T}}}$ is ignored because only leakage inductance is considered in the calculation of DM inductance. The leakage inductance mainly depends on the air gap inside the toroidal core and the core geometry. The leakage inductance of a CM choke can be calculated using the following equation [1],

$$L_{\text{leakage}} \cong 2.5\mu_0 N_{\text{T}}^2 \frac{A_{\text{e}}}{l_{\text{eff}}} \left(\frac{l_{\text{e}}}{2} \sqrt{\frac{\pi}{A_{\text{e}}}} \right)^{1.45} \tag{11}$$

Here, l_{eff} is the mean effective path length of the leakage flux and can be calculated from Equation (12). Here, d_{in}, d_{out} denote the inner and outer diameters of the toroidal core, respectively, and θ is the angle of the adjacent turns of the winding due to the curvature of the toroid [1].

$$l_{\text{eff}} = \sqrt{\frac{d_{\text{out}}^2}{\sqrt{2}} \left(\frac{\theta}{4} + 1 + \sin\frac{\theta}{2} \right) + d_{\text{in}}^2 \left(\frac{\theta}{4} - 1 + \sin\frac{\theta}{2} \right)} \tag{12}$$

Since the leakage inductance of a single winding in this new-design CM choke with a block core consists of the contributions of three windings, it can be stated as follows [5]:

$$L_{\text{DM\_New}} \cong L_{\text{leakage\_1}}\prime + L_{\text{leakage\_1}}\prime\prime + L_{\text{leakage\_B1}} + \frac{2(N_{\text{B}} + N_{\text{T}})^2}{\mathfrak{R}_{\text{C}}} \tag{13}$$

The additional factor accounts for the number of turns on the block core which aid in increasing the DM inductance of the structure. Usually, the leakage inductance of the winding on the block core $L_{\text{leakage\_B1}}$ is lower than the leakage inductance of the winding on the toroidal core. The reason for this is that the number of turns in the winding on the block core is lower. A lower number of turns is usually wound on the block core to avoid early core saturation. Since the two windings on the toroidal core count as a single winding, the equation written above can be simplified so that $L_{\text{leakage\_1}}\prime + L_{\text{leakage\_1}}\prime\prime + L_{\text{leakage\_B1}}$ is approximately equal to $2L_{\text{leakage}}$. In addition, the leakage inductance of the winding on the block core is negligible because its value is much smaller than the leakage inductance of the winding on the toroidal core. Therefore, Equation (13) can be expressed as follows [5]:

$$\begin{aligned} L_{\text{DM\_New}} &\cong 2L_{\text{leakage}} + \frac{2(N_{\text{B}} + N_{\text{T}})^2}{\mathfrak{R}_{\text{C}}} \\ &\cong 5\mu_0 N_{\text{T}}^2 \frac{A_{\text{e}}}{l_{\text{eff}}} \left(\frac{l_{\text{e}}}{2} \sqrt{\frac{\pi}{A_{\text{e}}}} \right)^{1.45} + \frac{2(N_{\text{B}} + N_{\text{T}})^2}{\mathfrak{R}_{\text{C}}} \end{aligned} \tag{14}$$

In this equation, N_{B} and N_{T} indicate the number of turns in each winding on the block core and the toroidal core, respectively. The total reluctance of the toroidal core and block core combined is denoted by $\mathfrak{R}_{\text{C}}$, which can be expressed as follows:

$$\mathfrak{R}_{\text{C}} = 2\mathfrak{R}_{\text{B}} + 2\mathfrak{R}_{\text{G}} + \mathfrak{R}_{\text{T}} \tag{15}$$

In this equation, $\mathfrak{R}_B$ represents the reluctance of either half of the block core and $\mathfrak{R}_G$ represents the reluctance of the air gap, where

$$\mathfrak{R}_B = \frac{l_{e\_B}}{\mu_{r\_B}\mu_0 A_{e\_B}} \text{ and } \mathfrak{R}_G = \frac{l_{e\_G}}{\mu_0 A_{e\_G}} \tag{16}$$

In the above equations, μ_{r_B}, l_{e_B}, and A_{e_B} denote the the relative permeability, effective path length, and effective cross-section of the block core, respectively.

III. Other Parameters of CM Choke with Block Core
A. Effect of the Number of Turns on the Block Core on the CM and DM Attenuation

To investigate the influence of the number of turns on the block core, we have performed an investigation with different numbers of turns on the block core ($L = 23.8\,\text{mm}$, $W = 5\,\text{mm}$, $H = 12.5\,\text{mm}$) and a fixed number of turns of 13 in each winding on the toroidal core ($d_{out} = 41.8\,\text{mm}$, $d_{in} = 26.2\,\text{mm}$, $h = 12.5\,\text{mm}$). Since $N_T = \frac{N}{2}$ is applicable, the attenuation of this CM choke with a block core is compared with a conventional CM choke with 26 turns. Fig. 4 shows a comparison of the DM inductance between the conventional CM choke and the CM choke with a block core. The relative permeability of the ferrite block core is $\mu_{r\_B} = 2200$, and the relative permeability of the ferrite toroidal core is $\mu_{r\_T} = 4300$. When a block core without turns is inserted into the air gap of the toroid, the DM inductance of the CM choke increases significantly, as mentioned in Tab. I. This unconnected block core could act as a shield to minimize the coupling between two windings on the toroidal core and results in a higher leakage or DM inductance. A comparison between the analytical and measured CM and DM inductance values is shown in Tab. I. The analytical values of the CM and DM inductance are very similar to the measured values, which confirms the correctness of the analytical formula for calculating the inductance of a CM choke with a block core.

As the number of turns on the block core increases, the DM inductance also increases, as shown in Fig. 4a. The effect of the increasing number of turns can be seen in the DM attenuation curves, as shown in Fig. 4b. However, as the number of turns on the block core increases, the winding resistance also increases. This winding resistance and the core loss of the block core also influence the total loss of the CM choke, which was not observed before in the investigation of the DM attenuation when the block core was unconnected. Moreover, when the block core is connected to the winding on the toroidal via different numbers of turns, the losses in the air gap between the toroid and the block core must also be taken into account, because the air gap also contributes to the total losses. To achieve optimum DM attenuation, it would make sense to use a block core without turns or with a low number of turns.

Figs. 4c and 4d show a comparison of the CM inductance and CM attenuation of the conventional CM choke and the single-phase CM choke with different numbers of turns on the block core. No substantial effect can be observed in the CM inductance and CM attenuation of a CM choke with different numbers of turns on the block core because the opposite winding direction on the block core mutually balances the magnetic fluxes for the CM current.

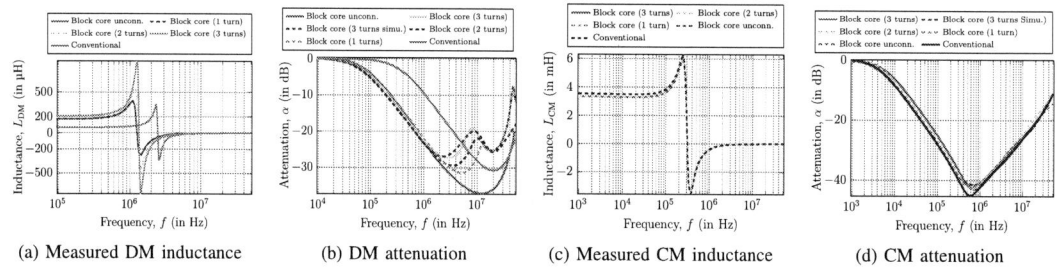

(a) Measured DM inductance   (b) DM attenuation   (c) Measured CM inductance   (d) CM attenuation

Fig. 4. A common-mode choke with a block core; (a, c) Measured DM and CM inductance; (b, d) Measured and simulated DM and CM attenuation

TABLE I
CM AND DM INDUCTANCE OF THE FERRITE CM CHOKE WITH THE FERRITE BLOCK CORE

Structure	N_T	N_B	CM inductance, L_{CM}		DM inductance, L_{DM}		
			Analytical	Measured	Analytical	Measured	% of increased L_{DM} from conventional L_{DM} (measured)
Conventional	$N = 26$	-	3.5 mH (7)	3.7 mH	67.8 µH (11)	69.381 µH	-
Unconnected block core	$N = 26$	-	3.5 mH (7)	3.7 mH	170.9 µH (14)	171.13 µH	146.89 %
Connected block core	13	1	3.5 mH (7)	3.7 mH	175.2 µH (14)	173.15 µH	149.56 %
Connected block core	13	2	3.5 mH (7)	3.6 mH	190.22 µH (14)	193.70 µH	179.183 %
Connected block core	13	3	3.5 mH (7)	3.6 mH	202.33 µH (14)	207.54 µH	199.131 %

B. Effect of the Air-Gap Length on the Single-Phase CM Choke with a Block Core

As the air gap is a high impact in the modified CM choke, the influence of the air gap is investigated for two distinct air gap lengths, 0.5 mm and 0.7 mm, as shown in Figs. 5 (a, b). The increased reluctance reduces the total DM inductance of the circuit, as shown in Tab. II.

The simulated CM and DM inductances are obtained by using the simulation model in ANSYS Q3D and the CM and DM inductances of the simulated model can be obtained using the following equations [11]: $L_{\mathrm{CM}} = \frac{L+M}{2}$, $L_{\mathrm{DM}} = 2(L-M)$. Here, L is the self-inductance of a winding and M is the mutual inductance between the windings. In Tab. II, it can be seen that the numerical results are in a good agreement with the analytical and measurement results.

(a) Air-gap length   (b) Air-gap length
= 0.45 mm   = 0.7 mm

Fig. 5. A CM choke with a block core (a) Simulation model with the air-gap length, $l_{\mathrm{e\_G}}$ = 0.5 mm (b) Simulation model with the air-gap length, $l_{\mathrm{e\_G}}$ = 0.7 mm

TABLE II

CM AND DM INDUCTANCES FOR DIFFERENT LENGTHS OF AIR GAP BETWEEN CM CHOKE AND BLOCK CORE

Length of air gap, $l_{\mathrm{e\_G}}$	N_{T}	N_{B}	CM inductance, L_{CM}			DM inductance, L_{DM}		
			Measured	Analytical	Simulated	Measured	Analytical	Simulated
0.5 mm	13	3	3.6 mH	3.48 mH (7)	3.5 mH	207.54 µH	202.33 µH (14)	201.65 µH
0.7 mm	13	3	3.6 mH	3.48 mH (7)	3.5 mH	204.68 µH	197.81 µH (14)	198.88 µH

C. Influence of the Relative Permeability of the Toroidal Core on Improvement of the DM Inductance

The relative permeability of the core plays one of the most important roles in achieving the required inductance. Two toroidal cores with the same dimensions but with different relative permeabilities of 2200 and 4300 are chosen to observe the effect on DM inductance when the permeability of the toroidal core varies. The block core's relative permeability is kept constant at 2200. When the relative permeability is the same for the block core and toroidal core, the improvement in the DM inductance after the block core is placed in the toroidal core is almost 56 % higher than for the high-permeability toroidal core, as shown in Fig. 6a and Tab. III.

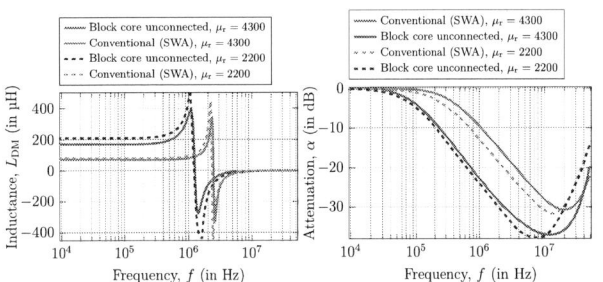

(a) Measured DM inductance   (b) Measured DM attenuation

Fig. 6. Comparison of the DM attenuation and DM inductance of two single-phase CM chokes with a ferrite core of different relative permeability and with an unconnected winding on the block core. Here, SWA refers to the small winding angle between two adjacent turns in each winding on the toroidal core (see Fig. 1)

TABLE III

COMPARISON OF THE MEASURED CM AND DM INDUCTANCES OF CM CHOKE MADE WITH A FERRITE CORE OF DIFFERENT RELATIVE PERMEABILITY AND WITH AN UNCONNECTED BLOCK CORE

μ_{r} of ferrite core	N	CM inductance, L_{CM}	DM inductance, L_{DM}		% of increase of L_{DM}
			Conventional	Block core	
$\mu_{\mathrm{r}} = 4300$	26	3.8 mH	67.8 µH	171.13 µH (14)	135.71 %
$\mu_{\mathrm{r}} = 2200$	26	1.7 mH	72.3 µH	210.9 µH (14)	191.70 %

D. Effect of Different Materials for the Toroidal Core and Block Core

This section contains a brief discussion about the performance of the newly proposed CM choke when the materials used for the toroidal core and block core are different. For this purpose, a nanocrystalline toroidal core and a ferrite block core are used. The block core's placement in the nanocrystalline choke also increases its DM inductance, which rises further in parallel with the numbers of turns. Tab. IV shows a comparison between the inductances of the unconnected block core in the conventional CM choke with a ferrite core and in the conventional CM choke with a nanocrystalline core. The rise in the DM inductance of the CM choke with a nanocrystalline core is somewhat lower compared to the conventional CM choke with a ferrite core. Due to the high permeability of the CM choke with a nanocrystalline core, more flux linkage and less leakage occur in the inner air gap, resulting in a higher coupling coefficient than for the choke with a ferrite core. Therefore, the rate of increase of the DM inductance in the CM choke with a nanocrystalline core and a ferrite block core is low. A nanocrystalline block core in the CM choke with a nanocrystalline core increases the DM inductance as expected. Including the block core in the CM choke with a nanocrystalline toroidal core improves the DM attenuation and shifts the DM self-resonance, both at low frequencies, as shown in Fig. 7c. Further increasing the number of turns on the block core improves the attenuation at low frequencies below the resonant frequency. However, above the resonant frequency, the attenuation is degraded. Since block core made with a ferrite has larger losses than the CM choke with a nanocrystalline core, the losses from the block core may dominate here, worsening the high-frequency attenuation.

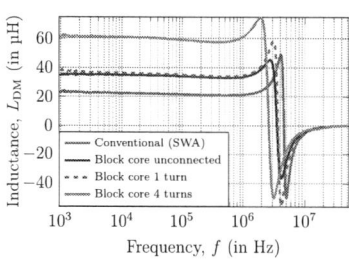
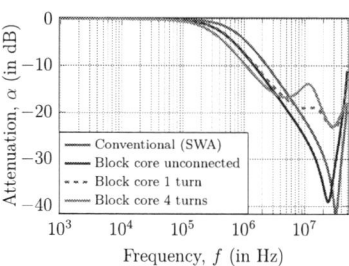

(a) Physical module   (b) Measured DM inductance   (c) Measured DM attenuation

Fig. 7. Nanocrystalline CM choke with 4 turns on block core: (a) Physical module; (b, c) Comparison of the measured DM inductance and DM attenuation of the CM choke with a nanocrystalline toroidal core with different numbers of turns on the ferrite block core

TABLE IV
COMPARISON OF MEASURED CM AND DM INDUCTANCES OF CM CHOKES (FERRITE AND NANOCRYSTALLINE TOROIDAL CORES) WITH AN UNCONNECTED FERRITE BLOCK CORE

Core type	N	CM inductance	DM inductance, L_{DM}		% of increase of L_{DM}
			Conventional	Block core	
Ferrite	26	3.8 mH	67.8 µH	171.13 µH	135.71 %
Nanocrystalline	14	12 mH	21.8 µH	35.9 µH	64.67 %

E. Core Saturation Analysis

Though it is considered that a high DM current and the associated leakage fluxes are the main reasons for the core saturation of a CM choke, the magnetic flux produced by the CM current can also cause the core saturation. Therefore, both CM and DM currents, and the associated CM and DM magnetic flux densities must be considered when analyzing core saturation. To investigate the core saturation of a CM choke using the CM and DM currents, we have conducted a measurement with the DPG10 1500B power choke tester. Fig. 8a shows a comparison of the DM saturation current for different winding strategies of the CM choke built with a ferrite core. As the leakage or DM inductance of the CM choke increases with different winding strategies, it can be seen that the core starts to saturate at a smaller current. However, the CM choke with a nanocrystalline core shows better performance than the choke with a ferrite core in the DM core saturation analysis, as shown in Fig. 8b. The nanocrystalline core has higher saturation flux density and smaller leakage inductance due to the high permeability and high coupling between windings. The ferrite block core with 4 turns in the CM choke with a nanocrystalline toroidal core provides higher DM inductance than the other configurations of the CM choke with a nanocrystalline core up to 35 A, and then starts to decrease. Therefore, it can be concluded that when improving the DM inductance, the number of turns on the block core must be carefully selected. In addition, when the DM current is high, it is better to choose a nanocrystalline core to avoid DM saturation of the core. The CM current that flows through a circuit is usually smaller than the DM current. Figs. 8c and 8d also verify that for both cores, the CM core saturation behavior of the choke with a block core is the same as the conventional CM choke without a block core.

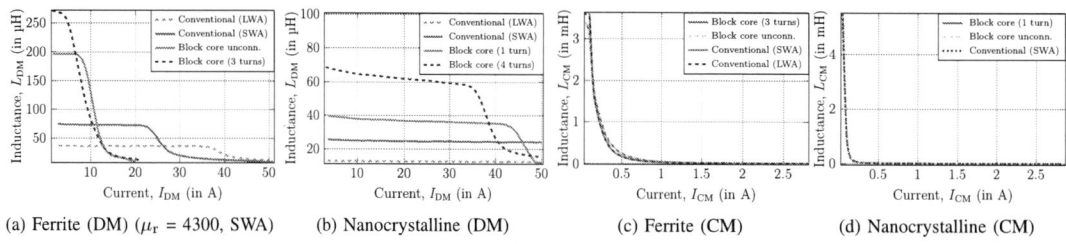

(a) Ferrite (DM) ($\mu_r = 4300$, SWA)   (b) Nanocrystalline (DM)   (c) Ferrite (CM)   (d) Nanocrystalline (CM)

Fig. 8. Comparison of the incremental inductance versus saturation current: (a) Ferrite CM choke (DM configuration); (b) Nanocrystalline CM choke (DM configuration); (c) Ferrite CM choke (CM configuration); (d) Nanocrystalline CM choke (CM configuration). Here, SWA and LWA mean small and large winding angles.

III. Performance of the CM Choke with a Block Core in an EMI Filter

To evaluate the performance in attenuating EMI noise, we inserted the CM choke with the block core into a ready built EMI filter. Fig. 9a shows the existing EMI filter, which is used to compare different winding strategies for a CM choke on the same PCB and under the same conditions. This filter works as a CLCL topology in the DM configuration and as an LCL topology in the CM configuration. Fig. 9d shows the equivalent electrical circuit diagram of the existing EMI filter. The magenta box in the circuit indicates where the first CM inductor is replaced for these measurements. Figs. 9b and 9e show that there is no significant difference in the CM attenuation of the EMI filter with the conventional CM choke and alternatively with the choke with the unconnected block core. Furthermore, it also proves that an increase in the number of turns on the block core does not have any remarkable effect on the CM attenuation of the filter. The DM attenuation of the filter shown in Figs. 9c and 9f indicates that the insertion of the block core serves its purpose quite well. Fig. 9c shows that for the ferrite choke with block core (3 turns), the first resonant frequency shifted to 6 kHz from 11 kHz. In addition, approximately 15 dB to 17 dB more attenuation can be achieved at the lower frequencies using a CM choke with a ferrite core (with 3 turns on the block core). Using the nanocrystalline CM choke with 4 turns on the block core, as shown in Fig. 9f, about 10 dB to 12 dB more attenuation can be obtained than when using the conventional CM choke, without any significant change in the high-frequency performance.

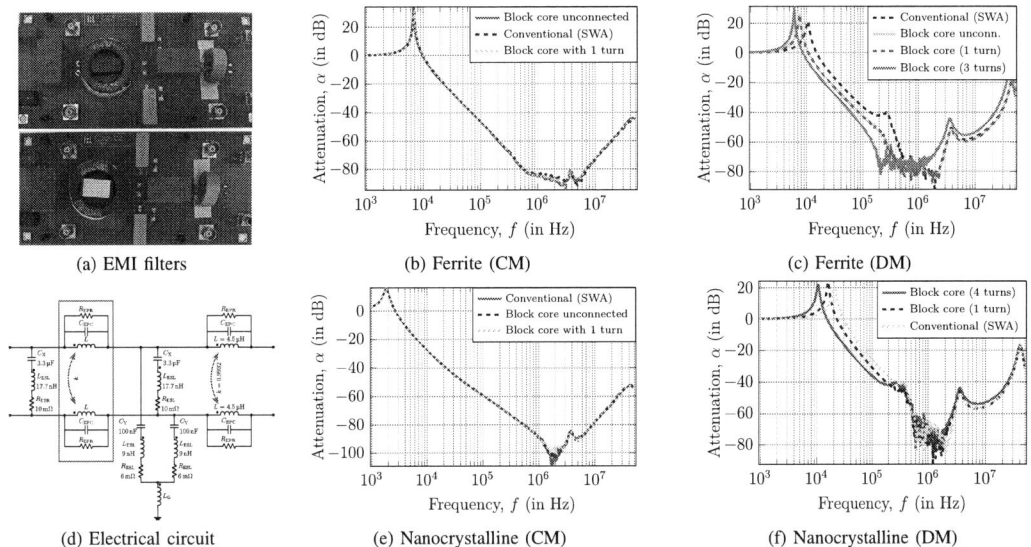

Fig. 9. Measurement of different CM chokes using an existing EMI filter: (a) EMI filter with the CM choke (made with ferrite and nanocrystalline cores + the ferrite block core); (b, e) Comparison of the CM attenuation with the conventional CM choke (made with a ferrite core) and the CM choke with a nanocrystalline core (with block core made from ferrite) in the existing EMI filter; (c, f) Comparison of the DM attenuation with the conventional CM choke (made from ferrite) and the CM choke with a nanocrystalline core (with block core made from ferrite) in the existing EMI filter; (d) An equivalent electrical circuit diagram of the existing EMI filter

IV. Possible Structure of a Three-Phase CM Choke with Integrated DM Inductance

For a three-phase CM choke, the block core must be Y-shaped to fit between the three windings of the conventional three-phase CM choke. The three legs of the Y-core should be spaced 120° apart to ensure symmetry. This newly proposed choke consists of six copper windings on the toroidal core and three on the Y-core, as shown in Fig. 10a. To maintain the symmetrical structure and symmetrical coupling between the windings, the six windings on the toroidal core must have an identical number of turns. The installation of a Y-core directly increases the leakage inductance by reducing the coupling between the windings.

A. Magnetic Flux Direction in a Three-Phase CM Choke with a Y-Core

The three-phase CM choke with Y-core works in the same way as the single-phase CM choke with block core, except that it has nine windings instead of six. The magnetic flux directions of the CM and DM configurations of this newly proposed three-phase choke are mentioned in Figs. 10b and 10c, respectively. In these figures, the violet, blue and green arrows denote the currents for phases U, V, W, respectively, and red arrows indicate the flux direction. As shown in Fig. 10b, the flux induced by the CM current is oriented anti-clockwise, and for the two windings on the toroid, the fluxes add to attenuate the CM noise. As seen in the single-phase CM choke with the block core, the induced fluxes in the Y-core also cancel each other out in the CM configuration, without affecting the CM performance of the three-phase CM choke. In contrast, in the DM configuration, the magnetic fluxes in the Y-core help to create magnetic paths through the air gap of the three-phase choke, as shown in Fig. 10c, thereby leading to more leakage flux and more DM inductance. However, fluxes φ_{B2} and φ_3 are in the opposite direction and can offset some of the flux, weakening the magnetic path through winding B2 for flux φ_3. Nevertheless, the other two windings on the Y-core help minimize this problem by keeping the magnetic path intact and providing the expected DM inductance.

B. Magnetic and Electrical Circuits of a Three-Phase CM Choke with a Y-Core

Based on the magnetic circuit discussed in Section (II-B), the CM and DM magnetic circuits are drawn in Figs. 11a and 11c for the three-phase CM choke with a Y-core, respectively. In the CM configuration, as shown in Fig. 11a, the induced fluxes φ_1, φ_1'', φ_2, φ_2'', φ_3, and φ_3'' in the toroidal core sum up to eliminate the CM noise. Furthermore, the induced fluxes φ_{B1}, φ_{B2}, and φ_{B3} in the Y-core cancel each other out without affecting the CM inductance of the choke. In the DM configuration shown in Fig. 11c, the fluxes φ_2, φ_2'', φ_3, and φ_3'' add up. But most of these fluxes are canceled out when they encounter the fluxes φ_1, φ_1'', and the fluxes that are not canceled out generate some leakage flux. However, as shown in Fig. 11c, the fluxes φ_{B1}, φ_{B2}, and φ_{B1}, φ_{B3} establish magnetic paths through the air gap for the leakage flux of the windings T_1, T_1'', T_2 and T_2''. These leakage fluxes improve the DM inductance. As discussed in Section (II-B) an equivalent electrical circuit for the three-phase CM choke with different numbers of turns on the Y-core can be easily obtained from the equivalent magnetic circuit shown in Fig. 11.

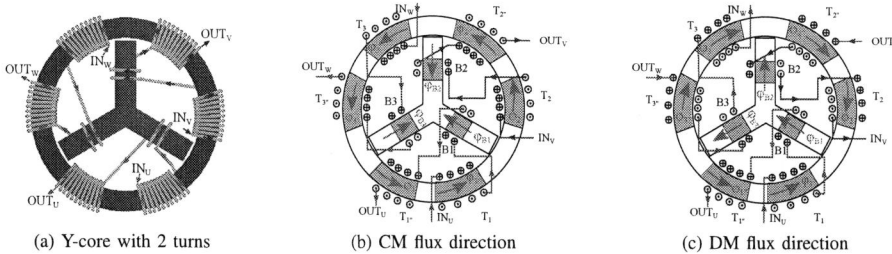

(a) Y-core with 2 turns   (b) CM flux direction   (c) DM flux direction

Fig. 10. Three-phase CM choke with a Y-core: (a) Simulation model of the three-phase CM choke with a Y-core; (b) CM flux direction; (c) DM flux direction

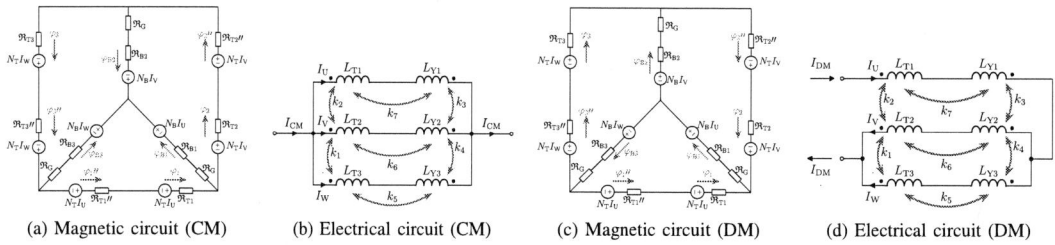

(a) Magnetic circuit (CM)   (b) Electrical circuit (CM)   (c) Magnetic circuit (DM)   (d) Electrical circuit (DM)

Fig. 11. Magnetic and electrical circuits for the three-phase CM choke with a Y-core

C. CM and DM Inductance of a Three-Phase Common-Mode Choke with a Y-core

The three-phase CM choke with the Y-core model shown in Fig. 10a is simulated in ANSYS Q3D to extract the self-inductance of the each winding and obtain the coupling between the windings. As with the newly designed single-phase choke $N_T = \frac{N}{2}$, which is discussed in Section (II-C), the newly proposed three-phase choke is built with 11 turns in each winding to fairly compare it with the conventional choke with 22 turns. Then the total DM and CM inductances of the choke using the extracted parameters from the 3D-FEM simulation model are obtained using Equations (9) and (10) as proposed in [7], respectively. Subsequently, to verify the simulated CM and DM inductances, the analytical values of the CM and DM inductances are calculated using Equations (8) and (14). The equations used for a single-phase CM choke also apply to the three-phase CM choke with a Y-core since each winding of the choke comprises two windings on the toroidal core and one winding on the Y-core. This winding structure is similar to that of the single-phase CM choke with a block core, and the only change is in the calculation of the total reluctance of the choke. In the three-phase CM choke with a Y-core, the Y-core consists of three parts, and there are three air gaps active between the toroid and the Y-core. Therefore, the total reluctance of this newly proposed choke can be expressed as follows:

$$\mathfrak{R}_C = 3\mathfrak{R}_B + 3\mathfrak{R}_G + \mathfrak{R}_T \tag{17}$$

A comparison between the analytical and simulated CM and DM inductances of the conventional three-phase CM choke and the three-phase CM choke with the Y-core is provided in Table V.

TABLE V

COMPARISON OF THE ANALYTICAL AND SIMULATED CM AND DM INDUCTANCES OF A CONVENTIONAL THREE-PHASE CM CHOKE AND THREE-PHASE CM CHOKE WITH A Y-CORE

Inductance	Conventional (N=22)		With Y-core (N_B=2, N_T=11)	
	Analytical	Simulation	Analytical	Simulation
k		0.9964		0.9931
CM inductance, L_{CM}	8.2 mH (7)	8.22 mH	8.2 mH (7)	8.16 mH
DM inductance, L_{DM}	43.79 μH (11)	44.02 μH	83.51 μH (14)	83.64 μH

When calculating the analytical DM inductance of the newly proposed choke, Equation (17) supplies the value of $\mathfrak{R}_C$ in Equation (14).

V. Performance of the Three-Phase CM Choke with a Y-core in a Filter

To investigate the performance of a three-phase CM choke with a Y-core in an EMI filter, we performed a simulation in ANSYS circuit simulator because the actual physical Y-core is not readily available on the market. The physical filter and the schematic diagram of the electrical circuit of the three-phase EMI filter are shown in Fig. 12a and Fig. 12b, respectively. In the beginning, the simulation results are validated using the measurement results from a conventional three-phase CM choke in a three-phase EMI filter, as shown in Fig. 12a. Fig. 12c shows that a Y-core in a three-phase CM choke does not affect the CM attenuation of the EMI filter, as with the block core in the single-phase CM choke. Nevertheless, the simulation result shown in Fig. 12d indicates that this newly proposed three-phase CM choke will also improve the DM attenuation of the EMI filter. So, it can be summarized that a three-phase CM choke with a Y-core could be an exciting new choice for a three-phase EMI filter.

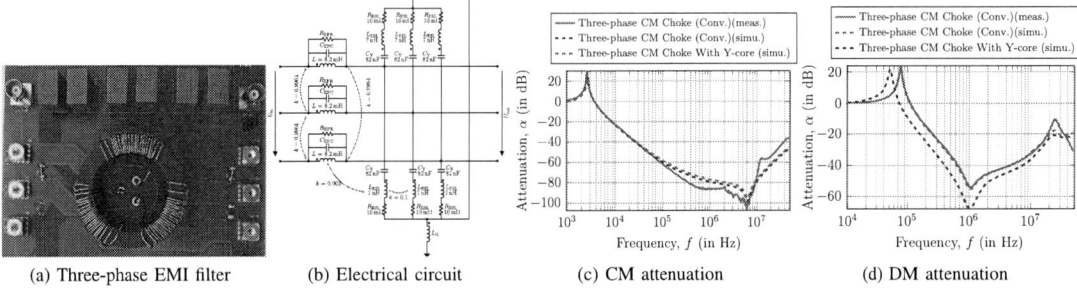

(a) Three-phase EMI filter   (b) Electrical circuit   (c) CM attenuation   (d) DM attenuation

Fig. 12. (a) Conventional three-phase CM choke in the three-phase EMI filter; (b) Circuit diagram of the electrical circuit of the three-phase EMI filter; (c, d) Measured and simulated CM and DM attenuation of the three-phase conventional CM choke and the three-phase CM choke with a Y-core

VI. Conclusion

A purpose-built common-mode choke designed well for the application is essential to achieve the desired attenuation from an EMI filter. To improve the DM inductance of a single-phase choke, the authors of [5] have proposed a CM choke with an integrated DM inductance, created with an additional block core. This configuration has been analyzed, designed and tested, and the performance has been evaluated as a standalone component and integrated into an EMI filter. The concept successfully increases the DM inductance up to a certain DM saturation current. Different core materials have been evaluated. It has been shown that the CM choke with a nanocrystalline core shows better performance than the choke with a ferrite core in the DM core saturation analysis. The DM attenuation of an EMI filter can be improved this way, until the core saturates. The numerical and experimental analysis of different parameters that affect the performance of the newly proposed choke has been discussed. It was noted that the numerical results are in a good agreement with the analytical and measurement results. Moreover, a proposal for a three-phase CM choke has been made. Its magnetic flux distribution, magnetic circuit and electrical circuit have been explained and discussed in this paper. The performance of a three-phase EMI filter with a three-phase CM choke incorporating a Y-core has also been discussed.

REFERENCES

[1] M.J. Nave, "On modeling the common mode inductor," *in IEEE International Symposium on Electromagnetic Compatibility*, 1991.

[2] L. Dehong, J. Xanguo, "High frequency model of common mode inductor for EMI analysis based on measurements," *in 3rd International Symposium on Electromagnetic Compatibility*, 2002.

[3] K. Kostov and J. Kyyra, "Common-mode choke coils characterization," *in 13th European Conference on Power Electronics and Applications*, 2009.

[4] A. Massarini, M. K. Kazimierczuk and G. Grandi, "Lumped parameter models for single- and multiple-layer inductors," *in PESC Record. 27th Annual IEEE Power Electronics Specialists Conference*, 1996.

[5] J. Borsalani, A. Dastfan, J. Ghalibafan, "An Integrated EMI Choke With Improved DM Inductance," *in IEEE Transactions on Power Electronics*, 2021.

[6] Fang Luo, Dushan Boroyevich, Paolo Mattevelli, Khai Ngo, David Gilham and Nicolas Gazel "An integrated common mode and differential mode choke for EMI suppression using magnetic epoxy mixture," *The 26th Annual IEEE Applied Power Electronics Conference and Exposition (APEC)*, 2011.

[7] M. Ali, T. Brinker, J. Friebe, A. Mertens, "Analysis of EMI Filter Attenuation under the Influence of Parasitic Elements of Components and their Mutual Couplings," *The 23rd European Conf. on Power Electronics and Applications (ECCE Europe 2021)*, September 2021.

[8] M. Ali, J. Friebe, A. Mertens, "Design and Optimization of Input and Output EMI Filters under the Influence of Parasitic Couplings," *The 23rd European Conf. on Power Electronics and Applications (ECCE Europe 2021)*, September 2021.

[9] H. Kim et al., "A new asymmetrical winding common mode choke capable of attenuating differential mode noise," *in 8th International Conference on Power Electronics - ECCE Asia*, 2011.

[10] D. Xu, C. K. Lee, S. Kiratipongvoot and W. M. Ng, "An Active EMI Choke for Both Common- and Differential-Mode Noise Suppression," *in IEEE Transactions on Industrial Electronics*, vol. 65, no. 6, pp. 4640-4649, June 2018, doi: 10.1109/TIE.2017.2764859.

[11] M. Ali, J. Friebe, A. Mertens, "Simplified Calculation of Parasitic Elements and Mutual Couplings of Wide-bandgap Power Semiconductor Modules," *in European Conference on Power Electronics and Applications (EPE'20 ECCE Europe)*, 2020.

Implementation options of a fully SiC Buck-CSI for advanced motor drive application

Yonghwa Lee and Alberto Castellazzi
Solid-State Power Processing (SP2) Lab
Kyoto University of Advanced Science
18, Gotanda-cho Yamanouchi, Ukyo Ward,
Kyoto, Japan, 615-8577
E-Mail: 2021md04@kuas.ac.jp

Keywords

«Current Source Inverter (CSI)», «Silicon Carbide (SiC)», «High-speed drive», «High frequency power converter», «High power density systems»

Abstract

This paper discusses the design option for a current source inverter for a 15kW high-speed machine with a wide-band-gap technology-based electric motor drive. Both control and modulation options are considered, as well as the integration of silicon carbide (SiC) bi-directional switches to yield high-performance bi-directional switches, as required by the topology. Recently, by taking advantage of the high-speed switching performance of SiC, a simple and robust control design method without voltage sensing and the Two-Third Modulation (TTM) method, which has the benefit of high efficiency, have been proposed. Based on this recent work, this paper proposes the simplified current-sensor-less controller of the CSI-fed PMSM. Also, this paper compared the characteristics of the TTM with the conventional PWM methods by simulation models.

Introduction

Research aiming at increasing efficiency and power density of power converters and machines is ubiquitous. A game-changing approach which has been receiving renewed and increasing attention in the very recent past is Current Source Inversion (CSI, or Current DC-Link Inversion). In this case, the square-wave PWM modulation of the output voltage followed by a filtering action is replaced by the direct synthesizing of a sinusoidal output voltage. In the pre-WBG era, application scenarios where CSIs could provide significant benefits were sparse and of limited relevance. Three-phase CSI topologies were mainly limited to boost (voltage step-up) and boost-buck (step-up step-down), but the proposed CSI concepts ended up mainly in obscurity or very niche applications. In comparison to VSI, the major benefits of CSI are:

■ removal of dV/dt stress, eddy currents and common mode noise on the load;
■ greatly improved harmonic signature, with virtual removal of any high-frequency harmonic components for greatly reduced motor losses;
■ extremely high switching frequency capability for a given power rating, due to major reduction in the switching losses resulting from the twofold ability of employing lower-voltage-rated devices and switching them on average at lower peak voltages (decreasing sinusoidally to zero at the fundamental frequency);
■ simpler circuit design and higher power densities.

So, the interest in CSI can be regarded as tightly related to the advent of WBG technology and, at present, the struggle for joint pursuit of increased efficiency and power density is renewing the momentum in CSI research [1-2]. Inherent advantageous features of CSI, Fig. 1 (a) such as joint inversion and voltage boosting capability, low output voltage harmonic distortion, removal of voltage DC-link and the related bulky capacitor technologies, are all very important for enabling disruptive progress beyond state-of-the-art in SiC-based power conversion, especially in the rapidly growing application domain of high-speed drives, as used increasingly in transportation.

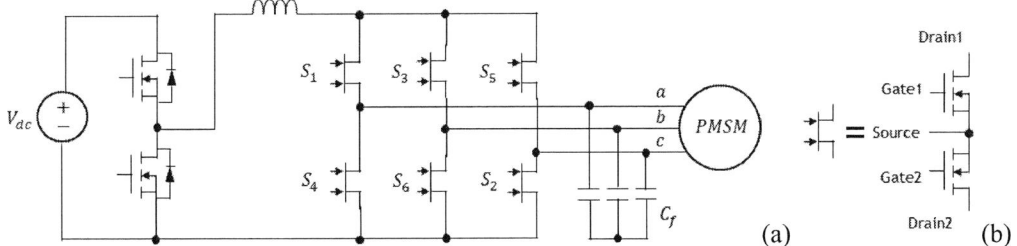

Fig. 1: Circuit schematic of Buck-CSI based motor-drive, with detail of bi-directional switch (BDS) implementation for the case of power MOSFETs (common source configuration).

Here, the aim is to go an important step further in enhancing efficiency and power density, with particular emphasis on future integrated electric drives: that will be pursued by original topology re-design and driving/control solutions and bespoke functional-structural integration. In particular, the converter requires a particular switch configuration: a bi-directional switch, capable of blocking voltage and conducting current in both directions (i.e., irrespective of the device terminals to the voltage and current are applied). The BDS schematic is shown in Fig. 1 (b); for the case of MOSFETs in common source configuration, by far the most practical implementation and favored solution. If built with discrete devices, the BDS cannot realistically be operated at the target high frequencies and thus enable disruptive progress in efficiency and power density: integrated SiC BDS`s are needed. Specifically, the BDS features a source-to-source connection between the two transistors. Whereas examples of monolithically integrated low-current GaN BDS`s have been shown, monolithic integration of SiC MOSFET is not viable, the difference being that GaN transistors are lateral devices, whereas SiC MOSFETs are vertical, with the source electrode corresponding to the semiconductor chip top surface. So, in view of the high integration aims of the project, design concepts and assembly solutions enabling the reliable top-to-top mounting of the devices are required, a world-first target in the framework of a technology development exercise targeting high-voltage large current capable industrial-grade solutions.

Pulse Width Modulation

In order to control gating signals for the switches of CSI, the space vector pulse width modulation (SVPWM) is developed [3]. The conventional SVPWM for VSIs are needed the dead time to avoid the arm short-circuit, which is the bottom, and the top switch turned on together at the same time. In contrast, the current source inverter requires overlap time during switching events when one device is being turned off while another is being turned on to ensure that a conducting path always exists for the dc-link inductor current to avoid the dangerous overvoltage from transient open circuits [4]. Therefore, the switching restraint for SVPWM of CSIs is that two switches should be conducted at any time of the current vectors.

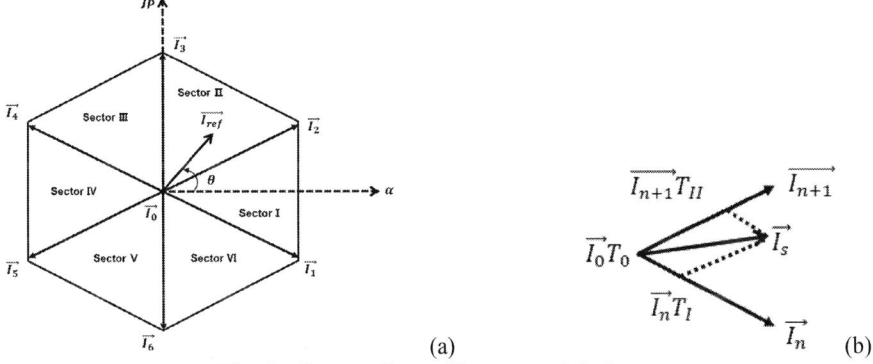

Fig. 2: Current Space Vector modulation.

As shown in Fig. 2(a), when operating CSI by SVPWM, the output current vector is divided into six sectors. A reference-current vector may be output by combining two active vectors and one zero vector for each vector, Fig. 2(b). Each active vector has a magnitude of a DC Link current, and if the DC-current may be controlled according to a target current, an optimized current vector may be output.

Table I: Switching state in CSIs and corresponding phase switching functions and switching space vectors

Three Zero Current Vectors					Six Active Current Vectors				
Vector	On Switches	$\vec{\iota_a}$	$\vec{\iota_b}$	$\vec{\iota_c}$	Vector	On Switches	$\vec{\iota_a}$	$\vec{\iota_b}$	$\vec{\iota_c}$
$\vec{I_0}$	S1, S4	0	0	0	$\vec{I_1}$	S6, S1	I_{dc}	$-I_{dc}$	0
$\vec{I_0}$	S2, S5	0	0	0	$\vec{I_2}$	S1, S2	I_{dc}	0	$-I_{dc}$
$\vec{I_0}$	S3, S6	0	0	0	$\vec{I_3}$	S2, S3	0	I_{dc}	$-I_{dc}$
					$\vec{I_4}$	S3, S4	$-I_{dc}$	I_{dc}	0
					$\vec{I_5}$	S4, S5	$-I_{dc}$	0	I_{dc}
					$\vec{I_6}$	S5, S6	0	$-I_{dc}$	I_{dc}

Looking at Table 1, CSIs have six active vectors and three zero vectors, and two switches must be operated in pairs to output them. The product of the current space vector and switching time is equated with the sum of the products of corresponding. The dwell time can be calculated with space vectors, and corresponding time intervals follow, where n is the number of the active vector. ($1 \leq n \leq 6$):

$$\vec{I_s}T_s = \vec{I_n}T_I + \vec{I_{n+1}}T_{II} + \vec{I_0}T_0 \tag{1}$$

$$T_s = T_I + T_{II} + T_0 \tag{2}$$

$$m_a = \frac{I_s^*}{I_{dc}}, \quad 0 \leq m_a \leq 1. \text{ where } I_s^* \text{ is desired current} \tag{3}$$

$$T_I = m_a \sin\left(\frac{\pi}{6} - \left(\theta - \frac{(k-1)\pi}{3}\right)\right)T_s, \tag{4}$$

$$T_{II} = m_a \sin\left(\frac{\pi}{6} + \left(\theta - \frac{(k-1)\pi}{3}\right)\right)T_s \tag{5}$$

$$T_0 = T_s - T_I - T_{II} \quad k \text{ is the sector number. } (1 \leq k \leq 6) \tag{6}$$

In order to output the active vector and the zero vector, a sequence with a specific order must be created, as shown in Fig. 3. CSIs need to generate a specific switching sequence to implement SVPWM, and the losses and quality of output current may change according to the number and order of the sequences. Usually, five or seven sequences are used, but it can be seen that it is advantageous to have five switching sequences in terms of switching loss through the study [5].

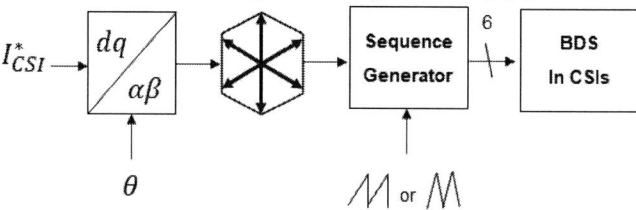

Fig.3: Space Vector PWM Generating Method of Current Source Inverters

Besides, according to a recent study [6], the six methods in Table 2 were discussed, and it can be seen that 8% of conduction loss and 86% of switching loss can be reduced when TTM is applied according to the modulation index. However, a DC-DC converter and synergetic control are required in this method.

Table 2: The state sequences

Modulation	State Sequence	Note
Asymmetric PWM	$[\overrightarrow{I_0}]\,[\overrightarrow{I_n}]\,[\overrightarrow{I_{n+1}}]$	CW-asymmetric pwm
	$[\overrightarrow{I_0}]\,[\overrightarrow{I_{n+1}}]\,[\overrightarrow{I_n}]$	CCW-asymmetric pwm
Symmetric PWM	$[\overrightarrow{I_n}]\,[\overrightarrow{I_0}]\,[\overrightarrow{I_{n+1}}]\,[\overrightarrow{I_0}]\,[\overrightarrow{I_n}]$	Zero Vector-symmetric pwm
	$[\overrightarrow{I_0}][\overrightarrow{I_n}][\overrightarrow{I_{n+1}}][\overrightarrow{I_n}][\overrightarrow{I_0}]$	In Vector-symmetric pwm
	$[\overrightarrow{I_0}]\,[\overrightarrow{I_{n+1}}]\,[\overrightarrow{I_n}]\,[\overrightarrow{I_{n+1}}]\,[\overrightarrow{I_0}]$	In+1 Vector-symmetric pwm
TTM	$[\overrightarrow{I_n}]\,[\overrightarrow{I_{n+1}}]$	without zero current vector

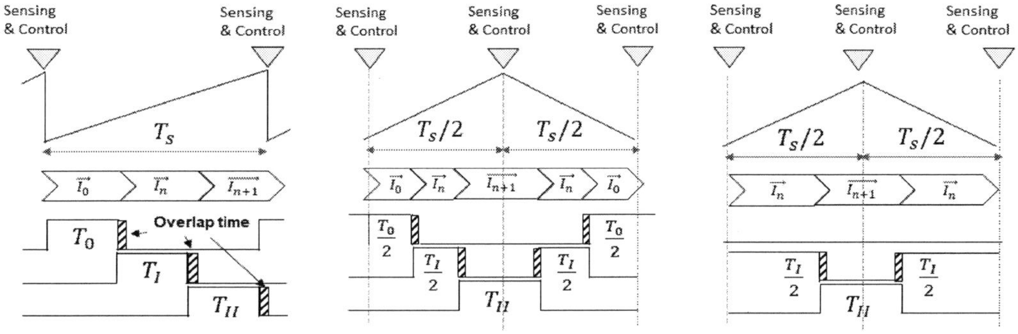

Asymmetric PWM (a)   Symmetric PWM (b)   Two-Third Modulation (c)

Fig. 4: (a) is shown that features of characteristic of Asymmetric PWM and (b) is indicated features of Symmetric PWM (c) is presented the characteristic of Two-Third Modulation.

As shown in the fig. 4, Symmetric PWM and TTM can be designed to facilitate the operation of sampling and control twice in one cycle period. In addition, since a switching operation of a power semiconductor can affect the sensing precise for current and voltages, which is essential in respect of the control stability in practice, the asymmetric PWM must find a way to avoid the switching noise. Therefore, it may be considered that the symmetric PWM and TTM are more suitable for implementing a high-speed machine in practice.

In general, the CSI needs the power converter, which can control the dc-link current, since it prevents overcurrent and damage from the failure of switches. Thus, the dc-link current in the CSI can be controlled using the power converter to associate with the PWM method. Due to the existence of the zero current vector in the symmetric PWM, the limitation of the maximum modulation index should be considered when the dc-link current controls by the power converter. That is why the controlled dc-link current by the power converter must be higher than the desired output current of the CSI. However, in TTM, the controlled dc-link current can close to the desired output current of the CSI since the TTM only needs to generate the active current vector.

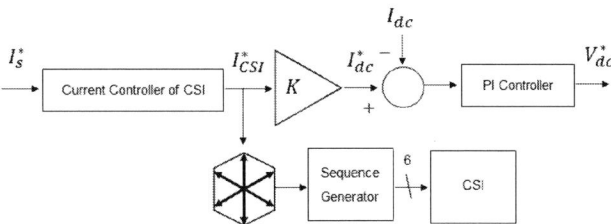

Fig. 5: The control concept for dc-link current controller; K is the margin for the limitation of pulse-width modulation method.

As a result of the simulation, Fig. 6 shows that the zero current vector was eliminated in the Two-Third Modulation compared to the conventional Symmetric PWM method. Additionally, due to the current of dc inductor ($i_{dc\ inductor}$) amplitude having a slight difference, the losses of using TTM might be decreased than the conventional PWM method.

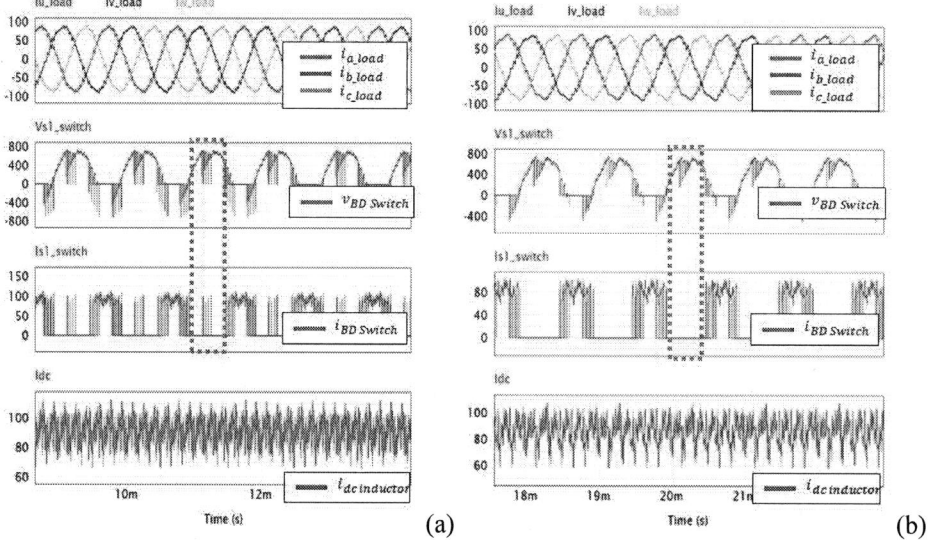

(a) (b)

Fig. 6: The results of the simulation model (a) is shown that features of control characteristic of Symmetric PWM and (b) is indicated control features of Two-Third Modulation.

Design of the current controller for the CSI-fed PMSM

Due to their higher efficiency and power density, permanent magnet synchronous motors (PMSM) have replaced induction motors in many industrial applications, such as home applications and electric vehicles. That is why this paper proposes the control algorithm for the CSI-fed PMSM system.

The Current Source Inverter causes resonant frequency due to the Filter Capacitor connected to the output terminal. Previously, the Proportional and Resonant (PR) controller or the multi-nested-loop controller was proposed to control the LC resonant frequency. Since the system parameter error heavily influences these controllers, the complex vector controller and Active Damping Control Methods using the two-stage modeling of CSI-fed PMSM system have been studied to improve response performance in previous studies [7].

Recently, as high-speed switching became possible through the application of the Wide-Band-Gap switch, a method of simplifying and robust controller was proposed [8]. In particular, the multi-loop-nested controller required the voltage sensor of the filter capacitor and ac current sensors. However, it was possible to design a controller using only current sensors without additional voltage sensors in recent research [8].

On the other hand, the current sensor is generally more expensive than the voltage sensor and inhabits a slightly larger volume in the power electronics. In the case of Voltage Source Inverter, the method of reducing the number of current sensors using DC-Link current sensing has been widely proposed. Therefore, this paper proposes a design method that can simplify the controller using only the voltage sensor of the filter capacitor instead of the load-side ac current sensors.

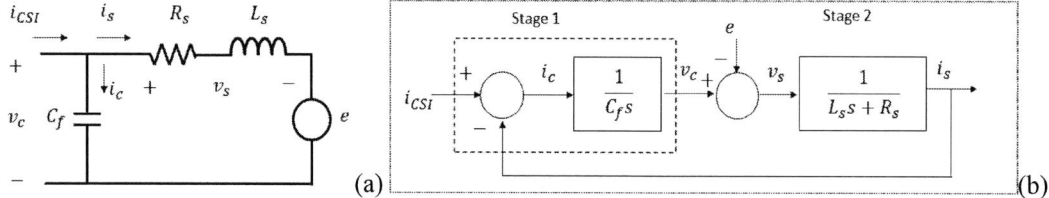

Fig. 7: (a) Equivalent circuit, and (b) Block diagram of CSI-fed PMSM in a stationary reference frame

First, for controller design, the system characteristics of the current source inverter with PMSM can be analyzed using two-stage modeling as in the previous study [7]. As illustrated in Fig, 7: Cf is the output filter capacitor, e is the machine electro-motive force, Ls the stator inductance, and Rs the stator resistance. Therefore, the plant of the system can represent by a second-order system (7).

$$G_p(s) = \frac{i_s(s)}{i_{CSI}(s)} = \frac{\dfrac{1}{L_s C_f}}{s^2 + s\dfrac{R_s}{L_s} + \dfrac{1}{L_s C_f}} \qquad (7)$$

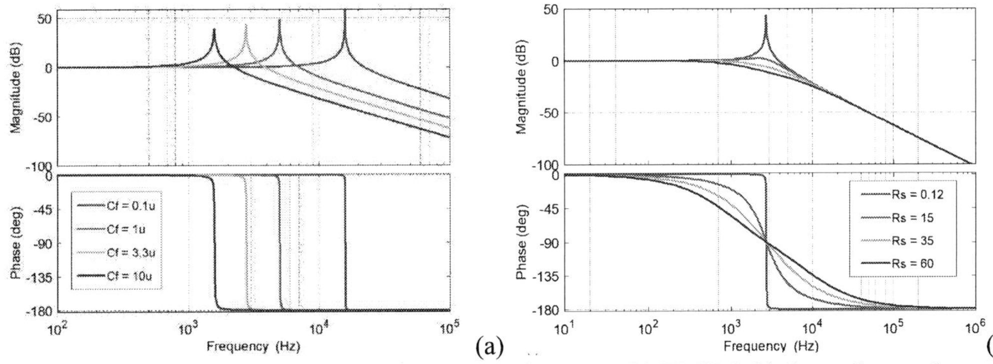

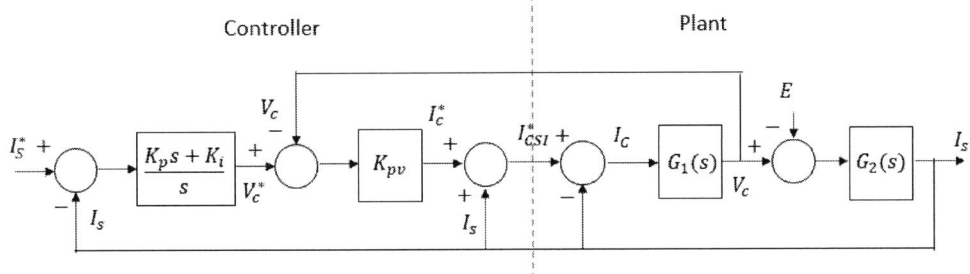

Fig. 8: The bode plot for the plant of current source inverter with PMSM, (a) depending on the output filter capacitance C_f, (b) depending on the stator resistance Rs.

Through the bode plot in Fig. 8, the resonance frequency can be adjusted according to the filter capacitance, and the resonance component can be controlled through the high-speed controller, so it is better to set it low enough compared to the switching frequency. Although, the gain and phase delay will be a problem if the resonant frequency is too low. In addition, if the stator resistance Rs can be increased, the resonance component can be damped, as shown in Fig. 8(b). By the way, in practice, increasing the resistance of the PM machine is not permitted because of the motor losses. Therefore, the controller could be designed to have the response characteristics of the desired 2nd-order system through the pole-zero cancellation technique as follows.

Fig. 9: The conventional multiple-nested-loop control design of CSI-fed PMSM

If the PI gain (K_p, K_i) are set as ($\widehat{L}_s\, \omega_{cc}$, $\widehat{R}_s\, \omega_{cc}$) respectively, and K_{pv} is $\widehat{C}_f \omega_f$, the transfer function can be induced as (10) where the estimated parameter(*hat*) are well matched.

$$G_1(s) = \frac{1}{sC_f} \qquad (8),$$

$$G_2(s) = \frac{1}{sL_s + R_s} \qquad (9)$$

$$CL(s) = \frac{I_s}{I_s^*} = \frac{\omega_f \omega_{cc}}{s^2 + \omega_f s + \omega_f \omega_{cc}} \qquad (10)$$

In recent research [8], a simple and robust voltage sensor-less controller was designed using the effect of the additional virtual resistance as follows in Fig. 10. C(s) is the proposed controller in the research [8].

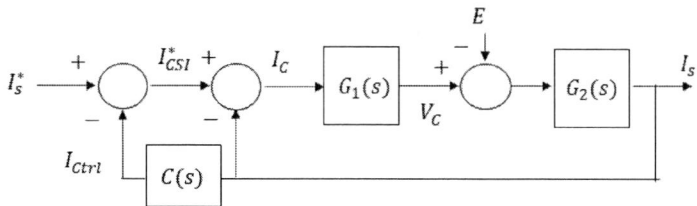

Fig. 10: The reference controller using only ac current sensor in the [8].

$$C(s) = \frac{I_{Ctrl}}{I_S} = sR_V\widehat{C_f} \quad (11)$$

$$CL(s) = \frac{I_S(s)}{I_S^*(s)} = \frac{G_p(s)}{1 + G_p(s)C(s)} = \frac{\dfrac{1}{L_sC_f}}{s^2 + s\dfrac{(R_s + R_V)}{L_s} + \dfrac{1}{L_sC_f}} \quad (12)$$

In order to obtain the fastest response corresponded to the critically damped system that contains no overshooting, the pole locations of the second-order system can be designed as:

$$Pole = -\frac{R_s}{2L_s} \pm \frac{1}{2}\sqrt{\frac{R_s^2}{L_s^2} - \frac{4}{L_sC_f}} \quad (13)$$

$$\frac{R_s^{*2}}{L_s^2} - \frac{4}{L_sC_f} = 0, \quad R_s^* = 2\sqrt{\frac{L_s}{C_f}} \cong 35.0\,\Omega \quad (14)$$

$$R_v = R_s^* - R_s = 34.88\,\Omega \quad (15)$$

As mentioned above, the use of voltage sensors may have advantages over the use of current sensors in the aspect of price and volume. Therefore, in this paper, the controller with the same response characteristics as discussed above [8] can be designed as follows.

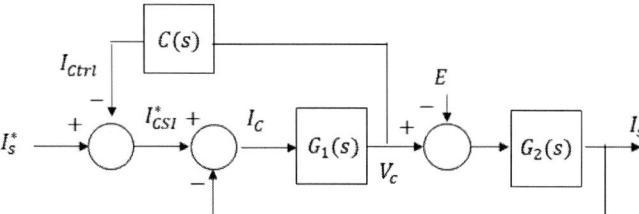

Fig. 11: The block diagram of proposed controller and system which has only voltage sensors.

$$C(s) = \frac{I_{Ctrl}}{V_C} = \frac{sR_V\widehat{C_f}}{s\widehat{L_s} + \widehat{R_s}}, \quad (16)$$

$$CL(s) = \frac{I_S(s)}{I_S^*(s)} = \frac{\dfrac{G_1(s)\,G_2(s)}{1 + G_1(s)\,C(s)}}{1 + \dfrac{G_1(s)G_2(s)}{1 + G_1(s)\,C(s)}} = \frac{\dfrac{1}{L_sC_f}}{s^2 + s\dfrac{(R_s + R_V)}{L_s} + \dfrac{1}{L_sC_f}} \quad (17)$$

There is a difference between the previous study [8] and the proposed controller. Since the reference controller has the derivative, the filtered discrete derivative is considered when expanding in the discrete-time domain because the filter can reduce the emitting disturbance, such as sensing noise. In contrast, the proposed controller in this paper basically has a feature like the High Pass Filter. However, due to the nature of the High Pass Filter, it may have the DC offset problem, so consideration of this should be sufficiently studied when expanding the controller in the discrete-time domain.

Table 3: The parameters of the simulation model

Parameters	Value	Parameters	Value
CSI output power	15kW	Rotor Type	SPM
Pole / Slots	12 / 18	Motor Phase inductance	1.0 mH
Input DC Voltage	500 Vdc	Motor Phase resistance	0.12 Ω
DC link Inductance	600uH	Switching Frequency in CSI	75kHz
Switching Frequency in DC-DC Converter	75kHz	Filter capacitor	3.3uF
Overlap Time	200ns	Fundamental frequency	540Hz (electrical)

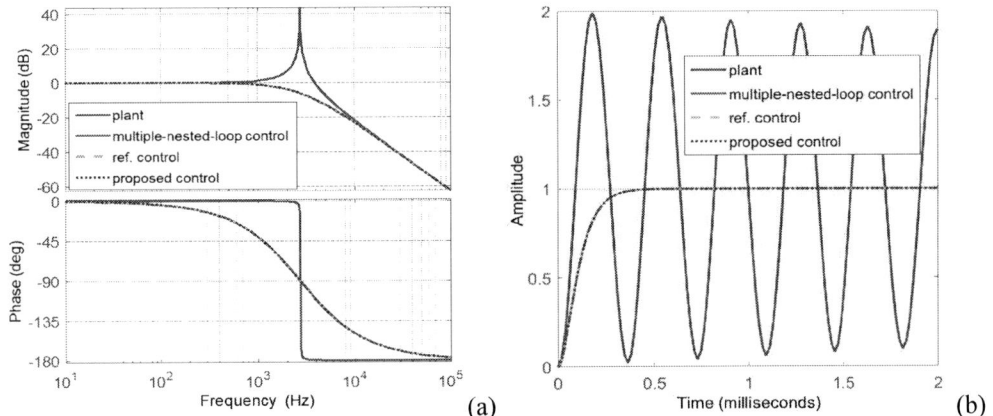

Fig. 12: (a) Bode plot and (b) unit step response for comparison the characteristic of system response depending on the plant with open-loop and the controller which are mentioned this paper.

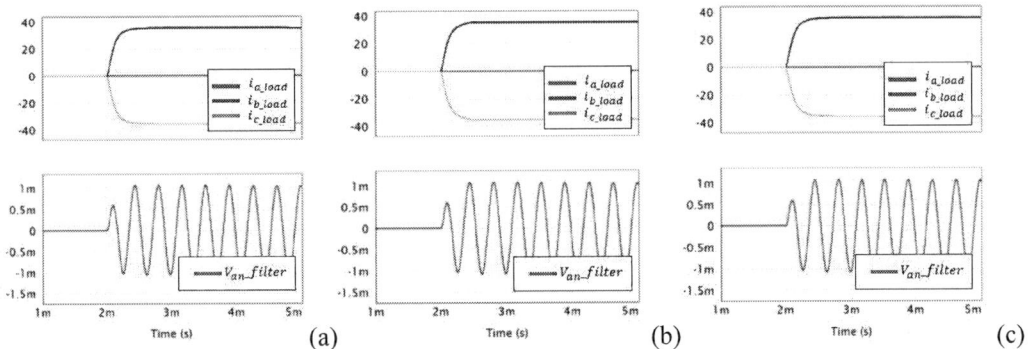

Fig. 13: Step Response depending on the controller design; (a) Conventional multiple-nested-loop controller, (b) Reference controller without voltage sensors, (c) Proposed controller without ac current sensors.

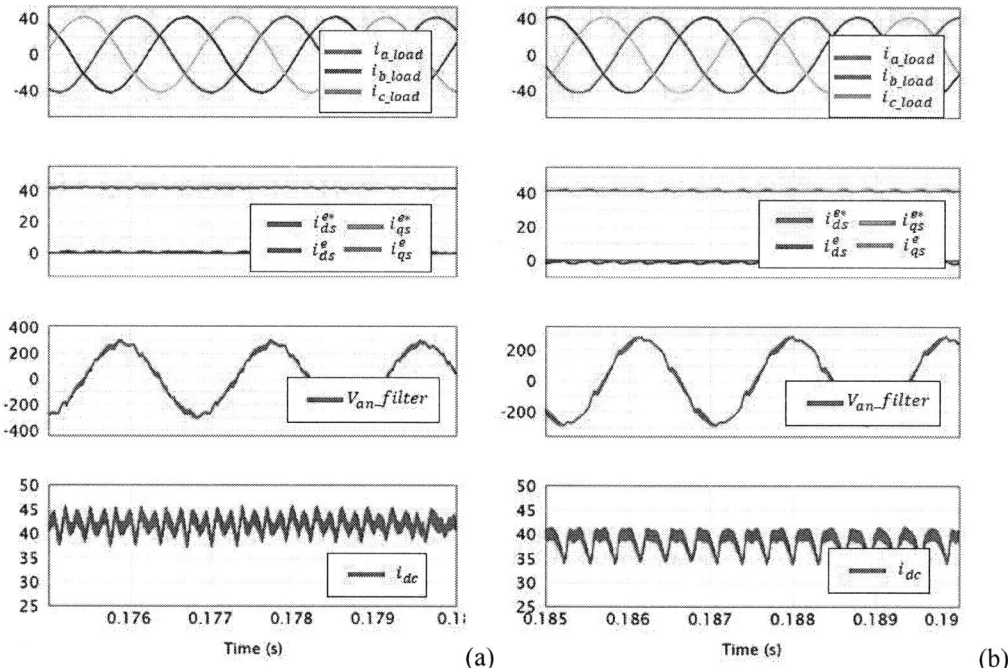

Fig. 14: Simulation Result of the proposed controller under the different PWM strategy between Symmetric PWM (a) and Two-Third PWM (b).

As can be seen from the bode plot in Fig. 12, it has almost the same response characteristics in continuous time domain. However, as seen in the recent paper [8], due to the derivative component, it can be a problem in stability, such as the disturbance. Therefore, further design is needed to allow the control design in the discrete time domain to have robust properties against possible disturbances such as harmonic components of EMF and sensing noise. This can be discussed through future research.

As shown in Fig. 13, the simulation results of 40A step responses below verified that each controller has similar response characteristics through RL load. The simulation result for the step response was performed by applying Symmetric PWM, and it was intended to verify the response characteristics of the current controller through DC Aline current. It can be said that the current controller designed through this has basically similar performance. However, as mentioned earlier, the impact on the error of the motor parameter and disturbance will be verified through further research.

In Fig 14, it could be verified through the simulation model that the proposed controller can be stably controlled at 15 kW load conditions by applying it under Symmetric PWM and Two-Third PWM to CSI-fed PMSM. Also, this result shows the control of the DC-DC converter varies depending on the Symmetric PWM and TTM methods. Although it has the same inverter output power current under the same load conditions, for TTM, it is essential to control the DC Link current in order to remove the Zero Current Vector. The harmonic of the current also increases depending on the shape of the DC current, so attention should be paid to the DC current control. On the other hand, in the case of Symmetric PWM, the DC Link current should be controlled in the DC constant with the Modulation Margin required by CSI so that the higher current should be controlled compared to the TTM. Accordingly, the losses of the power converter may be additionally generated compared to the TTM. However, since the harmonics of the output current control the DC current in a DC form, it may be said to be less than that of TTM. In other words, as shown in the previous study [6], in the case of TTM, attention to the syngeneic control between CSI and DC converter is required.

Finally, the simplified controller can reduce the mass on computational processors (e.g., microprocessor) because it has less computation and less input from sensors than the conventional controller [7,8]. That is, the high-speed controller operation through the high-speed switching power semiconductor may cause a cost increase mass of the control processor. However, the computational mass may be reduced using the simplified controller. It will be able to suppress the mass on the processor.

HW Development

Finally, for the hardware implementation, the creation of custom integrated high electro-magnetic and electro-thermal performance SiC MOSFET BDS`s is proposed, taking the moves from a concept developed and already successfully demonstrated for the case of a matrix converter [9, 10]. Here, Fig 15, the proposed concept of SiC MOSFET BDS differs from the previous one proposed for the Matrix Converter by horizontally connecting common sources to increase the heat dissipation area for application to Integrated Motor Drives with high temperatures and poor environmental conditions. The innovative packaging concept relies on a Power Overlay (POL) concept, capable of withstanding high temperatures, in excess of 200 °C [11]. In the future, the design will be verified through further research such as thermal analysis and experiments.

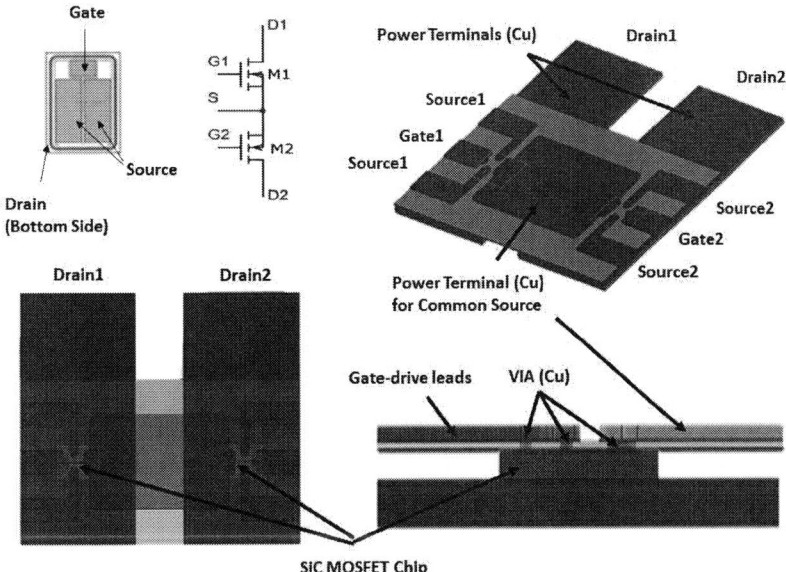

Fig. 15: The proposed design concept of the SiC Bi-directional Switch Power Module

Conclusion and Future work

Several pieces of research have recently been studied on the Current Source Inverter by taking advantage of the high switching frequency enabled by WBG devices. In this paper, the method to effectively control and implement these powerful functions has been discussed.

This research discussed the modulation technique and the design of the simplified current controller with respect to the new technologies of the current source inverter using the wide-band-gap power semiconductor, which is actively researched and proposes a new current controller to reduce the current sensors. Furthermore, we assess the PWM of CSI through Simulation Model and verify the proposed controllers.

Based on this study, we will continue to study various problems of the proposed controller through discrete-time analysis, i.e., current offset errors and so on. Accordingly, it is intended to extend from CSI to rotor position estimation technology using minimal sensors, such as Voltage Source Inverter, which can follow the rotor position of PMSM using only DC-Link voltage and current measurement. In addition, to develop future Integrated Motor Drives, we intend to verify and evaluate the Power Module design of SiC-based Bi-direction Switches suitable for CSI.

References

[1] J. W. Kolar and J. Huber, "Next-Generation SiC/GaN Three-Phase Variable-Speed Drive Inverter Concepts," PCIM Europe digital days 2021; International Exhibition and Conference for Power Electronics, Intelligent Motion, Renewable Energy and Energy Management, 2021, pp. 1-5.

[2] R. Amorim Torres, H. Dai, W. Lee, B. Sarlioglu and T. Jahns, "Current-Source Inverter Integrated Motor Drives Using Dual-Gate Four-Quadrant Wide-Bandgap Power Switches," in IEEE Transactions on Industry Applications, vol. 57, no. 5, pp. 5183-5198, Sept.-Oct. 2021, doi: 10.1109/TIA.2021.3096179.

[3] Y. W. Li, B. Wu, D. Xu and N. R. Zargari, "Space vector sequence investigation and synchronization methods for active front-end rectifiers in high-power current-source drives," IEEE Trans, Industrial Electronics, vol. 55, no. 3, pp. 1022-1034, Mar. 2008

[4] R. A. Torres, H. Dai, W. Lee, T. M. Jahns and B. Sarlioglu, "Current-Source Inverters for Integrated Motor Drives using Wide-Bandgap Power Switches," 2018 IEEE Transportation Electrification Conference and Expo (ITEC), 2018, pp. 1002-1008, doi: 10.1109/ITEC.2018.8450127.

[5] M. G. Sayed, O. Abdel-Rahim and M. Orabi, "Comparative Study to Investigate the Effect of Five VS Seven Segment Modulation Sequence on the Waveform Distortion Resulted by the Overlap Time in Current Source Inverter," 2019 International Conference on Innovative Trends in Computer Engineering (ITCE), 2019, pp. 576-580, doi: 10.1109/ITCE.2019.8646346.

[6] M. Guacci, M. Tatic, D. Bortis, J. W. Kolar, Y. Kinoshita and H. Ishida, "Novel Three-Phase Two-Third-Modulated Buck-Boost Current Source Inverter System Employing Dual-Gate Monolithic Bidirectional GaN e-FETs," 2019 IEEE 10th International Symposium on Power Electronics for Distributed Generation Systems (PEDG), 2019, pp. 674-683, doi: 10.1109/PEDG.2019.8807580.

[7] H. Lee, S. Jung and S. Sul, "A current controller design for current source inverter-fed PMSM drive system," 8th International Conference on Power Electronics - ECCE Asia, 2011, pp. 1364-1370, doi: 10.1109/ICPE.2011.5944414.

[8] R. A. Torres, H. Dai, W. Lee, T. M. Jahns and B. Sarlioglu, "A Simple and Robust Controller Design for High-Frequency WBG-Based Current-Source-Inverter-Fed.AC Motor Drive," 2020 IEEE Transportation Electrification Conference & Expo (ITEC), 2020, pp. 111-117.

[9] Aliyu, A. M., Castellazzi, A., Lasserre, P., & Delmonte, N. (2017, June). Modular integrated SiC MOSFET matrix converter. In 2017 IEEE 3rd International Future Energy Electronics Conference and ECCE Asia (IFEEC 2017-ECCE Asia) (pp. 1184-1188). IEEE.

[10] P. Lasserre, D. Lambert and A. Castellazzi, "Integrated Bi-directional SiC MOSFET power switches for efficient, power dense and reliable matrix converter assembly," 2016 IEEE 4th Workshop on Wide Bandgap Power Devices and Applications (WiPDA), 2016, pp. 188-193.

[11] https://www.shinko.co.jp/english/product/package/assembly/pol.php

Optimized Control Scheme to Achieve ZVS for the Complete Pre-Charging Phase of Supercapacitors with a 500 kHz SiC- and GaN-Based Dual Active Bridge

Patrick Lenzen, Martin Pfost
TU Dortmund University
Martin-Schmeisser-Weg 4
44227 Dortmund, Germany
Phone: +49 231-7557844
Email: patrick.lenzen@tu-dortmund.de
URL: https://ewa.etit.tu-dortmund.de/

Acknowledgments

This contribution was supported by the European Union and the Ministry of Economic Affairs, Innovation, Digitalization, and Energy of the State of North Rhine-Westphalia as part of the ERDF program under the funding code EFRE-0801621.

Keywords

≪Supercapacitor≫, ≪Dual Active Bridge (DAB) DC-DC converter≫, ≪Zero-voltage switching≫, ≪Energy storage≫, ≪High frequency power converter≫

Abstract

Large energy storage systems are used in many applications. If they are not in operation continuously, they will discharge themselves. To restart the operation, the supercapacitors should be pre-charged again to reach the nominal voltage range. Mostly, the capacitors are connected to a source via a DC/DC converter. For a fast reaction on load transients, fast-switching DC/DC converters are needed. Due to the high switching frequencies, soft switching is essential to keep the losses low. This is important for the normal operation but also for the pre-charging phase because pre-charging takes up to several minutes for large capacities. This work presents a zero-voltage switching (ZVS) analysis and extended-phase shift (EPS) modulation to reach ZVS switching for all semiconductors of the DC/DC converter during the complete pre-charge procedure. The analysis is done with a capacitance time-domain-based model and experimentally validated on a prototype.

Introduction

Energy-storage systems with supercapacitors become more and more attractive for many applications, e.g. for peak shaving in industrial applications like intralogistics. Supercapacitors have the advantage of a 500 times larger cycle life and higher power densities [1]. Nevertheless, supercapacitors have higher self-discharge than batteries [2]. If the system is not used continuously, the supercapacitor might be completely discharged and must be charged from 0 V before normal operation.

In industrial applications, AC and DC grids are used to supply all electrical loads. Lower losses can be reached with high voltage DC grids. In our case the nominal grid voltage is 650 V, much higher than the voltage of the supercapacitor modules, here 48 V. Therefore, the connection of supercapacitors to the DC grid is mostly handled by bidirectional DC/DC converters to overcome the voltage difference. Dual Active Bridges (DAB) and resonant converters, e.g CLLC converters, are commonly used bidirectional DC/DC converters with galvanic isolation, high conversion ratio possibility, and full bidirectional power flow. The DAB topology was first introduced in [3] and the CLLC converter in [4]. For high currents, the CLLC converters will need synchronous rectification to lower the losses. But for high switching frequencies, a realization of the synchronous rectification is challenging. Therefore and due to the multiple control options, the DAB is the more attractive choice. If using large supercapacitors, in this case 165 F, the pre-charging process to 48 V with constant 50 A takes around 2 min. For converters operating with high switching frequencies, which is necessary to reach a fast reaction on load transients, low switching losses are essential. Switching losses can be reduced by using zero-voltage switching. Nevertheless, a minimum dead time is required to reach zero-voltage switching (ZVS).

For the design process of the converter, the loss calculation for the starting process is as important as for the steady-state performance. To achieve soft switching during the pre-charging process, different charging techniques with soft switching are proposed in [5]-[6]. [5] presents a controlled trapezoidal current modulation. The authors calculate the switching points for zero-current switching (ZCS) from ideal current curves. However, the switching point is not controlled, the real current differs slightly from the ideal calculation. For high switching frequencies with high $\mathrm{d}i/\mathrm{d}t$, small time deviations result in large current changes. Therefore, switching at 0 A is not realizable and results in high switching losses.

[6] presents different pre-charging techniques. One solution uses a controlled primary and secondary full bridge. The source full bridge is under duty cycle control and the load full bridge has a constant duty cycle of 50 %. The two rectangular voltages are phase-shifted. The parameters are optimized with a current-based ZVS analysis. This work only considers if the current through the inductor flows in the right direction to reach ZVS. However, for fast-switching converters, the current-based ZVS analysis is not sufficient. This is because at low power and low currents, the dead time of the half bridge is often not large enough and the current changes during the transition. Moreover, the current direction can be reversed and it is impossible to reach full ZVS.

This paper presents an optimized extended phase-shift modulation scheme to reach ZVS for all switches during the pre-charging process for a fast-switching 500 kHz DAB. The ZVS area calculation relies on a capacitance time-domain-based model (CTD) as previously presented in [7]. The calculation is validated on a prototype.

Proposed System

The schematic of the proposed bidirectional DAB is shown in Fig. 1(a) and realized as a modular design with two LV and one HV full bridge, which was first introduced in [8]. For the LV and the HV full bridge, EPC GaN HEMTs and Wolfspeed SiC MOSFETs, respectively, are used. The practical realization is shown in Fig. 1(b). The normal operation voltage V_{SC} on the supercapacitor varies between 38 V and 48 V with a nominal value of 43 V and the DC bus voltage V_{bus} ranges from 580 V to 750 V with a nominal voltage of 650 V. The switching frequency is 500 kHz and the permissible maximal dead time is set to 100 ns. The DAB is in normal operation controlled by single phase-shift modulation.

Extended Phase-Shift Modulation

At the beginning of the pre-charging process of the supercapacitor, the voltage at the capacitor is usually 0 V. The modulation scheme should minimize the inrush current and enable soft switching also in this case. As previously described, for simplicity the converter is controlled by phase-shift modulation. Three different phase-shift modulation schemes are possible: Firstly, the single phase-shift scheme where the duty cycle of the primary and the secondary bridge voltage is $v_{\mathrm{AC,LV}} = 50\,\%$. The primary and secondary voltages are shifted by the angle φ. Secondly, the extended phase shift where the voltage of the primary HV bridge $v_{\mathrm{AC,HV}}$ has a duty cycle D_{p} which is lower than 50 % (see Fig. 2). Thirdly, the triple phase shift where the duty cycle D_{s} of the voltage of the secondary LV bridge $v_{\mathrm{AC,LV}}$ is lower than 50 % in addition to the previous shifts.

For fast-frequency converters, it is obvious that the modulation scheme should be as simple as possible. However, if using the single-phase-shift control, ZVS is not possible for all switching events. If the supercapacitor is discharged, the current is triangular. Consequently, ZVS will only be achieved for the

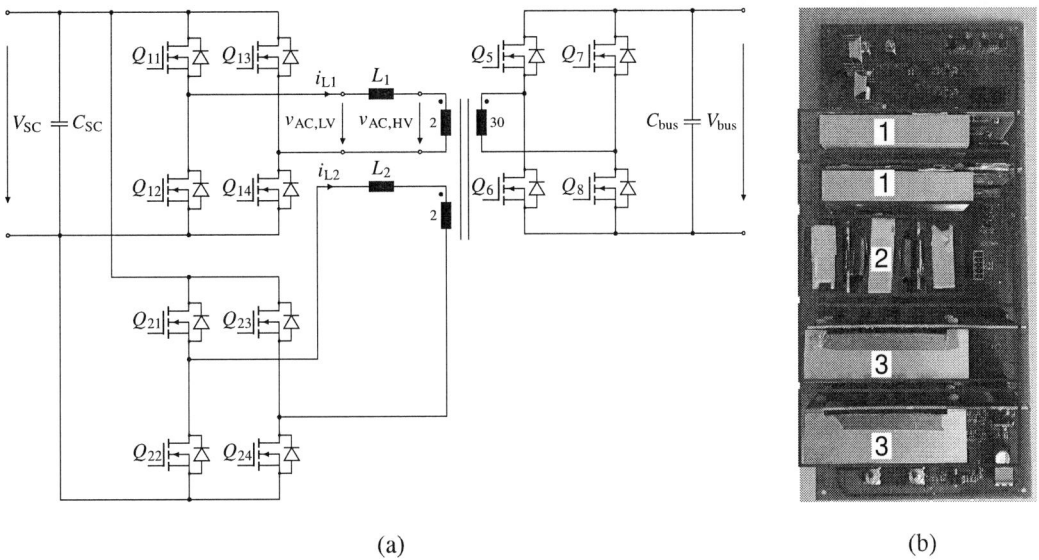

(a)   (b)

Fig. 1: (a) Schematic of the converter with two LV full bridges. The series inductors are placed on the LV side (L_1 and L_2). The supercapacitor is connected to V_{SC} and the DC bus to V_{bus}. (b) Modular realization of the converter. (1) HV half bridge (Q_5 - Q_6, Q_7 - Q_8), (2) Transformer, series inductors L_1, L_2, (3) LV full bridge (Q_{11} - Q_{14}, Q_{21} - Q_{24}). Transformer and full bridges are realized on pluggable PCBs.

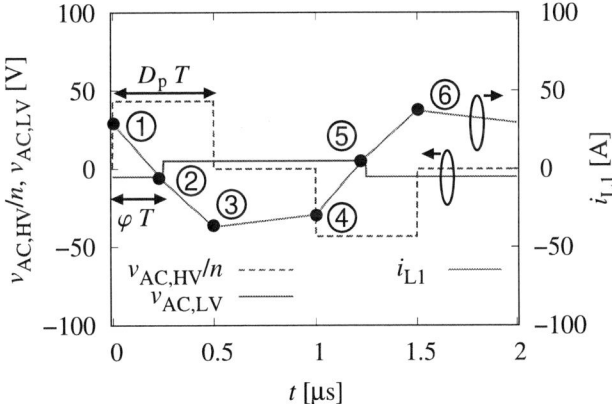

Fig. 2: Extended phase-shift control scenario. An outer phase shift φT is implemented between the low voltage $v_{AC,LV}$ and the high voltage $v_{AC,HV}/n$. The duty cycle of the high voltage is $D_p T$. The current i_{L1} flows through the inductor L_1. The same current flows also through the inductor L_2. All switching positions are marked with a specific number from ①-⑥.

HV switches and not for the LV switches. Therefore, this work investigates the extended phase-shift modulation to minimize the losses during the pre-charging procedure. In the following work, the phase shift φ and the duty cycle D_p are defined in degree, i.e. 360° corresponds to the PWM period T.

ZVS Switching Analysis with Equivalent Circuits

Clearly, low conduction losses and low switching losses are necessary to keep the converter in the specification, also during startup. The conduction losses can be minimized with a low current during startup. To reduce the switching losses, soft-switching methods are inevitable. As previously mentioned, zero-voltage switching (ZVS) is necessary. To identify which parameter combination of φ and D_p (see Fig. 2) enables ZVS for the complete pre-charging procedure, an accurate time-domain ZVS analysis as previously presented in [7] is used.

To analyze the switching behavior, equivalent circuits are constructed for each switching event, see Fig. 3. Therefore, the interaction of the output capacitance of the switches, the series inductor of the DAB, and the bulk capacitor is analyzed. The interwinding and the parasitic capacitance of the PCB are neglected in this work because they are much lower than the output capacitance of the switches. The source of the DAB is connected with a filter to the bulk capacitors (not shown in Fig. 1). Due to the low cut-off frequency of the filter, the energy of transient events will be provided by the bulk capacitors. Therefore, it is not necessary to add the input and output voltage source to the equivalent circuits. In the analysis, it is assumed that the switching operations of the LV full bridge and both HV half bridges are not simultaneously. This is achieved if at least the necessary dead time is between both events.

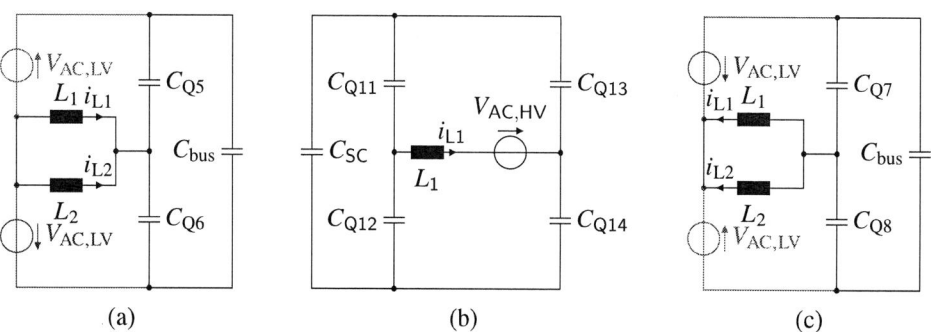

Fig. 3: Resonant commutation circuits for (a) the half bridge with switches Q_5, Q_6, (b) the full bridge with switches Q_{11}, Q_{12}, Q_{13}, Q_{14}, and (c) the half bridge with switches Q_7, Q_8.

For the ① switching event, see Fig. 2, the HV semiconductor Q_5 is turned on, Q_6 off, the equivalent circuit is constructed in Fig. 3(a). For the commutation only the output capacitances of the switches Q_5 and Q_6 are important. Therefore, the switches Q_5 and Q_6 are replaced in the equivalent circuit by the output capacitance C_{Q5} and C_{Q6}, respectively. Because the bridges do not switch simultaneously, the states on the other HV half bridge and on the LV full bridge does not change. Because of this, the LV full bridge can be replaced by the violet voltage source $V_{AC,LV}$. Moreover, Q_7 and Q_8 are replaced by a short and open, respectively. The teal voltage source is necessary for the ④ switching event and is removed for the ① switching event. To eliminate the transformer in the equivalent circuit, the capacitance of the switches and the C_{bus} capacitor are transformed to the low-voltage side with $n^2 C$. The assignment for the other switching events ②-⑤ is shown in Table I.

The effective capacitance $C_{oss,eff}(V_{DC})$ of the SiC HV and GaN LV switches is voltage-dependent. The effective output capacitance is calculated with the voltage-dependent stored energy from the datasheet as $C_{oss,eff} = 2E_{oss}/V_{DC}$. For the HV semiconductors only one constant capacitance $n^2 C_{oss,HV,eff} = 15.7$ nF is used because the bus voltage is fixed to 650 V. For the LV switches, the capacitance is calculated for each supercapacitor voltage. At the nominal voltage, here of the GaN LV switch, is $C_{oss,LV}(48 \text{ V}) = 1.7$ nF.

The ZVS analysis for all supercapacitor voltages, φ, D_p, and $v_{AC,LV}$ combinations is done with a state-space model of each equivalent switching circuit in Fig. 3. The assignment of the switching states in Fig. 2 to the mathematical models (1) or (2) with the state vectors $\mathbf{x_1} = [i_{Ls1} \ i_{Ls2} \ v_{C,SC} \ v_{Q,x} \ v_{Q,y}]^T$, $\mathbf{x_2} = [i_{Ls1} \ v_{C,SC} \ v_{Q,m1} \ v_{Q,m2} \ v_{Q,m3} \ v_{Q,m4}]^T$, the input scalars $u_1 = V_{AC,LV}$, $u_2 = V_{AC,HV}$, and the necessary additional parameters are shown in Table I. To prevent numerical problems, very small identical resistors R are added in series to the capacitors (not shown in Fig. 3). The input voltages and initial values on the capacitors and inductors at each switching state are calculated with an ideal analytical model of the DAB. The output capacitance of the switches is mapped to voltage-dependent capacitors in the circuit diagrams in Fig. 3.

Table I: Assignment of the different switching states to the equivalent circuits and the necessary additional model parameters.

Switching event	①	②	③	④	⑤	⑥
prior switch status	Q_6 on Q_5 off	Q_{x2}, Q_{x3} on Q_{x1}, Q_{x4} off	Q_8 on Q_7 off	Q_5 on Q_6 off	Q_{x1}, Q_{x4} on Q_{x2}, Q_{x3} off	Q_7 on Q_8 off
target switch status	Q_5 on Q_6 off	Q_{x1}, Q_{x4} on Q_{x2}, Q_{x3} off	Q_7 on Q_8 off	Q_6 on Q_5 off	Q_{x2}, Q_{x3} on Q_{x1}, Q_{x4} off	Q_8 on Q_7 off
Equivalent circuit	Fig. 3(a)	Fig. 3(b)	Fig. 3(c)	Fig. 3(a)	Fig. 3(b)	Fig. 3(c)
LV source	violet	-	violet	teal	-	teal
State-space model	Eq. (1)	Eq. (2)	Eq. (1)	Eq. (1)	Eq. (2)	Eq. (1)
model parameter z_1	1	–	1	1	–	−1
model parameter z_2	−1	–	1	1	–	1
model parameter $C_{Q,x}$	$x = 5$	–	$x = 7$	$x = 5$	–	$x = 7$
model parameter $C_{Q,y}$	$y = 6$	–	$y = 8$	$y = 6$	–	$y = 8$

$$\dot{\mathbf{x}}_1 = \begin{bmatrix} \dfrac{-R}{2L_{s1}} & \dfrac{-R}{2L_{s1}} & \dfrac{z_2}{2L_{s1}} & \dfrac{z_1}{2L_{s1}} & \dfrac{-z_1}{2L_{s1}} \\[2mm] \dfrac{-R}{2L_{s2}} & \dfrac{-R}{2L_{s2}} & \dfrac{z_2}{2L_{s2}} & \dfrac{z_1}{2L_{s2}} & \dfrac{-z_1}{2L_{s2}} \\[2mm] \dfrac{-z_2}{2C_{bus}} & \dfrac{-z_2}{2C_{bus}} & \dfrac{-1}{2RC_{bus}} & \dfrac{1}{2RC_{bus}} & \dfrac{1}{2RC_{bus}} \\[2mm] \dfrac{-1}{2C_{Q,x}} & \dfrac{-1}{2C_{Q,x}} & \dfrac{1}{2RC_{Q,x}} & \dfrac{-1}{2RC_{Q,x}} & \dfrac{-1}{2RC_{Q,x}} \\[2mm] \dfrac{z_1}{2C_{Q,y}} & \dfrac{z_1}{2C_{Q,y}} & \dfrac{1}{2RC_{Q,y}} & \dfrac{-1}{2RC_{Q,y}} & \dfrac{-1}{2RC_{Q,y}} \end{bmatrix} \cdot \mathbf{x}_1 + \begin{bmatrix} \dfrac{1}{L_{s1}} \\[2mm] \dfrac{1}{L_{s2}} \\[2mm] 0 \\[2mm] 0 \\[2mm] 0 \end{bmatrix} \cdot u_1 \tag{1}$$

$$\dot{\mathbf{x}}_2 = \begin{bmatrix} \dfrac{-R}{L_s} & 0 & \dfrac{-1}{2L_s} & \dfrac{1}{2L_s} & \dfrac{1}{2L_s} & \dfrac{-1}{2L_s} \\[2mm] 0 & \dfrac{-1}{RC_{SC}} & \dfrac{1}{2RC_{SC}} & \dfrac{1}{2RC_{SC}} & \dfrac{1}{2RC_{SC}} & \dfrac{1}{2RC_{bus}} \\[2mm] \dfrac{1}{2C_{Q,m1}} & \dfrac{1}{2RC_{Q,m1}} & \dfrac{-1}{2RC_{Q,m1}} & \dfrac{-1}{2RC_{Q,m1}} & 0 & 0 \\[2mm] \dfrac{-1}{2C_{Q,m2}} & \dfrac{1}{2RC_{Q,m2}} & \dfrac{-1}{2RC_{Q,m2}} & \dfrac{-1}{2RC_{Q,m2}} & 0 & 0 \\[2mm] \dfrac{-1}{2C_{Q,m3}} & \dfrac{1}{2RC_{Q,m3}} & 0 & 0 & \dfrac{-1}{2RC_{Q,m3}} & \dfrac{-1}{2RC_{Q,m3}} \\[2mm] \dfrac{1}{2C_{Q,m4}} & \dfrac{1}{2RC_{Q,m4}} & 0 & 0 & \dfrac{-1}{2RC_{Q,m4}} & \dfrac{-1}{2RC_{Q,m4}} \end{bmatrix} \cdot \mathbf{x}_2 + \begin{bmatrix} \dfrac{1}{L_s} \\[2mm] 0 \\[2mm] 0 \\[2mm] 0 \\[2mm] 0 \\[2mm] 0 \end{bmatrix} \cdot u_2 \tag{2}$$

An exemplary commutation at the switching event ① (Q_5 switches on) is shown in Fig. 4. The voltage before 0 ns is calculated with the ideal analytical model of the DAB (violet, solid). The transition is calculated with the state-space model (teal, solid). If the voltage at output capacitance of the switch Q_5 reaches the forward voltage of the body diode of Q_5, the body diode conducts and the voltage follows the violet solid line. Then, ZVS is achieved. If the body diode conducts, the solution of the state-space model is not valid anymore (teal, dashed). The minimum dead time t_c until the switch reaches ZVS can

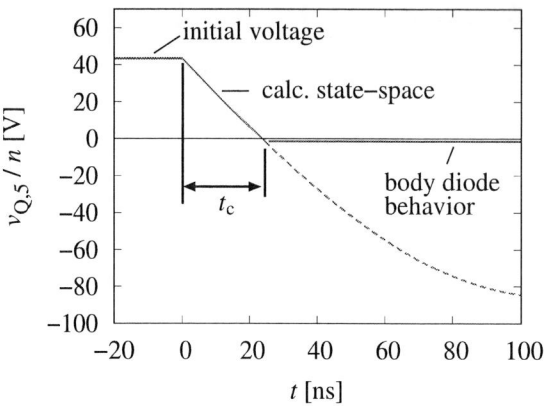

Fig. 4: Commutation of switch Q5. Violet: Initial voltage before the commutation and voltage if the body diode conducts. Teal (solid): Valid calculated commutation voltage of switch Q5 for switching event ①. Teal (dashed): Invalid calculated commutation voltage if the body diode conducts.

be extracted from the calculated resonant commutation. This time is also used during the calculations to ensure that the necessary dead time between the two HV half bridge switching events exists. A variable dead time that follows the calculated t_c could increase the efficiency. In our case, for simplicity, a 100 ns dead time for all operation points is implemented. After 100 ns, semiconductor Q_6 switches on.

As a result of the calculation of all operation points, a 3-dimensional ZVS boundary for different outer phase shifts φ and duty cycle D_p combinations is extracted, see. Fig. 5. For all operation points in the area, ZVS can be guaranteed for all switching events.

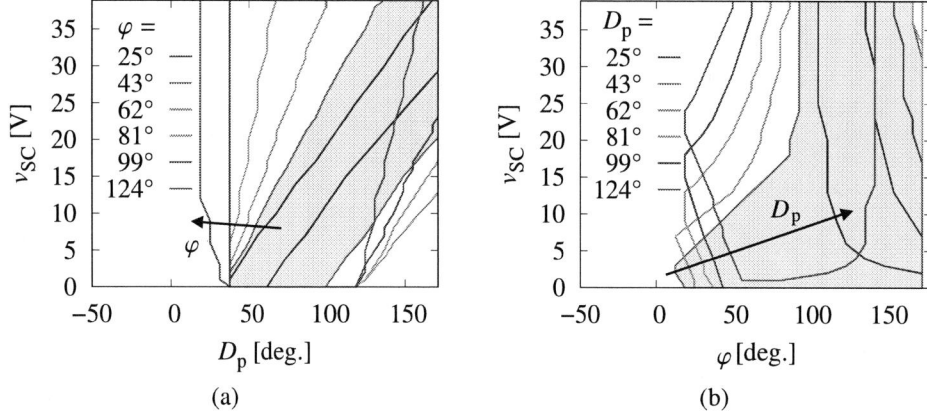

Fig. 5: The boundary of the inner ZVS switching area for all switching actions is shown for different combinations of supercapacitor voltage v_{SC}, outer phase shift φ, and primary duty cycle D_p. The different colors represent (a) the outer phase shift φ and (b) the primary duty cycle D_p. For clarification, an exemplary ZVS area for $\varphi = 43°$ and $D_p = 43°$, respectively, is filled.

As can be seen from Fig. 5, ZVS for the complete voltage range can be reached if both control parameters (φ, D_p) are varied, but also if one parameter is kept constant and only one parameter is varied. This is desirable to minimize the control expense. If a constant φ is chosen out of the range 19° and 174° and D_p varies, ZVS can be reached. Otherwise, if a constant D_p is chosen out of the range 37° and 112° and φ varies, ZVS can also be reached. Not all operation points are shown in Fig. 5. The optimal D_p, φ combination is achieved if 1) ZVS for all switches is reached, 2) the current through the inductors L_1 and L_2, respectively, does not exceed the maximum current of 60 A, and 3) only one parameter (either D_p or φ) is varied. Moreover, ZVS should be possible from 0 V to the minimum normal operating voltage of the converter, here 38 V.

Fig. 6(a) shows the ZVS operation area with its associated inductor currents for a constant $\varphi = 43°$, Fig. 6(b) for a constant $D_p = 37°$, respectively. The inductor currents for different operations points are shown with different colors to clarify that the inductor currents do not exceed the current limit. In Fig. 6(a) a possible control characteristic is shown in black. Nevertheless, also for lower φ even with lower inductor currents ZVS can be reached, but the area will be smaller, see Fig. 5(a).

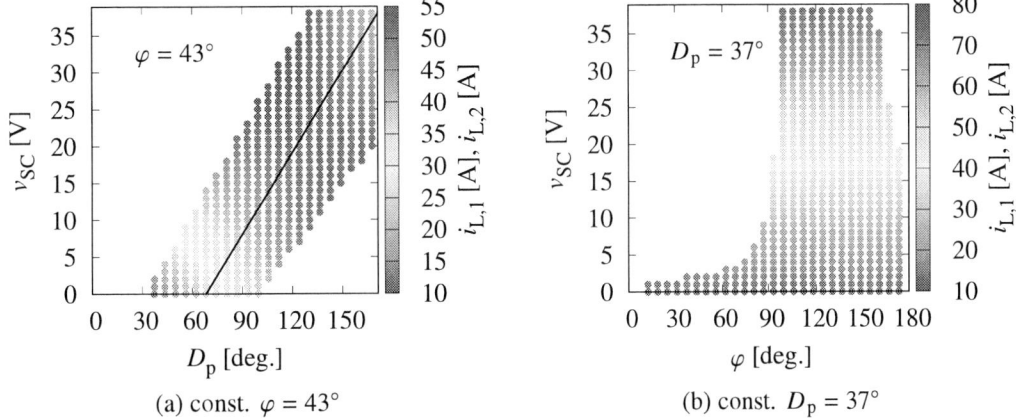

(a) const. $\varphi = 43°$ (b) const. $D_p = 37°$

Fig. 6: Possible ZVS operation points for supercapacitor voltages in the range $0\,\text{V} < v_{SC} < 39\,\text{V}$. Shown are (a) constant $\varphi = 43°$ with a variation of the primary duty cycle D_p and (b) constant $D_p = 37°$ with a variation of the phase shift φ. The different colors represent the inductor currents i_{L1} and i_{L2}. The black line displays a possible control characteristic.

For a small ZVS area and if slight variations occur, e.g. due to temperature changes, ZVS can not be guaranteed. Moreover, the pre-charging time will be extended due to the lower inductor currents. Therefore, the largest ZVS area which agrees with the current limitation is chosen. For a constant D_p, no ZVS area can be found which does not exceed the maximum current and simultaneously enables ZVS for the complete voltage range.

Keeping φ constant has the advantage that no additional DC offset current through the magnetizing inductance of the transformer occurs because of the phase shift during the starting procedure. Consequently, the modulation with a constant phase shift φ and a rising duty cycle D_p is chosen. Due to the minimal current variation around each operation point, all points are used to find the control characteristic. The duty cycle follows as

$$D_P = 2.64 v_{SC} + 68 \tag{3}$$

and is also shown in Fig. 6(a) in black.

Practical Realization

The previously presented extended phase-shift control for pre-charging the supercapacitor is implemented in the system shown in Fig. 1, also described in [8]. To verify that ZVS is possible for the previously presented area in Fig. 5(a), the switching behavior is tested for four different supercapacitor voltages 5 V, 10 V, 20 V, and 30 V. According to the previous calculations, the optimal outer phase shift should be $\varphi = 43°$.

Fig. 7 shows the voltages $v_{AC,HV}$, $v_{AC,LV}$ and the inductor current i_{L1}. From the voltage curves of $v_{AC,HV}$ and $v_{AC,LV}$ it is observed that at the end of the switching events the voltage increase slightly and the body diodes of the primary and secondary side conduct, see the small circles in Fig. 7. Consequently, ZVS is achieved for all switching events. The calculated ZVS area is verified by the four measurements.

The complete pre-charging procedure from 0 V to 38 V of a 165 F supercapacitor is shown in Fig. 8. It can be seen that the current through the inductor matches with the ideal calculations in Fig. 6(a). The maximal current of 60 A will not be exceeded.

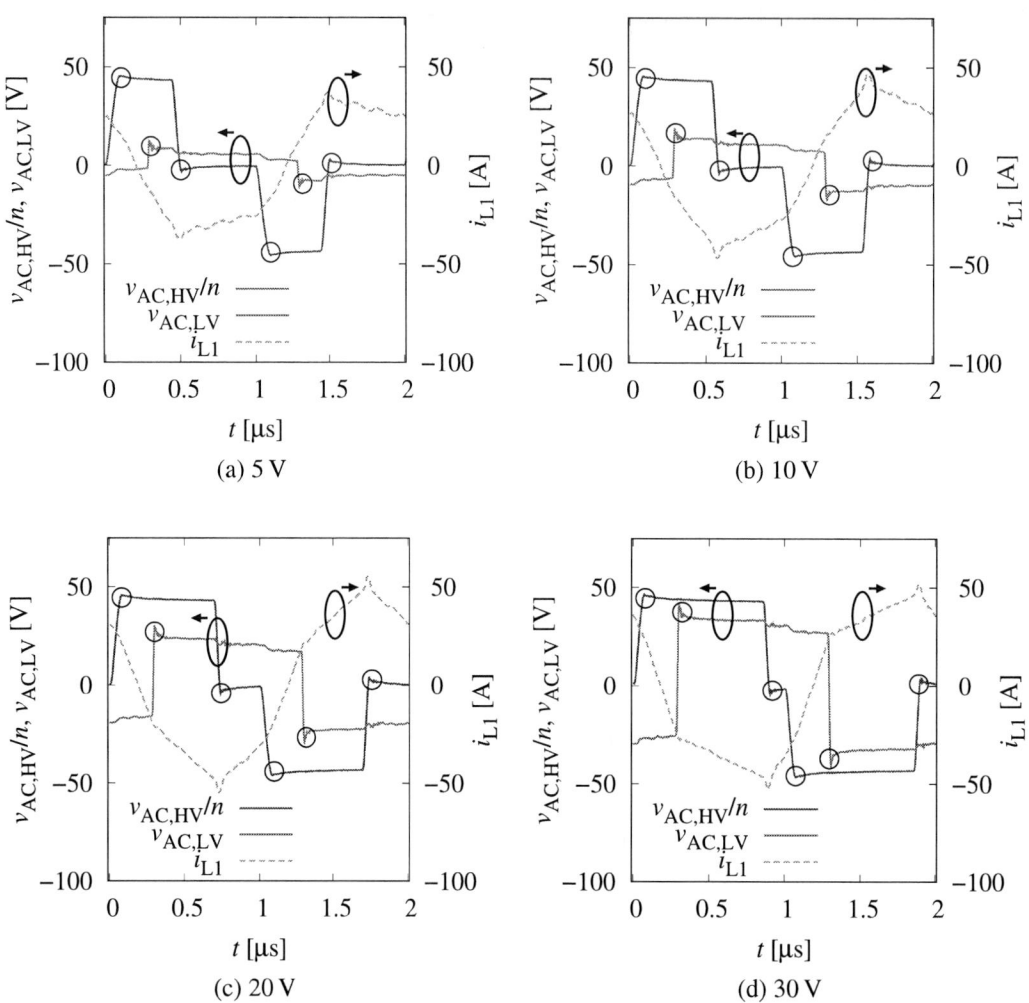

Fig. 7: Measured switching behavior for four different voltages (a) V_{SC} = 5 V, (b) 10 V, (c) 20 V and (d) 30 V. V_{bus} = 650 V and D_p is determined from (3). The small circles show the phases when the body diode is conducting.

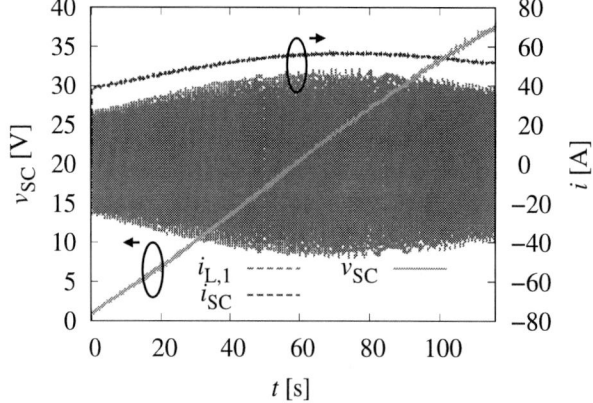

Fig. 8: Pre-charging procedure of a 165 F supercapacitor. The capacitor charging current i_{SC}, the capacitor voltage v_{SC}, and the inductor current $i_{L,1}$ are shown.

Conclusion

This paper presents an optimized control scheme to achieve ZVS for the complete pre-charging phase for a dual active bridge (DAB). A capacitance time-domain-based model is used to identify φ and D_p combinations to achieve full ZVS. Without a constant φ and rising D_p, full ZVS switching is possible for a supercapacitor voltage between 0 V and 38 V. Measurements confirm the correctness of the calculations. The presented extended phase-shift modulation is attractive for applications with supercapacitors that are often discharged completely and need to be pre-charged. With this modulation, an over-dimensioning of the DC/DC converter because of the pre-charging phase is not necessary.

References

[1] A. González, E. Goikolea, J. A. Barrena, and R. Mysyk, "Review on supercapacitors: Technologies and materials," *Renewable and Sustainable Energy Reviews*, vol. 58, pp. 1189–1206, 2016. [Online]. Available: https://www.sciencedirect.com/science/article/pii/S1364032115016329

[2] Y. Wu and R. Holze, *Electrochemical energy conversion and storage.* Wiley-VCH GmbH, 2022.

[3] R. De Doncker, D. M. Divan, and M. H. Kheraluwala, "A three-phase soft-switched high-power-density DC/DC converter for high-power applications," *IEEE Transactions on Industry Applications*, vol. 27, no. 1, pp. 63–73, Jan 1991.

[4] W. Chen, P. Rong, and Z. Lu, "Snubberless bidirectional DC-DC converter with new CLLC resonant tank featuring minimized switching loss," *IEEE Transactions on Industrial Electronics*, vol. 57, no. 9, pp. 3075–3086, 2010.

[5] J. Hu, S. Cui, and R. De Doncker, "Closed-loop black start-up of dual-active bridge converter with boosted dynamics and soft-switching operation," *IEEE Transactions on Power Electronics*, vol. 36, no. 10, pp. 11 009–11 013, 2021.

[6] P. Yao, X. Jiang, and F. F. Wang, "Soft starting strategy of cascaded dual active bridge converter for high power isolated DC-DC conversion," in *2020 IEEE Applied Power Electronics Conference and Exposition (APEC)*, 2020, pp. 1031–1037.

[7] F. Sommer, N. Menger, T. Merz, and M. Hiller, "Accurate time domain zero voltage switching analysis of a dual active bridge with triple phase shift," in *2021 23rd European Conference on Power Electronics and Applications (EPE'21 ECCE Europe)*, 2021, pp. 1–9.

[8] P. Lenzen and M. Pfost, "500 kHz SiC- and GaN-based dual active bridge with voltage conversion between 48 V and 650 V," in *PCIM Europe 2022; International Exhibition and Conference for Power Electronics, Intelligent Motion, Renewable Energy and Energy Management*, 2022.

Fault blocking capability in the DC-MMC with reduced number of sub-modules

J. D. Páez[1], F. Morel[1], S. Bacha[1,2], P. Dworakowski[1]

[1]SuperGrid Institute
23 Rue de Cyprian
69100 Villeurbanne, France
juan.paez@supergrid-institute.com
https://www.supergrid-institute.com/

[2]G2ELab
Univ. Grenoble Alpes, CNRS, Grenoble INP*
21 Avenue des Martyrs
38000 Grenoble, France
http://www.g2elab.grenoble-inp.fr/
*Institute of Engineering Univ. Grenoble Alpes

Acknowledgements

This work was supported by a grant overseen by the French National Research Agency (ANR) as part of the "Investissements d'Avenir" Program ANE-ITE-002-01.

Keywords

«HVDC», «Modular Multilevel Converter», «DC-DC converter», «Multi-terminal HVDC».

Abstract

The capability of DC-DC converters to block DC faults is an important issue for the development of HVDC grids. To include such feature, converters are generally oversized in terms of components by adding more semiconductors than needed in normal operation. This paper proposes to use a main switch instead of adding sub-modules in the topology to provide fault blocking capability. A converter control method is proposed to open the switch at zero current; thus, no breaking capability is required. An analysis on the impact on the converter design and a comparison with the classical solution are done. The operation of the proposed solution is verified through transient simulations. In the analysis, the requirements in terms of opening time for the switch were determined to be around 1 ms to 6 ms, which is feasible with a fast disconnector. However, the required control to keep the current on the switch at zero amps adds constraints to the lower arms of the topology in terms of installed capacitance and current withstanding.

Introduction

DC-DC converters are needed to enable the development of future HVDC grids. These structures allow the interconnection of DC systems with different characteristics like voltage rating or line configuration [1]. While enabling the interconnection of DC systems, DC-DC converters can also provide several functionalities to the system, like power flow control, DC voltage regulation and fault blocking capability (FBC) [2]. This last characteristic makes that DC-DC converters can play a role in the protection strategy of DC networks [3].

The FBC can be understood as the capability of the converter to prevent the apparition of anormal voltages or currents in one of the DC systems being interconnected by the converter when a fault appears in the second DC system. Anormal voltages and currents are defined as values that could lead to stop the operation of the network or to damage some of its components. This feature is achieved by liming the contribution of the healthy system to the fault current. The FBC can be achieved by a DC-DC converter integrated with an external protection device like a DC circuit breaker (DCCB), or by a DC-DC converter designed to provide the functionality by controlling the contribution of the heathy system to the fault current or being capable of interrupting this current.

Several DC-DC topologies adapted to HVDC applications have been proposed in literature. The common trend are modular multilevel converters that use chains of sub-modules (SMs) [2]. The SMs can be of different types, the half-bridge (HB) and the full-bridge (FB) being the most common. Two main categories of DC-DC converters are identified: isolated and non-isolated circuits. Non-isolated topologies seem to be more competitive in term of losses and component count [4], but they do not provide inherent FBC as the isolated circuits do. The non-isolated converters generally need some modifications in the structure to provide the FBC, like the use of FB-SMs instead of HB-SMs.

Among the non-isolated topologies, the DC-MMC has been identified as an interesting solution [4]–[8]. In this circuit, the FBC is provided by adding more SMs compared to those needed for normal operation [9]. This leads to an increase on the losses and costs, which can reduce the interest of the topology. This paper proposes an alternative solution to include the FBC.

The solution proposed in this paper relies in the use of a main high voltage switch (HV-Sw) instead of SMs to include the FBC against faults on the low-voltage side. A converter control method is proposed during faults to operate the HV-Sw at zero current. The aim is to reduce the constraints of the switch in terms of breaking capability.

The paper is organized in three sections. In the first section the DC-MMC topology is introduced and the constraints in terms of number of SMs to provide FBC are presented. The second section presents the proposed solution and the proposed control during faults. Finally, the third section presents the validation of the solution in simulation as well as a sensitivity analysis on the converter sizing and the constraints to the proposed HV-Sw. A comparison of the converter losses in the proposed solution and the classical DC-MMC is also done.

The DC Modular Multilevel Converter

The DC-MMC (also known as M2DC) (Fig. 1) is a non-isolated converter formed by several legs connected in parallel to the HV DC system. Each leg is formed by two arms made by a stack of SMs and an inductor. The legs have a middle point which interconnects both arms. On this middle point a filter interconnects the LV side DC system. The simplest filter is an inductance. In this paper a three-phase DC-MMC is studied.

The operation principle of the circuit is to use the SM stacks as controllable voltage sources by acting on the insertion and bypass of the SM capacitors. AC and DC voltages are produced on the SM stacks, and thanks to the voltage drop that is generated on the converter inductors, AC and DC currents are controlled. The DC currents control the power flow between DC systems while the AC currents are used to balance the energy in the converter, i.e. to control the voltage on the SM capacitors, by exchanging energy between upper arms and lower arms for instance. The balanced operation between the legs makes that the AC currents circulate between arms and legs but not into the DC ports.

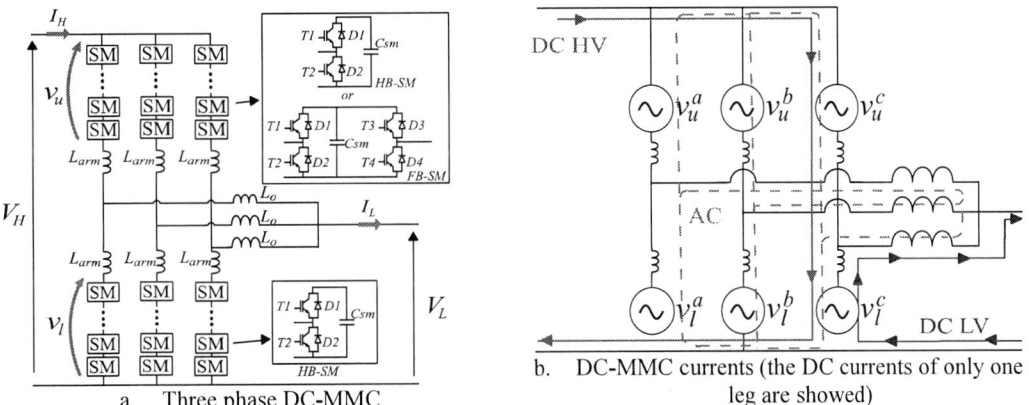

a. Three phase DC-MMC

b. DC-MMC currents (the DC currents of only one leg are showed)

Fig. 1: The DC-MMC and its operation principle with DC and AC circulating currents

In normal operation, each arm generates a maximum voltage described by Eq. (1) and Eq. (2) [9] which leads to the number of SMs per arm that would be required for normal operation which is shown on Eq. (3) and Eq. (4), where V_{SM} is the nominal voltage of one SM.

$$v_{u_{max}} = V_H - V_L + \min(V_L, V_H - V_L) \tag{1}$$

$$v_{l_{max}} = V_L + \min(V_L, V_H - V_L) \tag{2}$$

$$N_{SM_{u_{normal}}} = \text{ceil}\left(\frac{v_{u_{max}}}{V_{SM}}\right) \tag{3}$$

$$N_{SM_{l_{normal}}} = \text{ceil}\left(\frac{v_{l_{max}}}{V_{SM}}\right) \tag{4}$$

For HV side faults, to provide the FBC, the upper arms should generate a voltage equal to $-V_L$. For LV side faults, they should withstand V_H. Thus, the number of SMs required in upper arms to provide FBC are defined by Eq. (5) and Eq. (6) [9].

$$N_{SM_{FB_u}} \geq \text{ceil}\left(\frac{V_L}{V_{SM}}\right) \tag{5}$$

$$N_{SM_{HB_u}} \geq \text{ceil}\left(\frac{V_H}{V_{SM}}\right) - N_{SM_{FB_u}} \tag{6}$$

Comparing Eq. (3) and Eq. (4) with Eq. (5) and Eq. (6) it is seen how FBC leads to oversize the upper arms for transformation ratios less than 2 (defining the transformation ratio as the ratio between the HV and LV DC voltages, i.e . $n_{dc} = V_H/V_L$). This means that, that the upper arms require FB-SMs and more HB-SMs compared to the normal operation requirements. For the lower arm the number and kind of SMs to provide FBC is the same that for normal operation. Therefore, The DC-MMC has to be oversized on the upper arms to provide FBC [9]. This oversizing is presented in Fig. 2.

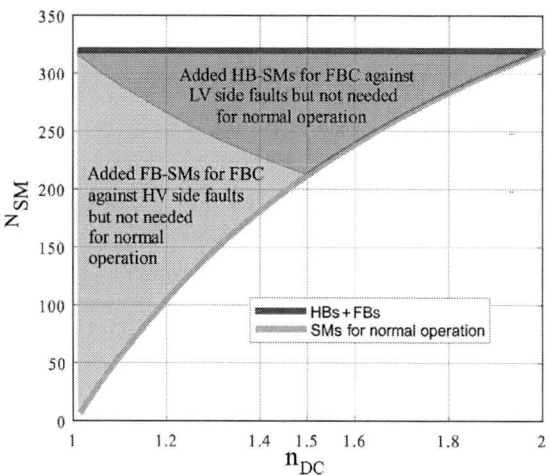

Fig. 2: Number of SMs required on each upper arm for normal operation and to provide FBC. A DC system voltage of $V_H = 640 \ kV$ and SM voltage of $V_{SM} = 2 \ kV$ were assumed as an example.

As an example, from Fig. 2, for $n_{dc} = 1.2$ ($V_L = 533$ kV) the required number of SMs for an operation in normal conditions is 107 per upper arm, all of them are HB-SMs. To include the FBC against HV side faults, the number of SMs per upper arm increases to 267 and must be of FB-SM type (Eq. (5)). To include the FBC against LV side faults, in addition to these SMs, 53 HB-SMs must be added (Eq. (6)). Thus, in total to include FBC for both types of faults the number of SMs per upper arm increased from 107 HB-SMs to 320 SMs (267 FBs+53 HBs). Considering that there are three upper arms, the oversizing is considerable. From Fig. 2, it is also observed that for $n_{dc} > 2$ there is no oversizing to include FBC.

The oversizing for the cases of $n_{dc} < 2$ degrades the performance indicators of the topology such as losses, number of switches and semiconductor utilization factor [9].

The use of a HV-Sw to avoid adding HB-SMs

To provide FBC without adding HB-SMs that are not necessary in normal operation on the upper arms, it is proposed to use an external switch HV-Sw between the HV side and the upper arms as shown in Fig. 3 [10]. This switch replaces the added HB-SMs on the upper arms to provide FBC against LV side faults. For HV side faults, the use of FB-SMs is still required as the proposed solution does not rely on a DCCB. During normal operation the switch is closed, and it is open only when there is a fault on the LV side.

Since breaking a DC current requires a circuit breaker, which is a costly solution, it is proposed to open the HV-Sw at zero current. To achieve this, a control method of the converter during the fault is proposed. With this approach it is expected that the requirements of breaking capability for the HV-Sw would be decreased and then a cost-effective solution like a disconnector could be used.

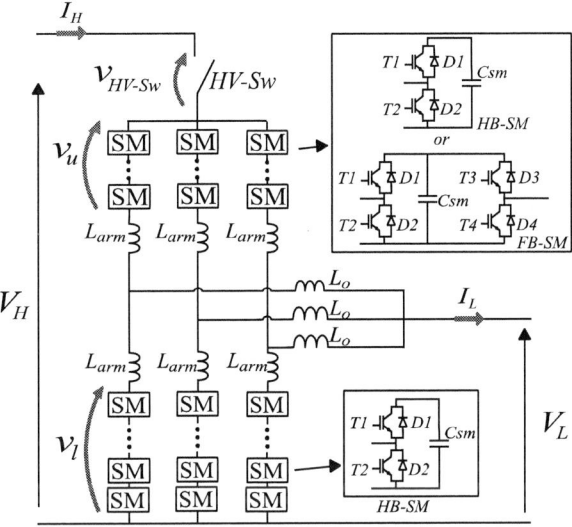

Fig. 3: DC-MMC with external HV-Sw to provide FBC against LV faults.

Reduction of number of SMs and losses

Fig. 4 shows the number of SMs (FBs and HBs) required on each upper arm in function of the transformation ratio for the classical DC-MMC as well as those needed with the proposed solution. It is observed how with the proposed solution, the number of HB-SMs needed in the topology to provide FBC is reduced. In the figure, the savings in SMs per upper arm are presented. For $n_{dc} = 1.5$ the solution gives the maximum reduction on SMs.

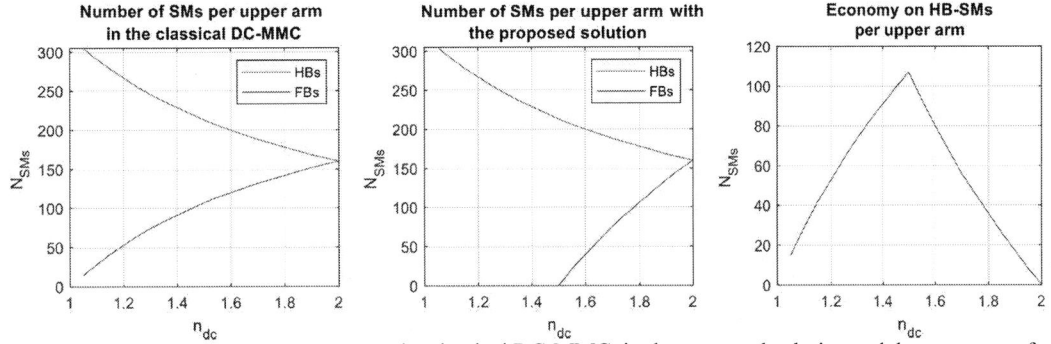

Fig. 4: Number of SMs per upper arm on the classical DC-MMC, in the proposed solution and the economy of HB-SMs for the proposed solution. A DC system voltage of $V_H = 640\ kV$ and SM voltage of $V_{SM} = 2\ kV$ were assumed as an example.

Since the conduction losses in a mechanical disconnector are lower than in the semiconductors of the removed SMs and because the current seen by the HV-Sw is only the DC current, contrary to the classical solution where the added SMs in the arms see the DC current and the circulating AC currents, it is expected that the proposed solution leads to an overall reduction on the conduction losses.

In addition, because the HV-Sw is kept closed during normal operation contrary to the classical solution where the added HB SMs are switching permanently to keep balanced the energy in the SM capacitors, it is expected that the proposed solution leads as well to a reduction on switching losses

In the section of validation of the proposed solution, the losses are quantified.

Proposed control

In order to open the HV-Sw at zero current it is proposed to use the lower arms to control the DC current on HV side after a LV side fault. The proposed control scheme is based on the following steps:

1. LV side fault detection by a measurement of the voltage drop, the rise of the DC current (di/dt) or by overcurrent.
2. Blocking of all SMs in the upper arms by turning OFF all IGBTs in the SMs.
3. Control of the lower arm SMs to generate the highest possible DC voltage by inserting all the SM capacitors.
4. When the current on each of the upper arms is zero, the corresponding lower arm in the same leg is controlled to maintain the current at zero amps.
5. When the current in all upper arms is zero (the DC current on the HV side is also zero), the HV-Sw is opened at zero current. The opening time depends on the HV-Sw technology.
6. When the HV-Sw is totally opened, (which can be verified by the voltage at its terminals), all the SMs in the lower arms are blocked.

The proposed control relies in the energy stored on the lower arms SMs to control the current. Since the lower arms still operate during the fault, the output inductances play an important role to decrease the fault current circulating by the arms.

In the following section the proposed control approach is validated, and a sensitivity analysis on the output inductor and SM capacitors is done.

Validation of the proposed solution in simulation

Control validation

To validate the proposed solution, simulations are done in Matlab/Simulink using the SimPowerSystems Toolbox. A case study based on the values of Table I is proposed. To model the circuit, an average model including the blocked state is used per arm as shown in Fig. 5. These models were proposed in [11], [12]. In these models, the SM stack is replaced by equivalent voltage and current sources, and all SM capacitors by an equivalent capacitor Ceq. IGBTs and diodes are added into the model to simulate the blocked state according to the blocking signal of the arm Blk.

The value of the equivalent capacitor depends on the number of SMs on the arm and the SM capacitance:

$$Ceq = \frac{C_{SM}}{N_{SM}} \tag{7}$$

The voltage at the terminals on this capacitor represents the sum of the voltages of all the SMs in the arm.

The blocking signal is a boolean and is calculated by the controller of the converter. The modulation index m is also used on the models. This variable represents the percentage of SMs being inserted at a given moment. It is calculated by the controller of the converter. The details of converter control during normal operation are omitted here for simplicity, but the reader can find more details about the control the converter in [5]. The focus of the paper is the proposed algorithm after the fault.

Table I: Circuit parameters for DC fault simulations

Parameter	Value
DC voltage	640 kV (HV side) and 500 kV (LV side)
Nominal Power	700 MW
Operating frequency	150 Hz

SM capacitance (sized to have an acceptable SM voltage ripple during normal operation at nominal power)	2.8 mF (Upper arms) and 10 mF (Lower arms)
Number of SMs per arm	313 FBs (Upper arms) and 400 HBs (Lower arms)
SM nominal voltage	2 kV
Arm inductance	25 mH
Filter inductance	250 mH

In the simulation, a LV side fault is done at the converter terminals at t=0.8 s, at nominal power, then the different stages of the proposed control are executed. It is assumed that there is no delay between the detection of the fault and the start of the control algorithm. The fault detection is done by undervoltage and overcurrent measurements at the converter terminals. The controller time step is 40 μs. Fig. 6 presents the simulation results. Only the variables related to one of the converter legs are presented for clarity (except for the voltage on the equivalent arm capacitance and the lower arms currents that are presented for the three legs). The figure highlights the different control steps described in the previous section. A delay of 1.5 ms is assumed on the opening of the HV-Sw, between the trigger signal and the effective switch opening.

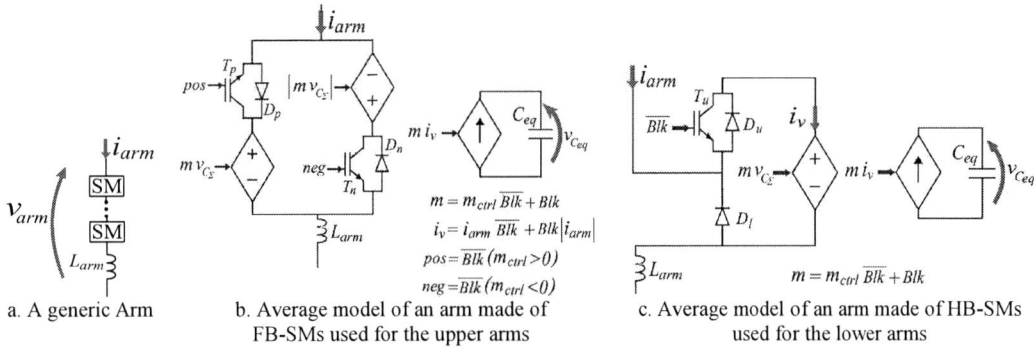

a. A generic Arm b. Average model of an arm made of c. Average model of an arm made of HB-SMs
        FB-SMs used for the upper arms   used for the lower arms

Fig. 5: Simulation models of an arm used to model the DC-MMC.

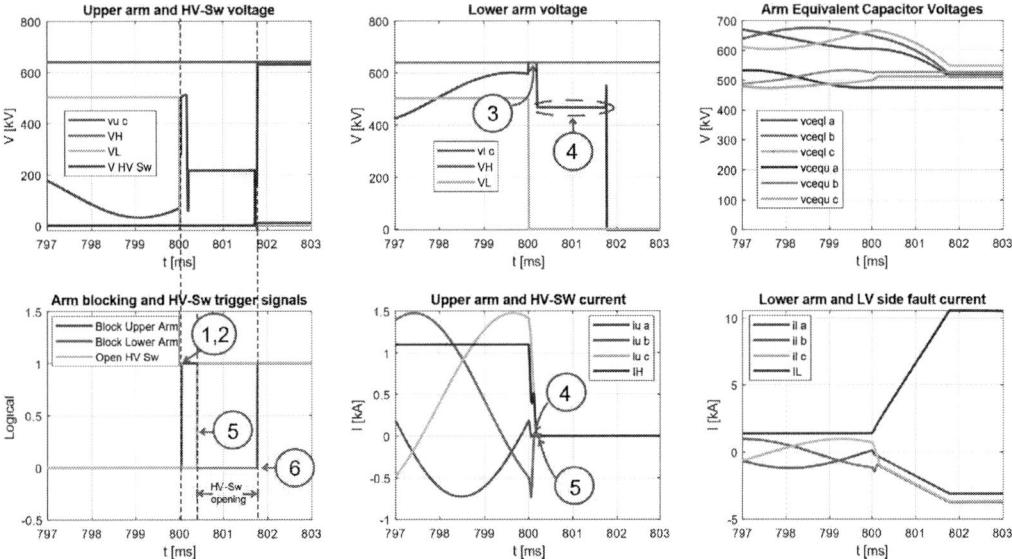

Fig. 6: Simulation results highlighting the control steps (1 to 6). The LV side fault is done at t=800ms

From the simulation results it is seen that effectively the opening of the HV-Sw (t=801.76 ms) is done at zero current. The control scheme achieves to maintain the I_H current on the switch at zero amps during all the opening. Once the switch is totally open, it is seen that the HV voltage is withstood by the HV-Sw and not by the FB-SMs on the upper arm.

According to the simulation results, during the transient, the high-side current I_H did not reached unacceptable values. Thus, the FBC is demonstrated.

In the figure it is also observed that, to keep controlled the HV current at zero during the opening of the HV-Sw, the lower arm should generate a voltage which feeds the LV fault. Thus, the SM capacitors are discharged into the LV side fault and the fault current increases. This is evidenced by the decrease of the capacitor voltage and the increase on the lower arm currents. This phenomenon poses two constraints to the proposed solution. From one side, if the capacitor voltages are discharged below the required voltage to control the upper arm currents, the control is lost and it is not anymore possible to regulate the HV current. From the other side, the current increase on the lower arms poses a constraint in the sizing of the semiconductor current rating. If the HV-Sw opening is too long, any of both situations will appear: the lower arm SM capacitors will be discharged below the control limit or the currents on the lower arms will reach unacceptable values.

Sensitivity analysis

An analysis of the maximum time to open the HV-Sw at zero current was done in function of the sizing of the converter. Two parameters were varied: the lower arm SM capacitance and the filter inductances Lo. The results are presented on Fig. 7. The maximum time before losing the control is presented as well as the current on the lower arms at that moment. For example, for a design with $Lo= 250$ mH and $Csml= 10$ mF the maximal time is around 3.5 ms and the lower arms must withstand a current of 6 kA.

The times presented on Fig. 7 are calculated as the time between the fault and when the current I_H cannot be longer controlled. Thus, this time represents the maximum time to open the switch including the fault detection, processing, triggering, and switch opening.

Fig. 7 shows that the technology of the HV-Sw must be capable of opening in some milliseconds, which can be done with a fast disconnector [13]. The increase on the SM capacitors can increase the available time to open the switch, but as a consequence the lower arms should withstand more current. Increasing the filter inductance can help to decrease the current rating and provide more time to open the HV-Sw.

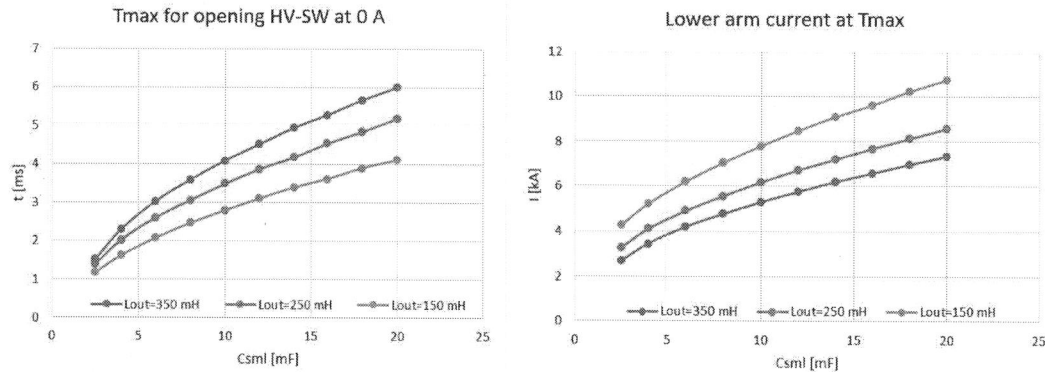

Fig. 7: Maximal time before losing the control at zero amps on the HV-Sw and current on the lower arms at that time in function of the lower arm SM capacitance and output filter inductance.

Power losses analysis

In this section, the losses of the proposed solution are compared with the losses of the classical DC-MMC. Only the semiconductor losses are considered. To calculate the conduction and switching losses, a semi-analytic detailed model of the converter is built. In this model the equivalent current source and equivalent capacitor used in the average models (Fig. 5) are replaced by a set of equations that represent the behavior of the individual SMs. The details of this modelling technique can be found in [14]. In these simulations, it assumed that the ac voltages are the same for a converter with or without FBC (even if a converter with FBC has additional SMs and then can generate different ac voltages).

From the simulation, the arm currents are extracted as well as the control signals of each SM. With this data, it is possible to establish which semiconductor is conducting at each time and to detect the switching actions (bypass or insertion of a SM). Then it is possible to establish the RMS and average currents per device as well as the switched current.

The conduction losses are estimated by Eq. (8) where V_o represents the saturation voltage if the switch is an IGBT or the threshold voltage if it is a diode, R_{ON} represents the device equivalent resistance in the ON state, I_{RMS} the device RMS current, I_{AVG} the device average current.

$$P_{conduction} = I_{RMS}^2 R_{ON} + I_{AVG} * V_o \tag{8}$$

The switching losses are calculated with Eq. (9), where $a_{on,off,rec}$, $b_{on,off,rec}$ and $c_{on,off,rec}$ are coefficients that approximate the energy loss at each switching of the device by a polynomial regression in function of the switched current I_{sw}. The power losses are calculated by adding all the energy losses and dividing by the elapsed time.

$$P_{switching} = \frac{1}{T_{total}} \sum \left(a_{on,off,rec} + b_{on,off,rec} * i_{switched} + c_{on,off,rec} * i_{switched}^2 \right) \tag{9}$$

The parameters R_{ON}, V_o, $a_{on,off,rec}$, $b_{on,off,rec}$ and $c_{on,off,rec}$ are obtained from the device datasheet. In this paper a 3.3 kV / 1500 A power module is assumed (FZ1500R33HL3).

For the simulation, a low-level controller was added to control the switching of each SM. The choice was a Nearest Level control modulation with a Balancing Control Algorithm (BCA) based in a tolerance band as described in [15]. The tolerance band was set to 180 V. The control of the FB-SMs in the structure does not consider a negative insertion.

The simulation is done during a long period of time (9 seconds) to average the "random" switching phenomena caused by the BCA. The data of one SM on the upper arms and one SM on the lower arms was recorded to do the post-processing. The losses are calculated for different scenarios. As base values, the parameters of Table I are used, Table II presents those parameters which are changed in comparison with the previous case.

Table II: Parameters for simulations for the calculation of power losses

Parameter	Value				
LV side DC voltages	533 kV	457 kV	427 kV	400 kV	356 kV
Ndc	1.2	1.4	1.5	1.6	1.8
Number of SMs (FBs) per upper arm	334	286	267	250	223
IGBT Losses parameters	$R_{ON} = 0.94$ mΩ , $V_o = 1.5$ V, $a_{on} = 0.5435, b_{on} = 0.9946x10^{-3}$ and $c_{on} = 5.7705x10^{-7}$ $a_{off} = 0.2929, b_{off} = 1.7079x10^{-3}$ and $c_{off} = 0.4486x10^{-7}$				
Diode Losses parameters	$R_{ON} = 0.43$ mΩ , $V_o = 1.2$ V, $a_{rec} = 0.5706, b_{rec} = 2.8654x10^{-3}$ and $c_{rec} = -7.2667x10^{-7}$				

Fig. 8 shows the power losses of the proposed solution compared with the classical DC-MMC. It is verified that the proposed scheme reduces the converter power losses. The reduction is mainly on the switching losses. The reduction on conduction losses is lower and increases with the transformation ratio n_{DC}. The conduction losses are still penalized by the high amount of FB SMs required on the upper arms to provide FBC against HV side faults.

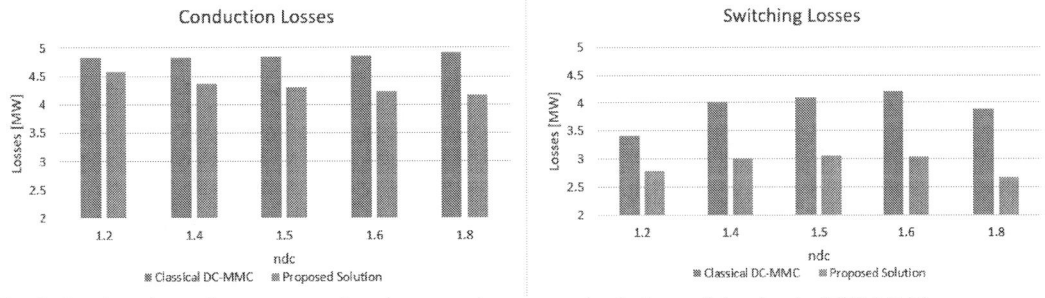

Fig. 8: Semiconductor losses comparison between the proposed solution and the classical DC-MMC.

Conclusions

An alternative method for providing FBC on the DC-MMC is proposed in this paper. The method is based on the use of an external switch. With the proposed solution, a decrease in the number of installed SMs is achieved and as well as a decrease in power losses. A control method was proposed to operate the switch at zero current. This avoids the need of breaking capability on the switch. Thus, it can be implemented with a fast disconnector. The operation of the switch and the control method was verified in simulation. A sensitive analysis on the circuit parameters was done to analyse the maximum opening time of the switch. It is seen that the HV-Sw must operate in some milliseconds and that the lower arms must withstand the fault current. The opening time of the HV-Sw is related to the maximal current that the semiconductors in the lower arms must be able to switch and to the minimal SM capacitance in lower arms. Thus, the proposed solution reduces the number of SM in the upper arms but, according to the HW-Sw performance, can lead to oversize the lower arm SMs. Finally, the proposed method can provide FBC against LV side faults but for HV side faults the DC-MMC is still oversized by the need of FB-SMs.

References

[1] D. Gómez, J. D. Páez, M. Cheah-Mane, J. Maneiro, P. Dworakowski, O. Gomis-Bellmunt, and F. Morel, "Requirements for interconnection of HVDC links with DC-DC converters," in *IEEE 45th Annual Conference of the Industrial Electronics Society (IECON 2019)*, 2019.

[2] J. D. Páez, D. Frey, J. Maneiro, S. Bacha, and P. Dworakowski, "Overview of DC–DC converters dedicated to HVdc grids," *IEEE Transactions on Power Delivery*, vol. 34, no. 1, pp. 119–128, 2019.

[3] J. D. Páez, J. Maneiro, D. Frey, S. Bacha, A. Bertinato, and P. Dworakowski, "Study of the impact of DC-DC converters on the protection strategy of HVDC grids," in *15th IET International Conference on AC and DC Power Transmission (ACDC 2019)*, 2019, pp. 1–6.

[4] J. D. Páez, J. Maneiro, S. Bacha, D. Frey, and P. Dworakowski, "Influence of the operating frequency on DC-DC converters for HVDC grids," in *2019 21th European Conference on Power Electronics and Applications (EPE'19 ECCE Europe)*, 2019.

[5] F. Gruson, Y. Li, P. Le Moigne, P. Delarue, F. Colas, and X. Guillaud, "Full State Regulation of the Modular Multilevel DC converter (M2DC) achieving minimization of circulating currents," *IEEE Transactions on Power Delivery*, 2019.

[6] D. Jovcic, P. Dworakowski, G. Kish, A. Jamshidi Far, A. Nami Abb, A. Darbandi, and X. Guillaud, "Case Study of Non-Isolated MMC DC-DC Converter in HVDC Grids," in *CIGRE Symposium 2019 - Aalborg, Denmark*, 2019.

[7] G. J. Kish, M. Ranjram, and P. W. Lehn, "A modular multilevel DC/DC converter with fault blocking capability for HVDC interconnects," *IEEE Transactions on Power Electronics*, vol. 30, no. 1, pp. 148–162, 2015.

[8] S. Norrga, L. Ängquist, and A. Antonopoulos, "The polyphase cascaded-cell DC/DC converter," in *Energy Conversion Congress and Exposition (ECCE), 2013 IEEE*, 2013, pp. 4082–4088.

[9] J. D. Paez, F. Morel, S. Bacha, P. Dworakowski, and D. Frey, "Impact of DC fault blocking capability on the sizing of the DC-DC Modular Multilevel Converter," in *2020 22nd European Conference on Power Electronics and Applications (EPE'20 ECCE Europe)*, 2020, pp. 1–10.

[10] J. D. Paez Alvarez, J. Maneiro, P. Dworakowski, S. Bacha, D. Frey, and F. Morel, "DC-DC VOLTAGE CONVERTER PROVIDED WITH A CIRCUIT BREAKER DEVICE," WO2021/1221872021.

[11] F. Xinkai, Z. Baohui, and W. Yanting, "Fast electromagnetic transient simulation models of full-bridge modular multilevel converter," in *2016 IEEE PES Asia-Pacific Power and Energy Engineering Conference (APPEEC)*, 2016, pp. 998–1002.

[12] H. Zhang, D. Jovcic, W. Lin, and A. J. Far, "Average value MMC model with accurate blocked state and cell charging/discharging dynamics," in *Environment Friendly Energies and Applications (EFEA), 2016 4th International Symposium on*, 2016, pp. 1–6.

[13] R. Derakhshanfar, T. Jonsson, U. Steiger, and M. Habert, "Hybrid HVDC breaker–Technology and applications in point-to-point connections and DC grids," in *CIGRE session*, 2014, pp. 1–11.

[14] A. Zama, S. Bacha, A. Benchaib, D. Frey, and S. Silvant, "A novel modular multilevel converter modelling technique based on semi-analytical models for HVDC application," *J. Elect. Syst.*, vol. 12, no. 4, pp. 649–659, 2016.

[15] A. Hassanpoor, L. Ängquist, S. Norrga, K. Ilves, and H.-P. Nee, "Tolerance band modulation methods for modular multilevel converters," *IEEE Transactions on power electronics*, vol. 30, no. 1, pp. 311–326, Jan. 2015.

An Open-Source FEM Magnetic Toolbox for Calculating Electric and Thermal Behavior of Power Electronic Magnetic Components

Nikolas Förster, Jonas Hölscher, Till Piepenbrock, Philipp Rehlaender,
Oliver Wallscheid, Frank Schafmeister, Joachim Böcker
Paderborn University
Warburger Str. 100
33098 Paderborn
Email: {foerster, piepenbrock}@lea.upb.de
URL: https://lea.upb.de

Keywords

≪Finite-element analysis≫, ≪Flux model≫, ≪Thermal model≫, ≪Cooling≫ ≪Simulation≫.

1 Abstract

Minimizing power losses and the thermal management are important factors in developing magnetics for power electronics. Both are very relevant to maintain the efficiency and the maximum operating temperatures of the core and conductors. To bring this calculation in the development workflow, the FEM (Finite Element Method) Magnetics Toolbox (FEMMT) will be continued. This open-source toolbox helps to automatically simulate magnetics in a guided, standardized format, requiring minimal effort. Based on the magnetoquasistatic simulation, a thermal simulation is set up. Conductor and core losses are taken from the magnetostatic simulation. The losses are homogenized inside the winding and inside the core. Thermal geometries and materials are set by script (e.g., for air gap filling and potting material), to automate the time-consuming drawing process inside the FEM tool. After simulation, the results are read back. With the help of this toolbox, which is publicly and collaboratively developed on GitHub, entire parameter sweeps can be easily performed and the simulation process is significantly speeded up.

2 Introduction

Losses of inductive components in winding and core can be calculated by different analytical or numerical methods [1, 2, 3, 4]. In this context, the question arises if the generated heat can be dissipated appropriately and if the given materials can be operated below the maximum temperature rating. Thus, it can be ensured that, for example, the core material keeps its magnetic properties (Curie temperature), or that the insulation of the conductors does not melt.

FEM simulations are suitable for calculating the local temperature distribution. Thus, hot spots are detected and it is ensured that the maximum temperatures of the different materials are maintained. For this purpose, there are excellent open-source programs [5, 6, 7], which provide interfaces for automation. Nevertheless, the model preparation is usually very time-consuming, as all parts of the magnetic component and the cooling structure have to be drawn manually, relationships and materials have to be established, and thermal boundary conditions have to be assigned. If a large number of setups (e.g., various cooling materials, different sizes of the inductive element and its housing) are to be tested, manual execution is practically infeasible. For that purpose, this paper presents the FEM magnetics toolbox (FEMMT), which can be used as the programming interface of the open-source tool ONELAB [5] to set up standardized magnetic components. The cooling structure is drawn by an abstracted code interface, using Python script language. Once the simulation has finished, the results for temperatures in conductors and core are read back.

The general workflow is shown in Fig. 1. First, a 2D-rotationally symmetric magnetoquasistatic simulation is parameterized according to [9], executed and the results are returned. Then, the thermal materials like air gap material, winding insulation or the potting material are set. After assigning the boundary conditions (ambient temperature and cooling method), the simulation is started. The simulation computes the power flow and temperatures. This paper focuses on the thermal simulation part, starting from the thermal geometry generation, since the magnetostatic part is already described in [9]. The geometry is described by code, so that different setups can be easily calculated and optimized.

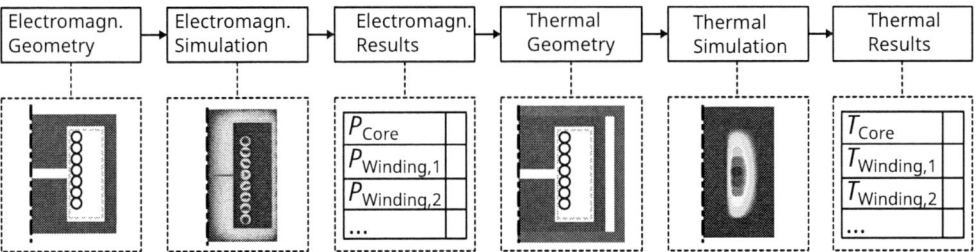

Fig. 1: Workflow to set up a thermal simulation using a magnetoquasistatic pre-simulation

3 Open-Source FEM-Magnetics Toolbox Using Thermal Simulations

This section describes the basics of the simulation tool, such as the abstract code interface for describing the cooling geometry.

FEM-Simulation Tool Details

The chosen FEM simulation software is ONELAB [5]. The tool has been chosen as it is controllable via Python, and is an open-source software that is actively developed. It is natively executable on the usual operating systems (Linux/MacOS/Windows). ONELAB contains a self-writable solver, so special features can be taken into account, such as litz wire approximation, eddy currents inside the core for the magnetoquasistatic simulation, and for the thermal simulation the heat conduction and convection can be considered. Based on ONELAB FEM software, the FEM magnetics toolbox (FEMMT) was developed. FEMMT is an interface to ONELAB and includes pre- and post-processing, to control external simulations. It should be noted that the FEMMT is continuously developed on GitHub [10] and welcomes open collaboration.

Abstract Code Interface for Thermal Simulations

The programming interface of ONELAB is used to automatically create geometries as well as to assign materials. In [9] this is shown how to draw the core including windings, assign the currents, start the simulation and read back the results. For this purpose, a separate code interface is designed, with which the geometries of the inductive elements are transferred with simple commands. This is where the current paper continues.

To be able to describe a thermal simulation, the geometries of the insulations as well as their material properties are necessary. A standardized model including thermal parameters is shown in Fig. 2a. To the rotationally symmetric structure, a winding body can be introduced, as well as insulation between primary and secondary winding. The winding window is filled with inner potting material, and the entire core can be filled with outer potting material. The boundary conditions are set at its outer surfaces. This corresponds, for example, to mounting the inductive element on a heat sink with water cooling as depicted in Fig. 2b.

The simulation results read back contain the power dissipation per turn as well as the power dissipation of the core. Both are now impressed into the thermal simulation. The power dissipation is assumed to be homogeneous for each individual turn as well as for the core. This simplifies the parameter transfer from magnetoquasistatic to thermal simulation.

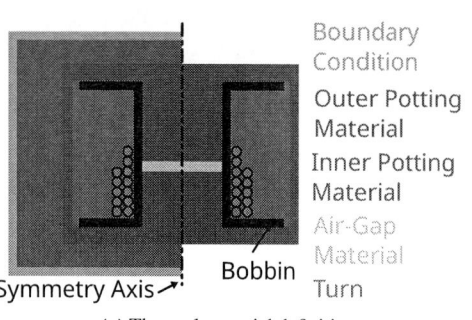

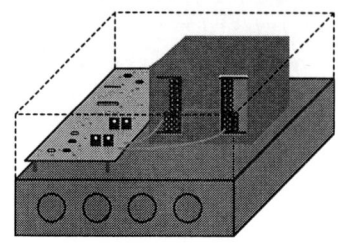

(a) Thermal material definitions

(b) Exemplary power converter based on a cooling plate with liquid cooling

Fig. 2: Cooling boundary condition options

The geometric parameterization is based on the previous work, published in [9]. The standard geometry for thermal parameterization is depicted in Fig. 3a. The following materials can be added: separate material for the air gaps, insulation inside the core (bobbin), inner potting material, and outer potting material. The thermal conductivities of the geometries generated there must be added (thermal conductivity of core, windings, insulations, and potting material). Outside of the core, the thickness of the potting material can be set individually for each side. Fig. 3b shows the parameterizable boundary conditions of

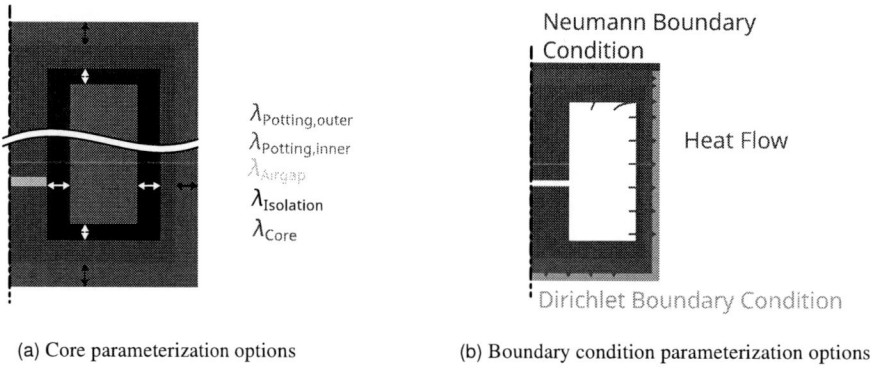

(a) Core parameterization options

(b) Boundary condition parameterization options

Fig. 3: Thermal parameterization options

the solver. The Dirichlet boundary condition sets the function values at the boundary, which corresponds to a fixed temperature. The Neumann boundary condition specifies an isolated environment which requires, e. g. no heat dissipation on that side. The type of boundary condition can be set via a flag in the program code. Possible examples are shown in Fig. 4.

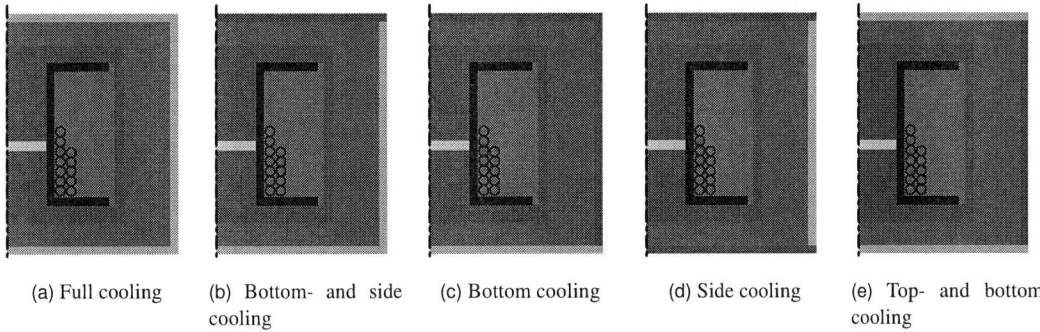

(a) Full cooling   (b) Bottom- and side cooling   (c) Bottom cooling   (d) Side cooling   (e) Top- and bottom cooling

Fig. 4: Cooling boundary condition options

4 Code Example

Listing 1 shows a code example to parameterize the thermal cooling structure. First, the thermal conductivities are described. Then the thickness of the potting material and the temperatures at the boundary conditions are set according to Fig. 3a and Fig. 3b. A flag can be used to determine whether the boundary condition is a Neumann boundary condition or a Dirichlet boundary condition. The inner insulation thickness is taken from the magnetoquasistatic simulation. Then the simulation is started. Please note that the code syntax may change as development progresses. The latest syntax can be found in [10].

This code generates the additional cooling structure inside the ONELAB simulation. The power loss is calculated for every single mesh cell.

```
thermal_conductivity_dict = {
    "air":          0.0263,# unit: W/(m*K)
    "case": {
        "top":    1.54, # unit: W/(m*K)
        "right": 1.54, # unit: W/(m*K)
        "bot":    1.54  # unit: W/(m*K)
    },
    "core": 5, # unit: W/(m*K)
    "winding":    400,  # unit: W/(m*K)
    "air_gaps":   180,  # unit: W/(m*K)
    "isolation": 0.42   # unit: W/(m*K)
}
# set distance from core to case
case_gap_top   =   0.0020 # unit: m
case_gap_right = 0.0025 # unit: m
case_gap_bot   =   0.0020 # unit: m

# set boundary temperatures
boundary_temperatures = {
    "value_boundary_top":      20, # unit: degree celcius
    "value_boundary_right":    20, # unit: degree celcius
    "value_boundary_bottom": 20   # unit: degree celcius
}

# 1: Dirichlet boundary condition
# 0: Neumann boundary condition
boundary_flags = {
    "flag_boundary_top":      1,
    "flag_boundary_right":    1,
    "flag_boundary_bottom": 1
}

# start thermal simulation (geo is the simulation object)
geo.thermal_simulation(thermal_conductivity_dict, boundary_temperatures,
    boundary_flags, case_gap_top, case_gap_right, case_gap_bot)
```

Thermal Simulation Details

In order to integrate the heat conduction in the 2D-rotationally symmetric case into the solver, the general heat conduction equation must be simplified and solved. The starting point is the fundamental law of heat conduction:

$$\dot{\vec{q}} = -\lambda(\rho, \vartheta, \vec{r}) \cdot \mathrm{grad}(\vartheta(\vec{r}, t)), \tag{1}$$

where ϑ describes the scalar temperature field which depends on time t and position $\vec{r}$, λ is the heat conductivity of the material and $\dot{\vec{q}}$ is the heat flow density.

In the following, the heat conduction equation derived from Fourier's law in [12] is assumed:

$$\rho \cdot c(\vartheta) \frac{\partial \vartheta}{\partial t} = \mathrm{div}[\lambda(\rho, \vartheta, \vec{r}) \cdot \mathrm{grad}(\vartheta)] + \dot{W}(\vartheta, \vec{r}, t). \tag{2}$$

This equation holds for an isotropic, incompressible material at rest, with material properties λ, c (specific heat capacity), and ρ (mass density of the material). Further, in an electro-magnetic context, the volumetric heat source density $\dot{W}$ equals the locally dissipated power density according to losses. For FEMMT, the following simplifications were made:

- Thermal conductivity λ depends on pressure only in gases and fluids (compressible materials) and thus, this value can be assumed as ρ-independent for solid materials [12].
- Thermal conductivity λ for typical materials (copper, aluminium, ferrite) is nearly temperature-independent as shown in [13], and thus, this value can be assumed as temperature-invariant for solid materials.
- This, thermal conductivity λ is partially homogeneous and can be extracted from the divergence operator.
- The heat source density $\dot{W}$ is assumed to be invariant to temperature and time.
- Only a steady-state temperature field is considered, since the internal heat sources are nearly time-independent due to the high thermal capacitances and the system does not change after settling. Thus, the left side of (2) is zero.

Thus, the heat conduction equation reduces to Poisson's equation:

$$\begin{aligned} 0 &= \lambda \cdot \mathrm{div}(\mathrm{grad}(\vartheta)) + \dot{W}(\vec{r}) \\ \Leftrightarrow 0 &= \lambda \cdot \Delta\vartheta + \dot{W}(\vec{r}) \\ \Leftrightarrow -\dot{W}(\vec{r}) &= \lambda \cdot \Delta\vartheta \end{aligned} \tag{3}$$

which is implemented in the solver. Finally, the starting point for the equations is the same as used in [7]. More details how to program the solver are shown in [8].

Conductor Insulation Simulation by an Equivalent Thermal Conductivity

Each conductor of an inductive component requires electrical insulation. Typically, the insulation has a much lower thermal conductivity compared to the electrical conductor. If this effect is to be modeled, a very fine mesh in each conductor insulation is necessary due to the small distances. To keep the mesh size, and thus the computation time low, an effective conductivity λ_{eff} of a thermally equivalent conductor is introduced. The equivalent conductor has the effective conductivity of a normal conductor, which consists of conductor and insulation, see Fig. 5a. The effective thermal conductivity is derived from the conductor radius r_{c}, the insulation radius r_{i}, the conductivity of the conductor λ_{c} and the conductivity of the insulation λ_{i} as (4):

$$\lambda_{\mathrm{eff}} = \frac{1}{2 \cdot \frac{\ln(r_{\mathrm{i}}/r_{\mathrm{c}})}{\lambda_{\mathrm{i}}} + \frac{1}{\lambda_{\mathrm{c}}}}. \tag{4}$$

The core temperature of the thermally equivalent conductor corresponds to the core temperature of the real conductor (see Fig. 5b). For this example, a conductor with radius of $1.0\,\mathrm{mm}$ and an insulation radius of $1.1\,\mathrm{mm}$ is used. The conductor heat conductivity is given with $\lambda_\mathrm{c} = 400\,\frac{\mathrm{W}}{\mathrm{mK}}$ and the insulation heat conductivity is $\lambda_\mathrm{i} = 0.42\,\frac{\mathrm{W}}{\mathrm{mK}}$. The outer temperature is $60\,^{\circ}\mathrm{C}$ and the dissipated electrical power density P_0 is given. For $10\,\mathrm{W}$ dissipated power in the conductor, the power density is $P_0 = \frac{P}{r_\mathrm{c}^2 \cdot \pi \cdot l} = \frac{10\,\mathrm{W}}{1\,\mathrm{mm}^2 \cdot \pi \cdot 1\,\mathrm{m}} = 3183\,\frac{\mathrm{kW}}{\mathrm{m}^3}$. The formula for the effective temperature is derived in [13]:

$$\vartheta_\mathrm{cond,eff}(\varrho) = -\frac{\varrho^2 P_0}{4\lambda_\mathrm{eff}} + \frac{r_\mathrm{c}^2 P_0}{2} \cdot \left(\frac{\ln(\frac{r_\mathrm{i}}{r_\mathrm{c}})}{\lambda_\mathrm{i}} + \frac{1}{2\lambda_\mathrm{c}}\right) + T_0. \tag{5}$$

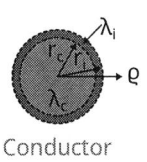
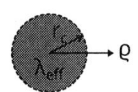

Conductor Insulation

Thermally Equivalent Conductor

(a) Conductor with insulation is summarized to an thermally equivalent conductor

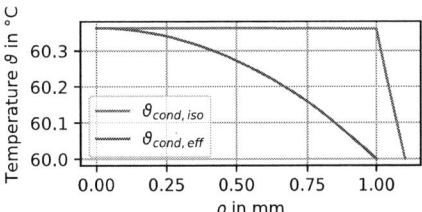

(b) Temperature distribution inside the conductor with insulation and the thermally equivalent conductor

Fig. 5: Description of an conductor with an effective thermally conductivity

A conductor made of stranded wires can be converted into a solid wire conductor beforehand with the help of the fill factor F. The conductor radius for stranded wires results to

$$r_\mathrm{c} = r_\mathrm{i} \cdot \sqrt{F}. \tag{6}$$

The core temperature of the stranded wire can then be calculated using (4) and (5).

This example shows, how little influence the insulation has on the temperature increase within the conductor $(0.37\,\mathrm{K})$. The influence is almost negligible. For a litz wire using the same insulation diameter and a fill factor $F = 0.5$, the temperature rise is $0.79\,\mathrm{K}$.

5 Verification

In a first step, the verification of the results is provided by a 2D rotationally symmetric reference simulation with the 2D FEM simulation tool FEMM to ensure the solver is implemented correctly. Further, the results are verified by measurements.

Verification by 2D Reference Simulation

To verify the automated simulation in FEMMT, a comparison simulation is set up in FEMM. A PQ40/40 N95 core is used for this purpose. The coil being investigated is set up by a winding with eight turns of solid wire with a radius of $1.5\,\mathrm{mm}$. The power dissipation of core and windings come from a magnetoquasistatic simulation performed before. A $3\,\mathrm{A}$ sinusodial peak current at $100\,\mathrm{kHz}$ flows through the winding. The boundary conditions on all sides are Dirichlet boundary conditions with a temperature of $20\,^{\circ}\mathrm{C}$. Fig. 6a shows the validation simulation for FEMM, Fig. 6b the simulation with FEMMT. The temperature distribution and peak temperatures match well.

Temperature Measurement Validation

In order to verify the results in practice, an experimental setup is constructed. This is shown schematically in Fig. 7a and in reality in Fig. 8a. In [9] the power dissipation of a coil is determined calorimetrically and compared with the simulation. The same coil with the same data is used, extended by some temperature sensors which are distributed inside the turns (see Fig. 7b. The coil is placed in potting material. The temperature of the container is approximately constant due to the thick aluminum

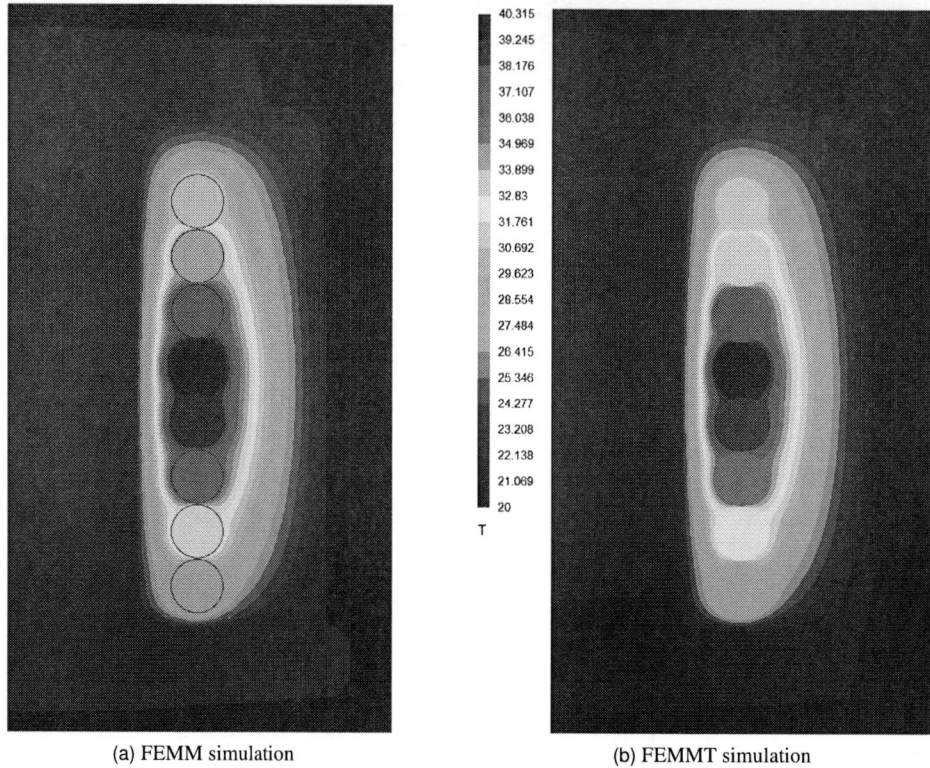

(a) FEMM simulation   (b) FEMMT simulation

Fig. 6: Temperature field calculation using FEMM and FEMMT

walls. Thus, a Dirichlet boundary condition can be assumed. Since only heat conduction, but not convection, was considered in the simulation, a styrofoam is put over the arrangement, which reproduces the Neumann boundary condition (perfect thermal insulation). Cooling is provided by an overdimensioned heat sink with a pressure chamber and a fan to enable an almost constant heat sink temperature. With the assistance of a resonant circuit, excited by a half-bridge, a sinusoidal current is injected into the coil. In the steady state, the measured temperature field is compared with the simulated temperature field.

Fig. 8b shows the comparison between the FEMMT simulation and the measurement results of the lab setup. The sensors are named according to Fig. 7b.

Even the simulation consideres a 2D rotationally symmetric inductor, what differs from the real PQ40/40 core, the results are matching well.

6   Using the Toolbox for Optimization

In the following, it is shown how the toolbox can be used to optimize magnetic components with regard to various criteria, see also Fig. 9. First, the user enters the possible geometry and current shapes for the magnetoquasistatic solver. The next step includes an analytical pre-calculation using a reluctance model to sort out parameter sets, which do not fit the goal-parameters (e.g., inductance values). The remaining geometries are built and simulated automatically with the help of the FEM module in ONELAB. Such an optimization is shown in [14]. In the next step, different cooling systems can be simulated and configurations with too high core- or conductor temperatures can be automatically sorted out.

7   Summary and Outlook

In order to easily create standard magnetic and thermal setups in FEM simulations, this paper presents an open-source FEM magnetics toolbox. The toolbox provides methods for drawing different cores, air

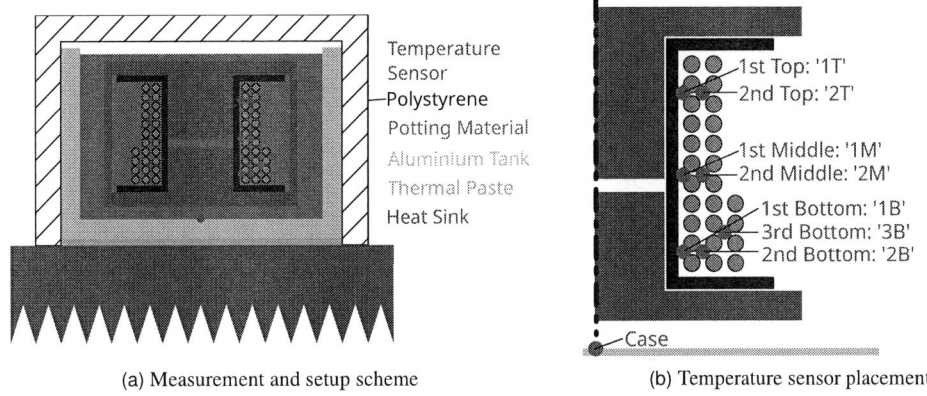

(a) Measurement and setup scheme

(b) Temperature sensor placement

Fig. 7: Inductor temperature verification setup

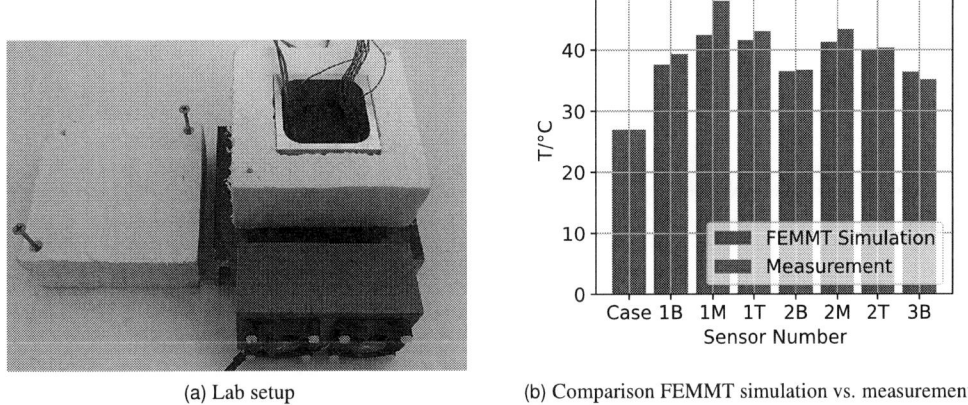

(a) Lab setup

(b) Comparison FEMMT simulation vs. measurement

Fig. 8: Inductor temperature verification measurement

gaps, winding schemes and winding types including its thermal behaviour. The built-in control of the FEM simulation tool is based on Python. The toolbox provides the basis to automate the design process of magnetic components with FEM simulations.

In the near future, the toolbox functions will be expanded in the direction of usability and optimization. Currently, work on a graphical user interface is in progress. An integrated material database will provide data on demand for certain standard materials, such as complex core parameters, electrical and thermal conductivities. A thermal equivalent circuit will be introduced according to [11]. Various optimization routines are to be integrated directly into the toolbox in order to be able to perform the design according to different criteria.

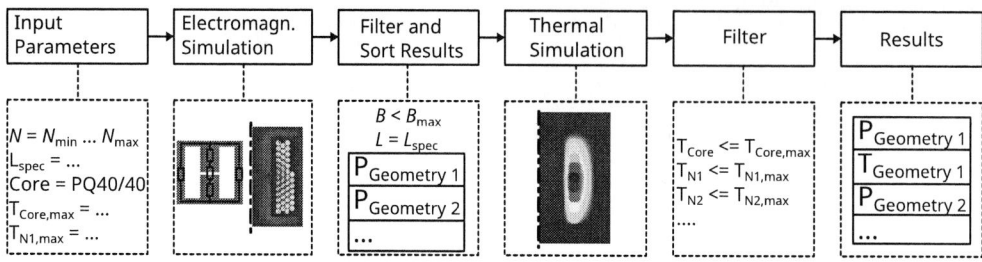

Fig. 9: Usage of FEMMT for an optimization

References

[1] R. Unruh, F. Schafmeister and J. Böcker, "11kW, 70kHz LLC Converter Design with Adaptive Input Voltage for 98 % Efficiency in an MMC," 2020 IEEE 21st Workshop on Control and Modeling for Power Electronics (COMPEL), 2020.

[2] M. J. Jacoboski, A. de Bastiani Lange and M. L. Heldwein, "Closed-Form Solution for Core Loss Calculation in Single-Phase Bridgeless PFC Rectifiers Based on the iGSE Method," in IEEE Transactions on Power Electronics, vol. 33, no. 6, pp. 4599-4604, June 2018, doi: 10.1109/TPEL.2017.2775106.

[3] J. B. Goodenough, "Summary of losses in magnetic materials," in IEEE Transactions on Magnetics, vol. 38, no. 5, pp. 3398-3408, Sept. 2002, doi: 10.1109/TMAG.2002.802741.

[4] J. Mühlethaler, "Modeling and multi-objective optimization of inductive power components", 2012.

[5] Christophe Geuzaine et al.: ONELAB, `https://onelab.info/`, Accessed Thursday 30th June, 2022.

[6] Alexandre Halbach: Sparselizard, `https://sparselizard.org/`, Accessed Thursday 30th June, 2022.

[7] David Meeker: FEMM, `https://www.femm.info`, Accessed Thursday 30th June, 2022.

[8] C. Geuzaine. „GetDP: a general finite-element solver for the de Rham complex". In: PAMM Volume 7 Issue 1. Special Issue: Sixth International Congress on Industrial Applied Mathematics (ICIAM07) and GAMM Annual Meeting, Zürich 2007.

[9] Nikolas Förster, Till Piepenbrock, Philipp Rehlaender, Oliver Wallscheid, Frank Schafmeister and Joachim Böcker. "An Open-Source FEM Magnetics Toolbox for Power Electronic Magnetic Components". PCIM Europe 2022; International Exhibition and Conference for Power Electronics, Intelligent Motion, Renewable Energy and Energy Management. 2022.

[10] LEA Power Electronics and Electrical Drives: FEM Magnetics Toolbox, `https://github.com/upb-lea/FEM_Magnetics_Toolbox`.

[11] Manfred Albach: Induktivitäten in der Leistungselektronik, ISBN: 978-3-658-15080-8, 2017.

[12] Hans Dieter Baehr and Karl Stephan: Wärme- und Stoffübertragung, ISBN: 978-3-662-58441-5, 2019.

[13] Jonas Hölscher. "Automatisierte thermische Analyse ausgewählter Transformatoren und Induktivitäten". Bachelor Thesis. Paderborn University 2022.

[14] Till Piepenbrock. „Automated FEM Transformer Design for a Dual Active Bridge". Master Thesis. Paderborn University 2022.

Comparison of Dual-Active-Bridge-based Topologies for single-phase single-stage EV On-board Chargers

Daniel Gaona, Denis Pauls, Eduardo Facanha de Oliveira
Huawei Technologies Co., Ltd.
Suedwestpark 48, 90449, Nuremberg, Germany
Corresponding Author's Email: daniel.gaona@huawei.com

Keywords

≪AC-DC≫, ≪Automotive application≫, ≪Dual-Active-Bridge (DAB)≫, ≪Electric Vehicle≫, ≪On-board charger≫, ≪Optimization method≫, ≪Single stage≫,

Abstract

Single-stage on-board chargers for EVs are an attractive solution as they reduce the number of converters stages leading to potential increments in power density and cost reduction. Dual-active-bridge topologies are particularly attractive for these applications as they offer a large voltage conversion range, isolation, and relatively simple controllability. Several DAB-based single-stage AC-DC topologies can be found in the literature; however, a comparison of said topologies has not been yet been presented. Filling the gap, this paper presents and compares the most promising DAB-based topologies used for single-stage OBC applications. For this purpose, a multi-objective optimization with a Pareto-front is used. Apart from the design of the main components, EMI filters are designed for each topology and their size is considered as part of the evaluation.

1 Introduction

Despite the continuously increasing deployment of DC-charging stations, onboard chargers (OBCs) are likely to remain the predominant EV charging solution in the next years [3]. The rate of transmission to DC-charging as well the permanence of OBCs as components of the EVs depends, not in a minor way, on them achieving high power density at minimum cost. In this context, single-stage AC-DC topologies have received particular attention in the last years. In contrast to two-stage converter topologies, they can lead to higher power density, lower costs, and reliability due to the non-existing DC-link capacitor and reduced component count.

Several single-stage concepts have been developed over the years [4] for both single-phase and three-phase grids. The DAB-based matrix converter and the T-type multi-port converter are particularly relevant for three-phase grids. However, they are not suitable for single-phase grids; moreover, they require a large number of components and complex control mechanisms. Single-phase converters, on the other hand, rely on simpler structures and control; however, they suffer from the second-harmonic power ripple, typical of single-phase systems.

Most single-stage single-phase OBCs are based on the Dual-Active-Bridge topology (DAB). The DAB offers a wide voltage conversion range, isolation, bi-directionality, and a wide range of control strategies. Its control strategy has a direct impact on the soft-switching range of the converter as well as on its size and efficiency. Different control strategies have been proposed in the last years [5, 6, 7]; nonetheless, some of them are highly complex and/or require auxiliary circuits which makes them less attractive for real applications.

To the best of the author's knowledge, a comparison of these DAB systems for the same design requirement and following simple control strategies has not been presented. To fill this gap, this paper presents a detailed analysis of the two most promising DAB-based topologies for single-stage single-phase OBCs: synchronous rectifier + DAB and an integrated DAB-Totem-Pole configuration. Section 2 discusses the operation of the DAB converter with two simple mechanisms of control: Single-Phase-Shift (SPS) control between the primary and secondary side, and Dual-Phase-Shift (DPS) control with an additional PWM control on the secondary side. The next section discusses the operation of the systems in the context of AC-DC converters. For each topology-control combination, a multi-objective optimization is performed following a Pareto-front-based approach of [8] to determine the effect of the control strategy on the converter design, as well as to facilitate the comparison of the topologies. Sections 4 and 5 cover the design process and results, respectively. Finally, the last section presents an estimation of the EMI filters for each topology in order to compare the overall converter size.

2 DAB Operation and Control Strategies

There are many control mechanisms for the DAB. The SPS and DPS modulation are the most commonly used due to their simplicity. These two strategies are considered in this work and adapted for AC-DC conversion.

2 Operation with Single-Phase-Shift Modulation

The simplest control method for a DAB is the SPS modulation (Fig. 1). Here the power transfer is controlled by adjusting the phase-shift (ϕ) between the two square-voltage waveforms (fixed duty-cycle of 50%) in the primary and secondary, respectively. The maximum power transfer is achieved at a phase-shift of $\pi/2$. The phase-shift (ϕ) required to transfer a required power (P) is given by (1):

$$\phi(P) = \frac{\pi}{2} \cdot \left[1 - \sqrt{1 - \frac{8 \cdot P \cdot f \cdot N \cdot L_{\text{DAB}}}{V_1 \cdot V_2}} \right];$$

(1)

where N : turn ratio, f_s : the switching frequency; the other variables are shown in Fig. 1. This modulation results in currents in the main inductor L_{DAB} similar to the one shown in Fig.1c). For the calculation of switching losses and the ZVS region, the commutation currents (I_0 and I_1 in Fig.1c)) are required. These can be calculated for each DC-DC operating point by means of (1)-(3):

$$I_{\text{o}} = -\frac{V_1}{2\pi \cdot f_s \cdot L_{\text{DAB}}} \cdot \left[\frac{V_2}{N \cdot V_1} \cdot \phi + \pi \cdot \frac{1 - V_2/(N \cdot V_1)}{2} \right];$$

(2)

$$I_1 = \frac{V_1}{2\pi \cdot f_s \cdot L_{\text{DAB}}} \cdot \left[\phi - \pi \cdot \frac{1 - V_2/(N \cdot V_1)}{2} \right].$$

(3)

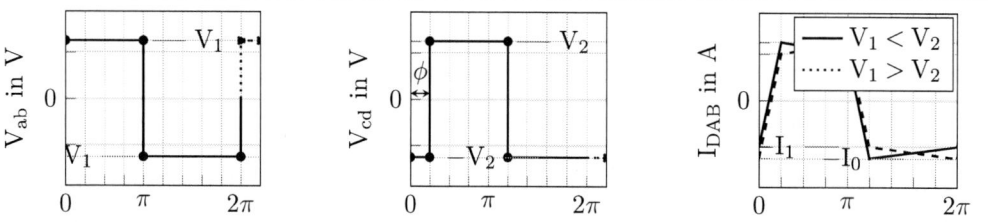

Fig. 1: a) Primary Voltage, b) Secondary Voltage, and c) Inductor Current of the DAB with a Single-Phase-Shift(SPS) Modulation.

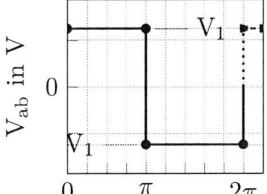 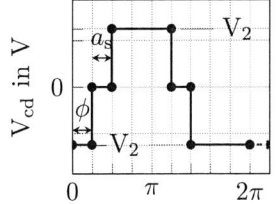 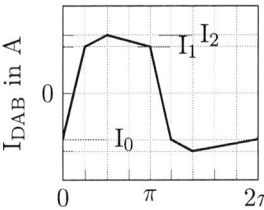

Fig. 2: Dual-Phase-Shift (DPS) Modulation

Based on these equations, the Zero-Voltage-Switching (ZVS) region can now be determined. For the primary side, ZVS occurs when $I_0 < 0$ and in the secondary, when $I_1 > 0$. The secondary side ZVS condition is almost fulfilled for the entire voltage range. The ZVS range can be dependent on the values of L_{DAB}, N, f_{s}. There is a correlation between the ZVS region and the inductor current. A compromise must be achieved during the design to ensure optimum performance.

2 Operation with Dual-Phase-Shift Modulation

Whereas the SPS modulation is fully based on one phase-shift the DPS modulation provides a further degree of freedom. The switches on the primary side are still switched with a fixed duty-cycle of 50%. On the secondary side, the duty-cycle and the phase-shift (ϕ) are both changed. The duty-cycle can be considered as a second phase-shift a_{s} as shown in Fig. 2. To simplify the power regulation, one can set this phase-shift to vary as: $a_{\mathrm{s}} = \pi - 2 \cdot \pi \cdot k \cdot \sin(2 \cdot \pi \cdot f_{grid} \cdot t)$. The value of k can be set to 0.5 to maximize the utilization of the converter [9].

The advantage of this control mechanism is to achieve a wider ZVS range and therefore potentially higher efficiency. It is worth noting that, contrary to what is stated in some literature (e.g [9]), there are two modes of operation when using this modulation strategy. Mode 1 is valid for high loads while Mode 2 dominates for partial loads. A failure to recognize these modes results in non-sinusoidal input currents. Since Mode 1 is used at full load, this mode is considered when sizing the main inductance L_{DAB}. This modulation results in currents such as the one shown in Fig. 2c). The equations for the commutation currents (I_0, I_1, and I_2 in Fig.2c)) as well as the equation for the phase-shift (ϕ) required to transfer a required power (P), are all shown below:

Mode 1

$$\phi_1(P) = \frac{\pi}{2} \cdot \left[1 - \sqrt{1 - \frac{8 \cdot P \cdot f \cdot N \cdot L}{V_1 \cdot V_2} - \left(\frac{a_{\mathrm{s}}}{\pi}\right)^2} \right]; \tag{4}$$

$$I_{0,1} = -\frac{V_1}{2 \cdot \pi \cdot f_{\mathrm{s}} \cdot L} \cdot \left[\frac{V_2}{N \cdot V_1} \cdot \phi + \frac{\pi}{2} \cdot \left(1 - \frac{V_2}{N \cdot V_1} \right) \right]; \tag{5}$$

$$I_{1,1} = \frac{V_1}{2 \cdot \pi \cdot f_{\mathrm{s}} \cdot L} \cdot \left[\phi - \frac{\pi}{2} \cdot \left(1 - \frac{V_2}{N \cdot V_1} \right) - \frac{a_{\mathrm{s}}}{2} \cdot \left(1 + \frac{V_2}{N \cdot V_1} \right) \right]; \tag{6}$$

$$I_{2,1} = \frac{V_2}{2 \cdot \pi \cdot f_{\mathrm{s}} \cdot} \cdot \left[\phi - \frac{1}{2} \cdot \left(1 - \frac{V_2}{N \cdot V_1} \right) \cdot (\pi - a_{\mathrm{s}}) \right]; \tag{7}$$

Mode 2

$$\phi_2(P) = \frac{2\pi^2 \cdot P \cdot f \cdot L \cdot N}{V_2 \cdot V_1 \left(\pi - a_s\right)}; \tag{8}$$

$$I_{0,2} = -\frac{V_1}{2 \cdot \pi \cdot f_s \cdot L} \cdot \left[\frac{a_s}{2} + \frac{1}{2} \cdot \left(1 - \frac{V_2}{N \cdot V_1}\right) \cdot (\pi - a_s)\right]; \tag{9}$$

$$I_{1,2} = \frac{V_1}{2 \cdot \pi \cdot f_s \cdot L} \cdot \left[\phi - \frac{1}{2} \cdot \left(1 - \frac{V_2}{N \cdot V_g}\right) \cdot (\pi - a_s)\right]; \tag{10}$$

$$I_{2,2} = \frac{V_1}{2 \cdot \pi \cdot f_s \cdot L} \cdot \left[\phi + \frac{1}{2} \cdot \left(1 - \frac{V_2}{N \cdot V_1}\right) \cdot (\pi - a_s)\right]; \tag{11}$$

3 DAB-based Single-Stage AC-DC Topologies

The two control strategies from the previous section are applied to two DAB-based topologies. The first one, Fig. 3, is a Synchronous Rectifier (SR) followed by a DAB. The second one, Fig. 5, is an Interleaved Totem-Pole DAB configuration. These are described in more detail in the following sections.

3 Synchronous Rectifier + DAB

The first topology corresponds to the SR DAB with SPS and DPS modulations. Here the grid voltage is rectified by the input bridge. The primary of the DAB imposes this voltage in a square waveform across the inductor and transformer of the DAB. The power flow and power factor correction are then regulated by the secondary side using the SPS or DPS modulations.

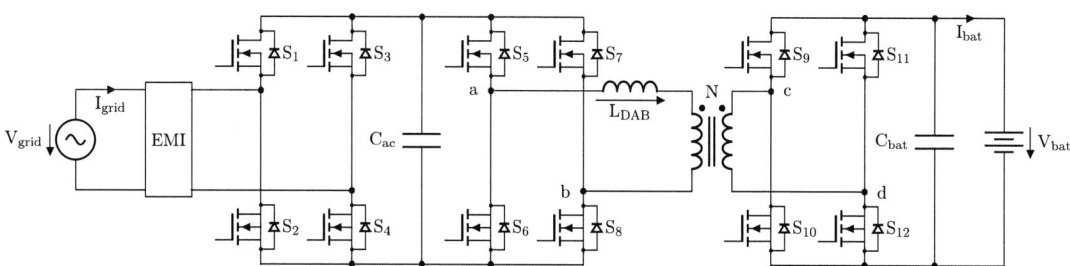

Fig. 3: Dual-Active-Bridge with Synchronous Bridge Rectification

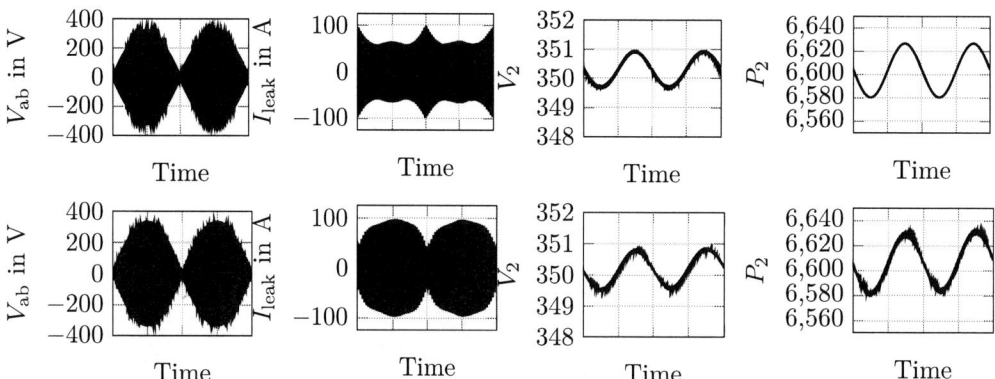

Fig. 4: Simulation Results: TOP ROW: SR DAB SPS, BOTTOM ROW: SR DAB DSP.

Single-Phase-Shift Modulation The use of this converter with SPS results in the waveforms shown in Fig. 4 (TOP ROW). The 100Hz oscillations seen in the secondary voltage (V_2) and power (P_2) are caused by the single-phase grid interconnection. It is worth noting that the current in the main inductor is high at the zero-crossing of the line voltage even though the power transferred to the load at this point is zero. These circulating currents are detrimental to the efficiency of the system.

Dual-Phase-Shift Modulation Although the voltage and power oscillation are similar, the current waveforms of the main inductor L_{DAB} are different when using DSP. As shown in Fig. 4 (BOTTOM ROW), lower currents can be seen at the zero-crossing of the line voltage, leading to better overall performance.

3 Interleaved Totem-Pole DAB

The second topology to be examined is the Interleaved Totem-Pole DAB (ITP) Topology using the DPS modulation previously described.

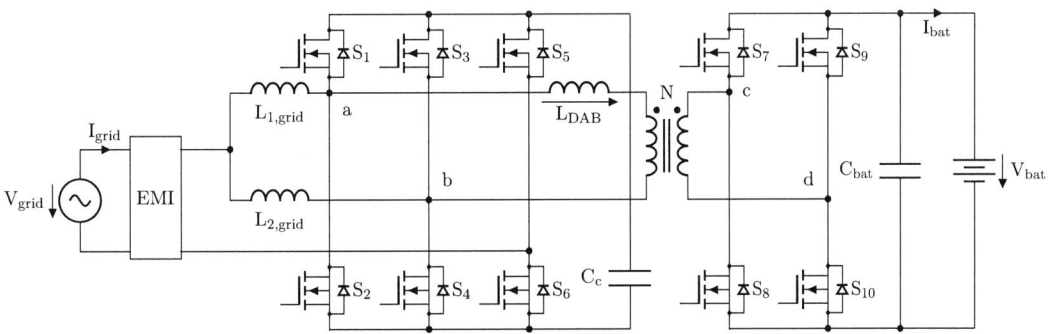

Fig. 5: Interleaved Totem-Pole Dual-Active-Bridge

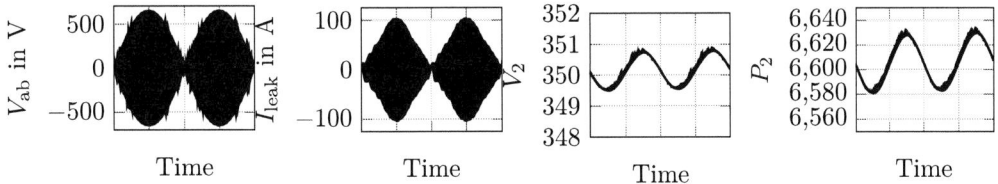

Fig. 6: Simulation Results - ITP DAB DPS

Dual-Phase-Shift Modulation The operation of the ITP DAB is as follows. Switches S_5-S_6 operate at the grid frequency while switches S_1-S_4 operated with 50% duty cycle synchronously - The same is true for S_2-S_3. Leg a and b are interleaved. As a result, the voltage in the capacitor C_c is twice the rectified grid voltage. This voltage is imposed in a square waveform at the primary of the DAB converter as shown in Fig. 6. The secondary is in charge of the power factor correction. This is achieved by making sure that the power flow follows the characteristic single-phase grid power: $P(t) = 2 \cdot P_{avg,out} \cdot \sin(2 \cdot \pi \cdot f_{grid} \cdot t)^2$.

4 Design Methodology

The number of design variables to be defined is large. Thus, a Pareto-front-based optimization was developed to consider all possible combinations of the turn ratio, switching frequency, output

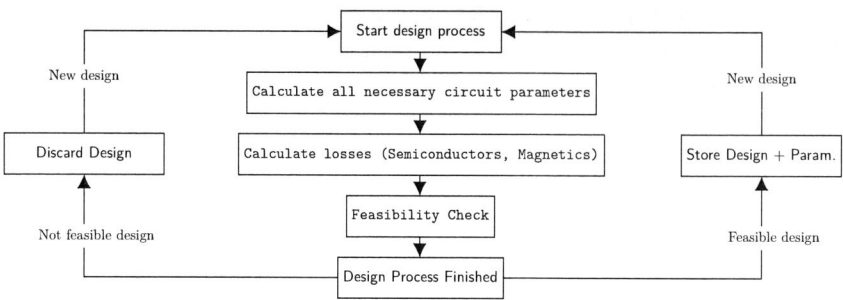

Fig. 7: Design Process developed in MATLAB

and input voltage, power switches, typical ferrite cores, core materials, etc. as shown in Fig. 7. The idea behind this design methodology is to evaluate all possible combinations within predetermined design constraints. Afterward, feasible designs are presented as a point in the Pareto-front plot from which the most appropriate design can be selected depending on the design objective.

5 Pareto-Front-based Design Results

The Pareto-front design is shown in the TOP ROW of Fig. 8 for the three topologies of interest. As expected, the average efficiency increases along with the volume of magnetic material (larger cores). The SR DAB SPS shows lower average efficiency than the SR DAB with DPS. This was expected as the SPS leads to higher circulating currents. Regarding switching frequency, larger values result in smaller cores, however, also in lower efficiencies, TOP ROW in Fig. 8. To achieve an average efficiency of 95%, the frequency is limited to 200kHz for the three topologies.

It can be seen that the ITP DAB DPS has a significantly higher volume of magnetic components for the same efficiency. This is due to the two grid inductors ($L_{1,grid}$ and $L_{2,grid}$ in Fig. 5). One can argue that this is an unfair comparison as the EMI filters have not been taken into consideration. The impact of the EMI filter is discussed in the next section.

The smallest designs that met the efficiency requirement of 95% were selected for each topology. The parameters for each of these designs are detailed in Table I. For each one of them, a loss-breakdown and efficiency are presented in the SECOND and THIRD ROWS of Fig. 8, respectively. A linear de-rating is considered for battery voltages greater than 450V or lower than 300V. The SR DAB DPS has the best performance in terms of average efficiency.

Table I: Benchmark of all three Systems

	SR DAB SPS	SR DAB DPS	ITP DAB DPS
Avg. Efficiency in %	95.17	96.39	95.24
Main Losses in W	$P_{cond,pri}$, $P_{off,pri}$, P_{rec}	$P_{off,sec}$, P_{rec}	$P_{con,pri}$, $P_{off,pri}$
Rectifier	STW78N65M5	STW78N65M5	No Rectifier
Switches	C3M0025065K	C3M0025065K	SCT3105KLHR
			C3M0025065K
Voltage Stress	650 V	650 V	900 V, 650 V
Estimated Cost of Switches (March 2022)	263€	263€	244€
Cores	T:PQ50/40	T:PQ50/40	T:PQ50/40
	L:PQ40/40	L:PQ40/40	L:3*PQ40/40
Core Material	N97 (100 °C)	N97 (100 °C)	N97 (100 °C)
DAB Box Volume in cm^3	147.2	147.2	265.6
EMI Filter Box Volume in cm^3	347	326	317
Overall system Volume in cm^3	494.2	473.2	582.6

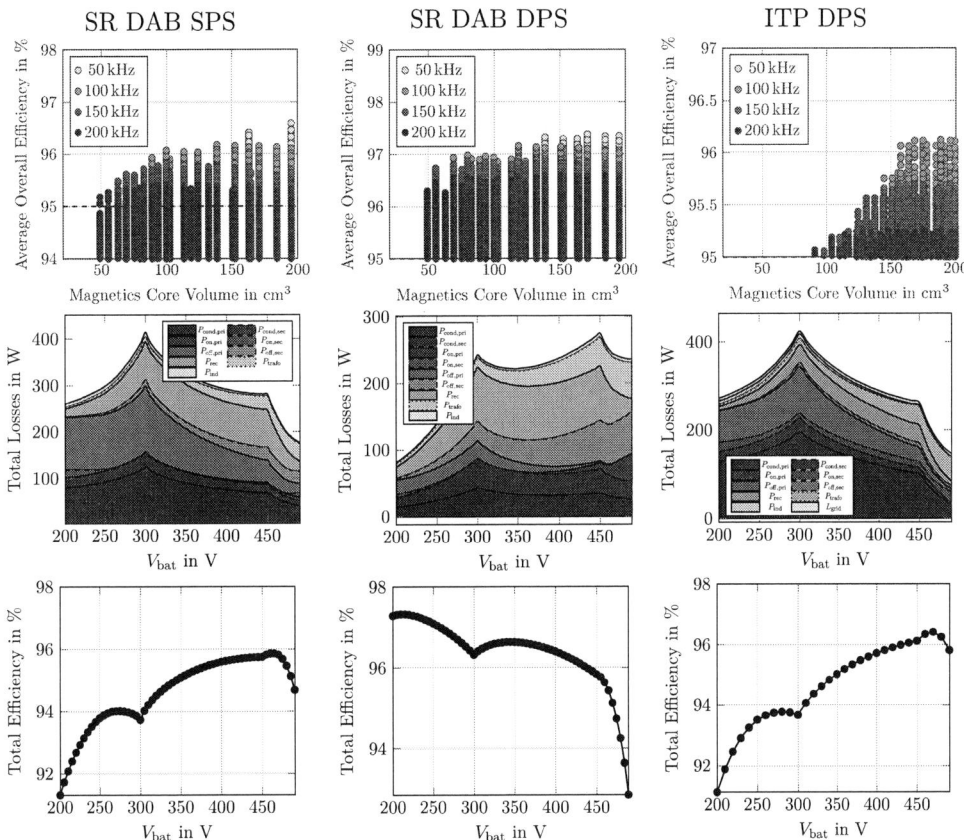

Fig. 8: Pareto-Front-Based Design. TOP ROW: Efficiency vs. Magnetic Core Volume vs. Frequency. The smallest designs that met the efficiency requirement of 95% were selected for each topology. For each one of them, a loss-breakdown and efficiency are presented in the SECOND and THIRD ROWS for different battery voltage and at rated power (6.6 kW).

6 EMI filter design

The EMI filter design followed the simplified approach presented in [10, 11, 12]. Thus, the LISN circuit shown in Fig.9b) was added to the simulation models in PLECS. Following the standards EN5502 and CISPR16, the voltage of the LISN was passed through a Quasi-Peak (QD) detector before being compared to the limits for Class B and A as shown in Fig.11d). From there, the most troublesome frequencies (f_D) are identified and their required attenuations (Att_{req}) are derived using (12):

$$Att_{req}(f_D)\,[dB] = U_{QP}(f_D)\,[\mathrm{dB}\cdot\mu\mathrm{V}] - \mathrm{Limit}_{Class\,B}(f_D)\,[\mathrm{dB}\cdot\mu\mathrm{V}] + \mathrm{Margin}\,[\mathrm{dB}\cdot\mu\mathrm{V}] \quad (12)$$

The filter components are then sized accordingly:

1. **Capacitor**: The capacitor of the LC filter imposes a reactive power into the system which is more severe at light loads. As a general requirement, the power factor (PF) should not be lower than $PF_{\min} = 0.9$ at 10% of the load. Thus, the maximum capacitor value can be calculated from (13):

$$C_{max} = \frac{\tan(\arccos(PF_{\min}))}{2\cdot\pi\cdot f_{grid}\cdot R_{load}} = 9.62\mu\mathrm{F}, \quad \text{with} \quad R_{load} = \frac{V_{grid,rms}}{0.1\cdot P_{rated}} \quad (13)$$

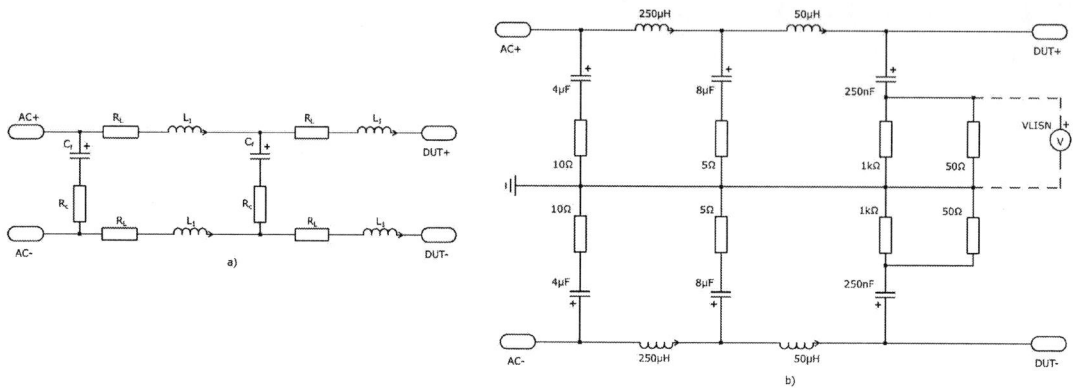

Fig. 9: *a)* Example of a DM EMI filter with 2-stages. *b)* LISN considered for the EMI filter design.

2. **Inductance:** With n_f as the number of stages, the inductance (L_1) required in each rail by the filter can be calculated from (14) for each troublesome frequency f_D. The selected inductor must achieve the required attenuation for all the troublesome frequencies.

$$L_1(f_D) = \frac{1}{2} \cdot \sqrt[n_f]{\frac{10^{Att_{req}(f_D)[\text{dB}]/20}}{(2\pi f_D)^{2 \cdot n_f} \times (C_{max}/n_f)^{nf}}}; \quad \text{and} \quad L_{1,final} = \max[L_1(f_{D}1), L_1(f_2), ...]$$
$$(14)$$

3. **Volume minimization**: The aforementioned process can be applied to different number of stages n_f. Their volume can be estimated following the method in [10]. The filter volume is given by (16) - (17):

$$\text{Vol}_{box,filter}[\text{cm}^3] = n_f \times (2 \cdot \text{Vol}_{box,L} + \text{Vol}_{box,C}); \tag{15}$$

$$\text{Vol}_{box,L}[\text{cm}^3] = k_{L1} \cdot L_1 \cdot i_{g,pk}^2 + k_{L2} \cdot L_1 + k_{L3} \cdot i_{g,pk}; \tag{16}$$

$$\text{Vol}_{box,C}[\text{cm}^3] = k_{C1} \cdot \frac{C_{max}}{n_f} \times U_{g,pk}^2 + k_{C2} \times U_{g,pk}; \tag{17}$$

where $i_{g,pk}$ and $U_{g,pk}$ are the maximum values of the grid current and voltage, respectively. Exemplary values are shown in Fig.10. With these, the number of stages is selected as the one which results in the smaller box-volume.

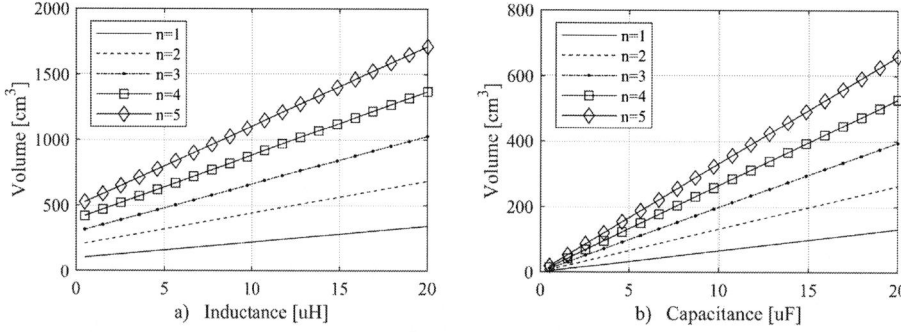

Fig. 10: Volume of the EMI *a)* Inductor and *b)* Capacitor for different number of stages. $i_{g,pk} = 40.6$A, $U_{g,pk} = 325$V. For the toroidal cores from *Magnetics*: $[k_{L1}, k_{L2}, k_{L3}] = [3, 8, 1.1]$; for the typical grid capacitors: $[k_{C1}, k_{C2}] = [62, 0.66]$ [4].

Table II: EMI Filter Design

	N° Stages	$L_1[\mu H]$	$\text{Vol}_{box,L}$ [3*] $[cm^3]$	$C_1[\mu F]$	$\text{Vol}_{box,C}$ $[cm^3]$	$\text{Vol}_{box,filter}$ $[cm^3]$
SPS[1*]	2	2.6	282	4.8	64	347
DPS[1*]	2	2.6	262	4.8	64	326
ITP[2*]	2	1.68	242	4.55/0.5	61.12/13.77	317

[1*]: The soft-DC link takes the role of one of the filter capacitors. [2*]: The $0.5\mu F$ is located in the soft DC-link.

[3*]: Note that here the *Box Volume* is used as oppose to the volume of the magnetic material as in Fig.8

Table II shows the EMI filter designs for the three studied topologies. $Fig.11c) - f)$ shows the LISN spectrum with and without the filter as well as the grid current and its harmonic content with the filter in place, for the three topologies of interest. The size of the ITP system is smaller than that of the other topologies. However, the difference in size is minimum and does not compensate for the larger size of the input grid inductors. The topology with the smaller filter is the SR DAB DPS.

7 Conclusions

The paper describes the operation of two promising DAB-based single-phase single-stage On-board chargers with two different modulation/control strategies. For each one of them, a design optimization is carried out with the objective to achieve an average efficiency of $\geq 95\%$ with the smaller possible volume. Different operating frequencies, switches, magnetic materials, and turn ratios were considered. The results show that it is possible to achieve average efficiencies of over 95% using single-stage topologies. Moreover, the optimization shows that the SR DAB DPS results in the highest efficiency and highest power density when compared to the SR DAB with SPS and the ITP DAB DPS. This opens an opportunity to research more complex modulation/control schemes that could further improve efficiency. It is also worth noting that, although the ITP DAB DPS requires a smaller EMI filter and fewer switches, its overall size is still larger than for the other topologies.

References

[1] Vanderkeyn Ralf W.: Example of fast switching component, EPE Journal Vol 20 no 5, pp. 48- 56
[2] Deboe B. D.: A novel type of grid converter, EPE 2013-ECCE Europe, paper 0321
[3] Yole. (2021).DC Charging for Plug in Electric Vehicles 2021. Market and Technology Report.
[4] J. Yuan, L. Dorn-Gomba, A. D. Callegaro, J. Reimers and A. Emadi, "A Review of Bidirectional On-Board Chargers for Electric Vehicles," in IEEE Access, vol. 9, pp. 51501-51518, 2021, doi: 10.1109/ACCESS.2021.3069448.
[5] Modeling and Optimization of Bidirectional Dual Active Bridge DC–DC Converter Topologies. (2010). Florian Krismer.
[6] Design and Operation Considerations of Three-Phase Dual Active Bridge Converters for Low-Power Applications with Wide Voltage Ranges (2016). Hauke van Hoek.
[7] Modeling and Optimization of Bidirectional Dual Active Bridge AC–DC Converter Topologies (2014). Jordi Everts.
[8] Modeling and Efficiency-Power Density-Pareto Optimization of Inductive Power Transfer Coils for Electric Vehicles Roman Bosshard, Johann Walter Kolar (2015). IEEE Journal of Emerging and Selected Topics in Power Electronics.
[9] Interleaved Totem-Pole ZVS Converter Operating in CCM for Single-Stage Bidirectional AC–DC Conversion With High-Frequency Isolation Hamza Belkamel, Hyungjin Kim (2021). IEEE Transaction of Power Electronics Vol. 36
[10] K. Raggl, T. Nussbaumer and J. W. Kolar: Guideline for a Simplified Differential-Mode EMI Filter Design, IEEE Transactions on Industrial Electronics, vol. 57, no. 3, pp. 1031-1040, March 2010, doi: 10.1109/TIE.2009.2028293.
[11] Marcelo Lobo: Heldwein: PhD Thesis, EMC filtering of Three-phase PWM Converters. ETH Zurich, 2008.
[12] T. Nussbaumer, M. L. Heldwein and J. W. Kolar: Differential Mode Input Filter Design for a Three-Phase Buck-Type PWM Rectifier Based on Modeling of the EMC Test Receiver, IEEE Transactions on Industrial Electronics, vol. 53, no. 5, pp. 1649-1661, Oct. 2006. doi: 10.1109/TIE.2006.881988

Comparison of Dual-Active-Bridge-based Topologies for single-phase single-stage EV On-board Chargers

GAONA Daniel

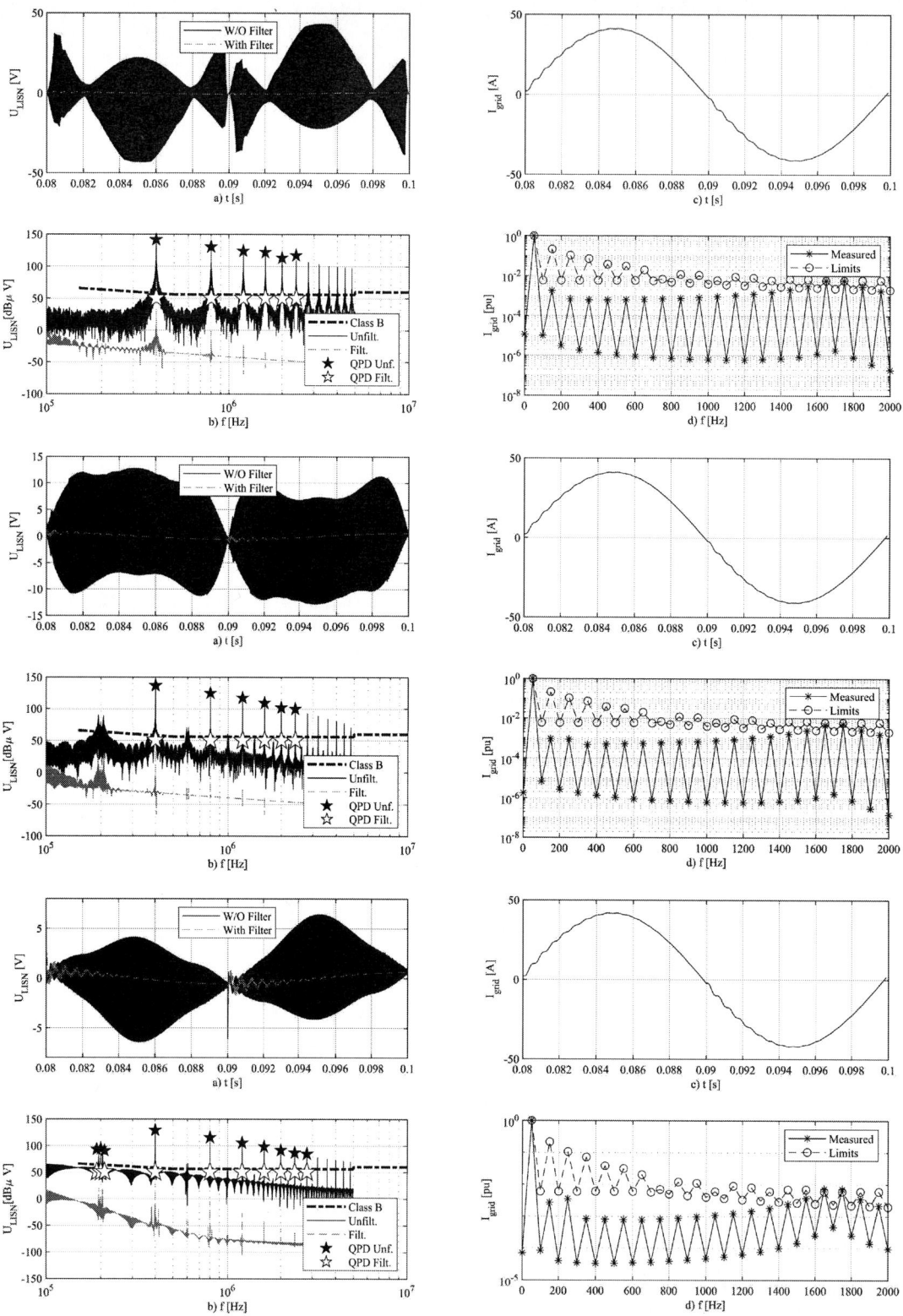

Fig. 11: For the DAB with SPS (top), DPS (middle) and the ITP with DPS (bottom): *a)* LISN measured voltage. *b)* Harmonic Spectrum and Limits (QP: Quasi-peak detector). *c)* Grid current with the EMI filter. *d)* Harmonic components and limits of grid current with the EMI filter.

Design Concepts for Medium Voltage DC Networks supplying the Future Circular Collider (FCC)

Manuel Colmenero*, Francisco R. Blanquez*, Ramon Blasco-Gimenez[+]
*CERN - European Organization for Nuclear Research
[+]Polytechnic University of Valencia
*1 Espl. des Particules, Meyrin, Switzerland
[+]Camino de Vera, s/n, Valencia, Spain
Email: mcolmene@cern.ch
URL: http://www.cern.ch

Keywords

≪Distribution of electrical energy≫, ≪Modular Multilevel Converters (MMC)≫, ≪Short circuit≫, ≪DC-DC≫, ≪Particle accelerators≫.

Abstract

In the frame of the feasibility studies for the construction of the Future Circular Collider (FCC), a new particle accelerator of 90 km of circumference under design, the use of Medium Voltage DC Grids (MVDC) for distribution has attracted attention considering their potential advantages in terms of transmission losses, power flow control and energy storage integration.

Traditionally, the main concerns of the DC networks have been their reliability and robustness. In this sense, a DC network where the AC/DC and DC/DC converters are based on the MMC (Modular Multilevel Converter) could offer advantages, thanks to the modularity, the easy maintenance, the possibility of integrating distributed energy storage and the redundancy. However, there is little knowledge about how a network for such a large power should be implemented.

Accordingly, this work describes the design aspects of a Medium Voltage DC network based on the MMC topology for the FCC, focusing on the specificities of non-isolated step-down DC/DC converters and examining how their control should be implemented to power a magnet chain and recover its energy. Additionally, a method to mitigate the impact of DC pole-to-pole and pole-to-ground faults is proposed.

Introduction

CERN is currently undergoing the feasibility studies for a new particle accelerator, the Future Circular Collider (FCC) [1]. This new ring of 90 km of circumference will overtake the CERN's Large Hadron Collider (27 km) as the world's biggest accelerator.

The technical challenges for the FCC are numerous. Considering the power consumption, the size of the facilities and the complexity of the machine, the electrical transmission and distribution networks shall be optimised to maximise the reliability, the efficiency and the robustness. Among the different research lines, the use of Medium Voltage DC Grids (MVDC) for the powering of some loads attracts much attention. MVDC grids could offer some advantages in terms of footprint reduction [2], mitigation of network perturbations [3], efficiency increase [4] and easy energy storage integration [5].

However, the implementation of MVDC networks is challenging. Compared with its AC counterpart, a DC distribution network, made of power converters, is less reliable due to their high sensitivity to overcurrents and overvoltages [6]. Besides, isolation of DC faults is difficult due to the lack of natural zero-crossings points on current [7]. Additionally, some faults in DC do not result in large overcurrents, but in large overvoltages [8], which must be cleared as soon as possible to avoid damages on insulation and reduce the risk of flashover. Finally, when energy storage (i.e. large capacitor or battery banks) is integrated, a DC fault can result in large fault currents [9] that need to be effectively suppressed to avoid damage or destruction of equipment. Therefore, a DC network requires the development of advanced control and protection strategies to achieve a level of reliability similar to a conventional AC network.

Among the different powering scenarios being considered, there is one where MMCs are used to build the MVDC network, using them as either AC/DC and DC/DC converters. Compared with a DC network based on other converter topologies [10],[11], MMCs could offer several advantages in terms of reliability and DC fault management: given their modular approach, a MMC can ride-though the failure of a single component without interrupting its operation [12]. Moreover, the submodules of the MMC allow to integrate a large amount of energy storage (i.e. batteries) in a distributed way [13], reducing or avoiding the damage in case of individual failure. In terms of protection, MMCs allow to significantly optimise the energy stored at the DC bus [14], reducing the impact of a DC fault. Finally, if certain submodule designs are employed (i.e. Full Bridge MMC (FB-MMCs)), the MMC is able to control the current in case of DC fault [15], allowing a faster restoration.

All these features have been extensively studied in case of transmission networks made of AC/DC converters [16], and have been partially addressed for networks with DC/DC converters [17]. In the case of the FCC MVDC network, if non-isolated DC/DC MMC converters (M2DC) are used to supply some accelerator loads, like magnets, then their control and protection aspects require further development considering their particularities. Therefore, this paper analyses and develops control methods to use these converters as magnet supplies and to cope with DC faults, either pole-to-pole fault or pole-to-ground. Then, the proposed methods are validated using simulation.

Medium Voltage DC Network for the FCC

For the considered scenario, the distribution network for one of the FCC technical sites would consist of an AC and a DC part. The AC part would be used to supply equipment not profiting from DC, whereas the MVDC network would power those loads that are inherently DC, such as power converters, drives, computing infrastructure, etc. Additionally, the DC network would include a certain amount of energy storage, which would be used to recover the energy from the magnets, as well as to support the network under contingencies. This network would have a radial structure following the principle of the existing AC network, with one or several rectifiers supplying a main busbar, and several DC feeders supplying the different loads. A simplified model of the network is shown in Fig. 1.

The total power of the loads to supply would be in the range of a few tens of MW. Therefore, a DC voltage in the 5-10 kV range would be suitable, considering the short distances between the rectifier and the load (usually less than a kilometer) and the difficulties in building DC/DC converters with high step-ratios.

As mentioned, building the MVDC network using MMCs offers several advantages, specially in terms of protection. If fault-blocking capabilities are employed, then there is no need of DC breakers [18], and the clearing and restoration of the DC voltage can be done within a short delay [19], minimizing the consequences of a fault. In this sense, it is well-known that AC/DC converters, both working as rectifiers or inverters (for the supply, for example, of large motors), can be designed as Full-Bridge MMCs to provide fault ride-though capabilities. However, for DC/DC converters based on the MMC, there are additional possibilities.

The main distinction that can be done is between isolated and non-isolated converters. Isolated converters employ a high-frequency transformer to step up and down the DC voltage [20], whereas in non-isolated converters this is achieved by means of a proper selection and arrangement of the submodules, as shown in [21]. The use of one topology or the other will depend on several factors.

Regarding isolated topologies, the Dual Active Bridge employing an MMC [22] on the medium voltage side allows to achieve high voltage conversion ratios while fulfilling the protection requirements. However, it is difficult to achieve a high modularity on the low voltage side of the converter (several windings

Design Concepts for Medium Voltage DC Networks supplying the Future Circular Collider (FCC)

COLMENERO MORATALLA Manuel

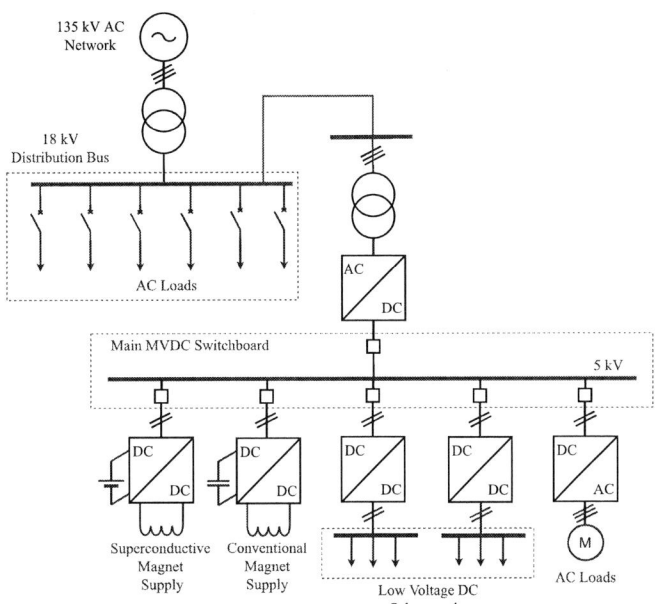

Fig. 1: Simplified electrical diagram of a technical site of the FCC including the MVDC network.

or HF transformers are required), and the integration of energy storage, specially batteries, becomes more complex and less reliable. Besides, in case of loads requiring four-quadrant operation, an additional conversion stage is required. On the other hand, non-isolated MMC topologies allows to supply directly the load with stepped-down voltages in the four-quadrants. However, there is not galvanic isolation between the high voltage network and the load and efficiency decreases for high voltage conversion ratios. Both converter topologies as shown in Fig. 2.

As seen, both topologies could be employed for the FCC. Isolated topologies would be convenient for supplying low voltage equipment where galvanic isolation is a requirement for safety reasons. On the other hand, non-isolated topologies would be suitable for supplying high-power loads, where safety requirements are lower and galvanic isolation is not critical.

Regarding protection, the behaviour of the DC network is highly dependent on the grounding scheme. If a high-impedance grounding scheme is used, as it is desired on the FCC to avoid stray currents, then only pole-to-pole faults will give rise to high fault currents. In the case of a pole-to-ground fault, an

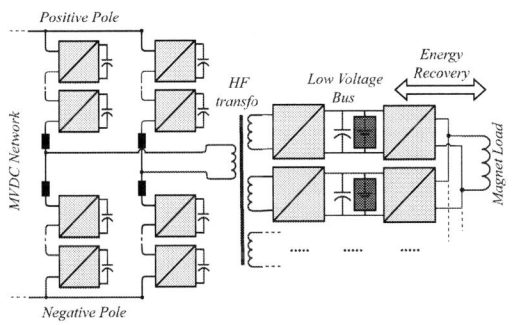
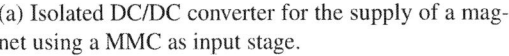

(a) Isolated DC/DC converter for the supply of a magnet using a MMC as input stage.

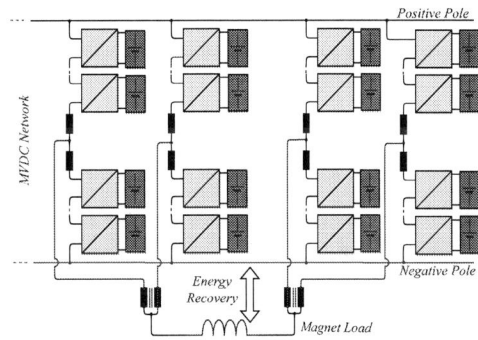

(b) Non-Isolated MMC DC/DC converter used to supply a magnet

Fig. 2: DC/DC converters that can be used to supply some magnets of the FCC.

EPE'22 ECCE Europe

overcurrent is not produced, but a shift of the DC voltages to ground happens: the voltage across the faulty pole drops to zero whereas on the other increases to 2 p.u.

If a protection scheme without DC breakers is desired, then, as stated, the isolated and non-isolated DC/DC converters need to be designed to be able to handle a DC fault. In case of pole-to-pole faults, by using full-bridge cells, it is possible to avoid the discharge of the converters and their contribution to the fault current by means of an appropriate control. On the other hand, in case of a pole-to-ground fault, there are significant differences between topologies. In isolated topologies, the overvoltage is not transferred to the low-voltage side of the converter. Instead, it appears between the HF transformer and ground, so a proper insulation must be foreseen to withstand it. On the contrary, in non-isolated converters, the voltage displacement of the poles immediately appears between the load and ground, leading to a dangerous situation where the risk of flashover is high. The explanation is as follows:

Considering the converter shown in Fig. 3, the voltages to ground of the load, V_+ and V_-, are kept centered around zero by controlling the upper and lower arm voltages during normal operation. Being V_{DC+} and V_{DC-} the voltages of the DC network, and $V_{arm,up}$ and $V_{arm,dw}$ the DC components of the MMC upper an lower arms, the voltages across the load, in this case a magnet, can be controlled according to (1) and (2).

$$V_+ = V_{DC+} - V_{arm,up} \tag{1}$$

$$V_- = V_{DC-} - V_{arm,dw} \tag{2}$$

However, if a pole-to-ground fault happens, for example, at the positive pole, and no control action is performed, then the voltages across the load will shift according to:

$$V_+ = -V_{arm,up} \tag{3}$$

$$V_- = 2 \cdot V_{DC-} - V_{arm,dw} \tag{4}$$

That is, the voltage between the load and ground increases by a magnitude close to half of the DC voltage. This situation could be very hazardous, specially if these converters were used to supply low voltage equipment, since a voltage of several kV would appear between the live parts and the enclosure. In the case of long chains of magnets, the situation is less critical since higher insulation levels are usually employed. Nonetheless, if the MVDC voltage is high, further increasing the insulation of the magnet coils might not be economically feasible. Therefore, other strategies are required to avoid or limit the overvoltage. In particular, it is possible to keep the load voltages within limits during the fault by means of a proper converter control. This and other control tasks will be developed in the following sections.

Modular Multilevel DC/DC Converter for magnet powering

Fig. 3 shows the circuit diagram of a DC/DC MMC converter used to power a magnet. The converter consists of two sub-converters, each composed of M legs whose outputs are connected together through a decoupling inductor L_m designed to filter out the AC components of the leg currents.

In particle accelerators, three quadrant operation is usually required to supply the magnets (positive current, positive, negative and zero voltage). This can be achieved by properly controlling the voltages generated by the arms. For proper converter operation, both a DC and an AC component need to be synthesized:

The DC components generated by the arms determine the input and output voltages. Considering V_{out} the desired output voltage across the magnet, either positive or negative, the lower arms of the two sub-converters must generate the DC components, $V_{arm,dw,dc}$ and $V_{arm,dw,dc}'$, according to (5).

$$V_{out} = V_{arm,dw,dc} - V_{arm,dw,dc}' \tag{5}$$

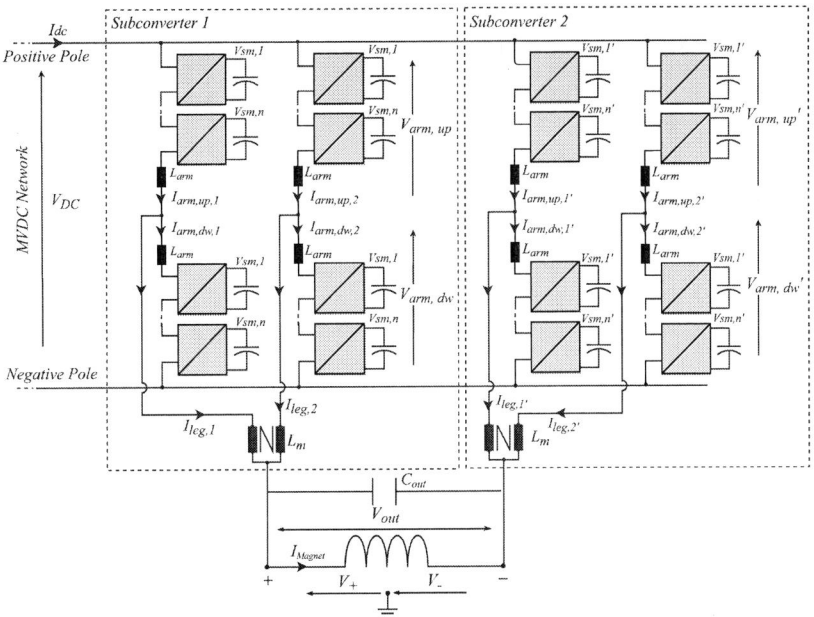

Fig. 3: Schematic of a M2DC converter supplying a magnet. Note that in this case, capacitors are used to recover the energy. Additionally, an output capacitor is added to filter the output ripple.

Which implies that the upper arms must generate their DC components, $V_{\text{arm,up,dc}}$ and $V_{\text{arm,up,dc}}$', according to (6) and (7) to match the voltage of the MVDC network:

$$V_{\text{DC}} = V_{\text{arm,up,dc}} + V_{\text{arm,dw,dc}} \tag{6}$$

$$V_{\text{DC}} = V_{\text{arm,up,dc}}' + V_{\text{arm,dw,dc}}' \tag{7}$$

To keep the load voltages centered around ground, the voltages generated at the magnet terminal V_+ and V_- must be equal with opposite signs.

$$V_+ = \frac{V_{\text{out}}}{2} \tag{8}$$

$$V_- = -\frac{V_{\text{out}}}{2} \tag{9}$$

Therefore, the DC components of the arm voltages must be set according to:

$$V_{\text{arm,up,dc}} = \frac{V_{\text{DC}} - V_{\text{out}}}{2} \qquad V_{\text{arm,dw,dc}} = \frac{V_{\text{DC}} + V_{\text{out}}}{2} \tag{10}$$

$$V_{\text{arm,up,dc}}' = \frac{V_{\text{DC}} + V_{\text{out}}}{2} \qquad V_{\text{arm,dw,dc}}' = \frac{V_{\text{DC}} - V_{\text{out}}}{2} \tag{11}$$

In this way, V_{out} can be generated with little variations of the produced arm voltages, even if the voltage required is small or zero. However, the converter cannot operate only with DC components. Depending on the direction of the power flow, the DC components flowing into the load will tend to charge or discharge the upper or lower arm submodules until the converter becomes unstable. Accordingly, to guarantee the proper operation of the converter, it is necessary to continuously rebalance the arms by transferring power between them. To do this, AC circulating currents, superimposed to the DC component, are generated,which allow to charge/discharge the submodules of the upper/lower arms.

Assuming L_m is sufficiently large, and considering $V_{\text{arm,up,ac}}$ and $V_{\text{arm,dw,ac}}$ the AC voltages generated by the upper and lower converter arms, a flow of power between them can be set by controlling the phase angle between these two AC components as given by (12) in order to keep the internal energy balance.

$$P_{ac,arms} = -P_{dc,arms} = \frac{V_{arm,up,ac} \, V_{arm,dw,ac}}{2 \, \omega \, L_{arm}} \, \sin(\varphi) \tag{12}$$

Where ω, the frequency of the circulating currents, can be arbitrarily chosen according to some optimization criteria. Note that the use of the decoupling inductor L_m, which is designed to cancel out the fluxes generated by AC components by using coupled coils, avoids that the AC currents flow into the load. On the other hand, if the shift between the currents generated by each leg is set to $\phi_M = 2\pi/M$, then these currents cancel out at the positive and negative rails of the converter and there is no flow of AC currents into the MVDC network.

The control diagram of the converter is shown in Fig. 4. Except for the total energy loop, the control is implemented in a per-leg basis. To control the current flowing though the magnet, the individual output current of each leg is regulated to I_{magnet}/M (opposite sign on the other sub-converter) by means of PI controllers.

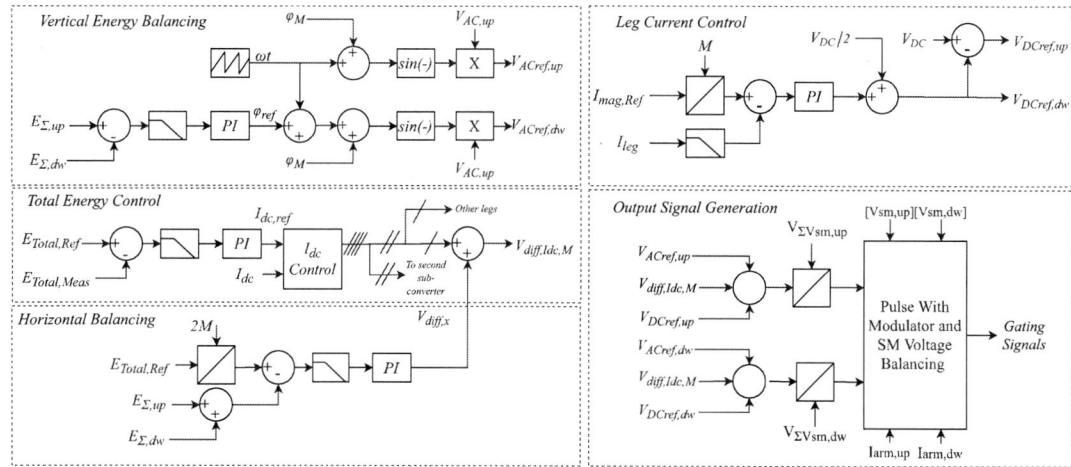

Fig. 4: Control schematics of the M2DC converter. The control corresponds to a single leg of the converter.

To control the differential energy (difference in energy between upper and lower arms), the angle φ, is regulated as stated in (12). The amplitudes of the AC components, $V_{arm,up,ac}$ and $V_{arm,dw,ac}$ are fixed and equal to the maximum voltage generation capability of the arms in order to minimize the magnitude of the AC circulating currents.

To recover of the energy of the magnets (the part corresponding to Li^2), a trajectory for the total energy stored by the converter, $E_{Total,Ref}$, is defined based on the magnet current reference, $I_{mag,ref}$ and on the magnet inductance L_{mag}. Known the initial energy stored by the converter, the trajectory that guarantees that the magnetic energy is supplied by the energy storage is:

$$E_{Total,Ref}(t) = E_{Total,o} - \frac{1}{2} L_{mag} \, I_{mag,ref}(t)^2 \tag{13}$$

To follow this trajectory, and ensure that only losses are supplied from the DC network, the input current, I_{DC}, is controlled by adding differential voltage components $V_{diff,Idc}$ to all the legs of the converter.

On the other hand, the energy stored by the different legs of the converter must be kept equal. For this purpose, an horizontal energy controller is used to control the legs energy to $E_{Total,Ref}(t)/2M$. A differential component, $V_{diff,x}$ is injected into each leg of the converter to keep them horizontally balanced.

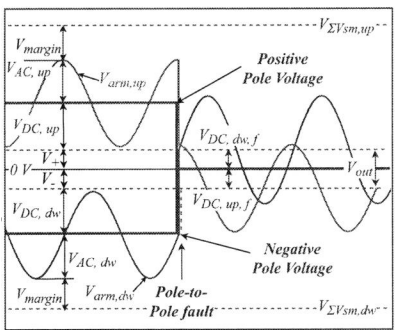
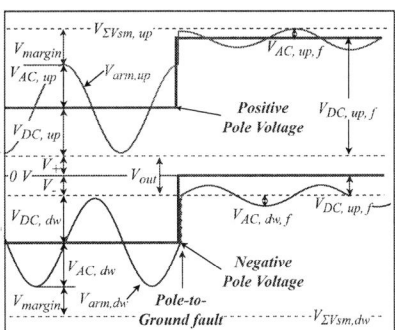

(a) Upper and lower arm voltages during a pole-to-pole fault.

(b) Upper and lower arm voltages during a pole-to-ground fault.

Fig. 5: Behaviour of the M2DC converter during a DC fault.

Finally, note that, in case of using batteries, the state-of-charge can be translated to energy references so that the control doesn't need modification. Note also that the protection functions, described in the following section, are implemented but not shown in Fig. 4. These are described next.

Protection Concepts using the M2DC converters

As described, the M2DC converter must avoid its discharge during a pole-to-pole fault, and the apparition of an overvoltage on the load during a pole-to-ground fault. Moreover, given the large amount of energy being transferred and the highly inductive character of the loads supplied, it is desired that the converter remains controllable during the fault so that the charge/discharge of the magnet is not interrupted. All this can be achieved by controlling the converter as follows:

Pole-to-Pole Fault

In case of a pole-to-pole fault, the DC voltage will drop abruptly to zero. To avoid the discharge of the converter and, thus, its blocking, the DC components generated by the converter arms, $V_{DC,up}$ and $V_{DC,dw}$ need to be replaced by new components as given by (14) for one sub-converter and by (15) for the other.

$$V_{DC,up,f} = V_- \qquad\qquad V_{DC,dw,f} = V_+ \qquad\qquad (14)$$

$$V_{DC,up,f'} = V_+ \qquad\qquad V_{DC,dw,f'} = V_- \qquad\qquad (15)$$

In this way, the DC voltage generated by the converter, as seen from the MVDC network, is zero, whereas from the load point of view it is still the imposed by the regulation. This is shown in Fig. 5a. On the other hand, even if the converter can still generate the AC components required for the vertical balancing, it can not compensate losses as the network is in short-circuit, so proper dimensioning is needed to avoid an excessive discharge during the clearing process.

Considering the protection strategy employed at the network level, when the fault is detected, all converters set their DC components as described (it is also valid for AC/DC converters) and the fault current is brought to zero. Then, the faulty line is identified and isolated by means of conventional switches. Finally, the DC voltage is restored (a process managed by the rectifiers) and operation is resumed.

Pole-to-Ground Fault

As explained, the fact that DC voltage remains during a pole-to-ground fault causes the voltage displacement that appears across the load. Depending on the pole at which the fault is produced, new references for the arm voltages needs to be generated to keep the load voltages centered. Considering a fault on the positive pole of the MVDC network, the arms connected to the negative pole have to increase their DC components by a value equal to half of the DC voltage whereas the arms connected to the faulty pole must still generate the low voltage output components (see (16) and (17)).

$$V_{DC,up,f} = -V_+ \qquad\qquad V_{DC,dw,f} = V_{DC} + V_+ \qquad\qquad (16)$$

$$V_{DC,up,f'} = -V_- \qquad\qquad V_{DC,dw,f'} = V_{DC} + V_- \qquad\qquad (17)$$

This operation is shown in Fig. 5b. However, since for MMCs the arms are usually charged to V_{DC}, the voltage generation capabilities of the arms might be exceeded if the voltage required by the load is high. One solution is to compensate partially the overvoltage so, even if a small increase of voltage across the load is produced, it will not cause damage. Another solution, used here, is to operate the submodules with a voltage higher than the one strictly required for the operation of the converter. By adding an additional component to the arms, V_{margin}, the maximum voltage is not exceeded and the load does not see any overvoltage. Note that, because of this limitation, the amplitude of the AC components needs to be reduced in order to avoid exceeding the voltage generation limit of the arms. Therefore, an imbalance between the upper and lower arms can be expected and properly dimensioning is needed to avoid any instability during the fault clearing process. Considering the duration of this is short, the impact of the imbalance is small and the operation is not affected.

From the point of view of the network, the fact that the DC voltage is kept during the fault means that power can still be transmitted to the loads during the clearing process, including to those connected to the faulty feeder. Therefore, to isolate the faulty line without using DC breakers, it is necessary to bring the current to zero on this feeder by sending a blocking order to the converters concerned, and then isolate it by opening a conventional breaker. If this cannot be done (for example, in case of a meshed configuration), then the DC voltage can be brought to zero to interrupt the current as it is done in case of a pole-to-pole fault.

Finally, note that by decreasing the grounding impedance of the network, the overvoltage can be reduced at expense of a higher fault current. In that case, both principles exposed can be combined to achieved proper protection of the network.

Simulation Results

To validate the presented concepts and design principles, a simulation model of a M2DC converter have been developed using MATLAB/Simulink. The model represents a converter used to supply the existing magnets of the Large Hadron Collider (LHC) which employs batteries to recover the magnetic energy. Under this scenario, the converter is connected to a 5 kV MVDC network and supplies a magnet with a peak current of 8 kA. During ramp-up and down of this current, the output voltage is ± 400 V whereas during the flat top the voltage is nearly zero.

In Fig. 6 the main operation curves of the converter during a magnetic cycle are shown. In Fig 6a), the magnet current and the output voltages are presented. In Fig 6b, the evolution of the energy stored in the converter is shown, demonstrating the proper behaviour of the energy management strategy. Finally, in Fig 6c, the current absorbed from the DC network is represented, showing how the energy storage contributes to reduce the power absorbed from the MVDC network.

For the MVDC protection, two scenarios are considered: one of a pole-to-pole fault and another one of a pole-to-ground fault. In both cases, the fault concerns a different feeder. On the first scenario, a pole-to-pole fault happens at $t = 40s$ during the ramp-up. As shown in Fig. 7, the fault leads to an overcurrent on the faulty feeder (Fig 7a) due to the discharge of the cable and a sudden drop of the DC voltages (Fig 7b). In reaction to this, the arms components of the healthy DC/DC converter are immediately modified as explained (see Fig 9a), allowing to keep the load voltages (Fig 7c) and avoiding the discharge of the converter into the fault (Fig 7d). Once the DC network has been de-energized, the faulty feeder is isolated by means of conventional breakers. Then, at $t = 40.5s$ the DC voltage is restored by the rectifier and and the power flow is resumed. During the whole process, the M2DC converter has been able to keep control of the output current, demonstrating the effectiveness of the strategy.

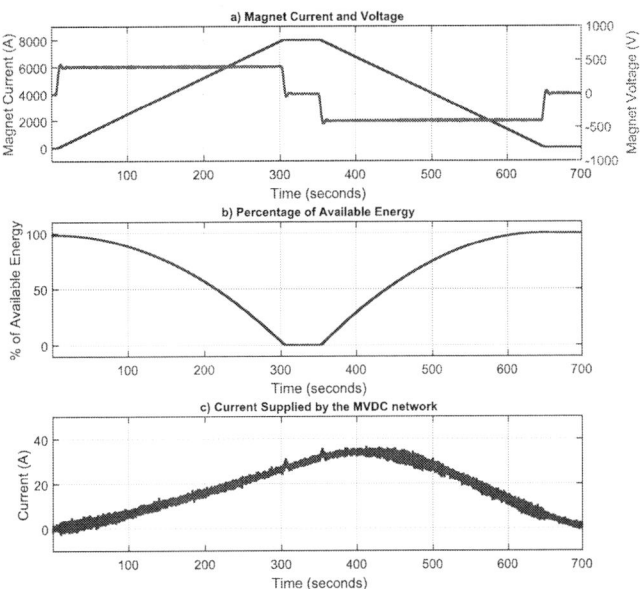

Fig. 6: Behaviour of the converter during an acceleration cycle. a) Current and voltage across the magnet. b) Energy available on the batteries of the converter. c) Current supplied by the MVDC network to cover the losses.

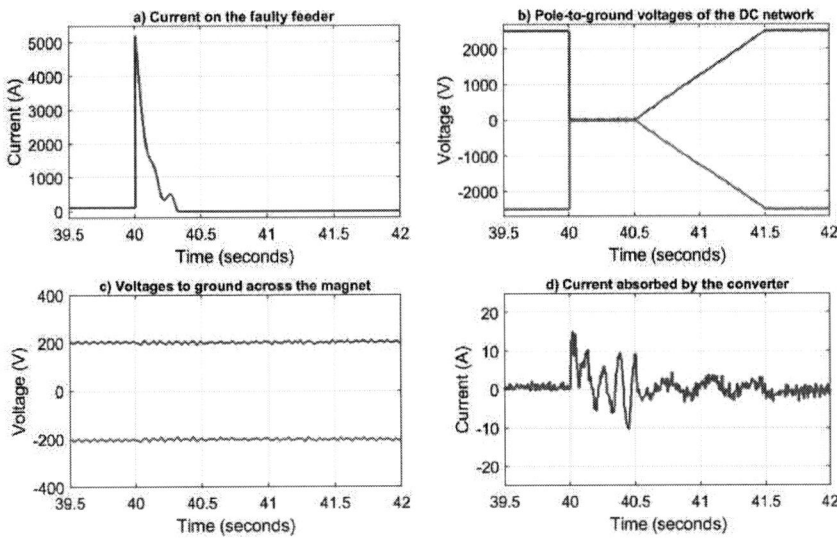

Fig. 7: Behaviour of the converter during a DC pole-to-pole fault.

Regarding the second scenario, the pole-to-ground fault happening at $t = 40s$ does not lead to an overcurrent (Fig 8a) but to a voltage displacement of the pole-to-ground voltages, producing a large overvoltage across the healthy pole (Fig 8b). However, the change of the voltages produced by the arms (Fig 9b) avoids the apparition of the overvoltage across the magnet (Fig 8c) without significantly impacting the current absorbed by the converter (Fig 8d). To isolate the fault, a blocking order is sent to the converters on the faulty feeder, which brings the current flowing through it to zero. This allows at $t = 40.5s$ to open the breaker on the faulty feeder, which clears the fault and permits the re-balancing of the poles. Again, the variation on the pole voltages is not transferred to the load during restoration.

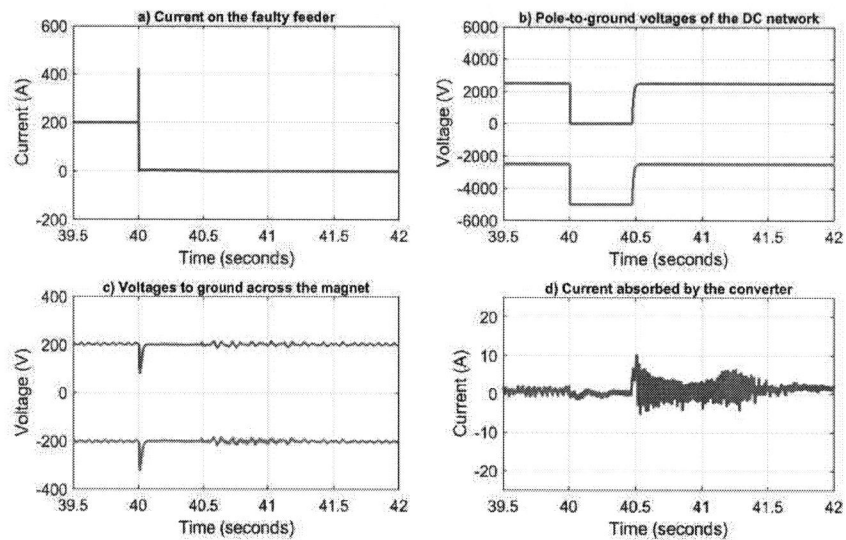

Fig. 8: Behaviour of the converter during a DC pole-to-ground fault.

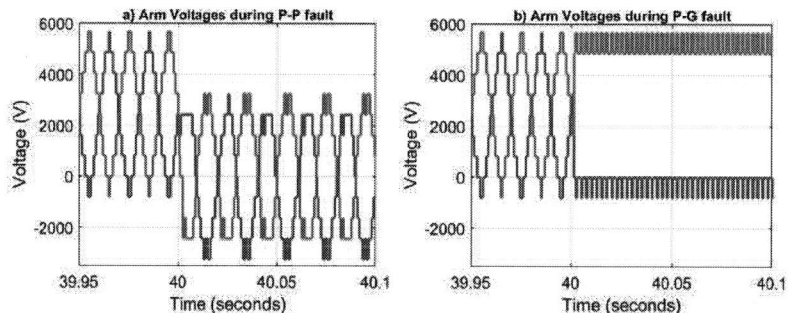

Fig. 9: Arm voltages during a DC fault. Left: Arm voltages during a pole-to-pole fault. Right: Arm voltages during a pole-to-ground fault.

Discussion and Conclusions

This paper analysed the design principles of a Medium Voltage DC network based on Modular Multi-level Converters. As explained, using this topology for building the AC/DC and DC/DC converters that constitute the MVDC network offers several advantages. However, there are also some drawbacks: since circulating currents are required to balance the DC/DC converters, the semiconductors need to cope with higher currents and there are extra losses. Nevertheless, the higher reliability of this solution, and the possibilities it offers in terms of protection compensate the disadvantages for this application.

Regarding the results, they show that the converter can be effectively used to recover the energy of the magnets and store it on its submodules. This reduces notably the power that needs to be supplied by the MVDC network. On the other hand, the behaviour of the converter during pole-to-pole and pole to ground faults have been analyzed. As shown, by properly manipulating the arm voltages of the converter, it is possible to avoid its discharge during a pole-to-pole fault and the apparition of an overvoltage across the load during a pole-to-ground fault. Moreover, in combination with a proper fault-clearing strategy, the supply of the magnet is not interrupted during the fault.

In summary, the use of MMCs for building the FCC MVDC network allows to fulfill several of the reliability requirements and simplifies its protection. Nonetheless, further research is required to understand the feasibility of the solution considering the higher ratings and the extra losses of the DC/DC converters.

References

[1] Benedikt, M., Blondel, A., Brunner, O., Capeans Garrido, M., Cerutti, F., Gutleber, J., Janot, P., Jimenez, J., Mertens, V., Milanese, A., Oide, K., Osborne, J., Otto, T., Papaphilippou, Y., Poole, J., Tavian, L. & Zimmermann, F. FCC-ee: The Lepton Collider: Future Circular Collider Conceptual Design Report Volume 2. Future Circular Collider. (CERN,2018,12), https://cds.cern.ch/record/2651299

[2] Bosich, D., Mastromauro, R. & Sulligoi, G. AC-DC interface converters for MW-scale MVDC distribution systems: A survey. *2017 IEEE Electric Ship Technologies Symposium, ESTS 2017.* pp. 44-49 (2017,10)

[3] Colmenero, M., Blanquez, F. & Kahle, K. Transient Voltage Dip Mitigation System Based On Hybrid Modular Multilevel Converters. *2020 22nd European Conference On Power Electronics And Applications (EPE'20 ECCE Europe).* pp. P.1-P.10.

[4] Zahedi, B. & Norum, L. Efficiency analysis of shipboard dc power systems. *IECON Proceedings (Industrial Electronics Conference).* pp. 689-694.

[5] Vu, T., Gonsoulin, D., Perkins, D., Diaz, F., Vahedi, H. & Edrington, C. Predictive energy management for MVDC all-electric ships. *2017 IEEE Electric Ship Technologies Symposium, ESTS 2017.* pp. 327-331.

[6] Abeynayake, G., Li, G., Joseph, T., Ming, W., Liang, J., Moon, A., Smith, K. & Yu, J. Reliability Evaluation of Voltage Source Converters for MVDC Applications. *2019 IEEE PES Innovative Smart Grid Technologies Asia, ISGT 2019.* pp. 2566-2570.

[7] Saat, J., Bleilevens, R., Mildt, D., Priebe, J., Wehbring, N. & Moser, A. Design Aspects of Medium and Low Voltage DC Distribution Grids-An Overview. *2020 5th IEEE Workshop On The Electronic Grid, EGRID 2020.*

[8] Fan, N., Shen, B., Huang, J. & Chen, J. Simulation Research on Grounding Mode of Modular Multilevel Converter Based Medium Voltage DC Distribution System. *2021 3rd Asia Energy And Electrical Engineering Symposium, AEEES 2021.* pp. 212-218.

[9] Monadi, M., Gavriluta, C., Candela, J. & Rodriguez, P. A communication-assisted protection for MVDC distribution systems with distributed generation. *IEEE Power And Energy Society General Meeting.*

[10] Freijedo, F., Rodriguez-Diaz, E. & Dujic, D. Stable and Passive High-Power Dual Active Bridge Converters Interfacing MVDC Grids. *IEEE Transactions On Industrial Electronics.* **65**, 9561-9570.

[11] Zhan, X., Xue, Z., Ning, G., Zhang, K., Chen, W. & Tao, Y. 250kW High-Frequency Transformer Design and Verification for MVDC Collection System for Renewable Energy Resources. *Proceedings Of The 15th IEEE Conference On Industrial Electronics And Applications, ICIEA 2020.* pp. 641-644.

[12] Abeynayake, G., Li, G., Joseph, T., Liang, J. & Ming, W. Reliability and Cost-oriented Analysis, Comparison and Selection of Multi-level MVdc Converters. *IEEE Transactions On Power Delivery.*

[13] Soong, T. & Lehn, P. Assessment of Fault Tolerance in Modular Multilevel Converters with Integrated Energy Storage. *IEEE Transactions On Power Electronics.* **31**, 4085-4095.

[14] Deng, F. & Chen, Z. Elimination of DC-link current ripple for modular multilevel converters with capacitor voltage-balancing pulse-shifted carrier PWM. *IEEE Transactions On Power Electronics.* **30**, 284-296.

[15] Wang, Y., Li, Q., Li, B., Wei, T., Zhu, Z., Li, W., Wen, W. & Wang, C. A Practical DC Fault Ride-Through Method for MMC Based MVDC Distribution Systems. *IEEE Transactions On Power Delivery.* **36**, 2510-2519.

[16] Alyami, H. & Mohamed, Y. Review and development of MMC employed in VSC-HVDC systems. *2017 IEEE 30th Canadian Conference On Electrical And Computer Engineering (CCECE).* pp. 1-6 (2017)

[17] Vidal-Albalate, R., Soto-Sanchez, D., Belenguer, E., Peña, R. & Blasco-Gimenez, R. Sizing and Short-Circuit Capability of a Transformerless HVDC DC–DC Converter. *IEEE Transactions On Power Delivery.* **35**, 2363-2377.

[18] Zhang, X., Kang, X. & Zhang, Y. A Strategy of DC Fault Ride Through and capacitor voltage balancing for Hybrid Modular Multilevel Converter (MMC). *2019 IEEE 8th International Conference On Advanced Power System Automation And Protection (APAP).* pp. 80-84.

[19] Soto, D., Sloderbeck, M., Ravindra, H. & Steurer, M. Advances to megawatt scale demonstrations of high speed fault clearing and power restoration in breakerless MVDC shipboard power systems. *2017 IEEE Electric Ship Technologies Symposium (ESTS).* pp. 312-315 (2017)

[20] Dincan, C., Kjaer, P., Chen, Y., Sarra-Macia, E., Munk-Nielsen, S., Bak, C. & Vaisambhayana, S. Design of a High-Power Resonant Converter for DC Wind Turbines. *IEEE Transactions On Power Electronics.* **34**, 6136-6154.

[21] Li, Y., Gruson, F., Delarue, P. & Le Moigne, P. Design and control of modular multilevel DC converter (M2DC). *2017 19th European Conference On Power Electronics And Applications, EPE 2017 ECCE Europe.* 2017-Janua pp. P1-P10.

[22] Mo, R., Li, R. & Li, H. Isolated modular multilevel (IMM) DC/DC converter with energy storage and active filter function for shipboard MVDC system applications. *2015 IEEE Electric Ship Technologies Symposium, ESTS 2015.* pp. 113-117.

A Novel Dual CC–CV Output Wireless EV Charger With Minimal Dependency on Both Coil Coupling and Load Variation

Subhranil Barman[*] and Kishore Chatterjee[†]
Indian Institute of Technology Bombay
Powai, Maharashtra - 400076
Mumbai, India
Email: subhrabarman@ee.iitb.ac.in[*]; kishore@iitb.ac.in[†]

Keywords

≪Electric Vehicle≫, ≪Battery charger≫, ≪Wireless power transmission≫, ≪Power converters for EV≫, ≪Compensation≫

Abstract

Robust and efficient wireless EV charging requires load and coupling independent CC and CV modes. Existing literature focus either on CC or CV mode. The characteristics of the schemes which cater to both the modes are affected by load and coupling coefficient. In this paper, an improved dual output charger topology is proposed that is maximally independent from variations in coil coupling and load. Relevant analysis and simulation results are shown to validate the working of the proposed topology.

Introduction

Moving forward, electric vehicles (EVs) are the way ahead for transportation. A key aspect in the development of EVs is the charging of their batteries. The most commonly used batteries in EVs are Lithium-ion (Li-ion) batteries. As they cannot withstand overvoltage, provision of the two standardised charging modes, constant current (CC) and constant voltage (CV), in an appropriate sequence is crucial [1], with smooth transition between them. This feature should be catered by wireless battery chargers (WBCs) in an efficient and robust manner, along with fulfilling other requisite characteristics.

WBCs generally employ inductive power transfer (IPT) using magnetically coupled coils, wherein the power transfer occurs in "non-radiative" form. The basic components of inductive WBCs are: isolated bridge converters, compensation circuits and optionally DC backend converters. The bridge converters usually involve single active bridges (SAB) for unidirectional power flow, or dual active bridges (DAB) for bidirectional power flow. Although WBCs provide contactless charging, their operation is dependent on the alignment between the transmitter and the receiver coils. For "true wireless" systems, the alignment between the aforesaid coils depends on factors such as ground level conditions, tire pressure, parking manoeuvres, etc. The change in the coupling coefficient of the two coils affects power transfer, output characteristics, operating frequency, and several other parameters. Coupling and load conditions also govern the charger operation in zero phase angle (ZPA) condition. Further, the cascade network consisting of inverter, coupled coils and rectifier reduces the overall efficiency. In view of these aspects, an ideal WBC should have the following features - (1) maximal operational independence from variations of both coil coupling and load, (2) minimized size and components, (3) ZPA operation with soft switching, and (4) maximum efficiency in both CC and CV modes [2].

The aforementioned features can be achieved by employing (1) appropriate modifications in the inversion and/or rectification stages, (2) DC front-end or back-end converters, (3) modified compensation circuits with or without additional switches, or (4) frequency or phase modulation. Literature survey in this regard reveals the following significant aspects. (1) A WBC configuration with a fully controlled input side inverter and a output diode rectifier is preferable to minimize component count and simplify the control at the vehicular end. (2) DC converters provide independent output regulation, but have extra switches, additional DC side magnetics, lower efficiency etc., which have led to their elimination with advanced topologies. (3) Extensive techniques that are reported in [1]– [4] are not fully sufficient, as they have either inherent CC output, or CV output with partial or no ZPA, and/or are operated at frequencies dependent on coupling coefficient, k. Besides this, topological modifications involving switched compensation networks [5], [6], wherein the compensation circuit is altered by applying appropriate switching states, have several drawbacks like - charge modes being dependent on k, higher component count, lower efficiency (η), etc. (4) Moreover, soft control methods involving frequency modulation and phase shift control can only partially fulfill the required specifications, and have poor wide load range operation and high complexity.

To improve upon all these aspects, a compensation strategy incorporating multiple and multi-resonant LC tanks has been proposed in this paper to achieve a re-configurable charger which can provide both CC and CV modes of operation. The proposed modifications have been targeted in the compensation section as it affects several parameters such as ZPA, output characteristics, power transfer capability, size, etc. Analysis and methods followed for the development of the proposed charging topology, as well as their features and benefits, are presented, along with the future scope of the topology. The salient features of the proposed topology are as follows : (1) it does not require additional switches to realize distinct compensation networks for both CC and CV modes; (2) it eliminates the dependency of operating frequencies from coil coupling; (3) constant output current and voltage are obtained during CC and CV modes respectively; (4) load and coupling coefficient independent ZPA is achieved in both the modes, thus fulfilling all desired features.

Passive Network Analysis

For obtaining both CC and CV modes from a single input source, different circuits are required. To achieve that, a single topology may be modified to have different structures, however, additional switches are undesirable. Hence, a passive network analysis is presented here as per [7] that describes the basic principle of obtaining constant current or voltage outputs from any given input source, which is - allocation of suitable passive networks between source and load. Accordingly, the requisite modifications for developing a dual output topology devoid of any additional switches are shown.

The passive networks required for obtaining load independent CC output from a voltage source, and CV output from a current source, are shown [7] in Fig. 1. There can be several combinations of z_a, z_b and z_c [7]. The CC and CV output expressions are also given. Similar networks can be derived for other possible combinations of input and output configurations [7]. Based on this principle, it is evident that the CC and CV modes require different network configurations. Hence, to implement them, resonance is used for reconfiguring the topology without utilizing any switches. Since in a single passive network stage, tunable parameters are less for the provision of ZPA independent of load resistance R_0 and coil coupling coefficient k [7], multiple network stages are employed with appropriate resonating conditions.

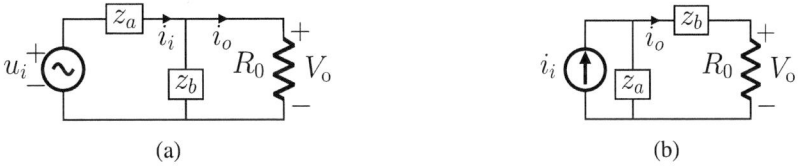

Fig. 1: Circuits for (a) CC output from a voltage source, and (b) CV output from a current source

$$\text{CC ouput from voltage source } i_o = \frac{u_i}{z_a} = \frac{u_i}{z_b}; \quad \text{CV output from current source } u_o = z_a i_i = z_b i_i \quad (1)$$

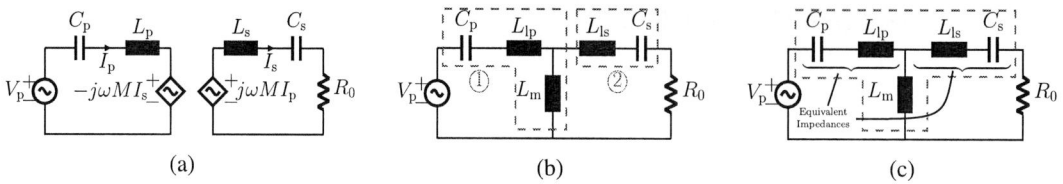

Fig. 2: (a) SS topology, and its constituent network stages for (b) CC mode (c) CV mode

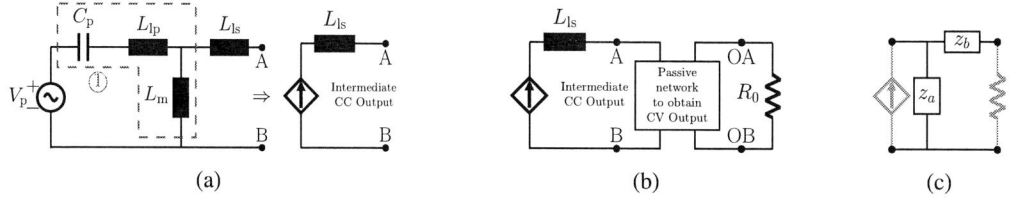

Fig. 3: (a), (b) Secondary side with intermediate current source ; (c) Network VC-LC/CL

Synthesis of the Proposed Topology

As a starting point, the existing series-series (SS) topology is chosen as a base circuit as it already provides CC mode with ZPA and has very few passive components [1]. This is subsequently modified to implement the CV mode of operation. The configuration of the SS topology is shown in Fig. 2a. As can be seen in Fig. 2, the CC mode circuit can be decomposed into passive network blocks [7] based on the input type - a voltage source, and output type - constant current. There are two cascaded stages in Fig. 2b - ① network CV-CL (nomenclature - 'C' : CC output, 'V' : voltage source input, 'C' : capacitor C_p, impedance as seen from the network input, 'L' - next impedance encountered - formed by L_{lp} and L_m), and ② L_{ls} and C_s which form a series impedance. Stage ① generates an intermediate CC output, which is then reflected to load R_0 via stage ②. Similar analysis for CV mode yields network type VV-xLx (Fig. 2c) - where 'x' signifies that the combination of C_p, L_{lp}, and C_s, L_{ls} are treated as equivalent inductors or capacitors. A single network stage is present in this case. In CC mode, as there are two stages, Z_p is independent of R_0 - the second stage facilitates tuning. In CV mode, Z_p becomes R_0 dependent, as the single stage does not present adequate tuning parameters. Besides, $\omega_0 \propto k$ for the CV mode because the resonance condition involves L_m (or L_{lp}). Thus it can be inferred that the synthesized topology must consist of multiple network stages for load independent ZPA provision. Also, resonance conditions must involve the entire self inductances of the coils for providing charging modes unaffected by k. Besides, for the proposed topology, the CC mode frequency ω_{0cc} is selected to be lower than the CV mode frequency ω_{0cv}. This is because R_0 assumes a low value in the CC mode and increases considerably in the CV mode [3]. To achieve optimal η for the entire load range, SS topologies are operated such that $\omega_{0cc} < \omega_{0cv}$ [2,3]. This can be supported by η plots for different frequency ratios.

Proposed Modifications for Achieving Desired Features

The secondary side structure is shown in Fig. 3. Using the intermediate current source as an input, as per Fig. 1b, a CV output can be obtained by placing network stages VC-CL or VC-LC [7] shown in Fig. 3c between 'A', 'B' and 'OA', 'OB' in Fig. 3b. As per the aforesaid nomenclature, for the network VC-CL, z_a is capacitive and z_b is inductive, and the opposite holds for VC-LC. Here only network VC-CL is feasible, as a contradiction occurs for VC-LC wherein CC and CV mode frequencies become equal. During CV mode, to implement VC-CL, two LC tanks - SLCA and SLCB, are used (as shown in Fig. 4), wherein SLCA forms an inductor L_{seq}, while SLCB forms a capacitor C_{seq}. During CC mode, SLCA forms a capacitor C_s and SLCB, which must remain disconnected, forms an open circuit. While shifting from ω_{0cc} to ω_{0cv}, the impedance of SLCA changes from being capacitive to inductive, while it is the opposite for SLCB (open circuit to capacitive). Thus a parallel tank is used for SLCB, and a series tank is used for SLCA. To enable ZPA operation, the primary side needs to resonate during both CC and

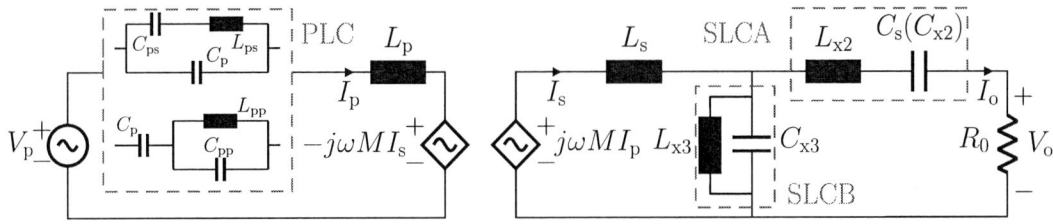

Fig. 4: Proposed topology

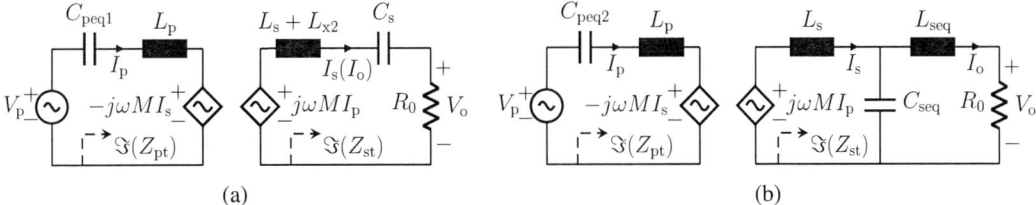

Fig. 5: Equivalent circuits of the proposed topology for (a) CC mode, and (b) CV mode

CV modes. Hence a multi resonant tank PLC involving capacitor C_p is used instead of just C_p being connected in series with L_p, (as is present in the primary of a conventional SS topology), since C_p and L_p can only resonate at one frequency. Two structures are possible in this case - a parallel LC tank in series with C_p, or a series LC tank in parallel with C_p, both of which are in series with L_p. PLC is designed to form C_{peq1} and C_{peq2} in CC and CV modes respectively. Combining the modified primary and secondary circuits, the final proposed topology is shown in Fig. 4.

Independence of ZPA from R_0, k in Both Modes

To achieve R_0 and k independent ZPA, the impedances $\Im(Z_{pt}), \Im(Z_{st})$ in Fig. 5 must be nullified in both modes. For $\Im(Z_{pt})$ and $\Im(Z_{st})$, given in (2) and (3), to be zero at both ω_{0cc} and ω_{0cv}, such that secondary to primary transferred impedance Z_{tr} is real ($=\frac{\omega^2 M^2}{R_0}$ [8], where $\omega = \omega_{0cc}, \omega_{0cv}$), we get $\omega_{0cc}^2 L_p C_{peq1} = \omega_{0cv}^2 L_p C_{peq2} = 1$, $\omega_{0cc}^2(L_s + L_{x2})C_s = 1$ and $\omega_{0cv}^2 L_s C_{seq} = \omega_{0cv}^2 L_{seq} C_{seq} = 1$ (implying $L_{seq} = L_s$).

$$\text{CC Mode}: \quad \Im(Z_{pt}) = \frac{\omega_{0cc}^2 L_p C_{peq1} - 1}{\omega_{0cc} C_{peq1}}, \quad \Im(Z_{st}) = \frac{\omega_{0cc}^2 (L_s + L_{x2})C_s - 1}{\omega_{0cc} C_s} \tag{2}$$

$$\text{CV Mode}: \quad \Im(Z_{pt}) = \frac{\omega_{0cv}^2 L_p C_{peq2} - 1}{\omega_{0cv} C_{peq2}}, \quad \Im(Z_{st}) = \frac{-R_0(\omega_{0cv}^2 L_s C_{seq} - 1)}{\omega_{0cv} C_s R_0 + j(\omega_{0cv}^2 L_{seq2} C_{seq} - 1)} \tag{3}$$

Load and Coupling Independent Resonant Frequencies

From (2) and (3), the CC and CV mode resonant frequencies are obtained as given below, which shows that both ω_{0cc} and ω_{0cv} (and hence ZPA as well) are independent of R_0 and k.

$$\omega_{0cc} = \frac{1}{\sqrt{(L_s + L_{x2})C_{sx2}}} = \frac{1}{L_p C_{peq1}}, \quad \omega_{0cv} = \frac{1}{\sqrt{L_{seq} C_{seq}}} = \frac{1}{L_p C_{peq2}} \tag{4}$$

Dual Output Characteristics

The dual output (CC and CV) characteristics can be derived by circuit analysis of the equivalent figures shown in Fig. 5a and Fig. 5b at the resonant frequencies ω_{0cc} and ω_{0cv} respectively. It can be seen that both the outputs are load independent as shown in (5), thereby fulfilling all the required features.

$$\text{CC mode} \rightarrow I_0 = \frac{V_p}{\omega_{0cc} M}, \quad \text{CV mode} \rightarrow V_0 = \frac{L_s}{M} V_p \tag{5}$$

Design Considerations for the Proposed Topology

The tanks PLC, SLCA and SLCB are designed as per their required functionality, such that the overall tank impedances are equal to the requisite equivalent impedances, as shown in Table I. For the design of coil parameters L_p and L_s, (5) is employed, wherein M is replaced by $k\sqrt{nL_pL_s}$, n being the turns ratio. Using the value of R_0 at the transition point between CC to CV mode, given as $R_{crit} = \frac{V_o}{I_o} = \frac{8}{\pi^2}(\frac{V_{Bat}}{I_{Bat}})$, we can get L_s and L_p. If $L_p = nL_s$ is chosen, then $V_p = kV_o$, where V_p has to be regulated if k varies; otherwise $L_p \propto \frac{nL_s}{k^2}$, wherein L_p has to be adjusted with variations in k. For this topology, the design approach keeping $L_p = nL_s$ is chosen. The coil and tank parameters are shown in Table II. Here, $\sigma_\omega > 1$ and $\omega_{0cc} < \omega_{0L_pC_p} < \omega_{0cv}$ is required ($\sigma_\omega = \frac{\omega_{0cv}}{\omega_{0cc}}$, and $\omega_{0L_pC_p}$ is the resonant frequency of L_p and C_p), which gives a design condition for $\omega_{0L_pC_p}$ as $\omega_{0cv} > \omega_{0cc}$. The frequencies f_{0cc} (ω_{0cc}), f_{0cv} (ω_{0cv}) and $f_{0L_pC_p}$ ($\omega_{0L_pC_p}$) are chosen as some factors of a base frequency f_{base}, such that $f_{0cc} = \beta_\omega f_{base}$, $f_{0cv} = \sigma_\omega f_{0cc}$, and $f_{0L_pC_p} = \lambda_\omega f_{0cc}$. Optimal values of the turns ratio n, and the frequency tuning parameters $\beta_\omega, \sigma_\omega$, and λ_ω are selected by analysis of the variation trends of three key aspects, mainly - (a) passive component values, (b) peak stresses of components, and (c) average efficiency η for the entire load range.

Table I: Primary and secondary tank design features

Tank	CC Mode Equivalent Impedance	CV Mode Equivalent Impedance
PLC	$\frac{-j}{\omega_{0cc}C_{peq1}}$	$\frac{-j}{\omega_{0cv}C_{peq2}}$
SLCA	$j\omega_{0cc}L_{x2} - \frac{j}{\omega_{0cc}C_{x2}}$	$j\omega_{0cv}L_{seq}$
SLCB	∞ (open circuit)	$\frac{-j}{\omega_{0cv}C_{seq}}$

Table II: Significant parameters of the proposed topology

$L_s = \frac{R_{crit}}{\omega_{0cc}}$	$L_p = nL_s$	$f_{0cc} = \beta_\omega f_{base}$, $f_{0L_pC_p} = \lambda_\omega f_{0cc}$, $f_{0cv} = \sigma_\omega f_{0cc}$
PLC → (C_p in series with parallel LC tank)	$L_{pp} = \frac{(\omega_{0cv}^2 L_pC_p-1)(1-\omega_{0cc}^2 L_pC_p)}{(\omega_{0cc}^2 C_p)(\omega_{0cv}^2 L_pC_p)}$ $C_{pp} = \frac{C_p}{(\omega_{0cv}^2 L_pC_p-1)(1-\omega_{0cc}^2 L_pC_p)}$	PLC → (C_p in parallel with series LC tank) $L_{ps} = \frac{L_p}{(\omega_{0cv}^2 L_pC_p-1)(1-\omega_{0cc}^2 L_pC_p)}$ $C_{ps} = \frac{(\omega_{0cv}^2 L_pC_p-1)(1-\omega_{0cc}^2 L_pC_p)}{(\omega_{0cc}^2 L_p)(\omega_{0cv}^2 L_pC_p)}$
C_p (for both forms of PLC) $= \frac{1}{\omega_{0L_pC_p}^2 L_p}$		
SLCA → $L_{x2} = \frac{\sigma_\omega^2+1}{\sigma_\omega^2-1}L_s$, $C_{x2} = C_s = \frac{\sigma_\omega^2-1}{2\omega_{0cv}^2 L_s}$	SLCB → $L_{x3} = \frac{1}{\omega_{0cc}^2 C_{x3}}$, $C_{x3} = \frac{\sigma_\omega^2 C_{seq}}{\sigma_\omega^2-1}$	

One thing to note is that using (4) and (5), and noting that $C_{sx2} = C_s$ (as C_{sx2} is the capacitor C_s from the base circuit of SS topology), we get (6). For a normal SS topology, with the same C_s, and secondary inductance denoted by L_s', the resonant frequency is $\omega' = 1/(L_s'C_s)$, and $|I_s| = (V_p/k)\sqrt{(C_s/L_p)}$. If for the sake of comparison $\omega' = \omega_{0cc}$ is considered, then I_s is increased by a factor $\sqrt{(1+L_{x2}/L_s)}$ in the proposed topology. Alternatively, it can be said that L_s can be reduced for obtaining the same I_o (or I_s), as is evident from the expressions of ω_{0cc} and ω'. If L_s can be reduced, then it not only reduces the coil size on the car side, but also helps reduce I_p, because noting that $L_{seq} = L_s$, we get (7). As per (7), the magnitude of I_p throughout the charging process can be quite high as $L_s > M$ by a large extent. $|I_p|$ will vary with load induced variations, but its peak will occur at the starting point of CV mode (or transition point from CC to CV mode) and is obtained by substituting the peak value of I_o during CC mode (a known constant) in (7). Hence I_p can be reduced by lowering the peak value of I_p by adjusting L_s in (7).

$$|I_s| = \frac{V_p}{\omega_{0cc}M} = \frac{V_p}{k}\sqrt{\frac{(L_s+L_{x2})C_s}{L_pL_s}} = \frac{V_p}{k}\sqrt{\frac{C_s}{L_p}}\sqrt{1+\frac{L_{x2}}{L_s}} \tag{6}$$

$$|I_p| = \frac{V_p(L_s+L_{x2})C_{sx2}R_0}{M^2} \text{ during CC mode;} \quad = \frac{L_sI_o}{M} \text{ during CV mode} \tag{7}$$

Table III: Simulation parameters {transition from CC to CV mode : $t = 0.7$ s (after $V_o = V_{Bat}$)}

$V_{Bat} = 300$ V	$I_{Bat} = 10$ A	$R_{Bat} = 30\ \Omega, R_{crit} = 24.3\ \Omega$
$L_s = 77.4\ \mu H$	$L_p = 344.1\ \mu H$	$n = 4.44$
$\beta_\omega = 1, f_{0cc} = 50$ kHz	$\lambda_\omega = 1.336, f_{0L_pC_p} = 66.8$ kHz	$\sigma_\omega = 1.6, f_{0cv} = 80$ kHz
$L_{pp} = 81.7\ \mu H, C_{pp} = 86.4$ nF	$L_{x2} = 176.6\ \mu H, C_{x2} = 39.9$ nF	$L_{x3} = 120.7\ \mu H, C_{x3} = 83.9$ nF
$k = 0.3, V_p = 170.8$ V		$k = 0.35, V_p = 199.3$ V

Simulated Performance

The proposed topology has been simulated in MATLAB-Simulink (where C_p is in series with a parallel LC tank in PLC) for two values of k : $k = 0.30$, and $k = 0.35$, with step variation of the load resistance to emulate the EV battery charging process. The simulated circuit is almost the same as in Fig. 4, with only V_p being replaced by a inverter with an appropriate DC voltage source input corresponding to V_p, and R_0 being replaced by a rectifier having a DC filter capacitor of 50 μF and a variable battery resistance $R_{Bat} = V_{Bat}/I_{Bat}$ at its output. Simulation parameters are given in Table III. The base frequency is $f_{base} = 50$ kHz. The input voltage V_p varies based on the value of k, as per the equation $V_p = (\sqrt{8}/\pi) \times kV_{Bat}\sqrt{L_p/L_s}$. Based on the designed parameter values, the frequency response characteristics of the circuit in Fig. 4, which is the fundamental harmonic equivalent circuit of the proposed topology, is shown in Fig. 6 for varying load values at k = 0.3. The plots show that constant current and voltage gain characteristics are obtained at the frequencies corresponding to ω_{0cc} and ω_{0cv} respectively. The system input phase angle is also zero at the CC and CV mode frequencies. Simulation results are shown in Fig. 7 and Table IV.

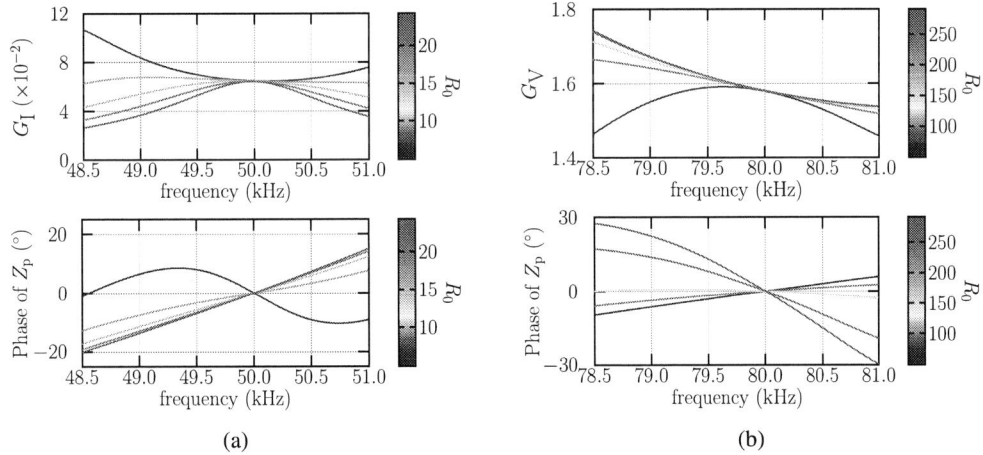

Fig. 6: (a) Transconductance gain and input phase angle at and around ω_{0cc}; (b) Voltage gain and input phase angle at and around ω_{0cv}. The characteristics are shown for the appropriate load ranges for the CC ($R_0 \leq R_{crit}$) and CV modes ($R_0 > R_{crit}$) respectively.

Table IV: Simulation results – output voltage V_o, current I_o and input phase angle at each load step

R_0 (Ω)		12	15	18	21	24	27	30	36	42	48	54	60
V_o (V)	$k = 0.30$	119.20	148.73	178.12	207.36	236.45	265.37	294.11	299.38	299.47	299.53	299.59	299.63
	$k = 0.35$	119.30	148.90	178.38	207.73	236.94	265.99	294.88	299.51	299.58	299.64	299.68	299.72
I_o (A)	$k = 0.30$	9.93	9.92	9.90	9.87	9.85	9.83	9.80	8.32	7.13	6.24	5.55	4.99
	$k = 0.35$	9.94	9.93	9.91	9.89	9.87	9.85	9.83	8.32	7.13	6.24	5.55	5.00
θ_p (°)	$k = 0.30$	1.98	2.85	3.66	4.43	5.18	5.91	6.63	-4.86	-5.68	-6.49	-7.29	-8.10
	$k = 0.35$	1.67	2.61	3.46	4.27	5.05	5.81	6.55	-4.97	-5.80	-6.63	-7.45	-8.27

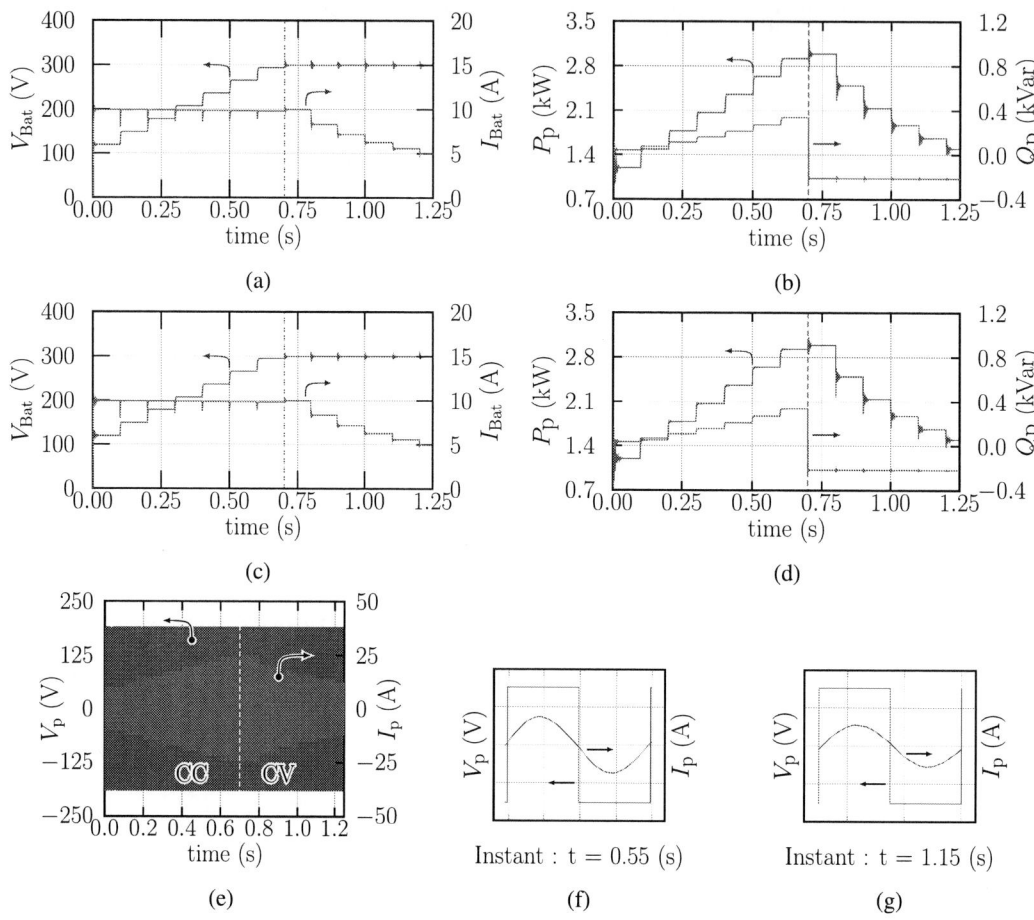

Fig. 7: Simulation results showing output voltage V_{Bat} and current I_{Bat} across load R_0, along with input side active and reactive power P_p and Q_p respectively. The plots for $k = 0.3$ are (a) V_{Bat} and I_{Bat}, and (b) P_p and Q_p; and for $k = 0.35$ are (c) V_{Bat} and I_{Bat}, and (d) P_p and Q_p. Plots of V_p and I_p for $k = 0.30$ are given in Fig. 7e. V_p and I_p for one switching cycle are also shown (for $k = 0.30$) in Fig. 7f (CC mode) and Fig. 7g (CV mode).

The results show that a constant current of 10 A in the CC mode, and voltage of 300 V in the CV mode is achieved for both $k = 0.30$ and $k = 0.35$ as given in Table IV, with very minor variations in the output current and voltage (due to stray resistances purposefully included in simulation). The spikes observed in all the plots are due to the step changes in the load which will not be present in a realistic system having the battery of the EV. The primary input phase angle θ_p in Table IV is small (though not negligible) throughout the CC and CV modes, irrespective of k. This is also evident in Fig. 7f and Fig. 7g, which show the plots of V_p and I_p for one instant during the CC and CV modes respectively for $k = 0.30$. The reason for this is the presence of the filter capacitor at the rectifier output. As a result, non-zero reactive power Q_p is drawn, which is not desired as the reactive power requirement of the circuit in Fig. 4 is $\simeq 0$. This occurrence can be eliminated by adjusting both ω_{0cc} and ω_{0cv} as per requirement so that θ_p can be nullified during the whole operation. Thus closed loop control of the system is required to counter the rectifier action and achieve ZPA. The phase difference can also be utilized to make the inverter output current I_p slightly lagging so as to enable soft switching (zero voltage switching or ZVS) operation. The output plots given in Fig. 7a and Fig. 7c indicate that there is a steady transfer of active power as per load requirement for both the coupling conditions and modes. It can also be seen that the CC to CV mode transition (at $t = 0.7$ s, wherein the operating frequency changes from ω_{0cc} to ω_{0cv}) is smooth and the fluctuations are not drastic. These findings validate the fulfillment of the desired targets in the proposed topology. Moreover, the simulated performance is also indicative of the robustness of the proposed topology as it can negotiate misalignment between the coils (which causes variations in the coupling coefficient) and provide the desired output current or voltage.

Conclusions

In this paper a dual output contactless charger with minimal dependence on both coil coupling coefficient and load variation has been proposed. This is achieved by utilizing passive network stages in the WBC which are used to reconfigure the circuit by employing resonance as per requirement. Proper analysis has been provided to describe the process of developing the proposed topology. Thereafter, in-depth explanation of the design methodology of the topology has also been given. Simulation results are then presented in detail to validate the effectiveness of the topology, wherein near ZPA enabled CC and CV outputs are obtained for a range of values of k at load and coupling independent frequencies.

References

[1] Vu V.B., Tran D.H. and Choi W.: Implementation of the constant current and constant voltage charge of inductive power transfer systems with the double-sided LCC compensation topology for electric vehicle battery charge applications, IEEE Transactions on Power Electronics Vol 33 no 9, pp.7398-7410.

[2] Zhang W., Wong S.C., Chi K.T. and Chen Q.: Analysis and comparison of secondary series-and parallel-compensated inductive power transfer systems operating for optimal efficiency and load-independent voltage-transfer ratio, IEEE Transactions on Power Electronics Vol 29 no 6, pp.2979-2990.

[3] Huang Z., Wong S.C. and Chi K.T.: Design of a single-stage inductive-power-transfer converter for efficient EV battery charging, IEEE Transactions on Vehicular Technology Vol 66 no 7, pp.5808-5821.

[4] Cai C., Wang J., Fang Z., Zhang P., Hu M., Zhang J., Li L. and Lin Z.: Design and optimization of load-independent magnetic resonant wireless charging system for electric vehicles, IEEE Access Vol 6, pp.17264-17274.

[5] Chen Y., Zhang H., Park S.J. and Kim D.H.: A switching hybrid LCC-S compensation topology for constant current/voltage EV wireless charging, IEEE Access Vol 7, pp.133924-133935.

[6] Qu X., Han H., Wong S.C., Chi K.T. and Chen W.: Hybrid IPT topologies with constant current or constant voltage output for battery charging applications, IEEE Transactions on Power Electronics Vol 30 no 11, pp.6329-6337.

[7] Zhang W. and Mi C.C., 2015.: Compensation topologies of high-power wireless power transfer systems, IEEE Transactions on Vehicular Technology Vol 65 no 6, pp.4768-4778.

[8] Wang C.S., Stielau O.H., and Covic G.A., 2005.: Design considerations for a contactless electric vehicle battery charger, IEEE Transactions on Industrial Electronics Vol 52 no 5, pp.1308-1314.

A high-performance EMI filter based on laminated ferrite ring cores

Marcin Kącki[1], Marek S. Ryłko[1], John G. Hayes[2], Charles R. Sullivan[3]

[1] Research and Development SMA Magnetics Sp. z o.o.
Modlniczka, Poland
e-mail: marcin.kacki@sma-magnetics.com / marek.rylko@sma-magnetics.com

[2] Power Electronics Research Laboratory School of Engineering University College Cork
Cork, Ireland
e-mail: john.hayes@ucc.ie

[3] Thayer School of Engineering Dartmouth College
Hanover, NH 03755, USA
e-mail: charles.r.sullivan@dartmouth.edu

Keywords

EMC/EMI, Passive filters, Ferrite, Optimization

Abstract

This paper presents a study on the flux distribution and impedance of common-mode chokes based on the solid and laminated ferrite ring cores. The analysis demonstrates the novel laminated core structure to improve the flux distribution across the core cross-section at high frequency, and so, also improves the overall EMI filter performance. Solid and laminated ferrite common-mode chokes are implemented in a two-stage single-phase EMI filter and compared experimentally.

Introduction

The continuous drive for improvement of energy conversion focuses on efficiency improvement combined with switching frequency increases, which result in size and cost reductions. This leads to new solutions and is fostering an interest in the wide-bandgap (WBG) silicon carbide (SiC) and gallium nitride (GaN) semiconductors, which have superior switching performance in comparison to the traditional Silicon (Si) based power semiconductors. The benefits of using such semiconductors, i.e., higher switching frequencies, higher rate of voltage (dv/dt) and current (di/dt) transients, and higher power density, create a new frontier for EMI filtering and magnetic component design. EMI generation is strongly related to the rate of voltage (dv/dt) transients and their frequencies. In order to compare EMI footprints, three spectra are compared to demonstrate the effect of switching frequency and fast voltage (dv/dt) transients. The EMI emission spectra for Si, SiC and GaN are shown in Fig. 1. The spectra are the Fourier analysis of trapezoidal switching signals with parameters as listed in Table I. The operating frequency of the semiconductors is taken as a most common for the respective type of semiconductors [1]-[6].

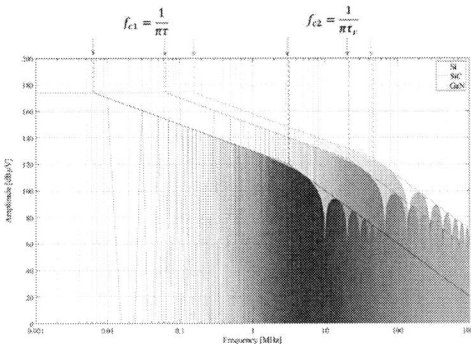

Fig. 1: EMI emission spectra for Si (black), SiC (blue) and GaN (orange).

Table I: Trapezoidal signal parameters

	Unit	Si	SiC	GaN
Amplitude	V	500	500	500
Switching Frequency	kHz	10	100	250
Rise/fall time $\tau_r = \tau_f$	ns	100	15	8

Application of WBG semiconductors allow for operation with higher frequencies, and thus, the EMI envelope is shifted to higher frequencies, which is challenging for magnetic component design. Effects that are otherwise negligible become significant, such as skin depth and dimensional resonance. The higher influence of filter parasitics and couplings must be addressed with specific models and more advanced designs to provide high-performance EMI filter designs. A laminated Mn-Zn ferrite ring core is proposed to improve the flux distribution, the core frequency characteristics, and the overall EMI filter performance, which is reduced by the skin and dimensional resonance effects in traditional designs. It has been known for over a century that laminating eliminates eddy-current losses in iron-based cores. Eddy-current losses in laminated cores are reduced by a factor $1/n^2$, where n is the number of laminations, as the path of the induced electric field is closed in each lamination [7][8]. However, ferrite is perceived as a high-impedance bulk body. However, the ferrite structure may develop conductive paths in certain conditions. Additionally, the ferrite core lamination helps to reduce one of the core cross-sectional dimensions, and therefore the frequency at which the dimensional resonance occurs can be significantly increased. FEA simulations were performed to quantify the effect of laminating on the flux distribution in the ferrite ring core. The FEA calculation uses a 3-D eddy-current field solver. The simulation results are shown in Fig. 3. Intuitively, in the solid core cross-section the magnetic flux is concentrated in the outer circumference, while the core center exhibits flux density weakening due to the skin effect. The laminated core, on the other hand, shows only a relatively marginal magnetic field change [9]-[12], with a resulting improvement in flux distribution and utilization.

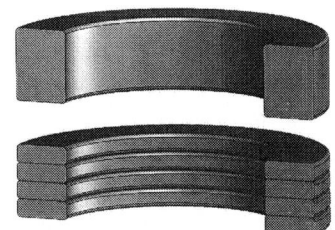

Fig. 2: Ferrite eddy current reduction with lamination.

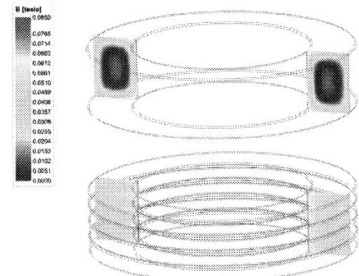

Fig. 3: FEA results of flux distribution in the solid and laminated ferrite ring core.

Core size and material selection effect on lamination

The laminated core structure divides the conduction path into sub-regions that reduce the high-frequency effects. For such structures, the upper end of the frequency range is limited by the minimum lamination size. The feasible lamination thickness directly impacts the core's complex characteristic at high frequencies. Thin laminations shift the skin and dimensional resonance effects to higher frequencies. In order to investigate how lamination performance depends on core size and material, we compare the complex permeability characteristics for three bulk core sizes, T29, T50, and T80, made of two materials, 3F36 and 3E10. The laminated cores are stacked together from smaller cores with a thickness of 3.5 mm. The T29 laminated core is stacked out of 3 laminations, while the T50 out of 4, and the T80 out of 5 laminations. Table II shows the cores parameters used for this test.

Table II: Core parameters used for the tests

Core type	Core material	Dimensions OD x ID x H	Core cross section dimension	Core cross section	Core volume
Unit	-	mm	mm	mm²	cm³
T29 Solid	3F36/3E10	29 x 19 x 10.5	5 x 10.5	53	3.96
T29 Laminated	3F36/3E10	29 x 19 x 3.5 x 3	5 x 3.5 x 3	53	3.96
T50 Solid	3F36/3E10	50 x 30 x 14	10 x 14	140	17.59
T50 Laminated	3F36/3E10	50 x 30 x 3.5 x 4	10 x 3.5 x 4	140	17.59
T80 Solid	3F36/3E10	80 x 45 x 17.5	17.5 x 17.5	306	60.13
T80 Laminated	3F36/3E10	80 x 45 x 3.5 x 5	17.5 x 3.5 x 5	306	60.13

The measured complex frequency characteristics are presented in Fig. 4 and Fig. 5 for the 3F36 material, and in Fig. 6 and Fig. 7 for the 3E10 material. Since the core size varies it requires a customized measurement fixture. Construction of the fixture provides one turn equally distributed around the toroidal core specimen. The fixture is in the shape of a cylindrical cup which forms a short-ended coaxial line over the tested core. This measurement method provides accurate results and allows for measurement up to 1 GHz. Measurements are performed with a Wayne Kerr 6550B analyzer.

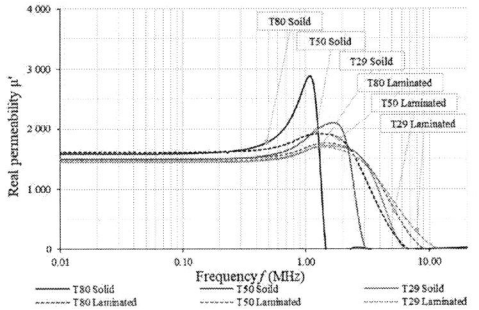

Fig. 4: 3F36 ferrite real permeability vs. frequency for various core sizes.

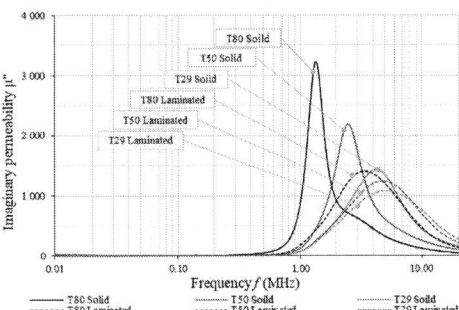

Fig. 5: 3F36 ferrite imaginary permeability vs. frequency for various core sizes.

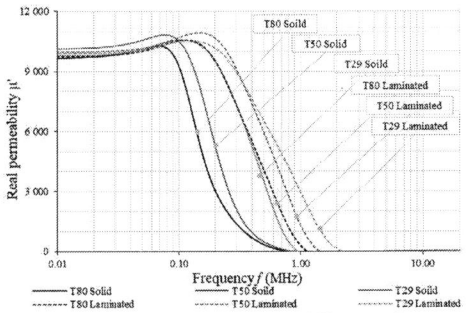

Fig. 6: 3E10 ferrite real permeability vs. frequency for various core sizes.

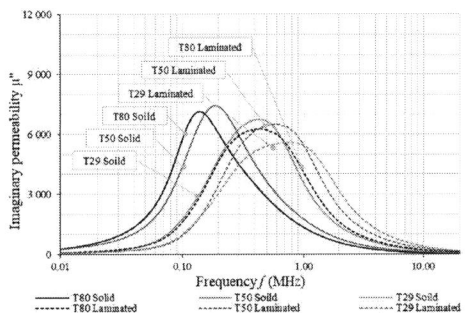

Fig. 7: 3E10 ferrite imaginary permeability vs. frequency for various core sizes.

As can be observed, the laminated cores, regardless of their size and material, show significant improvements in the frequency characteristic. The laminated T80 core maintains its performance up to 6 MHz for the 3F36 material, exactly the same as the smaller T29 solid core. The measured real permeability characteristic for the T29 bulk core, made of the 3F36 material, drops to zero at 6 MHz, while the laminated core keeps its characteristic up to 11 MHz. The biggest improvement in the imaginary permeability characteristic is visible in the T80 core. A significant loss increase, visible in the 3F36 material and caused by dimensional resonance, is reduced by half, and shifted to higher frequencies. Improvement is also visible in high permeability materials such as 3E10. The real permeability characteristic is extended from 1 MHz to 2 MHz, while the loss peak of the imaginary permeability characteristic for the T80 is reduced by 60%.

Single-phase EMC filter based on the laminated ferrite core

In order to show the differences between the solid and laminated cores in a potential application, four single-phase common-mode chokes are built. The basic construction parameters for the designed chokes are presented in Table III. The construction details are the identical for all the chokes, except for the core. The turns are wound in single layers to reduce the winding effects on the impedance characteristic

Table III: Common mode choke parameters

Parameter	Unit	Sample 1/ Sample 2	Sample 3/ Sample 4
Core material	-	Mn-Zn ferrite	Mn-Zn ferrite
Core dimensions OD x ID x H	mm	50 x 30 x 14	50 x 30 x 2.8 x 5 5 laminations
Material perm.	-	12 000	12 000
Number of turns	-	8/phase	8/phase
Wire type	-	1.8 mm Cu wire	1.8 mm Cu wire
CM Inductance	mH	1.19 / 1.15	1.05 / 1.11
Rated current	A	20	20

The measured choke inductance and impedance frequency characteristics are shown in Fig. 8 and Fig. 9, respectively. The laminated CM choke in comparison to the solid core has improved its performance in the frequency range between 200 kHz and 20 MHz. Compared to the CMC based on solid cores, the laminated choke impedance is doubled in the frequency range between 1 MHz and 20 MHz, while both chokes have the same size.

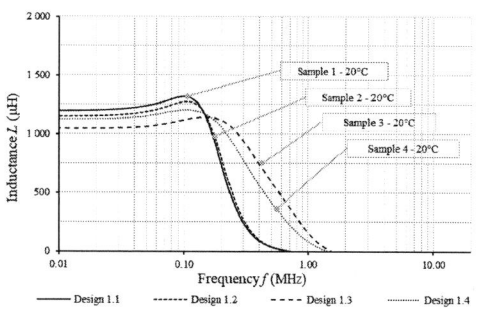

Fig. 8: Tested CMC inductance vs. frequency.

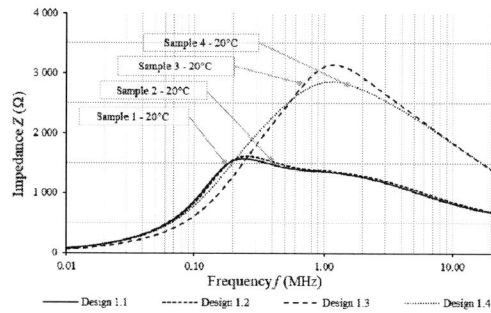

Fig. 9: Tested CMC impedance vs. frequency.

As a final step, the designed common mode chokes are integrated into the two-stage EMI filter to demonstrate the laminated core advantages in the filter structure. The EMI filter schematic is shown in Fig. 10, while the list of components is presented in Table IV.

Table IV: EMI filter components

Component	Description
CMC – solid core	Design 1.1 and Design 1.2
CMC – laminated core	Design 1.3 and Design 1.4
Capacitor C_{1X}	4.7μF - Kemet F863RL475M310ALW0L
Capacitor C_{2X}	2.2μF - Kemet F863FL225MK310ALW0L
Capacitor C_{1Y}, $C2_Y$, C_{3Y}, C_{4Y},	22nF - Epcos B32022A3223M

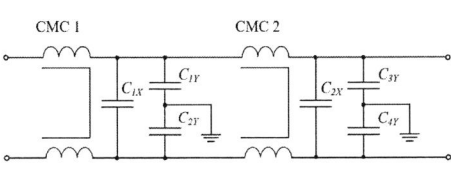

Fig. 10: Schematic of the two-stage EMI filter.

A photo of the tested EMI filters is shown in Fig. 11. Both PCBs used for the comparison are identical. The measured common-mode insertion loss characteristic for the two EMI filters is shown in Fig. 12. The evaluation shows that the proposed laminated ferrite structure provides significant attenuation improvement in the frequency range between 200 kHz and 20 MHz. Above 1 MHz, the difference is about 10 dB, while both filters maintain the same size.

Fig. 11: Complete EMI filters for CMC test.

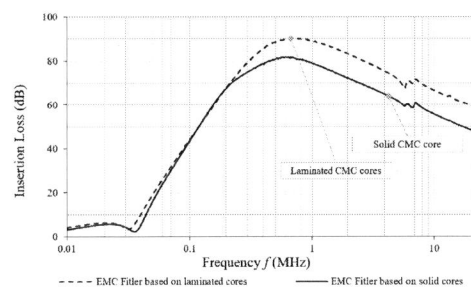

Fig. 12: EMI filters common-mode insertion loss.

Conclusion

Ferrite magnetic materials are subject to high-frequency effects which result in non-uniform frequency-dependent magnetic flux distributions. Laminated cores open up new horizons for EMI filter design. The proposed core structure can overcome the poor flux distribution caused by skin and dimensional-resonance effects. Laminated Mn-Zn ferrite ring cores show a significant improvement in the flux distribution resulting in improved frequency characteristics.

References

[1] E. Hoene, G. Deboy, C.R. Sullivan, G. Hurley, "Outlook on developments in power devices and integration: recent investigations and future requirements," *IEEE Power Electronics Magazine*, vol. 5, March 2018.

[2] J. Millan, P. Godignon, X. Perpina, A. Perez-Tomas, J. Rebollo, "A survey of wide bandgap power semiconductors devices," *IEEE Trans. Power Electron., vol.29 no.5 pp.2155-2163, 2014.*

[3] J. Biela, M. Schweizer, S. Waffler, B. Wrzecionko, J.W. Kolar, "SiC vs. Si evaluation of potentials for performance improvement of power electronics converter system by SiC power semiconductors," *Industrial Electronics, IEEE Transactions*, vol. 58, issue 7, pp. 2872–2882, July 2011.

[4] Di Han, J. Noppakunkajorn, B. Sarlioglu, "Comprehensive Efficiency, Weight, and Volume Comparison of SiC- and Si-Based Bidirectional DC–DC Converters for Hybrid Electric Vehicles," *IEEE Transactions on Vehicular Technology*, vol. 63, issue: 7, September 2014.

[5] J.W. Kolar, J.E. Huber, "Future Power Electronics 4.0 3-Phase SiC/GaN Converter Systems," *Tutorial at the 36th Applied Power Electronics Conference and Exposition (APEC2021),* June 2021

[6] N. Oswald, B.H. Stark, D. Holliday, C. Hargis, "Analysis of shaped pulse transitions in power electronics switching waveforms for reduced EMI," *IEEE Transactions on Industry Application,* vol. 47, issue: 5, October 2011.

[7] D.J. Griffiths, "Instruction to Electrodynamics" Price Hall, 1999.

[8] E.C. Snelling, Soft ferrites: properties and applications, Newnes-Butterworth; 1St Edition, 1969

[9] G.R. Skutt, "High-frequency dimensional effects in ferrite-core magnetic devices," Ph.D. dissertation, Virginia Polytechnic Institute, Blacksburg, Virginia, 1996

[10] G.R. Skutt, F.C. Lee, "Characterization of dimensional effects in ferrite-core magnetic devices," *IEEE Power Electronics Specialist Conference,* Jun. 1996

[11] M. Kącki, M.S. Ryłko, J.G Hayes, C.R. Sullivan, "Magnetic material selection for EMI filter," *IEEE Energy Conversion Congress and Exposition (ECCE)*, October 2017.

[12] M. Kącki, M.S. Ryłko, J.G Hayes, C.R. Sullivan, "A study of flux distribution and impedance in solid and laminar ferrite cores," *IEEE Applied Power Electronics Conference and Exposition (APEC)*, March 2019.

Investigation of the Static Performance and Avalanche Reliability of High Voltage 4H-SiC Merged-PiN-Schottky Diodes

Chengjun Shen, Saeed Jahdi, Phil Mellor, Juefei Yang
Erfan Bashar, Jose Ortiz-Gonzalez , Olayiwola Alatise

Department of Electrical Engineering
University of Bristol
Bristol, United Kingdom
Phone: +44 (0) 117 455 9492
Email: chengjun.shen@bristol.ac.uk
URL: http://www.bristol.ac.uk/engineering/research/em/

Acknowledgments

CT data were obtained at the XTM Facility, Palaeobiology Research Group, University of Bristol.

Keywords

≪Silicon Carbide≫, ≪Junction Barrier Schottky≫, ≪Merged-PiN-Schottky≫, ≪Schottky Diode≫

Abstract

A comprehensive range of static measurements and UIS tests have been conducted for Silicon PiN diodes, SiC JBS diodes and SiC MPS diodes with temperatures ranging to up to 175°C. The results shows that the forward voltage of Silicon PiN diode is lower at the on-state, even at high temperatures and at high currents. Higher forward voltage and positive temperature coefficient are observed for SiC devices during the static measurements, while they outperform the Silicon devices in terms of the electrothermal ruggedness, as validated by the UIS measurements and its subsequent calculated avalanche energy and die area as measured by means of CT-Scan imaging of the devices.

Introduction

Merged-PiN-Schottky (MPS) diode structure can be the compromise to exhibit the best attributes of both PiN and Schottky diode for power electronics applications [1]. This is because the significant reduction of electric field at Schottky contact suppress the leakage current [2] and carriers from the P$^+$ regions, as shown in Fig. 1, will be injected into the drift region [2, 3], allowing the occurrence of conductivity modulation to reduce the on-state voltage drop. A further reduction of leakage current is expected in SiC MPS diode as the peak electric field occurring at the SiC surface defects on Schottky contact [4] is reduced. In addition to performance metrics, reliability of power semiconductor devices and lifetime prediction has been a major topic of research in the last few decades [5, 6]. In some applications such as grid-level converters, power diodes can experience such high voltage transients that they may be led into the avalanche rating conduction, and potentially failure [7]. These diodes can also be used for high frequency and medium voltage applications as output diodes in Power Factor Correction (PFC) circuit and as clamping diode in high voltage DC transmission. Clamping diodes can experience such high voltage transients that they may be led into the avalanche rating conduction and potentially failure, while undetected grid failures in PFC circuit [8] may lead to overcurrent in output diodes. Previously, electrothermal ruggedness and avalanche robustness of SiC MPS diode have been assessed [9, 10, 11, 12] under Unclamped Inductive Switching (UIS) tests, though in absence of like-for-like comparison with similarly rated power rectifiers. Previous studies of static performance have also not dealt with the self-heating effects of the SiC MPS diode at different current levels while the high-level injection effect of MPS diode has not been discussed by means of experimental measurements.

This paper explores the static characteristic of commercially available 4H-SiC MPS diodes in contrast to Silicon PiN and 4H-SiC Junction Barrier Schottky (JBS) diodes in order to characterize the on-state limits. In addition, single event avalanche performance of these three devices is evaluated by means of experimental measurements. Section II of this paper provides the static performance analysis of the three devices while Section III explains the mechanism of UIS stressing and provides insights based on the experimental results. Conclusions are provided in Section IV.

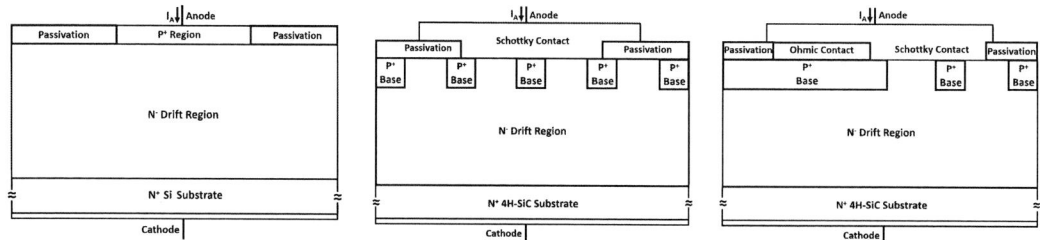

Fig. 1: From left: cross-section of Silicon PiN diode, SiC JBS diode and SiC MPS diode.

Static Performance

Table. I includes the key parameters of the three diodes to be used for further analysis. All the devices are in TO-220 packages. The static performance of these diodes have been characterized by using the static measurement test circuit shown in Fig. 2. The SiC power MOSFET SCT3160KL connected in series to diodes is used to accurately control the current conduction period through the diode. An ELC ALR3220 power supply provides the on-state current ranged from 5 to 20 A while the conducting period is set by an Agilent 33220A 20MHz arbitrary waveform generator to 3 seconds for all three diodes. The initial case temperature before the circuit is turned-on is controlled from 25°C to 175°C in 25-degree increments via ITC-100RL PID temperature controller. A Tektronix current probe model TCP312 in conjunction with a probe amplifier model TCPA300 was used for measuring the diode currents while a JAMECO P6100 100MHz voltage probe was used to measure the diode voltage. Both the captured current and voltage waveform are shown in a Keysight MSO7104 A 1-GHz 4 GSa/s oscilloscope. The I-V characteristic of all three diodes has also been measured by using B2901A Source/Measure Unit.

Table I: The key electrical parameters of the three Silicon and SiC rectifiers.

	Silicon PiN	4H-SiC JBS	4H-SiC MPS
Model	DSI30-12A	C4D20120A	GC20MPS12-220
Manufacture	IXYS	CREE	GeneSiC
Package	TO-220-2	TO-220-2	TO-220-2
Blocking Voltage	1200 V	1200 V	1200 V
Forward Current	30 A at 130°C	26 A at 135°C	30 A at 135°C
Leakage Current	40 μA	200 μA	10 μA

Fig. 3 and Fig. 4 show the IV characteristic of the three diodes. It is observed that the on-state voltage of SiC MPS diode is the highest followed by that of the SiC JBS diode and Silicon PiN diode. The higher on-state voltage of SiC devices is due to the larger junction voltage because of the much lower intrinsic carrier concentration. At high temperatures, the junction voltage is found to decrease by the increase of intrinsic carrier concentration, while the temperature stability of SiC device leads to the convergence of I-V curves as in Fig. 3. Fig. 5 highlights the conductivity modulation effect observed during the self-heating of Silicon PiN diode since a less voltage increment is observed at high currents. However, the conductivity modulation is not observed in SiC JBS and MPS diode as they are primarily functioning

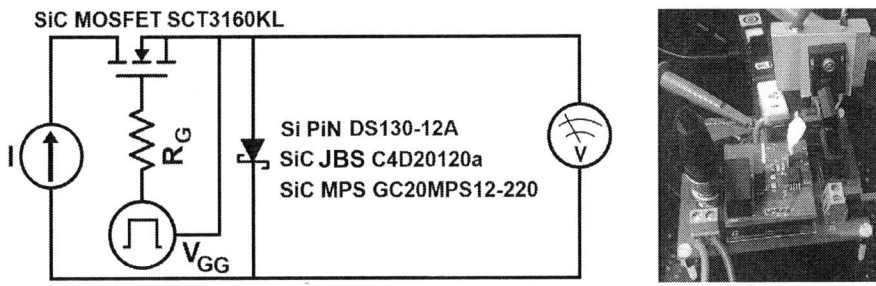

Fig. 2: (A) The equivalent circuit and (B) devices connection for On-State measurements.

as Unipolar devices with a large built-in voltage of the SiC P-N junction of about 3 V confirmed by the experiments.

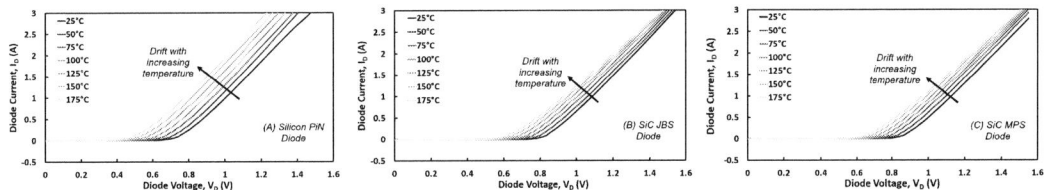

Fig. 3: The I-V characteristics of Silicon PiN, SiC JBS and SiC MPS diode.

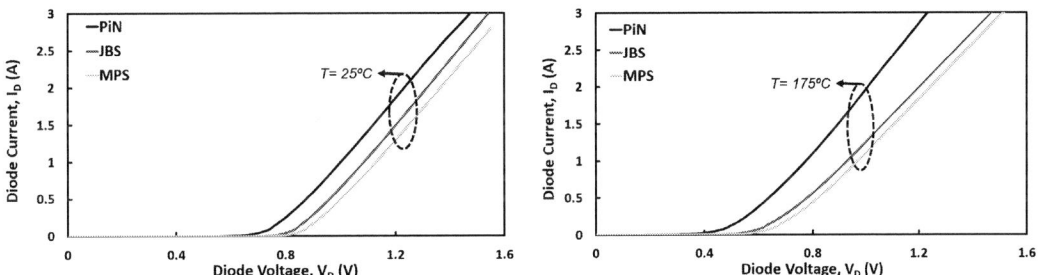

Fig. 4: The I-V characteristics of the 3 diode compared with each other at 25°C & 175°C.

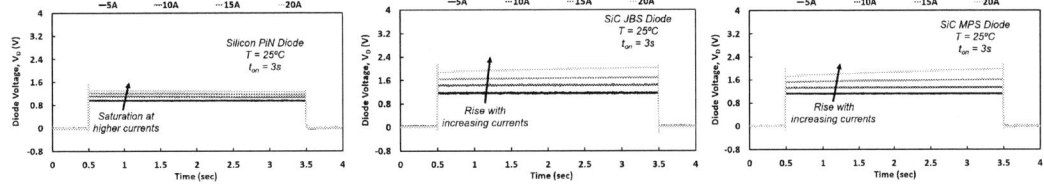

Fig. 5: On-state voltage during self-heating for Silicon PiN, SiC JBS and SiC MPS diode at 25°C.

Fig. 6 also shows the negative temperature dependence of the on-state voltage of Silicon PiN diode at the on-state. In contrary to the Silicon PiN diode, the positive temperature dependence for SiC devices, especially for SiC MPS diode, is also observed. This is because the increased current levels across the series resistance causes the forward voltage to have a positive temperature coefficient. The absence of conductivity modulation is beneficial for the JBS diode because it favours the unipolar conduction through the JBS structure which is designed to block high voltage while maintaining its unipolar conduction mode. However, this is a crucial disadvantage of MPS diodes which is expected to have high level injection. This property is suitable for high frequency application as the low stored charge and thus the fast-switching transient is maintained. Nevertheless, it is not beneficial for parallel connection of the Silicon PiN diodes with negative temperature dependence of on-state voltage [13, 14, 15]. A positive feedback loop between current and temperature is generated since the hotter diode with lower voltage can conduct more current which will continue to increase until failure. Fig. 5 and Fig. 6 also show larger on-state voltage at high currents and at high temperatures in SiC devices, further increases during the on-state. The on resistance is further increased due to the negative temperature coefficient of the carrier mobility.

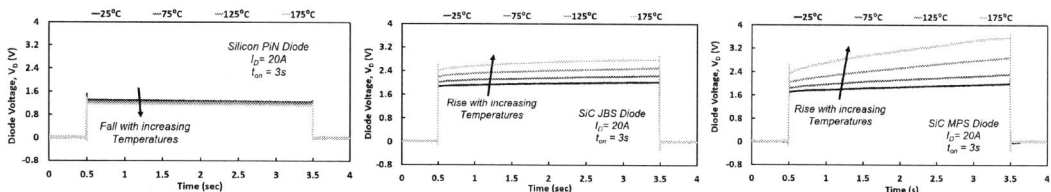

Fig. 6: On-state voltage during self-heating for Silicon PiN, SiC JBS and SiC MPS diode at 175°C.

Unclamped Inductive Switchings

The single event avalanche ruggedness of Silicon PiN, SiC JBS and SiC MPS diodes have been investigated through a wide scale of UIS measurements. All the devices are fabricated in a standard TO-220 package as shown in Table. I. The UIS testing board is shown in Fig. 7 with a high voltage IGBT (IXBX55N300) acting as the power switch [16]. The initial temperature of diodes before each UIS event is controlled in the same way as that of on-state measurement. A load inductor of 1.25 mH is charged to the peak avalanche current that is proportional to the length of the gate pulse L_P, ranging at 80 μs & 160 μs, and also proportional to the initial DC link voltage V_{DC} increased from 90 V to 360 V. Two GW-Instek GDP-100 100 MHz voltages probes and a CWT Ultra-mini 50 MHz Rogowski current coil (CWT1) are used to capture voltage and current waveforms while both are shown in the same oscilloscope as that for the on-state measurement.

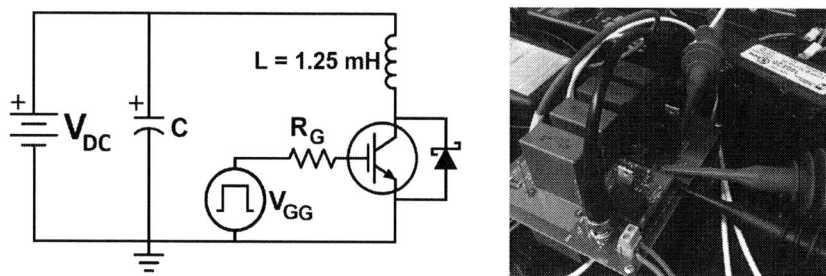

Fig. 7: The UIS test circuit schematic and the test board.

Typical current and voltage waveforms in a single UIS event is shown in Fig. 8. When the IGBT switches off, the current flowing through the inductor starts to decrease. Since a counter Electromagnetic Force (EMF) will be induced to resist the abruptly change of inductor current, the peak over-voltage transient voltage V_{PK} can be derived [17, 18] as:

$$V_{PK} = L \times \frac{dI_{off}}{dt} \tag{1}$$

Where L is the load inductance, $\frac{dI_{off}}{dt}$ is the rate of change of current at turn-off. Such surge voltage usually reaches the breakdown voltage [19] of the diode (V_{BR}), conducting the avalanche current and remain steady under higher DC link voltage. The resulting power dissipation cause the surge of junction temperature, degrades the diode breakdown ruggedness or destroys the device as the hotspot at junction termination with potential for melting of the anode metallization [9]. Unlike the power diodes which will suffer high electrothermal stress, the IGBT will stay safe due to the much higher voltage/current ratings (voltage of 3 kV & steady-state current of 55 A at 110°C). The increase of DC link voltage is used to increase the rate of current turn-on $(\dfrac{dI_{on}}{dt})$, because:

$$V_{DC} = L \times \frac{dI_{on}}{dt} \qquad (2)$$

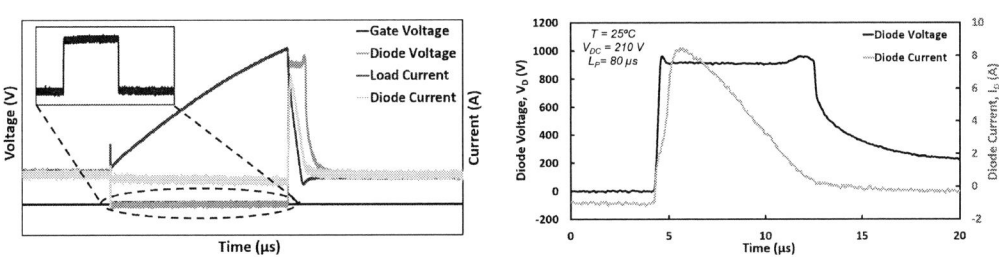

Fig. 8: Typical waveforms under UIS test and its zoomed-in version.

Therefore, the load current is proportional to the DC link voltage. To monitor the device degradation before & after the single UIS test, the I-V characteristic of all three diodes has also been measured. Fig. 9 shows the degradation of the IV characteristics of those diodes under test. Unlike the repetitive UIS test which imposes the same thermal stress for all UIS pulses, the single UIS tests aim to fail the device with just a few progressively prolonged pulses. Meanwhile, all three devices under test show stable behaviour with minor degradation at 25°C and at 175°C during the single UIS tests, when compared with the forward degradation in [11]. Therefore, this single UIS test methodology are shown to be reliable while the devices' degradation is also found to have a limited impact on their failure.

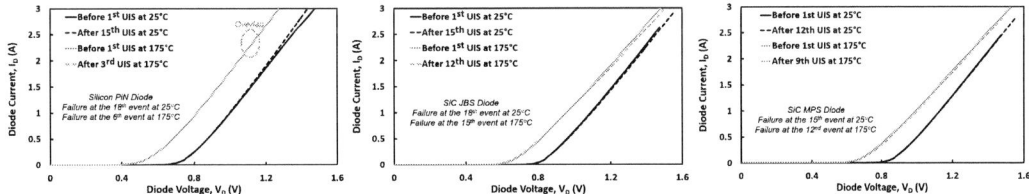

Fig. 9: Forward voltage degradation of the IV characteristics at 25°C and at 175°C for Silicon PiN, SiC JBS and SiC MPS diode.

Fig. 10 and Fig. 11 show the zoomed-in view of UIS waveforms for diodes with different technologies with load current increased until failure of devices. Although all three diodes are rated at 1.2 kV, a much higher breakdown voltage, especially for the Silicon PiN diode can be observed. This is mainly because the Schottky contact which causes higher leakage current when reverse biased is not present in the Silicon PiN diode structure. At load current of 5A, the diode voltage cannot reach the breakdown voltage in SiC devices. This is because the stored energy inside the load inductor is not sufficient to keep charging the parasitic capacitor inside the diode to the breakdown voltage, as expected by Equation 1. It can also be seen in Fig. 11 and in Fig. 12 that the higher effective breakdown voltage of Silicon PiN leads to much lower diode current compared to that of SiC devices. This is the leakage current instead of the avalanche current. The avalanche duration increases with increase of load current while the tail current indicates the process to discharging parasitic capacitor. Fig. 11 and Fig. 12 shows that the current decrease rate of SiC MPS diode is smaller than that of SiC JBS diode. This is because of the larger drift velocity in JBS diode due to the higher electric field applied. When the device failure occurs, the diode conducts in the reverse conduction with increasing current exceeding the preset load current levels because of the avalanche multiplication effect together with the thermal runaway effect, while the diode voltage drops to zero as the blocking capability is lost. Silicon PiN diode failed at lower load current compared with the SiC JBS & SiC MPS while its recovery process, as in Fig. 11, has been skipped as the device cannot handle such high induced avalanche current.

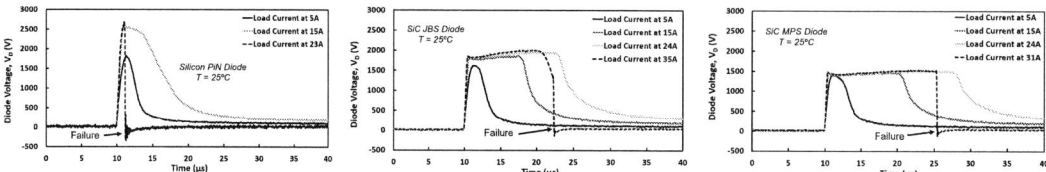

Fig. 10: Avalanche diode voltage as a function of time for different load currents until device failure for Silicon PiN diode, SiC JBS diode and SiC MPS diode.

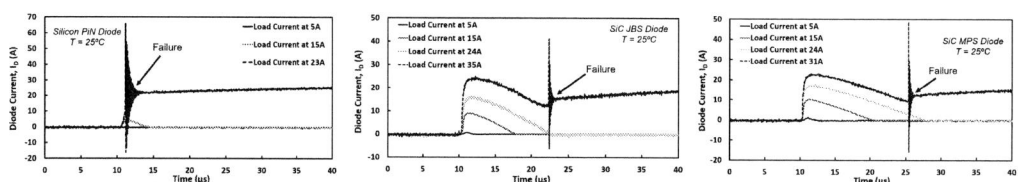

Fig. 11: Avalanche diode Current as a function of time for different load currents until device failure for Silicon PiN diode, SiC JBS diode and SiC MPS diode.

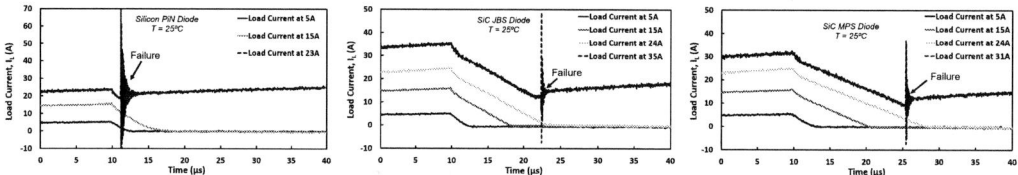

Fig. 12: Avalanche load current as a function of time for different load currents until device failure for Silicon PiN diode, SiC JBS diode and SiC MPS diode.

Fig. 13 and Fig. 14 emphasize the difference in avalanche ruggedness among three different diodes at failure mode. It is seen that the SiC devices can sustain the avalanche conduction for a longer time than that of Silicon device before the avalanche multiplication is triggered. In contrast with the diode current of Silicon devices which is increased to the preset load level immediately after the failure, the rate of current increase is found to be much smaller. This is because of the much smaller impact ionization coefficients in SiC devices which enables a slower generation process of electron-hole pairs [20]. It can also be seen that the highest breakdown voltage in Silicon PiN diode, followed by that of SiC JBS and that of MPS. At high temperatures, all devices is found to fail at lower currents with shorter recovery period. This can be explained by the fact that there is less headroom to dissipate power during the recovery process when the temperature of controller is increased even though the less avalanche energy is generated due to the smaller load current.

The avalanche energy is an important parameter since the avalanche breakdown mechanism of power rectifiers is avalanche energy breakdown as a high induced power increase the junction temperature to destroy the devices. The critical avalanche energy is determined as the maximum value before failure of the device during the progressive single UIS tests as highlighted in Fig. 15, can be derived as:

$$E_{ava} = \int_{0}^{t_{ava}} V_{diode}(t) \cdot I_{diode}(t) \cdot \mathrm{d}t \tag{3}$$

where t_{ava} is the time duration of device avalanche.

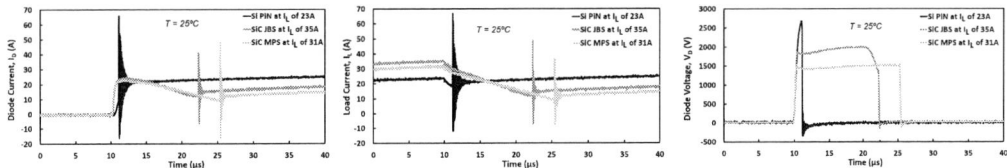

Fig. 13: The (left) diode current, (center) load current, and (right) diode voltage when failure occurs at 25°C.

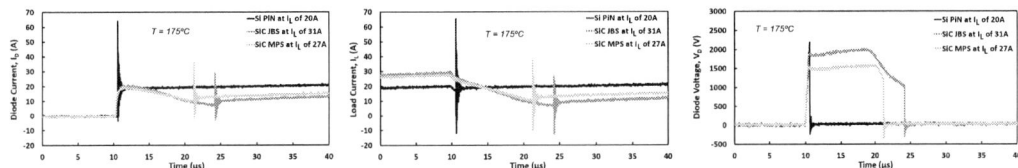

Fig. 14: The (left) diode current, (center) load current, and (right) diode voltage when failure occurs at 175°C.

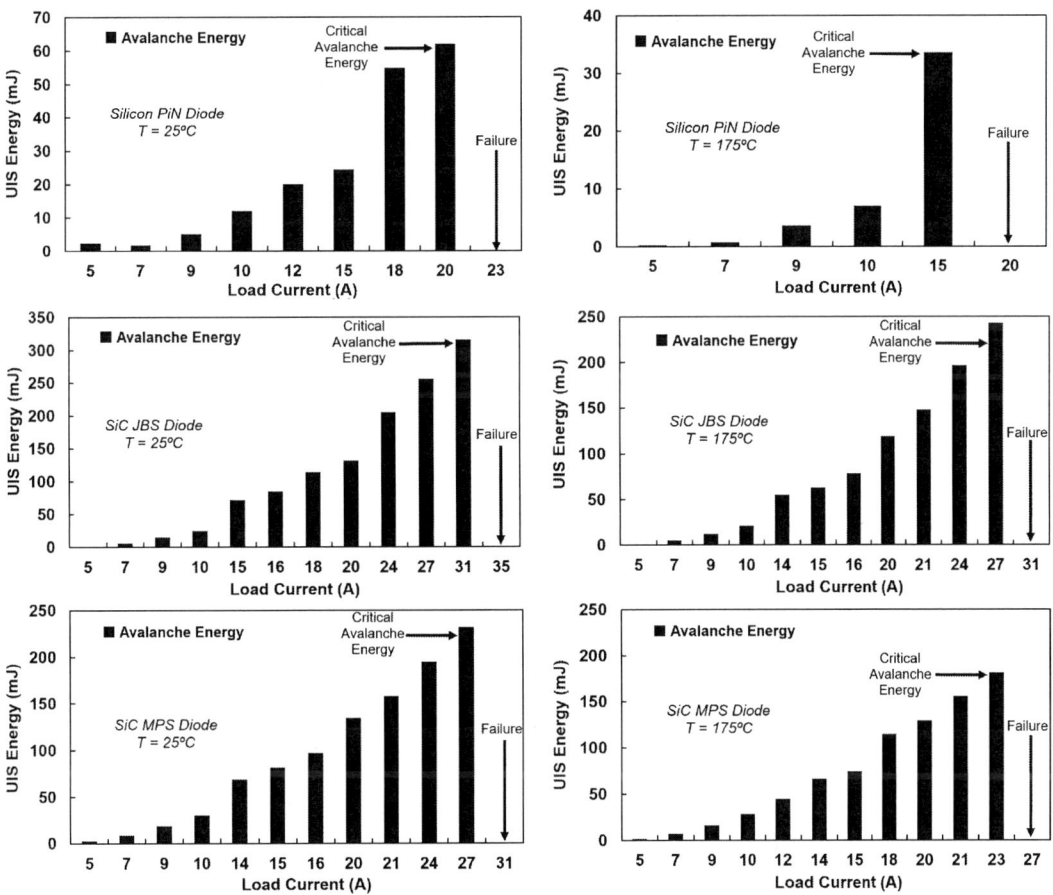

Fig. 15: Determination of critical avalanche energy from the UIS test at 25°C and at 175°C.

The critical avalanche energy of SiC JBS is the highest, followed by that of SiC MPS, and then that of Silicon PiN diode. This is expected from Fig. 11 and Fig. 10 as the recovery time after the event dominates the avalanche energy. The critical energy is found to be lower at higher temperatures. This is because of the shorter recovery period observed at high temperatures, primarily due to the smaller difference between the junction temperature and the preset temperature making the energy dissipation more difficult. Despite the Silicon PiN diode able to sustain a higher reverse voltage compared to SiC devices before the actual failure takes place, this device always failed at lower load currents during the progressive UIS tests. Fig. 16 shows the CT scan image of the three devices after failure. The three devices have similar packaging, though with different dies areas, and thus it is conceived that the packaging has limited impact on difference between the results of devices. The failure pattern of three test devices indicate that the molten anode metallization is likely to have happened, especially at elevated temperatures similar to that in [11], leading to failures of the die and subsequently wire-bonds.

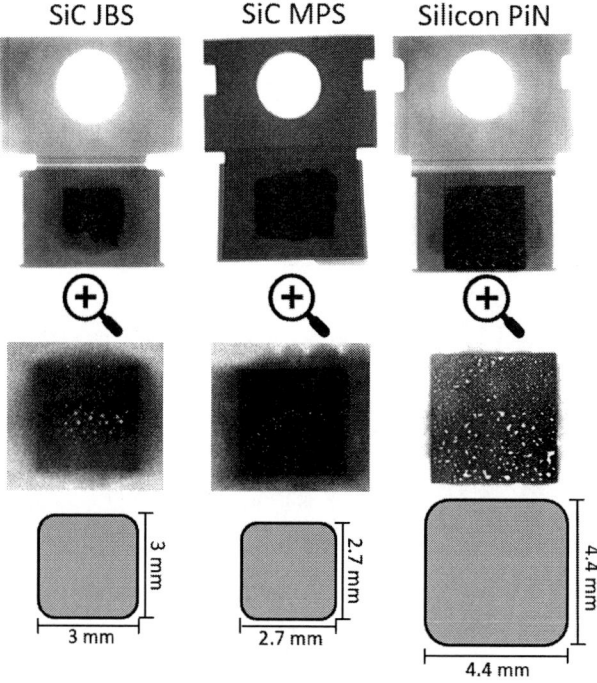

Fig. 16: Corresponding front side CT scan image of the devices from Fig. 13 for Silicon PiN diode, SiC JBS and SiC MPS diode.

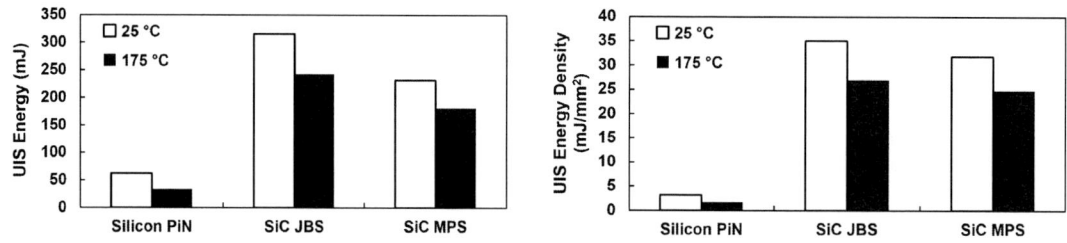

Fig. 17: Comparison of critical avalanche energy for Silicon PiN, SiC JBS & SiC MPS diode at different temperatures.

The factors that would influence the higher avalanche ruggedness of the SiC devices compared with the Silicon PiN, excluding the packaging, can be listed with the following factor. The thermal conductivity of SiC is two times higher than that of Silicon for a more effective heat dissipation. In addition, the wide-bandgap of SiC limits the generation of additional carriers due to the higher thermal energy [21, 22, 23]. The lower carrier lifetime of SiC also enables a faster recombination of the thermally generated carriers. Additionally, the impact ionization coefficient of SiC devices is smaller than Silicon, so less electron-hole pairs are generated at reverse bias. All of these would lead to a better performance of the SiC diodes than the Silicon counterparts as demonstrated in the measurements of this paper.

Conclusion

In this paper, the on-state performance and avalanche ruggedness of Silicon PiN diode, SiC JBS diode, and SiC MPS diode have been investigated by means of wide-scale experimental measurements in a range of temperatures to up to 175°C. It is shown that the Silicon PiN diodes have a lower on-state voltage compared with SiC JBS diode and SiC MPS diode, enabling a lower steady-state power dissipation while the negative temperature dependence of diode voltage further decreases the losses at the conduction mode and at high temperatures. On the other hand, SiC MPS diode and SiC JBS diode have shown positive temperature coefficient and a positive feedback loop is observed. The conductivity modulation effect is only observed for Silicon PiN diode at high currents, while the built-in voltage of P-N junction in SiC devices is larger which favors high-speed operation. In terms of the UIS measurements, the difference in device packaging have negligible influence to the single UIS tests. Although the Silicon PiN diode exhibits the highest breakdown voltage among these three similarly-rated power devices, the SiC JBS diode is the most electrothermally rugged devices, followed by that of the SiC MPS diode, and then the Silicon PiN diode, as is reflected by the measurement results and the calculated avalanche energy.

References

[1] S. Jahdi, et al., 'Renewable hybrids Grid-Connection using converter interferences' International Journal of Sustainable Energy Development (IJSED), vol. 2, no. 1, 2013.

[2] J. Baliga, 'Advanced Power Rectifier Concepts', Springer US, 2009.

[3] C. Shen, et al., 'Prospects and Challenges of 4H-SiC Thyristors in Protection of HB-MMC-VSC-HVDC Converters,' in IEEE Open Journal of Power Electronics, vol. 2, pp. 145-154, 2021.

[4] B. Thomas, 'SiC Schottky Diode Device Design: Characterizing, Performance & Reliability', Wolfspeed, A Cree Company, 2015.

[5] J. Liu, et al., 'Reliability consideration of low-power grid-tied inverter for photovoltaic application,' 24th European Photovoltaic Solar Energy Conference and Exhibition, Germany, Sep. 2009.

[6] S. Yang, et al., 'An industry based survey of reliability of power electronic converters,' IEEE Trans. Ind. Appl., vol. 47, no. 3, pp. 14411451, May 2011.

[7] S. Jahdi, et al., 'Electrothermal modeling and characterization of SiC Schottky and silicon PiN diodes switching transients', IEEE Energy Conversion Congress and Exposition (ECCE), 2014, pp. 2817-2823.

[8] J.W. Hancock, 'SiC Device Applications: Identifying and Developing Commercial Applications', 2006.

[9] T. Basler, et al., 'Avalanche Robustness of SiC MPS Diodes', PCIM Europe 2016; Int. Exhibition and Conf. for Power Electronics, Intelligent Motion, Renewable Energy and Energy Management, 2016, pp. 1-8.

[10] R. Rupp et al., 'Avalanche behaviour and its temperature dependence of commercial SiC MPS diodes: Influence of design and voltage class,' IEEE 26th International Symposium on Power Semiconductor Devices & IC's (ISPSD), 2014, pp. 67-70.

[11] S. Palanisamy, et al., 'Repetitive surge current test of SiC MPS diode with load in bipolar regime,' IEEE 30th International Symposium on Power Semiconductor Devices and ICs (ISPSD), 2018, pp. 367-370.

[12] J. Ortiz Gonzalez, et al., 'Dynamic characterization of SiC and GaN devices with BTI stresses', Microelectronics Reliability, vol. 100101, pp. 113389, 2019.

[13] R. Wu, et al., 'Performance of Parallel Connected SiC MOSFETs under Short Circuits Conditions', Energies, 2021, 14, 6834.

[14] R. Wu, et al., 'Measurement and simulation of short circuit current sharing under parallel connection: SiC MOSFETs and SiC Cascode JFETs', Microelectronics Reliability, vol. 126, 2021, 114271.

[15] R. Wu, et al., 'Current Sharing of Parallel SiC MOSFETs under Short Circuit Conditions', 23rd European Conference on Power Electronics and Applications (EPE'21 ECCE Europe), 2021, pp. 1-9.

[16] R. Bonyadi, et al., 'Physics-based modelling and experimental characterisation of parasitic turn-on in IG-BTs,' 17th European Conf. on Power Electronics and Applications (EPE'15 ECCE-Europe), 2015, pp. 1-9.

[17] Renesas, Unclamped Inductive Switching (UIS) Test and Rating Methodology Application note.

[18] Toshiba, MOSFET Avalanche Ruggedness Application note.

[19] P. Alexakis, et al., 'Improved Electrothermal Ruggedness in SiC MOSFETs Compared With Silicon IGBTs,' in IEEE Transactions on Electron Devices, vol. 61, no. 7, pp. 2278-2286, July 2014.

[20] J. Baliga, 'Fundamentals of Power Semiconductor Devices', Springer US, 2019.

[21] S. Jahdi, et al., 'The Impact of Temperature and Switching Rate on Dynamic Transients of High-Voltage Silicon and 4H-SiC NPN BJTs: A Technology Evaluation,' in IEEE Transactions on Industrial Electronics, vol. 67, no. 6, pp. 4556-4566, June 2020.

[22] S.N. Agbo, et al., 'UIS performance and ruggedness of stand-alone and cascode SiC JFETs', Microelectronics Reliability, Volume 114, pp. 113803, 2020.

[23] E. Bashar, et al., 'Comparison of Short Circuit Failure Modes in SiC Planar MOSFETs, SiC Trench MOSFETs and SiC Cascode JFETs', IEEE 8th Workshop on Wide Bandgap Power Devices and Applications (WiPDA), 2021, pp. 384-388.

On chain-link based multi-port converters able to connect HVDC and MVDC to AC transmission network

Daniele Falchi[1], Oriol Gomis-Bellmunt[1], Eduardo Prieto-Araujo[1], and Olivier Despouys[2]

[1]CITCEA-UPC, Polytechnic University of Catalonia.
Avinguda Diagonal, 647, 08028 Barcelona, Spain
Email: daniele.falchi@upc.edu, eduardo.prieto-araujo@upc.edu and oriol.gomis@upc.edu

[2]RTE (Réseau de Transport d'Électricité).
La Defense, 7C place du Dôme, Paris, France
Email olivier.despouys@rte-france.com

Keywords

≪Voltage Source Converters (VSCs)≫, ≪High voltage power converters≫, ≪HVDC≫, ≪Converter circuit≫, ≪Multi-port converters HVDC/MVDC/HVAC≫.

Abstract

Multi-port converters may offer an attractive solution to contribute in achieving a more compact, flexible and efficient network. Recently, in literature, a few converter topologies have been proposed to interconnect high-power and high-voltage systems. Most of converter topologies are based on chain-link or modular concept, which is considered a very beneficial way to scale different ranges of voltages and current keeping a high level of reliability, feasibility and power quality. The aim of this paper is to provide, first, a qualitative description of a few selected multi-port converter configurations (non-isolated and fully isolated arrangements) able to interconnect HVDC, MVDC to HVAC systems. And, second, to present a qualitative comparison among cost and footprint of each topology. Also, the paper introduces some practical implications to take into account abnormal conditions e.g fault and port disconnections for such converters.

Introduction

Within the last decade, the power system is facing the increasing amount of renewable and discontinuous energy sources together with the adoption of DC systems integrated in the conventional AC network. Compared to AC technology, DC technology offers improved features, performances, controlling and cost-effectiveness especially looking at long transmission energy systems. In addition to HVDC technology, medium voltage direct current (MVDC) technology has been recently considered for many high power applications and it has been attracting the interest of power distribution and transmission system experts in order to reinforce the grid. MVDC voltages range between 1.5 kV ($\pm$750 V) and 100 kV ($\pm$50 kV) approximately [1], [2]. Offshore renewable, distribution network interconnectors, transmission network interconnectors, rail electrification, shipping and harbors are some applications in which MVDC technology is applied [3] - [6]. Power electronic converters act as key-enablers for modern AC/DC network allowing such a systems integration. Nowadays, MVDC and HVDC energy provision is enabled by conventional AC/DC converters. In the industry, the main VSC reference for HVDC is the Modular Multilevel Converter (MMC) [7]. Regarding MVDC, some converter manufacturers have been proposing various AC/DC topologies, e.g. Two-Level-Converter (TLC) [8], cascade Neutral Point Clamped (NPC) [9] and modular multilevel converter (MMC) [10]. It is also noticed that DC/DC modular converters [11] are under research to provide an interconnection among different DC networks, HVDC and MVDC. However, in order to improve the overall cost, footprint and efficiency of the sub-station, recently, researchers and industries are looking at multi-port converters as another more feasible solution to interconnect HVDC, MVDC to HVAC systems instead of having several two-port converters (AC/DC and DC/DC) linked. Fig. 1 shows a possible application in which multi-port converter technology can be used in a generic system. Table I summarizes some general characteristics depending on the solution, 2-port converter solution and multi-port converter solution.

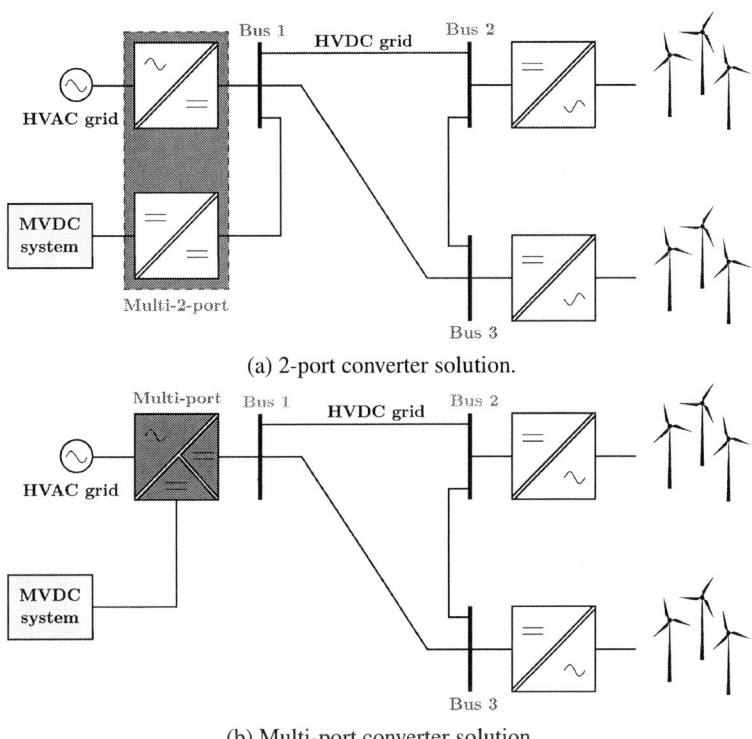

(a) 2-port converter solution.

(b) Multi-port converter solution.

Fig. 1: 2-port converter vs Multi-port converter solutions.

Table I: 2-port converter solution vs Multi-port converter solutions

	2-port converter based	Multi-port converter based
Cost/Footprint	High	Low
Protection system complexity	Medium	Medium
Load shedding complexity	Medium	High
Fault Ride Through complexity	Medium	High
Power density	Low	High
Energy stored	High	Low

The paper refers to chain-link based converters because of their significant benefits: greater modularity leads to scaling large voltage ranges while maintaining the same semiconductor parameters. Higher fault tolerance, compared with other multilevel converters such as NPCs or FCs (Flying Capacitors) is another key aspect leading to a more reliable solution; in addition chain-link converters achieve a high power quality that contributes to minimize the filtering equipment. The main drawbacks are: first, the complexity of the internal circulating currents, needed for balancing the internal energy stored in sub-module capacitors, second, the inability to perform soft-switching at sub-module level, which might increase overall losses. Anyway, in addition to those strictly related to HVDC, there are several applications in which chain-link based topologies are used: AC/AC as static frequency converter (SFC) [12], DC/AC as motor drive converter [13] and FACTS [14].

Various multi-port converter proposals have been presented in literature. For most of those, the modelling and control design were developed and tested [15]. However, an initial comparative analysis of different proposed concepts, looking at design implications and operations, might be needed to address the beginning approach to such converters.

Thus, the aim of the paper is to provide a qualitative comparison between four most relevant topologies that can be found in the literature which are based on four different concepts. Hence, two non-isolated and two fully-isolated based topologies were selected, all which are based on two different concepts. The purpose of the proposed work is to introduce the topic with initial insights on different chain-link based multi-port converters interconnecting HVDC, MVDC to HVAC systems. To do so, aspects such as footprint, cost, capabilities and limitation to deal with abnormal conditions (such as fault and port disconnection) were considered.

The following sections include a description among the selected topologies, a qualitative comparison in terms of cost and footprint, a qualitative comparison highlighting main concerns during abnormal conditions and finally, a concluding section.

Non-isolated multi-port topologies

As mentioned above, two non-isolated topologies based on two different concepts have been studied. Both are inspired by existing 2-port converters: the DC/DC Modular Multilevel Converter, M2DC [16], and the DC/DC autotransformer, AT-HVDC [17]. Fig. 2 shows a general representation of the two concepts. For both of them, MMC is the basic block, but each one suggests a different concept to get a multi-port converter. In literature, other similar topologies [18]-[19] based on the same two concepts have been suggested, so that the study focuses on the main two above mentioned references.

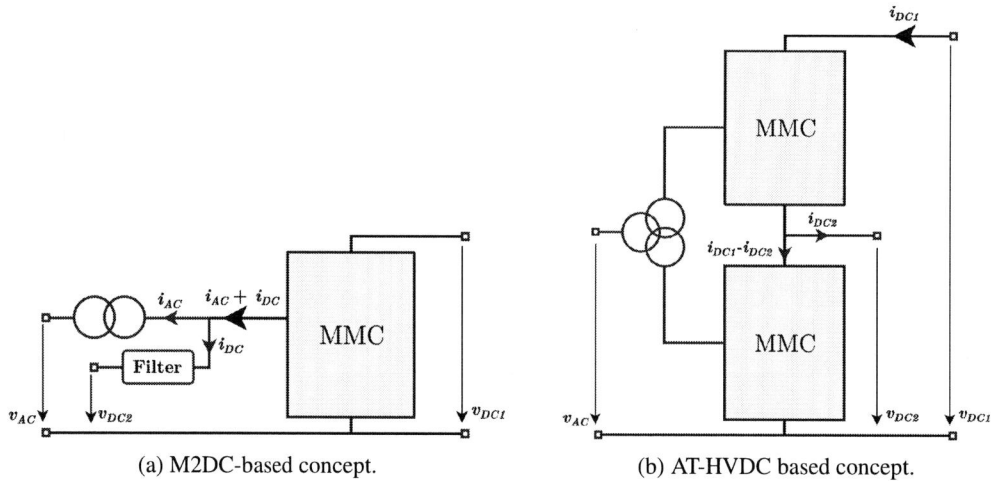

(a) M2DC-based concept. (b) AT-HVDC based concept.

Fig. 2: M2DC and AT-HVDC based concept for non-isolated multi-port converters.

Due to degrees of freedom of MMC at the intermediate point of connection (shunt connection to AC and to MVDC), the M2DC-based converter is able to impose AC and DC components simultaneously. The filter provides a pure DC component to the second DC port. On the other hand, the AT-HVDC converter makes it possible to derive another DC terminal (thanks to series-stacked interconnection of MMCs) to allow a second DC current circulation. The transformer has two main purposes: the first one is to provide energy circulation between upper and lower MMCs in order to achieve the internal energy balancing; the second purpose, by adding a third winding, is to offer a connection to the main AC network.

M2DC based multi-port converter

Asymmetrical and symmetrical monopole arrangements have been proposed in [20] and [21] respectively and Fig. 3 shows both converter arrangements. The first provides a shunt connection of the AC transformer with MVDC filter. The symmetrical monopole version provides the AC transformer connection at the converter midpoint.

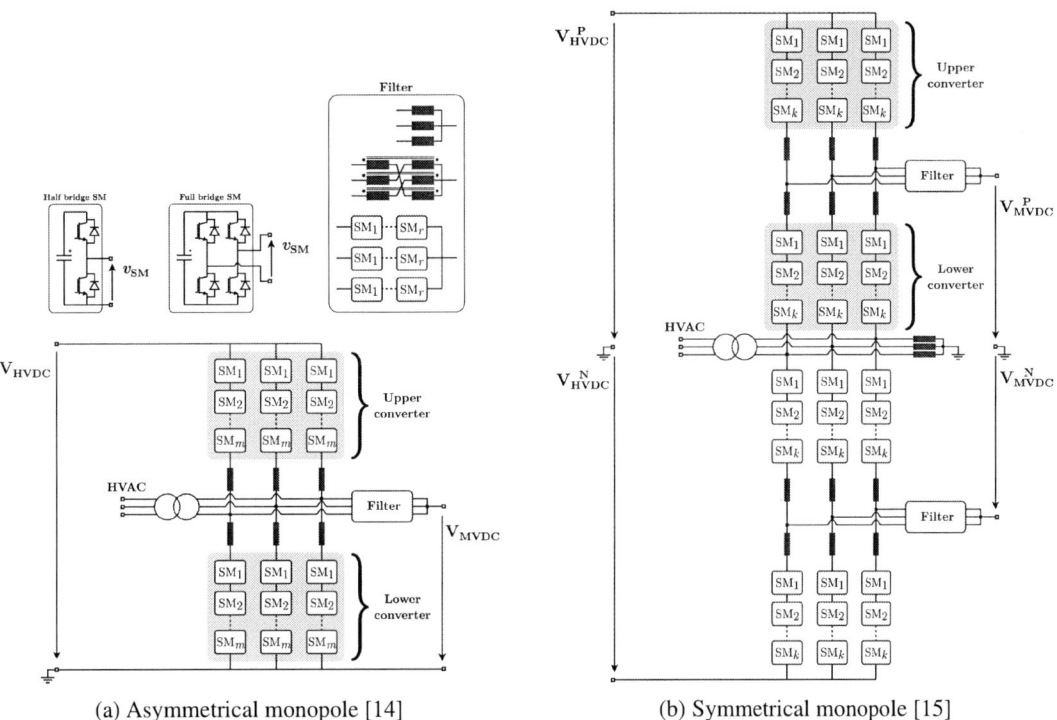

(a) Asymmetrical monopole [14]   (b) Symmetrical monopole [15]

Fig. 3: M2DC-based multi-port arrangements.

The configuration depicted in Fig. 3a requires a transformer sustaining DC voltage isolation between ground to neutral. This may have significant impact on the size and weight of this transformer. Internal current circulation between upper and lower converter do not affect the AC current through the AC port, so that AC power can be fully decoupled from internal power circulation. In the configuration depicted in Fig. 3b the transformer does not have to sustain any DC voltage isolation respect to ground but, in such a arrangement, the internal circulating current is not fully decoupled from power to AC terminals, the transformer impedance is in the loop of internal circulation between upper and lower converters, therefore this might lead to an AC-side power limitation or to transformer overrating.

The main purpose of the filter connected in shunt to mid-point of the converter and AC grid is to provide a pure DC component at MVDC terminals preventing any AC components. Depending on the design and functionalities, the type of the filter might be passive (L-C), magnetic (coupled reactors) or active (power electronics based).

In general, a main issue related on this topology concept regards the limited DC voltage step ratio between DC terminals, especially when the DC power step ratios are quite similar. Such limitation is due to the increased value of internal circulating current to provide a balanced operation of the converter and, semiconductor and losses constrains limit the capabilities of such a converter topology. The multi-port operation might help somehow to mitigate such DC voltage step ratio limitation since the rated power of both DC sides is not the same because of the AC-side power contribution.

AT-HVDC based multi-port converter

The AT-HVDC based multi-port converter depicted in Fig. 4 is the most relevant alternative to the M2DC-based converter. Asymmetrical and symmetrical monopole solutions have been proposed in [17] and [22]. In both solutions, due to the series-stacked sub-MMCs construction (upper MMC, lower MMC Fig. 4a), the transformer must sustain a large DC voltage isolation leading to increased size and design complexity.

The AC connection is provided by a multi-winding transformer. It is the same transformer used to

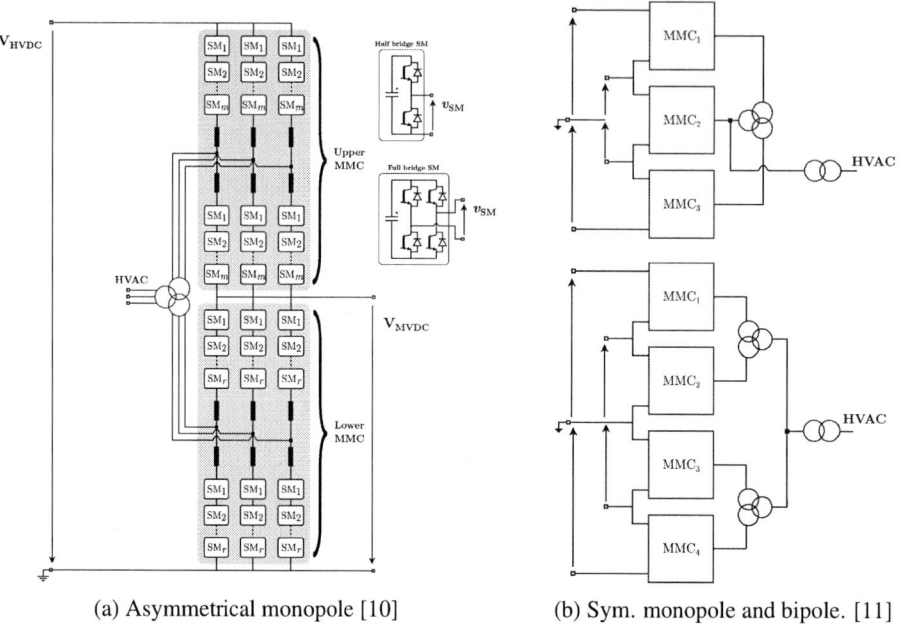

(a) Asymmetrical monopole [10]   (b) Sym. monopole and bipole. [11]

Fig. 4: AT-HVDC-based multi-port arrangements.

transfer power between upper and lower sub-MMCs. Therefore, the transformer must be rated for the nearly full power of the converter and not only rated for the partial power needed for internal energy balancing. Furthermore, in normal conditions, the transformer operates at a fixed frequency (50 Hz or 60 Hz) due to the AC network connection.

Compared to the prior M2DC-based multi-port converter, as a key difference, by acting on the winding turn ratio properly it is possible to minimize the circulating current by maximizing the voltages. Due to this, the stress on semiconductors and losses will be reduced and the converter is able to increase the step ratio capability.

Isolated topologies

Most of the time, flexible grounding solution at each port and improved safety level might be key requirements to select between multi-port topologies. Fully galvanic isolated multi-port topologies offer such kind of benefits. Nevertheless, compared to non-isolated converters, the cost, size and losses indicators of isolated arrangements will be negatively impacted. Fig. 5 shows the two main concepts used to implement galvanic isolated multi-port converters.

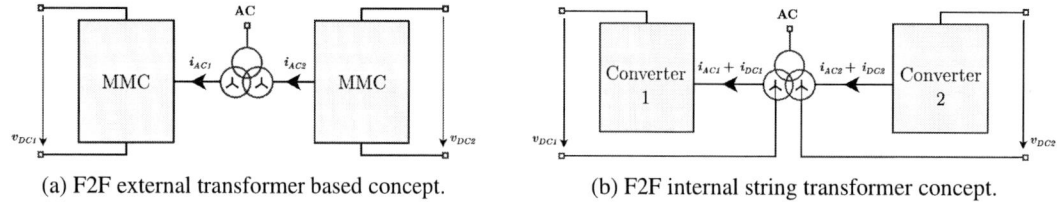

(a) F2F external transformer based concept.   (b) F2F internal string transformer concept.

Fig. 5: Galvanic isolated multi-port converter concepts.

The first solution in Fig. 6a is inspired by the most well known front-to-front (F2F) topology [23], which is also based on the DAB (Dual Active Bridge) concept. The transformer is connected in shunt to the AC terminals of MMC converters. The second solution in Fig. 6b is inspired by [24] and provides galvanic isolated ports by installing a series connected transformer to the converter string; still, it is possible to consider the topology based on F2F with incorporated string transformer. In literature, other proposals can be found in which the galvanic isolation is distributed by sub-modules or cells; in this case each cell (or some of them) are equipped by a medium frequency transformer (MFT) [25]. Such a solution is not part of this article, as their main disadvantages is the high voltage isolation requirements for those cell transformers and their low fault tolerance especially considering isolation lost in one sub-module. Hence, for HVDC applications, such a solution might not be adequate. Fig. 6 highlights the two isolated multi-port topologies and the transformer winding connections to the chain-link based converter arrangement.

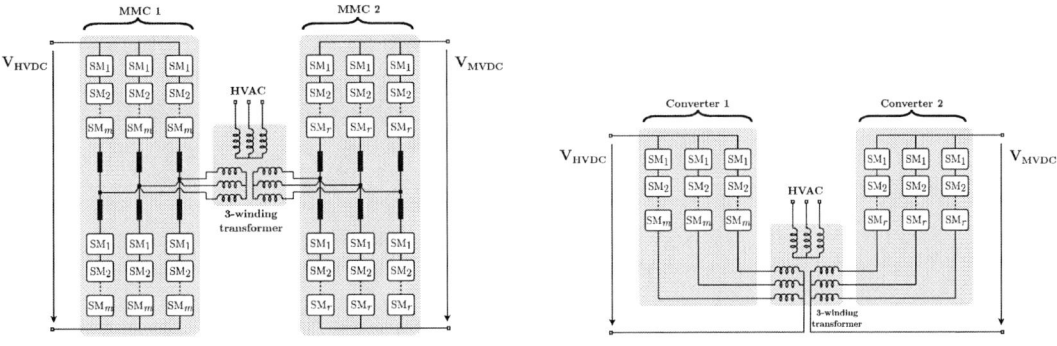

(a) External transformer multi-port converter.   (b) Series string transformer multi-port converter.

Fig. 6: Fully galvanic isolated multi-port converters.

F2F multi-port converter with external multi-winding transformer

F2F multi-port converter based on external multi-winding transformer is the most relevant reference suggested in literature. With Fig. 6a it is possible to note that two MMC converters share the same AC common coupling by means of the transformer and an additional winding provides the connection of the overall converter to the main AC grid. For such a topology the main advantage is the relatively low design complexity. The transformer has to be rated for the full power ratio of the converter and, in case of asymmetrical monopole at DC terminals, it has to comply to DC voltage isolation respect to the ground. Obviously, the transformer frequency must be equal to the AC grid frequency (50 Hz or 60 Hz). The high power coupling between the transformer windings brings to increase the complexity of the control especially in order to achieve good performances (e.g. AC-Fault Ride Through) during some grid contingencies, such as AC short circuit. Furthermore, since both MMCs are connected to the AC three-phase system, the topology needs full-bridge based sub-modules in both converters to achieve the full blocking capability.

F2F multi-port converter with series string multi-winding transformer

As it is possible to see from Fig. 6b, the main converter purpose is to reduce its overall footprint by incorporating in the transformer its leakage impedance and the impedance of string reactors [26]. The neutral of two wye winding connections provides the DC current circulation to the relative DC port. Then, transformer windings have to be sized to carry AC and DC components and this fact might result in substantial impact on the weight and cost. Furthermore, such a solution provides one degree of freedom less than the conventional MMC; the internal circulating current is not fully decoupled from AC current to the grid, overall performances and internal energy balancing may be limited especially during contingencies such as faults. In general, the transformer saturation due to DC bias has to be considered; although by a proper winding turn ratio and DC current ratio between two converters it is possible to ensure a zero net DC flux within the core in normal operation, this is not true when one of the two DC port is not demanding power, in the last case transformer core needs to be sized in order to prevent the saturation. To deal with the DC bias, it is possible to install different transformer topologies able to eliminate the DC bias within the core properly, for instance, one of these method is to use a zig-zag transformer [27].

General comparison between multi-port topologies

Multi-port converter solutions, compared to multi-2-port converter ones, offer relevant advantages in reducing the footprint, saving cost and materials. This is true for the multi-module non-isolated and isolated solutions. Table II summarizes some general characteristic that isolated arrangements offer with respect to non-isolated topologies in a generic system. As above mentioned, full galvanic isolated converters offer a more secure and flexible solution to interconnect HVDC, MVDC to HVAC, and most of the time these are required by the application. Anyway, such benefits lead to have higher costs and larger footprint compared with non-isolated converters.

Table II: General pros and cons on having a galvanically isolated converter.

Pros	Cons
High flexibility on grounding	Increased number of switches
High flexibility on DC voltage step ratio	Impact on the footprint
Improved security operating condition	Increased losses
Increased buck-boost capability	DC bias (saturated area)*

*For the transformer.

Since the galvanic isolation is a requirement from the application, the following sections provide a qualitative comparison between non-isolated topologies and isolated topologies separately.

Footprint and cost

Tables III and IV summarize the number of main components present in the selected topologies, non-isolated and isolated converters respectively. The comparison focuses on the number of sub-modules, reactors, transformers and filters. V_{SM} is the rated voltage of each sub-module, V_{HVdc} and V_{MVdc} the voltages at the DC terminals and n_{leg} the number of the converter phases. Special transformer or conventional transformer depends on the topology, special transformer has to be sized to sustain a large DC voltage isolation respect the ground (AT-HVDC based case) or in case to allow AC and DC current circulation (F2F series winding transformer based).

Table III: Non-isolated multi-port converters.

	SMs	String reactors	Transformer	Filter
M2DC based	$2n_{leg}V_{HVdc}/V_{SM}$	$2n_{leg}$	✓Conventional	✓
AT-HVDC based	$2n_{leg}V_{HVdc}/V_{SM}$	$4n_{leg}$	✓Special	-

Table IV: Isolated multi-port converters.

	SMs	String reactors	Transformer	Filters
F2F shunt based	$2n_{leg}(V_{HVdc}+V_{MVdc})/V_{SM}$	$4n_{leg}$	✓Conventional	-
F2F series based	$2n_{leg}(V_{HVdc}+V_{MVdc})/V_{SM}$	-	✓Special	-

In terms of footprint and cost the two non-isolated solutions might be quite equivalent; M2DC-based converter use a conventional transformer to connect to AC grid but a filter sized for the full AC voltage is needed. AT-HVDC based converter does not need a filter but a special three-winding transformer sustaining a large DC voltage isolation to ground is required. For isolated converters, although the F2F shunt based converter requires more string reactors than F2F series based converter, the complexity and footprint of transformer should be considerably less impacting on cost and footprint.

Fault and protection implications

The objective of this section is to give an initial evaluation on the main implications to consider depending on the topology in anomalous conditions such as fault. Fig. 7 shows three cases of fault selected for this analysis: C1 represents a symmetrical short circuit fault at AC port. C2 represents a pole to ground short circuit fault at HVDC port. C3 represents a pole to ground short circuit fault at MVDC port. Table V summarizes such observations for the four selected topologies.

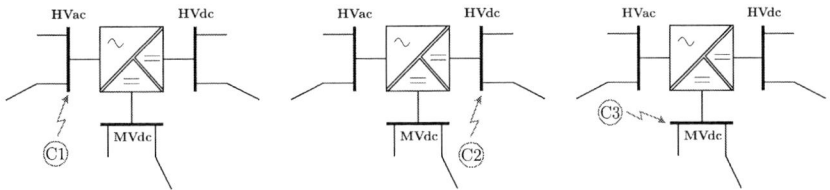

Fig. 7: Case of fault on multi-port converter.

Table V: Topologies implications in different fault conditions.

(a) M2DC-based converter.

	M2DC-based
C1	Converter is able to perform the FRT and DC/DC operation is still possible.
C2	Fault block: FB SMs to limit the short circuit current from AC and MVDC port.
C3	Fault block to HVDC: proper number of SM at the upper side converter.
	Fault block to HVAC: proper number of FB SMs at the lower side converter.

(b) AT-HVDC-based converter.

	AT-HVDC-based
C1	Converter is able to perform FRT. DC/DC operation limited by current.
C2	Fault block: FB SMs to limit the short circuit current from AC and MVDC port.
C3	Fault block to HVDC: proper number of SM at the upper MMC converter.
	Fault block to HVAC: proper number of FB SMs at the lower MMC converter.

(c) F2F shunt transformer based.

	Shunt multi-winding transformer based
C1	Converter is able to perform FRT to AC. DC/DC operation limited by current.
C2	Fault block to HVdc: FB SMs to HVdc side MMC to limit the short circuit current.
C3	Fault block to MVdc: FB SMs to MVdc side MMC to limit the short circuit current.

(d) F2F series string transformer based.

	Series multi-winding transformer based
C1	Converter is able to perform FRT to AC. DC/DC operation limited by current.
C2	Fault block to HVdc: FB SMs to HVdc side MMC to limit the short circuit current.
C3	Fault block to MVdc: FB SMs to MVdc side MMC to limit the short circuit current.

Port disconnection - load change implications

Each converter port of the converter is rated at a specific power. Here the assumption is that, in case of load change or port disconnection at the port, the overall converter has to continue to operate adjusting the power flow according with the new sets (P^*_{dc-dc} and P^*_{ac-dc}) of references between the remaining ports. Fig. 8 shows three cases of port disconnections (load/source): C1 represents the HVAC terminals disconnection, C2 represents the HVDC terminals disconnection and C3 represents the MVDC terminals disconnection.

For M2DC based topology, referring to table VIa, the worst case occurs when HVDC terminals are disconnected; high internal upper-lower energy unbalancing could occur leading high current stress in order to achieve the balancing among upper-lower side.

For AT-HVDC based topology, referring to table VIb, when AC is disconnected a DC/DC mode operation is possible and the transformer provides the energy transferring between upper MMC and lower MMC in order to achieve internal balancing. When the HVDC is disconnected the upper MMC should not provide any active power to the AC terminals and the lower MMC works to achieve the AC/DC power conversion. When the MVDC is disconnected the converter works similarly to a bipolar MMC, and the AC power is distributed between upper and lower MMC.

Referring to table VIc, F2F based on shunt connected transformer, the worst case occurs when AC port is disconnected; two MMCs have to change the mode of operation, keeping one them controlling the AC voltages and the other controlling the power. Finally, from table table VId, F2F based on series string connected transformer, the worst case occurs when one of two DC terminals is disconnected; DC flux cannot be compensated leading the transformer to work in saturated area, unless a special transformer is used (zig-zag transformer for instance).

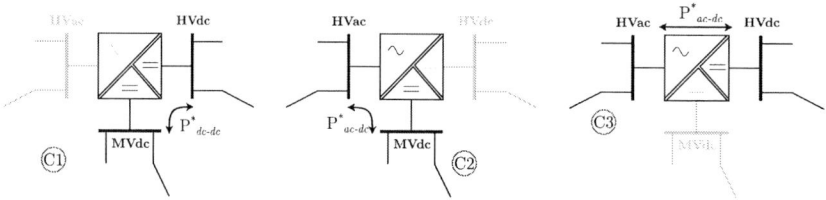

Fig. 8: Different load shedding cases in multi-port converter.

Table VI: Topologies implications in different port disconnection events.

(a) M2DC-based converter.

	M2DC based converter
C1	Pure DC/DC operation with considering a new P^*_{dc-dc}.
C2	Pure AC/DC operation with considering a new P^*_{ac-dc}. High upper-lower energy unbalancing.
C3	Pure AC/DC operation with considering a new P^*_{ac-dc}.

(b) AT-HVDC-based converter.

	AT-HVDC based converter
C1	Pure DC/DC operation with considering a new P^*_{dc-dc}.
C2	Pure AC/DC operation with considering a new P^*_{ac-dc}. Upper converter exchange zero active power.
C3	Pure AC/DC operation with considering a new P^*_{ac-dc}. AC/DC as a bipolar MMC

(c) F2F shunt transformer based.

	Shunt multi-winding transformer based converter
C1	Pure DC/DC operation with considering a new P^*_{dc-dc}. One converter form the grid the other controls the power.
C2	Pure AC/DC operation with considering a new P^*_{ac-dc}. HVdc side converter can provide just reactive power.
C3	Pure AC/DC operation with considering a new P^*_{ac-dc}. MVdc side converter can provide just reactive power.

(d) F2F series string transformer based.

	Series multi-winding transformer based converter
C1	Pure DC/DC operation with considering a new P^*_{dc-dc}. One converter form the grid the other controls the power.
C2	Pure AC/DC operation with considering a new P^*_{ac-dc}. DC biased transformer.
C3	Pure AC/DC operation with considering a new P^*_{ac-dc}. DC biased transformer.

Conclusion

After presenting the main four topologies based on chain-link concept, the paper provides a qualitative analysis of each of them highlighting the main implications in design, footprint and during anomalous grid contingencies depending on the four different concepts. M2DC based topology and AT-HVDC topology are very similar in footprint and cost, regarding performances in abnormal conditions, while M2DC based converter might offer a better response during AC FRT condition than AT-HVDC based converters. Regarding isolated topologies, F2F based on series string connected multi-winding transformer might offer a reduced footprint compared to F2F based on shunt connected multi-winding transformer, but it results in more limitations and complexities in sizing and control, especially during grid contingencies.

References

[1] CIGRE WG C6.31: Medium Voltage Direct Current (MVDC) Grid Feasibility Study
[2] Rentschler, A and Kuhn, G and Delzenne, M and Kuhn, O.: Medium voltage dc, challenges related to the building of long overhead lines, PES Transmission and Distribution Conference and Exposition (T&D). IEEE. 2018
[3] Hay S. et al.: MVDC Technology Study-Market Opportunities and Economic Impact, Scottish Enterprise, Report 9639-01 (2015), R0
[4] Dujic D.: Medium Voltage Direct Current-Technologies and Systems, KEYNOTE at Ee2017-19th International Symposium on Power Electronics. POST TALK. 2017.
[5] Stieneker M. and W De Doncker R.: Medium-voltage DC distribution grids in urban areas, 2016 IEEE 7th international symposium on power electronics for distributed generation systems (PEDG). 2016, pp. 1-7.
[6] Verdicchio A. et al.: New medium-voltage DC railway electrification system, IEEE Transactions on Transportation Electrification 4.2 (2018), pp. 591-604.
[7] R Marquardt.: A new modular voltage source inverter topology. Conf. Rec. EPE 2003.
[8] Skytt A.K., Holmberg P. and Juhlin L.E.: HVDC Light for connection of wind farms, (2001).
[9] Tibin J. et al.: Analysis of harmonic transfer through an MVDC Link, (2019).
[10] AG Siemens. MVDC PLUS: Medium Voltage Direct Current Managing the future grid. Whitepaper (2017).
[11] Kung S. H. and Kish G.J.: A Modular Multilevel HVDC Buck–Boost Converter Derived From Its Switched-Mode Counterpart, IEEE Transaction On Power Delivery, (2018).
[12] Bessegato L. et al.: Control and Admittance Modeling of an AC/AC Modular Multilevel Converter for Railway Supplies, IEEE Transaction On Power Electronics, (2019).
[13] Siemens: Principles of Modular Multilevel Converter Topology, Whitepaper, (2019).
[14] Fang Zheng Peng et al.: A Multilevel Voltage-Source Inverter with Separate DC Sources for Static Var Generation, IEEE Industry Applications Conference Thirtieth IAS Annual Meeting, (1995).
[15] Rouhani M.and Kish G.J.: Multiport DC–DC–AC Modular Multilevel Converters For Hybrid AC/DC Power Systems, IEEE Transaction On Power Delivery, (2020).
[16] Norrga S.: Bidirectional unisolated dc-dc converter based on cascaded cells. US Patent 9,484,808. DCDC Nov. 2016.
[17] Bakran M., Knaak H. and Schon A.: Modular multilevel DC/DC converter for HVDC applications. WO Patent 2014/056540 A1. Apr. 2014.
[18] Lin W., Wen J. and Cheng S.: Multiport DC–DC Autotransformer for Interconnecting Multiple High-Voltage DC Systems at Low Cost, IEEE Transaction On Power Delivery, (2015).
[19] Li Y. Liu D. and Kish G.J.: Generalized DC-DC-AC MMC Structure for MVDC and HVDC Applications, 20th Workshop on Control and Modeling for Power Electronics (COMPEL), (2019).
[20] Bakran M. and Schon A.: DC to DC Converter. WO Patent 2016/138949 A1. Sept. 2016.
[21] Papastergiou K. and Stamatiou G.: AC/DC multicell power converter for dual terminal HVDC connection. US Patent 9,065,328. June 2015.
[22] Schon A. and Bakran M.: High power HVDC-DC converters for the interconnection of HVDC lines with different line topologies, 2014 International Power Electronics Conference (IPEC-Hiroshima 2014-ECCE ASIA). IEEE. 2014, pp. 3255-3262.
[23] Kenzelmann S. et al: Isolated DC/DC Structure Based on Modular Multilevel Converter, IEEE Transaction On Power Electronics, (2015).
[24] Oates C.: A methodology for developing 'Chainlink' converters, 13th European Conference on Power Electronics and Applications, (2009).
[25] Ferreira Costa Levy et al: Comparative Analysis of Multiple Active Bridge Converters Configurations in Modular Smart Transformer, IEEE Transaction On Industrial Electronics, Vol. 66, No. 1,(2019).
[26] Tenca P. et al.: Patent: High voltage DC/DC converter with transformer driven by modular multilevel converters (MMC), WO 2013/075735 (2013).
[27] Serbia N. et al.: Half Wave Bridge AC/DC Converters – From diode rectifiers to PWM multilevel converters, PCIM Europe (2014).

Voltage Control Scheme for Multilevel Interfacing PV Application: Real-Time MRAC-Based Approach

Mohammad Sadegh Orfi Yeganeh
Technical University of Denmark
2800 Kgs. Lyngby, Denmark
morfi@dtu.dk

Mehdi Rahmani
Imam Khomeini International University
Iran, Qazvin
mrahmani@eng.ikiu.ac.ir

Nenad Mijatovic
Technical University of Denmark
2800 Kgs. Lyngby, Denmark
nm@dtu.dk

Tomislav Dragicevic
Technical University of Denmark
2800 Kgs. Lyngby, Denmark
tomdr@dtu.dk

Frede Blaabjerg
Aalborg University
Aalborg, Denmark
fbl@energy.aau.dk

Pooya Davari
Aalborg University
Aalborg, Denmark
pda@energy.aau.dk

Keywords

Harmonic mitigation, Model reference adaptive controller, Cascaded h-bridge multilevel inverter, Optimal switching angle, Total harmonic distortion.

Abstract

Cascaded H-bridge multilevel inverters (CHB-MLIs) have a proper structure and significant advantages like filter-less topology due to their desired output voltage levels and being suitable for employing isolated DC voltage sources that have been utilized in multi-sources applications such as photovoltaic (PV) power generation. However, unbalanced DC voltage sources of the DC/DC output section or fast voltage variations can decrease the output voltage quality and make some undesired fluctuations and harmonics, affecting the load behavior in different applications. To overcome these problems, a new real-time voltage control technique has been proposed to regulate the output voltage magnitude properly. This aim can be obtained by employing the model reference adaptive control (MRAC) technique for voltage regulation to improve total harmonic distortion (THD), and to decrease some harmonic orders (HOs) magnitude, especially lower harmonics (3rd, 5th, and 7th). Therefore, the proposed solution is designed to stabilize the dynamic error originating from the DC side. At the same time, the DC voltage sources are interfaced with variations, where they do not have equal magnitudes with each other. The effectiveness of the proposed control technique has been verified by simulation in MATLAB/Simulink.

1. Introduction

Inverters (DC/AC) are one of the integral parts of renewable energies such as photovoltaic (PV) and wind turbine systems. Multilevel inverters (MLIs) are introduced to provide better output voltage characteristics than the conventional two-level inverters. In recent years, MLIs have gained much attention in medium-voltage and high-power due to their various advantages such as lower electromagnetic interference, the lower voltage stress on power switches, an improvement in THD, and a reduction in massive filtering requirements. The classical MLIs are classified into diode clamped MLI (DC-MLI), flying capacitor (FC-MLI), and cascade H-Bridge MLI (CHB-MLI) [1]. In high-level inverter topologies, there is a limitation of complexity and number of clamping diodes for the DC-MLIs, and there is an increment in the number of required capacitors and complexities of considering DC-link balancing for FC-MLIs. Among these classical MLI topologies, the CHB-MLI has the least components for various levels. This topology consists of a series of H-bridge cells synthesized to the desired voltage from several isolated DC sources, which is suitable for PV applications and can be obtained from isolated PV modules. There is a growing interest in reduced component counts MLIs [2] and [3]. Still, the classical topologies have essential applications in most critical areas. Due to the overall efficiency of the entire system, modulation strategies are a crucial part of DC/AC converters. The purpose of a modulation signal is to control the output voltage/current characteristics, especially losses (both conduction and switching losses), THD, and individual harmonics content. There are different types of modulation strategies that are proposed for utilization in MLIs [4-6]. Three major types of modulation strategies are utilized in MLIs based on the switching frequency; low/fundamental, high, and variable switching frequency. In low/fundamental switching frequency, selective harmonic elimination (SHE), optimal switching angles (OSAs), space vector control (SVC), and nearest level control (NLC) are the major methods [7]. The OSA modulation strategy can minimize the harmonics in the inverter voltage waveform. This modulation strategy is a very flexible and perfect solution; by calculating the switching angles, it is possible to control the output voltage/current harmonics (to control the output voltage/current) and increase the output quality.

PI and fuzzy logic controllers are conventional controllers have some disadvantages like a slow dynamic response, high starting overshoot in some cases, and sensitivity to controller gains [8]. Model predictive control (MPC) is an advanced control technique, which is utilized in both industry and academia. Fast dynamic response and handling constraints are the benefits of this controller [9] and [10]. The main issues with the MPC controller are the weighting factor design process and real-time implementation [11]. The model reference adaptive control (MRAC) is one of the main techniques of adaptive control. It is a combination of a parameter estimator, which generates parameter estimates online [12]. These parameters are completely unknown and could change with time in an unpredictable manner. This controller is a way of adjusting the controller characteristics in response to changes in the plant and disturbance dynamics distinguishes one scheme from the other. This controller is a special type of nonlinear feedback control, which has useful capabilities and interesting properties that can be profitably incorporated into the design of control systems. Some benefits of adaptive control are as follows: utilizing time-varying systems uncertainty does not matter, the ability to eliminate noises and disturbances, and different types of control tools such as linear control [13]. There are many kinds of applications of the adaptive controller in converters [14] and [15].

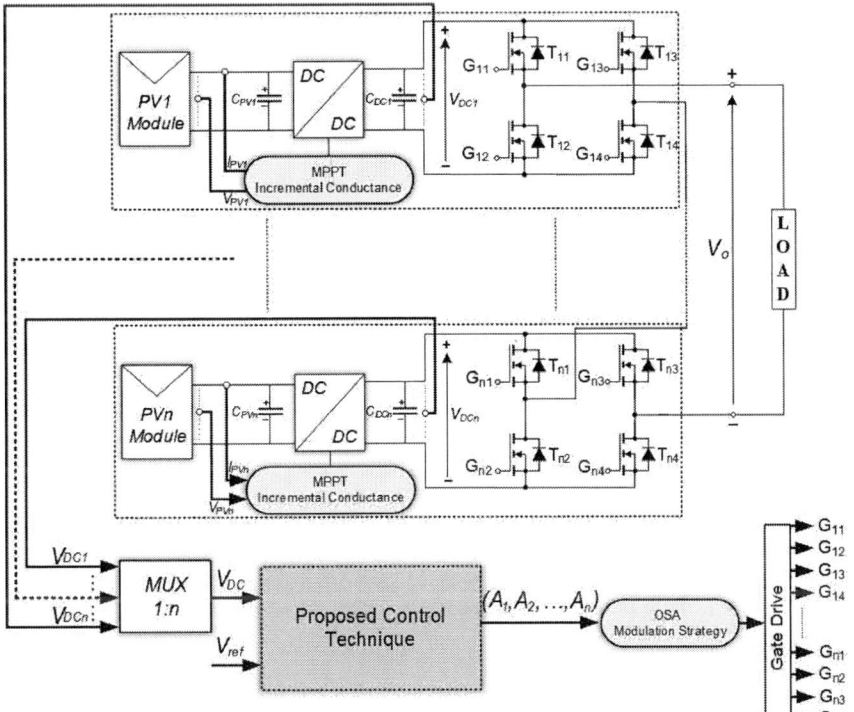

Fig. 1: System configuration consist of PV modules, DC/DC boost converter, CHB-MLI, OSA modulation strategy, and control block diagrams.

Motivated by the previous discussions, a new real-time modulation method is proposed to regulate the output voltage magnitude. In addition, it will improve the voltage THD and its harmonics content especially the lower ones (3rd, 5th, and 7th) by employing an adaptive controller in a closed-loop system for a single-phase MLI. By utilizing the proposed solution, an accurate control process can be applied to the harmonics, and it is possible to minimize their magnitude one by one. The proposed control technique is suitable for PV applications that have variable conditions such as changes in irradiation and temperature that cause variations in the DC link.

2. System configuration and voltage unbalanced problems

CHB-MLI configuration is a perfect topology in PV power generation due to its multiple isolated DC sources. Fig. 1 presents a general schematics of the utilized system configuration. At each pack, there are one PV module, a DC/DC boost converter, and a CHB inverter. PV modules generate power, and then the DC/DC link is used to increase the voltage magnitude and obtain the maximum power of the modules. An incremental inductance algorithm is employed for the maximum power point tracking (MPPT) section to control the switching states of the DC/DC converter. The output voltage of the DC/DC converters is considered as the input of the adaptive control system. The OSA modulation strategy organizes the output voltage levels and then sends the firing signals to the gate drivers. There are four ways to determine the value of the DC voltage sources: unary, binary, trinary, and proposed algorithms [3]. This study uses a unary method to value the DC sources. Therefore, the strategy is to generate the output voltage levels by equal DC sources. During this process, MRAC has an important role in determining the switching angles for each output voltage level. In this case, four packs of the PV modules, DC/DC converters, and DC/AC inverters (CHB-MLIs) are required to prepare a 9-level inverter.

The OSA strategy concept is organizing the switching angles at low switching frequencies, where it is possible to form the output voltage function. Therefore, there is an ability to control the output voltage characteristics such as root mean voltage square (V_{rms}), THD, and harmonic order magnitudes. Fig. 2 presents the output voltage waveform and the switching angles for a 9-level inverter and their positions at each level. For the first level, the output DC-link magnitude equals the DC voltage of pack one. For the second level, the output DC-link magnitude equals the DC voltage of pack two, and the same process for the rest of the levels. Based on the OSA concept, it is possible to control the output voltage magnitude by changing the switching angles at each level. It directly affects the on- and off-times for the power switches. In this modulation strategy, the output voltage waveform is symmetric, which means that it is possible to achieve the output voltage function of a 9-level inverter by considering four switching angles (a quarter of the waveform) and then form the Fourier series as follows:

$$V_{L1} = \sum_{i=1,3,5}^{h} \frac{4V_{DC1}}{i\pi} (cos\, iA_1 t) \tag{1}$$

$$V_{L2} = \sum_{i=1,3,5}^{h} \frac{4V_{DC2}}{i\pi} (cos\, i(A_1 + A_2)t) \tag{2}$$

$$V_{L3} = \sum_{i=1,3,5}^{h} \frac{4V_{DC3}}{i\pi} (cos\, i(A_1 + A_2 + A_3)t) \tag{3}$$

$$V_{L4} = \sum_{i=1,3,5}^{h} \frac{4V_{DC4}}{i\pi} (cos\, i(A_1 + A_2 + A_3 + A_4)t) \tag{4}$$

$$V_{out} = V_{L1} + V_{L2} + V_{L3} + V_{L4} \tag{5}$$

where $(V_{L1}, V_{L2}, V_{L3},$ and $V_{L4})$ are the output voltage function at each level, $(A_1, A_2, A_3,$ and $A_4)$ are the switching angles at each level, $(V_{DC1}, V_{DC2}, V_{DC3},$ and $V_{DC4})$ are the DC output voltage of the boost converter of each pack, and V_{out} is the output voltage function for a 9-level inverter.

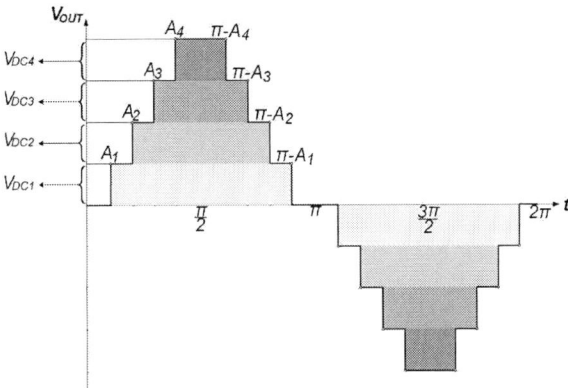

Fig. 2: Switching angles and DC voltage magnitudes of the CHB-MLI converter.

After forming the output voltage function, defining the cost functions to control the output voltage is in the process. Then, implementing the voltage regulation and controlling the output voltage characteristics such as minimizing the THD and setting constraints on the harmonic order magnitudes are achievable. Based on IEEE STD-519, there are several rules for the output voltage specifications; for example, the number of counted harmonics are equal to 49 and they need to be reduced. Therefore, the desired cost functions to cover all the restrictions and STDs items are as follows:

$$V_{rms} = \sqrt{\frac{1}{T}\int_0^T V_{out}^2 dt} \tag{6}$$

$$THD = \sqrt{\frac{\sum_{i=3,5,7}^h V_i^2}{V_1}} \tag{7}$$

$$HO = \sum_{i=1,3,5}^h \frac{4}{i\pi}(V_{DC1}\cos iA_1 t + V_{DC2}\cos i(A_1 + A_2)t + \cdots + V_{DCi}\cos i(A_1 + A_2 + \cdots + A_i)t) \tag{8}$$

where V_1 is the value of the fundamental term of the output voltage, V_i is the value of harmonics and h is the number of harmonics that are considered for minimization.

3. Voltage adaptive control strategy

The model-reference adaptive system is one of the well-known structures in adaptive control. The system has an ordinary feedback loop, and the error (e) of the systems is the difference between the output of the system and the reference model, which is a function of time and adapted parameters (θ). The purpose of an MRAC is to minimize this error such that the closed-loop system follows the reference model. To this end, a quadratic cost function (J) of the error signal (e) is considered as follows:

$$J(t,\theta) \triangleq \frac{1}{2}e^2(t,\theta) \tag{9}$$

In this way, adaption parameters are updated to minimize the cost function value. In the gradient descent method, adaption parameters are updated in the opposite direction of the gradient of the cost function (J). This adaptation rule can be presented as:

$$\frac{d\theta}{dt} = -\delta \frac{\partial}{\partial \theta} J(t,\theta) = -\delta e(t,\theta) \frac{\partial}{\partial \theta} e(t,\theta) \tag{10}$$

where δ is a real positive constant and called the adaption gain. The derivative $\partial e/\partial \theta$ is the sensitivity of the system. In the proposed approach, to apply the MRAC to a 9-level inverter, the following cost function is considered.

$$J(t,\theta) \triangleq \frac{1}{2} e^2(t,\theta) + \frac{1}{2} THD^2(t,\theta) + \frac{1}{2} \sum_{i=3,5,7}^{h} HO_i^2(t,\theta) \tag{11}$$

Minimizing the cost function (11) results in reducing the error signal, THD, and some specific harmonics. According to this cost function, the adaptation parameters are the switching angles given as

$$\theta = [A_1, A_2, A_3, A_4] \tag{12}$$

$$\delta = [\delta_1, \delta_2, \delta_3] \tag{13}$$

and the update rule of the parameters is given by

$$\frac{d\theta}{dt} = -\delta_1 e(t,\theta) \frac{\partial}{\partial \theta} e(t,\theta) - \delta_2 THD(t,\theta) \frac{\partial}{\partial \theta} THD(t,\theta) - \delta_3 \sum_{i=3,5,7}^{49} HO_i(t,\theta) \frac{\partial}{\partial \theta} HO_i(t,\theta) \tag{14}$$

Remark. The adaptation rule for the parameters (14) is obtained based on the gradient descent method. The gradient method often exhibits approximately linear convergence, i.e., the solution converges to an optimal point approximately as a geometric series [13].

In Fig. 3, the proposed adaptive controller is presented in block diagrams. The DC voltage magnitudes are inputs and switching angles of each level are the outputs for this algorithm. Replacing this controller in Fig. 1 completes our proposed control approach for the mentioned system.

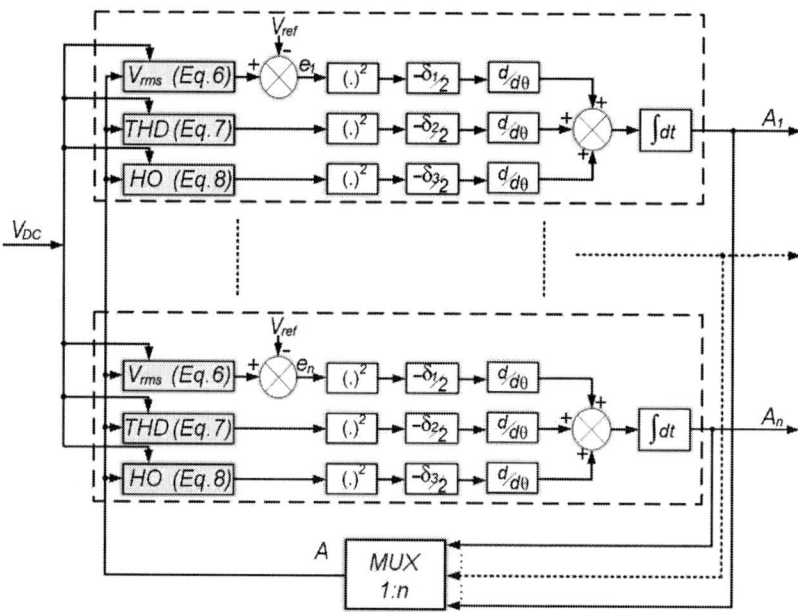

Fig. 3. Control block diagram of the proposed control technique.

4. Simulation results

To validate the proposed control technique, a single-phase 9-level CHB-MLI in MATLA/Simulink has been simulated. The proposed control technique has been tested for different values of the input voltage DC source with voltage ranges from 75 V to 95 V, and the sampling time ratio is equal to 50 µs. In this scenario, seven different conditions are applied to test the dynamic performance of the proposed solution. In the first step, all the DC sources are equal to $V_{DCi} = 75$ V. In the second to fifth steps, all the DC sources' values increase to $V_{DCi} = 95$ V for each step. In the sixth step, to analyze the dynamic response of the controller, all the DC source values are decreased to $V_{DCi} = 80$ immediately. For the final step, different values are referred to the DC sources randomly, they are equal to $V_{DC1} = 85\ V$, $V_{DC2} = 75\ V$, $V_{DC3} = 90\ V$, and $V_{DC4} = 80\ V$.

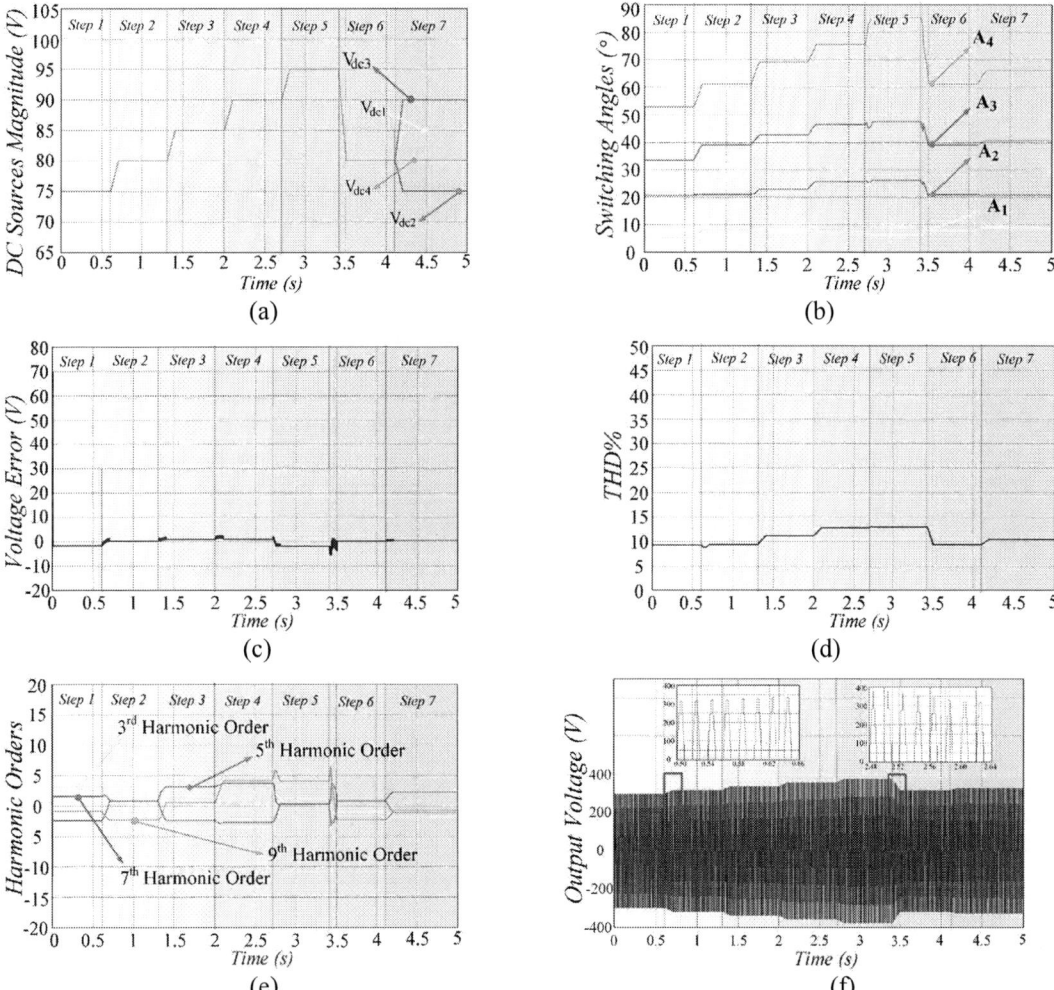

Fig. 4. Simulation results of the output characteristics and controllable parameters (a) Input DC voltage source, (b) Switching angles, (c) Error function of the output voltage, (d) THD% of the output voltage, (e) Different harmonic orders, and (f) Output voltage waveform.

Fig. 4 (a) depicts different steps for the DC voltage sources. In this study, the switching angles are the objective function utilized to optimize the cost functions. Therefore, switching angles take new positions as the DC voltage sources changes. Fig. 4 (b) presents the amplitude of different switching angles at each level of the 9-level inverter. It shows a fast dynamic response of the controller by playing with switching angles. The output voltage error is the main cost function optimized in real-time. The root-mean-square voltage is equal to 230 V all the time. Fig. 4 (c) presents the error magnitude of the output voltage for different steps of DC voltage sources. In addition to the voltage magnitude control, the THD function is optimized simultaneously in real-time. Fig. 4 (d) shows the THD magnitude where implementing the different scenarios for the input DC sources.

As it is a filter-less structure for the utilized topology, controlling the harmonics by employing the proposed control technique can be performed effectively. The magnitude of some specific harmonics are controlled, and they are limited based on the STDs by utilizing the HO's function (Eq. 11), which consists of 49 first orders of the output voltage harmonics. Fig. 4 (e) presents lower harmonic orders of 3rd, 5th, 7th, and 9th of the output voltage, whose amplitudes are lower than 5%. The output voltage waveform under the mentioned condition is illustrated in Fig. 4 (f) from 0 to 5 seconds. The variations of the DC voltage sources are clear at the peak of each voltage level. The controller aims to track the optimal switching angles under different DC voltage source conditions to improve the output voltage quality.

Conclusion

In this study, a new control scheme based on model-reference adaptive control (MRAC) is proposed by utilizing an optimal switching angle (OSA) modulation strategy in multilevel inverters (MLIs) having multiple isolated DC sources. The main cost function consists of the output voltage regulation, the total harmonic distortion (THD), and some low harmonic orders, which all are taken into account. The dynamic response of the proposed controller has been tested while the DC voltage sources have different magnitudes to keep the STDs. The proposed controller can determine the switching angles in real-time. In this online process, controlling the specified harmonic orders is possible. Besides, the proposed solution is a closed-loop system that makes it possible to control both the voltage magnitude and frequency. From a future work perspective, the proposed methodology can be extendable for higher output voltage levels. In addition, a higher number of switching angles can be considered at each level to obtain better results. Adding more cost functions based on IEEE STDs, testing the system under different load conditions and uncertainties, stability analysis of the whole system, and the experimental tests while MPPT section is operating could be suggested for future work.

References

[1] Dogga, R. and Pathak, M.K., "Recent trends in solar PV inverter topologies," Solar Energy, 183, pp.57-73, 2019.

[2] Ali, J.S.M. and Krishnaswamy, V., "An assessment of recent multilevel inverter topologies with reduced power electronics components for renewable applications," Renewable and Sustainable Energy Reviews, 82, pp.3379-3399, 2018.

[3] Yeganeh, M.S.O., Davari, P., Chub, A., Mijatovic, N., Dragičević, T. and Blaabjerg, F., "A single-phase reduced component count asymmetrical multilevel inverter topology," IEEE Journal of Emerging and Selected Topics in Power Electronics, 9(6), pp.6780-6790, 2021.

[4] Lashab, A., Sera, D., Hahn, F., Camurca, L., Terriche, Y., Liserre, M. and Guerrero, J.M., "Cascaded multilevel PV inverter with improved harmonic performance during power imbalance between power cells," IEEE Transactions on Industry Applications, 56(3), pp.2788-2798, 2020.

[5] Yegane, M.S.O. and Sarvi, M., "An improved harmonic injection PWM-frequency modulated triangular carrier method with multiobjective optimizations for inverters," Electric Power Systems Research, 160, pp.372-380, 2018.

[6] Yeganeh, M.S.O., Sarvi, M., Blaabjerg, F. and Davari, P., "Improved harmonic injection pulse-width modulation variable frequency triangular carrier scheme for multilevel inverters," IET Power Electronics, 13(14), pp.3146-3154, 2020.

[7] Haghdar, K., "Optimal DC source influence on selective harmonic elimination in multilevel inverters using teaching–learning-based optimization," IEEE Transactions on Industrial Electronics, 67(2), pp.942-949, 2019.

[8] Khezri, R., Oshnoei, A., Oshnoei, S., Bevrani, H., and Muyeen, S. M., "An intelligent coordinator design for GCSC and AGC in a two-area hybrid power system", Applied Soft Computing, 76, 491-504, 2019.

[9] Oshnoei, A., Kheradmandi, M., & Muyeen, S. M., "Robust control scheme for distributed battery energy storage systems in load frequency control", IEEE Transactions on Power Systems, 35(6), 4781-4791, 2020.

[10] Oshnoei, A., Kheradmandi, M., Muyeen, S. M., & Hatziargyriou, N. D., "Disturbance observer and tube-based model predictive controlled electric vehicles for frequency regulation of an isolated power grid," IEEE Transactions on Smart Grid, 12(5), 4351-4362, 2021.

[11] Yeganeh, M.S.O., Mijatovic, N. and Dragicevic, T., "Dynamic Performance Optimization of Single-Phase Inverter based on Model Predictive Control," In 2021 IEEE International Conference on Predictive Control of Electrical Drives and Power Electronics (PRECEDE) (pp. 235-240). IEEE, 2021.

[12] Åström, K.J. and Wittenmark, B., "Adaptive control," 1995.

[13] Boyd, S., Boyd, S.P. and Vandenberghe, L., "Convex optimization," Cambridge University Press, 2004.

[14] Kim, J., Choi, H.H. and Jung, J.W., "MRAC-based voltage controller for three-phase CVCF inverters to attenuate parameter uncertainties under critical load conditions," IEEE Transactions on Power Electronics, 35(1), pp.1002-1013, 2019.

[15] Zhao, T. and Chen, D., "A Power Adaptive Control Strategy for Further Extending the Operation Range of Single-Phase Cascaded H-Bridge Multilevel PV Inverter," IEEE Transactions on Industrial Electronics, 69(2), pp.1509-1520, 2021.

Control Principles for Island Operation and Black Start by Offshore Wind Farms integrating Grid-Forming Converters

Daniela Pagnani[*,♦], Łukasz Kocewiak[*], Jesper Hjerrild[*],
Frede Blaabjerg[♦], Claus Leth Bak[♦]

[*]ØRSTED
Gentofte, Denmark
{DAPAG, LUKKO, JESHJ}@orsted.com

[♦]AALBORG UNIVERSITY
Aalborg, Denmark
{fbl, clb}@energy.aau.dk

Keywords

«Wind energy», «Batteries», «Grid-forming converters», «Modelling», «Design».

Abstract

In this paper, control principles to perform black start services by offshore wind farms (OWFs) integrating grid-forming (GFM) control are presented. The strategy consists in exploiting a GFM battery energy storage system (BESS) to provide black start services by an OWF equipped by grid-following wind turbines. Controller modelling and operation methodology are explained. In order to show the proposed control and operation principles, the analysis is implemented on the CIGRE Working Group C4.49 benchmark, which could resemble modern large OWFs in the United Kingdom, such as Hornsea Project 1 and 2. Analysis simulations in the software PSCAD show the success of the proposed strategy.

Introduction

Today's power system is rapidly changing with the phaseout of synchronous generators (SGs) and the increasing deployment of renewable-based resources. SGs provide many grid services, in particular, the black start is of main interest for this study. Between renewable energy sources, a number of large offshore wind farms (OWFs) have been developed recently, especially in Europe and China and many more are under development worldwide, as seen in Fig. 1, where an overview of the total offshore wind installations by country is shown [1]. Despite the impact of COVID-19 pandemic challenging the normal development of operations, offshore wind had its second-best year ever in 2020 [1]. In the case of OWFs, black start represents a new service to be performed, as current wind turbines (WTs) are based on grid-following (GFL) converters that do not have the capability to work unless connected to the grid or an auxiliary SG. Black start has a specific grid code requirement to fulfil and some transmission system operators, e.g., National Grid Electricity System Operator (NGESO) in the United Kingdom, are investigating the implementation of this service by renewable energy sources [2, 3]. The most important requirement is the self-start capability (i.e., energisation without relying on the transmission grid). This implies equipment able to self-start, which can be achieved by grid-forming (GFM) converter control, to avoid relying on SGs and ultimately facilitate a 100% renewable-based black start strategy. Another requirement concerns the service availability (i.e., the ability to deliver black start services over a year), which must be a minimum of 90% for NGESO. This may be challenging for renewable energy, due to the intrinsic non-dispatchable nature of the source. Thus, the integration of an energy storage system seems advantageous.

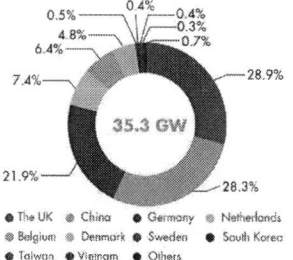

Fig. 1: Total offshore wind installations by country [1].

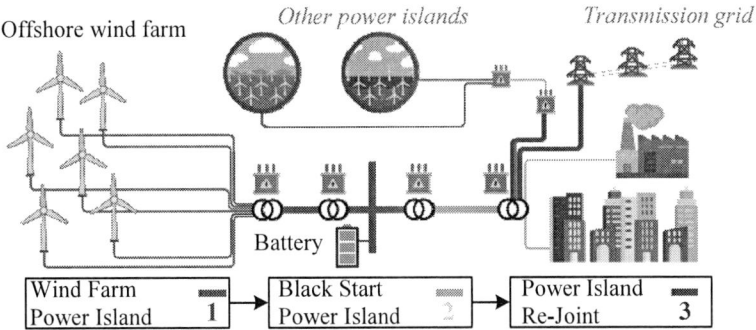

Fig. 2: Schematic of the proposed system consisting of an offshore wind farm equipped with a battery and the stages comprised in a black start process.

This paper primarily addresses the integration of a battery energy storage system (BESS) equipped with GFM control into an OWF to perform black start services, which has to work in parallel with GFL WTs. The system is shown in Fig. 2, where the BESS is a large, centralised unit located in the onshore substation of the OWF.

The integration of a BESS into renewable-based power plants and microgrids has been investigated in different studies, as able to increase flexibility by bidirectional active power exchange [4]. Moreover, such a storage system can provide black start, inertial and frequency regulation services [5]. A large STATCOM integrated with battery storage was initially proposed in [6], to enable the black start capability of OWFs. The STATCOM functionality provides fast and dynamic reactive power management and hence contributes to voltage regulation. These devices have been commercialised by the name of IBESS and or/ e-STATCOM [7-9].

The practical feasibility of successfully providing black start services from an OWF still has to be proven, as no OWF is currently able to restore a black grid. Therefore, more research needs to be conducted. Some literature around the main topic can be found in [7, 8, 10], where some discussions are made around OWF system configuration and control for a black start. However, little research is found, which discusses the whole black start procedure starting with an OWF with integrated BESS and addresses the control principles behind this hybrid power plant.

This paper analyses the implementation of a hybrid GFM-BESS+GFL-WT control system shown in Fig. 2. Furthermore, simulations in PSCAD are performed on an OWF model inspired by the CIGRE WG C4.49 benchmark, to validate the designed system and present the success of the black start procedure.

Grid-Forming Control Applied to Black Start and Island Operation

GFM converter control is a concept that has been developed extensively in the last decades and represents a novel way of integrating grid-connected converters [11]. For example, GFM converters in photovoltaic applications in microgrids have been available since around 2000 in the multi-kW range and around 2010 in high power applications (>MW) [12]. More recently, their application has been expanded to different technologies, i.e., BESSs and WTs, and several GFM strategies have been developed, e.g., droop-based control, virtual synchronous machine, power synchronisation control (PSC) [10, 13]. The designation *grid-forming* is used as an umbrella term for any inverter controller that regulates its internal voltage phasor and can be the only source in the grid to which it is connected, or it can coexist with other GFL and GFM inverters, as well as SGs [14]. In the literature, sometimes the latter converter (i.e., GFM) is defined as a grid-supporting voltage source [11, 15]. A universal definition is yet to be agreed upon. However, the main concept represents a controller for inverters in grid applications that do not necessarily need a voltage vector reference, e.g., as the phase-locked loop (PLL) does, to be the main loop to create its voltage phase reference and synchronise itself to the grid but it can create its own. Therefore, they are in contrast with GFL, where the need for an energised grid with a voltage vector to follow is indispensable for the controller. This is important with the operation of black start, as there is the need for a converter which can form the voltage without an existing grid, along the lines of a traditional SG. Additionally, it needs to have the capability to self-start. Thus, the concept of a GFM converter is a necessary but not sufficient condition when discussing black start by OWFs.

The interesting GFM converter concept for this research is applied to the converter that can both self-start and form its own voltage vector in the instance of the grid.

Selected Grid-Forming Control

In this paper, the GFM control strategy selected is implemented in the inverter of a large-scale BESS integrated into the onshore substation of the OWF as per Fig. 2. The GFM control applied is based on PSC with virtual inertia and damping, as shown in Fig. 3. This has been chosen as it is an established control topology, where the additional contribution of inertia and damping can be implemented.

For GFM purposes, the PSC was proposed in [16] to emulate the synchronisation mechanism based on transient power exchange seen in SGs. Thus, the resemblance with the swing equation of synchronous machines. PSC has been initially designed to be implemented in high-voltage direct current (HVDC) transmission to tackle weak grid conditions. In this GFM application, the frequency deviation $\Delta\omega$, corresponds to the deviation in active power ΔP, and it is given in Eq. 1.

$$\Delta\omega = \frac{1}{2H\,s + \dfrac{k_d\,s}{\omega_d + s} + k_g}\Delta P \tag{1}$$

where H indicates the inertia constant as defined in the conventional power system, which tends to resist the frequency deviations. The damping parameter k_d will be able to damp out the high-frequency oscillations if any in the power deviation. Furthermore, the reciprocal of the parameter k_g indicates the droop frequency response of the BESS. In this study, 1.6% frequency regulation droop is used as in [7].

Implemented Control Principles

In this section, the modelling and design of the system in this proposed OWF will be shown. Each controller will be presented, having the GFM controller first, and the GFL controller second. This will show the applied control principles for such a system and some key differences in the application of GFM and GFL control for black start purposes.

The converters are implemented using the space vector notation and an average model in the dq-frame rotating at rated frequency, in combination with proportional-integral (PI) controllers. These are universally known because of their flexibility combined with the relatively easy tuning. Due to the capabilities of the integral action to estimate load disturbance, a certain stability margin in the closed loop is obtained. Several techniques can be applied given the need for converter control modelling for design purposes. One of the most popular approaches is Laplace domain (s domain) analysis, which can be done using transfer functions. This method is based on the transfer function analysis, which relates the output to the input, which for this type of study are the measured variable and the reference variable. The analysis is based on the models in the s-domain, with an assumption that the controller sampling rate is fast enough and the impact from discretisation can be neglected. It is assumed that all the power needed during the operation can be extracted from the DC link, for simplification. This means that the DC-link voltages of the converter-based resources, i.e., BESS and WTs, are constant.

Grid-Forming Battery Energy Storage System Modelling and Design

The selected GFM controller for the BESS is the PSC with VSM variation, i.e., additional virtual inertia and damping terms, shown in Eq. 1, and the BESS parameters are adapted from [8].

A modular multilevel converter (MMC) type of structure is chosen for the large rating ($>$100 MVA) needed for this application. A single-line diagram of the BESS and its controllers is shown in Fig. 3. Thanks to the MMC internal filters, no large filters are needed at the inverter output, thus only an L filter is shown in the diagram. MMCs can reach high voltage levels, therefore the inverter output voltage is 33 kV. In order to connect the BESS to the onshore bus of the selected OWF, a step-up transformer is needed to reach the 220-kV level. The sizing of the BESS is not included in this study, where it is assumed that the BESS is large enough to perform the black start. However, the base values for the converter are needed, thus it is assumed that the active power rating of the BESS is 50 MW/100 MWh, while the reactive power rating (STATCOM capability) is 100 MVAr. Thus, the apparent power is assumed as 112 MVA. This has been inspired by [7].

The GFM controller is made of two parts: (i) the power synchronization loop (PSL) and (ii) the alternating voltage controller. Its principle is based on setting the converter voltage as $\boldsymbol{v}_c = V e^{j\theta}$.

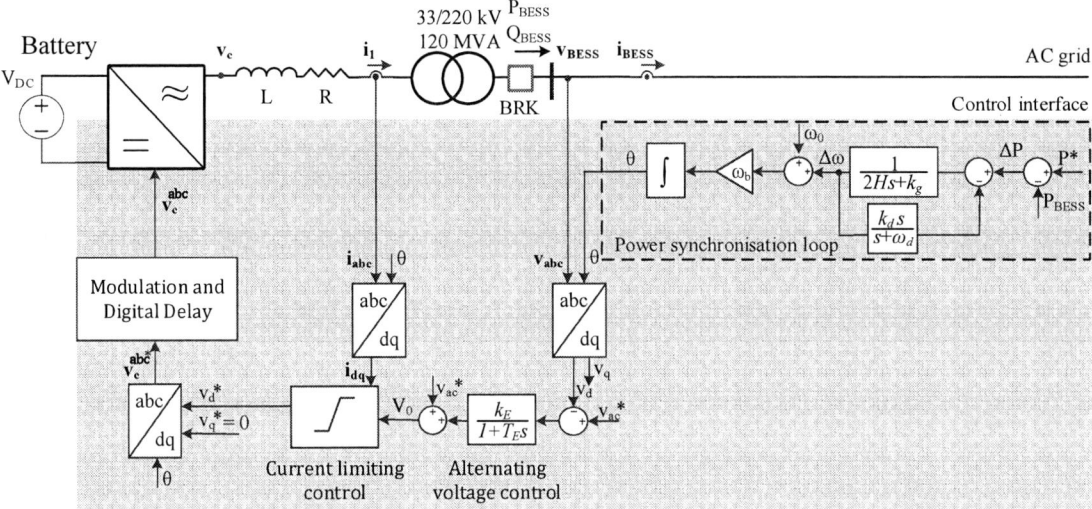

Fig. 3: System configuration and control structure for the grid-forming battery system implemented in the analysis.

Grid-Following Wind Turbine Modelling and Design

Fig. 4 shows the control strategy implemented for the GFL WTs. It is assumed that this is a typical two-level converter rated at 12 MW. The GFL control is using standard cascaded control loops, the inner loop controlling the converter side current, whereas the outer loop controls the active and reactive power generated by the WT. Since a GFL WT is considered, a PLL is applied to provide grid synchronisation. It is assumed an LCL filter is used, where the grid-side inductance is represented by the 0.69/66 kV transformer. Due to the resonance characteristic of this converter, an active damping term based on capacitive current is implemented in the current control loop. The system parameters are selected and adapted from [17].

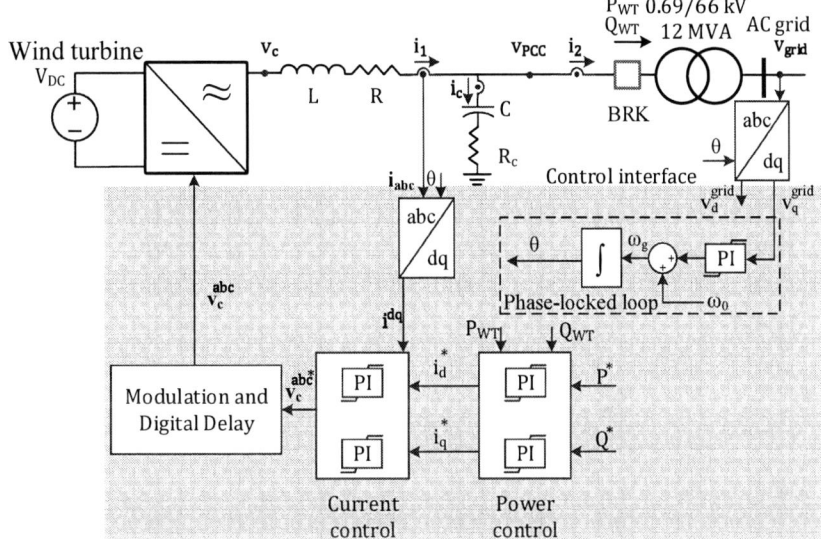

Fig. 4: System configuration and control structure of the grid-following wind turbine used in the analysis.

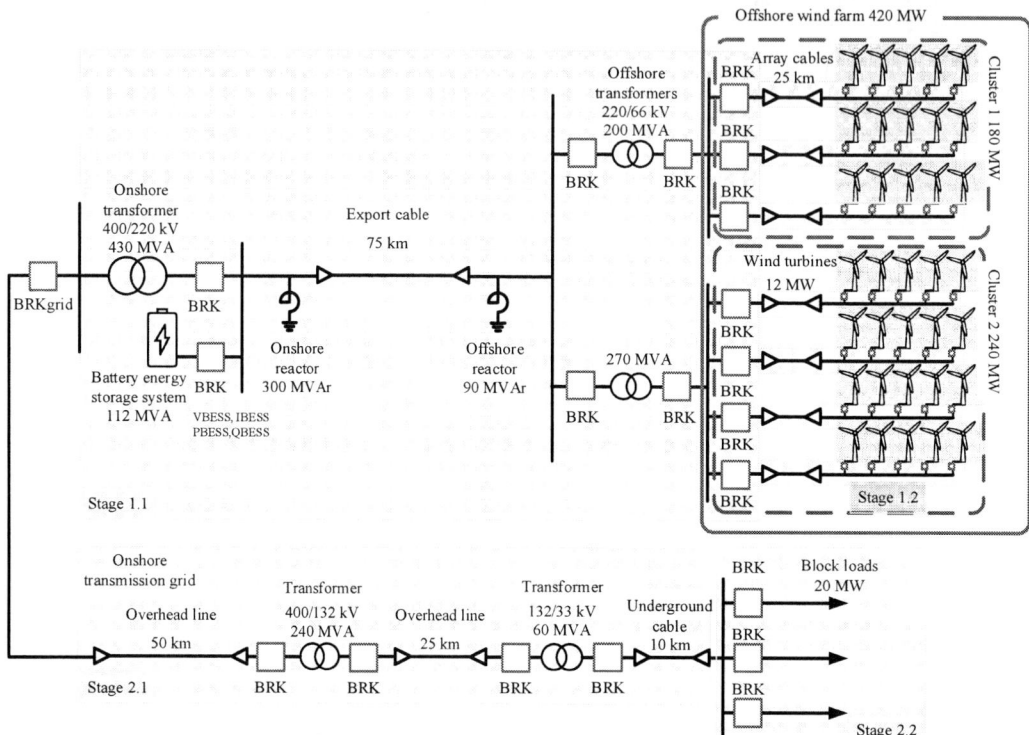

Fig. 5: Diagram of the study case based on the CIGRE WG C4.49 benchmark (BRK = circuit breaker) and onshore transmission network, where the energisation stages are highlited.

Simulation Results of Black Start and Island Operation

An OWF system inspired by the benchmark proposed by CIGRE WG C4.49 has been implemented in PSCAD, which is shown in Fig. 5.

A black start procedure implemented by an OWF can be said to be comprised of different stages, going from a state of no power to a state of restored normal operation for the power system, as shown in Fig. 2 [3]. It is assumed that the OWF is completely shut down following a blackout of the transmission grid or an event of similar impact. Moreover, the OWF is islanded, i.e., disconnected from the main grid. It is imagined that an assessment of the possible damage to the OWF following the blackout is performed, and the wind farm is able to continue without any issues. Once that is cleared, the next step for the OWF operator is to establish the energisation plan and capacity, which in this case means to assure the reserved BESS capacity for black start and the wind prediction. After that, the OWF is ready to receive the signal from the TSO to start the black start procedure. When using an OWF as a black start source, the first stage represents the energisation of the wind farm in island operation, i.e., working as a Wind Farm Power Island. Once the island is energised and stable, the energisation of the onshore transmission grid and the energising of the block loads can start and thus form a Black Start Power Island. This is also a challenging operation, as the transmission grid itself consists of many transmission lines and transformers with little or no load, determining a grid high harmonic impedance at low frequencies and little damping. Additionally, the main purpose is to energise large block loads (≥ 20 MW) in one single switching operation. After the energisation of a part of the system, the synchronisation with other power islands in the system has to take place until the full system is synchronised and restored. This final stage is referred to as Power Islands Re-Joint. These stages are shown in Fig. 2, and more in details in Fig. 6 [3]. As they have different characteristics, they will be analysed separately.

System Configuration

The OWF system is based on the CIGRE WG C4.49 benchmark, where a large BESS has been integrated into the study case [17]. This is depicted in Fig. 5.

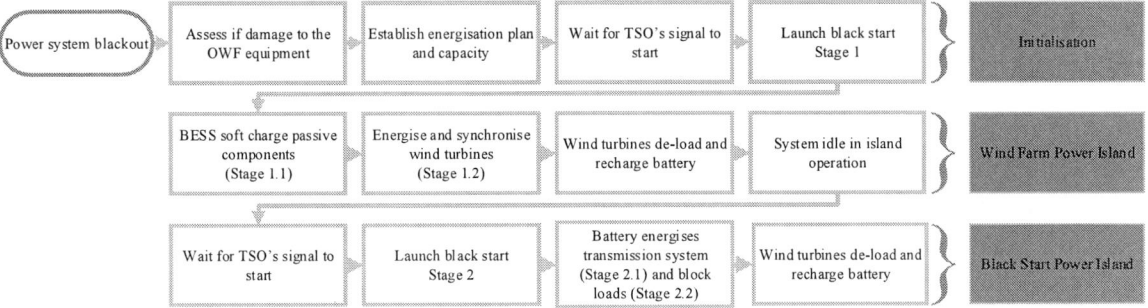

Fig. 6: Flowchart schematising the main steps of the implemented black start procedure.

The rated OWF active power is 420 MW and comprises 35 WTs rated at 12 MW. The export cable is a long, 220-kV high-voltage alternating current (HVAC) cable, which has 300 Mvar shunt compensation at the onshore end and 90 Mvar at the offshore end. The export cable length has been assumed as 75 km as it is considered a good scenario to represent modern, large OWFs which are far from shore such as the Hornsea projects in the UK, but without necessitating a midpoint reactive compensation station. The export cables comprise 35-km long land cables and 40-km long sea cables. These are modelled by a frequency-dependent model in PSCAD since it is useful for studies wherever the transient behaviour of the cable is relevant. The export cable is connected to the onshore substation where the BESS is located, together with a 400/220 kV, 420 MVA onshore transformer. In all the transformers, the saturation characteristics have been modelled, in order to show the impact of the phenomenon.

Offshore, there are 35 WTs rated at 12 MW and arranged in seven strings, each string having five WTs. These strings are collected in two WT clusters, one having three strings and the other having four strings. Two 220/66 kV transformers connect the export system to the collection system with a rated power equal to 200 and 270 MVA respectively for the two WT clusters. The 66-kV collection system is made of two different types of three-core cables, respectively 500 mm² and 150 mm² submarine cables. The three first WTs on the feeder are connected via the 500 mm², while the last two apply the 150 mm² cables. It is assumed that the cable length between WTs is constant and equal to 5 km. These two WT clusters have been aggregated in order to save computational time for the simulation, where they are modelled as seven individual feeders with five aggregated WTs each. The aggregated collection cables are modelled via the equivalent T model, while the WT ratings are correspondingly scaled up. An advantage of this aggregated model is lower computational time and speed up the simulation execution. However, this can cause a misrepresentation of the energisation phenomenon, since large groups of WTs are energised at the same time, while in reality it will be one at the time. Accordingly, it can be discussed that the success of this energisation in this simulation, therefore, is harder to achieve as a larger unit is energised at once. In contrast, the process is made of more and smaller steps in real life.

On the onshore side, it is assumed that the black start provider may have to energised block loads which are connected via a series of lines and transformers and at lower voltage level. Thus, the OWF+BESS system is connected onshore to a radial network which starts with a 50 km long overhead line at 400-kV level, which in turns is connected to a 400/132 kV transformer. This is in turn connected to a shorter overhead line (20 km) and another step-down transformer (132/33 kV). Finally, a 10 km long underground cable connects the system to three block loads each of 20 MW. This is assumed as a demonstration of the black start capabilities of the system, but it is assumed that a higher number of block loads can be covered by the OWF capacity when wind is available for the necessary period of time. This representation of the onshore grid, with both block loads and lines and transformers, is chosen to challenge the islanded network which has to comply with the requirements for black start, i.e., minimum 20 MW block loads capability and 100 Mvar reactive power capability [2, 3].

1st Stage: Wind Farm Power Island

The first step in the black start of the onshore transmission grid is the black start of the BESS and consequently of the OWF grid, in order to establish a Wind Farm Power Island. The steps for this to happen are: (i) energisation of the BESS and passive components, (ii) energisation and synchronisation

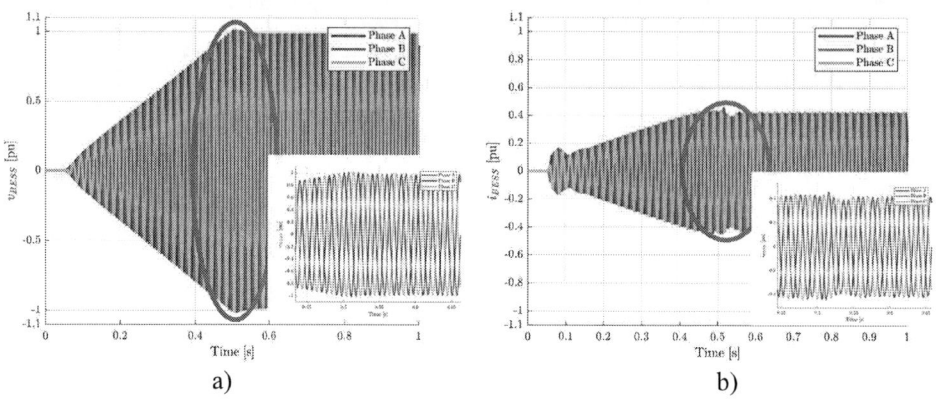

Fig. 7: Simulation results from the soft charge of the wind farm passive system in Fig. 5 by the grid-forming battery measured at its point of connection: a) Generated battery voltage, b) Generated battery current.

of the WTs, (iii) de-loading and recharging of the BESS by the WTs to compensate for the system losses. Being the BESS the largest active component in the system (50 MW against 12 MW), and being a dispatchable type of source, it is decided that this component will be the one used for the single energisation steps alone. In this way, we can ensure a more resilient system by avoiding multiple converter actions and their interactions. Thus, the BESS is always acting first and after a new steady state is achieved, it can be recharged by the WTs. This will allow the BESS to be fully recharged and idle till the next step can be performed. The procedure is illustrated in the flowchart in Fig. 6. The offshore grid can be energised in two ways: by hard switching or by soft charging (Stage 1.1). Hard switching is defined as performing the several switching operations for one component at the time after the BESS is self-energised and runs at 1 pu voltage. Inrush currents and temporary overvoltages when the BESS is connecting to the offshore grid are to be expected. On the other hand, by using soft-charging, all the circuit breakers in the OWF grid (without including the WTs) are connected and the voltage is slowly ramped up from 0 to 1 pu. Thus, the transient currents can be avoided to a large extent by decreasing the voltage ramp rate dV/dt. In this simulation study, the BESS voltage is ramped up from 0.0 to 1.0 pu in the period from 0.05 s to 0.5 s, while all the breakers shown in Fig. 5 are closed, with the exception of the breaker going to the onshore transmission network (BRKgrid in Fig. 5). Both the current and the voltage increase smoothly, and they remain balanced throughout the soft-charging process as shown in Fig. 7, where the voltage v_{BESS} and the current i_{BESS} are measured at the BESS converter terminals. This energises the onshore transformer, the export cable system with reactive power compensation, offshore transformers and the collection grid, including the WT transformers, and this is done in order to avoid a further switching operation. In general, the energisation of WTs starting from their transformers could likely cause sympathetic inrush due to their close electrical distance. Having

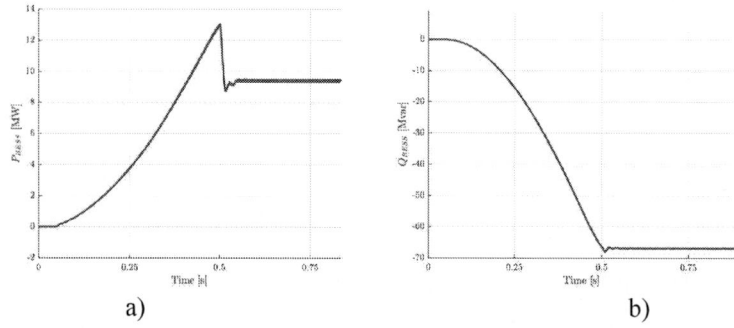

Fig. 8: Simulation results from the soft charge of the wind farm passive system in Fig. 5 by the grid-forming battery measured at its point of connection: a) Active power, b) Reactive power.

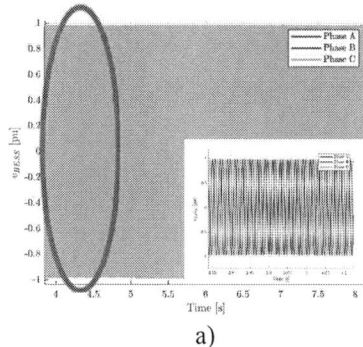 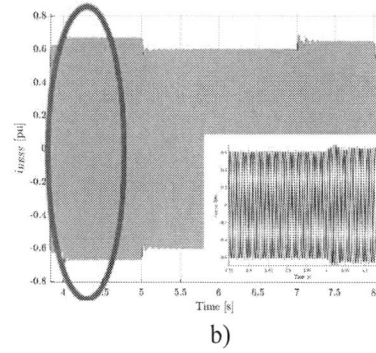

a)　　　　　　　　　　　　　　　　b)

Fig. 9: Simulation results from the energisation and synchronisation of the first (at 4 s) and second (at 7 s) group of wind turbines in Fig. 5 by the grid-forming battery measured at its point of connection: a) Generated battery voltage, b) Generated battery current.

the BESS soft charging these components will avoid this type of drawback. During Stage 1.1, shown in Fig. 8, the BESS supplies the maximum active power of 12.95 MW, while it absorbs the maximum reactive power of 67.55 Mvar, which is generated by the high voltage export cables. This is the excess reactive power considering that all the shunt reactors in the OWF are connected. In steady state, at 0.6 s, the BESS supplies 9.50 MW and absorbs 66.85 Mvar, as shown in Fig. 8. The BESS is controlled to regulate its terminals at the high-voltage side of its transformer, and this is 1.05 pu after the soft charging process. Thus, the BESS absorbs this large amount of reactive power, i.e., over 65 Mvar to decrease its voltage. It can be seen that the active power required for energising the offshore grid is not substantial (but at the same time not negligible), while the BESS unit has to handle a large amount of reactive power which fluctuates between the reactive components in the system, especially the export cable. Therefore, the STATCOM capability is implemented. The next step is the energisation and synchronisation of the WTs (Stage 1.2). In this simulation, the five WTs included in one string are modelled as a single unit, and thus energised as they were energised at the same time. However, in real operations, one WT will be energised at the time. From the electrical point of view, the energisation of the WT unit starts when the voltage is sensed at the low-voltage terminal of its transformer, as seen in Fig. 4. After this, it is assumed that the WT starts powering its auxiliaries and can start operation at zero power, deblocking its converter and working in standalone mode for a brief period of time before its breaker is closed. During this short time window, the converter is already synchronised to the grid via its PLL, and once the converter is deblocked and the breaker closed, the WT is fully synchronised to the main wind power island. The first WT group, which is an aggregate of five WT units, is synchronised at 4 s, while the second group is connected and synchronised at 7 s, as shown in Fig. 9. The connection of the WT strings is made in succession with an interval of 3 s between each other. This is sufficient time to see the new steady state before the connection of the new group of WTs. It can be seen that a transient disturbance arising due to the connection of the WT units. From the BESS side, there is a momentary increase of the injected current, which goes from 0.6 to 0.72 pu for 1 s, as seen in Fig. 9. This relatively small transient disturbance appears whenever a WT group is connected to the grid and are observed in the simulation study when the WT groups are connected as each group comprises an aggregation of multiple WT units. Thus, the resultant transient at 4 s is the consequence of the simultaneous connection of 5 WT units through a 60 MVA transformer, and an aggregated LC filter. At 22 s, the power reference of the first WT group is ramped up from 0.0 to 0.3 pu in 1 s. This is done assuming that the WTs can take on the active power losses of the system and re-charge the BESS. In response to the change in the reference, the output active power from the WT groups rises. The battery is thus charged by absorbing the active power produced by the WTs.

2nd Stage: Black Start Power Island

In this stage, the actual black start of the onshore transmission network is achieved. Once the Wind Farm Power Island is established and working stably, the OWF operator has to wait for instructions from the transmission system operator and be ready for energising block loads in the transmission grid, as shown

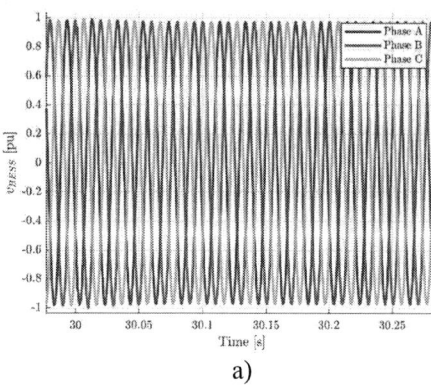 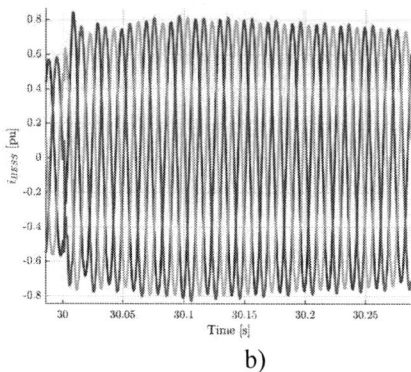

a)   b)

Fig. 10: Simulation results from the energisation of the first overhead line (at 30 s) in Fig. 5 by the grid-forming battery measured at its point of connection: a) Generated battery voltage, b) Generated battery current.

in the flowchart in Fig. 6. During this stage (Stage 2), the BESS, which is fully charged and de-loaded, is the one taking care of the extra active and reactive power loads of the transmission grid at first. It is only once the new components in the transmission grid are fully energised and stable, that the WTs will pick up the load and recharge the BESS. This cycle can be repeated for every block load to energise, and it is preferred in order to have the BESS as a single large unit to energise the large transmission system loads. To test the reactive power requirements for black start, the onshore grid has been modelled as a radial system where overhead lines, cables and transformers connect the block loads at 33 kV. This system is represented in Fig. 5 and it is made up of 50 km overhead line at 400 kV, connected to a 400/132 kV, 240 MVA transformer to another overhead line which is 25 km long. After that, there is a final 132/33 kV, 60 MVA transformer which connects to three 20 MW block loads.

The switching operation to energise the passive components of the transmission system (Stage 2.1) take place one component at the time and it is simulated by common switching operation, i.e., without advanced aids such as pre-insertion resistors or point of wave switching, and it causes heavy distortions to the BESS voltage and current waveforms. However, there are no overcurrents and overvoltages exceeding the 10% range, and the transients are all settled to steady state, thus the procedure is successful. The first overhead line is energised at 30 s, as seen in Fig. 11. This causes a large increase in the current (almost doubled), provided by the BESS, which goes from 0.5 to 0.9 pu, but this is still well below the limits. The transformer energisation is though the most problematic element to energise, as expected, and this is simulated at 33 s. The saturation phenomenon is known for causing inrush currents which are also seen here in Fig. 11, and last for several seconds. There is a rich harmonic component due to the switching and energisation of the transformer, which is relatively significant because the black start power island is very weak. However, it is damped due to the losses in the system.

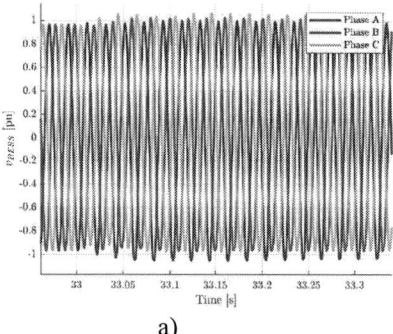 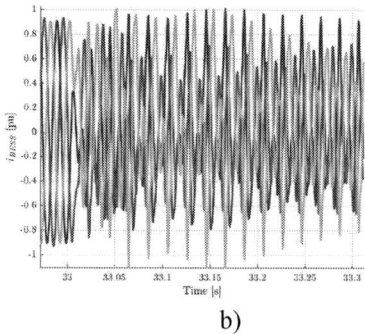

a)   b)

Fig. 11: Simulation results from the energisation of the first transformer at 33 s in Fig. 5 by the grid-forming battery measured at its point of connection: a) Generated battery voltage, b) Generated battery current.

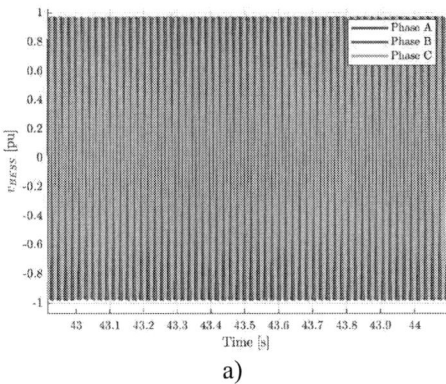
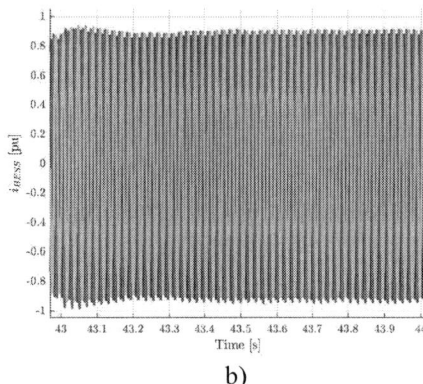

a)   b)

Fig. 12: Simulation results from the energisation of the block load at 43 in Fig. 5 by the grid-forming battery measured at its point of connection: a) Generated battery voltage, b) Generated battery current.

After connecting the remaining lines and transformer, the block loads are connected at 43 s, 45 s and 47 s, respectively, as shown in Fig. 12 (Stage 2.2). Compared to the energisation of the line, the block loads create negligible disturbance in the whole islanded grid. It is important for the WTs to support during the procedure as each block load draws a lot of energy and power.

The third and last stage of the black start procedure, i.e., the Power Island Re-Joint, is not addressed in this paper as it is directed, operated and controlled by the TSO. Thus, it is left as future work.

Discussion and Conclusion

In this paper, the discussion is mainly focused on the control implementation and on its performance during the black start provided by a complex OWF system equipped with GFM BESS and GFL WTs. This is especially challenging as the system is very weak since it is working in no grid conditions, i.e., as an islanded network. However, the system is able to form an islanded network and connect different transmission components and block loads in the transmission grid.

It can be argued that a lot of energy is necessary in the overall black start restoration, and that could be challenging for a single GFM unit. Thus, the implementation of GFM controllers in the WTs could be a favourable future work for in order to support the BESS and accelerate the black start procedure, especially during Stage 1.2.

The proposed strategy for black start by an OWF with integrated battery is designed, implemented and simulated. The designed GFM control allows the system to react to the transient phenomena caused during the system energisation with stable performance and shows the challenges in implementing a black start strategy from a converter-based network and the principles to overcome them, such as the soft charge of the passive OWF system. From the practical point of view, it can be seen that the energisation of the reactive components in the onshore transmission grid is the most challenging part for the operation, due to the high reactive power load and transient operations involved in switching, such as inrush currents and temporary overvoltages.

It is shown how the selected BESS with GFM control can successfully perform this service together with a large OWF based on a generic benchmark system. When a specific power plant is designed for black start, only a detailed study will enable an assessment of the risks and an evaluation of the efficacy of the additional solutions to be provided.

References

[1]   Global Wind Energy Council (GWEC), "Global Offshore Wind Report," Brussels, 2021.

[2]   National Grid Electrical System Operator, "Restoration Services," 21 January 2021. [Online]. Available: https://www.nationalgrideso.com/sites/eso/files/documents/Appendix%201%20-%20Tech%20Requirements%20and%20Assessment%20Criteria.pdf. [Accessed 1 June 2022].

[3] D. Pagnani, Ł. H. Kocewiak, J. Hjerrild, F. Blaabjerg and C. L. Bak, "Integrating Black Start Capabilities into Offshore Wind Farms by Grid-Forming Batteries," in *IET Renewable Power Generation*, pp. 1-12, submitted 2021.

[4] S. Saponara, R. Saletti and L. Mihet-Popa, "Hybrid Micro-Grids Exploiting Renewables Sources, Battery Energy Storages, and Bi-Directional Converters," in *Applied Sciences*, vol. 9, no. 22, pp. 1-18, 2019.

[5] M. P. S. Gryning, B. Berggren, Ł. H. Kocewiak and J. R. Svensson, "Delivery of Frequency Support and Black Start Services from Wind Power Combined with Battery Energy Storage," in *Proc. 19th Wind Integration Workshop*, Online, pp. 1-10, 2020.

[6] T. S. Sørensen, "Method for Black Starting an Electrical Grid". United States of America Patent 20200244070, 16 November 2018.

[7] S. K. Chaudhary, R. Teodorescu, J. R. Svensson, Ł. H. Kocewiak, P. Johnson and B. Berggren, "Islanded Operation of Offshore Wind Power Plant using IBESS," in *Proc. 2021 IEEE Power & Energy Society General Meeting (PESGM)*, 2021, pp. 1-5, doi: 10.1109/PESGM46819.2021.9638226.

[8] S. K. Chaudhary, R. Teodorescu, J. R. Svensson, Ł. H. Kocewiak, P. Johnson and B. Berggren, "Black Start Service from Offshore Wind Power Plant using IBESS," in *Proc. 2021 IEEE Madrid PowerTech*, 2021, pp. 1-6, doi: 10.1109/PowerTech46648.2021.9494851.

[9] M. R. A. Wara and A. H. M. A. Rahim, "Supercapacitor E-STATCOM for Power System Performance Enhancement," in *Proc. 2019 International Conference on Robotics, Electrical and Signal Processing Techniques (ICREST)*, 2019, pp. 69-73, doi: 10.1109/ICREST.2019.8644388.

[10] D. Pagnani, Ł. H. Kocewiak, J. Hjerrild, F. Blaabjerg and C. L. Bak, "Overview of Black Start Provision by Offshore Wind Farms," in *Proc. IECON 2020 The 46th Annual Conference of the IEEE Industrial Electronics Society*, 2020, pp. 1892-1898, doi: 10.1109/IECON43393.2020.9254743.

[11] J. Rocabert, A. Luna, F. Blaabjerg and P. Rodríguez, "Control of Power Converters in AC Microgrids," in *IEEE Transactions on Power Electronics*, vol. 27, no. 11, pp. 4734-4749, Nov. 2012, doi: 10.1109/TPEL.2012.2199334.

[12] E. Alegria, T. Brown, E. Minear and R. H. Lasseter, "CERTS Microgrid Demonstration with Large-Scale Energy Storage and Renewable Generation," in *IEEE Transactions on Smart Grid*, vol. 5, no. 2, pp. 937-943, March 2014, doi: 10.1109/TSG.2013.2286575.

[13] D. Pagnani, F. Blaabjerg, C. Bak, F. Faria da Silva, Ł. Kocewiak and J. Hjerrild, "Offshore Wind Farm Black Start Service Integration: Review and Outlook of Ongoing Research," in *Energies*, vol. 13, no. 23, pp. 6286, 2020.

[14] North American Electric Reliability Corporation (NERC), "Grid-Forming Technology," Atlanta, 2022.

[15] A. Narula, M. Bongiorno, M. Beza and P. Chen, "Tuning and evaluation of grid-forming converters for grid-support," in *Proc. 2021 23rd European Conference on Power Electronics and Applications (EPE'21 ECCE Europe)*, 2021, pp. 1-10, doi: 10.23919/EPE21ECCEEurope50061.2021.9570679.

[16] L. Zhang, "Modeling and Control of VSC-HVDC Links Connected to Weak AC Systems," Royal Institute of Technology School of Electrical Engineering, Electrical Machines and Power Electronics, PhD Thesis, Stockholm, 2010.

[17] Ł. Kocewiak, R. Blasco-Giménez, C. Buchhagen, J. B. Kwon, Y. Sun, A. Schwanka Trevisan, M. Larsson and X. Wang, "Overview, Status and Outline of Stability Analysis in Converter-based Power Systems," in *Proc. 19th Wind Integration Workshop*, Online, pp. 1-10, 2020.

Experimental study of the reduction and removal of turn-on snubber for IGCT based MMC submodule using fast silicon diodes

Arthur Boutry[1], Cyril Buttay[2], Besar Asllani[1], Bruno Lefebvre[1], Eric Vagnon[2], Dong Dong[3]

[1] SuperGrid Institute
23 rue Cyprian
F-69100 Villeurbanne,
France

[2] Univ Lyon, Ecole Centrale de Lyon,
INSA Lyon, Université Claude Bernard Lyon 1,
CNRS, Ampère, UMR5005,
69130 Ecully, France

[3] Center for Power Electronics Systems
The Bradley Dept. of Electrical and Computer Eng.
Virginia Polytechnic Institute and State University
Blacksburg, VA 24061, USA

E-mail: arthur.boutry@supergrid-institute.com

Acknowledgements

This work was supported by a grant overseen by the French National Research Agency (ANR) as part of the "Investissements d'Avenir" Program ANE-ITE-002-01.

Keywords

≪MMC≫, ≪IGCT≫, ≪Fast recovery diode≫, ≪Failure modes≫, ≪Reverse recovery≫

Abstract

IGCTs are attractive power semiconductor devices for HVDC applications. However, their switching speed at turn-on can only be controlled by the means of an external snubber circuit, which adds complexity and cost. This paper investigates experimentally the possibility to downsize and even completely remove this turn-on snubber/clamp in the case of an MMC submodule based on 6.5 kV IGCTs, using fast silicon diode modules (rated at up to $13\,\mathrm{kA\,\mu s^{-1}}$). The removal is found to be possible, although limiting phenomena (dynamic avalanche and snap-off) appear, reducing the actual operating range of the submodule.

1 Introduction – Relevance of IGCTs for HVDC MMCs, turn-on snubber

1.1 The Modular Multilevel Converters and the IGCT

The Modular Multilevel Converter (MMC) is a Voltage Source Converter (VSC) used for Medium or High Voltage Direct Current (MVDC or HVDC) applications. The MMC (Fig. 1a) is based on submodules (SMs), its elementary building blocks. Composed mainly of switches and a capacitor, a submodule can be seen as a small voltage source that can be inserted or not. The half-bridge submodule (HB, shown in Fig. 1b) is the most common submodule type. It consists of two switches and their freewheeling diode, one capacitor (and auxiliary systems, not shown here).

Most submodule designs, such as those described in [1], use IGBTs as controlled switches. These transistors offer high ratings (up to 6.5 kV and a few kA for commercially available IGBTs); their control circuitry is simple and requires low power. However, IGCTs are increasingly considered a potential competitor for IGBTs in HVDC MMCs, and several papers have shown that IGCTs may be a better alternative [2, 3, 4, 5, 6]. Indeed, they have the following advantages: lower losses (studied notably in [2, 3, 4, 5]) as well as higher power ratings than the IGBTs [5]. They also have a short-circuit failure mode, which is very desirable in HVDC MMCs as it permanently bypasses a failed submodule. IGCTs also exhibit some drawbacks, which may explain that the device was not considered for HVDC MMC submodules at first. Perhaps the main drawback is the need for a "turn-on snubber", described below.

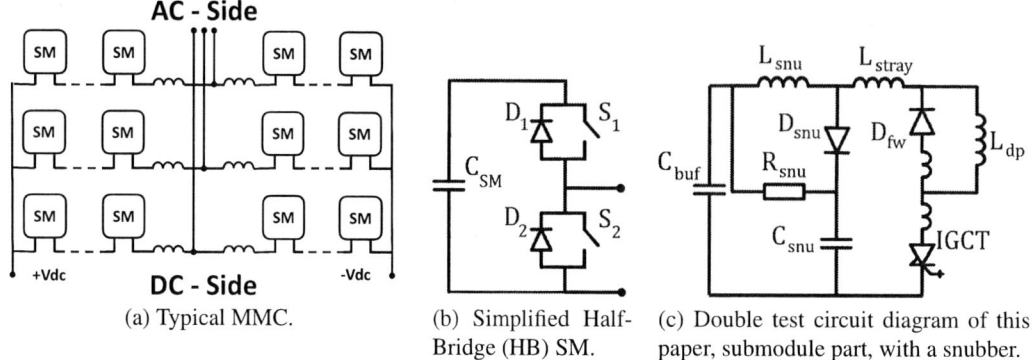

(a) Typical MMC.

(b) Simplified Half-Bridge (HB) SM.

(c) Double test circuit diagram of this paper, submodule part, with a snubber.

Fig. 1: Typical MMC and submodule, and the submodule double pulse circuit used for the tests in this paper.

1.2 Turn-on snubber requirement for IGCT MMC submodules

Based on a thyristor structure, IGCTs do not offer a way to control the speed of their turn-on transient, and in particular, the current slope (di/dt). This turn-on transient can be very fast (tens of $kA\,\mu s^{-1}$), and can cause an excessive burden on the opposite antiparallel diode [2, 3, 6, 7, 8, 9], which can be destroyed as shown in [6] and later in this paper. Indeed, typical silicon power diodes can only sustain di/dt of one or a few $kA\,\mu s^{-1}$ at turn-off (which corresponds to the opposite IGCT turning on). At a high di/dt, a diode faces the risk of destruction by dynamic avalanche. According to [10, 11], a high di/dt provokes a high peak power inside the diode at the end of the reverse recovery (when the diode voltage rises), with an indicative 'silicon limit' of 250kW/cm². Consequently, the current slope must be limited externally using a "turn-on snubber" circuit (presented in the figure 1c). The turn-on snubber comprises a snubber inductor (L_{snu} to set $di/dt = V_{bus}/L_{snu}$) and an RCD-clamp (resistor R_{snu}, capacitor C_{snu} and diode D_{snu}, to clamp the overvoltage produced by the snubber inductor at turn-off). The inductor is designed to limit turn-on speed to a value of around $1\,kA\,\mu s^{-1}$, consistent with the recommended diodes (the 5SDF20L4520 from Hitachi, intended for use in IGCT converters, is rated at $1.2\,kA\,\mu s^{-1}$).

It should be noted that the turn-on snubber has the advantage of improving the internal short-circuit capability of the submodule [12, 13, 14]: indeed, if both IGCTs of a SM happens to be ON simultaneously, the SM capacitor is short-circuited; the short-circuit current is then limited by the presence of the snubber inductor. This situation can occur with an erroneous triggering of the initially open IGCT (while the other IGCT is still on) or a failure to turn off an IGCT before turning-on the other. In [12], the snubber inductor is estimated to result in 5 to 10 times short-circuit current peak reduction compared to an IGBT-based SM, which does not contain such snubber.

However, this snubber represents added volume, added cost (4 more parts, including a high voltage diode, D_{snu}) and added switching losses (8 J for a $4\,\mu H$ snubber inductor and $2\,kA$ SM current for example). Consequently, it may be interesting to downsize it as much as possible. This is for example studied in [2, 6] for $2.2\,kV/1.5\,kA$ submodules using press-pack diodes. In these studies, the typical μH-range snubber inductor is reduced down to 320 nH, with the RCD-clamp remaining un-touched. A snubber inductor of 320 nH is comparable to the stray inductance encountered in standard IGCT circuits (L_{stray} in the figure 1c). These studies also demonstrate that the IGCTs themselves can withstand di/dt values higher than the recommended limit of $1\,kA\,\mu s^{-1}$ (according to the IGCT datasheet), which suggests that the main limitation to increasing the turn-on speed is the capability of the corresponding diode.

1.3 Aim of the study

In this paper, we investigate the possibility of removing entirely the snubber (smallest possible inductance and no RCD-clamp) in IGCT-based MMC submodules, or at least downsizing even further, compared with the studies in [2, 6], providing that fast enough freewheeling diodes are used. This dramatic simplification of the power circuit would have significant consequences on the cost and volume of an MMC submodule. It is studied experimentally with very fast silicon diodes (up to $13\,kA\,\mu s^{-1}$).

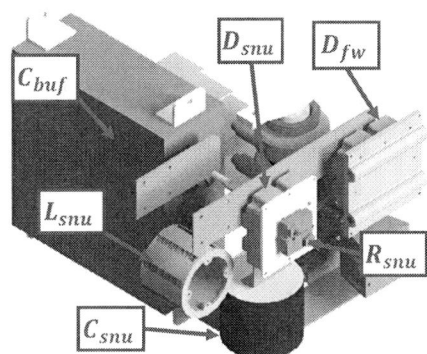

(a) IGCT side of the actual experimental setup.

(b) Snubber side, with the cable inductor as the snubber inductor, 3D rendering (AutoCAD).

Fig. 2: The experimental setup used for commutation tests (double pulse).

Table I: Overview of the different parts of the setup with their corresponding ratings, technologies, and part number (if applicable).

Part	Value/Ratings	Technology	Model
L_{snu}	0.8 to 3.4 µH, and 100 to 160 nH	- Air inductor (3D printed mandrel, Figs. 4a 4b) - Busbar connections (Figs. 4c 4d)	Custom
R_{snu}	0.5 Ω	Planar resistor	TAP800KR50E (Ohmite)
C_{snu}	5 µF/6 kV	Film capacitor	E51.S11-502R20 (UPE)
D_{snu}	6.5 kV/2×600 A	6.5 kV Si Diode	5SLD 0600J650100 (Hitachi)
C_{buf}	0.8 mF/3.8 kV	Film capacitor	Trafim (AVX)
L_{dp}	0.4 to 2.8 mH	Air inductor	Custom
L_{stray}	450 nH	*Non applicable*	*Non applicable*

In the next section, the design of the test bench is presented, with the selection of the semiconductors devices and the design of a configurable turn-on snubber. Then (in section 3), the results of the experiment are described and analysed for a reduced snubber inductance as well as for complete removal of the snubber circuit; finally, a particular focus is given to the limiting phenomena that appeared during the experiment, and their associated destructive events.

2 Design and characteristics of the test setup

To analyse the commutations of the IGCT, a "double-pulse test" circuit is used here. The experimental setup reproduces the possible layout of an HVDC MMC Half-Bridge submodule (although only one IGCT is mounted for the tests presented here) and shows similar behaviour. Therefore, conclusions can be extrapolated to the case of a real HVDC MMC submodule. The experimental setup is shown in figures 2a (IGCT side) and 2b (snubber side). The ratings and technologies of the components used for the test bench can be found in Tab. I (except for the IGCT and diodes, discussed further below). As a reminder, the test setup, with the corresponding components labels, is displayed in figure 1c. The temperature of the semiconductors is controlled by the radiator placed behind the diode pack and the heatsinks of the IGCT pressing clamp. They are connected to a Julabo A40 (thermostat) with EPDM Rubber pipes using HL60 cooling liquid to reach temperatures up to 120 °C.

2.1 Selection of power semiconductors devices

Various IGCT component references available on the market have been listed, without considering RC-IGCTs (as the diode they include is not rated for high di/dt). The list can be found in [15]. Here, we select a 6.5 kV device, as it allows higher voltage operation, a general trend in MMCs, and allow

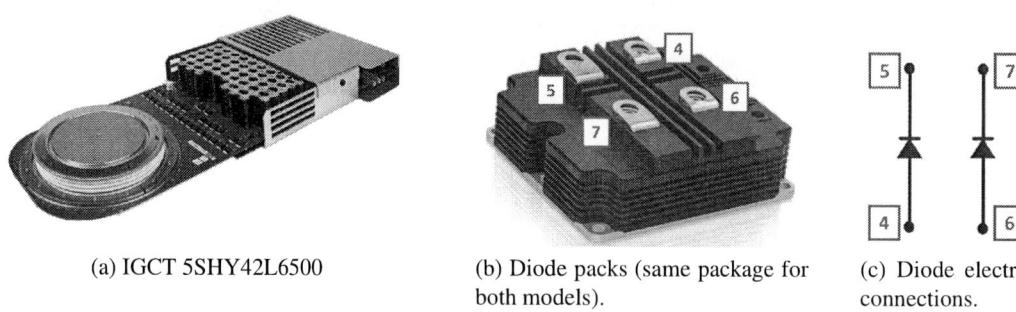

(a) IGCT 5SHY42L6500

(b) Diode packs (same package for both models).

(c) Diode electric connections.

Fig. 3: Photograph of the selected semiconductor devices.

Table II: Noticeable parameter values for the study. 3.6 kV corresponds to a typical bus voltage for a 6.5 kV device (55 % de-rating to ensure 100 FIT reliability). 2.8 kV corresponds to a typical bus voltage for a 4.5 kV device.

Conditions	di/dt Value	Voltage	Snubber inductor
Recommended di/dt for IGCT (w/ 6.5 kV diode)	1 kA µs^{-1}	3.6 kV	3.6 µH
Maximum di/dt (2×diode 6.5 kV)	8 kA µs^{-1}	3.6 kV	450 nH
Recommended di/dt for IGCT (w/ 4.5 kV diode)	1 kA µs^{-1}	2.8 kV	2.8 µH
Maximum di/dt (2×diode 4.5 kV)	13 kA µs^{-1}	2.8 kV	215 nH

testing of both 4.5 and 6.5 kV diodes. The selected IGCT is the 5SHY42L6500 from Hitachi, rated at 6.5 kV/3.8 kA). The current capability of this device is fully compatible with typical HVDC MMCs [5].

Regarding the selection of the freewheeling diode, the main criterion is its maximum allowed di/dt during turn-off, regardless of its package. Indeed, press-pack diodes are typically used together with IGCTs, which are all press pack devices, but most fast diodes, which are intended to be used with IGBTs, tend to be packaged in plastic module cases. For the sake of the study, all suitable diodes are investigated, and some diode modules are found to be the most promising devices: The 5SLD0600J650100 (6.5 kV, 2×600 A, 2×4 kA µs^{-1}) and 5SLD1200J450300 (4.5 kV, 2×1200 A, 2×6.5 kA µs^{-1}). In figure 3b, the diode package used here is displayed (both diodes have the same package). It contains two separate sets of diode chips, as shown in figure 3c. In this study, these two separate diodes are connected in parallel to double the current rating and di/dt. Note that for the sake of simplicity in the setup, the diode used in the RCD-clamp is one 5SLD0600J650100, although its current rating is much larger than required.

2.2 Snubber design

Suitable snubber inductance values depend on bus voltage (V_{bus}) and desired di/dt according to $L_{snu} = V_{bus}/(di/dt)$. Snubber inductor allowing to reach noticeable di/dt values recommended are displayed in Tab. II for the maximum voltage considered ($V_{bus} = 3.6$ kV for 6.5 kV devices and $V_{bus} = 2.8$ kV for 4.5 kV devices).

Custom snubber inductors ranging from 0.8 to 3.4 µH are formed by winding a cable around a 3D-printed mandrel, as shown in figures 4a and 4b. The different values are obtained by printing mandrels with different numbers of turns (2 turns for 800 nH, 3 for 1350 nH, etc.). The space between each turn ensures suitable clearance and creepage distance (low voltage cable is used so that insulation is ensured by the proper spacing). For lower inductance values, we take advantage of the stray inductance of the busbars as it is done in [2, 6], resulting in snubber inductance values of 160 and 100 nH (Figs. 4c and 4d, respectively). An extra stray inductance, evaluated at 450 nH by ANSYS Q3D finite elements calculations, must be added to all these values to obtain the total loop inductance. It corresponds to the unclamped stray inductance of the yellow parts in Fig. 2b.

The capacitor and resistor are selected after iterative circuit simulations using LTSpice, with two criteria: allowing pulses as short as the 40 µs (the minimum pulse time of the IGCT, as quoted in its datasheet) and the tolerated overvoltage on the commutation cell. Their ratings can be found in table I. More details

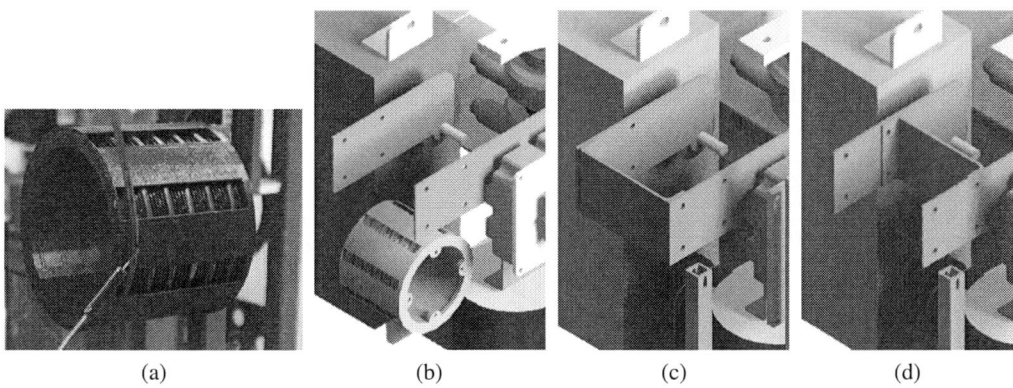

|(a)|(b)|(c)|(d)|

Fig. 4: Snubber inductor: technical choices to obtain different values from 100 nH to 3.4 µH.

on the selection process can be found in [15].

Although the snubber is not particularly optimised in size here (and no thermal analysis has been performed to allow continuous operation), its volume for this test setup is between 15L to 20L (calculated with AutoCad). This gives a good idea of typical volumes occupied by snubbers and confirms that this volume is significant. The entire snubber can be seen in Fig. 2b.

3 Experimental plan and results

3.1 Experimental plan

A "double pulse" test setup is used to explore a large range of switching current/voltage values, with moderate power dissipation (and therefore limited self-heating). In Fig. 1c the power circuit of the test setup is displayed and in [15] the whole circuit is detailed along with more explanations on the "double pulse". In the results presented below, we focus on the waveforms at the IGCT's turn-on, i.e. when the snubber circuit controls the current transient. The analysed physical quantities are the different reverse recovery parameters of the diode (E_{rr}, T_{rr}, Q_{rr}, I_{rr}), di/dt, the maximum voltage across the diode (and associated overvoltage), the total energy during turn-on ($E_{rr} + E_{on-IGCT} + E_{L-snubber}$) and the peak power during reverse recovery. The measurements are performed for the following conditions:

- Temperature: 25 °C, 60 °C, 90 °C, 120 °C for the heatsinks temperature.
- Voltage: 800 V, 1100 V, 1400 V, 1700 V, 2000 V, 2400 V, 2800 V, 3300 V*, 3600 V*. (* = 6.5 kV diode only)
- Current: 100 A, 300 A, 600 A, 900 A, 1200 A, 1800 A, 2400 A.
- The two diode modules described in section 2.1: the 6.5 kV PM diode and the 4.5 kV PM diode.
- Snubber inductor: 3.42 µH, 2.9 µH, 1.97 µH, 1.34 µH, 0.8 µH (wound inductors, see Fig. 4a), 160 nH, 100 nH (busbar connections). An unclamped stray inductance of 450 nH has to be added to the snubber inductance to obtain the total loop inductance of the circuit.

Due to undesirable phenomena (such as snap-off or dynamic avalanche), not all configurations of the parameters above are possible. This is especially true for the 6.5 kV diode, for which some failures are observed (this is described further below).

3.2 Typical waveforms and occurrence of snap-off

In this subsection, typical waveforms are presented at IGCT turn-on (see figure 5). Two types of reverse recovery can be observed: soft-recovery (figure 5a, in which the diode current $-I_{FWD}$ – smoothly reaches zero after recovery) and snappy-recovery (figure 5b, for which the diode current abruptly returns to zero at the end of the reverse recovery, causing oscillations). The chronology for the soft recovery (figure 5a) is the following:

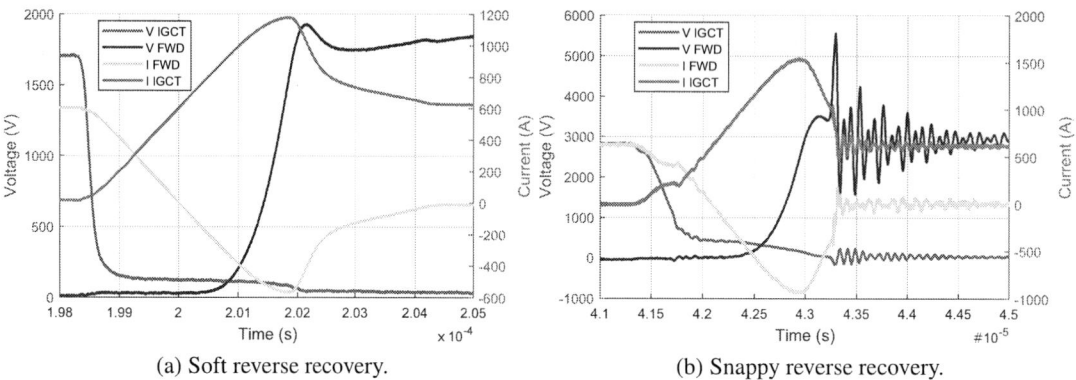

(a) Soft reverse recovery.   (b) Snappy reverse recovery.

Fig. 5: Typical waveforms obtained during the measures. Left: pulse at 1700 V/600 A/120 °C with the 4.5 kV diode and a 3.4 µH snubber inductor. Right: pulse at 2800 V/600 A/25 °C with the 6.5 kV diode and a 2.9 µH snubber inductor.

1. The IGCT receives the turn-on command and its voltage (V_{IGCT}, in red) starts decreasing ($t = 198.2\,\mu s$). The diode current (I_{FWD}, cyan) starts to transfer to the IGCT (I_{IGCT}, pink). This first phase ends when the IGCT voltage has finished its first decrease and is stable at a low voltage ($t = 199\,\mu s$).
2. Then, the current still transfers from the diode to the IGCT at a constant rate, determined by the loop inductance value; this loop inductor sustains most of the bus voltage during this phase. This second phase ends when the diode starts blocking voltage ($t = 200.5\,\mu s$).
3. The voltage across the diode (V_{FWD}, blue) increases as the diode enters reverse recovery. The reverse recovery current reaches its minimum (I_{rr}) at $t = 201.9\,\mu s$.
4. Starting from $t = 202.2\,\mu s$, a smooth recovery takes place: dI_{FWD}/dt slowly decreases until the recovery is over ($t = 204.3\,\mu s$).

In some cases, diodes can also display a "snappy recovery" behaviour, such as depicted in figure 5b. The abrupt diode current slope at $t = 43.2\,\mu s$ generates an overvoltage spike and oscillations that can be destructive [11, 16, 17]. Diode manufacturers optimise their devices to avoid snappy behaviour (among other criteria), adjusting doping levels and lifetime profile in the semiconductor [16], but, as described in [18], even soft diodes can present a snappy behaviour in certain conditions. In [16], operational conditions that favour snap-off are listed: low temperature, high voltage, high stray inductance, and low current.

3.3 Main results

This section starts with two subsections investigating the main stresses experienced by the diode (peak power density, overvoltage) and their evolution with current, voltage, L_{snu}, di/dt, temperature, and presence (or not) of an RCD clamp. Then, we analyse the impact of the snubber reduction and removal on the losses during turn-on. Finally, we focus on the impact of removing the RCD-clamp altogether.

In summary, it is possible to reduce the snubber inductance down to its lowest possible value (100 nH) for the 4.5 kV diode and even remove the RCD-clamp: proper operation is observed up to 2.4 kV and 1.8 kA at a temperature of 90 °C and 120 °C (up to 1.7 kV 2.4 kA for 25 °C and 60 °C). The 6.5 kV diode is found to work for a 800 nH inductor up to 3.6 kV/2.4 kA, but to fail at a lower snubber inductor (160 nH), as described in section 4.2.

3.3.1 Maximum power during reverse recovery

The maximum power densities measured for the 4.5 kV diode during reverse recovery are plotted in Fig. 6 over the voltage, current domain, for different snubber inductor values. This maximum power is calculated as the maximum of the $I_{FWD} \times V_{FWD}$ product divided by the total diode surface inside a package (active surface per die times the number of dies) assumed to be 19.44 cm^2 for the 4.5 kV diode

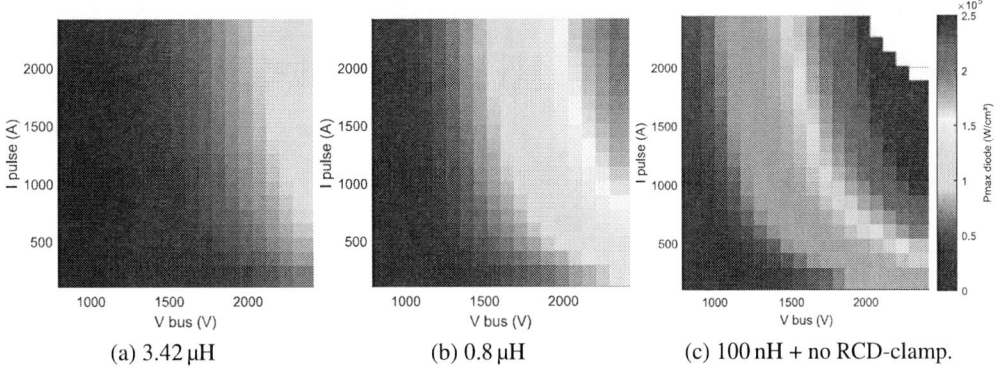

(a) 3.42 µH   (b) 0.8 µH   (c) 100 nH + no RCD-clamp.

Fig. 6: Maximum power density (W/cm²) in the 4.5 kV diode during reverse recovery, depending on the snubber inductor and structure, at 120 °C. The scale's maximum is the indicative 'silicon limit': 250kW/cm².

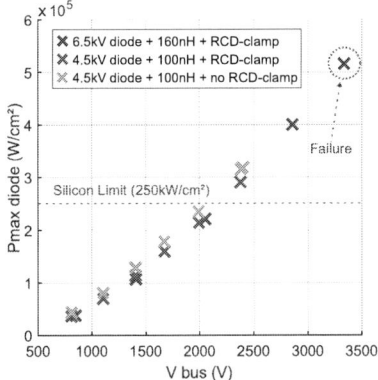

Fig. 7: Maximum power density (W/cm²) in the two diodes during the reverse recovery. In blue, for the 6.5 kV diode, 25 °C, 2400 A, 160 nH snubber inductance and RCD-clamp. In yellow and orange, for the 4.5 kV diode, 120 °C, 1800 A, 100 nH snubber inductance. In orange: RCD-clamp present, in yellow: RCD-clamp absent.

and 19.44 cm² as well for the 6.5 kV diode.

It can be seen that reducing the snubber inductance (and therefore increasing the di/dt at the IGCT turn-on) increases the peak power dissipation of the diode. In figure 6c, the indicative threshold of the "silicon limit" ($250\,\mathrm{kW\,cm^{-2}}$), described in [10], is even exceeded, but it did not lead to destruction. Fig. 7 presents the peak power density as a function of the bus voltage, for the most critical situations (i.e., lowest tested L_{snu}). The 6.5 kV diode is found to fail at a power level of $516\,\mathrm{kW\,cm^{-2}}$, after a successful test at $401\,\mathrm{kW\,cm^{-2}}$. This is in the order of magnitude of the indicative "silicon limit" [10, 11].

These results confirm that downsizing the snubber leads to higher stress on the diode during the reverse recovery, but they also demonstrate the possibility to minimise the snubber inductance value and to remove the RCD-clamp over a significant part of the (voltage,current) domain without causing diode failure.

3.3.2 Overvoltage across the diode due to snappy-recovery

As shown in Fig. 5b, the snap-off phenomenon causes a large overvoltage to appear across the diode (V_{FWD}) at the end of its recovery. This overvoltage spike may cause avalanche in the diode if it exceeds its blocking capability and eventually leads to the destruction of the device.

Fig. 8 presents the amplitude of the overvoltage (i.e. the difference between the overvoltage peak and the bus voltage). The overvoltage scale in the figure is blocked between 500V (when an overvoltage starts to be significant) and 1000V (when an overvoltage is too severe) in order to emphasise the phenomenon do-

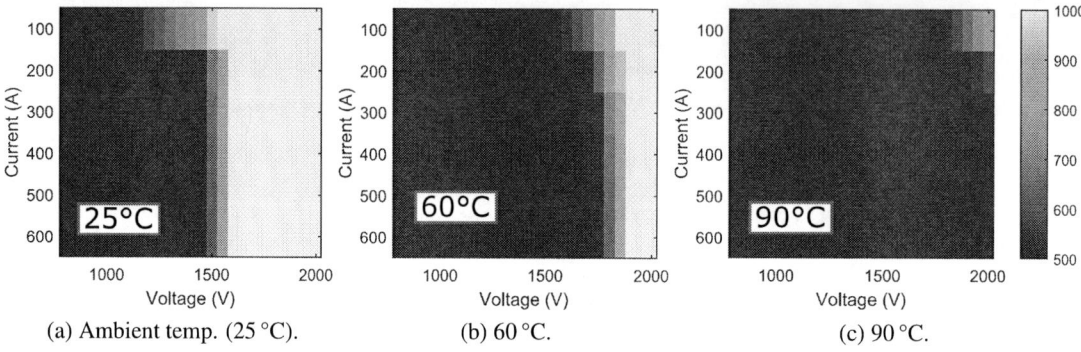

(a) Ambient temp. (25 °C).　　(b) 60 °C.　　(c) 90 °C.

Fig. 8: Heatmaps representing the overvoltage level depending on voltage, current, and temperature during snap-off events. Diode used here: 4.5 kV diode. L_{snu}=3.4 µH.

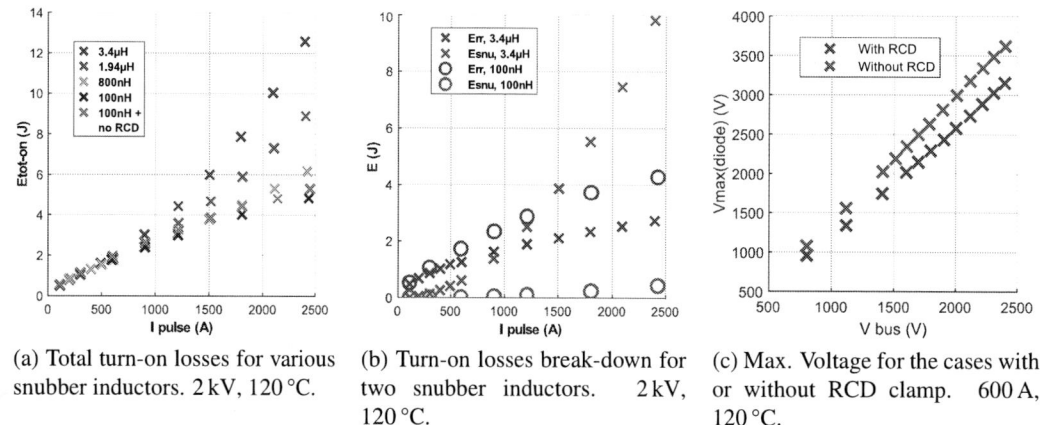

(a) Total turn-on losses for various snubber inductors. 2 kV, 120 °C.

(b) Turn-on losses break-down for two snubber inductors. 2 kV, 120 °C.

(c) Max. Voltage for the cases with or without RCD clamp. 600 A, 120 °C.

Fig. 9: Different results of the double pulse test (turn-on losses, maximum voltage).

main. It shows that, in the case presented here, significant snap-off occurs, with overvoltages exceeding 1000 V over a large fraction of the (I,V) domain, especially for the lowest junction temperature.

The trends are consistent with the literature [16]: diode snap-off is reported to depend on temperature, voltage, current, and stray inductance. Higher voltage, lower current, or/and lower temperature increase the snap-off overvoltage. In [15, 19], the effect of stray inductance on snap-off behaviour has been demonstrated. Here, the relatively high value of the stray inductance of our test setup (450 nH, in addition to the clamped snubber inductor) is suspected to be the reason why the snap-off occurs at these levels of voltage/current/temperature (as a comparison, datasheet values are quoted for a stray inductance value of 150 nH for the 4.5 kV diode). Reducing the stray inductance of the test setup is therefore required to reduce the area of the operational domain where snap-off occurs.

3.3.3 Snubber reduction impact on losses

The total switching energy losses ($E_{rr} + E_{on-IGCT} + E_{L-snubber}$) during IGCT turn-on are displayed in Fig. 9a, for 2 kV, 120 °C and various snubber inductors. This graph shows a clear reduction of the switching losses, especially for the highest current values, as the snubber inductor is reduced from 3.4 µH down to 100 nH.

Figure 9b explains this phenomenon, with the 3.4 µH snubber inductor situation marked with a cross ('x') and the 100 nH snubber inductor situation indicated with a circle ('o'): indeed, while the reverse recovery energy (in blue) is ≈ 50 % higher for the 100 nH snubber inductor than for the 3.4 µH, the energy dissipated by the snubber inductor (in orange) is drastically reduced (in the order of a 20-fold reduction). Downsizing the snubber inductor therefore offers a clear advantage regarding turn-on losses.

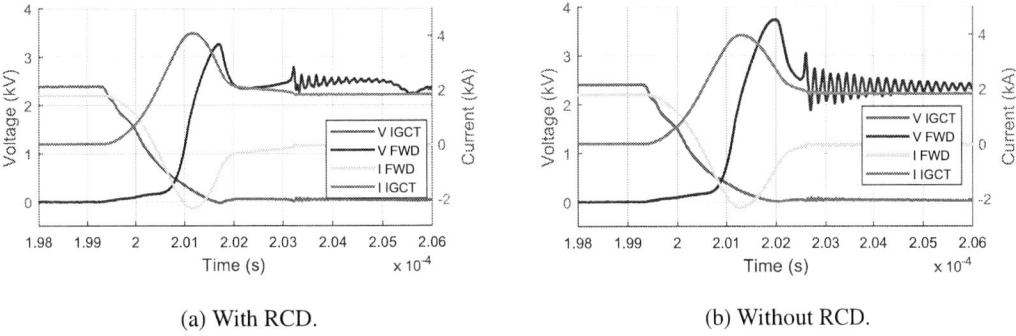

(a) With RCD.

(b) Without RCD.

Fig. 10: Waveforms of cases with and without RCD clamp. 2400 V, 1800 A, 120 °C, 4.5 kV diode, L_{snu}=100 nH.

3.3.4 Operation without snubber

Tests without RCD clamp (in the case of the 100 nH inductor) are completed up to 2.4 kV and 1.8 kA at 120 °C and 90 °C, and 1.7 kV and 600 A at 25 °C with the 4.5 kV diode.

Waveforms of tests with and without RCD clamp are displayed in Fig. 10. The resulting peak voltages are depicted in Fig. 9c at 120 °C and 600 A. Removing the RCD Clamp only results in a moderate increase of the turn-on losses (green and purple markers in Fig. 9a). This is confirmed in Fig. 9c, which shows that the peak voltage experienced by the diode increases by $\approx 15\%$. It is comparable with the increase in loop inductance: 450 nH in the clamped case (stray inductance only), $450 + 100$ nH in the un-clamped case (stray inductance + snubber inductor), i.e. a 22 % increase.

As shown in Fig. 9a, losses without RCD (green) are therefore increased compared to the case with RCD (purple) but are still lower than the losses with a 0.8 µH inductor (and RCD clamp). It must be noted that the increase in loop inductance caused by removing the RCD-clamp results in a slightly higher maximum power density inside the diode for the same voltage and current, as shown in Fig. 7 (at 2.4 kV/1.8kA/120 °C: 290 kW cm^{-2} with RCD, 315 kW cm^{-2} without RCD).

4 Limiting phenomena and failure events

Two destructive events are encountered during the tests. One is directly related to the snap-off of the diode, the other to dynamic avalanche. These two failure events are displayed in Fig. 11 and are explained in this section.

4.1 Snap-off

Snap-off can be destructive, as the resulting overvoltage can exceed the blocking capability of the diode. This is the case in the waveforms in Fig. 11a, for which a peak voltage of 5.8 kV is observed across the 4.5 kV diode. Analysis of the failed device shows that 3 of the diode chips in the diode package show clear signs of degradation (Fig. 11c), with "trenches" at the surface which are indicative of a focused phenomenon (filamentation, which causes localised melting of the silicon). More details about this failure mechanism can be found in [11, 15, 16, 17].

4.2 Dynamic avalanche

The dynamic avalanche occurs at a lower voltage than the static breakdown voltage of the diode, during reverse recovery. During turn-off, the diode blocks an increasing voltage, which is supported by the depletion region that grows within the drift region. But in certain conditions, especially at high di/dt, the electric field at the PN junction reaches a value higher than the limit (35 kV mm^{-1}), leading to the destruction of the device. The destructive dynamic avalanche (also called 3rd-degree dynamic avalanche) is described in more detail in [11, 15, 16, 20].

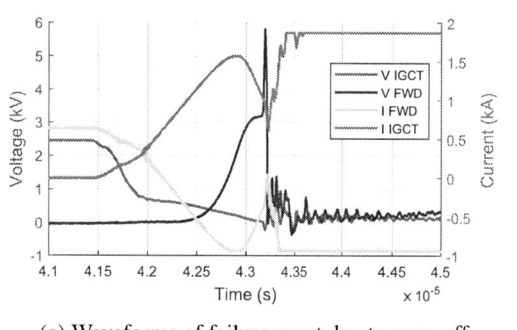 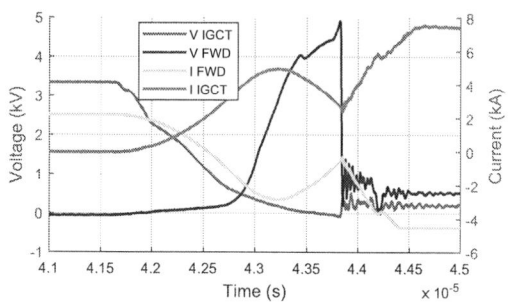

(a) Waveforms of failure event due to snap-off. (b) Waveforms of failure event due to dynamic avalanche.

(c) Dies after snap-off destruction. (d) Dies after dynamic avalanche destruction.

Fig. 11: Failure events during tests. (a-c-e) Snap-off destructive event: 2400 V, 600 A, di/dt of 1.7 kA μs^{-1}, 25 °C. Used diode for the pulse: the 4.5 kV. (b-d-f) Dynamic avalanche destructive event: 3300 V, 2200 A, di/dt of 4.5 kA μs^{-1}, 25 °C. Used diode for the pulse: the 6.5 kV.

Fig. 11b shows the waveforms of a failure event attributed to dynamic avalanche, for the 6.5 kV diode. Compared with the snap-off case (Fig. 11a), V_{FWD} does not exhibit a sharp peak and does not exceed the static blocking voltage of the diode. dI_{FWD}/dt, is relatively high, at 4.5 kA μs^{-1} (the diode module, which contains two separate groups of diode chips is rated at 2×4 kA μs^{-1}), and the test current is twice that of the package (2400 A vs. 2×600 A). This, together with the relatively large stray inductance (450 nH compared to 280 nH considered for the datasheet SOA figures) led to the diode failure.

Conclusion

Reducing the snubber inductor or removing the RCD clamp in an IGCT-based MMC submodule is found to be possible without failure at levels up to 2.4 kV/1.8 kA (using the 4.5 kV diode). A smaller snubber leads to lower losses and overall volume (15 L reduction). Removing the RCD clamp slightly increases switching losses, but the total losses (semiconductor devices and snubber circuit) remain lower than those observed in the standard case of 3.4 μH inductor and an RCD clamp (5.5 J and 13.5 J respectively, at 2 kV/2.4 kA/120 °C).

Limiting phenomena are observed (snap-off and dynamic avalanche). In real HVDC MMC implementations, they would be important problems as they occur within the normal operating range of the converter. Indeed, the observed overvoltage at ambient temperature is higher than 1 kV at 1700 V for the 4.5 kV diode. The relatively high stray inductance of the circuit could be significantly reduced in order to: (i) reduce the overvoltage during commutations and therefore the losses, (ii) reduce the occurrence and intensity of the snap-off, (iii) reduce the constraints due to high di/dt and reduce the risk of dynamic avalanche. It is estimated that re-arranging the various components of the system could yield a reduced loop inductance of less than 250 nH. Another solution to overcome these issues would be to use silicon carbide diodes, which have lower reverse recovery (or no reverse recovery at all in the case of Schottky diodes). The higher cost of these devices could be offset by the removal of the snubber circuit, the gains in efficiency and the simpler design of the structure.

References

[1] Kamran Sharifabadi, Lennart Harnefors, Hans-Peter Nee, Steffan Norrga, and Remus Teodorescu. *Design, Control, and Application of Modular Multilevel Converters for HVDC Transmission Systems*. John Wiley & Sons Inc, 2016.

[2] Rong Zeng, Biao Zhao, Tianyu Wei, Chaoqun Xu, Zhengyu Chen, Jiapeng Liu, Wenpeng Zhou, Qiang Song, and Zhanqing Yu. Integrated gate commutated thyristor-based modular multilevel converters: A promising solution for high-voltage dc applications. *IEEE Industrial Electronics Magazine*, 13(2):4–16, jun 2019.

[3] Davin Guédon, Philippe Ladoux, Mehdi Kanoun, and Sébastien Sanchez. Igcts in hvdc systems: Analysis and assessment of losses. In *PCIM Europe 2019*, 2019.

[4] Tomas Modeer, Hans-Peter Nee, and Staffan Norrga. Loss comparison of different sub-module implementations for modular multilevel converters in HVDC applications. *EPE Journal*, 22(3):32–38, sep 2012.

[5] Arthur Boutry, Cyril Buttay, Dong Dong, Rolando Burgos, Bruno Lefebvre, Florent Morel, and Colin Davidson. Figures-of-merit and current metric for the comparison of igcts and igbts in modular multilevel converters. In *2020 22nd European Conference on Power Electronics and Applications (EPE'20 ECCE Europe)*, pages P.1–P.10, 2020.

[6] Tianyu Wei, Qiang Song, Jianguo Li, Biao Zhao, Zhengyu Chen, and Rong Zeng. Experimental evaluation of IGCT converters with reduced di/dt limiting inductance. In *2018 IEEE Applied Power Electronics Conference and Exposition (APEC)*. IEEE, mar 2018.

[7] ABB. Applying igcts. Technical report, ABB, 2016.

[8] Silverio Alvarez-Hidalgo. *Characterisation of 3.3kV IGCTs for Medium Power Applications*. PhD thesis, ENSEEIHT, 2005.

[9] I. Etxeberria-Otadui, J. San-Sebastian, U. Viscarret, I. Perez de Arenaza, A. Lopez de Heredia, and J. M. Azurmendi. Analysis of IGCT current clamp design for single phase h-bridge converters. In *2008 IEEE Power Electronics Specialists Conference*. IEEE, jun 2008.

[10] Josef Lutz and Martin Domeij. Dynamic avalanche and reliability of high voltage diodes. *Microelectronics Reliability*, 43(4):529–536, apr 2003.

[11] Josef Lutz, Heinrich Schlangenotto, Uwe Scheuermann, and Rik De Doncker. *Semiconductor Power Devices*. Springer Berlin Heidelberg, 2011.

[12] David Weiss, Michail Vasiladiotis, Cosmin Banceanu, Noemi Drack, Bjorn Odegard, and Andrea Grondona. Igct based modular multilevel converter for an ac-ac rail power supply. In *PCIM Europe 2017; International Exhibition and Conference for Power Electronics, Intelligent Motion, Renewable Energy and Energy Management*, pages 1–8, 2017.

[13] Philippe Ladoux, Nicola Serbia, and Eric I. Carroll. On the potential of IGCTs in HVDC. *IEEE Journal of Emerging and Selected Topics in Power Electronics*, 3(3):780–793, sep 2015.

[14] Bjørn Ødegård, David Weiss, Tobias Wikström, and Remo Baumann. Rugged mmc converter cell for high power applications. In *2016 18th European Conference on Power Electronics and Applications (EPE'16 ECCE Europe)*, pages 1–10, 2016.

[15] Arthur Boutry. *Theoretical and experimental evaluation of the Integrated gate-commutated thyristor (IGCT) as a switch for Modular Multi Level Converters (MMC)*. Theses, INSA Lyon, December 2021.

[16] M. T. Rahimo and N. Y. A. Shammas. Freewheeling diode reverse-recovery failure modes in igbt applications. *IEEE Transactions on Industry Applications*, 37(2):661–670, 2001.

[17] Katsumi Nakamura, Fumihito Masuoka, Akito Nishii, Shin ichi Nishizawa, and Akihiko Furukawa. Freewheeling diode technology with low loss and high dynamic ruggedness in high-speed IGBT applications. *IEEE Transactions on Electron Devices*, 66(11):4842–4849, nov 2019.

[18] Peter Losee, Max-Josef Kell, Fabio Carastro, Jorge Mari, Matthias Menzel, Tobias Schuetz, and Thomas Zoels. Soft recovery diodes with snappy behavior. In *2015 17th European Conference on Power Electronics and Applications (EPE'15 ECCE-Europe)*. IEEE, sep 2015.

[19] N.Y.A. Shammas, M.T. Rahimo, and P.T. Hoban. Effects of external operating conditions on the reverse recovery behaviour of fast power diodes. *EPE Journal*, 8(1-2):11–18, jun 1999.

[20] B. Heinze, J. Lutz, H.P Felsl, and HA. Schulze. Ruggedness of high voltage diodes under very hard commutation conditons. In *2007 European Conference on Power Electronics and Applications*. IEEE, 2007.

Characterisation of a Ferrite-Polymer Based Magnetic Material

Johan Le Leslé, Guillaume Lefevre, Julien Morand
Rémi Perrin, Pierre-Yves Pichon, Guillaume Regnat
Mitsubishi Electric R&D Centre Europe
1 Allée de beaulieu
Rennes, France
Email: j.lelesle@fr.merce.mee.com

Keywords

≪Magnetic device≫, ≪Material≫, ≪Ferrite≫, ≪Polymer-Epoxy≫, ≪Iron Losses≫

Abstract

The design of highly integrated converter is more than ever a hot topic in the power electronic world. Integration also brings new challenges and opportunities for magnetic design, e.g. use of new magnetic material or innovative manufacturing process. This paper is presenting the manufacturing process and characterisation of a composite material applied to the design of a magnetic components which is fully compliant with the printed circuit board (PCB) process. This paper is presenting results of the characterisation on the ferrite powder and on the soft magnetic compound (SMC), some insight are given on the material manufacturing before presenting additional characterisation procedures, especially for core losses measurements.

Introduction

A major trend in power electronic is the high efficiency and power density design. With the advent of wide band gap (WBG) devices new challenges came along, and progress on packaging being definitely one of them. More recently, the design of converters using PCB embedding technique gains more interest [1]. Passives can also be integrated with, or close to the embedded power stage, such as resistors and/or capacitors leading to a size reduction of the active part. Nevertheless, magnetic components are still requiring magnetic core being difficult to embed into the PCB, especially for high operating power [2]. Therefore, magnetic components represent an important part of the overall volume of such converters. Ferrite cores, especially planar, are commonly implemented because of their capabilities to operate in the high frequency range with an acceptable amount of losses. Although several planar cores are compatible with PCB design, the mechanical possibilities remain limited. Hence, this paper introduces a soft magnetic compound and its preliminary characterisation. Finally, the tailored test bench dedicated to the material characterisations and the associated losses measurement procedures are presented.

Soft Magnetic Material Compound

Therefore, this paper is presenting investigations on a composite material made of ferrite powder and epoxy resin. The magnetic properties of the mixture widely depend on the relative proportion of ferrite powder and non-magnetic polymer binder. It is expected to take advantage of the low losses characteristic while the mechanical versatility should be increased due to infinite shape capabilities. Moreover, final magnetic component includes distributed air gaps which fully tackle the fringing flux related to copper losses and potential hotspots. The uncured magnetic mixture comes as a paste that can be moulded with moderate pressure in the desired shape. Curing can be performed at a low temperature. It makes this material compatible with PCB process and high volume production [3]. Defined portions of powder

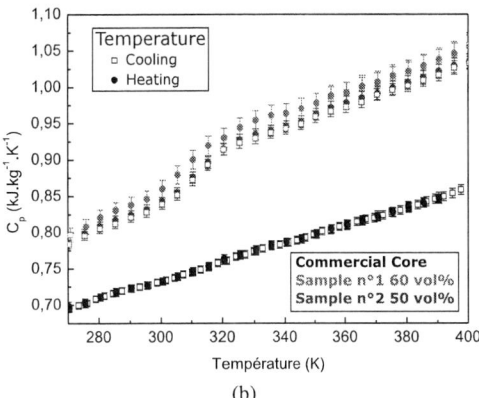

(a)   (b)

Figure 1: Results of the characterisation of the magnetic material. (a) SEM image using secondary electron detector, the powder is deposited on a carbon adhesive. (b) Comparison of specific heat coefficient for three samples, one being commercial tore and two being made of SMC with different volume portions.

and epoxy resin are mixed together, giving the paste. The moulding can be performed through several options. We choose to use a nested mould with polyurethane material having abrasion resistant properties in combination with a rigid backbone made of carbon steel. The polyurethane material can eventually be replaced by copper, or lubricated brass, as a liner between the mixture and the rigid mould.

An example of such a concept is the design of an isolated gate driver power supply having a specific transformer shape moulded on the PCB windings [4]. This type of soft magnetic compound is also very interesting for Electro-Magnetic Interferences (EMI) applications where multifunctional integration or conformal shielding is foreseen. Moreover, this type of material can be combined with any commercial cores. Indeed, by moulding the magnetic paste around a common mode choke the differential mode inductance is increased due the leakage flux concentration into the SMC [5]. In addition, the volume fraction of the magnetic powder and the binder can be adjusted to comply with some specific applications. For example, in the case of on-board wireless power transfer, the reception coil into the cars is also subject to mechanical stress, e.g. vibrations. Thus, a composite material with higher proportion of binder, epoxy or else, can take advantage of the plastic ruggedness compared to brittle ferrite core [6, 7].

Characterisation

Physical properties

First, it is important to highlight that the studied composite material is made of a by-product of milled Manganese-Zinc (Mn,Zn) ferrite and epoxy resin. As the selected magnetic powder comes from waste of the ferrite manufacturing process, some key data are missing. The manufacturer does not specify the distribution, composition and grain size of the material. Thus, the magnetic characterisation has to be performed. In order to investigate the material in detail, some experiments have been carried out in collaboration with the University of Rennes 1, Brittany. first, SEM analysis (second electrons detector) was performed to investigate grain size of the ferrite powder. The measurements concluded that the grain size ranges is from 10 μm to a few 100 μm, the results of the SEM analysis is presented in the Figure 1.(a). Another measurement protocol, being the Helium Pycnometer measurement [8], allowed to defined the powder density. The results gave a density of 4.95 g/cm^3, this value is quite important as it is a key parameter to calculate the mass fractions of ferrites versus epoxy in the mix.

Then, two samples based on T37 tore geometry have been manufactured with two different volume fraction of ferrites, 60 % and 50 %, it represents 89 % and 86 % of the weight, respectively. Having the value of the specific heat coefficient is important as it is useful for thermal design during transient, and for

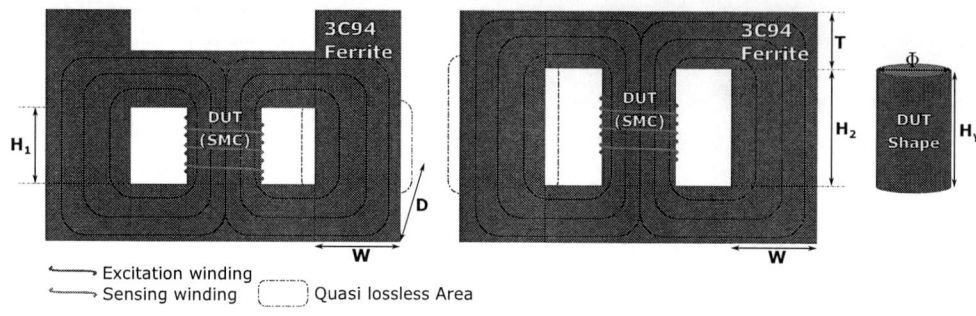

Figure 2: Magnetic frame principle applied to two SMC samples

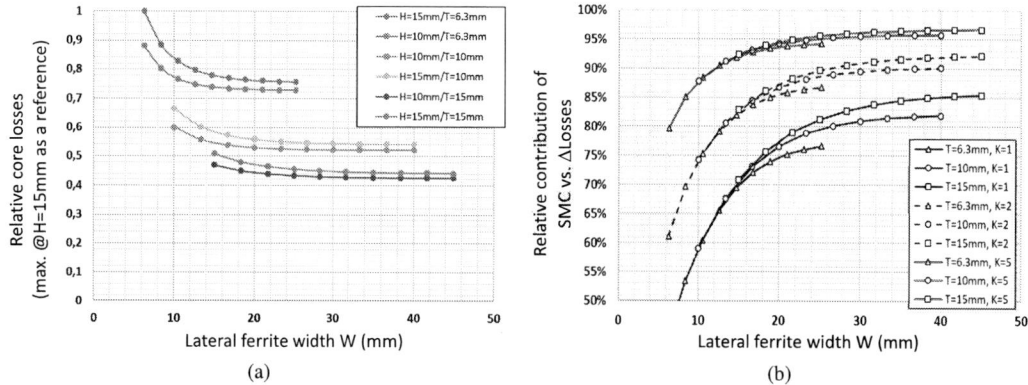

(a)

(b)

Figure 3: Investigation on the contribution of the SMC and the magnetic frame for different dimensions. (a) Impact of W for a given H and T. (b) Impact of T, K and W for H = 15 mm, K being the losses ratio between SMC and the ferrites.

material characterisation such as losses measurement using calorimetric method. Thus, the specific heat capacity of the composite material has been measured on these two samples and compared to commercial T37 ferrite core. The measurements have been performed with the relaxation method on a Physical Properties Measurement System. Prior to measurement, the magnetic core have been sawed and polished. The results are presented in the Figure 1.(b). As one can see, the specific heat coefficients of the two *in-situ* manufactured cores are higher than the one of commercial sintered cores. This is mostly due to the epoxy resin having a typical $Cp = 1.1 \text{ kJ.kg}^{-1}.\text{K}^{-1}$ while it is $0.7 \text{ kJ.kg}^{-1}.\text{K}^{-1}$ for the MnZn ferrite. The presence of the epoxy is also noticeable on the curve because of the bump occurring at 320 K (52°C) corresponding to the glass transition of the epoxy resin.

Core Losses Measurement

Experimental Set-up

In order to characterise the magnetic behaviour and the losses of the composite material, a dedicated test bench was designed to excite the core at different induction level and different frequency. It consists of placing one sample into a magnetic frame made of ferrite blocks. The motivation for using such a geometry is to ensure a homogeneous field distribution inside the DUT. The cylindrical shape has been chosen for the sample, being the simplest to manufacture with controllable dimensions. Excitation and sensing windings are located around the DUT, as illustrated in the Figure 2. Several FEA simulation campaigns were conducted on the magnetic frame. In a first time, the frame dimensions and how they affect the core losses distribution between the ferrite blocks and the DUT have been investigated. Indeed, the magnetic frame loss contribution must be minimised. In these simulations a perfect mechanical assembly and no residual air gap were assumed. Thus, the global losses are distributed between the

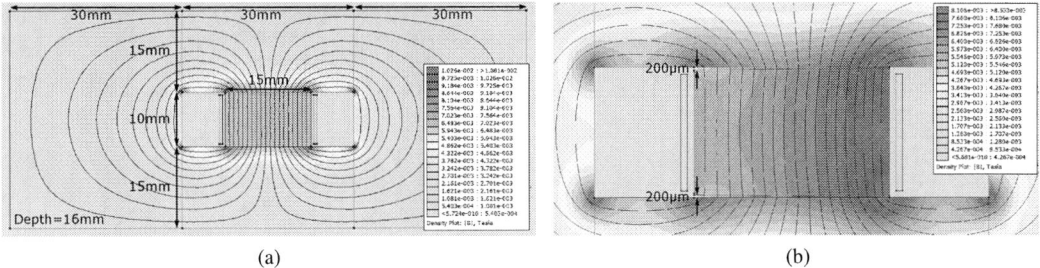

(a)   (b)

Figure 4: FEA simulation results investigating the lack of parallelism between the DUT and the magnetic frame. (a) Reference case with no residual air gap. (b) Zoom on the case having a presenting linearly increasing top and bottom air gaps between the DUT and the frame.

frame (P_F) and the SMC, both depending on the height, note that the sample diameter (Φ) being set to 15 mm.

First, the total losses were analysed assuming similar losses for the DUT and the frame (same Steinmetz coefficient). The losses are estimated using a Lua script applying the Steinmetz equation at the mesh level. The results are presented in the Figure 3.(a) in a relative manner for different dimensions, the case H = 15 mm/T = 6.3 mm being set as a reference. It highlights that an asymptotic value of the losses is reached for increasing width (W). This asymptote also decreases for thicker horizontal blocks which indicates that the magnetic flux efficiently spreads. This consequently leads to almost cancel the core losses in the lateral legs, reducing the contribution of the frame to the overall measured losses. The Figure 3.(b) is presenting the relative contribution of the SMC under test with regards to the losses difference (ΔP) when changing the thickness (T) and (W) values at a given height (H = 15 mm). The parameter K represents the loss factor between the DUT material and the frame, higher K meaning more lossy material. These results highlight that the main contributor is the DUT as the loss factor increases. Nevertheless, even with K = 1 an error below 15 % can be expected by properly selecting the width and the thickness of the constitutive blocks. According to the simulation results, a frame designed with a width of 30 mm and a thickness of 15 mm is a configuration that minimises the frame losses.

In a second time, simulations were carried out to investigate the impact of parallelism between the DUT and the frame. The dimensions are selected from the previous results. Two cases are simulated. One being the perfect assembly without any residual air gaps, and the second one including a linearly increasing air gap at the top and bottom junction of the DUT and the frame. The comparison is presented in the Figure 4. The results concluded that a few hundreds of residual air gap locally impact the induction filed distribution into the DUT, approximately $\pm$ 20 %, but can be easily tackled with a thin soft magnetic sheet. The mechanical deviations can also have an impact on the losses, in this case the analysis revealed that the deviation from the ideal case is less than 1 % . Therefore, these two FEA simulation campaigns validated the magnetic frame approach to characterise the soft magnetic compound.

The following is presenting the core losses evaluation of the soft magnetic compound through two different methods. First, the losses are measured using an improved electrical method based on magnetic voltage compensation technique. The second method is based on a calorimetric approach [9]. The methodology is simple to implement as no specific insulating materials nor complex temperature control loop are mandatory. The Table I is presenting the characteristic of the evaluated samples using electrical and thermal method, note that the SMC permeabilities are measured using the magnetic frame, and only the sample SMC 2 is characterised by the calorimetric method.

Electrical Measurements

Conventional electrical measurements assessing the core losses are performed by measuring the primary current flowing through the coil ($\propto$ to the magnetic field H [A/m]) and the secondary induced voltage ($\propto$ to induction B [T]). This classical 2-windings technique can lead to underestimated core losses due

Table I: Characteristics of the evaluated samples

	CK172060 (MF60)	CH172060 (HF60)	SMC 1	SMC 2
μ	60	60	≈ 22	≈ 24
Out Dia. [mm]	17.27	17.27	15.2	28
In Dia. [mm]	9.65	9.65	-	-
Height [mm]	2×6.35	2×6.35	10.4	23
Ae [cm^2]	1.61	1.61	1.81	6.16
Volume [cm^3]	2×1.0225	2×1.0225	1.886	14.7
Mass [g]	-	-	-	53.3
Cp [J.kg^{-1}.K^{-1}]	-	-	-	870 @298 K

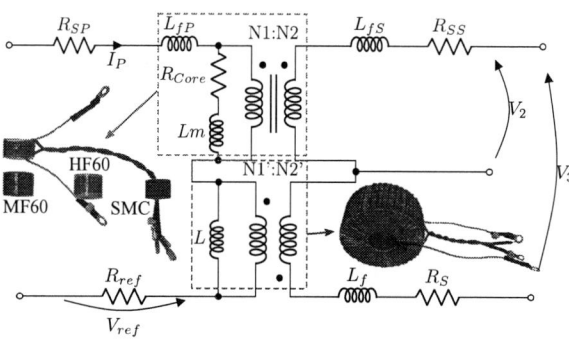

Figure 5: Compensation circuit used for electrical core losses measurements

to the voltage/current phase shift ($\varphi_{V,I}$) close to 90°. As presented in [10], compensation technique is mandatory when dealing with low permeability or low losses material characterisation. Series capacitive compensation can reduce $\varphi_{V,I}$ in an acceptable range, i.e. below 30°, suitable for common probes and associated deskew. However, inherent drawbacks remain. The compensation is only achieved for sinusoidal signal at a fixed frequency and the capacitor's ESR is included in the loss measurements, requiring special care for devices selection [11].

The selected method is using a two-windings air coil in series with the DUT, as presented in the Figure 5. The air coil is designed to have a similar coupling factor than the DUT and to compensate as much as possible the magnetic voltage, without adding any losses on the secondary side. Thus, by measuring the secondary compensated voltage (V_3) and the primary current it is possible to obtained the core losses for any type of induction.

$$V_3(t) = \left[R_{Core} \cdot I_P(t) + Lm\frac{dI_P(t)}{dt} - M\frac{dI_P(t)}{dt} \right] \rightarrow P_{Core}(t) = V_3(t) \cdot I_P(t)$$

This method has been applied on three different cores: two off-the-shelf cores and the proposed composite material. The reference samples have been selected to validate the compensation method and the associated probing [12]. Some waveforms obtained with the HF60 material at 25 kHz, and two induction levels are presented in the Figure 6. The primary current and the compensated secondary voltage are presented in the top figures, the middle figures are presenting the induced voltage across the DUT, as it would be classically obtained with the 2-windings method. The bottom figures are comparing the instantaneous powers with and without the compensation, in blue and in brown, respectively. On one hand, it is noticeable that without the compensation method the measured power is mainly reactive making

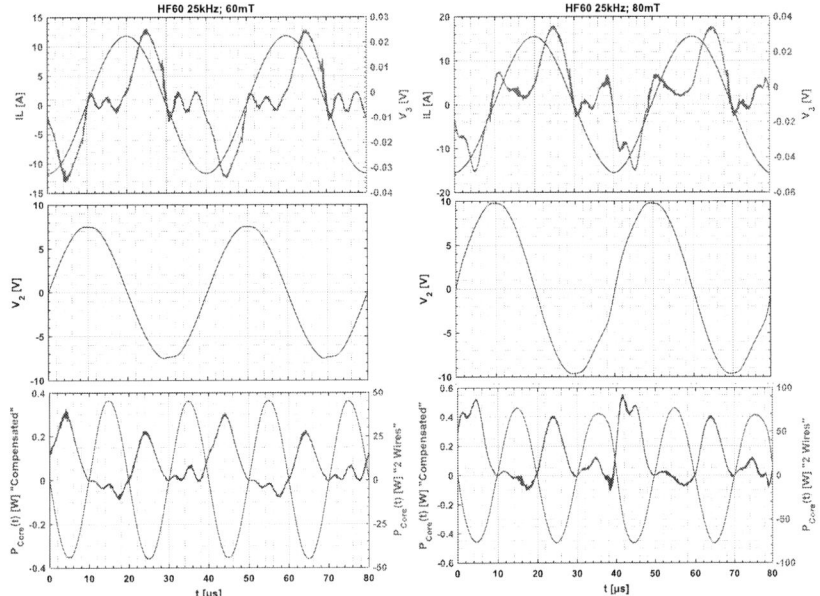

Figure 6: Voltage and current waveforms obtained with HF60 and the compensation method

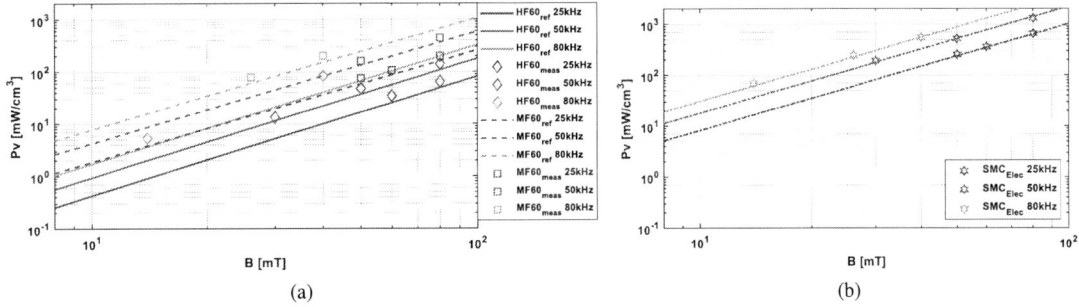

Figure 7: Core losses of the different materials, being High Flux 60, Mega Flux 60 and SMC. (a) Comparison between datasheet and electrical measurements for the HF60 and MF60. (b) Results obtained with the SMC.

the estimation of the core losses more complex and prone to significant errors. On the other hand, the compensation technique allows to measure only the active power related to the core losses. The measurement results are compared with the manufacturer data in the Figure 7.(a). Finally, the measurements on the reference material at different frequencies and induction levels are presenting a good agreement with the material datasheets, validating the experimental set-up. A similar experimental plan has been applied on the SMC, the DUT is evaluated for three induction values and frequencies. The results, shown in Figure 7, highlight much higher core losses, one to two orders of magnitude, with the SMC compared to the iron powder cores.

Calorimetric Measurements

To corroborate the results obtained with the electrical method, a second SMC sample was evaluated using a calorimetric approach. The magnetic frame is still used for this experiment, but no compensation coil is required facilitating the implementation and saving manufacturing time. As previously, the magnetic section of the different blocks is high enough to neglect the losses. The calorimetric approach is based on self-heating of the magnetic device. Assuming that the enclosure of the DUT is perfectly adiabatic the amount of losses generated by the sample is related to its temperature increase.

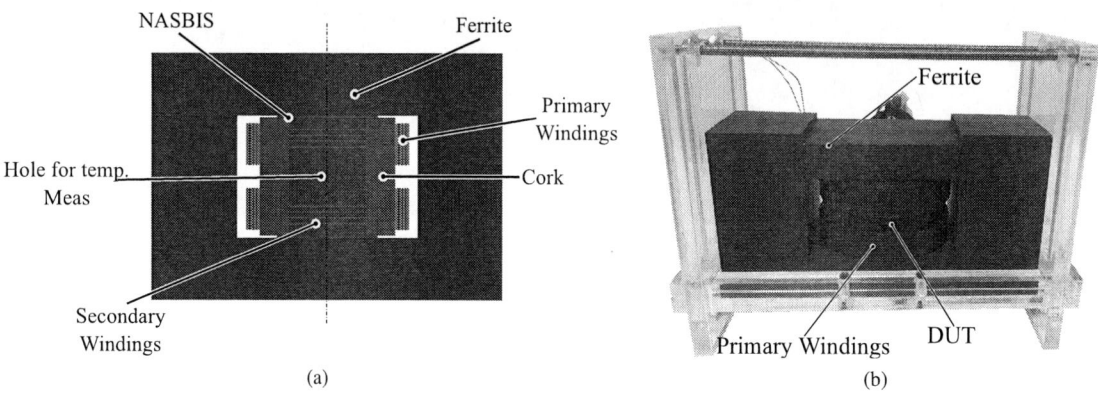

(a)

(b)

Figure 8: Set-up for calorimetric measurements. (a) Schematic of the DUT assembly. (b) Final assembly of the DUT and the magnetic frame.

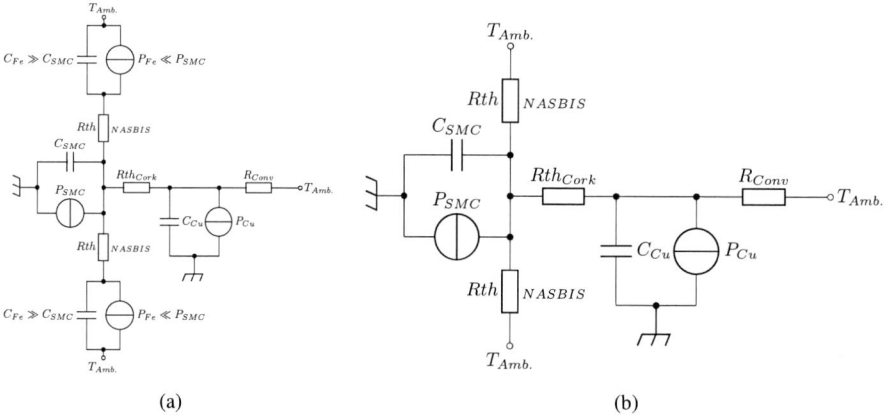

(a)

(b)

Figure 9: Thermal model of the calorimetric set-up. (a) Full model. (b) Simplified model.

$$P_{Core} = m_{Core} \cdot C_P \cdot \frac{dT_{Core}}{dt} \simeq m_{core} \cdot C_P \cdot \frac{\Delta T_{Core}}{\Delta t}$$

Additional heat leakages can be considered to compensate the thermal coupling with other elements (ambient air, copper, ferrite blocks, etc) as the set-up is definitely not perfect. To perform this experiment, the second SMC sample was manufactured, being ~8 times bigger than the first one. Nevertheless, the manufacturing process remains the same, as well as the volume fraction (60 %). The accuracy of such method relies on a proper insulation. In addition, a proper winding design is mandatory to avoid heating the DUT through the iron-to-copper thermal resistance. To do so, the DUT is thermally insulated by multiple layers of cork, 5 mm in total, having a thermal conductivity of $k \approx 0.043$ W.m^{-1}.K^{-1}, also acting as a coil former. The excitation winding was made of 4 stacked copper foils (6.5 mm width, 200 μm thick) to optimize both DC and AC resistances. A coil is wound around the core to sense the exact flux density. The DUT temperature is measured by an IR camera (FLIR A6752sc with a thermal sensitivity of 0.02 K) through a cavity punched into the cork. Nonetheless, the DUT has also to be thermally insulated from the top and bottom ferrite blocks without degrading the magnetic properties due to additional air gap. Inserting 100 μm thick NASBIS (Nano Silica Balloon InSulator) sheets in between the DUT and the ferrites was identified as the best trade-off. The thermal conductivity of such material is lower than the air $k \approx 0.018$ W.m^{-1}.K^{-1}. The black-painted set-up assembly is presented in the Figure 8. The complete, and simplified, 1D thermal models are presented in the Figure 9.

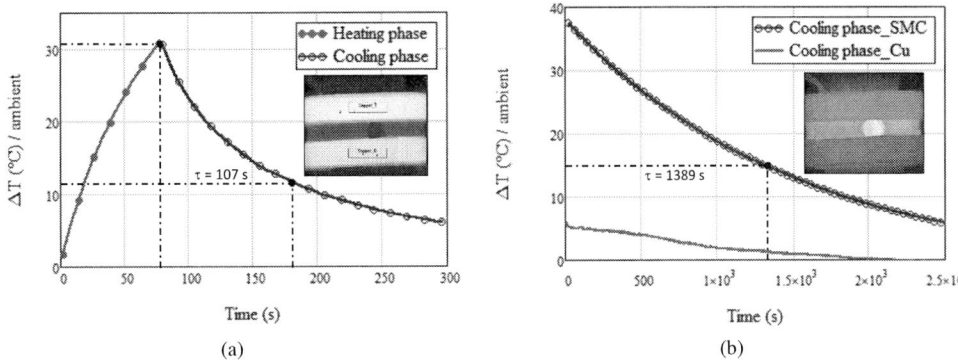

(a)   (b)

Figure 10: Temperature measurements performed to determine the thermal leakage path. (a) Heating and relaxation of copper to determine R_{Conv}. (b) Relaxation of the DUT to determine Rth_{Cork}.

The simplifications are made assuming the following assumptions :

- The ferrite losses are negligible and due to the high heat capacity the ambient temperature can be considered at the top/bottom surfaces of the NASBIS sheet.

- An equivalent copper layer can be defined by averaging the temperature of the top/bottom parts.

According to the simplified model, the thermal resistances of the NASBIS sheet, the cork, and the equivalent convection resistance have to be determined. The NASBIS thermal resistance has simply be calculated, $Rth_{NASBIS} = 216\,°C/W$. The other thermal resistances, Rth_{Cork} and R_{Conv}, have been experimentally determined through two separate procedures.

The convection resistance has been estimated by replacing the SMC sample by a polystyrene block, and the copper windings have been heated up with DC current. The R_{Conv} is calculated with the relaxation time constant and considering the copper heat capacity $C_{Cu} = 3.348\,J.K^{-1}$ for 8.7g of material. Thus, with 107 s time constant ($\tau = R_{Conv}.C_{Cu}$), R_{Conv} is equal to 41.8 °C/W. The determination of the cork thermal resistance is performed by heating the SMC block with a heating plate, and placing it into the cork enclosure. The top and bottom surfaces are insulated by polystyrene blocks. A forced air cooling is imposed on the copper winding to create a pseudo-isothermal boundary by reducing as much as possible the convection resistance ($R_{Conv} \approx 0$). Similarly to the previous procedure, the relaxation time constant of the hot SMC block through the cork thermal resistance is used. The heat capacity of the SMC is $C_{SMC} = 46.37\,J.K^{-1}$ for a 53.3 g block. Thus, with $\tau = Rth_{Cork}.C_{SMC} = 1389\,s$, we determine $Rth_{Cork} = 30\,°C/W$. Both temperature measurements obtained during the two identification procedures are presented in the Figure 10. Those value have also been confirmed through calculation. Hence, considering the simplified model and the identified thermal leakage paths, it is possible to derive the following equations, with the integral term being the correction function :

$$P_{Core} = m_{Core} \cdot C_P \cdot \frac{dT_{Core}}{dt} + 2 \cdot \frac{T_{Core} - T_{Amb}}{Rth_{NASBIS}} + \frac{T_{Core} - T_{Cu}}{Rth_{Cork}}$$

$$P_{Core} \approx \frac{1}{\Delta t} \cdot \left[m_{Core} \cdot C_P \cdot \Delta T_{Core} + \int_0^{\Delta T} 2 \cdot \frac{T_{Core} - T_{Amb}}{Rth_{NASBIS}} + \frac{T_{Core} - T_{Cu}}{Rth_{Cork}} dt \right]$$

The correction function can be applied to reduce the errors on the core loss estimation. To validate the proposed correction function, two sets of measurements have been performed, one with a high level of copper losses and one with reduced copper losses, in both cases the frequency and the induction amplitude remain the same, 150 kHz and 10 mT, respectively. The results are presented in the Figure 11.

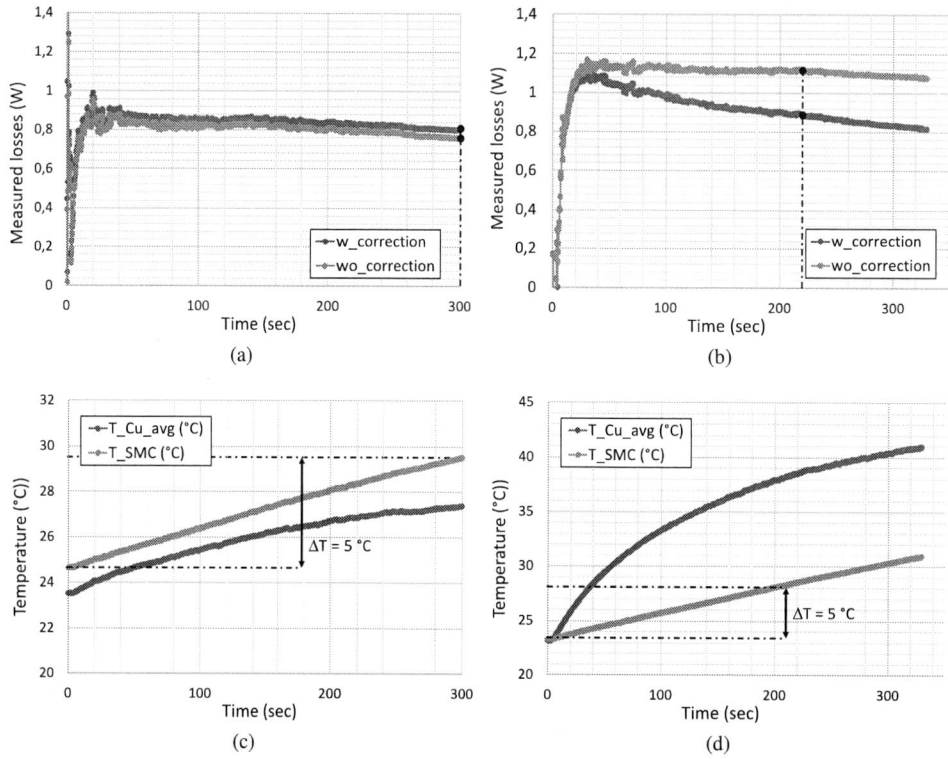

Figure 11: Experimental validation of the correction function. (a) Measured SMC losses with low copper losses. (b) Measured SMC losses with high copper losses. (c) Temperature variation with low copper losses. (d) Temperature variation with high copper losses.

As one can see, the correction function is mandatory to improve loss measurement accuracy and the operational range of the calorimetric method. Indeed, in the case of higher copper losses the windings contribute to the heating of the DUT, thus the losses can be overestimated, as in the example 1.11 W without correction against 0.88 W with the correction.

Finally new set of measurements have been carried out on the second SMC sample. The objective is to confirm, and complement the results obtained with the electrical method. The new operation points are at the same frequency as for electrical measurement but for different induction level. In total six new points have been added to loss graph and presented in the Figure 12.(a) by the red circled points. As one can see, the complementary results are in line with the previous electrical results, confirming the first employed method. According to this experimental results, electrical and thermal, the Steinmetz parameters are derived for the developed soft magnetic compound.

$$Pv_{SMC} = 0.00174 \cdot f^{1.1297} \cdot B^{2.0935}$$

The final comparison presented in the Figure 12.(b) highlights, and confirms, the higher volumic core losses of the manufactured composite material compared to iron powder and classical sintered ferrite. Therefore, the applicability of this material seems to be limited into high efficiency converters without further investigations on the material itself and optimisation on the manufacturing process. Further optimisation of the material properties (e.g. permeability and losses) will likely require to increase the volume portion of the ferrite material in the composite to levels significantly above 75 %[13]. This will require significant process adaptation from the approach proposed in this paper to overcome the practical limitations or highly packed powder-epoxy composites.

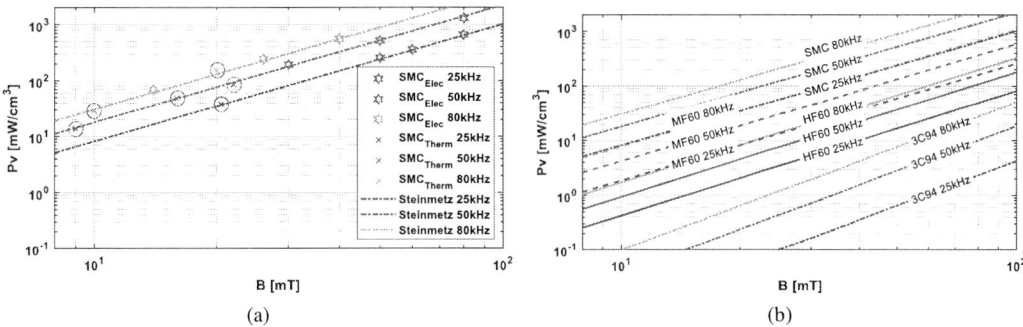

Figure 12: SMC losses abacus. (a) Addition of calorimetric measurements to the electrical measurements. (b) Comparison of the SMC with the reference materials.

Conclusion

This paper introduced manufacturing process of a composite magnetic material composed of a by-product of ferrite powder and epoxy resin. Some characterisations of the ferrite and the composite material have been presented. In addition, the design of a dedicated test bench aiming at ensuring a homogeneous flux density into the sample was also presented. This magnetic frame is used for several purposes. The first one is the magnetic characterisation, such as relative permeability as a function of the volume fraction. The second one being the evaluation of the material losses. Thus, the losses contribution of such magnetic frame has been investigated using FEA simulations, the presented results confirmed the applicability of the method. Continuing with the loss evaluation, two methods, one based on electrical measurements and one based on thermal measurements, have been presented. The electrical method includes a compensation technique ensuring a proper measurement of the core losses and has been validated using commercial core. The thermal method, being a quasi-adiabatic method, is simple to implement but requires preliminary experiment to extract model parameters, as detailed in this paper. Both methods are showing relevant results as the measurements from electrical and thermal methods are consistent with each other. A final comparison between the composite materials and commercial one highlighted the higher loss of the material. Nevertheless, the use of such material is still possible for applications with low stress conditions but requiring mechanically versatile material. Nevertheless, further investigations would be necessary at the material level and on the manufacturing process to deeply understand the root-cause of such results. The impact of the volume fraction between powder and binder should be one of the first investigated parameters along with process allowing to significantly increase the volume fraction of the magnetic powder.

References

[1] C. Buttay, C. Martin, F. Morel, R. Caillaud, J. Le Lesle, R. Mrad, N. Degrenne, and S. Mollov, "Application of the PCB-Embedding Technology in Power Electronics-State of the Art and Proposed Development," 3D-PEIM 2018 - 2nd International Symposium on 3D Power Electronics Integration and Manufacturing, 2018.

[2] R. Caillaud, C. Buttay, J. Le Leslé, F. Morel, R. Mrad, D. Nicolas, M. Stefan, and C. Combettes, "High power PCB-embedded inductors based on ferrite powder," in 5th Micro/Nano-Electronics Packaging and Assembly, Design and Manufacturing Forum (MiNaPAD), Grenoble,France, 2017.

[3] S. Egelkraut, M. März, and H. Ryssel, "Polymer bonded soft magnetic particles for planar inductive devices," CIPS 2008 - 5th International Conference on Integrated Power Electronics Systems, Proceedings, no. April, pp. 167–174, 2008.

[4] N. Degrenne, G. Lefevre, and M. Stefan, "A 2W, 5MHz, PCB-integration compatible 2.64cm^3 regulated and isolated power supply for gate driver," in 2016 18th European Conference on Power Electronics and Applications (EPE'16 ECCE Europe). IEEE, sep 2016, pp. 1–10. [Online]. Available: http://ieeexplore.ieee.org/document/7695517/

[5] F. Luo, D. Boroyevich, P. Mattevelli, K. Ngo, D. Gilham, and N. Gazel, "An integrated common mode and differential mode choke for EMI suppression using magnetic epoxy mixture," in 2011 Twenty-Sixth Annual IEEE Applied Power Electronics Conference and Exposition (APEC). IEEE, mar 2011, pp. 1715–1720. [Online]. Available: http://ieeexplore.ieee.org/document/5744827/

[6] D. Barth, G. Cortese, and T. Leibfried, "Evaluation of Soft Magnetic Composites for Inductive Wireless Power Transfer," in 2019 IEEE PELS Workshop on Emerging Technologies: Wireless Power Transfer (WoW). IEEE, jun 2019, pp. 7–10. [Online]. Available: https://ieeexplore.ieee.org/document/9030664/

[7] R. Perrin, G. Bueno Mariani, J. Morand, and S. Mollov, "PCB Embedded dies for low thickness Wireless rotary transformer," in CIPS 2020; 11th International Conference on Integrated Power Electronics Systems, Berlin, Germany, 2020.

[8] M. Viana, P. Jouannin, C. Pontier, and D. Chulia, "About pycnometric density measurements," Talanta, vol. 57, no. 3, pp. 583–593, may 2002. [Online]. Available: https://linkinghub.elsevier.com/retrieve/pii/S0039914002000589

[9] P. Papamanolis, T. Guillod, F. Krismer, and J. W. Kolar, "Transient Calorimetric Measurement of Ferrite Core Losses up to 50 MHz," IEEE Transactions on Power Electronics, vol. 36, no. 3, pp. 2548–2563, mar 2021. [Online]. Available: https://ieeexplore.ieee.org/document/9169835/

[10] N. F. Javidi and M. Nymand, "Error analysis of high frequency core loss measurement for low-permeability low-loss magnetic cores," in 2016 IEEE 2nd Annual Southern Power Electronics Conference (SPEC). IEEE, dec 2016, pp. 1–6. [Online]. Available: http://ieeexplore.ieee.org/document/7846098/

[11] M. Mu, Q. Li, D. J. Gilham, F. C. Lee, and K. D. Ngo, "New core loss measurement method for high-frequency magnetic materials," IEEE Transactions on Power Electronics, vol. 29, no. 8, pp. 4374–4381, 2014.

[12] C. S. Corporation, "SOFT MAGNETIC POWDER CORES," Tech. Rep.

[13] L. Siesing, "Development and Implementation of a Mouldable Soft Magnetic Composite," Ph.D. dissertation, 2016.

Model predictive-based control technique for fault ride-through capability of VSG-based grid-forming converter

Mobina Pouresmaeil[1], Amir Sepehr[1], Basit Ali Khan[1], Jafar Adabi[2], Edris Pouresmaeil[1]

Department of Electrical Engineering and Automation, Aalto University, Espoo, Finland[1]

Department of Electrical and Computer Engineering, Babol Noshirvani University of Technology, Babol, Iran[2]

mobina.pouresmaeil@aalto.fi, amir.sepehr@aalto.fi, basitali.khan@aalto.fi, j.adabi@nit.ac.ir, edris.pouresmaeil@aalto.fi

Keywords

≪Converter control≫, ≪Virtual Synchronous Generator≫, ≪Model predictive control≫, ≪Fault ride-through≫, ≪Power quality≫.

Abstract

Increasing integration of renewable energy resources emphasizes the importance of the grid-forming virtual synchronous generator (VSG)-based converters. An important issue in the control structure of these converters is the fault ride-through (FRT) capability under fault operating condition. In this paper a model predictive-based FRT control strategy is proposed to limit the converter current while ensures high power quality during fault situation. The proposed control method provides a fast dynamic response, high power quality, improved performance, and a simpler control structure. The effective performance of the proposed control method, as well as its superior performance in comparison with the conventional PI-based control and a model predictive-based control method, are validated through simulation results in MATLAB/Simulink.

Introduction

Renewable energy-based resources are expanding quickly to respond to the increasing demand for clean electric energy. However, increasing integration of these converter-based resources is a challenge to the modern power grid. Conventional synchronous generator (SG)-based power grid inherently possesses inertia and oscillation damping features, while converter-based generators do not possess these features inherently. This is a great challenge for the modern power grid dominated by renewable energy resources. To overcome this challenge, different control techniques have been proposed to virtually add important features of the SGs to the control structure of the converters as a virtual synchronous generator (VSG) [1]-[4]. Among these, grid-forming converters have a promising potential for a fully converter-based power grid [4], [5].

Although inertia and oscillation damping features of SGs are replicable in the control of converters, providing a large short circuit current is not possible for converters. This is because of semi-conducting devices of the converter which are not able to tolerate a very large overcurrent. So, fault ride-through (FRT) ability, i.e., the ability of the converter to continue connected to the grid with a reduced voltage level during fault conditions, is a vital issue of the converter that needs to be addressed [6]-[12].

Avoiding overcurrent by simply limiting the converter reference current to a certain level of its nominal value is proposed in [6]. However, this technique causes windup in the outer control loop of the converter when the converter current saturates. Anti-windup for the PI controller of the outer voltage control loop does not work well in the case of a large fault occurrence. An indirect current limiting control technique has been proposed in [7] in which virtual impedance is employed for limiting the reference voltage and indirectly current reference of the converter. However, this method is not fast enough in limiting the converter current. A control technique is proposed in [8] in which the converter current is directly limited, and the virtual impedance is employed to avoid windup in the voltage control loop. However, this technique is dependent on the fault size and location to determine the virtual impedance. A control technique for a grid-forming VSG-based converter is proposed in [9] that combines the direct and indirect current limiting strategies. In this paper, the reference current is directly limited, and the reference active and reactive power in fault operating mode are updated based on the voltage drop to provide the appropriate voltage reference compatible with the limited current, resulting in anti-windup in the outer control loop.

The aforementioned FRT control strategies are based on PI controller. Finite-set model predictive control (FS-MPC) has been introduced as an alternative to the conventional PI-based linear control techniques. The main advantages of FS-MPC over conventional PI-based control are the fast dynamic response, simple control structure, and multi-objective control possibility. For a cascaded voltage and current control loop, at least ten parameters need to be tuned, resulting in the complexity of the PI-based control schemes.

A simple FS-MPC is proposed in [10] to control a grid-forming converter working in the islanded mode of operation. Current reference tracking and overcurrent protection are two control objectives of the control scheme. In [11], a simultaneous voltage and current tracking based on FS-MPC has been proposed in which virtual voltage vectors are employed to reduce the harmonics. In [12] an FS-MPC strategy with three control objectives has been proposed for the control of a grid-forming VSG. Current and voltage reference tracking and current limitation are three control objectives followed by this control strategy. Although this control can provide a strict current limiter, the power quality during fault situations is still a vital issue that needs to be addressed.

Thereby, in this paper, an FS-MPC-based strategy is proposed for a grid-forming VSG-based converter taking into account both vital aspects of FRT ability: overcurrent protection and power quality. A reduced total harmonic distortion (THD) of voltage and current during fault condition verifies the power quality of the proposed control technique. The remaining of the paper is organized as follows. The model under study and the proposed control technique are described in Section 2 and Section 3, respectively. The simulation results in MATLAB/Simulink are presented and discussed in Section 4. Finally, Section 5 summarizes the main outcomes of the paper.

Description of the system model

Fig. 1 shows the general model and the proposed control, a VSG-based grid-forming converter connected to the grid through an LCL filter. According to Fig. 1, the equation representing the AC-side dynamics of the converter are expressed in $\alpha\beta$ reference frame as follows:

$$L_{fc}\frac{di_c}{dt} + R_{fc}i_c + v_c = u_c \tag{1}$$

$$C_f\frac{dv_c}{dt} + i_o = i_c \tag{2}$$

where L_{fc} and C_f are filter inductance and capacitance, $i_c = i_{c,\alpha} + ji_{c,\beta}$ and $v_c = v_{c,\alpha} + jv_{c,\beta}$ represent the inductor current and the capacitor voltage, respectively. $i_o = i_{o,\alpha} + ji_{o,\beta}$ and $u_c = u_{c,\alpha} + ju_{c,\beta}$ represent the output current and the converter terminal voltage, respectively.

Based on the AC-side dynamics given in (1),(2), the continuous-time state-space model of the system is defined in (3), considering i_c, v_c as state variables and u_c, i_o as input variables.

$$\begin{bmatrix} \frac{di_c}{dt} \\ \frac{dv_c}{dt} \end{bmatrix} = \underbrace{\begin{bmatrix} \frac{-R_{fc}}{L_{fc}} & \frac{-1}{L_{fc}} \\ \frac{-1}{C_f} & 0 \end{bmatrix}}_{A} \begin{bmatrix} i_c \\ v_c \end{bmatrix} + \underbrace{\begin{bmatrix} \frac{-1}{L_{fc}} & 0 \\ 0 & \frac{-1}{C_f} \end{bmatrix}}_{B} \begin{bmatrix} u_c \\ i_o \end{bmatrix} \tag{3}$$

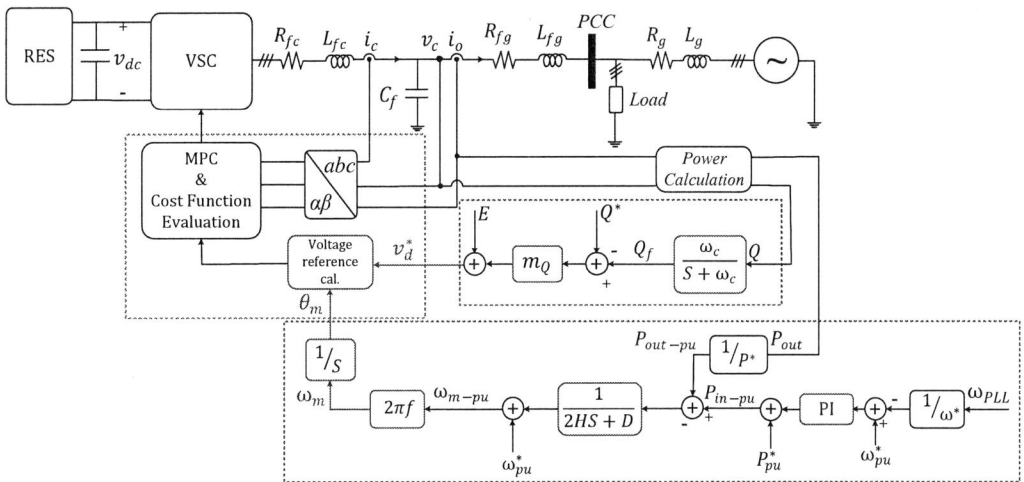

Fig. 1: The general model and the proposed control.

Description of the proposed control

The proposed control structure of the grid-forming converter includes two parts, the inner control part, and the outer control part. In the inner control part, a model predictive-based FRT control strategy is proposed to obtain desired performance in both normal and fault condition. In the outer control part, a VSG scheme is used to emulate the inertia and oscillation damping characteristics of SGs in the control structure of the converter.

Model predictive control

For the inner control part, a model predictive control is proposed instead of the conventional PI-based cascaded voltage and current control loop, resulting in a simpler control structure. In the conventional cascaded voltage and current control loop, at least ten parameters should be tuned, resulting in a complex control design. However, MPC can provide a simpler control structure with faster dynamic response. In principle MPC determines the performance of the system for the equivalent voltage vectors of the voltage source converter (VSC) in each of the eight possible switching states of the 2-level VSC. In fact, by employing a cost function, which follows some objectives, the cost function for each of these eight cases are estimated, and finally the optimal performance and switching state are predicted. These switching states are immediately used for the switching of the VSC without any modulator which is employed by PWM-based control. All the possible switching states and the equivalent voltage vectors of the 2-level voltage source converter are given in the Table 1.

Table I: Switching States and the Equivalent Voltage Vectors of the 2-level VSC

Index n	Switching state $\{S_{abc}\}$	Voltage Vectors
0	$\{000\}$	0
1	$\{100\}$	$\frac{2V_{dc}}{3}$
2	$\{110\}$	$\frac{V_{dc}}{3} + j\frac{\sqrt{3}V_{dc}}{3}$
3	$\{010\}$	$\frac{-V_{dc}}{3} + j\frac{\sqrt{3}V_{dc}}{3}$
4	$\{011\}$	$\frac{-2V_{dc}}{3}$
5	$\{001\}$	$\frac{V_{dc}}{3} - j\frac{\sqrt{3}V_{dc}}{3}$
6	$\{101\}$	$\frac{V_{dc}}{3} - j\frac{\sqrt{3}V_{dc}}{3}$
7	$\{111\}$	0

As model predictive control algorithm needs discrete model, the discrete-time equivalent of the model given in (3) considering a sampling period of T_s is represented as:

$$\begin{bmatrix} i_{c,k+1} \\ v_{c,k+1} \end{bmatrix} = \underbrace{\begin{bmatrix} a_{11} & a_{12} \\ a_{21} & a_{22} \end{bmatrix}}_{A_d} \begin{bmatrix} i_{c,k} \\ v_{c,k} \end{bmatrix} + \underbrace{\begin{bmatrix} b_{11} & b_{12} \\ b_{21} & b_{22} \end{bmatrix}}_{B_d} \begin{bmatrix} u_{c,k} \\ i_{o,k} \end{bmatrix} \tag{4}$$

where subscript k and $k+1$ define the current sampling instant and the next sampling instant in discrete mode, respectively and A_d and B_d matrices are discrete equivalent of matrices A and B, defined as:

$$A_d = e^{AT_s}, B_d = \int_0^{T_s} e^{A\tau} \cdot B \cdot d\tau \tag{5}$$

In order to make sure that all the state variables are immediately affected by the control input action of u_c, an algorithm with prediction horizon of two is employed. Therefore, sampling instant of $k+2$ is predicted using the following equation:

$$\begin{bmatrix} i_{c,k+2} \\ v_{c,k+2} \end{bmatrix} = \underbrace{\begin{bmatrix} a_{11} & a_{12} \\ a_{21} & a_{22} \end{bmatrix}}_{A_d} \begin{bmatrix} i_{c,k+1} \\ v_{c,k+1} \end{bmatrix} + \underbrace{\begin{bmatrix} b_{11} & b_{12} \\ b_{21} & b_{22} \end{bmatrix}}_{B_d} \begin{bmatrix} u_{c,k+1} \\ i_{o,k+1} \end{bmatrix} \tag{6}$$

Model predictive control algorithm employs a prediction process to generate switching signals of the converter. In this prediction process, a cost function is defined containing three control objectives, tracking reference voltage, tracking reference current, and limiting current as FRT ability. The cost function is defined in (7) and all the eight switching states are employed to predict the minimum value of the cost function, resulting in extracting the optimum switching state of the converter.

$$CF = \|v_{c,ref} - v_{c,k+2}\|^2 + \lambda \|i_{c,ref} - i_{c,k+2}\|^2 \tag{7}$$

where $v_{c,k+2}$ and $i_{c,k+2}$ are the predicted value of capacitor voltage and inductor current at time instant $k+2$, λ is the weighting factor of the cost function, and $v_{c,ref}$ and $i_{c,ref}$ are the reference of the capacitor voltage and inductor current. The voltage reference, $v_{c,ref}$, comes from the $Q-V$ droop control that is briefly described in the following subsection. The current reference, $i_{c,ref}$, is extracted from voltage reference provided by outer control part as follow:

$$i_{c,ref} = i_{o,k} + jC_f \omega_{m,k} v_{c,ref} \tag{8}$$

The proposed control includes a constraint for the time when the converter current exceeds its maximum allowable current, i.e., $I_{max} = 1.2 pu$. When the converter current exceeds $1.2 pu$, a new reference current, as expressed in (9), substitutes the previous reference current. This new reference current along with employing a new weighting factor for fault situation guarantee the FRT ability of the converter while THD of the voltage and current during fault are quite low which ensures high power quality.

$$i_{c,ref} = \begin{cases} I_{max} \frac{i_{c,ref}}{|i_{c,ref}|}, & \text{if } |i_{c,ref}| > I_{Max} \\ i_{c,ref} & \text{otherwise.} \end{cases} \tag{9}$$

Outer control loop

In the outer control part, as a VSG, the important features of SGs, i.e., inertia and oscillation daming features, are virtually emulated in the control structure of the VSC. To do so, the swing equation, expressed in (10), is implemented in the outer control part.

$$2H \frac{d(\omega_m - \omega_0)}{dt} = P_{in-pu} - P_{out-pu} - D(\omega_m - \omega_{PLL}), \frac{d\theta_m}{dt} = \omega_m \tag{10}$$

where, P_{in-pu} and P_{out-pu} are the input and output power of the converter in per unit and D, H represent the damping factor and inertia constant, respectively.

As shown in Fig. 1, the outer control part also includes $P - f$ droop control, expressed in (11), to provide frequency regulation. This outer control part defines the reference phase angle and the angular frequency for the inner control part. The $Q - V$ droop control, expressed in (12), is also employed to provide the reference voltage for the inner control part.

$$P^* = P_0 + \frac{1}{m_p}(\omega_0 - \omega_{PLL}) \tag{11}$$

$$v^* = E + m_Q(Q^* - Q_f) \tag{12}$$

where, m_p and m_Q represent the active and reactive power droop coefficients, Q^* and Q_f are the reference reactive power and the output reactive power of the converter, respectively.

Simulation results

To verify the performance of the proposed model predictive-based FRT control strategy simulation results are presented and discussed in this section. Parameters of the model and the proposed control technique, depicted in Fig. 1, are listed in Table II.

Table II: Parameters of the model and the proposed control method

Parameters	Values	Parameters	Values
Nominal voltage (v)	230 V	Damping factor (D)	240
Nominal frequency (f)	50 Hz	Inertia costant (H)	6 s
DC-Link voltage (V_{DC})	750 V	$P - f$ droop coefficient (m_p)	0.04
Reference power (P^*, Q^*)	12 kW	$Q - V$ droop coefficient (m_Q)	0.002
LCL filter (L_{fc}, C_f, L_{fg})	3.3 mH, 8.8 μF, 3 mH	Sampling time (T_s)	10 μs
Grid Impedance (R_g, L_g)	0.22 Ω, 7 mH	weighting factor (λ)	3/40

The grid-forming VSG-based converter is connected to the grid and is intended to supply a reference active power of 12 kW and reactive power of 0. At first, it operates in normal condition while after 1 s a three-phase fault with the duration of 0.15 s occurs at the line connecting the converter to the grid. Fig. 2 illustrates the capacitor voltage and the converter current during the normal and fault-mode conditions, comparing the proposed FS-MPC-based FRT with the control presented in [12], as well as the conventional PI-based cascaded controller equipped with anti-windup. As shown in Fig. 2 (e) and (f), the proposed controller in this paper provides the best performance during normal and fault conditions. It limits the converter current during fault condition while ensuring power quality with a low THD of 1.42% and 1.16% for voltage and current, respectively. Conventional PI-based cascaded controller has a more complex control structure in terms of parameter-tuning efforts and it also requires anti-windup to ensure the normal operation after fault clearance. Moreover, simulation results, shown in Fig. 2 (a) and (b), illustrate that the current limiting ability of this control technique is not reliable enough, as converter current during fault reaches $1.6 pu$ instead of limiting to $1.2 pu$. Fig. 2 (c) and (d) illustrate the performance of the FS-MPC-based control proposed in [12]. Although this control technique provides a strict current limiter during fault mode, power quality is very low with THD of 15.13% and 6.55% for voltage and current, respectively. THD of voltage and current waveform in all the three cases are shown in Fig. 3, comparing the power quality of the proposed controller with the other cases.

Conclusion

In this paper, an FS-MPC control is developed to improve the FRT ability of the grid-forming VSG-based converters. This control technique provides overcurrent protection for the converter while ensuring the power quality during normal operation and fault conditions. The proposed control technique is compared

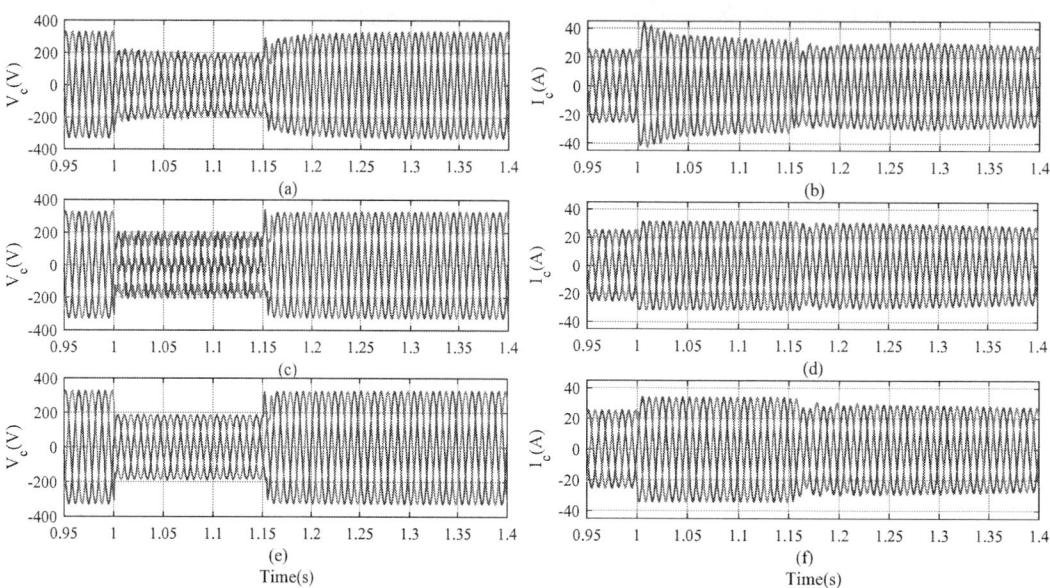

Fig. 2: Capacitor voltage and inductor current in (a),(b) the conventional PI-based cascaded controller equipped with anti-windup, (c),(d) the FS-MPC presented in [12], and (e),(f) the proposed FS-MPC-based FRT.

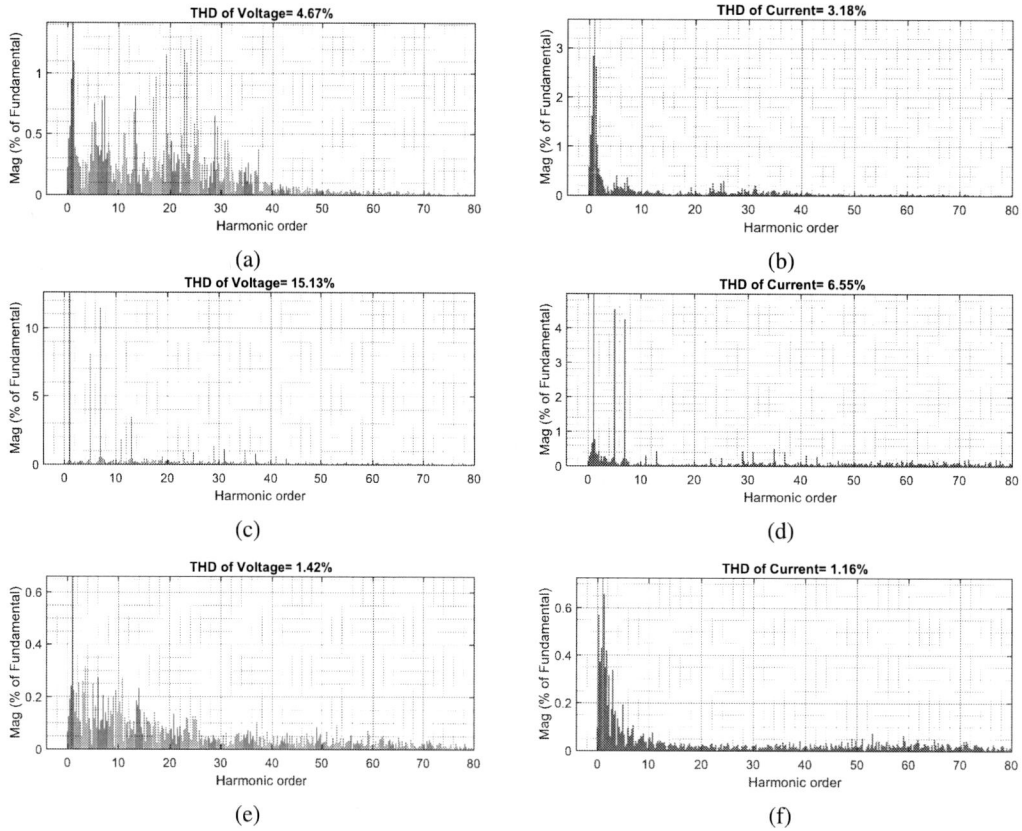

Fig. 3: THD of voltage and current waveforms in (a),(b) the conventional PI-based cascaded controller equipped with anti-windup, (c),(d) the FS-MPC presented in [12], and (e),(f) the proposed FS-MPC-based FRT.

with a PI-based control strategy as well as an MPC- based control technique with a defined current limiter as part of the cost function. Compared to the conventional PI-based cascaded controllers, it has a simple control structure without a modulator, anti-windup, and try and error for parameter-tuning. Besides, simulation results clearly confirmed that the proposed control technique is able to immediately avoid overcurrent when a fault occurs, while it is impossible for PI-based controller to avoid overcurrent so fast. In comparison with the MPC-based control with the defined current limiter, the proposed control technique shows a significant improvement in power quality.

References

[1] Alipoor J., Miura Y., and Ise T.,: Power system stabilization using virtual synchronous generator with alternating moment of inertia, IEEE J. Emerg. Sel. Topics Power Electron., vol. 3, no. 2, pp. 451-458, 2015.

[2] D'Arco S., Suul J. A., and Fosso O.B.: Small-Signal Modelling and Parametric Sensitivity of a Virtual Synchronous Machine, Power Systems Computation Conference, Wroclaw, Poland, pp. 1-9, Feb. 2015.

[3] Pouresmaeil M., Sangrody R., Taheri S., Pouresmaeil E.: An adaptive parameter-based control technique of virtual synchronous generator for smooth transient between islanded and grid-connected mode of operation, IEEE Access, Vol. 9, pp. 137322 - 137337, Oct. 2021.

[4] Tielens P., De Rijcke S., Srivastava K., Reza M., Marinopoulos A., and Driesen J.: Frequency Support by Wind Power Plants in Isolated Grids with Varying Generation Mix, 2012 IEEE Power & Energy Society General Meeting, San Diego, pp.1-8, July 2012.

[5] Rocabert J., Luna A., Blaabjerg F., and Rodr´ıguez P.: Control of Power Converters in AC Microgrids, IEEE Transactions on Power Electronics, vol. 27, no. 11, pp. 4734 - 4749, 2012.

[6] Huang L. , Xin H., et al.: Transient stability analysis and control design of droop-controlled voltage source converters considering current limitation, IEEE Trans. Smart Grid, vol. 10, no. 1, pp. 578–591, Jan. 2019.

[7] Gkountaras A., Dieckerhoff S., and Sezi T.: Evaluation of current limiting methods for grid forming inverters in medium voltage microgrids, IEEE ECCE, pp. 1223–1230, Sep. 2015.

[8] Zarei S. F., Mokhtari H., et al.: Reinforcing fault ride through capability of grid forming voltage source converters using an enhanced voltage control scheme, IEEE Trans. Power Del., Vol. 35 no 5, pp. 1827-1842, Oct. 2019.

[9] Pouresmaeil M., Saeedian M., Sepehr A., Sangrody R., Pouresmaeil E.: Fault-Ride-Through capability of VSG-Based grid-Forming converters, EPE 2021-ECCE Europe, Ghent, Belgium, pp. 1-7, Oct 2021.

[10] Young H.A., Marin V.A, Pesce C., Rodriguez J.: Simple finite-control-set model predictive control of grid-Forming inverters with LCL filters, IEEE Access Vol 8 , pp. 81246 - 81256, Apr. 2020.

[11] Jongudomkarn J., Liu J., Ise T.: Virtual synchronous generator control with reliable fault rde-through ability: A solution based on finite-set model predictive control, JESTPE, Vol. 8, no 4, pp. 3811- 3824, Dec. 2020.

[12] Zheng C., Dragicevic T., Blaabjerg F.: Model predictive control based virtual inertia emulator for an islanded AC microgrid, IEEE Trans. Ind. Electron. Vol. 68 no 8, pp. 7167-7177, Aug. 2021.

Grounding Points in
HV/MV Hybrid Transformer Auxiliary Converters

Adrian Wiemer and Jürgen Biela
Laboratory for High Power Electronic Systems, ETH Zurich
www.hpe.ee.ethz.ch

Keywords

<<Hybrid Transformer>>, <<Smart Grid>>, <<Distributed Generation>>, <<Power Control>>, <<Voltage Control>>, <<Imbalances Control>>

Abstract

The choice of grounding points in hybrid transformers is important to limit the overcurrents in case of a single line to ground fault. In this paper different grounding points for hybrid transformers are compared based on a controller with active damping for the filter resonances. Based on the comparison the best grounding point for limiting the single line to ground fault current for hybrid transformer auxiliary converters and the MV grid is identified.

1 Introduction

The increasing share of renewable energy generation requires a more flexible and smarter grid [1], [2], for controlling voltage and power fluctuations [3], [4]. To address this, various concepts as for example FACTS, grid expansions, and distribution transformers with controllable tap changer have been investigated [5], [6]. Another interesting solution for controllable transformers are hybrid transformers (HTs). HTs combine a line-frequency power transformer (LFT), which typically transfers the major power share, with an auxiliary AC-AC power electronic converter [7] for controlling the voltage and the power flow. The AC-AC auxiliary converter is typically connected to an auxiliary winding of the LFT as shown in the three-phase representation of a HT in Fig.1. The auxiliary converter system power rating is typically designed only for a relatively small fraction $\approx$ (10%) of the LFT power rating [7], [8]. In general, such power electronic converters as the HT auxiliary converter are relatively sensitive to overcurrents/-voltages, which could occur in the medium voltage (MV) grid in case of a fault, so that a protection system is required.

In the past, various concepts and topologies for HT [8], [9] as well as the protection against grid faults at the low voltage (LV) to medium voltage (MV) gird interface [7] have been investigated. A summary and

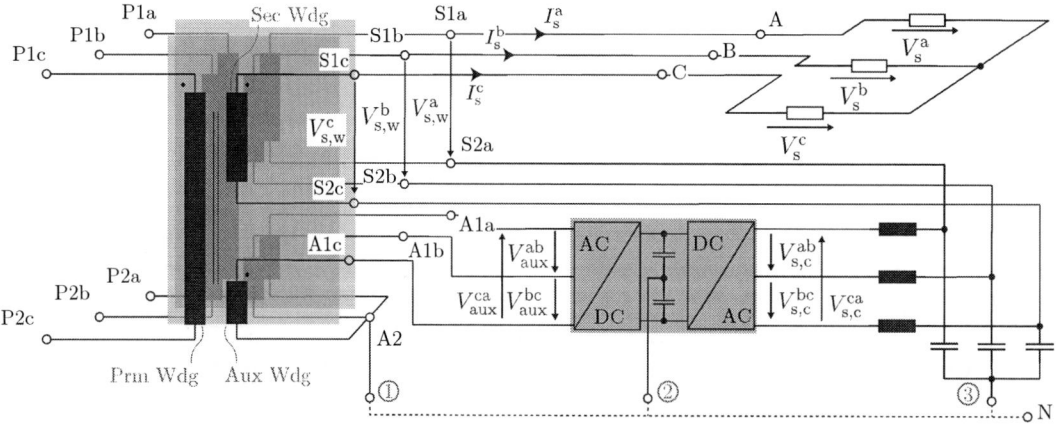

Figure 1: Three-phase HT with LFT, auxiliary converter, and grounding points ①-③, where a load is connected to the secondary winding.

discussion of the most interesting HT concepts can be found in [7]-[10]. Overvoltages and Overcurrents on the HT system level are invesitgated in [11], where it is shown, that the grid grounding method has a significant impact on the fault current through the converter in case of a single line to ground fault. The single line to ground fault is the most common fault in distribution grids and the fault current increases with the grid line length [11]. An increasing grid line length becomes problematic during a single line to ground fault in MV grids with isolated star point. In order to limit the single line to ground fault current for larger grid distances in the MV grid, the neutral point is connected to the ground via an inductance (denoted as resonant grounded grid [12]). The grounding inductance is designed to reduce the capacitive fault current to an acceptable level in case of a single line to ground fault [13], [14].

In general, there are different options for grounding points in HT. Three possible grounding points in HT are indicated by the red numbers ①-③ in Fig. 1. The impact of grounding points on the HT auxiliary converter design including for a resonant grounded grid has not been investigated yet, and is investigated in this paper. In case ①, a grounded star point of the star connected power transformer auxiliary winding is assumed. In case ②, the split DC-link of the HT auxiliary converter is grounded. In case ③, the star point of an assumed LC-filter is grounded. In the following grounding points ①-③ are investigated in terms of overcurrents flowing through the HT auxiliary converter and the reduction of the fault current by a grounding inductance is analyzed for each grounding point. Note that an investigation of the considered grounding points in terms of overvoltages is out of scope of this paper since the considered grounding points are primarily used to limit overcurrents.

As HT AC-AC auxiliary converter a three phase unidirectional AC/AC back-to-back converter is chosen in this paper. The unidirectional HT auxiliary converter has a low number of active switches, but some limiting constrains on the voltage and power controllability as discussed in [15].

In cases ② and ③, the single line to ground fault current flows either directly through the DC-Link capacitors or the LC-filter capacitors of the assumed LC-filter. In order to limit oscillations caused by the DC-Link or the LC-filter capacitors, damping methods are investigated.

In a first step, a simulation model of an unidirectional HT auxiliary converter is presented in **section 2**. Then, the possible grounding points are investigated in detail in **section 3**, and the preferred grounding point is selected. Results for the selected grounding point and a discussion are provided in **section 4**.

2 Simulation Model of the Three-Phase HT Auxiliary Converter

Before we compare grounding points ①-③ of the HT, we derive the simulation model and its parameters for the HT auxiliary converter. The investigation is based on the example of a unidirectional voltage source converter (UNI-VSC) as HT auxiliary converter shown in Fig. 2, where the line to ground capacitances C_{lg} with resistance R_c and the grounding inductance L_{co} with resistance R_{co} are neglected for now in order to develop the simulation model. The simulation model for the UNI-VSC is based on the power flow and voltage range considerations presented in [15] and on the exemplary voltage rating of the transformer primary/secondary winding as well as the number of primary N_{prm}, and secondary N_{sec} turns given in [11] and table I. The UNI-VSC allows a unidirectional active power flow, full reactive power flow and control of the voltage within the constrains of the unidirectional active power flow region as shown in [15]. Further advantages of the UNI-VSC are the low number of active switches due to the passive rectifier. With a passive rectifier also a simpler closed loop control strategy is possible [16]. A 2-level topology is assumed for the UNI-VSC as well as IGBTs are chosen as switches. With IGBTs the switching frequency f_{sw} is limited to a range of a few kHz, in order to keep the switching losses low. The current rating of the semiconductors is determined with the nominal current I_s of the MV grid, which is calculated based on the nominal power P_n of the power transformer and the nominal secondary winding voltage $V_{s,w}$.

The passive rectifier is connected to the power transformer auxiliary winding terminals A1a-A1c and the inverter is in series to the transformer secondary winding terminals S2a-S2c (see Fig. 1 & 2). The auxiliary and the secondary windings are modeled by the equivalent voltage sources $v_{th,aux/sec}$ and the inductances $L_{eq}^{aux/sec}$. To limit the harmonic injection of the inverter a LC-filter with filter elements L_f and C_f is included between the inverter and the secondary winding terminals S2a-S2c. The secondary winding terminals S1a-S1c are connected to the load R_L.

In a first step only, the inverter side of the UNI-VSC is considered and the DC-link capacitors in Fig. 2 are replaced by a constant voltage source $V_{dc,min}$. Based on the considered voltage source inverter (VSI) a controller and a plant model are derived in the following section. The controller model is used for controlling the filter capacitance voltages $v_{C,f}$ of the UNI VSC shown in Fig. 2 and to investigate the advantages of active damping to limit the fault current in case of a single line to ground fault.

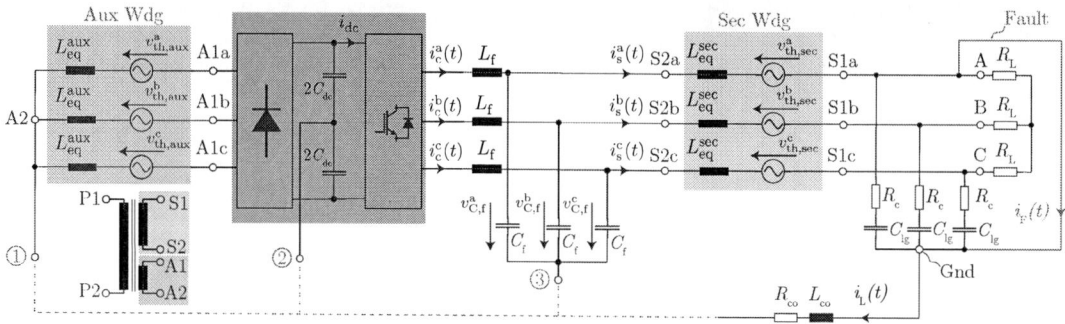

Figure 2: UNI-VSC with MV grid connected to load R_L and the considered grounding points ①-③, where a single line to ground fault is assumed in line "a".

2.1 Plant Model

The plant model for the VSI is derived to design the controller including active damping to limit the single line to ground fault current, by averaging the state space models of the VSI switching states. As PWM strategy space vector modulation (SVM) is chosen. Since the averaged state space model is transferred into dq-domain only a single sector of the SVM hexagon needs to be considered [17]. Within the considered sector the state space matrices are averaged to obtain the averaged state space model based on three phase quantities. Based on the obtained averaged state space model and states $i_c^a,...,i_c^c$, $i_s^a,...,i_s^c$, and $v_{C,f}^a,...,v_{C,f}^c$ (Fig. 2) the averaged state space model in dq-domain is derived [18]. Transferring the averaged state space model from the dq-domain into the Laplace domain results in the averaged state space model in (1).

$$
\underbrace{\begin{bmatrix}
s & -\omega & 0 & 0 & \frac{1}{L_f} & 0 \\
\omega & s & 0 & 0 & 0 & \frac{1}{L_f} \\
0 & 0 & s+\frac{R_L}{L_{eq}^{sec}} & -\omega & -\frac{1}{L_{eq}^{sec}} & 0 \\
0 & 0 & \omega & s+\frac{R_L}{L_{eq}^{sec}} & 0 & -\frac{1}{L_{eq}^{sec}} \\
-\frac{1}{C_f} & 0 & \frac{1}{C_f} & 0 & s & -\omega \\
0 & -\frac{1}{C_f} & 0 & \frac{1}{C_f} & \omega & s
\end{bmatrix}}_{A_{dq}}
\begin{bmatrix}
\underline{I}_c^d \\
\underline{I}_c^q \\
\underline{I}_s^d \\
\underline{I}_s^q \\
\underline{V}_{C,f}^d \\
\underline{V}_{C,f}^q
\end{bmatrix}
=
\begin{bmatrix}
\frac{\underline{V}_{ref}^d(s)}{L_f} \\
\frac{\underline{V}_{ref}^q(s)}{L_f} \\
\frac{\underline{V}_{th,sec}^d(s)}{L_{eq}^{sec}} \\
\frac{\underline{V}_{th,sec}^q(s)}{L_{eq}^{sec}} \\
0 \\
0
\end{bmatrix}
\tag{1}
$$

By inverting matrix A_{dq} the plant model is obtained to calculate the filter capacitance voltages $\underline{V}_{C,f}^d$ and $\underline{V}_{C,f}^q$ based on the secondary winding voltages $\underline{V}_{th,sec}^d$ and $\underline{V}_{th,sec}^q$ and the reference voltages $\underline{V}_{ref}^d$ and $\underline{V}_{ref}^q$. The plant model for the VSI is shown in Fig. 3a). The plant models to calculate the remaining states in (1) are derived in a similar way. The plant models are utilized to determine the controller variables and the feed forward to damp the filter resonance with active damping.

2.2 Power Transformer Model

For numerical simulations of the UNI-VSC, including the dynamic behavior of the control and plant model, the power transformer needs to be included in the simulation models. The power transformer's auxiliary and secondary windings are modeled by eq. circuits consisting of voltage sources $V_{th,i}$ and inductors L_{eq}^i with $i \in \{sec, aux\}$ in order to increase the simulation speed. To determine the voltage sources $V_{th,i}$ and the inductors L_{eq}^i, the turn numbers of primary and secondary winding N_{prm} and N_{sec} are assumed as described in [11]. The T-equivalent circuit from the primary to secondary winding to derive the eq. voltage source $V_{th,sec}$ and inductor L_{eq}^{sec} is shown in Fig. 3b). In the simulation model the transformer is then represented at the transformer secondary winding terminals S1 and S2 by the eq. voltage source $V_{th,sec}$ and inductor L_{eq}^{sec} shown in Fig. 3c). The calculation of the parameters for the T-equivalent circuits is only explained for the primary to secondary winding T-equivalent circuit since the T-equivalent circuit for the primary to auxiliary winding is calculated in the same way. The T-equivalent circuit is derived with the mutual inductances $L_{n_p n_s}$ of a single primary turn to a single secondary turn, the primary turns self inductance $L_{n_p n_p}$, and secondary turns self inductance $L_{n_s n_s}$ as well as the number of primary turns N_{prm}, and the secondary turns N_{sec} with (2)-(4). Note that inductances $L_{n_p n_s}$, $L_{n_p n_p}$, and $L_{n_s n_s}$ are calculated with FEM simulations as described in [11].

$$L_p = N_{prm}[L_{n_p n_p} + (N_{prm} - 1)L_{n_p n_p + 1}] - L_m^{ps} \tag{2}$$

$$L_m^{ps} = L_{n_p n_s} N_{prm} N_{sec} \tag{3}$$

$$L_s = N_{sec}[L_{n_s n_s} + (N_{sec} - 1)L_{n_s n_s + 1}] - L_m^{ps} \tag{4}$$

With inductances L_p, L_m^{ps}, and L_s, inductance L_{eq}^{sec} is calculated as given in (5). The value of the eq. voltage $V_{th,sec}$ is based on the assumption that the HT auxiliary converter operates at nominal power. Further it is assumed that the power at terminals S1a-S1c is limited to the nominal power P_n of the power transformer. Therefore, the eq. voltage $V_{th,sec}$ is given by $V_{th,sec} = V_{s,w} = V_s - V_{s,c}$. With voltage $V_{th,sec} = V_{s,w}$ the voltage $V_{th,aux}$ on the transformer's auxiliary winding side is calculated from the primary to secondary T-equivalent circuit and the primary to auxiliary winding T-equivalent circuit by (6). Inductances L_m^{pa} and L_a are calculated by replacing N_{sec} with N_{aux}, $L_{n_p n_s}$ with $L_{n_p n_a}$, and $L_{n_s n_s}$ with $L_{n_a n_a}$ in (2)-(4). Inductance L_{eq}^{aux} is then determined by replacing L_m^{ps} and L_s in (5) with inductances L_m^{pa} and L_a for the primary to auxiliary winding. Note that the number of auxiliary turns N_{aux} is determined in **section 2.5**.

$$L_{eq}^{sec} = L_s + \frac{L_p L_m^{ps}}{L_m^{ps} + L_p} \quad (5) \qquad V_{th,aux} = \frac{L_m^{pa}}{L_m^{pa} + L_p} \frac{L_m^{ps}}{L_m^{ps} + L_s} V_{s,w} \quad (6)$$

2.3 LC filter Model

With the *LC*-filter at the output of the VSI the harmonic content injected by the converter is limited. Due to the inductance L_{eq}^{sec} of the secondary winding, which is in series to the VSI/*LC*-filter, a simple *LC*-filter is sufficient for limiting the harmonics as shown in Fig. 2. The model of the *LC*-filter is based on the *LCL*-filter design scheme given in [19] with base impedance $Z_b = \frac{(V_{c,s})^2}{P_n}$ and base capacitance $C_b = \frac{1}{\omega_g Z_b}$. The filter capacitance C_f is calculated from the base capacitance C_b with $C_f = 0.05 C_b$. In order to determine the filter inductance L_f the maximum ripple current I_{rip} is defined with the nominal grid current I_s by $I_{rip} = 0.1 I_s$. The filter inductance is then calculated by $L_f = \frac{V_{dc,min}}{6 f_{sw} I_{rip}}$.

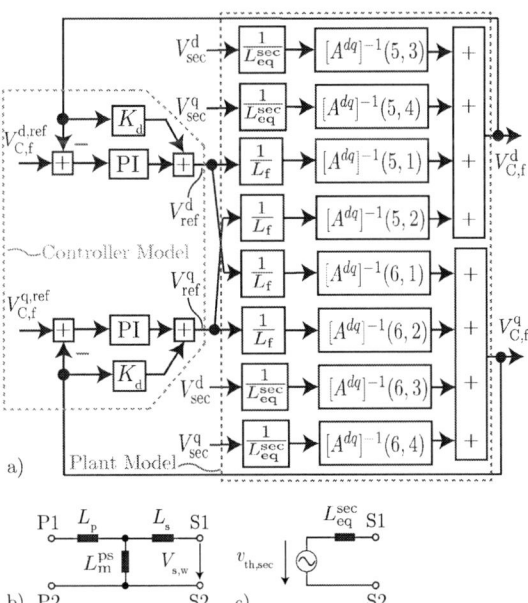

2.4 Controller Model

The controller model for the VSI in *dq*-domain is based on a PI controller with a feed forward of voltages $\underline{V}_{C,f}^d$ and $\underline{V}_{C,f}^q$. The feed forward of voltages $\underline{V}_{C,f}^d$ and $\underline{V}_{C,f}^q$ with parameter K_d actively damps the resonance in the *LC*-filter [20]. The controller model is shown in Fig. 3a). The PI controller parameters K_p and K_I and the feed forward parameter K_d are determined with the open loop control transfer function consisting of the *d*-axis PI controller with feed forward K_d and the transfer function given by $[A^{dq}]^{-1}(5,1)$. Note that the active damping in the controller model shown in Fig. 3a) is important for limiting the fault current with grounding point ③ as explained in **section 3**.

Figure 3: a) Controller and plant model for the VSI. b) T-equivalent circuit of the transformer primary and secondary winding. c) Eq. circuit representing the transformer at the secondary winding terminals S1 and S2.

2.5 DC-Link Model

To implement the simulation model including the passive rectifier and the auxiliary winding, the DC-link capacitance C_{dc} is determined based on the required DC-link voltage $V_{dc,min}$ of the VSI as discussed in the following. First the DC-link voltage $V_{\text{dc,min}}$ is determined with the converter current I_{c} at the nominal current I_{s} of the considered MV grid. There, the nominal current I_{s} is simply calculated with the rated power of the transformer P_{n} and the nominal secondary winding voltage $V_{\text{s,w}}$. The converter voltage $V_{\text{s,c}}$ is then determined so that, the chosen voltage range of $V_{\text{C,f}} \approx 13\%$ of voltage $V_{\text{s,w}}$ is achieved taking also the voltage drop across the LC-filter into account. The minimum DC-Link voltage $V_{\text{dc,min}}$ is then determined with voltage $\hat{v}_{\text{c,s}}$ including a small margin to avoid over modulation in practical cases.

With the determined DC-link voltage $V_{\text{dc,min}}$, the current i_{dc} through the DC-voltage source $V_{\text{dc,min}}$ of the VSI in Fig. 4a) is analyzed in a next step, in order to simplify determining the DC-link capacitance C_{dc}. For the design of C_{dc} the inverter is modeled by resistance R_{dc} connected to C_{dc} via switch S_{dc} as shown in Fig. 4b). The resistance value R_{dc} is calculated as follows. For determining R_{dc} the averaged DC-current $I_{\text{dc}}^{\text{avg}}$ of the VSI is required. This current is shown qualitatively in Fig. 4a). Resistor R_{dc} models the discharging of the DC-link during the time interval $d_0 T_{\text{s}} \le t \le T_{\text{s}}$ of the non zero space vectors. Therefore, the averaged DC-current $I_{\text{dc,act}}^{\text{avg}}$ in Fig. 4a) is determined in (7) with the duty cycles d_1 and d_2 of the non zero space vectors. The resistance R_{dc} is then obtained as $R_{\text{dc}} = \frac{V_{\text{dc,min}}}{I_{\text{dc,act}}^{\text{avg}}}$.

$$I_{\text{dc,act}}^{\text{avg}} = \frac{I_{\text{dc}}^{\text{avg}}}{d_1 + d_2} \quad (7) \qquad i_{\text{L}_{\text{dc}}} = \frac{V_{\text{rect,avg}} d}{R_{\text{dc}}} \quad (8) \qquad v_{\text{rip}} = \frac{1}{C_{\text{dc}}} \int_0^{(1-d)T_{\text{s}}} i_{\text{L}_{\text{dc}}} \, dt \quad (9)$$

With the value for R_{dc} the DC-link capacitor C_{dc} is determined with the equivalent circuit shown in Fig. 4b). There, the resistor R_{dc} is connected in parallel to the DC-link capacitor by switch S_{dc}, which is turned on for the duty cycle $d = d_1 + d_2$ shown in Fig. 4a). The passive rectifier of the UNI-VSC is represented by the voltage source $V_{\text{rect,avg}}$ and inductor $L_{\text{eq}}^{\text{dc}}$. Inductor $L_{\text{eq}}^{\text{dc}} = 2L_{\text{eq}}^{\text{aux}}$ is determined with inductors $L_{\text{eq}}^{\text{aux}}$ in Fig. 2. With voltage $V_{\text{rect,avg}}$, duty cycle d, and resistor R_{dc} the DC-link capacitance C_{dc} is calculated with current $i_{\text{L}_{\text{dc}}}$ given in (8). For the time interval $0 \le t \le d_0 T_{\text{s}}$ in Fig. 4a) the current through the inductor $L_{\text{eq}}^{\text{dc}}$ and the capacitance C_{dc} are identical. Therefore, the voltage ripple v_{rip} in voltage v_{dc} is determined by integrating the current $i_{\text{L}_{\text{dc}}}$ as shown in (9). Solving the integral and rearranging yields the DC-link capacitance C_{dc} in (10), which is similar to the expression for the input capacitance of a single phase buck regulator in [21]. In order to solve (10) for C_{dc} a maximum voltage ripple of v_{rip} and voltage $V_{\text{rect,avg}} = \frac{2\sqrt{3}}{\pi}(V_{\text{dc,min}} + v_{\text{rip}})$ are assumed. Note that the voltage v_{rip} only describes the ripple caused by the switching of the VSI. In the derivation of C_{dc}, the smoothing by C_{dc} of the passive rectifier voltage v_{rect} in Fig. 4c) is neglected for simplicity. Therefore, there is a margin between the minimum voltage shown in Fig. 4c) and the actual minimum voltage. Since the focus is on the selection of grounding points a more detailed modeling of the DC-link is out of scope of this paper.

$$C_{\text{dc}} = \frac{V_{\text{rect,avg}} d(1-d)T_{\text{s}}}{v_{\text{rip}} R_{\text{dc}}} \quad (10) \qquad V_{\text{aux}} = \frac{V_{\text{rect,avg}} \pi}{3\sqrt{2}} \quad (11) \qquad N_{\text{aux}} = \frac{N_{\text{prm}} V_{\text{aux}}}{V_{\text{p}}} \quad (12)$$

With the voltage $V_{\text{rect,avg}} = \frac{2\sqrt{3}}{\pi}(V_{\text{dc,min}} + v_{\text{rip}})$, the required RMS voltage of the voltage V_{aux} for the auxiliary winding side of the UNI-VSC is calculated as given in (11). Further, the number of turns N_{aux} for the auxiliary winding is given by (12).

With the auxiliary converter voltage $V_{\text{s,c}}$, the DC-link voltage $V_{\text{dc,min}}$, and the number of auxiliary winding turns N_{aux}, the UNI-VSC simulation model is defined for investigating the grounding points for the cases ①-③. The results are presented in **section 4**.

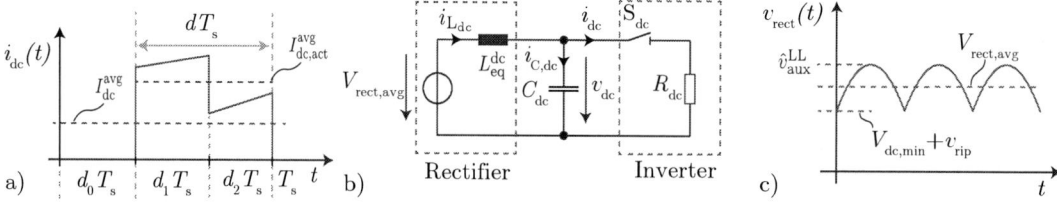

Figure 4: a) Wave form of the DC-link current i_{dc} in the VSI. b) Eq. circuit to design the DC-link capacitance of the UNI-VSC. c) Rectified voltage of the passive rectifier in the UNI-VSC.

3 Grounding Points in Unidirectional VSC

The grounding points ①-③ are investigated based on the developed simulation model in **section 2**. In order to evaluate the grounding points ①-③ the line to ground capacitances $C_{l,g}$ of a MV grid line are added to the UNI-VSC model in Fig. 2. The values of the line to ground capacitances $C_{l,g}$ are calculated from the per length capacitance $C'_{l,g}$ of the considered line and the grid line length l.

The single line to ground fault with short circuit current i_F is assumed to occur in phase "a" as shown in red in Fig. 2. The resistances R_c in series to capacitances $C_{l,g}$ are introduced to avoid numeric problems in the simulation due to the short circuit of the line to ground capacitance $C_{l,g}$ in phase "a" in an event of a single line to ground fault. Note that the star point of the load R_L needs to be isolated in order to limit the fault current to an acceptable level for the UNI-VSC in an event of a single line to ground fault. A load with isolated star point is typically given in case of a HV/MV HT connected to MV line, that is connected to a MV/LV distribution transformer with isolated neutral point [14]. To compensate the fault current i_F through the line to ground capacitors C_{lg}, a grounding inductor L_{co} is usually inserted. The grounding inductor L_{co} is designed dependent on the grid line length with $L_{co}(l) = \frac{1}{3\omega_g^2 C_{l,g}(l)}$ [14].

In a next step, the grounding points ①-③ are investigated for the grounding inductor L_{co} shown in Fig. 2.

① In case of grounding point ①, a star connected transformer auxiliary winding is assumed, where the grounding inductor L_{co} is connected to the star point A2 shown in Fig. 2. In case of a single line to ground fault, the converter currents $i_c^a, .., i_c^c$ as well as the rectifier currents become unbalanced. Inductances L_{eq}^{aux}, L_f and L_{eq}^{sec} limit the fault current i_F. In order to decrease the fault current i_F further, resistor R_{co} is determined, so that the fault current reaches a steady state value below the allowed residual current i_{res} for MV grids [14]. In case passive damping of the LC-filter is chosen with damping resistors in series to the LC-filter capacitances C_f, grounding point ① is preferred over grounding point ③. With passive damping and the damping resistors grounded with grounding point ③, a fault current above the allowed residual fault current i_{res} is reached as discussed below. A disadvantage of grounding point ① are the extra measures necessary to avoid an undesired charging of the DC-link capacitors. In case the filter capacitance C_f voltages $v_{C,f}$ are controlled close to zero, the current i_L starts to charge the DC-link capacitors even in normal operation. The path of i_L through the converter is shown in Fig. 5. In case voltage $v_{C,f} \approx 0$, approximately zero power is provided by the VSI and therefore only the passive rectifier conducts the current i_L with a single diode turned on. Due to $v_{C,f} \approx 0$ the VSI is mainly switched between the zero voltage space vectors (111) and (000) current i_L flows through the DC-link capacitors during the time the zero vector (000) is generated as shown in Fig. 5. In case a fault occurs at $v_{C,f} \approx 0$ the converter currents $i_c^a, .., i_c^c$ become unbalanced including a DC-offset. The DC-offset in converter currents $i_c^a, .., i_c^c$ further contributes to charge the DC-link. The charging of the DC-link is avoided with a sufficiently high reference voltage $V_{C,f}^{d,ref}$ so that active power is transferred, which allows to balance the charging of the DC-link by current i_L and the DC-offset in currents $i_c^a, .., i_c^c$. For the considered case study HT with the parameters in tables I-III, reference voltages of approximately $100V \le V_{C,f}^{d,ref}$ and $V_{C,f}^{q,ref} \approx 0$ are required to avoid a charging of the DC-link for grounding point ①.

To avoid an increase of the DC-link voltage v_{dc} for reference voltages $V_{C,f}^{d,ref} < 100V$ and $V_{C,f}^{q,ref} \approx 0$, a resistor can be added in parallel to the DC-link capacitors [22], which leads to additional losses.

② With the second grounding point ② the grounding inductor L_{co} is connected to the split DC-link. In case ②, inductances L_f and L_{eq}^{sec} limit the fault current i_F through the UNI-VSC. Since the fault current i_F flows through the DC-link capacitors and the grounding inductor L_{co} oscillations can occur. To damp these oscillations, the value of resistance R_{co} needs to be sufficiently high. Typically the value of R_{co}, for limiting the current i_F to a minimum residual fault current i_{res} within the permitted value for MV grids is sufficiently high to also damp the considered resonances. In contrast to grounding point ①, the DC-link voltage remains stable during normal operation with voltages $v_{C,f} \approx 0$ because the average of current i_L through the DC-link capacitors is zero. During a fault with voltages $v_{C,f} \approx 0$ the unbalanced currents $i_c^a, .., i_c^c$ lead to an increase of the DC-link voltage v_{dc} due to a DC-offset in $i_c^a, .., i_c^c$. The charging of the DC-link is avoided with a sufficiently high reference voltage $V_{C,f}^{d,ref}$ so that active power is transferred, what allows to balance the charging of the DC-link by the DC-offset in currents $i_c^a, .., i_c^c$. For the case study HT with the parameters in tables I-III, the DC-link voltage remains constant during the fault for reference voltages $5V \le V_{C,f}^{d,ref}$ and $V_{C,f}^{q,ref} \approx 0$. For reference voltages $V_{C,f}^{d,ref} < 5V$ and $V_{C,f}^{q,ref} \approx 0$ the increase of the DC-link voltage can be solved by adding parallel resistors to the DC-link capacitances as discussed in [22], which leads to additional losses.

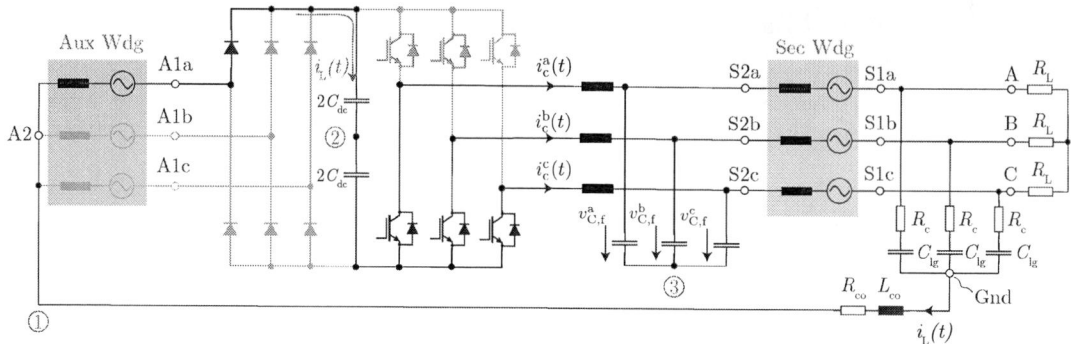

Figure 5: Flow of current i_L during zero space vector $\underline{V}_{(000)}$ with reference voltages $V_{C,f}^{d,ref} = V_{C,f}^{q,ref} = 0$ during normal operation and with grounding point ①

③ The third grounding point ③ is given by connecting the grounding inductor L_{co} to the star connected LC-filter capacitors C_f as shown in Fig. 2. Since the grounding inductor L_{co} is connected to the LC-filter no fault current i_F flows through the converter. However, the fault current i_F flows through the resonant tank consisting of the grounding inductor L_{co} and the filter capacitors C_f, which could lead to oscillations without sufficient damping. In case a passive damping strategy is chosen, the damping resistors in series to the filter capacitances C_f lead to a residual current i_{res} larger than the desired value in MV grids in case of a single line to ground fault with inductor L_{co} and resistor R_{co} connected to grounding point ③. The reason for the high residual current i_F is that the damping resistors lead to a current component that is not compensated by the grounding inductor L_{co}. This is also explained by considering the damping resistors in combination with the load resistors as an unbalanced resistive load, while the line to ground capacitances are compensated by inductor L_{co} during the fault. As a consequence, passive damping leads to unbalanced converter currents $i_c^a,...,i_c^c$ during a fault. Therefore active damping is investigated for grounding point ③, to eliminate the damping resistors of the LC-filter for passive damping. The controllers active damping leads to a residual fault current i_{res} well below the permitted value for MV grids. Furthermore, with active damping the unbalance of the converter currents $i_c^a,...,i_c^c$ is very limited since the value of R_{co} is low in order to achieve a short time for current i_F to reach its steady state value. Another advantage that comes with grounding location ③ is that no DC-link charging occurs due to the very limited unbalance of currents $i_c^a,...,i_c^c$. This allows to use a simple dq-domain control strategy for the normal and fault operation.

For the reasons mentioned above and as will be shown in **section 4**, the advantages of grounding point ③ compared to grounding points ① & ② are significant. Therefore, grounding points ① & ② are not considered in more detail in this paper.

4 Results and Discussion

Based on **section 3** only results for grounding point ③ are presented in the following. The advantages of active damping during a fault with grounding point ③ is shown for two operation points I & II in Fig. 6. Note that operation points I & II are normal non fault operation points for HT. At operation point I a lower voltage of the secondary winding voltage $V_{s,w}$ is assumed, where the HT auxiliary converter compensates, such that the auxiliary converter voltage $V_{s,c}$ and the secondary winding voltage $V_{s,w}$ add up to the nominal voltage $V_{s,N}$. At operation point I the auxiliary converter current I_c is at its nominal value $I_{c,N}$. Therefore, the highest over current is expected during a single line to ground fault at operation point I. At operation point II illustrated at the bottom of Fig. 6 the auxiliary converter voltage is $V_{s,c} = 0$ and the secondary winding voltage is $V_{s,w} = V_{s,N}$, while the auxiliary converter current I_c is close to its

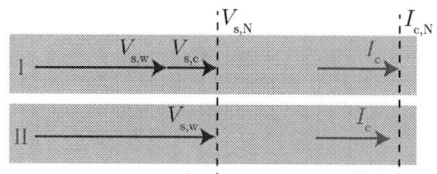

Figure 6: Voltages and currents at operation points I & II of the UNI-VSC for grounding point ③.

nominal value $I_{c,N}$. Operation point II is critical to maintain a stable DC-link voltage v_{dc} for grounding points ① & ② and is therefore investigated during a single line to ground fault with grounding point ③.

Grounding point ③ is investigated for a case study HV/MV HT with basic parameters listed in tables I and II. Since fault currents i_F in cable grids are significantly higher than in overhead line grids [11], only cable grids are investigated. The nominal load resistance R_L is calculated with the nominal current I_s and secondary winding voltage $V_{s,w}$ as $R_L=8.9\Omega$. Further simulation parameters are $L_{eq}^{sec} = 9.1mH$ and $L_{eq}^{aux} = 0.24mH$ for the inductances in the eq. circuits of the secondary and auxiliary winding calculated with (5). The values for the grid frequency f_g, switching frequency f_{sw}, the DC-link capacitor C_{dc}, the grounding inductor L_{co}, the LC-filter as well as the resistors R_{co} and R_c are given in tables I and II.

P_n	V_p	$V_{s,w}$	V_{aux}	I_s	L_{co}	R_{co}	$C'_{l,g}$	R_c	N_{prm}	N_{sec}	N_{aux}
45MW	110kV	20kV	3kV	1.3kA	0.27H	1.35Ω	250nF/km	0.1mΩ	1518	276	41

Table I: General design parameters for the case study HT and the reactance grounded grid.

The converter currents $i_c^a,...,i_c^c$ and fault current i_F for operation points I & II are determined for a cable line length of $l =50$km, where a single line to ground fault occurs after $t = 4$min $= 240$s. Note, that the converter currents $i_c^a,...,i_c^c$ increase linearly with increasing cable line length l. The reason for investigating the cable line length l of 50km is that the converter currents $i_c^a,...,i_c^c$ reach the nominal converter current $\hat{i}_{c,nom}$ at a line length of $l =50$km. According to [14], the steady state fault current i_F should not exceed the residual current i_{res} of 60A in MV grids. In cases I & II resistor R_{co} in table II is designed to insure a steady state fault current i_F below the residual current i_{res} of 60A with a settling time of $t_{set} \approx 1$s in case of a fault. Since the value of resistance R_{co} is dependent on the settling time t_{set} the chosen value for t_{set} is a compromise between a fast settling time and a low resistance R_{co} to obtain a low unbalance in the converter currents $i_c^a,...,i_c^c$ during a fault. The values of voltages $V_{th,sec}$ and $V_{th,aux}$ and the reference voltages $V_{C,f}^{d,ref}$ and $V_{C,f}^{q,ref}$ for the controller are listed in table III for operation points I & II.

$V_{s,c}$	f_g	f_{sw}	C_{dc}	$V_{dc,min}$	$V_{dc,max}$	L_f	C_f	v_{rip}
2.5kV	50Hz	3kHz	30.9mF	3.57kV	4.2kV	1.1mH	143.2μF	1V

Table II: Parameters for the UNI-VSC including the DC-link and the LC-filter of the case study HT.

4.1 Results

Results for operation point I are shown in Fig. 7. After the fault occurs, the UNI-VSC converter current i_c^a reaches a maximum value of 117% of the nominal peak current $\hat{i}_{c,nom}$ as shown in Fig. 7a) & b). From its maximum value current i_c^a drops to a value close to the nominal peak current $\hat{i}_{c,nom}$. After a settling time of $t_{set} \approx 1$s currents $i_c^a,..,i_c^c$ maintain a peak value very close to the peak value before the fault, with a negligible unbalance caused by resistor R_{co}. The fault currents i_F at operation points I & II are very similar in terms of shape and maximum peak values. Therefore, only the fault current i_F at operation point I is shown in Fig. 7c). With active damping the fault current decreases after the fault within the settling time of t_{set}. In steady state the fault current i_F reaches a peak value of $\hat{i}_F \approx 15.3$A.

For operation point II, the converter currents before and after the fault are shown in Fig. 8 a) & b). The converter currents i_c^a reaches a peak current of 113% of the nominal peak current $\hat{i}_{c,nom}$ after the fault. As with operation point I currents $i_c^a,..,i_c^c$ drop within a settling time of $t_{set} \approx 1$s to a peak value close to

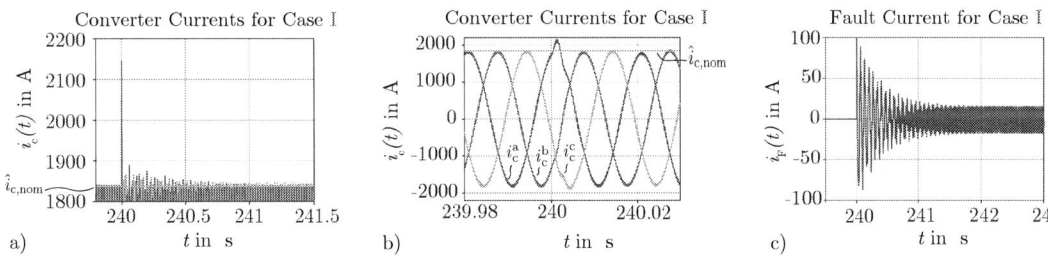

Figure 7: Operation point I: a) Converter currents $i_c^a,...,i_c^c$ during the fault. b) Converter currents $i_c^a,...,i_c^c$ during the fault in higher resolution. c) Fault current i_F.

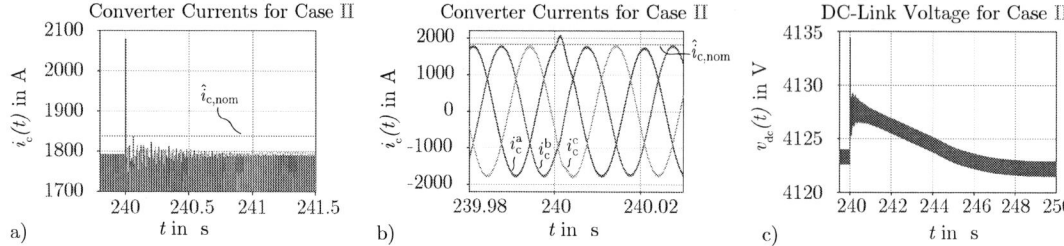

Figure 8: Operation point II: a) Converter currents $i_c^a,..,i_c^c$ during the fault. b) Converter currents $i_c^a,..,i_c^c$ during the fault in higher resolution. c) DC-link voltage during the fault.

the peak value before the fault, with a negligible unbalance caused by resistor R_{co}. Comparing currents $i_c^a,..,i_c^c$ at operation points I & II the shapes of the currents in Fig. 7b) are slightly different as in Fig. 8b) due to the difference in reference voltages $V_{C,f}^{d,ref}$ and $V_{C,f}^{q,ref}$ in table III.

	$V_{th,sec}$	$V_{th,aux}$	$V_{C,f}^{d,ref}$	$V_{C,f}^{q,ref}$
I	18kV	2.67kV	1.75kV	-1.2kV
II	20kV	3kV	0V	0V

Table III: Secondary winding voltage, auxiliary winding voltage, and reference voltages for the case study HT at operation points I & II.

Results for the DC-link voltage v_{dc} during a fault at operation point II with grounding point ③ are shown in Fig. 8c). The DC-link voltage v_{dc} increases to a maximum voltage of $v_{dc} \approx 4.14$kV at the time instant of the fault. After the fault the DC-link voltage v_{dc} settles to a stable voltage of $v_{dc} \approx 4.12$kV. Based on the results for the DC-link voltage v_{dc} in Fig. 8c) the maximum DC-link voltage $V_{dc,max}$ is given in table II.

4.2 Discussion

Due to active damping grounding point ③ is preferred. In case of passive damping only grounding points ① & ② limit the fault current to an acceptable level in MV grids. But additional measures are required to limit the UNI-VSC DC-link voltage v_{dc} for certain operation points with grounding points ①& ②. With grounding point ③ the DC-link voltage remains stable during a fault and normal operation. The advantage of grounding point ③ limiting the fault current and the DC-link voltage is shown for two operation points I & II. Since grounding point ③ is chosen, only a negligible unbalance remains in the converter currents after the fault, while the fault current i_F almost entirely flows through the transformer secondary winding and the filter capacitances C_f in contrast to the other grounding points. Furthermore, active damping allows the fault current to drop to a residual value i_{res} of less than 60A, which is required for MV grids.

5 Conclusion

This paper provides a comprehensive evaluation of possible grounding points in HT. Three possible grounding points are identified and compared for a HT with unidirectional auxiliary converter. Simulation models for the unidirectional HT auxiliary converter are derived based on state space averaging and equivalent circuits for the power transformer. Furthermore simulation results for the preferred grounding point ③ at different operation points are presented. With active damping, the single line to ground fault current for grounding point ③ is reduced below the allowed residual fault current for MV grids.

References

[1] P. Crossley and A. Beviz, "Smart Energy Systems: Transitioning Renewables onto the Grid," *Renewable Energy Focus*, pp. 54–59, 2010.

[2] M. Liserre, T. Sauter, and J. Y. Hung, "Future Energy Systems, Integrating Renewable Energy Sources into Smart Power Grid Through Industrial Electronics," *IEEE Industrial Electronics Magazine*, pp. 18–37, 2010.

[3] R. Yan, S. Roedinger, and T. K. Saha, "Impact of Photovoltaic Power Fluctuations by Moving Clouds on Network Voltage: A Case Study of an Urban Network," *Australasian Universities Power Engineering Conference (AUPEC)*, 2011.

[4] A. Woyte, V. V. Thong, R. Belmans, and J. Nijs, "Voltage Fluctuations on Distribution Level Introduced by Photovoltaic Systems," *IEEE Transactions on Energy Conversion*, vol. 21, no. 1, pp. 202–209, 2006.

[5] M. H. Oliver Brueckl, *Zukuenftige Bereitstellung von Blindleistung und anderen Massnahmen fuer die Netzsicherheit, Dienstleistungsauftrag fuer das Bundesministerium fuer Wirtschaft und Energie.* INA – Institut für Netz- und Anwendungstechnik GmbH, 2016.

[6] G. Glanzmann, "Flexible Alternating Current Transmission Systems," *ETH Zurich Research Collection*, 2005.

[7] J. Burkard and J. Biela, "Design of a Protection Concept for a 100-kVA Hybrid Transformer," *IEEE Transactions on Power Electronics*, vol. 35, no. 4, pp. 3543–3557, 2020.

[8] S. Bala, D. Das, E. Aeloiza, A. Maitra, and S. Rajagopalan, "Hybrid Distribution Transformer: Concept Development and Field Demonstration," in *IEEE Energy Conversion Congress and Exposition (ECCE)*, 2012.

[9] M. J. Mauger, P. Kandula, F. Lambert, and D. Divan, "Grounded Controllable Network Transformer for Cost-Effective Grid Control," in *IEEE Energy Conversion Congress and Exposition (ECCE)*, 2018.

[10] J. Kaniewski and Z. Fedyczak, "Modeling and Analysis of Dynamic Properties of the Hybrid Transformer with MRC," in *The International School on Nonsinusoidal Currents and Compensation*, 2010.

[11] A. Wiemer and J. Biela, "Overvoltages and Overcurrents in HV/MV Hybrid Transformers due to Grid Faults," in *European Conf. on Power Electronics and Applications (EPE ECCE Europe)*, 2021.

[12] S. P. D. Committee, *IEEE Guide for the Application of Neutral Grounding in Electrical Utility Systems.* IEEE, 2016.

[13] I.-S. S. Board, *IEEE Recommended Practice for Grounding of Industrial and Commercial Power Systems.* IEEE, 2007.

[14] I. Kasikci, *Short Circuits in Power Systems.* Wiley, 2002.

[15] A. Wiemer and J. Biela, "Comparison of Hybrid Transformers with Uni- and Bidirectional Auxiliary Converter," in *European Conference on Power Electronics and Applications (EPE ECCE Europe)*, 2019.

[16] L. Dongdong, T. Zhengyan, Y. Cikai, and S. S. Kumar, "Design and Implementation of Space Vector Modulated Three Phase Voltage Source Inverter," in *IEEE International Conference on Sustainable Energy Technologies and Systems (ICSETS)*, 2019.

[17] A. K. Kaviani and B. Mirafzal, "Dynamic Model of the Three-Phase Single-Stage Boost Inverter for Grid-Connected Applications," in *IEEE Energy Conversion Congress and Exposition (ECCE)*, 2012.

[18] R. Teodorescu, M. Liserre, and P. Rodriguez, *Grid Converters for Photovoltaic and Wind Power Systems.* Wiley, 2011.

[19] A. Reznik, M. G. Simoes, A. Al-Durra, and S. M. Muyeen, "LCL Filter Design and Performance Analysis for Grid-Interconnected Systems," *IEEE Transactions on Industry Applications*, vol. 50, no. 2, pp. 1225 – 1231, 2014.

[20] J. Dannehl, F. W. Fuchs, S. Hansen, and P. B. Thogersen, "Investigation of Active Damping Approaches for PI-Based Current Control of Grid-Connected Pulse Width Modulation Converters With LCL Filters," *Transactions on Industry Applications*, vol. 46, no. 4, 2010.

[21] J. Arrigo, "Input and Output Capacitor Selection," *Application Report*, no. SLTA055, 2006.

[22] Q.-C. Zhong, J. Liang, G. Weiss, C. Feng, and T. C. Green, "H∞ Control of the Neutral Point in Four-Wire Three-Phase DC–AC Converters," *IEEE Transactions on Industrial Electronics*, vol. 53, no. 5, 2006.

Non-parasitic induced transient overvoltage in ANPC topology due to critical switching sequences

Michael Geiss, Robert Kragl, Jürgen Thoma, Benjamin Volzer
FRAUNHOFER INSTITUTE FOR SOLAR ENERGY SYSTEMS ISE
Heidenhofstraße 2
79110 Freiburg, Germany
Tel.: +49 / (0) 761 4588-5069.
E-Mail: michael.geiss@ise.fraunhofer.de, robert.kragl@ise.fraunhofer.de,
juergen.thoma@ise.fraunhofer.de
URL: http://ise.link/hpe

Keywords

Multi-level inverters, Wide bandgap devices, MOSFET, IGBT, Converter control

Abstract

This paper describes a semiconductor overvoltage in an Active-Neutral-Point-Clamped Converter (ANPC). This overvoltage occurs in case of inductive load when the output voltage of the ANPC changes its polarity. In case of a grid inverter this occurs twice per grid period at the voltage zero crossing. It can be observed in most of the ANPC-based power electronics with classical PWM patterns and can reach the full DC-Link voltage. Although the ANPC is a well-known and widely spread topology there has been no particular concern in literature about this effect yet. From our point of view, the reason for this is a generous semiconductor dimensioning in terms of blocking voltage utilization and the limited energy due to the nature of the overvoltage. Nevertheless, this overvoltage could become a problem in modern designs when SiC MOSFETs are used, and their Safe Operating Areas (SOA) are pushed even further to the limits.

The shown overvoltage is not a switching overshoot due to parasitic inductances and high switching speeds. It cannot be explained by "hazardous" switching states either.

In the following, the emergence is described in detail and a theoretical model is introduced and evaluated by simulations and measurements. Afterwards, methods to avoid the overvoltage are shown and a risk estimation is performed.

Introduction

Multilevel topologies like ANPC, NPC or Flying-Cap are well-known and used in different applications for decades. They were first used in traction applications and afterwards they found their way to grid-connected inverters and even DC-DC-converters. Apart from the drawback of a higher number of semiconductors, they offer several benefits such as the use of semiconductors with lower blocking voltages, a better Total Harmonic Distortion (THD), and a smaller filtering effort.

In comparison to the Neutral Point Clamped Inverter (NPC), the active-NPC (ANPC) allows several additional zero-switching states. By corresponding implementation of these states, a better loss distribution [1] or a doubling of the apparent switching frequency [2] can be achieved, among others. Due to lower semiconductor blocking voltages the engineer has to ensure that under no circumstances is the full DC-Link voltage applied to a single switch. In addition to the "safe" switching states, there are "hazardous" states that can end up with one switch to be presented to the full DC-link voltage, and "destructive" states, which lead to a short circuit of half or the full DC-link [11]. Moreover, different cases like the emergency shutdown are known as critical in this context and therefore special switching sequences have to be followed [3].

This paper introduces another condition which leads to an overvoltage across one of the ouput switches (T2 or T3). In this case, the cause are not transient overvoltages in the switching moment or static "hazardous" switching states. The problem occurs with the use of - according to [11] - "safe" switching states in a critical sequence, when changing the polarity of the output voltage of the ANPC. The details will be explained in this paper.

The paper is structured as follows. In the first section the emergence of the overvoltage is described in detail and factors that favor its emergence are shown. Furthermore, a theoretical model is introduced and the influence of the output capacitance of the switches, the deadtime and the load current are discussed. Afterwards, the difference between full-SiC and hybrid (SiC MOSFET and Si IGBT) ANPC is explained. The theoretical model is compared with SPICE simulations and measured values. In the second section, possible strategies to avoid the overvoltage are given. The last section gives a risk assessment of the potential danger of the overvoltage.

Emergence of the overvoltage

In this first section, the emergence is explained in detail. All explanations are given for the upper half of the ANPC but apply in mirror image for the lower half as well.

The emergence of the overvoltage is explained based on one example switching sequence. Fig. 1 shows the different steps of this sequence. All switches are drawn as a combination of MOSFET, diode and capacitor. The diode represents the intrinsic diode, and the capacitance represents the output capacitance (C_{OSS}) of the MOSFET. Both is drawn for better understanding. The green rectangles show the actively switched on MOSFETs. The blue voltages show the present drain-source-voltage (V represents $V_{DC-link}/2$) and the trend (if it rises or falls). An ohmic-inductive load is assumed. Parasitic inductances were intentionally omitted because they have no influence on the occurrence of the overvoltage.

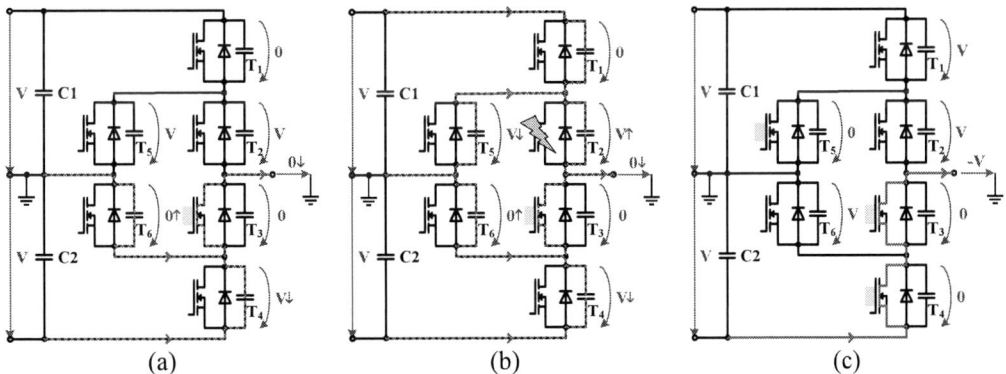

(a)   (b)   (c)

Fig. 1: Example „hazardous"-switching sequence for ohmic-inductive load

Starting point is the state "0L1" with positive output current as shown in Fig. 1 (a). The problem occurs when switching to the safe state "N". To get there, the MOSFETs T1 and T6 have to be switched off. This step is shown in Fig. 1 (b). At first, the output current charges $C_{OSS,T6}$. At the same time, $C_{OSS,T4}$ gets discharged to obey Kirchhoff's circuit laws. This leads to a falling output voltage. Due to the zero-current switch off of T1, $C_{OSS,T1}$ is not charged and so the problem occurs. Due to the falling output voltage, $C_{OSS,T1}$ has to be charged (Fig. 1 (b)). The only path for the charging current is through $C_{OSS,T2}$. Because this capacitance is already charged, the drain-source voltage of T2 rises above V. The overvoltage remains until the end of the deadtime when the state "N" is switched on by turning on the switches T4 and T5 (Fig. 1 (c)). With T5 on, T1 is clamped to V and $C_{OSS,T2}$ is discharged to V. This shows, that despite the use of only "safe" switching states an overvoltage occurs within the dead-time due to the missing charge of $C_{OSS,T1}$.

In summary, the problem always occurs when the outer switches (T1 / T4) are switched off at zero-current with the inner switches (T2 / T3) turned off, and a following change in polarity of the output voltage. As there is no current flowing through T2 just before the occurrence (Fig. 1 (a)), parasitic inductances have no influence on this effect.

Theoretical description of the overvoltage

In this Subsection, a simplified theoretical model was derived to calculate the maximum amplitude of the overvoltage. It is introduced shortly and evaluated using simulation and measurement data.

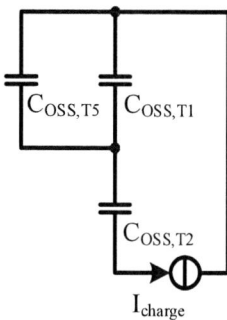

Fig. 2: Equivalent circuit diagram of the upper half of an ANPC to calculate the overvoltage.

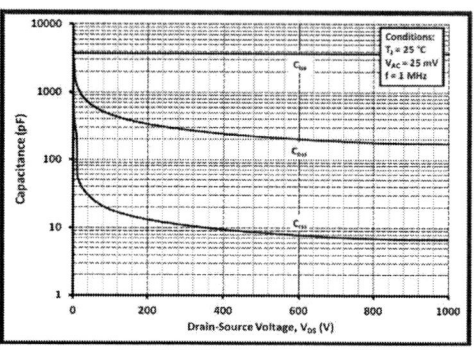

Fig. 3: Example datasheet [12] extract that shows the voltage dependency of C_{OSS} of a 1.7 kV SiC MOSFET.

During the occurrence of the overvoltage, there is no switching action in the upper half. The current flowing in the upper half occurs due to the decrease of the output voltage. Thus, the upper and the lower half of the ANPC can be considered separately. With these assumptions, the upper half can be transferred to an equivalent current source circuit diagram as shown in Fig. 2. This results in a capacitive divider. As charge current, a constant current is assumed as the interest of the calculation is the maximum overvoltage and not the rising speed. With this assumption, and the boundary condition that $V_{COSS,T1} + V_{COSS,T2} = V_{DCLink}/2$ at the end, the formula to calculate the overvoltage ΔV_{T2} can be derived as given in Eq. (1).

$$\Delta V_{T2} = \frac{V_{DCLink}}{2} * \frac{1}{\frac{C_{OSS,T2}(V_{DS,T2})}{C_{OSS,T1}(V_{DS,T1}) + C_{OSS,T5}(V_{DS,T5})} + 1} \tag{1}$$

Considering a full-SiC ANPC with 6 similar MOSFETs, a maximum overvoltage of 66 % could occur when assuming the same and constant C_{OSS} for all switches. This assumption is not correct since the output capacitance is highly dependent on V_{DS} as shown in Fig. 3.

At the starting point of the voltage rise, $C_{OSS,T2}$ is already charged to its rated voltage. Thus, the capacity will not change significantly with a further rising voltage. $C_{OSS,T1}$, on the other hand, is completely discharged and therefore has a much higher capacity at the beginning which becomes smaller with rising voltage. Assuming the same charge to be put into both the capacitors, the voltage of $C_{OSS,T2}$ would rise much faster than the one of $C_{OSS,T1}$. This makes the effect even worse. Therefore, the maximum voltage cannot be calculated directly with Eq. (1). Instead, it has to be calculated iteratively using theoretical equivalent capacities. This procedure is explained in the following.

Core of the iterative calculation is a charge-based replacement capacity. The stored charge at a certain voltage can be derived by the integral over the C_{OSS} curve as given in Eq. (2). From this charge, a constant capacity with the same charge stored at this voltage can be derived as shown in Eq. (3). The start and stop voltages are the drain-source-voltages at the beginning and at the end of the reloading process.

$$Q = \int C_{OSS}(V_{DS}) \, dV \tag{2}$$

$$C_{OSS,const} = \frac{\int_{V_{DS,Start}}^{V_{DS,Stop}} C_{OSS}(V_{DS}) \, dV}{V_{DS,Stop} - V_{DS,Start}} \tag{3}$$

This replacement capacity can be used in Eq. (1). The difficulty lies in the unknown voltage $V_{DS,Stop}$. Therefore, the calculation has to be done iteratively in the following steps:

1. The overvoltage is calculated by use of the rated output capacities.
2. The replacement capacities for T1, T2 and T5 are calculated using the calculated overvoltage and equation Eq. (3).
3. The new overvoltage is calculated using the replacement capacities.
4. Steps 2 and 3 are repeated until the calculated overvoltage reaches a fixed value

For a full-SiC ANPC with the MOSFETs shown in Fig. 3 and an assumed DC-link voltage of 2 kV, a maximum drain-source voltage across T2 of 1861 V is calculated.

There is no direct dependency between the output current and the resulting maximum overvoltage in case of full-SiC. But the output current determines how fast the reloading takes place and therefore how fast the overvoltage rises. With short deadtimes, small output currents, respectively big output capacities, and the fact that the overvoltage exists only during the deadtime, the theoretical maximum overvoltage value may not be reached.

Simulation and measurement results

The theoretical model was verified by simulations using LT-SPICE. A full-SiC ANPC was simulated using the manufacturers SPICE Models of the MOSFETs shown in Fig. 3. As load, a 15 A current source was used. Fig. 4 shows the simulation results of the switching sequence given in Fig. 1. In the simulation, the maximum drain-source voltage across T2 reached 1877 V. This fits very well into the iterative calculated value of 1861 V. The error can occur due to partly linearization and readout errors of the C_{OSS} curve.

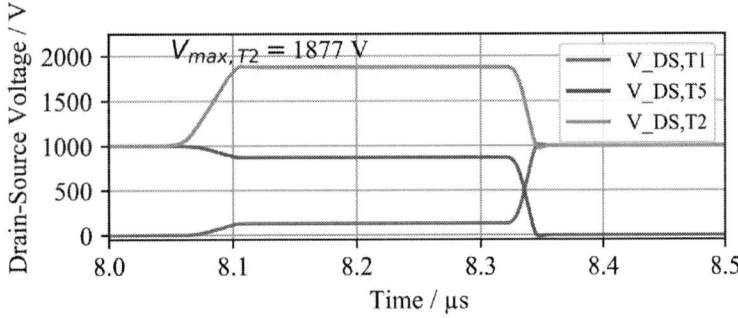

Fig. 4: LT Spice simulation results for the example switching sequence

The overvoltage was measured in different ANPC projects with different configurations. Thus, the formula was also evaluated using measurement data.

One example is given in Fig. 5. The curves were measured at an inverter based on a full-SiC ANPC consisting of 900 V, 120 mOhm MOSFETs. The inverter has a nominal power of 4 kW at an output voltage of 400 V and is driven in open-loop control with a fixed PWM pattern for doubling the output frequency as given in [2].

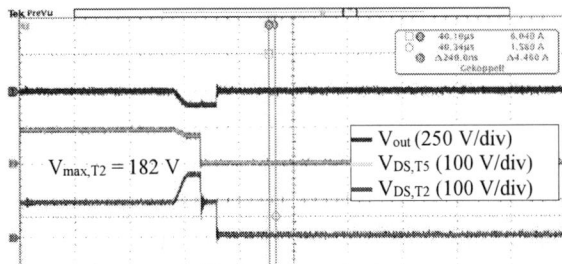

Fig. 5: Measured overvoltage during the deadtime in an ANPC with a DC-link voltage of 200 V

At the shown working point with a DC-link voltage of 200 V, a maximum drain-source-voltage of 182 V could be measured. The calculation for this configuration based on Eq. (1) and Eq. (3) gives a maximum of 185.8 V, a value that fits very well into the measurement data.

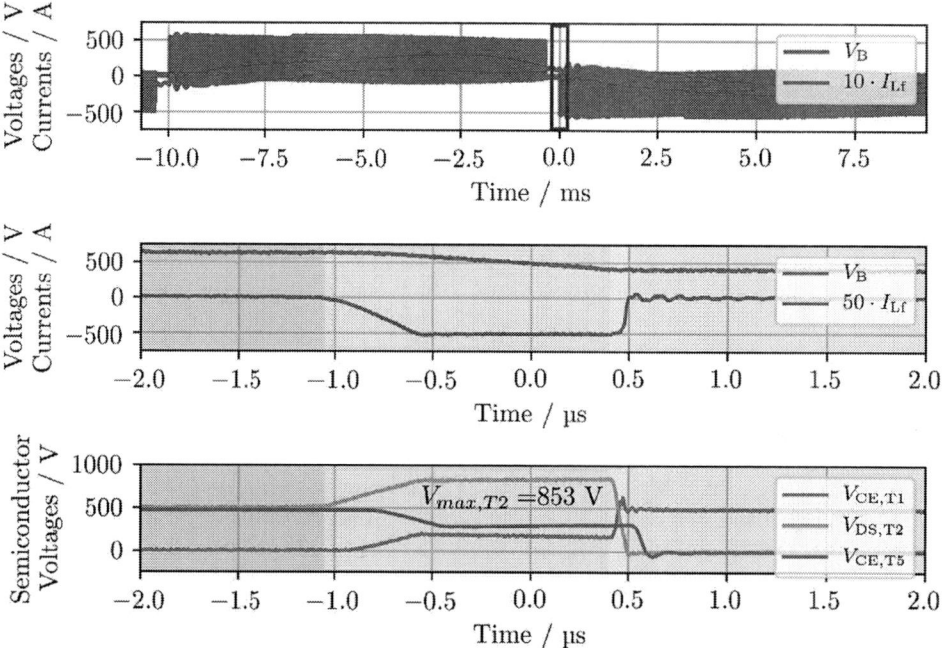

Fig. 6: Measured overvoltage during the deadtime in a hybrid ANPC with a DC-link voltage of 1000 V

Another measurement was performed on a 250 kW three-phase inverter based on a hybrid ANPC. In this case, the ANPC consists of two fast-switching 1.7 kV SiC MOSFETs for T2 and T3 and four slow-switching 1.7 kV Si IGBTs. The inverter is driven in open-loop control with a slightly adapted "ANPC-12" PWM pattern from [2]. The load is 5.16 kvar with a power factor of 0.8_{ind}. The measurement was performed with a lowered DC-link voltage of 1 kV and with an extremely long deadtime of 2 μs for a better visualization of the overvoltage.

Fig. 6 shows the measurement results. The first graph shows the ANPC output voltage and the line-filter current over one grid period. In the second graph, the same values are shown but with a zoom to the point where the output voltage changes its polarity. Graph 3 shows the drain-source respective collector-emitter voltages of the upper three switches at the same time interval.

Two things can be obtained here. First, the overvoltage at T2 of 71 % related to $V_{DC-Link}/2$. Secondly, the switching pattern is not the one used in Fig. 1 but also in this case overvoltages occur.

Unfortunately, the amplitude of the overvoltage cannot be compared to the calculated value because the given formula does not consider the forward recovery charge of IGBTs. The details are explained in the next section.

Comparison of full-SiC MOSFET and hybrid ANPC

In a 4 IGBT and 2 MOS hybrid version of the ANPC [8], the problem is getting worse. In addition to charging the output capacitor of an IGBT, there are carriers needed for the recovery of the junction. When the output voltage starts to fall as shown in Fig. 1 (b), a current flows through T1 an T2. At first, this current effects T1 only in the way of clearing out charge carriers, but already charges $C_{OSS,T2}$. Only then can $C_{OSS,T1}$ be charged. With this delayed charge of $C_{OSS,T1}$, the drain-source-voltage of T2 rises to a higher level and can reach the full DC-link voltage (or the avalanche voltage) even before the junction of T1 is fully recovered.

Due to the additional charge needed for the recovery of the junction, the simplified model behind Eq. (1) does not fit anymore. One option to approximate the forward recovery charge is to add an additional capacitor parallel to $C_{OSS,T1}$ that represents this additional charge. Unfortunately, it is complex to calculate the capacity because the recovery charge is highly dependent on the collector current, the switching speed and other values. Hence, a simple calculation of the overvoltage with the given formula is difficult for a hybrid ANPC. Furthermore, the current dependency of the forward recovery charge results in a load current dependency of the maximum overvoltage.

Another difference of the hybrid ANPC is the slower switching speed of the IGBTs and therefore the slow drop of the output voltage in Fig. 1 (b). Therefore, the rising speed of the overvoltage is more dependent on the switching speed of the IGBTs than on the load current.

As a conclusion it can be said that the overvoltage has a higher amplitude in a 4 IGBT and 2 MOS hybrid version of the ANPC, but the maximum cannot be precalculated with the given formula.

Further causes of occurrence

The direct use of a hazardous sequence as explained is just one possible trigger for the occurrence of the overvoltage. But there are other possibilities. The most important ones are described in the following.

Emergency shutdown

As explained in [3], there is a special switching sequence for an emergency shutdown, where the outer switches (T1/T4) always have to be switched off before the output switches (T2/T3) can be switched off. Otherwise, a difference in the switching speed of the semiconductors can lead to an overvoltage. This fixed shutdown sequence does not help to avoid the overvoltages explained in this Paper. If the emergency shutdown starts e.g. in the state shown in Fig. 1 (a) then this will lead to an overvoltage.

Missing gate signals due to deadtimes

Besides the direct use of a hazardous sequence, the effect can also occur because of missing gate signals due to deadtimes. One example is the sequence „P" → „0L1" → „P" → „0U1" → „N". Here, a save sequence is used for the transition from the upper to the lower half of the ANPC. In an ideal inverter without any deadtime everything works fine. In a real converter, however, there are "P" pulses missing around the zero crossing because the pulse duration given by PWM generator gets shorter than the deadtime. In this case, the given sequence is altered to „P" → „0L1" → „0U1" → „N". In the deadtime between "0L1" and "0U1", the overvoltage will occur.

Shadowing

Another trigger can be the shadowing function of the digital controller for the PWM. Shadowing means a sample-and-hold of the compare values at some defined point of the PWM period. This is mostly done in the highest and/or the lowest point of the triangular carrier value. Depending on which variant is chosen, the switching sequence can be alternated, and overvoltages can occur.

Occurrence in different PWM patterns

Table I: Overview of different ANPC SPWM

Name	Overvoltage without alternated implementation
ANPC-11-Sync [2]	No
ANPC-12 [2]	Yes
ANPC-DF [2]	Yes
ANPC-ALD [4]	No
ANPC-R2:1 [5]	Yes
ANPC-OOZS [6]	No
ANPC-SSLD [7]	Yes

With sine-triangle-PWM (SPWM) based modulations the used switching states are often fixed. When comparing different SPWM patterns for the ANPC, four out of seven investigated patterns showed the overvoltage in a straight forward implementation without any adjustments. An overview of the investigated PWM patterns is given in Table I.

Strategies to avoid the overvoltage

This section explains different options to solve the overvoltage problem.

Adaption of the PWM pattern

The most obvious way to solve this problem is to avoid the hazardous sequences of switching states. However, this could mean reducing or limiting the respective advantages of the chosen PWM pattern. For example, in a PWM pattern to double the apparent output frequency, all zero states have to be used in a fixed and alternating manner. This can lead to hazardous sequences. If the pattern is alternated in a way that there are no overvoltages, the doubling of the output frequency will no longer work.

Another option is to activate an alternative path to charge the output capacity of T1. In the case of the sequence shown in Fig. 1, a possible charge path would be via T5. Thus, in Fig. 1 (b) T1 and T6 must not be switched off at the same time. First, T1 has to be switched off and after the deadtime, T5 is switched on to charge $C_{OSS,T1}$. Then, T6 is switched off and then T4 is switched on after the deadtime. With this sequence, the overvoltage is avoided. The disadvantage of this sequence is that it takes longer because of two deadtimes. Also, the PWM generation becomes more complicated. Additionally, in case of emergency shutdown, a sequence where switches have to be switched on again is not acceptable.

Additional Switching Cell Capacitor

A very effective way to avoid the overvoltage is the use of a snubber capacitor across T2 and T3 as shown in Fig. 7. This capacitor creates a parallel path to T2 for charging T1. An additional advantage of the capacitor is a shorter commutation path between T2 and T3 as explained in [9], and therefore smaller voltage overshoots in the switching moment due to parasitic inductances. This can be interesting especially for the hybrid ANPC.

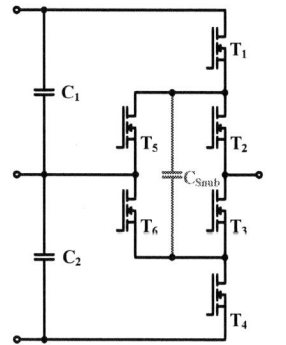

Fig. 7: Full-SiC ANPC with additional capacitor to avoid the overvoltage

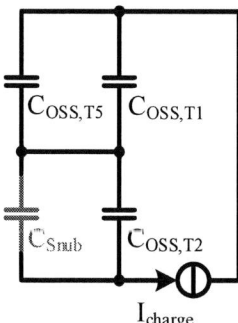

Fig. 8: Equivalent circuit diagram of the upper half of an ANPC to calculate the overvoltage with additional capacitor.

With the additional capacitor the equivalent circuit diagram in Fig. 2 can be extended as shown in Fig. 8. With the same assumptions as made before, a new formula to calculate the overvoltage can be derived and rearranged with the additional capacitor C_{Snub} as given in Eq. (4). With this formula, the size of the additional cap can be calculated for a chosen maximum overvoltage in a full-SiC ANPC. Again, charge-based replacement capacities have to be calculated with Eq. (3) for the chosen overvoltage.

$$C_{Snub} = \left(\left(\frac{\frac{U_{DC}}{2}}{\Delta U_{max}} - 1 \right) * \left(C_{OSS,T1} + C_{OSS,T5} \right) \right) - C_{OSS,T2} \qquad (4)$$

Fig. 9 shows the results of the same simulation that was used for Fig. 4 but with the additional capacitor C_{Snub}. The value was calculated with a chosen ΔU_{max} of 100 V. As can be seen, the calculation fits very well into the simulation results.

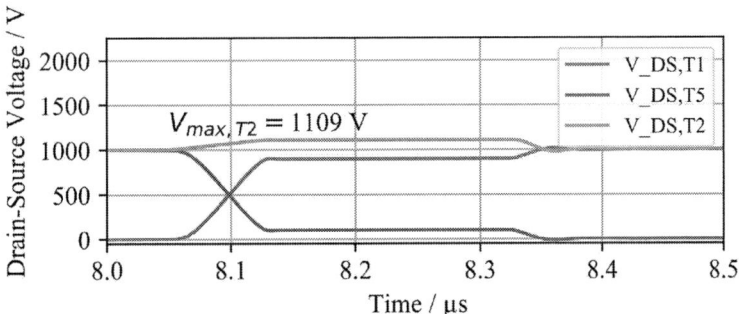

Fig. 9: LT Spice simulation results for the example switching sequence with additional capacitor

The additional capacitor was also successfully implemented in the hybrid ANPC as described before. Since the calculation of the overvoltage does not work for a hybrid ANPC as explained before, the calculation of the additional capacitor does not work either. Thus, the right value had to be found by simulation and was chosen as 100 nF. Fig. 10 shows the measurement results of the same inverter that was used for the measurement results in Fig. 6. But in this case the additional capacitor was added. As can be seen clearly, the capacitor works very well and there is no occurrence of overvoltage.

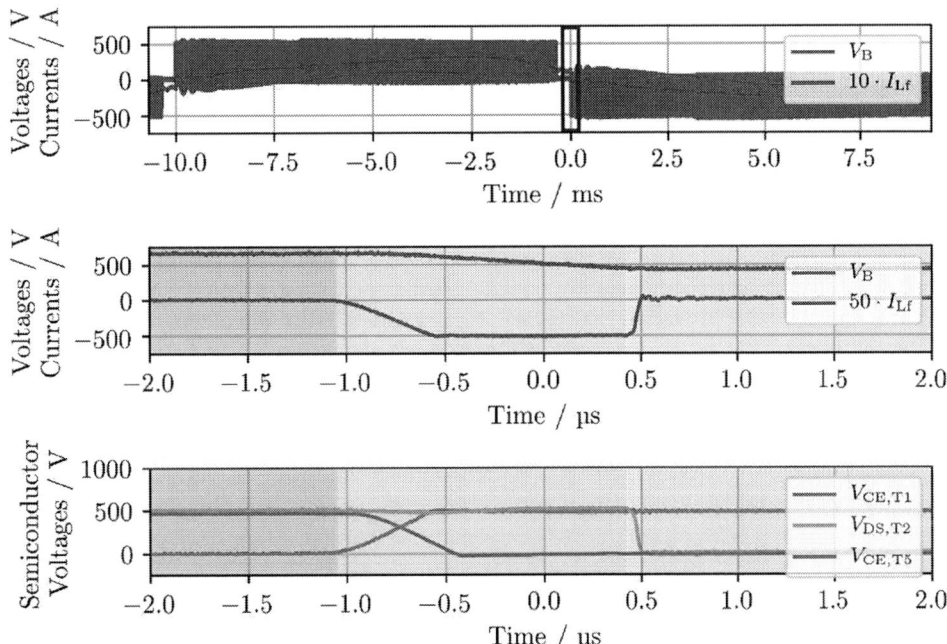

Fig. 10: Measured overvoltage during the deadtime in a hybrid ANPC with a DC-link voltage of 1000 V and additional capacitor

The drawback of this solution is the additional capacitor, which can lead to a ringing between itself, the parasitic inductances of the current path and the DC-Link when switching the outer transistors. This is described in [9] where the capacitor is used to shorten the commutation path of the output MOSFETs T2 and T3. No ringing problems occurred in the inverter presented.

Risk estimation

The ANPC is known and used for decades in different applications but there is no mentioning in the literature about the described overvoltage. So it is likely that the overvoltage does not lead to an immediate destruction of the semiconductors. It is difficult to judge how hazardous the overvoltage really is and depends on different factors. Two influences are discussed in the following.

Limited Energy and repetitive avalanche

One important point is that even if the overvoltage occurs, it will not have to be destructive for the semiconductors. This comes due to the limited energy that is brought into T2 during the overvoltage. The limit is given, on the one hand, by a time limit due to the deadtime. On the other hand, a possible avalanche current can only flow during the charge of $C_{OSS,T1}$, and therefore only until the summed up voltage of $V_{DS,T1} + V_{DS,T2}$ reaches the full DC-link voltage as can be seen in Fig. 1 (b). This limited energy is normally much lesser than the avalanche energy needed for destruction. In the hybrid case, the charge for the recovery of the junction adds on top.

Considering this, depending on the semiconductors, the overvoltage is expected to end with no or just a low energy avalanche and therefore with no immediate destruction. This does not mean that the effect will bring no harm. In the case of a grid inverter, this low energy avalanche occurs with the double grid frequency and in the case of a multilevel DCDC converter, it can occur with the switching frequency. As described in [10], a repetitive occurrence can impact the reliability of the device over its lifetime. Because the amount of energy is hard to predict and there is not much data from the manufacturers about repetitive avalanche withstand capability of the devices, an estimation is difficult for this case and will not be investigated further at this time.

Semiconductor utilization

Another important point is the semiconductor utilization in terms of blocking voltage, so what percent of the maximum blocking voltage is allowed by the developer to be seen by the semiconductors. If the ANPC is equipped with semiconductors that can block nearly the full DC-Link voltage, the additional overvoltage will not affect the semiconductors at all. There are different manufacturers that equip their inverters in such a way that the benefits of the ANPC in terms of smaller filter effort are fully taken but do not have to care about e.g. shutdown sequences.

In addition to a direct use of semiconductors that can block the full DC-Link voltage, there are cases where the semiconductors are utilized to a higher percentage of their blocking voltage for special operating points, but in the normal operating points the utilization is much lower. One example is a PV-inverter without DCDC converter. In this case, the maximum DC link voltage is the V_{OC} of the PV array. If the semiconductors are utilized by 80 % in this case, the utilization in the maximum power point operation will only be around 51 %.

In summary, the application and the chosen semiconductors will have a big influence when the overvoltage really is a destructive overvoltage from the semiconductors side of view. But through an increasing cost pressure in different markets, the semiconductor utilization is pushed further to its limits, so the problem could become more important in future.

Conclusion

In this paper, a drain-source overvoltage was investigated that occurs during normal switching operations in ANPC. It is caused due to critical switching sequences and can reach the full DC-link voltage. A theoretical model is introduced and validated by simulation and measurement results. The influence of the semiconductor output capacity and the load current is explained. Furthermore, strategies to avoid the overvoltage are discussed. Finally, a first risk estimation is given. Summarized, the maximum of the overvoltage has a strong dependency on the used semiconductors and the design of the power electronics. Furthermore, with the given considerations about the limited energy, the effect will most likely not lead to an immediate destruction of the semiconductors but could influence their lifetime.

References

[1] T. Bruckner and S. Bemet, "Loss balancing in three-level voltage source inverters applying active NPC switches," 2001 IEEE 32nd Annual Power Electronics Specialists Conference (IEEE Cat. No.01CH37230), 2001, pp. 1135-1140 vol.2, doi: 10.1109/PESC.2001.954272.

[2] D. Floricau, E. Floricau and M. Dumitrescu, "Natural doubling of the apparent switching frequency using three-level ANPC converter," 2008 International School on Nonsinusoidal Currents and Compensation, 2008, pp. 1-6, doi: 10.1109/ISNCC.2008.4627496.

[3] I. Staudt, "3L NPC & TNPC Topology", Semikron Application Note AN-11001, 2015, 2015-10-12 – Rev05

[4] Lin Ma, Xinmin Jin, T. Kerekes, M. Liserre, R. Teodorescu and P. Rodriguez, "The PWM strategies of grid-connected distributed generation active NPC inverters," 2009 IEEE Energy Conversion Congress and Exposition, 2009, pp. 920-927, doi: 10.1109/ECCE.2009.5316449.

[5] Bo Zhang, Qiongxuan Ge, Longcheng Tan, Xiaoxin Wang, Qiankun Chang and Jinxin Liu, "A new PWM strategy for three-level Active NPC converter," 2013 International Conference on Electrical Machines and Systems (ICEMS), 2013, pp. 1792-1795, doi: 10.1109/ICEMS.2013.6713292.

[6] E. Gurpinar, D. De, A. Castellazzi, D. Barater, G. Buticchi and G. Francheschini, "Performance analysis of SiC MOSFET based 3-level ANPC grid-connected inverter with novel modulation scheme," 2014 IEEE 15th Workshop on Control and Modeling for Power Electronics (COMPEL), 2014, pp. 1-7, doi: 10.1109/COMPEL.2014.6877124.

[7] G. Zhang, Y. Yang, F. Iannuzzo, K. Li, F. Blaabjerg and H. Xu, "Loss distribution analysis of three-level active neutral-point-clamped (3L-ANPC) converter with different PWM strategies," 2016 IEEE 2nd Annual Southern Power Electronics Conference (SPEC), 2016, pp. 1-6, doi: 10.1109/SPEC.2016.7846157.

[8] Q. -X. Guan et al., "An Extremely High Efficient Three-Level Active Neutral-Point-Clamped Converter Comprising SiC and Si Hybrid Power Stages," in IEEE Transactions on Power Electronics, vol. 33, no. 10, pp. 8341-8352, Oct. 2018, doi: 10.1109/TPEL.2017.2784821.

[9] D. Zhang, J. He and S. Madhusoodhanan, "Three-Level Two-Stage Decoupled Active NPC Converter With Si IGBT and SiC MOSFET," in IEEE Transactions on Industry Applications, vol. 54, no. 6, pp. 6169-6178, Nov.-Dec. 2018, doi: 10.1109/TIA.2018.2851561.

[10] Infineon, "Some key facts about avalanche", Application Note AN_201611_PL11_002, Version 1.0, 01.01.2017

[11] J. Dodge, "3L ANPC vs. 3L NPC Inverters", UnitedSiC Application Note UnitedSiC_AN0023, February 2020

[12] Cree, Inc, Datasheet of C2M0045170P, 04-2018

Open-Delta SBC: a New Converter Topology with Low Number of Sub-Modules for MV applications

D. Lanzarotto[1], P.B Steckler[1], K.Vershinin[1], F. Morel[1]
[1]Supergrid Institute
23 Rue Cyprian,
69100 Villeurbanne, France
E-Mail: damiano.lanzarotto@supergrid-institute.com, pierre-baptiste.steckler@supergrid-institute.com, konstantin.vershinin@supergrid-institute.com, florent.morel@supergrid-institute.com
URL: https://www.supergrid-institute.com/

Keywords

«Medium voltage converter», «AC-DC converter», «DC-AC converter», «Volume reduction», «Modular Multilevel Converters (MMC)»

Abstract

Medium voltage direct current (MVDC) technology has been experiencing a great boom of interest in recent years. This paper aims at giving a contribution to this field by proposing a new converter topology for MVDC applications. This topology is characterized by a low number of sub-modules (SMs) which is strongly related to the converter footprint and complexity. The new topology sizing is compared to the modular multilevel converter (MMC) for the same requirements to highlight advantages and disadvantages of the proposed solution.

Introduction

It is well established nowadays that high voltage direct current (HVDC) technology represents the most advantageous technical solution to problems such as long-distance energy transmission, asynchronous AC systems interconnection, interconnection of different regions requiring submarine and underground cables and transmission of offshore wind power to shore [1, 2]. In particular, the installation of HVDC lines is rapidly expanding in Europe and China. This is driven by the possibility of interconnecting large renewable energy sources located far away from where the main loads are [3-5].

The modular multilevel converter (MMC) represents the standard in terms of converter choice for recent HVDC applications mainly because its advantages compared to line-commuted converters (LCC): capability of realizing an independent P/Q control and small footprint due to the absence of any filter requirements. In addition, its modularity allows the use of low voltage building blocks to create high voltage stacks instead of using hundreds of high frequency switches connected in series. However, if compared to the traditional two-level voltage source converter (VSC), the MMC with half-bridge (HB) sub-modules (SM) requires twice the number of switches for the same output voltage [6] and a large overall converter capacitance leading to higher footprint and costs.

To address the issues mentioned above, researchers have been proposing new hybrid converter topologies with the aim of combining the advantages of the two-level VSC and the modular technology in a single architecture. Many solutions have been proposed following this line of research. The most significant ones consist of: the alternate arm converter (AAC) [7, 8], the series bridge converter (SBC) [9, 10], the hybrid series converter (HSC) [11] and the H-bridge hybrid multilevel converter [12].

Similar reasons to the ones that led to a rapid expansion of HVDC applications have recently started to drive the development of new medium voltage direct current (MVDC) concepts for power distribution [13, 14] with slightly different requirements (range of dc voltage and power of course, but also shape of required PQ domain). For applications with dc voltage of several tens of kV, MMC is an attractive solution thanks the converter modularity and low switching losses (if compared to two or three-level VSCs). Also, the know-how and technology transfer from the HVDC world allows a rapid growing of the number of applications of MMCs in MV domain. However, the above-mentioned weaknesses of MMCs in HVDC applications also arise MVDC then pushing research on alternative modular converters.

The goal of this paper is therefore to propose and analyze a new converter topology has the advantage of a reduced number of SMs in MVDC applications.

The paper is organized as follows. Section 1 deals with the description of the new converter topology, the strategies to balance the internal energy and its characteristic equations. Section 2 presents assumptions made to carry out the converter sizing, defines the key performance indicators (KPIs) and gives the sizing results themselves. Finally section 3 illustrates the comparison between the proposed topology and the MMC in a typical MVDC application.

Open-Delta SBC

General Description

The proposed converter (Figure 1) has been proposed in [15] and is named "open-delta SBC". This new topology consists of two phase-legs, or phase elements (PEs), similar to the ones found in conventional SBC, but there are only two single-phase transformers connected in open delta configuration. Each phase leg consists of a main H-bridge (MHB) and two stacks of SMs: Series Full Bridge (SFB) and Chain Link (CL). A SM stack is simply a series connection of a certain number of SMs which can be of the half-bridge (HB) or full-bridge (FB) type as shown in Figure 2. In particular, in this paper the SFB is a stack composed by HB and FB SMs while the CL in only composed by HB SMs. The voltage created by the stack is given instant by instant by the number of capacitors that are inserted in the circuit and it is characterized by the typical "staircase" profile. In order to create a symmetric and balanced load/generator from the grid standpoint, currents and voltages have to be properly controlled by both phase elements (PE1 and PE2 in Figure 1) as detailed in the following section.

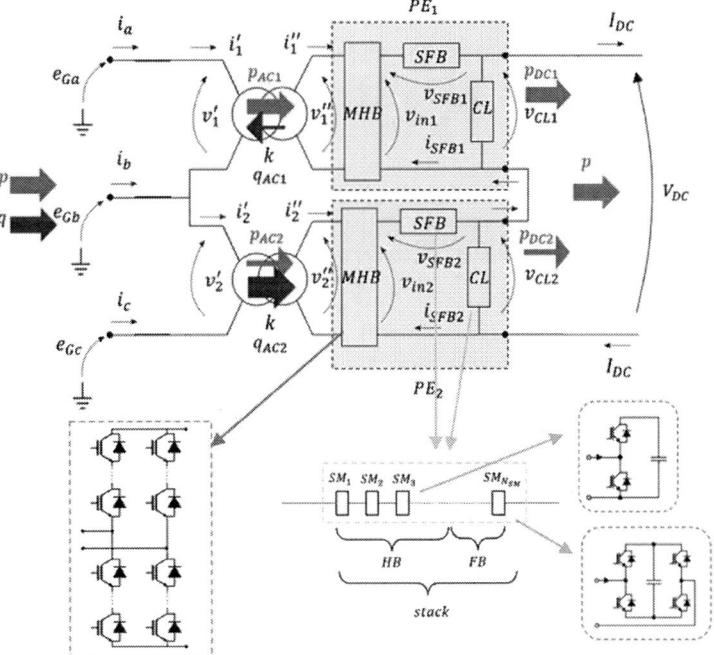

Figure 1 Open-delta SBC topology

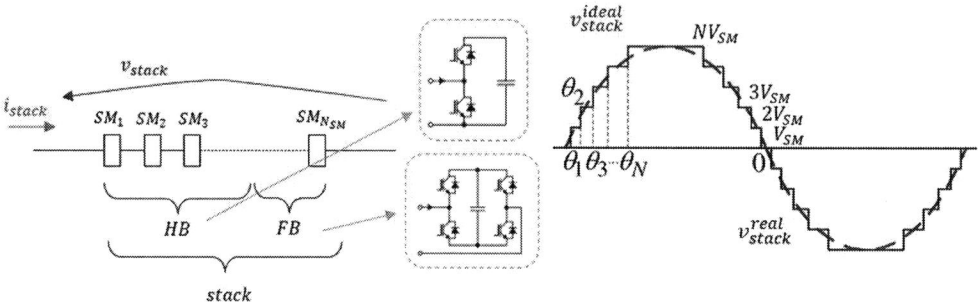

Figure 2 General sub-module stack consisting of both HB and FB SM

Operating Principle

In order to form a symmetric and balanced three-phase system, the open-delta current and voltage vector diagram must be the one of Figure 3 [15], In particular, the converter control must always ensure that:

$$\begin{cases} v_1' = e_{Ga} - e_{Gb} \\ v_2' = e_{Gb} - e_{Gc} \end{cases} \tag{1}$$

where the voltage drop due to the transformer leakage inductance and resistance is neglected. By satisfying (1) we obtain a 3-phase symmetric system, i.e.:

$$\begin{cases} i_1' = i_a \\ i_2' = -i_c \end{cases} \tag{2}$$

Hence, satisfying (2) in the normal operation implies that the two phases of the converter absorb currents which are always of the same amplitude and have a 60 degrees angle difference. Alternatively, if needed, currents with an inverse component can be created. It appears clear that the two phases carry the same amount of active power only when the reactive power is zero (i.e. when φ=0).

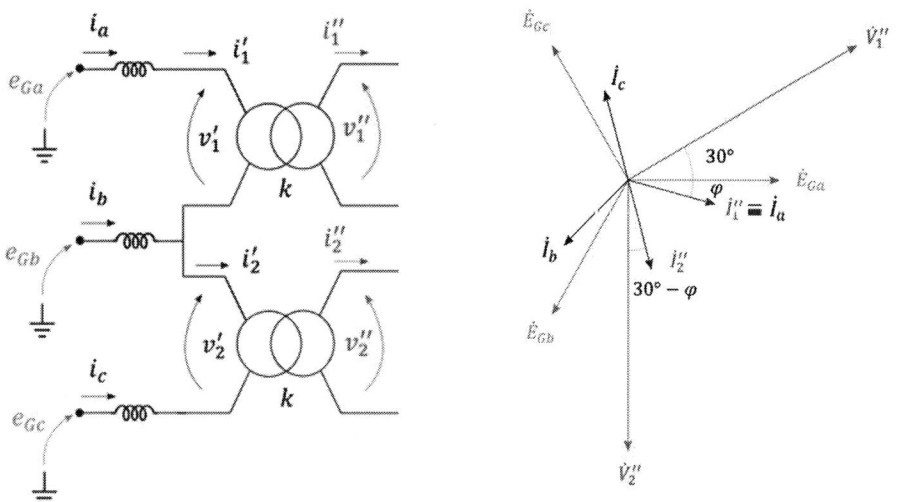

Figure 3 Open-delta connection voltage and current phasor diagram

Given the presence of SM stacks in the converter, the energy balance management issue acquires great importance. As a matter of fact, controlling and maintaining constant the average energy stored in the SM capacitors is crucial for converters with modular elements.

The basis V_b, I_b in Eq. (3) and $P_b = V_b I_b$ are used in the following in order to carry out calculations in per unit (pu).

$$V_b = V_{DC} \quad , \quad I_b = I_{DC,\max} \tag{3}$$

It can be easily verified that with reference to Figure 1 active and reactive powers in phase elements can be written as in Eq (4) where p and q are the active and reactive power absorbed from the grid in per unit (pu), $p_{AC1}, q_{AC1}, p_{AC2}$ and q_{AC2} are the active and reactive power flowing through phase 1 and 2 respectively, as shown in Figure 1.

$$p_{AC1} = \frac{3p - \sqrt{3}q}{6} \quad , \quad q_{AC1} = \frac{\sqrt{3}p + 3q}{6} \quad , \quad p_{AC2} = \frac{3p + \sqrt{3}q}{6} \quad , \quad q_{AC2} = \frac{-\sqrt{3}p + 3q}{6} \tag{4}$$

Energy management

Energy management of the converter and individual stacks is an essential aspect of the MMC converters. For individual stacks voltage and current waveforms observed shall result in zero power variation across fundamental cycle. For initial analysis of the topology presented in this paper we assume that appropriate control system is designed and ignore discrete nature of the voltage across the stack. Therefore, we can use analytical expressions for voltage and current across different elements of the converter.

There is a number of different ways how the converter can operate. In this paper we propose the following operation strategy for the converter which based on the minimizing of the circulating current, which is normally linked to the stack oversizing. In order to do so the CL1 and CL2 voltage are chosen to be time-constant equal to v_{DC1} and v_{DC2}. Considering that DC current is the same in both phase elements and $p_{DC1} = p_{AC1}$ and $p_{DC2} = p_{AC2}$, v_{CL1} and v_{CL2} can be expressed as:

$$v_{DC1} = \frac{p_{AC1}}{p} = \frac{3p - \sqrt{3}q}{6p} \quad , \quad v_{DC2} = \frac{p_{AC2}}{p} = \frac{3p + \sqrt{3}q}{6p} \tag{5}$$

In general, $p_{AC1} \neq p_{AC2}$ and v_{CL1} and v_{CL2} have to adapt to converter set point while their sum always has to be equal to v_{DC}. It is easy to verify from (5) that one has $v_{DC1} = v_{DC2} = 0.5$ at $\cos(\varphi) = 1$, whereas for $\cos(\varphi) = 0.95$ one obtains $v_{DC1} = 0.5 \pm 10\%$ and $v_{DC2} = 0.5 \mp 10\%$. Making the converter operate at lower power factors would further increase the max v_{DC1} and v_{DC2} determining a disadvantageous sizing. This is one of the reasons why the proposed converter is analyzed for MVDC applications, where the minimum $\cos(\varphi)$ is fixed to a certain value close to one (typically in the range 0.9-0.95) [14] leading to a cone-shaped domain in the PQ plane.

By choosing the phase element DC voltage according to (5), the overall energy of each phase element is balanced, therefore if the SFB energy is balanced than CL energy will also be balanced.

Equation (5) describes the CL steady state voltage and current (without loss of generality only the CL1 is considered).

$$\begin{cases} v_{CL1}(t) = v_{DC1} \\ i_{CL1}(t) = i_{SFB1}(t) - i_{DC} \end{cases} \tag{6}$$

By eliminating any harmonic voltage injection (i.e. by choosing that $v_{CL1} = v_{DC1}$ and $v_{CL2} = v_{DC}$), the only way to control the energy stored in the SFB is through the control of the average value of the SFB current. This can be obtained by the selection of an appropriate instant at which the MHB rectifies the transformer secondary side current and voltage. This particular instant depends on the operating condition, and it can be measured by an angle α quantifying its delay from the voltage zero crossing instant. The rectifying action above described is depicted in Figure 4. The part on the left of the figure shows the typical voltage and current waveforms of the MHB in the well-known SBC, where the voltage is rectified at its zero crossing.

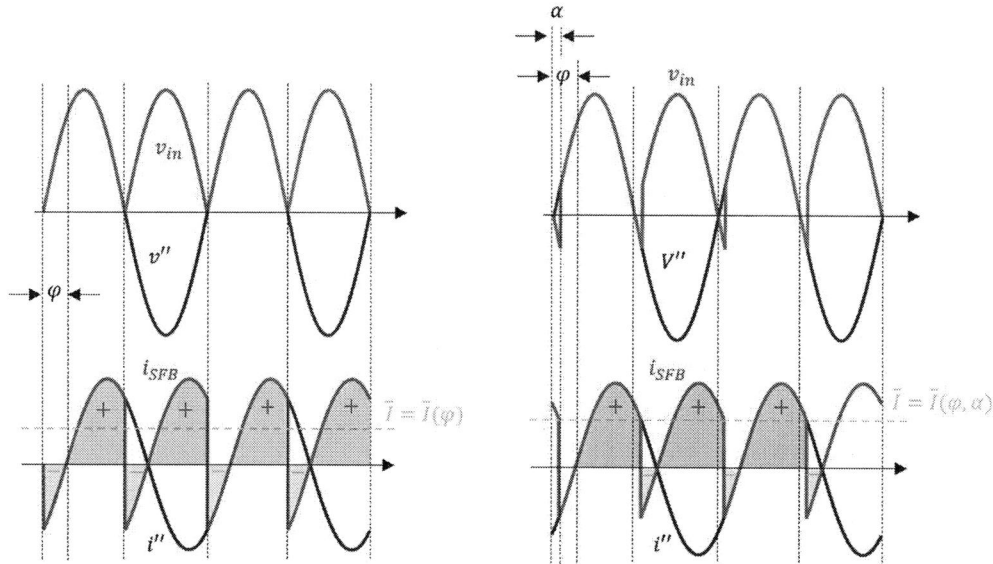

Figure 4 MHB average current control

The part on the right on the other hand shows that the voltage is rectified at an angle α measured from the zero crossing. Thanks to this technique the average value of the SFB current can be controlled. Considering PE1 only without the loss of generality voltage and current in SFB can be expressed as :

$$v_{in1} = \sqrt{2}V'' \sin\left(\omega t\right) \cdot sign\left[\sin\left(\omega t + \alpha_1\right)\right]$$
$$i_{SFB1} = \sqrt{2}I'' \sin\left(\omega t - \frac{\pi}{6} - \varphi\right) \cdot sign\left[\sin\left(\omega t + \alpha_1\right)\right] \tag{7}$$

$$p_{SFB1} = \left(v_{in1} - v_{DC1}\right)i_{SFB1} = p_{AC1} - v_{DC1}i_{SFB1} \tag{8}$$

where p_{SFB1} is the instantaneous power flowing into the SFB. Therefore, by averaging (8) over the period:

$$P_{SFB1} = \frac{1}{T}\int_0^T p_{SFB1}dt = P_{AC1} - v_{DC1}\overline{i_{SFB1}^{pu}}\left(\varphi, \alpha_1\right) \tag{9}$$

where P_{AC1} is the active power flowing into PE1 and $\overline{i_{SFB1}^{pu}}(\varphi, \alpha_1)$ is the mean value of the SFB current which depends on φ and α_1. Thus, in order to always maintain the SFB in energy balance, i.e. $P_{SFB1} = 0$, the following relation must always hold:

$$\overline{i_{SFB1}^{pu}} = \frac{P_{AC1}}{v_{DC1}} \tag{10}$$

It is important to remark that the MHB voltage rectified with angle $\alpha \neq 0$ becomes negative (see Figure 4), thus, a slight constructive modification has to be applied on the MHB itself in order to ne able to sustain it. Therefore, a certain amount of switches able to block a certain negative voltage are added as shown in Figure 5. In particular, it can be easily shown that the amount of the negative voltage to block depends on the minimum power factor ($\cos(\varphi)$) for which the converter has to be designed to operate (the lower the $\cos(\varphi)$ the higher is the absolute value of the negative voltage). For instance, with a minimal power factor of 0.95, the maximal negative voltage is roughly 25% of the maximal positive voltage. Thus, the number of the additional necessary switches in the MHB can be calculated. Limiting the power factor to a minimum value allows to not oversize the MHB for the reason explained above, the CL for reasons that can be found in equation (5) and consequently the SFB. This is another reason for the limitation of the converter work points in a cone-shaped PQ plane (MVDC applications) as already mentioned above.

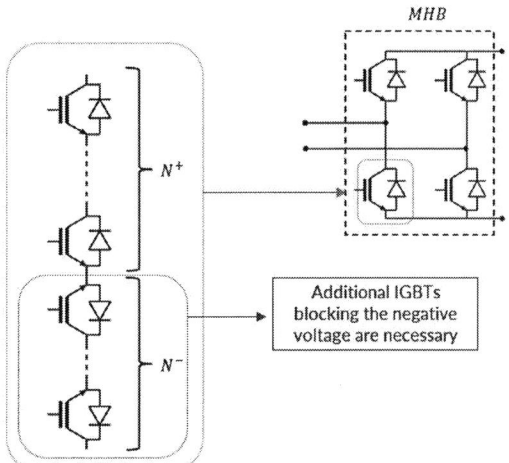

Figure 5 Additional IGBTs and diodes allowing the MHB to operate with negative v_{in} voltages.

Converter Steady State First Harmonic Equations

Summing up, the converter equations in p.u. for the MHB, the CL and the SFB can be expressed as:

$$\begin{cases} v_{in1} = R_v \sin(\omega t) \cdot sign\left[\sin(\omega t + \alpha_1)\right] \\ i_{in1} = \sqrt{2}I^{pu} \sin\left(\omega t - \dfrac{\pi}{6} - \varphi\right) \cdot sign\left[\sin(\omega t + \alpha_1)\right] \end{cases} \qquad MHB_1 \qquad (11)$$

$$\begin{cases} v_{CL1} = \dfrac{3p - \sqrt{3}q}{6p} \\ i_{CL1} = \sqrt{2}I^{pu} \sin\left(\omega t - \dfrac{\pi}{6} - \varphi\right) \cdot sign\left[\sin(\omega t + \alpha_1)\right] - p \end{cases} \qquad CL_1 \qquad (12)$$

$$\begin{cases} v_{SFB1} = v_{in1} - v_{CL1} \\ i_{SFB1} = i_{in1} \end{cases} \qquad SFB_1 \qquad (13)$$

where $R_v = \sqrt{2}V''/V_b$ in which V'' is the rms value of the secondary side voltage of the transformer and $I^{pu} = \sqrt{\dfrac{2}{3}\dfrac{\sqrt{p^2+q^2}}{R_v}}$. Voltage, current and stack energy waveforms are shown in the following figures for a particular operating point.

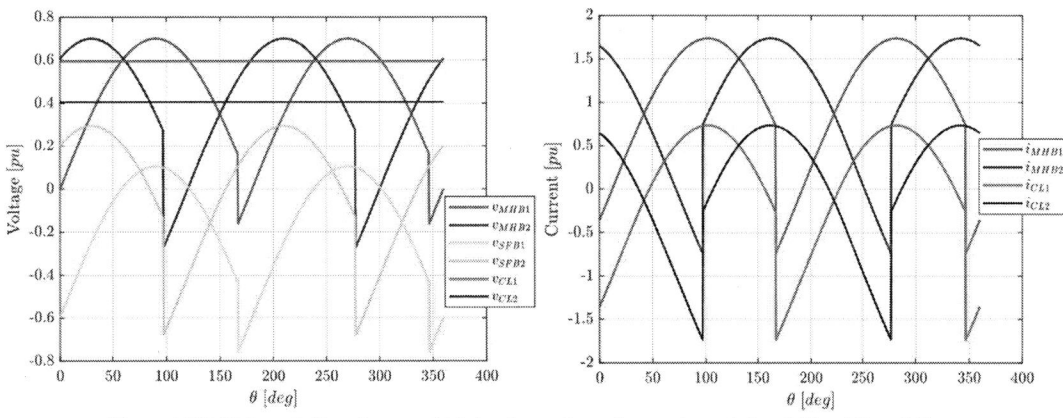

Figure 6 SFB Voltage (left) and current (right) and waveforms for $p = 1$, $cos(\varphi) = 0.95$ and $R_v = 0.7$

Please note that energy unit is time because it is calculated as Joules/Watts, i.e. it is normalized by P_b.

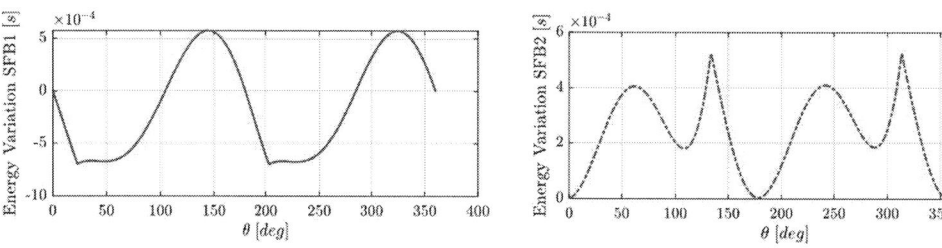

Figure 7 SFB 1 Energy Variation (left) and SFB 2 Energy Variation (right) for $p = 1$, $cos(\varphi) = 0.95$ and $R_v = 0.7$

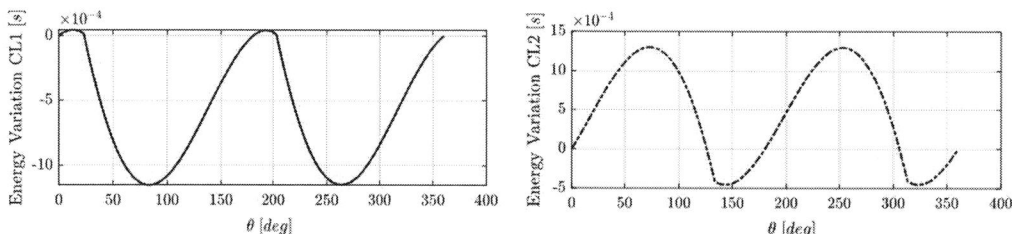

Figure 8 CL1 Energy Variation (left) and CL 2 Energy Variation (right) for $p = 1$, $cos(\varphi) = 0.95$ and $R_v = 0.7$

Sizing results comparison with the MMC

Study case presentation

It is well described in [17] how the interest in a particular converter configuration depends mainly on the converter cost, size and efficiency. It is therefore important to identify parameters able to suitably represent those aspects, consequently, the following key performance indicators (KPIs) are considered here:

- Transformer winding number and transformer sizing power: these parameters are related to the cost and volume of the transformers.
- Switch total sizing power (S_{SW}): it is related to the "quantity of silicon" (voltage to withstand by semiconductors and current passing through them) and therefore related to the converter cost. The total switch sizing power associated to one stack is defined by equation (14).

$$S_{SW}^{stack} = N_{SW}^{stack} V_{max}^{stack,pu} I_{max}^{stack,pu} \tag{14}$$

Where N_{SW}^{stack} is the stack switch number and:

$$I_{max}^{stack,pu} = \max\left\{i_{stack}(t)\right\} / I_{DC,max} \quad , \quad V_{max}^{stack,pu} = \max\left\{v_{stack(t)}\right\} / V_{DC} \tag{15}$$

The total converter switch sizing power is obtained by summing the sizing power associated to each stack.

- Stored Energy: this parameter quantifies the energy stored in the converter which is mainly due to SM capacitors which represents the major part of the SM volume. Therefore, this parameter is linked to the to the converter volume.
- Semiconductor switch total sizing power: This is related to the "quantity of silicon" (voltage the semiconductors must withstand and the current passing through them), and therefore is related to the converter cost. In simple terms, it represents the sum of the sizing power of all the switches.
- Sub-module number: it is related to the number of interconnexions between submodules and mechanical assemblies, the number of capacitor voltages to measure, number of discharge circuits (and number of by-pass circuits depending on the manufacturer's technical choices). It is then related to the converter cost.
- Sub-module cell capacitance: it is mainly related to the SM size (which is an important information on the ability to handle it during the construction phase and during replacing operations of faulty SMs). It is also related to the energy stored in an individual sub-module which is a constraint for the devices in the fault courant path in case of SM internal short-circuit. The calculation of the SM capacitance is conducted following the approach described in [18].
- Arm inductor number: it is linked to the converter volume and footprint.

Those KPIs are calculated for the MMC (which represent the state-of-the-art topology) and the open-delta SBC proposed here. The numerical values of the main parameters are listed in table 1 [20].
Thanks to the equations (11), (12) and (13) all the proposed KPIs can be evaluated for the most critical operating point.

Table 1 Numerical values of the main parameters

Name	Symbol	Value
Rated DC power	$P_{DC,max}$	30 MW
DC voltage	V_{DC}	54 kV
P.u. Secondary Voltage	R_v	0.7
Minimum $\cos(\varphi)$	$\cos(\varphi)_{min}$	0.95
Rated SM voltage	$V_{SM,N}$	3.6 kV
Max stack relative voltage variation	δV_{stack}	±10%

Result comparison

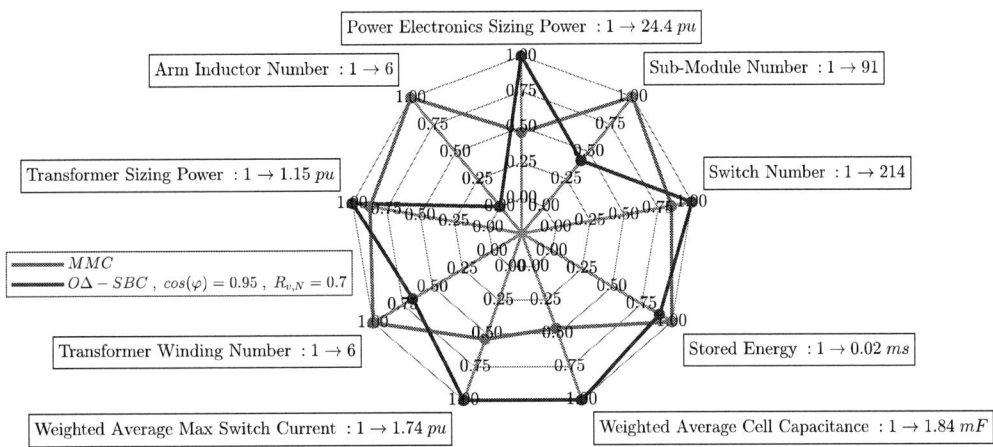

Figure 10 KPI comparison.

For each indicator, the lowest value, the better. Each separate type of KPI is normalized with respect to the maximum of the two to show a relative comparison between the topologies. The notation in the label text corresponding to each KPI: "1→x", means that 1 corresponds to the specified x value. The stored energy unit is time because it is calculated as Joules/Watts, i.e. it is normalized by P_b.

Based on the data presented in Figure 10 it can be seen that open-delta SBC does not have a distinct advantage of the standard configuration of HB MMC. Open-delta SBC has half the number of SM compared to MMC but these submodules are bigger (have larger capacitors and higher switch rated current). Although open delta SBC has slightly lower stored energy it shows much higher semiconductor sizing power.

An improvement in the trade off can be achieved for open-delta SBC if the "DC fault blocking capability" is required by application. In the scope of this paper we consider DC fault blocking capability is an ability of the converter to stay connected to AC grid under existence of the DC pole to pole fault. Indeed, for the MMC, at least 50% of the HB SMs must be replaced by FB SMs for the converter to block DC short-circuits [19]. With reference to for the open-delta SBC, as soon as the fault is detected, the switches in SMs (CLs and SFBs) can be blocked (all IGBTs in SMs controlled to be in off-state, letting only diodes as possible paths for the current) and two strategies can be adopted in order to block the DC fault current.

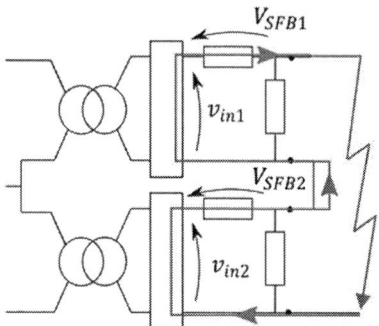

Figure 11 DC short-circuit current path in the open-delta SBC

1. Given the particular structure of the MHB (see Figure 5), it can be designed so that $N^+ = N^-$. Indeed, if the number of diodes pointing downward is the same as the number of diodes pointing upward the DC fault current can be blocked without applying an overvoltage to those devices and their IGBT in parallel.
2. The MHB can be controlled to reproduce the behavior of a diode rectifier when the DC fault occurs (in particular, only the IGBT blocking the negative voltage must be controlled). In this case the SFB voltage ($V_{SFB1,2}$) must be high enough to oppose the MHB output voltage ($v_{in1,2}$) so that the DC current can be taken to zero.

Let us consider the second option . The necessary amount of SFB voltage can be obtained as follows:

$$V_{SFB1} + V_{SFB2} \geq V_{in1} + V_{in2} \tag{16}$$

Assuming $V_{SFB1} = V_{SFB2} = V_{SFB}$ and $V_{in1} = V_{in2} = V_{in}$, then:

$$N_{SM,FB}^{SFB} V_{cell} \geq V_{in} = \sqrt{2}V'' \tag{17}$$

$$N_{SM,FB}^{SFB} \geq \frac{R_v}{V_{cell}^{pu}} \tag{18}$$

Thanks to (18) and to the values chosen in Table 1, one has that $N_{SM,FB}^{SFB} \geq 11$. Since already $N_{FB}^{SFB} = 5$, then 6 HB SMs must be replaced by FB SMs for each SFB. This means that in total $\frac{12}{35} * 100 \cong 34\%$ of the HB SMs must be replaced by FB SMs. Therefore, requiring the DC blocking capability from the open-delta SBC affects less its sizing than the MMC, where at least 50% of SMs must be replaced.

Conclusion

A new converter topology for MVDC applications is investigated. Converter structure principles of operations and energy balancing are briefly described. A set of KPIs are defined and used to assess the advantages and drawbacks of the proposed converter when compared to the MMC for an application case. Results show that the new topology is characterized by a significantly lower SM number with respect to the MMC. It can lead to an advantageous solution from complexity, cost and footprint points of view. Moreover, if the converter is required to able to block faults on the DC side, less switches should be added than in the MMC case.

References

[1] M. P. Bahrman and B. K. Johnson, "The ABCs of HVDC transmission technologies," *IEEE power and energy magazine,* vol. 5, pp. 32-44, 2007.
[2] P. Bresesti, W. L. Kling, R. L. Hendriks, and R. Vailati, "HVDC connection of offshore wind farms to the transmission system," *IEEE Transactions on energy conversion,* vol. 22, pp. 37-43, 2007.
[3] X. Qin, P. Zeng, Q. Zhou, Q. Dai, and J. Chen, "Study on the development and reliability of HVDC transmission systems in China," in *2016 IEEE International Conference on Power System Technology (POWERCON),* 2016, pp. 1-6.
[4] H. Xie, Z. Bie, and G. Li, "Reliability-oriented networking planning for meshed VSC-HVDC grids," *IEEE Transactions on Power Systems,* vol. 34, pp. 1342-1351, 2018.
[5] A. Orths, A. Hiorns, R. van Houtert, L. Fisher, and C. Fourment, "The European North seas countries' offshore grid initiative—The way forward," in *2012 IEEE Power and Energy Society General Meeting,* 2012, pp. 1-8.

[6]   M. B. Ghat, S. K. Patro, and A. Shukla, "The hybrid-legs bridge converter: A flexible and compact VSC-HVDC topology," *IEEE Transactions on Power Electronics,* vol. 36, pp. 2808-2822, 2020.

[7]   M. M. C. Merlin, D. Soto-Sanchez, P. D. Judge, G. Chaffey, P. Clemow, T. C. Green, *et al.,* "The extended overlap alternate arm converter: A voltage-source converter with DC fault ride-through capability and a compact design," *IEEE Transactions on Power Electronics,* vol. 33, pp. 3898-3910, 2017.

[8]   M. M. Merlin, T. C. Green, P. D. Mitcheson, D. R. Trainer, R. Critchley, W. Crookes, *et al.,* "The alternate arm converter: A new hybrid multilevel converter with dc-fault blocking capability," *IEEE transactions on power delivery,* vol. 29, pp. 310-317, 2013.

[9]   C. M. Diez, A. Costabeber, F. Tardelli, D. Trainer, and J. Clare, "Control and experimental validation of the series bridge modular multilevel converter for HVDC applications," *IEEE Transactions on Power Electronics,* vol. 35, pp. 2389-2401, 2019.

[10]   E. M. Farr, D. R. Trainer, O. E. Idehen, and K. Vershinin, "The series bridge converter (SBC): AC faults," *IEEE Transactions on Power Electronics,* vol. 35, pp. 4467-4471, 2019.

[11]   S. K. Patro, A. Shukla, and M. B. Ghat, "Hybrid series converter: a dc fault tolerant HVDC converter with wide operating range," *IEEE Journal of Emerging and Selected Topics in Power Electronics,* 2019.

[12]   M. B. Ghat and A. Shukla, "A new H-bridge hybrid modular converter (HBHMC) for HVDC application: operating modes, control, and voltage balancing," *IEEE Transactions on Power Electronics,* vol. 33, pp. 6537-6554, 2017.

[13]   P. Le Métayer, J. Paez, S. Touré, C. Buttay, D. Dujic, E. Lamard, *et al.,* "Break-even distance for MVDC electricity networks according to power loss criteria," in *2021 23rd European Conference on Power Electronics and Applications (EPE'21 ECCE Europe),* 2021, pp. 1-9.

[14]   "ANGLE-DC, 2015 Electricity Network Innovation Competition," 2015.

[15]   P.-B. Steckler, "Pierre-Baptiste Steckler, PATENT APPLICATION: Convertisseur de tension AC/DC triphasé comprenant uniquement deux modules de conversion électrique, Assignee: SUPERGRID INSTITUTE, FR3112042A1," 2020.

[16]   E. Amankwah, A. Costabeber, A. Watson, D. Trainer, O. Jasim, J. Chivite-Zabalza, *et al.,* "The series bridge converter (SBC): A hybrid modular multilevel converter for HVDC applications," in *2016 18th European Conference on Power Electronics and Applications (EPE'16 ECCE Europe),* 2016, pp. 1-9.

[17]   N. Evans, P. Dworakowski, M. Al-Kharaz, S. Hegde, E. Perez, and F. Morel, "Cost-performance framework for the assessment of Modular Multilevel Converter in HVDC transmission applications," in *IECON 2019-45th Annual Conference of the IEEE Industrial Electronics Society,* 2019, pp. 4793-4798.

[18]   M. Merlin, T. Green, P. Mitcheson, F. Moreno, K. Dyke, and D. Trainer, "Cell capacitor sizing in modular multilevel converters and hybrid topologies," in *2014 16th European Conference on Power Electronics and Applications,* 2014, pp. 1-10.

[19]   P.-B. Steckler, "Contribution à la conversion AC/DC en Haute Tension," Université de Lyon, 2020.

[20]   D. Lanzarotto, F. Morel, P.-B. Steckler, and K. Vershinin, "Rapid Evaluation Method for Modular Converter Topologies," Energies, vol. 15, no. 10, p. 3492, May 2022, doi: 10.3390/en15103492. [Online]. Available: http://dx.doi.org/10.3390/en15103492

Characterising the effect of an inverter on the regulation of the AC voltage using a frequency response identification technique

Mohamed Aldarmon, Joan Marc Rodriguez, Adria Junyent-ferre
Imperial Collage London
Exhibition Rd, South Kensington, SW7 2BX
London, UK
Phone:+44(0)20 7594 6290
Mohamed.Aldarmon18@Imperial.ac.uk
https://www.imperial.ac.uk

Keywords

≪Frequency-Domain Analysis≫, ≪Converter control≫, ≪Power quality≫.

Abstract

This paper presents a methodology to obtain a small-signal characteristics of a power electronic inverter in the frequency domain. The method is based on carrying out a series of simulations to observe the response of the inverter to a small voltage disturbance. The results obtained through the method provide information about the adequacy of the controller of the inverter and how it contributes to AC voltage regulation of the network. The paper describes the methodology and illustrates its use through a study case of a low-power three-phase inverter.

Introduction

The need to meet the future net-zero carbon targets requires the integration of more renewables to the power system using power electronic devices. Voltage source inverters (VSI) are commonly used to connect new low carbon technologies (LTCs). Depending on the grid code requirements, inverters use different high-level control modes such as unity power factor or AC voltage control. Looking at the specific implementation details, different manufacturers may choose to use different Phase Locked Loop (PLL) and internal reference calculation methods. Some controller formulations are more likely to trigger instability [1] or lead to different power quality at network level (eg voltage unbalance, harmonics). Several techniques have been proposed to study these problems using black-box models of the converters in simulation or through experimental testing. This paper studies the reaction of the inverter to a small-signal disturbance injected on the voltage of the grid. The characterisation is done through a series of simulations to create a frequency response plot for different operating conditions.

The disturbance injection method is a common tool used for measuring the impedance of devices. The disturbance injection can be either in series or in parallel with the tested device and use voltage or current injection. Choosing one of the two injection types depends on the system configuration and the characteristics of the point of injection (eg a stiff voltage or current source) and the device under test [2][3]. Different types of disturbances can be used, such as sinusoidal signal [3], chirp signal [2], and a multitone signal [4]. The measured response signals (eg current, voltage) are then processed using a discrete time Fourier (DFT) to transform the time domain signal into a phasor domain. This information can then be read in the form of Bode plots or used to obtain transfer functions and impedances for different types of assessments (eg impedance-based stability [5]).

The method can be used in simulation if accurate dynamic simulation models of the device are available. This is advantageous because it can be used with black-box manufacturer models where the internal equations of the model are not visible to the user. With adequate test equipment, the method can also be used for experimental system identification. These technique has been used to investigate and diagnose malfunctioning devices (such defected transformers and photovoltaic solar panels [6] [7] [8]). It can also be used to assess compliance of devices with power quality requirements.

In this paper we describe a method to characterise the response of an inverter based on voltage injection in series with a Thevenin equivalent of an AC network. The results shown across the paper are based on a study case for a low voltage inverter as described in the next section.

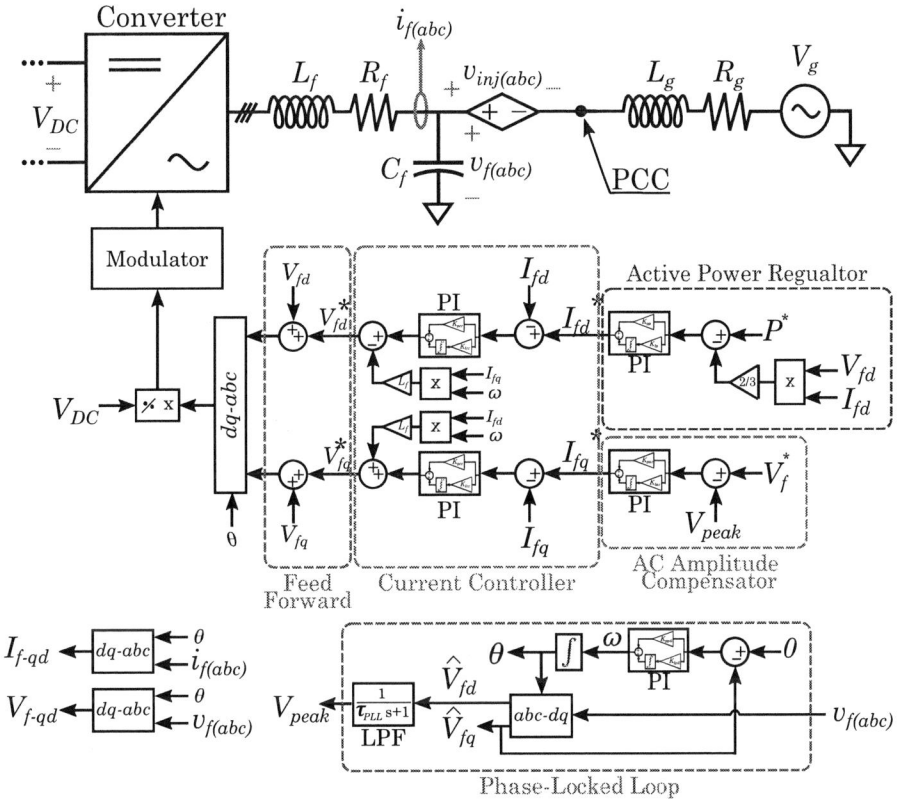

Fig. 1: Diagram of the system under test and its controller structure.

Study System

The system under the study is an inverter interfacing a DC power source (eg a solar photovoltaic DC/DC) to a distribution network. The inverter has a 2nd order LC pulse-width modulation (PWM) filter and the network is represented by a Thevenin equivalent. The injected disturbance is a small three-phase voltage in series with the Thevenin of the network. A diagram of the complete system is shown in Fig. 1. V_c is the converter voltage, v_f is the voltage of the point of common coupling (PCC), v_{inj} is the injected disturbance voltage and V_g is the grid voltage. The Thevenin equivalent of the grid has a series resistance R_g and inductance L_g. The inductance L_f of the filter has an effective series resistance (ESR) R_f, and the capacitance is C_f.

The controller of the inverter is based on a conventional synchronous reference frame vector control as shown in Fig. 1. It contains a decoupling feedback loop and a pair of Proportional Integral (PI) regulators [9]. This is nested within an active power control loop acting on the d component of the current reference and an AC voltage control loop acting on the q component both using PI controllers. A PLL fed with the voltage measured at the PCC provides information about the instantaneous modulus and angle of the positive sequence of the voltage. The angle of the positive sequence voltage is used as the angle for the Park reference frame.

The gains of the PI controllers were tuned based on the method explained in [9]. With an approximate bandwidth of 1 kHz for the current controller, 5 Hz for the PLL and 100 Hz for the power loops. All system parameters are listed in Table I.

Table I: System paramters and tuning values

Parameter	Description	Value
V_g	AC system voltage (RMS)	230 V
P	Converter rated power	50 kW
f	AC frequency	50 Hz
L_g	Grid side inductance	6.79 mH
R_g	Grid side resistance	0.2133 Ω
L_f	Inverter side inductance	8.48 mH
R_f	Inverter side resistance	53.3 mΩ
C_f	Inverter side capacitance	298 μF
τ_{cc}	Current controller time constant	0.5 ms
$K_{P_{cc}}$	Proportional gain current controller	1.6977
$K_{I_{cc}}$	Integral gain current controller	106.667
τ_{cc}	Time constant PLL loop	0.054
$K_{P_{pll}}$	Proportional gain PLL	0.1632
$K_{I_{pll}}$	Integral gain PLL	3.0219
K_{P_p}	Proportional gain power regulator	$1x10^{-3}$
K_{I_p}	Integral gain power regulator	0.042
$K_{P_{ac}}$	Proportional gain AC controller	0.5
$K_{I_{ac}}$	Integral gain AC controller	55

Furthermore, the pre-set power reference is compared with the calculated power to provide the d current reference to the inner control loop. The power is calculated from the measured current and voltage before the PCC. The voltage control loop is fed with the voltage measured at the PCC [10].

Frequency response identification methodology

In oder to obtain the system frequency response, a small-signal deviation of the steady-state operating point is caused by injecting a small voltage in series with the Thevenin equivalent of the grid. This voltage is a three-phase signal, ie for each tested frequency a voltage is injected in each of the three phases of the network with angles adjusted to have 120 degrees of difference between phases. Furthermore, the signal can be generated to contain negative or positive sequence (or both).

The frequency of the injected voltage is swept in the range between 10 Hz and 490 Hz with intermediate points generated using a logarithmic scale. For each value of frequency in the series, a frame of voltage and current data samples at 10 kHz is obtained for a duration of 0.5 s before the injection and 0.5 s in steady-state after injecting the disturbance. A Fast Fourier Transform (FFT) is applied to the data to transform the samples into phasors. The vector of *abc* voltages and currents corresponding to the specific frequency of test is multiplied by the Fortescue transformation matrix to obtain the phasors of the positive and negative sequence. The difference between the values obtained from the frame taken before and after the injection is used to eliminate the offset created on voltages and currents when operating at different power set-points.

An example of voltage at the PCC for an individual simulation is shown in Fig. 2. In this case, the voltage injected contained 1 V phase-to-phase rms of positive sequence and 1 V of negative sequence. Complete frequency response plots obtained using this method are presented for a series of scenarios of interest in the next section.

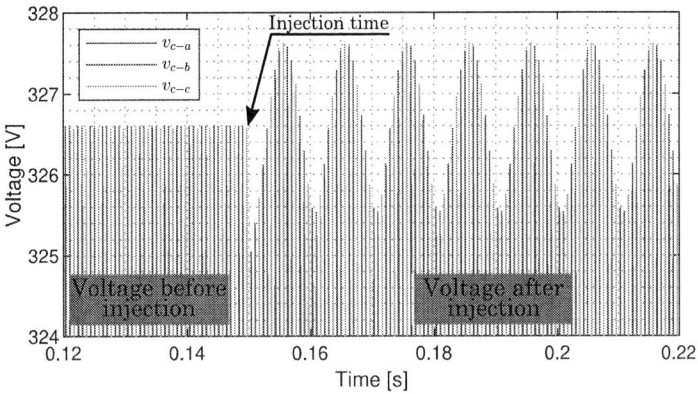

Fig. 2: Voltage at the point of common coupling upon injecting a small voltage disturbance

Case Study

The application of the proposed methodology is illustrated through a study case with three different simulation scenarios. First, the differences on the control performance observed when operating at different set-points is evaluated. The first scenario explores the effect of changing the power set-point of the converter. The second scenario investigates the operation at different AC voltage levels (see Fig.4a and 4b, respectively). The third scenario shows the impact of negative sequence voltage (see Fig. 5). The following subsections present the results obtained in detail. The values of the parameters considered in this study are summarised in Table I.

Scenario I: Varying the operating set-point

This section analyzes how the power set-point affects the system behaviour. Fig. 3b shows the performance of the *AC voltage control loop* at the base frequency of 50 Hz. This figure shows a detail of the peak of the AC voltages at the PCC before and after the disturbance. It can be seen that after the injection, the amplitude of the voltage goes through a transient excursion and finally returns to its nominal amplitude, which confirms the correct operation of the AC regulator. The diagram of the AC voltage regulator is shown in Fig. 1.

The effect of the injected voltage over the voltage measured at the PCC for different injected frequencies is shown in Fig. 3a. This plot is constructed by calculating the difference of the voltage amplitudes before and after the disturbance injection. Notice that the voltage difference at 50 Hz is zero, which means that the AC voltage regulator rejects the disturbance at nominal grid frequency as expected.

The trace in Fig. 3a also shows a resonant peak close to 350 Hz. This matches the resonant frequency of the LC filter ($f_{res} = \frac{1}{2\pi\sqrt{LC}}$), which gives 353Hz for the parameters shown in Table I.

Next, the frequency response for three different power set-points is obtained. The results are shown in Fig. 4a. This figure shows the module of the difference between the voltages before and after the injection of v_{inj}. This value would be zero if both the amplitude and the angle of the voltages before and after the injection were identical. This is not necessarily the goal of the AC voltage regulator, which only rejects the effect of the disturbance on the amplitude of the voltage. Furthermore, for all other frequencies different than zero, a value greater than one would imply that the inverter effectively amplified the effect of the disturbance on the voltage of the PCC.

It is noticed that the gain of the frequencies close to 50 Hz is not exactly 1 V. The design of the control loops (ie the PLL and the outer loops) is done such that they exhibit a low-pass behaviour. This helps neglecting disturbances above the design frequency (ie above 50 Hz). However, frequencies close to 50 Hz are not completely neglected. For instance, a 60 Hz disturbance is seen as a 10 Hz oscillation in the synchronous reference frame. Despite the low-bandwidth of the PIs in the outer loops, the low frequency ripple is still in the bandwidth of the controllers and causes the deviations observed in Fig. 4a. It is also

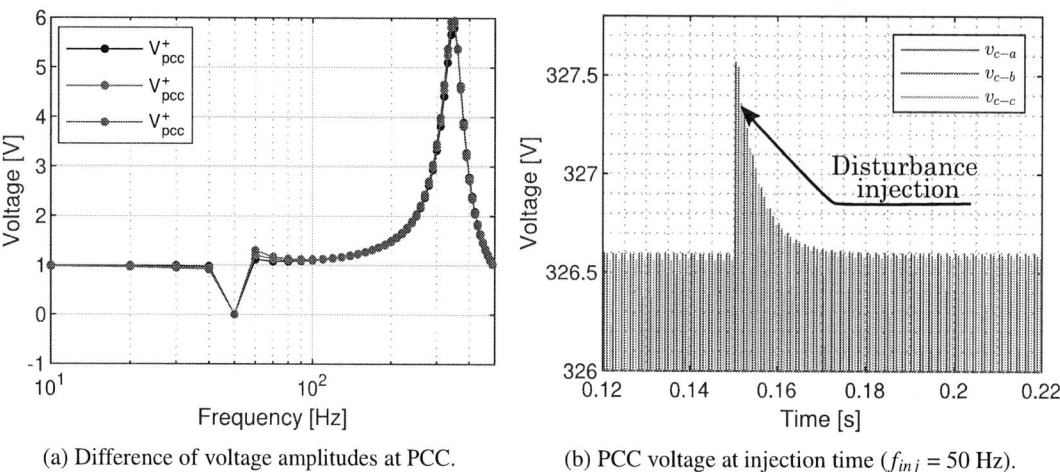

(a) Difference of voltage amplitudes at PCC.

(b) PCC voltage at injection time (f_{inj} = 50 Hz).

Fig. 3: Detail of system response at injection time (50 kW).

observed that the higher the power set-point of the converter, the larger the oscillation is seen in the $d - q$ components, which in turn impacts the voltage gain observed in the vicinity of 50 Hz.

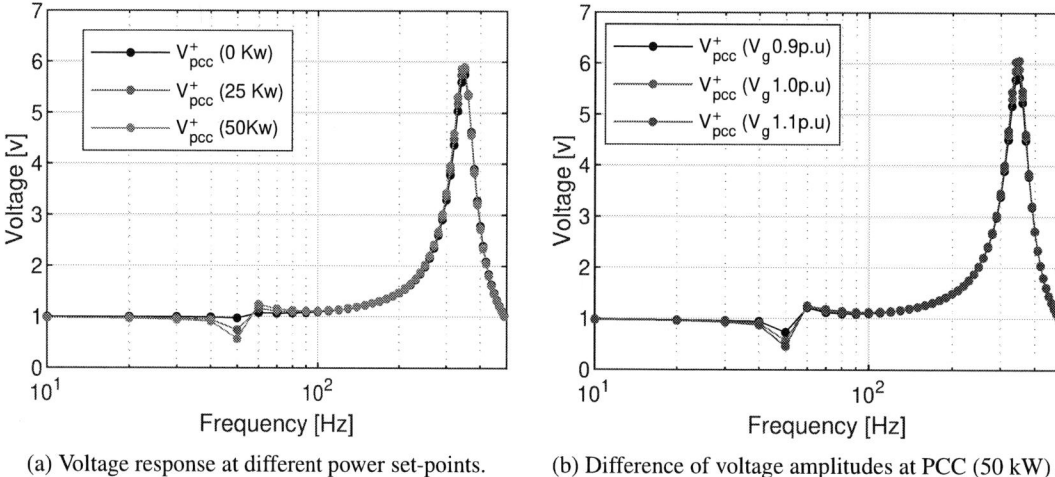

(a) Voltage response at different power set-points.

(b) Difference of voltage amplitudes at PCC (50 kW).

Fig. 4: Frequency response information of the system.

Scenario II: Different grid voltages

In this section, the grid voltage is changed in order to investigate its effect on the inverter's response. The result is shown in Fig. 4b. Even though the variation of the AC voltage is fairly large (specifically, from -0.1 pu to +0.1 pu), the effect on the response of the system is comparatively smaller than the change observed when varying the active power set-point, with slightly larger visible effect at 50 Hz the higher the AC voltage is.

Scenario III: Impact of the negative component on the control performance

The impact of negative sequence on the response is tested next. The results are shown in Fig. 5. Here the gain on the negative sequence component of the voltage is shown together with the gain of the positive sequence. Results are obtained for different power set-points: no active power, half rated power and full rated power.

The results show that there's no visible effect of the AC voltage regulator on the negative sequence voltage. This is because the AC voltage regulator uses the amplitude of the positive sequence voltage

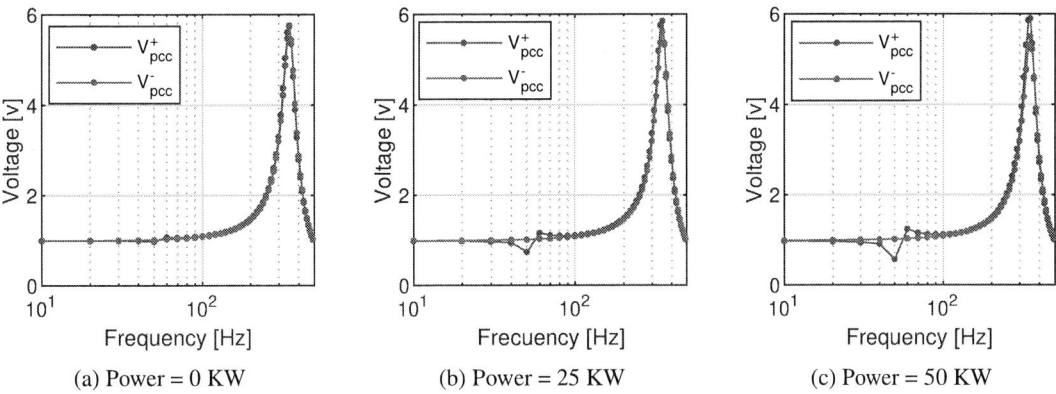

(a) Power = 0 KW (b) Power = 25 KW (c) Power = 50 KW

Fig. 5: Positive and negative voltage response

detected by the PLL, which effectively makes it ignore the negative sequence component of the voltage. Regarding the effect of changing the power set-point, here its effect on the positive sequence gain is seen as described in Scenario I, whereas no apparent change is observed on the negative sequence gain.

Conclusion

This paper has presented a method to identify how an inverter affects the voltage at its point of connection when voltages of different frequencies are injected in the equivalent model of the network. The results obtained through this method provide insights about the effectiveness of the AC voltage regulator, and the damping of the output filter of the inverter. This is useful when judging the design of the inverter hardware and its controller. The method also provides a way to judge how the converter contributes to the rejection of variations of the voltage of the network. This is desirable feature that can help improving the power quality of the voltage seen by loads connected in the vicinity of the inverter.

References

[1] M. Liserre, R. Teodorescu, and F. Blaabjerg, "Stability of photovoltaic and wind turbine grid-connected inverters for a large set of grid impedance values," *IEEE Transactions on Power Electronics*, vol. 21, no. 1, pp. 263–272, 2006.

[2] Z. Shen, M. Jaksic, P. Mattavelli, D. Boroyevich, J. Verhulst, and M. Belkhayat, "Three-phase ac system impedance measurement unit (imu) using chirp signal injection," in *2013 Twenty-Eighth Annual IEEE Applied Power Electronics Conference and Exposition (APEC)*, pp. 2666–2673, 2013.

[3] Y. Familiant, K. Corzine, J. Huang, and M. Belkhayat, "Ac impedance measurement techniques," in *IEEE International Conference on Electric Machines and Drives, 2005.*, pp. 1850–1857, 2005.

[4] M. Jaksic, Z. Shen, I. Cvetkovic, D. Boroyevich, R. Burgos, and P. Mattavelli, "Wide-bandwidth identification of small-signal dq impedances of ac power systems via single-phase series voltage injection," in *2015 17th European Conference on Power Electronics and Applications (EPE'15 ECCE-Europe)*, pp. 1–10, 2015.

[5] J. Sun, "Impedance-based stability criterion for grid-connected inverters," *IEEE Transactions on Power Electronics*, vol. 26, no. 11, pp. 3075–3078, 2011.

[6] X. Zhao, C. Yao, C. Zhang, and A. Abu-Siada, "Toward reliable interpretation of power transformer sweep frequency impedance signatures: experimental analysis," *IEEE Electrical Insulation Magazine*, vol. 34, no. 2, pp. 40–51, 2018.

[7] E. Al Murawwi and B. Barkat, "A new technique for a better sweep frequency response analysis interpretation," in *2012 IEEE International Symposium on Electrical Insulation*, pp. 366–370, 2012.

[8] K. Pourhossein and M. Asadi, "Identifying internal defects of photovoltaic panels using sweep frequency response analysis," in *2019 International Aegean Conference on Electrical Machines and Power Electronics (ACEMP) & 2019 International Conference on Optimization of Electrical and Electronic Equipment (OPTIM)*, pp. 481–485, 2019.

[9] A. Egea-Alvarez, S. Fekriasl, and O. Gomis-Bellmunt, "Advanced vector control for voltage source converters connected to weak grids," in *2016 IEEE Power and Energy Society General Meeting (PESGM)*, pp. 1–1, 2016.

[10] A. Egea-Alvarez, S. Fekriasl, F. Hassan, and O. Gomis-Bellmunt, "Advanced vector control for voltage source converters connected to weak grids," *IEEE Transactions on Power Systems*, vol. 30, no. 6, pp. 3072–3081, 2015.

Artificial-Intelligence based DC-DC Converter Efficiency Modelling and Parameters Optimization

Fanghao Tian, Diego Bernal Cobaleda and Wilmar Martinez
Electrical Engineering Department (ESAT)
KU Leuven - Energyville
Diepenbeek - Genk, Belgium
fanghao.tian@kuleuven.be

Keywords

≪DC-DC converter≫, ≪Design optimization≫, ≪Neural network≫, ≪Artificial intelligence≫

Abstract

This paper proposes a modeling and parameter design method for DC-DC converters based on artificial intelligence (AI). Initially, a database of switching losses is constructed using Spice simulation data with a single fixed semiconductor switch. Next, an artificial neural network (ANN) is trained by the database. Then Transfer Learning (TL) is implemented to train other ANNs for other switches with much less training data needed. Finally, under the restrictions of current and voltage ripples, a heuristic optimization algorithm is used to obtain the most efficient and optimal design. The results show that the ANN models give precise estimates of the converter properties.

Introduction

DC-DC power converters are widely used in many applications, such as electric vehicles [1], electric aircraft [2], and solar photovoltaic systems [3]. Additonally, the rising demand of these applications presents new challenges for power converter design, specifically on efficiency, volume, and mass. As a result, design automation has emerged as a novel research area in power electronics, with the goal of using artificial intelligence (AI) approaches to automate and accelerate the design process. However, the design of power converters is complicated and includes topology design, parameter selection, semiconductor and inductor modelling, power loss calculation, and optimization. Consequently, the design of power converters demands extensive simulation resources and time. However, AI has the ability to expedite the design process and reduce reliance on humans, hence enabling design automation.

Authors in [4] proposed an artificial neural network (ANN) based model named Mag-net, which is trained by data from practical experiment measurements, to calculate the magnetic core loss in a fast way. Furthermore, Transfer learning (TL), which is an idea of training a new ANN with much fewer data by adopting a trained ANN in a similar domain [5], is implemented to expand their database on more magnetic materials quickly [6]. Additionally, an ANN-based inductor evaluation model called AI-Mag, which is trained by data from Finite Element Methods (FEM) simulation, is proposed for accelerate the inductor design [7]. Besides, AI is also utilized to optimize the circuit parameters of power converters [8], and the ANN model is also trained to calculate efficiency under different combinations of frequency, inductance and capacitance values with the constraints of current and voltage ripples.

This paper is organized as follows: The data generation and neural network algorithm are explained in the first section. Following this, the first ANN is trained, and TL is utilized in training other ANNs with much less training data. Finally, based on all the ANN models, the particle swarm optimization (PSO) algorithm is used to optimize the efficiency, followed by results and conclusion.

Data Generation

Semiconductor switches are essential components in power converters, which usually work in high frequency and have unavoidable losses. In addition, It is difficult to derive an accurate mathematical power loss model of the switches under different working conditions. However, numerical simulation software such as LT Spice helps to evaluate the performance of circuits accurately. In this paper, a DC-DC synchronous buck converter is set as an example Fig. 1.

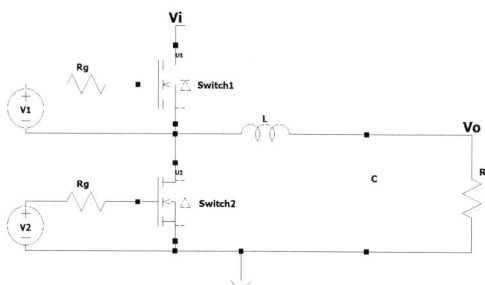

Fig. 1: Schematic of buck converter in LT Spice

The simulation parameters are shown in Table I. An input voltage of 30 [V] is converted to 12 [V] with an output power of 60 [W]. Frequency, inductance, and capacitance values are set as variables with a range. 20 frequencies, 50 inductance values, and 10 capacitance values are chosen to form $20 * 50 * 10 = 10000$ combinations of parameters for one specific type of switch. By simulating each case, measurement results of current and voltage on both sides are collected, based on which the efficiency, current ripple and voltage ripple are calculated to construct the training dataset.

Table I: Parameters setting for buck converter

Parameters	Values
Input Voltage (V)	30
Output Voltage (V)	12
Power (W)	60
Frequency(kHz)	10-1000
Inductance(μH)	10-1510
Capacitance(nF)	50-5050

In the simulation, Python is used to automate the data generation process. A list of different switches is created with their parameters, including component name, maximal drain-to-source voltage and current, rising and falling time. The external gate resistance is set as $5[\Omega]$, the dead time is set as $10[ns]$, and the rising and falling time are obtained based on the information from datasheets [10].

ANN Design and training

ANN is extensively used for supervised learning on a variety of classification and regression tasks, where labeled data is utilized to train the ANN, allowing it to identify the relationships between inputs and outputs. The fundamental ANN consists of layers of neural nodes. Nodes in adjacent layers are interconnected, forming a network. Every node is a function of all nodes from previous layer, using a weighted linear function and a nonlinear activation function. An example of ANN with 2 hidden layers is illustrated in Fig. 2.

For each node, the weighted linear transfer function and the nonlinear activation function are applied as shown in (1) and (2). The chosen activation function is Rectified Linear Unit (ReLU) function.

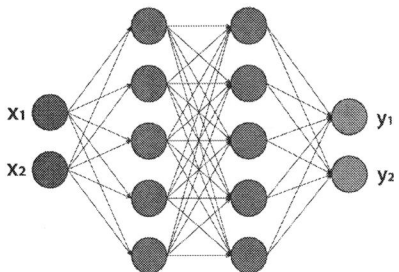

Fig. 2: Structure of an ANN

$$a = \mathbf{w}\mathbf{x} + b \tag{1}$$

$$y = max(0, a) \tag{2}$$

Where $\mathbf{x}$ is the vector of all values from the previous layer, $\mathbf{w}$ is the weight parameters from the nodes of previous layer to the current node, and b is the bias. In the ReLU function, positive a is kept while negative a is set as 0.

The ANN used to model the DC-DC buck converter is constructed as shown in Fig. 3. The ANN is trained with frequency, capacitance, and inductance data to model the outputs of efficiency, current ripple, and voltage ripple.

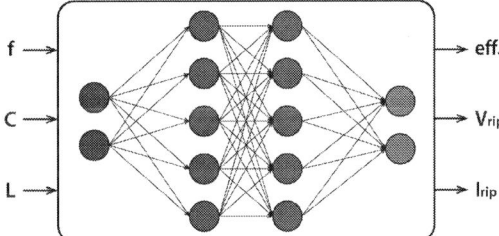

Fig. 3: The ANN model for modelling buck converter

In order to avoid training errors caused by parameters range differences, a logarithmic or linear normalization procedure is needed for the data as shown in (3) and (4) respectively.

$$y_{log} = log_{10}x \tag{3}$$

$$y_{linear} = \frac{x}{x_{max} - x_{min}} \tag{4}$$

Both of logarithmic and linear normalization are implemented on the switching frequency of the converter, and only linear normalization is implemented on the inductance and the capacitance values.

A benchmark switch is selected for the training of the benchmark ANN model. 10,000 sets of data are randomly divided into a training dataset (80%), validation dataset (10%), and testing dataset (10%). To obtain a precise ANN model, various ANN sizes with varying numbers of layers and nodes are evaluated. Eventually, a 4-layer ANN with 128 nodes per layer is selected for training because it has the least training error.

TL for other switches

TL is a technique for training a new ANN with significantly less data by adopting a previously trained ANN in a similar domain. The benchmark ANN model cannot be directly applied to other switches.

However, various switches share similar characteristics. As a result, TL can work as an effective way for training new ANN models for other switches, thereby constructing a database of ANN models for other switches that provide data for optimizing power converter design parameters. The workflow of TL is depicted in Fig. 4.

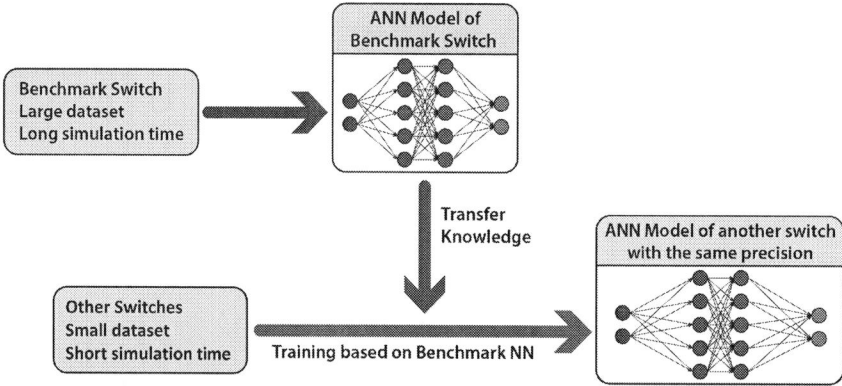

Fig. 4: Workflow of TL

Once the ANN model of the benchmark switch has been obtained, the data for another switch is also collected using a significantly smaller sample size of 2,000, which is only 20% of the size of the benchmark data set. Likewise, the 2000 data points are sorted into three categories. By using the benchmark ANN as the initial state, knowledge from the benchmark ANN model is transferred into the new ANN model, resulting in the need for far fewer data points to reach the same level of precision in the new ANN model as the benchmark ANN.

Parameters optimization by PSO Algorithm

Once the ANN models for several switches have been constructed. The efficiency and ripples can be easily calculated under various parameter combinations. PSO, a classic optimization algorithm, is applied to optimize the input parameters for maximizing the efficiency with satisfying ripples constraints. The optimization problem can be described as (5).

$$\begin{cases} \eta = f(switch, f, C, L) \\ switch \in switch\ list \\ Irip < Irip_{max}, Vrip < Vrip_{max} \end{cases} \tag{5}$$

where η is the efficiency and the optimization objective is to maximize the efficiency.

PSO is a classic bio-inspired optimization algorithm that imitates the swarming behaviors of birds flocking [10]. A group of searching agents work at the same time. After initializing the positions and velocity of the particle swarm, the particle with the best performance on the objective function is defined as global best *gbest*. In the meantime, the best position of the single particle during the whole iteration process is defined as particle best *pbest*. Each particle updates its speed and position as follows:

$$\begin{cases} v = v + c_1 \times r_1 \times (pbest - x) + c_2 \times r_2 \times (gbest - x) \\ x = x + v \end{cases} \tag{6}$$

where v is the updating speed and x is the position. c_1 and c_2 are the learning rate for individual and global updates. Additionally, r_1 and r_2 are two random number between $[0, 1]$, which increases the randomness and enhances the searching ability of the swarm. During the iteration process, the updating procedure is repeated until the optimal position converges to a stable value.

Finally, the whole design and optimization procedures is shown as Fig. 5

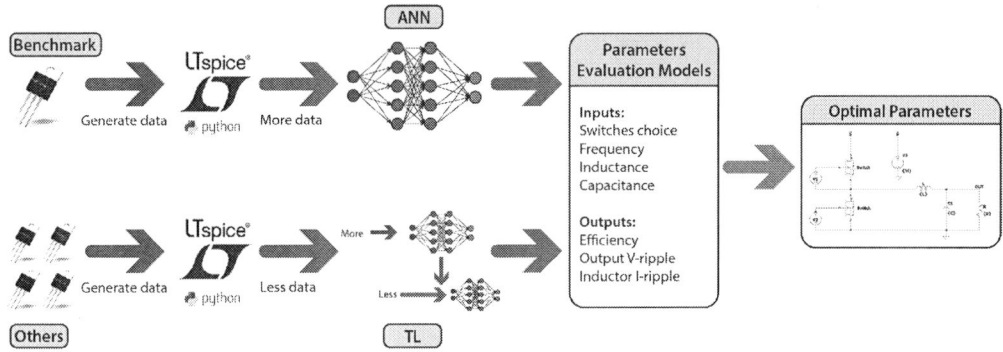

Fig. 5: Design and optimization workflow

Results

ANN training

The benchmark ANN was initially trained for 1000 epochs. As demonstrated in Fig. 6(a), the mean squared errors (MSE) of the training and validation datasets converged to small values. The decreasing errors in the validation dataset indicate that the ANN is not over-fitted. The error on the test dataset including 1000 sets of data is 2.67×10^{-6}. As a result, the trained ANN is effective at simulating efficiency and ripples. Fig. 6(b) depicts a comparison between the efficiency of the benchmark ANN and the Spice simulation. The curve shows the efficiency changes by frequency when capacitance is $2.5[\mu F]$ and inductance is $1[mH]$. The frequency varies from 10 [kHz] to 1 [MHz] on the logistic scale. The data represented by red dots, which were not part of the training data, are collected from Spice simulations. As depicted in Fig. 6, the benchmark ANN model is well trained and can compute the efficiency precisely under various scenarios.

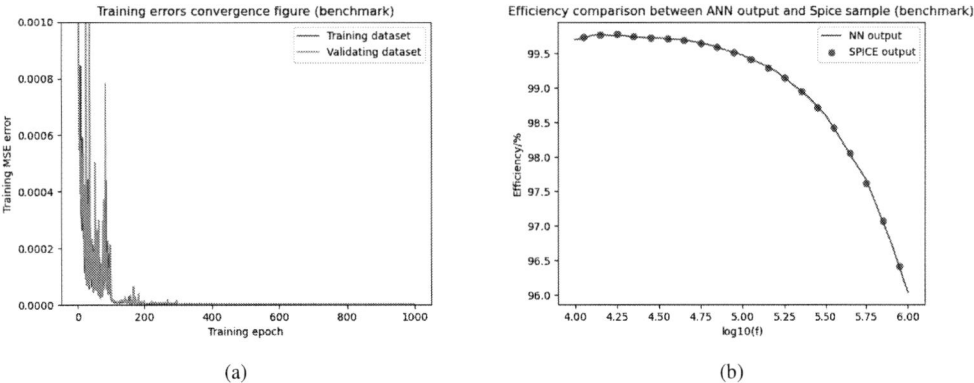

(a)   (b)

Fig. 6: Training loss (a) and efficiency validation (b) for the benchmark switch

After obtaining the benchmark ANN model, data from other switches is collected and used to train the other ANN models. On the basis of TL theory, the benchmark ANN can be utilized to efficiently train other models. For the generation of ANN models for other switches, only a modest database, whose size is only 10% of the benchmark database, is required. Fig. 7 shows the results of ANN models for three additional switches. Fig. 7(a)(b)(c) indicate the training errors which the comparisons between ANN outputs and Spice simulations are illustrated in Fig. 7(d)(e)(f). The testing loss for the 3 switches are 8.88×10^{-6}, 3.63×10^{-6} and 2.15×10^{-6} respectively, which proves that small database can train an ANN with high accuracy based on TL. The comparison figures shows the the ANN models provide an accurate estimation of efficiency calculation.

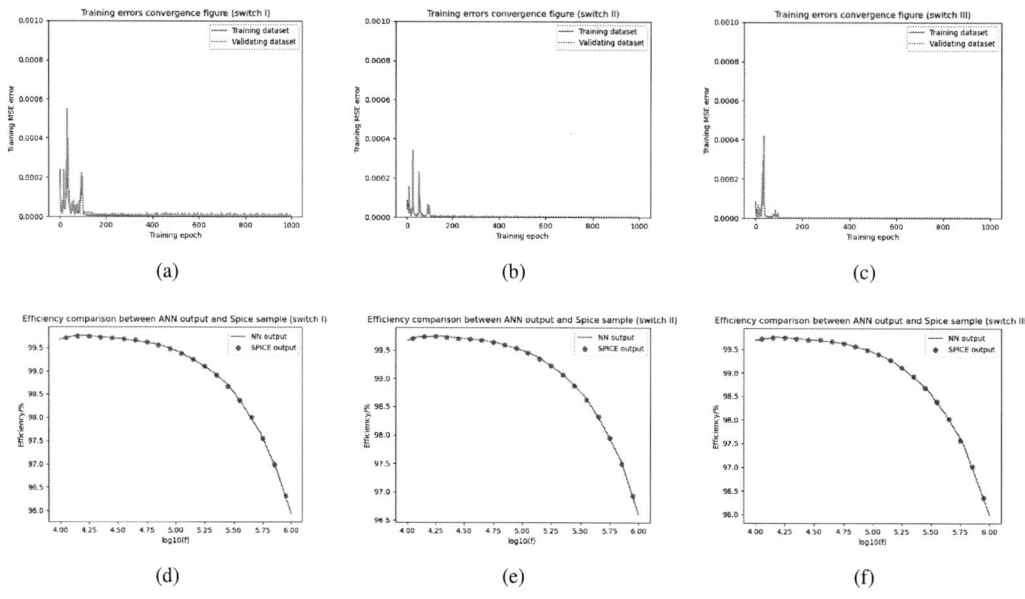

Fig. 7: Training loss (a)(b)(c) and efficiency validation (d)(e)(f) for the other switches

The results presented prove that TL is an effective technique to construct an ANN model efficiently. Much less data is needed to be collected for constructing the ANN models. Once all ANN models for the switches list are developed, heuristic algorithms, such as PSO, will be implemented to optimize the parameters.

Parameter optimization

The previous section shows that the ANN models can accurately represent the Spice simulation data. Much less simulation time is needed for Optimization by using the ANN models. A swarm group of 200 is chosen with the random initialization. 4 ANN models were used to be optimized respectively, the optimal values curves are shown in Fig. 8. The current ripple and voltage ripple limits are set as 10% and 1% respectively.

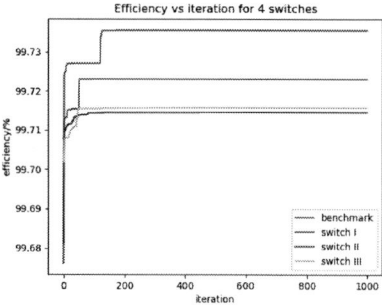

Fig. 8: Optimal efficiency via epoch in PSO algorithm

The best parameters selections are frequency of 30.367 $[kHz]$, inductance of 1.43 $[mH]$, and capacitance of 5.05 μF, with the benchmark switch, in which case the efficiency can reach 97.37% within the ripple constraints.

Conclusion

This paper proposed an AI-based DC-DC buck converter efficiency modelling and parameters optimization method. Spice simulations are used to generate data, which are collected to train an ANN model. TL is implemented to accelerate the process of modelling the power losses of different switches. Finally, the parameters are optimized to maximize the efficiency while considering the ripple constraints. Results indicate that TL is an effective method for training multiple ANNs in similar domains.

References

[1] Martinez W, Yamamoto M, Imaoka J, Velandia F and Cortes C. A.: Efficiency optimization of a two-phase interleaved boost DC-DC converter for electric vehicle applications, IPEMC-ECCE Asia 2016, pp. 2474-2480

[2] Benzaquen J, He J and Mirafzal B.: Toward more electric powertrains in aircraft: Technical challenges and advancements, CES Transactions on Electrical Machines and Systems vol. 5, no. 3, pp. 177-193, 2021

[3] Aguirre M and Yazdani A.: A Single-Phase dc-ac Dual-Active-Bridge Based Resonant Converter For Grid-Connected Photovoltaic (PV) Applications, EPE 2019, pp. 1-10

[4] Li H, Lee S. R, Luo M, Sullivan C. R, Chen Y and Chen M.: MagNet: A Machine Learning Framework for Magnetic Core Loss Modeling, 2020 IEEE 21st Workshop on Control and Modeling for Power Electronics (COMPEL), pp. 1-8

[5] Pan S. J and Yang Q,: A Survey on Transfer Learning, IEEE Transactions on Knowledge and Data Engineering, vol. 22, no. 10, pp. 1345-1359

[6] Dogariu E, Li H, Serrano López D, Wang S, Luo M and Chen M.: Transfer Learning Methods for Magnetic Core Loss Modeling, 2021 IEEE Workshop on Control and Modeling of Power Electronics (COMPEL)

[7] Guillod T, Papamanolis P and Kolar J. W.: Artificial Neural Network (ANN) Based Fast and Accurate Inductor Modeling and Design, IEEE Open Journal of Power Electronics, vol. 1, pp. 284-299, 2020

[8] Li X, Zhang X, Lin F and Blaabjerg F.: Artificial-Intelligence-Based Design (AI-D) for Circuit Parameters of Power Converters, IEEE Transactions on Industrial Electronics

[9] Vishay Siliconix, Appl. Note AN608A, pp.1-6. Available: https://www.vishay.com/docs/73217/an608a.pdf.

[10] del Valle Y, Venayagamoorthy G. K, Mohagheghi S, Hernandez J. -C and Harley R. G.: Particle Swarm Optimization: Basic Concepts, Variants and Applications in Power Systems, IEEE Transactions on Evolutionary Computation, vol. 12, no. 2, pp. 171-195, 2008

Analysis of the Loss Distribution of a 6 kW two Stage Power Supply for 600 V DC Applications

Lukas Fräger[1], Sascha Langfermann[1], Michael Owzareck[1], Dennis Kampen[1], Jens Friebe[2]

[1]BLOCK TRANSFORMATOREN-ELEKTRONIK GmbH
[2]Leibniz Universität Hannover, Institute for Drive Systems and Power Electronics
[1]Max-Planck-Straße 36-46 / [2]Welfengarten 1
[1]Verden / [2]Hannover, Germany
Tel.: +49 / (0) 4231 678 434
Fax: +49 / (0) 4231 678 277
E-Mail: lukas.fraeger@block.eu
URL: http://www.block.eu

Acknowledgements

Parts of this work were funded by the German Federal Ministry for Economic Affairs and Energy (BMWi) under grant number 03EN2010A-G in the project STIM. The authors are responsible for the content of this publication.

Keywords

Voltage Source Inverters (VSI), Modelling, Silicon Carbide (SiC), High frequency power converter, Switching Losses

Abstract

This paper presents a two-stage AC/DC power supply consisting of a silicon carbide power semiconductor based active front end and a dual active bridge DC/DC converter. The paper presents the design considerations, efficiency and loss analysis based on measurements. The focus is specially drawn to the influence of the variable DC-link voltage on the overall efficiency and loss distribution among the two stages.

1 Introduction

Current developments in industry and society lead to the need for high power AC/DC power supplies in low voltage grids (e.g. charging stations, energy storages, DC-grids for industry [1] [2] [3]). Additionally, for future applications, bidirectional supplies are favored to feed power back to the grid (e.g. braking energy, integration of renewables and batteries). This paper presents a 6 kW power supply designed for high power density and good efficiency. It gives a basis for optimization of the efficiency depending on input and output voltage and includes detailed measurement results.

2 System Specification

The power supply is designed to supply a fixed DC voltage of 600 V. This voltage can be used to supply off-the-shelf industry inverters that currently work with an uncontrolled rectified DC-link voltage which is at 563 V in a 400 V grid. For worldwide applicability, the power supply must be able to work in most low voltage grids. An overview of worldwide grid voltages is given in Table 1.

Table 1: Excerpt of international grid voltages [4]

Country / Region	Three Phase line to line grid voltage(s)
EU	400 V
USA/Canada	208 V, 480V, 600 V (only Canada)
Japan	200 V
China	380 V

Open DC-Grids as in [5] are one possible application for the presented power supply. Thus, a galvanic isolation is favored especially when the grids are supposed to have an independent earthing concept (e.g. grounded Δ-Grid to mid point grounded DC-Grid). Also, when connecting multiple supplies in parallel, an isolation can help reducing circular currents between the converters. The resulting specifications are given in Table 2.

Table 2: Specification of AC/DC-power supply

Specified Value	Value
Input voltage range (line to line)	$200\ V \dots 480\ V$
Output voltage (DC)	$600\ V\ (550\ V \dots 650\ V)$
Power	$6\ kW$
Isolation type	*Reinforced*

3 System description and design considerations

A block diagram of the built-up system is given in Figure 1. The system consists of two stages: A bidirectional active front end (AFE) and a bidirectional, galvanically isolated DC/DC converter.

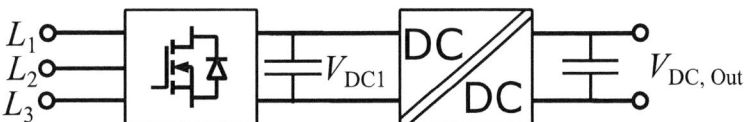

Figure 1: Block diagram of a two stage AC/DC power supply with a three phase input (L_1, L_2, L_3) and DC output ($V_{DC,Out}$)

3.1 Topology Considerations

Both, the AFE as well as the DC/DC converter must provide bidirectional power flow. Due to the recent developments in wide bandgap technologies, efficient converters can be built with SiC MOSFETs in a standard two-level topology.

For the AFE, a three leg, two stage inverter with an integrated All-Pole-LC filter is used. Due to the use of high switching frequencies ($\sim 100\ kHz$), it is possible to reduce the size of the LC filter compared to standard AFEs and integrate it in one housing with the AFE [6]. The specified values for the AFE are shown in Table 3. The losses of the AFE mainly depend on the DC-Link voltage (switching and AC-filter losses) and input current (switching and conduction losses) [7]. The highest losses thus appear for low input voltage (following high input currents) and high DC-Link voltage. While the input current in a certain grid for an intended power is fixed, the DC-Link voltage may be varied. For the bidirectional AFE, it is favorable to be operated close to the minimum DC-Link voltage which can be calculated with the line-to-line voltage $V_{AC,ll}$ and the control reserve factor $R_{C,\%}$ when neglecting the voltage drop across the inductor:

$$V_{DC1,min} = V_{AC,ll} \cdot \sqrt{2} \cdot (1 + R_{C,\%})$$

Table 3: Specification of bidirectional active rectifier

Specified Value	Value
Input voltage range (line to line)	$200\ V \dots 480\ V$
DC-Link voltage range	$300\ V \dots 750\ V$

The specification of the DC/DC stage is shown in Table 4. Many different DC/DC converter topologies have been published in recent years [8] [9] [10]. However, for resonant topologies in higher power applications ($> 2\ kW$), expensive capacitors are needed in addition to the transformer. Also, resonant topologies tend to have a limited bidirectional capability in terms of transfer ratio: In one direction, most resonant topologies have a maximum transfer-ratio equal to the transformer ratio. Thus, operated at a high AC input voltage $V_{AC,ll}$, energy transfer to the grid may not be possible at $V_{DC,Out} < V_{DC1,min}$ e.g.

$600\ V$ output voltage. To provide full bidirectionality, be able to vary the DC-Link voltage and to lower component cost, a dual active bridge (DAB) is chosen for the DC/DC converter. For first evaluations, the turns ratio of the medium frequency transformer is chosen to be 1:1. However, results show that depending on the preferred grids, a better optimum may be found [8].

Table 4: Specification of DCDC converter

Specified Value	Value
Input voltage range	$300\ V \dots 750\ V$
Output voltage	$600\ V$
Power	$6\ kW$
Power flow	*Bidirectional*
Isolation	*Reinforced*

3.2 System Description

A resulting overview schematic is given in Figure 2. All power semiconductors are Silicon Carbide MOSFETs, which allow a high switching frequency as well as a high efficiency. All half bridges are 2-level configurations to keep the component count and cost low.

To comply with electromagnetic compatibility (EMC) standards, a grid filter in front of the AFE is needed. The grid filter is integrated with the AFE. On one hand this way the power density can be optimized. Also, current paths for the filter currents are reduced, which reduces EMC issues.

The input and output stage of the DAB converter is comprised of full bridges to utilize the full voltage range at the input and output stage and thus reduce the currents. Also, this configuration enables the use of the third level and thus advanced modulation strategies [11] [12] [13].

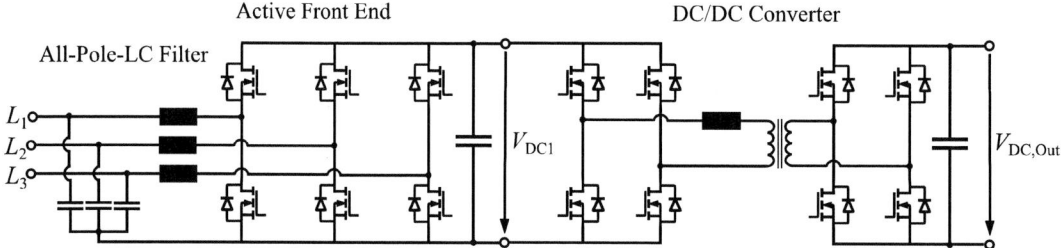

Figure 2: Schematic overview of the presented Power Supply with integrated all-pole-LC-filter, bidirectional AFE and DC/DC converter.

4 Hardware description

The test hardware is shown in Figure 3. For testing purposes, AFE and DC-DC converter are built up separately. However, they are designed to be integrated in a single housing.

Figure 3: Built up test hardware. Left: AFE with integrated LC filter and right: DC/DC converter with attached MF-Transformer.

Both AFE as well as DC/DC converter are built up with SiC power semiconductors. Lists of relevant components of the AFE and DC/DC converter are given in Table 5 and Table 6 respectively.

Table 5: AFE: List of relevant components and control settings

Component	Type
Semiconductors	Infineon IMZ120R060
DC-Link Capacitor	2x Vishay MKP1848C, 900 V 50 μF
Core Material	Sendust 60μ
Core Size	62x33x25 (Ring Core)
Number of Turns	35
Switching frequency	100 kHz

Table 6: DC/DC converter: List of relevant components and control settings

Component	Type
Semiconductors	Infineon IMZ120R090
DC-Link Capacitor	4x Vishay MKP1848C, 900 V 50 μF (2x primary, 2x secondary)
Core Material	EPCOS N87
Core Size	Stacked 3x EE55/25
Number of Turns Primary / Secondary	10 / 10
Litz Wire	1400x0.05 mm
Switching Frequency	100 kHz

5 Measurement Results

The aim of the presented measurements is to give a detailed picture of the influence of the DC-Link voltage V_{DC1} on the losses of both the AFE as well as the DC/DC converter. Thus, overall losses, losses of the DC/DC converter as well as losses of the AFE are measured and presented. Measurements were done separately with AFE and DC/DC converter. For all measurements, the measurement equipment described in Table 7 is used.

Table 7: Measurement Equipment

Equipment	Type
Oscilloscope	Tek MSO Series 4, 500MHz, 12 bit, 6Ch
Current Probe	TCP0030A
Voltage Probe	THDP0200
Power Analyzer	Yokogawa WT1800

The loss measurements include all losses including auxiliary supply losses for driving, control and fans. The auxiliary losses were analyzed independently and found to be below $4\,W$ for both the DC/DC converter as well as the AFE.

5.1 Loss Distribution

First, the losses of the AFE and the DC/DC converter are analyzed independently. Figure 4 shows the losses at $3.2\,kW$ output power and an input voltage $V_{AC,ll} = 208V$, Figure 5 at 6 kW output power and an input voltage $V_{AC,ll} = 400V$. Thermal measurements showed that the output power must be derated at low input voltages. Thus, measurements with nominal power are only shown for $V_{AC,ll} = 400V$.

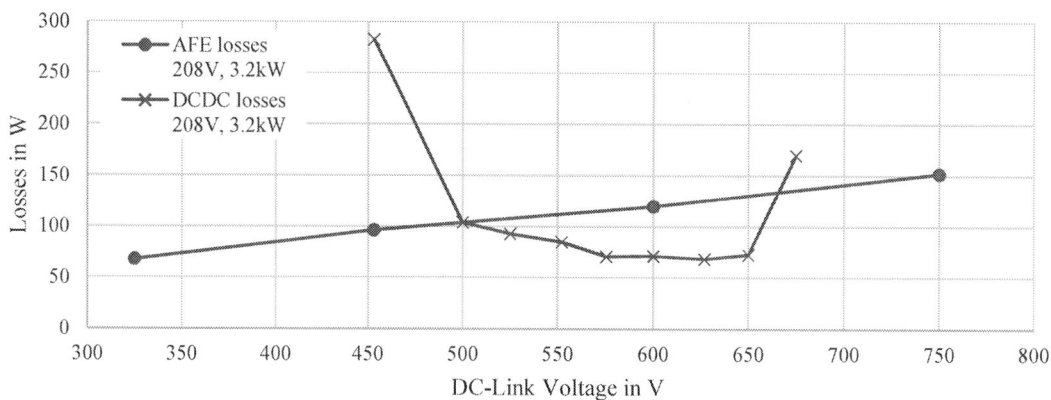

Figure 4: Measured and interpolated losses at 3.2 kW output power, 208 V input voltage and 600V output voltage of the AFE and the DC/DC converter.

Figure 4 shows the loss distribution among DC/DC converter and AFE at $P_{Out} = 3.2\ kW$ and $V_{AC,ll} = 208\ V$ for different DC-link voltages. Clearly, the loss distribution is influenced by the DC-Link voltage: The AFE-losses increase nearly linearly with a higher DC-Link voltage. The DC/DC converter losses increase strongly at transfer ratios far away from 1 (above 1.2 and below 0.9). It is assumed, that the converter leaves the soft switching region here. Between a DC-Link voltage of 500 V and 650 V, the DC/DC converter losses slightly decrease with a rising DC-link voltage. In this region, the losses can be distributed among DC/DC converter and AFE by means of the DC-link voltage. Minimum losses for the AFE can be found at $V_{DC1} = 500\ V$; minimum losses for the DC/DC converter at $V_{DC1} = 625\ V$. Further, for $500V \leq V_{DC1} \leq 650\ V$, the AFE losses dominate due to the low input voltage.

Figure 5 shows the losses of the DC/DC converter and AFE at $P_{Out} = 6\ kW$ and $V_{AC,ll} = 400\ V$ in dependence of the DC-link voltage. The soft switching region is increased compared to the measurements at 3.2 kW. Minimum losses are at $V_{DC1} = 500\ V$ for the DC/DC converter and at $V_{DC1} = 575\ V$ for the AFE respectively. It may be noted, that a and thus the system may operate at $V_{DC1,min} \geq 575\ V$.

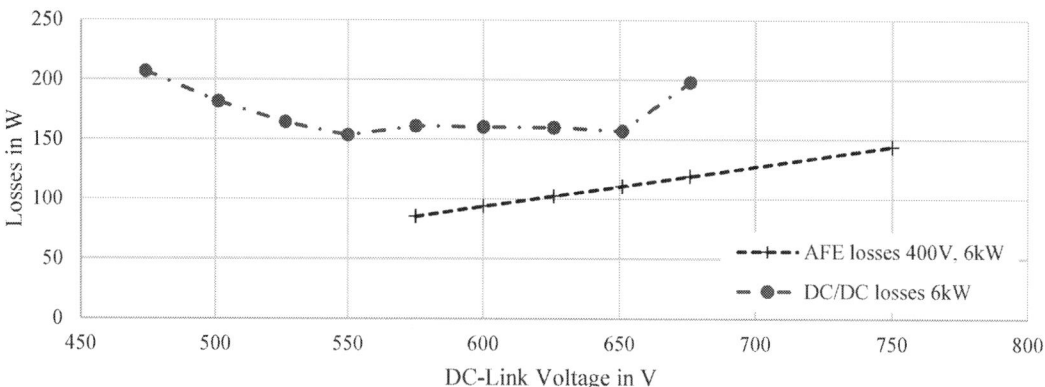

Figure 5: AFE and DC/DC converter losses at 6kW output power and 400V line to line input voltage in dependence on the DC-link voltage.

Figure 6 and Figure 7 show the overall losses for different power and input voltage levels. Figure 6 includes the system losses at $V_{AC,ll} = 208\ V$, Figure 7 at $V_{AC,ll} = 400\ V$.

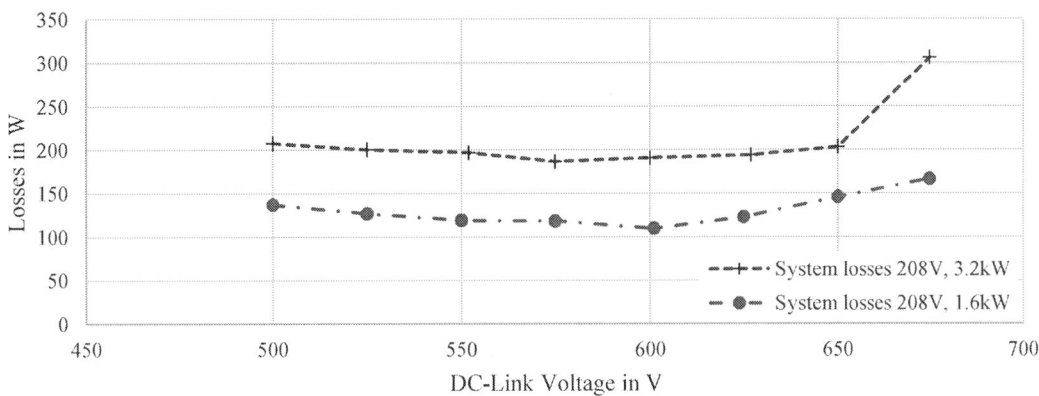

Figure 6: System losses at 208V input voltage, 600V output voltage in dependence of the DC-Link voltage for different output powers.

At $V_{AC,ll} = 208\ V$ the lowest overall losses for 1.6 kW and 3.2 kW are at $V_{DC} = 575\ V$ and $V_{DC1} = 600\ V$ respectively. However, losses of the AFE can be reduced by approximately 15% if operated at $V_{DC1} = 500\ V$ instead. This may be favorable as the AFE losses dominate for low input voltages.

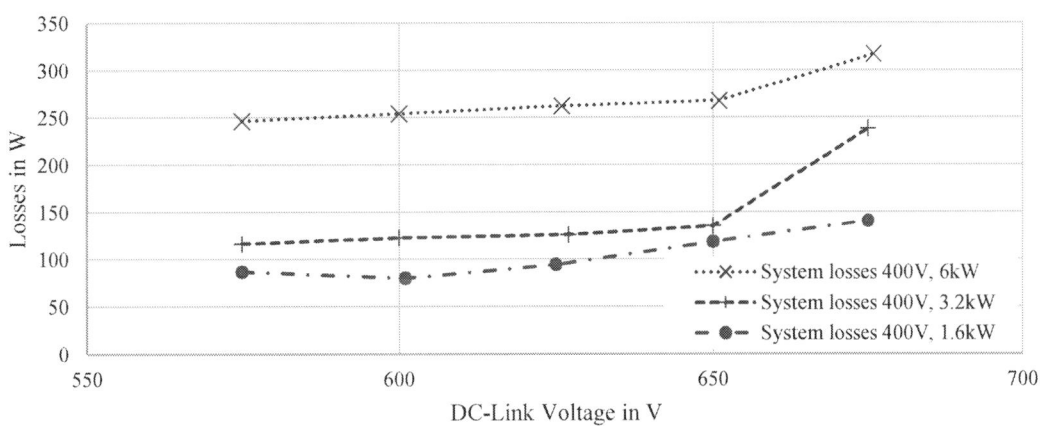

Figure 7: System losses at 400V input voltage, 600V output voltage in dependence of the DC-Link voltage for different output powers.

Figure 7 presents the system losses at $V_{AC,ll} = 400\ V$ at different output powers. At $P_{Out} = 6\ kW$ and $P_{Out} = 3.2\ kW$, the system losses are minimum at $V_{DC1} = V_{DC1,min} = 575\ V$. At $P_{Out} = 1.6\ kW$, losses become minimal at $V_{DC1} = 600\ V$.

At all presented operating points, the DC/DC converter losses would yet be lower for higher DC-link voltages (cmp. Figure 4 and Figure 5). It may be concluded that for the presented operating points a transformer turns ratio smaller than 1:1 would be favorable in terms of system efficiency.

5.2   System Efficiency

To complete the picture of the presented power supply, the overall system efficiency in dependence of the DC-Link voltage V_{DC1} is analyzed. The measurement results are given in Figure 8.

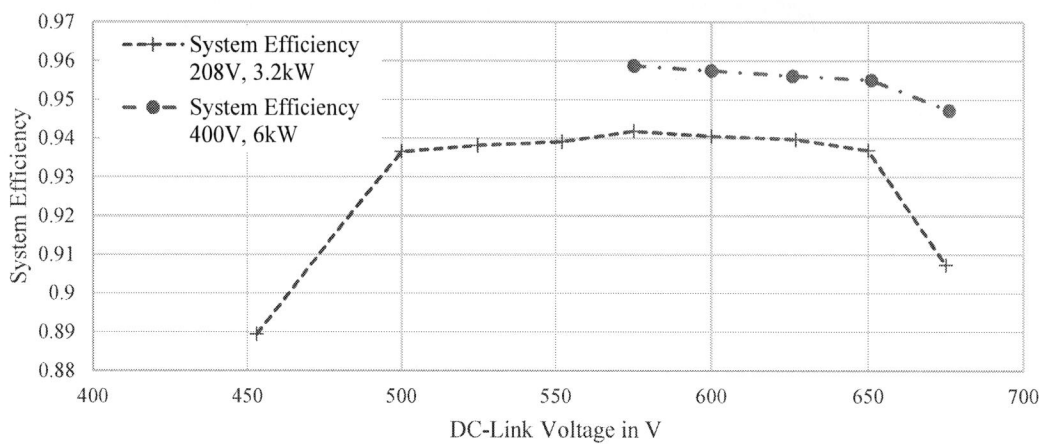

Figure 8: System Efficiency for an output power of 3.2kW and 6kW in dependence of the DC-Link voltage at an input voltage of $V_{AC,ll} = 208$ V and $V_{AC,ll} = 208$ V, respectively, and an output voltage $V_{DC,out} = 600$ V.

Maximum efficiency for both $V_{AC,ll} = 400\ V$ as well as $V_{AC,ll} = 208\ V$ is achieved for a DC-link voltage of $V_{DC1} = 575\ V$. Towards high / low transfer ratios between DC-link and output voltage, the efficiency drops clearly as it leaves the soft switching region of the DAB.

As the AFE-efficiency is much higher at larger input voltages $V_{AC,ll}$, the overall efficiency is higher at an input voltage of $V_{AC,ll} = 400\ V$.

6   Conclusion and Outlook

The paper analyzes the influence of the DC-Link voltage for a specific setup in detail. Results show, that by varying the DC-link voltage, losses can be distributed among the two stages of the AC/DC power supply. Depending on the grid voltage and the output power, different optimum DC-link voltages may be found. In future work, the influence of the DC-Link should be analyzed analytically for general setups. To optimize the DC-link voltage in systems during operation, analytic models are needed to be included in the control of power supplies. Further research should also focus on derating concepts depending on grid/input voltage and output voltage range.

7 References

[1] R. W. D. Doncker, *Fast Charging (350 kW) for Electric Vehicles - Possibilities and Issues,* Eindhoven, 2019.

[2] B. Sattler und S. Wiesner, „Research Project DC-Industrie - Energiewende meets Industrie 4.0,“ ZVEI - German Electrical and Eletronic Manufacturers' Association, Frankfurt, 2019.

[3] C. Meyer, M. Hoing, A. Peterson und R. W. D. Doncker, „Control and Design of DC-Grids for Offshore Wind Farms,“ *Conference Record of the 2006 IEEE Industry Applications Conference Forty-First IAS Annual Meeting,* 2006.

[4] "www.worldstandards.eu," 19 12 2021. [Online]. Available: https://www.worldstandards.eu/electricity/three-phase-electric-power/.

[5] B. Sattler and S. Wiesner, "Research Project DC-Industrie - Energiewende meets Industrie 4.0," ZVEI - German Electrical and Eletronic Manufacturers' Association, Frankfurt, 2019.

[6] S. Langfermann, M. Owzareck und L. Fräger, „Design Space Optimization of a SiC Drive Inverter with an Integrated All-Pole Sine Filter,“ *23rd European Conference on Power Electronics and Applications,* 2021.

[7] L. Fräger, S. Langfermann, M. Owzareck und J. Friebe, „An analytic inverter loss model for design and operation space optimization,“ *2021 23rd European Conference on Power Electronics and Applications,* 2021.

[8] P. Apte, S. Lin, L. Fräger und J. Friebe, „Design considerations for a 50 kW Dual-Bridge Series Resonant DC/DC Converter with Wide-Inpu Voltage Range for Solid-State Transformers,“ *2021 IEEE Energy Conversion Congress and Exposition (ECCE),* 2021.

[9] A. Hillers, D. Christen und J. Biela, „Design of a Highly Efficient Bidirectional Isolated LLC,“ *15th International Power Electronics and Motion Control Conference, EPE-PEMC,* 2012.

[10] B. Zhao, Q. Song, W. Liu und Y. Sun, „Overview of Dual-Active-Bridge Isolated Bidirectional DC–DC Converter for High-Frequency-Link Power-Conversion System,“ *IEEE Transactions on Power Electronics,* 2014.

[11] A. Jafari, M. S. Nikoo, F. Karakaya and E. Matioli, "Enhanced DAB for Efficiency Preservation Using Adjustable-Tap High-Frequency Transformer".*IEEE Transactions on Power Electronics.*

[12] V. Karthikeyan, S. Rajasekar, S. Pragaspathy und F. Blaabjerg, „A High Efficient DAB Converter under Heavy Load Conditions Using Inner Phase Shift Control,“ *IEEE International Conference on Power Electronics, Drives and Energy Systems (PEDES),* 2018.

[13] Fräger, Badenhop, Langfermann, Owzareck, Kampen and Friebe, "Dual Active Bridge Converter: Simple Peak Current Limitation by Dual Phase Shift Control," *PCIM 2022,* 2022.

[14] Langmaack et al, „Fast and Universal Semiconductor Loss Calculation Method".*IEEE PEDS 2019.*

[15] D. Christen und J. Biela, „Analytical Switching Loss Modeling Based on Datasheet Parameters for MOSFETs in a Half-Bridge,“ *IEEE TRANSACTIONS ON POWER ELECTRONICS,* 2019.

Study on the gate loop design and its impact on switching characteristics of GaN Transistors

Xiaomeng Geng[1], Carsten Kuring[1], Oliver Hilt[2], Mihaela Wolf[2], Joachim Würfl[2] and Sibylle Dieckerhoff[1]

[1] TECHNISCHE UNIVERSITÄT BERLIN
Einsteinufer 19
Berlin, Germany

[2] FERDINAND-BRAUN-INSTITUT
LEIBNIZ-INSTITUT FÜR HÖCHSTFREQUENZTECHNIK
Gustav-Kirchhoff-Str.4
Berlin, Germany
Tel.: +49 / (0)30 314-70024
Fax: +49 / (0)30 314-25526
E-Mail: xiaomeng.geng@tu-berlin.de
[1] URL: https://www.pe.tu-berlin.de
[2] URL: https://www.fbh-berlin.de

Keywords

«Parasitic inductance», «Gallium Nitride (GaN) », «Packaging», «Wide bandgap», «HEMT».

Abstract

This paper studies design parameters for the gate loop of GaN-based transistors. To achieve stable and fast switching of the GaN transistors aiming at MHz-range operation, different layout factors and their influence on gate loop inductances are investigated in simulations and measurements. Experimental results demonstrating the impact of driver ICs and their packages on the switching characteristics are presented as well.

Introduction

Wide bandgap gallium nitride (GaN)-based high-electron-mobility transistors (HEMTs) have low on-resistance and low parasitic capacitances compared with their Si-counterparts, which enable faster switching, lower losses, and smaller size of passive components [1]. These advantages make them increasingly preferred in high frequency and high efficiency converters. To achieve fast switching and take full advantage of GaN HEMTs aiming at MHz-range operation, both, the power loop and the gate loop have to be optimized. This paper focuses on the gate loop design. The gate loop inductance decreases the slew rate of the gate current and thus lowers the switching speed. Further, a large gate loop inductance forms a resonance circuit together with parasitic capacitances, causing overshoots and oscillations in the gate-source voltage. This degrades the electromagnetic interference (EMI) performance, efficiency, and device stability. The issue can be partially mitigated by large gate resistors damping critical oscillations, but this in turn limits the design of switching speed optimization [2, 3]. Generally, the gate loop inductance consists of the gate loop self-inductance and the common source inductance. The common source inductance quantifying the mutual interaction between the power and the gate loop has been studied in [4]. This paper focuses on the gate loop self-inductance L_g which quantifies the induced voltage in the gate loop due to the time-variant gate loop current.

To reduce the gate loop inductance and achieve a compact design, some manufacturers integrate the driver with GaN transistors [5, 6]. However, most of the commercial GaN transistors are still discrete devices. Several well-known "rule of thumb" recommendations for a proper gate loop layout to reduce L_g can be found in the driver IC datasheets or application notes [7, 8], e.g., placing the driver ICs close to power devices and reducing the length of current paths. However, to the authors' best knowledge, no quantitative investigations on the cause of the gate-loop inductance have been carried out to identify the most influential factors for GaN transistors. In order to optimize switching performance of GaN-based devices, further investigations are needed to establish a more systematic guiding for the gate loop design with discrete gate drivers.

As a key element in the gate loop circuit, the driver device itself has a large impact on the switching characteristics of GaN transistors. First, the parasitic inductances of the driver IC package directly contribute to the gate loop inductance. Second, additional parameters, such as the sink and source current capability of the driver ICs, internal parasitic resistances, and the different internal driver circuits, influence the switching performance. In recent years, many studies on the modeling and quantification of parasitic inductances of GaN transistors packages have been reported [9, 10]. Experimental tests were performed in [9] to extract the package parasitic inductances of the wide band-gap transistor in a through-hole technology (THT) package. In [10], the parasitic inductances of a cascode GaN transistor in a THT package are extracted using the 3D FEM simulation tool Ansoft Q3D Extractor and then used to build a lumped-element Spice simulation model. However, so far, the impact of different driver packages has not been studied in detail. Besides, these methods are difficult to adapt to the GaN-transistor drivers, since these are commonly surface-mounted devices (SMD) whose dimensions are much smaller than THT packages. Furthermore, assembly details of internal lead-frame and bonding geometry are usually not published. In addition, although it is well-known that the driver affects the switching behavior, it has not been studied in detail how large this effect is for GaN transistors, especially considering drivers with similar ratings stated in their datasheets. Thus, in this paper, we compare the switching behavior of one selected transistor type combined with different driver ICs while the PCB layout is kept widely identical, and use the experimental results in real operation conditions to directly show the impact of drivers and gain more straightforward insight into them.

The paper is organized as follows: The impact of different layout-related parameters on the gate loop inductance is evaluated with measurements and 3D FEM simulations in section 2. Section 3 demonstrates the impact of the driver ICs and their packages on the switching behavior of GaN transistors with experimental results. Conclusions are indicated at the end.

Parameter study of gate loop inductance

The gate loop self-inductance L_g is a result of the PCB layout as well as the incorporated passive and active device packages or chip dies. In order to optimize the converter design and achieve fast and stable switching transients, it is important to quantify the gate loop inductances and the contribution of different parameters, especially when a compromise between them is required. Therefore, the following different cases are compared (**Table I**) considering the *1EDN7511B* gate driver with a commonly used SOT-23 package [12]. Case 1 aims to achieve an optimal reference design case, in which the components are placed densely, and the 2nd layer of a 4-layer PCB is used as a ground return to reduce the vertical cross-section area of the gate loop enabling effective flux compensation. In case 2, the bottom layer is used as a ground return in order to compare a 2-layer design with a multi-layer design. In case 3, the 2nd layer is again used as a ground return path, but the bypass capacitors are moved 5 mm away from the driver. Lastly, in case 4, the load capacitor emulating the GaN transistors input capacitance is moved 10 mm away from the driver. Case 3 and case 4 serve for evaluating the influence of the position and distance of the components on L_g. Both, measurements and simulations are conducted to quantify L_g and to study the layout's impact.

Fig. 1 depicts the circuit diagram used for experimental investigation. In the measurement setup, the active GaN transistor is replaced by a fixed NP0 ceramic load capacitor C_{load}, which avoids the voltage-dependent nonlinearity of the transistor input capacitance and thus reduces complexity of the subsequent analysis. The load capacitor C_{load}=100 pF is in a similar range as the studied transistor's input capacitance C_{iss} (Fig. 7) and as for other commercial GaN transistors [13] to ensure a representative behavior. The transient gate current i_g is acquired through the voltage across the gate resistor R_{on} and R_{off}, and the voltage is measured by a 1 GHz high bandwidth voltage probe TPP1000.

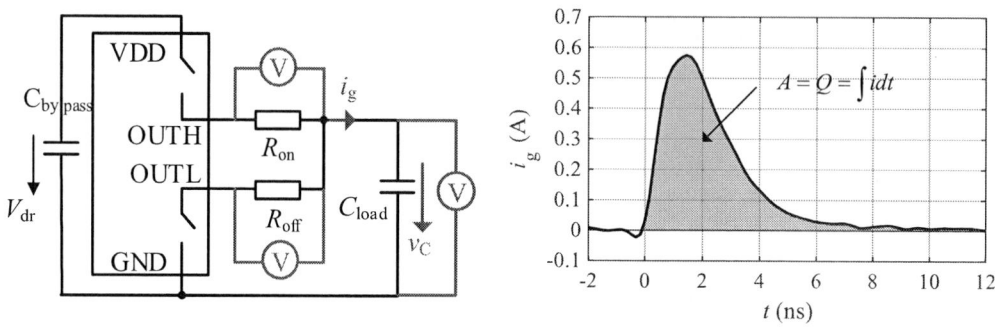

Fig. 1: Circuit diagram of the measurement setup

Fig. 2: Charge calculation using gate current at turn-on of case 1

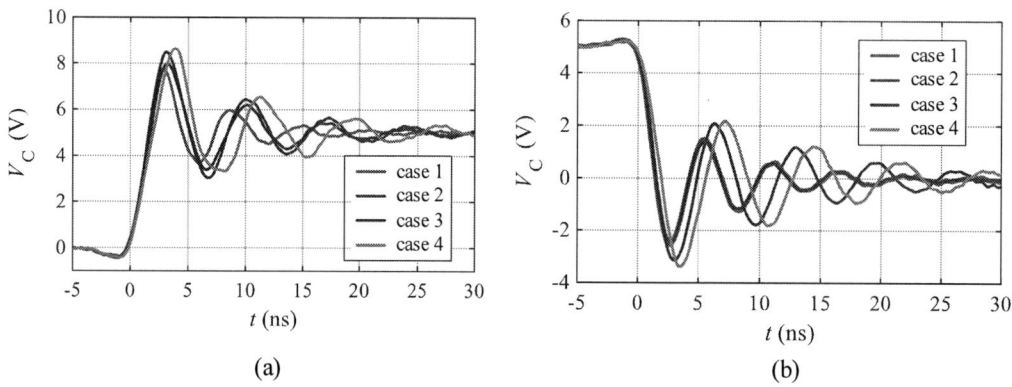

(a)

(b)

Fig. 3: Comparison of the experimental waveforms of v_C at (a) turn-on and (b) turn-off in the different studied cases with a driver supply voltage V_{dr}=5 V, gate resistors R_{on}=R_{off}=1 Ω, and load capacitor C_{load}=100 pF.

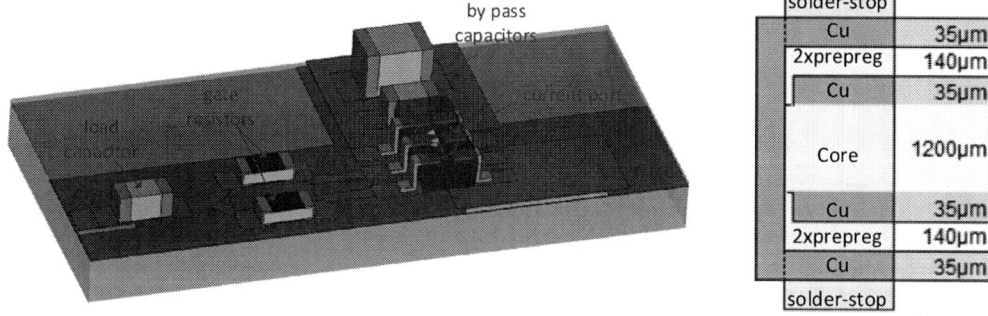

Fig. 4: 3D model of case 1 used in CST for parameter extraction

Fig. 5: Dimension of the used 4-layer PCB [15]

For an experimental characterization of the gate loop inductance, the voltage response (Fig. 3) to a turn-on and turn-off step signal at the driver output is measured. The gate loop forms a series RLC resonance network described by its transfer function

$$G(s) = \frac{Y(s)}{U(s)} = \frac{1}{s^2 LC + sRC + 1}.$$
(1)

where s is the Laplace transform frequency parameter. The R is the sum of the gate external resistor and the driver internal resistor which is unknown. Since R and L are both unknown parameters so far, L is determined by fitting the transient gate current response using time-domain data and a known C value. It should be noted here that C is the sum of the load capacitance C_{load}, the parasitic capacitance resulting from the PCB $C_{par,PCB}$, and the driver's parasitic output capacitance $C_{par,driver}$. For example, the extracted $C_{par,PCB}$ though FEM simulation equals 4 pF for case 1. The driver's $C_{par,driver}$ is measured with a Keithley parameter analyzer and equals 84 pF. Thus, the total C=188 pF can be calculated. To verify the results, the total charge is computed by integrating the transient gate current (Fig. 2). Using equation 2, C results in 185 pF. The good agreement of these two values suggests a satisfying reliability in the analysis and results.

$$C = \frac{Q}{V} = \frac{\int i dt}{V} \qquad (2)$$

Table I: Studied cases and extracted L_g

	Case 1	Case 2
Top view of layout and current paths		
Cross-section of layout and current paths		
$L_{g,meas}$ (turn on loop /turn off loop)	4.2 nH/3.8 nH	5.9 nH/5.5 nH
$L_{g,sim}$ (turn on/turn off, @100 MHz)	3.9 nH/ 3.4 nH	7.6 nH/6.6 nH
	Case 3	**Case 4**
Top view of layout and current paths		
Cross-section of layout and current paths		
$L_{g,meas}$ (turn on loop /turn off loop)	5.7 nH/3.8 nH	7.5 nH/6.7 nH
$L_{g,sim}$ (turn on/turn off, @100 MHz)	5.4 nH/3.4 nH	6.6 nH/6.1 nH

To verify the measurement results, the parasitic parameters are also characterized in simulation. The 3D model of the layout is imported to the 3D FEM field simulation tool CST Studio Suite [14] to extract the inductance values (Fig. 4). The vertical dimensions of the employed PCB are shown in Fig. 5 [15]. In this simulation, capacitors are built as solid copper blocks. The resistor models are built according to the datasheet specifications [16], consisting of a resistive layer on top (black), alumina ceramic substrate in the middle (white), and nickel terminations (silver) (Fig. 4). However, the accurate dimensions of each layer are not specified in the datasheet and thus they are only estimated, which potentially results in a slight inaccuracy of the 3D models. To reduce the meshing complexity, the cylindrical vias used to interconnect different copper layers are modeled by solid copper blocks. The structure inside the driver package and the signal path is not included in the model. The transistor die is assumed to be at the height of the lead frames and in the middle of package. A current port used for current excitation in the magnetoquasistatic simulation is therefore placed in the middle of the driver.

Based on this model, the impact of different factors can be studied. The simulation results agree well with the measurement results. The remaining difference can be attributed to the simplified driver-internal structure as well as the simplified models of the capacitors and resistors. Moreover, it is worth mentioning that L_g differs between turn-on and turn-off gate loop, and thus they should be considered separately in the converter design. Case 1 shows the lowest L_g as expected. Both, the experimental results $L_{g,meas}$ and simulation results $L_{g,sim}$ show that the multi-layer PCB can significantly reduce L_g for both, turn-on and turn-off gate loop. However, the reduction in L_g (approx. 30% in the measurement and approx. 50% in the simulation) is much smaller than the reduced distance between the power flow layer and the return ground layer (91%), which illustrates the contribution of the remaining active and passive circuit components to L_g. The bypass capacitors positioned further away results in an increased L_g for the turn-on gate loop, but for turn-off this is less relevant. The reason is that, assuming a unipolar gate-voltage supply as in our case-study, only the turn-on gate current flows through the bypass capacitors. An extended distance of 5 mm from the bypass capacitors to the driver IC in a 4-layer vertical design leads to an increase by approx. 1.5 nH of the turn-on gate loop inductance. In contrast, the farther positioned load capacitor (emulating the GaN transistor) increases L_g in both turn-on and turn off gate loop. A further extended distance of 10 mm between the load capacitor and the driver leads to an increase of L_g by approx. 3 nH.

Impact of the driver devices on switching characteristics

In this section, we show the impact of different driver ICs and their packages on the switching characteristics of GaN devices with experimental results in hard-switched double-pulse tests. The measurements are performed using 100 V/150 mΩ Schottky-gate type GaN transistors with −2.5 V threshold voltage fabricated by Ferdinand-Braun-Institute (FBH) [17]. To ensure comparability, driver ICs with similar ratings as given in the datasheets [7, 12, 18, 19] (**Table II**) are used in the measurements.

The investigated GaN transistors are fabricated with a Schottky gate, which means a parasitic Schottky diode is formed below the gate. Similar to the GaN Gate Injection Transistor (GIT), the non-insulating gate requires a continuous current sourced into the gate during on-state [20, 21]. The I-V curve of the GaN transistor's parasitic gate-source diode is shown in Fig. 6a. Considering commercially available driver ICs, the Panasonic driver *AN34092B* integrating a current source (Fig. 6b) or a conventional voltage source driver combined with an RC-type driver network as well as a NMOS aiming to discharge the boost C_s (Fig. 6c) could be used to drive non-insulating gate-type GaN transistors [22]. The gate loop parameters used in the tests are listed in **Table III**.

The driver type i.e. current source type driver or voltage source driver directly affects the switching behavior. Besides, even when using standard voltage source driver ICs with almost identical electrical ratings and assembled in the same package type, the switching transients appear to be notably different. Moreover, employing the same driver IC in different packages results in obvious differences in the switching transitions. To study the impact of these factors, the transient waveforms of the GaN transistor's gate-source voltage v_{gs}, drain-source voltage v_{ds}, and drain current i_d with different driver ICs and packages but almost identical PCB layout are shown in Fig. 8. The comparison of switching characteristics such as the slew rates of the drain-source voltage v_{ds} and drain-current i_d, as well as the peak value of v_{ds} during turn-off and i_d during turn-on are shown in Fig. 9, Fig. 10, and Fig. 11, respectively.

The current source driver IC *AN34092B* (Fig. 8b) achieves an obviously longer v_{gs} rise time compared with the voltage source driver IC *1EDN7550B* (Fig. 8a), although the given maximal sourcing current of *AN34092B* is much higher and sourcing resistance is lower than *1EDN7550B* (**Table II**). The GaN transistor's absolute slew rate values of drain-source-voltage and drain-current during turn-on with *AN34092B* are thus significantly lower compared with *1EDN7550B* (Fig. 9). As for turn-off, the v_{gs} with *AN34092B* shows more oscillations, which possibly results from the combination of two loops to turn the transistor off due to the active miller clamp function of the *AN34092B*, i.e., both OUT2 and OUT3 affect switching-off [19]. Consequently, v_{ds} and i_d exhibit stronger oscillations and higher peak values during turn-off transitions using *AN34092B*.

Table II: Parameters of the drivers

	1EDN7511	UCC27511	1EDN7550B	1EDN7550U	AN34092B
Package	SOT23-6	SOT23-6	SOT23-6	TSNP-6	HQFN-16
Sourcing current	4 A	4 A	4 A	4 A	6 A
Sinking current	-8 A	-8 A	-8 A	-8 A	-9 A (sum of I_{OUT2} and I_{OUT3})
Sourcing resistance	0.85 Ω	5 Ω	0.85 Ω	0.85 Ω	0.8 Ω
Sinking resistance	0.35 Ω	0.45 Ω	0.35 Ω	0.35 Ω	0.5 Ω (OUT2) 1 Ω (OUT3)

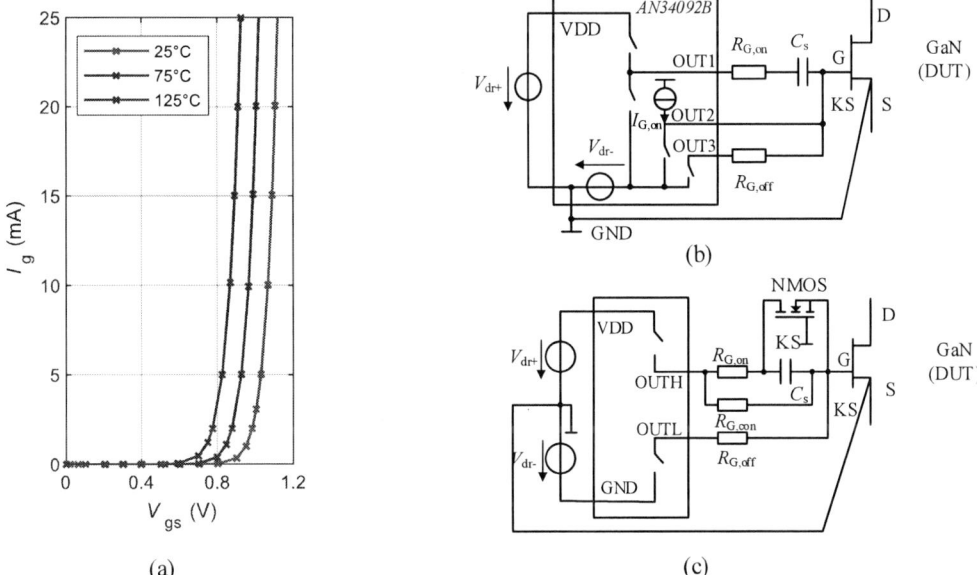

Fig. 6: (a) The I-V characteristics of the parasitic gate-source diode with open drain terminal for Schottky-type Gate GaN transistor fabricated by FBH. (b) Gate drive circuits for non-insulating gate GaN transistors with the current source driver *AN34092B*. (c) Gate drive circuits for non-insulating gate GaN transistors with standard voltage source driver combined with RC-type driver network and active boost-capacitor discharge [22].

Table III: Gate loop parameters

Parameters	Symbol	Value
Turn-on gate resistor	$R_{G,on}$	5.6 Ω
Turn-off gate resistor	$R_{G,off}$	2 Ω
Continuous gate resistor	$R_{G,con}$	3.9 Ω
Boost capacitor	C_s	330 pF
Continuous gate current turn-on	$I_{G,on}$	5 mA
Turn-on gate voltage	V_{dr+}	5 V
Turn-off gate voltage	V_{dr-}	−5 V

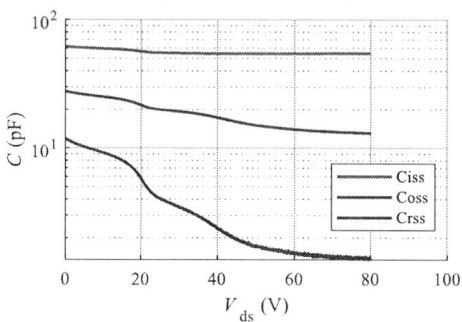

Fig. 7: Capacitance-voltage profile of the GaN transistor with $V_{GS}=-6$ V (100 kHz and 100 mV RMS small signal measurement with a Keithley parameter analyzer)

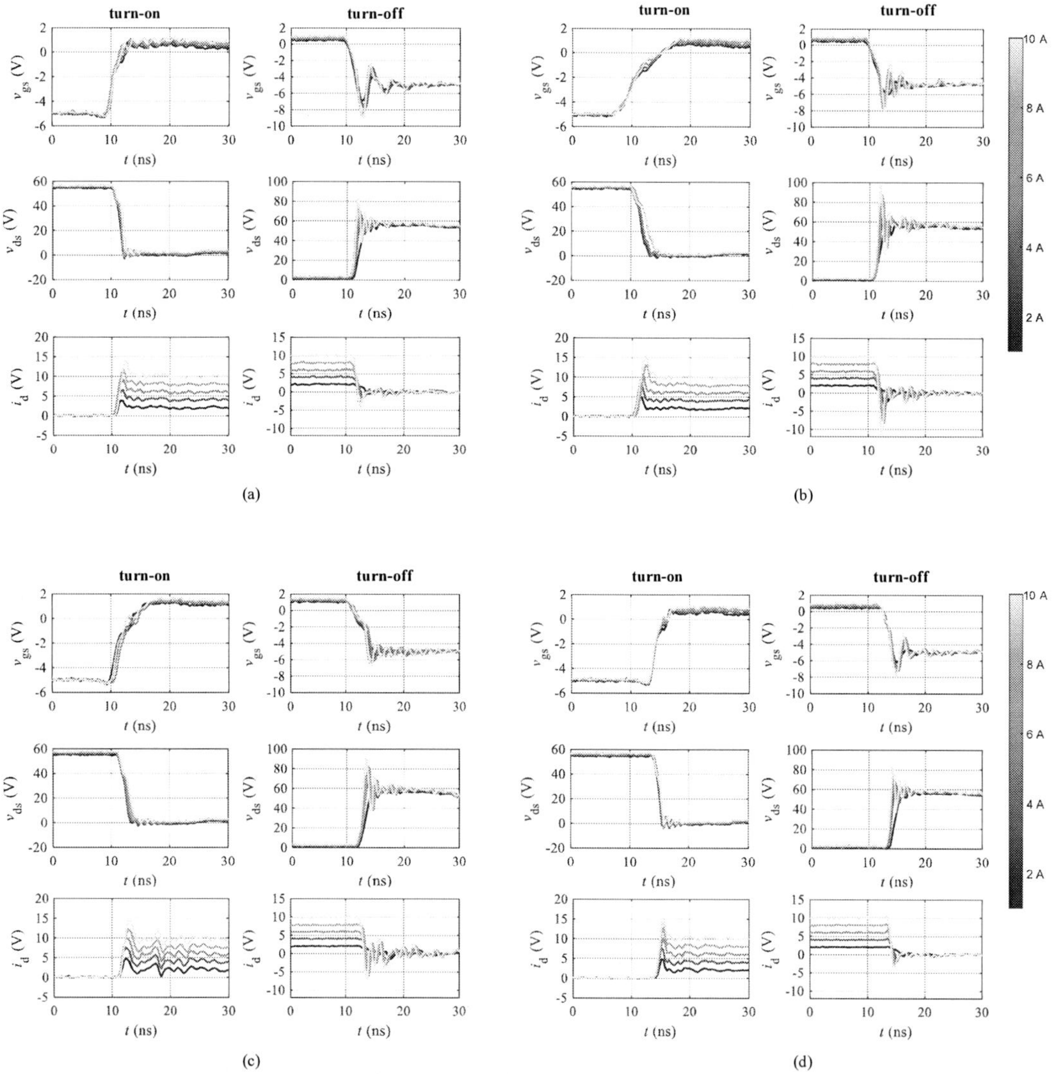

Fig. 8: Switching transitions of the GaN transistors at dc-link voltage $V_{dc}=50$ V, load current $I_L=2\ldots10$ A with (a) driver *1EDN7550B* with SOT23-6 package, (b) driver *AN34092B* with HQFN package, (c) driver *UCC27511* with SOT23-6 package and (d) driver *1EDN7550U* with TSNP-6 package.

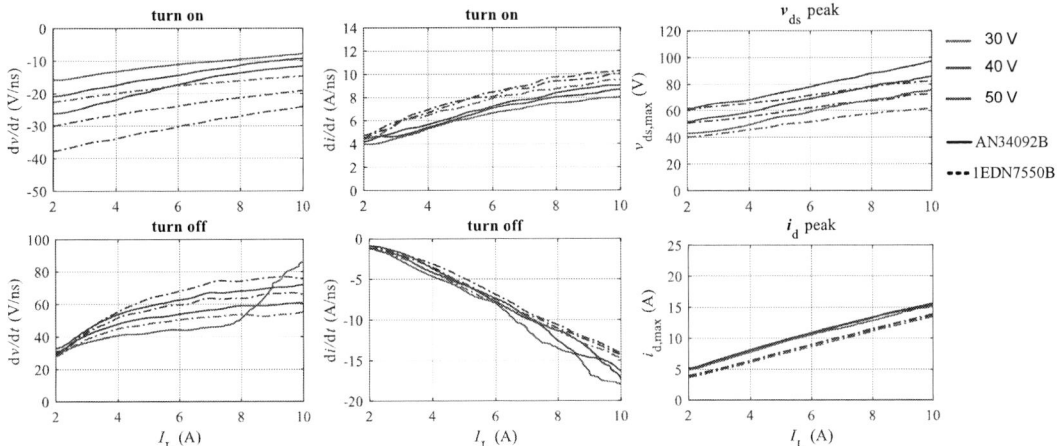

Fig. 9: Comparison of the switching characteristics between the current source driver and voltage source driver to drive non-insulating gate type GaN transistors for (a) slew rate of drain-source voltage v_{ds}, (b) slew rate of drain-current i_d and (c) peak value of v_{ds} and i_d.

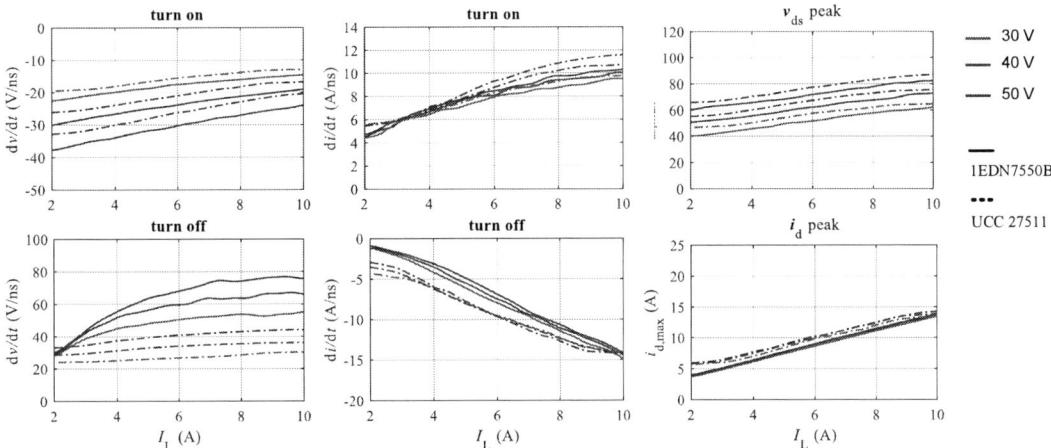

Fig. 10: Comparison of the switching characteristics between different voltage source driver ICs in the same package for (a) slew rate of drain-source voltage v_{ds}, (b) slew rate of drain current i_d and (c) peak value of v_{ds} and i_d.

Comparing the different standard voltage drivers with the same package, the slew rate of v_{gs} in both, turn-on and turn-off transitions is lower for driver IC *UCC27511* (Fig. 8c) than for driver IC *1EDN7550B* (Fig. 8a). The slew rate of the drain-source-voltage v_{ds} and drain current i_d is thus higher using *1EDN7550B* (Fig. 10), especially at turn-off. However, the turn-off voltage overshoot is still lower and the i_d in both turn-on and turn-off, as well as v_{ds} at turn-off, show fewer oscillations (Fig. 8) which is likely related to the lower parasitic inductances of *1EDN7550B*. Since the PCB layouts are almost identical for the two drivers, the influence from this layout is neglectable. Thus, the observed deviation in switching characteristics can be mainly attributed to the driver's internal structure.

The impact of different driver packages is investigated by comparing *1EDN7550B* (Fig. 8a) and *1EDN7550U* (Fig. 8d). Both drivers are based on the same IC chip but are embedded into different packages, a larger SOT package and a smaller TPSN package. The gate-source voltage v_{gs} for *1EDN7550U* in the smaller TPSN package shows slightly fewer oscillations at turn-on and an obvious lower negative peak value at turn-off (–8 V vs. –10 V). However, the drain-source voltage v_{ds} for *1EDN7550U* shows slightly stronger oscillations than v_{ds} for *1EDN7550B,* since the v_{ds} slew rates with

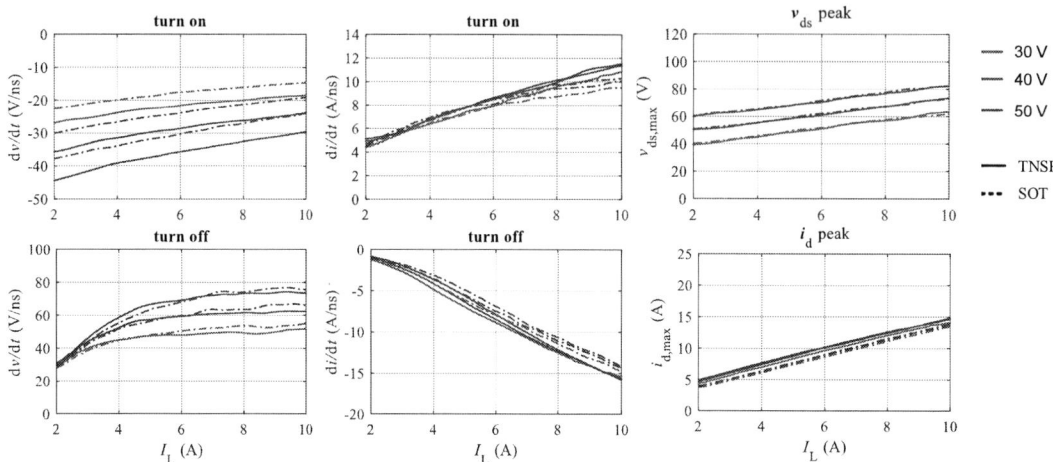

Fig. 11: Comparison of the switching characteristics between the same driver ICs in different packages: (a) slew rate of drain-source voltage v_{ds}, (b) slew rate of drain-current i_d and (c) peak value of v_{ds} and i_d.

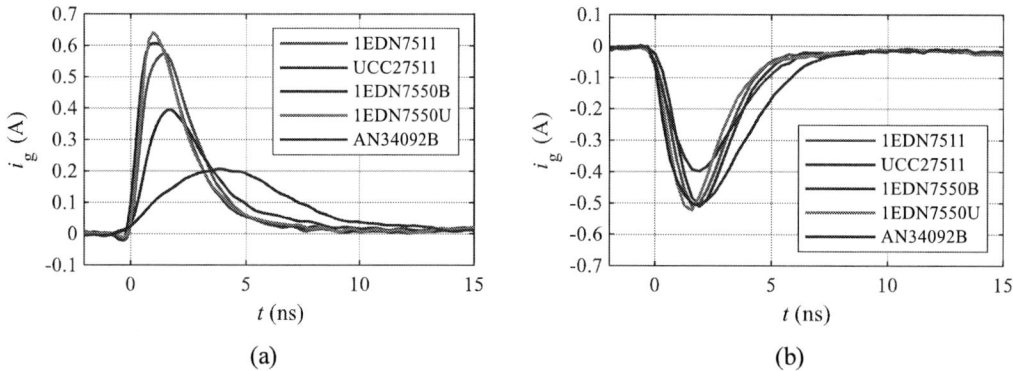

Fig. 12: Comparison of the (a) turn-on gate current and (b) turn-off gate current of the used drivers at a voltage difference $\Delta V = 10$ V, gate resistors $R_{on} = R_{off} = 10$ Ω and load capacitor $C_{load} = 100$ pF in the measurement setup shown in Fig. 1.

1EDN7550U are higher (Fig. 11). Overall, the smaller driver package is beneficial not only to achieve a more compact design but also to achieve a stable and less oscillating gate-source voltage, and a faster switching.

To study the origins of the differences caused by the driver ICs and their packages, the gate currents are measured during turn-on and turn-off transitions using the measurement setup shown in Fig. 1. Fig. 12 demonstrates that different driver ICs achieve different peak values and slew rates of the gate current i_g. The different peak values suggest that despite the same current ratings, the drivers have different current capabilities and parasitic resistances. Moreover, in all driver ICs the achieved gate current peak value is much lower than the datasheet rating. The expected value should be around 1 A, considering a driver voltage of 10 V and external gate resistors of 10 Ω, while the current source driver IC *AN34092B* achieves only a peak value of 0.2 A. The different slew rates of the gate current suggest that the drivers have different internal inductances and different switching speed, despite the fact that the package type of *1EDN7550B, 1EDN7511* and *UCC27511* is the same. The area below the i_g-curve equals to the charge of the total capacitance C. Hence, differences in the calculated charge indicate different parasitic output capacitances of the driver ICs, as both, the load capacitance and the parasitic layout capacitance remain unchanged.

At turn-on, driver *1EDN7550B* achieves a higher gate current peak value and slew rate than drivers *UCC27511* and *AN34092B*. This agrees with the previously presented switching tests in hard-switched double-pulse mode, where *1EDN7550B* exhibits faster turn-on switching and fewer oscillations. The differences of the gate current during turn-off are smaller among the driver ICs (Fig. 12b), which coincides with the switching test results, which show a more pronounced difference at turn-on than at turn off (Fig. 9, Fig. 10, and Fig. 11).

To achieve fast switching of GaN transistors, a fast and high gate current after switching signals are applied is desirable. The results indicate the importance of proper driver selection concerning both, current capability as well as driver package in order to exploit the fast switching potential of GaN transistors.

Conclusion

This work studies the gate loop design influencing parameters in order to achieve high switching performance of GaN transistors. The impact of a multi-layer PCB and the placement of devices and passive components on the gate inductance L_g are quantitatively evaluated to enable a straightforward insight to the resulting L_g-increase. The effects of the driver ICs and their packages on the switching characteristics are shown with experimental results. Our investigations prove the significant influence of electrical gate driver parameters including parasitic inductances and capacitances as well as peak current capability on the switching behaviors.

References

[1] E. A. Jones, F. F. Wang and D. Costinett, "Review of Commercial GaN Power Devices and GaN-Based Converter Design Challenges," *in IEEE Journal of Emerging and Selected Topics in Power Electronics*, vol. 4, no. 3, pp. 707-719, Sept. 2016, doi: 10.1109/JESTPE.2016.2582685.

[2] P. Nayak and K. Hatua, "Modeling of switching behavior of 1200 V SiC MOSFET in presence of layout parasitic inductance," *2016 IEEE International Conference on Power Electronics, Drives and Energy Systems (PEDES),* Trivandrum, 2016, pp. 1-6, doi: 10.1109/PEDES.2016.7914372.

[3] Y. Shen, J. Jiang, Y. Xiong, Y. Deng, X. He and Z. Zeng, "Parasitic Inductance Effects on the Switching Loss Measurement of Power Semiconductor Devices," *2006 IEEE International Symposium on Industrial Electronics*, Montreal, Que., 2006, pp. 847-852, doi: 10.1109/ISIE.2006.295745.

[4] X. Geng, C. Kuring, M. Wolf, O. Hilt, J. Würfl and S. Dieckerhoff, "Study on the Optimization of the Common Source Inductance for GaN Transistors," *2021 23rd European Conference on Power Electronics and Applications (EPE'21 ECCE Europe)*, 2021, pp. 1-10.

[5] Navitas, "GaNFast™ NV6113," NV6113 datasheet, Oct. 2020.

[6] STMicroelectronics, "High power density 600V half-bridge driver with two enhancement mode GaN HEMTs," MASTERGAN1 datasheet, Oct. 2020.

[7] Texas Instruments, "UCC2751x Single-Channel, High-Speed, Low-Side Gate Driver," UCC2751x datasheet, Oct. 2020.

[8] GaN Systems, "PCB Layout Considerations with GaN E-HEMT," July. 2021.

[9] B. Nelson, A. Lemmon, B. DeBoi, M. Olimmah and K. Olejniczak, "Measurement-Based Modeling of Power Module Parasitics with Increased Accuracy," *IEEE Applied Power Electronics Conference and Exposition (APEC)*, 2020, pp. 1430-1437, doi: 10.1109/APEC39645.2020.9124202.

[10] Z. Liu, X. Huang, F. C. Lee and Q. Li, "Package Parasitic Inductance Extraction and Simulation Model Development for the High-Voltage Cascode GaN HEMT," in *IEEE Transactions on Power Electronics*, vol. 29, no. 4, pp. 1977-1985, April 2014, doi: 10.1109/TPEL.2013.2264941.

[11] Paul, Clayton R. *Inductance: loop and partial*. John Wiley & Sons, 2011.

[12] Infineon, "EiceDRIVER™ 1EDN751x/1EDN851x," 1EDN751x/1EDN851x datasheet, Apr. 2018.

[13] GaN Systems, "GS66504B Bottom-side cooled 650 V E-mode GaN transistor," GS66504B datasheet, 2020.

[14] CST STUDIO SUITE®, CST AG, Germany, www.cst.com.

[15] Multi Circuit Board, "Multi-CB Defined layer buildup (4, 6, 8 layers) and layout examples for impedance." [Online]. Available: https://www.multi-circuit-boards.eu/fileadmin/pdf/leiterplatten_lagenaufbau/Multi-CB_Definierter_Lagenaufbau_Impedanzen_en.pdf.

[16] Yageo Phycomp, "Datasheet general purpose chip resistors RC0603." [Online]. Available: https://cdn-reichelt.de/documents/datenblatt/D150/PYu-RC0603_51_RoHS_L_4.pdf.

[17] R. Lossy, H. Blanck and J. Würfl, "Reliability studies on GaN HEMTs with sputtered Iridium gate module", Microelectronics Reliability, Vol. 52, pp. 2144-2148, 2012.

[18] Infineon, "Single-Channel EiceDRIVER™ Gate Driver IC with True Diffentential Inputs, " 1EDN7550 and 1EDN8550 datasheet, Dec. 2019.

[19] Panasonic, "Single-Channel GaN-Tr High-Speed Gate Driver," AN34092B datasheet, July. 2017.

[20] B. Zojer, "Driving 600V CoolGaN High Electron Mobility Transistors, " Infineon application note, May. 2018.

[21] J. Ao, D. Kikuta, N. Kubota, Y. Naoi and Y. Ohno, "Copper Gate AlGaN/GaN HEMT with Low Gate Leakage Current", *IEEE Electron Devices Letters*, Vol. 24, pp. 500-502, August 2003.

[22] X. Geng, C. Kuring, O. Hilt, M. Wolf, J. Würfl and S. Dieckerhoff, "Design and Optimization of the Driver Circuit for Non-Insulating Gate GaN-Transistors Enabling Fast Switching and High-Frequency Operation," *12th International Conference on Integrated Power Electronics Systems*, 2022.

Analysis of current sharing in the parallel connection of GaN transistors

Frederik Stalleicken, Sibylle Dieckerhoff[1], Karsten Handt[2], Sebastian Nielebock[2]

[1] TECHNISCHE UNIVERSITÄT BERLIN
Einsteinufer 19
Berlin, Germany
[2] SIEMENS AG
Frauenauracher Str. 80
Erlangen, Germany
Tel.: +49 / (0)30 314-70026
Fax: +49 / (0)30 314-25526
E-Mail: stalleicken@tu-berlin.de
[1] URL: https://www.pe.tu-berlin.de

Keywords

«Wide bandgap», « Gallium Nitride (GaN)», « Parallel Operation », «Design Optimization»

Abstract

Commercially available GaN transistors are connected in parallel in order to overcome thermal limitations prohibiting a further current increase for the single transistor. The driver concept and the circuit layout are presented. The parallel connection is examined in double-pulse tests, analyzing the stability of the gate-voltage and the current sharing between the parallel transistors. An oscillation of the measured drain currents of the individual transistors is identified and investigated in more detail.

Introduction

In order to overcome the thermal limitations of single GaN transistors and reach higher load currents, in this way enabling GaN applications in higher power classes, a parallel connection of multiple transistors is necessary [1], [2]. Assuming an equal current distribution among the transistors, the total load current can be scaled with the number of parallel transistors. Consequently, the converter power could ideally be scaled in the same way. Creating the basis for scaling the current carrying capacity should be the research topic of this paper. Another possible advantage of operating GaN tranistors in parallel is to achieve a higher converter efficiency if the current carrying capacity of the individual transistors is not fully utilized, so that each transistor is operated with lower load current [3]. However, the possible efficiency gain will not be the focus of this paper.

In order to ensure an even current distribution, some requirements need to be fulfilled so that the parallel connected transistors switch simultaneously. Apart from a synchronized control, layout symmetry is a key requirement [1], [3]. By a symmetrical layout, the influences of the parasitic layout elements such as the gate-loop- and common-source-inductances on the switching behavior and the ohmic influence on the conduction behavior of the parallel connected transistors are almost identical [3]. Regarding the need of a synchronized control, the literature and also commercial GaN manufacturers often recommend a common driver to switch two parallel transistors, so that the respective transistors receive the same control signal [1], [3], [4]. Since the layout presented here is designed for four parallel transistors, controlling them with a common driver would result in two problems. First, the power of a single driver might be insufficient to efficiently charge and discharge the input capacitances of four transistors at the same time. Second, the driver placement in the layout considering the necessary minimization of the gate loops is very complex in the case of a common driver. As a consequence, a driver concept using one driver as a synchronizing clock combined with an individual driver output stage per transistor will be presented in this paper, together with the entire layout.

The normally-off 600 V / 70 mΩ IGOT60R070D transistors from Infineon [5] are mounted step-by-step on the printed circuit board (PCB). First, a single half-bridge is equipped and analyzed in a double-pulse test. Afterwards, a second half-bridge is connected in parallel, and the setup is also characterized. When using the same gate circuit configuration in the single half-bridge and for the parallel circuit, unintended peaks in the gate voltage of the paralleled transistors occur. It is demonstrated that these peaks can be suppressed by a suitable gate configuration. Since an oscillation in the drain currents of the individual transistors appears in the first tests with two paralleled transistors, this effect is first studied in detail before the placement of the remaining transistors is completed.

Driver concept for synchronization of the parallel transistors

The driver concept is illustrated schematically in Fig. 1. For all individual drivers, the UCC27511 [6] from Texas Instruments is used, with a positive driver voltage of 9 V and a negative driver voltage of −5 V provided individually for each driver by low-dropout regulators (LDO). In order to switch the transistors simultaneously resulting in a symmetrical current distribution, it is useful to apply a synchronizing common driver, as suggested in literature [1], [3], [4]. Thus, one driver in the concept acts as a synchronizing clock getting the external control signal. The synchronized drive signal is then routed to the driver islands of the individual transistors, taking care to ensure identical conductor path lengths to prevent differences in propagation time (Fig. 4). The synchronizing driver cannot be used directly to drive the parallel transistors, since it would have to charge the input capacitances of all four transistors simultaneously, resulting in slow, or even not successful switching. Additionally, in the circuit design of four parallel transistors with only one common driver, very large gate loops would occur as the driver cannot be placed close to each transistor. The resulting feedback into the driver circuit could provoke strong gate oscillations [3]. To circumvent these problems, the proposed solution is placing a single driver output stage including the gate network directly next to each of the parallel connected transistors (Fig. 1). This enables an optimization of the gate-loops and sufficient driver power. One disadvantage that must be accepted hereby is the possible device-specific runtime difference between the four driver ICs. As there are few additional logic parts in the driver IC beside its output stage [6], this should be almost negligible.

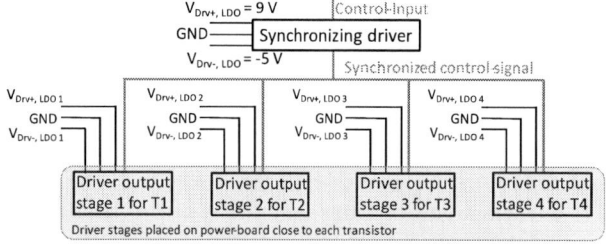

Fig. 1: Driver concept for synchronizing the transistors (T1-T4) switching events (identical for top and bottom)

Fig. 2: Gate network for controlling the Gate Injection Transistor (GIT)

Table I: Driver configuration for each transistor

Driver-input	R_{in} = 51 Ω	R_{in} = 4.7 kΩ	C_{in} = 100 pF		
Gate-circuit	R_{on} = 5.6 Ω	R_{off} = 1 Ω	R_{ss} = 390 kΩ	C_{couple} = 2.2 nF	R_{ks} = 0 Ω

The gate network and its initial configuration for each transistor are depicted in Fig. 2 and Table I. Since the GaN transistor from Infineon is a Gate Injection Transistor (GIT), an R-C network is necessary for proper driving [7]. Due to the presence of a non-insulated p-GaN gate to create a normally-off device, a pn diode is formed between gate and source [7]. As a result, there is a need for a continuous gate current during the transistor's on state [7] in addition to the classic gate peak currents for switching on and off. The transient gate peak currents are fed to the gate via the resistors R_{on} and R_{off} and the coupling capacitance C_{SS}, since for the transient currents C_{SS} has a low-impedance [7]. The continuous steady state current is provided by the high-impedance resistor R_{SS} [7] which loads the driver leading to a small DC current. Because all drivers share a common reference potential in their Kelvin sources, all gate circuits of the transistors are connected, and therefore all parallel transistors are also. Thus, the resulting path offers a potential way for equalizing currents. The resistor R_{ks} in the kelvin source path can be increased if necessary to damp resulting oscillations in the gate-voltage. At the input of the driver is a low pass filtering possible interfering signal.

Layout implementation of the parallel circuit

The whole setup is divided into two boards, the logic-board (Fig. 3, Fig. 4) and the power-board (Fig. 5, Fig. 6), which are plugged with connectors to obtain a compact arrangement. As indicated in the image, the main logic circuitry is placed on the logic-board frontside (Fig. 3). It contains the connection for a 12 V laboratory power supply to ensure the auxiliary voltage for the PCB. This voltage is transferred to the high voltage side via an isolating DC-DC-converter providing ±12 V at its dual output. These potentials are used to generate the driver voltages of +9 V / −5 V with the LDOs. The PWM control signals, marked on the right side, are fed via digital isolators to the synchronizing drivers located in the middle. A look at the backside reveals the conduction paths, optimized to the same length, of the synchronized control signal to the individual driver islands. Via the plugs on the back of the logic-board the +9 V / −5 V driver voltages, the ground reference potential and the control signal reach the power-board.

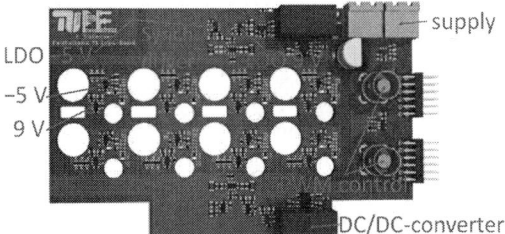

Fig. 3: Logic-board frontside   Fig. 4: Logic-board backside

The individual driver stages are located on the power-board (Fig. 5), to the left of the transistors as close as possible to their gate terminals optimizing the gate loop (Fig. 5). The drivers control the parallel transistors, which are defined from right to left for the bottom row as $T1_{bot} - T4_{bot}$, and for the top row as $T1_{top} - T4_{top}$. The transistors can be mounted step-by-step. The middle points of the half-bridges can be connected step-by-step via solder bridges (Fig. 5, green circle). This way, the switching behavior in the double pulse test is not always influenced by the whole layout, but only by the part relevant for the current parallel placement. To carry out a reference analysis in the first step, only a single half-bridge is equipped with the transistors $T1_{bot}$ and $T1_{top}$. In the second step, $T2_{bot}$ and $T2_{top}$ are added and the setup is further investigated. Following this analysis, the remaining transistors are added step-by-step, so that in the end the parallel connection of four transistors (total transistor number for top and bottom row: eight) will be analyzed. To the right of the transistors, the screw terminal for the external DC-link can be identified in Fig. 5. To monitor the drain currents of the transistors and thus the current sharing between the parallel connected transistors, an on-board current shunt [8] is used.

Fig. 5: Power-board frontside   Fig. 6: Power-board backside

The power-board backside in Fig. 6, contains the sockets for holding the logic-board, the various measuring points for observing the drain-source and the gate voltages of the transistors and the connection points for the load. These load connection points (LCP) are defined from right to left as LCP1 – LCP4. They are used to support the current injection. For example, when T1 and T2 are connected in parallel, both load connection points LCP1 and LCP2 are used accordingly to connect the air coil in the double pulse test.

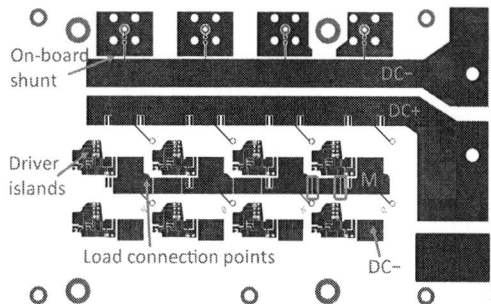 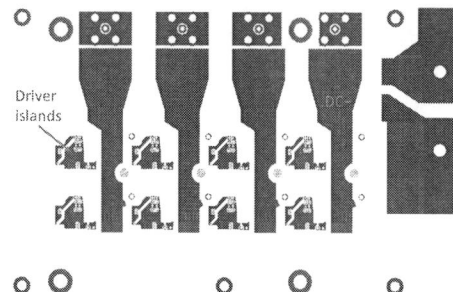

Fig. 7: First copper layer of power-board   Fig. 8: Power-board first inner copper layer

The first two copper layers of the power-board are shown in Fig. 7 and Fig. 8. Eight driver islands of the transistors can be identified, all routed identically to ensure layout symmetry. Furthermore, the connections of the DC-link and the switching node of the board are depicted. The green rectangles mark the mentioned separation of the switching node or middle points of the half-bridges, which can be connected according to the step-by-step assembly of the transistors. In Fig. 8, the DC− returns from the source terminals of the bottom transistors can be identified. In order to measure the separate drain currents using the on-board shunts, the DC− returns are designed without interconnections. As for the driver islands, the layout symmetry is given also for the commutation loops.

Double-pulse test with single half-bridge (T1$_{top}$, T1$_{bot}$) assembled on the PCB

To be able to assess the parallel circuit with regard to gate-voltage stability and current behavior and to evaluate the two-stage driver concept, a reference double-pulse test of the single half-bridge is carried out. Therefore, the board is equipped with a 728 μH air coil as load, connected to LCP1. Furthermore a large bulky external 1200 μF film capacitor combined with smaller film capacitors on a four-layer PCB to reach an overall better frequency behavior is used as DC-link (see Fig. 20). Since the transistor's maximum pulsed drain-current for a junction temperature of 25°C is given by 60 A [9], 40 A is chosen as reference value [10], providing sufficient safety margin. According to the voltage class of 600 V of the transistor, the DC-link voltage is set to 400 V.

As indicated in the diagram on the right (Fig. 9), the driver concept works satisfactory as it is stable, and the transistor shows clean switching of the drain-source voltage at the reference current of 40 A. The gate-source voltage slightly oscillates after the switching instant, but generally remains sta-

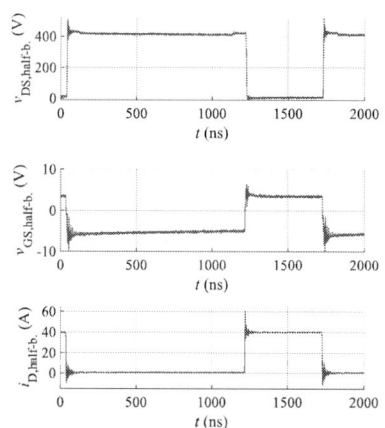

Fig. 9. Reference analysis of single half-bridge (signals of T1$_{bot}$)

ble. On the basis of this reference measurement, the parallel connection is analyzed starting with two transistors in parallel. Since the problems which are described in the next sections arose during the measurements of the parallel circuit, the following experiments are conducted with a smaller DC-link voltage of 200 V and a load-current of 30 A. Achieving a higher load current will be part of future work based on the detailed investigation in this paper.

Gate-voltage stability with two parallel transistors (T1$_{top}$, T2$_{top}$, T1$_{bot}$, T2$_{bot}$)

Following the reference analysis of the single half-bridge, the parallel connection of two transistors is analyzed focusing the stability of the gate-voltage and the synchronous switching of the transistors reflected in the respective drain current increase and decrease. The double-pulse measurements are conducted with the same load inductance and the same external DC-link capacitor as in the reference analysis. To connect the load, LCP 1 and 2 are both used to simplify current injection. To investigate the stability of the gate voltage at different points, the pulse time of the pause pulse is varied in the double pulse test as indicated in Fig. 10 – Fig. 12.

The stability analysis first starts with the initial driver configuration described in Fig. 2 and Table I. At a DC-link voltage of 200 V and a total load current of 30 A, measurements with different pause pulse times (500 – 2000 ns in 100 ns steps) are performed in the double pulse test (see Fig. 10). The drain-source (v_{DS}) and the gate-source voltage (v_{GS}) of the two bottom transistors are measured. Furthermore, the drain currents (i_D) are monitored. Since notable gate voltage peaks only occur in the gate voltage of the right transistor (T1$_{bot}$), only this voltage ($v_{GS,right}$) is shown in Fig. 10 – Fig. 12. As can be observed in the zoom-in in Fig. 10, some unintended gate voltage peaks occur which, in the most unfortunate case, could lead to the transistor switching on again and thus potentially destroy the circuit or at least endanger the stability of the parallel circuit. The peaks in the gate-voltage appear after its edge has already fallen and the switching process is already completed, suggesting electromagnetic coupling as the reason for the peaks.

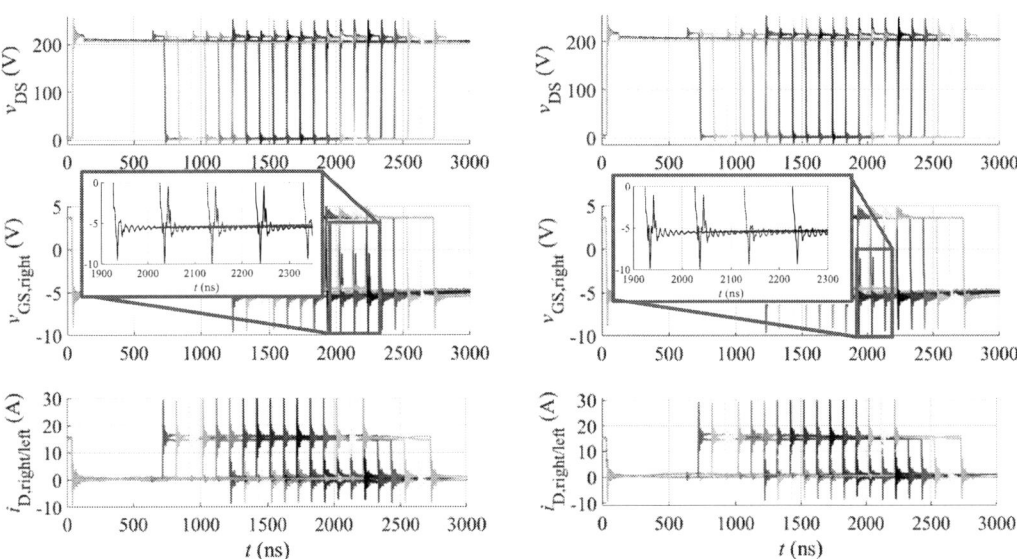

Fig. 10: Gate-voltage for different pause pulse lengths with initial gate-circuit configuration R_{in} = 51 Ω, R_{pd} = 4.7 kΩ, C_{in} = 100 pF, R_{on} = 5.6 Ω, R_{off} = 1 Ω, R_{ss} = 390 kΩ, C_{couple} = 2.2 nF, R_{ks} = 0 Ω

Fig. 11: Improved gate-voltage behavior with reduced pull-down resistor R_{in} = 51 Ω, R_{pd} = 200 Ω, C_{in} = 100 pF, R_{on} = 5.6 Ω, R_{off} = 1 Ω, R_{ss} = 390 kΩ, C_{couple} = 2.2 nF, R_{ks} = 0 Ω

Assuming electromagnetic coupling as cause of the peaks, the next step is to reduce the pull-down resistor (R_{pd}, Fig. 2) at the input of the driver to 200 Ω. The result of the measurement is depicted in Fig. 11. Most of the peaks are damped. Nevertheless, peaks are still present, even if they have been diminished compared to Fig. 10.

To eliminate the remaining peaks in the gate voltage, the resistor in the kelvin source path (R_{ks}) is increased from 0 Ω to 1 Ω. As illustrated in Fig. 12, the additional attenuation in the feedback path can successfully eliminate the peaks in the gate voltage, and the final gate configuration to ensure a stable gate voltage is found.

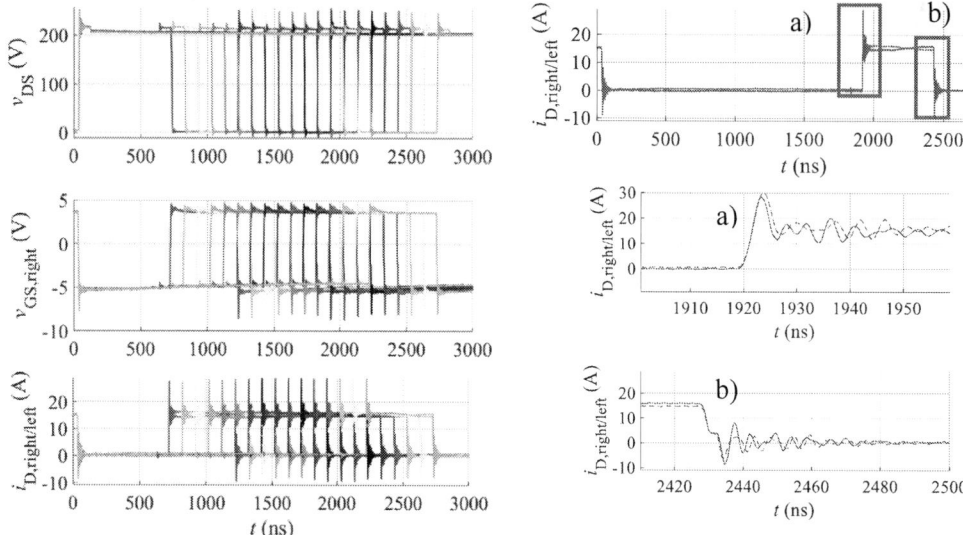

Fig. 12: Gate voltage without peaks due to additional damping by R_{ks}
R_{in} = 51 Ω, R_{pd} = 200 Ω, C_{in} = 100 pF, R_{on} = 5.6 Ω, R_{off} = 1 Ω, R_{ss} = 390 kΩ, C_{couple} = 2.2 nF, R_{ks} = 1 Ω

Fig. 13: Exemplary current transients verifying synchronous switching with the new driver concept a) rising edge, b) falling edge

Another important aspects of the parallel connection of GaN transistors is their synchronous switching. To analyze the synchronicity, an exemplary current pulse is examined in more detail in Fig. 13. As can be noticed when examining the enlarged rising and falling edges, the currents overlap very well in the transient region. Therefore, the proposed driver concept is well suited to drive transistors connected in parallel. The concept successfully combines the advantages of a common driver for all transistors with regard to synchronization of the control with the layout advantages of individual drivers per transistor. The layout advantage weights especially heavily for the parallel connection of more than two transistors.

Drain current behavior in double pulse test with two parallel transistors

The measurements in Fig. 10 – Fig. 12 already indicate the issue of opposing oscillations in the drain currents, which will be investigated in this section.

As a start, the layout configuration from the gate voltage stability analysis (see Fig. 15) is further examined. The measurement conditions in the double pulse test remain unchanged compared to the previous section. The drain currents of the lower transistors $T1_{bot}$ and $T2_{bot}$ measured with the on-board shunt and the summation of the two currents to $i_{D,total}$ is given in Fig. 14. In order to be able to assess more precisely the oscillation in the drain currents, the second pulse of the double pulse is varied, being stepwise extended to a maximum length of 8 µs. This allows to observe the oscillation over a longer period of time. It is apparent that the oscillation is independent of the length of the second pulse and is instead an oscillation occurring at a relatively low frequency over the entire measurement range. The estimated oscillation frequency of the measurement in Fig. 14 is in the range around 400 kHz. In the lower plot of Fig. 14, the sum of the individual drain currents results in the total load current, and it is obvious that no oscillations are visible. This leads to the conclusion that the total load current in the setup divides and oscillates.

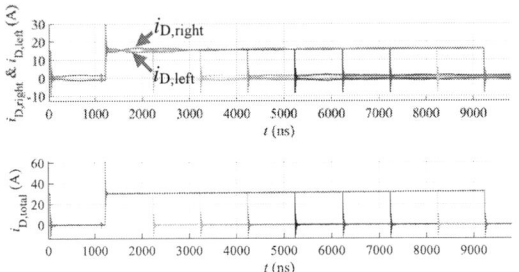

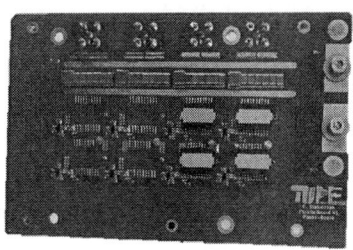

Fig. 14: Separate drain currents of $T1_{bot}$ and $T2_{bot}$ and their sum

Fig. 15: PCB configuration for current waveforms illustrated in Fig. 14 (LCP 1 and LCP 2 in combination)

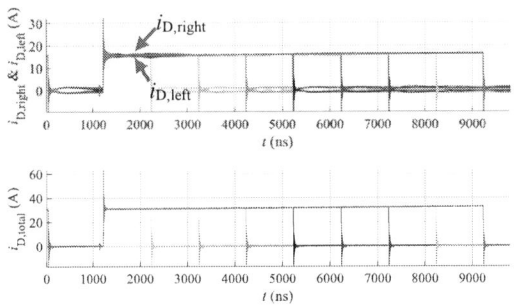

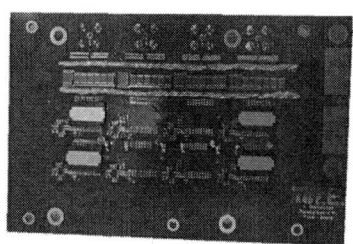

Fig. 16: Separate drain currents of $T1_{bot}$ and $T4_{bot}$ and their sum

Fig. 17: PCB configuration for current waveforms illustrated in Fig. 16 (LCP 1 and LCP 4 in combination)

To investigate possible layout influences due to the transistors placed differently close to the DC link on the PCB, the configuration illustrated in Fig. 17 is measured. Instead of transistors $T2_{top}$ and $T2_{bot}$, the leftmost transistors $T4_{top}$ und $T4_{bot}$ are placed on the board. As previously mentioned, the middle points of the two half-bridges are connected by solder bridges. Load connection points 1 and 4 are used in combination to connect the load. Due to the larger distance between the parallel transistors, the leakage inductance between them is higher compared to the configuration in Fig. 15, which can affect the oscillation. Another possible influence on the drain current is the slightly different ohmic resistance of this parallel connection, resulting from the longer current path for the left transistors ($T4_{top}$, $T4_{bot}$). The corresponding double pulse test results and the summation of the individual drain currents are given in Fig. 16.

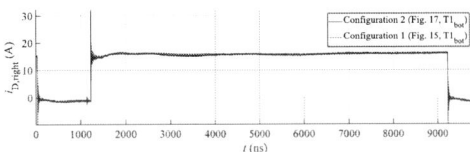

Fig. 18: Drain current comparison of the right transistors ($T1_{bot}$) in both PCB configurations

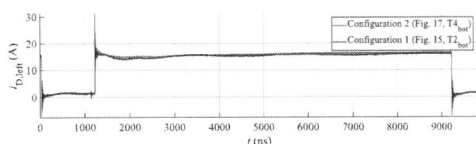

Fig. 19: Drain current comparison of the left transistors ($T2_{bot}$, $T4_{bot}$) in both PCB configurations

The drain currents exhibit the same oscillation behavior as the parallel circuit with $T2_{top}$ and $T2_{bot}$. Again, the oscillation occurs on the entire measurement range regardless of the length of the second pulse. Also, the sum of the individual currents results again in the total load current. For a direct comparison of the two PCB configurations, the drain currents are plotted in Fig. 18 and Fig. 19 for the 8 μs long second pulse. Fig. 18 compares the currents in the respective configuration through the right transistor, in both cases $T1_{bot}$, while Fig. 19 compares the currents through the left transistors, for configuration 1 $T2_{bot}$ and for configuration 2 $T4_{bot}$. The currents flowing through the transistor on the right and the currents flowing through the left transistor exhibit hardly any deviation from each other. This leads to the conclusion that the aforementioned layout differences between the configurations have no effect on the drain current oscillation.

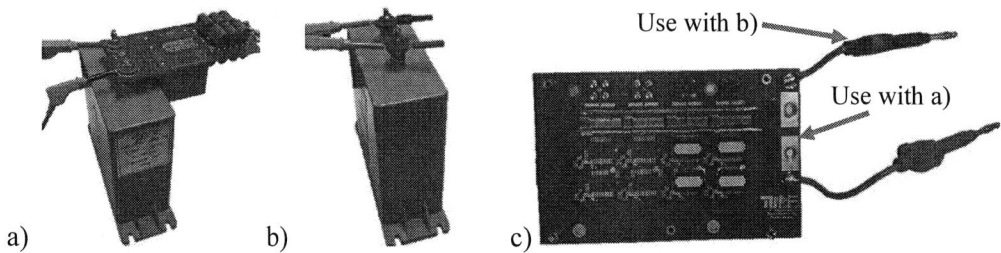

Fig. 20: a) DC link combined with additional PCB, b) only bulky film capacitor, c) connectors of the parallel board to the DC link

Another possible origin for the drain current oscillations is the DC link (capacitance and parasitic inductance), which together with the inductance of the power loop and the output capacitance of the transistors forms a resonant circuit. In the measurement results presented so far, the DC link combination illustrated in Fig. 20 a) was used. In order to investigate the influence of the additional DC link PCB on the drain current oscillations, the parallel board (Fig. 20 c)) is connected directly to the large capacitor (Fig. 20 b)) in the following measurement. The parallel connection consisting of the transistors $T1_{top,bot}$ and $T2_{top,bot}$ is examined again.

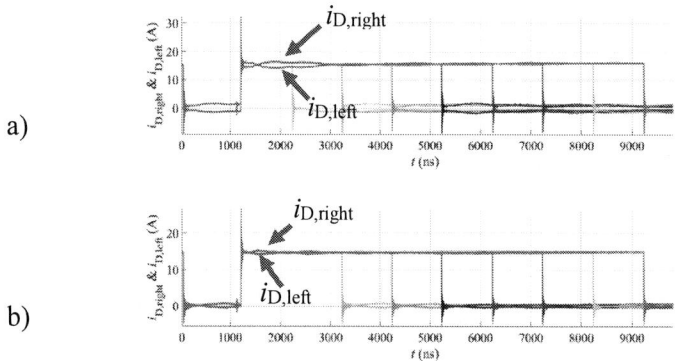

Fig. 21: a) Drain currents additional PCB used, b) Drain currents only bulky film capacitor used

The results of the comparison are presented in Fig. 21. From Fig. 21 a) it can be noticed that the removal of the additional DC link PCB has a positive influence on the opposing drain current oscillations since the deviation between the currents become significantly smaller. This indicates that further improvement of the DC link connection could potentially further improve the oscillation behavior.

In conclusion, it is apparent that the load current impressed in the circuit is divided in the layout, but exhibits oscillations. By investigating two different PCB configurations (Fig. 15, Fig. 17), it can be concluded that in the parallel connection, an increased leakage inductance between the two parallel transistors, as well as a slightly increased ohmic resistance in one of the current paths, does not significantly influence the oscillation behavior. However, the oscillation is significantly influenced by the examined variation of the DC link, so that a starting point is identified for further improvements by an optimized DC link connection. Therefore, for the further investigations in the next section, the additional DC link PCB will be dispensed.

Measurement of three and four transistors in parallel at 100 V

To get a complete picture of the parallel circuit characteristics, three- and four transistors in parallel are analysed in the following.

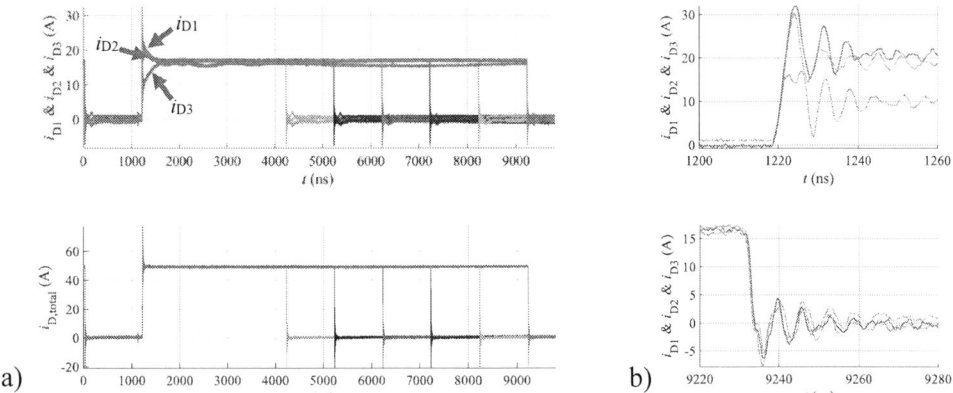

Fig. 22: a) Individual drain currents of $T1_{bot}$, $T2_{bot}$ and $T3_{bot}$ and their sum in the threefold parallel connection, b) synchronous current switching rising and falling edges

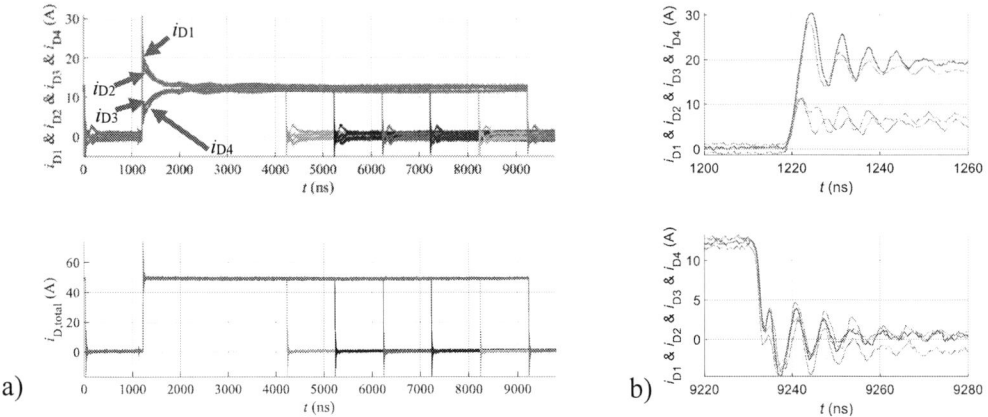

Fig. 23: a) Individual drain currents of $T1_{bot}$, $T2_{bot}$, $T3_{bot}$ and $T4_{bot}$ and their sum in the fourfold parallel connection, b) synchronous current switching rising and falling edges

The investigations start with the threefold parallel connection. To support an even current distribution, the load connection point 2, located in the center of the threefold parallel circuit, is used. In order to measure at safe operating points in the following analyses, three- and fourfold parallel circuits are investigated at a DC link voltage of 100 V and a load current of about 50 A. Fig. 22 a) depicts the individual drain currents and the sum of the three currents. The principle behavior of the drain currents is comparable to the twofold parallel connection. Again the sum of the individual currents results in the total load current, and opposing oscillations occur due to initial, unequal current distribution. The difference here is that now the current no longer oscillates between transistors 1 and 2, but between these two transistors and transistor three.

The measurement of the fourfold parallel connection, where load connection point 3 is used, depicted in Fig. 23 a) extends this finding, since the current in T_{4bot} behaves in the same way as the current in T_{3bot} previously. Hereby, the higher initial current flows in transistors 1 and 2, the lower current in transistors 3 and 4, with a resulting balancing process occurring during the pulse time. In summary, it seems that in comparison to the twofold parallel circuit, transistors 1 and 2 and transistors 3 and 4 behave equivalent to one transistor in there, exhibiting the corresponding current oscillation behavior.

Both, Fig. 22 b) and Fig. 23 b) underline the successful synchronization of the switching operations of the parallel transistors by the presented driver concept: the drain current rise and fall timings are well matched, leading to the conclusion that unequal switching instants of the transistors should not be the reason for the initial differences in the drain currents. Since the currents in the parallel connection of three and four transistors behave in the same way as in the twofold parallel connection, the conclusions found there, for example with regard to the DC link, also remain valid for the three- and fourfold parallel connection.

Conclusion

The paper presents design considerations and the resulting layout for the parallel connection of four GIT GaN transistors. A driver concept is proposed that synchronizes the transistors by a common driver and at the same time optimizes the gate loops by using individual drivers for each transistor. The general functionality of the driver concept is proved by a reference analysis with only a single half-bridge assembled. The driver concept is finally evaluated by the presented synchronism of the switching instants of the transistors in the parallel connections. Moreover, to solve the instabilities in the gate voltages most probably triggered by electromagnetic coupling, an appropriate gate circuit configuration is found so that a stable gate voltage can be guaranteed.

In addition to the evaluation of the gate driver concept, the drain current distribution is investigated in detail. The drain currents show opposing oscillations, which balance out after some time and do not drift apart. In the related investigations for two transistors in parallel, it can be demonstrated that the distance between the parallel transistors, and therefore the increased leakage inductance between the transistors has no noticeable effect on the oscillation. On the other hand, the variation of the DC link has a significant influence, which, as a result of the improvement presented, means that the DC link connection must be investigated again in detail in future work.

Connecting three and four transistors in parallel, the same effects appear as for two parallel transistors, whereby always the first two parallel transistors 1 and 2 behave like the first transistor in the twofold parallel circuit, and the remaining transistors 3 and 4 like the second transistor in the twofold parallel circuit.

Based on the results in this paper, future work will further improve the DC link connection, and also investigate the influence of selective heating of the transistors on the drain current behavior. In this context, operating the setup in a continuous test is also of interest. The current measurement is also examined with regard to possible influences of ground loops. In the end, the parallel circuit is to be operated at its actual operating point at 400 V and four times the load current compared to the single bridge.

References

[1] J. Burkard and J. Biela, "Paralleling GaN switches for low voltage high current half-bridges," 2019 IEEE Energy Conversion Congress and Exposition (ECCE), Baltimore, MD, USA, 2019, pp. 3245-3252, doi: 10.1109/ECCE.2019.8912830.

[2] I. Prakash, D. Klikic, N. Prabhakaran, G. Jagadanand and V. Pulakhandam, "Decoupled Layout Approach for Paralleling GaN Devices in Half Bridge Inverters," *2021 2nd Global Conference for Advancement in Technology (GCAT)*, 2021, pp. 1-5, doi: 10.1109/GCAT52182.2021.9587818.

[3] J. L. Lu and D. Chen, "Paralleling GaN E-HEMTs in 10kW–100kW systems," 2017 IEEE Applied Power Electronics Conference and Exposition (APEC), Tampa, FL, 2017, pp. 3049-3056, doi: 10.1109/APEC.2017.7931131.

[4] T. Kahl, F. Stalleicken, M. Schlüter and S. Dieckerhoff, "Reduction of the Turn-Off Overvoltage in an Active Full-Bridge Rectifier Stage by Paralleling GaN-HEMTs," *2021 23rd European Conference on Power Electronics and Applications (EPE'21 ECCE Europe)*, 2021, pp. P.1-P.10.

[5] Infineon, "IGOT60R070D1 600 V CoolGaN enhancement-mode Power Transistor" IGOT60R070D1 datasheet, October 2021.

[6] Texas Instruments, "UCC2751x Single-Channel, High-Speed, Low-Side Gate Driver (With 4-A Peak Source and 8-A Peak Sink)," UCC2751x datasheet, November 2014.

[7] Infineon, AN_201702_PL52_012 Application Note – "Driving CoolGaN 600 V high electron mobility transistors," 2018.

[8] J. Böcker, S. Schoos and S. Dieckerhoff, "Experimental Comparison and 3D FEM Based Optimization of Current Measurement Methods for GaN Switching Characterization," *2018 20th European Conference on Power Electronics and Applications (EPE'18 ECCE Europe)*, 2018, pp. P.1-P.9.

[9] Infineon, "IGOT60R070D1 600 V CoolGaN enhancement-mode Power Transistor," IGOT60R070D1 datasheet, 2021.

[10] X. Geng, C. Kuring, M. Tannhaeuser and S. Dieckerhoff, "Experimental Evaluation and Analysis of Dynamic On-Resistance in Hard- and Soft-switching Operation of a GaN GIT," PCIM Europe digital days 2020; International Exhibition and Conference for Power Electronics, Intelligent Motion, Renewable Energy and Energy Management, 2020, pp. 1-8.

Verification of GaN-HEMT Spice Models Using an S-parameters Approach

Alonso Gutierrez, Nasri Said, Emmanuel Marcault, Mathieu Gavelle
CEA Tech Occitanie
51 Rue de l'Innovation, Labege - France
Emails: alonso.gutierrez-galeano@cea.fr, nasri.said@u-bordeaux.fr,
emmanuel.marcault@cea.fr, Mathieu.Gavelle@cea.fr

Acknowledgments

The authors would like to thank to the Université Toulouse III - Paul Sabatier, Master in Electronics of Embedded Systems and Telecommunications (ESET) and the CEA Tech Occitanie in the framework of a successful end-of-study internship.

Keywords : « Gallium Nitride GaN», « Wide band gap », « Parasitic elements », « S-Parameters », « Transmission line », « Spice model», « De-embedding »

Abstract

This paper describes a complementary S-Parameters approach to verify the Spice model accuracy of power GaN-HEMT devices regarding their capacitive and inductive aspects represented in graphic Smith charts. This approach correlates the Smith charts of experimental and simulated results in order to provide insights to improve the model characteristics. The correlation is carried out by processing the experimental results and the Spice simulated data using the Smithplot Python library. Additionally, a complementary study considers the input and output reflection coefficients to stablish a connection between the measured and simulated parameters in frequency and voltage. Possibilities of dissociating both package parasitic elements and intrinsic GaN capacitances confirm the potential of S-parameters as a powerful tool for model verification and study of power GaN-HEMT devices using a radio-frequency approach.

1. Introduction

Current development of power GaN-HEMT devices has increased the performance of power transistors in the frequency domain [1]. Therefore, promising perspectives forecast a new generation of power converters with significant increase of the switching frequency at level of megahertz [2]. Therefore, the power electronics knowledge progressively requires complementary radio-frequency methodologies in order to facilitate the design of power converters based on GaN-HEMT [3]. In addition, given the high switching frequencies achieved by these power electronic devices, the parasitic elements associated to the package and the GaN-HEMT intrinsic capacitances are gradually fundamental in the design process [4]. In this context, verification of Spice models trends to be challenging given the high frequency phenomena and the interaction of the GaN-HEMT with their experimental setup [5]. Indeed, parasitic elements of test boards are not easily dissociable of the package parasitic elements and the device intrinsic capacitances [6][7]. Therefore, we propose through this paper a frequency domain methodology to partially verify GaN-HEMT Spice models by means of S-parameter simulations and measurements. Additionally, study results provide insights about the behavior of these power devices in a wide range of voltage and frequency.

The proposed approach begins with the design and characterization of the PCBs for the test fixtures, calibration standards, and Bias Tees. Then, a simplified model is implemented in Spice using a transmission line approach and the S-parameters measurements from test fixtures and the calibration standards. After, passive elements with accurate values are used to adjust the modeled test bench. Finally, considering several bias voltages, the experimental S-parameters from the GaN-HEMT test setup are compared with the associated Spice model. Given the S-parameters, a graphical comparison using Smith charts brings evidence about the Spice model performance regarding the capacitive, resistive and inductive aspects. These results highlight the potential of S-parameters to define and consequently improve the model accuracy using a methodology that requires a hardware and software low demanding to implement.

This document is organized as follows. Section 2 describes the proposed methodology based on S-Parameters. Section 3 explains the experimental setup and discusses the comparison results. Finally, conclusions and perspectives are reported.

2. Methodology for model verification using S-parameters

This section describes the proposed methodology to verify GaN-HEMT models using S-parameters. The focus of this approach is to use a conventional Spice simulation tool and measurements from a Vector Network Analyzer (VNA). Our proposed methodology based on S-parameters aims to provide insights about the model performance in the frequency domain using basic test boards modeled and characterized for further simulation in a well-known simulation tool in power electronics.

The S-parameters or Scattering parameters describe the electrical behavior of linear electrical networks when electrical signals propagate on it [8]. Two-port S-parameters are defined by considering a set of voltage traveling waves. When a voltage wave from a source is incident on a network, a portion of the voltage wave is transmitted through the network, and a portion is reflected back toward the source. Figure 1 shows a simplified diagram for S-parameters definition [9].

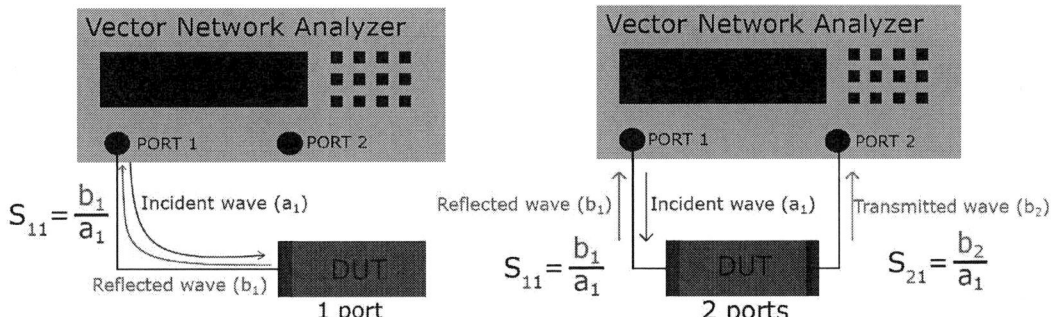

Fig. 1: S-parameters description

The relation between a_i and b_i (for i=1,2) can be written as a system of linear equations (see eq.(1)).

$$b_1 = S_{11}a_1 + S_{12}a_2$$
$$b_2 = S_{21}a_1 + S_{22}a_2 \tag{1}$$

Using VNA measurements, the physical meaning of S_{11} is the input reflection coefficient Γ_{in} with the output of the network terminated by a matched load ($a_2 = 0$). S_{21} is the forward transmission from port 1 to port 2. S_{12} is the reverse transmission (port 2 to port 1) and S_{22} is the output reflection coefficient Γ_{out}. From definition of the input reflection coefficient Γ_{in} and eq.(1), the relation between the S-parameter S_{11} and an unknown impedance Z_L connected to the generator of source and characteristic impedance Z_0 can be calculated using eq.(2).

$$S_{11} = \frac{b_1}{a_1}\bigg|_{a_2=0} = \Gamma_{in} = \frac{Z_L - Z_0}{Z_L + Z_0} = \frac{Z-1}{Z+1}, \;\; thus \;\; S_{11} = \frac{Z-1}{Z+1} \tag{2}$$

where $Z = Z_L/Z_0$ is the normalized impedance and Z_0 is the characteristic impedance, ie. $Z_0 = 50\Omega$

As deduced from eq.(2), S-parameters can be transformed directly into equivalent normalized impedances. Therefore, many electrical properties of the Device Under Test (DUT) (such as inductance, capacitance, and resistance) may be expressed and studied using these parameters. Indeed, two-port S-parameters and the associated impedances can be traced on a Smith chart using polar co-ordinates for graphical analysis. Additionally, a set of S-parameters covering a wide range of frequencies and depicted on a Smith chart can be used to visually represent how capacitive or how inductive is an electrical network across the defined frequency range. Next, we describe the proposed methodology to verify Spice models of power GaN-HEMTs using and S-parameters approach.

Figure 2 depicts the proposed steps for the verification of GaN-HEMT Spice models using the S-parameters approach. In the first step, the PCB layout of the test fixture boards are designed for a characteristic impedance Z_0=50Ω. In addition, two calibration standards [6], Open and Thru, are designed to extract the equivalent C and L of each PCB trace using the definition of characteristic impedance in eq.(3).

$$Z_0 = \sqrt{\frac{L}{C}} \qquad (3)$$

Second step aims to provide a simplified model of the PCBs traces by means of Pi networks to model them as transmission lines. This model is based on C and L parameters that are respectively measured from Open and Thru calibration boards. Furthermore, a resistive element R is added to compensate for an identified resistive offset measured by the VNA. The extraction of these parameters is a primordial step before applying the De-embedding on the device S-parameters.

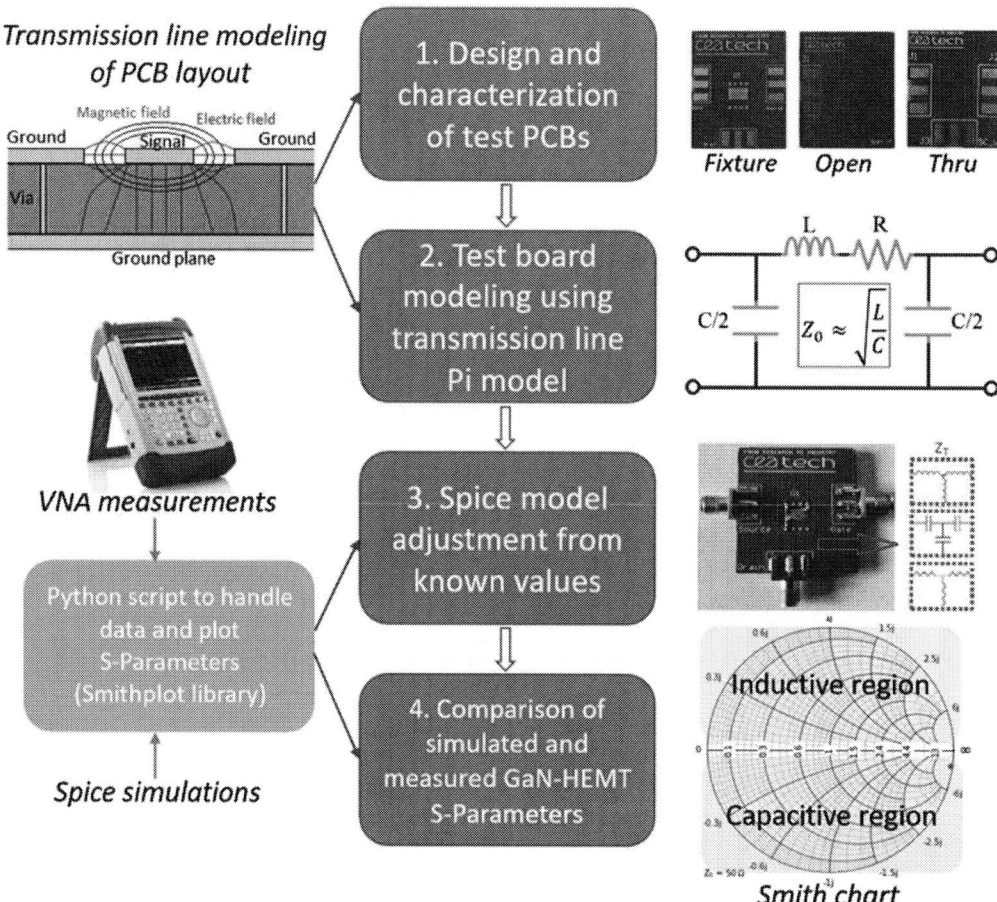

Fig. 2: Steps for power GaN-HEMT Spice model verification using S-parameters

The third step implements the simplified model of the fixture test board in a Spice simulation tool. Before measuring the GaN-HEMT S-parameters in the experimental setup, a calibration stage provides a fine adjustment of the Spice model using precision passive elements connected in a star configuration. Additionally, third step includes a characterized short circuit model to connect the Source terminal to ground. This configuration with unconnected Source terminal was selected given the flexibility to design several fixture test boards for different GaN-HEMT device footprints. Figure 3 shows details of the Pi networks used to model the PCB effect at the DUT accesses. The fixture test board of Figure 3 is designed according to guidelines proposed in [10].

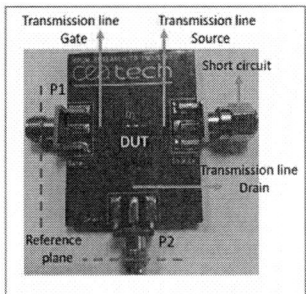 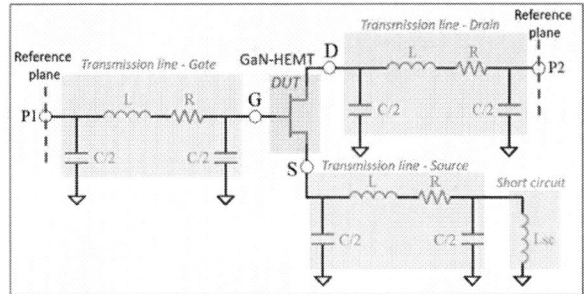

Fig. 3: a) Fixture test board   b) Simplified model of fixture text board using Tline Pi model

In the fourth step, the S-parameters of the GaN-HEMT under test are measured using the test fixture board connected to a VNA. To do the same measurements for different biasing voltages for V_{DS} from 0V to 400V, the designed Bias Tees are connected between the test fixtures and the VNA on gate and drain accesses (as shown in Figure 4) [9]. This stage provides measurements of S-parameters in the Touchstone format which are processed using a Python script. The Python script is used to apply the de-embedding techniques on the device measured data after measuring the parameters of the calibration standards. These data joined with the simulated data are plotted using the Python Smithplot library in the Smith chart for a graphical comparison [11]. The comparison results bring evidence about the Spice model performance in the frequency domain regarding the capacitive, resistive and inductive aspects for several biasing voltages. Finally, conclusions would potentially lead to model validation and/or improvements.

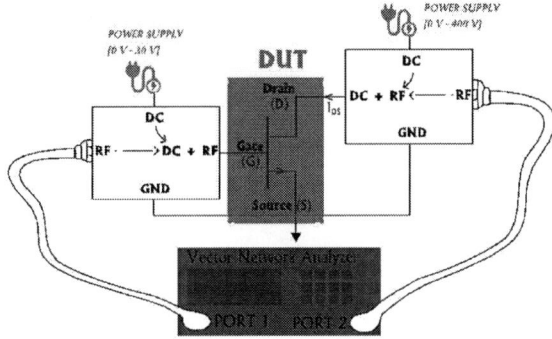

Fig. 4: Test bench schematic for S-parameters measurements under biasing voltages (V_{DS} from 0V to 400V)

3. Experimental results

This section shows the experimental and simulation results of comparing two GaN-HEMT devices from different commercial vendors using the proposed approach. The vendors are named Vendor 1 and Vendor 2. Additionally, the evaluated Spice models are available from the websites of each manufacturer. The experimental setup for this reported work uses as a main measurement equipment a VNA Rohde & Schwarz ZVH8 option ZVH-K42. Table I lists the configured setup parameters. In order to keep uniformity, the Spice simulation and the VNA are configured with similar setup parameters.

Table I: VNA parameters for all measurements

Parameter	Value
Start frequency	1MHz
Stop frequency	1GHz
Number of points	801
Tracking generator power	0 dBm
Resolution bandwidth	300Hz
Sweep time	4s
Characteristic impedance	50Ω

As a general description and following the connection ports, the parameter S_{11} is mainly associated to the loop around the Gate-Source terminals, the parameter S_{22} is mainly associated to the loop around the Drain-Source terminals, and the parameter S_{21} is mainly associated to the trajectory involved by the Gate-Drain terminals.

Figures 5 and 6 show that model provided by Vendor 1 is relatively close of experimental results in a wide range of frequencies and bias voltages of the transistor in OFF state (blocked canal: $V_{gs}<V_{th}$). Figure 7 confirms the tendency of intrinsic device capacitances to decrease when the V_{DS} voltage increases given the smaller curves into the Smith chart. However, the difference between experimental and simulation results in Figure 7 could be associated either to the model accuracy or potentially to the trapping effect caused by the sweep time not considered in the Spice approach. Figure 8 depicts the signal transmission through Gate-Drain terminals with a suitable impedance adaptation at high frequencies as expected. Therefore, this result agrees with a suitable modeling of the intrinsic capacitance C_{GD}.

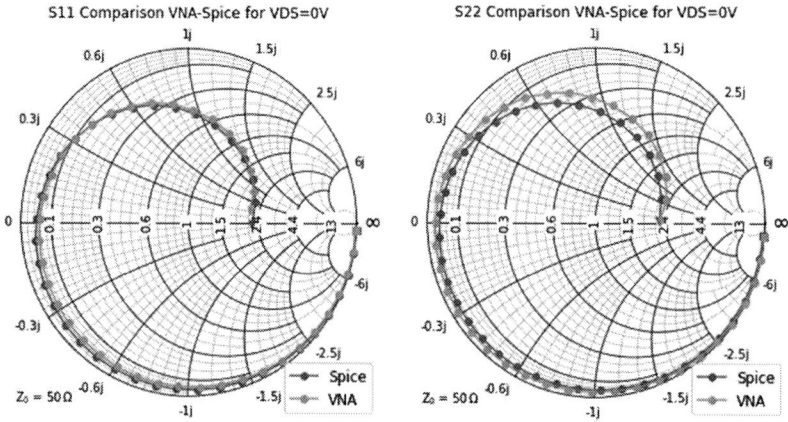

Fig. 5: S_{11} and S_{22} parameters for $V_{DS} = 0V$. Vendor 1.

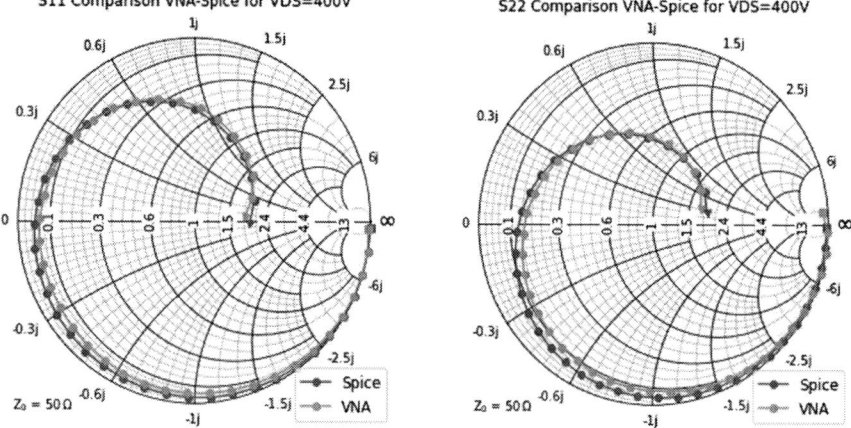

Fig. 6   : S_{11} and S_{22} parameters for $V_{DS} = 400V$. Vendor 1.

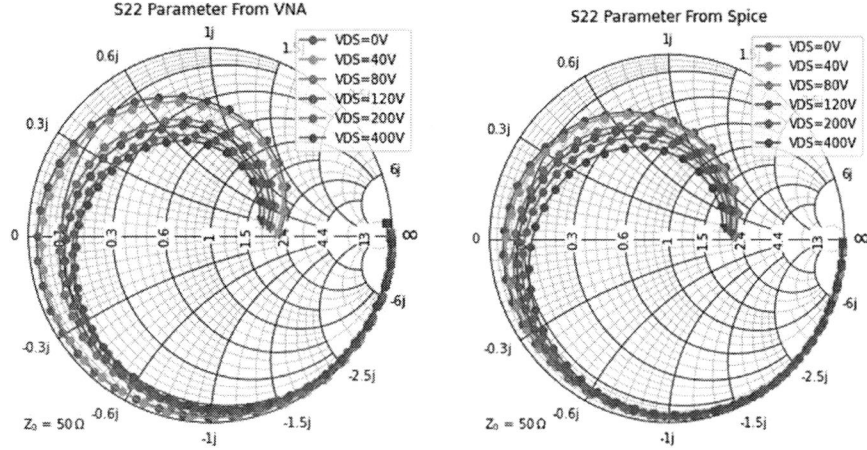

Fig. 7 : S_{22} parameter for several V_{DS} voltage. Vendor 1.

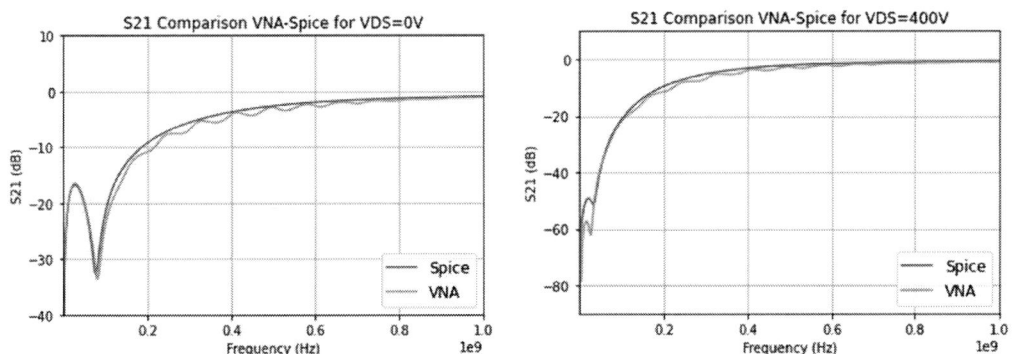

Fig. 8 : S_{22} parameter for $V_{DS} = 0V$ and $V_{DS} = 400V$. Vendor 1.

In contrast, experimental and simulated results for parameter S_{11} in Figure 9 and Figure 10 show some notorious differences in model provided by Vendor 2. Indeed, some capacitive aspects could be potentially improved in the associated C_{GS} capacitance. Furthermore, the inductive effects could be improved by reducing the length of the transmission lines from the SMA connectors to the DUT accesses to increase the test setup accuracy. Figure 11 illustrates the model performance for parameter S_{21} with a suitable behavior for the C_{GD} capacitance model.

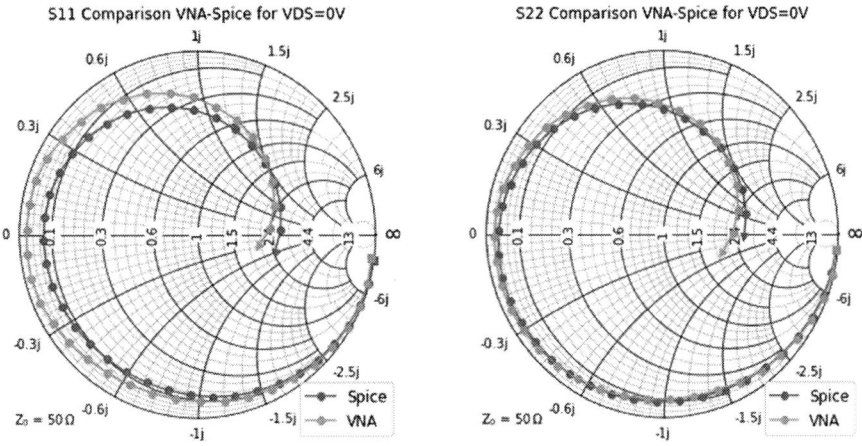

Fig. 9 : S_{11} and S_{22} parameters for $V_{DS} = 0V$. Vendor 2.

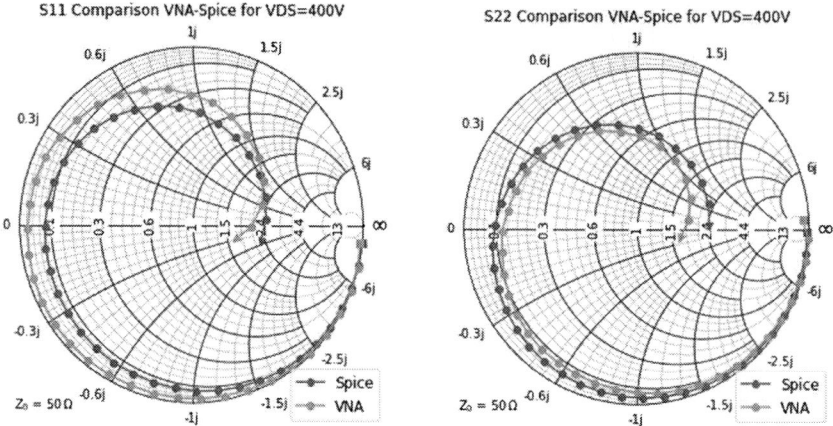

Fig. 10 : S_{11} and S_{22} parameters for $V_{DS} = 400V$. Vendor 2.

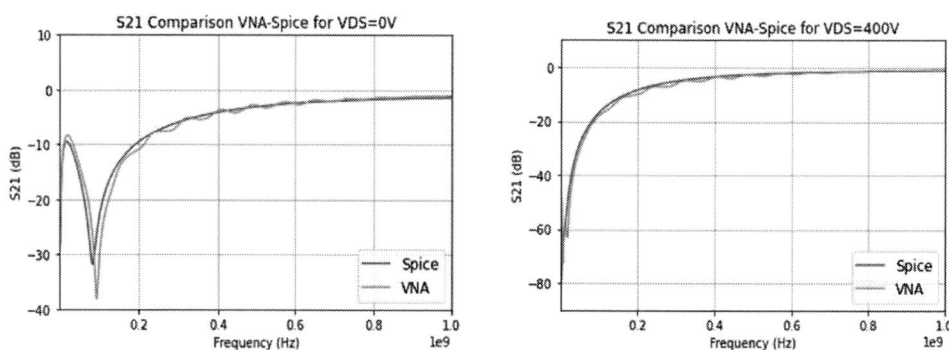

Fig. 11: S_{21} parameter for $V_{DS} = 0V$ and $V_{DS} = 400V$. Vendor 2.

Given that results for Vendor 2 have shown more divergence between Spice simulations and VNA measurements, a complementary analysis is developed as follows. In eq.(2), the input and output reflection coefficients are complex numbers that consider the magnitude and phase of the S-parameters S_{11} and S_{22} and their relation with the unknown impedance. Thus, the relation between the unknown impedance and the reflection coefficient is given by eq.(4).

$$Z_L = Z_0 \frac{1+\Gamma}{1-\Gamma} \tag{4}$$

Figure 12 depicts the magnitude of the input reflection coefficient $|\Gamma_{in}|$ for a bias voltage $V_{DS} = 400V$. This figure shows that the Spice model underestimates the input reflection coefficient from frequencies between 1MHz and 650MHz. Therefore, the model in this frequency range will underestimate the associated impedance between the Gate-Source terminals. On the other hand, the Gate-Source impedance will be overestimated by the model in frequencies higher that 650MHz. Figure 13 shows a suitable agreement between the phase of the input reflection coefficients for experimental and simulation results.

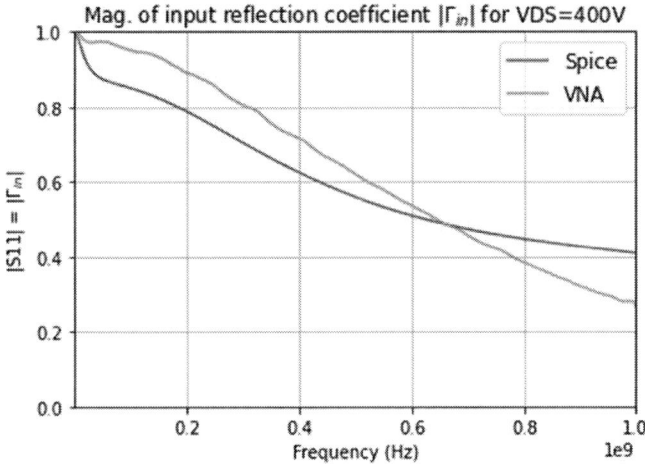

Fig. 12: Magnitude of input reflection coefficient $|\Gamma_{in}| = |S_{11}|$ for $V_{DS} = 400V$. Vendor 2.

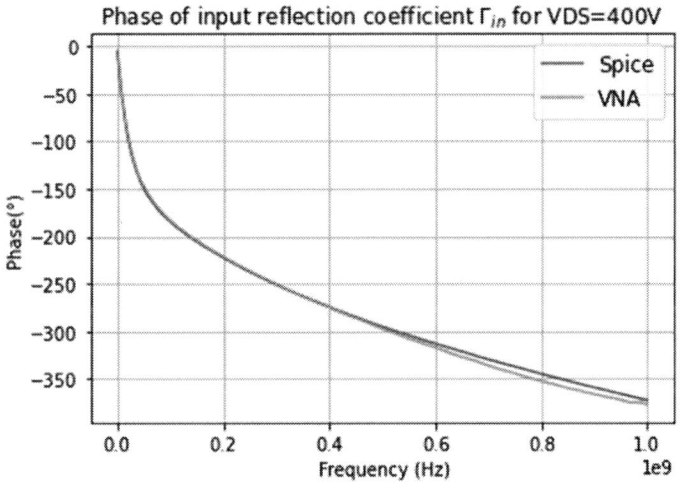

Fig. 13: Phase of input reflection coefficient $\angle\Gamma_{in} = \angle S_{11}$ for $V_{DS} = 400V$. Vendor 2.

Figure 14 illustrates the magnitude of the output reflection coefficient $|\Gamma_{out}|$ for a bias voltage $V_{DS} = 400V$. This figure highlights that the Spice model is higher than the VNA measurements for the output reflection coefficient in the frequencies range between 1MHz and 1GHz. Therefore, the model in this frequency range will overestimate the associated impedance between the Drain-Source terminals. Furthermore, Figure 15 shows a suitable agreement between the phase of the input reflection coefficients for experimental and simulation results.

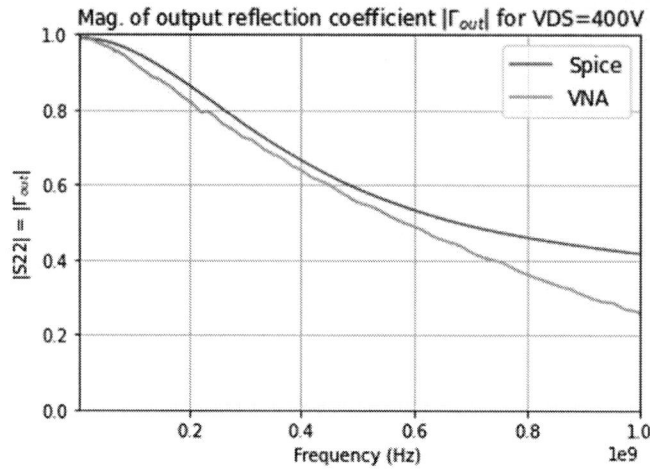

Fig. 14: Magnitude of output reflection coefficient $|\Gamma_{out}| = |S_{22}|$ for $V_{DS} = 400V$. Vendor 2.

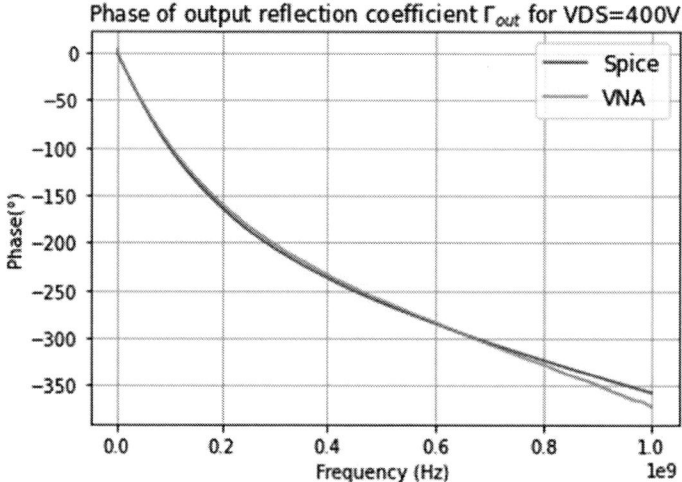

Fig. 15: Phase of input reflection coefficient $\angle\Gamma_{out} = \angle S_{22}$ for $V_{DS} = 400V$. Vendor 2.

To summarize, the results of this work highlight the potential of S-parameters to study the behavior of capacitive, inductive, and resistive elements of power GaN-HEMTs and their package in an extended span of frequencies and voltages. Indeed, the proposed approach has allowed the evaluation of power GaN models with an alternative perspective considering a radio-frequency approach and using an accessible power electronics simulator.

Conclusion

This work has described a radio-frequency technique based on S-parameters measurements and a Python library to verify and explore the characteristics of power GaN-HEMTs and their packaged in the frequency domain at several biasing voltages. The proposed approach has allowed the study of power GaN models with radio-frequency methodologies but using a well-known simulation tool in power electronics. This technique has required the implementation of a low complexity hardware setup to sweep the frequency and voltage at the same time. VNA measurements and Spice simulations were compared to evaluate Spice models from different commercial vendors suggesting potential improvements. The explored methodology confirmed the potential of S-parameters as a powerful tool to characterize power GaN-HEMTs and to build models including the parasitic effect of the package. Indeed, different elements of GaN-HEMT transistors could be extracted such as the parasitic aspects of the packaging in order to predict undesired oscillations and to improve the device reliability. Additionally, the device capacitive behavior could be determined in the frequency domain for different biasing voltages. The reflection coefficient allowed to verifying the Spice model accuracy in a wide range of frequencies.

The proposed methodology can be potentially extended to the study of more complex configurations such as Cascode or half-bridge interconnections. As a future work, a deeper result analysis will be discussed and their correlation with the time domain will be considered. Furthermore, pulsed DC and pulsed RF measurements can be done by adapting the same test bench in order to do reliability tests on power GaN devices.

References

[1] C. Bao and S. K. Mazumder, "GaN-HEMT Based Very-High-Frequency AC Power Supply for Electrosurgery," 2021 IEEE Applied Power Electronics Conference and Exposition (APEC), 2021, pp. 220-225.

[2] M. Baker, S. Jain and M. B. Shadmand, "GaN based High Frequency Power Electronic Interfaces: Challenges, Opportunities, and Research Roadmap," 2021 IEEE Power and Energy Conference at Illinois (PECI), 2021, doi: 10.1109/PECI51586.2021.9435203.

[3] R. Sun, J. Lai, W. Chen and B. Zhang, "GaN Power Integration for High Frequency and High Efficiency Power Applications: A Review," in IEEE Access, vol. 8, pp. 15529-15542, 2020, doi: 10.1109/ACCESS.2020.2967027.

[4] Z. Liu, X. Huang, F. C. Lee and Q. Li, "Package Parasitic Inductance Extraction and Simulation Model Development for the High-Voltage Cascode GaN HEMT," in IEEE Transactions on Power Electronics, 2014.

[5] S. Khandelwal, Y. S. Chauhan, J. Hodges and S. A. Albahrani, "Non-Linear RF Modeling of GaN HEMTs with Industry Standard ASM GaN Model," 2018 IEEE BiCMOS and Compound Semiconductor Integrated Circuits and Technology Symposium (BCICTS), 2018, pp. 93-97, doi: 10.1109/BCICTS.2018.8550974.

[6] L. Pace, " Caractérisation et Modélisation de Composants GaN pour la Conception des Convertisseurs Statiques Haute Fréquence," PhD thesis, Université de Lille, 2019.

[7] G. Curatola, G. Verzellesi, " Modelling of GaN HEMTs: From Device-Level Simulation to Virtual Prototyping, "In: Power GaN Devices, Power Electronics and Power Systems. Springer, Cham. 2017, doi.org/10.1007/978-3-319-43199-4_8

[8] A. J. Baden Fuller, " An Introduction to Microwave Theory and Techniques", Second Edition, Pergammon International Library, 1979, ISBN 0-08-024227-8.

[9] N. Said," Conception et Réalisation d'un Banc de Test pour la Caractérisation des Composants de Puissance GaN/Si et de Leur Packaging sous Stress Electrique," End-of-studies project, Master 2 - ESET, Université de Toulouse – CEA Tech Occitanie, 2021.

[10] L. Pace, N. Defrance, A. Videt, N. Idir, J.-C. De Jaeger, and V. Avramovic. "Extraction of Packaged GaN Power Transistors Parasitics using S-parameters". IEEE Transactions on Electron Devices, 2019.

[11] Pysmithplot project, repository at https://pypi.org/project/pysmithplot-fork/#files [Accessed: May 22, 2021]

Power Loss Modelling of GaN HEMT-based 3L-ANPC Three-Phase Inverter for different PWM Techniques

Salvatore Mita[1], Arjun Sujeeth[1], Giuseppe Aiello[1], Dario Patti[1], Francesco Gennaro[1], Giacomo Scelba[2], Mario Cacciato[2],

STMicroelectronics, Stradale Primo Sole 50, Catania, Italy [1]
University of Catania, Viale Andrea Doria 6, Catania, Italy [2]
E-Mail: giacomo.scelba@unict.it, francesco.gennaro@st.com

Acknowledgements

This work was carried out within the ECSEL-JU project GaN4AP (GaN for Advanced Power Applications), under grant agreement no. 101007310. This Joint Undertaking receives support from the European Union's Horizon 2020 research and innovation programme.

Keywords

« Gallium Nitride (GaN) », « PWM modulation techniques», « Power Converter Modelling », « Conduction losses», « Switching losses», « Active neutral point clamped converter», « Power converters for EV», «DC-AC converters».

Abstract

The paper presents a straightforward modelling approach to compute the power loss distribution in GaN HEMT-based three-phase and three-level (3L) active neutral point clamped (ANPC) inverters, for different pulse width modulated techniques. Conduction and switching losses averaged over each PWM switching period are analytically computed by starting from the operating conditions of the AC load and data of GaN power devices. The accuracy of the proposed analytical approach is evaluated through a circuit-based power electronics simulation tool, applied to different carrier-based PWM strategies.

Introduction

Nowadays, it is well recognized the key role of the electric traction as one of the prerequisites for the successful development of contemporary society, and the global electric vehicle market size is projected to a significant grow in the coming years even thanks to the policy objective set by many countries around the world of reducing greenhouse gas emissions from transport. Today, the rated power of the powertrain of electric vehicles generally varies from 60kW to 200kW, supplied by a lithium-ion battery pack whose capacity vary from 30kWh up to 90kWh, leading to optimistic estimated ranges lower than 450km. Although the recharging infrastructure is undergoing significant development in terms of charging points and installed power, most of the charging points are still limited to 50kW peak power, yielding to charging time not less than 45minutes ÷ 1h for reaching the 80% of full battery capacity when 50kW - 480V DC Fast Charging is used [1]-[3].

A shorter charge period could minimize the inconvenience of the driving range limitation from the consumer's point of view, and this goal could be achieved by using a higher charging power [1]-[5]. For instance, by increasing the fast-charging power level from 50 kW to 150 kW, the charging time is reduced by two-thirds. However, if the charging voltage level remains at the typical value of 400V, the current rating of the charging cable increases as well as the system power losses. Hence, some car makers are developing solutions to utilize an 800V DC bus beneficing in terms of shorter battery charge times [3]. This will however result in facing some technical issues related to the state of art technologies used in the electric powertrains, starting from the semiconductor technologies used to realize the electric traction inverter. Moreover, alternative power converter topologies could be considered to furtherly increase the efficiency, the reliability and the power density, while reduce the cost and weight of the power conversion processes.

Although still considered not suited for mostly motor drives, Gallium Nitride (GaN) power switches can potentially provide several benefits to the electric drives exploited in the traction, especially when high-speed electrical machines are suitably combined with GaN inverters topologies operating at high switching frequency [6]. In fact, GaN devices feature lower switching and conduction losses, higher power density, and higher-temperature operation compared to Si power switches. Moreover, GaN devices possess much less parasitic components and lower on-state resistance which make it more suitable than the SiC device in hard switched applications. Hence, this technology can furtherly contribute to realize electric drives for traction featuring extreme compactness, high efficiencies, robustness and reduced weight, all factors contributing to further extend the range of the vehicle. Moreover, the increase of DC bus voltage architectures (800V) in the next generation of EVs developed with the goal of enabling faster charging rates, will have direct consequence to increase the losses of the standard 2L VSIs based on Si and SiC devices, thus reducing the system efficiency. This makes the use of GaN multilevel inverters an attractive solution for designing very compact and high-efficiency traction inverters overcoming the breakdown voltage limits (650V) of the actual GaN technologies. Moreover, thanks to the reduced output voltage steps, the use of multilevel converters is an effective solution to reduce the voltage stress and thus the electrical aging in electrical traction machines. Several configurations of multilevel topologies have been presented in the technical literature, among them the neutral point clamped (NPC), the Active NPC (ANPC), and T-type NPC (TNPC) [7]-[12]. The 3L-ANPC inverter represents a good compromise between performances, compactness, efficiency, and investment cost and is considered one of the promising multilevel converter topologies for medium voltage application, even for electric traction application [10]-[12]. The 3L-ANPC topology is shown in Fig. 1. It consists of six switches for each leg, which allow to generate three voltage levels. In past literatures several modulation strategies have been presented with the main goal of minimizing conduction and switching power loss, but in most of the cases, sophisticated mathematical or circuital-based solutions have been used to carry out the loss distribution and they were specific only for a single modulation strategy [13]-[18].

This paper presents an accurate analytical modelling devoted to the power losses estimation for GaN HEMT-based three-phase 3L-ANPC inverters for different modulation techniques. The main goal of the proposed investigation is to evaluate, among a certain number of well-known PWM modulation strategies [13], [15], the power loss that allows to get a modulation technique with minimum power losses and better thermal distribution. The paper is structured to first provide the theoretical description of the ANPC operation, and the mathematical model used to compute the power loss. Then, the presentation of different modulation techniques is detailed provided, underlying pros and cons. Finally, some experimental tests are also provided to validate the accuracy of the theoretical modelling. Although the methodology presented in this paper has been applied to some modulation methods, it could be straightforwardly extended even to other PWM modulation strategies.

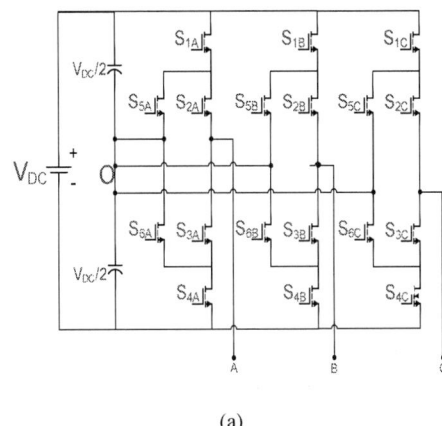

Table I:	
Parameter	**Value**
DC Bus Voltage [V_{DC}]	800V
Line-to-Line Voltage [V_{LL}]	400V
$\cos \varphi$ [PF]	0.9
Root Mean Square Current [I_{RMS}]	6.95A
Active Power [P]	4.5kW
Resistance Load [R_L]	25.65Ω
Inductance Load [L_L]	39.5mH
Fundamental Frequency [f_e]	50Hz
Switching Frequency [f_{sw}]	50kHz
Modulation Index [m]	0.7
Resistance Drain-Source [R_{DS}]	120mΩ
Junction Temperature [T_j]	80°C

(a)   (b)

Fig. 1: (a) Three-phase active neutral point clamped (ANPC) GaN-based topology (b) main power converter and load settings.

Three-Phase 3L-Active Neutral Point Clamped Inverter

Each leg of the 3L ANPC inverter consists of six bidirectional-conducting-unidirectional-blocking switches [4], which allow to generate three voltage levels. This converter can be considered an evolution of the standard neutral point clamped (NPC) inverter. The last name comes from the two diodes connected in anti-parallel that are used to "clamp" the output voltage to the neutral point of the DC circuit when the zero-voltage level is required. The output current direction determines whether the neutral point current flows through the upper or the lower current path.

Compared to the NPC, two additional switches S_5 and S_6 substitute the clamping diodes in the neutral point connection, allowing to actively clamp the output to the neutral point of the DC circuit. In this way, when the ANPC inverter outputs zero state, there are multiple redundant loops to choose from. By rationally selecting the zero-state loop, the loss balance of each device can be achieved [15]. Hereafter, some specific PWM strategies for the 3L ANPC inverter are described.

Carrier based Modulation Techniques

In the following analysis different PWM modulation techniques are considered [13], [15]: (a) DNPC modulation, (b) ANPC modulation with same-side clamping. (c) ANPC modulation with opposite-side clamping. (d) ANPC modulation with full-path clamping; these strategies can be applied to both sinusoidal and third harmonic injection SPWM.

- *DNPC Modulation (DNPC)*

The DNPC modulation keeps the two clamping switches S_5 and S_6 constantly OFF in the entire fundamental period, similarly to the diode clamped NPC. On the contrary, the gate signals for $S_1 - S_6$ are generated according to the phase disposition PWM, as shown in Fig. 2a. Note that S_1 is switched in complementary with S_3, and S_2 is complementary to S_4. The command signal applied to S_1 is generated by comparing the reference sinusoidal V_{ref} to V_{tri1}, whereas the switching pattern for S_4 is generated by comparing V_{ref} to V_{tri2}. The switches $S_1 - S_4$ are pulse width modulated at the switching frequency f_{sw} in half fundamental cycle while keep a constant state in the other half period.

- *ANPC Modulation With Same-Side Clamping (ANPC-SSCM)*

This PWM implementation forces the path of the load current to flow through the upper cell during the positive half cycle and vice-versa in the negative half; as shown in Fig. 2b, three pairs of complementary switches are used: S_1 and S_5, S_2 and S_3, S_4 and S_6. The command signal of S_1 is generated by comparing V_{ref} to V_{tri1} while S_4 is commutated by comparing V_{ref} to V_{tri2} and S_2 commutates for every half cycle by comparing V_{ref} to 0.

The switches S_1, S_4, S_5 and S_6 are modulated at f_{sw} in half of the fundamental period while remains at a constant state in the other half period. The inner switches S_2 and S_3 are commutated at the fundamental frequency f_e.

- *ANPC Modulation With Opposite-Side Clamping (ANPC-OSCM)*

An alternative PWM implementation can be achieved by using the lower neutral path for the top cell commutation and the upper neutral path for the bottom cell commutation, which is called "opposite-side clamping". As shown in Fig. 2c, the complementary switch pairs are the same of *ANPC-SSCM*, but the switching patterns are quite different. The command signals of S_1 and S_6 is carried out by comparing V_{ref} to 0 whereas S_4 and S_5 are the complementary. The command signal of S_2 is generated by comparing V_{ref} to V_{tri1} when V_{ref} is greater than 0 while V_{ref} is compared to V_{tri2} when it is less than 0. Only the inner switches S_2 and S_3 are modulated at f_{sw}, while the others are all switched at f_e.

- *ANPC Modulation With Full-Path Clamping: (ANPC-FPCM)*

Instead of exploiting only one current path during the neutral states, the upper path and the lower path can be used together; this strategy is referred in the literature as "full-path clamping". The two paralleled current paths can reduce the on-state resistance and thus the conduction loss during the neutral states. As shown in Fig. 2d, the devices S_3 and S_5 are simultaneously turned on and off and complementary to S_1, while the devices S_4 and S_6 simultaneously turned on and off and the command signals are complementary to S_2. This modulation is achieved by comparing V_{ref} to V_{tri1} for driving S_1 and V_{ref} to V_{tri2} for driving S_4 and S_6.

The above modulation methods can be also extended to the case of third harmonic injection PWM, where a third harmonic zero sequence voltage waveform set is added to the reference sinusoidal voltages [13]. The last approach allows to increase the DC bus voltage utilization of 15%.

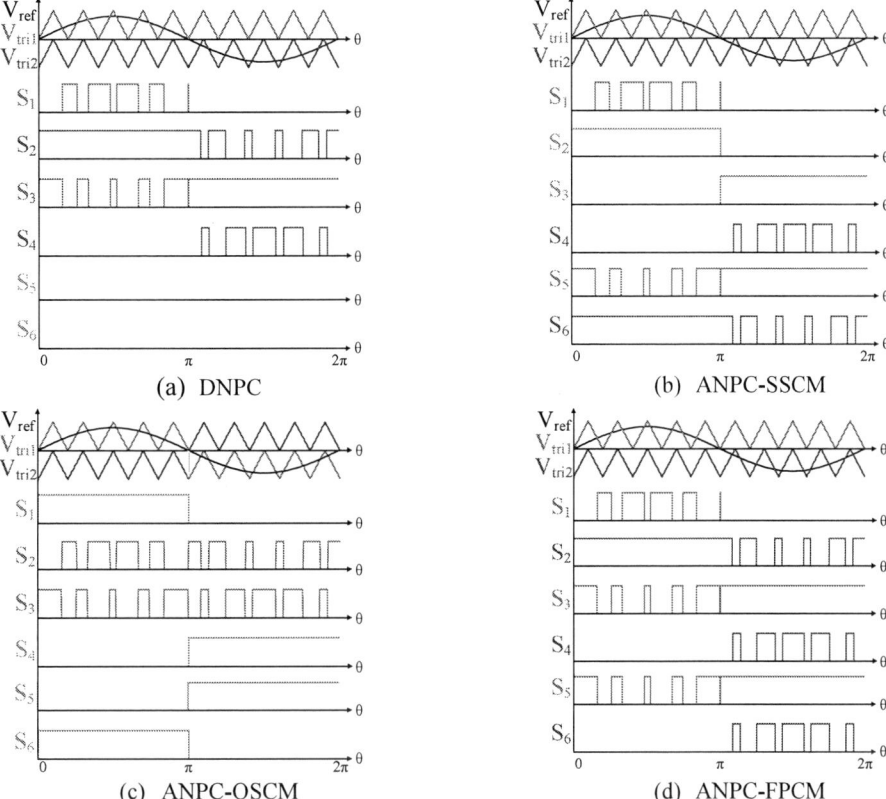

Fig. 2: Modulation strategies and corresponding switching patterns.

Power Loss Computation

Hereafter, the analytical approach to determine the conduction and switching losses of the devices composing the 3L-ANPC inverter is presented for the considered sinusoidal PWM carrier-based modulation strategies.

Conduction Losses

For each GaN power device the conduction losses are computed according to (1), where the RMS current is given by (3); we are assuming a sinusoidal load current (2) and the duty cycle $d(\theta)$ of one phase of the PWM voltage waveform is defined as a quantity variable with the modulation index m and power factor $cos(\varphi)$, [19]. The load current is identified by the peak value I_p and phase angle φ, while the angular position θ is given by $\omega_t t$, where ω_t is the fundamental voltage harmonic angular frequency.

$$P_{cond} = I_{RMS}^2 R_{DS(ON)} \quad (1) \qquad I(\theta) = I_p \, sin(\theta-\varphi) \quad (2) \qquad I_{RMS} = \sqrt{\frac{1}{\pi}\int_0^{2\pi}(I_p^2 sin^2(\theta - \varphi))d(\theta)} \quad (3)$$

The power device currents for the four type of sinusoidal carrier based modulation techniques considered in this study are shown in Fig. 2, with the power converter operating under the conditions indicated in Tab. I. The carrier-based modulation techniques feature a symmetrical behavior: $I_{rms(S1)}=I_{rms(S4)}$, $I_{rms(S2)}=I_{rms(S3)}$, $I_{rms(S5)}=I_{rms(S6)}$, thus the conduction losses computation can be limited to the top half GaN devices of the power converter i.e. S_1, S_2 and S_5. The computation of the RMS currents I_{RMS} (3) for each device are reported in Table. II. The computation has been carried out by considering the reverse conduction mechanism of GaN HEMT, thus considering the two current parallel paths for each device.

The state of the generic power switch S_j associated to forward and reverse current conductions is indicated with S_{jF} and S_{jR}, respectively. Same approach can be extended even to other modulation strategies, such as the third harmonic injection PWM techniques.

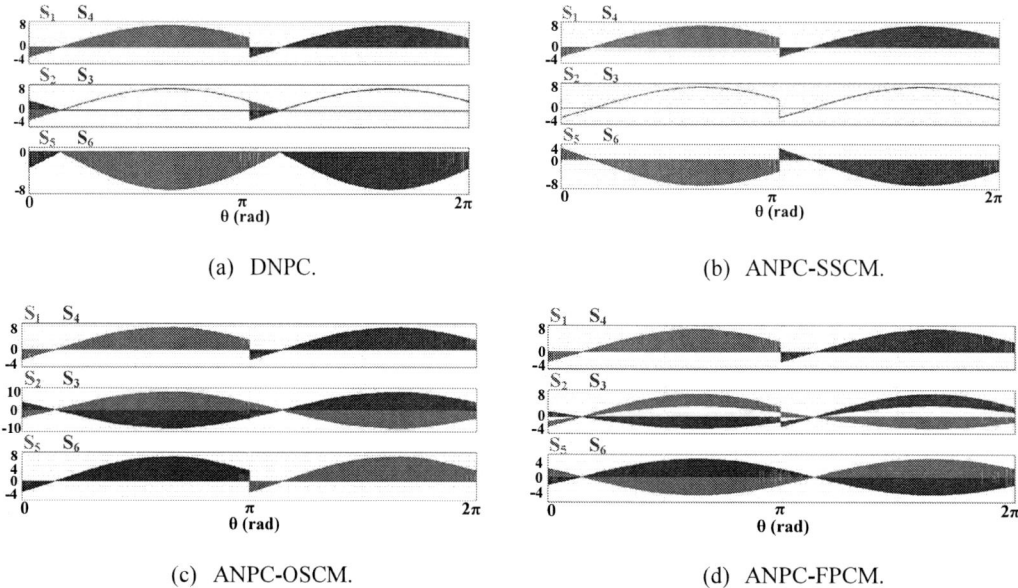

(a) DNPC.

(b) ANPC-SSCM.

(c) ANPC-OSCM.

(d) ANPC-FPCM.

Fig. 2: Currents waveforms flowing through the GaN power devices in the ANPC for sinusoidal PWM modulation techniques (cos $\varphi = 0.9$).

Switching Losses

An analytical approach for the estimation of switching losses is provided below. The input data required for switching losses estimation are given by the characteristic curves of the GaN energy losses (E_{ON} and E_{OFF}), expressed as a function of the current magnitude I_{ds} and gate resistor R_g (T=25°C), as shown in Fig. 3.

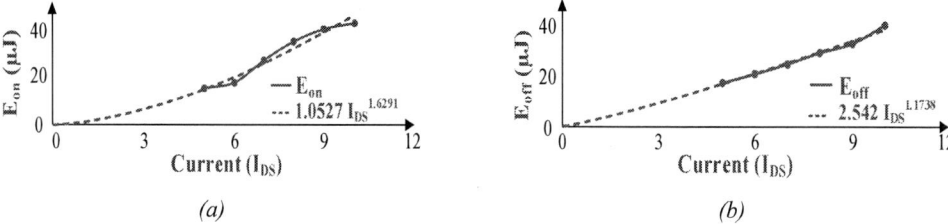

(a)

(b)

Fig. 3: Turn ON (a) and turn OFF (b) switching energy losses of the GaN HEMT under test.

It has been experienced that energy losses curves can be easily approximated by sampling and interpolating with a second order equation. For the considered device, the expression used to calculate the average value of the energy losses E_{ON} and E_{OFF} and powers P_{ON} and P_{OFF} for the generic GaN HEMT device S_j in each switching period are given by (4)-(7).

$$E_{ON} = (1.0527 * |I_{DS}|^{1.6291}) * 10^{-6} \text{ [J] (4)} \qquad E_{OFF} = (2.542 * |I_{DS}|^{1.1738}) * 10^{-6} \text{ [J] (5)}$$

$$P_{ON} = (1.0527 * |I_{DS}|^{1.6291}) * 10^{-6} * f_{sw} \text{ [W] (6)} \qquad P_{OFF} = (2.542 * |I_{DS}|^{1.1738})10^{-6} * f_{sw} \text{ [W] (7)}$$

By considering the number of commutations n of each power switch within the fundamental period, the average switching loss are given by:

$$P_{sw} = \frac{1}{n}\sum_{i=1}^{n}(P_{ONi} + P_{OFFi}) \qquad (8)$$

Table II: RMS currents for the considered sinusoidal modulation strategies

PWM strategy	S_j	$d(\theta)$; interval	I_{rms}
DNPC (a)	S_{1F}	$msin(\theta)$; $\varphi<\theta<\pi$	$\dfrac{I_P}{\sqrt{2\pi}}\sqrt{\dfrac{m(cos(\varphi)+1)^2}{3}}$
	S_{1R}	$msin(\theta)$; $0<\theta<\varphi$	$\dfrac{I_P}{\sqrt{2\pi}}\sqrt{\dfrac{m(cos(\varphi)-1)^2}{3}}$
	S_{2F}	1 ; $\varphi<\theta<\pi$ $1+msin(\theta)$; $\pi<\theta<\pi+\varphi$	$\dfrac{I_P}{\sqrt{2\pi}}\sqrt{\dfrac{\pi}{2}-\dfrac{(mcos(\varphi)-1)^2}{3}}$
	S_{2R}	$msin(\theta)$	$\dfrac{I_P}{\sqrt{2\pi}}\sqrt{\dfrac{m(cos(\varphi)-1)^2}{3}}$
	S_{5R}	$1-msin(\theta)$; $\varphi<\theta<\pi$ $1+msin(\theta)$; $\pi<\theta<\pi+\varphi$	$\dfrac{I_P}{\sqrt{2\pi}}\sqrt{\dfrac{\pi}{2}-\dfrac{2m}{3}(1+cos(\varphi)^2)}$
ANPC-SSCM (b)	S_{1F}	$msin(\theta)$; $\varphi<\theta<\pi$	$\dfrac{I_P}{\sqrt{2\pi}}\sqrt{\dfrac{m(cos(\varphi)+1)^2}{3}}$
	S_{1R}	$msin(\theta)$; $0<\theta<\varphi$	$\dfrac{I_P}{\sqrt{2\pi}}\sqrt{\dfrac{m(cos(\varphi)-1)^2}{3}}$
	S_{2F}	1 ; $\varphi<\theta<\pi$	$\dfrac{I_P}{\sqrt{2\pi}}\sqrt{\dfrac{\pi}{2}-\dfrac{\varphi}{2}-\dfrac{sin(2\varphi)}{4}}$
	S_{2R}	1 ; $0<\theta<\varphi$	$\dfrac{I_P}{\sqrt{2\pi}}\sqrt{\dfrac{\pi}{2}-\dfrac{sin(2\varphi)}{4}}$
	S_{5F}	$1-msin(\theta)$; $0<\theta<\varphi$	$\dfrac{I_P}{\sqrt{2\pi}}\sqrt{\dfrac{\varphi}{2}-\dfrac{sin(2\varphi)}{4}-\dfrac{m(cos(\varphi)-1)^2}{3}}$
	S_{5R}	$1-msin(\theta)$; $\varphi<\theta<\pi$	$\dfrac{I_P}{\sqrt{2\pi}}\sqrt{\dfrac{\pi}{2}-\dfrac{\varphi}{2}-\dfrac{sin(2\varphi)}{4}-\dfrac{m(cos(\varphi)-1)^2}{3}}$
ANPC-OSCM (c)	S_{1F}	$msin(\theta)$; $\varphi<\theta<\pi$	$\dfrac{I_P}{\sqrt{2\pi}}\sqrt{\dfrac{m(cos(\varphi)+1)^2}{3}}$
	S_{1R}	$msin(\theta)$; $0<\theta<\varphi$	$\dfrac{I_P}{\sqrt{2\pi}}\sqrt{\dfrac{m(cos(\varphi)-1)^2}{3}}$
	S_{2F}	$msin(\theta)$; $\varphi<\theta<\pi+\varphi$	$\dfrac{I_P}{\sqrt{2\pi}}\sqrt{\dfrac{4mcos\varphi}{3}}$
	S_{2R}	$msin(\theta+\pi)$; $0<\theta<\varphi$ $1-msin(\theta+\pi)$; $\pi+\varphi<\theta<2\pi$	$\dfrac{I_P}{\sqrt{2\pi}}\sqrt{\dfrac{\pi}{2}-\dfrac{\varphi}{2}+\dfrac{sin(2\varphi)}{4}-\dfrac{m(cos(\varphi)-1)^2}{3}-\dfrac{m(cos(\varphi)+1)^2}{3}}$
	S_{5F}	$1-msin(\theta)$; $0<\theta<\varphi$	$\dfrac{I_P}{\sqrt{2\pi}}\sqrt{\dfrac{\pi}{2}-\dfrac{\varphi}{2}+\dfrac{sin(2\varphi)}{4}-\dfrac{m(cos(\varphi)+1)^2}{3}}$
	S_{5R}	$1-msin(\theta)$; $\varphi<\theta<\pi$	$\dfrac{I_P}{\sqrt{2\pi}}\sqrt{\dfrac{\varphi}{2}-\dfrac{sin(2\varphi)}{4}-\dfrac{m(cos(\varphi)-1)^2}{3}}$
ANPC-FPCM (d)	S_{1F}	$msin(\theta)$; $\varphi<\theta<\pi$	$\dfrac{I_P}{\sqrt{2\pi}}\sqrt{\dfrac{m(cos(\varphi)+1)^2}{3}}$
	S_{1R}	$msin(\theta)$; $0<\theta<\varphi$	$\dfrac{I_P}{\sqrt{2\pi}}\sqrt{\dfrac{m(cos(\varphi)-1)^2}{3}}$
	S_{2F}	$msin(\theta)$; $\varphi<\theta<\varphi+\pi$ $1-msin(\theta)$; $\varphi<\theta<\varphi+\pi$	$\dfrac{I_P}{\sqrt{2\pi}}\sqrt{\dfrac{\pi}{8}+(mcos\varphi)}$
	S_{2R}	$1-2msin(\theta+\pi)$; $0<\theta<\varphi$ $1-msin(\theta+\pi)$; $\varphi<\theta<\varphi+\pi$	$\dfrac{I_P}{\sqrt{2\pi}}\sqrt{\dfrac{\pi}{8}-\dfrac{m(cos(\varphi)+1)^2}{12}}$
	S_{5F}	$1-msin(\theta+\pi)$; $\pi+\varphi<\theta<\varphi+2\pi$	$\dfrac{I_P}{\sqrt{2\pi}}\sqrt{\dfrac{\pi}{8}-\dfrac{mcos(\varphi)}{3}}$
	S_{5R}	$1-msin(\theta)$; $\varphi<\theta<\varphi+\pi$	$\dfrac{I_P}{\sqrt{2\pi}}\sqrt{\dfrac{\pi}{8}-\dfrac{mcos(\varphi)}{3}}$

Study Case

The proposed modelling has been tested under different operating conditions, and the following results are referred to the operating conditions listed in Tab. I. The budget of conduction and switching losses associated to each power switch for the considered PWM techniques are shown in Fig. 4, while the total losses for each device are displayed in Fig. 5. Note an unequal loss distribution among the devices depending on the modulation strategy, which is less evident in case of ANPC-FPCM.

The effectiveness and accuracy of the proposed approach has been initially evaluated by realizing the same power conversion system with PSIM, a circuit-based power electronics simulator containing a specific tool allowing to include the electrical and thermal characterization curves in the power device models. The circuital scheme and a picture of the device database editor are displayed in Fig. 7 and Fig. 8. Current and voltage acquisitions have been performed to compute the loss and delivered power. The error between the results obtained with the analytical approach and the circuit-based simulator are displayed in Fig. 6, confirming the feasibility of the method with satisfying achievements, i.e. differences lower than 3%.

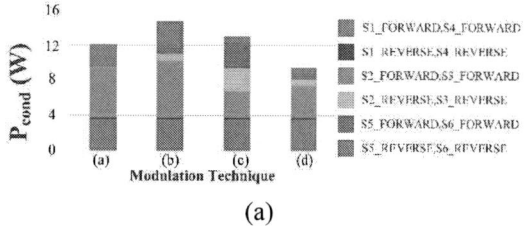

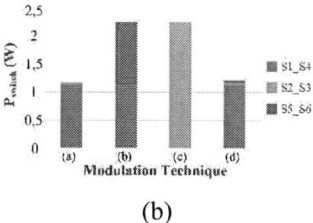

(a)   (b)

Fig. 4: (a) Conduction Losses and (b) Switching losses for the sinusoidal PWM techniques.

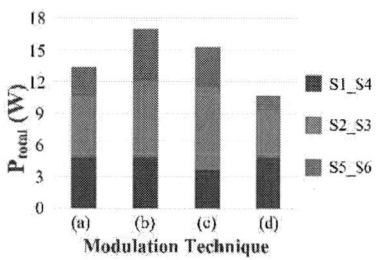

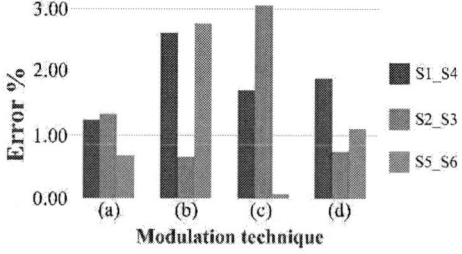

Fig. 5: Total losses distribution in the ANPC inverter leg for each modulation techniques.

Fig. 6: Difference between total losses computed according to the proposed approach and simulations.

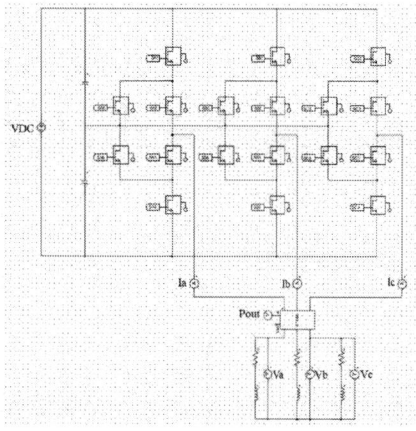

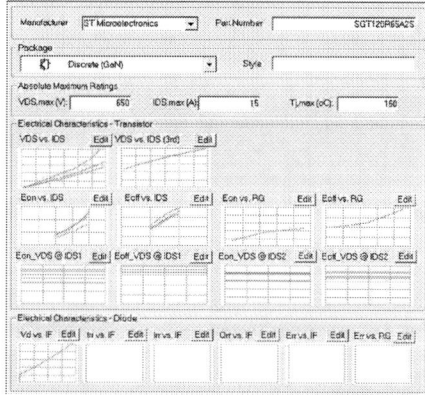

Fig. 7: Circuital modelling of the ANPC realized in PSIM for losses investigation.

Fig. 8: Device database editor used to set the electrical and thermal characteristics of the GaN devices.

Experimental Results

An experimental test bench has been arranged to carry out some preliminary results to be compared with that coming from the modelling. The single phase of an ANPC was realized by suitably connecting three half bridge converters, including GaN devices whose specifications are listed in Table III. The modulation strategies have been implemented in a STM32-G474RE control board, based on the high-performance Arm® Cortex®-M4 32-bit RISC core. The switching frequency of GaN devices is f_{sw}=50kHz, while the dead time is 100ns. The DC bus voltage has been set to 200V, while the inverter output has been connected to an ohmic-inductive load given by a series connection of R=20.65Ω and L=1mH. Fig. 9 displays the test rig, while Fig. 10 shows a zoomed in view of the single phase ANPC.

Fig. 9: Test Bench.   Fig. 10 GaN ANPC.

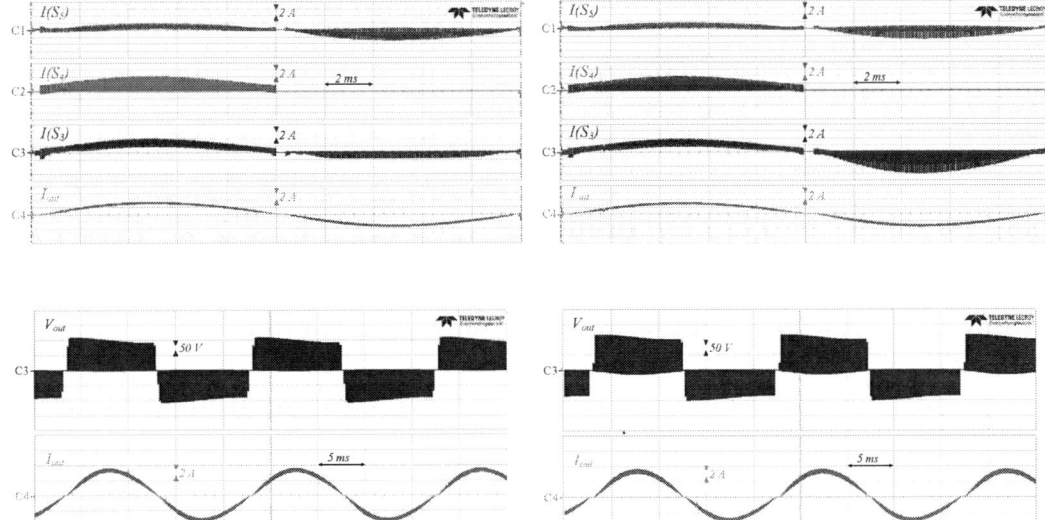

Fig. 11: Implementation of ANPC-SSCM:
V_{in}=200V I_{out}=3A m=0.7.

Fig. 12: Implementation of ANPC-FPCM:
V_{in}=200V I_{out}=3A m=0.7.

Preliminary experimental tests of Figs. 11 and 12 display the main electrical quantities of the power converter when the last is operated at a fundamental frequency of 50 Hz according to ANPC-SSCM and ANPC-FPCM, respectively. The power converter has been tested under different amplitude modulation indexes measuring the RMS values of the currents flowing through power devices. Such measurements have been compared to that coming from the analytical formulation of Tab. II. The results are graphically represented in Fig. 13, which underline a satisfying agreement between the estimated and measured RMS currents for most of the tests.

Tab. III –Technical specifications of GaN HEMT.

$V_{DS}(V_{GS} = 0V)$	650 V
I_D @ 100°C	10A
R_{dson}	65mΩ
V_{GS}	6V/-3V

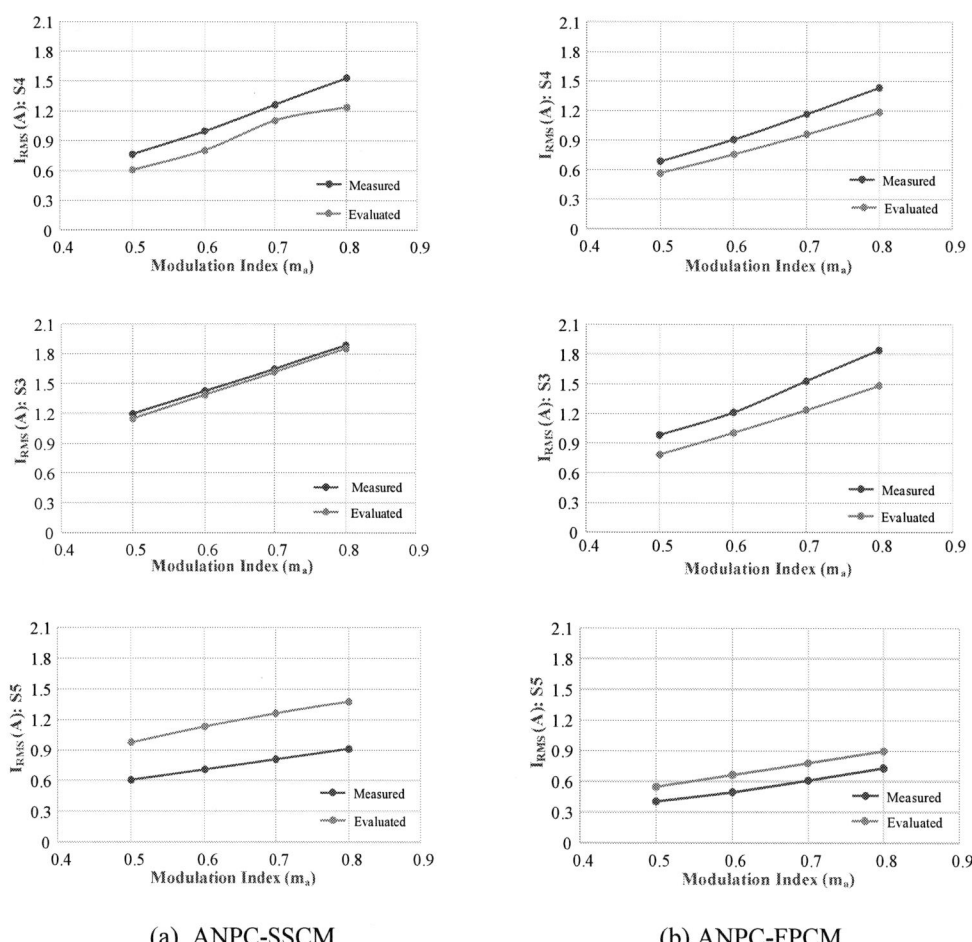

(a) ANPC-SSCM   (b) ANPC-FPCM

Fig. 13: RMS Currents determined with the analytical approach and with measurements

Conclusions

In this paper, a modelling approach to compute the power loss distribution in a three-phase three-level (3L) active neutral point clamped (ANPC) inverter has been investigated, evaluating the difference due to the modulation strategies. Simulations performed with a circuit-based power electronics tool have confirmed the good agreement with results carried out from the analytical model of Tab. II. The effectiveness of the proposed modelling has been also evaluated with some preliminary experimental tests comparing the measured and estimated RMS currents flowing through the power switches. As a main result of the comparison of the four carrier-based modulation techniques, it was confirmed that the modulation with full-path clamping technique allows to get lower total losses due to the effective current sharing strategy.

References

[1]. J. Reimers, L. Dorn-Gomba, C. Mak and A. Emadi, "Automotive Traction Inverters: Current Status and Future Trends," in IEEE Transactions on Vehicular Technology, vol. 68, no. 4, pp. 3337-3350, April 2019.

[2]. A. Morya, M. Moosavi, M. C. Gardner and H. A. Toliyat, "Applications of Wide Bandgap (WBG) devices in AC electric drives: A technology status review," 2017 IEEE International Electric Machines and Drives Conference (IEMDC), 2017, pp. 1-8.

[3]. I. Aghabali, J. Bauman, P. J. Kollmeyer, Y. Wang, B. Bilgin and A. Emadi, "800-V Electric Vehicle Powertrains: Review and Analysis of Benefits, Challenges, and Future Trends," in IEEE Transactions on Transportation Electrification, vol. 7, no. 3, pp. 927-948, Sept. 2021.

[4]. M. Galád, P. Špánik, M. Cacciato and G. Nobile, "Comparison of common and combined state of charge estimation methods for VRLA batteries," 2016 ELEKTRO, 2016, pp. 220-225.

[5]. S. De Caro et al., "THD and efficiency improvement in multi-level inverters through an open end winding configuration," 2016 IEEE Energy Conversion Congress and Exposition (ECCE), 2016, pp. 1-7.

[6]. Lidow, A., De Rooij, M., Strydom, J., Reusch, D. and Glaser, J., "GaN transistors for efficient power conversion," John Wiley & Sons, Second Edition, 2015.

[7]. G. Susinni, S.A. Rizzo, F. Iannuzzo, "Two Decades of Condition Monitoring Methods for Power Devices," Electronics 2021, 10, 683.

[8]. S. A. Rizzo, G. Susinni and F. Iannuzzo, "Intrusiveness of Power Device Condition Monitoring Methods: Introducing Figures of Merit for Condition Monitoring," in IEEE Industrial Electronics Magazine, vol. 16, no. 1, pp. 60-69, March 2022.

[9]. K. K. Gupta, A. Ranjan, P. Bhatnagar, L. K. Sahu and S. Jain, "Multilevel Inverter Topologies With Reduced Device Count: A Review," in IEEE Transactions on Power Electronics, vol. 31, no. 1, pp. 135-151, Jan. 2016.

[10]. D. Barater, C. Concari, G. Buticchi, E. Gurpinar, D. De and A. Castellazzi, "Performance Evaluation of a Three-Level ANPC Photovoltaic Grid-Connected Inverter With 650-V SiC Devices and Optimized PWM," in IEEE Transactions on Industry Applications, vol. 52, no. 3, pp. 2475-2485, May-June 2016.

[11]. M. Valente, F. Iannuzzo, Y. Yang and E. Gurpinar, "Performance Analysis of a Single-phase GaN-based 3L-ANPC Inverter for Photovoltaic Applications," 2018 IEEE 4th Southern Power Electronics Conference (SPEC), 2018, pp. 1-8.

[12]. Feng, Z.; Zhang, X.; Wang, J.; Yu, S. A High-Efficiency Three-Level ANPC Inverter Based on Hybrid SiC and Si Devices. Energies 2020, 13, 1159.

[13]. Holmes, D.G. and Lipo, T.A., "Pulse width modulation for power converters: principles and practice," John Wiley & Sons, IEEE Press Series on Power Engineering, 2003.

[14]. V. Jayakumar, B. Chokkalingam and J. L. Munda, "A Comprehensive Review on Space Vector Modulation Techniques for Neutral Point Clamped Multi-Level Inverters," in IEEE Access, vol. 9, pp. 112104-112144, 2021.

[15]. D. Pan, M. Chen, X. Wang, H. Wang, F. Blaabjerg and W. Wang, "EMI Modeling of Three-Level Active Neutral-Point-Clamped SiC Inverter Under Different Modulation Schemes," 2019 10th International Conference on Power Electronics and ECCE Asia (ICPE 2019 - ECCE Asia), 2019, pp. 1-6.

[16]. L. Ma, T. Kerekes, P. Rodriguez, X. Jin, R. Teodorescu and M. Liserre, "A New PWM Strategy for Grid-Connected Half-Bridge Active NPC Converters With Losses Distribution Balancing Mechanism," in IEEE Transactions on Power Electronics, vol. 30, no. 9, pp. 5331-5340, Sept. 2015.

[17]. C. Hu et al., "An Improved Virtual Space Vector Modulation Scheme for Three-Level Active Neutral-Point-Clamped Inverter," in IEEE Transactions on Power Electronics, vol. 32, no. 10, pp. 7419-7434, Oct. 2017.

[18]. M. Novak, V. Ferreira, F. Blaabjerg and M. Liserre, "Evaluation of carrier-based control strategies for balancing the thermal stress of a hybrid SiC ANPC converter," 2021 IEEE Applied Power Electronics Conference and Exposition (APEC), 2021, pp. 2077-2083.

[19]. K. Berringer, J. Marvin and P. Perruchoud, "Semiconductor power losses in AC inverters," IAS '95. Conference Record of the 1995 IEEE Industry Applications Conference Thirtieth IAS Annual Meeting, 1995, pp. 882-888, vol.1.

Generalized Core and Winding Area Ratio - Trends for Inductors and Transformers in Power Electronics with High Switching Frequencies

Siqi Lin[1], Leon Fauth[1], Wilmar Martnez[2], and Jens Friebe[1]

[1]Leibniz Universitat Hannover
[2]KU Leuven - Energyville
[1]Welfengarten - Hannover, Germany
[2]Diepenbeek - Genk, Belgium
Phone: +49 (0) 511 762 19477
Email: siqi.lin@ial.uni-hannover.de
URL: http://www.ial.uni-hannover.de

Keywords

≪Power Electronics≫, ≪Core area≫, ≪Winding area≫, ≪Inductors≫, ≪Transformers≫ .

Abstract

The paper describes the general trend for the design of core shapes for inductors and transformers in power electronic applications under the influence of increasing switching frequencies. The focus of the paper is on the ratio of the winding window and the core cross sectional area to identify the current trends and new shapes for magnetic circuits.

Introduction

Design of magnetic core components in power electronics is distinctly unique, which means that the magnetic designers need to design magnetic components specifically for each application. With the increasing use of wide-bandgap-semiconductors the high frequency losses of magnetic components become a major factor that limits the power density as well as the efficiency to be further improved. On the one hand coil losses have to be limited by different high frequency coil forms as well as winding arrangements [1] [2], on the other hand the size of the core should be constrained to avoid eddy current effects and dimensional resonance effects in the core caused by high frequencies and the material properties [3] [4]. This requires the magnetic designer to find a balance between the winding arrangement and the core cross section area. This paper shows the trend of core geometry in the case of high frequency power electronics by examining the ratio of winding window to core cross section, and giving a general approach for selecting or judging core geometry.

The most common method of core geometry selection is the A_p area product method. (1) shows the general A_p method for transformers [5] [6] with output power Pt, copper filling factor k_{cu} and waveform factor k_f etc. It can be seen by (1) that the A_p value is decreasing as the frequency increases, which is in accordance with the law of electromagnetic induction and also roughly demonstrates the ability of the core to handle power at different frequencies.

$$A_p = A_e \cdot A_w = \frac{Pt \cdot 10^4}{k_f \cdot k_{cu} \cdot J \cdot f \cdot B_m} \tag{1}$$

There might be many cores with similar A_p value, but this information is not enough to help determine the shape of a specific core. The proportion of A_e and A_w occupied in the power transfer during the frequency change still needs to be clarified. According to (1), and using the ferrite material N87 [7] as an example, the trend of the winding window area A_w over the core cross section A_e can be obtained while the relative core loss is kept at 100 mW/cm³. For the same frequency, the relationship between A_e and A_w can be calculated with constant A_p (Fig. 1). Fig. 1(a) shows that the core demand for the winding window area decreases as the frequency increases, which also means that the ratio of the core cross section area to the winding window area increases as the frequency increases (Fig. 1(b)). It can be inferred from Fig. 1 that as the frequency increases, a larger core cross section area and a smaller winding window area are required. In the next chapter the design of a transformer for a 10 kW series resonant converter will be used as an example to illustrate how the transformer loss varies with the A_e/A_w t different frequencies to determine whether the above inference holds true.

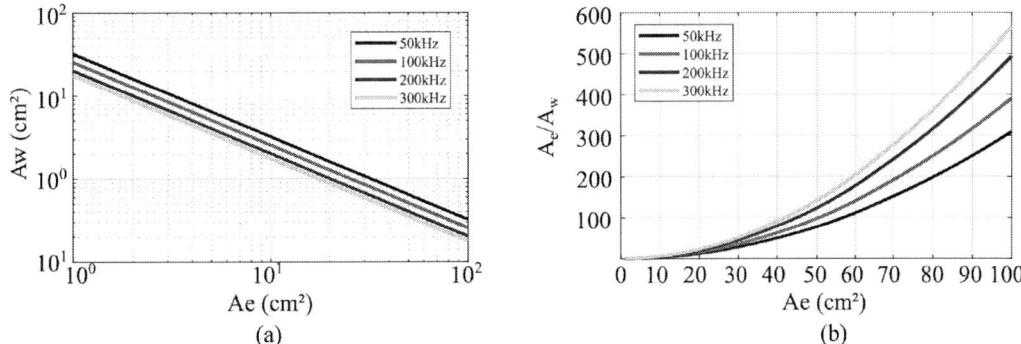

(a) (b)

Fig. 1: The trend of core geometry at different frequencies. (a) Trend of variation of window area A_w with core cross-sectional area A_e. (b) Trend of variation of ratio A_e/A_w with core cross section area A_e

Analytical calculation

Table I shows the specifications of the design requirements. It is worth noting that it is not intended to design and optimize a transformer here, but only to show the trend of the core geometry change through this example, thus the calculations are simplified accordingly. The core losses are calculated by the steinmetz equation [8] and the coil losses are calculated by the Dowell equation [9].

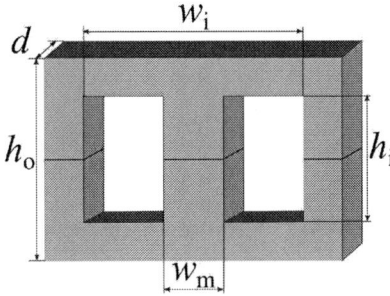

Fig. 2: Core geometry EE core as calculation example. To simplify the calculation, it is assumed that the core geometry satisfies the following relationship: w_i equals three times w_m.

$$A_e = d \cdot w_m \tag{2}$$

$$A_w = h_i \cdot w_m \tag{3}$$

(4) shows the actual geometric significance of A_e/A_w for EE-type cores.

$$A_p = A_e \cdot \frac{A_e}{A_w} = \frac{d}{h_i} \tag{4}$$

Table I: DESIGN SPECIFICATIONS FOR TRANSFORMERS

Output Power (W)	Primary Voltage (V)	Secondary Voltage (V)	Duty Cycle	Current Density (A/cm^2)	k_f	k_{cu}
10000	400	400	0.5	420	4	0.4

In Fig. 2, the size parameters of an EE core are shown as example. In the case of the EE type core, the individual side lengths of the core satisfy the following relationship ((2) and (3)) with the winding window area and cross section area.

Fig. 3 shows the commonality of the loss variation at all frequencies. As A_e/A_w increases, the coil loss decreases, but the decrease tends to be flat. The core loss increases as A_e/A_w increases. In order to fairly compare the degree of influence of A_e and A_w on the loss, the specific loss of the core and the type of coil (lize wire) are assumed the same at the same frequency. Since the core geometry is varied in three dimensions, the side length of the center column w_m must also be taken into account although it is known through (4) that w_m does not affect the ratio, but it does affect the coil losses and magnetic losses in different degrees.

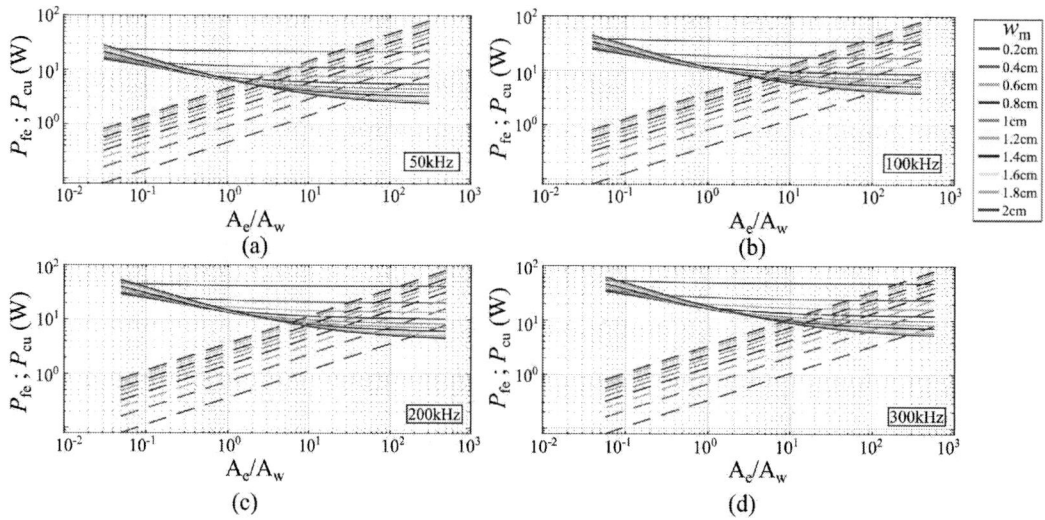

Fig. 3: The trend of core loss and copper loss with ratio A_e/A_w at different frequencies. The solid line is the coil loss and the dashed line is the core loss.

Fig. 4 shows the trend of the total transformer losses with frequency, ratio A_e/A_w and side length of the center column w_m. Each frequency corresponds to a case where the total loss is minimized. The ratios A_e/A_w are 0.78, 2, 2.46, and 4.6 for frequencies of 50 kHz, 100 kHz, 200 kHz, and 300 kHz, respectively.

For fixed frequency it is clear that the ratio A_e/A_w is not as large as it should be, the total loss decreases first as the ratio increases and then increases after reaching a minimum value. This is due to the fact that at first the coil loss decreases significantly as the ratio increases, and then tends to smooth out until the decrease in coil loss is compensated by the increase in core loss.

However, as the frequency increases, the trend in loss with ratio shows a tendency to gradually shift to the right side. This means that the core tends to have a larger cross section area and a smaller window area at high frequencies.

Numeric Verification

In order to verify the calculation results of the previous section, 2D finite element simulations have been used. Table II shows the simulation results of core losses. In order to avoid that the errors in the

Table II: 2D FEM TRANSFORMER LOSS SIMULATION RESULTS FOR DIFFERENT FREQUENCIES AND RATIOS.

	50 kHz			100 kHz			200 kHz			300 kHz		
	UO	OP	OO	UO	OP	OO	UO	OP	OO	UO	OP	OO
A_e/A_w	0.28	**0.78**	5.3	0.35	**2**	11.5	0.45	**2.46**	18	0.5	**4.6**	13
P_total/W	25	22	25	24.6	21	25.5	34	22	25	50	36	41

calculation and simulation affect the results of the comparison, the tolerance regions of the calculation are defined, as shown in the red areas of Fig. 4, and they represent the minimum regions of the total loss (range of 5% greater than the minimum value). UO, OP, and OO in Table II represent random cases located to the left side, inside, and right side of the loss-minimizing region (red region in Fig. 4), respectively. Meanwhile in Fig. 4 the ratios of PQ40/40 and EILP43 are shown as examples. As shown in the figure, as the frequency increases the distance between PQ40/40 and the loss minimum ratio gets increasingly farther, while the ratio of EILP43 gets closer.

As can be seen in Table II, the trends of 2D FEM simulation results are in agreement with the calculated results. In addition, Fig. 5 visualizes the 3D geometric model of the OP cases at different frequencies. As shown in the Fig. 5, the core decreases significantly in the z-direction and increases in the xy-plane as the frequency increases.

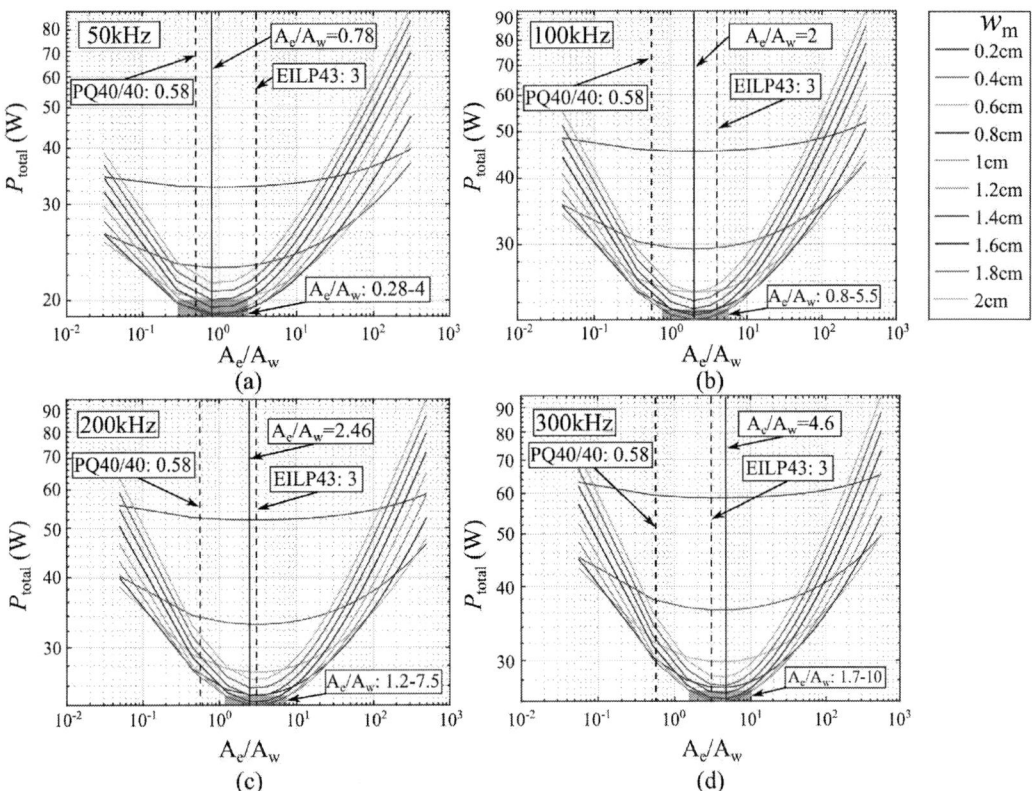

Fig. 4: Trend diagram of transformer losses with ratio A_e/A_w and side length w_m of the center column at different frequencies ((a), (b), (c), and (d) correspond to frequencies 50 kHz, 100 kHz, 200 kHz, and 300 kHz, respectively). The red area shows the range of the ratio A_e/A_w when the total losses are at minimum area (range of 5% greater than the minimum value).

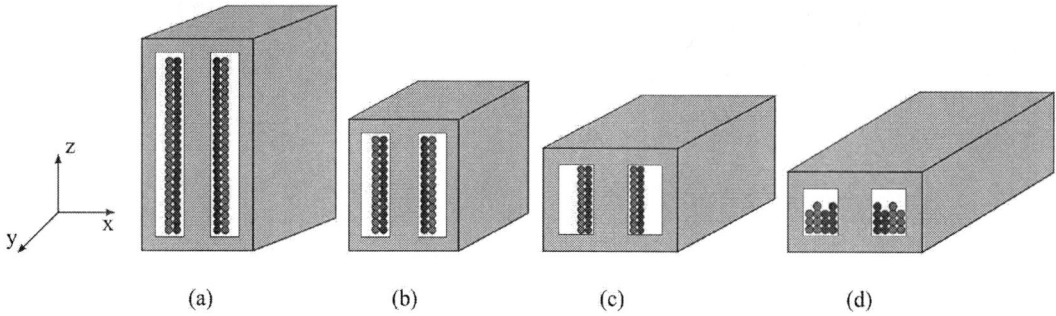

Fig. 5: Comparison of the height of the transformers. a, b, c, and d show the 3D schematic representation models of the variables with the lowest losses at different frequencies (50 kHz, 100 kHz, 200 kHz, and 300 kHz) according to Table I.

Measurements

The experimental data were obtained from two 600 W inductors of an ultrasonic converter at 600 kHz, and PQ 40/40 and EILP 43 shapes are used as the cores of the inductor, respectively. Their ratios A_e/A_w are: 0.58 and 3, respectively. The experimental results show that the loss of the inductor using PQ 40/40 is more than 4 W, while using EILP 43 is only about 1 W. In addition, as seen in Fig. 6 a, the new variant (EILP 43) has a significantly lower height and lighter weight. Fig. 6 b shows the inductors being used and the results of the temperature measurements. As shown in the figure, while the maximum temperature of the PQ 40 variant is 83 degrees Celsius, the maximum temperature of the EILP 43 variant is only 52 degrees Celsius. This is because, on the one hand, the EILP 43 variant requires a smaller number of turns due to the increased cross section area of the core, so the coil losses are reduced. On the other hand the flatter surface means better heat dissipation.

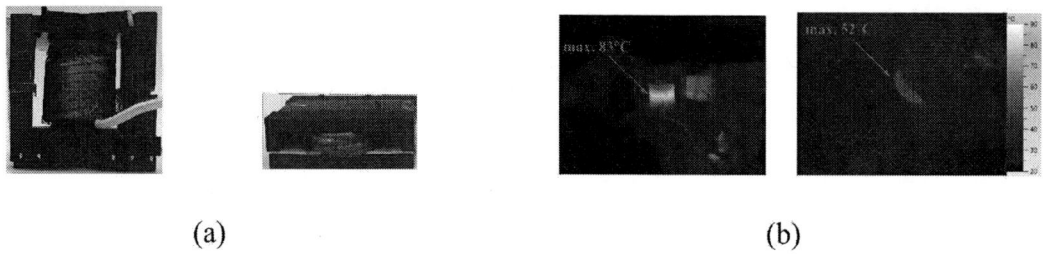

Fig. 6: Comparison of inductors for a 600 kHz ultrasonic converter. (a) Prototypes of PQ and EI cores are shown respectively. (b) Temperature measurement results are shown.

Conclusion

In this paper, for the first time, the ratio A_e/A_w is used as an indicator to show the trend of the E type core geometry with frequency. As the frequency increases, the larger the ratio, the more advantageous the core is in terms of efficiency, size and heat dissipation. This means that for high frequency magnetic designs, planar cores are preferred. It is because, on the one hand, the exponential growth of coil loss with the increase of frequency, especially for the multi-layer coil arrangement, the coil is severely affected by the proximity effect. Therefore, it is very uneconomical to increase the magnetic parameters by increasing the number of turns. On the other hand, a larger core area provides better heat dissipation. This helps the magnetic designers to determine the specific geometry parameters of the cores and clarify the design direction in the preliminary stage of the design. The research results can also be used as a reference for core manufacturers.

References

[1] A. Stadler and M. Albach, "The influence of the winding layout on the core losses and the leakage inductance in high frequency transformers," in IEEE Transactions on Magnetics, vol. 42, no. 4, pp. 735-738, April 2006, doi: 10.1109/TMAG.2006.871383.

[2] E. L. Barrios, A. Ursúa, L. Marroyo and P. Sanchis, "Analytical Design Methodology for Litz-Wired High-Frequency Power Transformers," in IEEE Transactions on Industrial Electronics, vol. 62, no. 4, pp. 2103-2113, April 2015, doi: 10.1109/TIE.2014.2351786.

[3] E. Cardelli, L. Fiorucci and E. Della Torre, "Estimation of MnZn ferrite core losses in magnetic components at high frequency," in IEEE Transactions on Magnetics, vol. 37, no. 4, pp. 2366-2368, July 2001, doi: 10.1109/20.951174.

[4] G. R. Skutt and F. C. Lee, "Characterization of dimensional effects in ferrite-core magnetic devices," PESC Record. 27th Annual IEEE Power Electronics Specialists Conference, Baveno, Italy, 1996, pp. 1435-1440 vol.2, doi: 10.1109/PESC.1996.548770.

[5] C. W. T. McLyman, Transformer and inductor design handbook. Place of publication not identified: CRC Press, 2017.

[6] C. P. Steinmetz, "The general equations of the electric circuit," in Proceedings of the American Institute of Electrical Engineers, vol. 27, no. 7, pp. 1121-1195, July 1908, doi: 10.1109/PAIEE.1908.6742132.

[7] EPCOS/TDK: Ferrites and accessories material N87, Datasheet, May 2017

[8] P. L. Dowell, "Effects of eddy currents in transformer windings," Proc. Inst. Elect. Eng., vol. 113, no. 8, pp. 1387–1394, Aug. 1966.

[9] R. L. Stoll, The Analysis of Eddy Currents. Oxford, U.K: Clarendon Press, 1974

Active substrate termination of discrete and monolithic bidirectional GaN HEMTs in a T-type inverter

Carsten Kuring[1], Yannic Lange[1], Xiaomeng Geng[1], Oliver Hilt[2], Mihaela Wolf[2], Joachim Würfl[2] and Sibylle Dieckerhoff[1]

[1] TECHNISCHE UNIVERSITÄT BERLIN
Einsteinufer 19
Berlin, Germany
[2] FERDINAND-BRAUN-INSTITUT
LEIBNIZ-INSTITUT FÜR HÖCHSTFREQUENZTECHNIK
Gustav-Kirchhoff-Str.4
Berlin, Germany
Tel.: +49 30 314 23404
Fax: +49 30 314 25526
E-Mail: carsten.kuring@tu-berlin.de
[1] URL: https://www.pe.tu-berlin.de
[2] URL: https://www.fbh-berlin.de

Keywords

«Gallium Nitride (GaN)», HEMT, Monolithic power integration, Multi-level inverters, DC-AC converter

Abstract

Monolithically integrated lateral Gallium-nitride bidirectional transistors can achieve symmetric conduction and blocking capability at a reduced on-state resistance compared to their discrete counterparts. An actively switched substrate effectively prevents back-gating effects in normally-off GaN-on-Si bidirectional transistors and enables symmetrical loss-minimized switching and on-state characteristics.

Introduction

Power electronic topologies such as the multi-level T-type inverter, matrix-inverter and current-sourced inverters require power semiconductor switches featuring bidirectional voltage-blocking capability. Lateral chip-design of Gallium-nitride high-electron-mobility transistors (GaN HEMTs) enables monolithic integration of several power transistors into a single chip. Monolithically integrated bidirectional GaN HEMTs allow to minimize the parasitic stray inductance in fast switching GaN-based power electronics. Bidirectional GaN HEMTs implementing a common drift region for both blocking voltage polarities can achieve an almost 50% lower on-resistance (R_{ON}) compared to discrete bidirectional switches [1, 2, 3]. Considering power electronic topologies such as the multilevel T-type inverter significantly reduced conduction losses are expected as long as stable bidirectional device characteristics are maintained and device degradation is prevented.

Normally-off GaN HEMTs fabricated on a conductive silicon (Si) substrate are prone to back-gating effects leading into current collapse and degraded on-state characteristics depending on the applied substrate bias. Therefore, an appropriate substrate termination is essential to ensure efficient and symmetric bidirectional switched operation. This paper discusses different implementation schemes of bidirectional GaN transistors in Sec. 1 including a discrete GaN-based bidirectional switch specifically designed to emulated substrate coupling-effects occurring in monolithic bidirectional GaN HEMTs. Sec. 2 demonstrates the degradation of static device characteristics due to negative substrate bias. Different substrate termination schemes suitable for switched operation are derived in Sec. 3. An experimental validation regarding switching characteristics and transient R_{ON} under hard- and soft switching conditions in double-pulse switching tests are presented in Sec. 4. Selected substrate termination schemes are further evaluated concerning converter efficiency and device temperature in continuous DC/DC buck operation. According to the obtained measurement results, an actively switched substrate enables efficient bidirectional operation of bidirectional GaN HEMTs.

1 Implementation of discrete and monolithic bidirectional GaN HEMTs

Bidirectional blocking capability can be implemented in either common drain (Fig. 1a) or common source configuration (Fig. 1b) [4, 5, 6]. In order to prevent device degradation in terms of reduced saturation current and increased R_{ON} induced by back-gating effect, the conductive Si-substrate in conventional unidirectional normally-off GaN HEMTs is shorted to the source node [7, 8]. This paradigm is satisfied for discrete bidirectional transistors with separate bulk node (Fig. 1a) as well as monolithic bidirectional transistors with common source and common bulk node (Fig. 1b). However, it is not feasible for monolithic bidirectional HEMTs with common drift region, which can achieve a significantly lower R_{ON} similar to a single unidirectional Gan HEMT as the same drift-region l_{GaGb} is utilized for positive as well as negative blocking voltage polarity (Fig. 1c) [1, 2, 3, 9, 10]. In case of the monolithic bidirectional HEMT presented in [2] saturation current degradation and increased R_{ON} by back-gating effects have been observed. The substrate must be terminated to the one of both source nodes S_a and S_b with the lower potential to avoid the observed back-gating effects. Since the applied blocking voltage polarity changes in device operation, switching of the backside termination between S_a and S_b may be feasible to prevent back-gating and enable minimized bidirectional conduction losses (TABLE I).

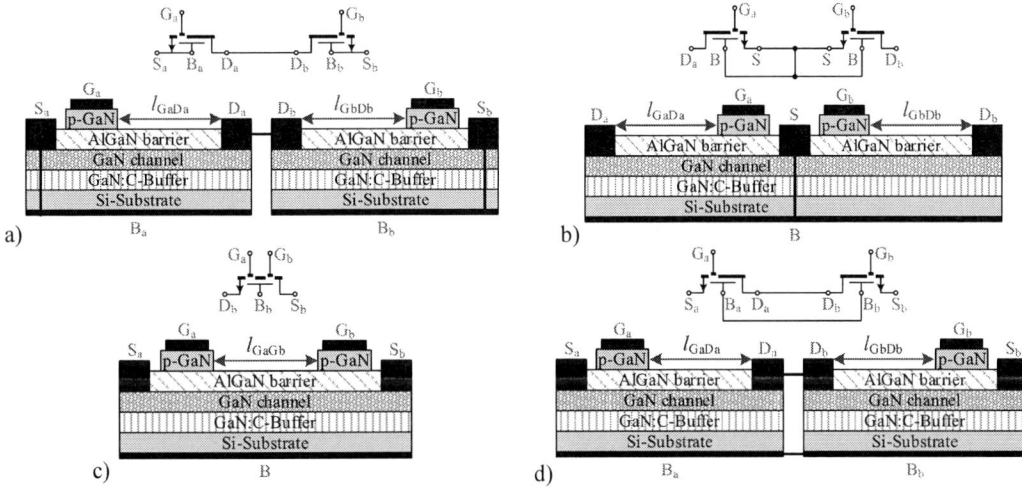

Fig. 1: Bidirectional switch implementation using lateral GaN HEMTs: (a) discrete switch in common drain configuration, (b) monolithic switch in common source configuration, (c) monolithic switch with common drift region and common substrate and (d) discrete switch with common bulk and common drain node

In this paper a monolithic bidirectional HEMT with common drift region is emulated using two commercially available discrete GaN-on-Si HEMTs with shorted bulk and drain terminals (Fig. 1d). This configuration reproduces substrate coupling and biasing effects of a bidirectional transistor with common substrate and common drift-region, while the R_{ON} is still doubled.

TABLE I. Optimum bulk termination scheme of a bidirectional GaN HEMT

Operation direction		Forward	Reverse
Off-state blocking voltage	$V_{SbSa,off}$	> 0 V	< 0 V
Operation mode of	S_a	actively switched	acting as Drain
	G_a		continuously turned on
	S_b	acting as Drain	actively switched
	G_b	continuously turned on	
Optimum bulk termination		bulk terminated to S_a	bulk terminated to S_b

2 Static device characteristics

The susceptibility to current collapse induced by back-gating is demonstrated by means of the static device characteristics extracted from pulsed IV measurements covering a single unidirectional GaN HEMT GS66508P as well as the emulated monolithic bidirectional (Fig. 1d).

A shorted termination of the substrate to source (V_{BS}=0 V) is recommended by the device manufacturers to achieve 'best device performance' [7, 8]. In case of the bidirectional transistor the second gate G_b is biased at V_{GbSb}=6 V. In this configuration both, the unidirectional and bidirectional GaN HEMT exhibit a good controllability by means of the gate-source bias. Thereby asymmetrical saturation currents of $I_{Sa,max,Q1}$=136 A and $I_{Sa,max,Q3}$=|−106 A| are observed for the bidirectional transistor compared to $I_{D,max,Q1}$=102 A in the unidirectional reference GaN HEMT (Fig. 2a vs. b). Despite the series interconnection of two separate unidirectional GaN HEMTs in the bidirectional transistor, the static R_{ON} is only increased from R_{on}=50.6 mΩ to R_{on}=89.6 mΩ (Fig. 2c). The observed deviations in device characteristics are related to a reduced threshold voltage in the bidirectional transistor (Fig. 2d) possibly caused by device degradation induced by the bulk-source voltage stress applied during the measurements.

Assuming an increasingly negative bulk-source bias V_{BS}<0 V the two-dimensional electron gas (2DEG) is depleted causing lower saturation currents and increased R_{ON} (Fig. 2e, f) in both, 1st- and 3rd-quadrant operation. At the lowest considered bulk-source bias $V_{BS,min}$=−300 V both devices are effectively blocking even at the highest considered gate-source bias $V_{GS,max}$=6 V. For positive bulk-source bias V_{BS}>0 V the output characteristic remains almost unaffected. In conclusion, the bulk acts as additional terminal controlling the GaN HEMT. In order to achieve a low R_{ON} and therefore efficient in-circuit switched operation the applied substrate termination schemes must prevent a negative bulk-source bias.

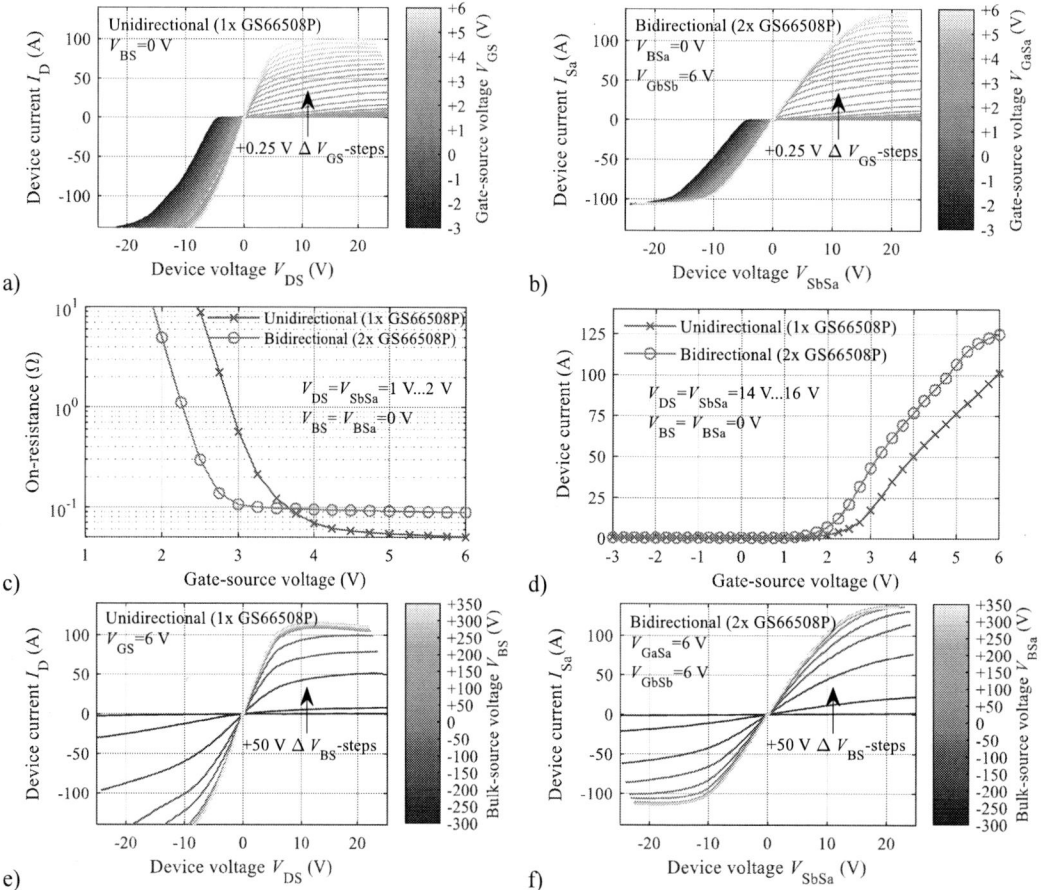

Fig. 2: Static IV characteristics of a unidirectional GaN HEMT (1x GS66508P) vs. an emulated bidirectional GaN HEMT (2x GS66508P) at (a...d) varied gate-source bias and (e...f) varied bulk-source bias

3 Substrate termination schemes

According to the obtained pulsed IV-measurement results a proper substrate termination is mandatory to ensure low on-state losses in monolithic bidirectional GaN HEMTs. Different substrate termination schemes have been proposed before [11, 12, 13]. However, measurement results validating the overall functionality especially concerning an actively switched substrate have not been published so far. Similar to [2] this paper covers two different fixed substrate terminations I and II , where the bulk is permanently shorted to one of both source nodes S_a, S_b (Fig. 3a, b). Assuming one of these substrate terminations, the other one will inevitably occur as soon as the blocking voltage polarity of the bidirectional

HEMT changes (TABLE I). According to the blocking voltage polarity the configurations I and II would results in asymmetric on-state characteristics of the bidirectional GaN HEMT as they correspond to a source- and drain-connected substrate of a unidirectional GaN HEMT. A floating substrate (Fig. 3c) can achieve symmetric, but still degraded on-state characteristics as previously demonstrated for a monolithic bidirectional GaN HEMT [2]. A self-managed or semi-floating substrate by means of diodes D_I, D_{II} (Fig. 3d, e) is proposed in [11, 12, 13] and experimentally evaluated in this paper as well. Both diode-based substrate termination schemes require anti-serial diode configuration to maintain bidirectional blocking capability of the bidirectional switch. Depending on the specific diode configuration IV and V the substrate bias is limited to negative values $V_{BSx} < 0$ V and positive values $V_{BSx} > 0$ V, respectively. Discrete high-voltage PiN-diodes D_I, D_{II} with small parasitic device capacitance are employed minimizing the additional parasitic switching node capacitance in the targeted T-type inverter (TABLE II). An actively switched substrate VI is proposed in [11, 12, 13, 2] and is herein implemented by means of discrete unidirectional GaN HEMTs T_I, T_{II} (Fig. 3f, TABLE II). The substrate switches are connected in common-source configuration enabling a dynamic termination of the bulk node according to the blocking voltage polarity of the bidirectional HEMT. This circuit design is especially suitable for future monolithic integration of the bidirectional transistor together with the active substrate switches on a common conductive Si-substrate as all four incorporated transistors share the same bulk potential. In this study, substrate termination scheme VIa, where T_I is in on-state and T_{II} in off-state is – apart from the $R_{ON} = 450$ mΩ of T_I – electrically equivalent to configuration I. Substrate termination scheme VIb, where both transistors T_I, T_{II} are in off-state, is electrically equivalent to configuration IV.

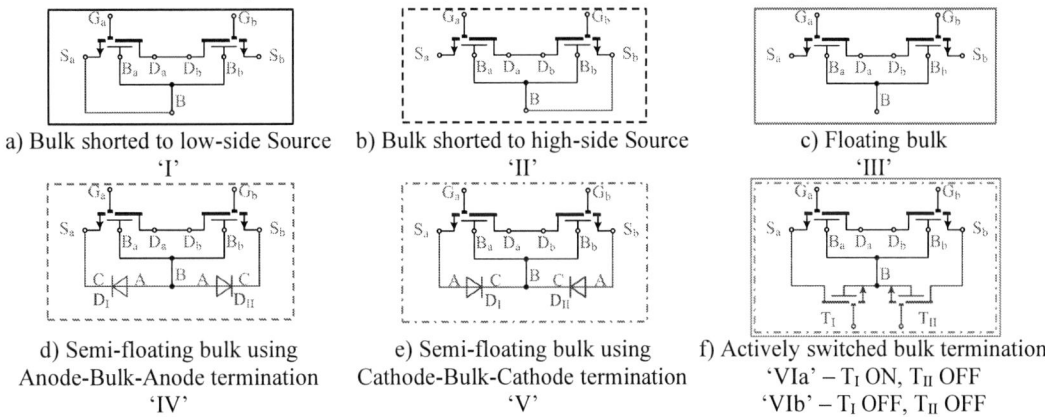

a) Bulk shorted to low-side Source 'I'
b) Bulk shorted to high-side Source 'II'
c) Floating bulk 'III'

d) Semi-floating bulk using Anode-Bulk-Anode termination 'IV'
e) Semi-floating bulk using Cathode-Bulk-Cathode termination 'V'
f) Actively switched bulk termination 'VIa' – T_I ON, T_{II} OFF 'VIb' – T_I OFF, T_{II} OFF

Fig. 3: Investigated fixed (a, b), (semi-) floating active (c…e) and active (f) substrate termination schemes

4 Switching characteristics in a T-type inverter

4.1 Test setup

The switching characteristics of the emulated bidirectional transistor T_2 formed by the unidirectional HEMTs T_{2a} and T_{2b} are investigated at DC-link voltages up to $V_{DC} = 500$ V in a T-type inverter providing three output voltage levels $[-0.5 \cdot V_{DC}, 0$ V$, +0.5 \cdot V_{DC}]$. The T-type PCB-layout implements a vertical commutation loop design enabling minimum parasitic stray inductances by magnetic flux compensation. The impact of six different substrate termination schemes (Fig. 3) on the switching transients, voltage slew rates, switching losses and dynamic R_{ON} is studied. The transient on-state voltage is extracted by an actively controlled clamping circuit enabling calculation of the R_{ON} is calculated by means of the load current [14]. The power transistors as well as substrate switches are implemented by commercial discrete GaN-on-Si HEMTs (Fig. 4, TABLE II).

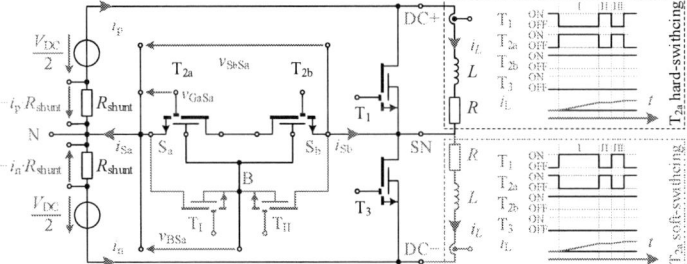

Fig. 4: Circuit diagram and control schemes of the emulated bidirectional GaN HEMT T_2 with active substrate termination in the T-type inverter

In hard-switched double-pulse tests of the bidirectional transistor T_2, the unidirectional GaN HEMTs T_{2a} and T_1 are operated in 1st-quadrant hard-switching and 3rd-quadrant zero-voltage-switching (ZVS) free-wheeling mode, respectively. Vice versa, soft-switching and 3rd-quadrant ZVS free-wheeling mode of the bidirectional transistor T_2 is achieved by reconnecting the load inductor and adjusting the control signals where T_1 is now operated in hard-switched mode (Fig. 4). Transistors T_{2b} and T_3 are hold in continuous on- and off-state respectively. In configuration VIa the substrate B is switched to source S_a by continuous on-state of T_I while T_{II} remains turned off.

TABLE II: Semiconductor devices and gate configuration

Device / Parameter		Rating / value
GaN HEMT	T_1, T_3	GS66508B, 650 V, 50 mΩ
GaN HEMT	T_{2a}, T_{2b}	GS66508P, 650 V, 50 mΩ
GaN HEMT	T_I, T_{II}	GS-065-004-1-L, 650 V, 450 mΩ
Diodes	D_I, D_{II}	MA4P7470F-1072T, 800 V, 0.7 pF, 0.8 Ω
Gate drive voltages	$V_{G,on}$ / $V_{G,off}$	+6.0 V / −2.0 V
Gate resistors	$R_{G,on}$ / $R_{G,off}$	20 Ω / 10 Ω

4.2 Hard-switching characteristics in double-pulse test

In hard-switched operation turn-on and turn-off switching speeds significantly depend on the implemented substrate termination scheme. At moderate load currents $I_L<8$ A, a floating (III) or semi-floating substrate (IV, V, VIb) leads to faster turn-off transitions including higher voltage slew rates and shorter current fall times (Fig. 5a and Fig. 6a). In contrast, at higher load currents $I_L>12$ A either a permanently shorted (I) or actively switched substrate termination (VIa) results in higher turn-off voltage slew rates. In consequence, minimum turn-off losses can be achieved for a floating or semi-substrate in light-load conditions while a termination of the substrate to source S_a (I, Via) yields into reduced switching losses at higher load currents (Fig. 6b). This effect may be further addressed and exploited by the proposed actively switched substrate VIa and VIb which differ only by the switching state of the substrate switches T_I and T_{II} (Fig. 3f). Note that the load current at which the terminated (VIa) and semi-floating substrate (VIb) achieve similar turn-off switching losses slightly decreases from 11.56 A at V_{DC}=300 V to 10.52 A at V_{DC}= 500 V due to non-linear voltage-dependent parasitic device capacitances. In substrate termination schemes II…V and VIb, where significant gradients of the substrate bias occur during turn-off transitions (Fig. 5a and b, left), the maximum turn-off voltage slew rate is limited to approx. +20 V/ns (Fig. 6a) due to device-internal substrate coupling effects.

The hard-switched turn-on transition is strongly affected by the applied substrate termination as well. Floating and semi-floating substrate configurations II-V and VIb result in slower turn-on transitions

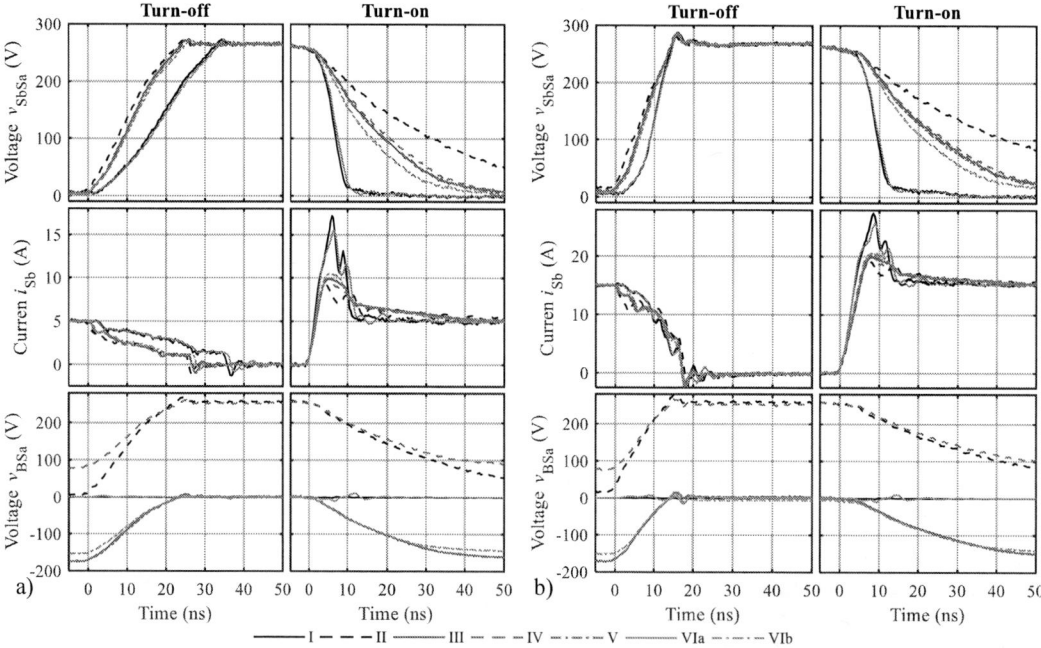

Fig. 5: Transients in hard-switched double-pulse test at load currents of (a) I_L=5 A and (b) I_L=15 A, V_{DC}=500 V

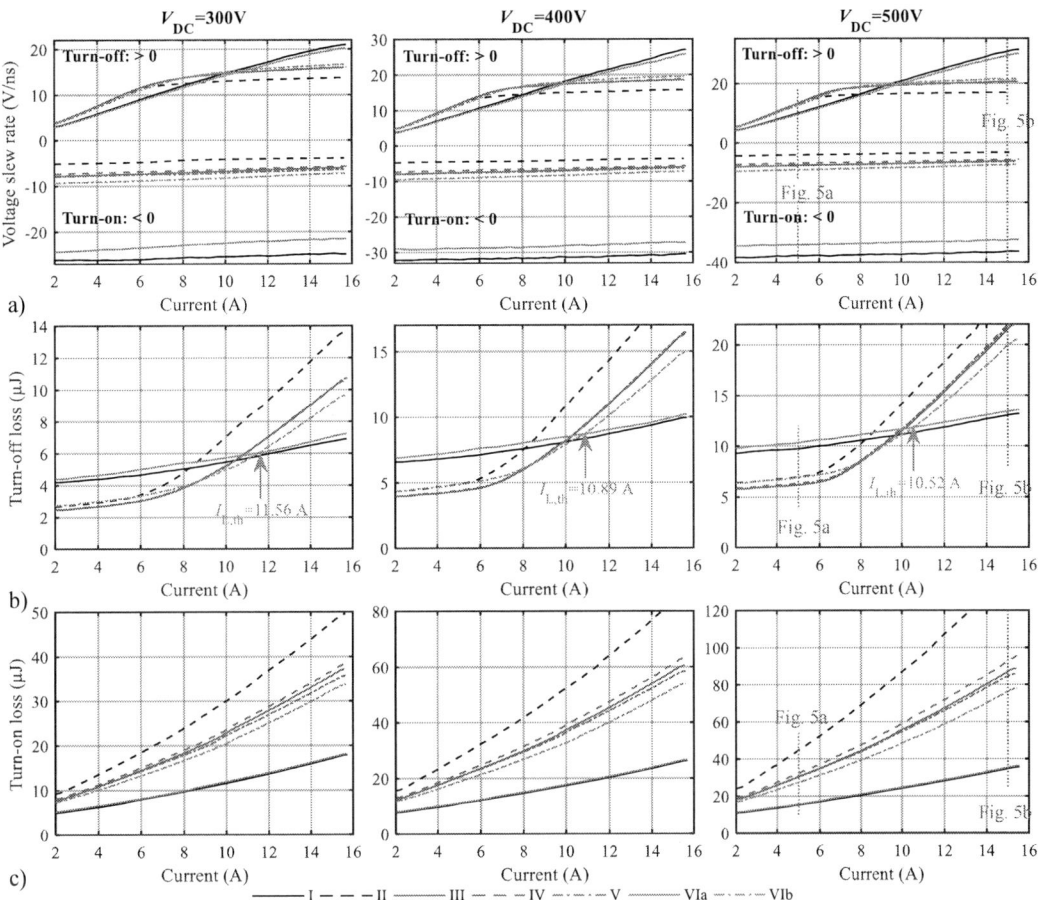

Fig. 6: Impact of the substrate termination on the (a) voltage slew rates (20%-80%), (b) turn-off and (c) turn-on switching losses

(Fig. 5a, b right) and severely increased turn-on losses (Fig. 6c). This effect occurs regardless of the substrate bias voltage prior turn-on (II, IV: $v_{BSa} \approx V_{DC}/2$ vs. III, V, VIb: $v_{BSa} \approx 0$ V), although device degradation in static measurements is limited to negative bulk-source bias (Sec. 2). However, a similar negative bulk-source voltage slew rate $d v_{BSa}/d t < 0$ is observed in all (semi-) floating substrate configurations (II-V, VIb). Despite the different substrate bias prior turn-on the bulk-source voltage drops by a similar magnitude during the hard-switched turn-on process for all the (semi-) floating substrate configurations (II-V, VIb). This is directly correlated to similar displacement currents charging and discharging the transistor's parasitic bulk capacitance. During GaN-on-Si device switching parasitic displacement currents aggravate short time charge trapping due to the continuous re-organization of the electron-hole equilibrium inside the GaN-based buffer and strain adaption layers of the transistor. Such semiconductor layer stack can be considered as an RC network with multiple time constants [15]. Partial depletion of the 2DEG and the observed slow turn-on process is a result of such dispersion phenomena (Fig. 6c). A fast and therefore loss-reduced turn-on transition is only achieved when the substrate is terminated to the low-side source S_a (Fig. 5, I and VIa). Additional parasitic device capacitances introduced by the active substrate switches (VIa) lead to marginally lower voltage slew rates as well as marginally increased switching losses in comparison to a permanently shorted substrate I. Considering the dominance of turn-on losses on the overall switching losses, an actively terminated substrate is mandatory to achieve loss-minimized hard-switched operation in bidirectional HEMTs. Consequently, subsequent dynamic R_{ON} measurements in double-pulse mode are limited to a comparison of the shorted substrate I with the switched substrate termination VIa.

The transient R_{ON} measurements of the bidirectional GaN HEMT in configuration I and VIa demonstrate a notable dynamic R_{ON}-increase when the R_{ON} before and after the blocking period are compared (Fig. 7a, $t < -1.1$ μs vs. $t > 0.15$ μs). The average R_{ON} before blocking achieves a minimum value at

V_{DC}=400 V while lower as well as higher DC-link voltages result in a slightly increased R_{ON} (Fig. 7a, b) as reported in [16]. In contrast the dynamic R_{ON}-increase is triggered by the maximum field strength during blocking stress [17] and becomes more extensive at higher DC-link-voltage and load current (Fig. 7c, dashed vs. dot plot lines). The actively switched substrate termination VIa achieves a similar or even marginally lower R_{ON} compared to the reference short circuit I (Fig. 7, black vs. green plot lines) which is in general agreement with [2].

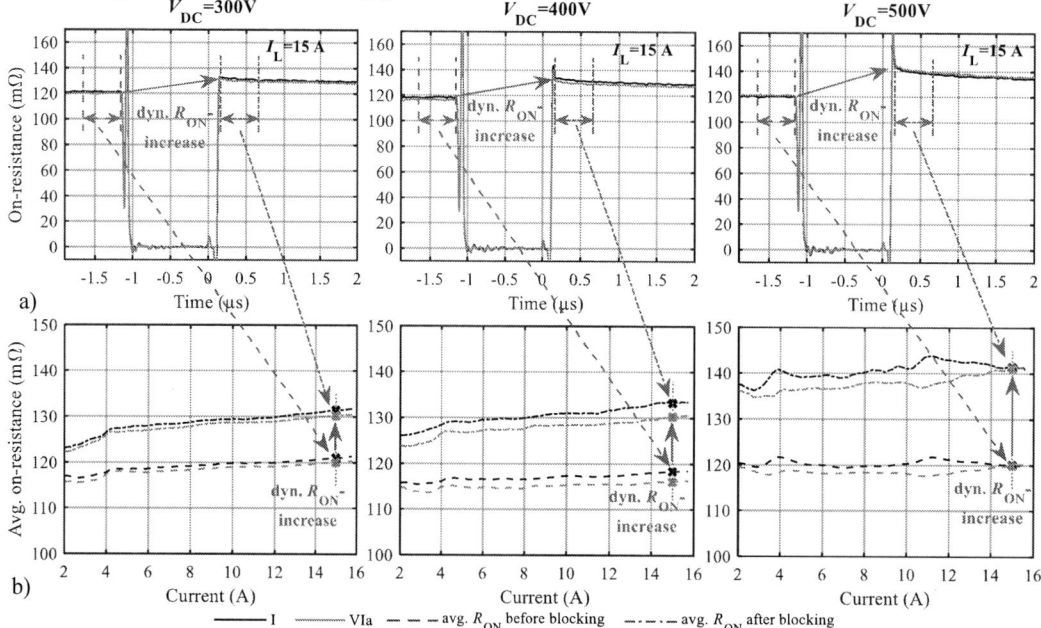

Fig. 7: Impact of the substrate termination on the (a) transient R_{ON} and (b) comparison of the R_{ON} before blocking vs. after blocking stress under hard-switched conditions. Note that the transistor off-state drain bias is $V_{DC}/2$.

4.3 Soft-switching characteristics in double-pulse test

In soft-switching conditions the emulated bidirectional GaN HEMT T_2 is operated in 3rd-quadrant reverse conduction mode achieving ZVS while the actively switched unidirectional GaN HEMT T_1 operated under hard-switched conditions. In consequence the impact of the employed substrate termination is much less pronounced, and the transient waveforms are inverted in comparison to the previous hard-switched characterization measurements.

In general, higher load currents result in shorter voltage fall-times during turn-on, as the charges in the parasitic switching node capacitance are displaced faster (Fig. 8a vs. b). The turn-off voltage slew rates decreases with the load current due to a longer current commutation. Non-linear voltage-dependent parasitic device capacitances result in higher absolute voltage slew rates during turn-on as well as turn off at increased DC-link voltage. Substrate configurations I and VIa, where the substrate is shorted and switched to source S_a respectively, exhibit almost identical switching transients. Marginally lower reduced voltage slew rate of VIa results from additional parasitic capacitances introduced by the active substrate switches T_I and T_{II} (Fig. 8, solid black vs. solid green plot lines). In (semi-)floating substrate termination schemes II-V and VIb a transient bulk-source bias ($v_{BSa} \neq 0$ V) occurs during switching transitions. In combination with the non-linear parasitic bulk capacitances, the overall output capacitance of the bidirectional GaN HEMT is reduced which finally results in faster discharging of the switching node capacitance and therefore steeper voltage gradients during turn-on.

The ZVS turn-off transition of the bidirectional GaN HEMT T_2 is again directly correlated with a hard-switched turn-on of the unidirectional GaN HEMT T_1. The transient current peak (Fig. 8a, b right, t=[5; 20] ns) is related to capacitive displacement currents as well as cross-conduction effects in the switching cell. Substrate configurations II-IV, where a significant positive bulk-source bias occurs, are correlated with a higher and/or longer transient current peak. Together with lower voltage slew rates dv_{SaSb}/dt increased switching losses of the unidirectional GaN HEMT T_1 must be considered. In contrast, a slightly lower transient current peak is observed in substrate termination schemes V and VIb, where a

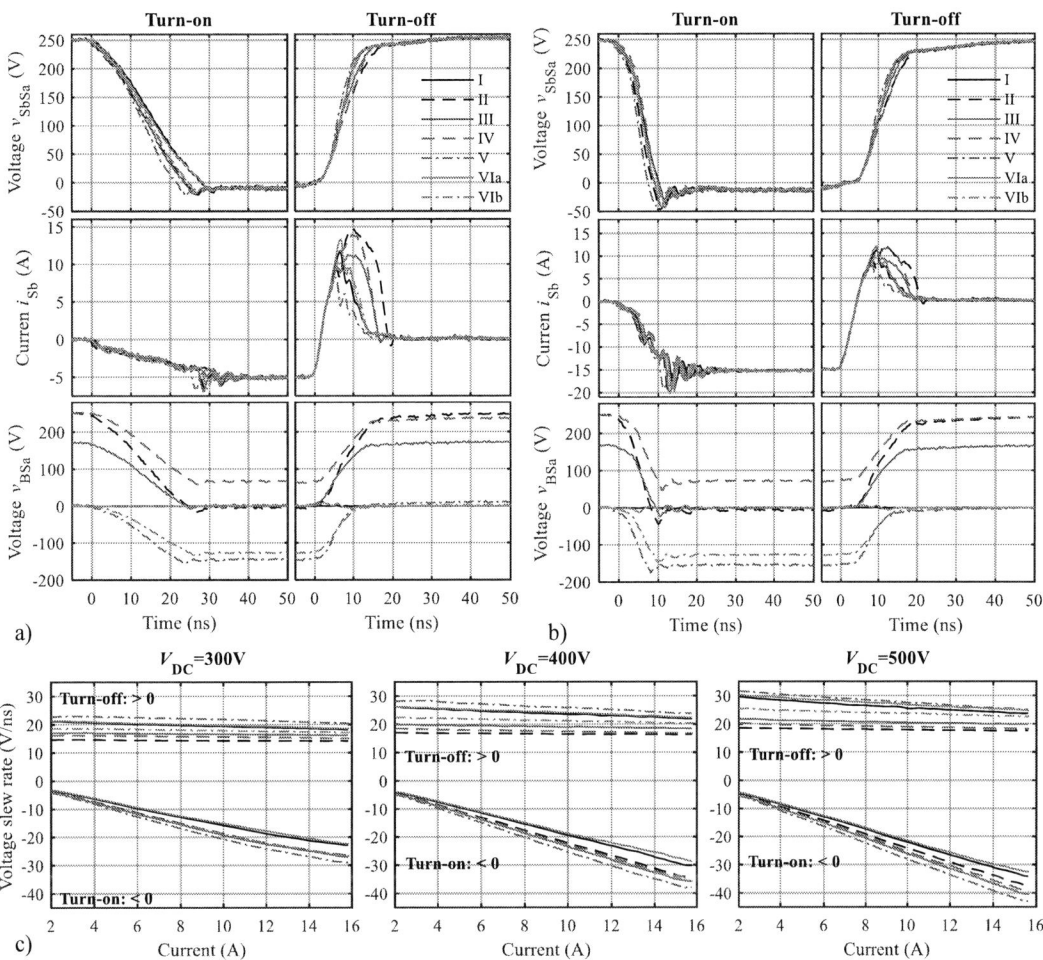

Fig. 8: Transient waveforms in soft-switched double-pulse test at V_{DC}=500 V and load currents of (a) I_L=5 A and (b) I_L=15 A and (c) average voltage slew rates (20%-80%)

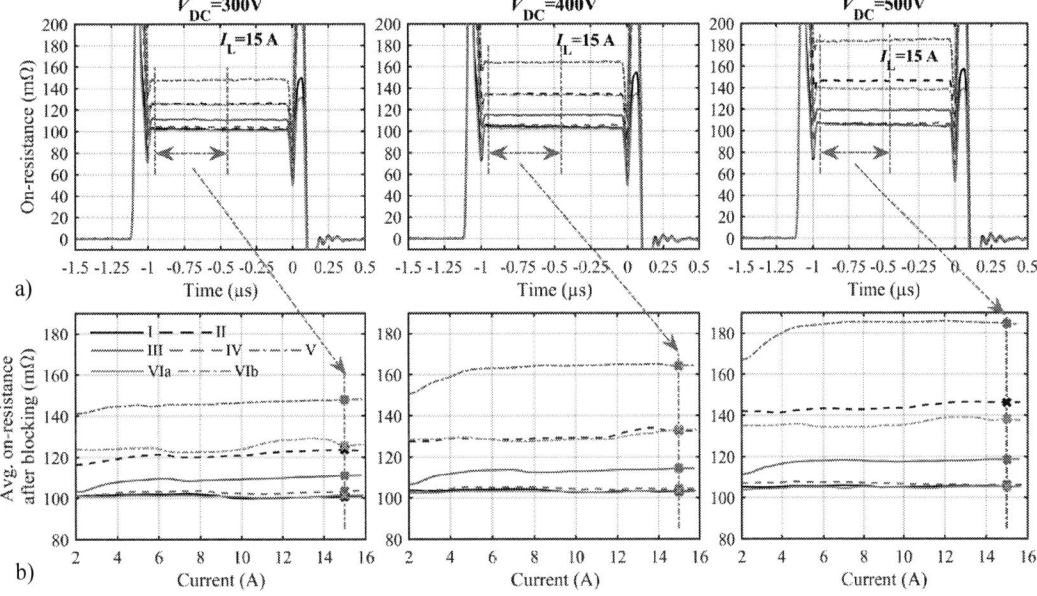

Fig. 9: Impact of the substrate termination on the (a) transient R_{ON} and (b) the average R_{ON} after blocking stress under soft-switched 3rd-quadrant ZVS conditions. Note that the transistor off-state drain bias is $V_{DC}/2$.

significant negative bulk-source bias occurs during on-state of the bidirectional transistor (Fig. 8a, b right, red and green dash-dot plot lines).

The transient R_{ON} in soft-switching double-pulse test is acquired during the 3rd-quadrant reverse conduction. In the timeframe of the preceding load inductor charging pulse the unidirectional GaN HEMT T_1 is conducting. Meanwhile the bidirectional GaN HEMT T_2 is exposed to off-state blocking stress and therefore prone to charge trapping causing dynamic R_{ON}-increase (Fig. 9). In soft-switching conditions the observed R_{ON} increases with the applied DC-link voltage (Fig. 9a). In comparison to hard-switched operation the absolute R_{ON} as well as its load-current dependency are notably reduced. The lowest R_{ON} is achieved using substrate terminations I and VIa, where a continuous substrate bias $v_{BSa} \approx 0$ V is guaranteed. Substrate termination IV exhibits a similar low R_{ON} (Fig. 9) although significant v_{BSa}-gradients occur during switching transitions while a positive bulk-source bias $v_{BSa} > 80$ V and $v_{BSa} > 250$ V is maintained during ON- and OFF-state respectively (Fig. 8a, b). The R_{ON}-degradation in soft-switching becomes especially extensive in configurations III, IV and VIb where a significant bulk-source bias gradient occurs combined with a temporarily bulk-source bias voltage which is close to zero or even negative (Fig. 8a and Fig. 9a).

4.4 Continuous operation in DC/DC buck-converter mode

The proposed active switched substrate termination scheme VIa has proven widely identical switching and dynamic on-state characteristics as a conventional fixed short of bulk and source (I) in both, hard- and soft-switched double-pulse characterization. The diode-based substrate termination scheme IV demonstrated similar on-state characteristics like configuration I and VIa at least under soft-switching conditions. Therefore, continuous operation measurements in DC/DC buck converter mode cover substrate termination schemes III, IV, VIa and VIb. The converter is operated in deep continuous conduction mode (CCM) and thermal steady state while a switching frequency $f_S = 50$ kHz, a duty cycle $D = 0.5$ and dead times $T_D = 80$ ns are applied. The employed PCB layout design is particularly optimized for minimum parasitic stray inductance and a small parasitic switching node capacitance while thermal design constraints are neglected. The maximum load current was limited to 6 A due to limited cooling capability of the PCB design. The bottom-side-cooled GaN HEMTs are exposed to forced air cooling. The T-type switching cell operation shows the highest efficiencies in both, hard- and soft-switching when using the active switches substrate termination scheme VIa, regardless of the DC-link voltage as well as load current (Fig. 10a vs. Fig. 11a). Other substrate termination schemes III, IV and VIb cause higher losses and device temperatures going along with reduced efficiency as well as lower maximum

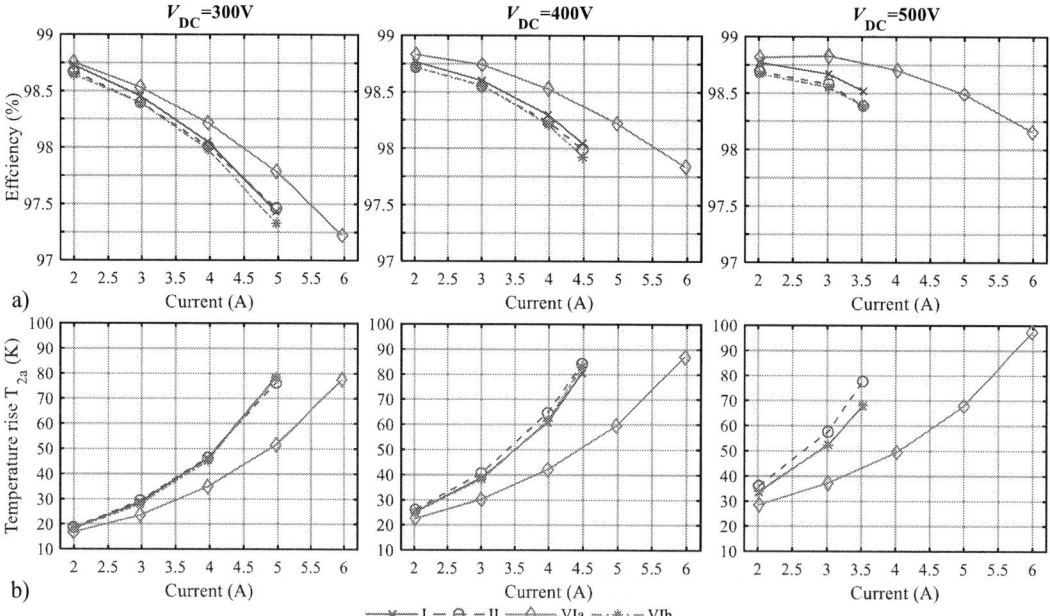

Fig. 10: Impact of the substrate termination on (a) the converter efficiency and (b) GaN HEMT package temperature rise vs. output current in hard-switched CCM DC/DC buck-converter operation

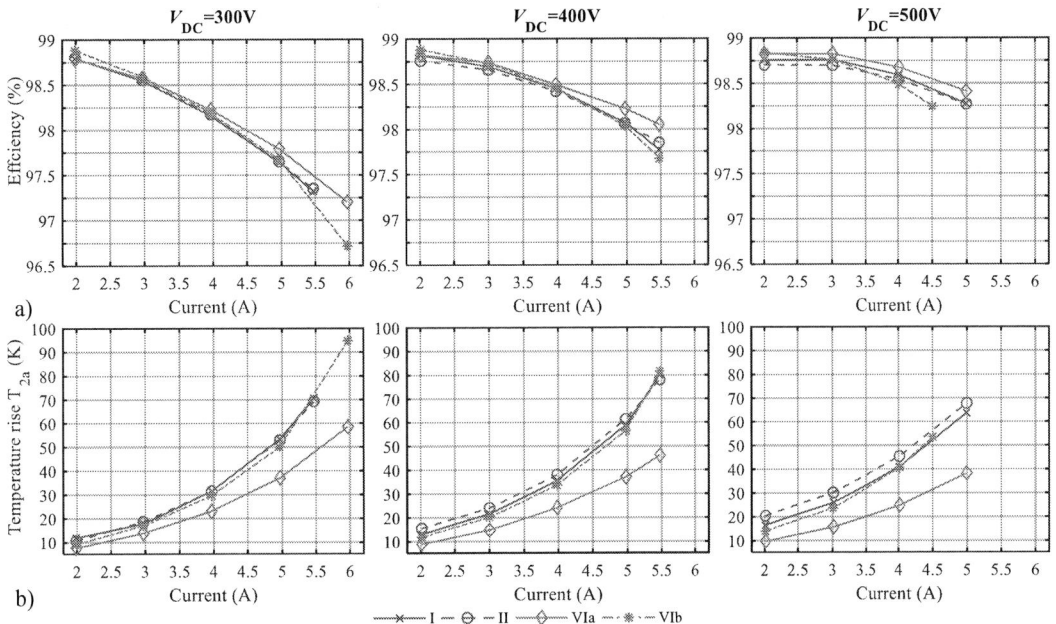

Fig. 11: Impact of the substrate termination on (a) the converter efficiency and (b) GaN HEMT package temperature vs. output current in soft-switched CCM DC/DC buck-converter operation

output currents. In agreement with the dynamic R_{ON}-characterization results in Sec. 4.2 and 4.3, the efficiency gains enabled by an active substrate termination are particularly pronounced in hard-switched operation. The significant impact on switching losses is further indicated when the temperature rises in Fig. 10 and Fig. 11 are compared. In 1st-quadrant hard-switching of the GaN HEMT T_{2a} which is part of the bidirectional transistor T2, the observed temperature rises are increased by up to 30 K in comparison to 3rd-quadrant ZVS operation when similar load conditions are considered (Fig. 10b vs. Fig. 11b).

Conclusion and future work

In this paper, different bidirectional switch implementations using lateral GaN HEMTs are presented. Monolithically integrated bidirectional GaN HEMTs with common drift region promise for significantly reduced R_{ON} compared to a discrete solution, but require a proper substrate termination preventing device degradation due to back-gating effects. Static and dynamic substrate coupling effects are studied for an electrically emulated monolithic bidirectional transistor which is based on discrete GaN-on-Si HEMTs. Several substrate termination schemes proposed before in literature are investigated with regard to their impact on switching and transient on-state characteristics in hard- and soft-switched operation of the device.

Based on the presented measurement results different degradation mechanisms due to substrate coupling are identified. Static device characteristics are degraded with respect to saturation current and R_{ON} when an increasingly negative substrate bias is applied, while a positive bias bulk-source voltage shows no effect. In contrast, in switched operation device degradation is mainly induced by gradients of the bulk-source bias and the related capacitive displacement currents occurring during switching transitions. An actively controlled substrate termination has demonstrated to overcome substrate coupling effects in switching tests of the emulated bidirectional GaN HEMT. Dynamic device characteristics in terms of switching speed and losses as well as the transient R_{ON} remain almost unaffected compared to a fixed substrate termination commonly used in unidirectional GaN HEMTs.

Efficiency measurements of the T-type switching cell in DC/DC operation with different substrate biasing configurations show the highest efficiencies for the proposed active substrate termination and thus support the findings from the static and switched device characterizations.

Future work will focus on the implementation and validation of the proposed active substrate termination in a monolithically integrated bidirectional GaN HEMT with common drift region in DC/AC operation of the T-type inverter. In addition, static and dynamic device degradation mechanisms will be analyzed by means of a physical transistor model.

References

[1] C. Kuring, O. Hilt, J. Böcker, M. Wolf, S. Dieckerhoff und J. Würfl, „Novel monolithically integrated bidirectional GaN HEMT," in *2018 IEEE Energy Conversion Congress and Exposition (ECCE)*, 2018.

[2] C. Kuring, N. Wieczorek, O. Hilt, M. Wolf, J. Böcker, J. Würfl und S. Dieckerhoff, „Impact of Substrate Termination on Dynamic On-State Characteristics of a Normally-off Monolithically Integrated Bidirectional GaN HEMT," in *2019 IEEE Energy Conversion Congress and Exposition (ECCE)*, 2019.

[3] M. Wolf, O. Hilt und J. Würfl, „Gate Control Scheme of Monolithically Integrated Normally OFF Bidirectional 600-V GaN HFETs," *IEEE Transactions on Electron Devices,* Bd. 65, pp. 3878-3883, 2018.

[4] EPC – Efficient Power Conversion Corporation, „eGaN® FET Datasheet EPC2110 – Dual Common-Source-Mode," 2017.

[5] S. Léo, F. Jean-Paul, F. David, J. Pierre-Olivier, P. Pierre und L. Othman, „Implementation of monolithic bidirectional switches in a AC/DC Dual Active Bridge in ZVS auto-switching mode," in *2018 IEEE International Conference on Industrial Technology (ICIT)*, 2018.

[6] Innoscience, „Datasheet 'INN40W08 - 40V Bi-GaN Enhancement-mode FET ',Rev.1.0 2021/11/26".

[7] „Datasheet 'GS66508P', Rev 190524".

[8] „Datasheet 'EPC2034C', revised June, 2020".

[9] H. Ueno, Y. Kinoshita, Y. Yamada, A. Suzuki, T. Ichiryu, M. Nomura, H. Fujiwara, H. Ishida und T. Hatsuda, „A 3-Phase T-type 3-Level Inverter using GaN Bidirectional Switch with Very Low On-State Resistance," in *PCIM Europe 2019; International Exhibition and Conference for Power Electronics, Intelligent Motion, Renewable Energy and Energy Management*, 2019.

[10] F. Vollmaier, N. Nain, J. Huber, J. W. Kolar, K. K. Leong und B. Pandya, „Performance Evaluation of Future T-Type PFC Rectifier and Inverter Systems with Monolithic Bidirectional 600 V GaN Switches," in *2021 IEEE Energy Conversion Congress and Exposition (ECCE)*, 2021.

[11] S. Bahl, M. Senesky, N. Tipirneni, D. Anderson and S. Pendharkar, "Bi-directional gallium nitride switch with self-managed substrate bias". Patent US20140374766A1 (20.06.2013).

[12] M. Imam, H. Kim, K. Leong, B. Pandya and G. Prechtl, "Semiconductor device having a bidirectional switch and discharge circuit". US / EU Patent US20190326280A1 (23.04.2018) / EP3562040A1 (15.04.2019).

[13] K. Leong, "Bidirectional switch with passive electrical network for substrate potential stabilization". Patent US10224924B1 (22.08.2017).

[14] C. Kuring, M. Tannhaeuser und S. Dieckerhoff, „Improvements on Dynamic On-State Resistance in Normally-off GaN HEMTs," in *PCIM Europe 2019; International Exhibition and Conference for Power Electronics, Intelligent Motion, Renewable Energy and Energy Management*, 2019.

[15] M. J. Uren, S. Karboyan, I. Chatterjee, A. Pooth, P. Moens, A. Banerjee und M. Kuball, „"Leaky Dielectric" Model for the Suppression of Dynamic R_{ON} in Carbon-Doped AlGaN/GaN HEMTs," *IEEE Transactions on Electron Devices,* Bd. 64, pp. 2826-2834, 2017.

[16] B. Kohlhepp, C. Kuring, S. Peller und D. Kübrich, „Measurement of Dynamic On-State Resistance of High-Voltage GaN-HEMTs under Real Application Conditions," in *2020 22nd European Conference on Power Electronics and Applications (EPE'20 ECCE Europe)*, 2020.

[17] A. Pozo, S. Zhang, G. Stecklein, RicardoGarcia, J. Glaser, Z. Tang, R. Strittmatter und A. Lidow, „GaN Reliability and Lifetime Projections," in *CIPS 2022; 12th International Conference on Integrated Power Electronics Systems*, 2022.

Transformer Design Optimization and Comparison for a DC-DC Converter used in PV Micro-Inverters

Tobias Manthey, Meriem Khader, Jens Friebe
Institute for Drive Systems and Power Electronics
Leibniz University Hannover
Email: tobias.manthey@ial.uni-hannover.de; friebe@ial.uni-hannover.de

Acknowledgement

Parts of this work were funded by the German Federal Ministry for Economic Affairs and Energy under Grant No. 03EE1057A (Voyager-PV) and also by the Ministry of Science and Culture of Lower Saxony and the Volkswagen Foundation. The authors are responsible for the content of this publication.

Keywords

≪Transformer≫, ≪Passive component≫, ≪Resonant converter≫, ≪ZVS converter≫, ≪Isolated converter≫

Abstract

The use of micro-inverters can lead to higher efficiency of photovoltaik systems, since shading problems and system failures have only a small impact. The most significant component in terms of losses and moreover extensive in the design process is the transformer which is necessary for the galvanic isolation and topology-dependent voltage boost. This paper compares several optimization options, both simulatively and through measurement-based verification.

Introduction

The availability of grid-connected micro-inverters has increased greatly in recent years and with it the demand for more and more efficient systems [1]. Since the semiconductor losses have an increasingly smaller proportion of the total losses due to soft switching, the transformer losses can have the greatest influence on the efficiency [2], [3]. However, transformers are required for the galvanic isolation between photovoltaic (PV) modules and the grid as well for voltage level transformation required for grid feeding. High currents, high switching frequencies and increasing power class of photovoltaic modules are challenging for the transformer design in terms of efficiency, temperature rise, volume and costs [4]. A fundamental design approach and an overview of magnetic materials as well as windings are given in [5] and [6].

In this paper, design criteria and parameter variations are discussed and performed based on simulations and measurements of a given hardware prototype of a series resonance converter (SRC) topology used in micro-inverters with an output power of 500 W and a switching frequency of 500 kHz serves as an application example. As a result of the extensive analysis of the topology and the operating point slightly below resonant frequency, the following section examines the influence of different core shapes as well as different winding materials and arrangements while keeping the core material and turn ratio constant. A major part of the simulations is performed with the artificial intelligence based software Frenetic, supported by investigations within a 2D FEM environment. In the last section, the measurement results are compared and analysed using efficiency curves of the entire set-up with the different transformers. For additional verification of the simulated core and winding losses, thermal images are taken with an IR camera.

Topology Description and Hardware Setup

First, a brief investigation of the influence of magnetizing inductance on the switching behavior is presented and the general conditions for the hardware prototype are listed in Table I. Fig. 1 shows the circuit diagram of the series resonant converter and the analyzed voltage and current characteristics. The objective is to discharge the output capacitance C_{OSS} of each MOSFET within the dead time to achieve zero voltage switching (ZVS) using the magnetizing current. Fig. 2 shows the simplified current and voltage waveforms for the possible operating points above and below resonance. The operating point $f_{sw} = f_{res}$ is not taken into account, as the operating point can shift under temperature and other external influences, since the proposed hardware prototype is an open loop system. In order to minimise the losses in the power semiconductors, the operating point shown in Fig. 2 (a) is preferred. In addition to ZVS in the primary-side MOSFETs, zero current switching (ZCS) at turn-off transition is also achieved in the diodes as the resonance is already completed before the switching transition. The challenge here is to design the transformer with a magnetizing inductance that is small enough to achieve a sufficiently high magnetizing current to discharge the output capacitances C_{OSS} in the switches within the deadtime. Fig. 9 shows the current and voltage waveforms for $L_m = 4\,\mu H$ and $L_m = 7\,\mu H$, in which ZVS is achieved only with the smaller inductance due to the higher magnetizing current. A further reduction of the magnetizing inductance enables a shorter dead time but also would lead to a higher current stress on the MOSFETs. An analysis of the trade off of dead time and ZVS was analyzed as an example in [7].

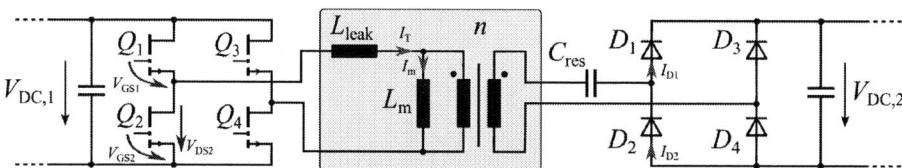

Fig. 1: Circuit description of the series resonant converter and color highlighting of the transformer.

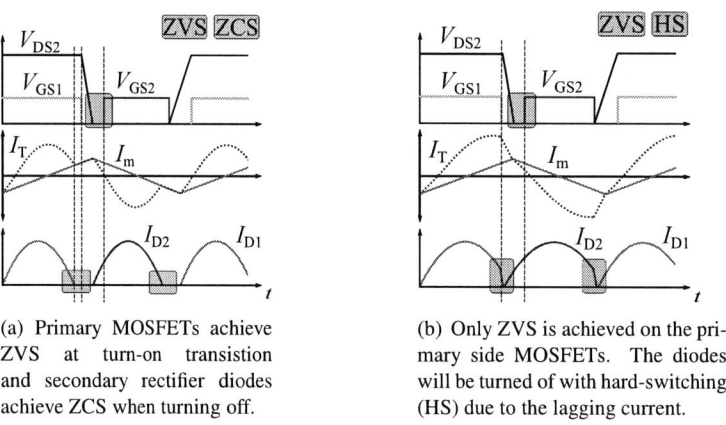

(a) Primary MOSFETs achieve ZVS at turn-on transistion and secondary rectifier diodes achieve ZCS when turning off.

(b) Only ZVS is achieved on the primary side MOSFETs. The diodes will be turned of with hard-switching (HS) due to the lagging current.

Fig. 2: Simplified waveforms of voltages and currents marked in Fig. 1. The operating point in (a) is preferred because of ZVS and ZCS.

To keep the magnetic components of the prototype as small as possible, the switching frequency is defined at 500 kHz, which has been tested to be acceptable for the switching losses of the primary and secondary side power semiconductors. In addition the leakage inductance of the transformer is used as the resonance inductance, which in combination with the resonance capacitance leads to the resonance frequency of the SRC defined in equation 1. The second resonance frequency, which results from the sum of the two inductances, is located far enough away from the operating point that it cannot be seen in

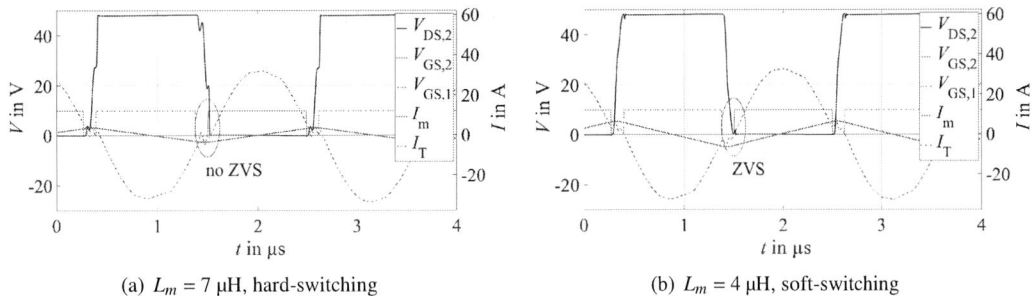

(a) $L_m = 7\,\mu\text{H}$, hard-switching

(b) $L_m = 4\,\mu\text{H}$, soft-switching

Fig. 3: Simulative switching behaviour of the primary MOSFETs with two different values of magnetizing inductance. The switching frequency is 500 kHz and the dead time between the gate signals is set to 130 ns.

the transfer function of the series resonance converter.

$$f_{\text{res},1} = \frac{1}{2\pi \cdot \sqrt{L_{\text{leak}} \cdot C_{\text{res}} \cdot n^2}} > f_{\text{sw}} \qquad f_{\text{res},2} = \frac{1}{2\pi \cdot \sqrt{(L_{\text{leak}} + L_{\text{m}}) \cdot C_{\text{res}} \cdot n^2}} << f_{\text{res},1} \qquad (1)$$

For the measurement of the transformers, a setup consisting of two PCBs, divided into primary and secondary side of the transformer, is designed, shown in Fig. 4. Consisting of two half-bridges equipped with Si MOFETs BSC025N08LS from Infineon, which are controlled via a gate driver and an external microcontroller for setting the switching frequency. A drain-source voltage measurement is implemented in the gate driver so that the dead time for reaching ZVS is adjusted by the driver itself. Large soldering pads for contacting the transformer offer the possibility to create a solder or screw connection. On the secondary side, next to the connection points, is the resonance capacitor which is realised with C0G capacitors to avoid a shift of the operating point because of voltage or temperature changes. Furthermore the different transformer designs result in different leakage inductances, which means that the resonance capacitance must be adjusted for each transformer. For this purpose, it is possible to connect several capacitors in parallel in order to be able to measure the design with a minimum leakage inductance of 22 nH. A full bridge diode rectifier with SiC Schottky diodes and an output capacitor provide the DC voltage on the output side. With the nominal input voltage of 48 V, the switching frequency of 500 kHz and a constant duty cycle of 50 %, an initial approximation of the magnetic flux density within the core can be estimated using equation 2, considering a triangular magnetizing current. The second DC-link voltage should be high enough to provide the negative and positive half-waves of the grid voltage using an inverter. A value of 384 V is chosen which can be realised with a transformer turn ratio of $n = 8$.

$$\hat{B} = \frac{V_{\text{DC},1}}{4 \cdot f_{\text{sw}} \cdot n_{\text{prim}} \cdot A_{\text{e}}} \qquad (2)$$

Transformer Design Process and Comparison

The idea of the design process is to change only one parameter at a time in order to identify the influence of the different degrees of freedom in the transformer design and focusing on methods to reduce losses and maximum temperatures as well as cost and volume optimization. For this approach, the core material and the secondary winding type are identical in each design and only the primary winding and its parameters as well as the core shape are varied. The N49 ferrite core material is the most suitable in the given frequency range [8] and is available for all cores considered in the simulation. With regard to the availability of core shapes on the market, the choice of cores is initially limited. Due to the high current on the primary side and the high switching frequency, the losses of the primary winding are dominant, so litz winding with a single strand diameter of 0.1 mm and 100 single wires can be selected for the secondary winding and is used in all designs for better comparability.

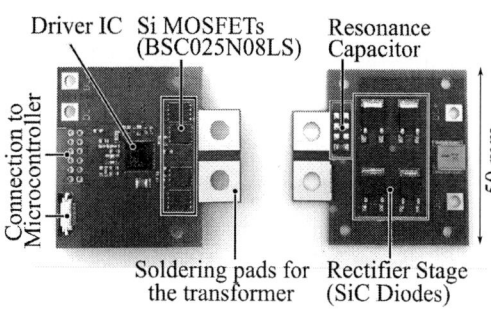

Fig. 4: Photograph of the 500 W hardware prototype without the transformer.

Table I: General conditions for the design of the transformer and fixed parameters.

Parameters	Symbol	Value
Input voltage	$V_{DC,1}$	48 V
Output voltage	$V_{DC,2}$	384 V
Rated output power	P_{out}	500 W
Switching frequency	f_{sw}	500 kHz
Resonance frequency	f_{res}	515 kHz
Turns ration	n	8
Core material		N49
Secondary winding		100 x 0.1 mm

Table II shows all the compared transformer designs as well as the design parameter changes. Overall, the influence on the core shape, primary turns, winding arrangement and winding type is investigated. A winding interleaving from PSS to SPS has only a small influence on the winding losses, but can be realised with little effort. Adding another turn on the primary side has a greater influence on the total losses. Although the winding losses increase, the core losses are reduced by a multiple. To reduce the increased winding losses, a litz wire winding with a smaller wire cross-section can be used. In this example a change from 50 μm (highlighted in green) to 30 μm (highlighted in blue) diameter leads to a reduction in winding losses of about 300 mW by keeping the total copper cross-section almost identical for both windings. The use of foil winding (highlighted in yellow) at first leads to higher losses simulatively, however, this will not show up in the measurements as it can be seen in the next section. In addition, foil winding is both less expensive and has a higher copper fill factor. Core shapes with larger core cross-sections and winding windows offer the possibility of further increasing the number of windings. However, since the winding losses increase more than the core losses decrease, this does not offer any optimization. The step towards larger cores offers no further advantages in this frequency and power range, and costs and construction volume can be saved.

Table II: Design parameters and simulation results of the individual transformers at 500 W and natural convection. Primary litz winding is marked in green and blue for 50 μm and 30 μm litz diameter as well as foil winding marked in yellow. Additional parameter changes are labeled on the side of the table.

Nr.	Core	w.a.	n_{prim}	P_c	P_w	P_{tot}	
1:	PQ 32/30	PSS	2	2.82 W	0.51 W	3.33 W	W. arrangement
2:	PQ 32/30	SPS	2	2.80 W	0.30 W	3.10 W	n_{prim}
3:	PQ 32/30	SPS	3	0.27 W	1.19 W	1.46 W	
4:	PQ 32/30	SPS	3	0.27 W	0.88 W	1.15 W	Litz parameters
5:	PQ 35/35	PSSS	2	1.79 W	1.19 W	2.98 W	W. arrangement
6:	PQ 35/35	SSPS	2	1.74 W	0.82 W	2.56 W	
7:	PQ 32/20	SPSS	2	3.02 W	0.69 W	3.71 W	Core & arragem.
8:	RM 14	SPS	2	2.50 W	0.50 W	3.00 W	Winding type and n_{prim}
9:	RM 14	SPS	3	0.68 W	3.24 W	3.92 W	
10:	PQ 40/40	SPSP	4	0.41 W	2.12 W	2.53 W	W. arrangement
11:	PQ 40/40	PSPS	4	0.38 W	2.24 W	2.62 W	

Transformer Prototypes and Measurement Verification

A part of the transformer prototypes are shown in Fig. 6. To achieve comparability of the overall system and to identify the transformer losses, the magnetizing inductance of each design is adjusted to about 4 µH by changing the air gap to ensure ZVS. The resonance frequency is set via the resonance capacitance, as each design has a different value of leakage inductance. Since only the overall system can be measured for the efficiency measurement, but the influence of the power semiconductors is not of interest, the difference in total power dissipation between the various designs is compared with the difference of transformer losses from the simulation results. The efficiency measurement of the four transformer prototypes is shown in Fig. 7 as well as voltage and current waveforms in Fig. 8 as an example resulting from the hardware setup with transformer No. 1 in which ZVS is achieved. The crosses in the efficiency measurement mark the measurement result after the steady state has been reached. Initial measurement investigations are relate to the arrangement of the foil winding, which in the PQ32/20 core has to be implemented in two layers due to its size, whereas the RM14 core offers the option of placing the windings on top of each other, as shown in Fig. 5 (b). After a 30-minute steady-state period, a maximum efficiency of 97.58 % was achieved with the PQ32/20. In order to compare only the difference in power loss, Table III include both, differences of the simulation results ΔP_{sim} as well as the differences of the measurement results ΔP_{meas}. It is observed that the transformers prototypes with the same primary winding, the simulations achieve well-matched results compared to the measurements. Since the simulation and measurement results are generally very close to each other at an output power of 500 W, Table II can be used as a good reference for the effect of changing the individual design parameters.

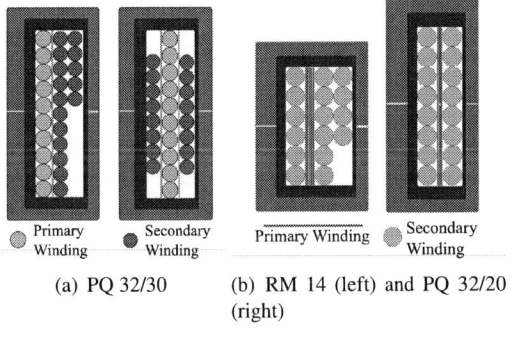

(a) PQ 32/30   (b) RM 14 (left) and PQ 32/20 (right)

Fig. 5: Schematic winding arrangement in one winding window of the core.

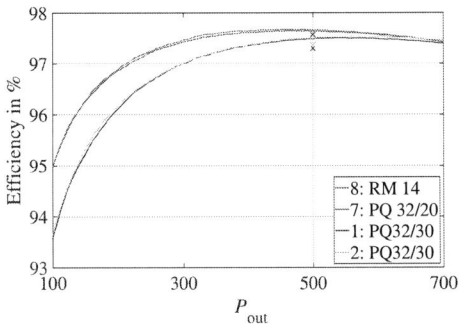

Fig. 7: Efficiency measurement of the two transformer prototypes RM 14 and PQ 32/20 with foil winding. The crosses mark the efficiency in steady state after about 30 minutes of operation.

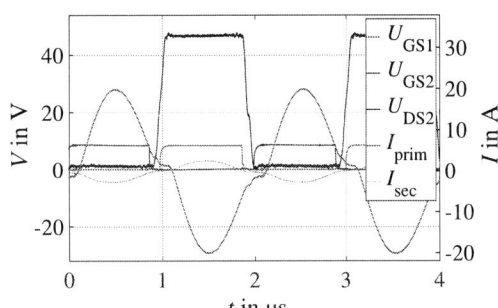

(a) PQ 32/20   (b) RM 14   (c) PQ 32/30

Fig. 6: Photographs of the first transformer prototypes.

Fig. 8: Resulting waveforms of transformer No. 1 with 500 W output power. ZVS is achieved with a dead time of 128 ns as it was expected from the simulation.

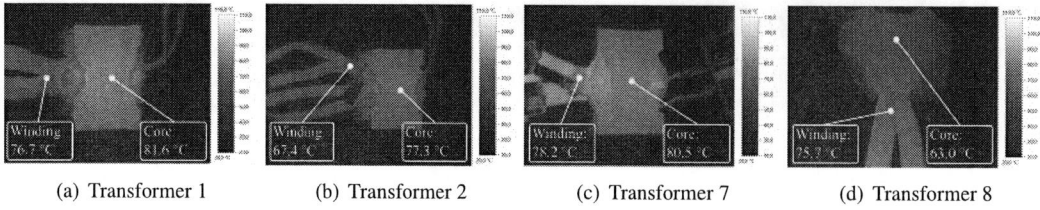

(a) Transformer 1   (b) Transformer 2   (c) Transformer 7   (d) Transformer 8

Fig. 9: Winding and core temperatures of the different transformers taken with a IR camera.

Table III: Comparison of simulation and measurement results.

Nr.	Core	w.a.	n_{prim}	P_c	P_w	P_{tot}		
1:	PQ 32/30	PSS	2	2.82 W	0.51 W	3.33 W	$\Delta P_{sim} = -0.23$ W	$\Delta P_{meas.} = -0.42$ W
2:	PQ 32/30	SPS	2	2.80 W	0.30 W	3.10 W	$\Delta P_{sim} = +0.61$ W	$\Delta P_{meas.} = -1.11$ W
7:	PQ 32/20	SPSS	2	3.02 W	0.69 W	3.71 W	$\Delta P_{sim} = -0.71$ W	$\Delta P_{meas.} = -0.11$ W
8:	RM 14	SPS	2	2.50 W	0.50 W	3.00 W		

Conclusion

This paper deals with the design optimization of transformers, especially for DC-DC converters for micro-inverters or other series resonant converter topologies. First, the switching behavior is simulated as a function of the magnetizing inductance. Furthermore, an analysis of operating point below and above the resonant frequency was made to show that both ZVS on the primary side and ZCS on the secondary side can be achieved. An extensive research of varying the different design parameters of the transformer to show the influence of winding and core losses was made in section . To make a comparison with the measurement results, only the difference of power loss between two transformer designs is considered in order to filter out the influence of the power semiconductors. This comparison shows that there is good correlation with the simulation results and ways of optimising the transformer design can be identified. A maximum efficiency of 97.58 % was achieved with a transformer prototype consisting out of a PQ 32/20 and two primary foil winding turns in two layers.

References

[1] J. M. A. Myrzik and M. Calais, "String and module integrated inverters for single-phase grid connected photovoltaic systems - a review," in IEEE Bologna Power Tech Conference Proceedings, pp. 8, vol. 2, 2003

[2] T. Manthey, T. Brinker and J. Friebe, "Design of an Isolated DC-DC Converter for PV Micro-Inverters with Planar Transformer and PCB Integrated Winding," 2021 IEEE Energy Conversion Congress and Exposition (ECCE), pp. 554-560, 2021

[3] H. Rezaei and A. Babaei, "Thermal analysis of inverters and high frequency transformers in the DC-DC converters," 2017 IEEE 4th International Conference on Knowledge-Based Engineering and Innovation (KBEI), pp. 0125-0130, 2017

[4] T. M. Undeland, J. Lode, R. Nilssen, W. P. Robbins and N. Mohan, "A single-pass design method for high-frequency inductors," in IEEE Industry Applications Magazine, vol. 2, no. 5, pp. 44-51, Sept.-Oct. 1996

[5] V. C. Valchev and V. A. D. Bossche, "Inductors and Transformers for Power Electronics (1st ed.)," CRC Press, 2005

[6] P. Zacharias, "Magnetische Bauelemente: Grundlagen und Anwendungen (1st ed.)," Springer Vieweg, ISBN 978-3-658-24741-6, 2020.

[7] GaN Systems Inc., "Gate Drive Circuit Design with GaN E-HEMTs," GN012 Application Note, 2020

[8] TDK, "Ferrites and accessories, SIFERRIT material N49," datasheet, 2017

Automated gate impedance network design for SiC MOSFETs using SPICE solver interfaced with MATLAB environment

Pawel Piotr Kubulus, Szymon Michal Beczkowski, Stig Munk-Nielsen, Asger Bjørn Jørgensen
AALBORG UNIVERSITY
Fredrik Bajers Vej 7K
Aalborg, Denmark
Email: ppk@energy.aau.dk

Keywords

≪Design optimization≫, ≪MOSFET≫, ≪Parasitic elements≫ ≪Silicon Carbide (SiC)≫ ≪Virtual prototyping≫

Abstract

In order to ensure proper switching of SiC devices gate impedance has to be carefully selected. Chosen topology and parameter values allow for damping the oscillations in poorly designed layouts, as well as adjusting dV/dt levels in cases where layout allows for too fast switching. Due to a wide choice of gate impedance topologies, some with multiple tunable parameters, experimental fine-tuning is a time-consuming process and analytical predictions do not take full effect of the parasitic elements into account. For this reason, an automated design process is developed using Matlab and LTSpice and the results are verified experimentally in a Double Pulse Test(DPT) setup, for the prediction accuracy assessment.

Introduction

Recently, wide band-gap (WBG) devices have been a subject of widespread integration in power converter design due to the significant improvement offered in the terms of switching speed, thermal performance and breakdown voltage in comparison with silicon devices [1]. However, the aforementioned improvements result in new design challenges due to high levels of voltage and current derivatives present [2]. High dV/dt and dI/dt tends to excite parasitic circuit elements, and degrades performance. Poorly designed layout results in an increased voltage overshoot, electromagnetic interference (EMI), gate loop stability issues and decreased reliability due to negative gate voltage spikes [3]. Those issues can be addressed in the early design phases by either designing the layout to mitigate the parasitics, or by proper choice of gate impedance network. The latter approach is investigated in this paper, as it can prove beneficial for both good and poor layouts.

Oftentimes gate impedance selection is performed manually and in the prototyping phase. Such a process is time-consuming and tends to omit the investigation of more advanced impedance networks than simple resistance, even though passive gate voltage waveform profiling has been presented in literature [4]. Additionally, manual selection rarely provides optimal performance and may not offer any compensation of layout shortcomings if different impedance network topologies are not considered. Due to aforementioned reasons, automation of gate impedance selection is desired as it is able to consider a higher variation of topologies, and improve parameter selection by applying optimisation algorithms.

In this paper, MATLAB environment has been interfaced with LTSpice circuit simulator, taking advantage of possibility to use manufacturer-provided device models and LTSpice accuracy combined with the extensive MATLAB function library. The possibility to freely manipulate SPICE netlists has been used for direct comparison of different gate impedance network topologies over a wide parameter range, and selection of optimal network using simple graphical optimisation. For accurate parasitic elements model, ANSYS Q3D has been used to extract full parasitics matrix as a SPICE subcircuit. Additionally, accuracy of simulation results has been verified experimentally.

Switching waveform prediction

The circuit considered is a standard double pulse test (DPT) setup, presented in Fig. 1. As it is a common practice to verify the switching performance using the DPT setup, it can be considered a good example for demonstrating the design methodology and interface proposed.

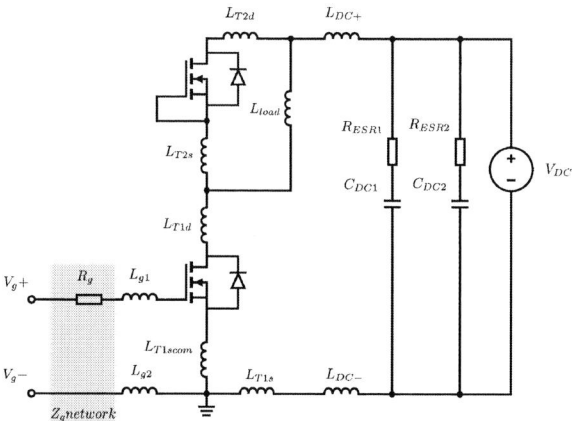

Fig. 1: Simplified Double Pulse Test circuit diagram.

The selected switches are CREE C3M0030090K. As an important mention, one of the SiC MOSFET selection criterion's was low internal gate resistance. This allows designer to have a better control over the switching waveform, as otherwise the gate loop would be over-damped and less responsive to the gate impedance modifications. Table I contains key circuit parameters and component types known at the start of design process. The rest of parameters is estimated using ANSYS Q3D or contained within manufacturer SPICE models.

Table I: Key circuit parameters

Parameter name	Symbol	Value
DC link bulk capacitance	C_{DC2}	4,7μF
DC link bypass capacitance	C_{DC1}	100nF
SiC MOSFET type	-	CREE C3M0030090K
DC link voltage	V_{DC}	400V
Gate voltage amplitude	$|V_{gd}|$	15V
Equivalent series resistance of C_{DC1}	R_{ESR1}	50mΩ
Equivalent series resistance of C_{DC2}	R_{ESR2}	7mΩ
Load inductance	L_{load}	107μH

Modelling and simulation

The modelling approach chosen assumes full digital design approach, and no experimental information available aside from the device models and datasheets provided by manufacturers. The reasoning for this assumption is, that a knowledge of digital design process accuracy without any case-specific modifications based on experimental work has not been extensively investigated in the literature, and is desirable for further design automation development in this area.

Therefore, additional emphasis has been put on circuit simulation accuracy and led to selection of LT-Spice. In the literature, this simulator has proven to be able to achieve closest match with experimental results [5]. However, a major drawback of Spice simulators are numerical stability issues, especially when the trapezoidal integration method is used [6]. Some of the physically viable circuits might consume too much computational power, return no or incorrect results, due to e.g. wrong solver settings, improper initial condition definition or an unfortunate combination of parameters. This is traditionally mitigated by manual fine-tuning, but in the case of investigating large variation of parameters and different topologies it can become a highly time-consuming process by itself, especially for sensitive circuits.

The above mentioned drawbacks have been mitigated by deploying the MATLAB-LTSpice interface. The possibility of monitoring the simulation progress has been used to stop and automatically correct the solver parameters whenever an excessive time consumption is detected. In case of problem persisting, parameter step size is adjusted in order to find a stable point in a close proximity. Operating directly on a netlist, new possibilities for analysis definitions are available. An example of this would investigating the modular structure for optimal number of modules, performing sweeps with a specified ratio of parameters or applying optimisation algorithms using the extensive MATLAB function libraries. Finally, connecting with multi-physics analysis tools is enabled.

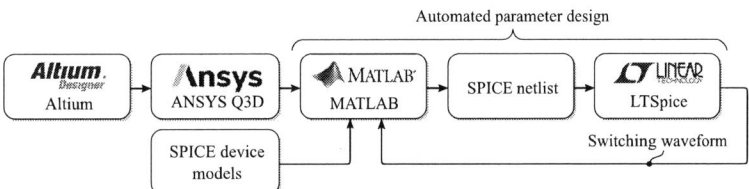

Fig. 2: Automated design flowchart

Fig. 2 depicts a workflow used for design automation. The geometry generated is exported from Altium Designer to Ansys Q3D, where parasitic element network is extracted. Then the network is used in MATLAB together with SPICE models in order to assemble a SPICE netlist for a given problem. The netlist is then simulated, and switching waveforms are fed back to MATLAB, where parameters or a structure of the netlist can be modified.

Extraction frequency selection

The choice of extraction frequency for lumped models of parasitic elements has been sparsely debated in the literature, due to broadband s-parameter based models being a default choice in electronics. Unfortunately, those models are not sufficient for power electronics, as frequency domain simulations are not an adequate mean of analysing the heavily non-linear behaviours. Even though attempts has been made in order to include non-linearities [7], time domain remain a standard. As the actual frequency present in the circuit is both a local and time dependent parameter in case of double pulse test, this selection is a non-trivial matter. There are three approaches used in practice.

Firstly, the highest frequency present can be used. While this can improve the model matching for the high frequencies, the response tends to get over-damped. Another approach may use the switching frequency, however this is hardly feasibly in the case of DPT circuit example with no particular switching frequency chosen. In this case, there is a risk of simulated response being under-damped, compared to the experiment. Last approach uses the rise time in order to approximate the signal bandwidth, using basic equation for the first-order step response. Once again, the response tends to be over-damped.

With those methods not being sufficient, the extraction frequency was selected by trial and error. The parasitic network values have been extracted at the frequency of 800kHz. A full RLGC model has been exported as a SPICE circuit and included in LTSpice circuit accordingly.

Gate impedance network design

The following gate impedance networks have been considered.

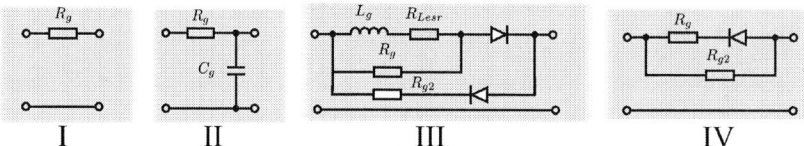

Fig. 3: Gate impedance networks considered

Networks I,II and IV are commonly used, while III is an alternative topology proposed in order to provide tunable frequency-varying gate impedance.

The optimisation criterion chosen is a sum voltage rise time(t_r) and fall time (t_f) further referred to as t_{tot}. The motivation behind this choice is simplifying the experimental verification, as switching energy measurement require meticulous adjustments in order to ensure current and voltage measurement to be synchronised in time. Additional constraint set is drain-source voltage overshoot lower than 10% of DC link voltage. The constraint value was set arbitrarily, as in practice it can vary with application. Standard definitions of t_r and t_f were used.

Simulation results

The impedance network topologies performance limitations have been investigated in simulations, with results present below. Firstly, the impedance network I has been investigated in the gate resistance range between 1 and 160Ω.

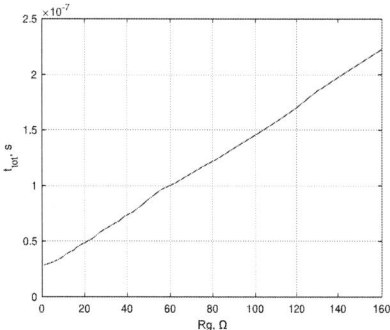

Fig. 4: Sum of drain-source voltage rise and fall times variation with gate resistance for topology I.

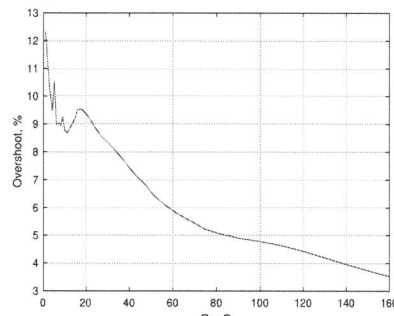

Fig. 5: Drain-source voltage percentage overshoot variation with gate resistance for topology I.

As expected, a linear increase of rise and fall time with increasing Rg is observed in Fig. 4. Slight deviations can be contributed to the oscillation presence for low values of gate resistance. The significant non-linearity present in Fig. 5 also originates from the drain-source voltage oscillations varying with gate resistance increase, until the response stops being heavily oscillatory.

The topology II is investigated next, within same Rg range and Cg ranging between 1 and 100nF.

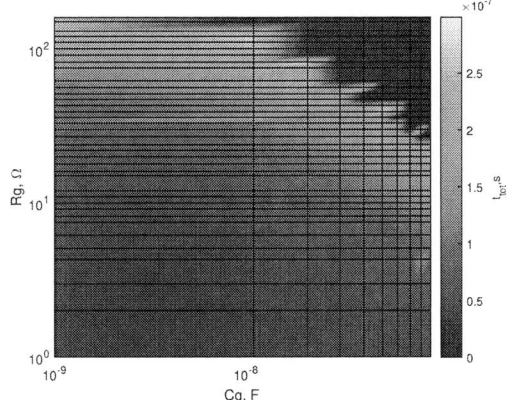 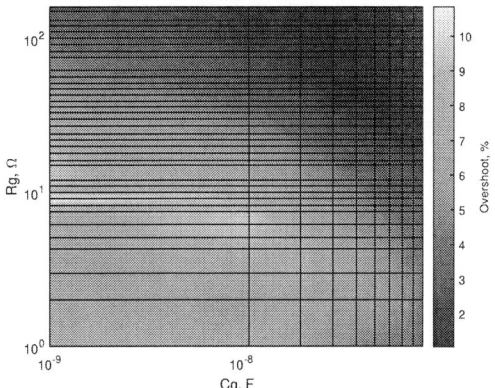

Fig. 6: Sum of rise and fall time variation with gate resistance (Rg) and capacitance (Cg) change for topology II.

Fig. 7: Overshoot variation with gate resistance (Rg) and capacitance (Cg) change for topology II.

In the Fig. 6, two regions divided with high t_{tot} region are visible. This is due to the high parameter range used, as both rise and fall time increase with gate resistance and capacitance. At some point, the switching process is slow enough for no switching to happen. Fig. 7 describes the overshoot variation with Rg and Cg. The overshoot does not seem to be linearly dependent on the design parameters, as high overshoot can be avoided by either keeping the Rg and Cg high enough, above the high overshoot zone, or by keeping Rg low and varying the Cg for fine-tuning.

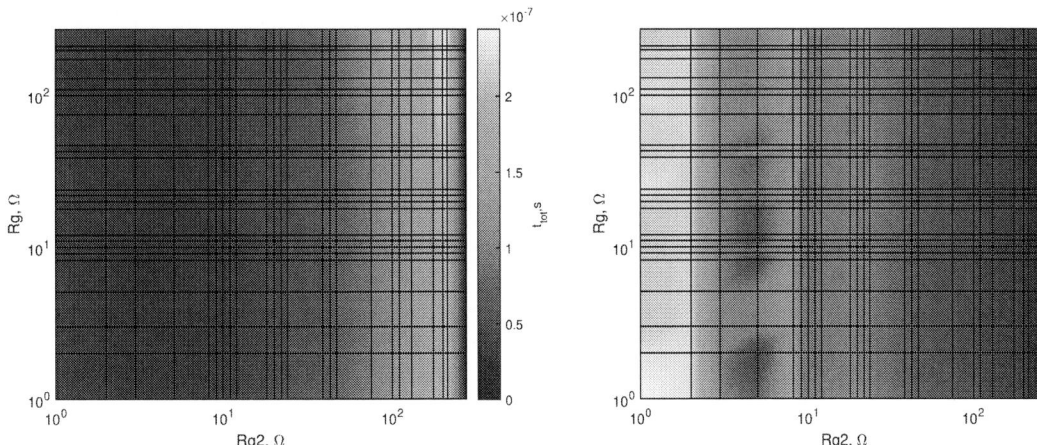

Fig. 8: The t_{tot} variation with two gate resistances (Rg and Rg2) change for topology IV.

Fig. 9: Overshoot variation with two gate resistances (Rg and Rg2) change for topology IV.

The topology IV behaves in a predictable, though not necessarily linear manner. The t_{tot} variation in Fig. 8 is dominated by Rg2. This leads to Rg2 being the main design parameter in terms of switching speed. Similar behaviour can be seen in the Fig. 9, however in this case oscillatory behaviour for the low values of Rg2 causes some non-linear dependencies.

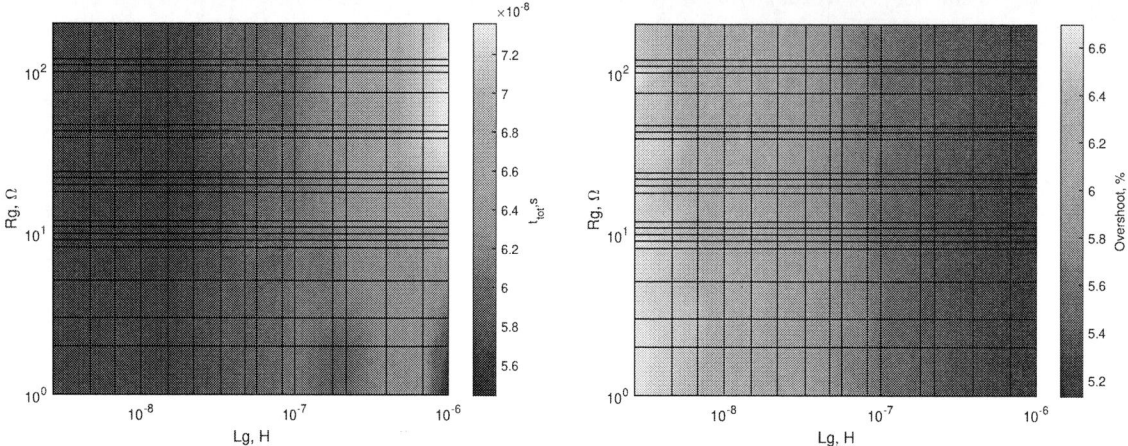

Fig. 10: Sum of rise and fall time variation with gate resistance (Rg) and gate inductance (Lg) change for topology III.

Fig. 11: Overshoot variation with gate resistance (Rg) and gate inductance (Lg) change for topology III.

Similarly to the topology IV, the behaviour is mainly controlled by one parameter, in this case L_g. The gate resistance sizing influence is visible mainly for very small values, due to the parallel connection. Surprisingly, there is a slight influence on the overshoot, even though the the main purpose of this topology is accelerating the turn-on. This is probably due to the oscillations originating from the turn-on undershoot.

Gate impedance network topology comparison

Finally, all of the topologies considered are compared in terms of the overshoot and sum of the rise and fall time.

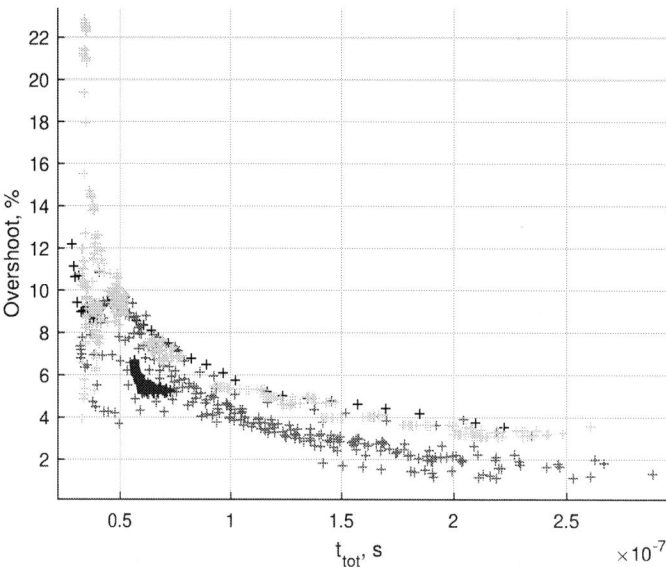

Fig. 12: Results of graphical optimisation for topologies I (black), II (red), III (blue), IV (green).

Fig. 12 shows the results of automated gate impedance design deploying graphical optimisation. Each of the points represents estimated voltage overshoot and a sum of rise and fall time for a single design variable point. The design variable combinations resulting in no switching, previously described in the

II topology simulation results, are removed from graphical optimisation. Finally, the minimal t_{tot} when subject to above mentioned overshoot limitation has been found to be 8 Ω gate resistance in topology I, closely followed by the topologies II and IV. This shows the importance of digital design for gate impedance network, as the topology IV would be a go-to in order to meet the overshoot requirements. However, in this case arresting the overshoot with special topology is not necessary, and topology I offers slightly better perfermonce in terms of t_{tot}. Topology III provides worse performance compared the other topologies investigated, which is most likely due to low impact of decreased turn-on fall time on t_{tot}.

Experimental validation

In order to validate the accuracy of switching waveform simulation and the automated design process results, the DPT board has been manufactured and tested in the setup presented in Fig. 14. Additionally, board layout exported from Altium Designer was depicted in Fig. 13.

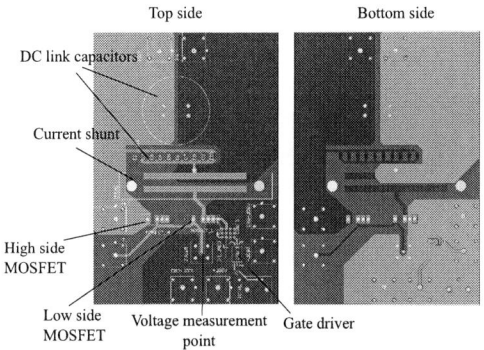

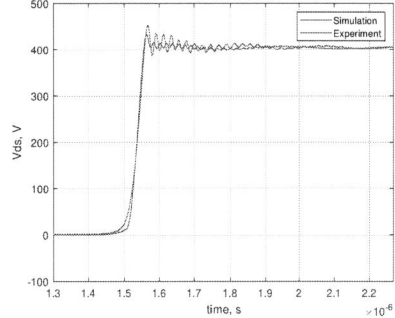

Fig. 13: DPT board layout.   Fig. 14: Experimental DPT setup.

The experimental test results for topology I have been presented in Fig. 15 and Fig. 16. Two data points with a significant parameter distance between them were chosen from previous chapter and verified experimentally. The verification was performed for all of the topologies aside from topology III, due to subpar performance.

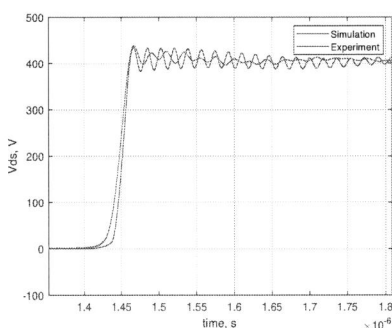

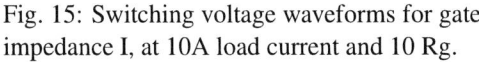

Fig. 15: Switching voltage waveforms for gate impedance I, at 10A load current and 10 Rg.

Fig. 16: Switching voltage waveforms for gate impedance I, at 10A load current and 30 Rg.

Firstly, the topology I has has been verified experimentally for the gate resistance values of 10 and 30 Ω, depicted in Fig. 15 and Fig. 16 respectively. A good matching in terms of rise time has been observed in both figures, as well as a very good match in terms of the overshoot in Fig. 15. In case of the second figure, overshoot predicted is lower than experimentally measured and this might be due to the too high extraction frequency. The extraction frequency issues are also visible in Fig. 15, as there is a lower damping in the simulation compared to the experimental results.

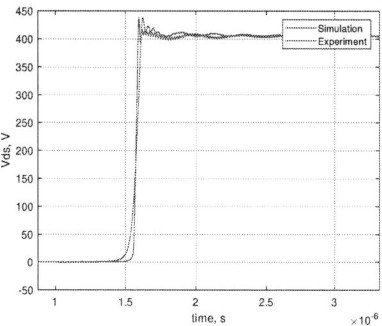
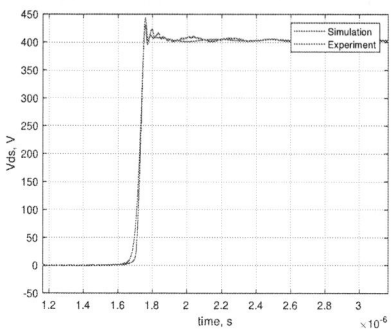

Fig. 17: Switching voltage waveforms for gate impedance II, at 10A load current, 3Ω Rg and 100nF Cg.

Fig. 18: Switching voltage waveforms for gate impedance II, at 10A load current, 10Ω Rg and 10nF Cg.

In the case of topology II, an overall good match between experimental and simulated waveforms can be observed in Fig. 17 and Fig. 18. In both case, oscillations and rise time is a close match, however in the Fig. 18 simulated overshoot is slightly lower than in the experimental waveform.

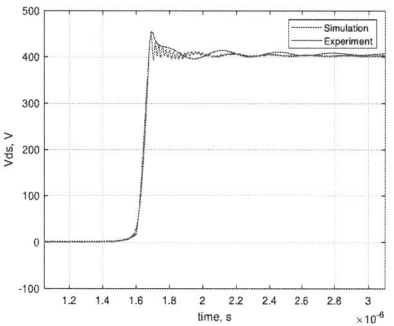
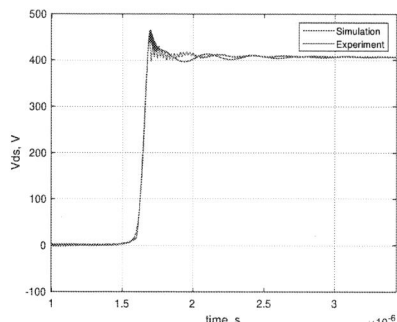

Fig. 19: Switching voltage waveforms for gate impedance IV, at 10A load current at 130Ω Rg and 75Ω Rg2

Fig. 20: Switching voltage waveforms for gate impedance IV, at 10A load current 75Ω Rg and 12Ω Rg2.

The last topology verfied experimentally was topology IV, once again with a good match between simulation and experimental results. Contrary to other topologies, the overshoot in Fig. 19 and Fig. 20 is slightly higher than in the experimental results.

Some of the simulation results were found to be mismatched with the experimental results. The main potential cause suspected is the limitation of frequency-dependent parameters used. Both capacitor ESR and parasitic network parameters used were chosen/extracted at one frequency. This assumes that the chosen frequency is valid for the whole sweep, which might not be the case.

Conclusion

The design framework based on LTSpice-Matlab interface and ANSYS Q3D extractor has been presented, along with the approach to automating the design of SiC MOSFET gate impedance design. Careful selection of SiC MOSFET gate impedance can increase the efficiency and reliability. Automation of this process allows to fully exploit its benefits and evaluate multiple impedance networks over a wide range of parameters. In future works, the passive gate voltage shaping circuits can be designed this way, further improving the performance.

The graphical optimisation has been used in order to determine the optimal gate impedance in analysed set, due to low number of optimisation parameters and possibility of easy visualisation. For cases requiring higher number of optimisation parameters, more advanced algorithms such as e.g. differential evolution should be used. The optimisation parameter definition is crucial to obtaining valid results. Therefore, a further investigation of available and more objective parameter definitions is needed in order to achieve fine-tuned gate impedances for specific applications.

Experimental setup for result verification has been developed and presented, showing a decent matching between simulations and experimental results. The model validity can possibly be improved by either adapting the frequency dependent parameters along the wide parameter sweeps or usage of broadband models, and should be investigated in the future works.

References

[1] J. Zhao, J. Zhang, H. Wu and Y. Zhang, "Design of Gate Driver and Power Device Evaluation Platform for SiC MOSFETS," 2018 1st Workshop on Wide Bandgap Power Devices and Applications in Asia (WiPDA Asia), 2018, pp. 1-6, doi: 10.1109/WiPDAAsia.2018.8734547.

[2] H. Zhou, C. Ye, X. Zhan and Z. Wang, "Designing a SiC MOSFETs Gate Driver with High dv/dt Immunity and Rapid Short Circuit Protection for xEV Drivetrain Inverter," 2019 22nd International Conference on Electrical Machines and Systems (ICEMS), 2019, pp. 1-5, doi: 10.1109/ICEMS.2019.8922429.

[3] S. Walder, X. Yuan and Q. Yan, "SiC MOSFET Switching Waveform Profiling Through Passive Networks," IECON 2018 - 44th Annual Conference of the IEEE Industrial Electronics Society, 2018, pp. 1483-1488, doi: 10.1109/IECON.2018.8591679.

[4] T. Shao et al., "Impact of Common Source Inductance on the Gate-Source Voltage Negative Spike of SiC MOSFET in Phase-Leg Configuration," 2020 IEEE 9th International Power Electronics and Motion Control Conference (IPEMC2020-ECCE Asia), 2020, pp. 3361-3365, doi: 10.1109/IPEMC-ECCEAsia48364.2020.9368245.

[5] I. Kovacevic-Badstübner, T. Ziemann, B. Kakarla and U. Grossner, "Highly accurate virtual dynamic characterization of discrete SiC power devices," 2017 29th International Symposium on Power Semiconductor Devices and IC's (ISPSD), 2017, pp. 383-386, doi: 10.23919/ISPSD.2017.7988984.

[6] P. Gubian and M. Zanella, "Stability properties of integration methods in SPICE transient analysis," 1991., IEEE International Sympoisum on Circuits and Systems, 1991, pp. 2701-2704 vol.5, doi: 10.1109/ISCAS.1991.176103.

[7] J. A. Jargon, K. C. Gupta and D. C. DeGroot, "Nonlinear large-signal scattering parameters: theory and applications," ARFTG 63rd Conference, Spring 2004, 2004, pp. 157-174, doi: 10.1109/ARFTG.2004.1387873.

An Improved Multi-loop Resonant and plug-in Repetitive Control Schemes for Three-Phase Stand-Alone PWM Inverter Supplying Non-Linear Loads

Ahmad Ali Nazeri, and Peter Zacharias
Centre of Competence for Distributed Electric Power Technology
Faculty of Electrical Engineering / Computer Science
University of Kassel, Kassel, Germany
Email: ahmad.nazeri@student.uni-kassel.de, peter.zacharias@uni-kassel.de

August 15, 2022

Acknowledgments

This work was financially supported by the German Academic Exchange Service (DAAD) Germany which provided a fully-funded scholarship to Ahmad Ali Nazeri for sponsoring his doctoral studies.

Keywords

≪Fractional delay≫, ≪Resonant control≫, ≪Repetitive control≫, ≪Total harmonic distortion≫

Abstract

This paper proposes an improved multi-loop control scheme for a three-phase voltage source inverter (VSI) for the island/microgrid operation. The constant voltage constant frequency (CVCF) pulse-width modulation (PWM) inverter can be used to regulate the output voltage with lower total harmonic distortion (THD). The output voltage is regulated under different load conditions, such as linear and rectifier loads for a CVCF for the uninterruptible power supply (UPS) inverter in stand-alone operation. An improved plug-in repetitive controller (RC) with the proportional-resonant (PR) control is used in the outer voltage loop to regulate the output AC voltage, and a simple proportional control is used in the inner current control loop for active damping and improving the transient performance. The instantaneous reference voltage of the converter is used as a feed-forward signal at the output of the converter to robust the system performance and simplify the controller design. This paper proposes a step-by-step design procedure of the voltage and current controllers, an analysis of the overall system stability from the frequency-response viewpoint, and the implementation of the PR with the improved plug-in RC for the three-phase stand-alone inverter supplying linear and nonlinear loads. The improved plug-in RC combined with PR control offers high-quality sinusoidal output voltages, robustness to the parameter uncertainties, fast response, and the need for fewer sensors. Moreover, the modified plug-in RC is effective and simple to be implemented on a digital signal processor (DSP). The three-phase VSI with the proposed multi-loop control is simulated in MATLAB/Simulink and experimentally implemented on a 7.5 kW system on TMDSDOCK28379D 32-bit floating-point DSP from Texas Instruments to validate the excellent steady-state, dynamic and transient performance of the proposed control scheme with better harmonic mitigation.

Introduction

The pulse-width-modulated (PWM) voltage source inverters (VSIs) are extensively integrated into different power conversion applications such as island microgrids, distributed generation, shunt active filters, and uninterruptible power supplies (UPS) [1, 2].

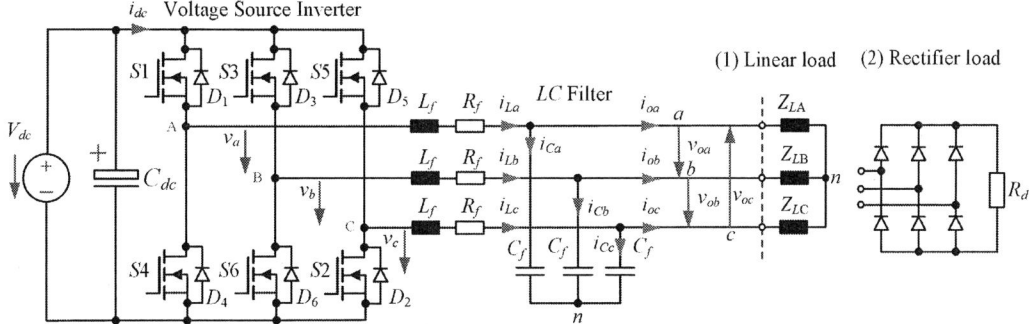

Fig. 1: Three-phase stand-alone inverter

Normally, the PWM VSI is connected to the rectifier loads, which introduces the disturbances into the inverter output voltage and results in the output voltage distortions [2]. Normally, a constant voltage constant frequency (CVCF) PWM inverter is used to achieve constant AC output with low total harmonic distortion (THD) and better transient response [3]. The set control objectives should be attained independent of the distortions caused by the connected load and the inverter. To achieve this, various control approaches have been reported such as deadbeat and model predictive control [4], resonant control [5, 6], and repetitive control (RC) [7, 8]. The RC based on the internal model principle [9], is capable of the perfect tracking of the periodic reference, eliminating the periodic error and has better periodic disturbance capability from the nonlinear load. Conventional RC (CRC) scheme has been extensively adopted in the CVCF PWM inverters [10]. The RC itself is capable of suppressing the odd and even harmonics, but typically it is used with another controller to improve system dynamic performance [11].

The parallel combination of the PR and RC control to mitigate the dead time harmonics with better transient performance in a single-phase PWM grid-connected inverter is analyzed in [12]. A modified RC with a finite impulse response filter (FIR) with the variation in the grid frequency is adopted in [13]. The multi-loop control strategies are normally adopted in the PWM VSIs to regulate the output voltage with nearly zero steady-state error and reduced THD supplying the nonlinear load. The inner current control is used to control the capacitor current or the inverter current for the inverter protection scheme, while the outer voltage control loop is used to track the periodic reference signal and compensate for the harmonics caused by the rectifier load and the nonlinearities of the inverter [14]. This paper proposes the step-by-step design procedure of the PR combined plug-in RC in the voltage loop and a proportional control in the current loop with the system stability analysis. An improved plug-in RC is designed in the discrete-time domain with the variation in the nominal grid frequency with fewer delays. Moreover, the proposed modified RC offers a simple implementation of the DSP with the implementation of the feed-forward loop at the fundamental frequency compared to the CRC. The phase lead compensator z^m is implemented to compensate for the phase lag of the CVCF PWM converter system [15]. A careful design of the improved RC with the lead step z^m is used to compensate for the system phase lag and is capable of eliminating the harmonics, which leads to higher tracking accuracy. Also, the phase lead compensator helps in stabilizing the inverter system with larger gains and results in achieving a fast convergence rate. Experimental results are provided to illustrate the mitigation of the harmonics caused by the nonlinear loads with lower THD and verify the effectiveness of the proposed multi-loop control.

Multi-loop Control System

Fig. 1 depicts the circuit diagram of a three-phase stand-alone CVCF PWM inverter. A constant DC voltage source V_{dc} connected to the three-phase SiC MOSFET with the DC-link capacitor C_{dc} and the current i_{dc}. The inverter is connected to an LC filter L_f, C_f, and the load Z_L. An equivalent resistor R_f presents the damping effects from the inverter losses, dead-time effects, and equivalent series resistance (ESR) of the filter inductance L_f. v_{inv} refers to the inverter output voltage, i_L, i_c and i_o are the inverter, capacitor and load currents respectively where v_o is the output voltage.

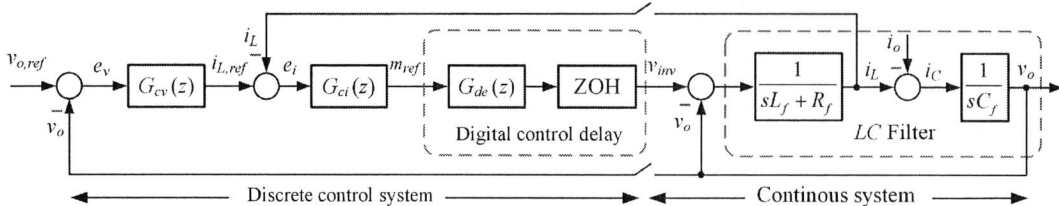

Fig. 2: Conventional multi-loop controller in the stationary reference frame

The VSI can be connected to the linear and/or the diode rectifier load with the output resistive load R_d as shown in Fig. 1. Fig. 2 illustrates the conventional multi-loop controllers for the CVCF PWM inverter in discrete domain [14]. The outer voltage loop $G_{cv}(z)$ regulates the capacitor voltage, $G_{ci}(z)$ is the inverter current feedback control with the inverter protection scheme. The load current i_o can be considered as a disturbance. The $G_{de}(z)$ is the computational and PWM delay. The computational delay is the time period between the sampling instance and the PWM duty cycle. The sampling instant is set to be at the middle of the PWM period, and the duty cycle will be updated at the next sampling instance [15]. The system parameters are listed in Table I. The computational delay is one sampling period T_s and is modeled as in (1) in the z-domain as

$$G_{de}(z) = z^{-1}. \tag{1}$$

The PWM inverter is modeled as a unity gain, and the PWM modulation is treated as a zero-order-hold (ZOH) block with the sampling period delay as shown in Fig. 2. Therefore, one and a half sampling period delay is introduced by the digital [16]. The effect of the digital delay has to be taken into consideration while designing the RC for the CVCF PWM inverter [17]. The open-loop transfer function of the inverter current loop (see Fig. 2) is given as

$$G_{\text{OL},ci}(z) = \frac{i_L}{e_i} = G_{ci}(z)G_{de}(z) \times Z_{\text{ZOH}} \left[\frac{sC_f}{(sL_f + R_f)\,sC_f + 1} \right]. \tag{2}$$

Table I: Inverter system parameters

Parameter	Symbol	Value
Nominal active power	$P_{\max}$	7.5 kW
Base voltage	V_B	325 V
Base current	I_B	15.3 A
Base impedance	Z_B	21.3 Ω
Base capacitance	C_B	150 μF
Output voltage (RMS)	V_{rms}	230 V
DC link voltage	V_{dc}	650 V
Rated output frequency	f_o	50 Hz
Switching frequency	f_{sw}	20 kHz
Sampling frequency	f_s	20 kHz
Computational time delay	T_d	75 μs
Filter inductance	L_f	2.6 mH
Equivalent resistance	R_f	0.1 Ω
Filter capacitance	C_f	6 μF
Connected load	R_d	60 Ω

The closed-loop transfer function of the current feedback control from i_L to $i_{L,ref}$ can be derived as

$$G_{CL,ci}(z) = \frac{i_L}{i_{L,ref}} = \frac{G_{OL,ci}(z)}{1 + G_{OL,ci}(z)}. \tag{3}$$

The inner current control $G_{ci}(z)$ is normally designed as a simple proportional gain [15]. A larger gain of the K_{pi} is needed to force the error to zero, but it should be carefully chosen not to restrict the bandwidth of the voltage control loop. The open-loop transfer function of the voltage control loop can be derived from Fig. 2 as [15]

$$G_{OL,cv}(z) = \frac{v_o}{e_v} = G_{cv}\left[1 - G_{CL,ci}(z)\right] G_{ci}(z) G_{de}(z) \times Z_{ZOH} \left[\frac{1}{(sL_f + Rf)sC_f + 1}\right]. \tag{4}$$

The closed-loop transfer function of the voltage feedback control v_o to $v_{o,ref}$ can be derived as

$$G_{CL,cv}(z) = H(z) = \frac{v_o}{v_{o,ref}} = \frac{G_{OL,cv}(z)}{1 + G_{OL,cv}(z)}. \tag{5}$$

Equation (5) depicts the precise tracking capability of the proposed multi-loop control system. The $H(z)$ can be derived based on (2)-(5) using the "ZOH" function in MATLAB [15].

Design of the Controllers

Current Control Loop

The inverter current control loop is designed in the stationary reference frame ($\alpha\beta$) using only a proportional gain of K_{pi}. A PI controller in a synchronous reference frame (dq) can also be used in the current control loop, but it introduces undesirable phase delay. The objective of using a proportional gain is to achieve proper resonance damping, and the reference voltage feed-forward path in the control scheme can be added at the output of the converter system to reduce the steady-state error, and the complexity of the controller design [18]. The open-loop transfer function of the current loop in the continuous domain, which is similar to the discrete domain in (2) and shown in Fig. 2 is given as

$$G_{OL,ci}^s(s) = G_{PR_{ci}}^s(s)G_{fi}(s) = \underbrace{\left(K_{pi} + \frac{K_i s}{s^2 + \omega_o^2}\right)}_{\text{PR current controller}} \underbrace{\left(\frac{V_{dc}}{sL_f + R_f}\right)}_{\text{Inverter}}. \tag{6}$$

where $'s'$ denotes the SRF, K_i is the integral gain of the current controller, and ω_o is the nominal frequency. The closed-loop transfer function of the current control is expressed as [16]

$$\begin{aligned} G_{CL,ci}^s(s) &= \frac{G_{PR_{ci}}^s(s)G_{fi}(s)}{1 + G_{PR_{ci}}^s(s)G_{fi}(s)} \\ &= \frac{\frac{V_{dc}}{L_f}\left(K_{pi}s^2 + K_i s + K_{pi}\omega_o^2\right)}{s^3 + \frac{1}{L_f}\left(K_{pi}V_{dc} + R_f\right)s^2 + \left(\omega_o^2 + \frac{K_i V_{dc}}{L_f}\right)s + \frac{\omega_o^2}{L_f}\left(K_{pi}V_{dc} + R_f\right)}. \end{aligned} \tag{7}$$

The third-order closed-loop characteristic equation is compared to the Naslin characteristic polynomial equation to extract the current control parameters as given as

$$P_N(s) = a_0 \left(1 + \frac{s}{w_0} + \frac{s^2}{\alpha w_0^2} + \frac{s^3}{\alpha^3 w_0^3} \right) = \frac{a_0}{\alpha^3 w_0^3} \left(\alpha^3 w_0^3 + \alpha^3 w_0^2 s + \alpha^2 w_0 s^2 + s^3 \right). \tag{8}$$

where w_0 is the pulsation, and the damping factor ζ is related to the value of α

$$\zeta = \frac{\alpha - 1}{2}. \tag{9}$$

The following relation holds after the comparison of the closed-loop current control (7) with the Naslin polynomial (8) as expressed as

$$\alpha^2 w_0 = \frac{1}{L_f} \left(K_{pi} V_{dc} + R_f \right), \qquad \alpha^3 w_0^2 = \omega_o^2 + \frac{K_i V_{dc}}{L_f},$$

$$\alpha^3 w_0^3 = \frac{\omega_o^2}{L_f} \left(K_{pi} V_{dc} + R_f \right), \qquad a_0 = \alpha^3 w_0^3. \tag{10}$$

The pulsation w_0 is derived from (10) as

$$w_0 = \frac{\omega_o}{\sqrt{\alpha}}. \tag{11}$$

which then leads to the parameters of the PR current control loop as [16]

$$K_{pi} = \frac{\alpha \sqrt{\alpha} \omega_o L_f - R_f}{V_{dc}}, \qquad K_i = \frac{\omega_o^2 L_f \left(\alpha^2 - 1 \right)}{V_{dc}}. \tag{12}$$

Voltage Control Loop

The fundamental component of the PR controller can be used in parallel with the resonant-harmonic compensators (RHCs) to mitigate the low-order harmonics [5, 6]. Fig. 3a shows the PR in parallel with the RHCs in the voltage control loop to reduce the voltage harmonics when the rectifier load is connected. The open-loop transfer function of the voltage control loop in $\alpha\beta$ frame can be determined as

$$G_{\text{PR},cv}^s(s) = \frac{i_{L,ref}(s)}{e_v(s)} = \underbrace{K_{pv} + \frac{K_{iv}s}{s^2 + \omega_o^2}}_{1^{\text{st}}\text{harmonic}} + \underbrace{\sum_n \frac{K_{ivn}s}{s^2 + (n\omega_o)^2}}_{n^{\text{th}}\text{harmonic}}. \tag{13}$$

where K_{pv} is proportional, and K_{ivn} is the integral gain of each RHCs with n being 3, 5, 7, etc. The PR controller is used for the capacitor voltage control design in the $\alpha\beta$ frame. The open-loop transfer function of the voltage control loop is given as [16]

$$G_{\text{OL},cv}^s(s) = G_{\text{PR}_{cv}}^s(s) G_{fv}(s) = \underbrace{\left(K_{pv} + \frac{K_{iv}s}{s^2 + \omega_o^2} \right)}_{\text{PR voltage controller}} \underbrace{\left(\frac{1}{sC_f} \right)}_{\text{Filter capacitor}}. \tag{14}$$

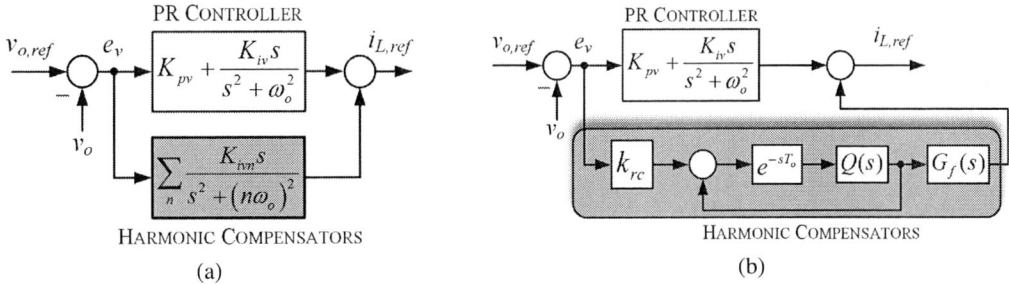

Fig. 3: Block diagram of the voltage control loop (a) PR controller with RHCs. (b) PR controller in parallel with RC

The closed-loop transfer function of the voltage loop is given as

$$G_{\text{CL},cv}^s(s) = \frac{G_{\text{PR}_{cv}}^s(s)G_{fv}(s)}{1 + G_{\text{PR}_{cv}}^s(s)G_{fv}(s)} = \frac{\frac{1}{C_f}\left(K_{pv}s^2 + K_{iv}s + K_{pv}\omega_o^2\right)}{s^3 + \frac{K_{pv}}{C_f}s^2 + \left(\frac{K_{iv}}{C_f} + \omega_o^2\right)s + \frac{K_{pv}\omega_o^2}{C_f}}. \tag{15}$$

Comparing the closed-loop transfer function $G_{\text{CL},cv}^s(s)$ in (15) to (8), the controller parameters for the voltage loop can be extracted in similar fashion as explained in (10) and (11), which are expressed as

$$\begin{aligned}
\alpha^2 w_0 &= \frac{K_{pv}}{C_f}, & \alpha^3 w_0^2 &= \frac{K_{iv}}{C_f} + \omega_o^2, \\
\alpha^3 w_0^3 &= \frac{K_{pv}\omega_o^2}{C_f}, & a_0 &= \alpha^3 w_0^3.
\end{aligned} \tag{16}$$

which then leads to the parameters of the PR voltage control loop as [16]

$$K_{pv} = \omega_o C_f \sqrt{\alpha^3}, \qquad\qquad K_{iv} = \omega_o^2 C_f\left(\alpha^2 - 1\right). \tag{17}$$

An Improved plug-in Repetitive Control Loop

According to the internal model principle [9], the output of the closed-loop system can track and reject the reference and disturbance signal respectively with zero steady-state error if the accurate model is included in the closed-loop system. Fig. 3b shows the PR controller in parallel with the repetitive controller for better transient performance and harmonic compensation [5, 8, 12]. The open-loop transfer function of the RC harmonic compensator $G_{\text{RC}}(s)$ for the voltage loop as shown in Fig. 3b in continuous domain can be expressed as

$$G_{\text{OL},cv}^s(s) = \frac{i_{L,ref}(s)}{e_v(s)} = G_{\text{PR}cv}^s(s) + \underbrace{k_{rc}\frac{e^{-sT_o}Q(s)G_f(s)}{1 - e^{-sT_o}Q(s)}}_{G_{\text{RC}}(s)} \tag{18}$$

where k_{rc} is the RC gain, which can be selected between $0 < k_{rc} < 2$, $\omega_o = 2\pi/T_o$ is the fundamental frequency, and T_o is the fundamental period. A low-pass filter LPF $Q(s)$ is included in the RC compensator to attenuate high-order harmonics and lead to the stability of the controller [11] but it might degrade the tracking capability of the controller. The $Q(s)$ can be presented in the z-domain as

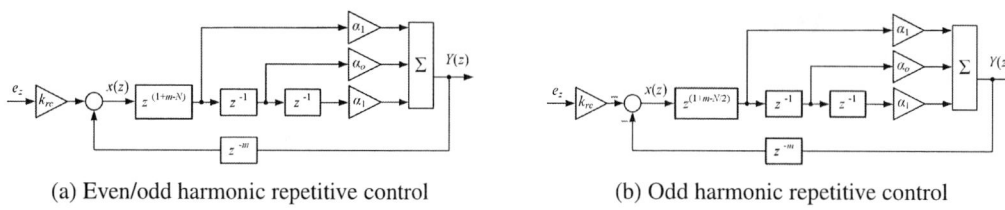

(a) Even/odd harmonic repetitive control   (b) Odd harmonic repetitive control

Fig. 4: Discrete-time implementation of the improved repetitive control

$$Q(z) = \alpha_1 z + \alpha_0 + \alpha_1 z^{-1}. \tag{19}$$

where $2\alpha_1 + \alpha_1 = 1$, $\alpha_1 > 0$, and $\alpha_0 > 0$. The design of the LPF can be done by examining the harmonic content of the voltage controlled by the PR controller. Moreover, $G_f(s)$ is a phase-lead compensator, which is also considered in the closed-loop system stability. The phase-lead step is given as

$$G_f(z) = z^m. \tag{20}$$

where m is the phase-lead number, determined using the trial and error method during the experiments. The RC with the LPF and the phase-lead compensator can be implemented in a digital microprocessor using cascaded delays. The ideal RC in (18) with $Q(s) = 1$ and $G_f(s) = 1$ can be expanded as

$$G_{RC}(s) = k_{rc}\left[-\frac{1}{2} + \frac{1}{T_o s} + \frac{2}{T_o}\left(\sum_{n}^{\infty}\frac{s}{s^2 + (n\omega_o)^2}\right)\right]. \tag{21}$$

where $n = 1, 2, 3, ..$ is the harmonic order including the fundamental component [12]. For practical implementation, the improved RC in (18) can be expressed in z-domain as

$$G_{RC}(z) = k_{rc}\frac{z^{-N}Q(z)G_f(z)}{1 - z^{-N}Q(z)}. \tag{22}$$

where $N = f_s T_o$ is the number of samples in one repetitive period, and f_s is the sampling frequency. Typically, $n \leq N_f$, where $N_f = f_s T_o/2$ is the Nyquist frequency. The RC scheme can compensate the harmonics up to the Nyquist frequency, since it contains N_f individual RHCs in parallel [12]. Fig. 4 shows the improved discrete implementation of the repetitive control for even/odd (N) harmonics (see Fig. 4a) and odd harmonics ($N/2$) (see Fig. 4b) realized on a DSP. The discrete transfer function of the RC output $Y(z)$ to the input $e(z)$ is given as

$$\frac{Y(z)}{e(z)} = \underbrace{k_{rc}\frac{z^{-N}Q(z)G_f(z)}{1 - z^{-N}Q(z)}}_{\text{Even/odd harmonics } G_{RC}(z)}, \qquad \frac{Y(z)}{e(z)} = \underbrace{-k_{rc}\frac{z^{-N/2}Q(z)G_f(z)}{1 + z^{-N/2}Q(z)}}_{\text{Odd harmonics } G_{RC}(z)}. \tag{23}$$

The improved repetitive controller in (22) and shown in Fig. 4a is used for the three-phase stand-alone VSI. Substituting (19) into (22) and rearranging results in

$$Y(z) = x(z)\left(\alpha_1 z^{1+m-N} + \alpha_0 z^{m-N} + \alpha_1 z^{m-N-1}\right). \tag{24}$$

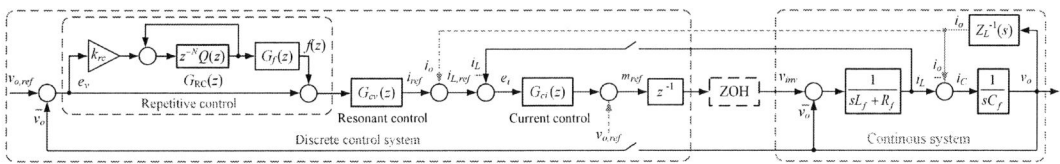

Fig. 5: The proposed improved multi-loop PR plug-in RC with the control scheme of the whole system in the stationary reference frame

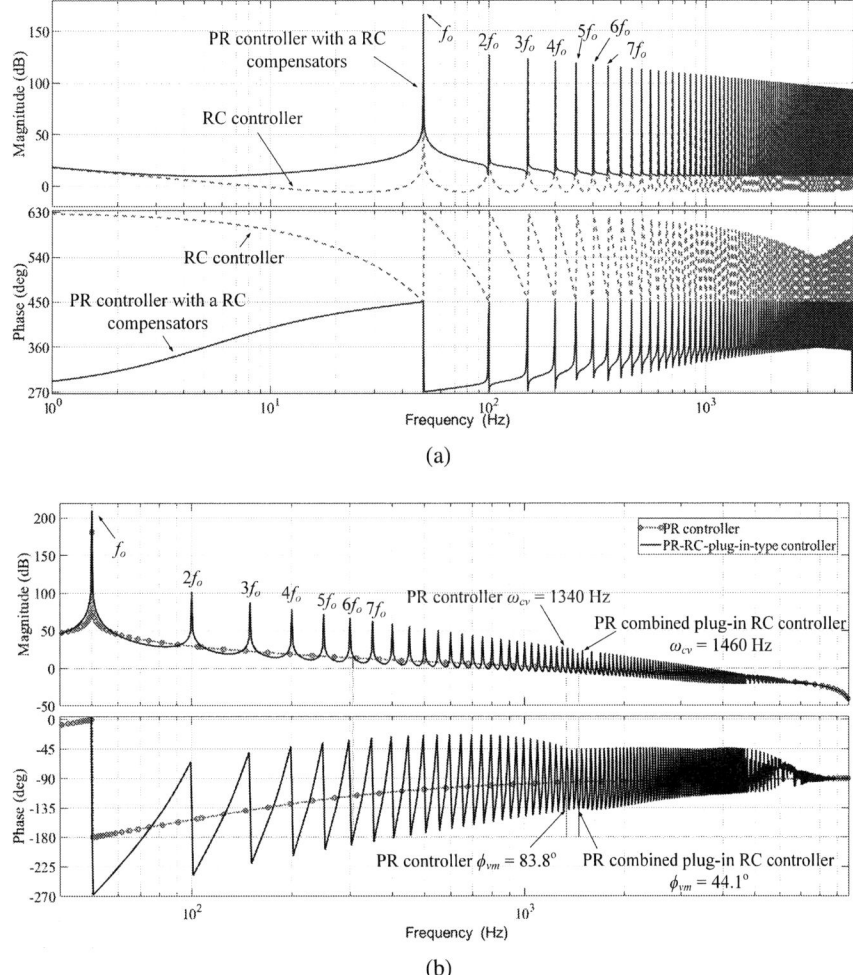

Fig. 6: Frequency response of the open-loop voltage control in stationary reference frame: (a) PR in parallel with RC controller as a harmonic compensator; (b) Improved PR combined plug-in RC controller

where $x(z)$ is given by

$$x(z) = e(z)k_{rc} + z^{-m}Y(z). \tag{25}$$

Equations (24) and (25) are the standard realization of the controller and shown in Fig. 4a. The overall improved multi-loop PR plug-in RC of the system is illustrated in Fig. 5. The plug-in type RC is given in purple dashed as shown in Fig. 5.

Table II: Parameters of the proposed control system

Parameter	Symbol	Value
Current loop proportional gain	K_{pi}	1.28
Voltage loop proportional gain	K_{pv}	0.91
Voltage loop integral gain	K_{iv}	795
Resonant controller frequency	ω_o	314
Controller value	α	3
Damping factor	ζ	0.95
Repetitive controller gain	k_{rc}	0.98
LPF $Q(z)$ coefficients	α_0, α_1	0.25, 0.5
Phase-lead compensator	$G_f(z)$	z^3
Delay line	$N = f_s/f_o$	400

The feed-forward reference voltage can also be added for better transient response and avoids the phase delay in the control system. Moreover, the load current i_o can also be used as a feedback control signal for a better transient and steady-state performance and provides better inverter protection as shown in Fig. 5 [6, 14]. The open-loop transfer function of the voltage control in z-domain for the parallel and plug-in type RC control with the PR control is given as

$$G_{\text{OL},cv}(z) = G_{\text{PR}cv}(z) + G_{\text{RC}}(z) \times G_{fv}(z) \qquad \rightarrow [\text{Parallel} - \text{type}] \qquad (26a)$$

$$G_{\text{OL},cv}(z) = G_{\text{PR}cv}(z) \times (1 + G_{\text{RC}}(z)) \times G_{fv}(z). \qquad \rightarrow [\text{Plug} - \text{in} - \text{type}] \qquad (26b)$$

Fig. 6 is extracted from (26) with the proposed control parameters given in Table II. Fig. 6a is the open-loop frequency response of the parallel combination of the RC and PR control (see 26a) where Fig. 6b is the improved plug-in RC combined with PR control (see 26b) in the $\alpha\beta$ frame. Fig. 6a and (21) shows that the RHCs can approach an infinite gain at the corresponding resonant frequency $n\omega_o$. Moreover, the PR control with the parallel and/or plug-in type combination of RC harmonic compensator can suppress all harmonics up to the Nyquist frequency as presented in Fig. 6. The cross-over frequency ω_{cv} for the PR and PR combined with the plug-in type RC control is 81490 rad/s and 91735 rad/s respectively. Moreover, the phase margin ϕ_{vm} for the PR and PR combined with the plug-in type RC control is 83.8° and 44.1° respectively as illustrated in Fig. 6b, which is still high enough for the closed-loop stability of the voltage control.

Experimental Verification

The simulation and experimental results were extracted by implementing the three-phase VSI system as illustrated in Fig. 1. The detailed system and proposed control parameters used for the simulation and experiments are listed in Table I and II respectively. The three-phase VSI system shown in Fig. 1 with the proposed control in Fig. 5 were implemented with the space vector modulation (SVM) in MATLAB/Simulink. The multi-loop control scheme shown in Fig. 5 with the improved PR combined plug-in RC in the voltage loop and a proportional control in the current loop was implemented. The base voltage V_B, base current I_B, base impedance Z_B, and the base capacitance C_B given in Table I were used to calculate the control parameters in the per-unit system, and the measured voltages and currents were scaled into the per-unit system for the simulation and experiments. The PR control was discretized and implemented using the forward and backward Euler method and the modified plug-in RC control depicted in Fig. 4a was implemented using (24) and (25) with the control parameters given in Table II.

Fig. 7 illustrates the laboratory setup for the implementation of the proposed PR combined plug-in RC control scheme. The prototype includes a constant DC source V_{dc} using the programmable DC power source from the Regatron, a three-phase inverter constructed from the CoolSiC MOSFET, an LC filter,

Fig. 7: Laboratory setup for the three-phase stand-alone inverter with an *LC* filter and connected linear and rectifier loads

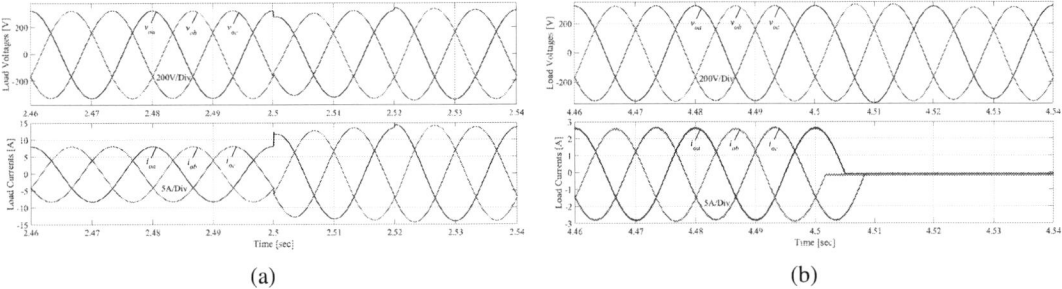

(a)   (b)

Fig. 8: Experimental transient results of the three-phase VSI with the step-change in the load with PR combined plug-in RC control; (a) half-load to full load, (b) half-load to no-load

a three-phase voltage and current sensors, and a DSP control board with sampling circuits. The voltage and current THDs and the results were extracted from the precision power analyzer ZES-Zimmer and an 8-channel MSO68B oscilloscope respectively as shown in Fig. 7. The reference signals $V_{\alpha,ref}$ and $V_{\beta,ref}$ are the input signals to the SVM to generate the control signals. The discrete control schemes were implemented on a TMDSDOCK28379D 32-bit floating-point DSP controller from Texas Instruments. The internal signals of the discrete control scheme implemented on the DSP were converted to 0-3.0 V analog signals via two embedded 12-bit digital-to-analog converters (DACs) and displayed on the oscilloscope. The analog signal sampling was performed in the middle of every switching period T_s. Fig. 8 illustrates the experimental results of the step-change in the resistive load using the PR combined plug-in RC control. Fig. 8a shows the step-change from half-load to full-load where Fig. 8b depicts the dynamic performance of the system from half-load to no-load. It can be seen that the proposed control is faster to the step-changes in the load and the output voltage is constant. Fig. 9 shows the experimental results of the improved PR combined plug-in RC control when highly nonlinear load is connected at the output. Fig. 9a and Fig. 9b illustrates the transient performance of the system with and without proposed plug-in RC control respectively. The proposed PR combined plug-in RC control has superior performance under rectifier load compared to the standard PR control. Also, the THD of the suggested PR combined RC control is 0.46 % compared to the 9.84 % when only PR control is implemented with the current THD of 30 %.

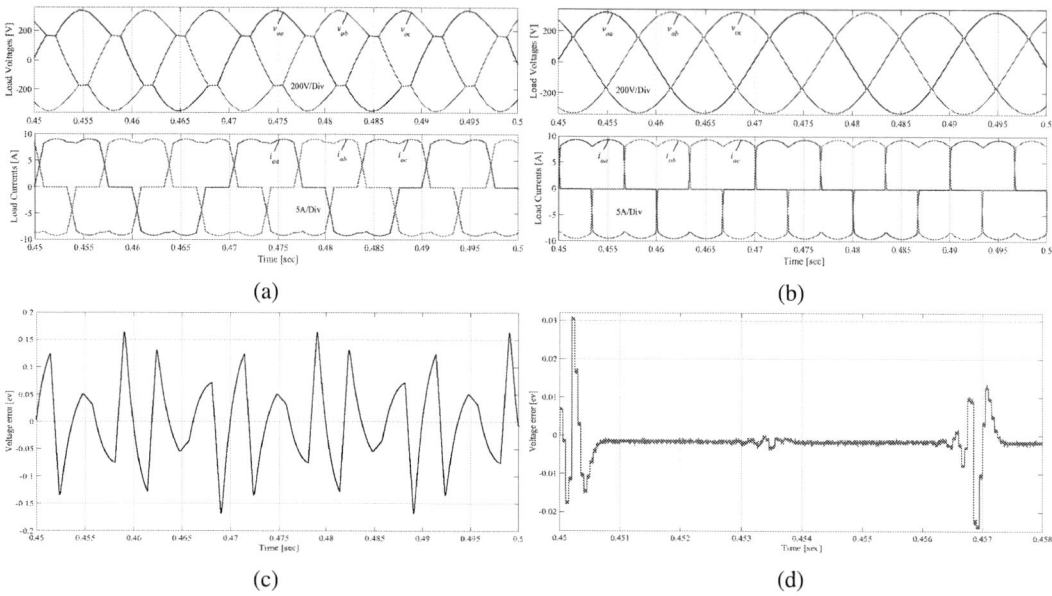

(a)   (b)

(c)   (d)

Fig. 9: Experimental dynamic performance of the three-phase VSI under rectifier load; (a) only PR control, (b) improved PR combined plug-in RC control (c) the voltage control loop error e_v with only PR control, (d) the voltage control loop error e_v with improved PR combined plug-in RC control

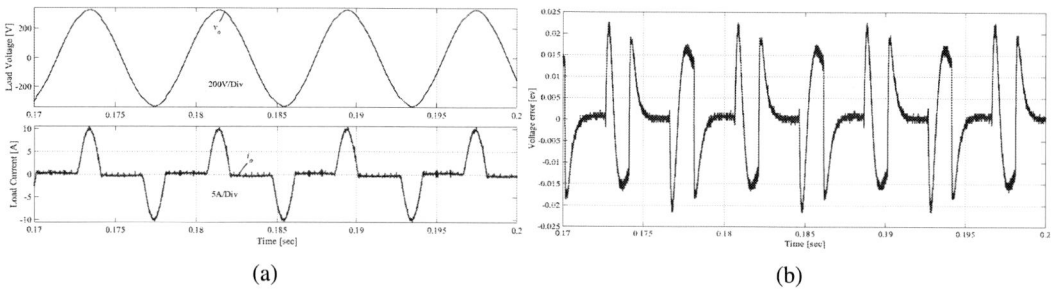

(a)   (b)

Fig. 10: Experimental performance of the single-phase VSI under nonlinear load with improved PR combined plug-in RC control; (a) output (v_o, i_o), (b) the voltage control loop error e_v

Moreover, the voltage control loop error e_v ($v_{o,ref} - v_o$) under nonlinear load with and without modified plug-in RC control can be seen in Fig. 9c and Fig. 9d respectively. Fig. 10 illustrates the experimental results of the single-phase inverter with the proposed improved PR combined plug-in RC control under highly nonlinear load conditions. Fig. 10a depicts the output voltage and current without the load current feedback (LCF) where Fig. 10b shows the voltage control loop error e_v.

Conclusion

This paper presents an improved PR combined plug-in repetitive control scheme capable to suppress the current harmonics caused by the rectifier loads at the output of the VSI-based islanded microgrid system. A step-by-step design procedure of the voltage and current control loop in $\alpha\beta$ frame is presented. The PR control combined plug-in RC control is implemented in the voltage control loop and the proportional control in the current loop considering the sampling and computational delay. The discrete PR combined RC control is simple in structure without the need for extra load current sensors. The proposed control scheme has higher system stability with maximum phase margin, better tracking accuracy, and better disturbance rejection capability. The output voltage presents a lower THD of 0.46% with the load current THD of 30% and without the need of the load current feedback when the nonlinear load is connected.

Also, the proposed PR combined plug-in RC control offers almost zero steady-state error, fast error convergence, and better dynamic and transient performance compared to the standard control. Moreover, the proposed control mechanism presents a simple implementation on a DSP system by implementing a feed-forward loop at the fundamental frequency that the voltage is bounded during the convergence period of the RC compared to the implementation of the traditional long delay feature of the RC. The load current feedback can also be used for better transient performance but it increases the cost of the overall system.

References

[1] R. Wai, C. Lin, Y. Huang and Y. Chang, "Design of High-Performance Stand-Alone and Grid-Connected Inverter for Distributed Generation Applications," IEEE Transactions on Industrial Electronics, vol. 60, no. 4, pp. 1542-1555, 2013.

[2] A. A. Nazeri, P. Zacharias, F. M. Ibanez and I. Idrisov, "Paralleled Modified Droop-Based Voltage Source Inverter for 100% Inverter-Based Microgrids," 2021 IEEE Industry Applications Society Annual Meeting (IAS), 2021, pp. 1-8, 2021.

[3] T. Yokoyama and A. Kawamura, "Disturbance observer based fully digital controlled PWM inverter for CVCF operation," IEEE Transactions on Power Electronics, vol. 9, no. 5, pp. 473-480, 1994.

[4] S. Kouro, P. Cortes, R. Vargas, U. Ammann and J. Rodriguez, "Model Predictive Control—A Simple and Powerful Method to Control Power Converters," IEEE Transactions on Industrial Electronics, vol. 56, no. 6, pp. 1826-1838, 2009.

[5] Teodorescu, Remus, Frede Blaabjerg, Marco Liserre, and P. Chiang Loh. "Proportional-resonant controllers and filters for grid-connected voltage-source converters." IEE Proceedings-Electric Power Applications, vol. 153, no. 5, 750-762, 2006.

[6] Somkun, Sakda. "Unbalanced synchronous reference frame control of singe-phase stand-alone inverter." International Journal of Electrical Power & Energy Systems, vol. 107, pp. 332-343, 2019.

[7] S. Jiang, D. Cao, Y. Li, J. Liu and F. Z. Peng, "Low-THD, Fast-Transient, and Cost-Effective Synchronous-Frame Repetitive Controller for Three-Phase UPS Inverters," IEEE Transactions on Power Electronics, vol. 27, no. 6, pp. 2994-3005, 2012.

[8] A. Lidozzi, C. Ji, L. Solero, P. Zanchetta and F. Crescimbini, "Resonant–Repetitive Combined Control for Stand-Alone Power Supply Units," IEEE Transactions on Industry Applications, vol. 51, no. 6, pp. 4653-4663, 2015.

[9] Francis, Bruce A., and William M. Wonham. "The internal model principle for linear multivariable regulators, Applied mathematics and optimization," vol. 2, no. 2, pp. 170-194, 1975.

[10] Kai Zhang, Yong Kang, Jian Xiong and Jian Chen, "Direct repetitive control of SPWM inverter for UPS purpose," IEEE Transactions on Power Electronics, vol. 18, no. 3, pp. 784-792, 2003.

[11] Nazir, R., Zhou, K., Watson, N. and Wood, A., "Analysis and synthesis of fractional order repetitive control for power converters." Electric Power Systems Research, vol. 124, pp. 110-119, 2015.

[12] Y. Yang, K. Zhou, H. Wang and F. Blaabjerg, "Analysis and Mitigation of Dead-Time Harmonics in the Single-Phase Full-Bridge PWM Converter With Repetitive Controllers," IEEE Transactions on Industry Applications, vol. 54, no. 5, pp. 5343-5354, 2018.

[13] D. Chen, J. Zhang and Z. Qian, "An Improved Repetitive Control Scheme for Grid-Connected Inverter With Frequency-Adaptive Capability," IEEE Transactions on Industrial Electronics, vol. 60, no. 2, pp. 814-823, 2013.

[14] Poh Chiang Loh, M. J. Newman, D. N. Zmood and D. G. Holmes, "A comparative analysis of multiloop voltage regulation strategies for single and three-phase UPS systems," IEEE Transactions on Power Electronics, vol. 18, no. 5, pp. 1176-1185, 2003.

[15] S. Yang, P. Wang, Y. Tang and L. Zhang, "Explicit Phase Lead Filter Design in Repetitive Control for Voltage Harmonic Mitigation of VSI-Based Islanded Microgrids," IEEE Transactions on Industrial Electronics, vol. 64, no. 1, pp. 817-826, 2017.

[16] Bacha, Seddik, Iulian Munteanu, and Antoneta Iuliana Bratcu. "Power electronic converters modeling and control." Advanced textbooks in control and signal processing, vol 454, no. 454, 2014.

[17] A. A. Nazeri, P. Zacharias, F. M. Ibanez and S. Somkun, "Design of Proportional-Resonant Controller with Zero Steady-State Error for a Single-Phase Grid-Connected Voltage Source Inverter with an LCL Output Filter," 2019 IEEE Milan PowerTech, pp. 1-6, 2019.

[18] M. Monfared, S. Golestan and J. M. Guerrero, "Analysis, Design, and Experimental Verification of a Synchronous Reference Frame Voltage Control for Single-Phase Inverters," IEEE Transactions on Industrial Electronics, vol. 61, no. 1, pp. 258-269, 2014.

High Switching Frequency Operation of a Single-Phase Five-Level Hybrid Active Neutral Point Clamped Inverter with a Model Predictive Control Approach

Mohammad Najjar[1], Mahdi Shahparasti[2], Rasool Heydari[3], Morten Nymand[3]

[1] Schneider Electric, Kolding, Denmark
[2] School of Technology and Innovations, University of Vaasa, 65200, Vaasa, Finland
[3] University of Southern Denmark, Odense, Denmark
E-Mail: mdn.najjar@gmail.com, mahdi.shahparasti@uwasa.fi,
rasool.heydari@gmail.com, mny@sdu.dk

Keywords

«Voltage Source Inverter (VSI)», «Model Predictive Control», «Multi-level converters», «Silicon Carbide (SiC)», « Capacitor voltage balancing».

Abstract

Wide bandgap (WBG) devices such as Silicon-Carbide (SiC) MOSFETs can be utilized to increase the switching frequency of power electronic converters. The size of passive components of an output filter can be reduced by increasing the switching frequency of converters or the number of output levels thorough the employment of multilevel topologies. Therefore, the combination of multilevel converters and WBG switches with a high switching frequency can improve the dynamic of converters. Meanwhile, a high bandwidth controller is also required to achieve a fast dynamic response of the system. In this paper, an advanced model predictive control (MPC) approach, based on the concept of hysteresis current control, is presented for a single-phase five-level hybrid active neutral point clamped (ANPC) inverter. A hybrid modulation technique with different switching frequencies is considered in this paper. As a result, different semiconductor technologies including SiC and Si are employed in the structure of the converter. Considering the AC and DC sides mathematical modeling of the converter, an MPC with the ability to control the neutral point (NP) voltage is designed. Finally, experimental results show that by utilizing the SiC MOSFETs and the proposed advanced MPC structure, the inverter's switching frequency is increased, with lower current ripple and fast dynamic performance.

Introduction

Multilevel converters offer lower dv/dt, common-mode voltage, current and voltage total harmonic distortion (THD), and smaller filter inductance compared to the conventional two-level converter [1]–[3]. Among the different utilizations of multilevel converters, single-phase power converters have been employed in photovoltaic (PV), electric railway traction, and power factor correction applications [4]–[9]. One interesting topology for these applications is the hybrid active neutral-point-clamped (H-ANPC) converter. In the converter structure, eight switches are utilized and can generate up to five output voltage levels [6], [10], [11].

Different control strategies have been proposed for single-phase applications. Meanwhile, with the development of advanced and fast microprocessors, the implementation of complex control algorithms such as model predictive control can be considered. The main concept of MPC control is based on the utilization of the discrete characteristics of power electronic converters. The system outputs' future behavior is predicted based on the mathematical model of the system and present values of converter outputs. In addition, system constraints and purposes are employed through a cost function. At each sampling time, the cost function is evaluated for different switching states, and the optimized solution will be utilized to determine the next switching state of the converter [12]–[14].

Among various MPC techniques, finite control set MPC (FCS-MPC) is a popular MPC strategy. In this strategy, the output voltage vectors of a converter are considered as the control inputs, which are applied to handle the optimal control problem [10], [15]–[17]. Fast-dynamic response, simple implementation, and the ability to integrate advanced control objectives in the design of its cost function can be counted as the features of FCS-MPC [13], [14], [18]. Considering the elimination of switching modulators and inner control loops and their delays by employing the FCS-MPC, the system can be operated with higher bandwidth [19].

In this paper, an MPC based on hysteresis current control is proposed for a single-phase five-level hybrid ANPC inverter. A hysteresis band for the current of the filter inductor is defined to limit the current ripple. The proposed method is implemented by utilizing both a microprocessor and an FPGA. Thus, the overall sampling frequency of the controller is set to 100 kHz. A prototype inverter with two different switch technologies, including Si and SiC MOSFETs, is selected to validate the controller.

Structure of single-phase five-level Hybrid ANPC Inverter

The structure of the single-phase hybrid ANPC is shown in Fig.1. The converter is comprised of eight switches and two dc-link capacitors. The hybrid switching states of this converter are listed in Table I. As it can be seen in Table I, five voltage levels at the output (v_{ab}) of the converter can be obtained. During the positive half cycle S_5 and S_8 are conducting, while S_6 and S_7 are in on-state condition at the negative half cycle. Thus, the switching frequency of S_5-S_8 is fundamental (low frequency switches). Meanwhile, during the positive and negative half cycles, the switching states of S_1-S_4 (high frequency switches) are altered to obtain different output voltage levels. Due to devices switching frequencies, two different switch technologies can be considered for low and high frequency switches. In this study, SiC and Si MOSFETs are employed in the structure of the converter.

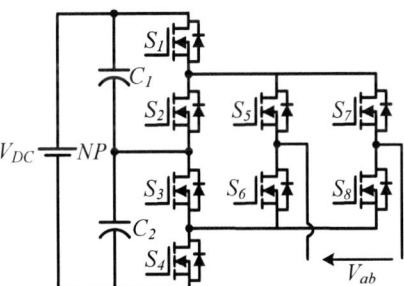

Fig. 1: The structure of the hybrid active neutral point clamped inverter.

Table I: The switching states of 5-level single-phase ANPC

Output Vector	Switching states								v_{ab}
	S_1	S_2	S_3	S_4	S_5	S_6	S_7	S_8	
V_1	1	0	0	1	1	0	0	1	V_{dc}
V_2	1	0	1	0	1	0	0	1	$V_{dc}/2$
V_3	0	1	0	1	1	0	0	1	$V_{dc}/2$
V_4	0	1	1	0	1	0	0	1	0
V_5	0	1	1	0	0	1	1	0	0
V_6	1	0	1	0	0	1	1	0	$-V_{dc}/2$
V_7	0	1	0	1	0	1	1	0	$-V_{dc}/2$
V_8	1	0	0	1	0	1	1	0	$-V_{dc}$

The general schematic of the system is shown in Fig. 2. As it can be seen, the load is supplied through an LC filter. To control the NP voltage, the effect of each vector on the NP voltage variation should be considered. The output vectors can be divided into three groups: large, small, and zero vectors. The large

(V_1, V_8) and zero (V_4, V_5) do not affect the NP voltage. Meanwhile, the small vectors (V_2, V_3, V_5, V_6) can change the NP voltage, in which the output current passes through one of the dc-link capacitors. Table II shows the effect of small vectors on the NP voltage [10].

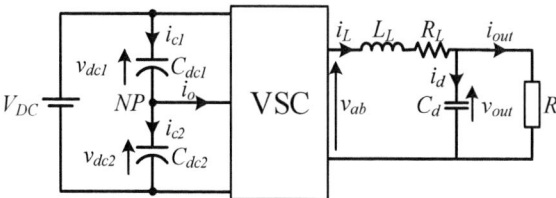

Fig. 2: The general schematic of the system.

Table II: The effect of small vectors on the NP voltage.

Vector	Output current	$v_{NP}=v_{dc1}-v_{dc2}$
V_2	$i_L>0$	↓
	$i_L<0$	↑
V_3	$i_L>0$	↑
	$i_L<0$	↓
V_6	$i_L>0$	↑
	$i_L<0$	↓
V_7	$i_L>0$	↓
	$i_L<0$	↑

Mathematical Modeling of the System

The mathematical model of the converter, including AC and DC sides, should be obtained to develop an MPC system. Considering the schematic of the system shown in Fig. 2, the voltage across the load (v_{out}) and filter inductor current (i_L) can be shown as state-space as follows [10]:

$$\frac{d}{dt}\begin{bmatrix} i_L \\ v_{out} \end{bmatrix} = A \begin{bmatrix} i_L \\ v_{out} \end{bmatrix} + B \begin{bmatrix} v_{ab} \\ i_{out} \end{bmatrix} \quad A = \begin{bmatrix} -\dfrac{R_L}{L_L} & -\dfrac{1}{L_L} \\ \dfrac{1}{C_d} & 0 \end{bmatrix}, B = \begin{bmatrix} \dfrac{1}{L_L} & 0 \\ 0 & -\dfrac{1}{C_d} \end{bmatrix} \tag{1}$$

where L_L, R_L are the inductance and resistance of filter inductor, and C_d is the value of the filter capacitor.

For the DC side of the converter, whenever the small vectors are applied the NP current (i_o) is equal to the filter inductor current. Thus, the voltage of dc-link capacitors can be obtained through the following equations [10]:

$$i_{c1} = C_{dc1}dv_{dc1}/dt \tag{2}$$

$$i_{c2} = C_{dc2}dv_{dc2}/dt \tag{3}$$

Moreover, the NP voltage is defined as:

$$v_{NP} = v_{dc1} - v_{dc2} \tag{4}$$

Assuming the sampling frequency is relatively high, the equations can be discretized based on the discrete step time of T_s. For example, the voltages of dc-link capacitors are discretized as follows:

$$v_{dc1}(k+1) = v_{dc1}(k) + \frac{T_s}{C_{dc1}} i_{c1}(k) \tag{5}$$

$$v_{dc2}(k+1) = v_{dc2}(k) + \frac{T_s}{C_{dc2}} i_{c2}(k) \tag{6}$$

which $v_{dc1}(k+1)$ and $v_{dc2}(k+1)$ are the predictive voltages of upper and lower dc-link capacitors at the $(k+1)^{th}$ instant.

MPC Algorithm

The cost function for MPC should be defined to achieve the desired purposes, such as tracking the reference voltage, keeping the current ripple of the filter inductor in the hysteresis band, and regulating the NP voltage. Consequently, the cost function is comprised of three subfunctions defined as follows:

$$CF: |v_{out}(n+1)^* - v_{out}(n+1)| + \lambda_1 (|v_{dc1}(n+1) - v_{dc2}(n+1)|) + hys(n+1) \tag{7}$$

where λ_1 is the weighting factor to balance the NP voltage. The first subfunction is used for tracking the reference voltage. The aim of the second subfunction is to keep the NP voltage constant and zero. The subfunction of $hys(n+1)$ is employed to reduce the current ripple in the filter inductor, which is defined as follows:

$$hys(n+1) = \begin{cases} \infty, & \text{if } |i_L(k+1)| > |i_L^*(k+1)| + lim \ \& \ |i_L(k+1)| < |i_L^*(k+1)| - lim \\ 0, & \text{if } |i_L(k+1)| < |i_L^*(k+1)| + lim \ \& \ |i_L(k+1)| > |i_L^*(k+1)| - lim \end{cases} \tag{8}$$

where lim is the hysteresis band as it is shown in Fig. 3.

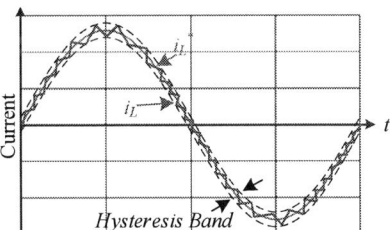

Fig. 3: The defined hysteresis band of the filter inductor current

Each of the output vectors (V_i, i=1-8) has a different effect on the value of the cost function (7) through (1), in which v_{ab} is replaced with the output voltage vectors. As mentioned earlier, the small vectors change the NP voltage and, consequently, the value of the cost function through (2) and (3). In each step, the value of the cost function is calculated for all the vectors, and a vector with the minimum related cost function is chosen to apply.

Experimental Results

To verify the performance of the proposed MPC, different experiments are performed. The prototype of the converter, including an LC filter, is shown in Fig. 4. To control the converter, dSPACE Micro-Lab Box ds1202 is used. The suggested method is implemented in the dSPACE microprocessor with a frequency of 100 kHz. Meanwhile, the dSPACE FPGA is employed to apply the selected vector with consideration of deadtimes between switches. The parameters of the built system are listed in Table III.

The steady-state experimental results of the system are shown in Fig. 5. Fig. 5(a) shows the output voltage (v_{out}), the inductor current (i_L), and the load current (i_{out}). In addition, Fig. 5(a) includes a zoomed view of the inductor current during the peak current.

The voltages of dc-link upper and lower capacitors (v_{dc1}, v_{dc2}) and the output of the converter (v_{ab}) are shown in Fig. 5 (b). It can be seen, the proposed MPC regulates the NP voltage and controls the output voltage of the converter.

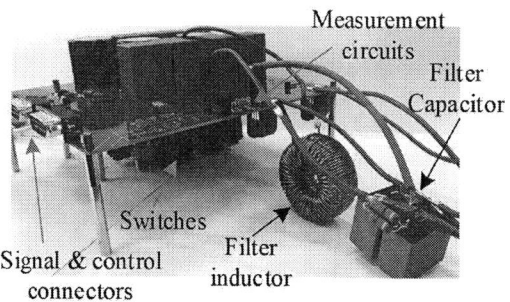

Fig. 4: The single-phase 5-level hybrid ANPC inverter with an LC filter.

Table III: The system parameters

Parameter	Definition	value
V_{DC}	DC supply voltage	200 V
V_{out}	The voltage across the load	110 V (rms)
C_1, C_2	DC-link capacitors	1 mF
L	Filter inductance	600 μH
R_L	The resistance of filter inductor	0.1 Ω
C_d	Filter capacitor	2 μF
R	Load	20 Ω
f	Reference output frequency	50 Hz
T_s	Sampling time	10 μs

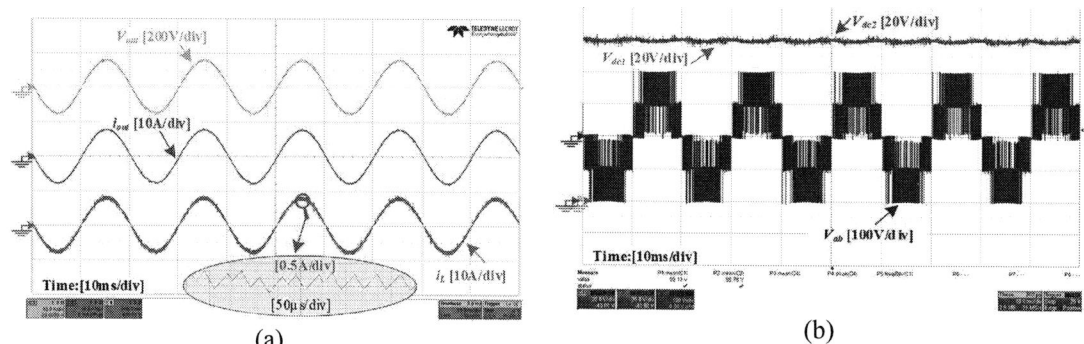

(a)　　　　　　　　　　　　　　　(b)

Fig. 5: The steady-state results of proposed MPC (a) The inductor current, output voltage, and load current, (b) the voltages of dc-link upper and lower capacitors and output voltage of the converter.

The harmonics spectrum of load current is shown in Fig. 6. The main harmonics are in the frequency range up to 30kHz.

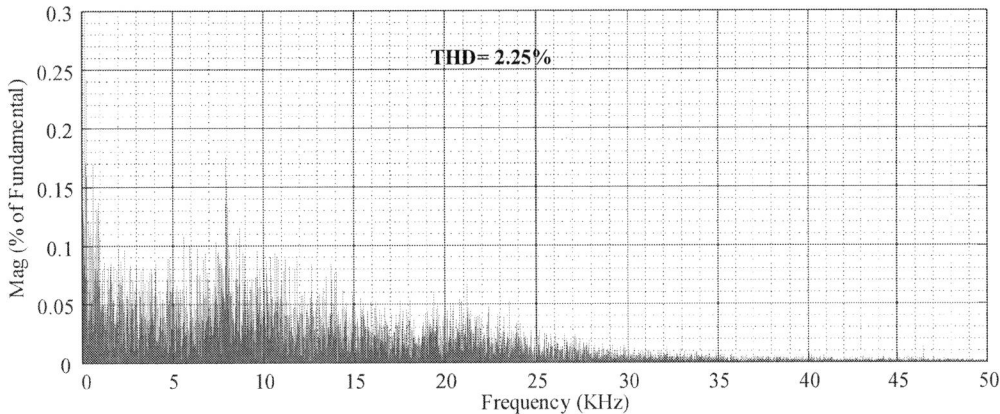

Fig. 6: The harmonics spectrum of output current.

To evaluate the performance of the suggested method during the transition, a step-change in the load value is introduced. The output voltage, the current of the filter inductor and the load are shown in Fig. 7. As it can be seen, the proposed MPC is able to keep the output voltage constant quickly and to follow the reference waveform.

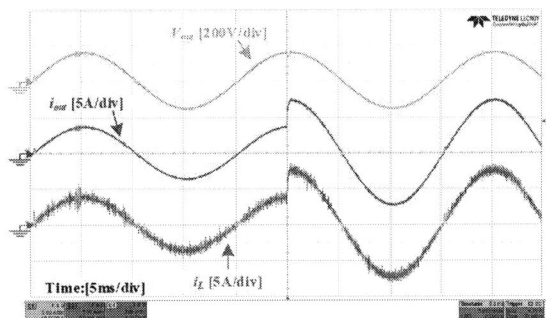

Fig. 7: The dynamic experimental results of proposed MPC when the load is changed.

Conclusion

In this paper, an MPC based on the hysteresis current control for a single-phase 5-level hybrid ANPC has been proposed. In the structure of the converter, SiC MOSFETs are utilized to increase the switching frequency of the converter in order to reduce the size of the filter and current ripples. The proposed MPC was implemented with an FPGA to have superior performance and fast dynamic response. Experimental results verify the performance of the proposed MPC both at steady-state and transition conditions.

References

[1]   Leon J. I., Vazquez S., Franquelo L. G., "Multilevel Converters: Control and Modulation Techniques for Their Operation and Industrial Applications," *Proc. IEEE*, vol. 105, no. 11, pp. 2066–2081, Nov. 2017.

[2]   Rodriguez J., Lai Jih-Sheng, Fang Zheng Peng, "Multilevel inverters: a survey of topologies, controls, and applications," *IEEE Trans. Ind. Electron.*, vol. 49, no. 4, pp. 724–738, Aug. 2002.

[3]   Najjar M., Kouchaki A., Nielsen J., Lazar R. Dan, Nymand M., "Design Procedure and Efficiency Analysis of a 99.3% Efficient 10 kW Three-Phase Three-Level Hybrid GaN/Si Active Neutral Point Clamped Converter," *IEEE Trans. Power Electron.*, vol. 37, no. 6, pp. 6698–6710, 2022.

[4]   Malinowski M., Leon J. I., Abu-Rub H., "Solar Photovoltaic and Thermal Energy Systems: Current Technology and Future Trends," *Proc. IEEE*, vol. 105, no. 11, pp. 2132–2146, Nov. 2017.

[5]   Kjaer S. B., Pedersen J. K., Blaabjerg F., "A review of single-phase grid-connected inverters for photovoltaic modules," *IEEE Trans. Ind. Appl.*, vol. 41, no. 5, pp. 1292–1306, 2005.

[6] Zhang L., Sun K., Feng L., Wu H., Xing Y., "A family of neutral point clamped full-bridge topologies for transformerless photovoltaic grid-tied inverters," *IEEE Trans. Power Electron.*, vol. 28, no. 2, pp. 730–739, 2013.

[7] Jacobina C. B., Dos Santos E. C., Rocha N., Fabrício E. L. L., "Single-phase to three-phase drive system using two parallel single-phase rectifiers," *IEEE Trans. Power Electron.*, vol. 25, no. 5, pp. 1285–1295, 2010.

[8] Song W., Feng X., Smedley K. M., "A carrier-based pwm strategy with the offset voltage injection for single-phase three-level neutral-point-clamped converters," *IEEE Trans. Power Electron.*, vol. 28, no. 3, pp. 1083–1095, 2013.

[9] Ortmann M. S., Mussa S. A., Heldwein M. L., "Generalized analysis of a multistate switching cells-based single-phase multilevel PFC rectifier," *IEEE Trans. Power Electron.*, vol. 27, no. 1, pp. 46–56, 2012.

[10] Najjar M., Shahparasti M., Heydari R., Nymand M., "Model Predictive Controllers With Capacitor Voltage Balancing for a Single-Phase Five-Level SiC/Si Based ANPC Inverter," *IEEE Open J. Power Electron.*, vol. 2, 2021.

[11] Najjar M., Shahparasti M., Kouchaki A., Nymand M., "Operation and Efficiency Analysis of a Five-Level Single-Phase Hybrid Si/SiC Active Neutral Point Clamped Converter," *IEEE J. Emerg. Sel. Top. Power Electron.*, vol. 10, no. 1, 2022.

[12] Cortés P., Kazmierkowski M. P., Kennel R. M., Quevedo D. E., Rodriguez J., "Predictive control in power electronics and drives," *IEEE Trans. Ind. Electron.*, vol. 55, no. 12, pp. 4312–4324, 2008.

[13] Yang Y., Wen H., Fan M., Xie M., Chen R., Wang Y., "A Constant Switching Frequency Model Predictive Control Without Weighting Factors for T-Type Single-Phase Three-Level Inverters," *IEEE Trans. Ind. Electron.*, vol. 66, no. 7, pp. 5153–5164, 2019.

[14] Vazquez S., Acuna P., Aguilera R. P., Pou J., Leon J. I., Franquelo L. G., "DC-Link Voltage-Balancing Strategy Based on Optimal Switching Sequence Model Predictive Control for Single-Phase H-NPC Converters," *IEEE Trans. Ind. Electron.*, vol. 67, no. 9, pp. 7410–7420, 2020.

[15] Zhang X., Tan G., Xia T., Wang Q., Wu X., "Optimized Switching Finite Control Set Model Predictive Control of NPC Single-Phase Three-Level Rectifiers," *IEEE Trans. Power Electron.*, vol. 35, no. 10, pp. 10097–10108, Oct. 2020.

[16] Ramírez R. O. *et al.*, "Finite-State Model Predictive Control With Integral Action Applied to a Single-Phase Z-Source Inverter," *IEEE J. Emerg. Sel. Top. Power Electron.*, vol. 7, no. 1, pp. 228–239, Mar. 2019.

[17] Novak M., Nyman U. M., Dragicevic T., Blaabjerg F., "Analytical Design and Performance Validation of Finite Set MPC Regulated Power Converters," *IEEE Trans. Ind. Electron.*, vol. 66, no. 3, pp. 2004–2014, Mar. 2019.

[18] Vazquez S. *et al.*, "Model Predictive Control for Single-Phase NPC Converters Based on Optimal Switching Sequences," *IEEE Trans. Ind. Electron.*, vol. 63, no. 12, pp. 7533–7541, 2016.

[19] Heydari R., Dragicevic T., Blaabjerg F., "High-Bandwidth Secondary Voltage and Frequency Control of VSC-Based AC Microgrid," *IEEE Trans. Power Electron.*, vol. 34, no. 11, pp. 11320–11331, Nov. 2019.

AUTHOR INDEX

Abdalrahman, Adil 2241, 3282, 3757
Abdullah, Ahmed.. 554
Abedini, Hossein... 865
Aceña, Javier Cañas... 484
Adabi, Jafar.. 2537
Addin, Ali Sharaf... 1824
Afonso, Luciana C. 4018
Aganza-Torres, Alejandro.................................... 1328
Agarwal, Ritika.. 3615
Agirrezabala, Eneko 3327
Aguglia, D. 1955
Ahmed, Emad M... 1015
Aiello, Giuseppe 2628
Aillerie, Michel... 315
Aizpuru, I... 2903
Aizpuru, Iosu 325, 3327, 3574, 3750
Akuru, Udochukwu B... 2958
Al-Haddad, Kamal 1025
Alaluss, Mohamed 1424
Alatise, Olayiwola 1497, 2477
Albrecht, Fabian 2726
Aldarmon, Mohamed 2574
Ali, Mohammad 2392, 3022
Ali, Ramy 390
Ali, Rana Asad 698
Allard, Bruno 169, 3862
Allioua, Abdelmoumin 2835
Alvarez, Asier 279
Alvarez-Herault, Marie-Cecile 1147
Alves, Wendell Da Cunha....................................... 1046
Alvi, Muhammad H 1692
Aly, Mokhtar 1015
Andersen, Michael A. E... 1561
Ando, Y. 1785
Andresen, Jan... 1684
Ansari, Sajad A. 3440
Antonopoulos, Antonios 297, 432
Anzola, J. 2903, 2967
Anzola, Jon 3574
Apostolidou, Nena 1796
Appel, Tobias... 1121
Apte, Pramod 2773
Arabsalmanabadi, Bita 1025
Arias, Manuel 152
Arrizabalaga, Antxon... 325
Arrozy, Juris 681
Arruti, Asier 3574, 3750
Artal-Sevil, J. S. 2903, 2967

Arza, Joseba 484, 2011
Asllani, Besar 2515
Asoodar, Mohsen 2843
Atzler, Frank 3391
Aunsborg, Thore Stig 825
Aviñó, Oriol 2715
Ayarzaguena, Ibán..................................... 1765, 3336
Aztiria, Jon 325
Baars, Nico 2788
Babin, Anthony 3696
Baburske, Roman 1424
Bacha, S. 2422, 3179
Bacha, Seddik.. 3140, 3928
Bacheti, Gabriel Gaburro 421
Bachmann, Matthias.. 3501
Badenhop, Niklas.......................................145, 1939
Baek, Seung-Hyuk 2877
Bagaber, Bakr.. 3037, 3711
Baimel, D. 3254
Baimel, N. 3254
Bak, Claus Leth 2504
Bakhos, Gianni 3928
Bakran, Mark-M.......................... 805, 1036, 2744
Bakri, Reda 1046
Balachandran, Arvind 1456
Balasubramanian, Sridhar 2030
Ballestín-Bernad, V.. 2903, 2967
Banana, Shady.. 1064
Banavath, Satish Naik 730
Banda, Joseph...187, 289
Barba, V. 1975
Barbi, Eli 3254
Barg, Sobhi 361
Barman, Subhranil.. 2462
Barón, Kevin Muñoz.. 2698
Bashar, Erfan 2477
Basic, Duro 125
Basler, Michael 242
Basler, Thomas.................... 1424, 1713, 1733, 3373
Bauer, Luca 971
Bauer, Pavol............................... 1319, 3607, 3729
Baumann, Michael 1167
Baumann, Timm Felix 2355
Bäumler, Christian 1733
Bayer, Markus 115
Bayhan, Sertac 3518
Bayram, Islam Safak 3518
Beck, Simon 1434, 2038

Beckemeier, Christian..2327
Beczkowski, Szymon Michal.......................................2661
Beineke, Stephan 3501
Beiranvand, Hamzeh.................... 833, 3092, 3846, 3966
Belhaouane, Mohamed Moez 582
Benchaib, Abdelkrim..3928
Bendfeld, Christian 1620
Benech, Philippe 169
Bensetti, Mohamed 3883
Bergmann, Lukas...1036
Bergveld, Henk Jan..3796
Bermejo, Jose Manuel..................................... 1765, 3336
Bernal, Carlos 3327
Bernal-Agustín, J. L...2967
Bernal-Ruiz, Carlos 3750
Bernichon, Thomas..3920
Bertilsson, Kent 361
Bertin, Matthieu 534
Beukes, Johan 3112
Beye, Mamadou Lamine...2736
Beza, Mebtu 1187
Bezerra, Vinicius Freire...2689
Bhatnagar, Pallavee 3804
Bhattacharya, Arghyadip 178
Bhoi, Sachin Kumar...3031
Biadene, Davide..865
Biela, Juergen 1402
Biela, Jürgen 651, 662, 933, 1391, 1434, 2038, 2544
Bieler, Arne 1121
Bier, Anthony 922, 2736
Billa, Laxma R...2301
Bimmel, Luc 2736
Binder, Andreas...2316
Bitsi, Konstantina 3246
Blaabjerg, Frede.............. 2110, 2182, 2496, 2504, 2939
Blanes, J. M. 3382, 3401
Blank, Thomas...232
Blanquez, Francisco R. 2189, 2451
Blasco-Gimenez, Ramon 2189, 2451
Blasuttigh, Nicola 3846
Blatsi, Zoe.. 2824, 3813
Blömeke, Alexander 4025
Böcker, Joachim 2276, 2432, 2754, 3625, 3686
Bockholt, Jan...1286
Boettcher, Norman...1128
Bohllaender, Marco 4016
Bohne, David 514
Boige, Francois 944
Boisson, Guillaume Piquet...960
Bolzoni, A...1371
Bongiorno, Massimo..1187
Bonten, Remco 634

Böorngen, Hannes 1754
Borcherding, Holger 2852
Börngen, Hannes 3362
Boroyevich, Dushan...2806
Bosch, Swen..2219
Bosga, Sjoerd G. 3246
Bouscayrol, Alain..2175
Boutleux, Emmanuel..251
Boutry, Arthur...2515
Brabetz, Ludwig...2383
Branco, Cesar Augusto Santana Castelo 2948
Braun, Gerrit 2205
Braz, Cesar 1445
Briff, Pablo 451
Bringezu, Thilo 662
Brinker, Tobias..2977
Brogioli, Doriano Constantino 833
Brommer, Volker 2726
Bronstein, S...3254
Brooks, Michael 279
Brückner, Thomas..1824
Brulin, Pierre-Yves..3831
Brunner, Andreas 593
Brunner, Frank 3775
Brüns, Michael 474
Bruyere, Antoine...1046
Bruyere, Paul 960
Bucarey, Victor 1074
Budo, Kohei 213, 351
Bueno, Emilio José 421
Bueno-Mariani, Guilherme 3272
Bugarski, Stevan 2334
Bünte, Andreas 380
Burgos, Rolando..1692, 2806
Burgos-Mellado, Claudio.................................1074, 3429
Burkart, Ralph M. 203
Burke, Richard 3696
Bushra, Rehnuma...2392, 3022
Busquets-Monge, Sergio 2715
Buticchi, Giampaolo 3014
Buttay, Cyril 2049, 2515
Byen, Byengjoo 1207
Caarls, Esin Ilhan 681
Cabrera, Michel 169
Cacciato, Mario 2628
Caillierez, Antoine 3883
Cajander, D. 1955
Cakal, Gokhan 3947
Caldognetto, Tommaso 865
Camargo, Renner Sartório...421
Camurca, Luis 3101
Can, Görkem 3092

Cano, Tania C. 335
Cao, Jingming 3215, 3225
Cao, Yongtao 2003
Cappelle, Jan 1300
Cárcamo, Alberto 1083
Carcouet, S. 843
Carpita, Mauro 1543
Carrasco, Miguel 370
Casado, P. 3382, 3401
Castellazzi, Alberto 689, 2156, 2285, 2402, 2893, 3084
Castelli-Dezza, Francesco 1476
Castro, Ignacio 335
Catalán, Pedro 2011
Catellani, Stéphane 922, 990
Ceccarelli, Lorenzo 681
Chakraborty, Sajib 2101, 3031
Chang, Che-Wei 1692
Charkaoui, Abdelmouneim 442
Chatterjee, Kishore 178, 2462
Chen, Zhe 2011
Chen, Zhu 3235
Chevalier, Florian 3582
Chida, Makoto 1580
Chinthavali, Madhu Sudhan 344
Chiumeo, Riccardo 3206
Choksi, Kushan 344
Choudhury, Soham 1966
Chub, Andrii 730
Cimetiere, Xavier 1046
Clerc, Guy 251
Clerici, Alessio 3206
Cobaleda, Diego Bernal 2581
Cogitore, Bruno 1216
Colmenero, Manuel 2189, 2451
Cosso, Simone 2919
Coumont, Martin 1966
Crovetti, Paolo 554
Cui, Yi 3986
Czerwenka, Philipp 593
Dahmen, Christopher 1824, 1855
Damian, Ioan Catalin 2266
Damm, Gilney 3590
Danielsson, Christer 2843
Dargahi, Vahid 2073
Davari, Pooya 2496
Davidson, Jonathan N. 3440
De Bernardinis, Alexandre 315
De Carne, Giovanni 3014
De Cesaris, Ivan 223
De Donato, Giulio 1569
De Doncker, Rik W. 709, 1266, 2119, 3599, 3676,
.... 3740, 3766, 3893

De Lillo, Liliana 3450
De Matos, Jose Gomes 2948
De Oliveira, Eduardo Facanha 2441
De, Dipankar 689
Deb, Arkadeep 1497
Deblecker, Olivier 504
Deboy, Gerald 3984
Deck, Patrick 514
Deckers, Martijn 2795
Degaa, Laid 3696
Delette, Gérard 922
Deng, Kai 3235
Dennetiere, Sébastien 582
Derammelaere, Stijn 3344
Despouys, Olivier 2486
Dick, Christian P. 514
Dickmann, Stefan 758
Dieckerhoff, Sibylle 1466, 2596, 2607, 2644, 3775
Dieng, A. 2092, 2930
Dierks, Rebecca 1533
Dietrich, Tim-Hendrik 1094
Disselkamp, Simon 2912
Domae, Shinichi 3084
Domes, Daniel 2744
Domes, Konrad 1137
Dong, Chaoyu 3215, 3225
Dong, Dong 1692, 2515
Dong, Jianning 1319
Dong, Tenghui 3084
Dorner, Oscar 1177
Dos Santos, Pedro Leal 604
Dragicevic, Tomislav 2496, 2939, 3429
Drexler, Christoph 411, 1167
Driesen, J. 3655
Driesen, Johan 2795
Drimizi, Youssef 2869
Drissi, Khalil El Khamlichi 3786
Duarte, Jorge L. 681, 798
Duarte, Jorge 2788
Duchamp, Jean-Marc 169
Dujic, Drazen 2049
Dumtzlaff, Jacob 1865
Duquesne, Thierry 3582
Dürbaum, Thomas 88, 307
Duun, Sune Bro 825
Dworakowski, P. 2422
Dworakowski, Piotr 2049
Ebel, Thomas 3130
Ebner, Kathrin 4015
Eckart, Martin 3646
Eckel, Hans-Guenter 3460

Eckel, Hans-Günter 11, 59, 70, 980, 1294, 1703,
.... 1744, 1885, 1895, 2308, 4003
Eckstein, Mattea 1277
Effenberger, Thomas 1754
Eggers, Malte 1466
Ehlich, Martin 2852
El Baghdadi, Mohamed 2101, 2293, 3031
El Sherif, Alaa 3796
El-Refaie, Ayman 719, 1692
Ellinger, Thomas 2885
Emmers, G. 3655
Emmers, Glenn 2795
Empringham, Lee 3450
Encarnação, Lucas Frizera 421
Endo, Yusuke 2285
Epping, Daniel 749
Erckrath, Tobias 1350, 1620
Eremia, Mircea 2266
Eriksson, Lars 1456
Erlbacher, Tobias 1128
Ernst, Alexander 3149, 3159
Es-Seghier, Hajar 922
Escoffier, René 990
Etoz, Burhan 1497
Faber, Samuel 307
Falchi, Daniele 2486
Faramehr, Soroush 3822
Farhangi, Shahrokh 787
Fauth, Leon 2003, 2638, 3838
Fayolle-Lecocq, Murielle 990
Fazli, Nastaran 11
Fehr, Hendrik 49, 3391
Felgemacher, Christian 442, 4004
Fernández, Arturo 152
Ferreyra, Fabio 554
Festerling, Tobias 1237
Finney, Stephen 80, 3470, 3813
Fischer, Katharina 1674, 1804
Fischer, Manuel 749
Fischer-Baeumer, Rico 1137
Fölkel, Lorandt 279
Formentini, Andrea 2919, 3975
Forouzesh, Mojtaba 1590, 1601
Forsstrom, Ville 3301
Förster, Nikolas 2432
Foster, Martin P. 3353, 3440
Foteinopoulos, Georgios 1985
Fräger, Lukas 145, 641, 1939, 2588, 2773
Frank, S. R. 3411
Franzki, Jonas 261
Frey, David 1147
Fricke, Tobias 1247

Fricke, Torben 1381
Friebe, Jens 1914, 2003, 2327, 2392, 2588, 2638,
.... 2655, 2689, 2773, 2977, 3022, 3059, 3545, 3838
Fritze, Eric 758
Fröhling, Sören 1674
Fuchs, Simon 1434, 2038
Fuhrmann, Jan 980
Fukunaga, Shuhei 108
Ganeshpure, Dhanashree Ashok 3729
Gao, Xiang 3014
Gaona, Daniel 2441
Garces, Santiago Ramos 3344
Garcia, Raul Murillo 2355
Garrigós, A. 3382, 3401
Gaubert, Jean-Paul 1525
Gauthier, Jean-Yves 3862
Gavelle, Mathieu 2618
Gehl, Adrian 2912
Geiss, Michael 2554
Gemma, Filippo 3975
Geng, Weiwei 3722
Geng, Xiaomeng 2596, 2644, 3775
Gennaro, Francesco 2628
Gensior, Albrecht 49, 370, 3391
Gerges, Tony 169
German, Ronan 2175
Germishuizen, J. J. 3318
Geury, Thomas 2101
Gholami, M. 3179
Gholami, Mehrdad 3140
Ghumman, Sukhjit S 2763
Gieraths, Antje 767
Gierschner, Magdalena 1294
Gierschner, Sidney 11
Gillon, Frédéric 1046
Girona-Badia, Jaume 3704
Glaser, Martin 4020
Gleissner, Michael 805, 1036
Gnärig, Lasse 370
Goetz, Stefan 1025, 1064, 1197, 3636, 3665
Gohler, Katherina 1804
Gohrmann, Kai 1137
Golev, Victor 1286
Goller, Maximilian 1733
Gomes, Lucas Vinícius De Araújo 3059
Gomes, Zariff Meira 3590
Gómez, Alexis A. 1765, 3336
Gomis-Bellmunt, Oriol 2486, 3704
Gonzalez, Jose Ortiz 1497
Gonzalez-Hernando, Fernando 3938
Gonzalez-Torres, Juan-Carlos 3928
Götz, Georg Tobias 709

Gräber, Hendrik	2977
Grabs, Volker	97
Gradinger, Thomas B.	203
Grant, Thomas	2301
Grass, Norbert	2366
Grau, Vivien	854
Gremme, Florian	4021
Griepentrog, Gerd	160, 2780, 2835
Grodnichev, Anton	624
Groke, Holger	3169
Groon, Fabian	3092
Groten, Jonas	279
Gruson, François	582
Guerrero, Bruno	944
Gui, Qiuye	49
Guillaud, Xavier	582
Günes, Ece Olcay	1361
Gupta, Kirti	2110
Gupta, Krishna Kumar	3615, 3804
Gutierrez, Alonso	2618
Haag, Felix	2726
Haake, Daniel	624
Haarer, Jörg	971, 1237, 1277
Habersetzer, Antoine	4015
Hably, A.	3179
Hably, Ahmad	3140
Hackl, Philipp	39
Haederli, Christoph	3282
Häfner, Ying-Jiang	2241, 3282, 3757
Hagedorn, Maximilian	1875
Hajar, K.	3179
Hajar, Khaled	3140
Hajian, Masood	468
Hakkila, Akseli	297
Hald, Alex	380
Hameyer, Kay	3005, 3235
Hammes, David	11
Handt, Karsten	2607
Hanf, Michael	3169
Hanisch, Lucas Vincent	261
Hanisch, Lucas	1094
Hänsel, Stefan	572
Hansen, Sandra	3966
Hanson, Alex J.	1722
Hanson, Jutta	1966
Hao, Chuantong	80, 3470
Hardan, Faysal	468
Harmand, Souad	2996
Hasan, Md. Mahamudul	3031
Hasler, J. P.	1371
Hassan, Tayssir	1466
Hatori, K.	1785

Hatori, Kenji	777
Hattori, Takato	739
Hauenschild, Philipp	1506
Haug, Martin	279, 698
Hayes, John G.	2470
Hegazy, Omar	2101, 2293, 3031
Heide, Daniel	3711
Heien, Christian	1294
Heimler, Patrick	1713
Hein, Yves	1294
Helmholdt-Zhu, Ting	97, 854
Hembel, Ahmed	3947
Henke, Markus	261, 1094, 2030
Henkenjohann, Jonas	1684
Henn, Jochen	3599
Henneberg, Dustin	2885, 3491
Herbold, Johannes	749
Hernando, Marta M.	1083, 1765, 3336
Herzog, Hans-Georg	952
Heydari, Rasool	2682
Hikihara, Takashi	108
Hiller, M.	3411
Hiller, Marc	115, 999
Hillmer, Hartmut	2383
Hilt, Oliver	2596, 2644, 3775
Himker, Niklas	1631
Himmelmann, Patrick	999
Hiraki, Eiji	2164
Hirning, David	971, 1237, 1277, 3536
Hissel, Daniel	315
Hjerrild, Jesper	2504
Hoerner, Michael	1754
Hofer, Heimo	1445
Hofer, Matthias	2251
Hoff, Bjarte	3198
Hoffmann, Klaus F.	758, 2726, 3188
Hoffmann, Madlen	3262
Hoffstadt, Thorben	1157
Hofmann, Viktor	195, 400
Hofmann, Wilfried	3957
Hofstetter, Patrick	195, 400
Hölscher, Jonas	2432
Holtje, Pauline	1665
Holzke, Wilfried	3149, 3159, 3169
Horn, Markus	2383
Hortans, Magnus	3309
Hoshi, Nobukazu	1776, 1844
Hosseinabadi, Farzad	3031
Hosseini, Elham	1025
Hou, Jingning	3722
Houwen, Simon	3344
Hridya, I	187

Hu, Anliang	651
Hu, Bin	2182
Hu, Xiaowei	3722
Huang, Jiasheng	1561
Huerta, Gabriel Ramos	1226
Huesgen, Till	2230
Huisman, Henk	634, 673, 681
Hutzler, Michael	1445
Idir, Nadir	2996, 3582, 3822
Igic, Petar	3822
Iida, Masaki	2164
Iman-Eini, Hossein	787
Imgart, Paul	1187
Incurvati, Maurizio	223, 268
Inoue, Michiko	3420
Iraola, Unai	3327
Ishihara, Mastaka	2164
Itoh, Jun-Ichi	902, 1104, 2127
Ittamveettil, Hridya	289
Izurza, Pedro	484
Jaber, Hamzeh J.	2156, 2285, 3084
Jacques, Dries	3344
Jagannath, Sriram	3362
Jahdi, Saeed	1497, 2477
Jain, Anekant	3615
Jain, Sanjay K.	3615, 3804
Jamal, Adeel	2780
Jaman, Shahid	3031
Jankovic, Marija	442
Jayathurathnage, Prasad	1947
Jena, Kasinath	3804
Jenhani, Firas	1343
Jeong, Byunghwang	1207
Jeschke, Sebina	3235
Jha, Kapil	187, 289
Jia, Hongjie	3215, 3225
Jia, Ming	1266
Joebges, Philipp	1266
Johansson, N.	1371
Johnson, C. Mark	3450
Jonsson, Tomas	1456
Jordà, Xavier	2715
Jørgensen, Asger Bjørn	825, 1641, 2661
Jöst, Dominik	4025
Jovanovic, Raka	3518
Juchem, Ralf	4023
Judge, Paul	80
Junemann, Lennart	1665
Jung, Marco	624, 1515, 1611, 1620
Junghans, Christoph	3460
Junyent-Ferre, Adria	2574
Kabbara, Wassim	3883

Kacetl, Jan	1197, 3636, 3665
Kacetl, Tomáš	1197, 3636, 3665
Kacki, Marcin	2470
Kadem, Karim	3590
Kaerst, Jens Peter	544
Kaiser, Jeremias	307
Kallfass, Ingmar	2698, 3565
Kamel, Tamer	468
Kaminski, Nando	2230, 3149, 3169
Kamm, Simon	2698
Kampen, Dennis	145, 1939, 2588
Kamper, Maarten J.	2958
Karakasli, Vefa	2835
Karamanakos, Petros	297, 1476, 1754
Karau, Fabian	3292
Karnehm, Dominic	767
Karwatzki, Dennis	195
Kasten, Henning	3501
Kayser, Felix	59, 4003
Keilmann, Robert	891
Kempchen, Malte	2912
Kemper, Philipp	749
Kennel, Ralph	1754, 2366, 3362
Kerekes, Tamas	1933
Keshavarzi, Davood	1064
Khader, Meriem	2655
Khan, Basit Ali	2537
Khan, Mohammed Ali	135
Khan, Nameer	3796
Khan, Siam Hasan	484
Khanzadeh, Babak	2344
Khenfri, Fouad	3831
Kiehnle, Philip	999
Kiffe, Axel	1157
Kikuchi, Naoto	1104
Kim, Dong-Uk	1207
Kim, Sungmin	1207, 2877
Kinzer, Dan	3987
Kirsch, Andreas	380
Kitagawa, Wataru	739
Kjærsgaard, Benjamin Futtrup	825
Klee, Matthias	1515
Klever, Severin	3676
Klötzer, Sebastian	4011
Knebusch, Benjamin	1665, 3048
Ko, Youngjong	3014
Kobayashi, Hiroyasu	1580
Kocewiak, Lukasz	2504
Koch, Jan-Niklas	2852
Koczy, Dawid	3149
Kohlhepp, Benedikt	88, 307
Kojima, Tetsuya	3740

Kondo, Keiichiro 1580
Kondratenko, Dmytro 1906
Kopp, Tobias 912
Kormska, Tomáš 1114
Körner, Patrick 2021
Korthauer, Bastian 3625
Kosesoy, Yusuf 634
Kostka, Benedikt 1649
Kostynski, Daniel 3855
Koteich, Mohamad 534
Kouro, Samir 1015
Koutroulis, Eftychios 1985
Kowal, Julia 4014
Kragl, Robert 2554
Krick, Alexander 3989
Krigar, Tim 2375
Krishnamoorthy, Harish Sarma 730
Krüger, Helge 3966
Krümpelmann, Marcel 1631
Kubulus, Pawel Piotr 2661
Kuder, Manuel 767
Kumar, Amit 451
Kumar, Kaushik Naresh 1486
Kumar, Manish 3511
Kuperman, A. 3254
Kuprat, Johannes 3067
Kuring, Carsten 2596, 2644, 3775
Kurrat, Michael 912
Kurukuru, V S Bharath 135
Kusaka, Keisuke 1104, 2127
Kusche, Stephan 3704
Kusebauch, Manuel 3491
Küster, Pierre 411
Kwak, Jaedon 2893
Kyyrä, Jorma 1947
La Mantia, Fabio 833
Labonne, A. 3179
Labonne, Antoine 3140
Labrousse, D. 843
Lacerda, Vinícius Albernaz 3704
Laclaverie, Julien 944
Laforet, David 1445
Lamar, Diego G. 335, 1083, 1765, 3336
Lange, Jarren 2276
Lange, Yannic 2644
Langfermann, Sascha 1939, 2588
Lanzarotto, D. 2564
Larrañaga, Uxue 3938
Larrazabal, Igor 1765, 3336
Larsson, Anders 1456
Lataire, Philippe 2293
Laumen, Michael 3766

Lauri, Andrea 865
Laza, Saioa Burutxaga 370
Lazkano, Markel Zubiaga 484
Le Leslé, Johan 2526
Le Métayer, Pierre 2049
Lee, Jaehong 2877
Lee, Seung-Hwan 2877
Lee, Yonghwa 2402
Lefebvre, Bruno 2515
Lefevre, Guillaume 2526
Legay, Florian 3529
Lehn, Peter W. 1995, 2084, 2145, 2763
Leifert, Torsten 4013
Lemaire-Semail, Betty 2175, 2996
Lembeye, Yves 1216
Lenz, Kevin 442
Lenzen, Patrick 2413
Leuer, Michael 3292
Leuzzi, Riccardo 3975
Lévy, PE 843
Lewicki, Arkadiusz 1906
Lexow, Daniel 1744
Li, Feifei 3235
Li, Ke 3822
Li, Marui 3215, 3225
Li, Qiang 3722
Li, Weihan 4025
Li, Xiang 2301
Li, Xupeng 3373
Li, Zheming 2744
Liang, Mincui 3786
Lichtenstein, Timo 1674
Liebfried, Oliver 2726
Liegmann, Eyke 1754, 3362
Lievre, Aurelien 2175
Lin, Siqi 1914, 2638
Lin-Shi, Xuefang 3862
Lindemann, Georg 3555
Linder, Stefan 3992
Lippold, Florian 1506
Liserre, Marco 421, 833, 3014, 3067, 3092, 3101,
.... 3846, 3966
Liu, Chao 1561
Liu, Steven 604
Liu, Xing 1733, 3373
Liu, Yan-Fei 1590, 1601
Liu, Yining 1947
Llanos, Jacqueline 3429
Löfgren, Jonas 3920
Lombard, Philippe 169
López, Abraham 152
Lorenz, Andreas 814

Lorenz, Erwin 1167
Lorenz, Malte 1875
Lorenz, Oscar 873
Loudot, Serge 3883
Lu, Xuyang 3822
Lu, Yizhou 883
Luan, Shaokang 3309
Luckert, Franz 2706
Luecke, Stefan 3075
Luh, Matthias 232
Luo, Fang 344, 2860
Lusardi, Federico 3975
Lutsch, Michael 88
Lutz, Josef 1713
Lutzen, Hauke 2230
Ma, Wenhao 80
Maamri, Nezha 1525
Maibach, Philippe 3282
Maier, Robert W. 2744
Maitra, Abhishek 1424
Mallwitz, Regine 891, 912, 1094, 1247, 1506
Mambetow, Arthur 145
Manthey, Tobias 2655, 2689, 3059
Marca, Ygor Pereira 798
Marcaide, Inko 3920
Marcault, Emmanuel 2618
Marchesoni, Mario 2919
Margreiter, Thomas 223
Margueron, Xavier 1046
Marks, Hendrik 2030
Marquardt, Rainer 1855
Marroquí, D. 3382, 3401
Martin, Jérémy 990, 2736
Martinez, Wilmar 1914, 2197, 2581
Martinez-Garcia, Herminio 1056
Martinez-Padron, Daniel S. 1256
Martnez, Wilmar 2638
Marx, Philipp 1237, 1277, 3536
März, Martin 493, 3262
Mashaly, Aly 442
Mashayekh, Ali 767
Mathúna, Cian Ó 4006
Mattavelli, Paolo 865
Matthies, David 3159
Maussion, Pascal 2869
Maynard, X. 843
Mazuela, Mikel 325, 3327, 3574
Meddour, Aissam Riad 3696
Mehran, Kamyar 614, 3353
Mehrasa, M. 3179
Mehrasa, Majid 3140
Meier, Hans 2021

Meinert, Janus Dybdahl 825
Meissner, Michael 758, 3188
Mellor, Phil 2477
Mendoza-Araya, Patricio 1177, 1226
Meng, Qingchao 933
Menzel, Steffen 3169
Merlin, Michael M. C. 2824, 3813
Merlin, Michael 80, 3470
Mersche, Stefan 115
Mertens, Axel 641, 1350, 1533, 1631, 1649, 1665,
..1684, 1865, 1875, 2003, 2066, 2392, 2706, 3022, 3037, 3048,
3075, 3555, 3711
Miaja, Pablo F. 152
Mijatovic, Nenad 2496, 2939
Miller, T. J. E. 3318
Minami, Masataka 2285
Mir, Tabish Nazir 468
Mirza, Abdul Basit 344
Mirzadeh, Mina 1350
Mirzaeva, Galina 3903
Miskiewicz, Rafal 1486
Mistretta, C. 1975
Mita, Salvatore 2628
Mo, Wai Keung 3130
Möckel, Andreas 3391
Moench, Stefan 242
Mogorovic, Marko 203
Mohanta, MK Kharabela 689
Möhlenkamp, Georg 3993
Mohsenzade, Sadegh 614, 3353
Moldenhauer, Deniz-Heinz 2205
Mondal, Gopal 572
Mondzik, Andrzej 3804
Monmasson, Eric 1256
Mönninghoff, Sebastian 3005
Montero, E. Rodriguez 1834
Morales-Paredes, Helmo K. 1074
Morand, Julien 2526
Morel, F. 2422, 2564
Morey, Philippe 1543
Morshed, Muhammad 2301
Motte-Michellon, Denis 1216
Mouselinos, Theodoros P. 1551
Moussa, Hassan 3590
Movagharnejad, Hedieh 3048
Mu, Yunfei 3215
Müller, Jonas 2230
Müller, Tankred 474
Munk-Nielsen, Stig 825, 1641, 2661, 3309
Muñoz-Carpintero, Diego 1074, 3429
Muruaga, Endika Bilbao 3529
Musolino, Francesco 554

Mustafeez-Ul-Hassan 2860
Musumeci, S. 1975
Muyllaert, Koenraad 2383
Mysore, Madhu Lakshman 1424
Naeve, Tomasz 1445
Nagayasu, Kiwa 2164
Naghibi, Javad 614, 3353
Nahalparvari, Mehrdad 2843
Najjar, Mohammad 2682
Nakamura, Keiichi 777
Nakamura, Taketsune 3084
Nami, Ashkan 2241, 3757
Nannen, Hauke 160
Nassurdine, B. Mohamed 843
Nayak, Khirod Kumar 2241, 3757
Nayampalli, Vishwas Acharya 1703
Nazeri, Ahmad Ali 1309, 1336, 1343, 2670, 3871
Neal, Harley 2301
Nee, Hans-Peter 2843
Nehmer, Dominik 1036
Neira, Sebastian 2824, 3813
Neuland, Tanja 3991
Neumann, Christian 1895
Neumann, Ingmar 1445
Neumeister, Matthias 572
Nguyen, Allen 1722
Nguyen, Khanh-Hung 562, 1309
Nguyen, Van-Sang 922, 990
Nguyen, Xuan Viet Linh 169
Nian, Heng 2182
Niasar, Mohamad Ghaffarian 3729
Nie, Shuang 2145
Niedernostheide, Franz-J. 2744
Niedernostheide, Franz-Josef 1424
Nielebock, Sebastian 493, 2607
Niemetz, Michael 2021
Niggemann, Oliver 3545
Nikowitz, Mario 2251
Nishio, Atsushi 351
Nishitani, Yota 3420
Nishizawa, Shin-Ichi 1128
Noboru, Wakana 777
Noisette, Philippe 3910
Nooshabadi, Morteza Tadbiri 787
Nordström, Lars 883, 1006
Nymand, Morten 2682
O'Donnell, Terence 390
O'Driscoll, Seamus 4006
Obernolte, Urs 854
Odeh, Charles 1906
Okada, Ryohei 1776, 1844
Olbrich, Markus 2912

Oliveira, Hercules Araujo 2948
Orbay, Raik 3920
Orchard, Marcos 3429
Orfanoudakis, Georgios I. 1985
Örgüt, Osman 1361
Orlik, Bernd 3149, 3159, 3169
Ortega, David 1765, 3336
Ortiz-Gonzalez, Jose 2477
Orts, C. 3382, 3401
Oshnoei, Arman 2939
Ota, Ryosuke 1776, 1844
Ouyang, Ziwei 1413, 1561
Owzareck, Michael 1939, 2588
Oyarbide, Estanis 3327
Paasch, Kasper M. 3130
Pace, Loris 3582
Páez, J. D. 2422
Pagnani, Daniela 2504
Panigrahi, Bijaya Ketan 2110, 3511
Papadopoulos, Georgios 1391
Papadopoulos, Theofilos 432
Papafotiou, George 2788
Papanikolaou, Nick 1796, 2257
Papastergiou, Konstantinos 2355
Pascal, Yoann 3067
Pasquier, Christophe 3786
Passalacqua, Massimiliano 2919
Passmore, Brandon 4005
Pathmanathan, Mehanathan 1995, 2084, 2145, 2763
Patin, Nicolas 1256
Patti, Dario 2628
Patzelt, Nikolaus 1923
Paul, Arup Ratan 178
Pauls, Denis 2441
Pavone, Mario 554
Pedroso, Douglas 335
Peftitsis, Dimosthenis 1486, 2355
Pelletier, Sebastien 223
Penczek, Adam 3804
Peng, Hujun 3235
Péra, Marie-Cécile 315
Pereda, Javier 2824
Pereira, Thiago 3014, 3092, 3101, 3846
Perez, Gaëtan 960
Perez-Cebolla, Francisco Jose 3574, 3750
Peroutka, Zdenek 1114
Perpiñá, Xavier 2715
Perrin, Rémi 2526
Perrin, Remi 3272
Petritz, Andreas 279
Petzoldt, Jürgen 2885, 3491
Peyghami, Saeed 2939

Pfeiffer, Jonas 411, 1167
Pfost, Martin 2375, 2413
Phanse, Ajinkya 1722
Phulpin, Tanguy 3883
Pichon, Pierre-Yves 2526
Pickert, Phil Leon 1381
Piepenbrock, Till 2432
Pietrzak-David, Maria 2869
Pigott, John 3796
Pinheiro, José Renes 3590
Piqué, Gerard Villar 3796
Piróg, Stanislaw 3804
Placzek, Julius M. 833
Plat, Arnaud 3862
Plötz, Till-Mathis 980
Pogulaguntla, Aditya 730
Pohlmann, Sebastian 767
Polezhaev, Vladimir 2230
Ponick, Bernd 1381, 1665, 3048, 3711
Poormohammadi, Fereshteh 2795
Pöschke, Florian 3704
Pouresmaeil, Edris 2537
Pouresmaeil, Kaveh 2788
Pouresmaeil, Mobina 2537
Pramanick, Sumit 1658, 3511
Pree, Elias 1445
Prenleloup, Pierre 3529
Prieto-Araujo, Eduardo 2486, 3704
Puls, Simon 2852
Puschmann, Frank 749
Qin, Zian 3607
Quabeck, Stefan 3893
Quade, Katharina Lilith 4025
Quay, Rüdiger 242
Rabkowski, Jacek 1486, 3938
Rädel, Uwe 2885, 3491
Radha, Krishna Moorthy 344
Rafiq, Aamir 1658
Raggini, Diego 3206
Raghavendra, I Venkata 730
Rahmani, Mehdi 2496
Raison, Bertrand 1147
Rajabian, Amir Azam 614
Ramdane, Brahim 1216
Ramirez, Fernando 289
Rasekh, Navid 3120
Rasool, Haaris 2101, 2293
Raßmann, Rando 1286
Rathjen, Kai-Uwe 758
Rault, Pierre 582
Ravyts, Simon 1300
Raya, Mariana 2715

Razi, R. 3179
Razi, Reza 3140
Regnat, Guillaume 2526
Rehlaender, Philipp 2432, 2754, 3625
Reimann, René 3159
Reincke-Collon, Carsten 370, 3391
Reindl, Andrea 2021
Reiner, Richard 242
Reißenweber, Lukas 525
Reitmeier, Dominik 2211
Remón, Daniel 1083
Rettner, Cornelius 4019
Reyes-Chamorro, Lorenzo 3429
Reynaud, Jean-François 3529
Ribeiro, Luiz Antonio De Souza 2948
Richard, Lucas 1147
Rickert, Kai 115
Rigbers, Klaus 4023
Rigogiannis, Nick 2257
Ringbeck, Florian 4025
Risch, Raffael 651
Rizoug, Nassim 3696, 3831
Robinson, Jonathan 572
Rocha, Gabriel Silva 2948
Roche, Jan-Philipp 3545
Rodríguez, Alberto 335, 1083, 1765, 3336
Rodriguez, Daniel C. 3893
Rodriguez, Joan Marc 2574
Rodriguez, José 1015
Roes, Maurice G. L. 798
Roes, Maurice 2788
Roß, Tilo 3391
Rossi, Mattia 1476
Rothenburger, Max 2383
Roth-Stielow, Jörg 971, 1237, 1277, 3536
Rouphael, Rosalie 1525
Rudolph, Christian 474
Rueß, Manuel 3565
Rufer, Alfred 30
Ruppert, Lukas A. 3766
Ruthardt, Johannes 971
Rylko, Marek S. 2470
Sadarnac, Daniel 3883
Saeidi, Mahmoud 1336, 1343, 3871
Safdarzadeh, Omid 2316
Sah, Gyanendra Kumar 1885
Sahan, Benjamin 1137
Sahin, Ilker 1361
Sahoo, Subham 2110, 2182
Sahu, Malaya Kumar 2241, 3757
Sahu, Silpashree 689
Said, Nasri 2618

Saito, Wataru	1128
Sakai, J.	1785
Salehi, Navid	1056
Samples, Ben	4005
Sanchez, Juan	873
Sanchez-Ruiz, Alain	484
Santos, Francisco	3101
Sanusi, Bima Nugraha	1413
Sanz-Alcaine, José Miguel	3750
Sarlioglu, Bulent	3947
Sato, Kota	1580
Sato, Takashi	3420
Sauer, Dirk Uwe	4012, 4025
Sauerland, Henning	3159
Sawicki, Jean–paul	315
Scarcella, Giuseppe	1569
Scelba, Giacomo	1569, 2628
Schäffner, Philipp	279
Schafmeister, Frank	2432, 2754, 3625, 3686
Schanen, Jean-Luc	787
Schanen, JL	843
Schefer, Hendrik	891, 912, 1094
Schellekens, Jan	634
Schierle, Guido	3188
Schiestl, Martin	223, 268
Schillinger, Tobias	3646
Schillingmann, Henning	2030
Schlegel, Christian	1923
Schlegel, Ludwig	3957
Schmid, Markus	268
Schmidhuber, Michael	411, 1167
Schmies, Dominik	2276
Schmitz, Laurids	3599
Schnabel, Fabian	624, 1515
Scholjegerdes, Moritz	3005
Schön, André	814
Schrödl, Manfred	2251
Schueltzke, Jens	1167
Schuerhuber, Robert	39
Schuhmann, Thomas	3646
Schullerus, Gernot	593, 2334
Schulte, Horst	3704
Schulz, D.	3411
Schulze, Gerold	2383
Schulze, Hans-Joachim	1424
Schumann, Christian	2058
Schumann, Sven	4022
Schümann, Ulf	1286
Schupp, Jan	3309
Schütt, Michael	1885, 2308
Schwarz, Babette	1381
Schwendemann, R.	3411

Scohier, Martin	504
Scrimizzi, F.	1975
Sebastián, Javier	1765, 3336
Seibel, Axel	1515, 1620
Seitz, Arne	4015
Seliger, Norbert	22
Semail, Eric	2996
Sen, Paresh C.	1590, 1601
Sepehr, Amir	2537
Serdyuk, Yuriy	2344
Sergentanis, Grigorios	3450
Serra, Amiron Wolff Dos Santos	2948
Seybold, Felix	3536
Shahparasti, Mahdi	2682
Sharma, Kanuj	2698
Shawky, Ahmed	1015
Shen, Chengjun	2477
Shen, Xiaobing	1914, 2197
Shinoda, Kosei	3928
Shintani, Michihiro	3420
Shousha, Mahmoud	279, 698
Shuqin, Wang	1815
Siala, Sami	125
Siemaszko, Daniel	3910
Siemieniec, Ralf	1445
Sievers, Markus	3855
Singh, Rupam	135
Singh, Shashank Shekhawat	279
Singh, Sukhjit	2084
Skala, Aleksander	3804
Skibin, Stanislav	3301
Soeiro, Thiago Batista	1319, 3729
Solomentsev, Michael	1722
Solovyov, Vyacheslav	2860
Soltau, N.	1785
Soltau, Nils	777
Sönmez, Ertugrul	593, 2334
Soundararajan, Ajeeth Phrassanna	3729
Soupremanien, Ulrich	922
Spieler, Matthias	1692
Sprunck, Sebastian	1611
Sreekanth, T	730
Stadler, Alexander	525
Stadlober, Barbara	279
Staiger, Jochen	2219
Stala, Robert	3804
Stalleicken, Frederik	2607
Stallmann, Frederik	641
Stärz, Ronald	223, 268
Stathis, Spyridon	1402
Staubach, Christian	1137
Steckler, P. B.	2564

Stefanski, L.	3411
Steffen, Jonas	1515
Steinhart, Heinrich	2219
Štengl, Josef	1114
Stevic, Marija	2985
Stewart, Joshua	2806
Steyn, Kyle	3112
Stille, Karl Stephan	2276
Stock, Alexander	1
Stöckl, Thomas	952
Stone, David A.	3440
Strunk, Robin	1350
Stul, Koen	1300
Stutz, Christian	493
Suberski, Martin	2885, 3491
Sujeeth, Arjun	2628
Sullivan, Charles R.	2470
Svensson, Jan R.	1187
Tabrizi, Gholamreza	1611
Takamori, Taro	1128
Takayama, Hajime	108
Takeshita, Takaharu	213, 351, 739
Talla, Jakub	1114
Tang, Chengjun	2813
Tang, Zhongting	1933
Tashakor, Nima	1025, 1064, 1197, 3636, 3665
Tatakis, Emmanuel C.	1551
Tegtmeier, Bernd	1674
Teske, Peter	1466
Thiringer, Torbjörn	2344, 2813, 3920
Thoma, Jürgen	2554
Thönelt, Nick	1713
Thönnessen, André	3676
Tian, Fanghao	2581
Tillmann, Philipp	3740
Tiwari, Arvind Kumar	289
Tiwari, Arvind	187
To, Pham Ha Trieu	59, 70, 4003
Tornello, Luigi Danilo	1569
Torres, C.	3382, 3401
Torrico, Grover	361, 1815
Tournez, Florian	2175
Tran, Dai Duong	2293
Tran, Manh Tuan	2101, 2293
Tresca, Giulia	3975
Trescases, Olivier	3796
Tricoli, Pietro	468
Trochimiuk, Przemyslaw	1486
Tschepp, Andreas	279
Turrisi, Gaetano	1569
Tzanakis, Athanasios	3920
Uicich, Simon	3862

Ulbing, Alexander	3855
Ulmer, Sabrina	593, 2334
Ulrich, Burkhard	459
Umetani, Kazuhiro	2164
Unruh, Peter	1620
Unruh, Roland	3686
Urkizu, June	325
Vaccaro, Luis	2919
Vaessen, Peter	3729
Vagg, Christopher	3696
Vagnon, Eric	2515
Vahid, Sina	719
Vala, Sama Salehi	344
Valderrama, Carlos	504
Valenzuela, Rodrigo Alonso Alvarez	814
Van Cappellen, Leander	2795
Van Mierlo, Joeri	2101
Van Oosterwyck, Nick	3344
Van Tuan, Mai	351
Vandenbussche, Thomas	1300
Vanfretti, Luigi	3928
Vanwalleghem, Bart	3344
Vasiladiotis, Michail	1923
Vatamanu, Lucian	1046
Vázquez, Aitor	1083
Vázquez, Francisco	1765, 3336
Velasco-Quesada, Guillermo	1056
Velazco, Diego	251
Vellvehi, Miquel	2715
Venkataramanan, Giri	3480
Venugopal, Ravinder	2985
Verdier, Jacques	169
Vermeerch, Pierre	582
Veroni, Alessandro	3206
Vershinin, K.	2564
Viana, Caniggia	1995, 2084
Viarouge, I.	1955
Viarouge, P.	1955
Vidal-Albalate, Ricardo	2189
Videau, Nicolas	944
Videt, Arnaud	3822
Villar, Irma	3529, 3938
Vitorino, Montiê Alves	2689, 3059
Vogelsberger, M.	1834
Volzer, Benjamin	2554
Von Hoegen, Anne	3740
Wada, Keiji	1128
Wagner, Valentin	514
Wakelin, Bruce	3309
Wallart, Francois	251
Wallscheid, Oliver	2276, 2432
Waltereit, Patrick	242

Wang, Chu	3722
Wang, Jun	2136, 3120
Wang, Kangan	3014
Wang, Rui	673, 1641
Wang, Xiaoya	3722
Wang, Xin	315
Wang, Yanbo	2011
Wang, Yangang	2301
Waradzyn, Zbigniew	3804
Watanabe, Hiroki	1104
Wattenberg, Martin	873
Weicker, Martin	2316
Weires, Jonas	604
Weiser, Mathias C. J.	3565
Weiss, Xavier	1006
Wenzel, Johannes C.	2066
Werlig, Christian	3966
Weyh, Thomas	767
Wicht, Bernhard	2912
Wieczorek, Nick	3775
Wiemer, Adrian	2544
Wiesemann, Julius	1865
Wiesner, E.	1785
Wiesner, Eugen	777
Wijnands, Korneel	673, 798, 2788
Wilkowski, Matt	4008
Willer, Felix	3838
Willich, Viktor	3555
Wohlrath, Fritz	525
Wolbank, T.	1834
Wolf, Mihaela	2596, 2644, 3775
Wolfstädter, Simon	4017
Wölk, Alexander	279
Wouters, Hans	2197
Woywode, Oliver	758
Wu, Weimin	1985
Wu, Xiangqiang	1933
Wu, Yuxuan	2860
Wunsch, Bernhard	3301
Würfl, Joachim	2596, 2644, 3775
Würsig, Andreas	3966
Xia, Peizhou	3470
Xiao, Qian	3215, 3225
Xiao, Xiong	1966
Xie, Jun	2885, 3491
Xie, Lihong	2136
Xu, Huihui	709
Xu, James	3796
Xu, Qianwen	883, 1006
Xu, Wei	2136
Xu, Zhongqing	912
Xu, Zixiao	2182

Yadav, Sachin	3607
Yamaguchi, Masamichi	2127
Yamashita, Shota	213
Yamauchi, Kohei	2119
Yang, Huoming	1466
Yang, Jiajun	3014
Yang, Juefei	2477
Yang, Yinghui	3993
Yang, Yongheng	2257
Yaqoob, M.	1815
Yasuda, Takumi	902
Yeganeh, Mohammad Sadegh Orfi	2496, 2939
Yu, Guangyao	1319
Yu, Xiao	562, 1309, 2383
Yu, Xiaodan	3225
Yuan, Xibo	2136, 3120
Zacharias, Peter	411, 562, 1309, 1328, 1336, 1343, 2383, 2670, 3871
Zacher, Benjamin H.	2058
Zampardi, Giorgia	833
Zanchetta, Pericle	3975
Zatocil, Heiko	160
Zdanowski, Mariusz	3938
Zhang, Bo	1733
Zhang, Shimin	709
Zhang, Yaqian	2182
Zhang, Zhe	1561
Zhang, Zhuoqi	1776
Zhang, Ziqian	39
Zhao, Hongbo	1641, 3309
Zheng, Zhixue	315
Zhetessov, Aidar	3480
Zhu, Zi-Qiang	2958
Ziani, Adel	944
Ziegler, Philipp	971, 1237, 1277, 3536
Zilic, Rufad	1336
Zocher, Markus	2366
Zolfi, Pouya	719
Zou, Zhixiang	3014
Zsurzsan, Tiberiu Gabriel	1561

IEEE
445 Hoes Lane
Piscataway, NJ 08854-4141

ISBN 978-1-6654-8700-9